# The Dictionary
# of Substances
# and their Effects

THE ROYAL SOCIETY OF CHEMISTRY

**Editorial advisers**

F S H Abram, *Hamilton Garrod, Consultant Biologists*
R Hunt, *Chiltern Water Management Services*

**Production team**

Janet Crombie (Staff Editor)
Rebecca Allen
Lynne Braybrook
Karin Crawford
David Evans
Gail Heward
Craig Mortimer
Robert Reid
Alan Skull
F John Taylor
Ghislaine Tibbs
Helly Whitfield
Christine Heading
Robert Hunt
Library and Information Centre

James Butler (Design)

A catalogue record for this book is available from the British Library.

ISBN 0-85186-371-X

© **The Royal Society of Chemistry 1994**

Published by The Royal Society of Chemistry.
Thomas Graham House, The Science Park, Cambridge, CB4 4WF

Photocomposed by Land & Unwin (Data Sciences) Ltd.
Bugbrooke, Northamptonshire

Printed in England by Clays Ltd, St Ives plc

# CONTENTS

# INTRODUCTION

The Royal Society of Chemistry's Dictionary of Substances and their Effects (DOSE) is a multi-volume work covering basic physico-chemical properties and hazard information on over 5000 chemicals. The chemicals have been selected because they are known to have an adverse effect on certain living organisms. DOSE contains, in summarised but in well referenced format, information on both physical/chemical profiles, bioaccumulation and degradation data, ecotoxicological, and mammalian data.

The information in DOSE has been collected and arranged in such a way as to provide users with all the relevant data necessary to identify the hazards associated with a chemical. The extensive referencing of each entry allows readers to obtain more detailed information on individual sections of specific interest.

DOSE is an invaluable source of data to scientists, regulators and information professionals alike. For scientists, toxicologists, and health, safety and environmental officers, DOSE provides easy and rapid access to information on the likely impact of chemicals on flora and fauna, including humans, and the ecosystem. DOSE meets the need of regulators for concise and accessible data on which to formulate the strategy for risk assessment and management. Research workers in chemistry and other disciplines can turn to DOSE for data during experimental research for new substances, modifications of existing substances and in assessing the potential hazards of substances they are handling. DOSE is also an important resource for librarians and other information scientists, who provide a service to a wide range of end users.

Although DOSE allows identification of the hazards associated with a particular chemical it cannot, on its own, provide risk assessment. Hazard identification is the determination of substances of concern, their adverse effects, target populations and conditions of exposure, taking into account toxicity data and knowledge of effects on human health, other organisms and their environment (1). Risk assessment is the identification and quantification of the risk resulting from a specific use or occurrence of a chemical, taking into account the possible harmful effects to individual people or society of using the chemical in the amount and manner proposed and by all possible routes of exposure. Quantification ideally requires the establishment of dose-effect and dose-response relationships in likely target individuals and populations (1,2). Risk evaluation involves the establishment of a qualitative or quantitative relationship between risks and benefits, involving the complex process of determining the significance of the identified risks and estimated risks to those organisms or people concerned with or affected by them (1,2). DOSE will help in hazard identification by providing the required information, where available, on the adverse effects of substances of concern. However, risk assessment requires knowledge of how the chemical is being used in a particular situation, i.e. knowledge of exposure levels, routes of exposure and any modifying effects of environmental factors on toxicity.

This information and the subsequent assessment must be provided by a trained specialist in risk assessment.

The Editorial Advisory Board decided that DOSE should contain only information available in the literature, from peer reviewed sources where possible. This gives the best knowledge available in the public domain at the time the data was collected. There are chemicals that have not been adequately tested to enable a complete DOSE item to be compiled, and others on regulatory lists that do not have sufficient reported information to enable a DOSE item to be compiled at all. In many cases, there is no toxicology data available for degradation products of well documented compounds. Therefore when carrying out risk assessments, the absence of information on the toxicology of a chemical can not be taken as an indication of its safety. It merely highlights the fact that either no studies have been undertaken or the findings are unpublished. It should be noted that a number of chemicals which were originally identified for inclusion in DOSE are not included because there are no relevant published data available.

The information in DOSE is brought together from a large number of disparate sources, sources which are constantly increasing in size and number as does the level of research into mammalian and ecotoxicity. The publishers of DOSE would therefore welcome readers' comments on any relevant information which could be included in future editions of DOSE.

1.  IUPAC, Glossary for Chemists and Terms Used in Toxicology, *Pure & Appl. Chem.* 1993, **65**(9), 2003-2122 (© 1993 IUPAC)
2.  IPCS, Training Module No. 1, Appendix 2, Chemical Safety – Fundamentals of Applied Toxicology, The Nature of Chemical Hazards, (WHO/IPCS 92.3, Geneva) (© 1992 IPCS)

# HOW TO USE DOSE

The data in DOSE are organised under the data headings summarised below:

**Identifiers**
Chemical name
Structure
CAS Registry Number
Synonyms
Molecular formula
Molecular weight
Uses
Occurrence

**Physical properties**
Melting point
Boiling point
Flash point
Specific gravity
Partition coefficient
Volatility
Solubility

**Occupational exposure**
Limit values
UN number
HAZCHEM code
Conveyance classification
Supply classification
Risk phrases
Safety phrases

**Ecotoxicity**
Fish toxicity
Invertebrate toxicity
Bioaccumulation

**Environmental fate**
Nitrification inhibition
Carbonaceous inhibition
Anaerobic effects
Degradation studies
Abiotic removal
Absorption

**Mammalian and avian toxicity**
Acute data
Sub-acute data
Carcinogenicity and long-term effects
Teratogenicity and reproductive effects
Metabolism and pharmacokinetics
Irritancy
Sensitisation
Genotoxicity
Any other adverse effects to man

**Any other adverse effects**

**Legislation**

**Any other comments**

**References**

These headings only appear in an item when data have been identified for that heading. The reader can, therefore, assume that the absence of a heading means that no relevant data were retrieved from the sources examined.

## Identifiers and basic chemistry

The identifiers section of the item provides data on synonyms and approved codes used to define the precise chemical structure of the compound.

### Chemical Abstracts Registry Number

The Chemical Abstracts Service Registry Number is a number sequence adopted by the Chemical Abstracts Service (American Chemical Society, Columbus, Ohio, USA) to uniquely identify specific chemical substances. The number contains no information relating to the chemical structure of a substance and is, in effect, a catalogue number relating to one of the 12 million or so unique chemical substances recorded in the Chemical Abstracts Registry. New numbers are assigned sequentially to each new compound identified by Chemical Abstracts Service. This information is also provided in the full index of CAS Registry Numbers available at the end of each volume.

### Synonyms

For common chemicals, several chemical names and numerous trade names may be applied to describe the chemical in question. In the synonyms field, many of these names are identified to aid users on the range of names which have been used to describe each substance. This information is also provided in the full index of alternative names available at the end of each volume.

### Molecular formula

This is the elemental composition of the compound. The elements appear alphabetically for inorganic compounds, i.e. $Ag_2CO_3$, $Cl_2Cr$, etc, but for organic compounds, carbon and hydrogen content are shown first followed by the other elements in alphabetical order, i.e. $C_6H_5Br$. This information is also provided in the molecular formula index at the end of each volume.

### Molecular weight

This is directly calculated from the molecular formula. No molecular weights are given for polymers.

### Uses

Principal uses of the substances are given, with information on other significant uses in industrial processes.

### Occurrence

Natural occurrences, whether in plants, animals or fungi are reported. Natural emissions, combustion and/or pyrolysis processes are also given.

## Physical properties

### Melting/Boiling point

These data are derived from various sources.

### Flash point

The flash point is the lowest temperature at which the vapours of a volatile combustible substance will sustain combustion in air when exposed to a flame. The flash point information is derived from various sources. Where possible the method of determination of the flash point is given.

## Specific gravity (density)
The specific gravity of each substance has been derived from a variety of sources. Where possible the data have been standardised.

## Partition coefficient
Partition coefficients, important for structure-activity relationship considerations, particularly in the aquatic environment, are indicated. Ideally the n-octanol/water partition coefficient is quoted. The major data source for this measurement is:

Sangster, J *J. Phys. Chem. Ref. Data* 1989, **18**(3), 1111-1229

Where no reference is quoted, it can be assumed that the information was derived from this source.

## Volatility
The vapour pressure and vapour density are quoted where available. Where possible, the data have been standardised.

## Solubility
Solubility data derived from several sources are quoted for both water and organic solvents.

# Occupational exposure

## Threshold limit values
The airborne limits of permitted concentrations of hazardous chemicals represent conditions under which it is believed that nearly all workers may be repeatedly exposed day after day without adverse effect. These limits are subject to periodic revision and vary between different countries. The term *threshold limit* relates primarily to the USA, but equivalent terms are available in most industrialised countries. In DOSE items, comparable values are available for the USA and United Kingdom. The data relate to concentrations of substances expressed in *parts per million (ppm)* and *milligrams per cubic metre (mg m$^{-3}$)*.

## USA threshold limits
The threshold limit values for the USA have been taken from the *Threshold Limit Values and Biological Exposure Indices, 1993-1994* produced by the American Conference of Governmental Industrial Hygienists, Cincinnati, USA. The limits relate to *Threshold Limit – Time Weighted Average, Threshold Limit – Short Term Exposure Limit* and *Threshold Limit – Ceiling Limit*. The Threshold Limit Value – Time Weighted Average (TLV-TWA) allows a time-weighted average concentration for a normal 8-hour working day and a 40-hour working week, to which nearly all workers may be repeatedly exposed day after day, without adverse effect. The Threshold Limit Value – Short Term Exposure Limit (TLV-STEL) is defined as a 15-minute, time-weighted average which should not be exceeded at any time during a working day, even if the 8-hour time-weighted average is within the TLV. It is designed to protect workers from chemicals which may cause irritancy, chronic or irreversible tissue damage, or narcosis which may increase the likelihood of accidental injury. Many STELs have been withdrawn pending further toxicological assessment. For Threshold Limit – Ceiling Values (TLV-C) the concentration should not be exceeded during any part of the working day.

### UK exposure limits

The occupational limits relating to airborne substances hazardous to health are published by the Health and Safety Executive annually in Guidance Note EH 40. The values in the DOSE items have been taken from the 1994 edition.

In the United Kingdom, there are Maximum Exposure Limits (MEL) which are subject to regulation and which should not normally be exceeded. They derive from Regulations, Approved Codes of Practice, European Community Directives, or from the Health and Safety Commission. In addition, there are Occupational Exposure Standards (OES) which are considered to represent good practice and realistic criteria for the control of exposure. In an analogous fashion to the USA Threshold Limits, there are long-term limits, expressed as time-weighted average concentrations over an 8-hour working day, designed to protect workers against the effects of long-term exposure. The short-term exposure limit is for a time-weighted average of 15 minutes. For those substances for which no short-term limit is listed, it is recommended that a figure of three times the long-term exposure limit averaged over a 15-minute period be used as a guideline for controlling exposure to short-term excursions.

### UN number

The United Nations Number is a four-figure code used to identify hazardous chemicals and is used for identification of chemicals transported internationally by road, rail, sea and air. In the UK this number is also called the ''Substance Identification Number'' or ''SI Number''.

### HAZCHEM code

The Hazchem Code is a code used to instruct United Kingdom emergency services on equipment, evacuation and other methods of dealing with transportation incidents. It is administered by the Chemical Industries Association.

### Conveyance classification

The information presented for the transportation of substances dangerous for conveyance by road is derived from the Approved Carriage List [Chemicals (Hazard Information and Packaging) Regulations 1993, Statutory Instrument No. 1746, amended by No. 3050] HSC, UK, (CHIP 1) and Approved Substance Identification Numbers, Emergency Action Codes and Classifications for Dangerous Substances in Road Tankers and Tank Containers, HSC, 1989. From 1 April 1994, the carriage requirements of CHIP 1 are transferred to the Carriage of Dangerous Goods by Road and Rail (Classification, Packaging and Labelling) Regulations 1994, Statutory Instrument No. 669.

### Supply classification

The information presented for the supply of substances is derived from the Approved Supply List: information approved for the classification and labelling of substances and preparations dangerous for supply [Chemicals (Hazard Information and Packaging) Regulations 1993, Statutory Instrument No. 1746, amended by No. 3050] HSC, UK. From January 1995, supply requirements will be updated and contained in the Chemicals (Hazard Information and Packaging for Supply) Regulations, also known as CHIP 2.

### Risk and safety phrases

Risk and safety phrases used in connection with DOSE items are approved phrases for describing the risks involved in the use of hazardous chemicals and have validity in the United Kingdom and throughout the countries of the European Community. The approved texts have designated R (Risk) and S (Safety) numbers from which it is possible to provide translations for all approved languages adopted by the European Community. The risk and safety phrases relate to the Approved Supply List: information approved for the classification and labelling of substances and preparations dangerous for supply [Chemicals (Hazard Information and Packaging) Regulations, 1993, Statutory Instrument No. 1746, amended by No. 3050] HSC, UK. The risk and safety phrases should be used to describe the hazards of chemicals on data sheets for use and supply; for labelling of containers, storage drums, tanks etc., and for labelling of articles specified as dangerous for conveyance by road. The Approved guide to the classification and labelling of substances and preparations dangerous for supply [Chemicals (Hazard Information and Packaging) Regulations 1993, Statutory Instrument No. 1746, amended by No. 3050] – Guidance on regulations, HSC, UK specifies the criteria necessary for assigning a particular R or S phrase to a substance which has not been assigned any phrases. They also now include ecotoxicity risk phrases, i.e.

R50 – Very toxic to aquatic organisms
R51 – Toxic to aquatic organisms
R52 – Harmful to aquatic organisms
R53 – May cause long-term adverse effects in the aquatic environment
R54 – Toxic to flora
R55 – Toxic to fauna
R56 – Toxic to soil organisms
R57 – Toxic to bees
R58 – May cause long-term adverse effects in the environment
R59 – Dangerous for the ozone layer

These risk phrases should be used to warn the user of the ecological damage that a chemical can cause.

## Ecotoxicity

Information is presented on the effects of chemicals on various ecosystems. Results of studies carried out on aquatic species, primarily fish and invertebrates, but also fresh water and marine microorganisms and plants are reported. Persistence and potential for accumulation in the environment and any available information on the harmful effects to non-target species, i.e. the unintentional exposure of terrestrial and/or aquatic species to a toxic substance is given.

### Fish Toxicity
$LC_{50}$, with duration of exposure, are quoted for two species of freshwater and one marine species if available. Any additional information on bioassay type (static or flow through) and water condition (pH, temperature, hardness or oxygen content) is reported.

### Invertebrate
$LC_{50}$, with duration of exposure, are quoted for molluscs and crustaceans. $EC_{50}$ values, i.e. concentrations which will immobilise 50% of an exposed population are given for microbes, algae and bacterium. Values which will inhibit microbial or algae growth are reported. Duration of exposure is given when available.

### Bioaccumulation
Bioaccumulation, biomagnification and bioconcentration data are quoted primarily for fish, invertebrates, bacteria and algae.

## Environmental Fate
Degradation data are used to assess the persistence of a chemical substance in the environment, in water, soil and air. If the substance does not persist, information on the degradation products is also desirable. Intermediates may be either harmless or toxic substances which may themselves persist. Degradation occurs via two major routes, microbial degradation utilising microorganisms from a variety of habitats and decomposition by chemical methods. Microbial degradation is associated with the production of elemental carbon, nitrogen and sulfur from complex molecules. Standard biodegradation tests estimate the importance of microbial biodegradation as a persistence factor. Most tests use relatively dense microbial populations adapted to the compound being studied. Rapid degradation results in these tests implies that the compound will degrade under most environmental conditions, although specialised environments where degradation would not occur can exist. Compounds which are not readily degradable are likely to persist over a wide range of environmental situations. Biodegradation in the environment is difficult to assess because the natural environment is highly variable and almost impossible to reproduce under laboratory conditions. This should be given due consideration when assessing the data in DOSE. Chemical degradation processes include photolysis, hydrolysis, oxidation and removal by reversible/irreversible binding to sediment. Factors which influence degradation rates: duration of exposure; temperature; pH; salinity; concentrations of test substance, microbial populations, and other nutrients; must also be taken into account.

### Nitrification inhibition
The nitrogen cycle is the major biogeochemical process in the production of nitrogen, an essential element contained in amino acids and proteins. Nitrogen is an essential element in microorganisms, higher plants and animals. Interference in nitrification from more complex molecules can be determined by standard tests using nitrogen-fixing bacteria. *Nitrosomonas* sp. detoxify ammonia to nitrite and *Nitrobacter* sp. detoxify nitrite to nitrate. The degree of nitrification inhibition can be used to estimate the environmental impact of the test chemical. There are a number of incidences where loss of nitrification because of dramatic reductions in populations of *Nitrobacter* sp. and *Nitrosomonas* sp. at sewage treatment works has resulted in the discharge of unacceptably high concentrations of ammonia.

### Carbonaceous inhibition
Another major biogeochemical process is the recycling of carbon via the decomposition of complex organic matter by bacteria and fungi. In nature the process is important in the cycling of elements and nutrients in ecosystems. The degradation sequence occurs in stages, cellulose→cellobiose→glucose→organic acids and carbon dioxide. Chemical inhibition of microbial processes at all or any of these stages is reported here.

### Anaerobic effects
Anaerobic microbial degradation of organic compounds occurs in the absence of oxygen and is an important degradation process in both the natural environment and in waste treatment plants. The importance of anaerobic processes in the degradation

of a particular substance is reported here. These include laboratory simulations of the natural processes occurring in flooded soils using flasks containing samples of the medium and radiolabelled substrate. The other important method uses anaerobic digestion tests which compare the production of methane and carbon dioxide by anaerobic microbes in a sludge sample with and without added test material. Methane production is at the end of the food chain process used by a wide range of anaerobic microorganisms.

**Degradation studies**
This section focuses on microbial degradation in both soil and water. The half-life of the chemical substance in the environment is reported with its degradation products where possible, giving an indication of the degree of its persistence. Water pollution factors: BOD, biochemical oxygen demand; COD, chemical oxygen demand; and ThOD, theoretical oxygen demand are stated, where available. BOD estimates the extent of natural purification which would occur if a substance is discharged into rivers, lakes or the sea. COD is a quicker chemical method for this determination which uses potassium dichromate or permanganate to establish the extent of oxidation likely to occur. ThOD indicates the amount of oxygen needed to oxidise hydrocarbons to carbon dioxide and water. When organic molecules contain other elements nitrogen, sulfur or phosphorus, the ThOD depends on the final oxidation stage of these elements.

**Abiotic removal**
Information on chemical decomposition processes is contained in this section. The energy from the sun is able to break carbon-carbon, and carbon-hydrogen bonds, cause photodissociation of nitrogen dioxide to nitric oxide and atomic oxygen and photolytically produce significant amounts of hydroxyl radicals. Hydrolysis occurs when a substance present in water is able to react with the hydrogen or hydroxyl ions of the water. Therefore the extent of photolytic and oxidative reactions occurring in the atmosphere and hydrolysis in water can be used as a measure of environmental pollution likely to arise from exposure to a substance.

**Absorption**
The environmental impact of a chemical substance is determined by its ability to move through the environment. This movement depends on the affinity of the chemical toward particulate matter; soil and sediment. Chemicals which have a high affinity for absorption are less readily transported in the gaseous phase or in solution, and therefore can accumulate in a particular medium. Chemical substances which are not readily absorbed are transported through soil, air and aquatic systems.

## Mammalian and avian toxicity

Studies on mammalian species are carried out to determine the potential toxicity of substances to humans. Avian species are studied primarily to assess the environmental impact on the ecosystem, however data from avian studies are also used for assessing human toxicity. This is specifically applied to pesticides, with neurotoxicology studies.

Procedures involve undertaking a series of established exposure studies on a particular substance using specific routes, oral, inhalation, dermal or injection for variable durations. Exposure durations include acute or single exposure to a given concentration of substance. Sub-acute or sub-chronic exposure, i.e. repeat doses increasing in concentration over an intermediate time period, up to 4 weeks for sub-acute and 90 day/13 week (in rodents) or 1 year (in dogs) for sub-chronic studies. Finally chronic/long-term studies involve exposure to specific concentrations of chemical for a duration of 18 month-2 years. A variety of species are used in toxicity testing, most commonly rodents, (rats, mice, hamsters) and rabbits, but tests can also be carried out on monkeys, domestic animals and birds, but rarely on wild animals which can be no less susceptible to toxic substances, e.g. pesticides.

**Acute Data**
Single exposure studies quoting $LD_{50}$ from at least two species.

**Sub-acute data**
Results of repeat doses, intermediate duration studies are quoted. Priority is given to reporting the adverse effects on the gastrointestinal, hepatic, circulatory, cardiopulmonary, immune, reproductive, renal and central nervous systems.

**Carcinogenicity and long-term effects**
Information on the carcinogenicity of substances unequivocally proven to cause cancer in humans and laboratory animals, together with equivocal data from carcinogenicity assays in laboratory animals are reported. Additionally treatment-related chronic adverse effects are reported. Criteria for inclusion required the study to report the species, duration of exposure, concentration and target organ(s); sex is also given where available.

**Teratogenicity and reproductive effects**
The results of studies carried out in intact animal and *in vitro* systems to determine the potential for teratogenic, foetotoxic and reproductive damage are reported here. Criteria for inclusion require the species, duration of exposure, concentration and details of the effect in relation to fertility to be stated. Adverse effects reported in this section include sexual organ dysfunction, developmental changes (to embryos and foetuses), malformations, increases in spontaneous abortions or stillbirths, impotence, menstrual disorders and neurotoxic effects on offspring.

**Metabolism and pharmacokinetics**
Data are quoted on the metabolic fate of the substance in mammals, and include absorption, distribution, storage and excretion. Mechanisms of anabolic or catabolic metabolism, enzyme activation and half-lives within the body are reported when available. Additionally findings from *in vitro* studies are reported.

**Irritancy**
Chemical substances which cause irritation (itching, inflammation) to skin, eye and mucous membranes on immediate contact in either humans or experimental animals is reported here. Exposure can be intentional in human or animal experiments, or unintentional via exposure at work or accident to humans.

**Sensitisation**
Sensitisation occurs where an initial accidental or intentional exposure to a large or small concentration of substance causes no reaction or irritant effects. However, repeat or prolonged exposure to even minute amounts of a sensitising chemical causes increasingly acute allergic reactions.

**Genotoxicity**
Genotoxicity testing is carried out to determine the mutagenic and/or carcinogenic potential of a chemical substance. A standard series of tests are carried out under

controlled laboratory conditions on an established set of test organisms. A hierarchical system using bacteria, yeasts, cultured human and mammalian cells, *in vivo* cytogenetic tests in mammals and plant genetics is used to assess the genotoxic potential of the substance under study. Bacteria, unlike mammals, lack the necessary oxidative enzyme systems for metabolising foreign compounds to the electrophilic metabolites capable of reacting with DNA. Therefore, bacteria are treated with the substance under study in the presence of a post-mitochondrial supernatant (S9) prepared from the livers of mammals (usually rats). This fraction is supplemented with essential co-factors to form the S9 mix necessary for activation. DOSE reports published studies: giving the test organisms, whether metabolic activation (S9) was required, and the result, positive or negative.

### Any other adverse effects to man
Adverse effects to humans from single or repeat exposures to a substance are given. The section includes results of epidemiological studies, smaller less comprehensive studies of people exposed through their work environment and accidental exposure of a single, few or many individuals.

### Any other adverse effects
Adverse effects to organisms or animals other than man are reported here.

### Legislation
Any form of legislation, medical (food and drugs) or environmental from European, American and worldwide sources is reported.

### Any other comments
All other relevant information including chemical instability and incompatibility is contained in this section. This includes reviews, phytotoxicity and toxic effects associated with impurities.

### References
Contains references to all data from above sections.

### Indexes
The most convenient means of accessing a chemical in DOSE is via one of the indexes at the back of the publication. DOSE contains three indexes: name and synonyms, CAS Registry Numbers and molecular formulae. Each index entry refers to an item number, not a page number.

### Index of names and synonyms
Contains the name of the chemical used in DOSE together with a number of synonyms for that chemical. All names are arranged alphabetically.

### Index of CAS Registry Numbers
Contains a list of the CAS Registry Numbers of the chemicals appearing within the volume of DOSE in ascending order.

### Index of molecular formulae
Contains a list of the molecular formulae of the chemicals appearing within the volume of DOSE in alphabetical order for inorganic compounds, i.e. $Ag_2CO_3$, $Cl_2Cr$, etc., but for organic compounds, carbon and hydrogen content are shown first followed by the other elements in alphabetical order, i.e. $C_6H_5Br$. However, salts are shown with the anion/cation incorporated into the principal structure.

**Note**

The Royal Society of Chemistry (RSC) can only assess published information when compiling the Dictionary of Substances and their Effects. However, the RSC would welcome any information not readily accessible, but in the public domain, on the chemicals covered so that information could be used when the items in DOSE are updated.

If you have any relevant information, please contact:

Hazards Production Department
The Royal Society of Chemistry
Thomas Graham House
The Science Park
Cambridge
CB4 4WF
UK
Telephone +44 (0) 223 420066
Fax: +44 (0) 223 423429

**Original articles and technical enquires**

The Royal Society of Chemistry, Library and Information Centre (LIC) can usually supply copies of articles cited in DOSE. In addition, LIC will answer technical enquires of all kinds including online chemical substructure searches.
LIC is at the Royal Society of Chemistry, Burlington House, Piccadilly, London W1V 0BN, UK
Telephone: +44 (0)71 437-8656, Fax: +44 (0)71 287-9798, Internet: LIBRARY@RSC.ORG

# I1   Ibenzmethyzin hydrochloride

$$CH_2NHNHCH_3$$

• HCl

$$C(O)NHCH(CH_3)_2$$

**CAS Registry No.** 366-70-1
**Synonyms** $N$-isopropyl-$\alpha$-(2-methylhydrazino)-$p$-toluamide monohydrochloride;
$N$-(1-methylethyl)-4-[(2-methylhydrazino)methyl]benzamide;IBZ;
$N$-4-isopropylcarbamoyl benzyl-$N'$-methylhydrazine hydrochloride;
Procarbazine hydrochloride; Natulan
**Mol. Formula** $C_{12}H_{20}ClN_3O$                    **Mol. Wt.** 257.77
**Uses** Chemotherapeutic agent.

## Physical properties
**M. Pt.** 223°C.

**Solubility**
Organic solvent: methanol, chloroform, diethyl ether

## Mammalian and avian toxicity

### Acute data
$LD_{50}$ oral mouse 1320 mg $kg^{-1}$ (1).
$LD_{50}$ oral rabbit 145 mg $kg^{-1}$ (1).
$LD_{50}$ intravenous rat, mouse 350-540 mg $kg^{-1}$ (2,3).
$LD_{50}$ subcutaneous rat, mouse 490-710 mg $kg^{-1}$ (3,4).
$LD_{50}$ intraperitoneal mouse 699 mg $kg^{-1}$ (5).

### Carcinogenicity and long-term effects
Sufficient evidence for the carcinogenicity in mice and rats. Limited evidence of its
carcinogenicity in monkeys. There is inadequate evidence of the carcinogenicity of
the substance alone in humans, IARC classification group 2A (6).
Mice were given (route unspecified) 300 mg $kg^{-1}$ once a wk for 8 wk. By 6-11 wk
9/14 had lung tumours and 2/14 had leukaemia and by 12-16 wk 21/21 had lung
tumours and 17/21 had leukaemia (7).
Ten ♀ Sprague-Dawley rats received single 50, 100 or 150 mg doses or 3 doses of
50 mg by gavage. All animals showed mammary tumours by 20 wk after treatment (8).
♂ and ♀ Sprague-Dawley rats received intraperitoneal doses of 15 or 30 mg $kg^{-1}$ in
buffered saline 3 × $wk^{-1}$ for 26 wk. Neoplasms, including those of the neuroepithelial
tissues, mammary gland, haematopoietic system and lymphoreticular tissue, were
observed in 19/30 ♂ and 27/30 ♀ of the low-dose group and in 30/33 ♂ and 30/31 ♀ of
the high-dose group. Other cancers detected were lymphomas, leukaemia, olfactory
neuroblastomas, adenocarcinomas of the mammary gland and oligodendroglioma (9).

Intravenous injections of 24 mg kg$^{-1}$ to ♂ BR46 rats once every wk for 52 wk. 1/34 and 14/34 rats died with respectively benign and malignant neoplasms. Malignant: 3 renal sarcomas, 1 adenocarcinoma, 2 intra-abdominal spindle-cell sarcomas, 1 rectal carcinoma, 1 testicular carcinoma, 1 squamous-cell carcinoma of the ear duct, 1 subaxillary sarcoma and 1 neurilemmoma. Benign: 1 prostatic adenoma and 1 subcutaneous fibromas (2).

### Teratogenicity and reproductive effects

Single intraperitoneal injection 5-550 mg kg$^{-1}$ to ♀ Wistar (CF) rats. After single injection of 100 mg kg$^{-1}$ on day 5-8 of pregnancy and a single dose >75 mg kg$^{-1}$ on day 9-14 or 17, 100% embryomortality was reported. Single doses of 25-75 mg kg$^{-1}$ on day 10, 11, 12, 14 or 17 were teratogenic with defects including tail and limb malformations, exencephaly, omphalocele, encephalocele and short maxilla or mandible (10).

0, 1, 5 or 15 mg kg$^{-1}$ administered to pregnant rats from gestation days 13 to 16. Offspring brains at age 12 to 15 wk showed cerebral atrophy, which increased with dose, and a reduction in neuronal numbers. Astrocyte and oligodendrocyte populations were unaffected (11).

♂ rats aged 10-90 days given a daily intraperitoneal (5 or 9 wk) 30 mg kg$^{-1}$ or gavage (9 wk) doses 5 or 50 mg kg$^{-1}$ showed substantial mortality in immature rats and at high oral doses. Body weight gain and the weights of testes and epididymis were reduced. Spermatogenic architecture was disrupted. The number of Sertoli cells was not affected, but a dysfunction was seen in cases of severe disruption of spermatogenesis. Leydig cells were moderately affected and epididymal sperm reserves were reduced. Foetus number was low in ♀ mated with treated ♂ (12).

### Metabolism and pharmacokinetics

Levels in the plasma and the cerebrospinal fluid equilibrated, following 100 mg kg$^{-1}$ intravenous dose, within 30 min in dog study (13).

After intraperitoneal injection of 20 or 200 mg kg$^{-1}$ of $^{14}$C-radiolabelled compound to rats 7-10% of the dose was exhaled as methane and 11-22% as $CO_2$ within 8 hr. It was suggested that metabolism proceeded via formation of methylhydrazine (14). Following human oral administration 70% appeared within 24 hr in urine. Most appeared as the metabolite N-isopropyl terephthalaminic acid with only 5% as the unmodified compound. Little was excreted in the faeces (15,16).

### Sensitisation

4/44 and 8/23 patients with Hodgkin's disease and non-Hodgkin's lymphoma, respectively, showed allergic skin reactions (17).

## Genotoxicity

*Salmonella typhimurium* TA98, TA100, TA1535, TA1537, with and without metabolic activation negative (18-20).
*Salmonella typhimurium* (strain unspecified) without metabolic activation positive (21).
*Saccharomyces cerevisiae* without metabolic activation mitotic crossover, gene conversions and reverse mutations positive (20).
*Drosophila melanogaster* recessive lethal mutations, total sex chromsome loss and dominant lethality positive (22).
*In vitro* mouse lymphoma L5178Y with metabolic activation positive (23).

*In vivo* mouse foetal liver and blood cells after transplacental treatment caused micronuclei (24)

## Any other adverse effects to man

Daily oral doses 250-300 mg caused dose-related reversible depression of peripheral leucocyte and platelet counts with a low 2-3 wk into the programme. Vomiting and nausea are common. Somnolence, hallucinations, agitation and lethargy are frequent with oral and intravenous injection, being more aggravated via the latter route (25). Peripheral neuropathy has been reported (26).

## Any other comments

Antitumour activity, metabolism, pharmacology and mode of action reviewed (1, 27).

## References

1. Reed, D. J. *Antineoplastic and Immunosuppresive Agents* 1975, Part 2, Springer, New York
2. Schmahl, D. et al *Arzneim.-Forsch.* 1970, **20**, 1461-1467
3. *Iyakuhin Kenkyu* 1973, **4**, 467
4. *Drugs in Japan. Ethical Drugs* 6th ed., 1982, Japan Pharmaceutical Information Centre, Yakugyo Jiho Co. Ltd., Tokyo
5. *Arch. Toxicol.* 1979, **41**, 287
6. *IARC Monographs 1987,* **Suppl. 7**, 327
7. Kelly, M. G. et al Cancer Chemother. Rep. 1964, **39**, 77-80
8. Henson, J. C. et al *Eur. J. Cancer* 1966, **2**, 385-386
9. National Cancer Institute *NCI Carcinog. Tech. Rep. Ser.* 1979, No. 19
10. Chaube, S. et al *Teratology* 1969, **2**, 23-32
11. Wright, J. A. et al *J. Comp. Pathol.* 1989, **101**(4), 421-427
12. Velez de la Calle, J. F. et al *J. Reprod. Fertil.* 1988, **84**(1), 51-61
13. Oliverio, V. T. et al *Cancer Chemother. Rep.* 1964, **42**, 1-7
14. Dost, F. N. et al *Biochem. Pharmacol.* 1967, **16**, 1741-1746
15. Bollag, W. *Natulan* 1965, Wright, Bristol
16. Schwartz, D. E. et al *Arzneim.- Forsch.* 1967, **17**, 1389-1393
17. Anderson, E. et al *Scand. J. Haematol.* 1980, **24**, 149-151
18. Heddle, J. A. et al *Progress in Genetic Toxicology* 1977, 265-274, Elsevier/North-Holland Biomedical Press, Amsterdam
19. Painter, R. B. et al *Mutat. Res.* 1978, **54**, 113-115
20. Bronzetti, G. et al *Mutat. Res.* 1979, **68**, 51-58
21. Pueyo, C. *Mutat. Res.* 1979, **67**, 189-192
22. Blijleven, W. G. H. et al *Mutat. Res.* 1977, **45**, 47-59
23. Clive, D. et al *Mutat. Res.* 1979, **59**, 61-108
24. Cole, R. J. et al *Nature (London)* 1979, **277**, 317-318
25. DeVita, V. T. et al *Clin. Pharmacol. Ther.* 1966, **7**, 542-546
26. *Martindale. The Extra Pharmacopoeia* 30th ed., 1993, The Pharmaceutical Press, London
27. Sartorelli, A. C. et al *Ann. Rev. Pharmacol.* 1969, **9**, 51-72

## I2    Ibuprofen

$$CH_3$$
$$CHCO_2H$$

$$CH_2CH(CH_3)_2$$

**CAS Registry No.** 15687-27-1
**Synonyms** α-methyl-4-(2-methylpropyl)benzeneacetic acid; *p*-isobutylhydratropic
acid; 2-(4-isobutylphenyl)propionic acid; Actifen; Brufen; Ibufen; Motrin
**Mol. Formula** $C_{13}H_{18}O_2$                                    **Mol. Wt.** 206.29
**Uses** Anti-inflammatory, analgesic and antipyretic drug.

### Physical properties
**M. Pt.** 75-77°C.

### Solubility
Organic solvent: ethanol, chloroform, diethyl ether, acetone

### Mammalian and avian toxicity

#### Acute data
$LD_{50}$ oral rat, mouse 636-740 mg $kg^{-1}$ (1,2).
$LD_{50}$ intraperitoneal mouse, rat 320-626 mg $kg^{-1}$ (3,4).

#### Teratogenicity and reproductive effects
Oral near-term rat 6 mg $kg^{-1}$ caused foetal ductal constriction of 70% within 1 to 8 hr,
60% dilation of both ventricles and 120% increase in pericardial fluid. These changes
partly disappeared at 24 hr. Cardiac failure was induced in some animals (5).

#### Metabolism and pharmacokinetics
Following ingestion by human, peak plasma concentrations occur at 1 to 2 hr. Bound
extensively to plasma proteins with $t_{1/2}$ of ≈2 hr. Rapidly excreted in urine mainly as
metabolites and their conjugates; 1% as unmodified compound and 14% as
conjugated compound (6).
After 400 mg single dose in healthy volunteers one preparation produced peak plasma
concentration of 30.0 μg $ml^{-1}$ at 1.6 hr and the other 23.2 μg $ml^{-1}$ at 2.3 hr (7).
90 and 120 min after dermal application to guinea pig detectable levels were observed
in blood and plasma. Concentrations of 13-228 μg, dependent on depth, were found
in muscle tissue below the application area (8).

#### Sensitisation
A fatal asthma attack occurred in a 65 yr old woman, adult-onset asthma, 30 min after
800 mg (9).
Hypersensitivity reactions may occur and include rashes (10).

## Any other adverse effects to man

Reversible amblyopia has been reported (11,12).
Can cause dyspepsia, nausea and vomiting, gastrointestinal bleeding, peptic ulcers and perforation (13, 14).
Acute renal failure (15-17).
In an evaluation of hepatic toxicity involving 1468 patients with rheumatoid arthritis and osteoarthritis no aspartate aminotransferase elevation was observed (18).

## Any other adverse effects

*In vitro* rat hepatocyte toxic effects at 10-fold therapeutic plasma concentration for 48 hr. Impaired gluconeogenesis from lactate after 6 hr at therapeutic level. 40% inhibition of albumin synthesis after 6 hr exposure to 5-fold therapeutic level (19).

## Any other comments

Central nervous system adverse effects reviewed (20).
Metabolism and pharmacokinetics reviewed (21-23).
No increase in δ-aminolaevulinic acid synthetase activity in rat suggesting the drug is safe to be used by porphyria sufferers (24).
Chemopreventive agent against carcinogenesis in *in vitro* mouse mammary gland (25).

## References

1. *Arzneim.-Forsch.* 1984, **34**, 280
2. *Pharm. Chem. J. (Engl. Transl.)* 1980, **14**, 119
3. *Toxicol. Appl. Pharmacol.* 1969, **15**, 310
4. *Arzneim.-Forsch.* 1977, **27**, 1006
5. Momma, K. et al *Am. J. Obstet. Gynecol.* 1990, **162**(5), 1304-1310
6. *Martindale. The Extra Pharmacopoeia* 30th ed., 1993, The Pharmaceutical Press, London
7. Karttunen, P. et al *Int. J. Clin. Pharmacol. Ther. Toxicol.* 1990, **28**(6), 251-255
8. Giese, U. *Arzneim.-Forsch.* 1990, **40**(1), 84-88
9. Ayres, J. G. et al *Lancet* 1987, **i**, 1082
10. Sternlieb, P. et al *Ann. Intern. Med.* 1980, **92**, 570
11. Collum, L. M. T. et al *Br. J. Ophthalmol.* 1971, **55**, 472-477
12. Palmer, C. A. L. *Br. Med. J.* 1972, **3**, 765
13. Ravi, S. et al *Postgrad. Med. J.* 1986, **62**, 773-776
14. Clements, D. et al *Br. Med. J.* 1990, **301**, 987
15. Brandstetter, R. D. et al *Br. Med. J.* 1978, **2**, 1194-1195
16. Kimberly, R. P. et al *Arthritis Rheum.* 1979, **22**, 281-285
17. Spierto, R. J. et al *Ann. Pharmacother.* 1992, **26**, 714
18. Freeland, G. R. et al *Clin. Pharmacol. Ther. (St. Louis)* 1988, **43**(5), 473-479
19. Castell, J. V. et al *Xenobiotica* 1988, **18**(6), 737-745
20. Hoppmann, R. A. et al *Arch. Intern. Med.* 1991, **151**, 1309-1313
21. Brune, K. et al *Dtsch. Apoth. Ztg.* 1989, **129**(5) (Suppl. 12), 3-15 (Ger.) (*Chem. Abstr.* **110**, 107468a)
22. Dinnendahl, V. *Pharm. Ztg.* 1989, **134**(1), 9-14 (Ger.) (*Chem. Abstr.* **110**, 107412c)
23. Brune, K. *Agents Actions Suppl.* 1988, **25**, 9-19
24. McColl, K. E. L. et al *Ann. Rheum. Dis.* 1987, **46**(7), 540-542
25. Mehta, R. G. et al *Anticancer Res.* 1991, **11**(2), 587-591

# I3  Imazalil

**CAS Registry No.** 35554-44-0
**Synonyms** 1-[2-(2,4-dichlorophenyl)-2-(2-propenyloxy)ethyl]-1$H$-imidazole;
Enilconazole; Florasan; Fungaflor; Fungazil; R23979
**Mol. Formula** $C_{14}H_{14}Cl_2N_2O$       **Mol. Wt.** 297.19
**Uses** Disinfectant of kennel and stable equipment. Fungicide.

## Physical properties

**M. Pt.** 50°C; **B. Pt.** Decomposes on distillation; **Specific gravity** $d^{23}$ 1.243; **Partition coefficient** log $P_{ow}$ 3.89; **Volatility** v.p. $1.18 \times 10^{-6}$ mmHg at 20°C.

## Solubility

Water: 1.4 g $l^{-1}$ at 20°C. Organic solvent: acetone, dichloromethane, methanol, propan-2-ol, toluene, hexane

## Occupational exposure

**Supply classification** harmful.
**Risk phrases** Harmful if swallowed – Irritating to eyes (R22, R36)

## Ecotoxicity

**Fish toxicity**
$LC_{50}$ (duration unspecified) rainbow trout 2.5 mg $l^{-1}$ (1).
$LC_{50}$ (duration unspecified) bluegill sunfish 3.2 mg $l^{-1}$ (1).

**Invertebrate toxicity**
$LC_{50}$ *Daphnia* 3.2 mg $l^{-1}$ (1).
$LC_{50}$ (48 hr) earthworm, contact 12.8 µg $cm^{-3}$ and (14 day) artificial soil test 541 µg $g^{-1}$. Survival in soil test >90% even at levels of 1000 µg $g^{-1}$ (2).

## Mammalian and avian toxicity

**Acute data**
$LD_{50}$ oral rat 227 mg $kg^{-1}$ (3).
$LC_{50}$ (4 hr) inhalation rat 16 mg $l^{-1}$ air (with 200 g $l^{-1}$ emulsifiable concentrate) (1).
$LD_{50}$ dermal rat 4200-4880 mg $kg^{-1}$ (1).
$LD_{50}$ intraperitoneal rat 155 mg $kg^{-1}$ (3).

**Sub-acute data**
$LC_{50}$ (8 day) mallard duck, bobwhite quail 5620-6290 mg $kg^{-1}$ diet (1).
Oral (8 wk) bobwhite quail 0-1000 mg $kg^{-1}$ in feed. Liver weight, hepatic microsomal protein content, induction of cytochrome $P_{450}$, NADPH-cytochrome c reductive activity were unaffected. 7-Ethoxyresorufin or 7-ethoxycoumarin $O$-deethylase activities in liver microsomes were neither induced nor inhibited. At high doses

increased aniline hydroxylase activity was observed, which normalised after a drug-free wk (4).

**Metabolism and pharmacokinetics**
Oral rat 90% eliminated in metabolised form within 4 days (5).

## Legislation
Limited under EC Directive on Drinking Water Quality 80/778/EEC. Pesticides: maximum admissible concentration 0.1 $\mu$g l$^{-1}$ (6).
Included in Schedule 6 (Release into Land: Prescribed Substances) Statutory Instrument No. 472, 1991 (7).
The log $P_{ow}$ value exceeds the European Community recommended level 3.0 (6th and 7th amendments) (8).

## Any other comments
WHO Class II; EPA Toxicity Class II (5).
TDI (human): 0.01 mg kg$^{-1}$ (5).

## References
1. *The Pesticide Manual* 9th ed., 1991, 482-483, British Crop Protection Council, Farnham
2. Van Laemput, L. et al *Ecotoxicol. Environ. Saf.* 1989, **18**(3), 313-320
3. *Arzneim.-Forsch.* 1981, **31**, 309
4. Lavrijsen, K. et al *Pestic. Sci.* 1990, **29**(1), 47-56
5. *The Agrochemicals Handbook* 3rd ed., 1991, RSC, London
6. *EC Directive Relating to the Quality of Water Intended for Human Consumption* 1982, 80/778/EEC, Office for Official Publications of the European Communities, 2 rue Mercier, L-2985 Luxembourg
7. *S. I. 1991 No. 472 The Environmental Protection (Prescribed Processes and Substances) Regulations* 1991, HMSO, London
8. *1967 Directive on Classification, Packaging, and Labelling of Dangerous Substances 67/548/EEC; 6th Amendment EEC Directive 79/831/EEC; 7th Amendment EEC Directive 91/32/EEC* 1991, HMSO, London

# I4    Imazamethabenz-methyl

**CAS Registry No.** 81405-85-8
**Synonyms** 2-[4,5-dihydro-4-methyl-4-(1-methylethyl)-5-oxo- 1*H*-imidazol-2-yl]-4 (or 5)-methylbenzoic acid, methyl ester; Imazamethabenz; AC222293; Assert; Dagger
**Mol. Formula** $C_{16}H_{20}N_2O_3$                    **Mol. Wt.** 288.35
**Uses** Herbicide.

## Physical properties

**M. Pt.** Ranges from 113-122 to 144-153°C depending on source; **Flash point** >93°C (closed cup); **Specific gravity** 0.22; **Partition coefficient** log $P_{ow}$ 1.54 (*p*-isomer), 1.82 (*m*-isomer) (1); **Volatility** v.p. $1.13 \times 10^{-10}$ mmHg at 25°C.

## Solubility

Water: 1.3 (*p*-isomer), 2.2 (*m*-isomer) g $l^{-1}$. Organic solvent: acetone, dimethyl sulfoxide, isopropyl alcohol, methanol, methylene chloride, toluene, dimethyl formamide

## Ecotoxicity

### Fish toxicity
$LC_{50}$ (96 hr) bluegill sunfish, rainbow trout >100 mg $l^{-1}$ (2).

### Invertebrate toxicity
$LC_{50}$ (48 hr) *Daphnia magna* >100 mg $l^{-1}$ (2).

## Environmental fate

### Anaerobic effects
Slowly degraded to free acids in sandy loam and clay loam soils under both aerobic and anaerobic conditions (3).

### Degradation studies
In tank experiments herbicide activity persisted longer in sandy loam than in clay soil, in fallow system. Under cropped conditions persistence was reduced in sandy loam and unaffected in clay soil (3).

### Abiotic removal
Photolytic degradation occurs in water and on soil surfaces. $t_{1/2}$ of residues in the soil is ca. 30-276 days with no accumulation of acid metabolites. Hydrolysis increases with increasing pH (3).

### Absorption
In arable-soil field experiments at recommended and 2 × recommended doses persistence times ranged from 4 to >10 months. Leaching down to 5-15 cm occurred in all trials and to 15-25 cm in several trials (4).

## Mammalian and avian toxicity

### Acute data
$LD_{50}$ oral rat >5000 mg $kg^{-1}$ (2).
$LD_{50}$ oral bobwhite quail, mallard duck >2150 mg $kg^{-1}$ (2).
$LC_{50}$ (duration unspecified) inhalation rat >5.8 mg $l^{-1}$ (2).
$LD_{50}$ dermal rabbit >2000 mg $kg^{-1}$ (2).

### Sub-acute data
$LC_{50}$ (8 day) oral bobwhite quail, mallard duck >5000 mg $kg^{-1}$ in diet (2).

### Metabolism and pharmacokinetics
Oral rat excreted 77% in urine and 13-19% in the faeces within 24 hr. After 48 hr blood and tissue residue levels were <0.05 mg $kg^{-1}$. Milk/tissue and egg/tissue residue levels were low in lactating goats and laying hens, respectively (2).

---

## Irritancy

Non-irritating to skin; reversible rabbit eye irritant (2).

## Legislation

Limited under EC Directive on Drinking Water Quality 80/778/EEC. Pesticides: maximum admissible concentration $0.1 \mu g \, l^{-1}$ (5).
Included in Schedule 6 (Release into Land: Prescribed Substances) Statutory Instrument No. 472, 1991 (6).

## Any other comments

EPA Toxicity Class IV (2).
$LD_{50}$ contact honeybee $>100 \mu g \, bee^{-1}$ (2).
Biology reviewed (7).
Tested against OECD species, wheat and oilseed rape, with the former tolerant (8).

## References

1.  *The Merck Index* 11th ed., 1989, Merck & Co., Inc., Rahway, NJ
2.  *The Agrochemicals Handbook* 3rd ed., 1991, RSC, London
3.  Allen, R. et al *Proc. Br. Crop. Prot. Conf. – Weeds* 1987, 569-576
4.  Nilsson, H. et al *Swed. Crop Prot. Sci.* 1989, **30**(2), 270-277
5.  *EC Directive Relating to the Quality of Water Intended for Human Consumption* 1982, 80/778/EEC, Office for Official Publications of the European Communities, 2 rue Mercier, L-2985 Luxembourg
6.  *S. I. 1991 No.472 The Environmental Protection (Prescribed Processes and Substances) Regulations* 1991, HMSO, London
7.  Los, M. *Pestic. Sci. Biochnol., Proc. Int. Congr. Pestic. Chem., 6th* 1987, 35-42
8.  Richardson, W. G. et al *Tests Agrochem. Cultiv.* 1985, **6**, 148-149

# I5   Imazapyr

**CAS Registry No.** 81334-34-1
**Synonyms** 2-(4-isopropyl-4-methyl-5-oxo-2-imidazolin-2-yl)nicotinic acid; 2-[4,5-dihydro-4-methyl-4-(1-methylethyl)-5-oxo-1*H*-imidazol-2-yl]-3-pyridinecarboxylic acid; AC243997; Arsenal; Assault; Chopper; Pivot
**Mol. Formula** $C_{13}H_{15}N_3O_3$         **Mol. Wt.** 261.28
**Uses** Herbicide.

## Physical properties

**M. Pt.** 169-173°C; **Partition coefficient** log $P_{ow}$ 0.113; **Volatility** v.p. $0.20 \times 10^{-6}$ mmHg at 45°C.

## Solubility

Water: 10-15 g $l^{-1}$ at 25°C.

## Ecotoxicity

### Fish toxicity

$LC_{50}$ (96 hr) rainbow trout, bluegill sunfish, channel catfish >100 mg $l^{-1}$ (1).
$LC_{50}$ (24, 48, 72, 96 hr) Nile tilapia 4.67, 4.63, 4.61, 4.36 ppm, respectively, in static bioassay (2).
$LC_{50}$ (96 hr) silver barb 2.71 ppm static bioassay (2).

### Invertebrate toxicity

$LC_{50}$ (48 hr) *Daphnia* >100 mg $l^{-1}$ (1).

## Environmental fate

### Degradation studies

Residual soil activity ranges from 6 months to 2 yr in temperate climates and 3 to 6 months in the tropics. The major soil residue is the parent compound (1).

### Abiotic removal

Photolytic degradation occurs in aqueous media in sunlight (1).

## Mammalian and avian toxicity

### Acute data

$LD_{50}$ oral bobwhite quail, mallard duck >2150 mg $kg^{-1}$ (1).
$LD_{50}$ oral ♀ mouse, ♂ ♀ rabbit, ♂ ♀ rat >2000->5000 mg $kg^{-1}$ (1).
$LC_{50}$ (duration unspecified) inhalation rat >1.3 mg $l^{-1}$ (1).
$LD_{50}$ dermal rat, rabbit >2000 mg $kg^{-1}$ (1).
$LD_{50}$ intraperitoneal ♂ rat 2500 mg $kg^{-1}$ (1).

### Sub-acute data

$LC_{50}$ (8 day) oral bobwhite quail, mallard duck >5000 mg $kg^{-1}$ in diet (1).
In 21 day dermal study with rabbits no indications at 400 mg $kg^{-1}$ $day^{-1}$ of systemic toxicity (1).
In 13 wk feeding study with rats no effect dose was 10,000 mg $kg^{-1}$ (highest tested) (1).

### Teratogenicity and reproductive effects

At 1000 mg $kg^{-1}$ in rats or 400 mg $kg^{-1}$ in rabbits (highest doses tested) no teratogenic effects observed (1).

### Metabolism and pharmacokinetics

Oral rats 87% excreted within 24 hr. Residual levels in liver and kidney were 0.03 and 0.02 mg $kg^{-1}$, respectively, at 24 hr and <0.01 ppm in both at 192 hr. In muscle, fat tissue and blood residues were <0.01 mg $kg^{-1}$ at both times (1).

### Irritancy

Rabbit eye irritant, mild rabbit skin irritant (doses and durations unspecified) (1).

## Legislation

Limited under EC Directive on Drinking Water Quality 80/778/EEC. Pesticides: maximum admissible concentration $0.1 \mu g \, l^{-1}$ (3).
Included in Schedule 6 (Release into Land: Prescribed Substances) Statutory Instrument No. 472, 1991 (4).

## Any other comments

$LD_{50}$ contact honeybee $>100 \mu g \, bee^{-1}$ (1).
WHO Class Table 5; EPA Toxicity Class IV (1).

## References

1. *The Agrochemicals Handbook* 3rd ed., 1991, RSC, London
2. Supamataya, K. et al *Warasan Songkhla Nakkharin* 1987, **9**(3), 309-313 (Thai.) (*Chem. Abstr.* **110**, 207524j)
3. *EC Directive Relating to the Quality of Water Intended for Human Consumption* 1982, 80/778/EEC, Office for Official Publications of the European Communities, 2 rue Mercier, L-2985 Luxembourg
4. *S. I. 1991 No. 472 The Environmental Protection (Prescribed Processes and Substances) Regulations* 1991, HMSO, London

# I6   Imazethapyr

**CAS Registry No.** 81335-77-5
**Synonyms** (*RS*)-5-ethyl-2-(4-isopropyl-4-methyl-5-oxo-2-imidazolin-2-yl)nicotinic acid; (±)-2-[4,5-dihydro-4-methyl-4-(1-methylethyl)-5-oxo-1*H*-imidazol-2-yl]-5-ethyl-3-pyridinecarboxylic acid; AC263,499; Event; Pivot; Pursuit
**Mol. Formula** $C_{15}H_{19}N_3O_3$          **Mol. Wt.** 289.34
**Uses** Herbicide.

## Physical properties

**M. Pt.** 172-175°C; **Partition coefficient** log $P_{ow}$ 1.04 at pH5;
**Volatility** v.p. $<1.0 \times 10^{-7}$ mmHg at 60°C.

## Solubility

Water: $1.4 \, g \, l^{-1}$ at 25°C. Organic solvent: acetone, methanol, toluene, dichloromethane, dimethyl sulfoxide, isopropanol

## Ecotoxicity

### Fish toxicity
$LC_{50}$ (96 hr) bluegill sunfish 420 mg $l^{-1}$ (1).
$LC_{50}$ (96 hr) rainbow trout 340 mg $l^{-1}$ (1).

## Environmental fate

### Degradation studies
$t_{1/2}$ in soil 1-3 months (1).
*In vitro* $^{14}$C-label experiments showed 95% unaltered after 12 wk incubation in sterilised soil. In unsterilised soil the major degradation product was $^{14}CO_2$ and adsorption was negatively correlated with degradation (2).

### Abiotic removal
$t_{1/2} \approx 3$ day in sunlight (1).
In $^{14}$C label experiments volatilisation losses for soil were <2%. Photodecomposition losses of up to 8% occurred from soil. The total $^{14}CO_2$ evolved from soils ranged from 2.4 to 3.6% of total applied. However, as detected by high performance liquid chromatography 62-82% had been degraded (3).

### Absorption
In laboratory and greenhouse experiments lower pH reduced effectiveness and mobility and increased adsorption. Adsorption was greatest in silty clay loam and least in sandy loam soil (4-6).

## Mammalian and avian toxicity

### Acute data
$LD_{50}$ oral rat, ♀ mouse >5000 mg $kg^{-1}$ (1).
$LD_{50}$ oral bobwhite quail, mallard duck >2150 mg $kg^{-1}$ (1).
$LD_{50}$ dermal rabbit >2000 mg $kg^{-1}$ (1).

### Sub-acute data
In 28 day feeding study with rats no effect dose was 10,000 mg $kg^{-1}$ (highest tested) (1).

### Metabolism and pharmacokinetics
Oral rat 92% excreted in urine and 5% in faeces within 24 hr. Residues in liver, kidney, muscle, fat tissue and blood were <0.01 ppm after 48 hr (1).

### Irritancy
Mild skin irritation, reversible eye irritation in rabbit (1).

## Legislation
Limited under EC Directive on Drinking Water Quality 80/778/EEC. Pesticides: maximum admissible concentration 0.1 µg $l^{-1}$ (7).
Included in Schedule 6 (Release into Land: Prescribed Substances) Statutory Instrument No. 472, 1991 (8).

## Any other comments
$LD_{50}$ contact honeybee >0.1 mg $bee^{-1}$ (1).
WHO Class Table 5; EPA Toxicity Class III (1).

# References

1. *The Agrochemicals Handbook* 3rd ed., 1991, RSC, London
2. Cantwell, J. R. et al *Weed Sci.* 1989, **37**(6), 815-819
3. Goetz, A. J. et al *Weed Sci.* 1990, **38**(4-5), 421-428
4. Stougaard, R. N. et al *Weed Sci.* 1990, **38**(1), 63-73
5. Loux, M. M. et al *Weed Sci.* 1989, **37**(5), 712-718
6. Loux, M. M. et al *Weed Sci.* 1989, **37**(2), 259-267
7. *EC Directive Relating to the Quality of Water Intended for Human Consumption* 1982, 80/778/EEC, Office for Official Publications of the European Communities, 2 rue Mercier, L-2985 Luxembourg
8. *S. I. 1991 No. 472 The Environmental Protection (Prescribed Processes and Substances) Regulations* 1991, HMSO, London

# I7   Imidazole

**CAS Registry No.** 288-32-4
**Synonyms** 1*H*-imidazole; 1,3-diazole; Imutex; Miazole; Glyoxaline
**Mol. Formula** $C_3H_4N_2$                    **Mol. Wt.** 68.08

## Physical properties

**M. Pt.** 90-91°C; **B. Pt.** 257°C; **Flash point** 145°C.

**Solubility**
Water: freely soluble. Organic solvent: ethanol, diethyl ether, chloroform, pyridine

## Ecotoxicity

**Invertebrate toxicity**
$EC_{50}$ (30 min) *Photobacterium phosphoreum* 231 ppm Microtox test (1).

## Mammalian and avian toxicity

**Acute data**
$LD_{50}$ oral mouse 1880 mg $kg^{-1}$ (2).
$LD_{50}$ subcutaneous mouse 817 mg $kg^{-1}$ (3).
$LD_{50}$ intraperitoneal mouse 610 mg $kg^{-1}$ (2).
$LD_{50}$ intravenous mouse 475 mg $kg^{-1}$ (4).

## Any other comments

*In vitro* 19-day-old Sprague-Dawley rat foetus brain cells incubated for 3 days then exposed for a further 3 days to 140 mg $l^{-1}$ showed no cytoxic effects (5).
Experimental toxicology, human health effects, workplace experience reviewed (6).

**References**

1. Kaiser, K. L. E. et al *Water Pollut. Res. J. Can.* 1991, **26**(3), 361-431
2. Nishie et al *Toxicol. Appl. Pharmacol.* 1969, **14**, 301
3. *J. Pharmacol. Exp. Ther.* 1957, **119**, 444
4. *Arzneim.-Forsch.* 1983, **33**, 716
5. Khera, K. S. et al *Toxicol. in Vitro* 1988, **2**(4), 257-273
6. *ECETOC Technical Report No. 30(4)* 1991, European Chemical Industry Ecology and Toxicology Centre, B-1160 Brussels

# I8   3,3′-Iminobis[propanenitrile]

## NH(CH₂CH₂CN)₂

**CAS Registry No.** 111-94-4
**Synonyms** 3,3′-iminodi-propionitrile; BBCE; bis(cyanoethyl)amine; IDPN; β,β′-iminodipropionitrile
**Mol. Formula** $C_6H_9N_3$ **Mol. Wt.** 123.16

### Physical properties

**M. Pt.** –5.5°C; **B. Pt.** 173°C; **Specific gravity** $d^{30}$ 1.0165; **Volatility** v. den. 3.3.

### Mammalian and avian toxicity

**Acute data**
$LD_{50}$ oral rat 2700 mg kg$^{-1}$ (1).
$LD_{50}$ dermal rabbit 2520 mg kg$^{-1}$ (2).

**Sub-acute data**
Gavage (5 wk) Sprague-Dawley rats 50 or 125 mg kg$^{-1}$ day$^{-1}$ for 7 days wk$^{-1}$ affected body weight, weight gain, food consumption, general activity, muscle tone, urination, motor reflexes, posture and gait. The ♂ in particular suffered significant sex effects. Minimal to moderately severe axonal swelling was observed in the brainstem and the cervical and lumbar regions of the spinal cord. This was most pronounced in the ♂. An increased incidence of chronic progressive nephropathy also occurred (3).
Mice that received 1 g kg$^{-1}$ (route unspecified) 3 × wk$^{-1}$ for 6 wk showed an increase in locomotor acitivity in wk 1 and 2, a decrease in acetylcholinesterase activity in the central nervous system, decreased diameter of neuronal cell bodies and an accumulation of bundles of 10 nm neurofilaments in the proximal axons (4).

### Any other adverse effects

Cat (5 wk) exhibited neurotoxic effects including swellings in intraparenchymal spinal axons and clumping of neurofilaments in some motor neurones (5).
A mechanism for neurotoxicity involving cyanoethenylation of amino groups is described (6).

### References
1. *J. Ind. Hyg. Toxicol.* 1949, **31**, 60
2. *Arch. Ind. Hyg. Occup. Med.* 1954, **10**, 61

3. Schulze, G. E. et al *Fundam. Appl. Toxicol.* 1991, **16**(3), 602-615
4. Yano, I. et al *Wakayama Med. Rep.* 1990, **32**(1), 1-9
5. Fiori, M. G. et al *Cytol. Pathol.* 1988, **20**(1), 137-146
6. Jacobson, A. R. et al *Mol. Toxicol.* 1987, **1**(1), 17-34

# I9   3,3′-Iminobis-1-propanol dimethanesulfonate (ester) hydrochloride

$(CH_3SO_3CH_2CH_2CH_2)_2N^+H_2Cl^-$

**CAS Registry No.** 3458-22-8
**Synonyms** Compound 864; Yoshi 864
**Mol. Formula** $C_8H_{20}ClNO_6S_2$                    **Mol. Wt.** 325.83
**Uses** Chemotherapeutic agent.

## Mammalian and avian toxicity

### Acute data

$LD_{50}$ intravenous rat 75 mg kg$^{-1}$ (1).
$LD_{10}$ intraperitoneal mouse 170 mg kg$^{-1}$ (2).

### Carcinogenicity and long-term effects

National Toxicology Program tested rats and mice via intraperitoneal injection. Equivocal evidence of carcinogenicity (3).

### References

1. *Arzneim.-Forsch.* 1974, **24**, 1139
2. *J. Med. Chem.* 1977, **20**, 515
3. *National Toxicology Program Research and Testing Division* 1992, Report No.TR-018, NIEHS, Research Triangle Park, NC

# I10   Iminoctadine

$H_2NC(=NH)NH(CH_2)_8NH(CH_2)_8NHC(=NH)NH_2$

**CAS Registry No.** 13516-27-3
**Synonyms** *N,N′′′*-(iminodi-8,1- octanediyl)bis-guanidine; bis(8-guanidinooctyl)amine; Mitrol; Panoctin; Panoctine
**Mol. Formula** $C_{18}H_{41}N_7$                    **Mol. Wt.** 355.57
**Uses** Fungicide.

## Physical properties

**M. Pt.** 143.0-144.2 (triacetate); **Volatility** v.p. $<3.0 \times 10^{-6}$ mmHg at 23°C.

## Solubility

Water: 764 g l$^{-1}$ (triacetate). Organic solvent: ethanol, methanol (triacetate)

## Occupational exposure

**Supply classification** harmful.
**Risk phrases** Harmful in contact with skin and if swallowed – Irritating to eyes and skin (R21/22, R36/38)
**Safety phrases** Wear suitable protective clothing and gloves (S36/37)

## Ecotoxicity

**Fish toxicity**
LC$_{50}$ (96 hr) carp 200 mg l$^{-1}$ (1).
LC$_{50}$ (96 hr) rainbow trout 36 mg l$^{-1}$ (1).

**Invertebrate toxicity**
EC$_{50}$ (48 hr) *Daphnia* 2.1 mg l$^{-1}$ (1).

## Environmental fate

**Nitrification inhibition**
Slight inhibitory effect on nitrification (2).

**Degradation studies**
t$_{1/2}$: diluvial sandy loam, 90 day; volcanic ash loamy upland soil, 122 day; colluvial clayey loamy upland soil, 75 day; volcanic ash loamy upland soil, 28 day (1).
Effects on soil microorganisms studied (2).

## Mammalian and avian toxicity

**Acute data**
LD$_{50}$ oral mallard duck 985 mg kg$^{-1}$ (1).
LD$_{50}$ oral rat, mouse 300-400 mg kg$^{-1}$ (1).
LC$_{50}$ (4 hr) inhalation rat 0.073 mg l air$^{-1}$ (for 25% liquid) (1).
LD$_{50}$ dermal rat 1500 mg kg$^{-1}$ (1).

**Teratogenicity and reproductive effects**
Non-toxic to embryos (species unspecified) at 8 mg kg$^{-1}$ (1).

**Irritancy**
Mild skin and eye irritant (rats) (1).

**Sensitisation**
Not a skin sensitiser (rats) (1).

## Any other adverse effects

Intravenous rabbit 3 mg kg$^{-1}$ transiently decreased blood pressure, followed by an elevation and then a continued depression. At ≤0.3 mg kg$^{-1}$ a mild transient hypotension was observed. Heart rate was dose dependently increased (3).

## Legislation

Limited under EC Directive on Drinking Water Quality 80/778/EEC. Pesticides: maximum admissible concentration 0.1 μg l$^{-1}$ (4).

---

Included in Schedule 6 (Release into Land: Prescribed Substances) Statutory Instrument No. 472, 1991 (5).

**Any other comments**

$LD_{50}$ oral honeybee >0.1 mg bee$^{-1}$ (1).
$LD_{50}$ contact honeybee >0.1 mg bee$^{-1}$ (1).

**References**

1.  *The Pesticide Manual* 9th ed., 1991, British Crop Protection Council, Farnham
2.  Torstensson, L. *Vaextskyddsrapp., Jordbruk* 1988, **49**, 165-172 (Swed.)
3.  Masaoka, T. et al *Azabu Daigaku Juigakubu Kenkyu Hokoku* 1985, **6**(2), 99-104
4.  *EC Directive Relating to the Quality of Water Intended for Human Consumption* 1982, 80/778/EEC, Office for Official Publications of the European Communities, 2 rue Mercier, L-2985 Luxembourg
5.  *S. I. 1991 No. 472 The Environmental Protection (Prescribed Processes and Substances) Regulations* 1991, HMSO, London

# I11  Iminodiacetic acid

$NH(CH_2CO_2H)_2$

**CAS Registry No.** 142-73-4
**Synonyms** *N*-(carboxymethyl)glycine; aminodiacetic acid; diglycine; diglykokoll; IDA; iminobis(acetic acid)
**Mol. Formula** $C_4H_7NO_4$          **Mol. Wt.** 133.10

**Physical properties**
**M. Pt.** 243°C (decomp.).

**Solubility**
Water: 2.43 g 100 ml$^{-1}$.

**Mammalian and avian toxicity**

**Acute data**
$LD_{50}$ intraperitoneal mouse 250 mg kg$^{-1}$ (1).

**Any other comments**
As a chelating agent for human serum albumin-bound copper(II) the reaction followed a process involving intermediate tertiary complexes (2).

**References**

1.  *NTIS Report* AD277-689, Natl. Tech. Inf. Ser. Springfield, VA
2.  Gao, L. et al *J. Inorg. Biochem.* 1989, **36**(2), 83-92

# I12    Iminodibenzyl

CAS Registry No. 494-19-9
Synonyms 10,11-dihydro-5*H*-dibenz[*b*,*f*]azepine; RP2 3669
Mol. Formula $C_{14}H_{13}N$                     Mol. Wt. 195.27

## Physical properties
M. Pt. 105-108°C.

## Mammalian and avian toxicity

**Acute data**
$LD_{50}$ intravenous mouse 320 mg $kg^{-1}$ (1).

**Irritancy**
100 mg instilled into rabbit eye caused well-defined erythema and slight oedema (2).

## References

1.   U.S. Army Armament Research and Development Command *Report NX No. 01352*,
     Chemicals Systems Laboratory, Aberdeen Proving Ground, MD
2.   *Food Chem. Toxicol.* 1982, **20**, 573

# I13    Imipramine

$(CH_2)_3N(CH_3)_2$

CAS Registry No. 50-49-7
Synonyms 10,11-dihydro-*N*,*N*-dimethyl-5*H*-dibenz[*b*,*f*] azepine-5-propanamine;
5-[3-(dimethylamino)propyl]-10,11-dihydro-5*H*-dibenz[*b*,*f*] azepine; Antideprin;
Prazepine
Mol. Formula $C_{19}H_{24}N_2$                     Mol. Wt. 280.42
Uses Antidepressant drug.

## Physical properties
M. Pt. 173-175°C; B. Pt. $_{0.1}$ 160°C.

## Mammalian and avian toxicity

### Acute data
$LD_{Lo}$ oral human 30-40 mg $kg^{-1}$ (1,2).
$LD_{50}$ oral mouse, rat 188-250 mg $kg^{-1}$ (3).
$LD_{50}$ intravenous rat, mouse 16-21 mg $kg^{-1}$ (4,5).
$LD_{50}$ intraperitoneal mouse, rat 52-79 mg $kg^{-1}$ (6-8).
$LD_{50}$ subcutaneous mouse 195 $\mu g$ $kg^{-1}$ (7).
$LD_{50}$ subcutaneous rat 250 mg $kg^{-1}$ (8).

### Teratogenicity and reproductive effects
Negative developmental toxicity for primate, rat and mouse. Equivocal in rabbit (9). Subcutaneous pregnant rats 3, 5, 10 mg $kg^{-1}$ $day^{-1}$ on days 8-20 of pregnancy. There were no dose-related differences in offspring body weights, but maternal weight gain declined in a dose-related fashion. The functional development of the central adrenergic systems was altered in a complicated way (10).

### Metabolism and pharmacokinetics
In a study of the biliary excretion of the compound and its metabolites, desipramine, 2-OH-imipramine and 2-OH-desipramine, rats were given an intraperitoneal daily dose of 10 mg $kg^{-1}$ over 13 days. An inhibition of imipramine demethylation and hydroxylation and acceleration of desipramine hydroxylation was suggested (11). Oral rat (2 wk) 20 mg $kg^{-1}$ $day^{-1}$ decreased imipramine hydroxylase activity and slightly reduced imipramine demethylase activity. Desipramine competitively inhibited imipramine metabolism (12).
Dermal hairless mouse 2 mg in distilled water followed by rapid evaporation of the water. After 1, 2, 4 or 6 hr the highest levels were recorded in the lung and the lowest in the heart and liver. Levels in blood were similar to low therapeutic to toxic concentrations in humans, whereas solid tissue levels were much lower than those observed in human overdose (13).

### Sensitisation
Urticaria, angioedema and other allergic skin reactions and photosensitisation reported in humans (14-16).
Agranulocytosis, particularly affecting the elderly, occurred 4 to 8 wk after starting treatment (17,18).

## Genotoxicity
*In vivo* polychromatic erythrocytes 1.87 and 2.81 mg $mouse^{-1}$ induced dose-related increase in frequency of micronuclei (19).
*Drosophila melanogaster* fed oral dose for 48 hr in Somatic Mutation and Recombination Test, positive for concentrations >280 mg (20).
*In vitro* human lymphocyte cells (24, 48, 72 hr) 25, 500 and 5000 ng $ml^{-1}$ gave a significant increase in chromosome damage at the upper plasma level and at higher concentrations. Only at levels higher than the plasma level (5 $\mu g$ $ml^{-1}$) were sister chromatid exchanges significantly increased. At no concentration was the mitotic index affected (21).

## Any other adverse effects to man
Volunteers given 100 and 40 mg, orally, on separate occasions displayed dose-related rises in blood pressure and resting heart rate. No appreciable effect was exhibited by

heart rate response to Valsalva's manoeuvre, respiratory sinus arrhythmia or the responses to exercise (22).

Patients without prior cardiovascular disease given therapeutic doses showed significant cardiovascular side effects of orthostatic hypotension and tachycardia (23,24).

Oral ♂ human 100 mg increased circulating levels of cortisol, prolactin and growth hormone. A dose of 40 mg had no effect on hormone release (25).

## Any other adverse effects

Intravenous rat decreased mean arterial blood pressure and heart rate and caused electrocardiogram rhythm and conduction changes (26).

## Any other comments

Sexual dysfunction reviewed (27-29).

Teratogenicity, metabolism and pharmacokinetics reviewed (30-32).

0.88 g is approximately equivalent to 1 g of imipramine hydrochloride (33).

## References

1.  *Handb. Exp. Pharmacol.* 1980, **55**, 527
2.  *Proc. Eur. Soc. Study Drug Toxicol.* 1965, **6**, 171
3.  *Pharm. Chem. J. (Engl. Transl.)* 1980, **14**, 773
4.  *Arzneim.-Forsch.* 1979, **29**, 193
5.  *Arch. Int. Pharmacodyn. Ther.* 1980, **245**, 283
6.  *British Patent Document No.1460700*
7.  *Pharm. Chem. J. (Engl. Transl.)* 1981, **15**, 412
8.  *Arch. Int. Pharmacodyn. Ther.* 1964, **148**, 560
9.  Jelovsek, F. R. et al *Obstet. Gynecol. (N. Y.)* 1989, **74**(4), 624-636
10. Ali, S. F. et al *Neurotoxicology* 1986, **7**(2), 365-380
11. Melzacka, M. et al *Pol. J. Pharmacol. Pharm.* 1987, **39**(4), 379-386
12. Wladyslawa, D. et al *Pol. J. Pharmacol. Pharm.* 1987, **39**(2), 135-141
13. Bailey, D. N. *J. Anal. Toxicol.* 1990, **14**(4), 217-218
14. Almeyda, J. *Br. J. Dermatol.* 1971, **84**, 298-299
15. Quitkin, F. *JAMA, J. Am. Med. Assoc.* 1979, **241**, 1625
16. Smith, A. G. *Adverse Drug React. Bull.* 1989, **136**, 508-511
17. Heimpel, H. *Med. Toxicol.* 1988, **3**, 449-462
18. Albertini, R. S. et al *J. Clin. Psychiatry* 1978, **39**, 483-485
19. Shaheen, S. et al *IRCS Med. Sci.* 1986, **14**(11), 1129-1130
20. Van Schaik, N. et al *Mutat. Res.* 1991, **260**(1), 99-104
21. Saxena, R. et al *Environ. Mol. Mutagen.* 1988, **12**(4), 421-430
22. Middleton, H. C. et al *Hum. Psychopharmacol.* 1988, **3**(3), 181-190
23. Glassman, A. H. *Annu. Rev. Med.* 1984, **35** 503-511
24. Mortensen, S. A. *Practitioner* 1984, 288, 1180-1183
25. Nutt, D. et al *Psychoneuroendocrinology (Oxford)* 1987, **12**(5), 367-375
26. Balcioglu, A. et al *Arch. Int. Pharmacodyn. Ther.* 1991, **309**, 64-74
27. Beeley, L. *Adverse Drug React. Poisoning Rev.* 1984, **3**, 23-42
28. *Med. Lett. Drugs Ther.* 1987, **29**, 65-70
29. Shen, W. W. et al *J. Reprod. Med.* 1983, **28**, 497-499
30. De Vane, C. L. et al *Psychopharmacol. Ser.* 1987, **3**, 174-178
31. Sallee, F. R. et al *Clin. Pharmacokinet.* 1990, **18**(5), 346-364
32. Gram, L. F. *Acta Psychiatr. Scand. Suppl.* 1988, **345**, 81-84
33. *Martindale. The Extra Pharmacopoeia* 30th ed., 1993, 255, The Pharmaceutical Press, London

# I14   Imipramine hydrochloride

$(CH_2)_3\overset{+}{N}H(CH_3)_2 \quad Cl^-$

**CAS Registry No.** 113-52-0
**Synonyms** Berkomine; Deprinol; Tofranil; Pryleugau; Janimine
**Mol. Formula** $C_{19}H_{25}ClN_2$            **Mol. Wt.** 316.88
**Uses** Antidepressant drug.

## Physical properties

**M. Pt.** 174-175°C.

**Solubility**
Water: freely soluble. Organic solvent: ethanol, chloroform, acetone

## Mammalian and avian toxicity

**Acute data**
$LD_{50}$ oral mouse, rat 275-305 mg kg$^{-1}$ (1,2).
$LD_{Lo}$ oral child 15 mg kg$^{-1}$ (3).
$LD_{50}$ intraperitoneal rat, guinea pig, mouse 72-104 mg kg$^{-1}$ (4-6).
$LD_{Lo}$ intravenous monkey 25 mg kg$^{-1}$ (7).

**Teratogenicity and reproductive effects**
Gavage ICR/SIM mouse on gestation days 8-12 non-teratogenic (8).

**Metabolism and pharmacokinetics**
Readily absorbed from the human small intestine. Considerably demethylated by
first-pass metabolism in the liver to desipramine, the primary active metabolite.
The metabolic pathways include hydroxylation and *N*-oxidation. Excretion occurs via
the urine, mainly as free or conjugated metabolites. Distributed throughout the body
and extensively bound to plasma and tissue protein. Elimination $t_{1/2}$ 9-28 hr (higher in
overdose). Can cross the blood-brain and placental barriers and is excreted in
breast-milk (9).
As a suppository, bioavailability in rabbit was 25% compared to intravenous
injection. When tested on 3 inpatients the plasma level and clinical efficiency were in
the same range as reported in literature for oral administration (10).

## Any other adverse effects to man

Agranulocytosis reported (11).
Orthostatic hypotension and tachycardia are the significant cardiovascular side
effects (12,13).
Sexual function effects, including painful ejaculation and anorgasmia or delayed
orgasm in women reported (14-16).
Some panic disorder patients exhibit symptoms of insomnia, jitteriness and
irritability (17).

Cold and blue hands and feet reported for a woman at dose of 150 mg daily (18).

## Any other comments
Pharmacokinetics reviewed (19).

## References

1. *Therapie* 1965, **20**, 67
2. *Toxicol. Appl. Pharmacol.* 1971, **18**, 185
3. *Br. Med. J.* 1974, **1**, 261
4. *Arzneim.-Forsch.* 1971, **21**, 391
5. *Pharmazie* 1983, **38**, 749
6. *Arzneim.-Forsch.* 1974, **24**, 166
7. *Indian J. Exp. Biol.* 1984, **22**, 539
8. Seidenberg, J. M. et al *Teratog., Carcinog., Mutagen.* 1987, **7**, 17-28
9. *Martindale. The Extra Pharmacopoeia* 30th ed., 1993, 255-256, The Pharmaceutical Press, London
10. Chaumeil, J. C. et al *Drug Dev. Ind. Pharm.* 1988, **14**(15-17), 2225-2239
11. Heimpel, H. *Med. Toxicol.* 1988, **3**, 449-462
12. Glassman, A. H. *Annu. Rev. Med.* 1984, **35**, 503-511
13. Mortensen, S. A. *Practitioner* 1984, **228**, 1180-1183
14. Beeley, L. *Adverse Drug React. Acute Poisoning Rev.* 1984, **3**, 23-42
15. *Med. Lett. Drugs Ther.* 1987, **29**, 65-70
16. Shen, W. W. et al *J. Report. Med.* 1983, **28**, 497-499
17. Yeragani, V. K. et al *Br. Med. J.* 1986, **292**, 1529
18. Appelbaum, P. S. et al *Am. J. Psychiatry* 1983, **140**, 913-915
19. Task Force on the Use of Laboratory Tests in Psychiatry *Am. J. Psychiatry* 1985, **142**, 155-162

# I15   Indan

**CAS Registry No.** 496-11-7
**Synonyms** 2,3-dihydro-1*H*-indene; benzocyclopentane; hydrindene; hydrindonaphthene
**Mol. Formula** $C_9H_{10}$                     **Mol. Wt.** 118.18
**Uses** Component of fuels, solvents and varnishes.
**Occurrence** In coal tar. Water pollutant (1).

## Physical properties

**M. Pt.** –51.4°C; **B. Pt.** 176.5°C; **Flash point** 50°C; **Specific gravity** $d_4^{20}$ 0.9639; **Partition coefficient** log $P_{ow}$ 3.33.

**Solubility**
Organic solvent: miscible in ethanol, diethyl ether

---

## Mammalian and avian toxicity

### Sub-acute data

Intragastric Fischer 344 rat (dose unspecified) alternate days for 14 days. In ♂ rats toxic injury shown by increased cytoplasmic hyaline droplets in proximal convoluted tubular epithelial cells. ♀ rats showed no renal damage (2).

### Metabolism and pharmacokinetics

In ♂ Fischer 344 rats metabolized to 1-, 2- and 5-indanol, 1- and 2-indanone, 2- and 3-hydroxy-1-indanone and *cis*- and *trans*-indan-1,2-diol (3,4).

## Any other adverse effects

♂, but not ♀, Fischer 344 rats, exhibit lesions to the kidney typical of cyclic hydrocarbons (3,4).

## References

1. Wang, Y. et al *Beijing Daxue Xuebao Ziran Kexueban* 1988, **24**(2), 163-166 (Ch.) (*Chem. Abstr.* **109**, 79296b)
2. Serve, M. P. *Gov. Rep. Announce. Index (U. S.)* 1989, **89**(19), Abstr. No.952, 011
3. Serve, M. P. et al *J. Toxicol. Environ. Health* 1990, **29**(4), 409-416
4. Yu, K. O. et al *Biomed. Environ. Mass Spectrom.* 1987, **14**(11), 649-651

# I16   1,3-Indandione

**CAS Registry No.** 606-23-5
**Synonyms** 1*H*-indene-1,3(2*H*)-dione; 1,3-diketohydrindene
**Mol. Formula** $C_9H_6O_2$                    **Mol. Wt.** 146.15

## Physical properties

**M. Pt.** 129-131°C; **Specific gravity** $d^{21}$ 1.37.

### Solubility

Organic solvent: hot ethanol, benzene

## Mammalian and avian toxicity

### Acute data

$LD_{Lo}$ intraperitoneal mouse 100 mg $kg^{-1}$ (1).

## References

1. *Arch. Toxicol.* 1975, **33**, 191

## I17  5-Indanol

HO. [structure]

**CAS Registry No.** 1470-94-6
**Synonyms** 2,3-dihydro-1*H*-inden-5-ol; 5-hydroxyhydrindene; 5-hydroxyindan
**Mol. Formula** $C_9H_{10}O$ **Mol. Wt.** 134.18

**Physical properties**
**M. Pt.** 51-53°C; **B. Pt.** 255°C; **Flash point** >110°C.

**Solubility**
Organic solvent: ethanol, diethyl ether

## Mammalian and avian toxicity

**Acute data**
$LD_{50}$ oral rat 3250 mg kg$^{-1}$ (1).
$LD_{50}$ dermal rabbit 450 mg kg$^{-1}$ (1).

**Any other comments**
A urinary metabolite of indan (2,3).

### References

1. *Am. Ind. Hyg. Assoc. J.* 1962, **23**, 95
2. Yu, K. et al *Biomed. Environ. Mass. Spectrom.* 1987, **14**(11), 649-651
3. Serve, M. P. et al *J. Toxicol. Environ. Health* 1990, **29**(4), 409-416

## I18  Indazole

[structure]

**CAS Registry No.** 271-44-3
**Synonyms** 1*H*-indazole; 2-azaindole; 1*H*-benzopyrazole; 1,2-diazaindene; isoindazole
**Mol. Formula** $C_7H_6N_2$ **Mol. Wt.** 118.14

**Physical properties**
**M. Pt.** 146.5°C; **B. Pt.** $_{743}$ 267-270°C.

**Solubility**
Organic solvent: ethanol, diethyl ether

## Mammalian and avian toxicity

**Acute data**
$LD_{50}$ intraperitoneal mouse 440 mg kg$^{-1}$ (1).

## References

1.  *J. Med. Chem.* 1963, **6**, 480.

# I19   1*H*-Indene

**CAS Registry No.** 95-13-6
**Synonyms** indonaphthene; indene; inden
**Mol. Formula** $C_9H_8$                                    **Mol. Wt.** 116.16
**Uses** In paint and coating manufacture. In tile manufacture. Chemicals synthesis
intermediate. In the production of coumarine-indene resins.
**Occurrence** From tars of coal, lignite and crude petroleum.

## Physical properties

**M. Pt.** −1.8°C; **B. Pt.** 181.6°C; **Specific gravity** $d_4^{20}$ 0.9968; **Partition coefficient**
log $P_{ow}$ 2.92.

## Occupational exposure

**US TLV (TWA)** 10 ppm (48 mg m$^{-3}$); **UK Long-term limit** 10 ppm (45 mg m$^{-3}$);
**UK Short-term limit** 15 ppm (70 mg m$^{-3}$).

## Ecotoxicity

**Fish toxicity**
$LC_{50}$ (1 hr) fathead minnow 39 mg l$^{-1}$ static bioassay in Lake Superior water at 18-22°C (1).
$LC_{50}$ (24-96 hr) fathead minnow 14 mg l$^{-1}$ static bioassay in Lake Superior water at
18-22°C (1).

## Legislation

Limited under EC Directive on Drinking Water Quality 80/778/EEC. Polycylic
aromatic hydrocarbons: maximum admissible concentration 0.2 µg l$^{-1}$ (2).

## Any other comments

Study of oil- and creosote-associated compounds under aerobic and anaerobic
biodegradation showed that polycyclic aromatic hydrocarbons with ≥ 4 aromatic rings
are degraded slowly (3).
Experimental toxicology and human health effects reviewed (4).

## References

1.  Mattson, V. R. et al *Acute Toxicity of Selected Organic Compounds to Fathead Minnows*
    1976, EPA-600/3-76-097
2.  *EC Directive Relating to the Quality of Water Intended for Human Consumption* 1982,
    80/778/EEC, Office for Official Publications of the European Communities, 2 rue Mercier,
    L-2985 Luxembourg

3.  Arvin, E. et al *Int. Conf. Physiochemical Biol. Detoxif. Hazard. Wastes* 1989, **2**, 828-847
4.  *ECETOC Technical Report No. 30(4)* 1991, European Chemical Industry Ecology and Toxicology Centre, B-1160 Brussels

# I20   Indeno[1,2,3-*cd*]pyrene

**CAS Registry No.** 193-39-5
**Synonyms** *o*-phenylenepyrene; 1,10-(*o*-phenylene)pyrene; 1,10-(*o*-phenylene)pyrene
**Mol. Formula** $C_{22}H_{12}$                           **Mol. Wt.** 276.34
**Occurrence** In fresh motor oil 0.03 mg $kg^{-1}$, used motor oil after 10,000 km
46.7-83.2 mg $kg^{-1}$ and petrol 0.04-0.18 mg $kg^{-1}$ (1).
In exhaust gases of petrol-engine cars 11-87 µg $m^{-3}$ (2).
In coke oven emissions 101.5 µg g $sample^{-1}$ (3).
Cigarette smoke 0.4 µg 100 $cigarettes^{-1}$ (4).
In groundwater 0.2-1.8 µg $m^{-3}$ and tapwater 0.9-3.0 µg $m^{-3}$ from man-made sources (5).

## Physical properties
**M. Pt.** 160-163°C; **B. Pt.** 536°C; **Volatility** v.p. $1.0 \times 10^{-10}$ mmHg at 20°C.

## Solubility
Water: 62 µg $l^{-1}$.

## Ecotoxicity

### Bioaccumulation
Levels of 160, 13 and 1.5 ng g dry $wt^{-1}$ in: seston; *Mytilus edulis* soft parts; and eider duck gallbladder, respectively, from the Baltic Sea. Concentration in eider duck: gall bladder >adipose tissue ≥ liver; on lipid weight basis: gall bladder >liver >adipose tissue (6).

## Environmental fate

### Degradation studies
$t_{1/2}$ 224-408 day in sandy loam soil (7).
$t_{1/2}$ at 10, 20, 30°C respectively 600, 730, 630 days in unacclimated agricultural sandy loam soil (8).
1.9 µg $ml^{-1}$ contaminated groundwater from creosote works site inoculated with indigenous adapted microorganisms sampled at 1, 3, 5, 8 and 14 days gave levels of 1.3, 1.4, 1.4, 1.2 and 0.9 µg $ml^{-1}$, respectively (9).

### Abiotic removal
Anaerobic sludge digestion, batch incubation with and without $NaN_3$ to arrest biological activity. Significant removal observed over 32 days. Removal was concluded to be non-biological (10).

After 48 hr incubation in sandy loam soil 11.5-13.5% removal by abiotic means. Volatilisation was negligible (7).

Following two-stages (powdered and granular) activated carbon treatment 98% removal from river water (60.4 ng l$^{-1}$) giving a concentration of 1.2 ng l$^{-1}$ in the drinking water (11).

## Mammalian and avian toxicity

### Acute data
Yolks of chicken eggs preincubated for 4 days were injected with 2.0 and 0.5 mg kg egg$^{-1}$ single dose and the mortality measured 2 wk later was 17/20 and 2/20, respectively (12).

### Carcinogenicity and long-term effects
No adequate evidence for carcinogenicity to humans, sufficient evidence for carcinogenicity to animals, IARC classification group 2B (13).

Dermal mice administered total initiating dose of 1 mg induced a 72% incidence of skin tumours (14).

At a total dose ranging from 0.14 to 0.58 mg mouse$^{-1}$ no tumorigenic effects were observed in newborn CD-1 mice (route unspecified) (15).

### Teratogenicity and reproductive effects
Single-dose injection of 2.0 mg kg egg$^{-1}$ caused abnormalities to the chicken embryo including degenerative hepatic lesions, pericardial oedema, subcutaneous oedema and microphthalmia (12).

### Metabolism and pharmacokinetics
Metabolised *in vivo* mouse skin the major metabolites were 8-hydroxyindeno[1,2,3-*cd*]pyrene; 9-hydroxyindeno[1,2,3-*cd*]pyrene; and *trans*-1,2-dihydro-1,2-dihydroxyindeno[1,2,3 -*cd*]pyrene. Minor metabolites identified included *trans*-1,2-dihydro-1,2,8-trihydroxyindeno[1, 2,3-*cd*]pyrene; *trans*-1,2-dihydro-1,2,9- trihydroxyindeno[1,2,3-*cd*]pyrene; indeno[1,2,3-*cd*]pyrene-1,2-dione; and 10-hydroxyindeno[1,2,3-*cd*]pyrene. Two metabolites, *trans*-1,2-dihydro-1,2- dihydroxyindeno[1,2,3- *cd*]pyrene and 1,2-dihydro-1,2-epoxyindeno[1,2,3- *cd*]pyrene both produced an 80% incidence of tumours at a total initiating dose of 1.0 mg. 8-Hydroxyindeno[1,2,3- *cd*]pyrene, which is mutagenic when assayed in the presence of a microsomal activation system, exhibited only weak tumour-initiating activity (16).

*In vitro* rodent liver metabolised to hydroxy, dihydrodiol and quinone metabolites (17).

## Genotoxicity

As component of PAH fraction of airborne pollutants *Salmonella typhimurium* TA100 with metabolic activation positive (18).

As component of a diesel exhaust fraction *Salmonella typhimurium* TA98 with and without metabolic activation positive (direct acting); TA100 with metabolic activation weakly positive and without metabolic activation negative (19).

*In vivo* ♀ CD-1 mice single topical application (unspecified), single major DNA adduct detected (20).

## Legislation

Limited under EC Directive on Drinking Water Quality 80/778/EEC. Polycyclic aromatic hydrocarbons: maximum admissible concentration 0.2 μg l$^{-1}$ (21).

## Any other comments

Effects of exposure to PAHs in road workers in New Zealand were investigated. In general the exposures were within permissible limits for these materials, and the health hazards were slight. The possible effects of exposure were mitigated by the work pattern (22).

PAHs pose a considerable occupational hazard to evaporative pattern casting process workers in metal industry (23).

PAHs are known to leach from pipe and storage tank linings. Potential cancer risks in drinking water discussed (24).

Environment monitoring of the mutagenic/carcinogenic hazard associated with bitumen fumes showed occupational exposures to these agents was low (25).

The relationship between exposure to environmental carcinogens (PAHs) and lung cancer discussed (26).

Median air concentration of 0.2 $\mu$g m$^{-3}$ (SD 144%) from tar bitumen in road works. Over an 8 hr working shift varied by a factor of 30. The study concludes that there is a considerable risk to health (27).

Experimental toxicology, environmental effects and human health effects reviewed (28).

## References

1. *Influence of engine/driving-conditions and operating time of engine oil on the exhaust emission of polycyclic aromatic hydrocarbons from gasoline passenger cars. Part 1. Emissions* 1977, German Society for Petroleum Sciences and Coal Chemistry, Hamburg
2. Schrodter, W. P. et al *Erdoel Kohle, Erdgas, Petrochem.* 1976, **294**
3. Coon, R. et al *Toxicol. Appl. Pharmacol.* 1970, **16**, 646
4. Wynder et al *Tobacco and Tobacco Smoke* 1967, Academic Press, New York
5. Borneff, J. et al *Arch. Hyg.* 1969, **153**(3), 220-229
6. Broman, D. et al *Environ. Toxicol. Chem.* 1990, **9**(4), 429-442
7. Park, K. S. et al *Environ. Chem.* 1990, **9**, 187-195
8. Coover, M. P. et al *Hazard. Waste Hazard. Mater.* 1987, **4**(1), 69-82
9. Mueller, J. G. et al *Appl. Environ. Microbiol.* 1991, **57**(5), 1277-1285
10. Kirk, P. W. W. et al *Environ. Technol.* 1991, **12**(1), 13-20
11. Harrison, R. et al *Environ. Sci. Technol.* 1976, **10**(12), 1151-1156
12. Brunstrom, B. et al *Environ. Pollut.* 1990, **67**(2), 133-143
13. *IARC Monograph* 1987, **Suppl. 7**, 64
14. Rice, J. E. et al *Carcinogenesis (London)* 1990, **11**(11), 1971-1974
15. Lavoie, E. J. et al *Cancer Lett. (Shannon, Irel.)* 1987, **34**(1), 15-20
16. Rice, J. E. et al *Carcinogenesis (London)* 1986, **7**(10), 1761-1764
17. Rice, J. E. et al *Polynuclear Aromatic Hydrocarbons (Pap. Int. Symp.)* 8th ed., 1985, Battelle Press, Columbus, OH
18. Weyand, E. H. et al *Cancer Lett. (Shannon, Irel.)* 1987, **37**(3), 257-266
19. Motykiewicz, G. et al *Mutat. Res.* 1989, **223**(2), 243-251
20. Ball, J. C. et al *Environ. Sci. Technol.* 1990, **24**, 890-894
21. *EC Directive Relating to the Quality of Water Intended for Human Consumption* 1982, 80/778/EEC, Office for Official Publications of the European Communities, 2 rue Mercier, L-2985 Luxembourg
22. Darby, F. W. et al *Ann. Occup. Hyg.* 1986, **30**(4), 445-454
23. Gressel, M. G. et al *Appl. Ind. Hyg.* 1988, **3**(1), 11-17
24. Olori, L. et al *Rapp. ISTISAN* 1989, ISTISAN 89/28 (Ital.) (*Chem. Abstr.* **113**, 120439y)
25. Monarca, S. et al *Int. Arch. Occup. Environ. Health* 1987, **59**(4), 393-402
26. Galassi, G. *AES* 1988, **10**(6), 9-10 (Ital.) (*Chem. Abstr.* **110**, 81633y)

27. Knecht, U. et al *Br. J. Ind. Med.* 1989, **46**(1), 24-30
28. *ECETOC Technical Report No. 30(4)* 1991, European Chemical Industry Ecology and Toxicology Centre, B-1160 Brussels

# I21   Indigo

**CAS Registry No.** 482-89-3
**Synonyms** 2-(1,3-dihydro-3-oxo-2*H*-indol-2-ylidene)-1,2-dihydro-3*H*-indol-3-one; C.I. 73000; indigo blue; Lithosol Deep Blue B; Vat Blue 1; Vynamon Blue A
**Mol. Formula** $C_{16}H_{10}N_2O_2$        **Mol. Wt.** 262.27
**Uses** Textile dye.
**Occurrence** *Indigofera* of the Leguminosae.

## Physical properties

**M. Pt.** 390-392°C. **Specific gravity** 1.35 **Volatility** v.p. 47 mmHg at 430°C

**Solubility**
Organic solvent: aniline, nitrobenzene, chloroform, glacial acetic acid

## Ecotoxicity

**Bioaccumulation**
Confirmed to be non-accumulative or low accumulative (1).

## Gentoxicity

*Salmonella typhimurium* TA98 with metabolic activation (3-methylcholanthrene induced) positive (2).

## Adverse effects to man

A man accidently splashed a 7% alkaline solution of the reduced form into his eyes. The conjunctiva appeared blue several hr later, this cleared within 10 days. The cornea was turbid but not stained (3).

**Sensitisation**
The allergic potential of 13 pharmaceutical dyes including indigo was investigated by sensitising guinea pig skin with a series of intradermal applications. Intradermally, all dyes elicited an allergenic response; epidermally only indigo gave an allergic response (4).

## References

1. *The list of the existing chemical substances tested on biodegradability by microorganisms or bioaccumulation in fish body* 1987, Chemicals Inspection and Testing Institute, Japan
2. Jongen, W.M.F. *Carcinogenesis (London)* 1982, **3**(11), 1321-1323
3. Grant, W.M. *Toxicology of the Eye* 2nd ed., 1974, 577, C.C. Thomas, Springfield, IL
4. Maurer, T. *Acta Pharm. Technol.* 1979, **Suppl. 8**, 37

## I22   Indigo carmine

**CAS Registry No.** 860-22-0
**Synonyms** 2-(1,3-dihydro-3-oxo-5-sulfo-2*H*-indol-2-ylidene)-2,3-dihydro-
3-oxo-1*H*-indole-5-sulfonic acid, disodium salt; Acid Blue 74; Amacid Brilliant Blue;
Intense Blue; C.I. 73015; C.I. 75781; FD & C Blue No. 2
**Mol. Formula** $C_{16}H_{10}N_2Na_2O_8S_2$            **Mol. Wt.** 468.37
**Uses** Dye. Reagent for detection of chlorate and nitrate. In testing milk. In test for
renal function. Clinical marker dye, particularly in urological procedures. Food colour.

## Physical properties

**Solubility**
Water: 1 g $\approx$ 100 ml$^{-1}$ at 25°C.

## Environmental fate

**Abiotic**
In a wastewater treatment plant using activated carbon with pore radius $<2 \times 10^{-8}$m; a
filling correlation between amount absorbed and pore size was obtained (1).
At 25°C adsorption rate by activated carbon was $\approx$ 1.7 to $6.5 \times 10^9$ cm$^2$s$^{-1}$(2).

**Absorption**
Uptake to give monolayer coverage for kaolinite is greater than that for chlorite (3).
Carbon-mineral adsorbents obtained from mixtures of waste carbonaceous materials
and montmorillonite, attapulgite and polygoroskite were more effective than activated
carbon in removal from wastewater (4).

## Mammalian and avian toxicity

**Acute data**
$LD_{50}$ oral rat, mouse 2000, 2500 mg kg$^{-1}$, respectively (5).
$LD_{50}$ intravenous rat 93 mg kg$^{-1}$ (6).
$LD_{50}$ subcutaneous mouse 405 mg kg$^{-1}$ (5).

**Teratogenicity and reproductive effects**
Gavage Charles River CD rats 25, 75 or 250 mg kg$^{-1}$ day$^{-1}$ on days 6-15 of gestation.
Dutch belted rabbits received the same dose on days 6-18 of gestation. No
teratogenicity was observed (7).

**Sensitisation**
Occasionally skin rash, pruritus and bronchoconstriction in humans (8).

## Genotoxicity

*Salmonella typhimurium* TA98, TA100, TA1535, TA1537, TA1538 with and without metabolic activation negative (9).

*In vitro* mouse lymphoma tk+/tk- with metabolic activation indeterminate (9).

Oral mouse in diet increased bone marrow chromosomal aberrations (10).

*Allium cepa* showed significant increase in polyploid cells and at high doses chromosome breaks and micronucleus formation occurred (11).

## Any other adverse effects to man

May cause vomiting, hypertension, nausea and bradycardia. Discoloured skin has occurred following a large parenteral dose (8).

Two elderly patients, both with a history of asthmatic bronchitis, suffered total cardiac arrest following 80 mg intravenously (12).

## Any other adverse effects

No inhibition of *in vitro* rabbit renal medulla prostaglandin synthetase activity at 117 mg l$^{-1}$ (13).

## Any other comments

Estimated acceptable daily intake: up to 2.5 mg kg body weight$^{-1}$ (8).

## References

1. Tamura, T. et al *Environ. Technol. Lett.* 1988, **9**(4), 281-286
2. Tamura, T. et al *Tokushima J. Exp. Med.* 1987, **34**(3-4), 97-100
3. Awal, K. P. *J. Nepal Chem. Soc.* 1988, **8**, 1-5
4. Tarasevich, Y. I. et al *Khim. Tekhnol. Vody* 1988, **10**(4), 315-317 (Russ.) (*Chem. Abstr.* **109**, 155525t)
5. *Scientia Pharmaceutica* 1979, **47**, 39
6. *Drugs in Japan. Ethical Drugs* 6th ed., 1982, Japan Pharmaceutical Information Centre, Yakugyo Jiho Co., Ltd., Tokyo
7. Borzelleca, J. F. et al *Food Chem. Toxicol.* 1987, **25**(7), 495-497
8. *Martindale. The Extra Pharmacopoeia* 30th ed., 1993, The Pharmaceutical Press, London
9. Cameron, T. P. et al *Mutat. Res.* 1987, **189**(3), 223-261
10. Das, S. K. et al *Cytobios* 1988, **53**(216), 25-29
11. Roychondhury, A. et al *Mutat. Res.* 1989, **223**(3), 313-319
12. Voiry, A. M. et al *Ann. Med. Nancy* 1976, **15**, 413-419
13. Yoshihira, K. et al *J. Pharmacobio-Dyn.* 1991, **14**(6), 327-334

# I23   Indium

In

CAS Registry No. 7440-74-6
**Mol. Formula** In                    **Mol. Wt.** 114.82
**Uses** In bearing and dental alloys; as a thin film on moving metallic surfaces. In nuclear reactor control rods. In semi-conductor research.
**Occurrence** In Earth's crust $1 \times 10^{-5}$%.

## Physical properties

**M. Pt.** 155°C; **B. Pt.** 2000°C; **Specific gravity** $d^{20}$ 7.3.

## Occupational exposure

**US TLV (TWA)** 0.1 mg m$^{-3}$ (as In); **UK Long-term limit** 0.1 mg m$^{-3}$ (as In); **UK Short-term limit** 0.3 mg m$^{-3}$ (as In).

## Mammalian and avian toxicity

### Acute data

LD$_{Lo}$ (calculated as indium) subcutaneous mouse 10 mg kg$^{-1}$ (1).

### Sub-acute data

Inhalation rat (9 month) 80 mg Pb-Zn dust containing trace amounts of indium, impaired liver function (2).

## Any other adverse effects

Intravenous Japanese quail (dose and duration unspecified) 32% [114]In radiolabel accumulates in kidney of estradiol-treated ♂ and 37% in growing oocytes plus ova (3). Injury to blood, heart, liver and kidneys. As sulfate relatively non-toxic when administered orally, but highly toxic when administered subcutaneously or intravenously as citrate (1).

## Legislation

Included in Schedule 4 (Release into the Air: Prescribed Substances) Statutory Instrument No. 472, 1991 (4).

## Any other comments

Uses, physical and chemical properties, acute/chronic toxicity, metabolism and exposure limits reviewed (5,6).
Experimental toxicology, human health effects and workplace experience reviewed (7).

## References

1. Browning, E. *Toxicity of Industrial Metals* 2nd ed., 1969, 164-168, Appleton-Century-Crofts, New York
2. Burkhanov, A. I. et al *Gig. Sanit.* 1987, (2), 90-91 (Russ.) (*Chem. Abstr.* **106**, 133362y)
3. Robinson, G. A. et al *Poult. Sci.* 1990, **69**(2), 300-306
4. *S. I. 1991 No. 472 The Environmental Protection (Prescribed Processes and Substances) Regulations* 1991, HMSO, London
5. Horiguchi, S. et al *Jinrui Idengaku Zasshi* 1988, **33**(4), 120-137 (Japan.) (*Chem. Abstr.* **110**, 226636p)
6. Horiguchi, S. et al *Sumitomo Sangyo Eisei* 1987, **23**(4), 120-137 (Japan.) (*Chem. Abstr.* **111**, 128332n)
7. *ECETOC Technical Report No. 30(4)* 1991, European Chemical Industry Ecology and Toxicology Centre, B-1160 Brussels

# I24   Indium trichloride

## InCl$_3$

**CAS Registry No.** 10025-82-8
**Synonyms** Indium chloride (InCl$_3$); trichloroindium
**Mol. Formula** InCl$_3$                    **Mol. Wt.** 221.18
**Uses** In electroplating.

## Physical properties

**M. Pt.** 586°C; **Specific gravity** 4.0.

## Mammalian and avian toxicity

### Acute data

LD$_{50}$ subcutaneous rabbit 2350 µg kg$^{-1}$ (1).
LD$_{Lo}$ subcutaneous rat, mouse 10, 60 mg kg$^{-1}$, respectively (1,2).
LD$_{50}$ intraperitoneal mouse 9500 µg kg$^{-1}$ (3).
LD$_{50}$ chicken egg 121 µg egg$^{-1}$ (4).

## Teratogenicity and reproductive effects

Chick egg injected with 1-100 µg egg$^{-1}$ on day 2 of incubation. Gross malformations observed included twisted limbs, haemorrhage, everted viscera and reduced body weight (4).

## Legislation

Limited under EC Directive on Drinking Water Quality 80/778/EEC. Chlorides: guide level 25 mg l$^{-1}$ (5).
Included in Schedule 4 (Release into the Air: Prescribed Substances) Statutory Instrument No. 472, 1991 (6).

## References

1. *Environ. Qual. Saf., Suppl.* 1975, **1**, 1
2. *J. Ind. Hyg. Toxicol.* 1942, **24**, 243
3. *C. R. Hebd. Seances Acad. Sci.* 1963, **256**, 1043
4. Gilani, S.H. et al *J. Toxicol. Environ. Health* 1990, **30**(1), 23-31
5. *EC Directive Relating to the Quality of Water Intended for Human Consumption* 1982, 80/778/EEC, Office for Official Publications of the European Communities, 2 rue Mercier, L-2985 Luxembourg
6. *S. I. 1991 No. 472 The Environmental Protection (Prescribed Processes and Substances) Regulations* 1991, HMSO, London

# I25 Indocyanine green

$$H_3C \quad H_3C \quad =CH-CH=CH-CH=CH-CH=CH \quad CH_3 \quad CH_3$$

(CH$_2$)$_4$SO$_3$Na

(H$_2$C)$_4$

$^-$O$_3$S

**CAS Registry No.** 3599-32-4
**Synonyms** 2-[7-[1,3-dihydro-1,1-dimethyl-3-(4-sulfobutyl)
-2$H$-benz[$e$]indol-2-ylindene]1,3,5-heptatrienyl]-1,1-dimethyl-3-(4-sulfobutyl)
1$H$-benz[$e$]indolium, hydroxide, inner salt, sodium salt;
4,5-benzoindotricarbocyanine; Cardio Green; Ujoviridin; Wofaverdin
**Mol. Formula** C$_{43}$H$_{48}$N$_2$NaO$_6$S$_2$              **Mol. Wt.** 775.99
**Uses** Dye in infrared photography. In preparation of Wratten filters. Aid for blood
volume determination, hepatic function and cardiac output.

## Physical properties

**Solubility**
Organic solvent: methanol

## Mammalian and avian toxicity

**Acute data**
LD$_{50}$ intraperitoneal mouse, rat 400-700 mg kg$^{-1}$ (1).
LD$_{50}$ intravenous mouse 60 mg kg$^{-1}$ (2).

**Metabolism and pharmacokinetics**
Bound to plasma protein following intravenous injection. Rapidly taken up by the
liver and is excreted unmodified into the bile (3).
Intravenous Sprague-Dawley rat 46 mg l$^{-1}$ kg$^{-1}$ maximum biliary secretory rate of 108
$\mu$g l$^{-1}$ kg$^{-1}$ min$^{-1}$ (metabolism by the liver was inhibited by administration of diethyl
maleate) (4).

## Any other adverse effects to man

Anaphylactoid reactions reported (5).

## Any other adverse effects

In Sprague-Dawley rats 14 mg$^{-1}$ kg$^{-1}$ inhibited the maximum biliary secretory rate of
intravenously administered amaranth (181 mg$^{-1}$ kg$^{-1}$) by 50% and 39 mg kg$^{-1}$
inhibited liver ADP-stimulated O$_2$ consumption by 20-30% (4).

## References

1. *Toxicol. Appl. Pharmacol.* 1973, **24**, 37
2. *Toxicol. Appl. Pharmacol.* 1978, **44**, 225
3. *Martindale. The Extra Pharmacopoeia* 30th ed., 1993, The Pharmaceutical Press, London
4. Varga, F. et al *Arch. Toxicol., Suppl.* 1989, **13**, 309-311
5. Speich, R. et al *Ann. Intern. Med.* 1988, **109**, 345-346

# I26 Indole

CAS Registry No. 120-72-9
Synonyms 1*H*-indole; 1-azaindene; ketole
Mol. Formula $C_8H_7N$                                   Mol. Wt. 117.15
Uses In perfume (in highly dilute solution).
Occurrence Coal tar. Faeces. Cigarette smoke.

## Physical properties

M. Pt. 52°C; B. Pt. 253°C; Flash point 110°C; Specific gravity 1.22; Partition coefficient log $P_{ow}$ 2.14.

### Solubility
Organic solvent: hot ethanol, diethyl ether, benzene

## Ecotoxicity

### Fish toxicity
5 ppm killed trout, bluegill sunfish and goldfish within 4, 5 and 5 hr, respectively. 1 ppm had no effect. Test conditions: pH 7.0; dissolved oxygen, 7.5 ppm; total hardness (soap method), 300 ppm; methyl orange alkalinity, 310 ppm; free carbon dioxide, 5 ppm; and 12.8°C (1).
At 10 mg $l^{-1}$ death occurred at 0-2 and 12-21 hr for the steelhead trout and bridgelip sucker, respectively. At 5 mg $l^{-1}$ the trout died between 2 and 6 hr. Test conditions: total hardness, 0-17 mg $l^{-1}$; methyl orange alkalinity, 14 mg $l^{-1}$; pH 7.6; or total hardness, 67-120 mg $l^{-1}$; methyl orange alkalinity, 151-183 mg $l^{-1}$; total dissolved solids, 160-175 mg $l^{-1}$; pH 7.1 (2).

### Invertebrate toxicity
$EC_{50}$ (30 min) *Photobacterium phosphoreum* 2.39 ppm Microtox test (3).

## Environmental fate

### Nitrification inhibition
Depending on soil type 1.2, 12 and 59 μg $l^{-1}$ $g^{-1}$ caused, respectively, 0, 1-5 and 3-6% inhibition of nitrification of ammonium in soil (4).

### Anaerobic effects
Oxindole accumulation under methanogenic, but not under denitrifying, conditions is caused by differences between relative rates of oxindole production and destruction (5). With an inoculum of sewage sludge and incubation under methanogenic conditions metabolism occurred within 10 days. Oxindole was the temporary intermediate formed. Almost complete mineralisation occurred (6).
The rate of transformation to oxindole and its subsequent disappearance was dependent on the concentration of inoculum and indole and the temperature. The dominant process was methanogenesis. Sulfate reducers predominated over

methanogens in the mixed culture of anaerobic bacteria. 75% of the substrate was mineralised to $CO_2$ (7).

**Degradation studies**
$BOD_5$ 2.07 mg $l^{-1}O_2$ using the standard dilution technique with normal sewage as seed material (8).
COD, 2.460 mg $l^{-1}$ $O_2$ (9).
ThOD, 2.46 mg$l^{-1}O_2$ (10).
$BOD_5$, 84% of ThOD (11).
Degradation time in groundwater from a gasoline-contaminated aquifer 310 hr. Initial concentration 0.2-1 mg $l^{-1}$, temperature 10°C (12,13).

## Mammalian and avian toxicity

**Acute data**
$LD_{50}$ oral rat 1000 mg $kg^{-1}$ (14).
$LD_{50}$ (estimated) redwing blackbird >100 mg $kg^{-1}$ (15).
$LD_{50}$ dermal rabbit 790 mg $kg^{-1}$ (14).
$LD_{50}$ intraperitoneal mouse 117 mg $kg^{-1}$ (16).
$LD_{50}$ subcutaneous mouse 225 mg $kg^{-1}$ (17).
Oral pony 0.1 or 0.2 g $kg^{-1}$ caused intravascular haemolysis and haemoglobinuria. Heinz body formation was observed in the higher dose animals. Plasma indole levels increased. Death of some high-dose ponies occurred between 24 and 72 hr after dose. At necropsy all body fat, elastic tissue and mucous membrane was stained yellow. The most prominent microscopic lesion was haemoglobinuric nephrosis (18).

## Genotoxicity

*Salmonella typhimurium* TA100 without metabolic activation positive (after nitrite treatment). Showed mutagenic precursor activity (19).

## Any other adverse effects

Intraperitoneal rat 5 mg $kg^{-1}$ reduced oxygen consumption significantly. *In vitro* rat diaphragm or liver showed no influence on respiration (20).

## Any other comments

Major metabolite of orally administered tryptophan in ponies (21).
0.25 µg $l^{-1}$ found in domestic sewage (22).
Metabolism and hazards reviewed (23,24).
Intense faecal odour.

## References

1.  US EPA *The Toxicity of 3400 Chemicals to Fish* 1987, EPA 560/6-87-002, Washington, DC
2.  US EPA *Fish Toxicity Screening Data* 1989, EPA 560/6-89-001, Washington, DC
3.  Kaiser, K. L. E. et al *Water Pollut. Res. J. Can.* 1991, **26**(3), 361-431
4.  McCarty, G. W. et al *Biol. Fertil. Soils* 1989, **8**, 204-211
5.  Madsen, E. L. et al *Appl. Environ. Microbiol.* 1988, **54**(1), 74-78
6.  Berry, D. F. et al *Appl. Environ. Microbiol.* 1987, **53**(1), 180-182
7.  Shanker, R. et al *Microb. Ecol.* 1990, **20**(2), 171-183
8.  Mills, E. J. et al *Proc. 8th Purdue Ind. Waste Conf.* 1953, 492
9.  Chudoba, J. et al *Water Res.* 1973, **7**, 663-668
10. Meinck, F. et al *Les Eaux Residuaires Industrielles* 1970

11. Wolters, N. *unterschiedliche Bestimmungsmethoden zur Erfassung der organischen Substanz in einer Verbindung* Lehrauftrag Wasserbiologie a.d., Technischen Hochschule, Darmstadt
12. Arvin, E. et al *Proc. Pap. Int. Symp. Groundwater Microbiol., Probl. Biol. Treat.* 1987, Kuopio, Finland
13. Arvin, E. et al *Water Sci. Technol.* 1988, **20**(3), 109-118
14. *Am. Ind. Hyg. Assoc. J.* 1962, **23**, 95
15. Schafer, E. W. et al *Arch. Environ. Contam. Toxicol.* 1983, **12**, 355-382
16. *Yakugaku Zasshi* 1974, **94**, 1620
17. *Klin. Wochenschr.* 1957, **35**, 504
18. Paradis, M. R. et al *Am. J. Vet. Res.* 1991, **52**(5), 748-753
19. Ochiai, M. et al *Mutat. Res.* 1986, **172**(3), 189-197
20. Hohenegger, M. et al *Nephron* 1988, **48**(2), 154-158
21. Paradis, M. R. et al *Am. J. Vet. Res.* 1991, **52**(5), 742-747
22. Faust, S. J. et al *Organic Compounds in Aquatic Environment* Marcel Dekker, New York
23. Subramanian, V. et al *Biol. Oxid. Syst., [Proc. Symp.] 1989* 1990, **1**, 417-429
24. *Dangerous Prop. Ind. Mater. Rep.* 1988, **8**(3), 63-67

# I27    Indole-3-acetonitrile

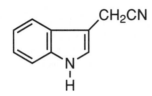

**CAS Registry No.** 771-51-7

**Synonyms** 1*H*-indole-3-acetonitrile; 3-(cyanomethyl)indole; IAN; indoleacetonitrile; 3-indolylacetonitrile

**Mol. Formula** $C_{10}H_8N_2$                          **Mol. Wt.** 156.19

## Physical properties

**M. Pt.** 35-37°C; **B. Pt.** $_{0.2}$ 157-160°C; **Flash point** >110°C.

## Mammalian and avian toxicity

### Acute data

$LD_{50}$ subcutaneous rat 255 mg kg$^{-1}$ (1).
$LD_{50}$ intraperitoneal mouse 200 mg kg$^{-1}$ (2).

## Any other comments

*In vitro* primary chick embryo hepatocytes 35 µg ml$^{-1}$ gave 1.6 fold increase in ethoxyresorufin-*O*-deethylase and a 2-fold increase in ethoxycoumarin-*O*-deethylase activities. Has a protective effect against benzo[*a*]pyrene genotoxicity (3,4).

## References

1. *Food Cosmet. Toxicol.* 1980, **18**, 159
2. *NTIS Report* AD277-689, Natl. Tech. Inf. Ser., Springfield, VA
3. Jongen, W. M. F. et al *Mutat. Res.* 1989, **222**(3), 263-269
4. Jongen, W. M. F. et al *Toxicol. in Vitro* 1987, **1**(2), 105-110

# I28 Indole-2,3-dione

**CAS Registry No.** 91-56-5
**Synonyms** 1*H*-indole-2,3-dione; *o*-aminobenzoylformic anhydride;
2,3-dioxoindoline; isatic acid lactam; isatin; isatinic acid anhydride
**Mol. Formula** $C_8H_5NO_2$                    **Mol. Wt.** 147.13
**Uses** In the manufacture of dyes. Analytical chemical reagent.

## Physical properties

**M. Pt.** 203.5°C.

### Solubility
Organic solvent: boiling ethanol, diethyl ether

## Mammalian and avian toxicity

### Acute data
$LD_{Lo}$ oral rat 5 g $kg^{-1}$ (1).
$LD_{50}$ (estimated) oral redwing blackbird >101 mg $kg^{-1}$ (2).
$LD_{50}$ intraperitoneal mouse 330 mg $kg^{-1}$ (3).
$LD_{50}$ intraperitoneal mouse 563 mg $kg^{-1}$ (4).

### Metabolism and pharmacokinetics
Showed a rapid rate of absorption and metabolism in rabbit study. Metabolites in
urine were anthranilic acid, tryptophan and nicotinic acid (5).

## References

1. *Indian J. Physiol. Pharmacol.* 1962, **6**, 145
2. Schafer, E. W. et al *Arch. Environ. Contam. Toxicol.* 1983, **12**, 355-382
3. *Nippon Yakurigaku Zasshi* 1959, **55**, 1514
4. *Pharm. Chem. J. (Engl. Transl.)* 1981, **15**, 858
5. Ahmad, M. et al *Proc. Pak. Acad. Sci.* 1990, **27**(1), 1-10

# I29 Indol-3-ylacetic acid

$$CH_2CO_2H$$

(structure of indole-3-ylacetic acid)

**CAS Registry No.** 87-51-4
**Synonyms** indole-3-acetic acid; indoleacetic acid; $1H$-indole-3-acetic acid; heteroauxin; IAA; Rhizopin; heteroauxin
**Mol. Formula** $C_{10}H_9NO_2$        **Mol. Wt.** 175.19
**Uses** Plant growth regulator.
**Occurrence** In all plants. In bacteria and fungi. In secondary domestic sewage plant effluent 0.013 mg $l^{-1}$ (1).

## Physical properties

**M. Pt.** 168-170°C; **Volatility** v.p. $0.15 \times 10^{-6}$ mmHg at 60°C.

**Solubility**
Water: 1.5 g $l^{-1}$ at 20°C. Organic solvent: ethanol, acetone, diethyl ether, chloroform

## Environmental fate

**Nitrification inhibition**
Growth, dinitrogen fixation and heterocyst frequency of *Anabaena* PCC7119 and *Nodularia* sp. were investigated. Concentrations higher than $1.75 \times 10^{-2}$ g $l^{-1}$ were inhibitory to growth. Lower concentrations did not alter the growth of *Anabaena* (2).

**Degradation studies**
Degraded by the fungi *Aspergillus niger*, *Paecilomyces varioti* and *Penicillium oxalicum*, but *Chaetomium cupreum* and *C. globosum* were totally inhibited. There was no positive correlation between degradation and vegetative growth or pH of the medium (3).
Degrades rapidly in soil (4).

**Abiotic removal**
Unstable to light (4).

## Mammalian and avian toxicity

**Acute data**
$LD_{50}$ dermal mouse 1000 mg $kg^{-1}$ (4).

## Genotoxicity

*Salmonella typhimurium* TA97, TA98, TA100, TA1535, TA1538 histidine revertants with and without metabolic activation negative (5).

## Any other comments

Metabolised in *Vicia faba* via biosynthetic pathway to indole-3-acetylaspartic acid,

3-hydroxy-2-indolone-3-acetylaspartic acid and 3-(*O*-β-glucosyl-2-indolone-3-acetylaspartic acid (6).
Plant biochemistry reviewed (7).

## References

1. Wilson Pitt, W. et al *Environ. Sci. Technol.* 1975, **9**(12), 1068-1073
2. Leganes, F. et al *Plant Cell Physiol.* 1987, **28**(3), 529-533
3. Reddy, V. K. et al *Indian J. Mycol. Plant Pathol.* 1987, **17**(1), 75-77
4. *The Agrochemicals Handbook* 3rd ed., 1991, RSC, London
5. Kappas, A. *Mutat. Res.* 1988, **204**(4), 615-621
6. Tsurumi, S. *Shokubutsu no Kagaku Chosetsu* 1986, **21**(2), 126-133 (Japan.) (*Chem. Abstr.* **107**, 112734b)
7. Nonhebel, H. M. et al *Plant Growth Subst. 1988, (International Conference on Plant Growth Substances)* 1990, 333-340, Springer, Berlin

# I30   4-(Indol-3-yl)butyric acid

**CAS Registry No.** 133-32-4
**Synonyms** 1*H*-indole-3-butanoic acid; indole-3-butyric acid; Hormex; Hormodiu; IBA; β-indolylbutyric acid; Seradix
**Mol. Formula** $C_{12}H_{13}NO_2$        **Mol. Wt.** 203.24
**Uses** Promoter/accelerator of roots from plant-cuttings.

## Physical properties

**M. Pt.** 123-125°C; **Volatility** v.p. $<8 \times 10^{-8}$ at 60°C.

### Solubility
Water: 250 mg $l^{-1}$ at 20°C. Organic solvent: benzene, ethanol, diethyl ether, acetone

## Ecotoxicity

### Fish toxicity
Trout, bluegill sunfish, goldfish (24 hr) 5 ppm non-toxic. Test conditions: pH 7; dissolved oxygen, 7.5 ppm; total hardness (soap method), 300 ppm; methyl orange alkalinity, 310 ppm; free carbon dioxide, 5 ppm; and 12.8°C (1).

## Environmental fate

### Degradation studies
Degrades rapidly in soil (2).

## Mammalian and avian toxicity

**Acute data**

$LD_{50}$ oral mouse 100 mg $kg^{-1}$ (3).
$LD_{50}$ intraperitoneal mouse 100 mg $kg^{-1}$ (4).

## Genotoxicity

*Salmonella typhimurium* TA97, TA98, TA100, TA1535, TA1538 histidine revertants
with and without metabolic activation negative (5).

## Any other comments

WHO Class III; EPA Toxicity Class III (2).

## References

1. Wood, E. M. *The Toxicity of 3400 chemicals to fish* 1987, EPA report 560/6-87-002, Washington, DC
2. *The Agrochemicals Handbook* 3rd ed., 1991, RSC, London
3. Thompson, W. T. *Agricultural Chemicals* Book 3, 1976/77, 76, Thompson Publications, Fresno, CA
4. Anderson et al *Proc. Soc. Exp. Biol. Med.* 1936, **34**, 138
5. Kappas, A. *Mutat. Res.* 1988, **204**(4), 615-621

# I31   Indomethacin

**CAS Registry No.** 53-86-1
**Synonyms** 1-(4-chlorobenzoyl)-5-methoxy-2-methyl-1*H*-indole-3-acetic acid; Dolovin; Indocid; Indomecol; Indren; Metacen ; Sadoreum
**Mol. Formula** $C_{19}H_{16}ClNO_4$                    **Mol. Wt.** 357.80
**Uses** Analgesic, antiflammatory and antihistaminic drug.

## Physical properties

**M. Pt.** $\approx$155 and $\approx$162°C (polymorphic crystals).

**Solubility**
Organic solvent: ethanol, diethyl ether, acetone, castor oil

## Mammalian and avian toxicity

### Acute data
$LD_{50}$ oral mouse 13 mg kg$^{-1}$ (1).
$LD_{50}$ oral dog 160 mg kg$^{-1}$ (2).
$LD_{50}$ intraperitoneal rat, mouse 13, 15 mg kg$^{-1}$, respectively (3,4).
$LD_{50}$ intravenous mouse, rat 30, 35 mg kg$^{-1}$, respectively (5,6).

### Sub-acute data
Oral marmoset (4 wk) 0.2, 6, 12 mg kg$^{-1}$ day$^{-1}$ 100% mortality at highest dose.
Intermediate dose induced severe gastrointestinal toxicity while low dose caused
functional and morphological renal alterations (7).
Rats 3.55 mg kg$^{-1}$ day$^{-1}$ (route unspecified) for 3 months caused kidney damage (8).
Three repeat administrations of 0.02 mg kg$^{-1}$ to rat (route unspecified) caused liver damage
manifested by increased activities of aminotransferases and alkaline phosphatase
activities, a decrease in glutathione levels and disturbance of liver functions (9).
Oral mice 5, 10, 20 mg l$^{-1}$ in drinking water hepatotoxicity was minimal at 5 mg l$^{-1}$
but higher doses killed (10).

### Teratogenicity and reproductive effects
Injection AKR mice on day 13.5 gestation no teratogenic effect observed (11).
Implant ♂ rat 50% rod adjacent to each epididymis. 1-4 wk after implant fertility was
reduced but returned to normal within 5-10 wk (12).
Oral rat maternal administration 1 and 2.5 mg kg$^{-1}$ day$^{-1}$ (days 18 and 21 gestation)
caused dose-dependent retardation in structural development in mesenteric lymph
nodes of foetuses (13).
Oral pregnant rat 0.7 mg kg$^{-1}$ caused constriction of foetal rat ductus and cardiac
failure. Effects were persistent (14).

### Metabolism and pharmacokinetics
Human single oral dose (concentration unspecified) peak concentrations in blood
were observed 1.5-3.5 hr (15).
Dermal rat single application 2.5 mg, plasma concentration peaked at 8-9 hr, muscle
concentration maximal levels after 8 hr. Dermal muscle/plasma concentration ratio
was higher than when administered orally (16).
99% bound to plasma proteins, it is distributed to the synovial fluid, central nervous
system, placenta and breast milk. Metabolised to its glucuronide conjugate and to
desmethyl- desbenzoylindomethacin; desbenzoylindomethacin;
desmethylindomethacin; and to their glucuronides. Some *N*-deacylation occurs.
Undergoes enterohepatic circulation. Metabolites excreted chiefly in the urine (17,18).

### Irritancy
100 mg placed into ♀ rabbit eye minimal irritation, recovery within 24 hr (19).

### Sensitisation
Hypersensitivity reactions, including acute asthma have occurred in patients with a
history of asthma or aspirin sensitivity (20-22).

## Genotoxicity
*In vivo* mice assays for bone marrow micronuclei, abnormal sperm formation,
meiotic chromosome abnormalities in spermatocysts positive in concentration range
12-36 mg kg$^{-1}$ (23).

## Any other adverse effects to man

Nephrotic syndrome, acute renal failure and renal papillary necrosis have been reported (24-26).
Gastrointestinal lesions, bleeding, ulceration and perforation as well as nausea, vomiting and dyspepsia recorded (27).

## Any other adverse effects

Potent prostaglandin synthesis inhibitor (28).
*In vitro* rat liver microsomes denatured cytochrome $P_{450}$ to cytochrome $P_{420}$ independent of NADPH and enzymes. Activities of NADH-cytochrome b5 reductase, NADPH cytochrome *C* reductase and epoxide hydratase decreased (29).

## Any other comments

Usually administered as the sodium salt (CAS RN 74252-25-8).
Potential use as antiovulatory drug (30).
Cyclo-oxygenase activity inhibitor, reported to be transient and not continuing beyond 20 hr (31,32).

## References

1. *Arzneim.-Forsch.* 1980, **30**, 1398
2. *Oyo Yakuri* 1968, **2**, 70
3. Klaassen, C. D. *Toxicol. Appl. Pharmacol.* 1976, **38**, 127
4. *Eur. J. Med. Chem.* 1989, **24**, 91
5. *Arzneim.-Forsch.* 1969, **19**, 1198
6. *Oyo Yakuri* 1973, **7**, 333
7. Oberto, G. et al *Fundam. Appl. Toxicol.* 1990, **15**(4), 800-813
8. Grom, V. V. et al *Eksp. Med. (Riga)* 1987, **24**, 67-71 (Russ.) (*Chem. Abstr.* **108**, 68518d)
9. Klimnyuk, E. V. *Farmakol. Toksikol. (Moscow)* 1989, **52**(2), 81-82 (Russ.) (*Chem. Abstr.* **110**, 165775j)
10. Sanjar, S. et al *Jpn. J. Pharmacol.* 1989, **51**(2), 151-160
11. Montenegro, M. A. et al *J. Craniofacial Genet. Dev. Biol.* 1990, **10**(1), 83-94
12. Ratnasooriya, W. D. et al *Int. J. Fertil.* 1987, **32**(2), 152-156
13. Morozova, E. V. *Arkh. Anat., Gistol. Embriol.* 1989, **96**(3), 48-55 (Russ.) (*Chem. Abstr.* **110**, 225204j)
14. *Am. J. Obstet. Gynecol.* 1990, **162**(5), 1304-1310
15. Wichlinski, L. M. et al *Acta Pol. Pharm.* 1986, **43**(4), 276-282
16. Yonemitsu, M. et al *Yakuri to Chiryo* 1988, **16**(5), 2027-2032 (Japan.) (*Chem. Abstr.* **109**, 115961a)
17. Lebeders, T. H. et al *Br. J. Clin. Pharmacol.* 1991, **32**, 751-754
18. Wiest, D. B. et al *Clin. Pharmacol. Ther.* 1991, **49**, 550-557
19. Sugai, S. et al *J. Toxicol. Sci.* 1990, **15**(4), 245-262
20. Timperman, J. A. *J. Forensic Med.* 1971, **18**, 30-32
21. Sheehan, G. J. et al *Ann. Intern. Med.* 1989, **111**, 337-338
22. Johnson, N. M. et al *Br. Med. J.* 1977, **2**, 1291
23. Shobha, D. P. et al *Mutat. Res.* 1987, **188**(4), 343-347
24. Chan, X. *Lancet* 1987, **ii**, 340
25. Boiskin, I. et al *Ann. Intern. Med.* 1987, **106**, 776-777
26. Mitchell, H. et al *Lancet* 1982, **ii**, 558-559
27. Stewart, J. T. et al *Br. Med. J.* 1985, **290**, 787-788
28. Nagasawa, H. et al *Breast Cancer Res. Treat.* 1986, **8**(3), 249-255
29. Falzon, M. et al *Biochem. Pharmacol.* 1986, **35**(22), 4019-4024

30. Wu, S. et al *Zhongguo Yike Daxue Xuebao* 1989, **18**(Zengkau), 18-20 (Ch.) (*Chem. Abstr.* **113**, 126692x)
31. Raud, J. et al *Proc. Natl. Acad. Sci. U. S. A.* 1988, **85**(7), 2315-2319
32. Patrick, J. et al *J. Dev. Physiol.* 1987, **9**(3), 295-300

# I32   Inosine

**CAS Registry No.** 58-63-9
**Synonyms** Atorel; hypoxanthine nucleoside; hypoxanthine riboside; hypoxanthosine; Ino; Inosie; Trophicardyl
**Mol. Formula** $C_{10}H_{12}N_4O_5$                **Mol. Wt.** 268.23
**Uses** Biochemical research. Pharmaceutical preparations and veterinary drugs.
**Occurrence** Meat. Sugar beet.

## Physical properties

**M. Pt.** 212-213°C; **Partition coefficient** log $P_{ow}$ −2.08 (1).

## Solubility

Water: 1.6 g 100 ml$^{-1}$ at 20°C. Organic solvent: ethanol

## Mammalian and avian toxicity

### Acute data

$LD_{50}$ intraperitoneal rat, mouse 2900, 3175 mg kg$^{-1}$ respectively (2,3).
$LD_{50}$ subcutaneous mouse 5000 mg kg$^{-1}$ (2).

### Carcinogenicity and long-term effects

Carcinogenicity under investigation in oral rat 3.0, 30 mg animal$^{-1}$ day$^{-1}$ and subcutaneous 5.0, 50 mg animal$^{-1}$ day$^{-1}$ in long-term study (N. N. Petrov Research Institute of Oncology, Ministry of Public Health, Leningrad) (4).

## Any other comments

Concentrations in primary and secondary domestic sewage plant effluent reported as 0.011-0.050 and 0.020 mg l$^{-1}$, respectively (5).

## References

1. Verschueren, K. *Handbook of Environmental Data on Organic Chemicals* 2nd ed., 1983, Van Nostrand Reinhold, New York
2. *Drugs in Japan. Ethical Drugs* 6th ed., 1982, Japan Pharmaceutical Information Centre, Yakugyo Jiho Co., Ltd., Tokyo

3. *Pharm. Chem. J. (Engl. Transl.)* 1986, **20**, 160
4. IARC *Directory of Agents being tested for Carcinogenicity* 1992, No.15, WHO/IARC, Lyon
5. Wilson Pitt, W. et al *Environ. Sci. Technol.* 1975, **9**(12), 1068-1073

# I33   Inositol

**CAS Registry No.** 87-89-8

**Synonyms** *myo*-inositol; *cis*-1,2,3,5-*trans*-4,6-cyclohexanehexol; Dambose; inosital; meat sugar; mesoinosite

**Mol. Formula** $C_6H_{12}O_6$  **Mol. Wt.** 180.16

**Occurrence** Plants and animals. Growth factor for animals and microorganisms.

## Physical properties

**M. Pt.** 225-227°C; **Specific gravity** 1.752.

### Solubility

Water: 14 g 100 ml$^{-1}$ at 25°C. Organic solvent: ethanol (slightly)

## Mammalian and avian toxicity

### Metabolism and pharmacokinetics

Initial take-up *in vitro* by 1210 mouse leukaemia cells is directly proportional to the extracellular concentration. Synthesised from glucose. Levels maintained by a combination of synthesis and uptake by either diffusion or a low-affinity carrier (1). Accumulation *in vitro* by lens epithelial, kidney endothelial and Chinese hamster ovary cells was stimulated by hypertonic stresses. Reduction in cellular activity and proliferation caused by hypertonic stress was exacerbated by removal from the culture medium. It was suggested that accumulation of the compound has an important role in the acute response to osmotic stress (2).

## Any other comments

Mutation lessening properties (3).
Metabolism reviewed (4-6).
Nine possible stereoisomers;the most common natural form is *cis*-1,2,3,5-*trans*-4,6-cyclohexanehexol.

## References

1. Moyer, J. D. et al *Biochem. J.* 1988, **254**(1), 95-100
2. Hohman, T. C. et al *Int. Congr. Ser. – Excerpta Med.* 1990, **913**, 31-41
3. Pak, C. H. et al *Han' guk Wonye Hakhoechi* 1987, **28**(3), 277-281 (Korean) (*Chem. Abstr.* **110**, 72647d)
4. Downes, C. P. *FIDIA Res. Ser.* 1989, **20**, 1-14
5. Lord, J. M. et al *Biochem. Soc. Trans.* 1991, **19**(2), 315-320
6. Handler, J. S. et al *Colloq. INSERM* 1991, **208**, 473-481

# I34   Inositol niacinate

**CAS Registry No.** 6556-11-2
**Synonyms** inositol nicotinate; *myo*-inositol hexa-3-pyridinecarboxylate; Dilcit; Esantene; Hexaniat; Linodil; Mesonex; Palohex
**Mol. Formula** $C_{42}H_{30}N_6O_{12}$                    **Mol. Wt.** 810.74
**Uses** Vasodilator.

## Physical properties
**M. Pt.** 254.3-254.9°C.

## Mammalian and avian toxicity

### Acute data
$LD_{50}$ intravenous rat, mouse 268, 345 mg kg$^{-1}$, respectively (1).
$LD_{50}$ intraperitoneal mouse 6400 mg kg$^{-1}$ (2).

## References
1. *Drugs in Japan. Ethical Drugs* 6th ed., 1982, Yakugyo Jiho Co., Ltd., Tokyo
2. *Oyo Yakuri* 1973, **7**, 149

# I35  Iodine

$I_2$

CAS Registry No. 7553-56-2
**Synonyms** Actomar; Diiodine; Eranol; Iosan Superdip
**Mol. Formula** $I_2$                                         **Mol. Wt.** 253.81
**Uses** Manufacture of iodine compounds. In germicides, antiseptics. Catalyst.
Analytical chemistry reagent. Antihyperthyroid. In lubricants. In dyestuffs.
**Occurrence** Igneous rocks. Brine lakes, seawater and seaweed.

## Physical properties

**M. Pt.** 113.60°C; **B. Pt.** 185.24°C; **Specific gravity** $d^{20}$ (solid) 4.93; **Partition coefficient** log $P_{ow}$ 2.49; **Volatility** v.p. 0.305 mmHg at 25°C.

## Solubility

Water: 0.16 g $l^{-1}$ at 25°C. Organic solvent: benzene, ethanol, carbon disulfide, cyclohexane, carbon tetrachloride, chloroform, glacial acetic acid

## Occupational exposure

**US TLV (TWA)** ceiling limit 0.1 ppm (1 mg m$^{-3}$); **UK Short-term limit** 0.1 ppm (1 mg m$^{-3}$); **Supply classification** harmful.
**Risk phrases** Harmful by inhalation and in contact with skin (R20/21)
**Safety phrases** Do not breathe vapour – Avoid contact with eyes (S23, S25)

## Ecotoxicity

**Fish toxicity**
$LC_{50}$ (24 hr) channel catfish 0.44 mg $l^{-1}$ (1).

**Bioaccumulation**
Soil microorganisms isolated from humus horizons of podzolic and sod-podzolic soils containing iodine at concentrations of 0.12-0.24 mg iodine kg$^{-1}$ dry matter could accumulate ≤315 mg iodine kg$^{-1}$ dry matter and thus possessed an iodine concentration function. Soil bacterial and fungal biomass could bind ≤3.24% of total iodine of the 0-20 cm soil layer. Some strains of *Penicillium chrysogenum* were able to accumulate 4.1% in a medium containing 1%; accumulation at this concentration was higher than at its threshold concentration by 9600 times (2).
$^{129}$I accumulates within marine algae at 740 times greater than background levels (3). The average atom ratio of $^{129}$I/$^{127}$I in algal samples was $2.5 \times 10^{-4}$, compared with $2.2 \times 10^{-12}$ in the ocean as the general background due to the natural production of $^{129}$I. Found in byssal threads of *Mytilus coruscus* (4).

## Mammalian and avian toxicity

**Acute data**
$LD_{50}$ oral rabbit, rat, mouse 10-22 g kg$^{-1}$ (5).
$LD_{Lo}$ oral dog 800 mg kg$^{-1}$ (6).
$LC_{Lo}$ (1 hr) inhalation rat 800 mg m$^{-3}$ (7).
$LD_{Lo}$ intravenous dog 40 mg kg$^{-1}$ (8).
$LD_{Lo}$ subcutaneous rabbit 175 mg kg$^{-1}$ (8).

## Teratogenicity and reproductive effects

C57 ♀ mice injected with 20 µl iodised oil were mated 8 days later. Urine content 24 hr$^{-1}$ was 20 times higher than control and 18 times higher 2 months later. Serum T4 values in the progeny were not affected, however T4 levels were depressed. There was no difference between treated animal and control progeny in brain protein, body weight, motility, DNA and learning ability (9).

## Metabolism and pharmacokinetics

Slight dermal absorption. When ingested converted to iodide and trapped by the thyroid gland. Iodides not trapped are excreted mainly in the urine, with lesser quantities released in the faeces, sweat and saliva. Iodides cross the placenta and also appear in milk (10).

## Irritancy

0.1-1.6 ppm reported to cause irritation to eyes with excessive flow of tears, chest tightness, sore throat and headache (11).

## Sensitisation

Occupationally exposed workers exhibit allergic dermatoses (12).
With a modified Beuhler's technique weak sensitiser (1/10 guinea pigs) (13).

## Genotoxicity

*Escherichia coli* PQ37 (uvrB$^{-1}$) SOS chromotest without metabolic activation negative (14).

## Any other adverse effects to man

Workers exposed to vapour had increased frequency of gingivitis, glossitis and stomatitis (15).
Can cause goitre, hypothyroidism and hyperthyroidism. At doses >2 mg day$^{-1}$ thyroid hormone production falls and may produce chronic inhibition of hormone synthesis (16-19).
Acute poisoning by ingestion may lead to death due to circulatory failure, oedema of the glottis resulting in asphyxia, pulmonary oedema or aspiration pneumonia (10).

## Legislation

Included in Schedules 4 and 6 (Release into Air/Land: Prescribed Substances)Statutory Instrument No. 472, 1991 (20).

## Any other comments

Accumulation in freshwater invertebrates compared with the soil and water from ecosystems of the forest-steppe zone and tundra studied. Concentrations were the same in different species. Distribution was habitat dependent (21).
Case studies of occupational exposure to vapour causing saturation are reported. Risk of dysthyroidism is discussed and preventive measures suggested (22).
Experimental toxicology, epidemiology, human health effects, physico-chemical properties and workplace experience reviewed (23).
Disease, environmental distribution, autoimmune thyroiditis, immunity, risks to infants from milk of iodine-exposed mothers, role in thyroid function and cretinism, toxicological and physiological effects and prophylaxis reviewed (24-32)

# References

1. Le Valley, M. J. *Bull. Environ. Contam. Toxicol.* 1982, **29**, 7
2. Letunova, S. V. et al *Mengen – Spurenelem., Arbeitstag.* 1987, **1**, 11-26 (Russ.) (*Chem. Abstr.* **109**, 145929p)
3. Doshi, G. R. et al *Indian J. Mar. Sci.* 1989, **18**(2), 87-90
4. Ikuta, K. *Kenkyu Hokoku – Miyazaki Daigaku Nogakubu* 1986, **33**(2), 255-264
5. *Drugs of the Future* 1979, **4**, 876, J. R. Prous SA, International Publishers, Barcelona
6. *Abdernalden's Handbuch der Biologischen Arbeitsmethoden* 1935, **4**, 1289
7. Izmerov, N. F. et al *Toxicometric Parameters of Industrial Toxic Chemicals Under Single Exposure* 1982, Centre of International Projects, GKNT, Moscow
8. *Handbook of Toxicology* 1959, 5, W. B. Saunders, Philadelphia
9. Liu, X. L. et al *J. Endocrinol. Invest.* 1988, **11**(6), 399-401
10. *Martindale. The Extra Pharmacopoeia* 30th ed., 1993, The Pharmaceutical Press, London
11. Health and Safety Executive *Occupational Exposure Limits: Criteria Document Summaries* 1993, HMSO, London
12. Budina, L. V. et al *Gig. Tr. Prof. Zabol.* 1991, (10), 32-34 (Russ.) (*Chem. Abstr.* **115**, 262347g)
13. Goh, C. L. *Contact Dermatitis* 1989, **21**(3), 166-171
14. Olivier, P. et al *Mutat. Res.* 1987, **189**(3), 263-269
15. Polyak, A. L. *Gig. Tr. Prof. Zabol.* 1988, (7), 26-28 (Russ.) (*Chem. Abstr.* **109**, 134321h)
16. Li, M. et al *Lancet* 1987, **ii**, 257-259
17. Murray, I. P. C. et al *Lancet* 1967, **i**, 922-926
18. Gomolin, I. H. *Drug Intell. Clin. Pharm.* 1987, **21**, 726-727
19. Smerdely, P. et al *Lancet* 1989, **ii**, 661-664
20. *S. I. 1991 No. 472 The Environmental Protection (Prescribed Processes and Substances) Regulations* 1991, HMSO, London
21. Zhulidov, A. V. et al *Gidrobiol. Zh.* 1989, **25**(4), 70-75 (Russ.) (*Chem. Abstr.* **111**, 191775z)
22. Proust, B. et al *Arch. Mal. Prof. Med. Trav. Secur. Soc.* 1987, **48**(3), 207-211 (Fr.) (*Chem. Abstr.* **107**, 120356v)
23. *ECETOC Technical Report No. 30(4)* 1991, European Chemical Industry Ecology and Toxicology Centre, B-1160 Brussels
24. Tan, J. et al *J. Environ. Sci. (China)* 1989, **1**(1), 107-114
25. Muramatsu, Y. et al *Hoshasen Kagaku (Tokyo)* 1988, **31**(8), 199-203 (Japan.) (*Chem. Abstr.* **110**, 100752e)
26. Sundick, R. S. *Immunol. Ser.* 1990, **52**, 213-228
27. Ye, J. *Zhongguo Mianyixue Zazhi* 1987, **3**(5), 312-314 (Ch.) (*Chem. Abstr.* **108**, 148355p)
28. Ythier, H. et al *Med. Infant.* 1987, **94**(1), 55-58 (Fr.) (*Chem. Abstr.* **107**, 32420f)
29. Wang, X. *Xibei Shifan Xueyuan Xuebao, Ziran Kexueban* 1987, (1), 80-85 (Ch.) (*Chem. Abstr.* **107**, 153217d)
30. Polyak, A. L. *Gig. Sanit.* 1988, (2), 67-70 (Russ.) (*Chem. Abstr.* **108**, 155577r)
31. Fuge, R. *Trace Subst. Environ. Health* 1987, **21**, 74-87
32. Pennington, J. A. T. *NTIS Report, FDA/CFSAN-89/54* order No.PB89-183016, 1989 Natl. Tech. Inf. Ser., Springfield, VA

# I36  Iodine monochloride

ICl

**CAS Registry No.** 7790-99-0
**Synonyms** iodine chloride; chlorine iodide; chlorine monoiodide; iodochlorine; Wijs' chloride
**Mol. Formula** ClI                    **Mol. Wt.** 162.36
**Uses** In Wijs' solutions used to determine iodine values of fats and oils. Topical anti-infective.

## Physical properties

**M. Pt.** 27.2°C (α-form), 13.9°C (β-form); **B. Pt.** 97°C; **Specific gravity** $d_4^{29}$ 310.

**Solubility**
Organic solvent: ethanol, diethyl ether, carbon disulfide, acetic acid

## Occupational exposure

**UN No.** 1792; **HAZCHEM Code** 2PE; **Conveyance classification** corrosive substance.

## Mammalian and avian toxicity

**Acute data**
$LD_{Lo}$ oral rat 50 mg kg$^{-1}$ (1).
$LD_{Lo}$ dermal rat 500 mg kg$^{-1}$ (1).

## Legislation

Limited under EC Directive on Drinking Water Quality 80/778/EEC. Chlorides: guide level 25 mg l$^{-1}$ (2).
Included in Schedules 4 and 6 (Release into Air/Land: Prescribed Substances) Statutory Instrument No. 472, 1991 (3).

## References

1. *Kodak Company Report,* 21 May 1971 Rochester, NY
2. *EC Directive Relating to the Quality of Water Intended for Human Consumption* 1982, 80/778/EEC, Office for Official Publications of the European Communities, 2 rue Mercier, L-2985 Luxembourg
3. *S. I. 1991 No. 472 The Environmental Protection (Prescribed Processes and Substances) Regulations* 1991, HMSO, London

# I37  Iodoacetic acid

ICH2CO2H

**CAS Registry No.** 64-69-7
**Synonyms** MIA; monoiodoacetic acid
**Mol. Formula** $C_2H_3IO_2$                    **Mol. Wt.** 185.95

## Physical properties

**M. Pt.** 82-83°C.

**Solubility**
Organic solvent: ethanol

## Occupational exposure

**Supply classification** toxic.
**Risk phrases** Toxic if swallowed – Causes severe burns (R25, R35)
**Safety phrases** Do not breathe dust – Wear suitable protective clothing, gloves and
eye/face protection – If you feel unwell, seek medical advice (show label where
possible) (S22, S36/37/39, S44)

## Environmental fate

**Degradation studies**
A stable methanogenic mixed culture was enriched from an industrial environment to
utilise iodoacetic acid as the sole carbon and energy source for growth.
Dehalogenation was identified as metabolic route and methane, $CO_2$ and iodide ions
as products of metabolism (1).

## Mammalian and avian toxicity

**Acute data**
$LD_{50}$ oral mouse 83 mg $kg^{-1}$ (2).
$LD_{Lo}$ (30 min) inhalation rat 94 g $m^{-3}$ (3).
$LD_{50}$ subcutaneous rat, rabbit 60 mg $kg^{-1}$ (4,5).
$LD_{50}$ intravenous dog 45 mg $kg^{-1}$ (6).

## Any other adverse effects

Hepatocytes isolated from 2 wk old and adult mice (8-10 wk) were exposed for
incubation times up to 24 hr; positive hepatotoxicity was indicated (7).
Intravenous administration destroyed visual cells in the retina of rabbits and caused an
initial increase, then decrease of total glucose-6-phosphatase activity in retinal
homogenate. Total glucose-6-phosphatase activity in the retinal microsomal fraction
was decreased after 5 days (8).

## Legislation

Included in Schedules 4 and 6 (Release into Air/Land: Prescribed Substances)
Statutory Instrument No. 472, 1991 (9).

## Any other comments

Experimental toxicology, human health effects and physico-chemical properties
reviewed (10).

## References

1.  Egli, C. et al *Arch. Microbiol.* 1989, **152**(3), 218-223
2.  *J. Pharmacol. Exp. Ther.* 1946, **86**, 336
3.  *Russ. Pharmacol. Toxicol. (Engl. Transl.)* 1978, **41**, 113
4.  *Toxicol. Appl. Pharmacol.* 1973, **26**, 93
5.  *J. Physiol. (London)* 1934, **80**, 360
6.  *J. Natl. Cancer Inst.* 1963, **31**, 297

7.  Adamson, G. M. et al *Biochem. Pharmacol.* 1988, **37**(21), 4183-4190
8.  Durlu, Y. K. et al *Prog. Clin. Biol. Res.* 1989, **314**, 585-600
9.  *S. I. 1991 No. 472 The Environmental Protection (Prescribed Processes and Substances) Regulations* 1991, HMSO, London
10. *ECETOC Technical Report No. 30(4)* 1991, European Industry Ecology and Toxicology Centre, B-1160 Brussels

# I38   3-Iodoaniline

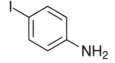

**CAS Registry No.** 626-01-7
**Synonyms** *m*-aminoiodobenzene; 3-aminoiodobenzene; *m*-iodoaniline; 3-iodobenzamine
**Mol. Formula** $C_6H_6IN$                     **Mol. Wt.** 219.03

## Physical properties

**M. Pt.** 25°C; **B. Pt.** $_{15}$ 145-146°C; **Flash point** >110°C; **Specific gravity** 1.821.

**Solubility**
Organic solvent: ethanol, chloroform

## Mammalian and avian toxicity

**Acute data**
$LD_{50}$ intravenous mouse 100 mg kg$^{-1}$ (1).

## Any other comments

Evaluated in 48 hr *Tetrahymena pyriformis* population growth impairment system (2).

## References

1.  US Army Armament Research and Development Command *Report NX #06766*, Chemical Systems Laboratory, NIOSH Exchange Chemicals, Aberdeen Proving Ground, MD 21010
2.  Schultz, T. W. et al *Sci. Total. Environ.* 1991, **109-110**, 569-580

# I39   4-Iodoaniline

**CAS Registry No.** 540-37-4
**Synonyms** *p*-iodoaniline; *p*-iodoaminobenzene; 4-aminoiodobenzene; 4-iodobenzamine
**Mol. Formula** $C_6H_6IN$                     **Mol. Wt.** 219.03

## Physical properties

**M. Pt.** 67-68°C.

**Solubility**
Organic solvent: ethanol, chloroform, petroleum ether

## Mammalian and avian toxicity

**Acute data**
LD$_{50}$ oral redwing blackbird, rat 100, 523 kg$^{-1}$, respectively (1,2).

**Any other comments**
Evaluated in 48 hr *Tetrahymena pyriformis* population growth impairment system (3).

## References

1. *Toxicol. Appl. Pharmacol.* 1972, **21**, 315
2. *Cesk. Hyg.* 1978, **23**, 168
3. Schultz, T. W. et al *Sci. Total Environ.* 1991, **109-110**, 569-580

# I40   Iodobenzene

**CAS Registry No.** 591-50-4
**Synonyms** benzene, iodo-; benzene iodide; phenyl iodide
**Mol. Formula** C$_6$H$_5$I                          **Mol. Wt.** 204.01

## Physical properties

**M. Pt.** –30°C; **B. Pt.** 188-189°C; **Flash point** 74°C; **Specific gravity** d$_4^{15}$ 1.838;
**Partition coefficient** log P$_{ow}$ 3.25.

**Solubility**
Organic solvent: miscible with ethanol, chloroform, diethyl ether

## Ecotoxicity

**Invertebrate toxicity**
EC$_{50}$ (30 min) *Photobacterium phosphoreum* 3.23 ppm Microtox test (1).

## Any other adverse effects

≈50% of cultured ♂ Sprague-Dawley rat hepatocytes were killed on addition of
40 g l$^{-1}$ iodobenzene in dimethyl sulfoxide (2).

## Legislation

Included in Schedule 4 and 6 (Release into Air/Land: Prescribed Substances) Statutory Instrument No. 472, 1991 (3).

## References

1.  Kaiser, K. L. E. et al *Water Pollut. Res. J. Can.* 1991, **26**(3), 361-431
2.  Coleman, J. B. et al *Toxicol. Appl. Pharmacol.* 1990, **105**, 393-402
3.  *S. I. 1991 No. 472 The Environmental Protection (Prescribed Processes and Substances) Regulations* 1991, HMSO, London

# I41   1-Iodobutane

## CH₃CH₂CH₂CH₂I

**CAS Registry No.** 542-69-8
**Synonyms** *n*-butyliodide; butyl iodide; butane, 1-iodo-
**Mol. Formula** $C_4H_9I$                          **Mol. Wt.** 184.02

## Physical properties

**M. Pt.** −103.0°C; **B. Pt.** 130.4°C; **Flash point** 36°C; **Specific gravity** $d_4^{20}$ 1.616.

## Solubility
Organic solvent: ethanol, diethyl ether

## Mammalian and avian toxicity

### Acute data
$LC_{50}$ (4 hr) inhalation rat 6100 mg m$^{-3}$ (1).
$LD_{50}$ intraperitoneal mouse, rat 101, 692 mg kg$^{-1}$, respectively (2).

### Carcinogenicity and long-term effects
A slight but significant increase in lung tumours was seen in Strain A mice injected with a total dose of 2.42 g kg$^{-1}$ (3).

## Legislation

Included in Schedule 4 and 6 (Release into Air/Land: Prescribed Substances) Statutory Instrument No. 472, 1991 (4).

## References

1.  Diechmann, W. B. *Toxicology of Drugs and Chemicals* 1969, Academic Press, New York
2.  Izmerov, N. F. et al *Toxicometric Parameters of Industrial Toxic Chemicals under Single Exposure* 1982, CIP, Moscow
3.  Poirier, L. A. et al *Cancer Res.* 1975, **35**, 1411-1415
4.  *S. I. 1991 No. 472 The Environmental Protection (Prescribed Processes and Substances) Regulations* 1991, HMSO, London

# I42   2-Iodobutane

## CH₃CH₂CHICH₃

$CH_3CH_2CHICH_3$

**CAS Registry No.** 513-48-4
**Synonyms** *sec*-butyl iodide; 2-butyl iodide; *sec*-iodobutane; butane, 2-iodo-
**Mol. Formula** $C_4H_9I$                      **Mol. Wt.** 184.02

## Physical properties

**M. Pt.** −104°C; **B. Pt.** 120°C; **Flash point** −10/21°C; **Specific gravity** $d_4^{20}$ 1.592.

### Solubility
Organic solvent: ethanol, diethyl ether

## Occupational exposure

**UN No.** 2390; **HAZCHEM Code** 2₩E; **Conveyance classification** flammable liquid.

## Ecotoxicity

### Invertebrate toxicity
$EC_{50}$ (4 hr) *Daphnia pulex* 10.93 mg $l^{-1}$ (1).

## Mammalian and avian toxicity

### Carcinogenicity and long-term effects
A slight but significant increase in the number of lung tumours was seen in ♂, ♀ mice injected with a total of 5.99 g $kg^{-1}$ (2).

## Genotoxicity

*Escherichia coli* P3478E without metabolic activation positive (3).

## Legislation

Included in Schedule 4 and 6 (Release into Air/Land: Prescribed Substances) Statutory Instrument No. 472, 1991 (4).

## References

1. Smith, S. B. et al *J. Great Lakes Res.* 1988, **14**(4), 394-404
2. Poirier, L. A. et al *Cancer Res.* 1975, **35**, 1411-1415
3. Fluck, E. R. et al *Chem.–Biol. Interact.* 1976, **15**, 219-231
4. *S. I. 1991 No. 472 The Environmental Protection (Prescribed Processes and Substances) Regulations* 1991, HMSO, London

# I43  5-Iododeoxyuridine

CAS Registry No. 54-42-2
Synonyms 2'-deoxy-5-iodouridine; 1-(2-deoxy-$\beta$-D-ribofuranosyl)-5-iodouracil;
Dendrid; Herplex; Iduridin, Ophthalmadine
Mol. Formula $C_9H_{11}IN_2O_5$                     Mol. Wt. 354.10
Uses Antiviral agent, used in treatment of *Herpes simplex* keratitis and cutaneous
forms of *Herpes simplex* and *zoster*.

## Physical properties

M. Pt. 194°C.

### Solubility
Water: 2.0 mg ml$^{-1}$ at 25°C. Organic solvent: methanol, ethanol, acetone, diethyl
ether, chloroform, ethyl acetate, 1,4-dioxane

## Occupational exposure

HAZCHEM Code 2⊠E; Conveyance classification flammable liquid.

## Mammalian and avian toxicity

### Acute data
$LD_{50}$ intraperitoneal mouse, rat 1000, 4000 mg kg$^{-1}$ respectively (1,2).

### Carcinogenicity and long-term effects
Squamous carcinoma observed in one patient associated with topical
iododeoxyuridine treatment (3).

### Teratogenicity and reproductive effects
Embryotoxic in chick embryotoxicity screening test at >0.3 µg embryo$^{-1}$.
Administration on day 2 was followed by various forms of the caudal regression
syndrome (4).

### Metabolism and pharmacokinetics
$^3$H-5-iodo-2'-deoxyuridine (IDU) was detected in rat epidermis but not in circulating
blood 60 min after local application of a 40% w/v solution of $^3$H-IDU in dimethyl
sulfoxide (5).
Obeys Michaelis-Menten equation elimination kinetics (6).

### Sensitisation

Guinea pig sensitisation studies by the Magnusson and Kligman method showed slight transient erythematous responses in some animals after the first and second challenges (5).

### Genotoxicity

*In vitro* human lymphocyte cells (48 hr) 50 mg l$^{-1}$ and *in vitro* human fibroblasts (72 hr) 50 mg l$^{-1}$ increased frequency of sister chromatid exchanges (7).

### Any other adverse effects to man

Cytotoxic to cultured human conjunctival cells, thus in clinical ophthalmological applications, concentrations below 100 µg ml$^{-1}$ are recommended (8).

### Legislation

Included in Schedule 4 and 6 (Release into Air/Land: Prescribed Substances) Statutory Instrument No. 472, 1991 (9).

### References

1. *J. Natl. Cancer. Inst.* 1979, **62**, 911
2. *Adv. Teratol.* 1968, **3**, 181
3. Koppang, H. S. et al *Br. J. Dermatol.* 1983, **108**, 501-503
4. Jelinek, R. et al *Folia Morphol. (Prague)* 1987, **35**(4), 374-380
5. Bravo, M. L. et al *Methods Find. Exp. Clin. Pharmacol.* 1988, **10**(11), 705-709
6. Wagner, J. G. et al *Sel. Cancer Ther.* 1989, **5**(4), 193-203
7. Cassiman, J. J. et al *Br. Med. J.* 1981, **283**, 817-818
8. Ikoma, N. *Kanazawa Ika Daigaku Zasshi* 1990, **15**(2), 121-129, (Japan.) (*Chem. Abstr.* **113**, 224235v)
9. *S. I. 1991 No. 472 The Environmental Protection (Prescribed Processes and Substances) Regulations* 1991, HMSO, London

# I44   Iodoethane

## CH3CH2I

**CAS Registry No.** 75-03-6
**Synonyms** ethane, iodo-; ethyl iodide; hydriodic ether; monoiodoethane
**Mol. Formula** $C_2H_5I$         **Mol. Wt.** 155.97

### Physical properties

**M. Pt.** −108°C; **B. Pt.** 72°C; **Flash point** >71°C; **Specific gravity** $d_{20}^{20}$ 1.950; **Partition coefficient** log $P_{ow}$ 2.00.

### Solubility

Organic solvent: ethanol

### Environmental fate

#### Degradation studies

Oxidised by *Nitrosomonas europaea* at a rate of 19 nmol min mg protein$^{-1}$. Major product detected was acetic acid (1).

## Mammalian and avian toxicity

### Acute data

$LC_{50}$ (30 min) inhalation rat 65,000 mg m$^{-3}$ (2).
$LD_{50}$ subcutaneous mouse 1000 mg kg$^{-1}$ (3).
$LD_{50}$ intraperitoneal guinea pig, rat, mouse 322-560 mg kg$^{-1}$ (4).

### Genotoxicity

*Escherichia coli* WP2 uvrA without metabolic activation positive (5).

### Legislation

Included in Schedule 4 and 6 (Release into Air/Land: Prescribed Substances)
Statutory Instrument No. 472, 1991 (6).

### References

1.  Rasche, M. E. et al *J. Bacteriol.* 1990, **172**(9), 5368-5373
2.  *Fiz. Akt. Vesh.* 1975, **7**, 35
3.  *Japan. J. Pharmacol.* 1954, **3**, 99
4.  Izmerov, N. F. et al *Toxicometric Parameters of Industrial Toxic Chemicals under Single Exposure* 1982, CIP, Moscow
5.  Hemminki, K. et al *Arch. Toxicol.* 1980, **46**, 277-285
6.  *S. I. 1991 No. 472 The Environmental Protection (Prescribed Processes and Substances) Regulations* 1991, HMSO, London

# I45   Iodofenphos

**CAS Registry No.** 18181-70-9
**Synonyms** phosphorothioic acid, *O*-(2,5-dichloro-4- iodophenyl) *O,O*-dimethyl ester; Alfacron; Ciba 9491; iodophos; Monocron 9491; Nuvanol N; OMS 1211; Jodfenphos
**Mol. Formula** $C_8H_8Cl_2IO_3PS$              **Mol. Wt.** 413.00
**Uses** Insecticide/acaricide used to control flies and ticks on cattle.

## Physical properties

**M. Pt.** 72-73°C.

### Solubility

Water: <2 mg l$^{-1}$ at 20°C. Organic solvent: acetone, xylene, dichloromethane, benzene, isopropanol

# Ecotoxicity

### Fish toxicity
$LC_{50}$ (96 hr) rainbow trout, goldfish, bluegill sunfish 0.06-0.75 mg $l^{-1}$ (1).
Toxic to carp at $\geq$0.01 mg $l^{-1}$. Changes were observed in the numbers of blood corpuscles, cardiovascular activity and growth rate. Juveniles were more sensitive than adults (2).

## Mammalian and avian toxicity

### Acute data
$LD_{50}$ oral rat, rabbit, mouse, dog 2-3 g $kg^{-1}$ (3,4).
$LC_{50}$ (6 hr) inhalation rat >0.246 mg $l^{-1}$ (1).
$LD_{50}$ dermal rabbit, rat 500, 2150 mg $kg^{-1}$, respectively (4,5).

### Sub-acute data
In 90 day feeding trials, no-effect level for rats was 5 mg $kg^{-1}$ diet and for dogs 15 mg $kg^{-1}$ diet (1).

### Metabolism and pharmacokinetics
92% of an oral dose in rats was eliminated within 24 hr (6).

## Legislation
Limited under EC Directive on Drinking Water Quality 80/778/EEC. Pesticides: maximum admissible concentration 0.1 $\mu$g $l^{-1}$ (7).
Included in Schedule 6 (Release into Land: Prescribed Substances) Statutory Instrument No. 472, 1991 (8).

## Any other comments
Non-toxic to birds. Toxic to bees (1).

## References
1. *The Agrochemicals Handbook* 3rd ed., 1991, RSC, London
2. Guseva, S. S. et al *Biol. Nauki (Moscow)* 1988, (1), 53-58 (Russ.) (*Chem. Abstr.* **108**, 145069n)
3. *Guide to the Chemicals Used In Crop Protection* 1968, **5**, 197, Information Canada, 171 Slater St., Ottawa, Ontario
4. *Wirksubstanzen der Pflanzenschutz- und Schaedlingsbekaempfungsmittel* 1971-1976, Verlag Paul Parey, Berlin
5. *Special Publication of the Entomological Society of America* 1974, **78-1**, 45, 4603 Calvert Rd., College Park, MD
6. Johanssen, F. R. et al *J. Econ. Entomol.* 1970, **63**, 693-697
7. *EC Directive Relating to the Quality of Water Intended for Human Consumption* 1982, 80/778/EEC, Office for Official Publications of the European Communities, 2 rue Mercier, L-2985 Luxembourg
8. *S. I. 1991 No. 472 The Environmental Protection (Prescribed Processes and Substances) Regulations* 1991, HMSO, London

# I46    Iodoform

## CHI₃

**CAS Registry No.** 75-47-8
**Synonyms** methane, triiodo-; carbon triiodide; triiodomethane
**Mol. Formula** $CHI_3$                          **Mol. Wt.** 393.73
**Uses** Topical anti-infective.

## Physical properties
**M. Pt.** 120-123°C; **Specific gravity** 4.008.

### Solubility
Organic solvent: benzene, acetone, ethanol, chloroform, diethyl ether, glycerol, carbon disulfide, olive oil

## Occupational exposure
**US TLV (TWA)** 0.6 ppm (10 mg m$^{-3}$); **UK Long-term limit** 0.6 ppm (10 mg m$^{-3}$); **UK Short-term limit** 1.0 ppm (20 mg m$^{-3}$).

## Ecotoxicity

### Fish toxicity
$LC_{50}$ (24 hr) minnow 1.2 mg l$^{-1}$ (1).

### Invertebrate toxicity
$LC_{50}$ (24 hr) *Daphnia* 0.1 mg l$^{-1}$ (1).

## Mammalian and avian toxicity

### Acute data
$LD_{50}$ oral rat, rabbit, mouse 355-810 mg kg$^{-1}$ (2).
$LC_{50}$ (7 hr) inhalation rat 165 ppm (3).
$LD_{50}$ dermal rat 1184 mg kg$^{-1}$ (2).
$LD_{50}$ subcutaneous mouse 630 mg kg$^{-1}$ (4).

### Carcinogenicity and long-term effects
Judged non-carcinogenic in studies of ♂ and ♀ Fischer 344 or Osborne-Mendel rats and B6C3F₁ mice by the National Cancer Institute and National Toxicology Program (5).

### Teratogenicity and reproductive effects
Minimum toxic concentration for rat spermatozoa exposed for 35 day was 2 mg kg$^{-1}$ and for rat embryos was 25 mg kg$^{-1}$. No mutagenic or teratogenic effects were observed in rats (1).

## Genotoxicity
*Salmonella typhimurium* TA98, TA100, TA1535 with and without metabolic activation positive, TA1537 with metabolic activation negative (5).
Induced unscheduled DNA synthesis. Exposure of Syrian hamster embryo cells for 18-20 hr produced a significant level of sister chromatid exchanges (6).

---

Treatment of V-79 cells with iodoform caused reduced growth at 30 μg ml$^{-1}$ for 24 or 48 hr, complete inhibition of growth at 100 μg ml$^{-1}$ for 24 or 48 hr and no inhibitory effect on survival at 10-30 μg ml$^{-1}$ for 2-24 hr (7).

## Any other adverse effects

Induced lipid peroxidation and inactivation of cytochrome $P_{450}$ in rat liver microsomes at oxygen partial pressures of 0-1 mmHg (8).

0.39 g l$^{-1}$ in dimethyl sulfoxide incubated with rat red blood cells for 2 hr did not significantly change the content of oxyhaemoglobin or methaemoglobin relative to untreated cells (9).

Morphological transformation was induced in Syrian hamster embryo cells by treatment with iodoform for 48 hr (6).

## Legislation

Included in Schedule 4 and 6 (Release into Air/Land: Prescribed Substances) Statutory Instrument No. 472, 1991 (10).

## Any other comments

Human health effects, experimental toxicology and workplace experience reviewed (11).

## References

1. Khristov, K. T. *Gig. Sanit.* 1987, (11), 74-75 (Russ.) (*Chem. Abstr.* **108**, 50831r)
2. *Zdravookhr. Kaz.* 1983, **27**(5), 9
3. *J. Toxicol. Environ. Health* 1981, **8**, 59
4. *Toxicol. Appl. Pharmacol.* 1962, **4**, 354
5. Zeiger, E. et al *Cancer Res.* 1987, **47**, 1287-1296
6. Suzuki, H. *Shigaku* 1987, **74**(6), 1385-1403
7. Bae, R. D. et al *Shigaku* 1987, **75**(5), 997-1004 (Japan.) (*Chem. Abstr.* **108** 126050h)
8. DeGroot, H. et al *Toxicol. Appl. Pharmacol.* 1989, **97**(3), 530-537
9. Hidalgo, F. J. et al *Toxicol. Lett.* 1990, **52**, 191-199
10. *S. I. 1991 No. 472 The Environmental Protection (Prescribed Processes and Substances) Regulations* 1991, HMSO, London
11. *ECETOC Technical Report No. 30(4)* 1991, European Chemical Industry Ecology and Toxicology Centre, B-1160 Brussels

# I47   Iodomethane

CH3I

**CAS Registry No.** 74-88-4
**Synonyms** methane, iodo-; methyl iodide
**Mol. Formula** CH$_3$I                    **Mol. Wt.** 141.94
**Uses** In methylations. Used in microscopy due to high refractive index. Used in testing for pyridine.
**Occurrence** Produced by various marine organisms.

## Physical properties

**M. Pt.** $-66.5°C$; **B. Pt.** $42.5°C$; **Specific gravity** $d_4^{20}$ 2.28; **Partition coefficient** log $P_{ow}$ 1.51.

**Solubility**
Water: $14 \text{ g l}^{-1}$ at 20°C. Organic solvent: ethanol, diethyl ether, benzene

## Occupational exposure

**US TLV (TWA)** 2 ppm ($12 \text{ mg m}^{-3}$); **UK Long-term limit** under review; **UK Short-term limit** under review; **UN No.** 2644; **HAZCHEM Code** 2XE; **Conveyance classification** toxic substance; **Supply classification** toxic.
**Risk phrases** Harmful in contact with skin – Toxic by inhalation and if swallowed – Irritating to respiratory system and skin – Possible risk of irreversible effects (R21, R23/25, R37/38, R40)
**Safety phrases** Wear suitable protective clothing and gloves – In case of insufficient ventilation, wear suitable respiratory equipment – If you feel unwell, seek medical advice (show label where possible) (S36/37, S38, S44)

## Ecotoxicity

**Bioaccumulation**
Concentrated in *Modialus modiolus*: $10 \text{ ng g}^{-1}$ in digestive tissue, $188 \text{ ng g}^{-1}$ in mantle; and in coalfish: $4 \text{ ng g}^{-1}$ in muscle and $166 \text{ ng g}^{-1}$ in brain. Enrichment was 2-25 times greater on dry weight basis compared with seawater concentrations (1).

## Mammalian and avian toxicity

**Acute data**
$LD_{50}$ oral rat $76 \text{ mg kg}^{-1}$ (2).
$LC_{50}$ (4 hr) inhalation rat $1300 \text{ mg kg}^{-1}$ (3).
$LD_{Lo}$ dermal rat $800 \text{ mg kg}^{-1}$ (4).
$LD_{50}$ subcutaneous mouse $110 \text{ mg kg}^{-1}$ (5).
$LD_{50}$ intraperitoneal guinea pig, rat, mouse $51\text{-}172 \text{ mg kg}^{-1}$ (6).

**Carcinogenicity and long-term effects**
No adequate evidence for carcinogenicity to humans or animals, IARC classification group 3 (7).
Subcutaneous sarcomas were observed in 9/12 and 6/6 rats injected with 10 and $20 \text{ mg kg}^{-1}$, respectively, for 1 yr and in 4/14 in rats injected once with $50 \text{ mg kg}^{-1}$. Local fibrosarcomas and sarcomas were seen more than 1 yr after the first injection (8).
Lung tumours were reported in 6/29, 4/19, 6/20 and 5/11 A/He mice treated with 0, 8.5, 21.3 and $44.0 \text{ mg kg}^{-1}$, respectively (9).

**Metabolism and pharmacokinetics**
Rats administered orally with $76 \text{ mg kg}^{-1}$ exhaled 1% unchanged within 30 min (2).
Urinary metabolites in rats injected with $50 \text{ mg kg}^{-1}$ originated from *S*-methyl glutathione and included *S*-methylcysteine, *N*-acetyl-*S*-methylcysteine, *S*-methyl thioacetic acid and *N*-(methythioacetyl)glycine (10).
Disappearance of methyl iodide in the presence of glutathione was catalysed by liver and kidney homogenates (11).

## Genotoxicity

*Salmonella typhimurium* TA100 with and without metabolic activation weakly positive (12).
*Salmonella typhimurium* TA1535, TA1538 with and without metabolic activation positive and negative, respectively (13).
*Escherichia coli pol A* without metabolic activation positive (13).
*In vitro* mouse lymphoma L5178Y tk$^+$/tk$^-$ positive (14).

## Any other adverse effects to man

A worker manufacturing methyl iodide exhibited severe neurological symptoms before death. Autopsy showed congested organs (15).
Workers poisoned non-fatally during the manufacture of methyl iodide suffered neurological disturbances such as vertigo, visual disturbances and weakness, followed by psychological disturbances and intellectual impairment (16).

## Legislation

Regulated under the OSHA Air Contaminants Standard 1989 (17).
Included in Schedule 4 and 6 (Release into Air/Land: Prescribed Substances) Statutory Instrument No. 472, 1991 (18).

## Any other comments

Induced chromatid aberrations in *Vicia fabia* root-tip meristems (19).
Human health effects, experimental toxicology, physico-chemical properties, epidemiology and workplace experience reviewed (20).

## References

1. Dickson, A. G. et al *Mar. Pollut. Bull.* 1976, **7**(9), Sept
2. Johnson, M. K. *Biochem. J.* 1966, **98**, 38-43
3. Diechmann, W. B. *Toxicology of Drugs and Chemicals* 1969, Academic Press, New York
4. *Kodak Company Reports* No.21, May 1971
5. *Toxicol. Appl. Pharmacol.* 1962, **4**, 354
6. Izmerov, N. F. et al *Toxicometric Parameters of Industrial Toxic Chemicals under Single Exposure* 1982, CIP, Moscow
7. *IARC Monograph* 1987, **Suppl. 7**, 66
8. Druckrey, H. et al *Z. Krebsforsch.* 1970, **74**, 241-273 (Ger.)
9. Poirier et al *Cancer Res.* 1975, **35**, 1411-1415
10. Barnsley, E. A. et al *Biochem. J.* 1965, **95**, 77-81
11. Johnson, M. K. *Biochem. J.* 1966, **98**, 44-56
12. McCann, J. et al *Proc. Natl. Acad. Sci. U. S. A.* 1975, **72**(12), 5135-5139
13. Rozenkranz, H. S. et al *J. Natl. Cancer Inst.* 1979, **62**, 873-892
14. Moore, M. M. et al *Mutat. Res.* 1985, **151**, 147-159
15. Garland, A. et al *Br. J. Ind. Med.* 1945, **2**, 209-211
16. Appel, G. B. et al *Ann. Intern. Med.* 1975, **82**, 534-536
17. Paxman, D. G. et al *Regul. Toxicol. Pharmacol.* 1990, **12**, (3, Pt.1), 296
18. *S. I. 1991 No. 472 The Environmental Protection (Prescribed Processes and Substances) Regulations* 1991, HMSO, London
19. Rieger, R. et al *Mutat. Res.* 1988, **208**(2), 101-104
20. *ECETOC Technical Report No. 30(4)* 1991, European Chemical Industry Ecology and Toxicology Centre, B-1160 Brussels

# I48    1-Iodo-2-methylpropane

$$CH_3CH(CH_3)CH_2I$$

**CAS Registry No.** 513-38-2
**Synonyms** propane, 1-iodo-2-methyl-; isobutyl iodide
**Mol. Formula** $C_4H_9I$                                   **Mol. Wt.** 184.02

## Physical properties
**M. Pt.** −93°C; **B. Pt.** 120°C; **Flash point** 22°C; **Specific gravity** $d^{20}$ 1.605.

## Solubility
Organic solvent: miscible with ethanol, diethyl ether

## Occupational exposure
**UN No.** 2391; **HAZCHEM Code** 2ME; **Conveyance classification** flammable liquid.

## Mammalian and avian toxicity

### Acute data
$LC_{50}$ (4 hr) inhalation rat 6700 mg m$^{-3}$ (1).
$LD_{50}$ intraperitoneal mouse 549 mg kg$^{-1}$ (1).
$LD_{50}$ intraperitoneal rat 1241 mg kg$^{-1}$ (1).

## Legislation
Included in Schedule 4 and 6 (Release into Air/Land: Prescribed Substances)
Statutory Instrument No. 472, 1991 (2).

## References
1.  Diechmann, W. B. *Toxicology of Drugs and Chemicals* 1969, Academic Press, New York
2.  *S. I. 1991 No. 472 The Environmental Protection (Prescribed Processes and Substances) Regulations* 1991, HMSO, London

# I49    2-Iodo-2-methylpropane

$$CH_3CI(CH_3)CH_3$$

**CAS Registry No.** 558-17-8
**Synonyms** propane, 2-iodo-2-methyl-; *tert*-butyl iodide; trimethyliodomethane
**Mol. Formula** $C_4H_9I$                                   **Mol. Wt.** 184.02

## Physical properties
**M. Pt.** −38°C; **B. Pt.** 99-100°C; **Flash point** 7°C; **Specific gravity** 1.544.

## Genotoxicity
*Escherichia coli* P3478 without metabolic activation positive (1).

## Legislation

Included in Schedule 4 and 6 (Release into Air/Land: Prescribed Substances) Statutory Instrument No. 472, 1991 (2).

## References

1. Fluck, E. R. et al *Chem.–Biol. Interact.* 1976, **15**, 219-231
2. *S. I. 1991 No. 472 The Environmental Protection (Prescribed Processes and Substances) Regulations* 1991, HMSO, London

# I50   4-Iodophenol

**CAS Registry No.** 540-38-5
**Synonyms** *p*-iodophenol
**Mol. Formula** $C_6H_5IO$                **Mol. Wt.** 220.01

## Physical properties

**M. Pt.** 93-94°C; **B. Pt.** $_5$ 138°C; **Specific gravity** 1.857; **Partition coefficient** log $P_{ow}$ 2.91 (1).

## Solubility

Organic solvent: ethanol, diethyl ether

## Mammalian and avian toxicity

### Acute data

$LD_{Lo}$ intraperitoneal mouse 700 mg $kg^{-1}$ (2).

### Carcinogenicity and long-term effects

The skin of 32 ♀ Sutter mice was treated with 25 µl of a solution of 0.3% 7,12-dimethylbenz[*a*]anthracene in benzene followed by 1 drop of 25 µl of a solution of 20% 4-iodophenol in benzene $2 \times wk^{-1}$ for 18 wk. Application to the back of the mouse resulted in an average of 0.47 papillomas $mouse^{-1}$ in 35% of the 31 survivors (3).

### Teratogenicity and reproductive effects

Chernoff/Kavlock assay performed on Sprague-Dawley rats exposed on day 11 of gestation to 0, 100, 333, 667, 1000 mg $kg^{-1}$. 333 mg $kg^{-1}$ decreased maternal body weight by 1 g at 24 hr. Calculated concentrations which would decrease the total litter weight by 60% day 6 post natal was >5000 mg $kg^{-1}$. Categorised as inactive in developmental potency category (4).

### Metabolism and pharmacokinetics

4 µg $cm^{-2}$ was applied to full-thickness hairless mouse skin. 46% was absorbed after 6 hr, 65% after 12 hr and 81% after 24 hr (5).

## Any other adverse effects

*p*-Halogenated phenols induced the release of $K^+$ from mitochondria decreasing the respiratory control index (6).

## References

1. Leo, A. et al *Chem. Res.* 1971, **71**, 525-616
2. *J. Pharm. Sci.* 1978, **67**, 1154
3. Boutwell, R. K. et al *Cancer Res.* 1959, **19**, 413-424
4. Kavlock, R. J. *Teratology* 1990, **41**, 43-59
5. Hinz, R. S. et al *Fundam. Appl. Toxicol.* 1991, **17**, 575-583
6. Izushi, F. et al *Acta. Med. Okayama* 1988, **42**(1), 7-14.

# I51   6,3-Ionene

$$[-(CH_3)_2N(CH_2)_3N(CH_3)_2(CH_2)_6-]_n^{2+}.2Br-$$

**CAS Registry No.** 28728-55-4

**Synonyms** poly(dimethyliminio) hexamethylene (dimethylimino) trimethylene dibromide; Hexadimethrine bromide; Polybrene; COP1

## Mammalian and avian toxicity

### Acute data

$LD_{50}$ intraperitoneal mouse 30 mg $kg^{-1}$ (1).
$LD_{50}$ intravenous rat, mouse 20-28 mg $kg^{-1}$ (2).

## Legislation

Included in Schedule 4 and 6 (Release into Air/Land: Prescribed Substances) Statutory Instrument No. 472, 1991 (3).

## References

1. *Russ. Pharmacol. Toxicol. (Engl. Transl.)* 1974, **37**, 267
2. *United States Patent Document* No.4013507, Commissioner of Patents and Trademarks, Washington, DC
3. *S. I. 1991 No. 472 The Environmental Protection (Prescribed Processes and Substances) Regulations* 1991, HMSO, London

# I52   α-Ionone

**CAS Registry No.** 127-41-3
**Synonyms** 3-buten-2-one, 4-(2,6,6-trimethyl-2-cyclohexen -1-yl)-, (*E*)-;
α-cyclocitrylideneacetone; (*E*)-α-ionone; *trans*-α-ionone
**Mol. Formula** $C_{13}H_{20}O$                                      **Mol. Wt.** 192.30

## Physical properties

**B. Pt.** $_{13}$ 131°C; **Flash point** >100/113°C; **Specific gravity** 0.930.

## Environmental fate

### Degradation studies

Converted by *Aspergillus niger* JTS191 to *cis*-3-hydroxy-α-ionone;
*trans*-3-hydroxy-α-ionone; 3-oxo-α-ionone; 2,3-dehydro-α-ionone;
3,4-dehydro-β-ionone; and 1-(6,6-dimethyl-2-methylene-
3-cyclohexenyl)-buten-3-one (1).

## Mammalian and avian toxicity

### Acute data

$LD_{50}$ oral rat 4590 mg kg$^{-1}$ (2).

## References

1.  Yamazaki, Y. et al *Appl. Environ. Microbiol.* 1988, **54**(10), 2354-2360
2.  *Food Cosmet. Toxicol.* 1964, **2**, 327

# I53   Ioxynil

**CAS Registry No.** 1689-83-4
**Synonyms** 4-hydroxy-3,5-diiodobenzonitrile; 2,6-diiodo-4-cyanophenol; Actvil;
Bentrol; CA69-15; Certrol
**Mol. Formula** $C_7H_3I_2NO$                                      **Mol. Wt.** 370.92
**Uses** Herbicide.

## Physical properties

**M. Pt.** 212-214°C; **Volatility** v.p. $<7.52 \times 10^{-6}$ mmHg.

### Solubility

Water: 50 mg $l^{-1}$. Organic solvent: acetone, ethanol, methanol, cyclohexanone, tetrahydrofuran, dimethylformamide, chloroform, carbon tetrachloride

## Occupational exposure

**Supply classification** toxic.

**Risk phrases** Toxic by inhalation, in contact with skin and if swallowed (R23/24/25)
**Safety phrases** Keep out of reach of children – Keep away from food, drink and
animal feeding stuffs – If you feel unwell, seek medical advice (show label where
possible) (S2, S13, S44)

## Ecotoxicity

### Fish toxicity

$LC_{50}$ (48 hr) harlequin fish 3.3 mg $l^{-1}$ (ioxynil-sodium) (1).

### Bioaccumulation

The sorption kinetics in the unicellular microalgae *Ankistrodesmus braunii* are
independent of the herbicide concentration. Sorption is influenced by environmental
factors including light, temperature, pH and oxygen concentration. Thus, the response
of the target cells to environmental factors is at least as important for sorption in the
cells and the prediction of accumulation can only be partially deduced from the
properties of the pesticide molecule (2).

## Environmental fate

### Degradation studies

In soil $t_{1/2} \approx 10$ days. Degraded to less toxic compounds such as hydroxybenzoic acid
by deiodination and hydrolysis (3).

### Abiotic removal

Decomposed by UV light (3).

### Absorption

Highly mobile and resisted degradation in groundwater under treated ploughed
land (4).

## Mammalian and avian toxicity

### Acute data

$LD_{50}$ oral rat, mouse, cat, rabbit, guinea pig, pheasant, chicken 75-230 mg $kg^{-1}$
(3,5,6).
$LD_{Lo}$ oral human 28 mg $kg^{-1}$ (7).
$LC_{50}$ (6 hr) inhalation rat >3 mg $l^{-1}$ (3).
$LD_{Lo}$ dermal rat 210 µg $kg^{-1}$ (8).
$LD_{50}$ intravenous mouse 56 mg $kg^{-1}$ (9).

### Sub-acute data

In 90 day feeding trials, no-effect level for rats was 5.5 mg $kg^{-1}$ $day^{-1}$ (sodium salt)
and 4 mg $kg^{-1}$ $day^{-1}$ (octanoate) (3).

## Any other adverse effects

*In vitro* $ID_{50}$ (72 hr) 3T3-L1 cell culture $47 \pm 1$ µg ml$^{-1}$ FRAME kenacid blue assay (10).

## Legislation

Included in Schedule 4 (Release into Air: Prescribed Substances) Statutory Instrument No. 472, 1991 (11).
Included in Schedule 6 (Release into Land: Prescribed Substances) Statutory Instrument No. 472, 1991 (11).
Limited under EC Directive on Drinking Water Quality 80/778/EEC. Pesticides: maximum admissible concentration 0.1 µg l$^{-1}$ (12).

## Any other comments

WHO Class II; EPA toxicity class II (3).
Experimental toxicology, human health effects, physico-chemical properties reviewed (13).

## References

1. *The Pesticide Manual* 9th ed., 1991, The British Crop Protection Council, Farnham
2. Neumann, W. et al *Pestic. Biochem. Physiol.* 1987, **27**(2), 189-200
3. *The Agrochemicals Handbook* 3rd ed., 1991, RSC, London
4. Vavra, J. *Agrochemia (Bratislava)* 1987, **27**(3), 82-84
5. *World Rev. Pest Control* 1970, **9**, 119
6. Klimmer, O. R. *Pflanzenschutz-und Schaedlingsbekaempfungsmittel: Abriss einer Toxikologie und Therape von Vergiftungen* 2nd ed., 1971, Hattingen, Germany (Ger.)
7. *Arch. Toxicol.* 1977, **37**, 241
8. Information Canada *Guide to the Chemicals used in Crop Protection* 1973, **6**, 304
9. *US Army Armament Research and Development Command, Chemical Systems Laboratory, NIOSH Exchange Chemicals*, Report No.02818, Aberdeen Proving Ground, MD
10. Riddell, R. J. et al *ATLA, Altern. Lab. Anim.* 1986, **14**(2), 86-92
11. *S. I. 1991 No. 472 The Environmental Protection (Prescribed Processes and Substances) Regulations* 1991, HMSO, London
12. *EC Directive Relating to the Quality of Water Intended for Human Consumption* 1982, 80/778/EEC, Office for Official Publications of the European Communities, 2 rue Mercier, L-2985 Luxembourg
13. *ECETOC Technical Report No. 30(4)* 1991, European Chemical Industry Ecology and Toxicology Centre, B-1160 Brussels

# I54    Ipecacuanha

**CAS Registry No.** 8012-96-2
**Synonyms** Cartagena ipecacuanha; Cephaelis ipecacuanha; ipecac, C. ipecacuanha; Uragoga ipecacuanha; ipecac syrup
**Uses** Emetic. Used as expectorant. Ruminatoric aid and antihistomonad for veterinary purposes.
**Occurrence** Dried rhizome and roots of Rio or Brazilian ipecac. Contains emetine, cephaeline, emetamine, ipecacuanic acid, protoemetrine psychotrine, methyl psychotrine and resin (1).

## Mammalian and avian toxicity

**Acute data**

$LD_{50}$ oral rat 7800 mg $kg^{-1}$ (2).

$LD_{Lo}$ oral human 70 mg $kg^{-1}$ (2).

$LD_{Lo}$ oral dog 5000 mg $kg^{-1}$ (3).

**Sensitisation**

Packers of ipecacuanha tablets suffered allergic reactions, characterised by rhinitis, conjunctivitus, and chest tightness due to inhalation (4).

**Any other adverse effects to man**

A 26 yr old woman experienced tachycardia, hypotension, dyspnoea and finally death after drinking 3-4 bottles of ipecac syrup a day for 3 months to induce vomiting in order to lose weight (5).

Humans dosed orally with 7.5 ml (syrup) experienced severe nausea, vomiting and headache (3).

### References

1. Lewis, R. J. *Sax's Dangerous Properties of Industrial Materials* 8th ed., 1992, Van Nostrand Reinhold, New York
2. Rumack, B. H. et al (Ed.) *Management of the Poisoned Patient* 1977, Science Press, Princetown, NJ
3. Goldenberg, M. M. et al *J. Pharm. Sci.* 1976, **65**, 1398-1400
4. Luczynska, C. M. et al *Clin. Allergy* 1984, **14**, 169-175
5. Adler, A. G. et al *JAMA, J. Am. Med. Soc.* 1980, **243**, 1927-1928

# I55 Iprobenfos

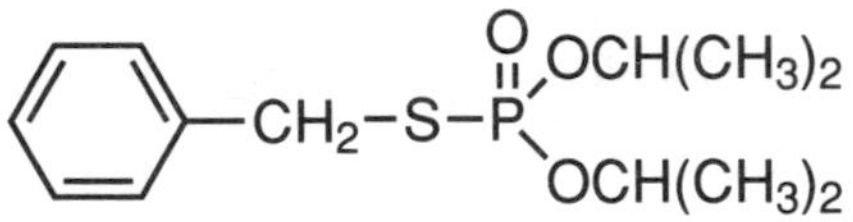

**CAS Registry No.** 26087-47-8

**Synonyms** phosphorothioic acid, *O,O*-bis(1-methylethyl) *S*-(phenylmethyl) ester; phosphorothioic acid, *S*-benzyl *O,O*-diisopropyl ester; IBP; Kitazin L; Ricid II

**Mol. Formula** $C_{13}H_{21}O_3PS$    **Mol. Wt.** 288.35

**Uses** Systemic fungicide used to control leaf and ear blast, stem rot and sheath blight.

### Physical properties

**B. Pt.** $_{0.04}$ 126°C; **Partition coefficient** log $P_{ow}$ 3.21 (1).

### Solubility

Water: 1 g $l^{-1}$ at 18°C. Organic solvent: methanol, acetone, xylene, acetonitrile

---

## Occupational exposure

**Supply classification** harmful.
**Risk phrases** Harmful if swallowed (R22)

## Ecotoxicity

**Fish toxicity**
$LC_{50}$ (24 hr) carp 3.7 mg $l^{-1}$ (2).

**Invertebrate toxicity**
$LC_{50}$ oral 2nd day of 5th instar of silkworm larvae ≈6.5 µg $g^{-1}$ (3).
$LC_{50}$ 3rd instar silkworm larvae 1.742 ppm in feed (3).

**Bioaccumulation**
Average bioconcentration factor in whole body of willow shiner was 33.2. The
excretion rate constant was 0.0017 g $ng^{-1}$ $hr^{-1}$ assuming 2nd-order kinetics (4).
Average bioconcentration factors in carp were 4.3-26.7 in various organs. The
excretion rate constants were between 0.002-0.024 for muscle, 0.001-0.020 for liver,
0.0004-0.004 for kidney and 0.002-0.023 for gallbladder (all g $ng^{-1}$ $hr^{-1}$) (5).

## Environmental fate

**Degradation studies**
Degradation in soil was affected by soil temperature, flooding, addition of rice straw
and treatment with mixed fertiliser. No degradation was observed in sterile soil
indicating degradation was due to microorganisms (6).

**Abiotic removal**
A 5000 mg $l^{-1}$ solution of NaClO was added to 10 mg $l^{-1}$ of iprobenfos in water
to give a $Cl^-$ concentration of 50 mg $l^{-1}$. Iprobenfos was degraded to various
products (7).

## Mammalian and avian toxicity

**Acute data**
$LD_{50}$ oral rat 490 mg $kg^{-1}$ (8).
$LD_{50}$ oral mouse 1760 mg $kg^{-1}$ (8).
$LC_{50}$ (4 hr) inhalation ♀, ♂ rat 0.34-1.12 mg $l^{-1}$ (1).
$LD_{50}$ dermal mouse 5000 mg $kg^{-1}$ (8).
Oral intoxication in guinea pigs, mice and rats (dose, duration unspecified) affected
the liver by inhibiting blood cholinesterase, causing changes to amino transferase and
alkaline phosphatase activities, weakening the antitoxic function of the liver and
increasing vital staining of the hepatic parenchyma (9).

**Carcinogenicity and long-term effects**
In 2 yr feeding trials, no effect level for ♂ rats 0.036, ♀ rats 0.45 mg $kg^{-1}$ $day^{-1}$ (1).

**Metabolism and pharmacokinetics**
Detected in heart, liver, spleen, lungs, kidney, fatty tissue and stomach wall of rats
administered 53 mg $kg^{-1}$ orally after 3 hr. After 1-2 days the heart, kidneys, stomach
wall and fat contained 0.47, 0.38, 0.225 and 2.7 mg $kg^{-1}$ respectively. $t_{1/2}$ in the body
was 2 days (10).

### Sensitisation

No sensitisation observed in guinea pigs treated orally but shock was observed after an intracardiac injection of $1: 1 \times 10^{-5}$ dilution (11).

### Legislation

Limited under EC Directive on Drinking Water Quality 80/778/EEC. Pesticides: maximum admissible concentration $0.1 \ \mu g \ l^{-1}$ (12).
Included in Schedule 6 (Release into Land: Prescribed Substances) Statutory Instrument No. 472, 1991 (13).

### References

1. *The Agrochemicals Handbook* 3rd ed., 1991, RSC, London
2. Hashimoto, E. et al *J. Pestic. Sci.* 1982, **7**, 457
3. Kuribayashi, S. *JARQ* 1988, **21**(4), 274-283
4. Tsuda, T. et al *Toxicol. Environ. Chem.* 1989, **24**(3), 185-190
5. Tsuda, T. et al *Comp. Biochem. Physiol., C: Comp. Pharmacol. Toxicol.* 1990, **96C**, (1), 23-26
6. Moon, Y. H. *Han'guk Nonghwa Hakhoechi* 1990, **33**(2), 133-137, (Korean) (*Chem. Abstr.* **114**, 160461b)
7. Imanishi, K. et al *Nara-ken Eisei Kenkyusho Nenpo* 1989 (Publ. 1990), (24), 52-55 (Japan.) (*Chem. Abstr.* **114**, 234671u)
8. *Farm Chemicals Handbook* 1991, Meister Publ. Co., Willoughby, OH
9. Chura, D. A. *Gig. Sanit.* 1988, (3), 70. (Russ.) (*Chem. Abstr.* **108**, 199834b)
10. Chura, D. A. et al *Gig. Tr. Prof. Zabol.* 1986, (11), 53-54 (Russ.) (*Chem. Abstr.* **106**, 28684g)
11. Petrus, V. S. et al *Gig. Tr. Prof. Zabol.* 1987, (4), 38-39 (Russ.) (*Chem. Abstr.* **107**, 34499n)
12. *EC Directive Relating to the Quality of Water Intended for Human Consumption* 1982, 80/778/EEC, Office for Official Publications of the European Communities, 2 rue Mercier, L-2985 Luxembourg
13. *S. I. 1991 No. 472 The Environmental Protection (Prescribed Processes and Substances) Regulations* 1991, HMSO, London

# I56   Iprodione

**CAS Registry No.** 36734-19-7
**Synonyms** 1-imadazolidinecarboxamide, 3-(3,5-dichlorophenyl)-*N*-(1-methylethyl)-2,4-dioxo-; Chipco 26019; Glycophen; Iprodial; Promidione; Rovral
**Mol. Formula** $C_{13}H_{13}Cl_2N_3O_3$          **Mol. Wt.** 330.17
**Uses** Fungicide used on pome and stone fruit, vegetables, ornamentals, cereals, potatoes, cotton, sunflowers, etc. Can be used as a post-harvest dip or seed treatment.

## Physical properties

**M. Pt.** ≈136°C.

**Solubility**
Water: 13 mg l$^{-1}$ at 20°C. Organic solvent: ethanol, methanol, acetone, dichloromethane

## Ecotoxicity

**Fish toxicity**
LC$_{50}$ (96 hr) bluegill sunfish, rainbow trout 2.25-6.7 mg l$^{-1}$ (1).

**Invertebrate toxicity**
9.9 mg l$^{-1}$ inhibited respiration in *Saccharomyces cerevisiae* by 9.6% with a glucose substrate and 8.3% with an ethanol substrate (2).

## Environmental fate

**Degradation studies**
25% was recovered in a plot treated with one application of 5 kg active ingredient ha$^{-1}$ 4 wk after treatment but none was recovered in plots treated with two and three applications of 5 kg active ingredient ha$^{-1}$ 16 and 10 days after treatment respectively implying rapid breakdown after repeated application (3).
Degradation occurred faster at pH 6.5 than pH 5.7 and little occurred at pH 4.3 (4).
Rapidly metabolised in soil forming carbon dioxide. t$_{1/2}$ 20-160 days (1).

## Mammalian and avian toxicity

**Acute data**
LD$_{50}$ oral mouse, rat 4000-4400 mg kg$^{-1}$ (5).
LC$_{50}$ (4 hr) inhalation rat >3.3 mg l$^{-1}$ (6).
LD$_{50}$ dermal rabbit, rat >1000-2500 mg kg$^{-1}$ (1).

**Carcinogenicity and long-term effects**
In 18 month feeding trials, rats receiving 1000 mg kg$^{-1}$ diet and dogs receiving 2400 mg kg$^{-1}$ day$^{-1}$ showed no ill effects (1).

## Any other adverse effects

♂ rats injected intraperitoneally with 0.13 or 0.33 g kg$^{-1}$ resulted in only minor or no alterations in the renal function parameters measured and renal morphology (7).

## Legislation

Limited under EC Directive on Drinking Water Quality 80/778/EEC. Pesticides: maximum admissible concentration 0.1 µg l$^{-1}$ (8).
Included in Schedule 6 (Release into Land: Prescribed Substances) Statutory Instrument No. 472, 1991 (9).

## References

1. *The Agrochemicals Handbook* 3rd ed., 1991, RSC, London
2. Chiba, M. et al *Can. J. Microbiol.* 1987, **33**(2), 157-161
3. Duah-Yentumi, S. et al *Soil Biol. Biochem.* 1986, **18**(6), 629-635
4. Walker, A. *Pestic Sci.* 1987, **21**(3), 219-231
5. *The Farm Chemicals Handbook* 1991, Meister Publ. Co., Willoughby, OH

6. *The Pesticide Manual* 9th ed., 1993, British Crop Protection Council, Farnham
7. Rankin, G. O. et al *Toxicology* 1989, **56**(3), 263-272
8. *EC Directive Relating to the Quality of Water Intended for Human Consumption* 1982, 80/778/EEC, Office for Official Publications of the European Communities, 2 rue Mercier, L-2985 Luxembourg
9. *S. I. 1991 No. 472 The Environmental Protection (Prescribed Processes and Substances) Regulations* 1991, HMSO, London

# I57   Iproniazid phosphate

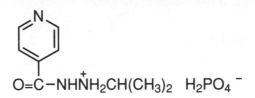

$$O=C-NH\overset{+}{N}H_2CH(CH_3)_2 \quad H_2PO_4^{-}$$

**CAS Registry No.** 305-33-9
**Synonyms** 4-pyridinecarboxylic acid, 2-(1-methylethyl)hydrazide, phosphate; isonicotinic acid, 2-isopropylhydrazide, phosphate; iproniazid dihydrogen phosphate; Marailid phosphate
**Mol. Formula** $C_9H_{16}N_3O_5P$                    **Mol. Wt** 277.22
**Uses** Antidepressant.

## Physical properties
**M. Pt.** 180-182°C.

## Mammalian and avian toxicity

**Acute data**
$LD_{50}$ dermal guinea pig 730 mg $kg^{-1}$ (1).
$LD_{50}$ intraperitoneal rat 442 mg $kg^{-1}$ (2).

**Teratogenicity and reproductive effects**
♀ Wistar rats single injection of 30 mg $kg^{-1}$ on day 15 of pregnancy or subchronically with 10 mg $kg^{-1}$ daily from day 10 to delivery. No change in neonatal reflexes of pups was observed after acute treatment; but increased numbers of pups showed active avoidance responses including cliff aversion, righting, barholding but not forelimb placing or grasping. Sexual activity was unimpaired (3).

## Genotoxicity
*Salmonella typhimurium* TA98, TA100, TA1535, TA1537, TA1538 with and without metabolic activation negative (4).
*In vivo* mouse bone marrow cells significant increase in sister chromatid exchange frequency (4).
*Escherichia coli* WP2, WP2 *uvrA*, WP67, CM871, TM1080 without metabolic activation negative (4).

## Legislation

Included in Schedule 4 (Release into Air: Prescribed Substances) Statutory Instrument No. 472, 1991 (5).

## References

1. *Arch. Int. Pharmacodyn. Ther.* 1962, **137**, 375
2. *Acta. Biol. Med. Germanica* 1967, **18**, 617
3. Drago, F. et al *Neurobehav. Toxicol. Teratol.* 1985, **7**, 493-497
4. Brambilla, G. et al *J. Toxicol. Environ. Health* 1982, **9**, 287-303
5. *S. I. 1991 No. 472 The Environmental Protection (Prescribed Processes and Substances) Regulations* 1991, HMSO, London

# I58   Iridium black

Ir

**CAS Registry No.** 7439-88-5
**Synonyms** iridium
**Mol. Formula** Ir                                   **Mol. Wt.** 192.20
**Occurrence** Occurs in earth's crust at 0.001 ppm. Found in nature combined with osmium, platinum or gold ores.

## Physical properties

**M. Pt.** 2450°C; **B. Pt.** 4500°C; **Specific gravity** $d_4^{20}$ 22.65.

## Ecotoxicity

**Invertebrate toxicity**
1.0-3.0 ppm iridium salts added to *Tetrahymena pyriformis* in the peptone-glucose culture medium stimulated cell proliferation but was inhibitory at >20 ppm (1).

**Bioaccumulation**
2.1 µg g$^{-1}$ dry weight detected in gut contents of yellow perch from an acidic lake (2).

## Legislation

Included in Schedule 4 (Release into Air: Prescribed Substances) Statutory Instrument No. 472, 1991 (3).

## References

1. Tang, R. et al *Yingyang Xuebao* 1986, **8**(3), 224-229 (Ch.) (*Chem. Abstr.* **106**, 15661x)
2. Heit, M. et al *Water, Air, Soil Pollut.* 1989, **44**, 9-30
3. *S. I. 1991 No. 472 The Environmental Protection (Prescribed Processes and Substances) Regulations* 1991, HMSO, London

# I59    Iridium tetrachloride

## IrCl$_4$

**CAS Registry No.** 10025-97-5
**Synonyms** iridium chloride
**Mol. Formula** Cl$_4$Ir                        **Mol. Wt.** 334.01

## Mammalian and avian toxicity

**Acute data**
LD$_{50}$ oral rat 8115 mg kg$^{-1}$ (1).

## Legislation

Limited under EC Directive on Drinking Water Quality 80/778/EEC. Chlorides guide level 25 mg l$^{-1}$ (2).
Included in Schedule 4 and 6 (Release into Air/Land: Prescribed Substances) Statutory Instrument No. 472, 1991 (3).

## References

1.  *Gig. Tr. Prof. Zabol.* 1977, **21**(7), 55
2.  *EC Directive Relating to the Quality of Water Intended for Human Consumption* 1982, 80/778/EEC, Office for Official Publications of the European Communities, 2 rue Mercier, L-2985 Luxembourg
3.  *S. I. 1991 No. 472 The Environmental Protection (Prescribed Processes and Substances) Regulations* 1991, HMSO, London

# I60    Iron

## Fe

**CAS Registry No.** 7439-89-6
**Mol. Formula** Fe                            **Mol. Wt.** 55.85
**Uses** Alloyed with carbon, manganese, chromium and nickel and other elements to form steels.
**Occurrence** About 5% of Earth's crust. Occurs in haematite, magnetite, limonite and siderite. Human body weight contains $\approx$60-70 µg g$^{-1}$ (1).

## Physical properties

**M. Pt.** 1535°C; **B. Pt.** 3000°C; **Specific gravity** d 7.86 (pure).

## Ecotoxicity

**Fish toxicity**
LC$_{50}$ (24 hr) starry flounder, bleak, perch 75-230 mg l$^{-1}$ (iron salts) (2).

**Invertebrate toxicity**
EC$_{50}$ (21 day) *Daphnia magna* 5.2 mg l$^{-1}$ (iron salts) (3).
LC$_{50}$ (48 hr, 96 hr) *Asellus aquaticus* Fe (III) 183, 124 mg l$^{-1}$ respectively (4).

## Bioaccumulation

Iron salts accumulate in fish (5).

Crayfish injected with 0.05-0.5 mg iron salts selectively stored iron in metal containing vacuoles of R- and F-cells in the hepatopancreatic cells at levels toxic to cells. High doses caused alterations in the ultrastructural morphology of the antennal gland cells although no accumulation was apparent (6).

## Environmental fate

### Nitrification inhibition

Carbon and nitrogen fixation increased fourfold in *Anabaena* with increased iron salt nutrition from $5.58 \times 10^{-7} - 5.58 \times 10^{-5}$ g l$^{-1}$. Iron salts may be the limiting factor in ocean nitrogen fixation (7).

### Anaerobic effects

Anaerobic bacteria in an iron-rich environment formed magnetite at high pHs and free $Fe^{2+}$ or siderite at low pHs (8).

### Degradation studies

Maximum concentration of iron not toxic to sludge microorganisms in wastewater treatment is 5.2 mg l$^{-1}$ (9).

## Mammalian and avian toxicity

### Acute data

$LD_{50}$ oral rat 30 g kg$^{-1}$ (form unspecified) (10).

$LD_{Lo}$ intraperitoneal rat 20 mg kg$^{-1}$ (form unspecified) (11).

Average human lethal oral dose is 200-300 mg elemental iron kg$^{-1}$ body weight. Toxic manifestations include severe haemorrhagic necrosis of the stomach and small intestine, metabolic derangements and effects on the liver, cardiovascular system and central nervous system (12).

### Sub-acute data

Iron overload causes inhibition of hepatic uroporphyrinogen decarboxylase and uroporphyria in C57BL/10ScSn but not DB A/2 mice. Studies of inbred mice 25 wk after treatment with 600 mg kg$^{-1}$ iron showed no correlation between Ah locus, microsomal enzyme activities associated with cytochrome $P_{450}$ or administration of 5-aminolevulinic acid in drinking water, with the propensity of strains to develop porphyria. Comparison with the congenic A2G-hr/+ srrain, which carries the recessive hr gene showed a modulating influence associated with the hr locus (13).

### Carcinogenicity and long-term effects

Anaemia may increase the risk of oral cancer (14).

Iron-depleted women but not men showed a lower risk of lung cancer (15).

Those occupationally exposed to iron or who work in iron foundries seem to be at increased risk of lung cancer (16).

Lack of body iron is common in cancer patients and is often associated with complications (17).

### Metabolism and pharmacokinetics

Absorbed from the gut and enters plasma attached to the protein transferrin. Transferrin enters cells by endocytosis and iron is released as pH drops and is used in synthesis of intracellular protein. Excess iron is stored in protein ferritin (18).

Incorporation into ferritin of bone marrow, liver and spleen mediated by ATP and ascorbic acid. Iron is stored in bone marrow as non-heme iron ($100 \, \mu g \, g^{-1}$ in humans) and to a lesser extent in liver and kidney (19).
$0.03-0.05 \, mg \, day^{-1}$ excreted from normal humans in bile, faeces and urine (1).

## Genotoxicity

*Salmonella typhimurium* TA98, TA102, TA1535, TA1537 with and without metabolic activation negative (19).
*In vitro* cultured mouse splenocytes dose dependent increase in micronuclei (20).

## Any other adverse effects to man

Cases of iron salt overload may occur in patients receiving repeated blood transfusions (21), oral preparations for real or supposed anaemia (22) and food fortified with iron salts (23).
South African Bantu siderosis caused by iron in cooking pots and home brewed beer stored in drums. 81% subjects autopsied both ♂ and ♀ had excessive haemosiderin-like deposits (24).
Acute oral poisoning from iron tablets common in children. The average lethal dose for a 2-year-old is 3 g (equivalent to ≈50 tablets). 600 mg reported fatal to one small child (25).

## Any other adverse effects

Animals (species unspecified) treated with a single intraperitoneal injection of $7.5 \, mg \, kg^{-1}$ ferric nitrilotriacetate showed transient changes in liver function and evidence for lipid peroxidation in liver homoganates (26).

## Legislation

Limited under EC Directive on Drinking Water Quality 80/778/EEC. Iron guide level $50 \, \mu g \, l^{-1}$, maximum admissible concentration $200 \, \mu g \, l^{-1}$ (27).
Included in Schedule 4 (Release into Air: Prescribed Substances) Statutory Instrument No. 472, 1991 (28).

## Any other comments

Toxic action reviewed (29).
Metabolism reviewed (30,31).
Magnetic.

## References

1. Abdel-Mageed, A. B. et al *Vet. Hum. Toxicol.* 1990, **32**(4), 324-328
2. Bagge et al *Ferrosulfaatin ja sitae sisaeltuevien jaeteresien akuutista myrkylisyydestae murtovedessae* 1975, **Meri 3**, 46-64
3. Biesinger, K. E. et al *J. Fish. Res. Board Can.* 1972, **29**, 1691-1700
4. Martin, T. R. et al *Water Res.* 1986, **20**(9), 1137-1147
5. Munshi, J. S. D. et al *Environ. Ecol.* 1989, **7**(4), 790-792
6. Roldan, B. M. et al *Comp. Biochem. Physiol., C. Comp. Pharmacol. Toxicol.* 1987, **86C**(1), 201-214
7. Rueter, J. G. *J. Phycol.* 1988, **24**(2), 249-254
8. Bell, P. E. et al *App. Environ. Microbiol.* 1987, **53**(11), 2610-2616
9. Boldyrev, A. I. et al *Vopr. Khim. Khim. Tekhnol.* 1986, **82**, 54-57 (Russ.) (*Chem. Abstr.* **108**, 43278n)

10. *Indian J. Pharm.* 1951, **13**, 240
11. *NTIS Report* PB158508, Natl. Tech. Inf. Ser., Springfield, VA
12. Aiser, P. et al *Int. Rev. Exp. Pathol.* 1990, **31**, 1-46
13. Smith, A. G. et al *Biochem. J.* 1993, **291**, 29-35
14. Altini, M. et al *S. Afr. Med. J.* 1989, Suppl., 6-10
15. Selby, J. V. et al *Int. J. Cancer* 1988, **41**, 677-682
16. Perera, F. F. et al *Cancer Res.* 1988, **48**, 2288-2291
17. Harju, E. *Met. Ions. Biol. Med., Proc. Int. Symp. 1st* 1990, 45-48
18. Halliwell, B. et al *ISI Atlas Sci.: Biochem.* 1988, **1**(1), 48-52
19. Wang, P. K. *Bull. Environ. Contam. Toxicol.* 1988, **40**(4), 597-603
20. Dreosti, I. E. et al *Mutat. Res.* 1990, **244**(4), 337-343
21. Cohen, A. et al *Ann. N. Y. Acad. Sci.* 1985, **445**, 274
22. Hennigar, G. R. et al *Am. J. Pathol.* 1979, **96**, 611
23. Crosby, W. H. *N. Engl. J. Med.* 1977, **297**, 543
24. Jacobs, A. et al (Eds.) *Iron in Biochemistry and Medicine* 1981, **2**, 427-459, Academic Press, London
25. Arena, J. M. (Ed.) *Poisoning* 1977, 431-436, Thomas, Springfield, IL
26. Yamanoi, Y. et al *Acta Haematol. Jpn.* 1982, **45**, 1229
27. *EC Directive Relating to the Quality of Water Intended for Human Consumption* 1982, 80/778/EEC, Office for Official Publications of the European Communities, 2 rue Mercier, L-2985 Luxembourg
28. *S. I. 1991 No. 472 The Environmental Protection (Prescribed Processes and Substances) Regulations* 1991, HMSO, London
29. Anke, M. et al *Trace Elem., Anal. Chem. Med. Biol., Proc. Int. Workshop, 4th* 1986 (Publ. 1987), 201-236
30. Lerner, A. et al *Front. Gastrointest. Res.* 1988, 14 (Prog. Diet Nutr.) 117-134
31. Flanagan, P. R. *Iron Transp. Storage* 1990, 247-261

# I61  Iron(II) ammonium sulfate

## $Fe(NH_4)_2(SO_4)_2$

**CAS Registry No.** 10045-89-3
**Synonyms** sulfuric acid, ammonium iron (2+) salt; ammonium ferrous sulfate; ferrous ammonium sulfate; ferrous diammonium disulfate; Mohr's Salt
**Mol. Formula** $FeH_8N_2O_8S_2$          **Mol. Wt.** 284.05

## Physical properties

**Specific gravity** $d_4^{20}$ 1.86.

## Occupational exposure

**US TLV (TWA)** 1 mg m$^{-3}$ (as Fe); **UK Long-term limit** 1 mg m$^{-3}$ (as Fe); **UK Short-term limit** 2 mg m$^{-3}$ (as Fe).

## Mammalian and avian toxicity

**Acute data**
LD$_{50}$ oral rat 3.25 g kg$^{-1}$ (1).

## Legislation

Limited under EC Directive on Drinking Water Quality 80/778/EEC. Sulfates guide level 25 mg $l^{-1}$ maximum admissible concentration 250 mg $l^{-1}$; ammonium guide level 0.05 mg $l^{-1}$ maximum admissible concentration 0.5 mg $l^{-1}$; iron guide level 50 μg $l^{-1}$ maximum admissible concentration 200 μg $l^{-1}$ (2).
Included in Schedule 4 (Release into Air: Prescribed Substances) Statutory Instrument No. 472, 1991 (3).

## References

1. Smyth, H. F. *Am. Ind. Hyg. Assoc. J.* 1969, **30**, 470
2. *EC Directive Relating to the Quality of Water Intended for Human Consumption* 1982, 80/778/EEC, Office for Official Publications of the European Communities, 2 rue Mercier, L-2985 Luxembourg
3. *S. I. 1991 No. 472 The Environmental Protection (Prescribed processes and Substances) Regulations* 1991, HMSO, London

# I62   Iron(II) chloride

## $FeCl_2$

**CAS Registry No.** 7758-94-3
**Synonyms** fero 66; ferrous chloride; ferrous dichloride; iron dichloride; iron protochloride
**Mol. Formula** $Cl_2Fe$                    **Mol. Wt.** 126.75
**Uses** Reducing agent in metallurgy. Used in pharmaceutical preparations and as mordant in dyeing.
**Occurrence** Occurs in nature as the mineral lawrencite.

## Physical properties

**M. Pt.** 674°C; **B. Pt.** 1023°C; **Specific gravity** $d^{25}$ 3.16.

## Occupational exposure

**US TLV (TWA)** 1 mg $m^{-3}$ (as Fe); **UK Long-term limit** 1 mg $m^{-3}$ (as Fe); **UK Short-term limit** 2 mg $m^{-3}$ (as Fe).

## Mammalian and avian toxicity

**Acute data**
$LD_{50}$ intraperitoneal mouse 59 mg $kg^{-1}$ (1).

## Genotoxicity

*Escherichia coli* WP2S (λ) without metabolic activation positive (2).

## Legislation

Limited under EC Directive on Drinking Water Quality 80/778/EEC. Iron guide level 50 μg $l^{-1}$ maximum admissible concentration 200 μg $l^{-1}$. Chlorides guide level 25 mg $l^{-1}$ (3).
Included in Schedule 4 and 6 (Release into Air/Land: Prescribed Substances) Statutory Instrument No. 472, 1991 (4).

## References

1. *Naunyn-Schmiedeberg's Arch. Exp. Pathol. Pharmakol.* 1962, **244**, 17
2. Rossman, T. G. et al *Mutat. Res.* 1991, **260**(4), 349-367
3. *EC Directive Relating to the Quality of Water Intended for Human Consumption* 1982, 80/778/EEC, Office for Official Publications of the European Communities, 2 rue Mercier, L-2985 Luxembourg
4. *S. I. 1991 No. 472 The Environmental Protection (Prescribed Processes and Substances) Regulations* 1991, HMSO, London

# I63   Iron(III) chloride

## $FeCl_3$

**CAS Registry No.** 7705-08-0
**Synonyms** ferric chloride; ferric trichloride; flores martis; iron chloride; iron perchloride; iron trichloride
**Mol. Formula** $Cl_3Fe$ **Mol. Wt.** 162.21
**Uses** Photography, ink. Catalyst for organic reactions. Purification and deodorisation of factory effluents and sewage.
**Occurrence** As mineral (molysite).

## Physical properties

**M. Pt.** ≈300°C (volatilises); **B. Pt.** ≈316°C; **Specific gravity** $d^{25}$ 2.90.

### Solubility

Organic solvent: ethanol, diethyl ether, acetone

## Occupational exposure

**US TLV (TWA)** 1 mg m$^{-3}$ (as Fe); **UK Long-term limit** 1 mg m$^{-3}$ (as Fe); **UK Short-term limit** 2 mg m$^{-3}$ (as Fe); **UN No.** 1773; 2582 (solution); **HAZCHEM Code** 2R; 2T (solution); **Conveyance classification** corrosive substance.

## Ecotoxicity

### Invertebrate toxicity

$EC_{50}$ (2 and 4 day) *Asellus aquaticus* 183 and 124 mg l$^{-1}$ respectively (1).
$EC_{50}$ (2 and 4 day) *Crangonyx pseudogracilis* 160 and 120 mg l$^{-1}$ respectively (1).

## Environmental fate

### Degradation studies

Respirometry data indicate that iron(III) chloride does not have toxic effects on the biomass of an activated sludge process when concentrations of <100 mg l$^{-1}$ are used; the pH does not vary significantly at these concentrations. Progressive inhibition of biomass occurs at iron(III) chloride concentrations of 500 mg l$^{-1}$, due to pH decrease (2).

## Mammalian and avian toxicity

### Acute data

$LD_{50}$ oral mouse 1280 mg kg$^{-1}$ (3).
$LD_{50}$ intraperitoneal mouse 68 mg kg$^{-1}$ respectively (3).
$LD_{50}$ intravenous mouse 142 mg kg$^{-1}$ (3).

**Carcinogenicity and long-term effects**
Oral Fischer 344 rat (2 yr) 0.25 or 0.5% in drinking water under investigation (4).

**Irritancy**
2% topical patch test: response graded at 48 hr, 72 hr and 5-7 day, erythema and oedema was reported in 3 of 17 subjects tested (5).

### Genotoxicity
*Escherichia coli* PQ37 SOS chromotest without metabolic activation negative (6).
*Escherichia coli* WP2s($\lambda$) without metabolic activation, weakly positive (7).
*Salmonella typhimurium* without metabolic activation positive (7).
*In vitro* mouse lymphoma L51787 $tk^+/tk^-$ cell forward mutation assay negative (8).
*In vitro* human cells (cell type unspecified), DNA damage negative (9).
*In vivo* mouse bone marrow induced micronuclei (10).

### Any other adverse effects
In fasting mice induced a dose-dependent increase in nuclear aberrations in the stomach tissue. In normal feeding animals no increase in the nuclear aberrations was seen. It is concluded that iron compounds have an intrinsic cellular toxicity when not administered with food but do not have any genotoxic potential for the gastrointestinal tract (11).

### Legislation
Limited under EC Directive on Drinking Water Quality 80/778/EEC. Iron guide level: 50 $\mu$g l$^{-1}$, maximum admissible concentration 200 $\mu$g l$^{-1}$; chloride guide level 25 mg l$^{-1}$ (approximate concentration at which effects might occur, 200 mg l$^{-1}$) (12).

### References
1. Martin, T. R. et al *Water Res.* 1986, **20**(9), 1137-1147
2. Broglio *Inquinamento* 1987, **29**(3), 112-114 (Ital.) (*Chem Abstr.* **107**, 140280u)
3. Lewis, M. A. et al *Registry of Toxic Effects of Chemical Substances* 1984, National Institute for Occupational Safety and Health, No. 83-107-4
4. *Directory of Agents Being Tested for Carcinogenicity No. 15* 1992, IARC, Lyon
5. Namikoshi, T. et al *J. Oral Rehabil.* 1990, **17**, 377-381
6. Olivier et al *Mutat. Res.* 1987, **189**(3), 263-269
7. Rossman, T. G. *Mutat. Res.* 1991, **260**(4), 349-367
8. McGregor et al *Environ. Mol. Mutagen.* 1988, **12**(1), 85-154
9. Sumiko *Cancer Res.* 1987, **47**(24), 6522-6527
10. Liao, M. et al *Weisheng Dulixe Zazhi* 1988, **2**(2), 83-86 (Ch.) (*Chem Abstr.* **112**, 134189h)
11. Bianchini et al *J. Appl. Toxicol.* 1988, **8**(3), 179-183
12. *EC Directive Relating to the Quality of Water Intended for Human Consumption* 1982, 80/778/EEC, Office for Official Publications of the European Communities, 2 rue Mercier, L-2985 Luxembourg

# I64   Iron(III) dextran

**CAS Registry No.** 9004-66-4
**Synonyms** Imferon; A100; Chinofer; ferric dextran; ferroglucin; Imposil; Myofer

## Physical properties
**M. Pt.** decomp.

## Mammalian and avian toxicity

### Acute data
$LD_{50}$ oral mouse 1 g(Fe) $kg^{-1}$ (1).
$LD_{50}$ intraperitoneal rat 3 g(Fe) $kg^{-1}$ (2).
$LD_{50}$ intravenous mouse 460 mg(Fe) $kg^{-1}$ (3).

### Carcinogenicity and long-term effects
No adequate evidence for carcinogenicity to humans, sufficient evidence for carcinogenicity to animals, IARC classification group 2B (4).

A woman who had multiple injections of iron-dextran complex developed an undifferentiated soft tissue sarcoma (5).

Only 1/9 malignancies in 5 reports from 1960-1977 was likely to be related to iron-dextran injections given 14 yrs before (a fibrosarcoma) (6).

Repeated intramuscular and subcutaneous injections in mice, rabbits and rats produced local tumours at the injection site (1,7).

Female F344 rats injected intraperitoneally with 600 mg Fe $kg^{-1}$ were fed 0.02% hexachlorobenzene (HCB) in diet after 1 wk for 65 wk. 8/8 rats receiving HCB after iron overload developed multiple hepatic nodules but only 3/8 receiving HCB alone had nodules. Iron overload potentiated the neoplastic process induced by HCB with both enhancing and depressing effects on various HCB-induced enzyme activities (8).

## Genotoxicity
L5178Y mouse lymphoma cell forward mutation assay with and without metabolic activation negative (9).

## Any other adverse effects
100-500 mg(Fe) $kg^{-1}$ given to anaesthetised cats lowered blood pressure inhibited noradrenaline and had curare-like effects but had no effect on acetylcholine, histamine or serotonin (10).

Small doses in guinea pigs protected 70% of animals from anaphylactic shock induced by horse serum (10).

Intramuscular administration to rats with dextran oedema had an antioedemic effect (10).

## Legislation
Limited under EC Directive on Drinking Water Quality 80/778/EEC. Iron guide level 50 µg $l^{-1}$ maxiumum admissible concentration 200 µg $l^{-1}$ (11).

Included in Schedule 4 (Release into the Air: Prescribed Substances) Statutory Instrument No. 472, 1991 (12).

## Any other comments
Full term infants injected with 150 mg iron-dextran at birth had a nutritional advantage in iron status up to 15 months compared to controls (13).

## References
1. *Br. J. Pharm. Chem.* 1965, **24**, 352
2. *Toxicol. Appl. Pharmacol.* 1971, **18**, 185
3. *Acta. Pol. Pharm.* 1961, **18**, 149

4.  *IARC Monograph* 1987, **Suppl. 7**, 226
5.  *IARC Monograph* 1973, **2**, 161-178
6.  Robertson, A. G. et al *Br. Med. J.* 1977, **i**, 946
7.  Hruby, M. et al *Sbornik Lek.* 1985, **87**, 114-120
8.  Smith, A. G. et al *Biochem. J.* 1993, **291**, 29-35
9.  McGregor, D. B. et al *Environ. Mol. Mutagen.* 1991, **17**(3), 196-219
10. Dilar, P. *Vet.-Med. Nauki* 1986, **23**(9), 35-44 (Bulg.) (*Chem. Abstr.* **106**, 168494v)
11. *EC Directive Relating to the Quality of Water Intended for Human Consumption* 1982, 80/778/EEC, Office for Official Publications of the European Communities, 2 rue Mercier, L-2985 Luxembourg
12. *S. I. 1991 No. 472 The Environmental Protection (Prescribed Processes and Substances) Regulations* 1991, HMSO, London
13. Olivares, M. et al *Nutr. Rep. Int.* 1989, **40**(3), 577-583.

# I65   Iron(III) fluoride

## FeF₃

**CAS Registry No.** 7783-50-8
**Synonyms** ferric fluoride; ferric trifluoride; iron trifluoride
**Mol. Formula** FeF₃                                            **Mol. Wt.** 112.84

### Physical properties

**M. Pt.** >1000°C (sublimes); **Specific gravity** 3.87.

### Occupational exposure

**US TLV (TWA)** 1 mg m$^{-3}$ (as Fe); **UK Long-term limit** 1 mg m$^{-3}$ (as Fe); **UK Short-term limit** 2 mg m$^{-3}$ (as Fe).

### Mammalian and avian toxicity

**Acute data**
$LD_{50}$ intravenous mouse 18 mg kg$^{-1}$ (1).

### Legislation

Limited under EC Directive on Drinking Water Quality 80/778/EEC. Iron guide level 50 µg l$^{-1}$ maximum admissible concentration 200 µg l$^{-1}$; fluoride maximum admissible concentration 1500 µg l$^{-1}$ at 8-12°C (700 µg l$^{-1}$ at 25-30°C) (2). Included in Schedule 4 and 6 (Release into Air/Land: Prescribed Substances) Statutory Instrument No. 472, 1991 (3).

### References

1.  *US Army Armament Research and Development Command* Report No.NX00135, Chemical Systems Laboratory, NIOSH Exchange Chemicals, Aberdeen Proving Ground, MD
2.  *EC Directive Relating to the Quality of Water Intended for Human Consumption* 1982, 80/778/EEC, Office for Official Publications of the European Communities, 2 rue Mercier, L-2985 Luxembourg
3.  *S. I. 1991 No. 472 The Environmental Protection (Prescribed Processes and Substances) Regulations* 1991, HMSO, London

# I66   Iron(II) fumarate

$$^-O_2C \underset{\phantom{E}}{\overset{E}{\diagdown\!=\!\diagup}} CO_2^- \quad .Fe^{2+}$$

**CAS Registry No.** 141-01-5
**Synonyms** 2-butenedioic acid, (*E*)-, iron (2+) salt (1:1); fumaric acid, iron (2+) salt (1:1); Cpiron; Erzoferro; Ferrone; Ircon; Toleron
**Mol. Formula** $C_4H_4O_4Fe$        **Mol. Wt.** 171.92

## Physical properties
**M. Pt.** >280°C; **Specific gravity** $d^{25}$ 2.435.

## Solubility
Water: 1.4 g $l^{-1}$ at 25°C. Organic solvent: ethanol

## Occupational exposure
**US TLV (TWA)** 1 mg $m^{-3}$ (as Fe); **UK Long-term limit** 1 mg $m^{-3}$ (as Fe); **UK Short-term limit** 2 mg $m^{-3}$ (as Fe).

## Mammalian and avian toxicity

### Acute data
$LD_{50}$ oral mouse, rat 1570-3850 mg $kg^{-1}$ (1,2).
$LD_{50}$ dermal rat 500 mg $kg^{-1}$ (2).
$LD_{50}$ intraperitoneal rat, mouse 185-450 mg $kg^{-1}$ (1,2).

## Legislation
Limited under EC Directive on Drinking Water Quality 80/778/EEC. Iron guide level 50 µg $l^{-1}$, maximum admissible concentration 200 µg $l^{-1}$ (3).
Included in Schedule 4 (Release into Air: Prescribed Substances) Statutory Instrument No. 472, 1991 (4).

## References
1. *Am. J. Med. Sci.* 1961, **241**, 296
2. *Drugs in Japan. Ethical Drugs* 6th ed., 1982, Japan Pharmaceutical Information Center, Tokyo
3. *EC Directive Relating to the Quality of Water Intended for Human Consumption* 1982, 80/778/EEC, Office for Official Publications of the European Communities, 2 rue Mercier, L-2985 Luxembourg
4. *S. I. 1991 No. 472 The Environmental Protection (Prescribed Processes and Substances) Regulations* 1991, HMSO, London

# I67    Iron(III) nitrate

## $Fe(NO_3)_3$

**CAS Registry No.** 10421-48-4
**Synonyms** nitric acid, iron (3+) salt; ferric nitrate; iron trinitrate
**Mol. Formula** $FeN_3O_9$                    **Mol. Wt.** 241.86

### Physical properties
**M. Pt.** 472°C (decomp.); **B. Pt.** 125°C.

### Solubility
Organic solvent: ethanol, acetone

### Occupational exposure
**US TLV (TWA)** 1 mg m$^{-3}$ (as Fe); **UK Long-term limit** 1 mg m$^{-3}$ (as Fe); **UK Short-term limit** 2 mg m$^{-3}$ (as Fe).

### Genotoxicity
*Eschericha coli* PQ37 SOS chromotest without metabolic activation negative (1).

### Legislation
Limited under EC Directive on Drinking Water Quality 80/778/EEC. Iron: guide level 50 µg l$^{-1}$, maximum admissible concentration 200 µg l$^{-1}$; nitrate: guide level 25 mg l$^{-1}$, maximum admissible concentration 50 mg l$^{-1}$ (2).

### References
1.  Olivier et al *Mutat. Res.* 1987, **189**(3), 263-269
2.  *EC Directive Relating to the Quality of Water Intended for Human Consumption* 1982, 80/778/EEC, Office for Official Publications of the European Communities, 2 rue Mercier, L-2985 Luxembourg

# I68    Iron oxide

## $Fe_2O_3$

**CAS Registry No.** 1309-37-1
**Synonyms** C.I. Pigment Red 101; ferric oxide; iron(III) oxide; Prussian red; Venetian red; ferric sesquioxide; α-iron oxide; α-ferric oxide; C.I. 77491
**Mol. Formula** $Fe_2O_3$                    **Mol. Wt.** 159.69
**Uses** As pigment. Polishing agent for glass, diamonds. In electrical resistors, semiconductors, magnets, magnetic tapes. Catalyst. In colloidal solutions as stain for polysaccharides.
**Occurrence** α form occurs as haematite; γ form occurs as maghaemite

### Physical properties
**Specific gravity** 5.24.

---

## Occupational exposure

US TLV (TWA) 2 ppm (5 mg m$^{-3}$) (as Fe); **UK Long-term limit** 5 mg m$^{-3}$ (as Fe); **UK Short-term limit** 10 mg m$^{-3}$ (as Fe).

## Ecotoxicity

**Fish toxicity**
There was a reduction in the percentage of carp eggs hatching, from 54 to 20%, when exposed to 1-500 ppm (1).

## Mammalian and avian toxicity

**Acute data**
LD$_{50}$ intraperitoneal mouse, rat 5400-5500 mg kg$^{-1}$ (2).
LD$_{Lo}$ subcutaneous dog 30 mg kg$^{-1}$ (3).

**Sub-acute data**
Syrian golden hamster 3 mg iron oxide in 0.2 ml saline instilled intralaryngeally 1 × wk for 5, 10, 15 wk had increased epithelial mitotic rates of the bronchioles after 5 instillations. Bronchioalveolar hyperplasia was also seen after 5 instillations but was less prominent after 10 or 15 treatments (4).

**Carcinogenicity and long-term effects**
No adequate evidence for carcinogenicity to humans or animals, IARC classification group 3 (5).
18.9% of rats dosed intraperitoneally with 21.6 g (5 injections) of α-ferric oxide hydrate had sarcomas, mesothelioma or carcinomas in the abdominal cavity (6).
Ferric oxide particles instilled intratracheally to hamsters induced interstitial fibrosis but benzo[a]pyrene administered bound to ferric oxide particles induced squamous-cell and anaplastic carcinomas. The ferric oxide particles act as cofactors, mainly carriers, in the system (7-9).

## Genotoxicity

Induced chromosomal aberrations in rat bone marrow cells (10).

## Legislation

Limited under EC Directive on Drinking Water Quality 80/778/EEC. Iron guide level 50 μg maximum admissible concentration 200 μg l$^{-1}$ (11).
Included in Schedule 4 (Release into Air: Prescribed Substances) Statutory Instrument No. 472, 1991 (12).

## Any other comments

Human health effects, experimental toxicity, epidemiology and workplace experience reviewed (13).

## References

1. Rao, K. S. et al *Geobios (Jodphur, India)* 1988, **15**(2-3), 111-113
2. *Gig. Tr. Prof. Zabol.* 1982, **26**(4), 23
3. *Abdernalden's Handbuch der Biologischen Arbeitsmethoden* 1935, **4**, 1289
4. Marshall, H. E. et al *Fundam. Appl. Toxicol.* 1987, **9**(4), 705-714
5. *IARC Monograph* 1987, **Suppl. 7**, 216-218
6. Pott, F. et al *Exp. Pathol.* 1987, **32**, 129-152

7. Saffiotti, U. *Prog. Exp. Tumor. Res.* 1969, **11**, 302-333
8. Stenbaeck, F. et al *Oncology* 1976, **33**, 29-34
9. Sellakumar, A. R. et al *J. Natl. Cancer Inst.* 1973, **50**, 507-510
10. He, W. et al *Dongwuxue Yanjuu* 1988, **9**(3), 263-268 (Ch.) (*Chem. Abstr.* **110**, 109561z)
11. *EC Directive Relating to the Quality of Water Intended for Human Consumption* 1982, 80/778/EEC, Office for Official Publications of the European Communities, 2 rue Mercier, L-2985 Luxembourg
12. *S. I. 1991 No. 472 The Environmental Protection (Prescribed Processes and Substances) Regulations* 1991, HMSO, London
13. *ECETOC Technical Report No. 30(4)* 1991, European Industry Ecology and Toxicology Centre, B-1160 Brussels

# I69   Iron pentacarbonyl

## Fe(CO)₅

**CAS Registry No.** 13463-40-6
**Synonyms** (*TB*-5-11)-iron carbonyl (Fe(CO)$_5$); iron carbonyl (Fe(CO)$_5$); pentacarbonyliron
**Mol. Formula** C$_5$FeO$_5$                                     **Mol. Wt.** 195.90
**Uses** To make finely divided iron ('carbonyl iron'). Anti knock agent in motor fuels. Catalyst in organic reactions.

## Physical properties

**M. Pt.** –20°C; **B. Pt.** 103°C; **Flash point** –15°C; **Specific gravity** d$_4^{25}$ 1.453; **Volatility** v.p. 40 mmHg at 30.3°C.

## Occupational exposure

**US TLV (TWA)** 0.1 ppm (0.23 mg m$^{-3}$) (as Fe); **US TLV (STEL)** 0.2 ppm (0.45 mg m$^{-3}$) (as Fe); **UK Short-term limit** 0.01 ppm (0.08 mg m$^{-3}$) (as Fe); **UN No.** 1994; **Conveyance classification** toxic substance, flammable liquid.

## Mammalian and avian toxicity

### Acute data

LD$_{50}$ oral rabbit, guinea pig 12-22 mg kg$^{-1}$ (1).
LC$_{50}$ (10 min) inhalation mouse 7 g m$^{-3}$ (2).
Inhalation ♂ rat (1, 3, 5 hr) 10-20 mg m$^{-3}$ (dust aerosol). Following 3 hr exposure the majority of carbonyl iron particles were located on alveolar duct bifurcations. Within 24 hr pulmonary macrophage had accumulated at site of deposition in the lung (3).

## Any other comments

Human health effects, epidemiology, work place experience and experimental toxicology reviewed (4).

## References

1. *J. Ind. Hyg. Toxicol.* 1943, **25**, 415
2. *NTIS Report 158/508*, Natl. Tech. Inf. Ser., Springfield, VA

3. Warheit, D. B. et al *Scanning Microsc.* 1988, **2**(2), 1069-1078
4. *ECETOC Technical Report No. 30(4)* 1991, European Chemical Industry Ecology and Toxicology Centre, B-1160 Brussels

# I70   Iron(II) sulfate

## FeSO₄

**CAS Registry No.** 7720-78-7
**Synonyms** sulfuric acid, iron (2+) salt (1:1); Feosol; ferrous sulphate; green vitriol; iron monosulfate; sulferrous
**Mol. Formula** FeO₄S                                   **Mol. Wt.** 151.91
**Uses** In the manufacture of iron and iron compounds. In fertiliser. Reducing agent. Weed killer. Wood preservative. In pesticides. In quantitative analysis.
**Occurrence** Hydrates occur in nature as minerals melanterite, siderotil, szomolnikite, tauriscite.

## Occupational exposure

**US TLV (TWA)** 1 mg m$^{-3}$ (as Fe); **UK Long-term limit** 1 mg m$^{-3}$ (as Fe); **UK Short-term limit** 2 mg m$^{-3}$ (as Fe).

## Ecotoxicity

### Invertebrate toxicity

$EC_{50}$ (2 day) *Crangonyx pseudogracilus* 143 mg l$^{-1}$ (1).
$LC_{50}$ (50 hr) *Asellus aquaticus* (calculated) 256-467 mg l$^{-1}$ at pH 4.5 (2).
200 µg cm$^{-2}$ on glass surface toxic to *Deroceras reticulatum* (3).

## Mammalian and avian toxicity

### Acute data

$LD_{50}$ oral rat, mouse 319-680 mg kg$^{-1}$ (4,5).
$TD_{Lo}$ oral woman 600 mg kg$^{-1}$ (6).
$LD_{50}$ subcutaneous mouse, rat 60-155 mg kg$^{-1}$ (7).
$LD_{50}$ intraperitoneal mouse 289 mg kg$^{-1}$ (8).
$LD_{50}$ intravenous dog, mouse 79-112 mg kg$^{-1}$ (9).

## Genotoxicity

*Saccharomyces cerevisiae* failed to induce diploid or disomic spores during meiosis (10).
*Escherichia coli* PQ37 without metabolic activation negative (11).
*In vivo* mouse gastrointestinal cells showed no increase in chromosomal aberrations in the stomach or duodenum of fasting or feeding mice but a dose related increase in nuclear aberrations in the colon for feeding and fasting mice (12).
*In vitro* mouse lymphoma L5178Y tk$^+$/tk$^-$ no significant result (13).

## Any other adverse effects to man

Gastro-intestinal disturbances, including colic, constipation and diarrhoea may occur in humans. In children, ingestion of large quantities can cause vomiting, haematemesis, hepatic damage, tachycardia and peripheral vascular collapse (14).

## Any other adverse effects

Increased rat liver peroxides and lipofuscins and reduced membrane fluidity and swelling of liver mitochondria (15).

15-106 mg $l^{-1}$ iron(II) sulfate shortened action potential duration, and decreased the contractile force of the guinea pig myocardium and decreased the action potential amplitude and maximum upstroke velocity of the myocardium, papillary muscles and $K^+$ depolarised papillary muscles (16).

Inhibition of transmembrane movement of $Ca^{2+}$ and $Na^+$ in myocardial cells by $Fe^{2+}$ may be one of the mechanisms of heart failure and circulatory collapse in acute iron poisoning (16).

## Legislation

Limited under EC Directive on Drinking Water Quality 80/778/EEC. Iron guide level 50 $\mu g\ l^{-1}$, maximum admissible concentration 200 $\mu g\ l^{-1}$; sulfates guide level 25 mg $l^{-1}$, maximum admissible concentration 250 mg $l^{-1}$ (17).

Included in Schedule 4 (Release into the Air: Prescribed Substances) Statutory Instrument No. 472, 1991 (18).

## Any other comments

Potentially explosive.

## References

1. Martin, T. R. et al *Water Res.* 1986, **20**(9), 1137-1147
2. Maltby, L. et al *Environ. Pollut.* 1987, **43**, 271-279
3. Henderson, I. F. et al *Monogr. – Br. Crop. Prot. Counc.* 1989, (Slugs, Snails World Agric.), 289-294
4. *J. Paediatrics* 1966, **69**, 663
5. *Br. J. Pharmacol. Chemother.* 1965, **24**, 352
6. *JAMA, J. Am. Med. Assoc.* 1974, **229**, 1333
7. *Drugs in Japan. Ethical Drugs* 6th ed., 1982, 888, Japan Pharmaceutical Information Center, Tokyo
8. *C. R. Hebd. Seances Acad. Sci. (Paris)* 1963, **256**, 1043
9. *Am. J. Med. Sci.* 1961, **241**, 296
10. Sora, S. et al *Mutat. Res.* 1988, **201**(2), 375-384
11. Olivier, P. et al *Mutat. Res.* 1987, **189**, 263-269
12. Bianchini, F. et al *J. Appl. Toxicol.* 1988, **8**(3), 179-183
13. McGregor, D. B. et al *Environ. Mol. Mutagen.* 1991, **17**(3), 196-219
14. Gosselin, K. E. et al (Eds). *Clinical Toxicology of Chemical Products*, 4th ed., 1976, Williams and Wilkins, Baltimore
15. Lu, X. *Shengwu Huaxue Yu Shengwu Wuli Jinzhan* 1989, **16**(5), 372-374, (Ch.) (*Chem. Abstr.* **112**, 153221d)
16. Luo, G. et al *Zhongguo Yaoli Xuebao* 1989, **10**(6), 523-525 (Ch.) (*Chem. Abstr.* **112**, 111802p)
17. *EC Directive Relating to the Quality of Water Intended for Human Consumption* 1982, 80/778/EEC, Office for Official Publications of the European Communities, 2 rue Mercier, L-2985 Luxembourg
18. *S. I. 1991 No. 472 The Environmental Protection (Prescribed Processes and Substances) Regulations* 1991, HMSO, London

# I71   Iron(III) sulfate

## Fe$_2$(SO$_4$)$_3$

**CAS Registry No.** 10028-22-5
**Synonyms** Iron(3+) sulfuric acid salt (3:2); diiron trisulfate; ferric sulfate; iron persulfate; iron sulfate (2:3); iron tersulfate
**Mol. Formula** Fe$_2$O$_{12}$S$_3$                              **Mol. Wt.** 399.88
**Uses** Agent for removing red tide plankton. Coagulant in water purification and sewage treatment. Iron alums, iron salts and pigment preparation. Soil conditioner. Catalyst in polymerisation. In textile dyeing and calico printing as a mordant.

## Physical properties

**M. Pt.** 480°C (decomp.); **Specific gravity** 3.097.

**Solubility**
Water: 440 g l$^{-1}$.

## Occupational exposure

**US TLV (TWA)** 1 mg m$^{-3}$ (as Fe); **UK Long-term limit** 1 mg m$^{-3}$ (as Fe); **UK Short-term limit** 2 mg m$^{-3}$ (as Fe).

## Ecotoxicity

**Fish toxicity**
LC$_{50}$ (24 hr) rabbit fish, striped goby 65 and 236 mg l$^{-1}$, respectively (1).

## Mammalian and avian toxicity

**Sub-acute data**
Inhalation rat (7 or 21 day) 7 component mixture including iron sulfate (concentration unspecified) for 4 hr day$^{-1}$. Significant effects observed on pulmonary macrophage function, more pronounced after 21 days. Effects persisted to 96 hr post-exposure (2).

## Any other adverse effects

Adverse effects to the respiratory system of acidic air pollutants including iron sulfate were studied in rats, exposure time ≤4 hr. Reflex breathing patterns, lung/nasal histopathological changes and particle clearance rate were examined. Acid components had no effect on the parameters examined (3).

## Legislation

Generally recommended as safe for direct use in food, including infant formula under US Federal Food, Drug and Cosmetic Act (4).
Limited under EC Directive on Drinking Water Quality 80/778/EEC. Iron: guide level 50 μg l$^{-1}$, maximum admissible level 200 μg l$^{-1}$; sulfate: guide level 25 mg l$^{-1}$, maximum admissible level 250 mg l$^{-1}$ (5).

## Any other comments

Mutagenicity, toxicity and bioavalibility of food additives reviewed (6, 7).

### References

1. Kanda et al *Suisan Zoshoku* 1989, **37**(3), 221-224 (Japan.) (*Chem. Abstr.* **112**, 113796p)
2. Phalen et al *Report* 1987, ARB-R-88/354, Order No. PB88-240437, NTIS
3. Mauts et al *Report* 1988, ARB-R-88/353, Order No. PB88-235718, NTIS
4. *Fed. Regist.* 12 May 1988, **53**(92), 16862-16867
5. *EC Directive Relating to the Quality of Water Intended for Human Consumption* 1982, 80/778/EEC, Office for Official Publications of the European Communities, 2 rue Mercier, L-2985 Luxembourg
6. *Fed. Regist.* 12 April 1987, **52**(76), 13086-13107
7. *Dangerous Prop. Ind. Mater. Rep.* 1987, **7**(2), 75-79

# I72   Isazofos

**CAS Registry No.** 42509-80-8
**Synonyms** $O$-[5-chloro-1(1-methylethyl)-1$H$- 1,2,4-triazol-3-yl]-$O,O$-diethyl phosphorothioic acid ester; Ciba-Geigy A 12223; Isazophos; Miral
**Mol. Formula** $C_9H_{17}ClN_3O_3PS$                **Mol. Wt.** 313.74
**Uses** Nematacide.

### Physical properties

**B. Pt.** $_{0.001}$ 100°C; **Specific gravity** $d^{20}$ 1.22; **Volatility** v.p. $3.225 \times 10^{-5}$ mmHg at 20°C.

### Solubility

Water: 250 mg $l^{-1}$ at 20°C. Organic solvent: miscible with chloroform, methanol, benzene

### Ecotoxicity

#### Fish toxicity

$LC_{50}$ (96 hr) bluegill sunfish, carp, trout 0.01, 0.22 and 0.008 mg $l^{-1}$, respectively (1).

### Environmental fate

#### Absorption

No leaching observed throughout a 21 wk rainfall study. Evaporation accounted for 44-67% (emulsifiable concentration) and 72-100% (granular formulation) (2).
Mobility and persistence in silt loam soils was studied at 1, 2, 4, 8, 12 and 21 wk, $t_{1/2}$ 3-4 wk. No mobility observed after 1 wk (3).

## Mammalian and avian toxicity

### Acute data
$LD_{50}$ oral rat 40-60 mg $kg^{-1}$ (4).
$LC_{50}$ (4 hr) inhalation rat 0.24 mg $l^{-1}$ in air (1).
$LD_{50}$ dermal ♂, ♀ rat 118, 290 mg $kg^{-1}$ respectively (5,6).

### Sub-acute data
No effect level (90 day) oral rat, dog 0.2 and 0.05 mg $kg^{-1}$ $day^{-1}$ (1).

### Irritancy
Mild irritant to skin and minimal to eyes of rabbits (no other details) (1).

## Any other comments
Toxic to honey bees (details not specified) (1).

## References

1. *The Agrochemical Handbook* 3rd ed., 1991, RSC, London
2. Bowman *Environ. Toxicol. Chem.* 1991, **10**(7), 873-879
3. Bowman *Environ. Toxicol. Chem.* 1990, **9**(4), 453-461
4. Thompson *Agricultural Chemicals* Volume 1, 1977, 131
5. *Farm Chemicals Handbook* 1980, Meister Publishing, Willoughby, OH
6. *The Pesticide Manual* 1968, **1**, British Crop Protection Council, Thornton Heath.

# I73   Isobenzan

**CAS Registry No.** 297-78-9
**Synonyms** 1,3,4,5,6,7,8,8-octachloro-1,3,3a,4,7,7a-hexahydro-
4,7-methanoisobenzofuran;
1,3,4,5,6,7,8,8-octachloro-3a,4,7,7a-tetrahydro-4,7-methanoisobenzofuran;Telodrin
**Mol. Formula** $C_9H_4Cl_8O$                **Mol. Wt.** 411.76
**Uses** Insecticide.

## Physical properties
**M. Pt.** 120-122°C; **Specific gravity** 1.87; **Volatility** $3.0 \times 10^{-6}$ mmHg at 20°C.

### Solubility
Organic solvent: acetone, benzene, toluene, diethyl ether, xylene, heavy aromatic
naptha

## Occupational exposure

**Supply classification** very toxic.

**Risk phrases** Very toxic in contact with skin and if swallowed (R27/28)

**Safety phrases** After contact with skin, wash immediately with plenty of soap and water – Wear suitable protective clothing and gloves – In case of accident or if you feel unwell, seek medical advice immediately (show label where possible) (S28, S36/37, S45)

## Environmental fate

### Degradation studies

95% disappearance from soils in 2-4 yr (1).

### Abiotic removal

Persistence in riverwater in sealed vessel kept in sunlight and artificial fluorescent light; initial concentraion 10 $\mu$g l$^{-1}$: after 1 hr, 1 wk, 2 wk, 4 wk the levels were respectively 100, 25, 10, 0% of original compound (2).

## Mammalian and avian toxicity

### Acute data

LD$_{50}$ oral dog, rabbit, rat, mouse 1-8.4 mg kg$^{-1}$ (3,4).

LD$_{50}$ oral housesparrow, common grackle, pigeon, 1.0, 1.33, 10.0 mg kg$^{-1}$ respectively (5).

LD$_{50}$ dermal guinea pig, rat 2-5 mg kg$^{-1}$ (3,6).

LD$_{50}$ dermal rabbit 12 mg kg$^{-1}$ (7).

LD$_{50}$ intravenous rat 1.8 mg kg$^{-1}$ (8).

LD$_{50}$ intraperitoneal rat, mouse 3.56-8.17 mg kg$^{-1}$ (3).

### Sub-acute data

Oral deer mouse (feeding study, 3 day) 87.5 mg kg$^{-1}$ killed 50% of the study population (9).

## Legislation

Included in Schedule 6 (Release into Land: Prescribed Substances) Statutory Instrument No. 472, 1991 (10).

## Any other comments

Included as one of a number of pesticides evaluated in epidemiological studies for haemotoxic effects. No leukaemogenic influence demonstrated (11).

Human health effects, environmental exposure, toxicity comprehensively reviewed (12).

## References

1. Edwards, C. A. *Residue Rev.* 1966, **13**, 83
2. Eichelberger, J. W. et al *Environ. Sci. Technol.* 1971, **5**, 430-435
3. Jager *Aldrin Dieldrin Endrin and Telodrin: An Epidemiological and Toxicological Study of Long-Term Occupational Exposure* 1970, Elsevier, New York
4. *Agricultural Research Service, USDA Information Memorandum* 1966, **20**, 13, Beltsville, MD
5. Schafer, E. W. et al *Vertebrate Pest Control and Management Materials* 1979, J. R. Beck (Ed.), ASTM STP 680, American Society for Testing and Materials, Philadelphia, PA
6. *World Rev. Pest Control* 1970, **9**, 119

7. *Pesticide Chemicals Official Compendium* 1966, 1099, Association of the American Pesticide Control Officials, Inc., Topeka, Kansas

8. Traberger *Arch. Exp. Pathol. Pharmakol.* 1957, **232**, 227

9. *Environmental Properties of Chemicals* 1990, 664, Research Report 91, Ministry of the Environment, VAPK- Publishing, Helsinki

10. *S. I. 1991 No. 472 The Environmental Protection (Prescribed Processes and Substances) Regulations* 1991, HMSO, London

11. David, A. *Prakt. Lek.* 1987, **39**(9), 421-426 (Czech.) (*Chem. Abstr.* **108**, 61698d)

12. *Environmental Health Criteria 129: Isobenzan* 1991, International Program on Chemical Safety, WHO, Geneva

# I74   Isobornyl acetate

**CAS Registry No.** 125-12-2
**Synonyms** 1,7,7-trimethylbicyclo[2.2.1]heptan-2-ol, acetate; Pichtosin; Pichtosine; isoborneol acetate
**Mol. Formula** $C_{12}H_{20}O_2$                **Mol. Wt.** 196.29

## Physical properties

**M. Pt.** >-50°C; **B. Pt.** 220-224°C; **Flash point** 88-90°C.

**Solubility**
Organic solvent: ethanol, acetone

## Ecotoxicity

**Fish toxicity**
$LC_{50}$ (calc.) bluegill sunfish, rainbow trout, fathead minnow 3.97-5.18 mg l$^{-1}$ (duration unspecified) (1).

## References

1. Fiedler, H. *Toxicol. Environ. Chem.* 1990, **28**(2-3), 167-188

# I75   Isobutane

$(CH_3)_2CHCH_3$

**CAS Registry No.** 75-28-5
**Synonyms** 2-methylpropane; *iso*-butane; 1,1-dimethylethane; trimethylmethane
**Mol. Formula** $C_4H_{10}$                **Mol. Wt.** 58.12

**Uses** Aerosol propellant in consumer products. Synthesis of polyurethane foams and resins. Component of gasoline.

**Occurrence** Emitted in waste gases from printing presses, paint booths, incinerators and car exhausts (1).

## Physical properties

**M. Pt.** −160°C; **B. Pt.** −12°C; **Flash point** −81°C; **Specific gravity** $d^{20}$ 0.5572; **Volatility** v. den. 2.0.

## Solubility

Water: 49 mg $l^{-1}$ at 20°C.

## Occupational exposure

**UN No.** 1969; **HAZCHEM Code** 2WE; **Conveyance classification** flammable gas.

## Environmental fate

### Degradation studies

*Scedosporium sp.* oxidised isobutane to *tert*-butanol but neither substrate was used for growth (2).

Incubation with natural flora in ground water (in the presence of other components of high-octane gasoline 100 μg $l^{-1}$) 0% remained after 192 hr at 13°C (3).

### Abiotic removal

Estimated $t_{1/2}$ 17 hr under photochemical smog conditions in S.E. England (4,5).

## Mammalian and avian toxicity

### Acute data

Inhalation human 250-1000 ppm (1 min to 8 hr) and 500 ppm (10 day) 1 to 8 hr day$^{-1}$ no deleterious effects (6).

$LC_{50}$ (1 hr) inhalation mouse 52 mg kg$^{-1}$ (7).

### Sub-acute data

Inhalation (13 wk) F344 rats 50:50 wt% mixture of isobutane and isopentane target concentration 4500 and 1000 ppm 6 hr day$^{-1}$ 5 day wk$^{-1}$. No evidence of hydrocarbon-induced nephropathy in either sex (8).

Inhalation (90 day) of aerosol (22%) rabbit 30 sec bursts twice daily 5 days wk$^{-1}$. No adverse effects (9).

### Metabolism and pharmacokinetics

*In vitro* rat liver microsomes oxidatively metabolised to its parent alcohol (10).

Blood levels of 20-90 ng ml$^{-1}$, humans exposed to 500 ppm, indicating isobutane can be absorbed through the lungs (11).

Inhalation rats detected in adipose, brain, liver and lungs (12).

## Genotoxicity

*Salmonella typhimurium* (strain unspecified) with and without metabolic activation negative (13).

## Any other adverse effects

Mice at near $LD_{50}$ concentrations showed depression, rapid and shallow breathing and apnoea (7,14).

## Any other comments

Human effects, epidemiology, workplace experience, experimental toxicology, safety test data and exposure conditions reviewed (15).

The gas will support growth of *Mycobacterium phlei* (16).

## References

1. Kzumikawa et al *Tokyo-to Kunkyo Kagaku Kenkyusho Nenpo* 1989, 36-43 (Japan.) (*Chem. Abstr.* **112**, 24938u)
2. Onodera *Agric. Biol. Chem.* 1990, **54**(9), 2413-16
3. Jamison, V. W. et al *Proceedings of the Third International Biodegradation Symposium* 1976, Applied Science Publishers
4. Brice, K. A. et al *Atmos. Environ.* 1978, **12**, 2045-2054
5. Greiner, N. R. *J. Chem. Phys.* 1967, **46**, 3389-3392
6. Sperling, F. et al *Environ. Res.* 1972, **5**(2), 164
7. Smyth, H. F. et al *Am. Ind. Hyg. Assoc. J.* 1969, **30**, 470
8. Aranyi et al *Toxicol. Ind. Health* 1986, **2**(1), 85-89
9. Moore, A. F. *J. Am. Coll. Toxicol.* 1982, **1**, 127-142
10. Brown, V. K. H. et al *Br. J. Ind. Med.* 1968, **25**, 75
11. Snyder, R. (Ed.) *Ethel Browning's Toxicity and Metabolism of Industrial Solvents* 2nd ed., 1987, Elsevier, Amsterdam
12. Shuyaev, B. B. *Arch. Environ. Health* 1969, **18**, 878-882
13. Kirwin, C. J. et al *J. Soc. Cosmet. Chem.* 1980, **31**, 367-370
14. Brown, V. K. H. et al *Ann. Occup. Hyg.* 1967, **10**, 123
15. *ECETOC Technical Report No. 30(4)* 1991, European Chemical Industry Ecology and Toxicology Centre, B-1160 Brussels
16. Tsuji M. et al *Oyo Yakuri* 1974, **8**(10), 1439

# I76   Isobutanol

## CH3CH(CH3)CH2OH

**CAS Registry No.** 78-83-1

**Synonyms** 2-methyl-1-propanol; isobutyl alcohol; isopropylcarbinol; 2-methylpropyl alcohol

**Mol. Formula** $C_4H_{10}O$          **Mol. Wt.** 74.12

**Uses** Solvent in varnish remover and paint. Manufacturing esters for fruit flavouring essences.

**Occurrence** Emissions from wastewater treatment plants (1).

Found in fusel oil.

Produced by fermentation of carbohydrates.

## Physical properties

**M. Pt.** –108°C; **B. Pt.** 108°C; **Flash point** 28°C (closed cup); **Specific gravity** $d^{15}$ 0.806; **Partition coefficient** log $P_{ow}$ 0.76; **Volatility** v.p. 10 mmHg at 21.7°C; v. den 2.6.

## Solubility

Water: 1 in 20. Organic solvent: miscible with ethanol, diethyl ether

## Occupational exposure

**US TLV (TWA)** 50 ppm (152 mg m$^{-3}$); **UK Long-term limit** 50 ppm (150 mg m$^{-3}$);
**UK Short-term limit** 75 ppm (225 mg m$^{-3}$); **UN No.** 1212; **HAZCHEM Code 3☒**;
**Conveyance classification** flammable liquid; **Supply classification** harmful.
**Risk phrases** Flammable – Harmful by inhalation (R10, R20)
**Safety phrases** Keep away from sources of ignition – No Smoking (S16)

## Ecotoxicity

### Fish toxicity

$LC_{50}$ (48 hr) ide 1520 mg l$^{-1}$ (2).
$LC_{50}$ (24 hr) goldfish 2600 mg l$^{-1}$ (3).
$LC_{50}$ (96 hr) bleak 1000-3000 mg l$^{-1}$ (4).

### Invertebrate toxicity

$EC_{50}$ (24, 48 hr) *Daphnia magna* 1463 and 1439 mg l$^{-1}$ respectively, $EC_{100}$ (24, 48 hr)
*Daphnia magna* both 2143 mg l$^{-1}$ as determined by acute *Daphnia* test (5).
21 day *Daphnia* sp. reproduction test to measure mortality of parent animals, the
reproduction rate and the appearance of the 1st offspring, 21 day NOEC 3.4 mg l$^{-1}$ with
the reproduction rate being the most affected (6).
$EC_{50}$ (0-48 hr) *Scenedesmus subspicatus* 2300 mg l$^{-1}$ cell multiplication test (7).
Threshold for narcosis, tadpole 4000 mg l$^{-1}$ (8).
NOEC on biomass *Uronema paraduczi* (20 hr), *Pseudomonas putida* (16 hr) and
*Entosiphon sulatum* (72 hr) 169-296 mg l$^{-1}$ (9).
NOEC of growth *Microcystis aeruginosa*, *Scenedesmus quadricauda* (8 day)
290-350 mg l$^{-1}$ (10).
$EC_{50}$ (5 min) *Photobacterium phosphoreum* 1659 ppm Microtox test (11).

### Bioaccumulation

Does not bioaccumulate (12).

## Environmental fate

### Degradation studies

Oxidised to corresponding acid by *Desulfovibrio vulgaris* (13).
5-day BOD determined using acclimated mixed microbial cultures was 3.92 mg l$^{-1}$ $O_2$
(14).
An oxygen requirement of 1.4 mg needed to oxidise 1 mg (15).
ThOD 37.4% (16).
Readily biodegradable. Degraded significantly within hours and does not persist
beyond a few days (17).

### Absorption

Low adsorption in soils and sediments. Liable to leaching (17).

## Mammalian and avian toxicity

### Acute data

$LD_{50}$ oral rat 2.46 g kg$^{-1}$ (18).
$LC_{Lo}$ (4 hr) inhalation rat 8000 ppm (18).
$LD_{50}$ dermal rabbit 3.4 g kg$^{-1}$ (19).
$LD_{50}$ intravenous rat 340 mg kg$^{-1}$ (20).
$LD_{50}$ intraperitoneal rabbit, rat 323-720 mg kg$^{-1}$ (20).

---

**Sub-acute data**

Oral (4 month) Wistar rat 74 g $l^{-1}$ as sole drinking liquid showed no adverse effect to liver. Oral (2 month) 148 g $l^{-1}$ as sole drinking liquid showed reduction in liver size and also a reduction in fat, glycogen and RNA content of the liver (21).

Inhalation (4 month) rat, continuous exposure 3 mg $m^{-3}$. Depression of leg withdrawal response to electrical stimulation, minor changes to formed elements of blood and serum enzymes. Estimated NOEC level 0.1 mg $m^{-3}$ (21).

**Carcinogenicity and long-term effects**

In a lifetime study 2 groups of rats were given subcutaneous 0.05 ml $kg^{-1}$ body weight twice a wk or oral 0.2 mg $kg^{-1}$ body weight twice a wk. Both groups exhibited liver damage ranging from steatosis to cirrhosis. Subcutaneous group showed 8 animals with malignant tumours, 3 in the oral group and none in the control group. The majority of treated animals also showed hyperplasia in blood-forming tissues (21).

**Teratogenicity and reproductive effects**

No adequate data are available to assess teratogenicity or effects on reproduction (21).

**Metabolism and pharmacokinetics**

Adsorbed through skin, lungs and gastrointestinal tract. Metabolised by alcohol dehydrogenase to isobutyric acid, via the aldehyde it may enter the tricarboxylic acid cycle. Small amounts are excreted unchanged or as glucuronide in urine. In rabbits metabolites in urine included acetaldehyde (22).

**Irritancy**

Eye rabbit (24 hr) 100 vol % severe irritation, 30 vol % moderate irritation 10-20 vol % mild irritation, Draize test (23).

Dermal rabbit (24 hr) 500 mg caused severe irritation (24).

Mildly irritating to eyes (details unspecified) (21).

## Genotoxicity

Single intragastric administration at equitoxic dose. 0.2 $LD_{50}$ induced chromosomal aberrations and polyploidy in rat bone marrow (25).

## Legislation

Included in Schedule 6 (Release into Land: Prescribed Substances) Statutory Instrument No. 472, 1991 (26).

## Any other comments

Included in Council of Europe (1981) list of flavouring substances that can be included in food and beverages at 25 mg $kg^{-1}$ without hazard to health (21).

Human health effects, experimental toxicology, physico-chemical properties, environmental effects, ecotoxicology, exposure levels and workplace experience reviewed (27).

Chemical properties, industrial uses and toxicity reviewed (28).

## References

1.  Hangartner, M. *Int. J. Environ. Anal. Chem.* 1979, **6**, 161-169
2.  Junkhe, J. et al *Z. Wasser-Abwasser Forsch.* 1978, **11**, 161-164
3.  Bridie, A. L. et al *Water Res.* 1979, **13**, 623
4.  Linden et al *Chemosphere* 1979, **11/12**, 843-851
5.  Kuoehn, R. et al *Water Res.* 1989, **23**(4), 495-499

6. Kuoehn, R. et al *Water Res.* 1989, **23**(4), 501-510
7. Kuoehn, R. et al *Water Res.* 1990, **24**(1), 31-38
8. Muoench, J. C. *Ind. Med.* 1972, **41**(4), 31-33
9. Bringmann, G. et al *GWF, Gas- Wasserfach: Wasser/Abwasser* 1981, **122**(7), 308-312
10. Bringmann, G. et al *Vom Wasser* 1978, **50**, 45-60
11. Curtis, C. A. et al *Aquatic Toxicology and Hazard Assessment: 5th Conference* 1982, 170-178, Pearson, J. G. et al (Eds), ASTM STP766, American Society for Testing of Materials, Philadelphia, PA
12. Chiou, C. T. et al *Environ. Sci. Technol.* 1977, **11**(5), 475-478
13. Tanaka *Arch. Microbiol.* 1990, **155**(1), 18-21
14. Babeu, L. et al *J. Ind. Microbiol.* 1987, **2**(2), 107-115
15. Nazarenko, I. O. *Sanit. Ochrana. Vodojemov at Zagrazen. Strochn. Vodami. M, Medgiz* 1969, **4**, 65-75
16. Vaisnav, D. D. et al *Chemosphere* 1986, **16**(4), 695-703
17. Howard, P. H. *Handbook of Environmental Fate and Exposure Data for Organic Chemicals* 1990, Lewis Publishers, Chelsea, MI
18. Smyth et al *Arch. Ind. Hyg. Occup. Med.* 1954, **10**, 61
19. *Raw Material Data Handbook: Organic Solvents* 1974, Volume 1, National Association of Printing Ink Research Institute, Francis McDonald Sinclair Memorial Laboratory, Lehigh University, Bethlehem, PA
20. *Environ. Health Perspect.* 1985, *61*, 321
21. WHO *Environmental Health Criteria 65* 1987, WHO, Geneva
22. *Environmental Properties of Chemicals*, 1990, Research Report 91, 658-662 Ministry of the Environment, VAPK-Publishing, Helsinki
23. Kennah, H. E. et al *Appl. Toxicol.* 1989, **12**(2), 258-268
24. Lewis, R. J. *Sax's Dangerous Properties of Industrial Materials* 8th ed., 1992, Van Nostrand Reinhold, New York
25. Burilyak *Tsitol. Genet.* 1988, **22**(2), 49-52, (Russ.) (*Chem. Abstr.* **109** 68639b)
26. *S. I. 1991 No. 472 The Environmental Protection (Prescribed Processes and Substances) Regulations* 1991, HMSO, London
27. *ECETOC Technical Report No. 30(4)* 1991, European Chemical Industry Ecology and Toxicology Centre, B-1160 Brussels
28. Sejbl *Rudy* 1986, **34**(12), 364-367, (Czech.) (*Chem. Abstr.* **106**, 181925f)

# I77   Isobutene

$$CH_3C(CH_3)=CH_2$$

**CAS Registry No.** 115-11-7
**Synonyms** γ-butylene; isobutylene; 2-methylpropene; 1-propene-2-methyl
**Mol. Formula** $C_4H_8$                    **Mol. Wt.** 56.11
**Uses** Producing polymers and antioxidants for foods, packaging, food supplements and plastics. Production of high octane aviation gasoline.
**Occurrence** As minor environmental contaminant in urban air.

## Physical properties

**M. Pt.** −140.3°C; **B. Pt.** −6.9°C; **Flash point** −76.1°C; **Specific gravity** $d_4^{20}$ 0.5942; **Partition coefficient** log $P_{ow}$ 2.35; **Volatility** v.p. 400 mmHg at 21.6°C; v. den. 2.01.

## Solubility
Water: 263 mg l$^{-1}$ at 20°C. Organic solvent: ethanol, acetone, diethyl ether, benzene

## Occupational exposure
**UN No.** 1055; **HAZCHEM Code** 2WE; **Conveyance classification** flammable gas; **Supply classification** highly flammable.
**Risk phrases** Extremely flammable liquefied gas (R13)
**Safety phrases** Keep container in a well ventilated place – Keep away from sources of ignition – No Smoking – Take precautionary measures against static discharges (S9, S16, S33)

## Mammalian and avian toxicity

### Acute data
LC$_{50}$ (4 hr) inhalation rat 620 g m$^{-3}$ (1).
LC$_{50}$ (2 hr) inhalation mouse 415 g m$^{-3}$ (1).
Mice (1-2 min) 50-70% isobutene resulted in immediate narcosis (2).
Inhalation mouse 30% no effect, 40% (7-8 min) excitement and narcosis (3).

### Carcinogenicity and long-term effects
A study of 62 ♂ and 29 ♀ workers; 50% with ≥ 11 yr service in a plant producing isobutylene and other hydrocarbons had symptomatic effects including anaemia, reduction of peroxidase activity in granulocytes and granulocytopenia (4).
Exposure of rubber workers to isobutylene, combined with isoprene and chloromethane depressed succinate dehydrogenase activity in the immunocompetent blood cells (5).
Inhalation rat (Fisher 344), mouse (B6C3F$_1$), prechronic study in progress (6).

### Metabolism and pharmacokinetics
*In Vitro* mice liver homogenates supplemented with NADPH metabolised to 2-methyl-1,2-epoxypropene. The epoxidation is cytochrome P$_{450}$ dependent, concentrations reach maximum after 20 min and is then converted to 2-methyl-1,2-propanediol and the glutathione conjugate by epoxide hydratase and glutathione *S*-transferase activities (7).
Inhalation-rats, mice steady state concentration was reached at 1200 ppm in rats and 1800 ppm in mice. At exposure concentrations of 500 ppm rate of metabolism was directly proportional to its concentration. Metabolite detected 1,1-dimethyloxirane (8).

## Genotoxicity
Butenes (isomer unspecified) in vapour phase *Salmonella typhimurium* TA97, TA98, TA100 with and without metabolic activation negative (9).

## Any other adverse to man
Acute exposure to organic chemicals in humans and chronic poisoning resulting from occupational exposure caused narcotic effects. Isobutylene caused disorders of the hypothalamo-hypophyseal system which were manifested in abnormally increased basal levels of immunoreactive insulin and somatotropin in blood serum (10).

## Any other adverse effects
Rubber which contains isobutylene may release anaesthetic and asphyxiant gases if isobutylene concentration is high (11).

Exposure of rats and mice to vapours of isobutylene showed concentrations found in brain, liver, kidney, spleen, perinephric fat, and hypodermic fat (12).

**Any other comments**

Human health effects, experimental toxicology, physico-chemical properties and work place experience reviewed (13).
Explosion and fire risk.
Butenes are weak anaesthetics and asphyxiant, and narcotic at high concentrations when heated to decomposition emits acrid smoke and irritating fumes.

**References**

1. Clayton, G. et al *Patty's Industrial Hygiene and Toxicology* 3rd ed., 1982, **2B**, 3175-3220, John Wiley, New York
2. von Oettingen, W. R. *Toxicity and Potential Dangers of Aliphatic and Aromatic Hydrocarbons* 1940, Public Health Bulletin No. 255
3. Foss, P. *Ann. N. Y. Acad. Sci.* 1971, **180**, 126
4. Khristeva, V. et al *Khig. Zdraveopaz.* 1985, **28**(6), 28-33 (Bulg.) (*Chem. Abstr.* **104**, 173617a).
5. Mamedov, A. M. et al *Azerb. Med. Zh.* 1985, **62**(1), 25-29 (Russ.) (*Chem. Abstr.* **102**, 225331h)
6. *Directory of Agents Being Tested for Carcinogenicity* 1992, IARC, Lyon
7. Cornet, M. et al *Arch. Toxicol.* 1991, **65**(4), 263-267
8. Csanady, G. A. *Arch. Toxicol.* 1991, **65**(2), 100-105
9. Hughes, T. J. et al *Gov. Rep. Announce. Index (U. S.)* 1984, **84**(10), 65 (*Chem. Abstr.* **101**, 85417t)
10. Danilin, V. A. *Gig. Tr. Prof. Zabol.* 1983, **12**, 29-32 (Russ.) (*Chem. Abstr.* **100**, 197080p)
11. Hamilton, A. *Industrial Toxicology* 3rd ed., 1974, 345-348, Publ. Sci. Group, MA
12. Shugaev, B. B. *Arch. Environ. Health* 1969, **18**(6), 78-882
13. *ECETOC Technical Report No*, **30**(4) 1991, European Chemical Industry Ecology and Toxicology Centre, B-1106 Brussels

# I78   Isobutyl acetate

$$CH_3CO_2CH_2CH(CH_3)CH_3$$

**CAS Registry No.** 110-19-0
**Synonyms** 2-methylpropyl acetic acid ester; acetic acid, isobutyl ester; 2-methylpropyl acetate; β-methylpropyl ethanoate
**Mol. Formula** $C_6H_{12}O_2$                     **Mol. Wt.** 116.16
**Uses** Solvent, flavouring.
**Occurrence** Wood rotting fungus. Animal (cattle) waste.

**Physical properties**

**M. Pt.** –99°C; **B. Pt.** 118°C; **Flash point** 18°C (closed cup); **Specific gravity** $d_4^{20}$ 0.871; **Partition coefficient** log $P_{ow}$ 1.60 (1); **Volatility** v.p. 10 mmHg at 12.8°C; v. den. 4.0.

**Solubility**

Water: 7.5 g l$^{-1}$. Organic solvent: ethanol

## Occupational exposure

**US TLV (TWA)** 150 ppm (713 mg m$^{-3}$); **UK Long-term limit** 150 ppm (700 mg m$^{-3}$); **UK Short-term limit** 187 ppm (875 mg m$^{-3}$); **UN No.** 1213; **HAZCHEM Code 3ME; Conveyance classification** flammable liquid; **Supply classification** highly flammable.

**Risk phrases** Highly flammable (R11)

**Safety phrases** Keep away from sources of ignition – No Smoking – Do not breathe vapour – Do not empty into drains – Take precautionary measures against static discharges (S16, S23, S29, S33)

## Ecotoxicity

**Invertebrate toxicity**

LOEC on reproduction (semichronic exposure) *Scenedemus quadricauda* 80 mg l$^{-1}$ (2).
Toxicity threshold (cell multiplication inhibition test) *Pseudomonas putida, Microcystis aeruginosa* 200-205 mg l$^{-1}$ (2,3).
Toxicity threshold (cell multiplication inhibition test) *Entosiphon sulcatum, Uronema parduczi* 411-727 mg l$^{-1}$ (2,4).

## Environmental fate

**Degradation studies**

Filtered sewage seed (5 and 20 day) theoretical BOD 60 and 81% respectively in fresh water and 23 and 37% respectively in salt water (5).
Significantly biodegradable suggesting soil microbial action (6).

**Abiotic removal**

$t_{1/2}$ for vapour phase reaction with photochemically produced hydroxyl radicals (average concentration $5 \times 10^5$ molecules cm$^{-3}$) in the atmosphere is about 6 hr (7).
Chemical hydrolysis may be important at ≥pH 9 (6).
Volatilisation $t_{1/2}$ from a 1 m deep river, flowing at 1 m sec$^{-1}$ with a wind velocity of 3 m sec$^{-1}$ estimated to be 5.3 hr (6).

**Absorption**

Moderate to high soil leaching based on $K_{oc}$ values of 36 and 177 (8,9).

## Mammalian and avian toxicity

**Acute data**

LD$_{50}$ oral rat 13,400 mg kg$^{-1}$ (10).
LD$_{50}$ oral rabbit 4763 mg kg$^{-1}$ (11).
LC$_{Lo}$ (4 hr) inhalation rat 8000 ppm (12).

**Irritancy**

Dermal (24 hr) rabbit 500 mg caused moderate to severe erythema and moderate oedema and 500 mg instilled into rabbit eye (24 hr) caused moderate irritation (13).

## Any other comments

Human health effects, epidemiology, workplace experience, experimental toxicology, safety test data and exposure conditions reviewed (14).

### References

1. *GEMS: Graphic Exposure Modeling System* CLOGP3 Program, Office of Toxic Substances, US EPA, 1989
2. Bringmann, G. et al *Water Res.* 1980, **14**, 231-241
3. Bringmann, G. et al *GWF, Gas Wasserfach: Wasser/Abwasser* 1976, **117**(9), (Ger.)
4. Bringmann, G. et al *Z. Wasser/Abwasser Forsch.* 1980, **1**, 26-31
5. Price, K. S. *J. -Water Pollut. Control Fed.* 1974, **46**, 63
6. Howard, P. H. *Handbook of Environmental Fate and Exposure Data for Organic Chemicals* 1990, **II**, 290-295, Lewis Publications, Chelsea, MI
7. Atkinson, R. *Int. J. Chem. Kinet.* 1987, **19**, 799-828
8. Lyman, W. K. et al *Handbook of Chemical Property Estimation Methods. Environmental Behaviour of Organic Compounds* 1982, McGraw-Hill, New York
9. Swann, R. L. et al *Res. Rev.* 1983, **85**, 17
10. *Raw Material Data Handbook: Organic Solvents* 1974, **1**, 8, National Association of Printing Ink Research Institute, Francis McDonald Sinclair Memorial Laboratory, Lehigh University, Bethlehem, PA
11. *Ind. Med. Surg.* 1972, **41**, 31
12. Smyth, H. F. et al *Am. Ind. Hyg. Assoc. J.* 1962, **23**, 95
13. *Food Cosmet. Toxicol.* 1978, **16**, 637
14. *ECETOC Technical Report No. 30(4)* 1991, European Chemical Industry Ecology and Toxicology Centre, B-1160 Brussels

## I79   Isobutyl acrylate

$$CH_2=CHCO_2CH_2CH(CH_3)CH_3$$

**CAS Registry No.** 106-63-8
**Synonyms** 2-methylpropyl 2-propenoic acid ester; acrylic acid, isobutyl ester; isobutyl 2-propenoate
**Mol. Formula** $C_7H_{12}O_2$                    **Mol. Wt.** 128.17

### Physical properties

**B. Pt.** $_{15}$ 61-63°C; **Flash point** 30°C (open cup); **Specific gravity** 0.9 (liquid); **Volatility** v. den. 0.9.

### Occupational exposure

**UN No.** 2527; **HAZCHEM Code 3☒**; **Conveyance classification** flammable liquid; **Supply classification** Xn (≥25%), Xi (≥1%).
**Risk phrases** ≥25% – Flammable – Harmful by inhalation and in contact with skin – Irritating to skin – May cause sensitisation by skin contact – ≥10%<25% – Flammable – Irritating to skin – May cause sensitisation by skin contact – ≥1%<10% – Flammable – May cause sensitisation by skin contact (R10, R20/21, R38, R43, R10, R38, R43, R10, R43)
**Safety phrases** Keep container in a well ventilated place – Avoid contact with skin – Wear suitable gloves (S9, S24, S37)

# Ecotoxicity

## Fish toxicity

$LC_{50}$ (96 hr) fathead minnow 2.09 mg $l^{-1}$ (1).
5 ppm caused death in 1, 7 or 23 hr in trout, bluegill sunfish and goldfish respectively.
Test conditions: pH7; dissolved oxygen content 7.5 ppm; total hardness (soap method) 300 ppm; methyl orange alkalinity 310 ppm; free carbon dioxide 5 ppm; temperature 12.8°C (2).
2-5 mg $l^{-1}$ steelhead trout, bridgelip sucker, stickleback caused death within 5 hr of exposure. Test conditions: artesian well water; total hardness 67-120 mg $l^{-1}$; methyl orange alkalinity 151-183 mg $l^{-1}$; total dissolved solids 160-175 mg $l^{-1}$; pH 7.1 (3).

## Mammalian and avian toxicity

### Acute data

$LD_{50}$ oral mouse, rat 6106-7070 mg $kg^{-1}$ (4, 5).
$LC_{Lo}$ (4 hr) inhalation rat 2000 ppm (6).
$LD_{50}$ dermal rabbit 890 mg $kg^{-1}$ (6).
$LD_{50}$ intraperitoneal rat, mouse 654-760 mg $kg^{-1}$ (7, 8).

### Irritancy

Dermal rabbit (duration unspecified) 500 mg open to atmosphere caused mild irritation (6).

## Genotoxicity

*Salmonella typhimurium* TA98, TA100, TA1535, TA1537 with and without metabolic activation negative (9).

## Any other comments

Human health effects, experimental toxicology, physico-chemical properties reviewed (10).
Toxicity and hazards reviewed (11).

## References

1. Russom, C. L. *Bull. Environ. Contam. Toxicol.* 1988, **4**, 589-596
2. Wood, E. M. *The Toxicity of 3400 Chemicals to Fish* 1987, EPA 560/6-87-002, PB 87-200-275, Washington, D.C
3. *Fish Toxicity Screening Data* 1989, EPA 560/6-89-001, PB 89-156715, Washington, DC
4. *Toxicol. Lett.* 1982, **11**, 125
5. *Toxicol. Appl. Pharmacol.* 1974, **28**, 313
6. *Union Carbide Data Sheet* 28-3-68, Industrial Medicine and Toxicology Department, Union Carbide Corp., New York
7. *Arch. Mal. Prof. Med. Trav. Secur. Soc.* 1975, **36**, 58
8. *J. Dent. Res.* 1975, **51**, 526
9. Zeiger, E. et al *Environ. Mutagen.* 1987, **9**(9), 1-109
10. *ECETOC Technical Report No. 30(4)* 1991, European Chemical Industry Ecology and Toxicology Centre, B-1160 Brussels
11. *Dangerous Prop. Ind. Mater. Rep.* 1987, **7**(6), 65-67

# I80　Isobutylamine

## CH₃CH(CH₃)CH₂NH₂

**CAS Registry No.** 78-81-9
**Synonyms** 2-methyl-1-propanamine; 2-methylpropylamine; monoisobutylamine; Valamine
**Mol. Formula** $C_4H_{11}N$ **Mol. Wt.** 73.14
**Uses** Organic synthesis. In the manufacture of insecticides.
**Occurrence** In various species of marine algae and Latakia tobacco leaves.

## Physical properties

**M. Pt.** –85°C; **B. Pt.** 68-69°C; **Flash point** –7°C; **Specific gravity** $d_4^{25}$ 0.724;
**Partition coefficient** log $P_{ow}$ 0.73; **Volatility** v.p. 100 mmHg at 20°C; v. den. 2.5.

### Solubility
Water: miscible. Organic solvent: ethanol, diethyl ether

## Occupational exposure

**UN No.** 1214; **HAZCHEM Code** 2WE; **Conveyance classification** flammable liquid.

## Ecotoxicity

**Fish toxicity**
$LC_0$ creek chub (24 hr) 20 mg $l^{-1}$, $LC_{100}$ creek chub (24 hr) 60 mg $l^{-1}$ (1).

## Environmental fate

### Degradation studies
Utilised by *Aspergillus versicolor* as a nitrogen source, however, only amines with alkyl chains ≥5 carbon atoms long supported significant growth in absence of a separate carbon substrate (2).
Confirmed to be biodegradable (3).

### Absorption
Not expected to be significantly adsorbed onto sediments or soils, therefore leaching is possible (4).

## Mammalian and avian toxicity

### Sub-acute data
$LD_{50}$ (14 day) oral ♂, ♀ rat 224 and 232 mg $kg^{-1}$, respectively (5).

### Irritancy
In humans skin contact can result in erythema and blistering (6).

## Any other adverse effects to man

Inhalation causes headache, dryness of nose and throat (6).

## Any other adverse effects

Upper airway irritation, $RD_{50}$ inhalation mouse 91 ppm. Pulmonary toxicity, $RD_{50}$ tracheally cannulated inhalation mouse 406 ppm (7).

---

## Any other comments

Human health effects, epidemiology, workplace experience and experimental toxicology reviewed (8).

Has been detected in air samples from ''sick-building'' syndrome workplaces. Pollution from isobutylamine has been attributed to biodegredation of casein by alkali-resistant *Clostridium* (9).

## References

1. Gillette, L. A. et al *Sewage Ind. Wastes* 1952, **24**(11), 1397-1401
2. Lindley, N. D. *Appl. Environ. Microbiol.* 1987, **53**(2), 246-248
3. *The list of the existing chemical substances tested on biodegradability by microorganisms of bioaccumulation in fish bodies* 1987, Chemicals Inspection and Testing Institute, Japan
4. Howard, P. H. *Handbook of Environmental Fate and Exposure Data for Organic Chemicals* 1990, Lewis Publishers, Chelsea, MI
5. Cheever, K. L. et al *Toxicol. Appl. Pharmacol.* 1982, **62** 150
6. *The Merck Index* 11th ed., 1989, Merck and Co., Rahway, NJ
7. Gagnair, F. et al *J. Appl. Toxicol.* 1989, **9**(5), 301-4
8. *ECETOC Technical Report No. 30(4)* 1991, European Chemical Industry Ecology and Toxicology Centre, B-1160 Brussels
9. Karlsson et al *Mater. Struct.* 1989, **22**(129), 163-169

# I81   Isobutyl formate

## HCO2CH2CH(CH3)CH3

**CAS Registry No.** 542-55-2

**Synonyms** 2-methylpropyl formic acid ester; formic acid, isobutyl ester; 2-methylpropyl formate; tetryl formate

**Mol. Formula** $C_5H_{10}O_2$                     **Mol. Wt.** 102.13

## Physical properties

**M. Pt.** –95°C; **B. Pt.** 98°C; **Flash point** 5°C; **Specific gravity** $d_4^{20}$ 0.885; **Volatility** v. den. 3.5.

**Solubility**
Water: 1 in 100. Organic solvent: miscible with ethanol, diethyl ether

## Occupational exposure

**UN No.** 2393; **HAZCHEM Code** 3☒E; **Conveyance classification** flammable liquid.

## Mammalian and avian toxicity

**Acute data**
$LD_{50}$ oral rat 3.06 g $kg^{-1}$ (1).

## References

1. *Ind. Med. Surg.* 1972, **41**, 31

# I82 Isobutyl isobutyrate

## CH₃CH(CH₃)CO₂CH₂CH(CH₃)CH₃

**CAS Registry No.** 97-85-8

**Synonyms** 2-methylpropanoic acid, 2-methylpropyl ester; isobutyric acid, isobutyl ester; isobutyl isobutanoate; 2-methylpropyl isobutyrate

**Mol. Formula** $C_8H_{16}O_2$ **Mol. Wt.** 144.22

## Physical properties

**M. Pt.** –81°C; **B. Pt.** 147°C; **Flash point** 38°C; **Specific gravity** $d_{20}^{20}$ 0.850-0.860; **Volatility** v.p. 10 mmHg at 39.9°C; v. den. 5.0.

**Solubility**
Organic solvent: miscible with ethanol

## Occupational exposure

**UN No.** 2528; **HAZCHEM Code** 3⚡; **Conveyance classification** flammable liquid.

## Mammalian and avian toxicity

**Acute data**
$LD_{50}$ oral rat 128 g $kg^{-1}$ (1).
$LD_{50}$ (6 hr) inhalation rat 5000 ppm (1).

## References

1. *Raw Material Data Handbook: Organic Solvents* 1974, Volume 1, National Association of Printing Ink Research Institute, Francis McDonald Memorial Laboratory, Lehigh University, Bethlehem, PA

# I83 Isobutyl methacrylate

## CH₂=C(CH₃)CO₂CH₂CH(CH₃)₂

**CAS Registry No.** 97-86-9

**Synonyms** 2-methyl-2-propenoic acid, 2-methylpropyl ester; methacrylic acid, isobutyl ester; isobutyl α-methylacrylate; 2-methylpropyl methacrylate

**Mol. Formula** $C_8H_{14}O_2$ **Mol. Wt.** 142.20

**Uses** Monomer for acrylic resins. To make concrete, water repellant. Manufacture of contact lenses.

## Physical properties

**B. Pt.** 155°C; **Flash point** 41°C (closed cup); **Specific gravity** $d^{20}$ 0.886; **Partition coefficient** log $P_{ow}$ 2.66; **Volatility** v.p. 1.8 mmHg at 20°C; v. den. 4.9.

**Solubility**
Organic solvent: ethanol, diethyl ether

## Occupational exposure

**UN No.** 2283; **HAZCHEM Code 3Ⅶ**; **Conveyance classification** flammable liquid; **Supply classification** irritant($\geq$1%).
**Risk phrases** $\geq$20% – Flammable – Irritating to eyes, respiratory system and skin – May cause sensitisation by skin contact – $\geq$1%<20% – Flammable – May cause sensitisation by skin contact (R10, R36/37/38, R43, R10, R43)
**Safety phrases** Avoid contact with skin – Wear suitable gloves (S24, S37)

## Ecotoxicity

**Fish toxicity**
$LC_{50}$ (96 hr) fathead minnow 38 mg l$^{-1}$ (1).

**Bioaccumulation**
A calculated bioconcentration factor of 62 indicates that bioaccumulation in aquatic organisms is unlikely (2).

## Environmental fate

**Degradation studies**
Will undergo significant biodegradation according to the MITI test (biodegradation test of the Japanese Ministry of International Trade and Industry) (3).

**Abiotic removal**
Based on the hydrolysis of methylmethacrylate it may be susceptible to hydrolysis especially in alkaline soils (4).
It will significantly volatilize from water $t_{1/2}$ 5.62 hr (estimate model river) (2).
Volatilization $t_{1/2}$ 4.2 day model pond (5).

**Absorption**
Has low mobility in soil, it is therefore possible that it may slowly leach into ground water (2,6,7).

## Mammalian and avian toxicity

**Acute data**
$LD_{50}$ oral mouse 12 g kg$^{-1}$ (8).
$LD_{50}$ intraperitoneal mouse 1.34 g kg$^{-1}$ (9).

**Teratogenicity and reproductive effects**
Has some developmental toxicity in rats (details not stated), calculated to have no human risk (10).
Pregnant rats on days 5, 10 and 15 of gestation, doses of $\frac{1}{10}$, $\frac{1}{5}$ or $\frac{1}{3}$ of their individual acute $LD_{50}$ values generally produced a dose-related increase in resorptions and gross and skeletal abnormalities, foetal birth weight also decreased (11).

**Irritancy**
Vapour mist is lachrymatory and irritating to the eyes and also irritating to the skin; it may be harmful if absorbed through the skin (species not specified) (12).

## Genotoxicity

*Salmonella typhimurium* TA98, TA100, TA1535, TA1537 with and without metabolic activation negative (13).

**Any other adverse effects to man**

Harmful if swallowed, inhaled or absorbed through the skin (12).

**Any other comments**

Human health effects, experimental toxicology and physico-chemical properties reviewed (14).

Occupational exposure occurs during production of synthetic fingernails (15).

**References**

1. Russom, C. L. et al *Bull. Environ. Contam. Toxicol.* 1988, **41**, 589-596
2. Lyman, W. J. et al *Handbook of Chemical Property Estimation Methods. Environmental Behaviour of Organic Compounds* 1982, 5-29, McGraw-Hill, New York
3. Susaki, S. *Aquatic Pollutants Transformation and Biological Effects* 1978, 283-298, Pergamon Press, Oxford
4. Ellington, J. J. et al *Measurement of Hydrolysis Rate Constants for Evaluation of Hazardous Wastes* US EPA-600/3-88/028, 1988, 18
5. *US EPA; EXAMS II Computer Simulations* 1987
6. Hansch, C. et al *Medchem Project Issue No. 26* 1985, Pomona College, Claremont, CA
7. Swann, R. L. et al *Res. Rev.* 1983, **85**, 17-28
8. *Toxicol. Lett.* 1982, **11**, 125
9. *J. Pharm. Sci.* 1973, **62**, 778
10. Jelovsek, F. R. et al *Obstet. Gynecol.* 1989, **74**(4), 624-636
11. Singh, A. R. et al *J. Dent. Res.* 1972, **51**(6), 1632-1638
12. *Chemical Safety Data Sheets* 1990, **3**, 39, RSC, London
13. Zeiger, E. et al *Environ. Mutagen.* 1987, **9**(9), 1-109
14. *ECETOC Technical Report No. 30(4)* 1991, European Chemical Industry Ecology and Toxicology Centre, B-1160 Brussels
15. Crable, J. V. et al *Appl. Ind. Hyg.* 1986, **1**(4), 200

# I84   Isobutyl methyl ketone

## CH3C(O)CH2CH(CH3)CH3

**CAS Registry No.** 108-10-1

**Synonyms** hexone; 4-methyl-2-pentanone; isopropylacetone; methyl isobutyl ketone; 2-methylpropyl methyl ketone

**Mol. Formula** $C_6H_{12}O$                     **Mol. Wt.** 100.16

**Uses** Solvent used in industrial processes, pharmaceuticals, pesticides, adhesives and coatings (paint, varnish etc.).

**Occurrence** Can be released into the environment from exhaust fumes, effluent and emissions from its manufacture and use, and from land disposal (1).

**Physical properties**

**M. Pt.** –84.7°C; **B. Pt.** 117-118°C; **Flash point** 23°C (closed cup); **Specific gravity** $d_4^{20}$ 0.801; **Partition coefficient** log $P_{ow}$ 1.31; **Volatility** v.p. 16 mmHg at 20°C; v. den. 3.5.

**Solubility**
Water: 20 g l$^{-1}$. Organic solvent: miscible with ethanol, benzene, diethyl ether

## Occupational exposure
**US TLV (TWA)** 50 ppm (205 mg m$^{-3}$); **US TLV (STEL)** 75 ppm (307 mg m$^{-3}$); **UK Long-term limit** 50 ppm (205 mg m$^{-3}$); **UK Short-term limit** 100 ppm (410 mg m$^{-3}$) change proposed; **Supply classification** highly flammable.
**Risk phrases** Highly flammable (R11)
**Safety phrases** Keep container in a well ventilated place – Keep away from sources of ignition – No Smoking – Do not breathe vapour – Take precautionary measures against static discharges (S9, S16, S23, S33)

## Ecotoxicity

**Fish toxicity**
LC$_{50}$ (96 hr) fathead minnow 505-537 mg l$^{-1}$ (2,3).
LC$_{50}$ (24 hr) goldfish 460 mg l$^{-1}$ (4).

**Invertebrate toxicity**
LOEC effect on reproduction (semichronic exposure) *Microcystis aeruginosa* 136 mg l$^{-1}$ (5).
EC$_{50}$ (5 min) *Photobacterium phosphoreum* 79.6 ppm Microtox test (6).
Threshold concentration for growth inhibition *Pseudomonas putida* (16 hr) 275 mg l$^{-1}$ (7).
Threshold concentration for growth inhibition *Uronema parduczi* (20 hr) 950 mg l$^{-1}$ (8).

**Bioaccumulation**
No data on bioaccumulation but it's moderate water solubility and low partition coefficient suggest low bioaccumulation potential (9).

## Environmental fate

**Degradation studies**
Microbial biodegradation can occur in soil (10, 11).
When incubated with settled domestic sewage as seed was found to have a 40 day ThOD of 64.8 mg l$^{-1}$ O$_2$ (12).
500 ppm incubated with 3 different activated sludge samples gave an average ThOD of 3% in 24 hr (13).
Standard dilution method with sludge from waste-treatment plant, BOD$_5$ to be 76% of ThOD (14).

**Abiotic removal**
In water primary removal mechanisms are volatilisation (t$_{1/2}$ 15-33 hr) and direct photolysis, chemical hydrolysis is not important (1,15).
In atmosphere degraded by hydroxy radicals; t$_{1/2}$ 0.57 days. It is also photodegraded with the major product being acetone (t$_{1/2}$ 16 days) (16).

**Absorption**
Estimated soil adsorption coefficent (K$_{oc}$) 19-106 (17).
Highly mobile in soil and will not be adsorbed significantly (18).

## Mammalian and avian toxicity

**Acute data**
LD$_{50}$ oral guinea pig, rat 1.6-2.8 g kg$^{-1}$ (19,20).

LD$_{50}$ oral redwing blackbird >100 mg kg$^{-1}$ (calc.) (21).
LC$_{50}$ inhalation mouse 23.3 g m$^{-3}$ (22).
LD$_{50}$ intraperitoneal mouse, rat 268-400 mg kg$^{-1}$ (19,23).

**Sub-acute data**
Inhalation (90 day) rat, dog, monkey 410 mg m$^{-3}$ continuous exposure under reduced oxygen tension and atmospheric pressure, increased liver and kidney weights. Histopathological changes in rat kidney epithelium (reversible 3-4 wk). No histopathological changes in dogs or monkeys (24).
Oral gavage (13 wk) Sprague-Dawley rats 0, 59, 250, 1000 mg kg$^{-1}$ day$^{-1}$ dose-related increase in liver and kidney weights no corresponding histopathological lesions in liver. No effects were observed at 50 mg kg$^{-1}$ (25).
Dermal rat (4 month) 300-600 mg kg$^{-1}$ day$^{-1}$, dose- and time-dependent morphological changes were observed in skin, liver, brain, adrenals, spleen and testis. Body temperature decreased and oxygen consumption increased (26).

**Carcinogenicity and long-term effects**
No studies reported (27).

**Teratogenicity and reproductive effects**
Pregnant Fischer-344 rats, CD-1 mice, 1230, 4100, 12,300 mg m$^{-3}$ (continuous) on days 6-15 (inclusive) of gestation. Animals killed on day 21 of gestation. Rats 12,300 mg m$^{-3}$, maternal toxicity included decreased body weight gain and decreased food consumption also reduced foetal body weight per litter and delays in skeletal ossification. No increase in foetal malformation. 1230, 4100 mg m$^{-3}$ showed no increase in maternal or foetal toxicity or malformations. Mice 12,300 mg m$^{-3}$ maternal effects included increased mortality, increased liver weight and foetotoxicity (increased dead foetuses, reduced body weight delayed or reduced ossification). No increase in malformations. No significant effects seen at 1230, 4100 mg m$^{-3}$ (28,29).

**Metabolism and pharmacokinetics**
Two metabolites were detected in serum, after intraperitoneal injection into guinea pig of 450 mg kg$^{-1}$ body weight. Major metabolite 4-hydroxy-4-methyl-2-pentanone formed by oxidation, and the minor metabolite 4-methyl-2-pentanol, formed by reduction. Serum t$_{1/2}$ and total clearance time are 66 min and 6 hr respectively (30).
Inhalation (4 hr) human studies 100 ppm caused steady-state levels in blood. Completely cleared from body by 90 min after exposure (31).

**Irritancy**
Eye rabbit (duration unspecified) 40 mg severe irritation (20).
Dermal rabbit (24 hr) 500 mg caused mild irritation (23).

**Sensitisation**
No studies reported (27).

## Genotoxicity
*Salmonella typhimurium* TA98, TA100, TA1537, TA1538 with and without metabolic activation negative (32).
*Eschericia coli* WP$_2$, WP$_2$ *uvr*A with metabolic activation negative (33).
*Saccharomyces cerevisae* JDI with and without metabolic activation mitotic gene conversion negative (33).

---

*In vitro* mouse lymphoma L51784 tk$^+$/tk$^-$ with and without metabolic activation negative (32).
*In vitro* cultured rat liver cells chromosomal damage assy RL$_4$ cells negative (33).
*In vitro* Balb/3T3 cell transformation assay with and without metabolic activation equivocol (32).
*In vivo* mouse micronucleus assay negative (32).

## Any other adverse effects to man

May depress the central nervous system at high concentrations, vapour may be irritating to mucous membranes (34).
Workers exposed to 2050 mg m$^{-3}$ for 20-30 min day$^{-1}$ and 328 mg m$^{-3}$ for the remainder of the day complained of weakness, loss of appetite, headache, eye irritation, stomach ache, nausea, vomiting and sore throat. After improvement in working conditions reduced levels to a maximum of 410-430 mg m$^{-3}$ most symptoms disappeared (35).

## Legislation

Listed in Schedule 6 (Release into Land: Prescribed Substances) Statutory Instrument No. 472, 1991 (36).

## Any other comments

Human health effects, experimental toxicology, environmental effects, ecotoxicology exposure levels, workplace experience and physico-chemical properties reviewed (37).

## References

1. Howard, P. H. *Handbook of Environmental Fate and Exposure Data for Organic Compounds* 1990, 341-348, Lewis Publishers, Chelsea, MI
2. Veith, G. et al *Aquatic Toxicology and Hazard Assessment: 6th Symposium* 1983, Bishop, W. E. et al (Ed.) ASTM STP 803, American Society for Testing and Materials, Philadelphia, PA
3. Broderius, S. et al *Aquat. Toxicol.* 1985, **6**, 307-322
4. Bridie, A. L. et al *Water Res.* 1979, **13**, 623
5. Bringmann, G. et al *GWF, Gas- Wasserfach: Wasser/Abwasser* 1976, **117** (9)
6. Curtis, C. A. et al *Aquatic Toxicology and Hazard Assessment: 5th Conference* 1982, Pearson, J. G. et al (Ed.) ASTM STP 766, American Society for Testing and Materials, Philadelphia, PA
7. Bringmann, G. et al *Z. Wasser Absasser Forsch.* 1977, **10**, 87-98 (Ger.)
8. Bringmann, G. et al *GWF, Gas- Wasserfach: Wasser/Abwasser* 1981, **122**, 308-318, (Ger.)
9. *The Assessment of Environmental Chemicals: Production Figures and Use Patterns for Some High Volume Chemicals* 1977, ENV/Chem./77.6, Organisation for Economic Cooperation and Development, Paris
10. Price, K. S. et al *J.- Water Pollut. Control Fed.* 1974, **46**, 63-77
11. Takemoto, S. et al *Suishitusu Odaku Kenkyu* 1981, **4**, 80-90
12. Ettinger, M. B. *Ind. Eng. Chem.* 1956, **48**, 256-259
13. Gerhold, R. M. et al *J.- Water Pollut. Control Fed.* 1966, **38**, 562-579
14. Bridie, A. L. et at *Water Res.* 1979, **13**, 627-630
15. Lyman, W. J. et al *Handbook of Chemical Property Estimation Methods. Environmental Behaviour of Organic Compounds* 1982, McGraw-Hill, New York
16. Cox, R. A. et al *Environ. Sci. Technol.* 1980, **14**, 57-61
17. Mill, T. et al *Science (Washington, D. C. 1883-)* 1980, **207**, 886-887
18. Swann, R. L. et al *Res. Rev.* 1983, **85**, 17-28

19. Clayton, G. D. et al *Patty's Industrial Hygiene and Toxicology* 3rd ed., 1985, John Wiley and Sons, New York
20. *Union Carbide Data Sheet 25/4/58*, Industrial Medicine and Toxicology Department, Union Carbide Corp., New York
21. Schafer, E. W. et al *Arch. Environ. Contam. Toxicol.* 1983, **12**, 355-382
22. *Gig. Tr. Prof. Zabol.* 1973, **17**(11), 52
23. Lewis, R. J. *Sax's Dangerous Properties of Industrial Materials* 8th ed., 1992, Van Nostrand Reinhold, New York
24. MacEwen, J. D. et al *Report No. AMRL TR-17-65* 1971, Aerospace Medical Research Laboratory, Wright-Patterson AFB, OH
25. Microbiological Associates *Preliminary Report* 1986, Study No. 5221.04, Research Triangle Park, NC
26. Malyscheva, M. V. et al *Gig. Sanit.* 1986, **2**, 90-91 (Russ.)
27. *Environmental Health Criteria: Methyl Isobutyl, Ketone* 1990, **No. 117**, International Programme on Chemical Safety, WHO, Geneva
28. Tyl, R. W. *Bushy Run Centre Report No. 47.505* 1984, Chemical Manufactures Association, Washington, DC
29. Tyl, R. W. et al *Fundam. Appl. Toxicol.* 1987, **8**, 319-327
30. DiVincenzo, G. D. et al *Toxicol. Appl. Pharmacol.* 1976, **36**, 511-522
31. Dick, R. et al *Toxicologist.* 1990, **10**, 122
32. O'Donoghue, J. L. et al *Mutat. Res.* 1988, **206**, 149-161
33. Brooks, T. M. et al *Mutagenesis* 1988, **3**, 227-232
34. *Martindale. The Extra Pharmacopoeia* 30th ed., 1993, Pharmaceutical Press, London
35. Armeli, G. et al *Lav. Um.* 1968, **20**, 418-424 (Ital.)
36. *S. I. 1991 No. 472 The Environmental Protection (Prescribed Processes and Substances) Regulations* 1991, HMSO, London
37. *ECETOC Technical Report No. 30(4)* 1991, European Chemical Industry Ecology and Toxicology Centre, B-1160 Brussels

# I85   Isobutylmethylxanthine

**CAS Registry No.** 28822-58-4
**Synonyms** 3,7-dihydro-1-methyl-3-(2-methylpropyl)-1*H*-purine-2,6-dione; 3-isobutyl-1-methylxanthine; methylisobutylxanthine
**Mol. Formula** $C_{10}H_{14}N_4O_2$                **Mol. Wt.** 222.25

## Physical properties

**M. Pt.** 200-201°C.

---

**Any other adverse effects**

Rat pups 7-10 days old transient exposure (dose, route not specified) can cause learning impairments and other undesirable behavioural consequences (1).

**References**

1. Neal, B. S. *Psychopharmacology (Berlin)* 1991, **103**(3), 388-397

# I86   Isobutyl-2-naphthyl ether

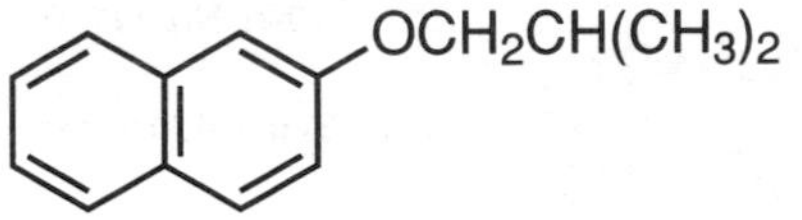

**CAS Registry No.** 2173-57-1
**Synonyms** 2-(2-methylpropoxy)naphthalene; 2-isobutoxynaphthalene
**Mol. Formula** $C_{14}H_{16}O$                     **Mol. Wt.** 200.28

**Physical properties**
**M. Pt.** 30-33.5°C; **B. Pt.** 304.5-307°C.

**Mammalian and avian toxicity**

**Acute data**
$LD_{Lo}$ oral rat 4.75 g kg$^{-1}$ (1).

**References**

1. *Food Cosmet. Toxicol.* 1964, **2**, 327

# I87   Isobutyl propionate

$$CH_3CH_2CO_2CH_2CH(CH_3)CH_3$$

**CAS Registry No.** 540-42-1
**Synonyms** 2-methylpropyl propanoic acid ester; propionic acid, isobutyl ester; isobutyl propanoate; 2-methylpropyl propanoate; 2-methylpropyl propinoate
**Mol. Formula** $C_7H_{14}O_2$                     **Mol. Wt.** 130.19
**Uses** Manufacture of fruit essences.

**Physical properties**
**M. Pt.** –71°C; **B. Pt.** 137°C; **Flash point** 18°C; **Specific gravity** $d_4^{20}$ 0.888.

**Solubility**
Water: miscible with ethanol.

## Occupational exposure

UN No. 2394; **HAZCHEM Code** 3◼E; **Conveyance classification** flammable liquid.
**Risk phrases** Flammable (R10)

## Mammalian and avian toxicity

**Acute data**
$LD_{50}$ oral rabbit 5.60 g $kg^{-1}$ (1).

## References

1. *Ind. Med. Surg.* 1972, **41**, 31

# I88   Isobutyraldehyde

## CH3CH(CH3)CHO

**CAS Registry No.** 78-84-2
**Synonyms** 2-methylpropanal; isobutanal; methylpropanal; isopropyl aldehyde;
isopropylformaldehyde; 2-methylpropionaldehyde
**Mol. Formula** $C_4H_8O$                                   **Mol. Wt.** 72.11
**Uses** Synthesis of pantothenic acid, plasticizers, resins, cellulose esters.

## Physical properties

**M. Pt.** –65.9°C; **B. Pt.** 64°C; **Flash point** <-6.67°C (open cup); **Specific gravity**
$d_4^{20}$ 0.7938; **Volatility** v.p. 98.3 mmHg at 12.9°C; v. den 2.5.

## Solubility

Water: 110 g $l^{-1}$. Organic solvent: miscible with ethanol, diethyl ether, carbon
disulfide, acetone, benzene, toluene, chloroform

## Occupational exposure

UN No. 2045; **HAZCHEM Code** 3WE; **Conveyance classification** flammable
liquid.

## Ecotoxicity

**Fish toxicity**
5 ppm (24 hr) no toxic effects on trout, bluegill sunfish, yellow perch or goldfish. Test
conditions: pH7; dissolved oxygen content 7.5 ppm; total hardness (soap method) 300
ppm; methyl orange alkalinity 310 ppm; free carbon dioxide 5 ppm; and temperature
12.8°C  (1).

## Environmental fate

**Degradation studies**
Confirmed to be biodegradable (2).

## Mammalian and avian toxicity

**Acute data**
$LD_{50}$ oral rat 3.7 g $kg^{-1}$ (3).

$LC_{50}$ (2 hr) inhalation mouse 39500 mg m$^{-3}$ (4).
$LD_{50}$ dermal rabbit 7.13 g kg$^{-1}$ (3).

### Carcinogenicity and long-term effects
Inhalation Fisher 344 rat, B6C3F$_1$ mouse 500, 1000 or 2000 ppm in air. Under investigation in the National Toxicology Program, USA (5).

### Teratogenicity and reproductive effects
Rat (13 wk) inhalation 500, 1000, 2000, 4000 ppm reduced body weight, epididymis and cauda epididymis weight. No effect on testis weight. Mice (13 wk) inhalation 500, 1000, 2000 ppm no effect on any of the above (6).

### Irritancy
Dermal rabbit (duration not stated) 397 mg open to atmosphere mild irritation (7).
Dermal rabbit (24 hr) 500 mg severe irritation (8).
Eye rabbit (24 hr) 100 mg moderate irritation (8).

### Sensitisation
May cause sensitisation (species not specified) (9).

## Genotoxicity
Mouse (13 wk) inhalation 500, 1000, 2000 ppm, rat (13 wk) inhalation 500, 1000, 2000, 4000 ppm no effects on sperm motility, density or morphology (6).
*Salmonella typhimurium* TA98, TA100, TA1535, TA1537 with and without metabolic activation negative (10).
*Salmonella typhimurium* G46, TA100, *Escherichia coli* WP2, WP2 *uvrA⁻* without metabolic activation, base pair substitution mutation (11).

## Any other adverse effects
Based on inhalation studies will cause central nervous system depression (as in alcohol intoxication) if swallowed (species not specified) (12).

## Any other comments
Recommended for subchronic toxicity testing under the US Federal Toxic Substances Control Act (13).
Autoignition temperature 223.33°C.

## References
1. Wood, E. M. *The Toxicity of 3400 Chemicals to Fish* 1987, EPA 560/6-87-002, PB 87-200-275, Washington, DC
2. *The list of the existing chemical substances tested on biodegradability by microorganisms or bioaccumulation in fish body.* 1987, Chemicals Inspection and Testing Institute, Japan
3. Smyth et al *Arch. Ind. Hyg. Occup. Med.* 1954, **10**, 61
4. Izmerov, N. F. et al *Toxicometric Parameters of Industrial Toxic Chemicals under Single Exposure* 1982, CIP, Moscow
5. *Directory of Agents Being Tested for Carcinogenicity* 1992, IARC, Lyon.
6. Morrisey, R. E. et al *Fundam. Appl. Toxicol.* 1988, **11**(2), 343-358
7. *Union Carbide Data Sheet* 3/11/71, Industrial Medicine and Toxicology Dept., Union Carbide Corp., New York, NY
8. Sax, N. I. et al *Dangerous Properties of Industrial Materials* 8th ed., 1992, Van Nostrand Reinhold, New York
9. *Chemical Safety Data Sheets* 1992, **5**, 159-161, RSC, London
10. Mortelmans, K. et al *Environ. Mutagen.* 1986, **8**(Suppl. 7), 1-119

11. McMahon, R. E. et al *Cancer Res.* 1979, **39**, 682-693
12. Gosselin, R. E. et al *Chemical Toxicology of Commercial Products* 5th ed., 1984, Williams and Wilkins, Baltimore, MD
13. *Fed. Regist.* 1991, **56**(44), 9534-9572

# I89   Isobutyric acid

## CH3CH(CH3)CO2H

**CAS Registry No.** 79-31-2
**Synonyms** 2-methylpropanoic acid; dimethylacetic acid; isobutanoic acid; isopropylformic acid; 2-methylpropionic acid
**Mol. Formula** $C_4H_8O_2$                                    **Mol. Wt.** 88.11

## Physical properties

**M. Pt.** –47°C; **B. Pt.** 152-155°C; **Flash point** 77°C (open cup); **Specific gravity** $d_4^{20}$ 0.950; **Volatility** v.p. 1 mmHg at 14.7°C; v. den. 3.0.

## Solubility

Water: 1 in 6. Organic solvent: miscible with ethanol, chloroform, diethyl ether

## Occupational exposure

**UN No.** 2529; **HAZCHEM Code** 2R; **Conveyance classification** flammable liquid; **Supply classification** harmful.
**Risk phrases** Harmful in contact with skin and if swallowed (R21/22)

## Ecotoxicity

**Invertebrate toxicity**
$EC_{50}$ *Chlorella pyrenoidosa* 345 mg $l^{-1}$ (1).

## Environmental fate

**Anaerobic effects**
*Desulfococcus multivorum* utilised isobutyrate as sole carbon source. In the presence of fresh water and marine sediments and sludge degraded to acetate and methane (2).

## Mammalian and avian toxicity

**Acute data**
$LD_{50}$ oral rat 280 mg $kg^{-1}$ (3).
$LD_{50}$ (estimated) oral redwing blackbird >100 mg $kg^{-1}$ (4).
$LD_{50}$ (duration unspecified) inhalation rat 400->800 mg $kg^{-1}$ (5).
$LD_{50}$ dermal rabbit 500 mg $kg^{-1}$ (3).

## Any other comments

Human health effects, experimental toxicology and physico-chemical properties reviewed (6).
Toxicology in relation to water pollution and waste water treatment reviewed (7).

# References

1. Jones, H. R. *Environmental Control in the Organic and Petrochemical Industries* 1971, Noyes Data Corporation
2. Stieb, M. et al *Arch. Microbiol.* 1989, **51**(2), 126-132
3. *Am. Ind. Hyg. Assoc. J.* 1962, **23**, 95
4. Schafer, E. W. et al *Arch. Environ. Contam. Toxicol.* 1983, **12**, 355-382
5. Patty, F. A. *Industrial Hygiene and Toxicology* 1967, Vol. 2, Interscience Publishers, New York
6. *ECETOC Technical Report No. 30(4)* 1991, European Chemical Industry Ecology and Toxicology Centre, B-1160 Brussels
7. Speece, R. E. et al *Int. Conf. Innovative Biol. Treat. Toxic. Wastewaters* 1986, 37-64, Scholze R. J. (Ed.), NTIS, Springfield, VA

# I90    Isobutyronitrile

## $CH_3CH(CH_3)CN$

**CAS Registry No.** 78-82-0
**Synonyms** 2-methylpropanitrile; 2-cyanopropane; dimethyl acetonitrile; isopropyl cyanide; isopropyl nitrile
**Mol. Formula** $C_4H_7N$                    **Mol. Wt.** 69.11

## Physical properties

**M. Pt.** –75°C; **B. Pt.** 101-102°C; **Flash point** 8°C; **Specific gravity** $d_{20}^{20}$ 0.773; **Partition coefficient** log $P_{ow}$ 0.46; **Volatility** v. den. 2.4.

**Solubility**
Organic solvent: ethanol, diethyl ether

## Occupational exposure

**UN No.** 2284; **HAZCHEM Code** 3WE; **Conveyance classification** flammable liquid, toxic substance.

## Environmental fate

**Degradation studies**
*Pseudomonas putida* utilised as sole source of carbon and nitrogen (1).

## Mammalian and avian toxicity

**Acute data**
$LD_{50}$ oral rat 102 mg $kg^{-1}$ (2).
$LC_{Lo}$ (4 hr) inhalation rat 1000 ppm (3).
$LD_{50}$ dermal rabbit 310 mg $kg^{-1}$ (3).
$LD_{Lo}$ subcutaneous frog 4800 mg $kg^{-1}$ (4).
$LD_{Lo}$ subcutaneous rabbit 9 mg $kg^{-1}$ (4).

**Sub-acute data**
Oral mouse, rat 6 month study (dose not specified) isobutyronitrile was moderately cumulative and affected blood indices, organ enzymes and nervous system (5).

## Any other comments

Human health effects, experimental toxicology and exposure reviewed (6).
Inhaled organonitriles are substantially detoxified by microsomal metabolism in the nasal cavity (7).

## References

1. Nawaz, M. S. *Appl. Environ. Microbiol.* 1989, **55**(9), 2267-2274
2. *Union Carbide Data Sheet* 29-10-59, Industrial Medicine and Toxicology Department, Union Carbide Corp., New York
3. *Am. Ind. Hyg. Assoc. J.* 1962, **23**, 95
4. *Arch. Int. Pharmacodyn. Ther.* 1899, **5**, 161
5. Fomochkin, I. P. et al *Gig. Sanit.* 1988, (10), 77-78 (Russ), (*Chem. Abstr.* **110**, 34923h)
6. *ECETOC Technical Report No. 30(4)* 1991, European Chemical Industry Ecology and Toxicology Centre, B-1160 Brussels
7. Dahl, A. R. *Xenobiotica* 1989, **19**(11), 1201-1205

# I91    Isocyanic acid, 3,4-dichlorophenyl ester

**CAS Registry No.** 102-36-3
**Synonyms** 3,4-dichlorophenyl isocyanate
**Mol. Formula** $C_7H_3Cl_2NO$                **Mol. Wt.** 188.01

## Physical properties

**M. Pt.** 42-44°C; **B. Pt.** $_{18}$ 118-120°C; **Flash point** >110°C.

## Ecotoxicity

**Invertebrate toxicity**
$EC_{50}$ (5-30 min) *Photobacterium phosphoreum* 0.964 ppm Microtox test (1).

## Mammalian and avian toxicity

**Acute data**
$LC_{Lo}$ (2 min, 4 min) inhalation mouse, rat 140 mg m$^{-3}$ (2).

## Any other comments

Lachrymator.

## References

1. Kaiser, K. L. E. et al *Water Pollut. Res. J. Can.* 1991, **26**(3), 361-431
2. *Gig. Tr. Prof. Zabol.* 1969, **13**(4), 50

# I92 Isodrin

CAS Registry No. 465-73-6
Synonyms 1,2,3,4,10,10-hexachloro-1,4,4a,5,8,8a-hexahydro-1,4,5,8-*endo,endo*-dimethanonaphthalene; ENT 19, 244; Compound 711; RCRA waste number P060
Mol. Formula $C_{12}H_8Cl_6$                                   Mol. Wt. 364.92
Uses Insecticide.

## Physical properties

M. Pt. 240-242°C.

## Occupational exposure

Supply classification very toxic.
Risk phrases Very toxic by inhalation, in contact with skin and if swallowed (R26/27/28)
Safety phrases Keep locked up – Keep away from food, drink and animal feeding stuffs – After contact with skin, wash immediately with plenty of soap and water – In case of accident or if you feel unwell, seek medical advice immediately (show label where possible) (S1, S13, S28, S45)

## Ecotoxicity

Fish toxicity
$LC_{50}$ fathead minnow, bluegill sunfish 0.006-0.012 mg $l^{-1}$ (1).

Invertebrate toxicity
Moderately toxic to grasshopper (details unspecified) (2).
$LD_{50}$ (duration unspecified) house fly 0.054 µg $fly^{-1}$ (isodrin), 0.113 µg $fly^{-1}$ (photoisodrin) (3).
$LD_{50}$ (duration unspecified) mosquito larvae 0.019 ppm (isodrin), 0.058 ppm (photoisodrin) (3).
$LC_{50}$ (48 hr) *Spodoptera litura* (Fibricius) 0.00524 g 100 $ml^{-1}$ (4).

Abiotic removal
Converted to photoisodrin by sunlight and UV light (5).

## Mammalian and avian toxicity

Acute data
$LD_{50}$ oral rat, mouse 7-8.8 mg $kg^{-1}$ (6,7).
$LD_{50}$ dermal rat 23 mg $kg^{-1}$ (6).
$LD_{Lo}$ intraperitoneal mouse 6.4 mg $kg^{-1}$ (8).
Chicken embryos injected with 10-500 ppm, isodrin was the most toxic of 25

organochloride compounds tested. The toxicity was greater in starved chicks than in fed chicks (9).

## Legislation

Land disposal prohibited under US Federal Resource Conservation and Recovery Act (10).

Limited under EC Directive on Drinking Water Quality 80/778/EEC. Pesticides: maximum admissible concentration 0.1 µg $l^{-1}$ (11).

Included in Schedule 6 (Release into Land: Prescribed Substances) Statutory Instrument No. 472, 1991 (12).

Reportable quantity regulated under the US Federal Comprehensive Environmental Response, Compensation and Liability Act (13).

Quality objective under EC Directives 86/280/EEC and 88/347/EEC 0.005 µg $l^{-1}$ for all waters. A 'standstill' provision applies to concentrations in sediments, molluscs, shellfish and for fish (14).

## Any other comments

Isodrin applied as a spray was the most toxic of a selection of cyclodiene insecticides tested against larvae of tobacco budworm (15).

Pretreatment with 2.5 µg $fly^{-1}$ sesamex synergised isodrin toxicity. This synergism appeared to be related to the inhibition of its detoxification by a system which can be stimulated by phenobarbital (16).

Isodrin is converted to photoisodrin by sunlight and UV light. Unlike photoisomers of aldrin, dieldrin and heptachlor, photoisodrin is less toxic to house flies and the mosquito *Aedes aegypti* due to its rapid detoxification. Photoisodrin is 3-5 times slower than isodrin in its toxicity to *Daphnia pulex*, *Asellus* sp., *Gammarus* sp., *Cambarus* sp., *Gambia affinis*, *Pimephalus promelas* and *Lepomis macrochirus* (5). Hazards reviewed (17).

The photoisomer is less toxic, partly due to its more rapid metabolism (18,19).

## References

1. Khan, M. A. Q. et al *Arch. Environ. Contam. Toxicol.* 1973, **1**(2), 159-169
2. Olaifa, J. I. *Insect Sci. Its Appl.* 1986, **7**(2), 135-138
3. Rosen, J. D. et al *J. Agric. Food Chem.* 1969, **17**(2), 404-405
4. Mukherjee, A. B. et al *Indian J. Entomol.* 1970, **32**(3), 251-255
5. Georgackakis, E. et al *Bull. Environ. Contam. Toxicol.* 1971, **6**(6), 535-538
6. *World Rev. Pest Control* 1970, **9**, 119
7. *Gig. Tr. Prof. Zabol.* 1964, **8**(4), 30
8. *Toxicol. Appl. Pharmacol.* 1972, **23**, 288
9. Dunachie, J. F. et al *Ann. Appl. Biol.* 1969, **64**(3), 409-423
10. US EPA *Fed. Regist.* 1991, **56**(21), 3864-3928
11. *EC Directive Relating to the Quality of Water Intended for Human Consumption* 1982, 80/778/EEC, Office for Official Publications of the European Communities, 2 rue Mercier, L-2985 Luxembourg
12. *S. I. 1991 No. 472 The Environmental Protection (Prescribed Processes and Substances) Regulations* 1991, HMSO, London
13. *Fed. Regist.* 14 Aug 1989, **54**(155), 33462-33484
14. *DoE Circular 7/89: Water and the environment. The implementation of EC directives on pollution caused by certain dangerous substances discharged into the aquatic environment* 1989, HMSO, London
15. Wolfenbarger, D. A. et al *J. Econ. Entomol.* 1970, **63**(5), 1563-1573

16. Khan, M. A. Q. et al *J. Econ. Entomol.* 1970, **63**(2), 470-475
17. *Dangerous Prop. Ind. Mater. Rep.* 1987, **7**(6), 72-75
18. Rosen, J. D. et al *J. Agric. Food Chem.* 1969, **17**(2), 404-405
19. Georgacakis, E. et al *Nature (London)* 1971, **233**(5315), 120-1211

# I93   Isoeugenol

**CAS Registry No.** 97-54-1
**Synonyms** 2-methoxy-4-(1-propenyl)phenol; 4-hydroxy-3-methoxy-1-propenylbenzene; 4-propenylguaiacol; NCI-C6 0979; FEMA No.2468
**Mol. Formula** $C_{10}H_{12}O_2$                         **Mol. Wt.** 164.21
**Uses** In vanillin manufacture.
**Occurrence** In ylang ylang and other essential oils. In cigarette smoke condensate.

## Physical properties

**M. Pt.** –10°C; **B. Pt.** 266°C; **Specific gravity** $d_4^{25}$ 1.08.

### Solubility
Water: slightly soluble. Organic solvent: miscible with ethanol, diethyl ether

## Mammalian and avian toxicity

### Acute data
$LD_{50}$ oral rat 1560 mg $kg^{-1}$ (1).
$LD_{50}$ oral guinea pig 1410 mg $kg^{-1}$ (2).

### Sensitisation
Induces allergic contact dermatitis (species unspecified) (3).
Positive in guinea pig maximisation test (4).

## Genotoxicity

*Salmonella typhimurium* TA98, TA100, TA1535, TA1537 with or without metabolic activation negative (5).
Induced sister chromatid exchanges in human lymphocytes *in vitro* (6).

## Any other comments

Selected for general toxicology study by US National Toxicology Program (7).

## References

1.  Jenner, P. M. et al *Food Cosmet. Toxicol.* 1964, **2**, 327

2. *Toxicol. Appl. Pharmacol.* 1964, **6**, 378
3. Sekine, M. et al *Nippon Hifuka Gakkai Zasshi* 1988, **98**(5), 513-519 (Japan.) (*Chem. Abstr.* **109**, 228065e)
4. Maurer, T. et al *Food Chem. Toxicol.* 1989, **27**(12), 807-811
5. Mortelmans, K. et al *Environ. Mutagen.* 1986, **8**(Suppl.7), 1-119
6. Jansson, T. et al *Mutat. Res.* 1986, **169**, 129-139
7. *National Toxicology Program Research and Testing Division Management Status Report* 1992, NIEHS, Research Triangle Park, NC

# I94   Isofenphos

**CAS Registry No.** 25311-71-1
**Synonyms** 2-[[ethoxy[(1-methylethyl)amino]phosphinothioyl]oxy]benzoic acid, 1-methylethyl ester; Amaze; Oftanol; isophenphos
**Mol. Formula** $C_{15}H_{24}NO_4PS$          **Mol. Wt.** 345.40
**Uses** Systemic insecticide.

## Physical properties

**M. Pt.** <-12°C; **B. Pt.** 120°C; **Specific gravity** $d_4^{20}$ 1.13; **Partition coefficient** log $P_{ow}$ 4.12; **Volatility** v.p. $4 \times 10^{-6}$ mmHg at 20°C.

**Solubility**
Water: 23.8 mg $kg^{-1}$ at 20°C. Organic solvent: dichloromethane, acetone, diethyl ether, benzene, cyclohexanone, ethanol

## Occupational exposure

**Supply classification** toxic.
**Risk phrases** Toxic in contact with skin and if swallowed (R24/25)
**Safety phrases** Wear suitable protective clothing and gloves – If you feel unwell, seek medical advice (show label where possible) (S36/37, S44)

## Ecotoxicity

**Fish toxicity**
$LC_{50}$ (96 hr) rudd, orfe, goldfish, carp 1-4 mg $l^{-1}$ (1).

**Invertebrate toxicity**
$EC_{50}$ (5 min) *Photobacterium phosphoreum* 97.3 ppm Microtox test (2).

## Environmental fate

**Degradation studies**
Persistence and degradation strongly affected by interaction of pH, soil temperature

and moisture; degradation is greatest at higher temperatures (35>25>15°C) (except at alkaline pH), medium moisture (25>20>15%) and in both acidic (pH 6) and alkaline (pH8) compared with neutral soils (3).

An *Arthrobacter* isolated from soil with a history of isofenphos use rapidly metabolised isofenphos in pure culture (4).

Treatment of soil with 10 ppm enhanced microbial activity within 6 wk by up to 100×; time to non-detectability of enhanced activity was >164 wk (5).

Treatment with manure increases persistence in soil due to binding effect of organic matter decreasing availability for microbial degradation (6).

### Absorption
Insecticidal activity of isofenphos detectable 2 wk after initial application to clay loam in May and was still present the following spring (7).

Soil mobility in relation to organic matter, clay, cation exchange capacity, pH, water-holding capacity and log $P_{ow}$ reviewed. Soil mobility expressed as $R_f$ varied from 0.09 for clay loam and silty clay loam, to 0.16 for silt loam and sandy loam (8).

## Mammalian and avian toxicity

### Acute data
$LD_{50}$ oral rat 28 mg $kg^{-1}$ (9).
$LD_{50}$ oral chicken, quail 3-13 mg $kg^{-1}$ (10,11).
$LC_{50}$ (4 hr) inhalation rat 0.144-0.210 mg $l^{-1}$ (9).
$LD_{50}$ dermal rabbit 162 mg $kg^{-1}$ (12).
Oral toxicity in grey partridge greater in birds fed on insect-poor diet as juveniles (13).

### Sub-acute data
$LC_{50}$ (8 day) oral Japanese quail 5-12.5 mg $kg^{-1}$ (9).

### Carcinogenicity and long-term effects
No effect level in 2 yr feeding trials in rats 1 mg $kg^{-1}$ diet (9).

### Metabolism and pharmacokinetics
After single oral administration to rats it was distributed throughout the body within 6 hr, then gradually eliminated mainly via urine. Major metabolites were *O*-ethyl-(isopropylcarboxyphenyl)phosphoric acid; amino, *O*-ethyl phosphorous acid and (isopropylamino)phosphorothioic acid, *O*-ethyl ester (14).

## Legislation
EEC maximum residue limit for bananas 0.1 ppm (9).
WHO Class 1b; EPA Toxicity Class 1; tolerable daily intake in humans 0.001 mg $kg^{-1}$ (9).
Limited under EC Directive on Drinking Water Quality 80/778/EEC. Pesticides: maximum admissible concentration 0.1 µg $l^{-1}$ (15).
Included in Schedule 6 (Release into Land: Prescribed Substances) of Statutory Instrument No. 472, 1991 (16).
The log $P_{ow}$ value exceeds the European Community recommended level 3.0 (6th and 7th amendments) (17).

## Any other comments
Metabolism reviewed (18).
Toxicity of organophosphate pesticides reviewed (19).

## References

1. *The Agrochemicals Handbook* 3rd ed., 1991, RSC, London
2. Kaiser, K. L. E. et al *Water Pollut. Res. J. Can.* 1991, **26**(3), 361-431
3. Abou-Assaf, N. et al *J. Environ. Sci. Health, Part B* 1987, **B22**(3), 285-301
4. Racke, K. D. et al *J. Agric. Food Chem.* 1988, **36**(1), 193-199
5. Chapman, R. A. et al *ACS Symp. Ser.* 1990, **426**(Enhanced Biodegrad. Pestic. Environ.), 82-96
6. Somasundaram, L. et al *Bull. Environ. Contam. Toxicol.* 1987, **39**(4), 579-586
7. Harris, C. R. et al *J. Environ. Sci. Health, Part B* 1988, **B23**(1), 1-32
8. Somasundaram, L. et al *Environ. Toxicol. Chem.* 1991, **10**(2), 185-194
9. *Agricultural Chemicals* 1976, Thomson Publications, Fresno, CA
10. *Bull. Environ. Contam. Toxicol.* 1984, **33**, 386
11. *Ecotoxicol. Environ. Saf.* 1984, **8**, 551
12. *Farm Chemicals Handbook* 1983, Meister Publishing Co., Willoughby, OH
13. Dahlgren, J. *Environ. Pollut.* 1988, **49**(3), 177-181
14. Ueji, M. et al *Nippon Noyaku Gakkaishi* 1987, **12**(2), 245-251
15. *EC Directive Relating to the Quality of Water Intended for Human Consumption* 1982, 80/778/EEC, Office for Official Publications of the European Communities, 2 rue Mercier, L-2985 Luxembourg
16. *S. I. 1991 No. 472 The Environmental Protection (Prescribed Processes and Substances) Regulations* 1991, HMSO, London
17. *1967 Directive on Classification, Packaging, and Labelling of Dangerous Substances 67/548/EEC; 6th Amendment EEC Directive 79/831/EEC; 7th Amendment EEC Directive 91/32/EEC* 1991, HMSO, London
18. Ueji, M. et al *J. Pestic. Sci.* 1987, **12**, 245-269
19. *Environmental Health Criteria 63* 1986, IPCS/WHO, Geneva

# I95   Isofluorphate

## [(CH3)2CHO]2P(O)F

**CAS Registry No.** 55-91-4

**Synonyms** isopropyl phosphorofluoridate; diisopropoxylphosphoryl fluoride; diisopropyl fluorophosphonate; isopropyl fluophosphate; diisopropyl fluorophosphate; diisopropyl phosphofluoridate; Dyflos

**Mol. Formula** $C_6H_{14}FO_3P$ **Mol. Wt.** 184.15

**Uses** FDA proprietary drug, meiotic agent; cholinergic. It is an irreversible cholinesterase inhibitor similar in action to ecothiopate. Used mainly in treatment of open-angle glaucoma and in diagnosis and management of accommodative convergent strabismus. It is administered locally as an ointment.

## Physical properties

**M. Pt.** $-82°C$; **B. Pt.** $_5$ 46°C; **Specific gravity** 1.055; **Volatility** v.p. 0.579 mmHg at 20°C; v. den. 5.24.

## Solubility

Water: 1.54% (w/w) at 25°C. Organic solvent: ethanol

## Mammalian and avian toxicity

### Acute data
$LD_{50}$ oral mouse, rat 2-5 mg $kg^{-1}$ (1).
$LC_{50}$ (10 min) inhalation rat, mouse 360-440 mg $m^{-3}$ (2,3).
$LD_{50}$ dermal mouse 72 mg $kg^{-1}$ (4).
$LD_{50}$ intraperitoneal rat, mouse 1280-2450 $\mu$g $kg^{-1}$ (5,6).
$LD_{50}$ subcutaneous mouse 3 mg $kg^{-1}$ (7).

### Teratogenicity and reproductive effects
0.920 mg $l^{-1}$ had no effect on human sperm motility, but inhibited penetration of zona-free hamster ova by human sperm; it is capable of inhibiting sperm function and associates with the proacrosin-acrosin system in live, motile sperm (8).

### Metabolism and pharmacokinetics
Readily absorbed from the human gut, skin, mucous membranes and lungs. Interacts with cholinesterases producing stable phosphonylated and phosphorylated derivatives which are then hydrolysed by phosphorylphosphatases. These hydrolysis products are excreted mainly in the urine (9).

### Irritancy
Vapour is extremely irritating to eye and mucous membranes in humans (9). Prolonged exposure to the eye can cause slowly reversible depigmentation of lid margins of dark-skinned patients (9).

## Legislation
Land disposal prohibited under US Federal Resource Conservation and Recovery Act (10).

## Any other comments
Isofluorophate-contaminated material should be immersed in 2% aqueous sodium hydroxide for several hours (9).
Long-term effects on memory reviewed (11,12).
Pharmacokinetics and use in Alzheimer's therapy reviewed (13).

## References
1. *NTIS Report PB158-508* Natl. Tech. Inf. Ser., Springfield, VA
2. *J. Chem. Soc.* 1948, 695
3. *Nature (London)* 1946, **157**, 287
4. *J. Pharmacol. Exp. Ther.* 1946, **87**, 414
5. *Arzneim.-Forsch.* 1964, **14**, 85
6. *Arch. Immunol. Ther. Exp.* 1975, **23**, 769
7. *J. Pharm. Pharmacol.* 1982, **34**, 603
8. Odem, R. R. et al *Biol. Reprod.* 1990, **42**(2), 329-336
9. *Martindale. The Extra Pharmacopoeia* 30th ed., 1993, The Pharmaceutical Press, London
10. U.S. EPA *Fed. Regist.* 1991, **56**(21), 3864-3928
11. Olton, D. S. et al *NIDA Res. Monogr.* 1990, **97** (*Neurobiol. Drug Abuse: Learn. Mem.*), 37-47
12. Raffaele, K. C. et al *Neurotoxicology* 1990, **11**(2), 237-256
13. Waser, P. G. et al *Curr. Res. Alzheimer Ther.: Cholinesterase Inhib.* 1988, 31-42

# I96 Isoflurane

## CF$_3$CH(Cl)OCHF$_2$

**CAS Registry No.** 26675-46-7
**Synonyms** 2-chloro-2-(difluoromethoxy)-1,1,1-trifluoroethane;
1-chloro-2,2,2-trifluoroethyl difluoromethyl ether; Compound 469
**Mol. Formula** C$_3$H$_2$ClF$_5$O                    **Mol. Wt.** 184.49
**Uses** Anaesthetic. Used for maintenance when anaesthesia induced by an intravenous agent; if used for induction it is given with oxygen or oxygen and nitrous oxide. Solvent and dispersant for fluorinated compounds.

## Physical properties

**B. Pt.** 49°C; **Specific gravity** 1.45; **Volatility** v.p. 330 mmHg at 25°C.

### Solubility
Organic solvent: miscible with organic solvents including oils and fats

## Occupational exposure

**UK Long-term limit** 50 ppm (380 mg m$^{-3}$) (proposed).

## Mammalian and avian toxicity

### Acute data
LD$_{50}$ oral rat, mouse 4770-5080 mg kg$^{-1}$ (1).
LD$_{50}$ (3 hr) inhalation rat, mouse 15,300-16,800 ppm (1).
LD$_{50}$ intraperitoneal mouse, rat 3030-4280 mg kg$^{-1}$ (1).

### Carcinogenicity and long-term effects
No increased incidence of tumours in mice exposed to 0.1 or 0.4% 4 hr day$^{-1}$, 5 days wk$^{-1}$ for 78 wk (2).

### Teratogenicity and reproductive effects
Isofluorane at concentrations similar to those used during human oocyte recovery for *in vitro* fertilisation inhibited mouse embryo development *in vitro* (3).
No major or minor teratogenic effects reported in rats exposed to 1.05%, 6 hr day$^{-1}$, on days 14-16, 11-13 or 8-10 of pregnancy (4).

### Metabolism and pharmacokinetics
Absorbed on inhalation by humans; blood/gas coefficient <enflurane or halothane. 0.2% metabolised mainly to inorganic fluoride (5).
In patients sedated for 24 hr plasma inorganic fluoride increased from 0.79 to 2.49 mg l$^{-1}$, a level too low to cause renal dysfunction (6).

## Genotoxicity

Did not induce sister chromatid exchanges in Chinese hamster lung cells (7).
Sex-linked recessive lethal assay in *Drosophila melanogaster* negative for a mixture of isofluorane and nitrous oxide (8).

## Any other adverse effects to man

Hepatic function was unaffected by exposure for 64 min (9).

---

Levels of urinary D-glucaric acid excretion was increased in operating theatre personnel exposed to 1 ppm plus <100 ppm nitrous oxide (10).

Isofluorane was not considered the likely cause of post-operative liver impairment in 45 cases of isofluorane-associated hepatotoxicity reported to the US FDA between 1981 and 1984 (11).

A subsequent case of post-operative hepatic necrosis and death may be attributable to isofluorane (12).

Respiratory depression, hypotension, arrhythmias and malignant hyperthermia have been reported. Post-operative shivering, nausea and vomiting may occur and coughing and laryngospasm on induction (5).

### Any other adverse effects

No pulmonary or renal injury, but slight liver injury, reported in organ specimens taken from enzyme-induced, hypoxic rats 24 hr after exposure to 1.2 min alveolar concentration (13).

### Any other comments

Metabolism in humans and laboratory animals reviewed (14).

Pharmacokinetics reviewed (15).

Metabolism and possible effects on liver, kidneys and nervous system of operating room personnel reviewed (16).

### References

1. *Kiso to Rinsho* 1987, **21**, 3031
2. Baden, J. M. et al *Anesthesiology* 1988, **69**(5), 750-753
3. Chetkowski, R. J. et al *J. Periodontal Res.* 1987, **22**(6), 171-173
4. Mazze, R. I. et al *Anesthesiology* 1986, **64**, 339-344
5. *Martindale. The Extra Pharmacopoeia* 30th ed., 1993, The Pharmaceutical Press, London
6. Kong, K. L. et al *Br. J. Anaesth.* 1990, **64**, 159-162
7. Trudnowski, R. J. et al *J. Med. (Westbury, N. Y.)* 1987, **18**(1), 55-60
8. Baden, J. M. et al *Br. J. Anaesth.* 1987, **59**(6), 772-775
9. Chulia, V. et al *Anaesthesiol. Intensivmed. (Berlin)* 1986, **182**(Isoflurane), 45-51
10. Franco, G. et al *Appl. Occup. Environ. Hyg.* 1992, **7**(10), 677-681
11. Stoelting, R. K. et al *Anesth. Analg. (N. Y.)* 1987, **66**, 147-153
12. Carrigan, T. W. et al *Anaesthesiology* 1987, **67**, 581-583
13. Eger, E. I. et al *Anesth. Analg. (N. Y.)* 1987, **66**(12), 1227-1229
14. Hempel, V. *Anaesthesiol. Intensivmed. (Berlin)* 1986, **182**(Isoflurane), 29-31
15. Narlander, O. *Anaesthesiol. Intensivmed. (Berlin)* 1986, **182**(Isoflurane), 18-23
16. Efthymion, M. L. *Arch. Mal. Prof. Med. Trav. Secur. Soc.* 1988, **49**(4), 236-247 (Fr.) (*Chem. Abstr.* **110**, 44014b)

# I97   Isohexanol

$$CH_3CH(CH_3)CH_2CH_2CH_2OH$$

**CAS Registry No.** 626-89-1

**Synonyms** 1-pentanol, 4-methyl-; isohexyl alcohol

**Mol. Formula** $C_6H_{14}O$ **Mol. Wt.** 102.18

## Physical properties

**B. Pt.** 136°C; **Flash point** 41°C; **Specific gravity** 0.810; **Partition coefficient** log $P_{ow}$ 1.25.

## Ecotoxicity

**Fish toxicity**

$LC_{50}$ (96 hr) zebra fish 340 mg $l^{-1}$ (1).
$LC_{50}$ (48 hr) ide 287-332 mg $l^{-1}$ (1).

**Invertebrate toxicity**

LOEC *Microcystis aeruginosa* and *Scenedesmus quadricauda* 32-72 mg $l^{-1}$ (2,3).
$ICG_{50}$ (50% growth inhibitory concentration) *Tetrahymena pyriformis* 0.542 g $l^{-1}$ (4).

## Environmental fate

**Degradation studies**

Biodegradability of higher alcohols in the activated sludge process decreases with increasing number of carbon atoms (5).
Oxidised by a thermophilic obligate methane-oxidising bacterium H-2 (type I) (6).

## Any other comments

US EPA have recommended testing for environmental and health effects (7).

## References

1. Wellens, H. *Z. Wasser/Abwasser Forsch.* 1982, **15**, 49
2. Bringmann, G. et al *GWF, Gas- Wasserfach: Wasser/Abwasser* 1976, **117**(9)
3. Bringmann, G. et al *Water Res.* 1980, **14**, 231-241
4. Schultz, T. W. et al *Bull. Environ. Contam. Toxicol.* 1993, **51**, 681-688
5. Niitsuma, T. et al *Tohoku Gakuin Daigaku Kogakubu Kenkyu Hokoku* 1988, **23**(1), 57-60 (Japan.) (*Chem. Abstr.* **110**, 140816k)
6. Imai, T. et al *Appl. Environ. Microbiol.* 1986, **52**(6), 1403-1406
7. *Fed. Regist.* 6 Mar. 1991, **56**(44), 9534-9572

# I98   Isolan

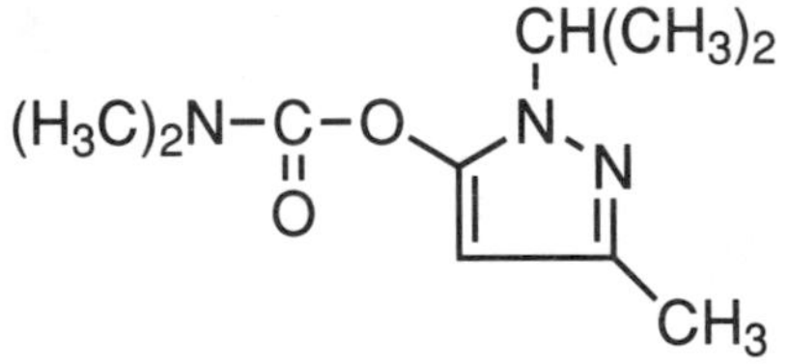

**CAS Registry No.** 119-38-0
**Synonyms** dimethylcarbamic acid, 3-methyl-1-(methylethyl)-1*H*-pyrazol-5-yl ester; 1-isopropyl-3-methyl-5-pyrazolyl  dimethylcarbamate
**Mol. Formula** $C_{10}H_{17}N_3O_2$                    **Mol. Wt.** 211.27

**Uses** Insecticide.

## Physical properties

**B. Pt.** $_{0.7}$ 103°C; **Specific gravity** 1.07; **Volatility** v.p 0.001 mmHg at 20°C.

## Occupational exposure

**Supply classification** very toxic.
**Risk phrases** Very toxic in contact with skin and if swallowed (R27/28)
**Safety phrases** After contact with skin, wash immediately with plenty of soap and water – Wear suitable protective clothing, gloves and eye/face protection – In case of accident or if you feel unwell, seek medical advice immediately (show label where possible) (S28, S36/37/39, S45)

## Mammalian and avian toxicity

### Acute data

$LD_{50}$ oral mouse, rat 5600-9800 $\mu g\ kg^{-1}$ (1).
$LD_{50}$ oral starling 7.94 mg $kg^{-1}$ (2).
$LD_{50}$ dermal rat 5600 $\mu g\ kg^{-1}$ (3).
$LD_{50}$ intraperitoneal mouse 1 mg $kg^{-1}$ (4).

## Legislation

Limited under EC Directive on Drinking Water Quality 80/778/EEC. Pesticides: maximum admissible concentration 0.1 $\mu g\ l^{-1}$ (5).
Included in Schedule 6 (Release into Land: Prescribed Substances) Statutory Instrument No. 472, 1991 (6).

## References

1.  *Proc. Eur. Soc. Toxicol.* 1976, **17**, 351
2.  Schafer, E. W. et al *Arch. Environ. Contam. Toxicol.* 1983, **12**, 355-382
3.  *Wirksubstanzen der Pflanzenschutz und Shadlingsbekampfungsmittel* 1971-76, Verlag, Berlin
4.  *Toxicol. Appl. Pharmacol.* 1964, **6**, 402
5.  *EC Directive Relating to the Quality of Water Intended for Human Consumption* 1982, 80/778/EEC, Office for Official Publications of the European Communities, 2 rue Mercier, L-2985 Luxembourg
6.  *S. I. 1991 No. 472 The Environmental Protection (Prescribed Processes and Substances) Regulations* 1991, HMSO, London

# I99   D-Isoleucine

$$CH_3CH_2CH(CH_3)CH(NH_2)CO_2H$$

**CAS Registry No.** 319-78-8
**Synonyms** *d*-2-amino-3-methylpentanoic acid; *d*-α-amino-β-methylvaleric acid
**Mol. Formula** $C_6H_{13}NO_2$                    **Mol. Wt.** 131.18
**Occurrence** Dietary amino acid.

## Physical properties

**M. Pt.** 283°C (decomp.).

**Solubility**
Water: 41,200 mg l$^{-1}$ at 25°C.

## Environmental fate

**Degradation studies**
2.4, 5.3 and 15.8% of ThOD after 6, 12 and 24 hr respectively in bench scale
activated sludge, fill and draw operations (1).
*Halobacterium halobium* R1mR utilised D-isoleucine after an induction period (2).

## References

1. Malaney, G. W. et al *J.- Water Pollut. Control Fed.* 1969, **41**(2), Part 2, R18-R33
2. Tanaka, M. et al *Viva Origino* 1991, **18**(3), 109-117 (Japan.) (*Chem. Abstr.* **115**, 154696m)

# I100   Isoniazid

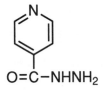

$O=\overset{\|}{C}-NHNH_2$

**CAS Registry No.** 54-85-3
**Synonyms** isonicotinic acid hydrazide; isonicotinoyl hydrazide; Chemiazid; Isonin;
Percin; Tyvid; 4-pyridinecarboxylic acid hydrazide
**Mol. Formula** $C_6H_7N_3O$ **Mol. Wt.** 137.14
**Uses** Antitubercular, antibacterial and antiactinomycotic agent.

## Physical properties

**M. Pt.** 171.4°C.

**Solubility**
Water: 14% at 25°C. Organic solvent: ethanol, chloroform

## Mammalian and avian toxicity

**Acute data**
LD$_{50}$ oral mouse 133 mg kg$^{-1}$ (1).
LD$_{50}$ oral rat 1250 mg kg$^{-1}$ (2).
LD$_{50}$ intravenous mouse, rat 150, 365 mg kg$^{-1}$ respectively (2,3).

**Carcinogenicity and long-term effects**
Inadequate evidence for carcinogenicity to humans, limited evidence for
carcinogenicity in animals, IARC classification group 3 (4).
Carcinogenic in mice, causing lung tumours, after oral, subcutaneous and
intraperitoneal administration. Did not induce tumours in hamsters, and evidence in
rats is inconclusive, after oral administration (5).

## Teratogenicity and reproductive effects

95% of 1480 pregnancies in which isoniazid was used resulted in normal term infants; 1% of foetuses/infants had abnormalities, mainly of the central nervous system. It is recognised as suitable for use in pregnant women and neonates (6).

Embryotoxic, inducing resorption and neonatal death, in mice given 2.2 mg mouse$^{-1}$ day$^{-1}$ intragastrically at day 1-4 or 10-13 of pregnancy (7).

Did not inhibit chondrogenesis in rat embryo limb bud cell culture teratogen screening (8).

Isoniazid had been used to evaluate the embryo malformation endpoint of the aquatic FETAX (frog embryo teratogenesis assay: *Xenopus*) test system (9).

## Metabolism and pharmacokinetics

Readily absorbed from the gut after intramuscular injection. Plasma $t_{1/2}$ 1-4 hr; rate of acetylation and hence $t_{1/2}$ is genetically determined. Acetylation to acetylisoniazid by *N*-acetyltransferase in liver and small intestine is the major metabolic route. This is then hydrolysed to isonicotinic acid, which is further conjugated with glycine to isonicotinyl glycine, and monoacetylhydrazine, which is further acetylated to diacetylhydrazine. Some unmetabolised isoniazid is conjugated to hydrazones. Metabolites are less toxic and not tuberculostatic (6,10).

Major urinary metabolites in humans are 1-acetyl-2-isonicotinoylhydrazine, *N*-acetyl-*N'*-isonicotinic acid, isonicotinylglycine, pyruvic acid isonicotinylhydrazone, and α-oxoglutaric acid isonicotinylhydrazone (5).

Liver injury may be caused by the metabolite acetylhydrazine formed by metabolic hydrolysis of acetylisoniazid by cytochrome $P_{450}$ (11).

Distributed in breast milk and although adverse effects in breast feeding infants have not been reported, they should be monitored (11).

## Genotoxicity

*Salmonella typhimurium* TA100 (12), TA1530 and TA1535 (13) with metabolic activation positive. Negative results have also been reported in TA100 (14).

*Salmonella typhimurium* TA98 (15) and TA1538 (16) with metabolic activation negative; TA1537 without metabolic activation negative (17).

*Salmonella typhimurium* TA1530, TA1535 without metabolic activation positive in plate test; TA1530, TA1535, *his*G46, and *Escherichia coli* TA85, TA86 and WP2 *uvr*A with or without metabolic activation positive in fluctuation assays (13).

*Escherichia coli* with and without metabolic activation negative (13).

Did not induce unscheduled DNA synthesis in rat primary hepatocytes *in vitro* (11,18,19).

Induced sister chromatid exchanges (20,21) and chromosomal aberrations in Chinese hamster cells *in vitro* (22).

Did not induce unscheduled DNA synthesis in human fibroblasts *in vitro* (22).

Inhibited recovery of DNA synthesis in human fibroblasts but did not inhibit protein synthesis (23).

Did not induce sister chromatid exchanges or chromosome aberrations in human lymphocytes *in vitro* (24).

Negative (25) and equivocal results reported for induction of chromosome aberrations in animal bone marrow cells *in vivo* (26).

Dominant lethal test in mice negative (7).

## Any other adverse effects to man

Side effects include peripheral neuritis (especially in poorly nourished patients), psychotic reactions, convulsions, optic neuritis, nausea and vomiting. Incidence of hepatic damage is higher in patients >35 who are slow inactivators and those who consume alcohol (6,27).

Haematological effects include anaemias, agranulocyctosis, thrombocytopenia, and eosinophilia. Hypersensitivity reactions occur infrequently including skin eruptions, fever, lymphadenopathy and vasculitis. Other effects include hyperglycaemia, metabolic acidosis, lupus-like syndrome, rheumatoid syndrome, urinary retention and gynaecomastia (6).

Symptoms of overdose include slurred speech, metabolic acidosis, hypoglycaemia, respiratory and central nervous system depression, convulsions, coma and can be fatal (6,28,29).

Non-acetylator populations are more susceptible to isoniazid toxicity (6).

## Any other comments

Isoniazid-induced liver injury reviewed (30).

The human embryonic palatal mesenchymal cell growth inhibition and the mouse ovarian tumour cell attachment inhibition assay has also been evaluated (31).

## References

1. *Am. Rev. Tuberc.* 1952, **65**, 376
2. *Arzneim.-Forsch.* 1976, **26**, 409
3. *J. Pharmacol. Exp. Ther.* 1958, **122**, 110
4. *IARC Monograph* 1987, **Suppl. 7**, 65, 227
5. *IARC Monograph* 1974, **4**, 159-172
6. *Martindale. The Extra Pharmacopoeia* 30th ed., 1993, The Pharmaceutical Press, London
7. Menon, M. M. et al *Indian J. Exp. Biol.* 1980, **18**, 1104-1106
8. Renault, J. Y. et al *Teratog., Carcinog., Mutagen.* 1989, **9**, 83-96
9. Dawson, D. A. et al *Ecotoxicol. Environ. Saf.* 1991, **21**(2), 215-226
10. Peretti, E. et al *Eur. J. Clin. Pharmacol.* 1987, **33**(3), 283-286
11. Neis, J. M. et al *Cancer Lett. (Shannon, Irel.)* 1986, **30**, 103-111
12. Wade, D. R. et al *Mutat. Res.* 1981, **89**, 9-20
13. Gatehouse, D. et al *Carcinogenesis (London)* 1984, **5**(3), 391-397
14. Parodi, S. et al *Cancer Res.* 1981, **41**, 1469-1482
15. Menon, M. M. et al *Indian J. Exp. Biol.* 1981, **19**, 939-942
16. Malca-Mor, L. et al *Appl. Environ. Microbiol.* 1982, **44**, 801-808
17. Herbold, B. A. et al *Mutat. Res.* 1976, **40**, 73-84
18. Probst, G. S. et al *Environ. Mol. Mutagen.* 1981, **3**, 11-32
19. Mori, H. et al *Jpn. J. Cancer Res.* 1988, **79**(2), 204-211
20. Speit, G. et al *Hum. Genet.* 1980, **54**, 155-158
21. MacRae, W. D. et al *Mutat. Res.* 1979, **68**, 351-365
22. Whiting, R. F. et al *Mutat. Res.* 1979, **62**, 505-515
23. Yanagisawa, K. et al *Mutat. Res.* 1987, **183**, 89-94
24. Schroeder, T. M. et al *Hum. Genet.* 1979, **52**, 309-321
25. Basler, A. *Hum. Genet.* 1978, **42**, 26-27
26. Manna, G. K. et al *Nucleus (Quezon City, Philipp.)* 1979, **22**, 96-103
27. Moss, J. D. et al *Am. Rev. Respir. Dis.* 1972, **106**, 849-856
28. Rubin, D. H. et al *Clin. Pediatr. (Philadelphia)* 1983, **22**, 518-519
29. Terman, D. S. et al *Neurology* 1970, **20**, 299-304
30. Yamamoto, T. *Tan, Sui* 1989, **19**(4), 813-818
31. Steele, V. E. et al *Fundam. Appl. Toxicol.* 1988, **11**(4), 673-684

# I101    Isooctane

## (CH3)2CHCH2C(CH3)3

**CAS Registry No.** 540-84-1
**Synonyms** 2,2,4-trimethylpentane; isobutyltrimethylethane
**Mol. Formula** $C_8H_{18}$                              **Mol. Wt.** 114.23
**Uses** Organic synthesis. Solvent. In determining octane number of motor fuel.
Spectrophotometric analysis.

## Physical properties

**M. Pt.** −107.4°C; **B. Pt.** 99.3°C; **Flash point** −12°C (closed cup); **Specific gravity**
$d_4^{20}$ 0.692; **Volatility** v.p. 40.6 mmHg at 21°C; v. den. 3.93.

## Solubility

Water: 0.56 mg $l^{-1}$ at 25°C. Organic solvent: benzene, toluene, xylene, chloroform,
diethyl ether, carbon tetrachloride

## Occupational exposure

**UN No.** 1262; **HAZCHEM Code** 3☒E; **Conveyance classification** flammable liquid.

## Environmental fate

### Degradation studies

13% biodegradation of an initial concentration of 3.47 μl $l^{-1}$ after 192 hr at 13°C
incubated with natural flora in groundwater in presence of other components of high
octane petrol (1).
Degraded by *Pseudomonas putida*, *Pseudomonas stutzeri* and *Micrococcus* sp.
isolated from contaminated well water (2).

### Abiotic removal

$t_{1/2}$ (calc.) in water at 25°C and 1 m depth 5.55 hr; evaporation rate 0.124 m $hr^{-1}$ (3).

## Mammalian and avian toxicity

### Metabolism and pharmacokinetics

In rats exposed to 1 or 350 ppm radiolabelled $^{14}$C-isoctane for 2 hr excretion was
almost entirely via kidneys over the entire 70 hr post-exposure period. After this time
1-2% remained in the body (4).
Urinary metabolites in rats after oral administration included:
1-hydroxy-2,3,4-trimethylpentane; 2,3,4-trimethyl-1-pentanoic acid; and
2,3,4-trimethyl-5-hydroxy-1-pentanoic acid (5).

### Irritancy

Eye irritation reactivity in humans 0.9 (6).

## Any other adverse effects

Hepatotoxic as well as nephrotoxic in rats dosed with 2 ml $kg^{-1}$ in corn oil by gavage (7).

## References

1.  Jamison, V. W. et al *Proc. Third Int. Biodeg. Symp.* 1976, Applied Science Publishers
2.  Ridgeway, H. F. et al *Appl. Environ. Microbiol.* 1990, **56**(11), 3565-3575

3. Mackay, D. et al *Environ. Sci. Technol.* 1975, **9**(13), 1178-1180
4. Dahl, A. R. *Toxicol. Appl. Pharmacol.* 1989, **100**(2), 334-341
5. Olson, C. T. et al *Toxicol. Lett.* 1987, **37**(3), 199-202
6. Yeung, C. K. K. et al *Atmos. Environ.* 1973, **7**, 551
7. Fowlie, A. J. et al *J. Appl. Toxicol.* 1987, **7**(5), 335-341

# I102   Isooctyl alcohol

## (CH3)2CH(CH2)4CH2OH

**CAS Registry No.** 26952-21-6
**Synonyms** isooctanol, 1-heptanol, 6-methyl-
**Mol. Formula** $C_8H_{18}O$                    **Mol. Wt.** 130.23

### Physical properties
**M. Pt.** <100°C; **B. Pt.** $_1$ 186°C; **Specific gravity** $d_{20}^{20}$ 0.832.

### Solubility
Water: 640 mg $l^{-1}$ at 25°C.

### Occupational exposure
**US TLV (TWA)** 50 ppm (266 mg m$^{-3}$); **UK Long-term limit** 50 ppm (270 mg m$^{-3}$).

### Ecotoxicity

**Invertebrate toxicity**
$EC_{50}$ (24 hr) *Daphnia magna* 115 mg $l^{-1}$; $EC_0$ (24 hr) *Daphnia magna* 77 mg $l^{-1}$ (1).
NOEC (21 day) *Daphnia* reproduction test 2.3 mg $l^{-1}$ (nominal value), 1.6 mg $l^{-1}$ (minimum value) (1).

### Mammalian and avian toxicity

**Acute data**
$LD_{50}$ oral rat, mouse 1480-1670 mg kg$^{-1}$ (2,3).
$LD_{50}$ dermal rabbit 2520 mg kg$^{-1}$ (4).

**Irritancy**
100 mg instilled into rabbits' eyes caused severe irritation. 2600 mg kg$^{-1}$ applied to rabbits skin for 24 hr caused moderate irritation (2).

### Any other comments
Human health effects, experimental toxicology and workplace experience reviewed (5).

### References
1. Kuhn, R. et al *Water Res.* 1989, **23**(4), 501-510
2. *Am. Ind. Hyg. Assoc. J.* 1973, **34**, 493
3. Izmerov, N. F. et al *Toxicometric Parameters of Industrial Toxic Chemicals under Single Exposure* 1982, CIP, Moscow
4. Monick, J. A. *Alcohols and their Chemistry* 1968, Reinhold Books, New York
5. *ECETOC Technical Report No. 30(4)* 1991, European Chemical Industry Ecology and Toxicology Centre, B-1160 Brussels

# I103   Isopentane

CH3CH(CH3)CH2CH3

**CAS Registry No.** 78-78-4
**Synonyms** ethyldimethylmethane; isoamyl hydride; 2-methylbutane
**Mol. Formula** $C_5H_{12}$                                 **Mol. Wt.** 72.15

## Physical properties

**M. Pt.** −160.5°C; **B. Pt.** 27.8°C; **Flash point** <-50°C (closed cup); **Specific gravity** $d^{19}$ 0.62; **Volatility** v.p. 595 mmHg at 21.1°C; v. den. 2.48.

**Solubility**
Water: 48 mg $l^{-1}$ at 20°C.

## Occupational exposure

**UN No.** 1265; **HAZCHEM Code** 3ME; **Conveyance classification** flammable liquid; **Supply classification** highly flammable.
**Risk phrases** Highly flammable (R11)
**Safety phrases** Keep container in a well ventilated place – Keep away from sources of ignition – No Smoking – Do not empty into drains – Take precautionary measures against static discharges (S9, S16, S29, S33)

## Environmental fate

### Degradation studies

Zero biodegradation of an initial concentration of 3.29 μl $l^{-1}$ after 192 hr at 13°C incubation with natural flora in groundwater in presence of other components of high octane petrol (1).

## Any other comments

Human health effects, environmental toxicity and physicochemical properties reviewed (2).

## References

1.  Jamison, V. W. et al *Proc. Third Int. Biodeg. Symp.* 1976, Applied Science Publishers
2.  *ECETOC Technical Report No. 30(4)* 1991, European Chemical Industry Ecology and Toxicology Centre, B-1160 Brussels

# I104   Isopentyl acetate

CH3CH(CH3)CH2CH2OC(O)CH3

**CAS Registry No.** 123-92-2
**Synonyms** isoamyl acetate; isoamyl ethanoate; isopentyl alcohol acetate; 3-methylbutyl acetate; pear oil; banana oil
**Mol. Formula** $C_7H_{14}O_2$                                 **Mol. Wt.** 130.19
**Uses** Solvent. Flavouring. In textile manufacture and manufacture of photographic

film, artificial pearls and leather, celluloid cements, waterproof varnishes, bronzing liquids and metallic paints.

## Physical properties

**M. Pt.** –78˚C; **B. Pt.** $_{756}$ 142˚C; **Flash point** 25˚C; **Specific gravity** $d_4^{15}$ 0.876.

### Solubility
Water: 1600 mg $l^{-1}$ at 20˚C. Organic solvent: miscible with ethanol, diethyl ether, ethyl acetate, amyl alcohol

### Occupational exposure
**US TLV (TWA)** 100 ppm (532 mg m$^{-3}$); **UK Long-term limit** 100 ppm (525 mg m$^{-3}$); **UK Short-term limit** 125 ppm (655 mg m$^{-3}$).

## Mammalian and avian toxicity

### Acute data
$LD_{50}$ oral rabbit, rat 7420-16,600 mg kg$^{-1}$ respectively (1,2).
$LC_{Lo}$ (duration unspecified) inhalation cat 35,000 mg m$^{-3}$ (3).
$LD_{Lo}$ subcutaneous guinea pig 5000 mg kg$^{-1}$ (3).

## Any other comments

Human health effects, experimental toxicity, ecotoxicology and workplace experience reviewed (4).

## References

1. *Ind. Med. Surg.* 1972, **41**, 31
2. *Yakkyoku* 1981, **32**, 1241
3. *Arch. Gewerbepathol. Gewerbehyg.* 1933, **5**, 1
4. *ECETOC Technical Report No. 30(4)* 1991, European Chemical Industry Ecology and Toxicology Centre, B-1160 Brussels

# I105   Isophorone

**CAS Registry No.** 78-59-1
**Synonyms** isoacetophorone; 1,1,3-trimethyl-3-cyclohexen-5-one; isooctaphenone; 3,5,5-trimethyl-2-cyclohexen-1-one
**Mol. Formula** $C_9H_{14}O$                    **Mol. Wt.** 138.21
**Uses** Solvent. Intermediate for synthesis of alcohols. Raw material for 3,5-dimethylaniline. In pesticide lacquers and finishes manufacture.

## Physical properties
**M. Pt.** –8°C; **B. Pt.** 215.2°C; **Flash point** 84°C (open cup); **Specific gravity** 0.9229; **Volatility** v.p. 1 mmHg at 38°C; v. den. 4.77.

### Solubility
Water: 12,000 mg $l^{-1}$.

## Occupational exposure
**US TLV (TWA)** ceiling limit 5 ppm (28 mg $m^{-3}$); **UK Short-term limit** 5 ppm (25 mg $m^{-3}$); **Supply classification** irritant.
**Risk phrases** Irritating to eyes, respiratory system and skin (R36/37/38)
**Safety phrases** In case of contact with eyes, rinse immediately with plenty of water and seek medical advice (S26)

## Ecotoxicity

### Fish toxicity
$LC_{50}$ (96 hr) fathead minnow 145-255 mg $l^{-1}$ (1).
$LC_{50}$ (96 hr) bluegill sunfish 220 mg $l^{-1}$ (2).

### Invertebrate toxicity
$LC_{50}$ (48 hr) *Daphnia magna* 120 mg $l^{-1}$ (3).
$EC_{50}$ (24 hr) *Tetrahymena pyriformis* 420 mg $l^{-1}$ (4).
$LC_{50}$ (96 hr) mysid shrimp 12.9 mg $l^{-1}$ (5).
$EC_{50}$ (96 hr) *Selenestrum capricornutum* 126 mg $l^{-1}$ (5).

## Environmental fate

### Degradation studies
Activated sludge adsorbability 0.193 g $g^{-1}$ C; 96.6% reduction, influent 1000 mg $l^{-1}$, effluent 34 mg $l^{-1}$ (6).
Activated sludge $EC_{50}$ (3 hr) 100 mg $l^{-1}$; $EC_{50}$ (24 hr) *Tetrahymena pyriformis* 420 mg $l^{-1}$ (7).

### Absorption
Adsorbability 0.193 g $g^{-1}$ carbon 96.6% reduction (8).

## Mammalian and avian toxicity

### Acute data
$LD_{50}$ oral rat 2330 mg $kg^{-1}$ (8).
$LC_{Lo}$ (4 hr) inhalation rat 1840 ppm (9).
$LD_{50}$ dermal rabbit 1500 mg $kg^{-1}$ (10).

### Sub-acute data
Kidney tubule injury not reported in ♀ rats exposed to 5 intraperitoneal injections wkly for 2 wk of 5-20% the $LD_{50}$ (11).

### Carcinogenicity and long-term effects
National Toxicology Program tested rats and mice via gavage. Some evidence of carcinogenicity in ♂ rats, equivocal evidence in ♂ mice, no evidence in ♀ mice or rats. Increased incidence of kidney tubular cell adenoma and adenocarcinoma occurred in ♂ rats fed 250-500 mg $kg^{-1}$, 5 days $wk^{-1}$ for 103 wk; evidence for preputial gland carcinoma in ♂ rats was equivocal as was liver adenoma/carcinoma, integumentary

system tumours and haematopoeitic system malignant tumours in ♂ mice (12,13). Neither isophorone nor its metabolites showed covalent binding to DNA in liver and kidneys of rats and mice, hence tumours reported in ♂ rats and mice after long-term oral administration are not likely to be caused by genotoxic mechanisms (14).

**Metabolism and pharmacokinetics**
After oral administration to rabbits and rats, the urinary metabolites included: isophorol; *cis/trans*-3,5,5-trimethyl-2-cyclohexen-1-one; dihydroisophorone; diisophorone glucuronide; and 5,5-dimethyl-3-oxocyclohex-1-enecarboxylic acid (15-17).

**Irritancy**
100 mg applied to rabbits' skin for 24 hr caused mild irritation (18).
920 μg applied to rabbits' eyes caused severe irritation (10).
25 ppm caused eye, nose and throat irritation in humans (19).

## Genotoxicity
*Salmonella typhimurium* TA98, TA100, TA1535, TA1537 with or without metabolic activation negative (20).
Induced sister chromatid exchanges but not chromosome aberrations in Chinese hamster ovary cells *in vitro* (21).
L5178Y $tk^+/tk^-$ mouse lymphoma cell forward mutation assay without metabolic activation positive (22).

## Any other comments
Human health effects, experimental toxicity, environmental effects, ecotoxicology, exposure levels, epidemiology, workplace experience, hazard assessment and physico-chemical properties reviewed (23).
Physico-chemical properties, toxicity, hazards and French regulations of isophorone reviewed (24).
Toxicology reviewed (25,26).

## References
1. Cairns, M. A. et al *Arch. Environ. Contam. Toxicol.* 1982, **11**, 703-707
2. Buccafusco, R. J. et al *Bull. Environ. Contam. Toxicol.* 1981, **26**, 446
3. LeBlanc, G. A. *Bull. Environ. Contam. Toxicol.* 1980, **24**, 684-691
4. Yoshioka, Y. et al *Sci. Total Environ.* 1985, **43**, 149
5. *US EPA Contract No. 68-01-4646* 1978
6. Guisti, D. M. et al *J.- Water Pollut. Control. Fed.* 1974, **46**(5), 947-965
7. Yoshioka, Y. et al *Ecotoxicol. Environ. Saf.* 1986, **12**, 206-212
8. *Toxicol. Appl. Pharmacol.* 1970, **17**, 498
9. *J. Ind. Hyg. Toxicol.* 1940, **22**, 477
10. *Union Carbide Data Sheet* 1971, Union Carbide Corp., New York
11. Bernard, A. M. et al *Toxicol. Lett.* 1989, **45**(2-3), 271-280
12. *National Toxicology Program Research and Testing Division* 1992, Report No.TR-291, NIEHS, Research Triangle Park, NC
13. Bucher, J. R. et al *Toxicology* 1986, **39**, 207-219
14. Thier, R. et al *Arch. Toxicol.* 1990, **64**(8), 684-685
15. *Chemical Safety Data Sheets 1. Solvents* 1988, RSC, London
16. Dutertue, L. *Toxicol. Eur. Res.* 1978, **1**(4), 209
17. Truhort, R. *C. R. Acad. Sci. Ser. 7* 1970, **271**, 1333
18. *J. Eur. Toxicol.* 1972, **5**, 31

19. Patty, F. A. *Industrial Hygiene and Toxicology* 1967, Interscience Publishers, New York
20. Mortelmans, K. et al *Environ. Mutagen.* 1986, **8**(Suppl. 7), 1-119
21. Buzzi, R. et al *Mutat. Res.* 1990, **234**, 269-288
22. McGregor, D. B. et al *Environ. Mol. Mutagen.* 1988, **12**, 85-154
23. *ECETOC Technical Report No. 30(4)* 1991, European Chemical Industry Ecology and Toxicology Centre, B-1160 Brussels
24. *Cah. Notes Doc.* 1989, **134**, 175-178
25. *Gov. Rep. Announce. Index (U. S.)* 1990, **90**(11), Abstr. No.028,423
26. *Gov. Rep. Announce. Index (U. S.)* 1988, **88**(13), Abstr. No.833,883

# I106   Isophorone diisocyanate

**CAS Registry No.** 4098-71-9
**Synonyms** isophorone diamine diisocyanate; 3-isocyanatomethyl-3,5,5-trimethyl-cyclohexylisocyanate
**Mol. Formula** $C_{12}H_{18}N_2O_2$                          **Mol. Wt.** 222.29
**Uses** Crosslinking and curing agent.

## Physical properties

**B. Pt.** $_{15}$ 158-159°C; **Flash point** >110°C; **Specific gravity** 1.049.

## Occupational exposure

**US TLV (TWA)** 0.005 ppm (0.045 mg m$^{-3}$); **UK Long-term limit** MEL 0.02 mg m$^{-3}$ (as –NCO); **UK Short-term limit** MEL 0.07 mg m$^{-3}$ (as –NCO); **UN No.** 2290; **HAZCHEM Code** 2X; **Conveyance classification** harmful substance; **Supply classification** T (≥2%), Xn (≥0.5%).
**Risk phrases** ≥20% – Toxic by inhalation – Irritating to eyes, respiratory system and skin – May cause sensitisation by inhalation and skin contact – ≥2%<20% – Toxic by inhalation – May cause sensitisation by inhalation and skin contact – ≥0.5%<2% – Harmful by inhalation – May cause sensitisation by inhalation and skin contact (R23, R36/37/38, R42/43, R23, R42/43, R20, R42/43)
**Safety phrases** In case of contact with eyes, rinse immediately with plenty of water and seek medical advice – After contact with skin, wash immediately with plenty of soap and water – In case of insufficient ventilation, wear suitable respiratory equipment – In case of accident or if you feel unwell, seek medical advice immediately (show label where possible) (S26, S28, S38, S45)

## Mammalian and avian toxicity

### Acute data

$LC_{50}$ (4 hr) inhalation rat 260 mg m$^{-3}$ (1).
$LD_{50}$ dermal rabbit 1060 mg kg$^{-1}$ (1).

## Any other comments

Human health effects, experimental toxicity, physicochemical effects, epidemiology, workplace experience and ecotoxicology reviewed (2).

## References

1. *Documentation of Threshold Limit Values in Workroom Air* 1980, ACGIH, Cincinnati, OH
2. *ECETOC Technical Report No. 30(4)* 1991, European Chemical Industry Ecology and Toxicology Centre, B-1160 Brussels

# I107   Isophosphamide

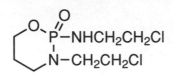

**CAS Registry No.** 3778-73-2

**Synonyms** *N,N*-bis(β-chloroethyl)amino-*N'*-*O*-propylenephosporic acid ester diamide; *N*,3-bis(2-chloroethyl)tetrahydro-2*H*-1,3,2-oxazaphosphorin-2-amine, 2-oxide; 3-(2-chloroethyl)-2-[(2-chloroethyl)amino]tetrahydro-2*H*-1,3,2-oxazaphosphorin-2-oxide

**Mol. Formula** $C_7H_{15}Cl_2N_2O_2P$          **Mol. Wt.** 261.09

**Uses** Antineoplastic agent used to treat solid tumours of the lung, ovary and testis, and for sarcomas and lymphomas. Some suggest its superiority over its congener cyclophosphamide is unsubstantiated, particularly in childhood neoplasms and because of its toxicity argue its use may be unjustified.

## Physical properties

**M. Pt.** 40°C.

## Solubility

Organic solvent: methanol, isopropanol

## Mammalian and avian toxicity

### Acute data

$LD_{50}$ oral rat 143 mg kg$^{-1}$ (1).
$LD_{50}$ oral mouse 1005 mg kg$^{-1}$ (1).
$LD_{50}$ intraperitoneal rat, mouse 140-397 mg kg$^{-1}$ (2,3).

### Carcinogenicity and long-term effects

No adequate data for carcinogenicity in humans, limited evidence for carcinogenicty in animals, IARC group 3 (4).
National Toxicology Program tested rats and mice by intraperitoneal injection. Designated carcinogen in ♀ rats and mice, non-carcinogen in ♂ rats and mice (5). Significantly increased incidence of lung tumours in ♂, ♀ mice after intraperitoneal injection over 8 wk of 5 doses totalling 1.3 g kg$^{-1}$ or 24 doses totalling 0.45-1.125 g kg$^{-1}$ (6).

Increased incidence of malignant lymphoma in ♀ mice given 20 mg kg$^{-1}$ intraperitoneally 3 × wk$^{-1}$ for 52 wk (7).
In 2 studies involving subcutaneous injection of 0.2-2.0 mg mouse$^{-1}$ to ♀ NZB/NZW mice one found significantly increased incidence of neoplasms (8), whereas the other study produced negative results (9).

**Teratogenicity and reproductive effects**
Increased rate of resorption, growth retardation and incidence of hydrocephalus, micromelia, adactyly, syndactyly, kidney ectopia and delayed ossification in mice given 20 mg kg$^{-1}$ intraperitoneally on day 11 of pregnancy; growth retardation alone occurred at 10 mg kg$^{-1}$ (10).

**Metabolism and pharmacokinetics**
In humans clearance after a single intravenous bolus is biphasic, with mean terminal elimination $t_{1/2}$ 15 hr. Clearance from plasma after repeated, lower doses is monoexponential with $t_{1/2}$ 7 hr. It is extensively metabolised in the liver to active metabolites; metabolism appears saturated at higher doses. It is excreted largely in urine as unchanged drug and metabolites (11).
It is rapidly metabolised in many animal species, in a similar way to cyclophosphamide, to acrolein and, in dogs, the carboxy derivative and 4-ketoisophosphamide (12).
Metabolic activation in humans may be like cyclophosphamide, with ring hydroxylation to isophosphomide mustard and acrolein. An intravenous dose is excreted in urine as one or both dechloroethylated metabolite (50%), intact drug (20%) and carboxyisophosphamide (2%), there is wide interindividual variation (13,14).
In humans plasma pharmacokinetics of high doses is biphasic with a secondary $t_{1/2}$ 15 hr, but monophasic with $t_{1/2}$ 7 hr with lower doses (15).

# Genotoxicity

*Salmonella typhimurium* TA100, TA1535 with metabolic activation positive (16,17).
Dose-dependent increase in chromosomal aberrations in Chinese hamster bone marrow cells after intraperitoneal administration (18).
*Drosophila melanogaster* sex-linked recessive lethal test positive (19).

**Any other adverse effects to man**
Toxic effects of treatment involve urinary tract, kidney and central nervous system. Kidney damage is irreversible. Central nervous system effects include encephalopathy, with EEG abnormalities, disorientation, confusion, catatonia, coma and occasionally death from central nervous system depression and circulatory collapse. Incidence of encephalopathy is greater after oral than intravenous administration, and the contributory role of mesna, with which it is given, is unclear but may be due to chelating properties (11).
Urinary tract toxicity, characterised by signs of cystitis, limits therapeutic doses to 50 mg kg$^{-1}$. Less important side effects are myelosuppression, nausea, vomiting, alopecia, lethargy, confusion and reversible glycosuria (20).

**Any other adverse effects**
Results in rats suggest renal damage may be due to formation of toxic metabolites in the kidney. Central nervous system toxicity may also be due to a metabolite (11).

## Any other comments

Pharmacology and toxicity reviewed (21).

## References

1. *Kiso to Rinsho* 1982, **16**, 431
2. *US Patent* 3732340 U. S. Patent Office, Box 9, Washington, DC
3. *Arzneim.-Forsch.* 1974, **24**, 1149
4. *IARC Monograph* 1987, **Suppl. 7**, 65
5. *National Toxicology Program Research and Testing Division* 1992, Report No. TR-032, NIEHS, Research Triangle Park, NC
6. Stoner, G. D. et al *Cancer Res.* 1973, **33**, 3069-3085
7. *NCI Carcinog. Tech. Rep. Ser.* 1977, No.32
8. Mitron, P. S. et al *Arzneim.-Forsch.* 1979, **29**, 483-488
9. Mitron, P. S. et al *Arzneim.-Forsch.* 1979, **29**, 662-667
10. Bus, J. S. et al *Proc. Soc. Exp. Biol. Med.* 1973, **143**, 965-970
11. *Martindale. The Extra Pharmacopoeia* 30th ed., 1993, The Pharmaceutical Press, London
12. Hill, D. L. et al *Cancer Res.* 1973, **33**, 1016-1022
13. Norporth, K. *Cancer Treat. Rep.* 1976, **60**, 437-443
14. Goren, M. P. *J. Chromatogr.* 1991, **570**(2), 351-359
15. Nelson, R. L. et al *Clin. Pharmacol. Ther.* 1976, **19**, 365-370
16. Benedict, W. F. et al *Cancer Res.* 1977, **37**, 2209-2213
17. Matney, T. S. et al *Teratog., Carcinog., Mutagen.* 1985, **5**, 319-328
18. Rohrborn, G. et al *Arch. Toxicol.* 1977, **38**, 35-43
19. *Environ. Mol. Mutagen.* 1986, **8**(6), 849-865
20. van Dyk, J. et al *Cancer Res.* 1972, **32**, 921-924
21. Zalupski, M. et al *J. Natl. Cancer Inst.* 1988, **80**(8), 556-566

# I108   Isophthalic acid

**CAS Registry No.** 121-91-5

**Synonyms** *m*-benzenedicarboxylic acid; *m*-phthalic acid.

**Mol. Formula** $C_8H_6O_4$                    **Mol. Wt.** 166.13

## Physical properties

**M. Pt.** 345-348°C; **B. Pt.** sublimes.

## Solubility

Water: 130 mg $l^{-1}$ at 25°C. Organic solvent: acetic acid, ethanol

## Environmental fate

### Anaerobic effects

Anaerobically metabolised by *Pseudomonas* sp. strain P136 (1).

### Degradation studies
Biodegradable (2).
Decomposed by a soil microflora in 8 days (3).
95% COD removal at 76 mg COD $g^{-1}$ $O_2$ dry inoculum $hr^{-1}$; adapted activated sludge at 20°C, product is sole carbon source (4).

## Mammalian and avian toxicity

### Acute data
$LD_{50}$ oral rat 10400 mg $kg^{-1}$ (5).
$LD_{50}$ intraperitoneal mouse 4200 mg $kg^{-1}$ (6).

### Irritancy
100 mg instilled into rabbits' eyes for 24 hr caused mild irritation (5).

## References

1. Nozawa, T. et al *J. Bacteriol.* 1988, **170**(12), 5778-5784
2. *List of the existing chemical substances tested on biodegradability by microorganisms or bioaccumulation in fish body* 1987, Chemicals Inspection and Testing Institute, Japan
3. Alexander, M. et al *J. Agric. Food Chem.* 1966, **14**, 410
4. Pitter, P. *Water Res.* 1976, **10**, 231-235
5. Marhold, J. V. *Sbornik Vysledku Toxixologickeho Vysetreni Latek A Pripravku* 1972, Prague
6. *C. R. Seances Acad. Sci., Ser. 3* 1958, **246**, 851

# I109   Isophthalonitrile

**CAS Registry No.** 626-17-5
**Synonyms** *m*-phthalodinitrile; 1,3-benzenedicarbonitrile; *m*-dicyanobenzene; 1,3-dicyanobenzene
**Mol. Formula** $C_8H_4N_2$                    **Mol. Wt.** 128.13

## Physical properties
**M. Pt.** 160-162°C; **B. Pt.** Sublimes; **Volatility** ; v. den. 4.42.

## Ecotoxicity

### Invertebrate toxicity
$EC_{50}$ (5-30 min) *Photobacterium phosphoreum* 107 ppm Microtox test (1).

## Mammalian and avian toxicity

### Acute data
$LD_{50}$ oral mouse 180 mg $kg^{-1}$ (2).

LD$_{50}$ oral rat 1860 mg kg$^{-1}$ (3).
LD$_{50}$ oral rabbit, guinea pig 350-370 mg kg$^{-1}$ (4).
LD$_{50}$ intraperitoneal mouse 480 mg kg$^{-1}$ (5).

**Irritancy**
500 mg instilled into rabbits' eyes for 24 hr caused mild irritation (3).

**Any other comments**
Human health effects, experimental toxicity, environmental effects, ecotoxicology, exposure levels and workplace experience reviewed (6).

**References**

1. Kaiser, K. L. E. et al *Water Pollut. Res. J. Can.* 1991, **26**(3), 361-431
2. *J. Med. Chem.* 1978, **21**, 906
3. Marhold, J. V. *Sbornik Vysledku Toxixologickeho Vysetreni Latek A Pripravku* 1972, Prague
4. Izmerov, N. F. et al *Toxicometric Parameters of Industrial Toxic Chemicals under Single Exposure* 1982, CIP, Moscow
5. *Ind. Health* 1966, **4**, 11
6. *ECETOC Technical Report No. 30(4)* 1991, European Chemical Industry Ecology and Toxicology Centre, B-1160 Brussels

# I110 Isoprene

$$CH_2=C(CH_3)CH=CH_2$$

**CAS Registry No.** 78-79-5
**Synonyms** 2-methylbivinyl; 2-methylbutadiene; 2-methyl-1,3-butadiene
**Mol. Formula** C$_5$H$_8$          **Mol. Wt.** 68.12
**Uses** Monomer for polyisoprene manufacture; manufacture of butyl rubber.
**Occurrence** Isolated from pyrolysis products of natural rubber.

## Physical properties

**M. Pt.** −146.7°C; **B. Pt.** 34°C; **Flash point** −54°C; **Specific gravity** d$_4^{20}$ 0.681; **Volatility** v.p. 400 mmHg at 15.4°C; v. den. 2.35.

**Solubility**
Organic solvent: miscible with ethanol, diethyl ether

## Occupational exposure

**UN No.** 1218; **HAZCHEM Code** 3$\blacksquare$E; **Conveyance classification** flammable liquid; **Supply classification** extremely flammable.
**Risk phrases** Extremely flammable (R12)
**Safety phrases** Keep container in a well ventilated place – Keep away from sources of ignition – No Smoking – Do not empty into drains – Take precautionary measures against static discharges (S9, S16, S29, S33)

## Ecotoxicity

### Fish toxicity
$LC_{50}$ (96 hr) bluegill sunfish, fathead minnow and goldfish 43, 74, 180 mg l$^{-1}$, respectively (1).

## Environmental fate

### Degradation studies
Degraded by *Nocardia* strains (2).

### Abiotic removal
Activated carbon 10 ×, 89% reduction, influent 1000 ppm, effluent 110 ppm; 78% reduction, influent 500 ppm, effluent 110 ppm (3).

## Mammalian and avian toxicity

### Acute data
$LC_{50}$ (2 hr) inhalation mouse 140 g m$^{-3}$ (4).
$LC_{50}$ (4 hr) inhalation rat 180 g m$^{-3}$ (5).

### Sub-acute data
No effect level inhalation rat 1670 ppm 16 × 6 hr (1).
Short term inhalation study in rats and mice scheduled for peer review (6).
Haematological changes and microscopic lesions including testicular atrophy, olfactory epithelial degeneration, and forestomach epithelial hyperplasia observed in mice but not rats exposed to 7000 ppm for 2 wk (7).

### Metabolism and pharmacokinetics
In ♂ mice and rats rate of metabolism was directly proportional to concentration below 300 ppm; small amounts were exhaled unchanged (25% and 15% respectively) and $t_{1/2}$ was 4.4 and 6.8 min respectively (8).
In ♂ rats exposed to 8-8200 ppm 75% excretion of total metabolites was via urine (9).
Metabolised *in vitro* by rabbit, mouse and rat liver microsomes to 3,4-epoxy-3-methyl-1-butene and 3,4-epoxy-2-methyl-1-butene which are not mutagenic. Formation of the mutagenic and presumably carcinogenic isoprene diepoxide is possible, hence genotoxicity in rodents or other species cannot be eliminated (10).

### Irritancy
Irritating to skin and mucous membranes (species unspecified) (11).

## Genotoxicity
Induced sister chromatid exchanges in bone marrow cells, increased levels of micronucleated polychromatic and normochromatic erythrocytes, but did not induce chromosomal aberrations or alter mitotic index in ♂ mice exposed to 438, 1750 or 7000 ppm for 12 days (12,13).

## Any other adverse effects
Narcotic in high concentrations (species unspecified) (11).

## Legislation
Reportable quantity listed in US Federal Comprehensive Environmental Response, Compensation and Liability Act (14).

## Any other comments

Human health effects, experimental toxicity, ecotoxicology, workplace experience and physicochemical effects reviewed (15).

Future directions in toxicology research on isoprene reviewed (16).

Slight inhibition of microbial growth after 24 hr at saturation concentration (17).

## References

1. Cage, J. C. *Br. J. Ind. Med.* 1970, **27**
2. Van Ginkel, C. G. et al *FEMS Microbiol. Ecol.* 1987, **45**(5), 275-279
3. Dahm, D. B. et al *Environ. Sci. Technol.* 1974, **8**(13)
4. *Gig. Tr. Prof. Zabol.* 1965, **9**(1), 36
5. *Russ. Pharmacol. Toxicol.* 1968, **31**, 162
6. *National Toxicology Program Research and Testing Division* 1992, Management Status Report, NIEHS, Research Triangle Park, NC
7. Melnick, R. L. et al *Environ. Health Perspect.* 1990, **86**, 93-98
8. Peter, H. et al *Toxicol. Lett.* 1987, **36**(1), 9-14
9. Dahl, A. R. et al *Toxicol. Appl. Pharmacol.* 1987, **89**(2), 237-248
10. Gervasi, P. G. et al *Environ. Health Perspect.* 1990, **86**, 85-87
11. *Merck Index* 11th ed., 1989, Merck & Co., Rahway, NJ
12. Tice. R. R. et al *Mutagenesis* 1988, **3**(2), 141-146
13. Shelby, M. D. et al *Environ. Health Perspect.* 1990, **86**, 71-73
14. US EPA *Fed. Regist.* 1989, **54**(155), 33426-33484
15. *ECETOC Technical Report No. 30(4)* 1991, European Chemical Industry Ecology and Toxicology Centre, B-1160 Brussels
16. Bird, M. G. *Environ. Health Perspect.* 1990, **86**, 99-102
17. Cabridenc. R. et al *Etude des possibilités de biodegradation d'un effluent de fabrication d'isoprene* 1968, IRCHA, France

# I111   Isoprocarb

**CAS Registry No.** 2631-40-5

**Synonyms** 2-isopropylphenyl methylcarbamate; 2-(1-methylethyl)phenyl methylcarbamate; MIPC; MIPCIN; MIPSIN; *o*-cumenyl methylcarbamate

**Mol. Formula** $C_{11}H_{15}NO_2$                **Mol. Wt.** 193.25

**Uses** Contact insecticide.

## Physical properties

**M. Pt.** 88-93°C; **Volatility** v.p. 2.85 mmHg at 20°C.

## Solubility
Organic solvent: acetone, methanol

## Occupational exposure
**Supply classification** harmful.
**Risk phrases** Harmful if swallowed (R22)

## Ecotoxicity
**Fish toxicity**
$LC_{50}$ (48 hr) carp 4.2 mg l$^{-1}$ (1).

**Invertebrate toxicity**
Inhibited ammonium oxidisers, nitrate reducers and denitrifying soil bacteria (2).

## Mammalian and avian toxicity
**Acute data**
$LD_{50}$ oral mouse, rat 150-180 mg kg$^{-1}$ (3,4).
$LD_{50}$ dermal mouse 7600 mg kg$^{-1}$ (3).
$LD_{50}$ intraperitoneal rat 142 mg kg$^{-1}$ (5).

**Sub-acute data**
No effect level in 90 day feeding studies in rats 300 mg kg$^{-1}$ (1).

**Irritancy**
Non-irritating to rabbits' eyes and skin (dose and duration unspecified) (1).

## Genotoxicity
Questionable positive results for induction of sister chromatid exchanges and chromosome aberrations in Chinese hamster ovary cells (6).

## Legislation
WHO Class II; EPA Toxicity Class II (1).
Limited under EC Directive on Drinking Water Quality 80/778/EEC. Pesticides: maximum admissible concentration 0.1 µg l$^{-1}$ (7).
Included in Schedule 6 (Release into Land: Prescribed Substances) Statutory Instrument No. 472, 1991 (8).

## References
1. *The Agrochemicals Handbook* 3rd ed., 1991, RSC, London
2. Lee, K. B. et al *Han' guk T' oyang Piryo Hakhoechi* 1988, **21**(2), 149-159 (Kor.) (*Chem. Abstr.* **110**, 2520h)
3. *Guide to the Chemicals used in Crop Protection* 1968, Information Canada, Ottawa, Ontario
4. *Special Publication of the Entomological Society of America* 1978, College Park, MD
5. *Bull. W H O* 1971, **44**(1-3), 241
6. Wang, T. C. et al *Bull. Inst. Zool. Academia Sinica* 1987, **26**(4), 317-329
7. *EC Directive Relating to the Quality of Water Intended for Human Consumption* 1982, 80/778/EEC, Office for Official Publications of the European Communities, 2 rue Mercier, L-2985 Luxembourg
8. *S. I. 1991 No. 472 The Environmental Protection (Prescribed Processes and Substances) Regulations* 1991, HMSO, London

# I112   Isopropalin

$$N(CH_2CH_2CH_3)_2$$

$$O_2N \quad\quad NO_2$$

$$CH(CH_3)_2$$

**CAS Registry No.** 33820-53-0
**Synonyms** 4-(1-methylethyl)-2,6-dinitro-*N,N*-dipropylbenzenamine;
2,6-dinitro-*N,N*-dipropylcumidine; 4-isopropyl-2,6-dinitro-*N,N*-dipropylaniline
**Mol. Formula** $C_{15}H_{23}N_3O_4$          **Mol. Wt.** 309.37
**Uses** Selective pre-plant herbicide.

## Physical properties

**Flash point** 40.6°C; **Volatility** v.p. $1.4 \times 10^4$ mmHg at 30°C.

**Solubility**
Water: 0.1 mg $l^{-1}$ at 25°C. Organic solvent: acetone, *n*-hexane, benzene, chloroform, diethyl ether, methanol, acetonitrile

## Ecotoxicity

**Fish toxicity**
$LC_{50}$ (96 hr) fathead minnow, goldfish 0.1-0.15 mg $l^{-1}$ (1).

## Environmental fate

**Degradation studies**
Microorganisms may play a role in degradation as well as loss from volatilisation and photodecomposition (2).

**Abiotic removal**
Decomposed by UV light (3).

**Absorption**
Strongly adsorbed in soil with negligible leaching (1).

## Mammalian and avian toxicity

**Acute data**
$LD_{50}$ oral rat 5000 mg $kg^{-1}$ (3).
$LD_{50}$ oral quail, mallard duck, chicken 1000-2000 mg $kg^{-1}$ (1).
$LD_{50}$ dermal rabbit >2000 mg $kg^{-1}$ (1).

**Sub-acute data**
No effect level in 90 day feeding trials in dogs and rats >250 mg $kg^{-1}$ diet (1).

**Irritancy**
Slightly irritating to rabbits' skin and eyes (1).

## Any other comments

Hazards reviewed (4).

## References

1. *The Agrochemicals Handbook* 3rd ed., 1991, RSC, London
2. Romanowski, R. *Weed Sci.* 1978, **26**, 258
3. *The Merck Index* 11th ed., 1989, Merck & Co., Rahway, NJ
4. *Dangerous Prop. Ind. Mater. Rep.* 1990, **10**(4), 68-71

# I113   Isopropenyl acetate

$$CH_3CO_2C(CH_3)=CH_2$$

**CAS Registry No.** 108-22-5
**Synonyms** 1-propen-2-ol acetate; acetic acid, isopropenyl ester; 2-acetoxypropene; isopropenyl acetate; 1-methylvinyl acetate; propen-2-yl acetate
**Mol. Formula** $C_5H_8O_2$                              **Mol. Wt.** 100.12
**Uses** Reagent for acylation in organic syntheses.

## Physical properties

**M. Pt.** −92.9°C; **B. Pt.** $_{746}$ 96.6°C; **Flash point** 18°C (closed cup); **Specific gravity** $d^{20}$ 0.9; **Volatility** v. den. 3.45.

## Occupational exposure

**UN No.** 2403; **HAZCHEM Code** 3☒E; **Conveyance classification** flammable liquid.

## Mammalian and avian toxicity

**Acute data**
$LD_{50}$ oral rat 3000 mg $kg^{-1}$ (1).

## References

1. Smyth et al *J. Ind. Hyg. Toxicol.* 1949, **31**, 60

# I114   Isopropenylbenzene

**CAS Registry No.** 98-83-9
**Synonyms** (1-methylethenyl-)benzene; α-methylstyrene ; isopropenylbenzene; α-methylstyrol; 2-phenylpropene
**Mol. Formula** $C_9H_{10}$                              **Mol. Wt.** 118.18
**Uses** Copolymer in speciality resin systems.

## Physical properties

**M. Pt.** $-96°C$; **B. Pt.** $152.4°C$; **Flash point** $45°C$ (closed cup); **Specific gravity** $d_4^{20}$ 0.862; **Volatility** v. den. 4.08.

## Solubility

Organic solvent: ethanol, diethyl ether

## Occupational exposure

**US TLV (TWA)** 50 ppm (242 mg m$^{-3}$); **US TLV (STEL)** 100 ppm (483 mg m$^{-3}$); **UK Short-term limit** 100 ppm (480 mg m$^{-3}$); **UN No.** 2303; **HAZCHEM Code 3☒**; **Conveyance classification** flammable liquid; **Supply classification** irritant.
**Risk phrases** ≥25% – Flammable – Irritating to eyes and respiratory system – <25% – Flammable (R10, R36/37, R10)

## Mammalian and avian toxicity

### Acute data

TC$_{Lo}$ (duration unspecified) inhalation human 600 ppm (1).
LC$_{Lo}$ (duration unspecified) inhalation rat 3000 ppm (2).
LC$_{Lo}$ (duration unspecified) inhalation guinea pig 3000 ppm (1).

### Sub-acute data

Inhalation rats, guinea pigs, rabbits, mice and monkeys (6 month) 200 ppm, no adverse effects were reported (3).
Inhalation rats, guinea pigs (27 day) 800 ppm, slight changes in liver and kidney weight and growth retardation occurred. Inhalation rats (3-4 day), 3000 ppm, fatal (3).

### Irritancy

Dermal rabbit 100% caused moderate irritation, and 91 mg instilled into rabbit eye caused mild irritation (1).
Strong eye irritation was reported in humans briefly exposed to 600 ppm, slight irritation occurred after 2 min exposure to 200 ppm (3).

## Genotoxicity

Cultured human lymphocytes positive effect with sister chromatid exchanges (metabolic activation unspecified) (4).

## Any other adverse effects to man

Effects on blood clotting (5) and liver function have been reported (6).
Isopropenylbenzene is irritating to the upper respiratory tract and prolonged exposure may cause central nervous system depression (7).

## Any other adverse effects

The liquid caused slight conjunctival irritation in rabbit's eye. The liquid caused erythema after application to rabbit's skin (3).
Reversible skin damage was reported in an inhalation study on animals (8).

## Any other comments

Epidemiology, experimental toxicology, human health effects, physico-chemical properties and workplace experience reviewed (9).
Toxicity has been reviewed (10).

---

# References

1. *Arch. Ind. Health* 1956, **14**, 387
2. Gerarde, H. *Toxicology and Biochemistry of Aromatic Hydrocarbons* 1960, 131, Elsevier, New York
3. Wolf, M. A. et al *Arch. Ind. Health* 1956, **14**, 387-398
4. Norppa, H. et al *Mutat. Res.* 1983, **116**(3-4), 379-387
5. Netesa, V. A. et al *HSE Transl.* Jul-Sep 1985, **3**, 1
6. Putalova, T. V. *HSE Transl.* Oct-Nov 1984, 1
7. Proctor, N. H.et al *Chemical Hazards of the Workplace* 1991, Van Nostrand Reinhold, New York
8. Nikiforova, A. A. *HSE Transl.* Jul-Sep 1984, 3
9. *ECETOC Technical Report No. 30(4)* 1991, European Chemical Industry Ecology and Toxicology Centre, B-1160 Brussels
10. *Toxicity Profile: α-Methylstyrene* 1989, BIBRA, Carshalton

# I115   Isopropyl alcohol

## CH3CH(OH)CH3

**CAS Registry No.** 67-63-0

**Synonyms** 2-propanol; isopropyl alcohol; Avantin; isopropanol; Propol; *sec*-propyl alcohol

**Mol. Formula** $C_3H_8O$ **Mol. Wt.** 60.10

**Uses** Solvents. Cosmetics and Pharmaceutical aid (1).
Acetone manufacture (2).

**Occurrence** Plant volatile, animal waste and emissions from volcanoes and microbes (3). Leachate from landfills (4), emissions from petroleum storage, car exhaust and plastics combustion (3).

## Physical properties

**M. Pt.** −88.5 to −89.5°C; **B. Pt.** 82.5°C; **Flash point** 22°C (closed cup); **Specific gravity** $d_4^{20}$ 0.7854; **Partition coefficient** log $P_{ow}$ 0.05 (5); **Volatility** ; v. den. 2.07.

## Solubility

Water: freely soluble. Organic solvent: miscible with ethanol, diethyl ether, chloroform

## Occupational exposure

**US TLV (TWA)** 400 ppm (983 mg m$^{-3}$); **US TLV (STEL)** 500 ppm (1230 mg m$^{-3}$); **UK Long-term limit** 400 ppm (980 mg m$^{-3}$); **UK Short-term limit** 500 ppm (1225 mg m$^{-3}$); **UN No.** 1219; **HAZCHEM Code** 2SE; **Conveyance classification** flammable liquid; **Supply classification** highly flammable.

**Risk phrases** Highly flammable (R11)

**Safety phrases** Keep container tightly closed – Keep away from sources of ignition – No Smoking (S7, S16)

## Ecotoxicity

### Fish toxicity

$LC_{50}$ (96 hr) fathead minnow 10400 mg $l^{-1}$ (6).
$LC_{50}$ (7 day) guppy 7060 ppm (6).

### Invertebrate toxicity

$LC_{50}$ (5 min) *Photobacterium phosphoreum* 35390 ppm Microtox test (7).
LOEC semichronic *Microcystis aeruginosa* 1000 mg $l^{-1}$ (8).
Cell multiplication inhibition test, toxicity threshold, *Pseudomonas putida* 1050 mg
$l^{-1}$ (9), *Microcystis aeruginosa* 1000 mg $l^{-1}$ (8), *Entosiphon sulcatum* 4930 mg $l^{-1}$ (9).

## Environmental fate

### Degradation studies

Degradation with sewage at 20°C for 5 days resulted in 58% ThOD (10).
Filtered sewage seed resulted in 49% ThOD (11).
It was 99% degraded with acclimated activated sludge at 20°C (52 mg COD g-hr
rate) (12).
$BOD_5$ at 1000 ppm, warburg/sewage (13).
ThOD in fresh dilution water 28% and in salt dilution water 13% (14).

### Abiotic removal

Reacts with hydroxyl ions in the atmosphere, the estimated $t_{1/2}$ ranges from 33 hrs to
3.5 days, depending on the hydroxyl ion concentration (3,15).

## Mammalian and avian toxicity

### Acute data

$LD_{50}$ oral dog 4797 mg $kg^{-1}$ (16).
$LD_{50}$ oral mouse 3600 mg $kg^{-1}$ (17).
$LD_{50}$ oral rat 5045 mg $kg^{-1}$ (17).
$LD_{50}$ intraperitoneal rat 2735 mg $kg^{-1}$ (18).
$LD_{50}$ intraperitoneal mouse 4477 mg $kg^{-1}$ (18).
$LD_{Lo}$ oral man 250 ml (19).

### Sub-acute data

Oral humans (6 wk) daily dose of 2.6-6.4 mg $kg^{-1}$, no adverse effects (20).
Combined exposure to organic solvents including isopropyl alcohol caused liver
injury and hospitalisation in 3 workers within 2-4 months of starting work in a
chemical plant (21).

### Carcinogenicity and long-term effects

Insufficient evidence for carcinogenicity to humans or animals, IARC classification
group 3 (22).
Inhalation C3H, ABC or C57BL mice (5-8 month) 7700 mg $m^{-3}$. Animals were killed
at 8-12 months of age. The incidence of lung tumours was not increased, compared
with that in controls (23).
Inhalation rat, mouse, 500; 2500; 5000 ppm, experiment in progress (24).

### Teratogenicity and reproductive effects

Inhalation rats (1-19 day gestation) 5000 ppm 7 hr $day^{-1}$, congenital malformations
were reported at maternally toxic levels (25).

Increased incidence of malformations, resorptions and reduced foetal weight were reported following inhalation exposure of pregnant rats to maternally toxic levels of 10,000 ppm (duration unspecified) (26).

**Metabolism and pharmacokinetics**
It is readily absorbed from the gastro-intestinal tract but there is little absorption through intact skin. Lungs may absorb the vapour. 15% of an ingested dose is metabolised to acetone, Isopropyl alcohol is metabolised slower than ethanol (19). Some of the compound conjugates with glucuronic acid. Acetone is excreted in expired air and urine (27).

**Irritancy**
Dermal (duration unspecified) rabbit 500 mg caused mild irritation, and 10 mg instilled into rabbit eye caused moderate irritation (28, 29).
Application to human skin may cause dryness and irritation (30).

**Sensitisation**
Following 2 cases of contact dermititis in laboratory workers handling pre-injection swabs (70% isopropyl alcohol, 1% propylene oxide), one was found to be allergic to isopropyl alcohol (31).

## Genotoxicity
*Salmonella typhimurium* TA98, TA100, TA1535, TA1537, TA1538 with or without metabolic activation negative (32).
*In vitro* V79 Chinese hamster lung fibroblasts with or without metabolic activation, no increase in sister chromatid exchange frequencies (33).

## Any other adverse effects to man
Ingestion or inhalation of large quantities of the vapour may cause flushing, headache, dizziness, nausea, mental depression, narcosis and coma (26).
Isopropyl alcohol is more toxic than ethyl alcohol. Due to the presence of the main metabolite, acetone, in the circulation, ketoacidosis and ketonuria commonly occur (19).
Second and third degree chemical skin burns have been reported, caused by topical application in 2 premature infants of very low birth weight (34).
Cases of acute hepatitis, renal failure and pulmonary oedema in colour printing factory workers were attributed to combined use of carbon tetrachloride and isopropyl alcohol in cleaning operations (35).
Several lethal cases have been described as a result of lethal ingestion of 0.47 l of 70% isopropyl alcohol, death was preceded by deep coma and shock and resulted from respiratory arrest (36).

## Any other adverse effects
High levels of vapour inhalation cause ataxia, prostration and deep narcosis in mice (37).

## Legislation
Included in Schedule 6 (Release into Land: Prescribed Substances) Statutory Instrument No. 472, 1991 (38).

## Any other comments
Ecotoxicology, environmental effects, epidemology, experimental toxicology,

exposure, exposure levels, human health effects, physico-chemical properties and workplace experience reviewed (39).

Volatisation and leeching into ground water are important in the removal from soil. It is not expected to absorb onto aquatic sediments (37).

Metabolism has been reviewed (27).

Carcinogenicity has been reviewed (40).

Early reports associating excess incidence of paranasal cancer with isopropyl alcohol manufacture has since been attributed to the intermediate 'isopropyl oil' (22, 41).

## References

1. Budvari, S. (Ed.) *Merck Index* 11 ed., 1989, Merck & Co. Inc., Rahway NJ, USA
2. *Chemical Safety Data Sheets: Flammable Chemicals* 1992, **5**, Royal Society of Chemistry, London
3. Graedel, T. E. *Chemical Compounds in the Atmosphere* 1978, 244, Academic Press, New York
4. Sawkney, B. L. et al *J. Environ. Qual.* 1984, **13**, 349-352
5. Hansch, C. et al *Medchem Project Issue No. 26* 1985, Claremont, CA, Pomona College
6. Konemann, W. H. *Quantiative structure-activity relationships for kinetics and toxicity of aquatic pollutants and their mixtures in fish* 1979, Univ. Utrecht, Netherlands
7. Kaiser, K. L. E. et al *Water Pollut. Res. J. Can.* 1991, **26**(3), 361-431
8. Bringman, G. et al *Gwf-Wasser-Abwasser* 1976, **117**(9), (Ger.)
9. Bringmann, G. et al *Water Res.* 1980, **14**, 231-241
10. Heukelekian, H. et al *J. Water Pollut. Control Assoc.* 1955, **29**, 1040-1053
11. Bridie, A. L. et al *Water Res.* 1979, **8**, 391-414
12. Pitter, P. *Water Res.* 1976, **10**, 231-235
13. Gellman, I. *Studies on Biochemical Oxidation of Sewage, Industrial Waters and Organic Compounds* 1952, Ph.D. Thesis, Rutgers Univ
14. Price, K. S. et al *J. Water Pollut. Control Fed.* 1974, **46**
15. Atkinson, R. et al *Adv. Photochem* 1979, **11** 375-488
16. *J. Lab. Clin. Med.* 1944, **29**, 561
17. *Gig. Sanit.* 1978, **43**(1), 8
18. *Environ. Health Perspect.* 1985, **61**, 321
19. Reynolds, J. E. F. (Ed.) *Martindale. The Extra Pharmacopoeia* 30th ed., 1993, The Pharmaceutical Press, London
20. Wills, J. et al *Toxicol. Appl. Pharmacol.* 1969, **15**(3), 560
21. Dossing, M. et al *L. Br. J. Ind. Med.* 1984, **41**(1), 142-144
22. Weil, C. S. et al *Arch. Ind. Hyg. Occup. Med.* 1952, **5**, 535-547
23. *IARC Monograph* 1987, **Suppl. 7**, 229
24. *Directory of Agents being Tested for Carcinogenicity* 1992, **No. 15**, 224
25. Nelson, B. K. et al *Toxicol. Ind. Health* 1990, **6**(3/4), 373-387
26. Nelson, B. K. et al *Food. Chem. Toxicol.* 1988, **26**(3), 247-254
27. Nordman, R. *Colloq. Inst. Natl. Sante Rech. Med.* 1980, **95** (Alcohol Tractus Dig.), 187-205
28. *NTIS Report No. AD-A106-994* Natl. Tech. Inf. Ser., Springfield, VA
29. *Toxicol. App. Pharmacol.* 1980, **55**, 501
30. Browning, E. *Toxicity and Metabolism of Industrial Solvents* 1965, 335-341, Elsevier, New York
31. Jensen, O. *Contact Dermititis* 1981, **7**(3), 148-150
32. Shimizu, H. et al *Sangyo Igaku* 1985, **27**(6), 400-419
33. von der Hude, W. *Environ. Mutagen.* 1987, **9**, 401-410
34. Schick, J. B. et al *Pediatrics* 1981, **68**, 587-8
35. Deng, J. F. et al *Am. J. Ind. Med.* 1987, **12**(1), 11-19
36. Adelson, L. *Am. J. Clin. Path.* 1962, **38**, 144-151

37. Henning, H. *Solvent Safety Sheets: A compendium for the working chemist* 1993, Royal Society of Chemistry, London
38. *S. I. 1991 No. 472 The Environmental Protection (Prescribed Processes and Substances) Regulations* 1991, HMSO, London
39. *ECETOC Technical Report No. 30(4)* 1991, European Chemical Industry Ecology and Toxicology Centre, B-1160 Brussels
40. Henschler, D. *Dtsch. Mediz. Wehschr.* 1984, **109**(19), 763-764 (HSE Translation No. Glaxo T2539)
41. Hueper, W. C. *Occupational and Environmental Cancers of the Respiratory System* 1968, Springer, New York

# I116    Isopropyl butyrate

$$CH_3CH_2CH_2CO_2CH(CH_3)_2$$

**CAS Registry No.** 638-11-9
**Synonyms** butanoic acid,1-methylethyl ester; butyric acid, isopropylester; isopropyl butanoate; isopropyl butyrate
**Mol. Formula** $C_7H_{14}O_2$            **Mol. Wt.** 130.19

## Physical properties

**M. Pt.** –95°C; **B. Pt.** 130-131°C; **Flash point** 30°C; **Specific gravity** $d_4^{15}$ 0.859.

## Solubility
Organic solvent: miscible with ethanol, diethyl ether

## Occupational exposure

**UN No.** 2405; **HAZCHEM Code** 3☒; **Conveyance classification** flammable liquid.

## Mammalian and avian toxicity

**Acute data**
$LD_{50}$ oral rat 15000 mg $kg^{-1}$ (1).

## References

1.  Jenner, P. M. et al *Food Cosmet. Toxicol.* 1964, **2**, 327

# I117   Isopropyl chloroformate

$$ClCO_2CH(CH_3)_2$$

**CAS Registry No.** 108-23-6

**Synonyms** carbonochloridic acid,1-methylethyl ester; formic acid, chloroisopropyl ester; chloroformic acid isopropyl ester; isopropyl chlorocarbonate; isopropyl chloroformate

**Mol. Formula** $ClC_4H_7O_2$                                      **Mol. Wt.** 122.55

## Physical properties

**B. Pt.** 105°C; **Flash point** 20°C (closed cup); **Specific gravity** $d_4^{20}$ 1.078; **Volatility** 20°C, v. den. 4.2.

## Solubility

Organic solvent: ethyl ether, acetone, chloroform

## Occupational exposure

**UK Long-term limit** 1 ppm (5 mg m$^{-3}$); **UN No.** 2407; **Conveyance classification** flammable liquid, corrosive substance.

## Mammalian and avian toxicity

### Acute data

$LD_{50}$ oral rat 1070 mg kg$^{-1}$ (1).
$LD_{50}$ oral mouse 178 mg kg$^{-1}$ (2).
$LD_{50}$ dermal mouse 12 mg kg$^{-1}$ (2).
$LD_{50}$ dermal rabbit 11300 mg kg$^{-1}$ (1).

### Irritancy

500 mg (duration unspecified) instilled into rabbit eye caused severe irritation (1).

## Any other adverse effects

Limited under EC Directive on Drinking Water Quality 80/778/EEC. Organochlorine compounds: guide level 1 μg l$^{-1}$. Haloform concentrations must be as low as possible (3).

Included in Schedule 4 (Release Into Air: Prescribed Substances) Statutory Instrument No. 472 1991 (4).

## References

1. *Ind. Hyg. Found. Am., Chem. Toxicol. Series, Bull.* 1967, **6**, 1, Pittsburgh, PA
2. Eckroth, D. & Grayson, M. (Eds.) *Kirk-Othmer Encyclopedia of Chemical Technology* 3rd ed., 1978, Wiley, New York
3. *EC Directive Relating to the Quality of Water Intended for Human Consumption* 1982, 80/778/EEC, Office for Official Publications of the European Communities, 2 rue Mercier, L-2985 Luxembourg
4. *S. I. 1991 No. 472 The Environmental Protection (Prescribed Processes and Substances) Regulations* 1991, HMSO, London

# I118   Isopropylcyclohexane

CH(CH₃)₂

CAS Registry No. 696-29-7
Synonyms cyclohexane, 1-methylethyl-; cyclohexane, isopropyl-
Mol. Formula $C_9H_{18}$        Mol. Wt. 126.24

### Physical properties
B. Pt. 155°C; Flash point 35°C; Specific gravity $d^{20}$ 0.802.

### Mammalian and avian toxicity

#### Metabolism and pharmacokinetics
Oral gavage rats, (concentration unspecified). The rats experienced moderate kidney
damage similar to that produced by acyclic, branched chain hydrocarbons. The
urinary metabolites identified included: *cis*-4-isopropylcyclohexanol;
*trans*-4-isopropylcyclohexanol; 2-cyclohexylpropanoic acid;
2 cyclohexyl-1,3-propanediol; 2-*trans*-hydroxy-4- *trans*-isopropylcyclohexanol;
2-*cis*-hydroxy-4- *cis*-isopropylcyclohexanol; and
2-*cis*-hydroxy-4-*trans*-isopropylcyclohexanol (1).

### References
1.   Henningsen, G. M. *J. Toxicol. Environ. Health* 1988, **24**(1), 19-25

# I119   Isopropyl formate

HCO₂CH(CH₃)₂

CAS Registry No. 625-55-8
Synonyms formic acid, 1-methylethyl ester; formic acid, isopropyl ester
Mol. Formula $C_4H_8O_2$        Mol. Wt. 88.11

### Physical properties
B. Pt. 68.3°C; Flash point 12.2°C (closed cup); Specific gravity $d^{20}$ 0.873;
Volatility v. den. 3.03.

### Occupational exposure
UN No. 1281; HAZCHEM Code 3▣E; Conveyance classification flammable
liquid; Supply classification highly flammable.
Risk phrases Highly flammable (R11)
Safety phrases Keep container in a well ventilated place – Keep away from sources
of ignition – No Smoking – Take precautionary measures against static discharges
(S9, S16, S33)

## Mammalian and avian toxicity

### Acute data
$LD_{50}$ oral guinea pig 1400 µg kg$^{-1}$ (1).

### Any other comments
Experimental toxicology, human health effects and physico-chemical properties reviewed (2).

### References

1.  Frear, E. H. (Ed.) *Pesticide Index* 1969, **4**, 256, College Science Publication, State College, PA
2.  *ECETOC Technical Report No. 30(4)* 1991, European Chemical Industry Ecology and Toxicology Centre, B-1160 Brussels

# I120   4,4′-Isopropylidenediphenol

**CAS Registry No.** 80-05-7
**Synonyms** 4,4′-(1-methylethylidene)bisphenol; bisphenol A; Dian; *p,p′*-dihydroxydiphenylpropane; diphenylolpropane; Ipognox 88; Rickabanol
**Mol. Formula** $C_{15}H_{16}O_2$                    **Mol. Wt.** 228.29
**Uses** Fungicide. Manufacture of epoxy resins and polycarbonates.

## Physical properties
**M. Pt.** 150-155°C; **B. Pt.** $_4$ 220°C; **Partition coefficient** log $P_{ow}$ 3.32.

### Solubility
Water: 120 mg l$^{-1}$. Organic solvent: ethanol, acetone

## Ecotoxicity

### Fish toxicity
$LC_{50}$ (96 hr) fathead minnow 4.7 mg l$^{-1}$ static test and 4.6 mg l$^{-1}$ flow through test (1).
$LC_{50}$ (96 hr) Atlantic silverside 9.4 mg l$^{-1}$ (1).

### Invertebrate toxicity
$EC_{50}$ (96 hr) *Selenastrum capricornutum* 2.7 mg l$^{-1}$ (1).
$EC_{50}$ (96 hr) *Skeletonema costatum* 1.0 mg l$^{-1}$ (1).
$LC_{50}$ (96 hr) *Mysidopsis bahia* 1.1 mg l$^{-1}$ (1).
$EC_{50}$ (48 hr) *Daphnia magna* 10 mg l$^{-1}$ (1).

### Bioaccumulation
Bioconcentration factors of 42 and 196 were estimated, based on the water solubility and log $K_{ow}$, respectively (2).

## Environmental fate

### Nitrification inhibition
At 50 mg $l^{-1}$ does not inhibit nitrification (activated sludge) (3).

### Degradation studies
Acclimated activated sludge innocula, 72% COD removal in 24 hr, initial concentration 105 mg $l^{-1}$ (4).
Acclimated activated sludge, 72% COD and 57% TOC removal in 24 hr, initial concentration 58 mg $l^{-1}$ (5).

### Abiotic removal
The $t_{1/2}$ for 4,4'-isopropylidenediphenol vapour reacting with photochemically generated hydroxyl radicals is estimated as 4 hr (6).
It is not expected to undergo chemical hydrolysis under environmental conditions, since it contains no hydrolyzable functional groups (2).
Photodecomposition products of the vapour are phenol; 4-isopropylphenol; and a semiquinone derivative (7).

### Absorption
Estimated soil adsorption coefficients of 314 and 1524 indicate moderate to low mobility in soil and moderate to extensive mobility in suspended solids (2,8).

## Mammalian and avian toxicity

### Acute data
$LD_{50}$ oral rat 3250 mg $kg^{-1}$ (9).
$LD_{50}$ oral mouse 2500 mg $kg^{-1}$ (9).
$LD_{50}$ oral rabbit 2230 mg $kg^{-1}$ (9).
$LC_{50}$ inhalation rat 200 ppm (duration unspecified) (10).
$LD_{50}$ dermal rabbit 3000 mg $kg^{-1}$ (11).

### Sub-acute data
Oral ♂ or ♀ mice (14 day) 0.0, 0.31, 0.62, 1.25, 2.5 and 5.0%. No clinical signs of toxicity were observed in either ♂ or ♀ (12).

### Carcinogenicity and long-term effects
Oral feed ♂ or ♀ rats and mice (103 wk) 2000-10,000 ppm. ♂ rat equivocal, ♀ rat, ♂ and ♀ mice negative. Tumour site was in the hematopoietic system (13-15).

### Teratogenicity and reproductive effects
CD rats (6-15 day gestation) 0, 160, 320 or 640 mg $kg^{-1}$ $day^{-1}$, CD-1 mice (6-15 day gestation) 0, 500, 750, 1000 or 1250 mg $kg^{-1}$ $day^{-1}$. In rats gravid uterine weight and average foetal body weight $litter^{-1}$ was not affected, no increase in percentage resorptions $litter^{-1}$ or percentage foetuses malformed $litter^{-1}$. In mice, there was a reduction in gravid uterine weight and average foetal body weight and increase in the percentage of resorptions per litter with 1250 mg $kg^{-1}$ $day^{-1}$ dose. No alteration in the incidence of malformations were observed in mice. (16).

### Irritancy
Dermal (duration unspecified) rabbit 250 mg, exposed to atmosphere, caused mild irritation and 20 mg instilled into rabbit eye (24 hr) caused severe irritation (17,18).

### Sensitisation
Reported induction of contact and allergic dermatitis and sensitisation in humans (19).

Dermal mice, photoallergic contact dermatitis was induced; sites of photochallenge reactions flared when tested with UV-A. Attempt to induce photoallergy in guinea pigs failed (20).

## Genotoxicity

*Salmonella typhimirium* TA98, TA100, TA1535, TA97/TA1537 with and without metabolic activation negative (14,21).

*In vitro* mouse lymphoma L5178Y cells with and without metabolic activation negative (22).

*In vitro* Chinese hamster ovary cells, sister chromatid exchange with and without metabolic activation, negative (23).

*In vitro* Chinese hamster ovary cells, chromosomal aberration assays, with or without metabolic activation, negative (23).

*In vivo* human leukocytes chromosomal aberrations positive in occupationally exposed workers (24).

## Legislation

Limited under EC Directive on Drinking Water Quality 80/778/EEC. Phenol compounds: maximum admissible concentration 0.5 $C_6H_5OH$ mg l$^{-1}$, excluding natural phenols which do not react to chlorine (25).

Included in Schedule 6 (Release into Land: Prescribed Substances) Statutory Instrument No. 472, 1991 (26).

The log $P_{ow}$ value exceeds the European Community recommended level 3.0 (6th and 7th amendments) (27).

## Any other comments

Experimental toxicology, human health effects and exposure reviewed (28). Toxicity reviewed (29).

## References

1. Alexander, H. C. et al *Environ. Toxicol. Chem.* 1988, **7**(1), 19-26
2. Lyman, W. K. et al *Handbook of Chemical Property Estimation Methods. Environmental Behaviour of Organic Compounds* 1982, McGraw-Hill, New York
3. Wood, L. B. et al, *Water Res.* 1981, **15**, 543-551
4. Matsui, S. et al *Prog. Water Technol.* 1975, **7**, 645-659
5. Matsui, S. et al *Water Sci. Technol.* 1988, **20**(10), 201-210
6. *GEMS: Graphical Exposure Modeling System. Fate of Atmospheric pollutants* 1987, Washington, DC.
7. Peltonen, K. et al *Photochem. Photobiol.* 1986, **43**, 481-484
8. Swann, R. L. et al *Residue Rev.* 1983, **85**, 17-28
9. *Am. Ind. Hyg. Assoc. J.* 1967, **28** 301
10. Marhold, J. V. *Prehled Prumyslove Toxikologie: Organicke Latky* 1986, 238, Prague (Czech.)
11. *Arch. Ind. Hyg. Occup. Med.* 1951, **119**, 51
12. Reel, J. R. et al *Bisphenol A: reproduction and fertility assessment in CD-1 mice when administered in the feed* Report, 1985, RTI-149, NTP-85-192
13. *National Toxicology Program Research and Testing Division* 1992, Report No. TR-215, NIEHS, Research Triangle Park, NC 27709
14. Ashby, J. et al *Mutat. Res.* 1988, **204**(1), 17-115
15. Hansch, C. et al *Medchem Project Issue No. 26* 1985, Pomona College, Claremont, CA
16. Morrisey, R. E. et al *Fundam. Appl. Toxicol.* 1987, **8**(4), 571-582
17. *Union Carbide Data Sheet* 1965, **7**, 14

18. Marhold, J. V. *Sbornik Vysledku Toxixologickeho Vysetreni Latek A Pripravku* 1972, 58 Prague (Czech.)
19. Yardley-Jones, A. et al *Br. J. Ind. Med.* 1988, **45**(10), 694-700
20. Maguire, H. C. *Acta Derm. Venereol.* 1988, **68**(5), 408-412
21. Ashby, J. et al *Mutat. Res.* 1991, **257**(3), 229-306
22. Myhr, B. C. et al *Environ. Mol. Mutat.* 1991, **18**, 51-83
23. Ivett, J. L. et al *Environ. Mol. Mutagen.* 1989, **14**, 165-187
24. Chernykh, L. V. et al *Gig. Tr. Prof. Zabol.* 1990, **3**, 51-52
25. *EC Directive Relating to the Quality of Water Intended for Human Consumption* 1982, 80/778/EEC, Office for Official Publications of the European Communities, 2 rue Mercier, L-2985 Luxembourg
26. *S. I. 1991 No. 472 The Environmental Protection (Prescribed Processes and Substances) Regulations* 1991, HMSO, London
27. *1967 Directive on Classification, Packaging, and Labelling of Dangerous Substances 67/548/EEC; 6th Amendment EEC Directive 79/831/EEC; 7th Amendment EEC Directive 91/32/EEC* 1991, HMSO, London.
28. *ECETOC Technical Report No. 30(4)* 1991, European Chemical Industry Ecology and Toxicology Centre, B-1160 Brussels
29. *BIBRA Toxicity Profile* 1991, British Biological Industrial Research Association, Carshalton

# I121    Isopropylidene glycerol

**CAS Registry No.** 100-79-8
**Synonyms** 1,3-dioxolane-4-methanol, 2,2-dimethyl-; acetone glycerin ketal; Dioxolan; glycerol acetonide; glycerol dimethylketal; 1,2-isopropylideneglycerol; 2,3-isopropylideneglycerol; acetone ketal of glycerine
**Mol. Formula** $C_6H_{12}O_3$                     **Mol. Wt.** 132.16
**Uses** Versatile solvent and plasticizer. Pharmaceutical aid.

## Physical properties

**M. Pt.** −26.4°C; **B. Pt.** $_{10}$ 82°C; **Flash point** 80°C; **Specific gravity** $d^{20}$ 1.065; **Volatility** v. den. 2.6.

## Mammalian and avian toxicity

### Acute data
$LD_{50}$ oral rat 7 g $kg^{-1}$ (1).
$LD_{50}$ interperitoneal rat 3 g $kg^{-1}$ (1).

## Legislation

Included in Schedule 6 (Release into Air: Prescribed Substance) Statutory Instrument No. 472, 1991 (2).

## References

1. *Merck Index* 11th ed., 1989, Merck & Co. Inc., Rahway, NJ
2. *S. I. 1991 No. 472 The Environmental Protection (Prescribed Processes and Substances) Regulations* 1991, HMSO, London

# I122   1-Isopropyl-4-methylcyclohexane

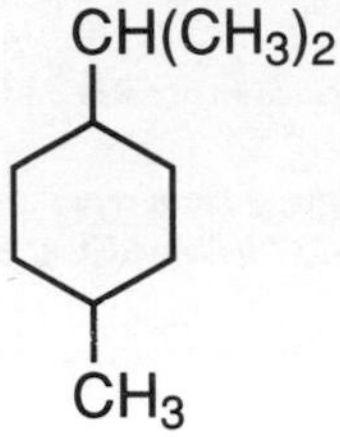

CH(CH$_3$)$_2$

CH$_3$

**CAS Registry No.** 1678-82-6
**Synonyms** *trans*-cyclohexane, 1-methyl-4-(1-methylethyl)-; *trans-p*-menthane
**Mol. Formula** C$_{10}$H$_{20}$                    **Mol. Wt.** 140.27

## Mammalian and avian toxicity

### Irritancy
At 2% and 5% in ethanol recovery from oedema and erythema occurred within 48 hr in rats and rabbits (1).

## References

1. Okabe, H. et al *Drug Des. Del.* 1990, **6**, 229-238

# I123   2-Isopropylnaphthalene

CH(CH$_3$)$_2$

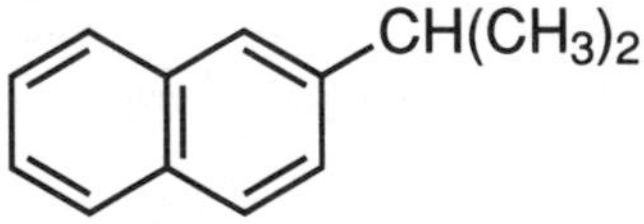

**CAS Registry No.** 2027-17-0
**Synonyms** 2-(1-methylethyl)naphthalene; β-isopropylnaphthalene
**Mol. Formula** C$_{13}$H$_{14}$                    **Mol. Wt.** 170.26

## Physical properties
**M. Pt.** 10.5-11°C; **B. Pt.** 268°C; **Specific gravity** d$^{20}$ 0.9762;
**Volatility** $5.18 \times 10^{-3}$ mmHg at 25°C.

### Solubility
Water: 0.9 mg l$^{-1}$.

**Abiotic removal**

Photodegradibility in aquatic systems studied, $t_{1/2}$ 22.3 hr in distilled water. Photolysis rate (4hr) 93.7% in 14 g l$^{-1}$ NaCl solution (1).

**Mammalian and avian toxicity**

**Acute data**

Intraperitoneal (24 hr) mice 3000 mg kg$^{-1}$ no pulmonary damage. No affect to the lipid peroxidation or phospholipid contents in the lung. Maximum levels detected in the lung, liver and kidney 6 hr after administration (2,3).

**References**

1. Fukuda, K. *Chemosphere* 1988, **17**(4), 651-659
2. Honda, T. et al *Chem. Pharm. Bull.* 1990, **30**(11), 3130-3135
3. Honda, T. et al *Eisei Kagaku* 1990, **37**(4), 300-306 (Japan.) (*Chem. Abstr.* **115**, 225651t)

# I124   Isopropyl nitrate

### CH3CH(NO3)CH3

**CAS Registry No.** 1712-64-7

**Synonyms** nitric acid, 1-methylethyl ester; isopropyl nitrate; propane-2-nitrate

**Mol. Formula** $C_3H_7NO_3$     **Mol. Wt.** 105.09

**Uses** Fuel ignition promoter in rocket fuel. Organic intermediate.

**Physical properties**

**B. Pt.** 102°C; **Flash point** 12°C (closed cup); **Specific gravity** $d^{20}$ 1.036.

**Solubility**

Organic solvent: ethanol, diethyl ether

**Occupational exposure**

**UN No.** 1222; **Conveyance classification** flammable liquid.

**Mammalian and avian toxicity**

**Acute data**

LC$_{50}$ (2 hr) inhalation mouse 65 g m$^{-3}$ (1).

**Metabolism and pharmacokinetics**

4 human volunteers were exposed to 45.8 mg m$^{-3}$ for 60 min. Expired air, blood and urine of the volunteers contained isopropyl nitrate. Respiratory $t_{1/2}$ 98 min. 0.02% of the estimated intake was excreted in the urine within 6 hr post exposure (2).

**Legislation**

Included in Schedule 6 (Release into Air: Prescribed Substances) Statutory Instrument No. 472, 1991 (3).

**References**

1. Izmerov, N. F. et al *Toxicometric Parameters of Industrial Toxic Chemicals under Single Exposure* 1982, 78, CIP, Moscow
2. Ahonen, I. *Toxicol. Lett.* 1989, **47**(2), 205-211

3. *S. I. 1991 No. 472 The Environmental Protection (Prescribed Processes and Substances) Regulations* 1991, HMSO, London

# I125   *p*-Isopropylphenol

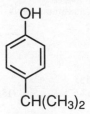

**CAS Registry No.** 99-89-8
**Synonyms** 4-(1-methylethyl)phenol; Australol; *p*-cumenol; 4-isopropylphenol;
**Mol. Formula** $C_9H_{12}O$                     **Mol. Wt.** 136.20

## Physical properties
**M. Pt.** 59-61°C; **B. Pt.** 212-212.5°C; **Specific gravity** $d^{20}$ 0.990.

### Solubility
Organic solvent: ethanol, diethyl ether at 25°C

## Ecotoxicity

### Fish toxicity
Steelhead trout 3 mg $l^{-1}$ caused death within 4 hr (1).

### Invertebrate toxicity
$EC_{50}$ (5 min) *Photobacterium phosphoreum* 0.25 mg $l^{-1}$, Microtox test (2).

### Bioaccumulation
Accumulates in *Acinetobacter calcoacetius anitratus* grown on media containing 26-105 µg $ml^{-1}$ *p*-isopropylphenol. There was a direct relationship between bioaccumulation and incubation time (3).

### Abiotic removal
Photooxidation $t_{1/2}$ of alkyl phenols in middle latitude shallow surface waters ranged from 1 day to several months (4).

## Mammalian and avian toxicity

### Acute data
$LD_{50}$ oral mouse 875 mg $kg^{-1}$ (5).
$LD_{50}$ intravenous mouse 40 mg $kg^{-1}$ (6).
$LD_{Lo}$ intraperitoneal mouse 250 mg $kg^{-1}$ (7).

## Any other comments
A potential method for treating chemical pollution with phenoloxidases has been tried on *p*-isopropylphenol. Phenoloxidase activity was differently inhibited by the presence of clays and clay-humus complexes; pH and temperature also influenced

activity (8).
Air pollution emission factors reviewed (9).

## References

1. MacPhee, C. et al *Lethal Effects of 2014 Chemicals Upon Sockeye Salmon, Steelhead Trout and Three-Spine Stickleback* 1989, 77, EPA 560/6-89-001, PB890156715, Washington, DC
2. Heil, T. P. et al *J. Environ. Sci. Health, Part B* 1989, **B24**(4), 349-360
3. Zaim, K. K. *Doga: Turk Mehendislik Cevre Bilimleri Derg* 1991, **15**(1), 39-53
4. Faust, B. C. et al *Environ. Sci. Technol.* 1987, **21**(10), 957-964
5. *Hyg. Sanit. (USSR)* 1980, **46**(1), 94
6. *J. Med. Chem.* 1980, **23**, 1350
7. *Summary of Tables of Biological Tests* 1953, **5**, 339, National Research Council Chemical-Biological Coordination Center, Washington, DC
8. Claus, H. et al *Water Sci. Technol.* 1990, **22**(26), 69-77
9. Pope, A. A. et al *U.S. Environ. Prot. Agency Off. Air. Qual. Plann. Strand. [Tech. Rep.]* 1988, EPA-450/2-88-006a, EPA, Research Triangle Park, NC

# I126   Isoprothiolane

**CAS Registry No.** 50512-35-1
**Synonyms** 1,3-dithiolan-2-ylidenepropanedioic acid, bis(1-methylethyl) ester; Fuji-one
**Mol. Formula** $C_{12}H_{18}O_4S_2$                **Mol. Wt.** 290.40
**Uses** Fungicide. Control of rice blast (*Pyricularia oryzae*), rice stem rot and *Fusarium* leaf spot on rice. Also reduces insect population on rice.
**Occurrence** Residues have been detected in river waters and sediments (1).

## Physical properties

**M. Pt.** 54-54.5°C; **B. Pt.** $_{0.5}$ 167-169°C; **Volatility** v.p. $1.4 \times 10^{-4}$ mmHg at 25°C.

### Solubility

Water: 48 mg $l^{-1}$ at 20°C. Organic solvent: methanol, dimethyl sulfoxide, acetone, chloroform, benzene, xylene, hexane

## Ecotoxicity

### Fish toxicity

$LC_{50}$ (48 hr) carp 6.7 mg $l^{-1}$ (2).
Medaka fry (HO5 strain) sensitive to isoprothiolane toxicity (3).
*In vitro* GF-scale cells (derived from goldfish scales) midpoint cytotoxicity value 253 mg $kg^{-1}$ (4).

### Invertebrate toxicity

$EC_{50}$ *Chlamydomonas reinhardti* growth 3.4 mg $l^{-1}$ (5).

## Mammalian and avian toxicity

**Acute data**

$LD_{50}$ oral ♂ rat, ♂ mouse 1190-1340 mg $kg^{-1}$ (2).
$LD_{50}$ dermal rat >10.25 g $kg^{-1}$ (2).

## Genotoxicity

Not toxic and not mutagenic in *Salmonella typhimurium* test strains, metabolic activation not specified (6).
*Salmonella typhimurium* TA98, TA100, TA102, TA1535, TA1538 with and without metabolic activation negative, *Saccharomyces cerevisiae* without metabolic activation negative (7).
*In vitro* Syrian hamster embryo morphological transformation negative. High concentrations were cytotoxic (8).
*In vitro* Chinese hamster ovary cells sister chromatid exchanges and chromosomal aberrations equivocal (9).

## Legislation

Included in Schedule 6 (Release into Land: Prescribed Substances) Statutory Instrument No. 472, 1991 (10).
Limited under EC Directive on Drinking Water Quality 80/778/EEC. Pesticides: maximum admissible concentration 0.1 μg $l^{-1}$ (11).

## Any other comments

WHO Class III; EPA Toxicity Class III (2).

## References

1. Iwakuma, T. et al *Kokuritsu Kogai Kenkyysha Kenkyu Hokoku* 1988, **114**, 73-83 (Japan.) (*Chem. Abstr.* **109**, 27350v)
2. *The Agrochemicals Handbook* 3rd ed., 1991, RSC, London
3. Miyashita, M. et al *Kokuritsu Kogai Kenkyusho Kenkyu Hokoku* 1988, **114**, 137-145 (Japan) (*Chem. Abstr.* **109**, 33309e)
4. Saito, H. et al *Chemosphere* 1991, **23**(4), 525-527
5. Lee, E. K. *Sikmul Hakhoe Chi* 1982, **25**, 113 (Korea.), (*Chem. Abstr.* **109**, 157691x)
6. Rusina, O. Y. et al *Gig. Sanit.* 1990, (3), 59-60 (Russ.)
7. Choi, E. J. et al *Environ. Mutagen. Carcinog.* 1985, **5**(1), 11-18
8. Zhany, Z. et al *Huaxi Yike Daxue Xuebao* 1989, **20**(1), 96-98 (Ch.) (*Chem. Abstr.* **110** 226951n
9. Wang, T. C. et al *Bull. Inst. Zool., Acad. Sin.* 1987, **26**(4), 317-329
10. *S. I. 1991 No. 472 The Environmental Protection (Prescribed Processes and Substances) Regulations* 1991, HMSO, London
11. *EC Directive Relating to the Quality of Water Intended for Human Consumption* 1982, 80/778/EEC, Office for Official Publications of the European Communities, 2 rue Mercier, L-2985 Luxembourg

# I127   Isoproturon

$$(H_3C)CH-\!\!\!\bigcirc\!\!\!-NH\overset{\overset{O}{\|}}{C}-N(CH_3)_2$$

**CAS Registry No.** 34123-59-6
**Synonyms** *N,N*-dimethyl-*N'*-[4-(1- methylethyl)phenyl]urea; Arelon; Graminon;
Ipuron; Tolkan
**Mol. Formula** $C_{12}H_{18}N_2O$                    **Mol. Wt.** 206.29
**Uses** Herbicide, control of annual grasses and broad-leaved weeds.

## Physical properties

**M. Pt.** 155-156°C; **Specific gravity** $d^{20}$ 1.16; **Partition coefficient** log $P_{ow}$ 2.25 (1).;
**Volatility** v.p. $2.33 \times 10^{-8}$ mmHg at 20°C.

## Solubility
Water: 72 mg $l^{-1}$ at 20°C. Organic solvent: benzene, dichloromethane, hexane,
methanol

## Occupational exposure

**Supply classification** harmful.
**Risk phrases** Harmful if swallowed – Possible risk of irreversible effects (R22, R40)
**Safety phrases** Wear suitable protective clothing and gloves (S36/37)

## Ecotoxicity

**Fish toxicity**
$LC_{50}$ (96 hr) guppy, goldfish, bluegill sunfish, rainbow trout 90-240 mg $l^{-1}$ (1).

## Environmental fate

### Nitrification inhibition
Effect of 0.0001-0.1% isoproturon on the nitrification of ammonium nitrate.
0.0001-0.001% had no inhibitory effect; nitrification was 85-90% in the first 16 day
and the process was completed by 16-24 days. 0.01% depressed nitrification; 27-38%
in 32 days, completed over 32-64 days; 0.1% only 36% nitrification was observed by
64 days (2).
Nitrification by 2 *Azotobacter* strains was not influenced by 15 and 300 ppm
concentrations (3).

### Degradation studies
Undergoes enzymic and microbial demethylation at the nitrogen and hydrolysis of the
phenylurea to 4-isopropylaniline. $t_{1/2}$ 12-29 days in soil (1).
Laboratory incubation with soil from a field that had recieved 4 doses over a
12 month period. Degradation rates of isoproturon were the same in pretreated and
control soil samples (4).
*Arthrobacter* sp., *Bacillus* sp. and *Pseudomonas* sp. were unable to utilize isoproturon
as carbon or nitrogen sources. Had no significant effect on bacterial, actinomycetes
and fungal populations (5).

Aerobic soil bacteria, *Pseudomonas convexa* and *Bacillus* sp. and nitrogen fixers *Rhizobium phaseoli* and *Azotobacter chroococcum* exposed 0-50 µg ml$^{-1}$ in shake cultures did not show any adverse effect on growth, higher concentrations were inhibitory to growth. 10- 20 µg ml$^{-1}$ caused stimulation of growth in *Rhizobium phaseoli* and *Azotobacter chroococcum* (6).

In both sandy and clay soil exposed to isoproturon the metabolite 4-(2-hydroxyisopropylphenyl) urea was identified (7).

**Absorption**

Degradation of isoproturon in sand clay and clay loam soils demonstrated an apparent increase in the strength of adsorption with time (8).

Studied on homoionic clays at 10°C and 26°C containing $Cu^{2+}$, $Ca^{2+}$ and $K^+$ cations. Cationic adsorption pattern was $Cu^{2+}$, $Ca^{2+}$ and $K^+$ in decreasing order in all 3 clays. Adsorption increased at lower temperatures (9).

## Mammalian and avian toxicity

### Acute data

$LD_{50}$ oral mouse, rat 3350, 3600 mg kg$^{-1}$, respectively (10).
$LD_{50}$ oral Japanese quail, pigeon >3000, >5000 mg kg$^{-1}$, respectively (1).
$LC_{50}$ (4 hr) inhalation ♀ rat >0.67 mg 1 air$^{-1}$ (1).
$LD_{50}$ intraperitoneal sheep (death after 48-56 hr) 3.6 g kg$^{-1}$ heart rate increased above controls and body temperature and respiration rate decreased below controls 1-24 hr after administration (11).

### Sub-acute data

Oral (90 day) dog, rat no effect level 50, 400 mg kg$^{-1}$, respectively (1).
Dermal (21 days) ♂ ♀ rats repeated application (unspecified dose) of technical grade and powder formulation caused mild to moderate toxic effects. Technical grade was more toxic to ♂ rats as evidenced by animal mortality and haematological enzymic changes (12).

### Metabolism and pharmacokinetics

Oral rat 50% is eliminated within 8 hr, predominantly in urine (1).

## Genotoxicity

*In vivo* mammalian chromosomal aberration, micronucleus and sperm-shape abnormality assays. Dose dependent mutagenic effect in chromosomal aberrations and sperm-shape abnormality tests, micronucleus assay only significant effects seen at highest dose (200 mg kg$^{-1}$) (species unspecified) (13).

## Legislation

Limited under EC Directive on Drinking Water Quality 80/778/EEC.Pesticides: maximum admissible concentration 0.1 µg l$^{-1}$ (14).

## Any other comments

Human health effects, experimental toxicology, physico-chemical properties reviewed (15).
0.125 kg ha$^{-1}$ sprayed onto black grass at the 2-3 leaf stage resulted in a decrease in the mean fresh weight of plants (16).
Biodegradation reviewed (17).
Stable to light, alkali and acid.

# References

1. *The Agrochemicals Handbook* 3rd ed., 1991, RSC, London
2. Haleemi, M. A. et al *Pak. J. Sci. Ind. Res.* 1988, **31**(11), 798-800
3. Gudkari, D. *Zentralbl. Mikrobiol.* 1987, **142**(4), 283-291
4. Walker, A. et al *Weed Res.* 1991, **31**(1), 49-57
5. Kapur, R. et al *Asian Environ.* 1986, **8**(4), 24-26
6. Gupta, R. et al *J. Res. (Punjab Agric. Univ.)* 1987, **24**(3), 464-468
7. Berger, B. et al *Fresenius' Z. Anal. Chem.* 1989, **334**(4), 360-362 (*Chem. Abstr.* **111**, 110898s)
8. Blair, A. M. et al *Crop Prot.* 1990, **9**(4), 289-294
9. Kumar, Y. et al *J. Indian Soc. Soil. Sci.* 1987, **35**(3), 394-399
10. Thizy, A. et al *C. R. Ser. D* 1972, **274**, 2053
11. Mikhailov, G. et al *Vet.-Med. Nauki* 1987, **24**(2), 30-35 (Bulg.) (*Chem. Abstr.* **108**, 162978a)
12. Dikshith, T. S. et al *Vet. Hum. Toxicol.* 1990, **32**(5), 432-434
13. Behera, B. C. et al *Indian J. Exp. Biol.* 1990, **28**(9), 826-827
14. *EC Directive Relating to the Quality of Water Intended for Human Consumption* 1982, 80/778/EEC, Office for Official Publications of the European Communities, 2 rue Mercier, L-2985 Luxembourg
15. *ECETOC Technical Report No. 30(4)* 1991, European Chemical Industry Ecology and Toxicology Centre, B-1160 Brussels
16. Blair, A. M. *Weed Res.* 1978, **18**(6), 381-387
17. Frimmel, F. H. *Vom Wasser* 1989, **73**, 273-286 (Ger.) (*Chem. Abstr.* **112**, 104455z)

# I128   Isopulegol

**CAS Registry No.** 89-79-2
**Synonyms** 5-methyl-2-(1-methylethenyl)cyclohexanol, [1*R*- (1α,2 β,5α)]-, (1*R*,3*R*,4*S*)-(–)-*p*-menth-8-en-3-ol
**Mol. Formula** $C_{10}H_{18}O$                                **Mol. Wt.** 154.25

## Physical properties
**B. Pt.** 268°C; **Flash point** 78°C; **Specific gravity** $d_4^{20}$ 0.911.

## Solubility
Organic solvent: ethanol, diethyl ether

## Ecotoxicity

### Fish toxicity
5 ppm (24 hr) caused no toxic effects to trout, bluegill sunfish, yellow perch and goldfish (1).

Stickleback 10 mg l$^{-1}$ caused the death of ½ of the fish within 1 hr. The remaining fish remained alive for the duration of the experiment (48 hr) (2).

## Environmental fate

### Degradation studies

Bacterium (species unspecified) from sewage isolated by enrichment on (–)-methanol will utilize (–)-isopulegol as a sole carbon source, but not (+)-isopulegol (3).

## Mammalian and avian toxicity

### Metabolism and pharmacokinetics

*In vitro* rat liver microsomes, major metabolite is the allylic alcohol with menthofuran as a minor metabolite (4).

## References

1. Wood, E. M. *The Toxicity of 3400 Chemicals to Fish* 1987, EPA 560/6-87-002, PB87-200-275, Washington, DC
2. MacPhee, C. et al *Lethal Effects of 2014 Chemicals upon Sockeye Salmon, Steelhead Trout and Three-Spine Stickleback* 1989, EPA 560/6-89-001, PB89-156715, Washington, DC
3. Shukia, O. P. et al *Can. J. Microbiol.* 1987, **33**(6), 489-497
4. Madyastha, K. M. *Biochem. Biophys. Res. Commun.* 1990, **173**, 1086

# I129    Isoquinoline

**CAS Registry No.** 119-65-3
**Synonyms** 2-azanaphthalene; 2-benzazine; benzopyridine; benzo[*c*]pyridine; leucoline; β-quinoline
**Mol. Formula** $C_9H_7N$                                    **Mol. Wt.** 129.16
**Uses** In the synthesis of insecticides, antimalarials and dyes.
**Occurrence** In coal tar. Component of creosote.

## Physical properties

**M. Pt.** 26-28°C; **B. Pt.** 242°C; **Flash point** >107°C; **Specific gravity** $d_4^{30}$ 1.099; **Partition coefficient** log $P_{ow}$ 2.08.

### Solubility

Organic solvent: miscible, diethyl ether, benzene; soluble, ethanol, acetone, chloroform

## Ecotoxicity

### Fish toxicity

5 ppm (24 hr) caused no toxic effects on trout, bluegill sunfish, yellow perch and goldfish (1).

**Invertebrate toxicity**

$LC_{50}$ (24 hr) *Daphnia pulex* 39.9 mg $l^{-1}$ (2).

$LC_{50}$ (24 hr) *Tetrahymen pyriformis* 0.8 g $l^{-1}$ (3).

$EC_{50}$ (5, 15 min) *Photobacterium phosphoreum* 1.7, 2.2 mg $l^{-1}$ respectively, Microtox test (4).

## Environmental fate

**Anaerobic effects**

Anaerobically degradable to methane and carbon dioxide (5).

**Degradation studies**

Transformed by microbial action to oxygenated analogue (6).

Microbial community of Gram-negative rods isolated from sewage utilized isoquinoline as sole nitrogen and carbon-source. 1-Hydroxyisoquinoline was identified as the major transformation-product. Transformation via dihydroxyisoquinoline was suggested by the accumulation of a pink product in the reaction mixture (7).

*Acinetobacter* sp. isolated from oil– and creosote-contaminated soils utilized isoquinoline as sole carbon and nitrogen source. Degradation was associated with the build up of the pigment, 1-hydroxyisoquinoline which was further degraded to unknown intermediate ring cleavage products and carbon dioxide (8).

*Alcaligenes faecalis* and *Pseudomonas diminuta* will utilize isoquinoline as sole carbon source. Both excreted the metabolite 1-oxo-1,2-dihydroisoquinoline (9). 1.5 µg $ml^{-1}$ mixed with ground water and microorganisms from creosote contaminated soil (27 µg bacterial protein added) incubated 30°C for 14 days. Samples at 1, 3, 5, 8, and 14 days contained 0.9, 0.3, 0.3, 0.1, and undetectable µg $ml^{-1}$ respectively. The sterile control contained 1.4 µg after 14 days (10).

## Mammalian and avian toxicity

**Acute data**

$LD_{50}$ oral rat 360 mg $kg^{-1}$ (11).

$LD_{50}$ dermal rabbit 590 mg $kg^{-1}$ (11).

$LD_{Lo}$ intraperitoneal mouse 128 mg $kg^{-1}$ (12).

**Irritancy**

Dermal rabbit (24 hr) 10 mg (uncovered) caused severe irritation and 250 µg instilled into the eye (duration unspecified) caused severe irritation (11).

## Genotoxicity

A CASE study reported isoquinoline to have marginal genotoxic potential (13).

## References

1. Wood, E. M. *The Toxicity of 3400 Chemicals to Fish* 1987, EPA 560/6-87-002, PB87-200-275, Washington, DC
2. Southwold, G. R. et al *Environm. Sci. Technol.* 1978, **12**(9), 1062-1066
3. Schultz, T. W. et al *Arch. Environ. Contam. Toxicol.* 1978, **7**, 457-463
4. Birkholz, D. A. et al *Water Res.* 1990, **24**(1), 67-73
5. Pereira, W. E. et al *Water Sci. Technol.* 1988, **20**(11-12), 17-23
6. Gobsy, E. M. et al *U.S. Environ. Prot. Agency Res. Dev.* [Rep.] EPA 1990, EPA-600/9-90/401, 39-42
7. Shukla, O. P. *Microbiol. Lett.* 1990, **45**(178), 89-96

8.  Aislabie, J. et al *Appl. Environ. Microbiol.* 1989, **55**(12), 3247-3249
9.  Roeger, P. et al *Biol. Chem. Hoppe-Seyler* 1990, **371**(6), 511-13
10. Muller, J. K. et al *Appl. Environ. Microbiol.* 1991, **57**(5), 1277-1286
11. Smyth et al *Arch. Ind. Occup. Med.* 1951, **4**, 119
12. *Summary Tables of Biological Tests* 1950, **2**, 189, National Research Council Chemical-Biological Coordination Centre, Washington, DC
13. Klopman, G. et al *Mutat. Res.* 1990, **228**(1), 1-50

# I130   Isosafrole

**CAS Registry No.** 120-58-1
**Synonyms** 5-(1-propenyl)-1,3-benzodioxole;
1,2-(methylenedioxy)-4-propenylbenzene
**Mol. Formula** $C_{10}H_{10}O_2$                        **Mol. Wt.** 162.19
**Uses** Modification and strengthening of perfumes. Manufacturing of heliotropin.

## Physical properties
**M. Pt.** 8.2°C; **B. Pt.** 253°C; **Flash point** 104°C; **Specific gravity** $d_4^{20}$ 1.1206.

## Solubility
Organic solvent: miscible in ethanol, diethyl ether, benzene.

## Mammalian and avian toxicity

### Acute data
$LD_{50}$ oral rat, mouse 1340, 2470 mg $kg^{-1}$, respectively (1,2).
$LD_{50}$ oral redwing blackbird >1000 mg $kg^{-1}$ (3).
$LC_{Lo}$ intraperitoneal, intravenous mouse, rabbit 256, 300 mg $kg^{-1}$, respectively (4,5)
$LD_{Lo}$ subcutaneous cat 2 g $kg^{-1}$ (5).

### Carcinogenicity and long-term effects
No adequate evidence for carcinogenicity to humans, limited evidence for carcinogenicity to animals, IARC classification group 3. Degree of evidence not previously categorised; evaluation based on criteria in monograph 10, 1976 (6).
Oral ♂ ♀ rat, mouse maximum tolerated dose continuous administration from 7 days of age caused no neoplasms of the forestomach (7).
Oral (82 wk) mouse 215 mg $kg^{-1}$ daily in distilled water induced lung and liver tumours (8).

### Irritancy
Dermal rabbit (24 hr) caused moderate irritation (9).

## Genotoxicity
*Salmonella typhimurium* TA100, TA1535 with and without metabolic activation negative (10).

CASE structural activity methodology, based on qualitative chemical structural features to predict mutagenic potential, classified isosafrole as being marginally active (11).
*In vitro* rat hepatocytes unscheduled DNA synthesis negative. Cytotoxic at concentrations between $10^{-3}$ and $10^{-2}$ M (12).

## Any other adverse effects

*In vitro* (72 hr) rat hepatocytes induced several cytochrome $P_{450}$ isoenzymes demonstrated by increased catalytic activity by western blotting and by immunocytochemistry (13).

## Any other comments

Human health effects and experimental toxicology reviewed (14).
♀ CD-1 mice injected intraperitoneally with [$^{32}$P]-isosafrole showed a low level of raidolabel binding to liver-DNA (15).

## References

1. *Toxicol. Appl. Pharmacol.* 1965, **7**, 18
2. *Food Cosmet. Toxicol.* 1964, **2**, 327
3. Schafer, E. W. et al *Arch. Environ. Contam. Toxicol.* 1983, **12**, 355-382
4. *Summary Tables of Biological Tests* 1949, **1**, 45, National Research Council Chemical-Biological Coordination Centre
5. *Arch. Exp. Pathol. Pharmakol.* 1895, **35**, 342 (Ger.)
6. *IARC Monograph* 1987, **Suppl. 7**, 65
7. Reuber, M. D. et al *Digestion* 1979, **19**(1), 42-47
8. Innes, J. R. M. *J. Natl. Cancer Inst.* 1969, **42**, 1101-1114
9. *Food Cosmet. Toxicol.* 1976, **14**, 307
10. Wislocki, P.G. et al *Cancer Res.* 1977, **37**(6), 1883-1891
11. Klopman, G. et al *Mutat. Res.* 1990, **228**(1), 1-50
12. Howes, A. J. *Food Chem. Toxicol.* 1990, **28**(8), 537-542
13. Bars, R. G. *Biochem. J.* 1989, **262**(1), 151-158
14. *ECETOC Technical Report No. 30(4)* 1991, European Chemical Industry Ecology and Toxicology Centre, B-1160 Brussels
15. Randerath, K. et al *Carcinogenesis (London)* 1984, **5**(12), 1613-1622

# I131   Isosorbide

**CAS Registry No.** 652-67-5
**Synonyms** 1,4:3,6-dianhydro-D-glucitol; Devicorun; Hydronol; Isobide, 1,4:3,6-dianhydrosorbitol
**Mol. Formula** $C_6H_{10}O_4$                    **Mol. Wt.** 146.14
**Uses** As a diuretic

**Physical properties**

**M. Pt.** 61-64°C.

**Mammalian and avian toxicity**

**Acute data**

$LD_{50}$ oral rat 24,150 mg $kg^{-1}$ (1).
$LD_{50}$ intravenous mouse, rat 6870 and 11,300 mg $kg^{-1}$ respectively (1,2).
$LD_{50}$ intraperitoneal mouse 13,600 mg $kg^{-1}$ (3).

**References**

1. *Pharmacometrics* 1969, **3**, 15
2. *Jpn. J. Pharmacol.* 1982, **6**, 71
3. *Drugs in Japan. Ethical Drugs* 6th ed., 1982, Edo Japan Pharmaceutical Information Centre, Yukugyo Jiho Co. Ltd., Tokyo

# I132   Isotretinoin

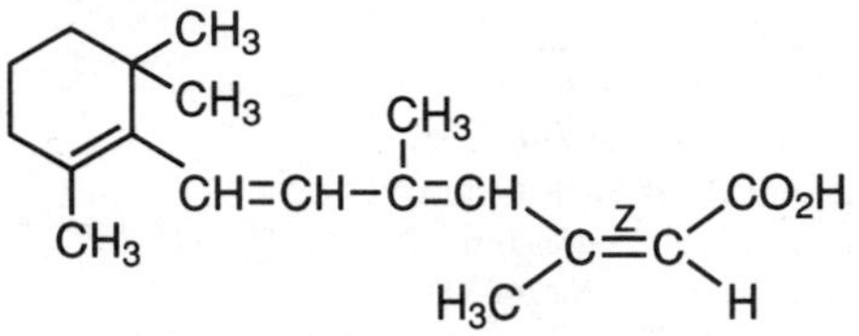

**CAS Registry No.** 4759-48-2
**Synonyms** 13-*cis*-retinoic acid; neovitamin A acid; 13-*cis*-vitamin A acid; Accutane
**Mol. Formula** $C_{20}H_{28}O_2$                     **Mol. Wt.** 300.44
**Uses** Treatment of severe acne and other skin disorders.

**Physical properties**

**M. Pt.** 174-175°C.

**Solubility**
Organic solvent: ethanol

**Ecotoxicity**

**Invertebrate toxicity**
Interferes with final differentiation of hydra buds, *in vitro* Hydra Assay (duration, dose unspecified) (1).
American sea urchin, isotretinoin induced a dose related delay in development. Isotretinoin metabolites, 4-oxoisotretinoin and 4-oxotretinoin and the isomer tretinoin induced strikingly dysomorphic development. It is suggested that it is the metabolites rather than the parent compound which are responsible for foetal abnormalities observed in higher animals (2).
Frog embryo teratogenesis assay, small cell frog blastulae exposed 96 hr. Scored as having strong teratogenic potential (3).

## Mammalian and avian toxicity

### Acute data
$LD_{50}$ oral rabbit, mouse 1960, 3389 mg kg$^{-1}$, respectively (4).
$LD_{50}$ intraperitoneal mouse 138 mg kg$^{-1}$ (4).

### Teratogenicity and reproductive effects
Teratogenicity of single oral dose 200 mg kg$^{-1}$ to pregnant ICR mice on day 7, 8 or 9
of gestation on craniofacial development was tested. Day 8 of gestation was the most
sensitive period; resorption rates were 46%; 88% of embryos and 100% of litters
developed malformations. Treatment on day 9 of gestation, resorption rates normal
but 9/10 litters and 75% of embryos had malformations mostly of their palate and ear
auricle. This indicates a strong developmental stage-dependent susceptibility to the
drug and stage-specific pattern of malformations. The target of the drug appears to be
the neural crest and developing neuroepithelium (5).
Post natal mouse screening test 85 mg kg$^{-1}$ per day (duration not specified). Caused
no effect on maternal mortality, the number of viable litters, litter size or pup birth
weight (6).
Pregant cynomolgus macaque oral, 4 dose regimes: dose A: 2, 10, 25 mg kg$^{-1}$ on
gestation days (GD) 18-28; dose B: 5 mg kg$^{-1}$ as an equally divided dose twice daily,
GD 21-24; dose C: 5 mg kg$^{-1}$ as an equally divided dose twice daily GD 25-27 and
dose D: 2.5 mg kg$^{-1}$ daily GD 10-25 and then twice ($2 \times 2.5$ mg kg$^{-1}$) daily GD 26-27.
Dose A caused maternal death and toxicity indicated by a reduction in weight and
food consumption and diarrhoea at all dose levels. No significant maternal toxicity
observed in dose regimes B-D. Dose regimes A and B were embryo lethal, embryo
death was not significant in dose C. Examination of GD 100 embryos showed no
malformations in doses A-C, dose D resulted embryo malformations of 71% with
malformed external ears, 57% with hypo- or aplasia of the thymus and 29% with
malformations of the heart (7).
Chick embryo injected via the yolk sac single 15 μl doses of either 1.5 μg, 15 μg or
150 μg on embryonic days 2, 3, 4, 5 or 6. Examination on day 14 of incubation
showed mortality and total malformations were dose and developmental-stage
responsive. Defects occurred in mesenchymal tissues derived from the cranial neural
crest ectomesenchyme which are analogous to those observed in animal models and
in human foetuses exposed during maternal therapy for cystic acne. The greatest
incidence of malformations occurred when isotretinoin was given after cranial neural
crest cell migration was complete (8).
*In vitro* 9.5 day rat embryo grown in culture for 48 hr, serum concentration of
500 ng ml$^{-1}$ induced defects in visceral arch development. Embryos must be exposed
for a minimum period of time regardless of the concentration of isotretinion above the
500 ng ml$^{-1}$ threshold (9).
Oral 40-80 mg day$^{-1}$ during the 1st month of human pregnancy can induce severe
congenital malformation. The human Accutane dysmorphic syndrome is characterised
by rudimentary external ears, absent or imperforate auditory canals, cleft palate,
depressed midface, abnormalities of brain jaw and heart. Humans are $\approx 16 \times$ more
sensitive to the teratogenic effects of oral isotretinoin than hamsters (10).

### Metabolism and pharmacokinetics
Percutaneous absorption *in vivo* monkey. Exposure to light caused 60% degradation
on the surface of the skin but did not change the amount penetrating the epidermis.

On human skin the amount penetrating the epidermis did not increase over a 25-fold range of dose (11).

Oral pregnant hamster single dose 35 μg kg$^{-1}$ administered during primitive streak stage of embryo development. Found distributed in all tissues sampled (including placenta and foetus) the largest accumulation was in the liver with the least in fat. Isotretinoin had a slower clearance and the longest elimination ($t_{1/2}$) time of the acidic retinoids tested (12).

Human oral 80 mg (single dose), excreted 17- 23% of drug by 4 days. Major metabolites in bile were glucuronide conjugates of 4-oxoisotretinoin and 16-hydroxyisotretinoin. Minor metabolites were glucuronide conjugates of isotretinoin and 18-hydroxyisotretinoin (13).

Single 100 mg kg$^{-1}$ and multiple $3 \times 100$ mg kg$^{-1}$ 4 hr apart, given to NMRI mice on day 10 of gestation. Major plasma metabolite was 13-*cis*-retinoyl-β-glucuronide followed by 4-oxo-metabolites and all-*trans*-retinoic acid. Transfer to the mouse embryo was very efficient for all-*trans*-retinoic acid and 10-fold less efficient for 13-*cis*-retioic acid (14).

*In vitro* mice liver microsomes metabolised to *cis*- and *trans*-4-hydroxy-13-retinoic acid and 4-oxo-13- retinoic acid as major metabolites but not in mice skin microsomes (15).

## Genotoxicity

*In vitro* human embryonic palatal mesenchymal (HEPM) cells with and without metabolic activation interferes with DNA synthesis and decreases HEPM cell proliferation (16).

## Any other adverse effects to man

Healthy human volunteers dermal 1.5% solution, dose related exfoliative effects, expressed by an increase of cell counts (number of corneocytes cm$^{-2}$ skin surface) and a decrease of corneocyte area (17).

*In vitro* human teratocarcinoma derived cell line, PA-1, 10$^{-6}$ to 10$^{-8}$ M did not significantly alter the log-phase growth rate but did decrease the saturation cell density and mitotic indexes, also induced changes in cell morphology which appear to be related to reorganization of microtubules and microfilaments (18).

## Any other adverse effects

Systemic administration (dose unspecified) hamster caused a reduction of acinar tissue in the Meibomiam gland. Clinical observations included alopecia and weight loss. Ocular complications included crusting of the eyelid margin and the external surface of the lid and erythema of the conjunctiva (19).

Oral mice (dose, duration unspecified), mice killed and examined at 4, 7 and 12 wk. Effects on *in vivo* immune regulation were characterised by expansion of the splenic marginal zone and the paracortical region of the lymph nodes (20).

Oral (in food) rat 45 mg kg$^{-1}$ increased total serum triacylglycerol and cholesterol concentrations after 3 days. Higher doses ($\geq$450 mg kg$^{-1}$) did not increase triacylglycerol earlier (21).

## Any other comments

Effects on epidermal proliferation, mechanisms of action, anti-tumour pharmacology and anti-inflammatory action reviewed (22-24).

Rats pretreated with isotretinoin, 300 mg kg$^{-1}$ in diet, 1 wk prior to tumour induction with 1,2-dimethylhydrazine (20 mg kg$^{-1}$ each wk for 20 wk). A significant delay in tumour onset was observed (25).

## References

1. Johnson, E. M. et al *Teratology* 1989, **39**(4), 349-361
2. Kahn, T. A. et al *Fundam. Appl. Toxicol.* 1988, **11**(3), 511-518
3. DeYoung, D. J. et al *Drug Chem. Toxicol.* (1977) 1991, **14**(1-2), 127-141
4. Sporn, M. B. et al (Ed.) *The Retinoids* Volume 2, 1984, Academic Press, New York
5. Park, S. H. et al *Koryo Taehakkyo Uikwa Taehalz Nonmunjip* 1986, **23**(3), 145-156 (Korean) (*Chem. Abstr.* **107**, 51983h
6. *Environmental Health Research and Testing Inc., Report* 1987, Order No.PB89-139075, Avail. NTIS. (*Chem. Abstr.*, **111**, 1483269
7. Hummler, H. et al *Teratology* 1990, **42**(3), 263-272
8. Hart, R. C. et al *Teratology* 1990, **41**(4), 463-472
9. Ritchie, H. et al *Teratology* 1991, **43**(1), 71-81
10. Willhite, C. G. et al *J. Craniofacial Genet. Dev. Biol.* 1986 (Suppl.2), 193-209
11. Lehman, P. A. et al *J. Invest. Dermatol.* 1988, **91**(1), 56-61
12. Howard, W. B. et al *Arch. Toxicol.* 1989, **63**(2), 112-120
13. Vane, F. M. et al *Xenobiotica* 1990, **20**(2), 193-207
14. Kraft, J. C. et al *Teratog., Carcinog., Mutagen.* 1991, **11**(1), 21-30
15. Oldfield, N. *Drug Metab. Dispos.* 1990, **18**(6), 1105-1107
16. Watanabe, T. et al *J. Nutr. Sci. Vitaminol.* 1990, **36**(4), 311-325
17. Haidl, G. et al *J. Soc. Cosmet. Chem.* 1988, **39**(1), 53-67
18. Taylor, D. D. et al *Differentiation* 1990, **43**(2), 123-130
19. Lumbert, R. W. et al *J. Invest. Dermatol.* 1989, **92**(3), 321-325
20. Kutz, D. R. et al *Br. J. Exp. Pathol.* 1987, **68**(3), 343-350
21. McMaster, J. et al *Arch. Dermatol. Res.* 1989, **281**(2), 116-118
22. Lever, L. R. et al *Retinoids: 10 years on, Retinoid Symp. 1990* Pub. 1991, Ed., Saurat, J. H., Karger, Basel
23. Chomienne, C. *Pathol. Biol.* 1989, **37**(8), 865-867 (Fr.) (*Chem. Abstr.* **111**, 224695d)
24. Coffey, J. W. et al *Int. Congr. Ser. – Excerpta Med.* 1987, **750**, 705-708
25. O'Dwyer, P. J. et al *Surg. Forum* 1986, **37**, 436-438

# I133   Isouron

**CAS Registry No.** 55861-78-4
**Synonyms** $N'$-[5-(1,1-dimethylethyl)-3-isoxazolyl]-$N$,$N$-dimethylurea; isuron
**Mol. Formula** $C_{10}H_{17}N_3O_2$                    **Mol. Wt.** 211.27
**Uses** Pre- and post-emergence control of broad-leaved and grass weeds in sugar cane and pineapples. Also used for total weed control on non-crop land.

## Physical properties

**M. Pt.** 119-120°C; **Specific gravity** $d^{20}$ 1.23; **Partition coefficient** log $P_{ow}$ 1.98 (1).

## Solubility

Water: 790 mg l$^{-1}$ at 25°C. Organic solvent: ethanol, acetone, xylene

## Ecotoxicity

**Fish toxicity**

$LC_{50}$ (48 hr) carp, Japanese killifish 78.7, 173.0 mg $l^{-1}$, respectively (1).
$LC_{50}$ (96 hr) rainbow trout, bluegill sunfish 110-140 mg $l^{-1}$ (1).

## Environmental fate

**Degradation studies**

In soil $t_{1/2}$ 22 days, some microbial degradation can occur (1).
Biodegradation in sewage and river water, incubation period of 120 days
0.01, 0.1 ppm added to samples, in sewage 31 and 25%, respectively were
mineralized whereas only 7% was mineralized in river water although the amount of
isouron decreased to 66-68% the original amount. Metabolites identified in the
aqueous environment were:
$N,N$-dimethyl-$N'$-[5-(1,1-dimethyl-2-formylethyl)-3-isoxazolyl]urea;
$N$-methyl-$N'$-[5-(1,1- dimethylpropyl)-3-isoxazolyl]urea; N-[5-(1,1-
dimethylpropyl)-3-isoxazolyl]urea; $N,N$-dimetyl- $N'$-[5-(1,1-dimethyl-
3-hydroxypropyl)-3- isoxazoly]urea; and,
$N$-methyl-$N'$-[5-(1,1-dimethyl-3-hydroxypropyl)-3-isoxazolyl]urea (2).
Degradation of 4 ppm applied to soils followed 1st order kinetics. $t_{1/2}$ 42-203 days at
10-40°C with water contents adjusted to 20-90% of the field capacity. The rate of
degradation was correlated with temperature between 20-40°C. The same metabolites
as above were formed (3).

## Mammalian and avian toxicity

**Acute data**

$LD_{50}$ oral mouse, rat 520-760 mg $kg^{-1}$ (1).
$LD_{50}$ oral bobwhite quail >2000 mg $kg^{-1}$ (1).
$LC_{50}$ (8 hr) inhalation rat >0.415 mg 1 $air^{-1}$ (1).
$LD_{50}$ dermal rat 5000 mg $kg^{-1}$.
$LD_{50}$ intraperitoneal rat 270 mg $kg^{-1}$ (1).

**Carcinogenicity and long-term effects**

Oral (2 yr) rat no effect level 7.26-8.77 mg $kg^{-1}$ $day^{-1}$, mice 3.42-16.6 mg $kg^{-1}$ $day^{-1}$ (1).

## Any other comments

Limited under EC Directive on Drinking Water Quality 80/778/EEC.Pesticides:
maximum admissible concentration 0.1 µg $l^{-1}$ (4).
Included in Schedule 6 (Release into Land: Prescribed Substances) Statutory
Instrument No. 472, 1991 (5).

## References

1. *The Agrochemicals Handbook* 3rd ed., 1991, RSC, London
2. Wang, Y. S. et al *Nippon Noyaku Gakkaishi* 1991, **16**(1), 19-25
3. Wu, T. C. et al *Nippon Noyaku Gakkaishi* 1991, **16**(2), 195-200
4. *EC Directive Relating to the Quality of Water Intended for Human Consumption* 1982,
   80/778/EEC, Office for Official Publications of the European Communities, 2 rue Mercier,
   L-2985 Luxembourg
5. *S. I. 1991 No. 472 The Environmental Protection (Prescribed Processes and Substances)
   Regulations* 1991, HMSO, London

# I134　Isovaleraldehyde

## (CH3)2CHCH2CHO

CAS Registry No. 590-86-3
Synonyms 3-methylbutanal; 3-methylbutyraldehyde; isoamyl aldehyde; isopentanal; isovaleric aldehyde; 3-methylbutyraldehyde
Mol. Formula $C_5H_{10}O$　　　　　　　　　　　　Mol. Wt. 86.13
Uses Flavour/perfume manufacture, pharmaceuticals and synthetic resins.
Occurrence Orange, peppermint, lemon, eucalyptus and other oils.

## Physical properties

M. Pt. –51°C; B. Pt. 92-93°C; Flash point –5°C; Specific gravity $d_{20}^{20}$ 0.785.

### Solubility
Water: miscible with ethanol, diethyl ether.

## Occupational exposure

UN No. 2058; HAZCHEM Code 3▨E; Conveyance classification flammable liquid.

## Ecotoxicity

### Fish toxicity
$LC_{50}$ (14 day) guppy 2.19 µg $l^{-1}$ (1).

## Environmental fate

### Degradation studies
Waste water treatment (activated sludge): 6 hr, 9.2% of ThOD; 12 hr, 14.2% of ThOD; 24 hr, 16.1% of ThOD (2).

## Mammalian and avian toxicity

### Acute data
$LD_{50}$ oral rat 8.91 g $kg^{-1}$ (3).
$LC_{Lo}$ (4 hr) inhalation rat 16,000 ppm (3).
$LD_{50}$ dermal rabbit 3.18 g $kg^{-1}$ (3).

### Metabolism and pharmacokinetics
Metabolised *in vitro* by cytochrome $P_{450}$ rabbit liver isoenzymes by oxidative cleavage and olefin formation (4).

## Genotoxicity

*Bacillus subtilis* H17 with and without metabolic activation positive (5).

## Any other comments

Recommended for testing for health effects, environmental and ecological fate by the US Federal Toxic Substances Control Act March 1991 (6).

## References

1.　Deneer, J. W. *Aquat. Toxicol.* 1988, **12**(2), 185-192
2.　Gerhold, R. M. et al *J. -Water Pollut. Control Fed.* 1966, **38**(4), 562
3.　*Toxicol. Appl. Pharmacol.* 1974, **28**, 313

4.  Roberts et al *Proc. Nat. Acad. Sci. U.S.A.* 1991, **88**(20), 8963-8966
5.  Matsui, S. et al *Water Sci. Technol.* 1989, **21**(8-9), 875-887
6.  *Fed. Regist.* 06 Mar 1991, **56**(44), 9534-9572

# I135   Isoxathion

**CAS Registry No.** 18854-01-8
**Synonyms** phosphorothioic acid *O,O*-diethyl-*O*-(5-phenyl-3-isoxazolyl) ester;
Karphos
**Mol. Formula** $C_{13}H_{16}NO_4PS$                          **Mol. Wt.** 313.31
**Uses** Control of scale insects, aphids, hoppers, beetles and caterpillars on citrus fruit,
tea, tobacco, garden trees, rice, sugarcane and vegetables.

## Physical properties

**B. Pt.** $_{0.15}$ 160°C; **Partition coefficient** log $P_{ow}$ 0.589 (1);
**Volatility** $1.0 \times 10^{-6}$ mmHg at 25°C.

## Solubility
Water: 1.9 mg $l^{-1}$ at 20°C.

## Occupational exposure

**Supply classification** toxic.
**Risk phrases** Toxic in contact with skin and if swallowed (R24/25)
**Safety phrases** After contact with skin, wash immediately with plenty of soap and
water – Wear suitable protective clothing and gloves – If you feel unwell, seek
medical advice (show label where possible) (S28, S36/37, S44)

## Ecotoxicity

**Fish toxicity**
*In vitro* GF-scale cells (derived from goldfish scales) midpoint cytotoxicity value
14.7 mg $l^{-1}$ (2).
$LC_{50}$ (48 hr) carp 2.1 mg $l^{-1}$ (2).

## Environmental fate

**Degradation studies**
$t_{1/2}$ 9-40 days in soil (1).

## Mammalian and avian toxicity

**Acute data**
$LD_{50}$ oral chicken, mouse 21.6, 79.1 mg $kg^{-1}$, respectively (3).
$LD_{50}$ oral rat 112 mg $kg^{-1}$ (4).
$LD_{50}$ dermal mouse, rat 193, 450 mg $kg^{-1}$, respectively (4).
$LD_{50}$ intraperitoneal mouse 105 mg $kg^{-1}$ (3).

LD$_{50}$ subcutaneous mouse 720 mg kg$^{-1}$ (3).

**Carcinogenicity and long-term effects**
Oral (2 yr) rat no-effect level 1.2 mg kg$^{-1}$ (1).

## Legislation

Included in Schedule 6 (Release into Land: Prescribed Substances)Statutory Instrument No. 472, 1991 (5).
Limited under EC Directive on Drinking Water Quality 80/778/EEC. Pesticides: maximum admissible concentration 0.1 µg l$^{-1}$ (6).

## Any other comments

Unstable in alkalis. Decomposes at 160°C.

## References

1. *The Agrochemicals Handbook* 3rd ed., 1991, RSC, London
2. Saito, H. et al *Chemosphere* 1991, **23**(4), 525-537
3. *Annual Report of Sankyo Research Laboratories* 1977, **29** 1, Tokyo 140
4. *Bull. Entomol. Soc. Am.* 1969, **15**, 121
5. *S. I. 1991 No. 472 The Environmental Protection (Prescribed Processes and Substances) Regulations* 1991, HMSO, London
6. *EC Directive Relating to the Quality of Water Intended for Human Consumption* 1982, 80/778/EEC, Office for Official Publications of the European Communities, 2 rue Mercier, L-2985 Luxembourg

# J1    Jasmolin I

**CAS Registry No.** 4466-14-2
**Synonyms** cyclopropanecarboxylic acid, 2,2-dimethyl-3-(2-methyl-1-propenyl)-, 2-methyl-4-oxo-3-(2-pentenyl)-2-cyclopenten-1–yl ester, [1*R*-[1α[*S*\*(*Z*)], 3β(*E*)]]-; Jasmoline I
**Mol. Formula** C$_{21}$H$_{30}$O$_3$                    **Mol. Wt.** 330.47
**Uses** Insecticide and acaricide.
**Occurrence** An active insecticidal constituant of pyrethram flowers.
One of the family of pyrethrin insecticides (1).

**Abiotic removal**
Oxidation and loss of insecticidal activity can occur in air and light. In the prescence of alkalis is rapidly hydrolysed causing loss of activity. Synergists have a stabilizing effect (1).
For pyrethrins in general, degradation is promoted by sunlight and UV light.

Degradation begins at the alcohol group and involves the formation of numerous cleavage products (1).

## Mammalian and avian toxicity

### Metabolism and pharmacokinetics
In mammals, pyrethrins are rapidly degraded in the stomach by hydrolysis of the ester bonds to form harmless metabolites (1).

### Irritancy
Pyrethrins may cause dermatitis in sensitized persons. Slightly irritating to skin and eyes (1).

## Legislation
Included in Schedule 6 (Release into Land: Prescribed Substances)Statutory Instrument No. 472, 1991 (2).
Limited under EC Directive on Drinking Water Quality 80/778/EEC. Pesticides: maximum admissible concentration 0.1 $\mu$g l$^{-1}$ (3).

## Any other comments
Toxicity and metabolism reviewed (4).

## References
1. *The Agrochemicals Handbook* 3rd ed., 1991, RSC, London
2. *S. I. 1991 No. 472 The Environmental Protection (Prescribed Processes and Substances) Regulations* 1991, HMSO, London
3. *EC Directive Relating to the Quality of Water Intended for Human Consumption* 1982, 80/778/EEC, Office for Official Publications of the European Communities, 2 rue Mercier, L-2985 Luxembourg
4. Timofiyevskaya, L. A. et al *Scientific Literature in Russian on Selected Hazardous Chemicals. Pyrethroids* 1993, **119**, UNEP/IRPTC, Eng. Transl. (Ed.) Richardson, M. L., Geneva

# J2   Jasmolin II

**CAS Registry No.** 1172-63-0
**Synonyms** cyclopropanecarboxylic acid, 3-(3-methoxy-2-methyl-3-oxo-1-propenyl)-2,2-dimethyl-, 2-methyl-4-oxo-3-(2-pentenyl)-2-cyclopenten-1-yl ester, [1R-[1α[S*(Z)],3β(E)]]-
**Mol. Formula** $C_{22}H_{30}O_5$              **Mol. Wt.** 374.48
**Uses** Insecticide and acaricide.

**Occurrence** In extracts from the plant *Pyrethrum (Chrysanthemum) cinerariaeflium* and its flowers.
One of the family of Pyrethrin insecticides (1).

## Physical properties

### Solubility
Water: Practically insoluble. Organic solvent: readily soluble in organic solvents.

## Environmental fate

### Abiotic removal
Oxidation and loss of insecticidal activity can occur in air and light. In the prescence of alkalis is rapidly hydrolysed causing loss of activity. Synergists have a stabilizing effect (1).
For pyrethrins in general, degradation is promoted by sunlight and UV light. Degradation begins at the alcohol group and involves the formation of numerous cleavage products (1).

## Mammalian and avian toxicity

### Metabolism and pharmacokinetics
In mammals, pyrethrins are rapidly degraded in the stomach by hydrolysis of the ester bonds to harmless metabolites (1).

### Irritancy
Pyrethrins in senstized persons dermatitis is possible. Slightly irritating to skin and eyes (1).

## Legislation
Included in Schedule 6 (Release into Land: Prescribed Substances)Statutory Instrument No. 472, 1991 (2).
Limited under EC Directive on Drinking Water Quality 80/778/EEC. Pesticides: maximum admissible concentration 0.1 $\mu g \, l^{-1}$ (3).

## Any other comments
Toxicity and metabolism of pyrethroids reviewed (4).

## References
1.  *The Agrochemicals Handbook* 3rd ed., 1991, RSC, London
2.  *S. I. 1991 No. 472 The Environmental Protection (Prescribed Processes and Substances) Regulations* 1991, HMSO, London
3.  *EC Directive Relating to the Quality of Water Intended for Human Consumption* 1982, 80/778/EEC, Office for Official Publications of the European Communities, 2 rue Mercier, L-2985 Luxembourg
4.  Timofiyevskaya, L. A. et al *Scientific Literature in Russian on Selected Hazardous Chemicals. Pyrethroids* 1993, **119**, UNEP/IRPTC, Eng. Transl. (Ed.) Richardson, M. L., Geneva

# J3  Jasmone

**CAS Registry No.** 488-10-8
**Synonyms** 3-methyl-2-(2-pentenyl)-(Z)-2-cyclopenten-1-one; *cis*-jasmone
**Mol. Formula** $C_{11}H_{16}O$                    **Mol. Wt.** 164.25
**Uses** In purfumery.
**Occurrence** Volatile portion of jasmin flower oil.

## Physical properties

**B. Pt.** $_{27}$ 146°C; **Flash point** 107°C; **Specific gravity** 0.904.

## Genotoxicity

*In vitro* Chinese hamster ovary K-1 cells, sister chromatid exchanges induced with mitomycin C were investigated for modifying effects by jasmone. Post treatment with jasmone significantly inceased the frequency of sister chromatid exchanges (1).

## References

1.   Sasaki, Y. F. et al *Mutat. Res.* 1989, 226 (2), 103-110

# K1   Kadethrin

**CAS Registry No.** 58769-20-3
**Synonyms** cyclopropanecarboxylic acid, 3-[(dihydro-2-oxo-3(2H)-thienylidene)methyl]-2,2-dimethyl-, [5-(phenylmethyl)- 3-furanyl]methyl ester, [1R-[1α,3α(E)]]-
**Mol. Formula** $C_{24}H_{26}O_3S$                    **Mol. Wt.** 396.51
**Uses** Insecticide used in aerosols and sprays in combination with other insecticides.

## Physical properties

**M. Pt.** 31°C; **Volatility** v.p. $7.52 \times 10^{-7}$ mmHg at 20°C.

## Solubility

Organic solvent: dichloromethane, ethanol, benzene, toluene, xylene, acetone, piperonyl butoxide

## Mammalian and avian toxicity

### Acute data
$LD_{50}$ oral ♀ rat 650 mg $kg^{-1}$, dog, ♂ rat >1000-1324 mg $kg^{-1}$ (1).
$LD_{50}$ percutaneous ♀ rat >3200 mg $kg^{-1}$ (1).

### Sub-acute data
Oral (90 day) no effect level dog, rat 15, 25 mg $kg^{-1}$, respectively (1).
Inhalation rat, guinea pig 200 × normal aerosol dose (exposure unspecified) caused no adverse effects (2).

## Legislation
Included in Schedule 6 (Release into Land: Prescribed Substances) Statutory Instrument No. 472, 1991 (3).
Limited under EC Directive on Drinking Water Quality 80/778/EEC.Pesticides: maximum admissible concentration 0.1 μg $l^{-1}$ (4).

## Any other comments
Unstable in heat. Hydrolysed by aqueous alkalis. Rapidly decomposes in light (1).

## References
1. *The Agrochemicals Handbook* 3rd ed., 1991, RSC, London
2. *The Pesticide Manual* 9th ed., 1991, British Crop Protection Council, Farnham
3. *S. I. 1991 No. 472 The Environmental Protection (Prescribed Processes and Substances) Regulations* 1991, HMSO, London
4. *EC Directive Relating to the Quality of Water Intended for Human Consumption* 1982, 80/778/EEC, Office for Official Publications of the European Communities, 2 rue Mercier, L-2985 Luxembourg

# K2   Kaolin

**CAS Registry No.** 1332-58-7
**Synonyms** Argiflex; Hydrogloss; Lustra; Porcelain clay; Satintone; Suprex clay; aluminium silicate hydroxide
**Mol. Formula** $Al_2H_4O_9Si_2$               **Mol. Wt.** 258.16
**Uses** Treatment for diarrhoea and for reducing inflammation and pain when applied in a poultice. Manufacture of porcelain, pottery and bricks.

## Occupational exposure
**US TLV (TWA)** 2 mg $m^{-3}$ (respirable dust); **UK Long-term limit** 2.5 mg $m^{-3}$ proposed.

## Mammalian and avian toxicity

### Acute data
$TD_{Lo}$ oral rat 590 g $kg^{-1}$ (1).

## References
1. *J. Nutr.* 1977, **107**, 2027

# K3　Karaya gum

**CAS Registry No.** 9000-36-6

**Synonyms** Indian tragacanth gum; katilo gum; Lame gum; Muccira; Siltex gum; Tab gum

**Uses** Denture adhesive. Binder in paper manufacture. Emulsifier and stabiliser in food. Laxative.

**Occurrence** As a dried exudate from tree *Sterculia urens* Roxbo., *Sterculiqceae* which grows in S.E. Asia.

## Physical properties

**Solubility**
Water: swells to form gel.

## Mammalian and avian toxicity

**Acute data**
$LD_{Lo}$ oral rat 30 g $kg^{-1}$ (1).

## Genotoxicity

*Salmonella typhimurium* TA98, TA100, TA1535, TA1537, TA1538 with and without metabolic activation negative (2).

## Any other comments

DNA damaging activity reviewed (3).

## References

1. *Food Res.* 1948, **13**, 29
2. Prival, M. J. et al *Mutat. Res.* 1991, **260**(4), 321-329
3. Ishizaki, M. et al *Shokuhin Eiseigaku Zasshi* 1987, **28**(6), 498-501 (Japan.) (*Chem. Abstr.* **109**, 5417v.)

# K4　Karbutilate

**CAS Registry No.** 4849-32-5

**Synonyms** Carbamic acid, (1,1-dimethylethyl)-, 3-[[(dimethylamino)carbonyl]amino]phenyl ester; Tandex

**Mol. Formula** $C_{14}H_{21}N_3O_3$　　　　　　　　　　　**Mol. Wt.** 279.34

**Uses** Herbicide, non-selective used for residual control of most annual and perennial broad-leaved weeds and grasses.

## Physical properties
**M. Pt.** 169-169.5°C.

**Solubility**
Water: 325 mg $l^{-1}$. Organic solvent: acetone, propan-2-ol, xylene, dimethyl sulfoxide

**Abiotic removal**
Degraded in soil, $t_{1/2}$ 20-120 day (1).

## Mammalian and avian toxicity

**Acute data**
$LD_{50}$ oral rat 3000 mg $kg^{-1}$ (2).
$LD_{50}$ intravenous mouse 320 mg $kg^{-1}$ (3).

## Legislation
Limited under EC Directive on Drinking Water Quality 80/778/EEC.Pesticides: maximum admissible concentration 0.1 µg $l^{-1}$ (4).
Included in Schedule 6 (Release into Land: Prescribed Substances) Statutory Instrument No. 472, 1991 (5).

## Any other comments
Stable in acid media.

## References

1. *The Pesticide Manual* 9th ed., 1993, British Crop Protection Council, Farnham, Surrey
2. *Guide to Chemicals Used in Crop Protection* 1972, Information Canada, Ottawa
3. U.S. Army Armament Research and Development Command, *Report: NX 03896*, Chemical Systems Laboratory, NIOSH Exchange Chemicals
4. *EC Directive Relating to the Quality of Water Intended for Human Consumption* 1982, 80/778/EEC, Office for Official Publications of the European Communities, 2 rue Mercier, L-2985 Luxembourg
5. *S. I. 1991 No. 472 The Environmental Protection (Prescribed Processes and Substances) Regulations* 1991, HMSO, London

# K5    Kasugamycin

**CAS Registry No.** 6980-18-3

**Synonyms** D-*chiro*-inositol, 3-*O*-[2-amino-4-[(carboxyiminomethyl)amino]-2,3,4,6-tetradeoxy-α-D-*arabino*-hexopyranosyl]-; Kasumin L

**Mol. Formula** $C_{14}H_{25}N_3O_9$                    **Mol. Wt.** 379.37

**Uses** Control of diseases in rice especially rice blast. Also the control of other plant diseases such as leaf mould, leaf spot and scab on apples and pears.

**Occurrence** Produced by the fermentation of *Streptomyces kasugaensis*.

## Physical properties

**M. Pt.** 202-204°C (decomp.).

**Solubility**
Water: 125 g $l^{-1}$. Organic solvent: methanol, acetone, xylene

## Ecotoxicity

**Fish toxicity**
$LC_{50}$ (48 hr) carp, goldfish >40 mg $l^{-1}$ (1).

**Invertebrate toxicity**
$LC_{50}$ (6 hr) *Daphnia pulex* >40 mg $l^{-1}$ (1).
$LC_{50}$ frog tadpoles (duration unspecified) >100 ppm (2).

**Abiotic removal**
At 50°C, $t_{1/2}$ 47 day at pH 5 and $t_{1/2}$ 14 day at pH 9 (3).

## Mammalian and avian toxicity

**Acute data**
$LD_{50}$ oral ♂ rat, ♀ mouse 20000-20500 mg $kg^{-1}$ (3).
$LD_{50}$ oral Japanese quail >4000 mg $kg^{-1}$ (3).
$LD_{50}$ dermal rat, mouse 4000, 10000 mg $kg^{-1}$, respectively (3).

**Carcinogenicity and long-term effects**
Oral (2 yr) rat, dog no effect level 1000, 800 mg $kg^{-1}$, respectively (3).

## Genotoxicity

*In vitro* Chinese hamster ovary cells (metabolic activation unspecified) in combination with carbendazim sister chromatid exchanges negative, chromosomal aberrations positive (4).

---

## Legislation

Included in Schedule 6 (Release into Land: Prescribed Substances)Statutory Instrument No. 472, 1991 (5).
Limited under EC Directive on Drinking Water Quality 80/778/EEC. Pesticides: maximum admissible concentration 0.1 $\mu$g l$^{-1}$ (6).

## Any other comments

No adverse effects observed in adult fire flies (dose duration unspecified) (7).
Levels of 0.2 g kg$^{-1}$ daily for 7 days to carp with bacterial infection resulted in 98% cure during this period (8).
WHO Class Table 5; EPA Toxicity Class IV (3).
Very stable at room temperature. Stable in weak acids, but is unstable in strong acids and alkalis.

## References

1. Hashimoto, Y. et al *J. Pestic. Sci.* 1981, **6**, 257
2. Nishiuch, Y. *Seitai Kayaku* 1989, **9**(4), 23-26 (Japan.) (*Chem. Abstr.* **113**, 72754y)
3. *The Agrochemicals Handbook* 3rd ed., 1991, RSC, London
4. Wang, T. C. et al *Bull. Inst. Zool. Acad. Sin.* 1987,**26**(4), 317-329
5. *S. I. 1991 No. 472 The Environmental Protection (Prescribed Processes and Substances) Regulations* 1991, HMSO, London
6. *EC Directive Relating to the Quality of Water Intended for Human Consumption* 1982, 80/778/EEC, Office for Official Publications of the European Communities, 2 rue Mercier, L-2985 Luxembourg
7. Miyashita, M. *Nippon Koshu Eisei Zasshi* 1988, **35**(3), 125-132 (Japan.) (*Chem. Abstr.* **109**, 68550d)
8. Okamoto, H. *Jpn. Kokai Tokkyo Koho JP 61, 267, 521 [86, 267, 521]* 1986, 1-3, Hokko Chemical Industry Co. Ltd., (*Chem. Abstr.* **106**, 149469z)

# K6   Kelevan

**CAS Registry No.** 4234-79-1
**Synonyms** 1,3,4-metheno-1*H*-cyclobuta[*cd*]pentalene-2-pentanoic acid, 1,1a,3,3a,4,5,5a,6-decachlorooctahydro-2-hydroxy-$\gamma$-oxoethyl ester; Despirol
**Mol. Formula** $C_{17}H_{12}Cl_{10}O_4$                **Mol. Wt.** 634.81

## Physical properties

**M. Pt.** 89-90°C.

## Occupational exposure

**Supply classification** toxic.
**Risk phrases** Harmful if swallowed – Toxic in contact with skin (R22, R24)
**Safety phrases** Wear suitable protective clothing and gloves – If you feel unwell, seek medical advice (show label where possible) (S36/37, S44)

## Environmental fate

**Degradation studies**
Supported growth of 3 *Pseudomonas* sp. (1).

**Abiotic removal**
Photoxidation by UV light in aqueous medium at 90-95°C, time for the formation of $CO_2$ (% of theoretical); 25%: 1.2 hr, 50%: 9.6 hr, 75%: 19 hr (2).

## Mammalian and avian toxicity

**Acute data**
$LD_{50}$ oral redwing blackbird >104 mg $kg^{-1}$ (3).

## Any other comments

Toxicity, human health effects, environmental effects comprehensively reviewed (4).

## References

1. George, S. et al *Xenobiotica* 1988, **18**(4), 407-416
2. Knoevenagel *Arch. Environ. Contam. Toxicol.* 1976, **4**, 324-333
3. Schafer, E. W. et al *Arch. Environ. Toxicol.* 1983, **12**, 355-382
4. *Environmental Health Criteria: Kelevan* No. **66**, 1986, World Health Organization, Geneva

# K7    Kerosene

**CAS Registry No.** 8008-20-6
**Synonyms** kerosine; jet fuel
**Uses** Fuel in lamps, stoves. Cleaner and degreaser. Jet fuel.
**Occurrence** Mixture of petroleum hydrocarbons, it constitutes the fifth fraction in petroleum distillation.

## Physical properties

**B. Pt.** 175-325°C; **Flash point** 81°C; **Specific gravity** d ≈0.80.

**Solubility**
Organic solvent: miscible with other petroleum solvents

## Occupational exposure

**UN No.** 1223; **HAZCHEM Code** 3▥; **Conveyance classification** flammable liquid.

## Mammalian and avian toxicity

### Acute data
$LD_{50}$ oral rabbit 28 ml $kg^{-1}$ (1).
$LD_{50}$ oral guinea pig 20 g $kg^{-1}$ (2).
$LD_{50}$ oral ♂ rat >60 ml $kg^{-1}$ (3).
$LD_{50}$ intravenous, intratracheal rabbit 180, 200 mg $kg^{-1}$ respectively (4,5).
$LD_{50}$ intraperitoneal rabbit 6600 mg $kg^{-1}$ (4).
Gavage ♂ rat 24 ml $kg^{-1}$ showed moderate renal and hepatic functional alterations 1-3 days later (3).

### Sub-acute data
Inhalation (90 days) ♂ Fischer 344 rats continuous exposure to 150 and 750 mg $m^{-3}$ developed dose related kidney damage with cytoplasmic hyaline droplets, necrosis of proximal tubular cells and accumulation of intratubular necrotic debris (6,7).
Dermal (60 wk) mice 5 µl 3 times a week developed atrophied and degenerating nephrons as well as papillary necrosis (8).

### Carcinogenicity and long-term effects
Dermal (2 yr) B6C3F1 mice 250 or 500 mg $kg^{-1}$ no evidence of carcinogenicity (9).

### Teratogenicity and reproductive effects
Charles River CD rats, 6-15 days of gestation inhalation 6 hr $day^{-1}$ 100, 400 ppm, no embryotoxic, foetotoxic or teratogenic effects observed (10).
Application to the shell surface of duck embryos, day 6 of incubation, 1-20 µl of weathered or unweathered aviation kerosene no toxic effects observed (11).

### Metabolism and pharmacokinetics
Baboons given radiolabelled kerosene via a nasogastric tube after tracheostomy. Radiolabel was found localised in the kidney, brain, liver, lungs and spleen (12).

### Irritancy
Dermal rabbit (duration not specified) 500 mg caused severe irritation (2).
Dermal (90 wk) ♀ B6C3F1 mice 500 mg $kg^{-1}$ caused excessive irritation and ulceration at site of application (9).
Dermal B6C3F1 mice 250 or 500 mg $kg^{-1}$ (unspecified duration) caused dose related increased incidence of chronic dermatitis identified by acanthosis, hyperkeratosis, necrosis and ulceration of the overlying epidermis. Dermal changes frequently included fibrosis, increased amounts of melanin and acute and chronic inflammatory cell infiltrates (9).

## Genotoxicity
*Salmonella typhimurium* TA97, TA98, TA100, TA1535 with and without metabolic activation negative (9).

## Any other adverse effects to man
A group of human ♂ exposed to levels in air >350 mg $m^{-3}$ were studied for 8 yrs; no increase in incidences of cancer were seen (13).
Case-control study of cancer in Canada revealed an increased incidence of kidney cancer in human ♂ with occupational exposure (14).

## Any other comments

Human health effects and experimental toxicology reviewed (15).
Autoignition temperature 210°C.

## References

1. Deichmann, W. B. et al *Ann. Int. Med.* 1944, **21**, 803
2. *J. Am. Coll. Toxicol.* 1990, **1**, 30
3. Paker, G. A. et al *Toxicol. Appl. Pharmacol.* 1981, **57**, 302-317
4. *Ann. Intern. Med.* 1944, **21**, 803
5. *Toxicol. Appl. Pharmacol.* 1961, **3**, 689
6. Bruner, R. H. *Advances in Modern Environmental Toxicology, Vol.III, Renal effects of Petroleum Hydrocarbons* 1984, 133-140 Ed. Mehlman, M. A. et al, Princeton Scientific Publishers, Princeton, NJ
7. Gaworsk, C. L. et al *Advances in Modern Environmental Toxicology Vol.VI, Applied Toxicology of Petroleum Hydrocarbons* 1984, 33-47, Ed. MacFarland, H. N. et al, Princeton Scientific Publishers, Princeton
8. Easley, J. R. et al *Toxicol. Appl. Pharmacol.* 1982, **65**, 84-91
9. *National Toxicology Program* 1986, Report No.TR-310 Toxicology and Carcinogenesis Studies of JP-5 7m Navy Fuel, NIEHS Research Triangle Park, NC 27709
10. Beliles, R. P. et al *Proceedings of a Symposium. The Toxicity of Petroleum Hydrocarbons* 1982, 233-238, Ed. MacFarland, H. N. et al, American Petroleum Institute, Washington, DC
11. Albers, P. H. et al *Bull. Environ. Contam. Toxicol.* 1982, **28**, 430-434
12. Mann, et al *J. Pediatr.* 1977, **91**(3), 495
13. *IARC Monograph* 1989, **10**, 232-218
14. Siemiatyck, J. et al *Scand. J. Work Environ. Health.* 1987, **13**, 493-504
15. *ECETOC Technical Report No. 30(4)* 1991, European Chemical Industry Ecology and Toxicology Centre, B-1160 Brussels

# K8   Ketene

$$CH_2=C=O$$

**CAS Registry No.** 463-51-4
**Synonyms** ethenone; carbomethene
**Mol. Formula** $C_2H_2O$ **Mol. Wt.** 42.04
**Uses** For acetylation in the manufacture of cellulose acetate and aspirin.

## Physical properties

**M. Pt.** –150°C; **B. Pt.** –56°C; **Flash point** –107°C; **Volatility** v. den. 1.45.

## Solubility

Organic solvent: diethyl ether, acetone (decomposes in ethanol)

## Occupational exposure

**US TLV (TWA)** 0.5 ppm (0.86 mg m$^{-3}$); **US TLV (STEL)** 1.5 ppm (2.6 mg m$^{-3}$);
**UK Long-term limit** 0.5 ppm (0.9 mg m$^{-3}$); **UK Short-term limit** 1.5 ppm
(3 mg m$^{-3}$).

## Mammalian and avian toxicity

### Acute data
$LD_{50}$ oral rat 1300 mg $kg^{-1}$ (1).
$LC_{Lo}$ (10 min) inhalation monkey 200 ppm (2).
$LC_{Lo}$ (10 min) inhalation cat 750 ppm (2).
$LC_{Lo}$ (30 min) inhalation mouse 23 ppm (2).
$LC_{Lo}$ (100 min) inhalation rat, guinea pig 53 ppm (2).

### Any other comments
Human health effects, epidemiology and experimental toxicology reviewed (3).
Autoignition temperature 528°C.

### References
1. *Union Carbide Data Sheet* Industrial Medicine and Toxicology Department, Union Carbide Corp., 270 Park Avenue, New York, NY10017
2. *J. Ind. Hyg. Toxicol.* 1949, **31**, 209
3. *ECETOC Technical Report No. 30(4)* 1991, European Chemical Industry Ecology and Toxicology Centre, B-1160 Brussels

# K9   Ketoconazole

**CAS Registry No.** 65277-42-1
**Synonyms** Piperazine, 1-acetyl-4-[4-[[2-(2,4-dichlorophenyl)-2-(1*H*-imidazol-1-ylmethyl)-1,3-dioxolan-4-yl]methoxy]phenyl]-, *cis*-
**Mol. Formula** $C_{26}H_{28}Cl_2N_4O_4$                    **Mol. Wt.** 531.44
**Uses** Oral broad-spectrum antimycotic. Possible treatment for prostrate carcinoma (1).

## Physical properties
**M. Pt.** 146°C.

## Mammalian and avian toxicity

### Acute data
$LD_{50}$ oral guinea pig, mouse, rat, dog 202-780 mg $kg^{-1}$ (2).
$LD_{50}$ intravenous guinea pig, mouse, rat, dog 28-86 mg $kg^{-1}$ (2).

### Teratogenicity and reproductive effects
Teratogenic potential assessed using post-implantation rat embryo culture system,

malformations at concentrations well below those affecting embryonic growth and differention were observed (3).

Healthy human ♂, oral doses caused a transitory decrease in circulating levels of both total and free testosterone without affecting oestradiol concentrations. The time of maximum effect was dose related, being maximum at 4 hr after 200 or 400 mg and 8 hr after 600 mg. No effect on circulating levels of testosterone or oestradiol in women (4). Oral (72 hr) ♂ rats 200 mg kg$^{-1}$ reduced fertility compared to the controls. 400 mg kg$^{-1}$ caused complete loss of fertility. Sperm motility and forward progression was reduced but there was no change in testicular weight, epididymal sperm concentrations or epididymal weight (5).

*In vitro* rat embryos (48 hr) with metabolic activation, ketoconazole was determined to have relatively high teratogenic potential (6).

## Metabolism and pharmacokinetics

Elimination is reported to be biphasic in humans, with an initial $t_{1/2}$ 2 hr and a terminal $t_{1/2}$ 8 hr (7).

Rats, rabbits, absorbed rapidly by the skin and then gradually distributed throughout the body. Excretion in urine and faeces was 0.4 and 2.1% of applied dose, respectively, after skin absorption (8).

Oral (1-6 month) humans 200 mg daily. Mean elimination $t_{1/2}$ 3.3 hr. 0.22% excreted in urine unchanged, suggesting almost complete metabolism (9).

Human adults 2% cream single application (5 g). Skin absorption rate was calculated to be 2.5-12.5%. Haematological and biochemical parameters remained normal, no unchanged drug was detected in blood or urine. Ketoconazole may be accumulated mainly in cutaneous layer and has no clinical safety problem (10).

## Irritancy

Healthy adult subjects 2% cream single application (5 g) did not cause skin irritation (10).

## Any other adverse effects to man

Nausea and vomiting reported in 3-10% of patients administered orally, and topical administration has resulted in irritation or dermatitis (11).

Demonstrated to inhibit testosterone biosynthesis in human ♂ (12).

Shown to have immunosuppressive effects (13,14).

## Any other adverse effects

Rat testes testosterone formation inhibited by the inhibition of cytochrome $P_{450}$ dependent C17,20-lyase (15).

Intraperitoneal administration caused a rapid dose-dependent reduction of bile acid synthesis in eight-day bile diverted rats, single dose 50 mg kg$^{-1}$ reduced bile synthesis to 5% of the control value (16).

## Any other comments

Toxicity and pharmacokinetics reviewed (17).

Has dose/time related cytotoxic effects against malignant cell lines *in vitro* (18).

## References

1. Savbensty, M. *Cas. Lek. Cesk.* 1986, **125**(48), 1488-1489 (Czech.) (*Chem. Abstr.* **106**, 131063c)

2.  Heel, R. C. *Drugs* 1982, **23**, 1-36
3.  Bechter, R. et al *Food Chem. Toxicol.* 1986, **24**(6-7) 641-642
4.  Mangas-Rojas, A. et al *Rev. Clin. Esp.* 1988, **183**(7), 358-364 (Span.) (*Chem. Abstr.* **110** 2250611z)
5.  Waller, D. P. et al *Contraception* 1990 **41**(4), 411-417
6.  Bechter, R. et al *Toxicol. In Vitro* 1987, **1**(1), 11-15
7.  Daneshmend. T. K. *Clin. Pharmacokinet.* 1988, **14**, 13-34
8.  Fujita, H. et al *Yakuri to Chiryo* 1991, **19**(5), 1845-1855, (Japan.) (*Chem. Abstr.*, **115**, 126341y)
9.  Babcock, N. R. et al *Eur. J. Clin. Pharmacol.* 1987, **33**(5), 531-534
10. Kobayashi, T. et al *Yakuri to Chiryo* 1991, **19**(5), 1857-1861 (Japan) (*Chem. Abstr.* **115**, 126342z)
11. *Martindale. The Extra Pharmacopoeia* 30th ed., 1993, 326-328, The Pharmaceutical Press, London
12. Morita, K. et al *J. Pharmacobio-Dyn* 1990, **13**(6), 336-343
13. Pawelec, G. et al *Int. J. Inmmunopharmacol.* 1991, **13**(2-3), 299-304
14. Senior, D. S. et al *Int. J. Immunopharmacol.* 1988, **10**(2), 169-173
15. Vanden Bossche, H. et al *Microsomes Drug Oxid. Proc. Int. Symp. 6th 1984* Publ. 1985, 63-73, Boobis, A. R. Ed., Taylor & Francis, London
16. Kuipers, F. et al *Lipids* 1989, **24**(9), 759-764
17. Liu, G. et al *Zhongguo Yaoxue Zazhi* 1989, **24**(5), 267-270 (Ch.) (*Chem. Abstr.* **112** 48105w)
18. Rochlitz, C. F. et al *Cancer Chemother. Pharmacol.* 1988, 21(4), 319-322

# K10   Ketoprofen

**CAS Registry No.** 22071-15-4

**Synonyms** benzeneacetic acid, 3-benzoyl-α-methyl-; Alrheumun; Orudis; hydratopic acid, *m*-benzoyl-; Profenid; 3-benzoylhydratropic acid

**Mol. Formula** $C_{16}H_{14}O_3$                           **Mol. Wt.** 254.29

**Uses** Analgesic, anti-inflammatory and antipyretic. Inhibitor of cyclo-oxygenase activity. Used in musculoskeletal and joint disorders.

## Physical properties
**M. Pt.** 94°C.

## Solubility
Organic solvent: diethyl ether, ethanol, acetone, chloroform, dimethyl formamide, ethyl acetate

## Mammalian and avian toxicity

### Acute data
$LD_{50}$ oral rat 101 mg kg$^{-1}$ (1).

### Metabolism and pharmacokinetics
Readily absorbed from the gastrointestinal tract in humans; peak plasma concentrations occur about 0.5-2 hr after a dose. When taken with food total bioavailability is not altered but rate of absorption is slowed. It is well absorbed from intramuscular and rectal routes though only a small amount is absorbed via topical application. Extensively bound to plasma proteins and substantial concentrations are found in synovial fluid. Plasma $t_{1/2}$ 2-4 hr. Metabolised mainly by conjugation with glucuronic acid and excreted mainly in urine (2).

Seven elderly subjects were administered a single rectal dose of 75 mg. No abnormalities were seen in clinical and physical findings. Maximum plasma concentration was reached at 0.5-2.0 hr. Urinary excretion of total ketoprofen during 72 hr after administration was 35-82% of dose, and 97% of the urinary total ketoprofen was in the form of glucuronide (3).

Ten healthy volunteers received daily 15 g of 2.5% ketoprofen topical gel, corresponding to 375 mg of ketoprofen on skin. The peak plasma concentration was 144 mg ml$^{-1}$ after the first administration with apparent absorption and elimination $t_{1/2}$ 3.2 and 27.7 hr respectively. The total quantities eliminated in the urine represented about 2.6% of the first dose applied. Apparent $t_{1/2}$ of ketoprofen was 17.1 hr and there was no accumulation (4).

### Irritancy
0.3%, adhesive agent was tested in ♂ rabbits for both primary and cumulative skin irritations. In the primary test hardly any irritation occurred and no irritation occurred in the cumulative test (5).

### Sensitisation
In photopatch tests for skin photosensitisation to UV-A plus UV-B light 3.8% of the subjects gave a positive reaction with ketoprofen (6).

## Any other adverse effects to man

Life threatening asthma, urticaria and angioedema developed in 2 aspirin sensitive patients after taking ketoprofen, 50 mg by mouth (7).
Cardiac and respiratory arrest occurred in an asthmatic patient (8).
Reported to cause photosensitivity reactions (9).

## Any other adverse effects

Oral rat (concentration unspecified) increased incidence of gastric ulcers (10).
DNA damage, single strand breaks was photoinduced by ketoprofen. No particular base specificity observed (11).

## Legislation
Included in Schedule 4 (Release into Air: Prescribed Substances) Statutory Instrument No. 472, 1991 (12).

## References
1. Veno, K. et al *J. Med. Chem.* 1976, **19**, 941
2. *Martindale. The Extra Pharmacopoeia* 30th ed., 1993, The Pharmaceutical Press, London

3.  Kobayashi, M. *Rinsho Yakuri* 1990, **21**(4), 675-681 (Japan.) (*Chem. Abstr.* **114** 239831f)
4.  Flourat, B. *Arzneim.-Forsch.* 1989, **39**(7), 812-815
5.  Saita, M. et al *Oyo Yakuri* 1986, **32**(5), 991-999 (Japan.) (*Chem. Abstr.* **106** 78445a)
6.  Przybilla, B. et al *Hautarzt* 1987, **38**(1), 18-25 (Ger.) (*Chem. Abstr.* **106**, 188635v)
7.  Frith, P. et al *Lancet* 1978, **ii** 847-848
8.  Schreuder, G. *Med. J. Aust.* 1990, **152**, 332-333
9.  *Med. Lett. Drugs Ther.* 1986, **28**, 51-52
10. Kojima, T. et al *Yakuri to Chirgo.* 1988, **16**(2), 611-619 (Japan.) (*Chem. Abstr.* **109**, 427p)
11. Artuso, T. et al *Photochem. Photobiol.* 1991, **54**(2), 205-13
12. *S. I. 1991 No. 472 The Environmental Protection (Prescribed Processes and Substances) Regulations* 1991, HMSO, London

# K11   Kinetin

**CAS Registry No.** 525-79-1

**Synonyms** 1*H*-purin-6-amine, *N*-(2- furanylmethyl)-; adenine, $N^6$-furfuryl-; FAP; *N*-furfuryladenine; 6-(furfurylamino)purine

**Mol. Formula** $C_{10}H_9N_5O$                               **Mol. Wt.** 215.22

**Uses** Plant growth regulator. To augment growth of microbial cultures.

**Occurrence** A cell division factor found in various plant parts and in yeasts. Isolation from autoclaved water slurries of deoxyribonucleic acid (1).

## Physical properties

**M. Pt.** 266-267°C (sealed tube)

## Environmental fate

### Absorption

Activated carbon was effective in adsorbing plant growth regulators including kinetin (2).

### Genotoxicity

*In vitro* wheat cell cultures, no significant effect on sister chromatid exchange induction (3).

### Legislation

Included in Schedule 4 (Release into Air: Prescribed Substances) Statutory Instrument No. 472, 1991 (4).

## Any other comments

Sublimes at 220°C.

## References

1. Miller, et al *J. Am. Chem. Soc.* 1955, **77**, 1392
2. Bu, X. et al *Zhiwa Shengli Xuebao* 1988, **14**(4), 401-405 (Ch.) (*Chem. Abstr.* **110**, 18812b)
3. Murata, M. *Theor. Appl. Genet.* 1989, **78**(4), 521-4
4. *S. I. 1991 No. 472 The Environmental Protection (Prescribed Processes and Substances) Regulations* 1991, HMSO, London

# L1   Lactic acid

## CH3CH(OH)CO2H

**CAS Registry No.** 50-21-5
**Synonyms** propanoic acid, 2-hydroxy-; Milk acid; 2-hydroxypropionic acid
**Mol. Formula** $C_3H_6O_3$                                                **Mol. Wt.** 90.08
**Uses** In the treatment of infective skin and vaginal disorders. Infusions to provide a source of bicarbonate for treatment of metabolic acidosis. Food preservatives and cosmetics.
**Occurrence** In sour milk as a result of lactic acid bacteria

## Physical properties

**M. Pt.** 16.8°C; **B. Pt.** $_{15}$ 122°C; **Flash point** 110°C (closed cup);
**Specific gravity** $d^{20}$ 1.249 at 15°C.

### Solubility
Water: miscible. Organic solvent: ethanol, furfurol

## Ecotoxicity

### Fish toxicity
$LC_{50}$ (18 hr) trout 100 mg $l^{-1}$ (1).

### Invertebrate toxicity
$LC_0$ (26-72 hr) *Daphnia* 170 mg $l^{-1}$ (2).

## Environmental fate

### Anaerobic effects
Methanogenic bacteria *Methanosarcina* sp. and *Methanobacterium* sp. were capable of metabolising lactate under methanogenic conditions (3).
Biodegradation occurs in anaerobic environments with transformation rate constants decreased as the food chain ascends acetate >lactate >glucose (4).
Anaerobic degradation by a mixed microbial culture, without sulfate and with both sulfate and molybdate, lactate was rapidly consumed and propionate and acetate were produced; whereas with sulfate alone, only acetate accumulated. Propionate oxidation was strongly accelerated by the presence of sulfate (5).

### Degradation studies
$BOD_5$ 0.63 mg $l^{-1}$ $O_2$ Warburg Sewage (6).
$BOD_{10}$ 0.88 mg $l^{-1}$ $O_2$ standard dilution sp. culture (7).
COD 100% ThOD (7).
*Tubifex tubifex* (0-2 hr), 80-85% of [14]C lactate was found in the intermediary products,

with glutamate and malate as the main constituents. During aerobic long-term incubation (12-24 hr) the largest proportion of label was incorporated into proteins. In the absence of oxygen most of the radioactivity remained in the intermediaries, mainly in alanine and succinate, during initial period of aerobic incubation, high amounts of [14]C-carbon dioxide were released (8).

100 ppm was treated by activated sludge process in a rotary cylinder type biological treatment at 20°C for 120 hr. 70-100% of lactic acid was removed (rotating tube) (9).

## Mammalian and avian toxicity

### Acute data
$LD_{50}$ oral rat 3730 mg $kg^{-1}$ (10).
$LD_{50}$ oral mouse 4875 mg $kg^{-1}$ (11).
$LD_{50}$ oral guinea pig 1.81 g $kg^{-1}$ (1).
$LD_{Lo}$ oral rabbit 500 mg $kg^{-1}$ (12)
$LD_{50}$ subcutaneous mouse 4500 mg $kg^{-1}$ (13).
Injection of 22.5 mg $l^{-1}$ into jugular vein of anesthetised rabbits initially increased breathing rate but no change or decrease in tidal volume. This was followed by deep and fast respiration (14).

### Carcinogenicity and long-term effects
Subcutaneous 16 mice (18 month) 125 mg, 2 lymphomas, 1 sarcoma and 1 pulmonary tumour were observed (15).

### Metabolism and pharmacokinetics
Bacterial metabolites (lactate) in the gut contents and the blood in relation to the faecal excretory cycle were studied in anaesthetised rabbits. The level of organic acids in the alimentary tract varied cyclically with the faecal excretion pattern. Lactate originates from the stomach; it was available for extrahepatic tissue metabolism (16).
Mechanism of hepatic lactate uptake was studied in the gulf toadfish by following the accumulation of [14]C lactate by isolated hepatocytes *in vitro*. Lactate uptake is by passive diffusion in toadfish hepatocytes. Lactate uptake by toadfish hepatocytes further differed from lactate uptake by mammalian tissues in that rates were not altered by changes in either extracellular pH or extracellular sodium ion concentration. Rates of lactate conversion to glucose and carbon dioxide were measured and compared to uptake rates; it appears the rates of lactate metabolism are not limited by passive diffusion (17).
The effect of glucose concentration (0-20 ml) on lactate uptake at low lactate concentration were studied in perfused livers from 48 hr starved rats with perfusate pH values of 7.4 and 6.8. Lactate uptake was independent of glucose concentration (0.18-1.80 mg $l^{-1}$), but was slightly inhibited with time at 3.60 mg $l^{-1}$ glucose (18).

### Irritancy
Dermal rabbit 500 mg (4 hr) caused severe irritation and 750 mg (24 hr) caused severe irritation (19).

## Genotoxicity
*Saccharomyces cerevisiae* and *Salmonella typhimurium* with and without metabolic activation, negative (15).
*Salmonella typhimurium* TA97, TA98, TA100, TA104 with and without metabolic activation, negative (20).

*In vitro* Chinese hamster ovary K1 cells 900-1261 mg l$^{-1}$ induced chromosomal aberrations at the initial pH of ≈6.0 with and without metabolic activation. No clastogenic activity was observed when the culture medium was first acidified with each of the acids and then neutralised with sodium hydroxide (21).

### Any other adverse effects

Lactic acidosis has been proposed to be one factor promoting cell death following cerebral ischemia. It has been demonstrated that cultured neurons and glia are killed by relatively brief (10 min) exposure to acidic solutions of <pH 5. The onset of death after exposure to moderately acidic solutions were delayed in some cells, such that death of the entire cell population became evident only 48 hr after acid exposure (22).

### Legislation

Included in Schedule 4 (Release into Air: Prescribed Substances) Statutory Instrument No. 472, 1991 (23).

### Any other comments

Toxicity reviewed (24).
Experimental toxicity and human health effects reviewed (25).
Analysis of the concentration of low molecular weight organic acids in soil, in the unsaturated zone, and in groundwater was undertaken. An unexpectedly high concentration of organic substances in natural surroundings that had significant dependence on the extractable amount of organic acids and on the pH value of the extract was observed. A clear connection between the spectrum of extractable organic acids and the microbiological activity was found in a depth profile from the soil down to the groundwater. In general, the spectrum of organic acids reduces to increasingly simpler substances with increasing depth (26).
Metabolism reviewed (27).

### References

1. Verschueren, K. *Handbook of Environmental Data on Organic Chemicals* 2nd ed., 1983, Van Nostrand Reinhold, New York
2. Meinck, F. et al *Les eaux residuaires industrielles* 1970
3. Soubes, M. *J. Appl. Microbiol. Biotechnol.* 1989, **5**(2), 193-198
4. Phelps, T. J. *J. Microbiol. Methods* 1991, **13**(4), 243-254
5. Qatibi, A. I. *Antonie van Leeuwenhoek* 1990, **58**(4), 241-248
6. Meissner, B. *Wasserwirtsch.-Wassertech., WWT* 1954, **4**, 166
7. Zobell, C. E. *Biol. Bull.* 1940, **78**, 388
8. Bock, S. et al *Comp. Biochem. Physiol., B: Comp. Biochem.* 1989, **92B**(1), 35-40
9. Muto, N. et al *Henkyu Hokoku – Kanto Gakuin Daigaku Kogakubu* 1987, **31**(2), 257-266 (Japan.) (*Chem. Abstr.* **112**, 83347y)
10. *J. Ind. Hyg. Toxicol.* 1941, **23**, 259
11. *Food Agri. Organ. U. N.* Report Series, 1967, **40**, 146
12. *Ind. Eng. Chem.* 1923, **15**, 628
13. *Z. Gesammte. Exp. Med.* 1944, **113**, 536 (Ger.)
14. Ducros, G. et al *Pediatr. Res.* 1991, **29**(6), 548-552
15. Fawell, J. K. et al *Environmental Toxicology: Organic Pollutants* 1988, Ellis Horwood Ltd., Chichester, England
16. Vernay, M. *Br. J. Nutr.* 1987, **57**(3), 371-381
17. Walsh, P. J. *J. Exp. Biol.* 1987, **130**, 295-304
18. Sestoft, L. et al *Clin. Sci.* 1988, **74**(4), 403-406

19. Marhold, J. V. *Sbornik Vysledku Toxixologickeho Vysetreni Latek A Pripravku* 1972, Prague
20. Al-Ani, F. Y. et al *Mutat. Res.* 1988, **206**, 467-470
21. Morita, T. *Mutat. Res.* 1990, **240**(3), 195-202
22. Nedergaard, M. *J. Neurosci.* 1991, **11**(8), 2489-2497
23. *S. I. 1991 No. 472 The Environmental Protection (Prescribed Processes and Substances) Regulations* 1991, HMSO, London
24. *BIBRA Toxicity Profile* 1991, British Biological Industrial Research Association, Carlshalton
25. *ECETOC Technical Report No. 30(4)* 1991, European Chemical Industry Ecology and Toxicology Centre, B-1160 Brussels
26. Cordt, T. et al *Vom Wasser* 1990, **74**, 287-298 (Ger.) (*Chem. Abstr.* **113**, 103034v)
27. Baessler, K. H. *Ernaehr.-Umsch.* 1988, **35**(3), 71-74 (Ger.) (*Chem. Abstr.* **109**, 52147s)

# L2   Lactic acid, butyl ester

## $CH_3CH(OH)CO_2CH_2CH_2CH_2CH_3$

**CAS Registry No.** 138-22-7
**Synonyms** 2-hydroxypropanoic acid, butyl ester; butyl $\alpha$-hydroxypropionate; butyl lactate; *n*-butyl lactate
**Mol. Formula** $C_7H_{14}O_3$                                **Mol. Wt.** 146.19

## Physical properties

**M. Pt.** –28°C; **B. Pt.** 183-187°C; **Flash point** 71°C (closed cup);
**Specific gravity** 0.984; **Volatility** v.p. 0.4 mmHg at 20°C; v. den. 5.04.

**Solubility**
Organic solvent: miscible in ethanol, diethyl ether

## Mammalian and avian toxicity

**Acute data**
$LD_{50}$ subcutaneous mouse, rat 11-12 g $kg^{-1}$ (1,2).
$LD_{Lo}$ intraperitoneal mouse 200 mg $kg^{-1}$ (3).

**Irritancy**
Dermal rabbit (24 hr) 500 mg caused moderate to severe erythema and moderate oedema (1).

## Legislation

Included in Schedule 4 (Release into Air: Prescribed Substances) Statutory Instrument No. 472, 1991 (4).

## Any other comments

Experimental toxicology, human health effects, workplace experience reviewed (5).
Affinity in biological systems described using solubility parameter techniques (6).
Autoignition temperature 382°C.

## References

1. *Food Cosmet. Toxicol.* 1979, **17**, 727
2. *Raw Material Data Handbook, Vol.1 Organic Solvents* 1974, National Association of Printing Ink Research Institute, Francis McDonald Sinclair Memorial Laboratory, Lehigh University, Bethlehem, PA
3. *Summary Tables of Biological Tests* 1955, No.7, National Research Council Chemical – Biological Coordination Center, Washington, DC
4. *S. I. 1991 No. 472 The Environmental Protection (Prescribed Processes and Substances) Regulations* 1991, HMSO, London
5. *ECETOC Technical Report No. 30(4)* 1991, European Chemical Industry Ecology and Toxicology Centre, B-1160 Brussels
6. Hansen, C. M. et al *Am. Ind. Hyg. Assoc. J.* 1988, **49**(6), 301-308

# L3   Lactonitrile

## CH3CH(OH)CN

**CAS Registry No.** 78-97-7
**Synonyms** propanenitrile, 2-hydroxy-; acetocyanohydrin; 2-hydroxypropionitrile
**Mol. Formula** $C_3H_5NO$                    **Mol. Wt.** 71.08
**Uses** Solvent, intermediate in production of lactate and lactic acid.

## Physical properties

**M. Pt.** –40°C; **B. Pt.** 183°C; **Flash point** 76.6°C; **Specific gravity** $d_{25}^{20}$ 0.9834;
**Volatility** v. den. 2.45.

## Solubility
Organic solvent: ethanol

## Ecotoxicity

**Fish toxicity**
$LC_{50}$ (24 hr) pinperch 0.215 mg kg$^{-1}$ (1).
$LC_{50}$ (96 hr) fathead minnow 0.9 mg kg$^{-1}$ (2).
$LC_{50}$ (96 hr) guppy 1.37 mg kg$^{-1}$ (3).
Median threshold limit (24-96 hr) bluegill sunfish 4.0 mg l$^{-1}$ (3).

## Environmental fate

**Degradation studies**
60% ThOD after 8 days at 20°C, 10 mg l$^{-1}$ of feed (4).

## Mammalian and avian toxicity

**Acute data**
$LD_{50}$ oral rat 87 mg kg$^{-1}$ (5).
$LD_{50}$ dermal rabbit 20 mg kg$^{-1}$ (5).
$LD_{Lo}$ subcutaneous rabbit 5 mg kg$^{-1}$ (6).
$LD_{Lo}$ subcutaneous frog 200 mg kg$^{-1}$ (6).

## Legislation

Included in Schedule 4 (Release into Air: Prescribed Substances) Statutory Instrument No. 472, 1991 (7).

## Any other comments

A retrospective structure-activity relationship (SAR) comparison has been reported of acute, subchronic toxicity, teratogenicity and biochemical mechanism studies of a series of structurally similar aliphatic nitriles including lactonitrile (8).

## References

1.  Garrett, J. T. et al *Texas J. Sci.* 1951, **3**, 391-396
2.  Jones, H. R. *Environmental Control in the Organic and Petrochemical Indutries* 1971, Noyes Data Corporation
3.  Nikunen, E. et al *Environmental Properties of Chemicals, Research Report 91/1990* 1991, VAPK-Publishing, Helsinki
4.  Ludzack, F. J. et al *J. Water Pollut. Control Fed.* 1960, **32**, 1173
5.  *Am. Ind. Hyg. Assoc. J.* 1969, **30**, 470
6.  *Arch. Int. Pharmacodyn. Ther.* 1899, **5**, 161
7.  *S. I. 1991 No. 472 The Environmental Protection (Prescribed Processes and Substances) Regulations* 1991, HMSO, London
8.  Johnson, F. R. *Fundam. Appl. Toxicol.* 1986, **7**(4), 690-7

# L4   β-Lactose

CAS Registry No. 5965-66-2
Synonyms β-D-glucopyranose,4- *O*-β-D-galactopyranosyl; lactose, β-; β-D-lactose
Mol. Formula $C_{12}H_{20}O_{11}$                                  Mol. Wt. 340.29

## Physical properties

M. Pt. 253-255°C.

## Any other adverse effects

It is a potent inhibitor of the lytic activity of natural killer cells as well as of the cytotoxic T-lymphocytes activated in mixed lymphocyte cultures (1).

## Legislation

Included in Schedule 4 (Release into Air: Prescribed Substances) Statutory Instrument No. 472, 1991 (2).

## References

1. Kornbluth, J. et al *Cell. Immunol.* 1984, **88**(1), 162-173
2. *S. I. 1991 No. 472 The Environmental Protection (Prescribed Processes and Substances) Regulations* 1991, HMSO, London

# L5    Lanthanum

## La

**CAS Registry No.** 7439-91-0

**Mol. Formula** La                                          **Mol. Wt.** 138.91

**Uses** $La^{3+}$ used in experimental biology as a specific antagonist of $Ca^{2+}$ (1).
Oxide in glass to improve optical properties.

**Occurrence** Estimated abundance in earth's crust: 5-18 ppm in ores monazite, bastnaesite and cerite.

## Physical properties

**M. Pt.** 920°C; **B. Pt.** 3454°C; **Specific gravity** 6.166 at 25°C.

## Ecotoxicity

**Fish toxicity**

$LC_{50}$ (28 day) rainbow trout 0.02 mg $kg^{-1}$ as lanthanum salt (2).

**Bioaccumulation**

Adult rainbow trout were exposed to various unspecified concentrations of $^{140}La^{3+}$. The dissociation constant for gill metal binding was 43 mg $l^{-1}$ for $La^{3+}$ (3). Accumulation within mussels and limpets was 0.904 and 1.14 µg $g^{-1}$ $La^{3+}$, respectively. The highest concentration within the mussel was 6.89 µg $g^{-1}$ $La^{3+}$ in the digestive gland (4).

## Genotoxicity

*Drosophila melanogaster* yw and mus $302^{D1}$ strain at 10, 20 or 30 Gy, chromosome loss 0.31, 0.89 or 0.46% respectively; recessive lethals 0.40, 1.3 or 2.4% respectively and translocations 0.69% at 10 Gy and 0% at 20 and 30 Gy (5).

## Any other adverse effects to man

High levels of lanthanum, antimony, arsenic, cadmium and lead and low levels of selenium salts found in the lung tissues of smelter workers who died from lung cancer, compared to control groups, suggesting a multifactorial genesis for the development of lung cancer and a protective effect of selenium in occupational exposure to certain carcinogens (6).

## Legislation

Included in Schedule 4 (Release into Air: Prescribed Substances) Statutory Instrument No. 472, 1991 (7).

## Any other comments

Cytoxicity reviewed (8).

Four field studies detailing application of lanthanum salts in ecological research are reported. Studies covered:

(I) Accretion and erosion of sediment in wetlands;

(II) Fire ant behaviour in a crop;

(III) Adsorption of chelated and nonchelated ions to aquatic roots;

(IV) Sorption by aquatic insect larvae (9).

Effects on cell growth and reproduction were studied in the unicellular cellular organism *Tetrahymena pyriformis*, elements with higher molecular weights showed lower nutritive values and higher toxicities (10).

If radioactive lanthanum was released in sufficiently large quantities from nuclear power stations it could contribute significantly to early bone marrow damage in humans (11).

## References

1.  Weiss *Ann. Rev. Phramacol.* 1974, **14**, 343
2.  Birge, W. J. et al *Aquatic toxicity tests on inorganic elements occurring in oil shale* 1980, EPA 600/9-80-022. NTIS, Springfield, VA
3.  Reid, S. D. et al *Can. J. Fish Aquat. Sci.* 1991, **48**(6), 1061-1068
4.  Smith, D. R. et al *Mar. Biol.* 1989, **102**, 127-133
5.  Fritz-Niggel, H. et al *Radiat. Environ. Biophys.* 1988, **27**, 133-141
6.  Gerhardsson, L. *Acta. Pharmacol. Toxicol. Suppl.* 1986, **59**(7), 256-259
7.  *S. I. 1991 No. 472 The Environmental Protection (Prescribed Processes and Substances) Regulations* 1991, HMSO, London
8.  Das, T. et al *Biol. Trace Elem. Res.* 1988, **18**, 201-28
9.  Knaus, R. M. *Eur. J. Solid State Inorg. Chem.* 1991, **28** (Suppl.), 379-382
10. Liv, Y. et al *Beijing Daxue Xuebao, Ziran Kexueban* 1986, (3), 101-104 (Ch.) (*Chem. Abstr.* **106** 2044y)
11. Alpert, D. J. et al *Environ. Effects* 1987, **28**(1), 77-86

# L6    Lanthanum trichloride

## LaCl$_3$

**CAS Registry No.** 10099-58-8

**Synonyms** lanthanum chloride (LaCl$_3$); lanthanium chloride (La$_2$Cl$_6$)

**Mol. Formula** Cl$_3$La                    **Mol. Wt.** 245.27

**Uses** Reagent for the conversion of carbonyls to thioacetal under mild conditions.

## Physical properties

**M. Pt.** 907°C.

## Solubility

Organic solvent: ethanol

## Mammalian and avian toxicity

### Acute data

LD$_{50}$ oral rat 4180 mg kg$^{-1}$ (1).

LD$_{50}$ intraperitoneal rat 106 mg kg$^{-1}$ (2).
LD$_{50}$ subcutaneous mouse 2420 mg kg$^{-1}$ (1).
LD$_{50}$ intravenous rabbit 148 mg kg$^{-1}$ (1).

### Any other adverse effects

Intravenous mice (concentration unspecified) as lanthanum chloride caused significant elevation in alanine transaminase activity and serum lipid peroxide 24 hr later. Degradation of hepatocytes and disappearance of glycogen in the cytoplasm was observed (3).

### Legislation

Included in Schedule 4 (Release into Air: Prescribed Substances) Statutory Instrument No. 472, 1991 (4).
Limited under EC Directive on Drinking Water Quality 80/778/EEC, guide level 25 mg chloride l$^{-1}$. Approximate concentration above which effects might occur 200 mg l$^{-1}$ (5).

### References

1. *Environ. Qual. Safety, Supp.* 1975, **1**, 1
2. *Arch. Ind. Health* 1957, **16**, 475
3. Chen, Q. et al *Zhongguo Yaolixue Yu Dulixue Zazhi* 1990, **4**(2), 131-133
4. *S. I. 1991 No. 472 The Environmental Protection (Prescribed Processes and Substances) Regulations* 1991, HMSO, London
5. *EC Directive Relating to the Quality of Water Intended for Human Consumption* 1982, 80/778/EEC, Office for Official Publications of the European Communities, 2 rue Mercier, L-2985 Luxembourg

# L7   Lasiocarpine

**CAS Registry No.** 303-34-4
**Synonyms** 2-butenoic acid, 2-methyl-, 7-[[2,3-dihydroxy-2-(1-methoxyethyl)-3-methyl-1-oxobutoxy]methyl]-2,3,5,7a-tetrahydro-1*H*-pyrrolizin-1-yl ester, [1*S*-[1a(*Z*),7(2*S*; 3*R*),7aα]]-
**Mol. Formula** C$_{21}$H$_{33}$NO$_7$     **Mol. Wt.** 411.50
**Uses** As an emetic and in the treatment of snake bites in south east Asia.

**Occurrence** Isolated only from plant species of the family Boraginaceae; *Heliotropium europaeum, Heliotropium lasiocarpum* and *Symphytum caucasicums*.

## Physical properties
**M. Pt.** 94-95.5°C.

**Solubility**
Organic solvent: diethyl ether, ethanol, benzene

## Mammalian and avian toxicity

**Acute data**
$LD_{50}$ oral rat 150 mg $kg^{-1}$ (1).
$LD_{Lo}$ intravenous mouse 85 mg $kg^{-1}$ (2).
$LD_{Lo}$ intravenous guinea pig 50 mg $kg^{-1}$ (3)
$LD_{Lo}$ intravenous monkey 20 mg $kg^{-1}$ (2).

**Carcinogenicity and long-term effects**
No adequate evidence for carcinogenicity to humans, sufficient evidence for carcinogenicity to animals, IARC classification group 2B (4).
Oral (104 wk) ♂, ♀ Fischer 344 rats at 7, 15 or 30 ppm in diet. Mean body weights of the high dose ♂, ♀ rats were lower than the control groups, weights of mid dose rats were lower only in second year and weights of low dose rats were unaffected. All surviving ♂ rats developed tumours except for one low dose and one high dose animal. Among ♀ rats 23 low dose and 22 high dose animals developed tumours. It is carcinogenic in Fischer 344 rats producing hepatocellular tumours and angiosarcomas of the liver in both sexes and hematopoietic tumours in ♀ animals (5).
Intraperitoneal injections (dose unspecified) were given to rats fed aflatoxin $B_1$ in the diet and also pretreated with lasiocarpine to produce an antimitotic effect, liver tumours developed after a similar time (18 wk) and in similar numbers to those in rats given aflatoxin alone. The tumours were associated with post necrotic cirrhosis or advanced portal scarring not seen in rats receiving aflatoxin alone (6).

**Teratogenicity and reproductive effects**
Suckling rats showed toxic signs and died with severe liver lesions when their mothers were given total doses of 125 mg $kg^{-1}$ body weight twice wkly. The mothers showed no outward ill effects (7).

**Metabolism and pharmacokinetics**
*In vivo* rats, metabolised via pyrrole formation (8).
Most of the toxic effects appeared to be mediated via the very reactive dehydroalkaloid metabolites that are produced by the liver mixed function oxidases. The toxicity is not related to the level of activity of this enzyme system (9,10). Dehydroheliotridine has been isolated and identified as a product of microsomal oxidation (11).

## Genotoxicity
*Salmonella typhimurium* TA100 with metabolic activation positive (12).
*Salmonella typhimurium* TA98, TA1535, TA1537 with and without metabolic activation negative (12).
*Salmonella typhimurium* TA1535 with and without metabolic activation negative (13).

## Any other adverse effects

*In vitro* rat hepatocytes significant toxicity observed (14).
Low doses produce severe haemorrhagic necrosis of the liver, gastro-intestinal
haemorrhage, sometimes congestion and oedema of the lungs, congestion of the
adrenals and sometimes pyloric, duodenal and rectal ulceration (species unspecified)
(15-17).

## Legislation

Included in Schedule 4 (Release into Air: Prescribed Substances)Statutory Instrument
No. 472, 1991 (18).

## Any other comments

Environmental toxicology and human health effects reviewed (19).
It is a pyrrolizidine alkaloid.

## References

1.  *Toxicol. App. Pharmacol.* 1970, **17**, 290
2.  *J. Pharmacol. Exp. Ther.* 1959, **126**, 179
3.  Chen, K. K. et al *J. Pharmacol. Exp. Ther.* 1940, **68**, 123-129
4.  *IARC Monograph* 1987, **Suppl. 7**, Lyon
5.  *National Toxicology Program Research and Testing Division* 1992, Report No. TR-039,
    NIEHS, Research Triangle Park, NC
6.  Reddy, J. K. et al *Arch. Pathol.* 1972, **93**, 55-60
7.  Schoental, R. *J. Pathol. Bact.* 1959, **77**, 485-495
8.  Buhler, D. R. et al *Adv. Exp. Med. Biol.* 1986, **197**(3), 611-620
9.  Mattocks, A. R. *Chem.-Biol. Interact.* 1972, **5**, 227-242
10. Mattocks, A. R. et al *Chem.-Biol. Interact.* 1971, **3**, 383-396
11. Jago, M. V. et al *Mol. Pharmacol.* 1970, **6**, 402-406
12. Zeiger, E. *Cancer Res.* 1987, **47**, 1287-1296
13. Shinada, T. et al *Cancer Res.* 1989, **49**, 3218-3228
14. Moore, D. J. *Toxicol. Appl. Pharmacol.* 1989, **101**(2), 271-284
15. Bull, L. B. et al *J. Pathol. Bact.* 1958, **75**, 17-25
16. Bull, L. B. et al *The Pyrrolizidine Alkaloids* 1968, Elsevier, Amsterdam
17. Schoental, R. et al *J. Pathol. Bact.* 1957, **74**, 305-319
18. *S. I. 1991 No. 472 The Environmental Protection (Prescribed Processes and Substances)
    Regulations* 1991, HMSO, London
19. *ECETOC Technical Report No. 30(4)* 1991, European Chemical Industry Ecology and
    Toxicology Centre, B-1160 Brussels

# L8   Lauric acid

$$CH_3(CH_2)_{10}CO_2H$$

**CAS Registry No.** 143-07-7
**Synonyms** dodecanoic acid; ABL; Aliphat No.4; Vulvic acid; Laurostearic acid;
Neo-Fat 12
**Mol. Formula** $C_{12}H_{24}O_2$                                            **Mol. Wt.** 200.32

---

## Physical properties

**M. Pt.** 48°C; **B. Pt.** $_{100}$ 225°C; **Flash point** 110°C (closed cup);
**Specific gravity** $d_4^{50}$ 0.869; **Partition coefficient** log $P_{ow}$ 4.6.

## Solubility

Organic solvent: chloroform, benzene, diethyl ether, ethanol, petroleum ether

## Ecotoxicity

### Fish toxicity

$LC_{50}$ red killifish (seawater) no mortality occurred at saturation (unspecified duration) (1).
$LC_{50}$ red killifish (freshwater) 20 mg $l^{-1}$ (unspecified duration) (1).

### Invertebrate toxicity

Untreated retort waters were highly toxic to *Nitzschia closterium*, growth inhibited by concentration as low as 0.01% of retort water in sea water. Aliphatic acids are one of the many toxic components of retort waters (2).

## Environmental fate

### Anaerobic effects

The effect of lauric acid on the microbial formation of methane was investigated using *Methanothrix* sp. Inhibition commenced at 320 and 861 mg $l^{-1}$, the maximum specific acetoclastic methanogenic activity was reduced to 50%. Synergistic toxicity was observed with capric acid and myristic acid (3).
Under laboratory conditions 60-90% degradation occurred. Results obtained by measuring the amount of fermentation gas and methane concentration (4).

### Degradation studies

Activated sludge (6 hr) 4.1% ThOD; (12 hr) 4.3% ThOD; (24 hr) 6.1% ThOD (5).

## Mammalian and avian toxicity

### Acute data

$LD_{50}$ oral rat 12 g $kg^{-1}$ (6).
$LD_{50}$ intravenous mouse 131 mg $kg^{-1}$ (7).

### Irritancy

Based on the available data for studies in animals and humans it was concluded that lauric acid is non-irritant and is safe in present practice of use and concentration in cosmetics (8).

## Genotoxicity

*Salmonella typhimurium* TA97, TA98, TA100, TA1535, TA1537, with and without metabolic activation, negative (9).

## Legislation

Included in Schedule 4 (Release into Air: Prescribed Substances) Statutory Instrument No. 472, 1991 (10).

## References

1. Onitsuka, S. et al *Chemosphere* 1989, **18**(7-8), 1621-1631
2. Mann, K. et al *Fuel* 1987, **66**(3), 404-407
3. Koster, I. W. et al *Appl. Environ. Microbiol.* 1987, **53**(2), 403-409

4. Petzi, S. *Seifen, Oele, Fette, Wasche* 1989, **115**(7), 229-232 (Ger.) (*Chem. Abstr.* **111**, 83456n)
5. Maloney, G. W. et al *J. Water Poll. Control Fed.*, 1966, **41**(2), R18-R83
6. *Acta. Pharmacol. Toxicol.* 1961, **18**, 141
7. *Food Drug Res. Labs., Papers* 1976, 123
8. *J. Am. Coll. Toxicol.* 1987, **6**(3), 321-401
9. Zeiger, E. et al *Environ. Mol. Mutagen.* 1988, **11**(Suppl.12), 1-158
10. *S. I. 1991 No. 472 The Environmental Protection (Prescribed Processes and Substances) Regulations* 1991, HMSO, London

# L9   Lauryl pyridinium chloride

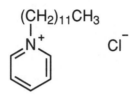

**CAS Registry No.** 104-74-5
**Synonyms** pyridinium, 1-dodecyl-, chloride; Dehyquart C; DPC; Eltren; 1-dodecylpyridinum chloride; Quaternario LPC
**Mol. Formula** $C_{17}H_{30}ClN$                     **Mol. Wt.** 283.89

## Physical properties
**M. Pt.** 66-70°C.

## Environmental fate

### Absorption
Adsorption of lauryl pyridinium chloride on sodium kaolinite was studied (1).
The adsorption of lauryl pyridinium chloride onto different adsorbents was examined at equilibrium concentrations from 10 to 300 $\mu$mol $l^{-1}$. The adsorption maximum for soils and compost was less than 10 $\mu$mol $g^{-1}$. There is no significant correlation between adsorption behaviour and organic matter contents in soils and composts (2). The reaction of the cationic surfactant, lauryl pyridinium chloride, with model humic substances was studied. The compound was bound to the humic substance, whereby the products contained between 25-70% of the surfactant. Reaction with *in situ* generated model humic substance gave products containing higher lauryl pyridinium chloride contents than those prepared by precipitation reactions with the various humic substances, indicative of multilayer binding. Both the former and the sodium hydroxide extract revealed a drastic reduction of free carboxylic groups following reaction with lauryl pyridinium chloride (3).

## Mammalian and avian toxicity

### Irritancy
Eye guinea pig 5% (w/v) was found to be extremely irritating (4).

## Legislation

Included in Schedule 4 (Release into Air: Prescribed Substances) Statutory Instrument No. 472, 1991 (5).

Limited under EC Directive on Drinking Water Quality 80/778/EEC. Guide level 1 μg l$^{-1}$ haloform concentrations must be as low as possible (6).

## Any other comments

Cationic surfactant biodegradability was studied in river waters (7).

Human mixed lymphocyte reaction 94±% inhibition and toxic to cells (8).

With increase in the ionic strength of the medium, which contained 2 ml sheep red blood cells in a buffer, haemolytic activity of lauryl pyridinium chloride decreased (9).

## References

1. Mehrian, T. et al *Langmuir* 1991, **7**(12), 3097-3098
2. Gerke, J. et al *Chem. Erde* 1990, **50**(4), 247-253 (Ger.) (*Chem. Abstr.* **114**, 145850j)
3. Gerke, J. et al *Chem. Erde* 1991, **51**(1), 23-8 (Ger.)
4. Bracher, M. et al *Mol. Toxicol.* 1987, **1**(4), 561-570
5. *S. I. 1991 No. 472 The Environmental Protection (Prescribed Processes and Substances) Regulations* 1991, HMSO, London
6. *EC Directive Relating to the Quality of Water Intended for Human Consumption* 1982, 80/778/EEC, Office for Official Publications of the European Communities, 2 rue Mercier, L-2985 Luxembourg
7. Ruiz Cruz, J. *Grasas Aceites (Seville)* 1987, **38**(6), 383-388 (Span.) (*Chem. Abstr.* **109**, 98379x)
8. Coy, E. A. et al *Int. J. Immunopharmacol.* 1990, **12**(8), 871-881
9. Murmatsuo, N. *J. Jpn. Oil Chem. Soc.* 1990, **39**(8), 555-559

# L10   Lead

## Pb

**CAS Registry No.** 7439-92-1

**Synonyms** Lead; C.I. 77575; C.I. Pigment Metal 4; Lead Flake; Pb-S 100; SO

**Mol. Formula** Pb                                    **Mol. Wt.** 207.19

**Uses** Construction material for tank linings, piping. For x-ray and atomic radiation protection. Pigments for paints. Bearing metal and alloys. Storage batteries. Solder and other lead alloys.

**Occurrence** Occurs in galena, galenite, lead sulphide, cerussite, anglesite, lancarksite, massicot and matlockite ores. Contaminant in cigarette smoke (1). Environmental pollutant through surface runoff and airborne lead. Extent of occurrence in earth's crust ≈15 g ton$^{-1}$, or 0.002%. Residue in soils (2). Lead, in the form of simple salts is acutely toxic to aquatic invertebrates at concentration between 0.1->40 mg l$^{-1}$ (freshwater) and 2.5->500 mg l$^{-1}$ (marine). However lead salts are poorly soluble in water and the presence of other salts reduces the bioavailability because of precipitation. Concentrations of lead in water are nominal in most studies; the contribution to toxicity of factors such as pH, water hardness, anions and complexing agents cannot be fully evaluated.

## Physical properties

**M. Pt.** 327.4°C; **B. Pt.** 1740°C; **Specific gravity** $d_4^{20}$ 11.34.

## Occupational exposure

**US TLV (TWA)** 0.05 mg m$^{-3}$ (as Pb) intended change; **UK Long-term limit** (MEL) 0.15 mg m$^{-3}$ (as Pb) under review.

## Ecotoxicity

### Fish toxicity

$LC_{50}$ (28 day) rainbow trout 0.22 mg kg$^{-1}$ (Pb salt) (3).
$EC_{50}$ (96 hr) giant gourami 26 mg kg$^{-1}$ (Pb salt) (4).
LOEC chronic survival rainbow trout 0.0007 mg kg$^{-1}$ (Pb salt) (5).

### Invertebrate toxicity

$EC_{50}$ (96 hr) reproduction *Navicula incerta* 11 mg kg$^{-1}$ (Pb salt) (6).
$LC_{50}$ (48 hr) *Daphnia magna* 0.3 mg kg$^{-1}$ (Pb salt) (7).
$LC_{50}$ (48 hr) *Bufo arenarum* 0.47-0.90 mg l$^{-1}$. Concentrations $\geq$0.25 mg Pb$^{2+}$ l$^{-1}$ interfered with normal embryo development (8).
$EC_{50}$ (96 hr) *Perna viridie* 8820 µg l$^{-1}$ (Pb salt) (9).
$EC_{50}$ (48 hr) *Daphnia magna* 3.61 ppm (Pb salt) (10).

### Bioaccumulation

Lead uptake is slow and reaches equilibrium only after prolonged exposure.
*Crassostrea virginica* (49 day) exposed to 25, 50, 100, 200 mg Pb$^{2+}$ l$^{-1}$ in water, final concentrations in soft tissues were 17, 35, 75 and 200 mg kg$^{-1}$ lead respectively (11).
Uptake by mallard ducks (14 day) exposure to 178 g m$^{-2}$ of lead shot. The liver contained 28.4 mg kg$^{-1}$ wet weight and bones contained 176 mg kg$^{-1}$ dry weight (12).
A bioconcentration factor of less than 1 was observed in earthworms in soils with 15-50 mg lead kg$^{-1}$ soil (13,14).
*Arca granosa* accumulated lead in soft tissue and blood. Distribution order: internal organs >gill >mantle >blood. Positive correlation with lead concentration in seawater (15).
*Artemia salina* exposed to concentrations of 5 µg l$^{-1}$ accumulated 250 µg l$^{-1}$. Lead had a synergistic effect on heavy metal uptake in combined heavy metal element solution (16).
Several species of fish, molluscs and crustacea accumulate lead. Main storage tissues are digestive tract and exoskeleton. Spherocrystals and lysosomes are prominent in accumulation and elimination (17).

## Environmental fate

### Nitrification inhibition

20 mg (Pb salt) l$^{-1}$ inhibited denitrification in rotating disc and 20 mg (Pb salt) l$^{-1}$ inhibited nitrification and denitrification in activated sludge (18).
0.5 mg l$^{-1}$ was toxic to *Nitrosomonas* (19).
*Azotobacter* exposed to high concentrations showed inhibited urease activity. Inhibition threshold concentration range 30-500 ppm (20).

## Absorption

Cores of marine sediment were used as model systems to examine the degradation of digested sewage sludge in the marine environment at the sediment-water interface. Lead added to the model systems in the sludge were immobilised by the sediment and not exported from the model (21).

Adsorption to sediment occurs rapidly and almost quantitatively (22).

Inorganic lead tends to form highly insoluble salts and complexes with various anions and is tightly bound to soil, therefore availability to terrestrial plants is drastically reduced and effects to plants are normally only observed at very high environmental concentrations 100-1000 mg kg$^{-1}$ soil (23).

## Mammalian and avian toxicity

### Acute data

$TD_{Lo}$ oral woman 450 mg kg$^{-1}$ (24).
$TD_{Lo}$ oral mouse 4800 mg kg$^{-1}$ (25).
$LD_{Lo}$ oral pigeon 160 mg kg$^{-1}$ (26).
$LD_{Lo}$ intraperitoneal rat 1000 mg kg$^{-1}$ (27).

### Sub-acute data

American kestrel (7 month) 0, 10 or 50 mg kg$^{-1}$. Lead levels increased in liver and bones though no adverse effects were observed (28).

Oral American kestrel (10 day) 125 or 625 mg kg$^{-1}$ body weight. Reduced haematocrit, haemoglobin levels and plasma creatine phosphorylase activity was observed (29).

Oral ♂ bobwhite quail (4 wk) 10 lead shot. Increased mortality was observed, >90% of ♂ dosed with 30 lead shots wk$^{-1}$ died within 4 wk (30).

Oral rat (6 month) single concentration lead or combined heavy metals in non-toxic, minimal or toxic concentrations. Every single metal affected the concentrations of some of the other metals in the liver, brain, femoral bone, spleen and kidney. The changes are due to compensatory mechanism and/or toxic effects of the metal (31).

### Carcinogenicity

Renal tumours have been induced in rats following administration of large doses (32). Gavage (12 month) ♂, ♀ Fischer 344 rat, 10 mg twice a month. One lymphoma and 4 leukaemias in 47 treated rats. No significant difference from control rate (33).

### Teratogenicity and reproductive effects

Lead salts are gonadotoxic and embryotoxic (34,35).

Caused a reduction in the number of pregnancies in successfully mated mice compared with controls (36).

Reduced foetal birth weight, neonatal body weight and motor activity and induced skeletal deformities in mice (37).

Lead concentrations (Pb salts) were above normal in women with spontaneous abortions and were above normal in mothers of newborn with intrauterine growth retardation (38).

Developmental toxicity in humans not determined; in rabbits negative; and in rats mice and hamsters positive (39).

### Metabolism and pharmacokinetics

Lead crosses placental barrier (40).

Inhalation (different periods) rats 0.05, 77, 249 and 1546 µg (Pb salt) m$^{-3}$. Blood Pb levels declined with a $t_{1/2}$ of between 3 and 5 days. Baseline blood concentration was reached by 25 and 50 days for the 2 smallest doses of lead. The most heavily contaminated groups had not returned to initial concentration after 180 days (41). Single oral dose (unspecified) of $^{203}$Pb radiolabel to pregnant mice. Effects differed depending on which gestation period the animals were treated, retention was markedly increased in later stages of pregnancy. Lead accumulated in bone and kidney (42).

Absorbed from the gastro-intestinal tract, also absorbed by the lungs from dust particles. Inorganic lead is not absorbed through intact skin whilst organic lead is absorbed rapidly. Distributed in soft tissues, higher concentrations in liver and kidneys. Associated with erythrocytes in blood. Accumulates in body, deposited in calcified bone, hair and teeth. It crosses placental barrier. Excreted in faeces, urine, sweat and milk (43).

A level of 0.20 mg l$^{-1}$ was associated with a 75% decrease in δ-aminolevulinic acid dehydratase activity (44).

Single oral dose of 0.5 and 1.5 mg $^{204}$Pb, kinetic data was derived to estimate degree of human intestinal absorption from the empty stomach which varied from 10 to 80%, $t_{1/2}$ of lead in blood was in the range of 39 to 53 days (45).

75 volunteers from plastic industry had lead blood levels of 15-20 mg 100 ml$^{-1}$. The δ-aminolevulinic acid excretion in urine was 2-8.4 mg in 24 hr (46).

## Genotoxicity

*Salmonella typhimurium* TA98, TA102, TA1535, TA1537 with metabolic activation and *Salmonella typhimurium* TA102 without metabolic activation positive (47).

*In vitro* Chinese hamster cells induced an increased number of UV induced mutations and sister chromatid exchange (48).

*In vitro* mice bone marrow induced chromosomal aberrations and increased frequency of sister chromatid exchanges (49).

*Monopterus albus*, *Cyprinus carpio* and *Aristichthys nobilis* micronucleus test positive (50).

*In vitro* human peripheral lymphocytes (72 hr) 0.05 µg l$^{-1}$ increased number of sister chromatid exchanges (51).

## Any other adverse effects to man

90 ♂ occupationally exposed, the lead exposure levels may have resulted in a subclinical increase in follicle-stimulating hormone, which was related to blood lead levels. This suggests that lead may be causing some subclinical primary damage to the seminiferous tubules in the testes. However at blood levels of >47 µg ml$^{-1}$ this effect on serum follicle-stimulating hormone was not apparent (52).

Acute toxicity-anorexia, vomiting, malaise, convulsions due to increased intercranial pressure, common in young children with a history of pica. Children with chronic toxicity show weight loss, weakness, anaemia. Lead poisoning in adults is usually occupational, due mainly to inhalation of lead dust or fumes (53).

Organic lead poisoning produces mainly central nervous system symptoms, there can be gastro-intestinal and cardiovascular effects, renal and hepatic damage (43).

Higher mortality from malignant neoplasms has been reported in smelter workers. In battery workers no association was found between lead exposure and cancer (54,55). Chromosomal abberrations have been reported in occupationally exposed workers (56).

Lung cancer mortality was examined in 4393 men employed in a lead smelter. There was an excess of lung cancer which was particularly evident for those employed for >20yr. Lung cancer mortality was associated with estimates of cumulative exposure to arsenic and lead. It was not possible to determine whether the increased risk might be due to arsenic, lead or other contaminants in the smelter (57).

A case study has shown an increased risk for glassworks employees to die from stomach cancer, lung cancer and cardiovascular disorders. A follow up study covering the entire glass-producing industry of Sweden confirmed the earlier results and, furthermore, an excess risk for colon cancer was also identified. The grouping of glassworks employees according to type of metal contamination showed an excess risk of stomach cancer, colon cancer and cardiovascular deaths related to glassworks with a high consumption of lead, arsenic, antimony and manganese. Their exposure might be oral, involving the glassblowers pipe as a vector for the exposure to various metals (58).

A study of working conditions at all stages of lead production showed that the worker morbidity was affected by chemical factors involving complex polymetallic dust and sulfur dioxide (59).

The relationship between blood lead concentration and blood pressure was examined in a survey of 7371 men aged 40 to 59 in UK. There exists a very weak but statistically significant positive association between blood lead and both systolic and diastolic blood pressure. After 6 years of follow-up, 316 of these men had major ischemic heart disease and 66 had had a stroke. There is no evidence that blood lead is a risk factor for these cardiovascular events. This and other surveys provides reasonably consistent evidence of the relationship between lead and blood pressure (60).

## Any other adverse effects

The effects of atmospheric lead concentrations were investigated in Madrid city pigeons. Renal tissues contained >30 $\mu$g g$^{-1}$ (dry weight). Inclusions present in proximal convoluted tubules and hepatic haemosiderosis were observed (61).

## Legislation

Included in Schedule 4 (Release into Air: Prescribed Substances) Statutory Instrument No. 472, 1991 (62).

Limited under EC Directive on Drinking Water Quality 80/778/EEC. Maximum admissible concentration 50 $\mu$g Pb l$^{-1}$ (in running water). Where lead pipes are present, content should not exceed 50 $\mu$g l$^{-1}$ in a sample after flushing. If the sample is taken either directly or after flushing and lead content exceeds 100 $\mu$g l$^{-1}$, measure must be taken to reduce the exposure (63).

A maximum admissible concentration of 10 $\mu$g dl$^{-1}$ was established for Canadian drinking water in Autumn 1989. Lead was classified as a possible carcinogen and the concentration evaluated using threshold concentrations (64).

## Any other comments

Carcinogenicity, mutagenicity, teratogenicity and neurotoxiocology reviewed (65-68). Ecotoxicity, epidemiology, environmental effects, environmental toxicity, exposure,

exposure levels, human health effects, physico-chemical properties and work exposure reviewed (69).

Killing of 34 simian primates and 3 fruit bats in Washington Zoo by lead in paint on their cages has been reported (22).

Samples of topsoil were taken and the total content of 19 elements determined. Mean lead values for uncontaminated soil in the 30-40 mg kg$^{-1}$ range were calculated (70). Calculated concentration below which biological effects were minimal in sediment quality criteria was 50 μg Pb g$^{-1}$ (dry weight sediment) (71).

Carcinogenicity, sources, environmental levels, human exposure and absorption reviewed (72-75).

Environmental exposure to humans and related blood effects reviewed (76).

The biogeochemical availability to aquatic and semiaquatic wildlife reviewed (77).

Lead ingestion and inhibition studies in humans 1937-1971 reviewed (78).

Environmental fate, toxicity and genotoxicity reviewed (79).

Toxicity to terrestial animals reviewed (80).

Health hazards of airborne particulate matter reviewed (81).

Toxicology, metabolism, teratogenicity, neurotoxicity and carcinogenicity reviewed (82-87).

Toxicity profile lead (88).

Lead and lead compounds comprehensively reviewed (89).

# References

1. UNEP/WHO *Lead (Environmental Health Criteria 3)*, United Nations Environment Programmes/World Health Organization, Geneva
2. Davies, B. E. *Sci. Total. Environ.* 1978, **9**, 243-262
3. Birge, W. J. et al *Aquatic toxicity tests on inorganic elements occurring in oil shale* 1980 EPA600/9-80-022. NTIS, Department of Commerce, Springfield, VA
4. Saxena, O. P. et al *J. Environ. Biol.* 1983, **4**, 91
5. Nikunen, E. et al *Environmental Properties of Chemicals, Research Report 91/1990* 1991, VAPK-Publishing, Helsinki
6. Rachlin, J. W. et al *Bull. Torrey Bot. Club.* 1983, **110**, 217
7. Biesinger, K. E. et al *J. Fish. Res. Board Can.* 1972, **29**, 1691-1700
8. Perez-Coll, C. S. et al *Bull. Environ. Contam. Toxicol.* 1988, **41**(2), 247-252
9. Chan, H. M. *Mar. Ecol.: Prog. Ser.* 1988, **48**(3), 295-303
10. Khangarot, B. S. et al *Bull. Environ. Contam. Toxicol.* 1987, **38**(4), 722-726
11. Pringle, B. H. *J. Sanit. Eng. Div. Proc. Am. Soc. Civil Eng.* 1968, **94**(SA3) 455-475
12. Irwin, J. C. et al *J. Wildl. Dis.* 1972, **8**, 149-154
13. Beyer, W. N. et al *J. Environ. Qual.* 1982, **11**, 381-385
14. Van Hook, R. I. *Bull. Environ. Contam. Toxicol.* 1974, **12**, 509-512
15. Wang, C. et al *Haiyang Xuebao* 1986, **8**(6), 724-728 (Ch.) (*Chem. Abstr.* **109**, 185126f)
16. Chen, J. C. et al *J. World Aquacult. Soc.* 1987, **18**(2), 84-93
17. Chassard-Bouchaud, C. *Anal. Chim. Acta.* 1987, **195**, 307-315
18. Knoetze, C. *Water Res.* 1979, (9), 5-6
19. Martin, G. et al *Water Sci. Technol.* 1982, **14**, 781-784
20. Liu, Q. et al *Huanjing Kexue Xuebao* 1986, **6**(4), 385-394 (Ch.) (*Chem. Abstr.* **106**, 170531y)
21. Nedwell, D. B. et al *Mar. Pollut. Bull.* 1990, **21**(2), 87-91
22. Zook, B. C. et al *J. Wildl. Dis.* 1972, **8**, 264-272
23. *Environmental Health Criteria 85: Lead Environmental Aspects* 1989, WHO, Geneva
24. *J. Am. Med. Assoc.* 1977, **237**, 2627
25. *Bull. Environ. Contam. Toxicol.* 1977, **18**, 271

26. *Aber. Hand. Biol. Arbeit.*, 1935, **4**, 1289
27. *Environ. Qual. Safety, Supp.* 1975, **1**, 1
28. Pattee, O. H. *Arch. Environ. Contam. Toxicol.* 1984, **13**, 29-34
29. Hoffman, D. J. et al *Comp. Biochem. Physiol.* 1985, **80**, 431-439
30. Damron, B. L. et al *Bull. Environ. Contam. Toxicol.* 1975, **14**, 489-496
31. Wadeenko, V. G. et al *Gig. Sanit.* 1990, **6**, 24-26 (Russ.)
32. Mao, P. et al *Am. J. Pathol.* 1967, **50**, 571
33. Furst, A. et al *Cancer Res.* 1976, **36**, 1779-1783
34. Krassowski, G. N. et al *Z. Gesammte Hyg. Ihre Grenzgeb.* 1980, **26**(5), 336-338
35. Charyev, O. G. *Gigi. Aspekty Okhr. Okruzh. Sredy* 1978, (6), 106-108
36. Jacquet, P. *Arch. Pathol. Lab. Med.* 1977, **101**(12), 641-643
37. Bederka, J. P., Jr et al *Survival Toxic. Environ. (Paper. Symp.)* 1973, 515-520
38. Kharkwal, S. et al *J. Obstet. Gynaecol. India* 1988, **38**(5), 555-558
39. Jelovesk, F. R. et al *Obstet. Gynecol.* 1989, **74**(4), 624-636
40. Carpenter, S. J. *Environ. Health Perspect.* 1974, **7**, 129-131
41. Grobler, S. R. et al *S. Afr. J. Sci.* 1988, **84**(4), 260-262
42. Donald, J. M. et al *Environ. Res.* 1986, **41**(2), 420-431
43. *Martindale. The Extra Pharmacopoeia* 30th ed., 1993, The Pharmaceutical Press, London
44. Dieter, M. P. *Arch. Environ. Contam. Toxicol.* 1976, **5**, 1-13
45. Heusler-Bitschy, S. et al *Trace. Elem. Anal. Chem. Med. Biol. Proc. Int. Workshop, 5th* 1988, 627-634
46. Tabaku, A. et al *Rev. Mjekesore* 1988, (5), 21-24 (Albanian) (*Chem. Abstr.* **110**, 218237u)
47. Wong, P. K. *Bull. Environ. Contam. Toxicol.* 1988, **40**(4), 597-603
48. Hartwig, A. et al *Mutat. Res.* 1990, **241**(1), 75-82
49. Nayak, B. N. et al *Exp. Pathol.* 1989, **36**(2), 65-73.
50. Chen, Y. et al *Hunan Shifan Daxue Xuebao, Ziran Kexueban* 1986, **9**, (2, Suppl.), 85-91 (Ch.) (*Chem. Abstr.* **106**, 97652g)
51. Bassendowska-Karska, E. et al *Rocz. Panstw. Zakl. Hig.* 1986, **37**(4), 278-285
52. McGregor, A. J. et al *Hum. Exp. Toxicol.* 1990, **9**(6), 371-376
53. *The Merck Index* 11th ed., 1989, Merck & Co. Inc., Rahway, NJ
54. Cooper, W. C. et al *J. Occup. Med.* 1975, **17**, 100
55. Dingwall-Fordyce, I. et al *Br. J. Ind. Med.* 1963, **20**, 313
56. Calugar, A. et al *Rev. Med. Chir.* 1977, **81**(1), 87-92
57. Ades, A. E. et al *Br. J. Ind. Med.* 1988, **45**(7), 435-442
58. Wingren, G. et al *Scand. J. Work, Environ. Health* 1987, **13**(5), 421-416
59. Abdasimov, A. D. *Gig. Tr. Prof. Zabol.* 1989, (5), 18-21 (Russ.) (*Chem. Abstr.* **111**, 120121x)
60. Pocock, S. J. et al *Environ. Health Perspect.* 1988, **78**, 23-30
61. Antonio Garcia, M. et al *Environ. Technol. Lett.* 1988, **9**(3), 227-238 (Fr.) (*Chem. Abstr.* **108**, 226093t)
62. *S. I. 1991 No. 472 The Environmental Protection (Prescribed Processes and Substances) Regulations* 1991, HMSO, London
63. *EC Directive Relating to the Quality of Water Intended for Human Consumption* 1982, 80/778/EEC, Office for Official Publications of the European Communities, 2 rue Mercier, L-2985 Luxembourg
64. Sitwell, J. et al *Trace. Subst. Environ. Health* 1990, **23**, 51-62
65. Kazantizis, G. *Environ. Health Perspect.* 1981, **40**, 143-161
66. Gerber, G. B. et al *Mutat. Res.* 1980, **76**(2), 115-141
67. Needleman, M. L. *Prog. Clin. Biol. Res.* 1988, **281**, 279-287
68. Cavanagh, J. B. *NATO ASI Ser., Ser. A* 1988, **100**, 177-202
69. *ECETOC Technical Report No. 30(4)* 1991, European Chemical Industry Ecology and Toxicology Centre, B-1160 Brussels
70. McGrath, S. P. *Trace. Subst. Environ. Health* 1986, **20**, 242-252

71. Chapman, P. M. *Environ. Toxicol. Chem.* 1986, **5**(11), 957-964
72. O'Neil, I. K. et al *IARC Sci. Publ.* 1986, **71**, 3-13
73. Kazantis, G. *IARC Sci. Publ.* 1986, **71**, 103-111
74. Kazantiz, G. *Trace Subst. Environ. Health* 1990, **23**, 63-79
75. Wojtczak-Jaroszowa, J. et al *Med. Hypotheses* 1989, **30**(2), 141-150
76. Myzlak, Z. W. et al *Zentralbl. Arbeitsmed., Arbeitsschutz, Propyl. Ergon.* 1987, **2**(2), 87-92
77. Scheuhammer, A. M. *Environ. Pollut.* 1991, **71**(2-4), 329-375
78. Kehoe, R. A. *Food Chem. Toxicol.* 1987, **25**(6), 425-493
79. Schlag, R. D. *Adv. Mod. Environ. Toxicol.* 1987, **11**, 211-243
80. Newman, J. R. et al *Environ. Toxicol. Chem.* 1988, **7**(5), 381-390
81. Viola, A. et al *Pollut. Atmos.* 1988, **120**, 389-394 (Fr.) (*Chem. Abstr.* **110** 120293j)
82. Skerfuing, S. *Curr. Top. Nutr. Dis.* 1986, **18**, 611-630
83. Winder, C. *Reprod. Toxicol.* 1989, **3**(4), 221-233
84. Schweinberg, F. et al *Comp. Biochem. Physiol., C: Comp. Pharmacol. Toxicol.* 1990, **95**(2), 117-123
85. Taylor, A. *Rev. Environ. Health* 1986, **6**(1-4), 1-83
86. Humphreys, D. J. *Br. Vet. J.* 1991, **147**(1), 18-30
87. Marcus, W. L. *Adv. Mod. Environ. Toxicol.* 1990, **17**, 69-113
88. Syracuse Research Corp. *Report 1990*. From *Gov. Rep. Announce. Index (US)* 1990, **90**(23)
89. *IARC Monograph* 1980, **23**, 325-415

# L11   Lead arsenate

## $PbHAsO_4$

**CAS Registry No.** 7784-40-9
**Synonyms** arsenic acid ($H_3AsO_4$), lead(2+) salt; acid lead arsenate; lead hydrogen arsenate; arsenic acid, lead salt
**Mol. Formula** $AsHO_4Pb$                                   **Mol. Wt.** 347.12
**Uses** Constituent of various insecticides.
**Occurrence** in the mineral schultenite.

## Physical properties

**Specific gravity** 5.79.

## Occupational exposure

**US TLV (TWA)** 0.15 mg m$^{-3}$ (as Pb); **UK Long-term limit** (MEL) 0.15 mg m$^{-3}$ (as Pb) under review; **UN No.** 1617; **HAZCHEM Code** 2Z;
**Conveyance classification** toxic substance.

## Ecotoxicity

**Fish toxicity**
$LC_{50}$ (24, 96 hr) channel catfish >100 mg l$^{-1}$ static bioassay, 18°C, hardness 44 mg l$^{-1}$ ($CaCO_3$) and pH 7.1 (1).

## Mammalian and avian toxicity

**Acute data**
$LD_{50}$ oral rat 100 mg $kg^{-1}$ (2).
$LD_{50}$ oral chicken 450 mg $kg^{-1}$ (2).

**Sub-acute data**
$LC_{50}$ (5 day) oral Japanese quail 2760 mg $kg^{-1}$ (3).

**Carcinogenicity and long-term effects**
Oral (2 yr) ♂ rat 0.1%. Enlargement of cells, vesiculation of nuclei and accumulation of brown granules but no tumours were reported (4).

## Any other adverse effects

Acute and sub-acute testing in animals (species unspecified) showed lead arsenate was moderately toxic. Skin resorption effects were observed. The safe exposure reference level was calculated at 0.01 mg $m^{-3}$ (As) (5).

## Legislation

Limited under EC Directive on Drinking Water Quality 80/778/EEC. Arsenic: maximum admissible concentration 50 μg $l^{-1}$; lead: maximum admissible concentration 50 μg $l^{-1}$ (6).
Included in Schedule 4 and 6 (Release into Air/Land: Prescribed Substances) Statutory Instrument No. 472, 1991 (7).

## Any other comments

Experimental toxicology and human health effects reviewed (8).
The risk of developing Non-Hodgkins lymphoma in farmworkers using lead arsenates in combination with phenoxy herbicides had an odds ratio of 1.89 and a 95% confidence interval of 1.1-3.3 (9).

## References

1. Mayer, F. L. et al *Manual of acute toxicity: interpretation and data base for 410 chemicals and 66 species of freshwater animals* 1986, 1-506, US Department of the Interior, Fish and Wildlife Service, Report No.160, Washington, DC
2. *Pesticide Chemicals Official Compendium* 1966, 653, Topeka, Kansas
3. Hill, E. F. et al *Lethal dietary toxicities of environmental contaminants and pesticides to Cotornix* 1986, 86-88, U.S. Department of the Interior, Fish and Wildlife Service Report No.2, Washington, DC
4. Fairhall, L. T. et al *Publ. Health Rep. (Wash.)* 1941, **28**, 98-100
5. Neizvestnova, E. M. et al *Aktual Probl. Gig. Metall. Gornodobyuayushcei Prom-sti* 1985, 28-32 (Russ.) (*Chem. Abstr.* **107**, 91374s)
6. *EC Directive Relating to the Quality of Water Intended for Human Consumption* 1982, 80/778/EEC, Office for Official Publications of the European Communities, 2 rue Mercier, L-2985 Luxembourg
7. *S. I. 1991 No. 472 The Environmental Protection (Prescribed Processes and Substances) Regulations* 1991, HMSO, London
8. *ECETOC Technical Report No. 30(4)* 1991, European Chemical Industry Ecology and Toxicology Centre, B-1160 Brussels
9. Woods, J. S. *Chemosphere* 1989, **18**, 401-406

# L12   Lead arsenite

## $Pb(AsO_2)_2$

**CAS Registry No.** 10031-13-7
**Synonyms** arsenious acid, lead(2+) salt; arsenious acid $(HAsO_2)$, lead (2+) salt
**Mol. Formula** $As_2O_4Pb$                              **Mol. Wt.** 421.03
**Uses** Insecticide.

## Physical properties

**Specific gravity** 5.85.

## Occupational exposure

**US TLV (TWA)** 0.05 mg m$^{-3}$ (as Pb) intended change; **UK Long-term limit** (MEL)
0.5 mg m$^{-3}$ (as Pb) under review; **UN No.** 1618; **HAZCHEM Code** 2Z; **Conveyance
classification** toxic substance.

## Environmental fate

### Absorption

Inorganic lead tends to form highly insoluble salts and complexes with various anions
and is tightly bound to soil, therefore availability to terrestrial plants is drastically
reduced and effects on plants are normally only observed at very high environmental
concentrations 100-1000 mg kg$^{-1}$ soil (1).

## Mammalian and avian toxicity

### Carcinogenicity and long-term effects

No adequate evidence for carcinogenicity to humans, sufficient evidence for
carcinogenicity to animals, IARC classification group 2B (2).

## Legislation

Limited under EC Directive on Drinking Water Quality 80/778/EEC. Arsenic:
maximum admissible concentration 50 μg l$^{-1}$; lead: maximum admissible
concentration 50 μg l$^{-1}$ (3).
Included in Schedule 4 and 6 (Release into Air/Land: Prescribed Substances)
Statutory Instrument No. 472, 1991 (4).

## References

1.  *IPCS Environmental Health Criteria 85: Lead-Environmental Aspects* 1989, WHO, Geneva
2.  *IARC Monograph* 1987, **Suppl. 7**, 230-232
3.  *S. I. 1991 No. 472 The Environmental Protection (Prescribed Processes and Substances)
    Regulations* 1991, HMSO, London
4.  *EC Directive Relating to the Quality of Water Intended for Human Consumption* 1982,
    80/778/EEC, Office for Official Publications of the European Communities, 2 rue Mercier,
    L-2985 Luxembourg

# L13   Lead chloride

## PbCl$_2$

**CAS Registry No.** 7758-95-4
**Synonyms** lead chloride (PbCl$_2$); lead dichloride; plumbous chloride
**Mol. Formula** Cl$_2$Pb                                **Mol. Wt.** 278.10
**Uses** Catalyst. Solder. Flux for steel galvanising. Cathode in magnesium-lead chloride seawater batteries. Chemical scrubber. Flame retardant. Pigment and organolead manufacture.

## Physical properties
**M. Pt.** 501°C; **B. Pt.** 950°C; **Specific gravity** 5.85.

## Solubility
Water: 9.9 g l$^{-1}$ at 20°C.

## Occupational exposure
**US TLV (TWA)** 0.05 mg m$^{-3}$ (as Pb) intended change; **UK Long-term limit** (MEL) 0.15 mg m$^{-3}$ (as Pb) under review.

## Ecotoxicity

### Fish toxicity
Preliminary static acute toxicity tests using fathead minnows, show lead salts are less toxic in hard alkaline waters. 3000 mg l$^{-1}$ as the Cl$^-$ ion did not cause mortality of the newly-hatched fathead minnows. Mortality increased at 3500-4000 mg Pb l$^{-1}$ and appears to be partly due to the removal of protective carbonates initially present in the hard water (1).
LC$_{50}$ (24, 48, 96 hr) fathead minnow, bluegill sunfish, goldfish and guppy 8-6 mg l$^{-1}$, 26-24 mg l$^{-1}$, 45-31 mg l$^{-1}$, 24-20 mg l$^{-1}$, respectively. Static bioassay, alkalinity 18 mg l$^{-1}$ (CaCO$_3$), hardness 20 mg l$^{-1}$ (CaCO$_3$), 25°C and pH 7.5. Increasing alkalinity and hardness increased lethal concentration to 470-480 mg l$^{-1}$ (2).

### Invertebrate toxicity
EC$_{50}$ (4 hr) *Scenedesmus quadricauda* 2.7 mg kg$^{-1}$ (3).
EC$_{50}$ (48 hr) *Daphnia magna* 0.45 mg l$^{-1}$ static bioassay at pH 7.4, alkalinity 41-50 mg l$^{-1}$ (CaCO$_3$), and hardness 44-53 mg l$^{-1}$ (CaCO$_3$) (4).
LC$_{50}$ (96 hr, 30 day) crayfish 2.6, 1.5 mg l$^{-1}$ flowthrough bioassay, temperature 15-17°C and pH 7 (5).
EC$_{50}$ (3 wk) *Daphnia magna* 100 µg l$^{-1}$ (lead), reproductive impairment more sensitive measure for toxicity than survival (4).
Soil bacteria, 2500 ppm lead as lead chloride depressed bacterial growth at pH 4 and 6, but not in alkaline soil pH 8 (6).

## Environmental fate

### Absorption
Inorganic lead tends to form highly insoluble salts and complexes with various anions and is tightly bound to soil, therefore availability to terrestrial plants is drastically

reduced and effects on plants are normally only observed at very high environmental concentrations 100-1000 mg kg$^{-1}$ soil (7).

## Mammalian and avian toxicity

### Acute data
$TD_{Lo}$ oral rat 570 mg kg$^{-1}$ (8).
$TD_{Lo}$ intravenous mouse 20 mg kg$^{-1}$ (9).
$LD_{Lo}$ oral guinea pig 1500 mg kg$^{-1}$ (10).

### Sub-acute data
Oral dog (24 day) 5 mg kg$^{-1}$ day$^{-1}$ lead salt mixture containing unspecified concentrations of lead chloride, lead bromide and lead sulfate caused histopathological lesions consisting of acid fast lead inclusions were seen in osteoclasts and renal tubular epithelial cells. Tissue lead concentrations (liver, kidney and bones) were greatly elevated (11).

### Carcinogenicity and long-term effects
No adequate evidence for carcinogenicity to humans, sufficient evidence for carcinogenicity to animals, IARC classification group 2B (12).

### Teratogenicity and reproductive effects
Intravenous mice 40 mg kg$^{-1}$ on day 3, 4 or 6 of pregnancy. Attachment of the blastocyst was observed on day 5, invasion of the trophoblast was seen on day 6 and formation of the primitive streak was seen on day 6, all three developmental stages studied were affected by the action of lead (13).
Embryotoxic and teratogenic to trout and chick embryos (14).
Intravenous mouse 75 ppm on day 4 of pregnancy was found to interfere in embryo implantation (15).
Caused anophthalamia, fused ribs, spina bifida and exencephaly in golden hamster embryos (16).

### Irritancy
Oral guinea pig 1.5-2.0 g kg$^{-1}$ caused mild irritation (17).

## Genotoxicity
*Allium cepa* 1.0 mg l$^{-1}$ slight but significant clastogenic effects without disturbing the mitotic activity (18).
The fidelity and rate of synthesis of avian mycloblastosis virus DNA polymerase mediated DNA synthesis was decreased by 1.11 g kg$^{-1}$ of lead chloride (19).
In systems containing *Escherichia coli* RNA polymerase and either calf-thymus DNA or T4 DNA, stimulation of chain initation of RNA synthesis at a concentrations of 28 mg kg$^{-1}$ inhibited overall RNA synthesis (20).

## Any other adverse effects to man
Symptoms of poisoning include fatigue, headache, sleep disturbance, constipation, aching bones and muscles, gastrointestineal tract disturbances and reduced appetite (21).
Symptoms often precipitated by alcohol or exercise (22).

## Any other adverse effects
Intravenous rat 5 mg kg$^{-1}$ decreased hepatic cytochrome $P_{450}$ levels, with associated

decreases in microsomal oxidative demethylation and hydroxylation enzyme activities (23).

*In vitro* chick neurons and freshwater snail neurons incubated for 3-4 days with lead chloride reduced the percentage of cells that grew neurites; however, the mechanism of action for both species was different (24).

## Legislation

Limited under EC Directive on Drinking Water Quality 80/778/EEC. Guide level 25 mg chloride $l^{-1}$; approximate chloride concentration above which effects might occur, 200 mg $l^{-1}$ and maximum admissible concentration 50 µg lead $l^{-1}$ (25).

Included in Schedule 4 (Release into Air: Prescribed Substances) Statutory Instrument No. 472, 1991 (26).

## Any other comments

Causes algicidal and herbicidal effects (27).

## References

1. Erten, M. Z. et al *Proc. Ind. Waste Conf.* 1988, **43**, 617-629
2. Pickering, Q. H. et al *Air. Water Pollut. Int. J.* 1966, **10**, 453-463
3. Starodub, M. E. et al *Sci. Total Environ.* 1987, **63**, 101
4. Biesinger, K. E. et al *J. Fish. Res. Board Can.* 1972, **29**, 1691-1700
5. Boutet, C. et al *C. R. Soc. Biol. (Paris)* 1973, **167**, 1933-1938
6. Badura, L. et al *Arch. Ochr. Srodowiska* 1987, (3-4), 129-39 (Pol.) (*Chem. Abstr.* **110**, 70746m)
7. *IPCS Environmental Health Criteria 85: Lead-Environmental Aspects* 1989, WHO, Geneva
8. *Pharmacol. Biochem. Behav.* 1975, **11**, 95
9. *Teratol. J. Abnorm. Develop.* 1986, **34**, 207
10. *The Merck Index* 11th ed., 1989, Merck & Co. Inc., Rahway, NJ
11. Hamir, A. N. *Br. Vet. J.* 1988, **144**(3), 240-245
12. *IARC Monograph* 1987, **Suppl. 7**, 230-232
13. Wide, M. et al *Teratology* 1977, **16**, 273-276
14. Birge, W. J. et al *Trace Contam. Environ., Proc. Ann. NSF-RANN Trace Contam. Conf.* 2nd., 1974, (LBL-3217), 308-315
15. Wide, M. et al *Teratology* 1979, **20**(1), 101-103
16. Ferm, V. H. et al *Exp. Mol. Pathol.* 1967, **7**, 208-213
17. Spector, S. (Ed.), *Handbook of Toxicology* 1956, **1**, 176-177, Saunders, Philidelphia
18. Wierzbicka, M. *Caryologia* 1988, **41**(2), 143-160
19. Sirover, M. A. et al *Science* 1976, **194**, 1434-1436
20. Hoffman, D. J. et al *Science* 1977, **198**, 513-514
21. Kehoe, R. A. *J. Occup. Med.* 1972, **14**, 298
22. *Chemial Safety Data Sheets; Volume 2: Main Group Metals & their Compounds* 1989, RSC, London
23. Alvares, A. P. et al *J. Exp. Med.* 1972, **135**, 1406-1409
24. Audesirck, G. *In Vitro Cell. Dev. Biol.* 1989, **25**(12), 1121-1128
25. *EC Directive Relating to the Quality of Water Intended for Human Consumption* 1982, 80/778/EEC, Office for Official Publications of the European Communities, 2 rue Mercier, L-2985 Luxembourg
26. *S. I. 1991 No. 472 The Environmental Protection (Prescribed Processes and Substances) Regulations* 1991, HMSO, London
27. Roderer, G. *Arch. Environ. Contam. Toxicol.* 1987, **16**, 291

# L14   Lead chromate

## PbCrO₄

CAS Registry No. 7758-97-6

**Synonyms** chromic acid ($H_2CrO_4$), lead(2+) salt(I:I); lead chromium oxide; plumbous chromate

**Mol. Formula** PbCrO₄                    **Mol. Wt.** 323.18

**Uses** Pigment in oil and water colours. Chemical analysis of organic substances. In traffic paints.

**Occurrence** In the minerals (crocite, phoenicochroite) (1).

## Physical properties

**M. Pt.** 844°C; **B. Pt.** decomposes; **Specific gravity** 6.30.

**Solubility**
Water: 0.58 mg $l^{-1}$ at 25°C.

## Occupational exposure

**US TLV (TWA)** 0.05 mg m$^{-3}$ (as Pb) 0.012 mg m$^{-3}$ (as Cr); **UK Long-term limit** (MEL) 0.15 mg m$^{-3}$ (as Pb) under review 0.05 mg m$^{-3}$ (as Cr); **Supply classification** harmful.

**Risk phrases** Danger of cumulative effects – Possible risk of irreversible effects (R33, R40)

**Safety phrases** Do not breathe dust (S22)

## Environmental fate

**Absorption**

Inorganic lead tends to form highly insoluble salts and complexes with various anions and is tightly bound to soil, therefore availability to terrestrial plants is drastically reduced and effects on plants are normally only observed at very high environmental concentrations 100-1000 mg kg$^{-1}$ soil (2).

## Mammalian and avian toxicity

**Acute data**

$LD_{50}$ oral mouse 12 g kg$^{-1}$ (3).

$LD_{Lo}$ oral guinea pig 2000 mg kg$^{-1}$ (4).

$LD_{50}$ intraperitoneal guinea pig 156 mg kg$^{-1}$ (1).

**Carcinogenicity and long-term effects**

Subcutaneous (117-150 wk) ♂ and ♀ Sprague Dawley rats 30 mg in water. Sarcomas developed at the injection site in 26/40 and 27/40 animals, respectively (5).

Intramuscular (16 month) 25 ♀ NIH-Swiss weaning mice 3 mg injection every 4 month incidence of carcinomas was 2/22 (6).

Intramuscular (9 month) 25 ♂ and ♀ weaning Fischer 344 rat 8 mg month$^{-1}$ induced 14 fibrosarcomas and 17 rhabydosarcomas at the site of injection in 31/47 rats. Renal carcinomas were observed in 3/23 ♂ rats in 24 months (6).

## Genotoxicity

*Salmonella typhimurium* TA100 with metabolic activation was positive in the presence of nitrilotriacetic acid (7).

*Eschericha coli* PQ37 without metabolic activation negative (8).

*In vitro Vicia faba* root tips lead chromate significantly enhances the frequency of micronucleated cells only in the presence of nitrilotriacetic acid (9).

Chinese hamster V79 cells, with and without metabolic activation negative due to the slow uptake by the cells. When nitrilotriacetic acid was present, positive dose dependent mutagenicity was observed (10).

Chinese hamster ovary cells and C3H/10T½ mouse embryo cells, non-mutagenic in either system. Induced a dose dependent, low frequency of focus formation; and transformation (11).

Cultured Chinese hamster cells, frequency of sister chromatid exchanges significantly increased (12).

*In vitro* Syrian hamster embryo cells morphological transformations positive (13).

*In vivo Drosophila melanogaster* no significant increase of the mutation frequency. The addition of nitrilotriacetic acid caused significant increases of the frequency of sex linked lethal mutations, with a significant dose-effect relationship with respect to lead chromate, as a result of the interaction of the compounds and subsequent release of the genotoxic heavy metal chromium (IV) ions (14).

It increased the frequency of micronuclei in polychromatic erythrocyte ratio in bone marrow cells of intraperitoneally treated (57B1/6N mice (15).

## Any other adverse effects to man

Irritating to eyes, may cause conjunctivitis and lachrymation (4).
May cause ulceration and eczematous dermatitis on skin (5).

## Legislation

Limited under EC Directive on Drinking Water Quality 80/778/EEC. Chromium: maximum admissible concentration 50 $\mu$g l$^{-1}$ (16).
Included in Schedule 4 (Release into Air: Prescribed Substances) of Statutory Instrument No. 472, 1991 (17).

## Any other comments

Toxicity reviewed (18).
Acute toxicity is greater than other inorganic lead compounds due to contributory toxicity of chromate portion (4).
Experimental toxicology, human health effects and physico-chemical properties reviewed (19).
Lead chromate chemistry reviewed (20).

## References

1. *The Merck Index* 11th ed., 1989, Merck & Co. Inc., Rahway, NJ
2. *IPCS Environmental Health Criteria 85: Lead-Environmental Aspects* 1989, WHO, Geneva
3. Oyo, Y. *Pharmacometics* 1968, **2**, 76
4. *Chemical Safety Data Sheets* 1989, **2**, 205, RSC, London
5. Furst, A. et al *Cancer Res.* 1976, **36**, 1779-1783
6. Maltoni, C. et al *Adv. Med. Environ. Toxicol.* 1982, **2**, 77-92
7. Venier, P. et al *Toxicol. Environ. Chem.* 1987, **14**(3), 201-218
8. Venier, P. *Mutagenesis* 1989, **4**(1), 51-57
9. De Marco, A. et al *Mutat. Res.* 1988, **206**(3), 311-315
10. Celotti, L. *Mutat. Res.* 1987, **190**(1), 35-39
11. Patierno, S. R. et al *Cancer Res.* 1988, **48**(18), 5280-5288

12. Montaldi, A. *J. Toxicol. Environ. Health* 1987, **21**(3), 387-394
13. Elias, Z. et al *Carcinogenesis (London)* 1991, **12**(10), 1811-1816
14. Costa, R. et al *Mutat. Res.* 1988, **204**(2), 257-261
15. Watanabe, M. et al *Tohoku J. Exp. Med.* 1985, **146**, 373-374
16. *EC Directive Relating to the Quality of Water Intended for Human Consumption* 1982, 80/778/EEC, Office for Official Publications of the European Communities, 2 rue Mercier, L-2985 Luxembourg
17. *S. I. 1991 No. 472 The Environmental Protection (Prescribed Processes and Substances) Regulations* 1991, HMSO, London
18. Gardner, J. S. et al *Double Liason-Chim. Peint.* 1981, **28**(307), 111-118
19. *ECETOC Technical Report No. 30(4)* 1991, European Chemical Industry Ecology and Toxicology Centre, B-1160 Brussels
20. Algra, G. P. *J. Oil Colour Chem. Assoc.* 1988, **71**(3), 71-77

# L15   Lead cyanide

## $Pb(CN)_2$

**CAS Registry No.** 592-05-2
**Synonyms** lead cyanide; C.I. 77610; C.I. Pigment Yellow 48
**Mol. Formula** $C_2N_2Pb$                                **Mol. Wt.** 259.23

## Occupational exposure

**US TLV (TWA)** 0.05 mg m$^{-3}$ (as Pb) intended change; **UK Long-term limit** (MEL) 0.15 mg m$^{-3}$ (as Pb) under review; **UN No.** 1620; **Conveyance classification** toxic substance.

## Mammalian and avian toxicity

**Acute data**
$LD_{Lo}$ intraperitoneal rat 100 mg kg$^{-1}$ (1).

**Carcinogenicity and long-term effects**
No adequate evidence for carcinogenicity to humans, sufficient evidence for carcinogenicity to animals, IARC classification group 2B (2).

## Legislation

Limited under EC Directive on Drinking Water Quality 80/778/EEC. Lead: maximum admissible concentration 50 µg l$^{-1}$; cyanides maximum admissible concentration 50 µg l$^{-1}$ (3).
Included in Schedule 4 (Release into Air: Prescribed Substances) Statutory Instrument No. 472, 1991 (4).

## Any other comments

Lead and its compounds have been comprehensively reviewed (5).

---

## References

1. *National Academy of Sciences, National Reseach Council, Chemical-Biological Coordination Center, Review* 1953, **5**, 27
2. *IARC Monograph* 1987, **Suppl. 7**, 230-232
3. *EC Directive Relating to the Quality of Water Intended for Human Consumption* 1982, 80/778/EEC, Office for Official Publications of the European Communities, 2 rue Mercier, L-2985 Luxembourg
4. *S. I. 1991 No. 472 The Environmental Protection (Prescribed Processes and Substances) Regulations* 1991, HMSO, London
5. *IPCS Environmental Health Criteria 85: Lead-Environmental Aspects* 1989, WHO, Geneva

# L16    Lead dimethyldithiocarbamate

$$(H_3C)_2N-\overset{S}{\underset{S}{<}}\overset{S}{\underset{S}{>}}-N(CH_3)_2$$

**CAS Registry No.** 19010-66-3
**Synonyms** lead, bis(dimethylcarbamodithioato- $S,S'$)-,(T-4)-;
bis (dimethyldithiocarbamato)lead; Ledate
**Mol. Formula** $C_6H_{12}N_2PbS_4$         **Mol. Wt.** 447.62
**Uses** Accelerator for vulcanisation. Compounding natural, styrene-butadiene, isobutylene-isopropene, isoprene and butadiene rubbers.

## Physical properties
**M. Pt.** 258°C; **Specific gravity** 2.50.

## Occupational exposure
**UK Long-term limit** (MEL) 0.15 mg m$^{-3}$ (as Pb) under review.

## Mammalian and avian toxicity

### Acute data
TD$_{Lo}$ subcutaneous mouse 1000 mg kg$^{-1}$ (1).

### Carcinogenicity and long-term effects
Oral (104 or 105 wk) ♂, ♀ Fisher 344 rats and B6C3F1 mice 25 or 50 ppm in feed. No tumours occurred in rats or mice of either sex at incidences that were significantly higher than the control groups (2,3).
Oral (76 wk) ♂, ♀ mice, 46.4 mg kg$^{-1}$ bodyweight for first 4 wk then 130 mg kg$^{-1}$ diet. Incidence of reticulum cell sarcomas in ♂ was significantly different from that in controls (4).

## Genotoxicity
*Salmonella typhimurium* TA97, TA98, TA100, TA1536, TA1537 with and without metabolic activation positive (3).

## Legislation

Included in Schedule 4 and 6 (Release into Air/Land: Prescribed Substances) Statutory Instrument No. 472, 1991 (5).

## References

1. *NTIS Report No. PB223-1561*, Natl. Tech. Inf. Ser., Springfield, VA
2. *National Toxicology Program Research and Testing Division* 1992, Report No. TR151, NIEHS, Research Triangle Park, NC27709
3. Zeiger, E. *Cancer Res.* 1987, **47**, 1287-1296
4. Innes, J. R. M. et al *J. Natl. Cancer Inst.* 1969, **42**, 1101-1114
5. *S. I. 1991 No. 472 The Environmental Protection (Prescribed Processes and Substances) Regulations* 1991, HMSO, London

# L17   Lead dioxide

## $PbO_2$

**CAS Registry No.** 1309-60-0
**Synonyms** lead oxide brown; lead peroxide; lead superoxide; plumbic oxide; C.I.77580
**Mol. Formula** $O_2Pb$                                 **Mol. Wt.** 239.19
**Uses** Electrodes in lead-acid batteries. Paint, rubber and ceramics industry. Manufacture of chemicals, dyes, pyrotechnics and liquid polysulphides polymers.
**Occurrence** mineral plattnerite (1).

## Physical properties

**M. Pt.** decomp. at 290°C; **Specific gravity** 9.375.

## Occupational exposure

**US TLV (TWA)** 0.05 mg m$^{-3}$ (as Pb) intended change; **UK Long-term limit** (MEL) 0.15 mg m$^{-3}$ (as Pb) under review; **UN No.** 1872; **HAZCHEM Code** 2Z; **Conveyance classification** oxidizing substance.

## Mammalian and avian toxicity

### Acute data
LD$_{50}$ intraperitoneal guinea pig 220 mg kg$^{-1}$ (2).

### Carcinogenicity and long-term effects
No adequate evidence for carcinogenicity to humans, sufficient evidence for carcinogenicity to animals, IARC classification group 2B (3).

### Metabolism and pharmacokinetics
Oral mice (30 day) ≈100-1000 ppm in drinking water. Blood lead levels at the end of the period were dose-dependent. Lead dioxide provided the lowest level of blood level concentrations when compared to a series of lead salts (4).

## Any other adverse effects to man

Symptoms of poisoning include fatigue, headache, sleep disturbances, constipation, aching bones and muscles, gastrointestinal tract disturbances and reduced appetite. Later anaemia, lead-line on the gums and lead colic may occur. Large doses affect the central nervous system, causing severe headaches, convulsions, coma and possibly death, kidney damage may result from chronic exposure (5).
Causes irritation to eyes and skin (5).

## Legislation

Limited under EC Directive on Drinking Water Quality 80/778/EEC. Lead: maximum admissible concentration 50 $\mu$g $l^{-1}$ (6).
Included in Schedule 4 (Release into Air: Prescribed Substances) Statutory Instrument No. 472, 1991 (7).

## Any other comments

Lead and its compounds have been comprehensively reviewed (8).

## References

1. *The Merck Index* 11th ed., 1989, Merck & Co. Inc., Rahway, NJ
2. *Environmental Quality and Safety, Supplement*, 1975, **1**, 1
3. *IARC Monograph* 1987, **Suppl. 7**, 230-232
4. Cheng, Y. L. et al *J. Hazard. Mater.* 1991, **27**(2), 137-147
5. *Chemical Safety Data Sheets, Vol 2: Main Group Metals & their Compounds* 1989, RSC, London
6. *EC Directive Relating to the Quality of Water Intended for Human Consumption* 1982, 80/778/EEC, Office for Official Publications of the European Communities, 2 rue Mercier, L-2985 Luxembourg
7. *S. I. 1991 No. 472 The Environmental Protection (Prescribed Processes and Substances) Regulations* 1991, HMSO, London
8. *IPCS Environmental Health Criteria 85: Lead-Environmental Aspects* 1989, WHO, Geneva

# L18   Lead fluoborate

## $Pb(BF_4)_2$

**CAS Registry No.** 13814-96-5
**Synonyms** borate(1- ),tetrafluro-, lead(II); lead boron fluoride; lead tetrafluoroborate
**Mol. Formula** $B_2F_8Pb$                    **Mol. Wt.** 380.80

## Mammalian and avian toxicity

### Acute data
$LD_{50}$ oral rat 50 mg $kg^{-1}$ (1).

### Carcinogenicity and long-term effects
No adequate evidence for carcinogenicity to humans, sufficient evidence for carcinogenicity to animals, IARC classification group 2B (2).

## Legislation

Included in Schedule 4 (Release into Air: Prescribed Substances) Statutory Instrument No. 472, 1991 (3).

## Any other comments

Lead and its compounds have been comprehensively reviewed (4).

## References

1. *Kodak Company Reports* 1971
2. *IARC Monograph* 1987, **Suppl. 7**, 230-232
3. *S. I. 1991 No. 472 The Environmental Protection (Prescribed Processes and Substances) Regulations* 1991, HMSO, London
4. *IPCS Environmental Health Criteria 85: Lead-Environmental Aspects* 1989, WHO, Geneva

# L19   Lead fluoride

## $PbF_2$

**CAS Registry No.** 7783-46-2
**Synonyms** lead fluoride ($PbF_2$); lead difluoride; Plumbous fluoride
**Mol. Formula** $F_2Pb$                      **Mol. Wt.** 245.19

## Physical properties

**M. Pt.** 824°C; **B. Pt.** 1293°C; **Specific gravity** 7.750 (cubic); 8.445 (orthorhombic).

**Solubility**
Water: 0.57 g $l^{-1}$ at 0°C, 0.65 g $l^{-1}$ at 20°C.

## Mammalian and avian toxicity

**Acute data**
$LD_{Lo}$ subcutaneous guinea pig 2800 mg $kg^{-1}$ (1).

**Carcinogenicity and long-term effects**
No adequate evidence for carcinogenicity to humans, sufficient evidence for carcinogenicity to animals, IARC classification group 2B (2).

## Legislation

Limited under EC Directive on Drinking Water Quality 80/778/EEC. Lead: maximum admissible concentration 50 µg $l^{-1}$ (3).
Included in Schedule 4 (Release into Air: Prescribed Substances) Statutory Instrument No. 472, 1991 (4).

## Any other comments

Lead and its compounds have been comprehensively reviewed (5).

## References

1. *C. R. Seances Soc. Biol. Ses Fil.* 1937, **124**, 137 (Fr.)
2. *IARC Monograph* 1987, **Suppl. 7**, 230-232
3. *EC Directive Relating to the Quality of Water Intended for Human Consumption* 1982, 80/778/EEC, Office for Official Publications of the European Communities, 2 rue Mercier, L-2985 Luxembourg
4. *S. I. 1991 No. 472 The Environmental Protection (Prescribed Processes and Substances) Regulations* 1991, HMSO, London
5. *IPCS Environmental Health Criteria 85: Lead-Environmental Aspects* 1989, WHO, Geneva

# L20    Lead hexafluorosilicate

## $PbSiF_6$

**CAS Registry No.** 25808-74-6
**Synonyms** silicate(2-), hexafluoro-, lead(2+)(1:1); lead fluorosilicate; lead silicon fluoride
**Mol. Formula** $F_6SiPb$                                   **Mol. Wt.** 349.27
**Uses** In refining lead by electrolytic methods.

## Physical properties

**M. Pt.** (decomp.).

## Occupational exposure

**UK Long-term limit** (MEL) 0.15 mg m$^{-3}$ (as Pb) under review.

## Mammalian and avian toxicity

**Acute data**
$LD_{Lo}$ oral rat 250 mg kg$^{-1}$ (1).

**Carcinogenicity and long-term effects**
No adequate evidence for carcinogenicity to humans, sufficient evidence for carcinogenicity to animals, IARC classification group 2B for inorganic lead compounds (2).

## Any other comments

Lead and its compounds have been comprehensively reviewed (3).

## References

1. *Nat. Acad. Sci., Nat. Res. Coun., Chem. Biol. Coord. Cent.* 1953, **5**, 27
2. *IARC Monograph* 1987, **Suppl. 7**, 65
3. *IPCS Environmental Health Criteria 85: Lead – Environmental Aspects* 1989, WHO, Geneva

# L21   Lead iodide

## $PbI_2$

**CAS Registry No.** 10101-63-0
**Synonyms** C.I. 77613; lead diiodide; lead(II) iodide; plumbous iodide
**Mol. Formula** $I_2Pb$                                          **Mol. Wt.** 461.00
**Uses** Bronzing, gold pencils, mosaic gold, printing, photography.

### Physical properties
**M. Pt.** 402°C; **B. Pt.** 954°C; **Specific gravity** 6.16.

### Solubility
Water: 0.74 g $l^{-1}$ cold, 4.34 g $l^{-1}$ hot.

### Occupational exposure
**US TLV (TWA)** 0.05 mg $m^{-3}$ (as Pb) intended change; **UK Long-term limit** (MEL) 0.15 mg $m^{-3}$ (as Pb) under review.

### Mammalian and avian toxicity

#### Carcinogenicity and long-term effects
No adequate evidence for carcinogenicity to humans, sufficient evidence for carcinogenicity to animals, IARC classification group 2B for inorganic lead compounds (1).

### Any other comments

Lead and its components have been comprehensively reviewed (2).

### References

1. *IARC Monograph* 1987, **Suppl. 7**, 65
2. *IPCS Environmental Health Criteria 85: Lead – Environmental Aspects* 1989, WHO, Geneva

# L22   Lead nitrate

## $Pb(NO_3)_2$

**CAS Registry No.** 10099-74-8
**Synonyms** nitric acid, lead (2+)salt; lead dinitrate; plumbous nitrate
**Mol. Formula** $N_2O_6Pb$                                        **Mol. Wt.** 331.20
**Uses** Manufacture of matches and explosives. As mordant in dyeing and printing on textiles. Oxidiser in dye industry. Sensitiser in photography. Has been used as a caustic in equine canker.

## Physical properties

**M. Pt.** 470°C (decomp.); **Specific gravity** $d^{20}$ 4.53.

### Solubility

Water: 500 g $l^{-1}$ cold, 1333 g $l^{-1}$ hot. Organic solvent: ethanol, methanol

## Occupational exposure

**US TLV (TWA)** 0.05 mg m$^{-3}$ (as Pb) intended change; **UK Long-term limit** (MEL) 0.15 mg m$^{-3}$ (as Pb) under review; **UN No.** 1469; **HAZCHEM Code** 2Y; **Conveyance classification** oxidizing substance, toxic substance.

## Ecotoxicity

### Fish toxicity

$LC_{50}$ (96 hr) snakehead fish 13.2 mg $l^{-1}$ (1).
$LC_{50}$ (96 hr) airsac catfish 12.0 mg $l^{-1}$ (1).
$LC_{50}$ (7 day) goldfish 6.6 mg $l^{-1}$ (2).
Red barb (30 day) 0.0474 mg $l^{-1}$ caused changes in physiochemical processes including glycogen uptake, cholesterol levels and lipid analysis (3).
Giant gourami (4 day) 9.5 mg $l^{-1}$ caused physical damage to tissue (4).

### Invertebrate toxicity

Clawed toad continuously exposed for 72 and 48 hr from the blastula and gastrula stage, respectively, to the larval stage. Lethality and abnormality in embryos and larvae; dose dependency was noted. The main abnormality observed was neural tube defects (5).
*Chara vulgaris* (7 day) 2.5 mg $l^{-1}$, 100% mortality including algicidal and herbicidal effects (6).
$LC_{Lo}$ (exposure unspecified) *Peranema gracilis, Euglena gracilis* 1000 mg lead $l^{-1}$; *Blepharisma* 42 mg $l^{-1}$; *Tetrahymena, Paramecium multimicronucleatum* 24 mg $l^{-1}$; and *Chilomanas* 5.6 mg $l^{-1}$ (7).
$EC_{50}$ (2 day) *Asellus aquaticus* (mobility) 120 mg $l^{-1}$ (8).
$EC_{50}$ (2 day) *Crangonyx pseudogracilis* (mobility) 43.8 mg $l^{-1}$ (8).
$LC_{50}$ (96 hr) *Barytelphues guerini* 25 ppm, haemolymph pH, $CO_2$ and lactate levels increased and haemolymph oxygen levels decreased (9).

### Bioaccumulation

Accumulation in earthworms 4.3-756 ppm. Accumulation of lead from soil with added lead nitrate was faster in soils with low (3%) organic carbon than in soils with high (17- 42%) organic carbon. Earthworms accumulated 1000 ppm without any ill effects (10).
*Procambarus clarkii* (96 hr) exposure to 10, 50 or 100 µg $l^{-1}$. Concentrations of lead accumulated in gills >midgut glands >muscle (11).
*Selenastrum capricornutum* whole plant accumulation after exposure to 4.5 µg $l^{-1}$ was 70,000 (7 day) and 102,000 (28 day), and after exposure to 40.1 µg $l^{-1}$ was 27,000 (7 day) and 32,000 (28 day), at temperatures 21-24°C and pH 7.2-7.8 (12).
*Mytilus edulis* (13 day) 100 µg $l^{-1}$ at 15°C, bioconcentration factor 3000 accumulation in kidney (13).
*Daphnia magna* (7, 28 day) 4.5 µg $l^{-1}$ bioconcentration factors 2900 and 5140, respectively (12).

Carp exposed to 10,000 μg l⁻¹ for 2 days had bioconcentration factors of 4200 for the viscera and 304 for the gills (14).

Guppys exposed to 4.6 μg l⁻¹ for 7 or 28 days had whole body bioconcentration factors of 654 and 3459, respectively (12).

## Environmental fate

### Absorption

Podzolic soil spiked with lead nitrate, content of water solute lead 0.5-2.22 mg kg⁻¹. Nitrates acidified the 3-40 cm soil layer by adsorption and displacement of $H^+$ and $Al^{3+}$ (15).

Inorganic lead tends to form highly insoluble salts and complexes with various anions and is tightly bound to soil, therefore availability to terrestial plants is drastically reduced and effects to plants are normally only observed at very high environmental concentrations 100-1000 mg kg⁻¹ soil (16).

## Mammalian and avian toxicity

### Acute data

$LD_{Lo}$ oral guinea pig 500 mg kg⁻¹ (17).
$LD_{Lo}$ intraperitoneal rat 270 mg kg⁻¹ (18).
$LD_{50}$ intraperitoneal mouse 74 mg kg⁻¹ (19).
$LD_{50}$ intravenous rat 93 mg kg⁻¹ (20).
$LD_{50}$ injection (4 day) chicken egg 0.10 mg egg⁻¹ (21).

### Sub-acute data

$LC_{50}$ (>100 day) mallard duck >50 mg kg⁻¹ diet (22).
$LC_{50}$ (5 day) Japanese quail >5000 mg kg⁻¹ diet (23).

### Carcinogenicity and long-term effects

No adequate evidence for carcinogenicity to humans, sufficient evidence for carcinogenicity to animals, IARC classification group 2B for inorganic lead (24).

### Teratogenicity and reproductive effects

Maternal exposure to 12.5, 25, 50 mg kg⁻¹ on day 9 of gestation did not cause embryonic resorption nor foetal lethality, it induced chromosomal aberrations in foetal liver and maternal bone marrow cells (25).

Intravenous golden hamster 25 or 50 mg kg⁻¹ body weight on days 8 or 9 of pregnancy. Malformations of the sacral and tail region and a few cases of rib fusion were observed, suggesting lead interferes with a specific enzymic event during pregnancy (26).

Intravenous golden hamster 50 mg kg⁻¹ body weight on day 8 of pregnancy produced hyperplasia. Disorientation of neuroepithelial cells of the dorsal region of the caudal neural tube were observed in embryos on day 9. Haematomas and extensive necrosis occurred throughout the dorsal region of the compressed neural tube on day 10 of gestation (27).

Intravenous Simonsen Sprague-Dawley rats single dose 25-70 mg kg⁻¹ body weight on days 8-17 of pregnancy. With 50 and 70 mg kg⁻¹ body weight, malformations of the urogenital and intestinal tracts and abnormalities of the posterior extremities were produced when doses were given on day 9 of pregnancy. Lead was increasingly lethal to foetuses when given on days 10-15 of pregnancy. Hydrocephalus and haemorrhage of the central nervous system occurred with treatment on day 16 of pregnancy (28).

## Genotoxicity

*In vivo* mice bone marrow increased aneuploidy and chromosomal aberrations (29).
*In vivo* mouse bone marrow did not induce micronuclei (30).
*Drosophila melanogaster* induced sex linked recessive lethal mutations (31).

## Any other adverse effects

Single concentration (unspecified) to ♂ Wistar rats caused increased activity of liver enzyme, inhibited ATPase and adenylate cyclase activity at 48-72 hr and 24 hr, respectively (32).

## Any other comments

The aquatic toxicity of lead is dependent on the free ionic concentration, which affects the bioavailability of lead to organisms. Inorganic lead toxicity is strongly dependent on environmental conditions such as water hardness, pH, and salinity. The presence of other salts reduces the bioavailability of lead to organisms because of precipitation (16).
*Scenedesmus obliquus* showed tolerance/resistance to concentrations of ≤300 ppm $Pb^{2+}$. Inhibited nitrate reductase activity (33).

## References

1. Sastry, K. V. et al, P. K. *Ecotoxicol. Environ. Safety 4* 1980, 232
2. Kemp, H. T. et al *Water Quality Data Book Vol. 5. Effects of Chemicals on Aquatic Life* 1973, EPA, Water Pollution Control Series 09/73
3. Teusari, H. et al *Bull. Environ. Contam. Toxicol.* 1987, **38**, 778
4. Srivastava, A. K. *J. Environ. Biol.* 1987, **8**, 329
5. Sakamato, M. et al *Kank. Kaga. Kenkyu. Kenk. Hok. (Kin. Daig.)* 1986, **14**, 185-188 (Japan.) (*Chem. Abstr.* **106**, 115027p)
6. Heumann, H. G. *Protoplasma* 1987, **136**, 37
7. Ruthren, J. A. et al *J. Protozool.* 1973, **20**, 127-135
8. Martin, T. R. et al *Water Res.* 1986, **20**(9), 1137-1147
9. Tulasi, S. J. et al *Bull. Environ. Contam. Toxicol.* 1988, **40**(2), 198-203
10. Streit, B. et al *Umweltwiss. Schadst.-Forsch.* 1990, **2**(1), 10-13 (Ger.) (*Chem. Abstr.* **113**, 93015y)
11. Paster, A. et al *Bull. Environ. Contam. Toxicol.* 1988, **41**(3), 412-418
12. Vighi, M. *Ecotoxicol. Environ. Saf.* 1981, **5**, 177-193
13. Coombs, T. L. *Proc. Anal. Div. Chem. Soc.* 1977, **14**, 219-222
14. Muramoto, S. *Bull. Environ. Contam. Toxicol.* 1980, **25**, 941-946
15. Obukhov, A. I. et al *Vestn. Mosk. Univ., Ser. 17: Pochvoved* 1990, (3), 39-44 (Russ.) (*Chem. Abstr.* **114**, 61047q)
16. *IPCS Environmental Health Criteria 85: Lead – Environmental Aspects* 1989, WHO, Geneva
17. *Arch. Hyg. Bakteriol.* 1941, **125**, 273 (Ger.)
18. *Environ. Qual. Saf., Suppl.* 1975, **1**, 1
19. *Bull. Environ. Contam. Toxicol.* 1973, **9**, 80
20. *Proc. Soc. Exp. Biol. Med.* 1956, **92**, 331
21. Ridgeway, L. P. et al *Ann. N.Y. Acad. Sci.* 1952, **55**, 203-215
22. Dewitt, J. B. et al *Pesticide-Wildlife Studies. A review of Fish and Wildlife Service investigations during 1961 and 1962* 1963, 71-96, US Dept. Int. Fish & Wildlife Ser., Circular No. 167, Washington, DC
23. Hill, E. F. et al *Lethal Dietary Toxicities of Environmental Contaminants and Pesticides to Coturnix* 1986, 86-88, US Dept. Int. Fish & Wildlife Ser. Report No. 2, Washington, DC

24. *IARC Monograph* 1987, **Suppl. 7**, 65
25. Vayak, B. N. et al *Acta Anat.* 1989, **135**(2), 185-188
26. Ferm, V. H. et al *Life Sci.* 1971, **10**, 35-39
27. Carpenter, S. J. et al *Anat. Rec.* 1974, **178**, 323
28. Mclain, R. M. et al *Toxicol. Appl. Pharmacol.* 1975, **31**, 72-82
29. Savic, G. et al *Acta. Biol. Med. Exp.* 1987, **12**(1), 39-45
30. Liao, M. et al *Weish. Duli. Zashi.* 1988, **2**(2), 83-86 (Ch.) (*Chem. Abstr.* **112**, 134189h)
31. Rodriguez-Arnaiz, R. et al *Contam. Ambiental* 1986, **2**(1), 57-61
32. Columbano, A. et al *Basic Appl. Histochem.* 1988, **32**(4), 501-510
33. Adam, M. S. et al *Acta. Hydrobiol.* 1990, **32**(1-2), 93-99

# L23   Lead perchlorate

## $Pb(ClO_4)_2$

**CAS Registry No.** 13637-76-8
**Synonyms** perchloric acid, lead(2+) salt; lead diperchlorate; lead(2+) perchlorate
**Mol. Formula** $Cl_2O_8Pb$                              **Mol. Wt.** 406.09

## Physical properties

**M. Pt.** (decomp.).

## Occupational exposure

**US TLV (TWA)** 0.05 mg m$^{-3}$ (as Pb) intended change; **UK Long-term limit** (MEL)
0.15 mg m$^{-3}$ (as Pb) under review; **UN No.** 1470; **HAZCHEM Code** 2Y;
**Conveyance classification** oxidizing substance, toxic substance.

## Mammalian and avian toxicity

### Carcinogenicity and long-term effects

No adequate evidence for carcinogenicity to humans, sufficient evidence for
carcinogenicity to animals, IARC classification group 2B (1).

## Any other comments

Lead and its compounds have been comprehensively reviewed (2).

## References

1. *IARC Monograph* 1987, **Suppl. 7**, 65
2. *IPCS Environmental Health Criteria 85: Lead – Environmental Aspects* 1989, WHO,
   Geneva

# L24    Lead phosphate

$Pb_3(PO_4)_2$

**CAS Registry No.** 7446-27-7
**Synonyms** phosphoric acid, lead (2+)salt (2:3); C.I. 77622; lead orthophosphate;
lead diphosphate; Perlex Paste 500; trilead phosphate
**Mol. Formula** $O_8P_2Pb_3$                                    **Mol. Wt.** 811.51
**Uses** Stabiliser for plastics.

## Physical properties
**M. Pt.** 1014°C; **Specific gravity** 6.9-7.3.

**Solubility**
Water: 0.00014 g $l^{-1}$ at 20°C.

## Occupational exposure
**US TLV (TWA)** 0.05 mg m$^{-3}$ (as Pb) intended change; **UK Long-term limit** (MEL)
0.15 mg m$^{-3}$ (as Pb) under review; **Supply classification** toxic.
**Risk phrases** May cause birth defects – Danger of cumulative effects – Harmful:
danger of serious damage to health by prolonged exposure if swallowed (R47, R33,
R48/22)
**Safety phrases** Avoid exposure-obtain special instruction before use – If you feel
unwell, seek medical advice (show label where possible) (S53, S44)

## Environmental fate

### Absorption
In adsorption studies on meadow brown soil, lead phosphate was the dominant
species. Desorption of lead occurred at pH <5 and at pHs >12.5. Distribution
coefficient in soil-water system was 0.5-6.0 × 10$^3$ (1).

## Mammalian and avian toxicity

### Carcinogenicity and long-term effects
No adequate evidence for carcinogenicity to humans, sufficient evidence for
carcinogenicity to animals, IARC classification group 2B (2).
Subcutaneous 270 albino rats (strain and sex unspecified) (16 month) 40-760 mg kg$^{-1}$
b.w. 19/29 rats that survived for 10 or more months developed renal tumours. The
tumours included adenomas, papillomas, cystadenomas and 3 carcinomas of the renal
cortex (3).
Subcutaneous (18 month) 80 albino rats wkly/fortnightly injections of 20 mg kg$^{-1}$.
Renal adenomas developed in 29 rats, no malignant tumours were found (3).

## Any other comments
Lead and its compounds have been comprehensively reviewed (4).

## References
1.  Li, S. et al *Huanjing Kexue Xuebao* 1987, **7** (4), 493-497 (Ch.) (*Chem. Abstr.* **109**, 210154y)
2.  *IARC Monograph* 1987, **Suppl. 7**, 65

3. *Fourth Annual Report on Carcinogens (NTP 85-002)* 1985, 121
4. *IPCS Environmental Health Criteria 85: Lead – Environmental Aspects* 1989, WHO, Geneva

# L25   Lead phosphite dibasic

**CAS Registry No.** 12141-20-7
**Synonyms** lead oxide phosphonate; dibasic lead phosphite
**Mol. Formula** $HO_5PPb_3$                                     **Mol. Wt.** 733.55

## Occupational exposure

**US TLV (TWA)** 0.05 mg m$^{-3}$ (as Pb) intended change; **UK Long-term limit** (MEL) 0.15 mg m$^{-3}$ (as Pb) under review; **UN No.** 2989; **Conveyance classification** flammable solid.

## Mammalian and avian toxicity

### Carcinogenicity and long-term effects
No adequate evidence for carcinogenicity to humans, sufficient evidence for carcinogenicity to animals, IARC classification group 2B (1).

## Any other comments

Lead and its compounds have been comprehensively reviewed (2).

## References
1. *IARC Monograph* 1987, **Suppl. 7**, 65
2. *IPCS Environmental Health Criteria 85: Lead – Environmental Aspects* 1989, WHO, Geneva

# L26   Lead stearate

$$Pb(C_{18}H_{35}O_2)_2$$

**CAS Registry No.** 1072-35-1
**Synonyms** octadecanoic acid, lead(2+) salt; lead distearate; normal lead stearate; Stabinex NC$_{18}$
**Mol. Formula** $C_{36}H_{70}O_4Pb$                             **Mol. Wt.** 774.15
**Uses** In extreme pressure lubricants. As drier in varnishes.

## Physical properties
**M. Pt.** 125°C.

## Occupational exposure
**UK Long-term limit** (MEL) 0.15 mg m$^{-3}$ (as Pb) under review.

240     **L26. Lead stearate**

## Mammalian and avian toxicity

### Carcinogenicity and long-term effects
No adequate evidence for carcinogenicity to humans or animals, IARC classification group 3 (1).

### Metabolism and pharmacokinetics
In workers occupationally exposed to lead stearate plasma (Pb) concentrations were significantly higher ($1.0 \pm 5.7 \mu g\ l^{-1}$) than those workers exposed to inorganic lead ($0.42 \pm 3 \mu g\ l^{-1}$) (2,3).

### Any other comments
Lead and its compounds have been comprehensively reviewed (4).

### References
1. *IARC Monograph* 1987, **Suppl. 7**, 65
2. Ong, C. N. et al *J. Appl. Toxicol.* 1990, **10**(1), 65-68
3. Cavalleri, A. et al, C. *Scand. J. Work, Environ. Health* 1987, **13**(3), 218-220
4. *IPCS Environmental Health Criteria 85: Lead – Environmental Aspects* 1989, WHO, Geneva

# L27   Lead subacetate

## $Pb(CH_3CO_2)_2 \cdot 2Pb(OH)_2$

**CAS Registry No.** 1335-32-6
**Synonyms** lead, bis(acetato-*O*)tetrahydroxytri-; lead acetate,basic; monobasic lead acetate; bis(acetato)dihydroxytrilead
**Mol. Formula** $C_4H_{10}O_8Pb_3$                     **Mol. Wt.** 807.69
**Uses** In sugar analysis to remove colouring matters, for clarifying and decolourising other solutions of organic substances.

## Physical properties

**Solubility**
Water: 62.5 g $l^{-1}$ cold, 250 g $l^{-1}$ hot.

## Occupational exposure

**US TLV (TWA)** 0.05 mg m$^{-3}$ (as Pb) intended change; **UK Long-term limit** (MEL) 0.15 mg m$^{-3}$ (as Pb) under review; **UN No.** 1616; **HAZCHEM Code** 2Z;
**Conveyance classification** harmful substance; **Supply classification** toxic.
**Risk phrases** May cause birth defects – Danger of cumulative effects – Possible risk of irreversible effects – Harmful: danger of serious damage to health by prolonged exposure if swallowed (R47, R33, R40, R48/22)
**Safety phrases** Avoid exposure-obtain special instruction before use – If you feel unwell, seek medical advice (show label where possible) (S53, S44)

## Mammalian and avian toxicity

**Sub-acute data**
$LC_{50}$ (5 day) oral Japanese quail >5000 mg $kg^{-1}$ (diet) (1).

**Carcinogenicity and long-term effects**
No adequate evidence for carcinogenicity to humans, insufficient evidence for carcinogenicity to animals, IARC classification group 3 (2).
Oral 50 ♂, ♀ Swiss mice (2 yr) 0.1% or 1.0%/0.5% in diet induced 3 adenomas and 4 carcinomas in 0.1% dose. Most of the mice receiving 1.0%/0.5% dose died and only 1 carcinoma was found (3).
Oral, 40 Wistar rats (29 or 24 month) 0.1% or 1.0%. Incidence of renal tumours occurred in 11/32 animals administered 0.1% dose, including 3 carcinomas; for 1.0% dose renal tumours were induced in 13/24 animals, including 6 carcinomas (4).
Oral 10 ♂ Wistar rats (48 wk) 1.5%. All rats developed renal tumours, 6 adenomas and 4 carcinomas, no extrarenal tumours were found (5).

**Teratogenicity and reproductive effects**
$TD_{Lo}$ oral mouse (28 day) 258 g $kg^{-1}$ caused reproductive effects (6).

## Any other comments

Lead and its compounds have been comprehensively reviewed (7).

## References

1.  Hill, E. F. et al *Lethal dietary toxicities of environmental contaminants and pesticides to Coturnix* 1986, 86-88, U.S. Dept. Int. Fish & Wildlife Ser. Report No. 2, Washington, DC
2.  *IARC Monograph* 1987, **Suppl. 7**, 65
3.  van Esch, G. J. et al *Br. J. Cancer* 1969, **23** 765-771
4.  van Esch, G. J. et al *Br. J. Cancer* 1962, **16**, 289-297
5.  Ito, N. et al *Acta Pathol. Jpn.* 1973, **23**, 87-109
6.  *Experientia* 1974, **30**, 486
7.  *IPCS Environmental Health Criteria 85: Lead – Environmental Aspects* 1989, WHO, Geneva

# L28    Lead sulfate

## $PbSO_4$

**CAS Registry No.** 7446-14-2
**Synonyms** sulfuric acid, lead (2+) salt (1:1); Anglislite; C.I. 77630;
C.I. Pigment White 3; Freemans white lead; Lead Bottoms
**Mol. Formula** $O_4SPb$                              **Mol. Wt.** 303.25
**Uses** Replaces white lead as pigment. Used with zinc in galvanic batteries. In lithography. Preparing rapidly drying oil varnishes. Weighting fabrics.

## Physical properties

**M. Pt.** 1170°C(decomp.); **Specific gravity** 6.20.

## Occupational exposure

**US TLV (TWA)** 0.05 mg m$^{-3}$ (as Pb) intended change; **UK Long-term limit** (MEL) 0.15 mg m$^{-3}$ (as Pb) under review; **UN No.** 1794; **HAZCHEM Code** 2X; **Conveyance classification** corrosive substance.

## Ecotoxicity

**Fish toxicity**

Preliminary static, acute toxicity tests using fathead minnows, >3000 mg (Pb) l$^{-1}$ introduced as sulfate did not cause morbidity to newly hatched minnows (1).

## Environmental fate

**Absorption**

Increased pH and humic acid concentration favoured complexation. At low pH values particulate lead compounds adsorbed humic acid inhibiting lead dissolution (2). Inorganic lead tends to form highly insoluble salts and complexes with various anions and is tightly bound to soil, therefore availability to terrestial plants is drastically reduced and effects to plants are normally only observed at very high environmental concentrations 100-1000 mg kg$^{-1}$ soil (3).

## Mammalian and avian toxicity

**Acute data**

LD$_{Lo}$ oral dog 2 g kg$^{-1}$ (4).
LD$_{Lo}$ oral guinea pig 30 g kg$^{-1}$ (5).
LD$_{Lo}$ intraperitoneal guinea pig 290 mg kg$^{-1}$ (6).

**Carcinogenicity and long-term effects**

No adequate evidence for carcinogenicity to humans, sufficient evidence for carcinogenicity to animals, IARC classification group 2B (7).

## Genotoxicity

*Salmonella typhimurium* TA100 with and without and metabolic activation in presence and absence of nitrilotriacetic acid. No mutagenic activity observed with nitrilotriacetic acid to solubilize. Encapsulation of lead pigments with amorphous silica rendered compounds non-mutagenic and not-toxic indicating active moieties were biologically unavailable (8).
*In vitro* human lymphocytes incidence of sister chromatid exchanges increased (9).
*In vitro* Chinese hamster cells significant increased frequency of sister chromatid exchanges (10).

## Any other adverse effects to man

It is highly irritating and destructive to the eyes and skin (11).

## Any other comments

Occupational exposure to pottery workers and toxicity discussed (12).
Soluble in cold water, hot water or ammonium salts; nitric acid, sodium hydroxide and concentrated hydriodic acid.

## References

1.  Erten, M. Z. et al *Proc. Ind. Waste Conf.* 1988, **43**, 617-629
2.  Law, K. J. et al *Min. Miner. Process. Wastes, Proc. West. Reg. Symp.* 1990, 61-67

3.   *IPCS Environmental Health Criteria 85: Lead – Environmental Aspects* 1989, WHO, Geneva
4.   *Abder. Handb. Biol. Arbeit.* 1935, **4**, 1289 (Ger.)
5.   *Arch. Hyg. Bakteriol.* 1941, **125**, 273
6.   *Merck Index* 1983, **10**, 779, Rahway, NJ
7.   *IARC Monograph* 1987, **Suppl. 7**, 65
8.   Connor, T. H. & Pier, S. M. *Mutat. Res.* 1990, **245**(2), 129-133
9.   Wulf, H. C. *Danish Med. Bull.* 1980, **27**(1), 40-42
10.  Montaldi, A. et al *J. Toxicol. Environ. Health* 1987, **21**(3), 387-394
11.  *Chemical Safety Data Sheets Vol. 2: Main Group Metals & their Compounds* 1990, RSC, London
12.  Kieser, J. et al *Z. Gesamte Hyg. Ihre Grenzeb.* 1990, **36**(3), 166-168 (Ger.) (*Chem. Abstr.* **112**, 222575m)

# L29   Lead sulfide

## PbS

**CAS Registry No.** 1314-87-0
**Synonyms** lead sulfide; C.I. 77640; Natural lead sulfide; P37; P128; plumbous sulfide
**Mol. Formula** PbS                                   **Mol. Wt.** 239.25
**Uses** Glazing earthenware.
**Occurrence** In the mineral galena.

## Physical properties

**M. Pt.** 1114°C; **B. Pt.** 1281°C; **Specific gravity** 7.50.

## Occupational exposure

**US TLV (TWA)** 0.05 mg m$^{-3}$ (as Pb) intended change; **UK Long-term limit** (MEL) 0.15 mg m$^{-3}$ (as Pb) under review.

## Ecotoxicity

**Fish toxicity**
Preliminary static, acute toxicity test using fathead minnow, >3000 mg (Pb) l$^{-1}$ introduced as sulfide did not cause mortality to newly hatched minnows (1).

## Environmental fate

**Absorption**
Inorganic lead tends to form highly insoluble salts and complexes with various anions and is tightly bound to soil, therefore availability to terrestial plants is drastically reduced and effects to plants are normally only observed at very high environmental concentrations 100-1000 mg kg$^{-1}$ soil (2).

## Mammalian and avian toxicity

**Acute data**
LD$_{Lo}$ oral guinea pig 10 g kg$^{-1}$ (3).
LD$_{Lo}$ intraperitoneal rat 1847 mg kg$^{-1}$ (4).

---

**Carcinogenicity and long-term effects**
No adequate evidence for carcinogenicity to humans, sufficient evidence for
carcinogenicity to animals, IARC classification group 2B (5).
Oral Fischer 344 rat prechronic experiments planned (6).

**References**

1. Erten, M. Z. et al *Proc. Ind. Waste Conf.* 1988, **43**, 617-629
2. *IPCS Environmental Health Criteria 85: Lead – Environmental Aspects* 1989, WHO,
   Geneva
3. *Ind. Med.* 1941, **10**(2), 15
4. *Arch. Hyg. Bakteriol.* 1941, **125**, 273
5. *IARC Monograph* 1987, **Suppl. 7**, 65
6. *Directory of Agents Being Tested for Carcinogenicity, No 15* 1992, IARC, WHO

# L30  Lead tetraacetate

## $(CH_3CO_2)_4Pb$

**CAS Registry No.** 546-67-8
**Synonyms** acetic acid, lead(IV) salt; lead(IV) acetate; plumbic acetate; lead acetate
**Mol. Formula** $C_8H_{12}O_8Pb$                                **Mol. Wt.** 443.37

**Physical properties**
**M. Pt.** 175-180°C; **Specific gravity** $d_4^{17}$ 2.23.

**Solubility**
Organic solvent: hot glacial acetic acid, benzene, chloroform, nitrobenzene

**Occupational exposure**
**US TLV (TWA)** 0.05 mg m$^{-3}$ (as Pb) under review; **UK Long-term limit** (MEL)
0.15 mg m$^{-3}$ (as Pb) intended change; **UN No.** 1616; **HAZCHEM Code** 2Z;
**Conveyance classification** harmful substance.

**Ecotoxicity**

**Invertebrate toxicity**
Cell multiplication inhibition test *Uronema parduczi Chatton-Lwoff* 0.07 mg l$^{-1}$
(Threshold concentration) (1).
*Pseudomonas putida*, cell multiplication inhibition starts at 1.8 mg l$^{-1}$ (2).
*Microcystis aeruginosa*, cell multiplication inhibition starts at 0.45 mg l$^{-1}$ (2).

**Mammalian and avian toxicity**

**Acute data**
$TD_{Lo}$ oral rat 18.9 mg kg$^{-1}$ (3).

### Carcinogenicity and long-term effects

No adequate evidence for carcinogenicity to humans, sufficient evidence for carcinogenicity to animals, IARC classification group 2B (4).

### Genotoxicity

*In vitro* Chinese hamster ovary cells sister chromatid exchange (metabolic activation unspecified) positive (5).

*Tradescentia* micronucleus assay, repeated tests for clastogenicity with lead acetate yielded the minimum ED for clastogenicity of 0.44 ppm (6).

### Legislation

Limited under EC Directive on Drinking Water Quality 80/778/EEC. Lead: maximum admissible concentration 50 $\mu$g l$^{-1}$ (7).

Included in Schedule 4 and 6 (Release into Air/Land: Prescribed Substances) Statutory Instrument No. 472, 1991 (8).

### Any other comments

Lead and its compounds have been comprehensively reviewed (9).

### References

1. Bringmann, G. et al *Water Res.* 1980, **14**, 231-241
2. Bringmann, G. et al *GWF, Gas-Wasserfach: Wasser/Abwasser* 1976, **117**(9)
3. *Andrologia* 1989, **21**, 161
4. *IARC Monograph* 1987, **Suppl. 7**, 230-232
5. Tai, E. C. H. et al *Bull. Inst. Zool., Acad. Sin.* 1990, **29**(2), 121-125
6. Sandhu, S. S. et al *Mutat. Res.* 1989, **224**(4), 437-45
7. *EC Directive Relating to the Quality of Water Intended for Human Consumption* 1982, 80/778/EEC, Office for Official Publications of the European Communities, 2 rue Mercier, L-2985 Luxembourg
8. *S. I. 1991 No. 472 The Environmental Protection (Prescribed Processes and Substances) Regulations* 1991, HMSO, London
9. *IPCS Environmental Health Criteria 85: Lead – Environmental Aspects* 1989, WHO, Geneva

# L31   Lead tetraethyl

## Pb(CH2CH3)4

**CAS Registry No.** 78-00-2

**Synonyms** plumbane, tetraethyl-; lead, tetraethyl-; tetraethylplumbane; TEL

**Mol. Formula** $C_8H_{20}Pb$                              **Mol. Wt.** 323.44

**Uses** As a gasoline additive to prevent 'knocking' in motors.

**Occurrence** In soils and organisms close to roads with high traffic density.

### Physical properties

**M. Pt.** –130°C; **B. Pt.** 227.7°C (decomp.); **Flash point** 93.3°C;
**Specific gravity** $d^{20}$ 1.653; **Volatility** v.p. 0.2 mmHg at 20°C; v. den. 8.6.

## Solubility

Water: 0.29 mg l$^{-1}$ at 25°C. Organic solvent: benzene, petroleum ether, ethanol, diethyl ether

## Occupational exposure

**US TLV (TWA)** 0.1 mg m$^{-3}$ (as Pb); **UK Long-term limit** (MEL) 0.15 mg m$^{-3}$ (as Pb) under review; **UN No.** 1649; **HAZCHEM Code** 2WE; **Conveyance classification** toxic substance.

## Ecotoxicity

### Fish toxicity

LC$_{50}$ (48 hr) bass 0.065 mg l$^{-1}$ static bioassay at 20°C (1).
LC$_{50}$ (24, 48, 96 hr) bluegill sunfish 2.0-0.02 mg l$^{-1}$ static bioassay at pH 6.9-7.5, 20°C, alkalinity 33-81 mg l$^{-1}$ (CaCO$_3$), hardness 84-163 mg l$^{-1}$ (CaCO$_3$) (2).
LC$_{50}$ (96 hr) plaice 0.23 mg l$^{-1}$ flow through bioassay at 15°C and 34.9 mg l$^{-1}$ (CaCO$_3$) salinity (3).
LC$_{50}$ (48 hr) stickleback, coho salmon 14 μg l$^{-1}$ for effluent discharges from tetraethyllead production plants. No attempt was made to assess the indirect hazard caused to birds by food organisms concentrating the lead (4).

### Invertebrate toxicity

EC$_{50}$ (4 hr) *Ankistrodesmus falcatus* < 0.3 mg l$^{-1}$ (5).
*Poteriouchromonas malhamensis* (3 day) 100 mg l$^{-1}$ culture in dark no toxic effects, in light all cells were killed at 80 mg l$^{-1}$ . In light, <80 mg l$^{-1}$ caused dose- related effects on growth, mitosis and cytokinesis. Conversion to a highly toxic derivative (triethyllead) took place within 3-6 hr, maximum concentration 24-32 hr (6).
Compounds used to alleviate lead poisoning in humans (including EDTA salts, dimercaprol) increased the effects of triethyllead on *Poteriouchromonas malhamensis* (7).
LC$_{50}$ (96 hr) mussel, brown shrimp 0.1 mg l$^{-1}$ and 0.02 mg l$^{-1}$, respectively (flowthrough), at 15°C and 34.9% salinity (3).
It is nontoxic to freshwater and marine algae (duration and concentration unspecified); it is the trialkyllead degradation product which is responsible for the apparent toxicity of tetraethyllead (8).

## Bioaccumulation

In Canada, northern pike and redhorse sucker living in rivers near alkyllead production plants contained 0.17 and 5.04 μg g$^{-1}$ respectively (9).
Caged clams concentrated tetraethyllead in the muscle and visceral tissues (9).
Accumulation (96 hr) in shrimp at 0.02 mg l$^{-1}$ gave a concentration factor of 650 (5).
Accumulation (96 hr) *Mytilus edulis* at 10 mg l$^{-1}$ gave a concentration factor of 120 mean of digestive gland, foot, gill and gonad (10).
Accumulation (96 hr) plaice at 0.23 mg l$^{-1}$ gave a concentration factor of 130 (10).

## Environmental fate

### Degradation studies

BOD using mixed coastal marine bacteria, concentrations < 0.16 mg l$^{-1}$ O$_2$ no significant effect on lag phase. EC$_{50}$ (48 hr) 0.2 mg l$^{-1}$ (1).
EC$_{50}$ (48 hr) *Dunialiella tertiolecta* 0.15 mg l$^{-1}$ (1).
Tetraalkyllead compounds are the predominant alkyllead compounds in the

atmosphere. They are removed from the air in rainwater and transferred to surface and highway drainage water, where sequential breakdown of tetraalkyllead to tri- and dialkyllead compounds occurs (11).

## Mammalian and avian toxicity

### Acute data
$LD_{50}$ oral Japanese quail 24.6 mg $kg^{-1}$ (12).
$LD_{50}$ oral mallard duck 107 mg $kg^{-1}$ (12).
$LD_{50}$ oral rat 12.3 mg $kg^{-1}$ (13).
$LD_{Lo}$ oral rabbit 30 mg $kg^{-1}$ (14).
$LD_{Lo}$ dermal dog, guinea pig 547, 995 mg $kg^{-1}$, respectively (14).
$LD_{50}$ intravenous rat, rabbit 15, 22 mg $kg^{-1}$, respectively (15,16).
$LD_{Lo}$ subcutaneous rabbit 32 mg $kg^{-1}$ (16).

### Sub-acute data
Oral starling (11 day) 0, 200, 2000 µg $l^{-1}$. All birds receiving low doses survived treatment. Lead accumulated in the brain, kidney and liver (17).
Mallard ducks, Japanese quail (6 day) 6 mg $kg^{-1}$ bw, no effect on eggshell thickness (18).

### Carcinogenicity and long-term effects
No adequate evidence for carcinogenicity to humans or animals, IARC classification group 3 (19).

### Teratogenicity and reproductive effects
Oral Simonsen Sprague-Dawley rats 7.5, 15 or 30 mg $kg^{-1}$ body weight on 3 consecutive days (9, 10, 11 or 12, 13, 14) of organogenesis. Doses up to those lethal to some mother animals were essentially non-teratogenic to offspring (20).
Transplacental effect of tetraethyllead on tissue plasminogen activator activity, plasminogen activator inhibition and plasma inhibition in rat. Changes in these processes were observed in lungs, liver, heart, brain and kidneys. Foetal tissue plasminogen activator activity, plasminogen activator inhibition and plasma inhibition can be affected transplacentally by tetraethyllead (21).

### Metabolism and pharmacokinetics
Tissue distribution studies of lead in rats and dogs exposed by inhalation revealed lead levels of 7-100 mg $kg^{-1}$ tissue in lung, brain, liver, kidney, spleen and heart (22).
Oral, intravenous (duration unspecified) or inhalation (5 hr) rabbits at 3 mg, 12 mg or 200 µg $m^{-3}$ air, respectively. Correlation was observed between diethyllead excretion in urine and the dose and method of administration (23).

## Any other adverse effects to man

7 occupationally exposed workers had altered plasma renin activity. The changes are discussed with regard to neurogenic, hormonal, humoral and electrolyte factors influencing the release of renin (24).

## Any other adverse effects

In rats with cerebral oedema induced by tetraethyllead, brain mitochondrial monophosphoinositides, sphingomyelins and phosphatidylserines, phosphatidylethanolamines and cardiolipins increased. Decrease in monophosphoinositides, phosphatidylcholines and phosphatidylethanolamines in

brain microsomes. Increases in lysophophatidylcholines and sphingomyelins were observed. Increased lipid peroxidation also accompanied cerebral oedema (25).
In 1979, 2400 birds were found dead or dying in the Mersey estuary, UK, the majority being waders. Smaller numbers of dead birds were found in 1980 and 1981. A plant manufacturing petrol additives was in the vicinity. Affected birds contained elevated lead levels mostly as alkyllead (26).

## Legislation

Included in Schedule 6 (Release into Land: Prescribed Substances) Statutory Instrument No. 472, 1991 (27).

## Any other comments

Generally tetraalkyllead compounds are of higher toxicity to microorganisms than inorganic lead compounds. Organolead compounds are generally 10-100 × more toxic to aquatic organisms than inorganic leads. Tetraalkyllead becomes toxic by conversion to trialkyllead (28).
Acute or chronic poisoning may occur if inhaled or absorbed through skin (29).
Dilute solution in water decomposes to give triethyl salt, then diethyl salt and finally inorganic lead (30).
Lead from tetraethyllead in soil and vegetation decreases exponentially with the distance from the road. Levels in plants and animals increase in areas close to roads, the levels are positively correlated with traffic volume and proximity of roads (28).
Hazardous properties reviewed (31).
*Scrobicularia plana* did not suffer any siphonal contraction when pure tetraethyllead was applied to the preparation. It has a low toxicity or is non-toxic in pure form (32).

## References

1. Marchetti, R. *Mar. Pollut. Bull.* 1978, **9**, 206-207
2. Turnbull, H. et al *Ind. Eng. Chem.* 1954, **46**, 324-333
3. Maddock, B. G. et al *Lead in the Marine Environment* 1980, 233-261, Pergamon Press, Oxford
4. Gill, J. M. et al *J. Water Pollut. Control Fed.* 1960, **32**, 858-867
5. Silverberg, B. A. et al *Arch. Environ. Contam. Toxicol.* 1977, **5**
6. Roderer, G. *Environ. Res.* 1980, **23**, 371-384
7. Roderer, G. *Chem.-biol. Interact.* 1983, **48**[A, 247-254
8. Jarvie, A. W. P. *Appl. Organomet. Chem.* 1987, **1**(1), 29-38
9. Chau, Y. K. et al *Chem. Environ. Proc. Int. Conf.* 1986, 77-82
10. Grove, J. R. *Investigation into the formation and behaviour of aqueous solutions of lead alkyls*, The Associated Octel Co. Ltd., Ellesmere Port
11. Radojevic, M. *Heavy Met. Environ., Int. Conf. 5th* 1985, **1**, 82-84
12. Hudson, R. H. et al *Handbook of toxicity of pesticides to wildlife* 2nd ed., 1984, US Dept. Int., Fish & Wildlife Ser., Report No. 153, Washington, DC
13. Schroeder et al *Experientia* 1972, **28**, 923
14. *Jpn. J. Ind. Health* 1973, **15**, 3
15. *Ind. Med.* 1963, **54**, 486
16. *Environ. Qual. Saf., Suppl.* 1975, **1**, 1
17. Osborn, D. et al *Environ. Pollut.* 1983, **31**, 261-275
18. Haegele, M. A. et al *Bull. Environ. Contam. Toxicol.* 1974, **11**, 98-102
19. *IARC Monograph* 1987, **Suppl. 7**, 230-232
20. McClain, R. M. et al *Toxicol. Appl. Pharmacol.* 1972, **21**, 265-274
21. Smokovitis, A. et al *Biol. Neonate* 1990, **58**(1), 41-49

22. Davies, B. E. et al *Arch. Environ. Health* 1963, **6**, 473-479
23. Kozarzewska, Z. et al *Br. J. Ind. Med.* 1987, **44**(6), 417-421
24. Carmignani, M. et al *Acta Med. Rom.* 1988, **26**(3), 277-286
25. Pogosyan, A. Y. et al *Biol. Zh. Arm.* 1989, **42**(1), 37-41 (Russ.)
26. Head, P. C. et al *The Mersey Estuary bird mortality Autumn-Winter 1979 – preliminary report* 1980, Directorate of Scientific Services, Report No. DSS-EST-80-1, Warrington
27. *S. I. 1991 No. 472 The Environmental Protection (Prescribed Processes and Substances) Regulations* 1991, HMSO, London
28. *IPCS Environmental Health Criteria 85: Lead – Environmental Aspects* 1989, WHO, Geneva
29. Browning, E. *Toxicity of Industrial Metals* 2nd ed., 1969, 192-199, Appleton-Century-Crofts, London
30. Harrison, G. F. *The Cavtat Incident* 1977, presented at the International Experts Discussion Meeting on Lead – Occurrence, Fate and Pollution in the Marine Environment, Rovinj
31. *Dangerous Prop. Ind. Mater. Rep.* 1989, **9**(4), 77-87
32. Marshall, S. J. et al *Appl. Organomet. Chem.* 1988, **2**(2), 143-149

# L32   Lead tetramethyl

## $Pb(CH_3)_4$

**CAS Registry No.** 75-74-1
**Synonyms** plumbane, tetramethyl-; lead, tetramethyl-; tetramethyllead; tetramethylplumbane
**Mol. Formula** $C_4H_{12}Pb$                              **Mol. Wt.** 267.33
**Uses** Component of anti-knock mixes for motor petrol.
**Occurrence** In soils and organisms close to roads with high traffic density.

## Physical properties

**M. Pt.** –27.8°C; **B. Pt.** 110°C (decomp.); **Flash point** 37.8°C; **Specific gravity** 1.99; **Volatility** v. den. 9.2.

**Solubility**
Water: 15 mg $l^{-1}$ seawater. Organic solvent: benzene, ethanol, diethyl ether

## Occupational exposure

**US TLV (TWA)** 0.15 mg m$^{-3}$ (as Pb); **UK Long-term limit** (MEL) 0.15 mg m$^{-3}$ (as Pb) under review; **UN No.** 1649; **HAZCHEM Code** 2WE; **Conveyance classification** toxic substance.

## Ecotoxicity

**Fish toxicity**
$LC_{50}$ (48 hr) bass 0.10 mg $l^{-1}$ static bioassay at 20°C (1).
$LC_{50}$ (96 hr) tidewater silverside, bluegill sunfish, 13.5-84 (static bioassay) pH 7.6-7.9, hardness 55 mg $l^{-1}$ ($CaCO_3$) and temperature 20-23°C (2).
$LC_{50}$ (96 hr) plaice 0.05 mg $l^{-1}$ flowthrough bioassay of 15°C and 34.9 mg $l^{-1}$ ($CaCO_3$) salinity (3).

**Invertebrate toxicity**
*Scenedesmus quadricauda, Ankistrodesmus falcatus*, and *Chlorella pyrenoidosa*

(7 day) < 0.5 mg decrease in growth 32, 32 and 74%, respectively. Tetramethyllead is 2 × as toxic as trimethyllead acetate and 20 × as toxic as lead nitrate for the same organisms (4).

$LC_{50}$ (96 hr) mussel, brown shrimp 0.27 mg $l^{-1}$ and 0.11 mg $l^{-1}$, respectively (flow through bioassay) at 15°C and 34.9% salinity (3).

It is non-toxic to freshwater and marine algae (duration and concentration unspecified); it is the trialkyllead degradation product which is responsible for the apparent toxicity of tetramethyllead (5).

### Bioaccumulation

In Canada, northern pike and redhorse sucker living in rivers near alkyllead production plants contained 0.17 and 5 μg $g^{-1}$, respectively (6).

Accumulation (96 hr) in shrimp at 0.11 mg $l^{-1}$ gave a concentration factor of 20 (7).

Accumulation (96 hr) *Mytilus edulis* at 0.27 mg $l^{-1}$ gave a concentration factor 170, mean of digestive gland, foot, gill and gonads (7).

Accumulation (96 hr) plaice at 0.05 mg $l^{-1}$ gave a concentration factor of 60 (7).

## Environmental fate

### Degradation studies

BOD using mixed coastal marine bacteria, concentrations < 3.2 mg $l^{-1}$ $O_2$ no significant effect on lag phase. $EC_{50}$ (48 hr) 1.9 mg $l^{-1}$ (1).

$EC_{50}$ (48 hr) *Dunialiella tertiolecta* 1.65 mg $l^{-1}$ (1).

Tetramethyllead was produced from inorganic lead salts using biologically active sediments and estuarine waters. Tetramethyllead formation was a 2 stage process involving an initial lag of ≈100 hr followed by the exponential appearance of tetramethyllead which amounted to ≈0.03% of total added lead (8).

Tetraalkyllead compounds are the predominant alkyllead compounds in the atmosphere. They are removed from the air in rainwater and transferred to surface and highway drainage water, sequential breakdown of tetraalkyllead to tri- and dialkyllead compounds occurs (9).

## Mammalian and avian toxicity

### Acute data

$LD_{50}$ oral rabbit, rat 24, 105 mg $kg^{-1}$, respectively (10,11).

$LD_{Lo}$ dermal rabbit 3391 mg $kg^{-1}$ (11).

$LD_{50}$ intraperitoneal rat 90 mg $kg^{-1}$ (10).

$LD_{50}$ intravenous rat, rabbit 88-90 mg $kg^{-1}$ (10,12).

### Carcinogenicity and long-term effects

No adequate evidence for carcinogenicity to humans or animals, IARC classification group 3 (13).

### Teratogenicity and reproductive effects

Oral Simonsen Sprague-Dawley rats 40, 80, 112 or 160 mg $kg^{-1}$ on 3 consecutive days (9, 10, 11 or 12, 13, 14) of organogenesis. Doses up to those lethal to some mother animals were 'essentially non-teratogenic' to offspring (7).

### Metabolism and pharmacokinetics

Intravenous (7 day) rabbit 9.9 or 39.7 mg $kg^{-1}$. Urinary total lead excretion for 9.9 mg $kg^{-1}$ was 73% dimethyllead, 19% trimethyllead, 6% inorganic lead and 2% tetramethyllead on the day following the injection and 100% trimethyllead 7 days after the injection. For 39.7 mg kg$^{-1}$ the urinary total lead excretion was 67% dimethyllead, 14% trimethyllead, 17% inorganic lead and 2% tetramethyllead on the day following the injection and 8% dimethyllead, 74% trimethyllead, 17% inorganic lead and 1% tetramethyllead 7 days after the injection. Faecal total lead excretion for the 7 days was composed of 100% inorganic lead. During the 7 days 1-3% of either administered dose was excreted in the urine and 7-19% in the faeces (14).

## Legislation

Included in Schedule 6 (Release into Land: Prescribed Substances) Statutory Instrument No. 472, 1991 (15).

## Any other comments

Organolead compounds are generally 10-100 × more toxic to aquatic organisms than inorganic leads. Tetraalklyllead becomes toxic by conversion to trialkyllead (16). Lead from tetramethyllead in soil and vegetation decreases exponentially with the distance from the road. Levels in plants and animals increase in areas close to roads, the levels are positively correlated with traffic volume and proximity of roads (16). Tetramethyllead is not soluble in water and is volatile.

## References

1. Marchetti, R. *Mar. Pollut. Bull.* 1978, **9**, 206-207
2. Dawson, G. W. et al *J. Hazard Mater.* 1977, **1**, 303-318
3. Maddock, B. G. et al *Lead in the Marine Environment* 1980, 233-261, Pergamon Press, Oxford
4. Silverberg, B. A. et al *Arch. Environ. Contam. Toxicol.* 1977, **5**, 305-313
5. Jarvie, A. W. P. *Appl. Organomet. Chem.* 1987, **1**(1), 29-38
6. Chau, Y. K. et al *Chem. Environ. Proc. Int. Conf.* 1986, 77-82
7. McClain, R. M. et al *Toxicol. Appl. Pharmacol.* 1972, **21**, 265-274
8. Walton, A. P. et al *Appl. Organomet. Chem.* 1988, **2**(1), 87-90
9. Radojevic, M. *Heavy Met. Environ., Int. Conf. 5th* 1985, **1**, 82-84
10. *Ind. Med.* 1963, **54**, 486
11. *Jpn. J. Ind. Health* 1973, **15**, 3
12. *J. Pharmacol. Exp. Ther.* 1930, **38**, 161
13. *IARC Monograph* 1987, **Suppl. 7**, 230-232
14. Arai, F. et al *Ind. Health* 1990, **28**(2), 63-76
15. *S. I. 1991 No. 472 The Environmental Protection (Prescribed Processes and Substances) Regulations* 1991, HMSO, London
16. *IPCS Environmental Health Criteria 85: Lead – Environmental Aspects* 1989, WHO, Geneva

# L33   Lead thiocyanate

## Pb(SCN)2

CAS Registry No. 592-87-0
**Synonyms** thiocyanic acid, lead(2+) salt; lead bis(thiocyanate); lead dithiocyanate; lead(II) thiocyanate
**Mol. Formula** $C_2N_2S_2Pb$                                **Mol. Wt.** 323.35
**Uses** Reverse dyeing with aniline black. Manufacture of safety matches and cartridges.

## Physical properties

**M. Pt.** 190°C (decomp.); **Specific gravity** 3.82.

## Occupational exposure

**US TLV (TWA)** 0.05 mg m$^{-3}$ (as Pb) intended change; **UK Long-term limit (MEL)** 0.15 mg m$^{-3}$ (as Pb) under review.

## Mammalian and avian toxicity

### Carcinogenicity and long-term effects

No adequate evidence for carcinogenicity to humans, sufficient evidence for carcinogenicity to animals, IARC classification group 2B (1).

## Any other comments

Lead and its compounds have been comprehensively reviewed (2).

## References

1. *IARC Monograph* 1987, **Suppl. 7**, 65
2. *IPCS Environmental Health Criteria 85: Lead – Environmental Aspects* 1989, WHO, Geneva

# L34   Lenacil

**CAS Registry No.** 2164-08-1
**Synonyms** 1*H*-cyclopentapyrimidine- 2,4(3*H*,5*H*)-dione, 3-cyclohexyl-6,7-dihydro-; Adol 80WP; Elbatan; Herbicide 634; Hexilure; Venzar
**Mol. Formula** $C_{13}H_{18}N_2O_2$                    **Mol. Wt.** 234.30
**Uses** Herbicide.

## Physical properties

**M. Pt.** solid 290°C, commercial product 316-317°C; **Specific gravity** 1.32.

## Solubility

Water: 6 mg l$^{-1}$ 25°C. Organic solvent: ethanol, acetone, benzene

## Ecotoxicity

### Fish toxicity

$LC_{50}$ (96 hr) carp 10 mg l$^{-1}$ (1).

## Environmental fate

### Degradation studies

Under laboratory conditions microbial activity in clay loam soil stimulated heterotrophic $N_2$ fixation and production of carbon dioxide. Effects on heterotrophic $N_2$ fixation and soil respiration were only exhibited at high rates of application $100\ \mu g\ g^{-1}$ soil (2).

### Absorption

Adsorption on montmorillonite and soil $0.028\ g\ l^{-1}$ per mol $kg^{-1}$ soil. At a concentration $<0.14\ g\ l^{-1}$ sorption linearly increased with concentration and was complete, $\geq 0.14\ g\ l^{-1}$ sorption was non-linear and decreased to 16% of triazines (3).

## Mammalian and avian toxicity

### Acute data

$LD_{50}$ rat $>11\ g\ kg^{-1}$ (4,5).
$LD_{50}$ dermal rabbit $>5000\ mg\ kg^{-1}$ (5).

### Sub-acute data

$LC_{50}$ (8 day) oral bobwhite quail $2300\ mg\ kg^{-1}$ (diet) (6).
Oral rats (22 wk) $200\ mg\ kg^{-1}\ day^{-1}$ diet caused increased glycolysis in rat liver, while phosphofructokinase and glucose-6-phosphatase enzyme activities were slightly inhibited (7).

### Carcinogenicity and long-term effects

Oral rats (lifetime study) 130 or $200\ mg\ kg^{-1}\ day^{-1}$. In rats which died a natural death hyperplastic changes of flat multilayer gastric epithelium were larger and more frequent than in controls. Symptoms of carcinogenic activity were not found (7).

### Teratogenicity and reproductive effects

Oral rats (3 generations) 500 ppm, no adverse effects were found on reproductive indices or lactation and there was no pathology in the $F_3$ generation (8).
Oral (2 yr) dogs 10,000 ppm. 2 ♀ dogs produced healthy offspring (8).

### Irritancy

Caused mild irritation to eyes of rabbits (5).

## Genotoxicity

*Salmonella typhimurium* TA97, TA98, TA100, TA102 (metabolic activation unspecified) negative (9).

## Legislation

Included in Schedule 4 (Release into Air: Prescribed Substances) Statutory Instrument No. 472, 1991 (10).
Limited under EC Directive on Drinking Water Quality 80/778/EEC. Pesticides: maximum admissible concentration $0.1\ \mu g\ l^{-1}$ (11).

## Any other comments

In soil, 100% microbial degradation occurred within 5-6 months (5).
Adsorption, decomposition dynamics and environmental fate studied (12).

# References

1. Nikunen, E. et al (Ed.) *Enviromental Properties of Chemicals* 1990, 672 VAPK-Publishing, Helsinki
2. Magallanes, M. et al *An. Edafol. Agrobiol.* 1985, **44**(11-12), 1693-1700 (Span.) (*Chem. Abstr.* **106**, 14582k)
3. Zima, J. et al *Agrochemia (Bratislavia)* 1987, **27**(6), 177-182 (Czech.) (*Chem. Abstr.* **107**, 193000j)
4. *Merck Index* 11th ed., 1989, Merck & Co. Inc., Rahway, NJ
5. *The Agrochemicals Handbook* 3rd ed., 1991, RSC, London
6. *The Pesticide Manual* 9th ed., 1991, British Crop Protection Council, Farnham
7. Pliss, M. B. et al *Ratsion. Pitan.* 1977, **12**, 90-92, 98-102, (Russ.) (*Chem. Abstr.* **93**, 108483)
8. Clayton, G. D. et al (Ed.) *Patty's Industrial Hygiene and Toxicology* 3rd ed., 1981, **2A**, John Wiley & Sons, New York
9. Grancharov, K. et al *Arzneim.-Forsch.* 1986, **36**(11), 1660-1663
10. *S. I. 1991 No. 472 The Environmental Protection (Prescribed Processes and Substances) Regulations* 1991, HMSO, London
11. *EC Directive Relating to the Quality of Water Intended for Human Consumption* 1982, 80/778/EEC, Office for Official Publications of the European Communities, 2 rue Mercier, L-2985 Luxembourg
12. Karoly, G. *Novenyvedelen (Budapest)* 1986, **22**(10), 438-444 (Hung.) (*Chem. Abstr.* **108**, 145360a)

# L35   Leptophos

**CAS Registry No.** 21609-90-5
**Synonyms** *O*-(4-bromo-2,5-dichlororophenyl) *O*-methyl phenylphosphonothioate; Phosvel; MBCP; NK 711; *O*-(2,5-dichloro-4-bromophenyl) *O*-methyl phenylthiophosphonate
**Mol. Formula** $C_{13}H_{10}BrCl_2O_2PS$         **Mol. Wt.** 412.08
**Uses** Insecticide (superseded).

## Physical properties
**M. Pt.** 70°C; **Specific gravity** $d_4^{25}$ 1.53; **Partition coefficient** log $P_{ow}$ 5.881.

## Solubility
Water: 4.7 mg $l^{-1}$ at 20°C. Organic solvent: benzene, xylene, acetone, cyclohexane, heptane, isopropyl alcohol

## Occupational exposure

**Supply classification** toxic.

**Risk phrases** Harmful in contact with skin – Toxic if swallowed – Toxic: danger of very serious irreversible effects if swallowed (R21, R25, R39/25)

**Safety phrases** Avoid contact with eyes – Wear suitable protective clothing, gloves and eye/face protection – If you feel unwell, seek medical advice (show label where possible) (S25, S36/37/39, S44)

## Ecotoxicity

**Fish toxicity**

$LC_{50}$ (96 hr) *Leiostomus xanthurus*, rainbow trout 4-10 µg l$^{-1}$ (1,2).
$LC_{50}$ (48 hr) carp >40 mg l$^{-1}$ (2).

**Invertebrate toxicity**

$LC_{50}$ (96 hr) mysid shrimp, marine shrimp 1.88-3.16 µg l$^{-1}$ (1).
$LC_{50}$ (3 hr) *Daphnia sp.* 2 µg l$^{-1}$ (2).
$LC_{50}$ (96 hr) *Saccobranchus fossilis* 22.6 mg l$^{-1}$ at 18°C (3).

**Bioaccumulation**

Bioconcentration factor (4 day) *Leiostomus xanthurus* 68 (1).
Bioconcentration factor (14 day) *Pseudorasbora parva* 0.019 (4).

## Environmental fate

**Anaerobic effects**

Nitrogen fixing cyanophytes growing on rice plants showed tolerance limits of: *Aulosira fertilissima* 300 ppm; *Anabaena doliulum* 175 ppm; and *Nostoc sp.* 200 ppm. Reduced nitrogen content of the algae (20 day): *Aulosira fertilissima* by ≈30%; *Anabaena doliulum* by ≈4%; and *Nostoc sp.* by ≈28% of initial (control) nitrogen value (5).

**Abiotic removal**

Photodegradation of thin dry film by UV irradiation after 24 hr and by sunlight after 65 days, 47 and 37% of initial concentration, respectively (6).

## Mammalian and avian toxicity

**Acute data**

$LD_{50}$ oral rat, mouse 30-65 mg kg$^{-1}$ (7-11).
$LD_{50}$ oral rat 19 mg kg$^{-1}$ (12).
$LD_{50}$ oral rabbit 124 mg kg$^{-1}$ (13).
$LD_{50}$ dermal ♂ rat 103 mg kg$^{-1}$ (10).
$LD_{50}$ dermal rat 44 mg kg$^{-1}$ (12).
$LD_{50}$ dermal rabbit 800 mg kg$^{-1}$ (13).
$LD_{50}$ intraperitoneal ♂ rat 175 mg kg$^{-1}$ (11).
$LD_{50}$ subcutaneous mouse 120 mg kg$^{-1}$ (11).

**Sub-acute data**

Oral (30 day) domestic chicken 1200 ppm (diet), fatal to 64% of birds (14).
Oral dog (30 day) 40 mg kg$^{-1}$ day$^{-1}$ caused cholinergic poisoning in 100% of animals treated (15).
Peking duck (30 day) 100 mg kg$^{-1}$ day$^{-1}$ caused delayed neuropathy in 100% of treated birds (16).

Oral mallard duck (30, 60, 90 day) 60 or 270 mg kg$^{-1}$ day$^{-1}$ caused neuropathy
and ataxia, while 10 mg kg$^{-1}$ failed to produce ataxia (17-19).
Intravenous (3 day) hens 5 mg kg$^{-1}$ day$^{-1}$ delayed ataxia was observed in 44% of test
population. Transfer from blood to affinitive tissues such as sciatic nerve or leg
muscles occurred. Accumulated in nerves and muscles at initial stages and caused
enhancement of neuropathy (20).

### Teratogenicity and reproductive effects
Pregnant rats (8-20 days gestation) were fed 125, 50 or 12.5 ppm. 2 cases of gross
foetal malformation were found in 50 ppm group, wavy rib and abnormal
construction of nasal cavity. Survival rate of neonates was decreased (11).

### Metabolism and pharmacokinetics
Following oral administration to domestic chicken 250 mg kg$^{-1}$ as a single dose,
levels in adipose tissue were 25, 9, 1.5 and 0 ppm after 1, 3, 7 and 38 days,
respectively. Blood plasma levels peaked 3 hr after dose (21).
Oral administration to domestic chicken 400 mg kg$^{-1}$. After 24 hr levels in the brain
and blood plasma were 0.66 and 0.15 ppm, respectively. After 72 hr these decreased
to 0.06 ppm in the brain and 0.11 ppm in the spinal cord. Complete metabolism
occurred after 8 days (brain) and 15 days (spinal cord) (22).
The disappearance rate in hens was 70% at 6 hr and 93% at 96 hr. The $t_{1/2}$ was 1.37 hr
for the early phase and 45 hr for the late phase. Only 0.1% of the administered dose
was detected in excreta, its disappearance from the hen's body was due to its
metabolism (23).
In mammals, monooxygenase system enzymes in the liver metabolically converted
leptophos into more active substances, i.e. oxonium leptophos, desmethyl leptophos,
desbromoleptophos. It is rapidly metabolised in non-sensitive species, e.g. mice and
rats. Eliminated in the urine in the form of polar degradation products with an
elimination $t_{1/2}$ of 9 hr (24).
A single intragastric dose of [14]C leptophos was given to rats as a 15 mg kg$^{-1}$ dose,
the [14]C in the liver, kidneys and fatty tissues did not exceed 2-5% of the [14]C dose. In
9 days 46-75% of the dose was eliminated in the urine, 12-29% in the faeces and
1.35-5.38% in exhaled air. The following metabolic products of leptophos were
detected in the urine: oxonium leptophos; $O$-methylphenylthiophosphoric acid;
phenylphosphoric acid; and 4-bromo-2,5-dichlorophenol (25).
Cholinesterase activity in pregnant rats following leptophos treatment decreased as
follows; maternal brain, liver, placenta, foetus and maternal serum (7).
The uncoupling action on liver mitochondrial oxidative phosphorylation was not
marked (7).

## Genotoxicity
*In vitro* Chinese hamster ovary cells without metabolic activation induced a
significant increase in sister chromatid exchanges (26).

## Any other adverse effects to man
Oral, inhalation and/or dermal, exposure at conditions of local agriculture usage
(4 day). Epigastric pain, diarrhoea, nausea, coughing, bradycardia, insomnia,
headaches and weariness were observed in subjects. Erythrocyte cholinesterase
activity was moderately inhibited (27).

## Legislation

The log $P_{ow}$ value exceeds the European Community recommended level 3.0 (6th and 7th amendments) (28).

## Any other comments

No neuropathy detected in Japanese quails (29).

Denied registration as pesticide in USA (30).

Residues detected in foods (0.12), fruits (0.005 ppm), potatoes (0.1 ppm) and water in USA (31-33).

Highly toxic to sporulation in *Metarhizium anisopliae* var. *anisopliae* isolate $E_9$ (34).

Toxicology and environmental fate reviewed (35).

## References

1.  Schimmel, S. C. et al *Contrib. Mar. Sci.* 1979, **22**, 193-203
2.  Kanazawa, J. *Pestic. Sci.* 1981, **12**(4), 417-424
3.  Verma, S. R. et al *Water Res.* 1982, **16**(5), 525-529
4.  Kanazawa, J. *Seitai Kagaku* 1983, **6**(3), 37-44 (Japan.) (*Chem. Abstr.* **100**, 186961a)
5.  Sharma, V. K. et al *Int. J. Ecol. Environ. Sci.* 1981, **7**, 117-122
6.  Riskallah, M. R. et al *Bull. Environ. Contam. Toxicol.* 1979, **23**, 630-641
7.  Hollinghaus, J. G. et al *J. Agric. Food Chem.* 1979, **27**, 1197
8.  Francis, B. M. et al *J. Environ. Sci. Health, Part B* 1982, **17**(6), 611-633
9.  Kanoh, S. et al *Teratol. J. Abn. Develop.* 1978, **18**, 146
10. Gaines, T. B. et al *Fundam. Appl. Toxicol.* 1986, **7**, 299-308
11. Kanoh, S. et al *Oyo Yahon* 1981, **112**(3), 373-380 (Japan.) (*Chem. Abstr.* **97**, 67476j)
12. *Fundam. Appl. Toxicol.* 1986, **7**, 299
13. *J. Eur. Toxicol.* 1973, **6**, 70
14. Belal, M. et al *Egypt. J. Anim. Prod.* 1982, **2**(1-2), 127-131
15. Soliman, S. A. *Neurotoxicology* 1983, **4**(41), 107-115
16. Soliman, S. A. et al *Toxicol. Appl. Pharmacol.* 1983, **69**(3), 417-431
17. Soliman, S. A. et al *J. Toxicol. Environ. Health* 1984, **14**(5-6), 789-801
18. Soliman, S. A. et al *J. Environ. Sci. Health, Part B* 1986, **21**(5), 401-411
19. Hoffman, D. J. et al *Toxicol. Appl. Pharmacol.* 1984, **75**(1), 128-136
20. Yamauchi, T. et al *J. Toxicol. Sci.* 1989, **14**(1), 11-21
21. Konno, N. et al *Proc. Int. Congr. Rural Med.* 7th ed., 1978, 296-298
22. Konno, N. et al *Nippon Koshu Eisei Zasshi* 1980, **27**(8), 367-372 (Japan.) (*Chem. Abstr.* **96**, 1757u)
23. Yamaguchi, Y. et al *Nippon Eiseigaku Zasshi* 1989, **43**(6), 1159-1168 (Japan.) (*Chem. Abstr.* **111** 52057d)
24. Makhayeva, G. F. et al *Agrokhimiya* 1987, **12**, 103-104 (Russ.)
25. Kagan, Y. S. *Leptophos in 'Harmful Substances in Industry' A Handbook* 1985, 276-277, Khimiya, Leningrad
26. Nishio, A. et al *J. Toxicol. Environ. Health* 1981, **8**, 939
27. Hassan, A. et al *Chemosphere* 1978, **7**(3), 283-290
28. *1967 Directive on Classification, Packaging, and Labelling of Dangerous Substances 67/548/EEC; 6th Amendment EEC Directive 79/831/EEC; 7th Amendment EEC Directive 91/32/EEC* 1991, HMSO, London
29. Kinebuchi, H. et al *Proc. ICMR Semin.* 1988, **8**, 541-545
30. *Pesticide Index* US EPA, 1985
31. Johnson, R. D. et al *Pestic. Monit. J.* 1979, **13**(3), 97-98
32. Luke, M. A. et al *J. Assoc. Off. Anal. Chem.* 1981, **64**(5), 1187-1195
33. *J. Assoc. Off. Anal. Chem.* 1984, **67**(1), 154-166

34.  Mohamed, A. K. et al *Mycopatholagia* 1987, **99**(2), 99-105
35.  Kagan, Yu. S. et al *Scientific Reviews of Soviet Literature of Toxicity and Hazards of Chemicals: Leptophos 124* 1990, UNEP IRPTC, Richardson, M. L. (Ed.), CIP, Moscow

# L36   D-Leucine

$$CH_3CH(CH_3)CH_2CH(NH_2)CO_2H$$

**CAS Registry No.** 328-38-1
**Synonyms** D-leucine
**Mol. Formula** $C_6H_{13}NO_2$                    **Mol. Wt.** 131.18

## Physical properties

**M. Pt.** 290°C (decomp.); **Specific gravity** $d_{18}$ 1.293.

**Solubility**
Water: 7.97 g $l^{-1}$ 0°C.

## Mammalian and avian toxicity

**Acute data**
$LD_{50}$ intraperitoneal rat 6430 mg $kg^{-1}$ (1).

## Legislation

Included in Schedule 4 (Release into Air: Prescribed Substances) Statutory Instrument No. 472, 1991 (2).

## Any other comments

*Halobacterium halobium* can utilise D-leucine for growth (3,4).

## References

1.  *Arch. Biochem. Biophys.* 1956, **64**, 319
2.  *S. I. 1991 No. 472 The Environmental Protection (Prescribed Processes and Substances) Regulations* 1991, HMSO, London
3.  Tanaka, M. et al *Viva Origino* 1991, **18**(3), 109-117 (Japan.) (*Chem. Abstr.* **115**, 154696m)
4.  Tanaka, M. et al *Can. J. Microbiol.* 1989, **35**(4), 508-511

# L37   Levonorgestrel

CAS Registry No. 797-63-7
Synonyms 18,19-dinorpregn-4-en-20-yn-3-one, 13-ethyl-17- hydroxy-, (17 α)-;
Norplant
Mol. Formula $C_{21}H_{28}O_2$                                Mol. Wt. 312.46
Uses Inhibitor of ovulation, oral contraceptive.

## Physical properties
M. Pt. 239-241°C.

### Solubility
Organic solvent: ethanol, acetone, diethyl ether, chloroform, dioxane

## Mammalian and avian toxicity

### Acute data
$LD_{50}$ oral ♂, ♀ rat, mouse 1140-2130 mg $kg^{-1}$ (50:30 ratio
levonorgesterol/ethinylestradiol mix) (1).
$LD_{50}$ intraperitoneal ♂, ♀ mouse 700-800 mg $kg^{-1}$ (1).

### Sub-acute data
$TD_{Lo}$ implant woman (52 wk pregnant) 1600 μg $kg^{-1}$ caused reproductive effects (2).
Oral ♀ rat (3 month) 0.0025, 0.025, 0.25, 2.5 and 25 mg $kg^{-1}$. Highest dose increased
liver weight, blood α1-globulin, decreased blood cholesterol. No histological changes
observed as all changes were reversible (3).
Rabbits administered 500 μg $day^{-1}$ 6 day $wk^{-1}$ for 3 months, no influence on growth
or body weight, enzyme activity, cholesterol levels and no organ damage (4).

### Carcinogenicity and long-term effects
$TD_{Lo}$ oral mouse (69 wk) 29 mg $kg^{-1}$ caused neoplastic effects (5).

### Metabolism and pharmacokinetics
Intravenous rabbits (unspecified dose) absorbed bioavailability was 40%. Oral rabbit
0.3 mg (initial dose) blood concentration peaked after 30 min. Rapid absorption
followed with blood concentration of <0.2 ng $ml^{-1}$ detected (6).

## Any other adverse effects to man

A 3 yr study of users of Norplant-2 rods for contraception indicated a possible
increased predisposition to thrombosis, evidenced by significant increase in platelet
count and aggregability (7).

---

## Any other comments

American women are filing a lawsuit against Wyeth-Ayerest the manufacturer of Norplant, implantable contraceptive, because of the pain caused on removal and the potential for scarring. Wyeth-Ayerest state if properly inserted removal is painless, with fewer than 6% of women experiencing difficulties. Norplant is approved for use in 39 countries (8).

## References

1. Makino, M. et al *Oyo Yakuri* 1991, **41**(3), 295-304 (Japan.) (*Chem. Abstr.* **115** 85610q)
2. *Contraception* 1975, **12**, 615
3. Usui, T. et al *Oyo Yakuri* 1991, **41**(4), 389-409 (Japan.) (*Chem. Abstr.* **115**, 85614u)
4. Cao, L. *Tongji Yike Daxue Xuebao* 1987, **16**(2), 93-96 (Ch.) (*Chem. Abstr.* **107**, 191164s)
5. *C. R. Seances. Soc. Biol. Ses Fil.* 1974, **168**, 1190
6. Wang, P. et al *Beijing Yike Daxue Xuebao* 1985, **17**(4), 300-302 (Ch.) (*Chem. Abstr.* **106**, 701m)
7. Singh, K. et al *Adv. Contracept.* 1990, **6**(2), 81-89
8. Mestel, R. *New Scientist* 1994, **1935**, 5

# L38  Levopropoxyphene

**CAS Registry No.** 2338-37-6

**Synonyms** benzeneethanol, α-[2-(dimethylamino)-1-methylethyl]-α-phenyl-, propanoate (ester),[R-(R'S')]; 2-butanol, 4-(dimethylethylamino)-3-methyl-1,2-diphenyl-, propionate (ester), (-)-; Leropropoxyphene; *l*-propoxyphene

**Mol. Formula** $C_{22}H_{29}NO_2$          **Mol. Wt.** 339.48

**Uses** Antitussive.

## Physical properties

**M. Pt.** 75-76°C.

## Mammalian and avian toxicity

### Acute data

$LD_{50}$ oral ♀ rat 1455 mg $kg^{-1}$ (1).

## References

1. Goldenthal, E. I. *Toxicol. Appl. Pharmacol.* 1971, **18**, 185

# L39   Levulinic acid

## $CH_3C(O)CH_2CH_2CO_2H$

**CAS Registry No.** 123-76-2
**Synonyms** pentanoic acid, 4-oxo-; β-acetylpropionic acid; 4-ketovaleric acid;
levulic acid; 4-oxovaleric acid
**Mol. Formula** $C_5H_8O_3$                                         **Mol. Wt.** 116.12
**Uses** In organic synthesis; in the manufacture of nylon, synthetic rubber, plastics and
medicinals.

## Physical properties

**M. Pt.** 33-35°C; **B. Pt.** 245-246°C; **Flash point** 137°C; **Specific gravity** 1.144.

**Solubility**
Water: freely soluble. Organic solvent: ethanol, diethyl ether

## Mammalian and avian toxicity

**Acute data**
$LD_{50}$ oral rat 1850 mg $kg^{-1}$ (1).
$LD_{50}$ intraperitoneal mouse 450 mg $kg^{-1}$ (2).

**Irritancy**
Dermal rabbit (24 hr) 500 mg caused moderate irritation (1).

## Legislation

Included in Schedule 4 (Release into Air: Prescribed Substances) Statutory Instrument
No. 472, 1991 (3).

## References

1.  *Food Cosmet. Toxicol.* 1979, **17**, 695
2.  *NTIS Report AD 607-952*, Natl. Tech. Info. Ser., Springfield, VA
3.  *S. I. 1991 No. 472 The Environmental Protection (Prescribed Processes and Substances)
    Regulations* 1991, HMSO, London

# L40   Lewisite

## ClCH=CHAsCl₂

**CAS Registry No.** 541-25-3
**Synonyms** arsonous dichloride, (2-chloroethenyl)-; chlorovinylarsine dichloride;
dichloro(2-chlorovinyl)arsine
**Mol. Formula** $C_2H_2AsCl_3$                                    **Mol. Wt.** 207.32
**Uses** Formerly used in chemical warfare.

## Physical properties

**M. Pt.** 0.1°C (solidifies at –13°C); **B. Pt.** 190°C (decomp.); **Flash point** –13°C;
**Specific gravity** $d_4^{20}$ 1.88; **Volatility** v. den. 7.15.

## Mammalian and avian toxicity

### Acute data
$LD_{Lo}$ dermal human 38 mg kg$^{-1}$ (1).
$LD_{50}$ dermal rat, mouse, dog 15 mg kg$^{-1}$ (2,3).
$LD_{50}$ subcutaneous rat, dog 1-2 mg kg$^{-1}$ (3).

### Sub-acute data
Gavage rats (13 wk) 0, 0.01, 0.5, 1.0 or 2.0 mg kg$^{-1}$ in sesame oil 5 day wk$^{-1}$. A treatment-related lesion was detected in the forestomach of both sexes at 2.0 mg kg$^{-1}$. Lesions were characterised by necrosis of the stratified squamous epithelium accompanied by infiltration of neutrophils and macrophages, proliferation of neocapillaries, haemorrhage, oedema and fibroblast proliferation. Mild acute inflammation of glandular stomach also observed with 1.0 and 2.0 mg kg$^{-1}$ (4).

### Teratogenicity and reproductive effects
Gavage ♂, ♀ rats (42 wk) 2 generation study 0, 0.10, 0.25 or 0.60 mg kg$^{-1}$ day$^{-1}$ 5 day wk$^{-1}$. No adverse effect on reproduction performance, fertility or reproductive organ weights through 2 consecutive generations. Severe inflammation of the lung was observed at necropsy (5).
Rats (unspecified route) 1.5 mg kg$^{-1}$ did not cause toxic or teratogenic effects in maternal animals or foetuses, however 10% maternal mortality occurred at 2 mg kg$^{-1}$. In rabbits 13 and 100% mortality were observed after doses of 0.07 and 1.5 mg kg$^{-1}$, respectively (6).

## Genotoxicity
*Salmonella typhimurium* TA97, TA98, TA100, TA102 with and without metabolic activation negative (7).

## Any other adverse effects to man
0.5 ml may give rise to sufficient absorption to produce severe systemic effects, 2 ml may cause death (8).

## Any other adverse effects
Dermal pig (high dose) caused blistering of skin. Exposure to vapour produced diffuse lesions of less severity (9).
*In vitro* isolated perfused porcine skin flap topical exposure to 6 concentrations ranging from 0.07–5.0 mg ml$^{-1}$. Lesions were characterised 1, 3, 5 and 8 hr after exposure, increased activation of lacatate dehydrogenase and vascular resistance, and mild increases in glucose utilisation were observed (10).

## Legislation
Included in Schedules 4 and 6 (Release into Air/Land: Prescribed Substances) Statutory Instrument No. 472, 1991 (11).

## Any other comments
Antidote: British Anti Lewisite (dimercaptopropanol) (8).
Neutralised and inactivated by sodium hypochlorite (8).
Toxicity amd physical and chemical properties reviewed (12).

## References

1. Sollman, T. *Manual of Pharmacology and Its Application to Therapeutics and Toxicology* 8th ed., 1957, 192, W. B. Saunders, Philadelphia, PA
2. *NTIS Report PB 158-508*, Natl. Tech. Inf. Ser., Springfield VA
3. *J. Path. Bacteriol.* 1946, **58**, 411
4. Sasser, L. B. et al *Gov. Rep. Announce. Index (U. S.)* 1990, **90**.(12), 031,315
5. Sasser, L. B. et al *Report, 1989 From Gov. Rep. Announce. Index (U. S.)* 1990, **90**(5), Abstr. No. 010.031
6. Hackett, P. L. et al *Report 1987, Energy Res. Abstr.* 1988, **13**(17), Abstr. No. 41594
7. Stewart, D. L. et al *Report* 1989, PNL-6872; Order No. AD-A213146, from *Gov. Rep. Announce. Index (U. S.)* 1990, **90**(3), Abstr. No. 006.319
8. *Merck Index* 11th ed., 1989, Merck & Co. Inc., Rahway, NJ
9. Mitcheltree, L. W. et al *J. Toxicol., Cutaneous Ocul. Toxicol.* 1989, **8**(3), 307-317
10. King, J. R. et al *Toxicol. Appl. Pharmacol.* 1992, **116**(2), 189-201
11. *S. I. 1991 No. 472 The Environmental Protection (Prescribed Processes and Substances) Regulations* 1991, HMSO, London
12. Goldman, M. et al *Rev. Environ. Contam. Toxicol.* 1989, **110**, 75-115

# L41   Lignin alkali

**CAS Registry No.** 8068-05-1

**Synonyms** alkali lignin; Indulin A; Kraft lignin; Meadol MRM; sulfate lignin; Tomlinite

## Environmental fate

### Anaerobic effects

*Desulfovibrio desulfuricans*, lignin alkali could not substitute for lactate or sulfate when added to the culture medium, though it did enhance the viability of the cells. After biological treatment it bound larger quantities of heavy metals (1).

### Degradation studies

During a study of metabolism of various lignins, lignin alkali was found to be most refactory (2).

*Heterobasidion annosum, Pleurotus ostreatus* and *Stereum purpureum* caused the degradation of lignin alkali, shown by spectophotometric studies at 280 nm. *Stereum purpureum* had the highest lignolytic activity (3).

## Any other comments

Physico-chemical properties and uses reviewed (4).

## References

1. Ziomek, E. et al *Appl. Environ. Microbiol.* 1989, **55**(9), 2262-2266
2. Pellinen, J. et al *Holzforschung* 1987, **41**(5), 271-276
3. Stevanovic-Janezic, T. et al *Bull. Liaison-Groupe Polyphenola* 1988, **14**, 161-164
4. Lin, S. Y. et al *Zhongguo Zaozhi* 1990, **9**(2), 45-53, (*Chem. Abstr.* **114**, 45239y)

# L42   Lignocaine

$$(H_3CCH_2)_2NCH_2\overset{\displaystyle O}{\overset{\displaystyle \|}{C}}-NH$$

(structure with 2,6-dimethylphenyl ring, $H_3C$ and $CH_3$ substituents)

**CAS Registry No.** 137-58-6
**Synonyms** acetamide, 2-(diethylamino)-$N$-(2,6-dimethylphenyl)-; Anestacon; Duncaine; Xyline; Lidocaine; Xylocitin
**Mol. Formula** $C_{14}H_{22}N_2O$           **Mol. Wt.** 234.34
**Uses** Local anaesthetic. Antiarrhythmic.

## Physical properties

**M. Pt.** 68-69°C; **B. Pt.** $_4$ 180-182°C.

### Solubility
Organic solvent: ethanol, diethyl ether, benzene, chloroform

## Ecotoxicity

### Invertebrate toxicity
*Scaphechinus mirabilis* exposed for 10 min to 234-2343 mg $kg^{-1}$ increased intracellular pH in unfertilised egg and decreased the pH rise associated with fertilisation (1).

## Mammalian and avian toxicity

### Acute data
$LD_{50}$ oral mouse, rat 220, 317 mg $kg^{-1}$, respectively (2,3).
$LD_{50}$ subcutaneous mouse, rat 238, 335 mg $kg^{-1}$, respectively (4,5).
$LD_{Lo}$ intravenous rat 25 mg $kg^{-1}$ (6).
$LD_{50}$ intraperitoneal mouse 102 mg $kg^{-1}$ (4).
$TD_{Lo}$ intravenous human 23 mg $kg^{-1}$ (7).
$LD_{100}$ intravenous sheep 1450 mg. Death occurred from respiratory depression with bradycardia and hypotension without arrhythmias (8).

### Irritancy
Eye rabbit 100 mg in 0.1 ml caused moderate irritation, persisting for more than 24 hr but recovery occurred within 21 days after treatment (9).

## Genotoxicity
*In vitro* Syrian hamster embryo cells induced sister chromatid exchanges but not chromosomal aberrations at 30-300 µg $l^{-1}$ for 18-20 hr (10).

## Any other adverse effects
Extraneural injection rat (concentration unspecified) produced concentration-

dependent direct cellular toxicity including oedema, lipid inclusions and fibre injury (11).

Intravenous dog 8 mg kg$^{-1}$ min$^{-1}$, seizure occurred at dose level of 20.8 mg kg$^{-1}$ and arterial plasma concentration of 47.2 µg ml$^{-1}$. 2 animals died of progressive hypotension, respiratory arrest and cardiovascular collapse (12).

## References

1.  Bozhkova, V. P. et al *Ontogenez* 1988, **19**(1), 73-81 (Russ.) (*Chem. Abstr.* **108**, 165031x)
2.  *Boll. Chim. Farmaceut.* 1971, **110**, 330
3.  *Arzneim. Forsch.* 1966, **16**, 1275
4.  *Eur. J. Microbiol.* 1974, **9**, 188
5.  *J. Med. Chem.* 1981, **24**, 1059
6.  *Br. J. Anaesth.* 1951, **23**, 153
7.  *Arch. Toxicol.* 1971, **28**, 72 (Ger.)
8.  Nancarrow, C. et al *Anesth. Analg. (N. Y.)* 1989, **69**(3), 276-283
9.  Sugai, S. et al *J. Toxicol. Sci.* 1990, **15**, 245-262
10. Kayukwa, E. et al *Shigaku* 1988, **76**(5), 941-962 (Japan.) (*Chem. Abstr.* **110**, 88535b)
11. Kalichman, M. W. et al *J. Pharmacol. Exp. Ther.* 1989, **250**(1), 406-413
12. Feldman, H. S. et al *Anesth. Analg. (N. Y.)* 1989, **69**(6), 794-801

# L43   Limonene

**CAS Registry No.** 138-86-3

**Synonyms** 1-methyl-4-(1-methylethenyl)cyclohexene; *p*-mentha-1,8-diene; α-limonene; Dipanol

**Mol. Formula** $C_{10}H_{16}$                    **Mol. Wt.** 136.24

**Uses** In air-fresheners and perfumes. In the manufacture of cleaning compounds. Flavouring agent. Lubricating oil additive. Solvent.

**Occurrence** In the essential oil of many plants. In the defensive secretions of termite species.

## Physical properties

**M. Pt.** −96.9°C; **B. Pt.** 170-180°C; **Flash point** 48°C (open cup);
**Specific gravity** $d_4^{21}$ 0.8402; **Volatility** v. den. 4.7.

## Solubility

Organic solvent: diethyl ether, dimethyl sulfoxide, dimethylformamide, ethanol

## Occupational exposure

**UN No.** 2052; **HAZCHEM Code 3ᴹ**; **Conveyance classification** Flammable liquid;

**Supply classification** irritant.

**Risk phrases** ≥25% – Flammable – Irritating to skin – <25% – Flammable
(R10, R38, R10)

**Safety phrases** After contact with skin, wash immediately with plenty of soap and
water (S28)

## Ecotoxicity

**Invertebrate toxicity**

$EC_{50}$ (48 hr) *Daphnia pulex* 70 mg $l^{-1}$ (1).

## Environmental fate

**Degradation studies**

Removed from water to below the limit of detection in activated sludge system and by
air stripping (2).

## Mammalian and avian toxicity

**Carcinogenicity and long-term effects**

A CASE study predicted marginal carcinogenicity (3).

**Irritancy**

Dermal rabbit (24 hr) 500 mg caused moderate irritation (4).

## Any other adverse effects

Inhibited oxidative phosphorylation in mitochondria *in vitro* through depletion of the
transmembrane potential and inhibition of oxygen consumption (species and
concentration not specified) (5).

## Legislation

Organic solvents are included in Schedule 6 (Release into Land: Prescribed
Substances) Statutory Instrument No. 472, 1991 (6).

## Any other comments

Toxicity reviews cited (7).
Autoignition temperature 237°C.

## References

1. May-Passino, D. R. et al *Environ. Toxicol. Chem.* 1987, **6**(11), 901-907
2. Schwartz, S. M. et al *Chemosphere* 1990, **21**(10-11), 1153-1160
3. Rosenkranz, H. S. et al *Carcinogenesis (London)* 1990, **11**(2), 349-353
4. *Food Cosmet. Toxicol.* 1974, **12**, 703
5. Uribe, S. et al *Xenobiotica* 1991, **21**(5), 679-688
6. Included in Schedule 6 (Release into Land: Prescribed Substances) Statutory Instrument
   No. 472, 1991
7. *ECETOC Technical Report No. 30(4)* 1991, European Chemical Industry Ecology and
   Toxicology Centre, B-1160 Brussels

# L44   *R*-Limonene

CAS Registry No. 5989-27-5
Synonyms cyclohexene, 1-methyl-4-(1-methylethenyl)-, (*R*)-; Carvene; (+)-limonene;
*d*-limonene
Mol. Formula $C_{10}H_{16}$                     Mol. Wt. 136.24
Uses Flavour and fragrance additive for food and domestic products.
Industrial solvent. Naturally occurring insecticide.
Occurrence Volatile oils, especially citrus oils.

## Physical properties

B. Pt. 175.5-176°C; Flash point 48°C; Specific gravity $d_4^{25}$ 0.8402.

## Solubility
Organic solvent: ethanol

## Occupational exposure

UN No. 2052; HAZCHEM Code 3☒; Conveyance classification flammable liquid.

## Ecotoxicity

### Invertebrate toxicity
$LD_{50}$ earthworm 6 ppm. Chronic and acute intoxication involved a rapid and
predictable cascade of behavioural and morphological symptoms, including increased
mucus secretion, writhing, clitellar swelling and elongation of the body (1).

## Environmental fate

### Nitrification inhibition
Limits nitrogen mineralisation and nitrification, enhances immobilisation of $NO_3^--N$
relative to $NH_4^+-N$ and stimulates overall net immobilisation of nitrogen during
breakdown of high carbon content material (2).

## Mammalian and avian toxicity

### Acute data
$LD_{50}$ oral rat 4400 mg $kg^{-1}$ (3).
$LD_{50}$ oral mouse 5600 mg $kg^{-1}$ (3).
$LD_{50}$ oral redwing blackbird >111 mg $kg^{-1}$ (4).
$LD_{50}$ intravenous rat 110 mg $kg^{-1}$ (3).
$LD_{50}$ intraperitoneal mouse 600 mg $kg^{-1}$ (5).

### Sub-acute data
Gavage ♂, ♀ dog (6 month) 100 or 1000 mg $kg^{-1}$ $day^{-1}$ induced no histopathological
changes in kidney (6).

---

Gavage ♂ rat (1, 4 wk) 75, 150, 300 mg kg$^{-1}$ single daily dose 5 day wk$^{-1}$ kidney damage observed (7).

**Carcinogenicity and long-term effects**

Gavage ♂, ♀ rat (103 wk) 0, 75, 150 mg kg$^{-1}$ and 0, 300, 600 mg kg$^{-1}$ in corn oil, respectively; ♂, ♀ mouse 0, 250, 500 mg kg$^{-1}$ and 0, 500, 1000 mg kg$^{-1}$ in corn oil, respectively. No evidence for carcinogenicity in ♀ rats, ♂, ♀ mice. Clear evidence for carcinogenicity, kidney tubular cell adenoma and adenocarcinoma in ♂ rats (8-12).

## Genotoxicity

*Salmonella typhimurium* TA97, TA98, TA100, TA1535, TA1537 with or without metabolic activation negative (11-13).

Mouse L5178Ytk$^+$/tk$^-$ assay, did not significantly increase the number of trifluorothymidine (Tft)-resistant cells (11).

*In vitro* Chinese hamster ovary cells did not induce chromosomal aberrations or sister chromatid exchanges (11).

## Legislation

Included in Schedule 4 (Release into Air: Prescribed Substances) Statutory Instrument No. 472, 1991 (14).

Limited under EC Directive on Drinking Water Quality 80/778/EEC. Pesticides: Maximum admissible concentration 0.1 µg l$^{-1}$ (15).

## Any other comments

Insect repellency, fumigant activity, acute toxicity (LD$_{50}$: ♀ house fly, ♂ German cockroach 90 and 700 µg insect$^{-1}$, respectively), reproductive toxicity and neurotoxicity investigated. The study concluded the substance has considerable potential as a naturally occurring insecticide (16).

Nephrotoxicity and nephrocarcinogenicity reviewed (17).

## References

1. Karr, L. L. et al *Pestic. Biochem. Physiol.* 1990, **36**(2), 175-186
2. White, C. S. *Biogeochemistry* 1991, **12**, 43-68
3. *Drugs in Japan. Ethical Drugs* 6th ed., 1982, 887
4. Schafer, E. W. *Arch. Environ. Contam. Toxicol.* 1983, **12**, 355-382
5. *Pharmacomet.* 1974, **8**, 1439
6. Webb, D. R. et al *Food Chem. Toxicol.* 1990, **28**(10), 669-675
7. Kanerva, R. L. et al *Food Chem. Toxicol.* 1987, **25**(5), 345-353
8. Jameson. C. W. *Report 1990* from *Gov. Rep. Announce. Index (U.S.)* 1990, **90**(18), Abstr. No. 045,992
9. Hasseman, J. K. et al *Environ. Mol. Mutagen.* 1990, **16**(Suppl. 18), 15-31
10. Tennant, R. W. et al *Mutat. Res.* 1991, **257**(3), 209-227
11. *National Toxicology Program Research and Testing Division* 1992, Report No. TR-347, NIEHS, Research Triangle Park, NC
12. Zeiger, E. *Environ. Mol. Mutat.* 1990, **16**(Suppl. 18), 32-54
13. Myhr, B. et al *Environ. Mol. Mutat.* 1990, **16**(Suppl. 18), 138-167
14. *S. I. 1991 No. 472 The Environmental Protection (Prescribed Processes and Substances) Regulations* 1991, HMSO, London
15. *EC Directive Relating to the Quality of Water Intended for Human Consumption* 1982, 80/778/EEC, Office for Official Publications of the European Communities, 2 rue Mercier, L-2985 Luxembourg

16. Coats, J. R. et al *ACS Symp. Ser.* 1991, **449**, 305-316
17. Flamm, W. G. & Lehman-Mckeeman, L. D. *Regul. Toxicol. Pharmacol.* 1991, **13**(1), 70-86

# L45   Lindane

**CAS Registry No.** 58-89-9
**Synonyms** cyclohexane, 1,2,3,4,5,6-hexachloro-, (1α,2α,3β,4α,5α,6β)-; 666; γ-BHC; γ-HCH; TAP85; benzene hexachloride
**Mol. Formula** $C_6H_6Cl_6$                                    **Mol. Wt.** 290.83
**Uses** Insecticide.
**Occurrence** Environmental contamination occurred following application of lindane containing pesticide formulations. Emissions can cross national boundaries in water and air.

## Physical properties

**M. Pt.** 112.8°C (γ-isomer); **B. Pt.** 288°C; **Specific gravity** 1.85; **Partition coefficient** log $P_{ow}$ 3.29-3.72 (1); **Volatility** v.p. $3.26 \times 10^{-5}$ mmHg at 20°C.

### Solubility
Water: 10 mg $l^{-1}$ at 20°C. Organic solvent: ethanol, acetone, aromatic and chlorinated solvents

## Occupational exposure

**US TLV (TWA)** 0.5 mg m$^{-3}$; **UK Long-term limit** 0.5 mg m$^{-3}$ under review;
**UK Short-term limit** 1.5 mg m$^{-3}$ under review; **Supply classification** toxic.
**Risk phrases** Toxic by inhalation, in contact with skin and if swallowed – Irritating to eyes and skin (R23/24/25, R36/38)
**Safety phrases** Keep out of reach of children – Keep away from food, drink and animal feeding stuffs – If you feel unwell, seek medical advice (show label where possible) (S2, S13, S44)

## Ecotoxicity

### Fish toxicity
$LC_{50}$ (48 hr) goldfish, guppy 0.12-0.3 mg $l^{-1}$ (2,3).
$LC_{50}$ (96 hr) rainbow trout, perch, bluegill sunfish 0.027-0.057 mg $l^{-1}$ (4-6).
$LC_{50}$ (96 hr) Atlantic silverside 9 ppb static bioassay (7).
Bluegill sunfish (18 month) 0.6-9.1 μg $l^{-1}$ no adverse effect was observed at any of the tested concentrations. Fathead minnow (43 wk) 1.4-23.5 μg $l^{-1}$, a statistically significant increase in mortality was observed at the highest dose. Growth of surviving fish was not adversely affected and spawning was normal in all test groups (8).

In fish symptoms of acute poisoning are gross irritability, loss of equilibrium, changes in pigmentation and localised peripheral haemorrhage (9).

### Invertebrate toxicity

$EC_{50}$ (15 min) *Photobacterium phosphoreum* 5.67 ppm Microtox test (10).
*Microcystis aeruginosa, Cylindrospermum licheniforme* 0.3-2 mg $l^{-1}$ no inhibition of growth. Both concentrations are higher than those that would result in adverse effects to fish or crustaceans (11, 12).
*Scendesmus acutus* (5 day) 5 mg $l^{-1}$ lethal dose, concentrations of 1-5 mg $l^{-1}$ caused a 50% reduction in growth (13).
A range of 10 common European freshwater macroinvertebrates in a continuous flow system (96 hr), $LC_{50}$ concentration 25-430 µg $l^{-1}$, test conditions water temperature 11°C, pH 7.5-8.0, dissolved oxygen >96%, water hardness 93 mg $l^{-1}$ ($CaCO_3$), *Baetis rhodani, Leuctra moselyi*, and *Protonemura meyeri* were the most sensitive species, *Physa fontinalis* and *Polycelis tenius* were the most tolerant (14).
$EC_{50}$ (96 hr) *Daphnia magna* 516 µg $l^{-1}$ static bioassay at 20°C (15).
$LC_{50}$ (96 hr) *Gammarus lacustris* 48 µg $l^{-1}$ static bioassay at 21°C (16).
$LC_{50}$ (96 hr) *Penaeus duorarum* 0.17 µg $l^{-1}$ flow-through bioassay 24-26°C, grass shrimp 10 µg $l^{-1}$ static bioassay at 20°C (17).

### Bioaccumulation

The bioconcentration factor is dependent on the exposure concentration, the highest concentration factors were seen with the lowest exposure concentrations (18).
Bioconcentration factors for fish 167-727; mussels (*Mytilus edulis*) 100; crustacean 25-143 (19).
Exposure via water or food in gudgeon, 0.22-142 µg $l^{-1}$ in water, bioconcentration factors in brain, liver and muscle were 600, 200 and 100, respectively, at higher concentrations decreased to 50. Exposure via food led to accumulation 2.5 times higher (18, 20).

## Environmental fate

### Nitrification inhibition

*Nitrobacter, Nitrosomonas* showed inhibited nitrification when exposed to 100 and 10 mg $kg^{-1}$. Nitrification was restored within 35 days (21).

### Carbonaceous inhibition

100 mg $kg^{-1}$ applied to soil inhibited carbon dioxide evolution, application of 1 and 10 mg $kg^{-1}$ incubated for 120 days caused no or slight inhibition to $CO_2$ evolution (21).

### Anaerobic effects

Anaerobic bacteria degraded ≤90% within 4 days, transformation to chlorine-free metabolites. $CO_2$ is not formed under anaerobic conditions (22).
Soils under flooded conditions at 30°C, $t_{1/2}$ 10-30 days, degradation occurred faster at higher temperatures (23).

### Degradation studies

An extensive study of sandy loam, silt loam and muck soils were treated in 1954 at application rates of 1.1, 11.2. 112.1 kg $ha^{-1}$ to a depth of 15 cm. Follow-up in 1959 found no lindane on low-dose application but ≈36% remained at higher application rates. Persistence was affected by the amount of organic matter in the soil and climatic conditions (24).

Degraded by *Pseudomonas* sp. in soil. Bacterial degradation led to the accumulation of a transitory, and eventual release of covalently linked Cl⁻ in stoichiometric amounts (25). Under aerobic conditions bacteria degrade lindane via dehydrochlorination of the polychlorinated cyclohexane moiety to intermediates which include chlorobenzenes and the end product carbon dioxide (26,27).

**Abiotic removal**
Photooxidation by UV light in aqueous medium at 90-95°C: time for formation of carbon dioxide (% theoretical): 25% in 3 hr; 50% in ≈17 hr; and 75% in ≈46 hr (19). Evaporation is an important process in the removal of lindane (28).

**Absorption**
Strongly absorbed onto organic soil material and weakly adsorbed on to inorganic material (29-35).
30-40% of application adsorbed onto aquifer sand at 5°C after 3-100 hr equilibrium time (19).

## Mammalian and avian toxicity

**Acute data**
$LD_{50}$ oral dog, mouse, rabbit, rat 40-76 mg kg$^{-1}$ (36-38).
$LD_{50}$ oral starling, bobwhite quail 100-130 mg kg$^{-1}$ (39,2).
$LC_{50}$ (4 hr) inhalation rat 1600 mg m$^{-3}$ (9).
$LD_{50}$ dermal rabbit 200-300 mg kg$^{-1}$ (40).
$LD_{50}$ intraperitoneal mouse 125 mg kg$^{-1}$ (41).
Oral ♂ rat 30, 40 or 50 mg kg$^{-1}$ in oil caused reduced food intake and hypothermia but death did not occur even at the highest dose. Intraperitoneal 4, 6 or 8 mg kg$^{-1}$ in dimethylsulfoxide caused hyperthermia and convulsions. 43% of animals administered the highest dose died (42).

**Sub-acute data**
Oral rat (3 month) 0, 0.2, 0.8, 4, 20 or 100 mg kg$^{-1}$ diet, no effect on mortality, food consumption or haematological parameters (9).
Oral ♂ rat (2 wk) 0-800 mg kg$^{-1}$ diet, high doses caused glucosuria and increased excretion of creatinine and urea and hypertrophy of the liver (43).
Inhalation mice (14 wk) 0, 0.3, 1.0, 5 or 10 g m$^{-3}$ 6 hr day$^{-1}$ 5 day wk$^{-1}$, significant mortality observed in ♀ mice (9).
Dermal rat (13 wk) 60 and 400 mg kg$^{-1}$ caused reversible hypertrophy of the liver and non-reversible degeneration of the kidneys (9).

**Carcinogenicity and long-term effects**
Inadequate evidence for carcinogenicity to humans, limited evidence for carcinogenicity to animals, IARC classification group 2B (44).
National Toxicology Program (2 yr) tested rats and mice via feed. No evidence of carcinogenicity in either species (45).
Oral ♂, ♀ CF1 mice (110 wk) 400 mg kg$^{-1}$ diet. Did not induce hepatocellular carcinoma in animals of either sex but increased the incidence of hepatocellular adenoma and hyperplastic nodules in ♂ mouse only (46).

**Teratogenicity and reproductive effects**
Oral ♀ mouse 0, 12, 30 or 60 mg kg$^{-1}$ on days 6-15 or 11-12 of gestation. Foetal mortality increased and foetal weights decreased, and increased maternal mortality in

high-dose animals. No effect on implantation, resorption or malformations were observed (9).

Gavage ♀ rat 5, 10 or 20 mg kg$^{-1}$ (days 6-15 gestation). 10 and 20 mg kg$^{-1}$ caused maternal toxicity but no evidence of embryo or foetal toxicity (47).

Gavage rabbit (days 6-18 gestation) 5, 10 and 20 mg kg$^{-1}$ caused slight tachypnoea and lethargy. Post implantation loss and the incidence of resorptions were increased at 5 and 20 mg kg$^{-1}$ (47).

**Metabolism and pharmacokinetics**

After uptake, lindane is distributed to all organs and tissues in the body of laboratory animals, at measurable concentrations within a few hours. Oral rat (56 day) 1, 10 or 100 mg kg$^{-1}$, the highest concentrations were found in adipose tissue after 2-3 wk (48). Inhalation rat (90 day) 0.02, 0.1, 0.5 and 5 mg m$^{-3}$ highest concentrations found in fatty tissues, after 4 wk concentrations in all organs reached control values (9). Metabolism of lindane is initiated by one of four possible reactions: dehydrogenation leads to the formation of γ-HCH; dehydrochlorination leads to the formation of γ-PCCH; dechlorination to γ-tetrachlorohexene; and hydroxylation to hexachlorocyclohexanol. These compounds are considered intermediates, metabolism continues by a series of further dehydrogenating, dechlorinating, dehydrochlorinating and hydroxylating steps (49-51).

**Irritancy**

Dermal rabbit 0.5 g, 4 hr occlusive application, did not cause irritation (9).

Reported to cause sensory irritation in humans (52).

652 patients were patch tested with 1% lindane no irritant or allergic reactions observed (53).

**Sensitisation**

Magnusson-Kligman maximisation test, guinea pigs unspecified dose 99.6% lindane. Animals were challenged at 24 and 48 hr. No skin sensitisation potential observed (9).

## Genotoxicity

*Salmonella typhimurium* TA98, TA100, TA1535, TA1537 with and without metabolic activation negative (54).

*Bacillus subtilis* H17 rec$^+$, M45 rec$^{-1}$ without metabolic activation negative (55).

*Escherichia coli* WP2, WP2 *uvr*A$^{-1}$ with metabolic activation negative (56).

*In vitro* Chinese hamster V79 cells with metabolic activation negative (9).

*In vitro* Chinese hamster fibroblasts chromatid gaps, chromatid and chromosomal breaks equivocal (57).

*Drosophila melanogaster* dominant lethal mutation positive (58).

*In vivo* mouse bone marrow sister chromatid exchanges negative (9).

## Any other adverse effects to man

Acute symptoms include dizziness, headache, nausea, vomiting, diarrhoea, tremors, weakness, convulsions, dyspnoea and circulatory collapse (59).

No indication of a cause-effect relationship between exposure to lindane and blood dysergasias (60).

## Any other adverse effects

May be absorbed through the skin and produce central nervous system effects including motor excitability, muscle twitching, myoclonic jerking and convulsive seizures (61).

## Legislation

The log $P_{ow}$ value exceeds the European Community recommended level 3.0 (6th and 7th amendments) (62).

Limited under EC Directive on Drinking Water Quality 80/778/EEC. Pesticides: maximum admissible concentration 0.1 $\mu$g l$^{-1}$ (63).

Included in Schedule 5 (Release into Water: Prescribed Substances) Statutory Instrument No. 472, 1991 (64).

Included in Schedule 6 (Release into Land: Prescribed Substances) Statutory Instrument No. 472, 1991 (64).

Recommended guideline for drinking water quality 2 $\mu$g l$^{-1}$ (65).

## Any other comments

>90% exposure in non-occupationally exposed humans results from contaminated food (66).

Carcinogenicity, toxicity, toxicology and health effects reviewed (67,68).

## References

1. Geyer, H. et al *Chemosphere* 1984, **13**(2), 269-284
2. *The Agrochemicals Handbook* 3rd ed., 1991, RSC, London
3. Hashimoto, Y. A. et al *J. Pestic. Sci.* 1981, **6**, 257
4. Verschueren, K. *Handbook of Environmental Data on Organic Chemicals* 2nd ed., 1983, Van Nostrand Reinhold, New York
5. Macek, K. J. et al *Trans. Am. Fish Soc.* 1979, **99**(1), 20-27
6. Randall, W. F. et al *Bull. Environ. Contam. Toxicol.* 1979, **21**, 849-854
7. Eisler, R. *Acute Toxicities of Organochlorine and Organophosphorus Insecticides to Estuarine Fishes* 1970, Bureau of Sport Fisheries & Wildlife Tech. Paper 45, Gov. Print. Off., Washington, DC
8. Macek, K. J. et al *Bull. Environ. Contam. Toxicol.* 1969, **4**(3), 174-183
9. *IPCS Environmental Health Criteria 124: Lindane* 1991, World Health Organization, Geneva
10. Kaiser, K. L. E. et al *Water Pollut. Res. J. Can.* 1991, **26**(3), 361-431
11. Bringmann, G. et al *Jahresher Wasser* 1978, **50**, 45-60
12. Palmer, C. M. et al *Ohio J. Sci.* 1955, **55**(1), 1-8
13. Krishnakumari, M. K. *Life Sci.* 1977, **20**, 1525-1532
14. Green, D. W. J. et al *Arch. Hydrobiol.* 1986, **106**(2), 263-273
15. Randall, W. F. et al *Bull. Environ. Contam. Toxicol.* 1979, **21**, 849-854
16. Sanders, H. O. *Toxicity of Pesticides to the Crustacean Gammarus lacustris* 1969, 2-18, Bureau of Sport Fisheries and Wildlife Tech. Paper No. 25, Washington, DC
17. Eisler, R. *Crustaceana Int. J. Crustacean Res.* 1969, **16**(3), 302-310
18. Marcelle, C. et al *Bull. Environ. Contam. Toxicol.* 1983, **31**, 453-458
19. Schimmel, S. C. et al *Arch. Environ. Contam. Toxicol.* 1977, **6**, 355
20. Marcelle, C. et al *Bull. Environ. Contam. Toxicol.* 1984, **33**, 423-429
21. Gaur, A. C. et al *Plant. Soil.* 1977, **46**, 5-15
22. Haider, K. et al *Arch. Microbiol.* 1975, **104** (Ger.)
23. Yoshida, T. et al *Soil Sci. Soc. Am. Proc.* 1970, **34**, 440-442
24. Lichtenstein, E. P. et al *J. Econ. Entomol.* 1960, **53**(1), 136-142
25. Sahu, S. K. et al *Appl. Environ. Microbiol.* 1990, **56**(11), 3620-3622
26. Vonk, J. W. et al *Pest. Biochem. Physiol.* 1979, **12**, 68-74
27. Mathur, S. P. et al *Soil Sci.* 1975, **120**(4), 301-307
28. MacKay, D. et al *Environ. Sci. Technol.* 1973, **7**, 611
29. Baluji, G. et al *Environ. Qual. Saf. Suppl.* 1975, **3**, 243-249
30. Kay, B. D. et al *Soil Sci.* 1967, **104**(5), 314-322

31. Mills, A. C. et al *Soil Sci. Soc. Am. Proc.* 1969, **33**, 210-216
32. Portmann, J. E. *Evaluation of the Impact on the Aquatic Environment of HCH Isomers* 1979, Commission of the European Communities, Brussels
33. Wahid, P. A. et al *J. Agric. Food Chem.* 1979, **27**(5), 1050-1053
34. Wahid, P. A. et al *J. Agric. Food Chem.* 1980, **28**, 623-625
35. Wirth, H. *GKSS Research Inst. Report No. GKSS 85/E/30* 1985
36. *Spec. Publ. Entomol. Soc. Am.* 1974, **74**, 1
37. *J. Econ. Entomol.* 1972, **65**, 632
38. *J. Hyg. Epidemol. Microbiol. Immunol.* 1978, **22**, 115
39. Schafer, E. W. *Arch. Environ. Contam. Toxicol.* 1983, **12**, 355-382
40. Izmerov, N. F. *IRPTC Scientific Review of Soviet Literature on Toxicity and Hazards of Chemicals* 1983, No. 40, CIP, Moscow
41. *Sov. Genet.* 1966, **2**(1), 80
42. Wolley, D. E. et al *Pharmacol. Biochem. Behav.* 1989, **33**(4), 787-792
43. Srinivasan, K. *J. Environ. Sci. Health* 1988, **B23**(4), 367-386
44. *IARC Monograph* 1987, **Suppl. 7**, 64
45. *National Toxicology Program Research and Testing Division* 1992, Report No. TR-104, NIEHS, Research Triangle Park, NC
46. Vesselinovitch, S. D. *Toxicol. Pathol.* 1983, **11**(1), 12-22
47. Palmer, A. K. et al *Toxicology* 1978b, **9**, 239-247
48. Oshiba, K. *Osaka Shiritsu Daigaku Igaku Zasshi* 1972, **21**(3), 1-19
49. Engst, R. et al *Residue Rev.* 1970, **68**, 59-84
50. Engst, R. et al *Residue Rev.* 1979, **75** (Part II), 71-95.
51. Engst, R. et al *J. Environ. Sci. Health* 1976, **B11**(2), 95-117
52. Grasso, G. et al *J. Environ. Sci. Health* 1987, **A22**(6), 543-547
53. Lisi, P. et al *Contact Dermatitis* 1987, **17**, 212-218
54. Haworth, S. et al *Environ. Mutagen.* 1983, (Suppl. 1), 3-142
55. Shirasu, Y. et al *Mutat. Res.* 1976, **40**, 19-30
56. Probst, G. S. et al *Environ. Mutagen.* 1981, **3**, 11-32
57. Ishidate, M. et al *Mutat. Res.* 1977, **48**, 337-354
58. Sinha, R. R. P. et al *Comp. Physiol. Ecol.* 1983, **8**(2), 87-89
59. Gosselin, R. E. et al (Eds.) *Clinical Toxicology of Commercial Products* 5th ed., 1984, Williams & Wilkins, Baltimore, MD
60. *Health Effects Assessment for Lindane* 1984, US EPA 540/1-86/056, Cincinnatti, OH
61. Moody, R. P. et al *J. Toxicol. Environ. Health* 1989, **28**(2), 161-169
62. *1967 Directive on Classification, Packaging and Labelling of Dangerous Substances 67/548/EEC; 6th Amendment EEC Directive 79/831/EEC; 7th Amendment EEC Directive 91/32/EEC* 1991, HMSO, London
63. *EC Directive Relating to the Quality of Water Intended for Human Consumption* 1982, 80/778/EEC, Office for Official Publications of the European Communities, 2 rue Mercier, L-2985 Luxembourg
64. *S. I. 1991 No. 472 The Environmental Protection (Prescribed Processes and Substances) Regulations* 1991, HMSO, London
65. *Guidelines for Drinking Water Quality* 2nd ed., 1993, **1**, WHO, Geneva
66. Geyer, H. et al *Regul. Toxicol. Pharmacol.* 1986, **6**, 313-347
67. *Gov. Rep. Announce. Index (U.S.)* 1990, **90**(10), Abstr. No. 025089
68. *Gov. Rep. Announce. Index (U.S.)* 1987, **87**(18), Abstr. No. 740936

# L46  ε-Lindane

**CAS Registry No.** 6108-10-7
**Synonyms** (1α,2α,3α,4β,5β,6β)- 1,2,3,4,5,6-hexachlorocyclohexane;
ε-benzene hexachloride; ε-BHC; ε-HCH; ε-hexachlorocyclohexane;
ε-1,2,3,4,5,6-hexachlorocyclohexane
**Mol. Formula** $C_6H_6Cl_6$       **Mol. Wt.** 290.83
**Uses** Pesticide.

## Physical properties
**M. Pt.** 219.3°C.

## Occupational exposure
**Supply classification** toxic, irritant.
**Risk phrases** Toxic by inhalation, in contact with skin and if swallowed – Irritating
to eyes and skin (R23/24/25, R36/38)
**Safety phrases** Keep out of reach of children – Keep away from food, drink and
animal feeding stuffs – If you feel unwell, seek medical advice (show label where
possible) (S2, S13, S44)

### Abiotic removal
40-80% removal from contaminated groundwater by biological trickling filter and
rotating biocontactor (1).

## Legislation
Limited under EC Directive on Drinking Water Quality 80/778/EEC. Pesticides:
maximum admissible concentration 0.1 μg l$^{-1}$ (2).
Included in Schedule 6 (Release into Land: Prescribed Substances) Statutory
Instrument No. 472, 1991 (3).

## References
1. van Comper, A. L. B. M. et al *Environ. Technol. Proc. Eur. Conf. 2nd* 1982, 624-626
2. *EC Directive Relating to the Quality of Water Intended for Human Consumption* 1982,
   80/778/EEC, Office for Official Publications of the European Communities, 2 rue Mercier,
   L-2985 Luxembourg
3. *S. I. 1991 No. 472 The Environmental Protection (Prescribed Processes and Substances)
   Regulations* 1991, HMSO, London

# L47   Linoleic acid

$$H_3C(CH_2)_3 \underset{E}{=} \underset{E}{=} (CH_2)_6CO_2H$$

**CAS Registry No.** 60-33-3
**Synonyms** 9,12-octadecadienoic acid, (Z,Z)-; linolic acid; Polylin 515; Telfairic acid; Uifac 6550
**Mol. Formula** $C_{18}H_{32}O_2$                    **Mol. Wt.** 280.45
**Uses** Manufacture of paints, coatings, emulsifiers, vitamins. Aluminium salt used to manufacture lacquers. Nutrient.
**Occurrence** Major constituent of many vegetable oils (1).

## Physical properties

**M. Pt.** −12°C; **B. Pt.** $_{16}$ 230°C; **Flash point** > 66°C (closed cup); **Specific gravity** $d_4^{18}$ 0.9038; **Partition coefficient** log $P_{ow}$ 7.05 (2).

## Solubility

Organic solvent: freely soluble diethyl ether, miscible dimethylformamide, soluble ethanol, petroleum ether

## Ecotoxicity

**Invertebrate toxicity**
$EC_{50}$ (duration unspecified) purple sea urchin 0.28-1.07 mg $kg^{-1}$ inhibited fertilisation (3).

## Environmental fate

**Anaerobic effects**
Readily biodegradable in batch bioassay innoculated with anaerobic granular sludge at 30°C and 35-200 mg $l^{-1}$ linoleic acid (4).

**Degradation studies**
The correlation between BOD and COD for fatty acids was studied. COD values were 1.2-29.6% ThOD averaging 8.38%. BOD values were 12.7-98.8% ThOD averaging 71% BOD/COD ratio 2.76-52.9% (5).

## Mammalian and avian toxicity

**Metabolism and pharmacokinetics**
Oral rat (unspecified concentration) as cream accumulates in triacylglycerols of the small intestine and plasma (6).
Intraduodenal infusion rats radiolabelled linoleic acid after 1 hr 64% of radiolabel disappearing from mucosa was recovered in blood (7).
Oral hamster, mouse, rat, rabbit, dog studied for 3 months after lactation, accumulates in adipose tissue (8).

**Irritancy**
Dermal human (3 day) 75 mg caused moderate irritation (9).

### Genotoxicity

*Salmonella typhimurium* TA98, TA100, TA1535, TA1537 with or without metabolic activation negative (10).

### Any other adverse effects

Administration to dogs of 10 mg kg$^{-1}$ (route unspecified) causes no haemodynamic changes, while 50 mg kg$^{-1}$ caused cardiotoxic effects (11).

### Legislation

The log $P_{ow}$ exceeds the European Community recommended level 3.0 (6th and 7th amendments) (12).

### Any other comments

The role of fatty acids in the development of cancers discussed (13).
Hazards, metabolism reviewed (14-16).

### References

1. *Merck Index* 11th ed., 1989, Merck & Co. Inc., Rahway, NJ
2. Sangster, J. *J. Phys. Chem. Ref. Data* 1989, **18**(3), 1111-1229
3. Cherr, G. N. et al *Environ. Toxicol. Chem.* 1987, **6**(7), 561-569
4. Sierra-Alvarrez, R. et al *Environ. Technol.* 1990, **11**(10), 891-898
5. Matsuoka, C. et al *Kenkyu Kiyo-Tokushima Bunri Daigaku* 1987, **34**, 193-199 (Japan.) (*Chem. Abstr.* **107**, 222653g)
6. Nilsson, A. & Melin, T. *Am. J. Physiol.* 1988, **255**(5, Pt. 1), G612-G618
7. Bernard, A. & Carlier, H. *Exp. Physiol.* 1991, **76**(3), 445-455
8. Martinez-Conde, A. et al *Rev. Esp. Fisiol.* 1986, **42**(3), 365-370 (Span.) (*Chem. Abstr.* **106**, 99856a)
9. Drill, V. A. & Lazar, P. (Eds.) *Cutaneous Toxicity* 1977, 127, Academic Press, New York
10. Zeiger, E. et al *Environ. Mutagen.* 1987, **9**(Suppl. 9), 1-109
11. Fukushima, A. et al *Cardiovasc. Res.* 1988, **22**(3), 213-218
12. *1967 Directive on Classification, Packaging, and Labelling of Dangerous Substances 67/548/EEC; 6th Amendment EEC Directive 79/831/EEC; 7th Amendment EEC Directive 91/32/EEC* 1991, HMSO, London
13. Hietanen, E. et al *Carcinogenesis (London)* 1986, **7**(12), 1965-1969
14. *Dangerous Prop. Ind. Mater. Rep.* 1988, **8**(2), 63-66
15. Horrobin, D. F. *NATO ASI Ser., Ser. A* 1989, **171**, 297-307
16. Watanabe, S. et al *Tanpakushitsu Kakusan Koso* 1991, **36**(3), 584-588 (Japan.) (*Chem. Abstr.* **114**, 162825x)

# L48   Lipase

**CAS Registry No.** 9001-62-1
**Synonyms** glycerol ester hydrolase; butyrinase; triacylglycerol hydrolase; triglyceridase; triolein hydrolase; tweenase
**Uses** In detergent formulations.
**Occurrence** Digestive enzyme, widely distributed in plants, moulds, bacteria and mammalian tissues.

## Mammalian and avian toxicity

### Metabolism and pharmacokinetics

Undergoes renal filtration and reabsorption followed by intrarenal degradation to amino acids. Rats were injected intravenously with a single dose of $^{125}$I-labelled homologous lipase. <1% of the lipase activity and >10% of the radioactivity was found in the urine after 120 min (1).

### Any other adverse effects

Elevated serum lipase levels in an acute pancreatitis episode may potentiate the autodigestion of the pancreas (2).

### References

1.  Malyusz, M. et al *J. Clin. Chem. Clin. Biochem.* 1988, **26**(10), 611-615
2.  Farooqui, A. A. et al *Neurochem. Pathol.* 1987, **7**(2), 99-128

# L49   Lithium

Li

CAS Registry No. 7439-93-2
**Mol. Formula** Li                                          **Mol. Wt.** 6.94
**Uses** Used in alloys, catalysts, aircraft fuels.
**Occurrence** Occurs in minerals spodumene; lepidolite; petalite; amblyganite; triphylite. Found naturally in food, air and drinking water. Occurrence in Earth's crust: 0.005% by wt.

## Physical properties

**M. Pt.** 180.54°C; **B. Pt.** 1336 ± 5°C; **Specific gravity** 0.584.

## Occupational exposure

**UN No.** 1415; **Conveyance classification** substance which in contact with water emits flammable gas; **Supply classification** highly flammable, corrosive.
**Risk phrases** Reacts violently with water, liberating highly flammable gases – Causes burns (R14/15, R34)
**Safety phrases** Keep container dry – In case of fire, use dry powder or sand – never use water (S8, S43)

## Ecotoxicity

### Fish toxicity

$LC_{50}$ (28, 35 day) rainbow trout 9.28, 1.4 mg $l^{-1}$ (salt), respectively (1,2).

### Invertebrate toxicity

Microinjection of lithium chloride into prospective ventral blastomeres of the 32-cell *Xenopus laevis* embryo gives rise to duplication of dorsanterior structures such as the notochord, neural tube and eyes (3).

### Bioaccumulation
Accumulates in several species of fish, molluscs and crustaceans. Stored in the digestive tract and exoskeleton (4).

## Environmental fate

### Anaerobic effects
Methanogenesis of granular anaerobic sludge at 35°C with initial COD of 5750 mg $l^{-1}$ $O_2$ at pH 7.2 was stimulated at lithium ion concentration 10-20 mg $l^{-1}$, slightly inhibited at lithium ion 350 mg $l^{-1}$ and seriously inhibited at lithium ion concentration ≥ 500 mg $l^{-1}$ (5).

## Mammalian and avian toxicity

### Sub-acute data
Adult ♂ rat, 2 or 4 mg lithium chloride 100 $g^{-1}$ $day^{-1}$ for 21 days reduced red blood cell count, haemoglobin concentration, haematocrit and mean corpuscular Hb on the 22nd day. Chronic $Li^+$ treatment can cause anaemia (6).
Newborn rats administered with lithium salts orally for 8-16 wk had chronic renal failure (7).
Lungs of rabbits exposed to 0.6 and 1.9 mg $m^{-3}$ lithium chloride for 4-8 wk, 5 day $wk^{-1}$, 6 hr $day^{-1}$ showed no significant effects (8).
2 mg $m^{-3}$ combustion products of metallic lithium via inhalation in rats caused damage of nervous and cardiovascular systems, liver and kidney (9).

### Teratogenicity and reproductive effects
♀ $F_1$ rats were treated with elemental lithium for 3 wk before mating and throughout pregnancy. Treatment related effects were detected in the endocrine system of ♀ and their pups. Reproductive parameters of the ♀ offspring were not greatly affected (10).
Inhibits human sperm motility *in vitro* at concentrations comparable with those in semen after oral administration (11).
≥ 100 μM administered to rats had no adverse effect on cell viability or hormone induced steroidogenesis in adrenal and Leydig cells of the testis (12).
Reduced litter weights, increased numbers of resorption, wavy ribs, short or deformed limb bones or increased incidence of incomplete ossification of sternebrae and wide bone separation in the skull observed in ♀ pregnant Wistar rats administered orally with 100 mg $kg^{-1}$ lithium carbonate (13).

### Metabolism and pharmacokinetics
Lithium is absorbed rapidly from the gastrointestinal tract and excreted unchanged, primarily by the kidneys (14).
30-67% of a single oral dose is excreted within 6-8 hr (15).
Rat plasma clearance of $Li^+$ was 169 ml hr $kg^{-1}$ with a terminal $t_{1/2}$ 5.0 hr (16).
Lithium absorption from guinea pig jejunum is unaffected by metabolic inhibition and appears to be a passive process occurring primarily through the paracellular pathway (17).

## Genotoxicity
Trilithium citrate was the only compound tested in the following study. *Salmonella typhimurium* TA98, TA100, TA1535, TA1537, TA1538 with and without metabolic activation negative (18).
*Escherichia coli* K12/343/113 with and without metabolic activation negative (18).
*Drosophila melanogaster* sex linked recessives negative (18).

*In vivo* intraperitoneal mice 2 injections of 27 mg kg$^{-1}$ did not induce micronuclei (18).

## Any other adverse effects to man

Compared with an age-matched control group the conduction velocity of the sural nerve was reduced in patients receiving long-term lithium therapy (60.8 and 57.8 m sec$^{-1}$, respectively) (19).

20-30% of patients receiving lithium therapy experienced gastrointestinal symptoms including anorexia, nausea, vomiting, diarrhoea and abdominal pain (20).

Absorption of lithium from workplace exposure is usually insufficient to alter lithium body burden or cause systemic toxicity although some upper respiratory symptoms are associated with high levels of dusts containing lithium carbonate and hydride. The 8-hour time-weighted average workplace exposure limit of 0.025 mg m$^{-3}$ for lithium hydride is based on nasal irritation (21).

A retrospective study of patients treated for 1 month-15yr with lithium chromate (LiCrO$_3$) indicated no correlation betweeen lithium treatment and chronic nephrotoxicity (22).

## Legislation

Included in Schedule 4 (Release into Air: Prescribed Substances) Statutory Instrument No. 472, 1991 (23).

## Any other comments

Reproductive effects reviewed (24).

Human health effects, experimental toxicity, environmental effects and physico-chemical properties reviewed (25).

## References

1. Birge, W. J. et al *EPA Report* No.6009-80-022, 1980, Natl. Tech. Inf. Ser., Springfield, VA
2. Downing, A. L. 1970 *Report of the Director of Water Pollution Research* 1970, 58-60, HMSO, London
3. Busa, W. B. et al *Dev. Biol.* 1989, **132**(2), 315-324
4. Chassard-Bouchaud, C. *Anal. Chim. Acta* 1987, **197**, 307-315
5. Weimin, W. et al *Zhongguo Huanjing Kexue* 1988, **8**(1), 68-72 (Ch.) (*Chem. Abstr.* **109**, 98187h)
6. Ghosh, P. K. *Indian J. Physiol. Allied Sci.* 1988, **42**(1), 29-33
7. Marcussen, N. et al *Lab. Invest.* 1989, **61**(3), 295-302
8. Johansson, A. et al *J. Appl. Toxicol.* 1988, **8**(5), 373-375
9. Kushneva, V. S. *Gig. Sanit.* 1990, (5), 36-39 (Russ.) (*Chem. Abstr.* **114**, 76505m)
10. Gloeckner, R. et al *Wiss. Z. Ernst-Moritz-Arndt-Univ. Greifsw., Med. Reihe* 1990, **39**(2), 39-40
11. Raoof, N. T. et al *Br. J. Clin. Pharmacol.* 1989, **28**(6), 715-717
12. Ng, T. B. et al *In Vitro Cell. Dev. Biol.* 1990, **26**(1), 24-28
13. Marathe, M. R. et al *Toxicol. Lett.* 1986, **34**(1), 113-120
14. Baldessarini, R. J. *The Pharmacological Basis of Therapeutics* 7th ed., 1985, Macmillan, New York
15. Trautner, E. M. et al *Med. J. Aust.* 1955, 5th August, 280-291
16. Karlsson, M. O. et al *Pharm. Res.* 1989, **6**(9), 817-821
17. Davie, R. J. et al *Lithium: Inorg. Pharmacol. Psychiatr. Use, Proc. Br. Lithium Congr., 2nd* 1987 (Publ.1988), 107-111
18. King, M-T. et al *Mutat. Res.* 1979, **66**, 33-43
19. Eisenstaedter, A. et al *Wein. Klin. Wochenschr.* 1989, **101**(5), 166-168 (Ger.) (*Chem. Abstr.* **110**, 205470q)

20. Groleau, G. et al *Am. J. Emerg. Med.* 1987, **5**, 572-532
21. *Threshold Limit Values and Biological Exposure Indices* 1989-1990 American Conference of Government Industrial Hygienists
22. Bismuth, C. et al *Arch. Toxicol., Suppl.* 1988, **12**(Target Organ Toxic Process), 179-185
23. *S. I. 1991 No. 472 The Environmental Protection (Prescribed processes and Substances) Regulations* 1991, HMSO, London
24. Johnson, E. M. *Lithium Biol. Med. [Int. Symp.]* 1990 (Publ.1991), 101-112
25. *ECETOC Technical Report No. 30(4)* 1991, European Chemical Industry Ecology and Toxicology Centre, B-1160 Brussels

# L50   Lithium hydride

## LiH

**CAS Registry No.** 7580-67-8
**Synonyms** lithium monohydride
**Mol. Formula** LiH                              **Mol. Wt.** 7.95
**Uses** Reducing agent. Dessicant. Hydrogen generator. Condensing agent with ketones and acid esters.

## Physical properties

**M. Pt.** 680°C; **Specific gravity** 0.76-0.77.

## Occupational exposure

**US TLV (TWA)** 0.025 mg m$^{-3}$; **UK Long-term limit** 0.025 mg m$^{-3}$; **UN No.** 1414, 2805 (fused solid); **Conveyance classification** substance which in contact with water emits flammable gas.

## Mammalian and avian toxicity

**Acute data**
$LC_{Lo}$ (4 hr) inhalation rat 10 mg m$^{-3}$ (1).

**Irritancy**
Rabbit eye exposed to 5 mg m$^{-3}$ (72 hr) caused irritation (1).

## Legislation

Included in Schedule 4 (Release into Air: Prescribed Substances) Statutory Instrument No. 472, 1991 (2).

## Any other comments

Human health effects, experimental toxicity, workplace experience, environmental effects and epidemiology reviewed (3).

## References

1. *AMA Arch. Ind. Health* 1956, **14**, 468
2. *S. I. 1991 No. 472 The Environmental Protection (Prescribed Processes and Substances) Regulations* 1991, HMSO, London
3. *ECETOC Technical Report No. 30(4)* 1991, European Chemical Industry Ecology and Toxicology Centre, B-1160 Brussels

# L51    Lithium hypochlorite

## LiOCl

**CAS Registry No.** 13840-33-0
**Synonyms** hypochlorous acid, lithium salt; lithium chloride oxide;
lithium oxychloride
**Mol. Formula** ClLiO                                **Mol. Wt.** 58.39
**Uses** Swimming pool sanitiser.

## Occupational exposure

**UN No.** 1471; **Conveyance classification** oxidizing substance.

## Mammalian and avian toxicity

### Teratogenicity and reproductive effects
Maternal and developmental no observable adverse effect level for lithium
hypochlorite in rats 100 mg kg$^{-1}$. Effects are consistent with chlorine toxicity (1).

## Genotoxicity

*Salmonella typhimurium* TA98, TA100, TA1535, TA1537, TA1538 with and without
metabolic activation negative (2).
*In vitro* Chinese hamster ovary cells chromosomal aberrations positive at 12 and 18 hr
without metabolic activation, at 22 hr with metabolic activation (2).
*In vivo* rat bone marrow cells showed no increased incidence in chromosomal
aberrations 6, 24, 48 hr after a single oral dose of 100, 500, 1000 mg kg$^{-1}$ to ♂ and 50,
250, 500 mg kg$^{-1}$ to ♀ (2).

## Any other adverse effects to man

Ionic lithium (up to 40 mg Li l$^{-1}$) did not affect serum levels of bathers in hot spas
(101°F) under repetitive exposure, 20 min day$^{-1}$, 4 days wk$^{-1}$ for 2 wk (3).

## Legislation

Included in Schedule 4 (Release into Air: Prescribed Substances) Statutory Instrument
No. 472, 1991 (4).

## References

1.  Hoberman, A. M. et al *J. Am. Coll. Toxicol.* 1990, **9**(3), 367-379
2.  Weiner, M. L. et al *Toxicology* 1990, **65**(1-2), 1-22
3.  McCarthy, J. D. et al *Hum. Exp. Toxicol.* 1994, **13**(5), 315-319
4.  *S. I. 1991 No. 472 The Environmental Protection (Prescribed Processes and Substances) Regulations* 1991, HMSO, London

# L52   Lithocholic Acid

**CAS Registry No.** 434-13-9
**Synonyms** cholan-24-oic acid, 3-hydroxy-, $(3\alpha,5\beta)$-; $3\alpha$-hydroxycholanic acid; $5\beta$-cholan-24-oic acid, $3\alpha$-hydroxy-
**Mol. Formula** $C_{24}H_{40}O_3$                    **Mol. Wt.** 376.58
**Occurrence** In ox, rabbit and human bile and in ox and pig gallstones.

## Physical properties
**M. Pt.** 184-186°C.

## Solubility
Organic solvent: ethanol, diethyl ether, benzene, glacial acetic acid

## Mammalian and avian toxicity

### Acute data
$LD_{50}$ oral mouse 3900 mg $kg^{-1}$ (1).

### Carcinogenicity and long-term effects
National Toxicology Program tested mice and rats via gavage. No evidence of carcinogenicity in ♂ or ♀ rats or mice (2).

### Teratogenicity and reproductive effects
Addition of 36 mg $l^{-1}$ lithocholic acid to *in vitro* 10.5 day old rat embryos resulted in 9% growth-retarded and 12% malformed embryos. Abnormalities in the development of the neural tube and exencephaly were the most common malformations (3).

### Metabolism and pharmacokinetics
Four hydroxylation reactions of lithocholic acid occur in rat liver microsomes resulting in 5 major metabolites: $3\alpha,6\beta$-dihydroxy-$5\beta$-cholanic acid (80%); $3\alpha,6\alpha$-dihydroxy-$5\beta$-cholanic acid; $3\alpha,7\alpha$-dihydroxy-$5\beta$-cholanic acid; $3\alpha,6\beta,7\beta$-trihydroxy-5-cholanic acid; and $3\alpha$-hydroxy-6(3)-oxo-$5\beta$-cholanic acid (4). Organs of rats sacrificed hrly for 8 hr after intrarectal instillation of 0.5 ml of a peanut oil solution containing 183.7 μg showed ≈45% of the instilled acid was absorbed (5).

## Genotoxicity
*Salmonella typhimurium* TA97, TA98 with and without metabolic activation negative (6).
*Saccharomyces cerevisiae* chromosome loss induced but no mitotic recombination or mutation (7).

---

**Any other adverse effects**

50% cytotoxic concentration to isolated rat hepatocytes was 18.37 mg $l^{-1}$ (8).

**Any other comments**

The additive effects of lithocholic acid in oxygen radical generation may overcome the antioxidant defense mechanism of the stem cell leading ultimately to semiquinone and hydroxyl radical mediated DNA damage and tumour induction (9).

**References**

1. *Progress Report for Contract No. NIH-NCI-E-C-72-3252* submitted to the Natl. Cancer Inst. by Litton Bionetics Ltd. (Bethesda, MD)
2. *National Toxicology Program Research and Testing Division* 1992, Report No. PB2884761AS, NIEHS, Research Triangle Park, NC
3. Zusman, I. et al *Acta Anat.* 1990, **138**(2), 144-149
4. Zimniak, P. et al *J. Lipid. Res.* 1989, **30**(6), 907-918
5. Sawada, K. et al *Tokushima J. Exp. Med.* 1986, **33**(3-4), 29-37
6. Zeiger, E. et al *Environ. Mol. Mutagen.* 1988, **11**(Suppl.12) 1-157
7. Albertini, S. et al *Environ. Mol. Mutagen.* 1988, **11**(4), 497-508
8. Takikawa, H. et al *Biochem. Biophys. Acta* 1991, **1091**(2), 173-178
9. Blakeborough, M. H. et al *Free Radical Res. Commun.* 1989, **6**(6), 359-367

# L53   Locust Bean Gum

**CAS Registry No.** 9000-40-2

**Synonyms** carob gum; algaroba; fructoline; Indalca PR90; lupogum; tragacol

**Uses** Stabiliser and thickener in foods and cosmetics. Sizing and finishing agent in textiles, and bonding agent in paper manufacture.

**Occurrence** Ground kernal endosperms of tree pods of *Ceratonia siliqua*

## Environmental fate

**Degradation studies**

Degraded to reducing sugars by *Pseudomonas* sp. (1).

## Mammalian and avian toxicity

**Acute data**

$LD_{50}$ oral rabbit, hamster, rat, mouse 9-13 g $kg^{-1}$ (2).

**Carcinogenicity and long-term effects**

National Toxicology Program investigated locust bean gum carcinogenicity in rat and mouse. Designated non-carcinogen in rat and mouse (3).

## Genotoxicity

*Salmonella typhimurium* TA98, TA100, TA1535, TA97/TA1537 with and without metabolic activation negative (4).

## References

1. Yoshida, O. et al *Seikatsu Eisei* 1990, **34**(3), 119-123 (Japan.) (*Chem. Abstr.* **113**, 168718r)

2. *Food and Drug Research Laboratories Papers* 1976, **124**, Waverly, New York
3. National Toxicology Program Research and Testing Division 1992, Report No. PB82163320, NIEHS, Research Triangle Park, NC
4. Ashby, J. et al *Mutat. Res.* 1988, **204**, 17-115

# L54   Lomustine

**CAS Registry No.** 13010-47-4
**Synonyms** Belustine; chloroethylcyclohexylnitrosourea; urea, *N*-(2-chloroethyl)-*N'*-cyclohexyl-*N*-nitroso-; 1-(2-chloroethyl)-3-cyclohexyl-1-nitrosourea
**Mol. Formula** $C_9H_{16}ClN_3O_2$                     **Mol. Wt.** 233.70
**Uses** Antineoplastic agent. Direct acting bifunctional alkylating agent.

## Physical properties

**M. Pt.** 90°C.

**Solubility**
Water: <0.05 mg $ml^{-1}$. Organic solvent: ethanol

## Mammalian and avian toxicity

**Acute data**
$LD_{50}$ oral rat 70 mg $kg^{-1}$ (1).
$LD_{50}$ subcutaneous mouse 54 mg $kg^{-1}$ (1).

**Carcinogenicity and long-term effects**
Sufficient evidence of carcinogenicity to humans and animals, IARC classification group 2A (2).
Produces lung tumours in rats after intraperitoneal/intravenous injection, slight increase in incidents of lymphomas in mice by intraperitoneal injection (3,4).
Intravenous rats (life study) 1.5, 4, 8 or 20 mg $kg^{-1}$ every 6 wk; treatment was discontinued when severe lethal toxicity became apparent (7 applications highest dose or 10 applications of other doses). Tumours induced included lung adenomas and papillomas of the forestomach (5).
Intravenous rats exerted a weak tumorigenic response to nervous system, lung and forestomach (6).

**Teratogenicity and reproductive effects**
Intraperitoneal rats (days 6-9 pregnancy) 2, 4 or 8 mg $kg^{-1}$ $day^{-1}$ caused omphalocell, ectopia cordis and aortic arch anomalies, syndactyly and anophthalmia occurred most frequently (7).
Intravenous/intraperitoneal rabbits (days 6-18 gestation) 1.5 or 3 mg $kg^{-1}$ $day^{-1}$ caused no teratogenic effect, but a high incidence of abortion was observed with the highest dose (8).

**Metabolism and pharmacokinetics**

Decomposition in physiological conditions to yield alkylating species and organic isocyanates is discussed. Its antitumour activity has been attributed to the alkylation of DNA and the carbamoylation of intracellular proteins by isocyanates having pharmacological and toxicological relevance (9).

Undergoes spontaneous decomposition under physiological conditions to release alkylating and carbamoylating entities (10).

Disappears from plasma within 5 min of administration but the antitumour effects of its metabolites may persist for 15 min (11).

Oral mouse (24 hr), 80% excreted in urine (12).

May be converted by microsomal metabolism to 6 isomeric hydroxylated derivatives (13).

Intraperitoneal ♂ Fischer rats 50 mg kg$^{-1}$ caused a peak plasma concentration of 3 µg ml$^{-1}$. Lomustine was eliminated with biphasic kinetics with terminal $t_{1/2}$ 47 min (14).

Binds to proteins and nucleic acids in cells (15).

Sensitive to oxidation and hydrolysis; forms alkylating and carbamoylating intermediates, $t_{1/2}$ 117 min 25°C at neutral pH (16).

## Genotoxicity

*Salmonella typhimurium* TA100 (metabolic activation unspecified) positive (17).
*Escherichia coli* indused *alkA* gene expression (18).
*In vitro* Chinese hamster lung V79 cells induced DNA single strand breaks (19).
*In vitro* Chinese hamster cells without metabolic activation positive sister chromatid exchanges (20).
*Neurospora crassa* slim mutant *in vitro* DNA synthesis caused increased strand separation. *Neurospora crassa* had excessive and unscheduled replication, indicating a destabilised nature of its DNA (21).
Human peripheral lymphocytes induced dose dependent increases in sister chromatid exchanges through DNA interstrand cross-links (22).
Rat liver microsomes catalysed biotransformation to alkylating metabolites that bound covalently to microsomal protein and to DNA (23).
*In vivo* rats with metabolic activation induced dominant lethal mutations and DNA damage (8).

## Any other adverse effects to man

Human systemic effects by ingestion include anorexia, nausea, vomiting, thrombocytopenia, leukopenia and nephrotoxicity. A characteristic feature of the antitumour nitro-sources is their delayed bone marrow toxicity (24).

## Any other adverse effects

In rats lesions of cerebal blood vessels with accompanying disfunction to the blood brain barrier and changes in brain function (25).

Single dose rats 20 or 50 mg kg$^{-1}$ caused irreversible liver lesions which were histologically detectable after 1 month. At the lower dose, they remained stable for 3 months, but at the 50 mg kg$^{-1}$ the lesions continued to develop, with progression toward cholangiolitic adenomatous transformation of the parenchyma or binary cirrhosis (26).

Morphological alterations were studied in the common bile duct and interlobular bile duct in rats given a single oral dose of 50 mg kg$^{-1}$. The results indicate that early injury appears to be localised in the large bile duct and reflects inflammatory oedema, bile stasis and degeneration of epithelial cells (27).

In dog and monkey, a single intravenous infusion caused toxic changes in bone marrow and lymphoid tissue with neutropenia and lymphopenia, also causes cardiopulmonary and gastrointestinal toxicity in dogs (28).

## Any other comments

Toxicity and hazards reviewed (29-31).

## References

1. *Toxicol. Appl. Pharmacol.*, 1972, **21**, 405
2. *IARC Monograph* 1987, **Suppl. 7**, 150-152
3. Cantor, K. P. et al *J. Natl. Cancer Inst.* 1978, **61**, 979-985
4. Cantor, K. P. et al *Environ. Health Perspect.* 1982, **46**, 187-195
5. Habs, M. *Experimental Studies on the Carcinogenic Effect of Cytostatic Drugs* 1980, Thesis Medical Faculty, Heidelberg University
6. Eisenbrand, G. et al *Dev. Toxicol. Environ. Sci.* 1980, **8**, 273-278
7. Thompson, D. J. et al *Teratology* 1975, **11**, 36A
8. Thompson, D. J. et al *Toxicol. Appl. Pharmacol.* 1975, **34**, 456-466
9. Dive, C. et al *Biochem. Pharmacol.* 1988, **37**(20), 3987-3993
10. Read, D. J. et al *Life Sci.* 1975, **16**, 1263-1270
11. Oiverio, V.T. et al *Cancer Treat. Rep.* 1976, **60**, 703-707
12. Oliverio, V. T. et al *Cancer Res.* 1970, **30**, 1330-1377
13. Wheeler, G. P. et al *Biochem. Pharmacol.* 1977, **26**, 2331-2336
14. Wong, K. H. et al *Int. J. Radiat. Oncol., Biol., Phys.* 1990, **18**(5) 1043-1050
15. Connors, T. A. *Br. J. Cancer* 1974, **30**, 477-480
16. Schein, P. S. et al *Fundam. Cancer Chemother. Antibiot. Chemother.* 1978, **23**, 64-75
17. Frenza, B. R. et al *J. Natl. Cancer Inst.* 1980, **65**, 149-179
18. Fran, R. J. et al *Cancer Res.* 1988, **48**(17), 4823-4827
19. Bradley, M. O. et al *Cancer Res.* 1980, **40**(8/1), 2719-2725
20. Singh, G. et al *Cancer Res.* 1983, **43**, 577-584
21. Beljanski, H. et al *Oncology* 1987, **44**(5), 327-330
22. Best, R. G. et al *Teratog., Carcinog., Mutagen.* 1988, **8**(6), 339-346
23. Kramer, R. A. *Biochem. Pharmacol.* 1989, **38**(19), 3185-3192
24. Szczech, J. et al *Neuropathol. Pol.* 1986, **24**(4), 573-584
25. Viotte, G. et al *Pathol. Biol.* 1987, **35**(2), 139-144 (*Chem. Abstr.* **106**, 2073380)
26. Kretschmer, N. W. et al *Cancer Chemother. Pharmacol.* 1987, **19**(2), 109-117
27. Young, R. C. et al *New Engl. J. Med.* 1971, **285**, 475-479
28. Olivero, V. T. et al *Cancer Chemother.* 1973, **4**(3), 13-20
29. *IARC Monograph* 1981, **26**, 79-95, 137-149
30. Katz, M. E. et al *Cancer Clin. Treat.* 1979, **2**, 297-316
31. *Dangerous Prop. Ind. Mater. Rep.* 1993, **13**(1), 63-66

# L55   LPG (liquified petroleum gas)

**CAS Registry No.** 68476-85-7
**Synonyms** petroleum gases, liquified; Burstane; Pyrofax

## Occupational exposure

**US TLV (TWA)** 1000 ppm (1800 mg m$^{-3}$); **UK Long-term limit** 1000 ppm (1800 mg m$^{-3}$); **UK Short-term limit** 1250 ppm (2250 mg m$^{-3}$); **UN No.** 1075.

## Any other adverse effects to man

LPG containing mixed $C_3$ and $C_4$ compounds (TLV 1000 mg m$^{-3}$) caused necrosis (sites unspecified) (1).

## Any other comments

Human health effects, experimental toxicity and workplace experience reviewed (2).

## References

1. Sunshine, I. *Handbook of Analytical Toxicology* 1969, The Chemical Rubber Co., Cleveland, OH
2. *ECETOC Technical Report No. 30(4)* 1991, European Chemical Industry Ecology and Toxicology Centre, B-1160 Brussels

# L56   Lutetium

## Lu

**CAS Registry No.** 7439-94-3
**Synonyms** lutecium
**Mol. Formula** Lu                                    **Mol. Wt.** 174.97
**Occurrence** 0.8-1.7 ppm in Earth's crust, occurs in xenotime, gadolinite and other rare earth minerals.

## Physical properties

**M. Pt.** 1652°C; **Specific gravity** 9.842.

## Ecotoxicity

### Bioaccumulation

370 ng g$^{-1}$ dry weight found in *Euglena* sp. in tailings drainage at a Canadian mining camp (1).

### Genotoxicity

$Lu^{3+}$ weakly inhibited the proliferative response of human lymphocytes to various polyclonal mitogens and the purified protein derivative of the tuberculin antigen (2).

## References

1. Mann, H. et al *Toxic. Assess.* 1988, **3**(1), 1-16
2. Yamage, M. et al *Experientia* 1989, **45**(11-12), 1129-1131

# L57   2,3-Lutidine

**CAS Registry No.** 583-61-9
**Synonyms** pyridine, 2,3-dimethyl-; 2,3-dimethylpyridine
**Mol. Formula** $C_7H_9N$                    **Mol. Wt.** 107.16

## Physical properties
**M. Pt.** −15°C; **B. Pt.** 162-164°C; **Flash point** 50°C; **Specific gravity** 0.945.

## Mammalian and avian toxicity

### Carcinogenicity and long-term effects
No reports of carcinogenicity; predicted to be non-carcinogenic (1).

## References
1.   Sakamoto, Y. et al *Bull. Chem. Soc. Jpn.* 1989, **62**, 330-332

# L58   2,4-Lutidine

**CAS Registry No.** 108-47-4
**Synonyms** pyridine, 2,4-dimethyl-; α,γ-dimethylpyridine; 2,4-dimethylpyridine
**Mol. Formula** $C_7H_9N$                    **Mol. Wt.** 107.16

## Physical properties
**M. Pt.** −60°C; **B. Pt.** 159°C; **Flash point** 46°C; **Specific gravity** $d_4^{20}$ 0.9309.

## Solubility
Organic solvent: ethanol, diethyl ether, acetone

## Ecotoxicity

### Invertebrate toxicity
$EC_{50}$ (30 min) *Photobacterium phosphoreum* 18.6 ppm Microtox test (1).

---

## Mammalian and avian toxicity

### Carcinogenicity and long-term effects
Predicted non-carcinogenic (2).

### Any other comments
Pyridine bases (unspecified) at concentrations of 1000 mg $l^{-1}$ kill fish in 1 hour (3).
Pyridine bases (unspecified) at concentrations of 1000 mg $l^{-1}$ kill *Tubifex tubifex* in 24 hr (3).

### References

1. Kaiser, K. L. E. et al *Water Pollut. Res. J. Can.* 1991, **26**(3), 361-431
2. Sakamoto, Y. et al *Bull. Chem. Soc. Jpn.* 1989, **62**(1), 330-332
3. Roubickova, J. *Vodni Hospod.: B* 1986, **36**(10), 271-277 (Czech.) (*Chem. Abstr.* **106**, 55222m)

# L59   2,5-Lutidine

**CAS Registry No.** 589-93-5
**Synonyms** Pyridine, 2,5-dimethyl-; 2,5-dimethylpyridine
**Mol. Formula** $C_7H_9N$                                **Mol. Wt.** 107.16

## Physical properties
**M. Pt.** −15°C; **B. Pt.** 157°C; **Flash point** 46°C; **Specific gravity** 0.926.

## Mammalian and avian toxicity

### Acute data
$LD_{50}$ oral mouse, rat, guinea pig 670-827 mg $kg^{-1}$ (1).

### Carcinogenicity and long-term effects
No studies of carcinogenicity reported; predicted to be non-carcinogenic (2).

## References

1. *Hyg. Sanit. (USSR)* 1968, **33**, 341
2. Sakamato, Y. et al *Bull. Chem. Soc. Jpn.* 1989, **62**, 333-335

# L60   Lycopene

**CAS Registry No.** 502-65-8
**Synonyms** ψ,ψ-carotene; lycopene, *all-trans*-; *trans*-lycopene; C.I. 75125;
lycopene 7; C.I. Natural Yellow 27
**Mol. Formula** $C_{40}H_{56}$                                    **Mol. Wt.** 536.89
**Occurrence** In ripe fruit, especially tomatoes.

## Physical properties

**M. Pt.** 172-173°C.

**Solubility**
Organic solvent: chloroform, benzene

## Mammalian and avian toxicity

### Metabolism and pharmacokinetics
Oral administration to rats and rhesus monkeys of [$^{14}$C]lycopene in olive oil resulted
in a peak accumulation of radioactivity in plasma at 4-8 hr in rats and 8-48 hr in
monkeys. The liver contained the largest amount of radioactive pigment in rats and
monkeys compared with other organs tested (1).

## Any other comments

Human health effects and experimental toxicology reviewed (2).

## References

1.  Mathews-Roth, M. M. et al *J. Nutr.* 1990, **120**(10), 1205-1213
2.  *ECETOC Technical Report No. 30(4)* 1991, European Chemical Industry Ecology and
    Toxicology Centre, B-1160 Brussels

# L61   Lymecycline

CAS Registry No. 992-21-2
Synonyms L-lysine, N[6]-[[[[4-(dimethylamino)-1,4,4a,5,5a,6,11,12a-octahydro-
3,6,10,12,12a-pentahydroxy-6-methyl-1,11-dioxo-2-naphthacenyl]carbonyl]amino]-
methyl]-, [4S-(4α,4aα,5aα,6β,12aα)]-; Armyl; Ciclisin; cidolysal; tetracycline
methylenelysine
Mol. Formula $C_{29}H_{38}N_4O_{10}$                Mol. Wt. 602.65
Uses Antibacterial.

## Physical properties

### Solubility
Organic solvent: ethanol

## Mammalian and avian toxicity

### Acute data
$LD_{50}$ intravenous mouse 181 mg kg$^{-1}$ (1).

## References
1.  *Kalikasan* 1970, **2**, 333

# L62   Lynestrenol

CAS Registry No. 52-76-6
Synonyms 19-norpregn-4-en-20-yn-17-ol, (17α)-;
19-nor-17α-pregn-4-en-20-yn-17-ol; 3-deoxynorlutin; ethynylestrenol; exluton;
orgametril
Mol. Formula $C_{20}H_{28}O$                Mol. Wt. 284.45
Uses Progestogen. Used in combination with oestrogen as oral contraceptive.

---

**Occurrence** Predicted riverine concentration 0.09 μg l$^{-1}$ (1).

## Physical properties
**M. Pt.** 158-160°C.

**Solubility**
Organic solvent: ethanol, acetone, chloroform, diethyl ether

## Mammalian and avian toxicity

### Carcinogenicity and long-term effects
Oral ♂ ♀ rats, mice received 2-5, 50-150 and 200-400 × the human contraceptive dose of lynestrenol or lynestrenol and mestranol (33:1) in diet for 80 wk (dose unspecified). Benign liver tumours occurred in 7% ♂ mice fed lynestrenol only compared with 2% in controls. Malignant mammary tumours occurred in 4% ♀ mice fed lynestrenol only, 6% fed the combination (0% in controls) and 3% in rats fed lynstrenol alone (0% in controls) (2).

### Teratogenicity and reproductive effects
No effects observed on rat embryos on day 7 after subcutaneous injection of 0.1 mg lynestrenol on days 2, 3, 4, 5 gestation. 1 mg dose terminated pregacy (3).
No effects on fertility, sexual behaviour or fecundity of offspring of next 2 generation of female golden hamsters receiving 0.6 mg lynestrenol and 18.7 μg mestranol daily for 4.5-8 months (route unspecified) (4).

### Metabolism and pharmacokinetics
Metabolised in rabbits via ethynodiol to norethisterone (5,6).

## Genotoxicity
Dominant lethal mutations observed in ♀ mice dosed 3 days prior to mating with 420 μg lynestrenol and 12 μg kg$^{-1}$ mestranol (7).

## Any other adverse effects to man
No significant difference in sex ratio or frequency of abnormal karyotypes in 124 abortuses of women who had taken oral contraceptives including lynestrenol compared with 122 controls (8).

## References
1. Richardson, M. L. et al *J. Pharm. Pharmacol.* 1985, **37**, 1-12
2. Committee on Safety of Medicines *Carcinogenicity Tests of Oral Contraceptives* 1972, HMSO, London
3. Castro-Vazquez, A. et al *Fert. Steril.* 1971, **22**, 741-744
4. Cottinet, D. et al *C. R. Seances Soc. Biol. Ses Fil* 1974, **168**, 517-520 (Fr.)
5. Okada, H. et al *Nippon Naibunpi Gakkai Zasshi* 1964, **40**, 1095-1098
6. Mazaheri, A. et al *J. Endocrinol.* 1970, **47**, 251-252
7. Badr, F. M. et al *Mutat. Res.* 1974, **26**, 529-534
8. Launtsen, J. A. *Acta Obstet. Gynecol. Scand.* 1975, **54**, 261-264

# L63  D-Lyxose

**CAS Registry No.** 1114-34-7
**Synonyms** lyxose, D-; α-D-lyxose
**Mol. Formula** $C_5H_{10}O_5$                    **Mol. Wt.** 150.13

## Physical properties
**M. Pt.** 106-107°C; **Specific gravity** $d^{20}$ 1.545.

## Solubility
Water: freely soluble. Organic solvent: ethanol

## Environmental fate

### Degradation studies
Degraded by soil microbes more slowly than natural pentoses (1).

## References

1.  Izumori, K. et al *Kagawa Daigaku Nogakubu Gakujutsu Hokoku* 1987, **39**(1), 83-86
    (Japan.) (*Chem. Abstr.* **111**, 170854j)

# M1  Magenta I

**CAS Registry No.** 632-99-5
**Synonyms** benzenamine, 4-[(4-aminophenyl)(4-imino-3-methyl-2,5-cyclohexadien-1-ylidene)methyl]-2-methyl-, monohydrochloride;
C.I. Basic Violet 14 mono hydrochloride
**Mol. Formula** $C_{20}H_{20}N_3Cl$                    **Mol. Wt.** 337.86

Uses Antifungal. Used as a dye or in dye manufacture.

## Physical properties
**M. Pt.** 200°C (decomp.).

**Solubility**
Organic solvent: ethanol

## Mammalian and avian toxicity

### Carcinogenicity and long-term effects
Magenta: No adequate evidence for carcinogenicity to humans or animals, IARC classification group 3. Magenta manufacturing process, sufficient evidence for carcinogenicity to humans and animals, IARC classification group 1 (1).

12 mg wk$^{-1}$ in arachis oil administered orally to mice for 52 wk by gavage caused stained tissues at autopsy. Four lymphomas and one hepatoma were seen in the 21 test mice compared with 5 lymphomas in controls (2).

10 mg p-magenta, a component of commercial magenta, subcutaneously or intramuscularly injected into rats as a 1% aqueous solution $1 \times$ wk$^{-1}$. One sarcoma appeared after 300 days at a dose of 370 mg. 7 sarcomas were seen in 12 rats surviving after the appearance of the first tumour (3).

## Genotoxicity
*Salmonella typhimurium* TA98, TA1538 with metabolic activation positive (4).

## Any other adverse effects to man
5/85 men manufacturing magenta but not exposed to 1- or 2-naphthylamine or benzidine had bladder cancer showing a higher risk than expected. Modern manufacturing replaces aniline with *o*-toluidine and it is possible the *o*-toluidine is implicated in the aetiology of the magenta tumours (5).

## Legislation
Listed as a controlled substance in the UK Carcinogenic Substances Regulations 1967 (6).

## Any other comments
Technical grade magenta consists of a mixture of magenta-I, *p*-magenta and related compounds (1).

## References
1. *IARC Monograph* 1987, **Suppl.7**, 65
2. Bonser, G. M. et al *Br. J. Cancer* 1956, **10**, 653
3. Druckrey, H. et al *Naturwissenschaften* 1956, **43**, 543
4. Bonin, A. M. et al *Mutat. Res.* 1981, **89**, 21-34
5. Case, R. A. M. et al *Br. J. Ind. Med.* 1954, **11**, 213
6. *S. I. 1967 No. 879 The Environmental Protection (Prescribed Processes and Substances) Regulations* 1967, HMSO, London

# M2   Magnesium arsenate

$Mg_x \cdot AsH_3O_4$

**CAS Registry No.** 10103-50-1
**Synonyms** arsenic acid, magnesium salt
**Mol. Formula** $AsH_3O_4Mg_x$

**Physical properties**
**Specific gravity** 2.60-2.61.

**Occupational exposure**
**UN No.** 1622; **HAZCHEM Code** 2Z; **Conveyance classification** toxic substance.

**Mammalian and avian toxicity**

**Acute data**
$LD_{50}$ oral mouse 315 mg $kg^{-1}$ (1).

**References**

1. *Int. Arch. Gewerbepathol. Gewerbehyg.* 1963, **20**, 21


# M3   Magnesium chlorate

$Mg(ClO_3)_2$

**CAS Registry No.** 10326-21-3
**Synonyms** chloric acid, magnesium salt; magnesium dichlorate; magron
**Mol. Formula** $Cl_2MgO_6$                 **Mol. Wt.** 191.21

**Physical properties**
**M. Pt.** ≈35°C; **B. Pt.** 120°C (decomp.); **Specific gravity** $d^{25}$ 1.80.

**Solubility**
Water: soluble in 0.01 parts water. Organic solvent: slightly soluble ethanol

**Occupational exposure**
**UN No.** 2723; **HAZCHEM Code** 1SE; **Conveyance classification** oxidizing substance.

**Mammalian and avian toxicity**

**Acute data**
$LD_{50}$ oral mouse, rat, rabbit 5.2-8.6 g $kg^{-1}$ (1,2).
$LD_{Lo}$ intraperitoneal rat 1100 mg $kg^{-1}$ (1).

**Carcinogenicity and long-term effects**
Oral (156 day) rat 24.7 mg $kg^{-1}$ $day^{-1}$ increased incidence of lung, stomach, skin, liver tumours (3).

**Teratogenicity and reproductive effects**

Oral administration of 127 mg kg$^{-1}$ day$^{-1}$ to pregnant rats increased preimplantation mortality of embryos, decreased number of progeny dam$^{-1}$, and decreased body weight of progeny. No teratogenic effects were observed (4).

**Genotoxicity**

*In vivo* intraperitoneal mice 1500 mg kg$^{-1}$ caused a 5-fold increase in chromosomal aberration frequency (5).

**Any other adverse effects**

Increased levels of methaemoglobin and sulfhaemoglobin reported in rats following oral administration of up to 10 mg kg$^{-1}$ (6).

**Legislation**

Limited under EC Directive on Drinking Water Quality 80/778/EEC magnesium guide level and maximum admissible concentration 50 mg l$^{-1}$ chloride guide level 25 mg l$^{-1}$ (7).

**Any other comments**

Health hazards evaluated (8).

**References**

1. Ulrich *J. Pharmacol. Exp. Ther.* 1929, **35**, 1
2. *Gig. Sanit.* 1983, **48**(4), 68
3. Ponomareva, L. A. et al *Med. Zh. Uzb.* 1990, (5), 44-46 (Russ.) (*Chem. Abstr.* **113**, 147075p)
4. Bairammuradova, M. K. et al *Zdravookhur. Turkm.* 1987 (6), 27-29 (Russ.) (*Chem. Abstr.* **108**, 33334p)
5. Tashkhodzhaev, P. I. et al *Uzb. Biol. Zh.* 1989, (3), 69-72 (Russ.) (*Chem. Abstr.* **112**, 17599f)
6. Bairammuradova, M. K. *Zdravookhr. Turkm.* 1987, (7), 15-17 (*Chem. Abstr.* **107**, 53827j)
7. *EC Directive Relating to the Quality of Water Intended for Human Consumption* 1982, 80/778/EEC, Office for Official Publications of the European Communities, 2 rue Mercier, L-2985 Luxembourg
8. Perepelento, S. D. et al *Dokl. Akad. Nauk. UzSSR* 1989, (10), 56-58 (Russ.) (*Chem. Abstr.* **112**, 193477y)

# M4    Magnesium fluorosilicate

## MgSiF$_6$

**CAS Registry No.** 16949-65-8

**Synonyms** hexafluoromagnesium(1:1) silicate (2-); magnesium hexafluorosilicate; magnesium silicofluoride; silicon fluoride magnesium salt

**Mol. Formula** F$_6$SiMg                                **Mol. Wt.** 166.39

**Uses** Mothproofing of textiles.

**Physical properties**

**Specific gravity** 1.788 (hexahydrate).

**Solubility**
Water: $650 \text{ g l}^{-1}$.

**Occupational exposure**
**UN No.** 2853; **HAZCHEM Code** 1Z; **Conveyance classification** harmful substance.

**Mammalian and avian toxicity**

**Acute data**
$LD_{50}$ oral guinea pig 200 mg $kg^{-1}$ (1).

**Legislation**
Limited under EC Directive on Drinking Water Quality 80/778/EEC. Magnesium: guide level and maximum admissible concentration 50 mg $l^{-1}$ (2).

**References**

1. Spector, W. S. Ed. *Handbook of Toxicology* 1956, **1**, 180-81, Saunders, Philadelphia
2. *EC Directive Relating to the Quality of Water Intended for Human Consumption* 1982, 80/778/EEC, Office for Official Publications of the European Communities, 2 rue Mercier, L-2985 Luxembourg

# M5   Magnesium nitrate

## $Mg(NO_3)_2$

**CAS Registry No.** 10377-60-3
**Synonyms** magnesium salt nitric acid; magnesium dinitrate; magniosan
**Mol. Formula** $MgN_2O_6$                          **Mol. Wt.** 148.32
**Uses** In pyrotechnics. In the concentration of nitric acid.
**Occurrence** Hydrated form occurs in the mineral nitromagnesite.

**Physical properties**
**M. Pt.** 89°C (hexahydrate); **Specific gravity** 1.636.

**Solubility**
Water: 1 in 0.8 parts water. Organic solvent: ethanol

**Occupational exposure**
**UN No.** 1474; **HAZCHEM Code** 1■; **Conveyance classification** oxidizing substance.

**Any other adverse effects to man**
Increases human amniotic membrane stability on maternal side but decreases it on foetal side (1).

**References**

1. Bara, U. et al *Magnesium Res.* 1988, **1**(2), 23-27

# M6　Magnesium oxide

## MgO

**CAS Registry No.** 1309-48-4
**Synonyms** animag; maglite; magnesa preprata; magnesium monoxide; magox; seasorb
**Mol. Formula** MgO　　　　　　　　　　　　　**Mol. Wt.** 40.31
**Uses** Manufacture of fire bricks, magnesia cements. Reflector in optical instruments. Antacid. Food additive. White colour standard.
**Occurrence** Occurs as the mineral periclase.

## Physical properties

**M. Pt.** 2500-2800°C; **B. Pt.** 3600°C; **Specific gravity** 3.65-3.75.

### Solubility
Water: slightly soluble in pure water, solubility increased by carbon dioxide.

## Occupational exposure

**US TLV (TWA)** 10 mg m$^{-3}$ (fume); **UK Long-term limit** 10 mg m$^{-3}$ (as Mg) (total inhalable dust) 5 mg m$^{-3}$ (as Mg) (fume and respirable dust); **UK Short-term limit** 10 mg m$^{-3}$ (as Mg) (fume and respirable dust).

## Mammalian and avian toxicity

### Acute data
$TC_{Lo}$ (duration unspecified) inhalation human 400 mg m$^{-3}$ (1).
Inhalation human (5-9 min) 15-29 mg of freshly generated magnesium oxide. 2/4 experienced mild body temperature rise followed by fever and a rise in white blood cell count from 7000 to 15500 after 6 hr. Recovery was complete by the next day (2).

### Carcinogenicity and long-term effects
In a cohort study of 2391 ♂ workers producing magnesium metal (work experience >1 yr) 152 cases of cancer were recorded (20 more than expected); cancers included cancer of lip, stomach and lung (3).

## Any other adverse effects to man

95 men exposed to magnesium oxide dust had slight irritation of eyes and nose. The magnesium level in the serum was above the normal level of 3.5 mg percent (4).
Can cause diarrhoea (5).
Occupational exposure causes higher than normal incidence of digestive disorders and gastric or duodenal ulcers (6).

## References

1. *Documentation of Threshold Limit Values for Substances in Workroom Air* 1971, **3**, 147
2. Drinker, D. et al *J. Ind. Hyg.* 1927, **9**, 187
3. Heldaas, S. S. et al *Br. J. Ind. Med.* 1989, **46**(9), 617-623
4. Pleschitzer, A. *Arch. Gewerbepathol. Gewerbehyg.* 1936, **7**, 8
5. *Martindale. The Extra Pharmacopoeia* 30th ed., 1993, The Pharmaceutical Press, London
6. Gwizdelz, E. et al *Arch. Mal. Prof., Med. Trav. Secur. Soc.* 1967, **28**(6), 531-534

# M7　Magnesium perchlorate

## Mg(ClO4)2

**CAS Registry No.** 10034-81-8

**Synonyms** magnesium salt perchloric acid; anhydrous magnesium perchlorate; anhydrone; dehydrite; magnesium diperchlorate

**Mol. Formula** $Cl_2MgO_8$　　　　　　　　　　**Mol. Wt.** 223.21

**Uses** Drying agent for gases. Growth stimulant in animal feed.

## Physical properties

**B. Pt.** >250°C (decomp.); **Specific gravity** $d^{25}$ 2.6.

## Solubility

Water: 993 g $l^{-1}$ at 25°C. Organic solvent: ethanol

## Occupational exposure

**UN No.** 1475; **HAZCHEM Code** 1Y; **Conveyance classification** oxidizing substance.

## Mammalian and avian toxicity

### Acute data

$LD_{50}$ intraperitoneal mouse 1500 mg $kg^{-1}$ (1).

## Any other comments

Limited under EC Directive on Drinking Water Quality 80/778/EEC. Magnesium: guide level 50 mg $l^{-1}$; maximum admissible concentration 50 mg $l^{-1}$; chloride: guide level 25 mg $l^{-1}$ (2).

## References

1. *J. Agric. Food. Chem.* 1966, **14**, 512
2. *EC Directive Relating to the Quality of Water Intended for Human Consumption* 1982, 80/778/EEC, Office for Official Publications of the European Communities, 2 rue Mercier, L-2985 Luxembourg

# M8　Magnesium phosphide

## Mg3P2

**CAS Registry No.** 12057-74-8

**Mol. Formula** $Mg_3P_2$　　　　　　　　　　**Mol. Wt.** 134.88

**Uses** Fumigant. Insecticide.

## Physical properties

**Specific gravity** 2.055.

## Solubility

Water: decomp.

## Occupational exposure

**UN No.** 2011; **Conveyance classification** substance which in contact with water emits flammable gas, toxic substance; **Supply classification** highly flammable, very toxic.
**Risk phrases** Contact with water liberates toxic, highly flammable gas – Very toxic if swallowed (R15/29, R28)
**Safety phrases** Keep locked up and out of reach of children – Do not breathe dust – In case of fire, use combustibles and powdered metals – In case of accident or if you feel unwell, seek medical advice immediately (show label where possible) (S1/2, S22, S43, S45)

## Mammalian and avian toxicity

### Acute data

$LC_{Lo}$ inhalation rat (1 hr) 580 ppm; cat, guinea pig (2 hr) 173, 288 ppm, respectively (1). $LC_{50}$ inhalation (1 hr) rat 580 ppm (2).

## Any other adverse effects

Is a potent acute mammalian poison but feeding trials with fumigated feedstuffs have shown no chronic effects in rats (3).

## Legislation

Limited under EC Directive on Drinking Water Quality 80/778/EEC. Pesticides: maximum admissible concentration 0.1 $\mu g \, l^{-1}$ (4).

## Any other comments

Stable when dry. Reacts with atmosphere moisture, and reacts violently with acids producing phosphine.

## References

1. *Z. Gesundheitstech. Staedtehyg.* 1933, **25**, 279
2. *Chemical Safety Data Sheets* 1991, RSC, London
3. Hackenberg, U. *Appl. Pharmacol.* 1972, **23**, 147
4. *EC Directive Relating to the Quality of Water Intended for Human Consumption* 1982, 80/778/EEC, Office for Official Publications of the European Communities, 2 rue Mercier, L-2985 Luxembourg

# M9   Malaoxon

$$CO_2CH_2CH_3$$
$$|$$
$$CH_2 \quad O$$
$$| \qquad \| \quad OCH_3$$
$$CH-S-P$$
$$| \qquad \quad OCH_3$$
$$CH_2CO_2CH_2CH_3$$

**CAS Registry No.** 1634-78-2
**Synonyms** butanedioic acid, [(dimethoxyphosphinyl)thio]-, diethyl ester; succinic acid, mercapto-, diethyl ester, *S*-ester with *O,O*-dimethyl phosphorothioate; Liromat; malathion-*O*-analog; oxycarbophos
**Mol. Formula** $C_{10}H_{19}O_7PS$                    **Mol. Wt.** 314.30

## Ecotoxicity

### Invertebrate toxicity
$LC_{50}$ (24 hr) *Chironomus riparius* larva 5.4 mg $l^{-1}$ (1).

## Mammalian and avian toxicity

### Acute data
$LD_{50}$ oral rat 158 mg $kg^{-1}$ (2).
$LD_{50}$ intraperitoneal rat 17.5 mg $kg^{-1}$ (3).

### Carcinogenicity and long-term effects
National Toxicology Program tested (103 wk) rats and mice orally 500 or 1000 ppm via feed. No evidence for carcinogenicity in rats or mice (4).
Oral rat, mice 1000 ppm in diet (duration unspecified) no increased incidence of tumours compared to controls (5).

### Teratogenicity and reproductive effects
Clawed frog embryos, dose dependent severe developmental effects, reduced growth, notochordal defects, abnormal pigmentation (6).

## Genotoxicity
*Salmonella typhimurium* TA97, TA98 with and without metabolic activation negative (7).
*In vitro* Chinese hamster ovary cells with metabolic activation weakly positive, without metabolic activation positive for sister chromatid exchanges, and with and without metabolic activation negative for chromosomal aberrations (8).
*In vitro* mouse lymphoma $tk^{+}/tk^{-}$ with metabolic activation equivocal and without metabolic activation positive (9).

## Any other adverse effects
Reduced $NAD^{+}$ levels in clawed frog (6).
Brain acetylcholinesterase activity inactivation: mice and rats were more sensitive than minnows or trout (10).

## Any other comments
In mammals it is the degradation metabolite of the insecticide malathion (11).

## References

1. Estenik, J. F. et al in *Pesticide and Xenobiotic Metabolism in Aquatic Organisms* 1979, Am. Chem. Soc., Symposium Series **99**
2. *Toxicol. Appl. Pharmacol.* 1966, **4**, 408
3. *Can. J. Physiol. Pharmacol.* 1967, **45**, 621
4. *National Toxicology Program Research and Testing Division* 1992, Report No.TR-135, Research Triangle Park, NC
5. Haseman, J. K. et al *Environ. Mol. Mutagen.* 1990, **16**(Suppl.18) 15-31
6. Snawder, J. E. et al *J. Environ. Sci. Health, Part B* **B24**(3), 205-218
7. Zeiger, E. et al *Environ. Mol. Mutagen.* 1988, **11**(Suppl. 12) 1-157
8. Ivett, J. L. et al *Environ. Mol. Mutagen.* 1989, **14**, 165-187
9. Myhr, B. C. *Environ. Mol. Mutagen.* 1991, **18**(1), 51-83
10. Johnson, J. A. et al *Toxicol. Appl. Pharmacol.* 1987, **88**(2), 234-241
11. *The Agrochemicals Handbook* 3rd ed., 1991, RSC, London

# M10 Malathion

$$CO_2CH_2CH_3$$
$$|$$
$$CH_2 \quad S$$
$$| \quad \quad || \quad OCH_3$$
$$CH-S-P$$
$$| \quad \quad OCH_3$$
$$CO_2CH_2CH_3$$

**CAS Registry No.** 121-75-5
**Synonyms** $S$-[1,2-bis(ethoxycarbonyl) ethyl $O,O$-dimethyl phosphorodithioate; carbofos; mercaptothion; phosphothion; moldison; sodophos; [(dimethoxyphosphinothioyl)thio]-butanedioic acid, diethyl ester; mercaptosuccinic acid diethyl ester, $S$-ester with $O,O$-dimethyl phosphorodithioate; diethyl (dimethoxythiophosphorylthio)succinate
**Mol. Formula** $C_{10}H_{19}O_6PS_2$        **Mol. Wt.** 330.36
**Uses** Insecticide. Pediculicide. Veterinary ectoparasiticide.
**Occurrence** Residues have been found in water, air, soil, crops and animal and fish tissues.

## Physical properties
**M. Pt.** 2.9°C; **B. Pt.** $_{0.7}$ 156-157°C; **Specific gravity** $d_4^{25}$ 1.2315;
**Partition coefficient** log $P_{ow}$ 2.36 (1); **Volatility** v.p. $4 \times 10^{-5}$ mmHg at 30°C.

## Solubility
Water: 145 mg $l^{-1}$ at 20°C. Organic solvent: esters, alcohols, ethers, ketones, aromatic and alkylated aromatic hydrocarbons, vegetable oils

## Occupational exposure
**US TLV (TWA)** 10 mg m$^{-3}$; **UK Long-term limit** 10 mg m$^{-3}$; **Supply classification** harmful.
**Risk phrases** Harmful if swallowed (R22)
**Safety phrases** Avoid contact with skin (S24)

## Ecotoxicity

**Fish toxicity**
$LD_{50}$ (96 hr) bluegill sunfish, bass, brown trout, rainbow trout 0.1-0.29 mg $l^{-1}$ (2,3).
$LC_{50}$ (48 hr) carp 9.80-10.04 mg $l^{-1}$ (4).
Snakehead fish (6 month resting to spawning phase) 0.20 mg $l^{-1}$. ♂ inhibition of gonadal development and gonadosomatic index. ♀ immature oocytes exhibited cytoplasmic protinaceous inclusion bodies which ultimately lead to their degeneration (5). *In vivo* spinally transected rainbow trout exposed to toxic aqueous concentrations, caused immediate reduction in oxygen utilisation and heart rate and an increase in ventilation volume (6).

**Invertebrate toxicity**
$LC_{50}$ (24, 48, 72, 96 hr) freshwater crab 8.9, 6.5, 5.6, 3.8 ppm, respectively (7).
*Anabaena oryzae* and *Phormidium fragile* 93.5% and 85.7% of applied dose recovered in respiration the metabolites identified were mono- and dicarboxylic acid, mercaptoethyl succinate, mono- and di-ethylsuccinate (8).

*Anabaena* survived up to 0.5 mg l$^{-1}$ (9).
LC$_{50}$ (48 hr) *Daphnia pulex* 1.8 µg l$^{-1}$ (10).

**Bioaccumulation**
Bioconcentration factor for carp in muscle, liver and kidney 2.7-17.3 (11).
Bioconcentration factor coho salmon 29.3 (12).

## Environmental fate

**Nitrification inhibition**
Threshold for inhibition of denitrification and nitrification (rotating disc and activated sludge) 10.0 mg l$^{-1}$ (13).

**Degradation studies**
The time required for complete biodegradation in a model river water were 8, 12 and 18 days for 5, 10 and 15 mg dm$^{-3}$, respectively (14).
Rate of degradation in 10 days 81-92% in various non-sterile loam soils and 5-19% in various sterile loam soils (15).
Unsterile seawater and sedimented cores under laboratory light at 20°C, pH 8 t$_{1/2}$ 2.6 and 20 days, respectively (16).
Field application, 1 ml in 40 ml distilled water, tropical grassland soil initially adversely affected microfungal and microbial biomass. Statistical analysis over a 15 month period showed no significant difference from controls (17).

**Abiotic removal**
Removal rates by activated carbon from waste water; COD 50-55% and organic phosphorous moiety 90% (18).
The adsorption capacity of activated charcoal for malathion in saline water was 117 mg g$^{-1}$ (19).
The rate of removal with aqueous oxidisers was up to 5 times as rapid when exposed to UV irradiation (280-320 nm) (20).
t$_{1/2}$ for thin film of 0.67 µg cm$^{-2}$ irradiated at environmentally important wavelengths 2.1 days (21).
Alkaline hydroysis is the most important pathway of degradation in the Indian River Estuary, biological and photolytic degradation only play a small role (22).

**Absorption**
Of clay minerals, maximum absorption occurred with calcium saturated Trancos montmorillonite (23).
Model to establish the potential to reach ground water. The model assumes steady waterflow, equilibrium linear adsorption and depth-dependent first-order biodegradation. K$_{oc}$, groundwater travel times and residual concentrations are determined (24).
Mobility through Sakit sandy loam and Itwa loam soils and the effect of soil organic matter (including humic and fulvic acids), clay fractions, free aluminium oxide and ferric oxide, soil pH and and exchangeable cations were studied. All parameters decreased mobility except for the addition of humic acid which increased mobility (25).

## Mammalian and avian toxicity

**Acute data**
LD$_{50}$ oral mouse, rat 190, 290 mg kg$^{-1}$, respectively (26).
LC$_{50}$ oral redwing blackbird 400 mg kg$^{-1}$ diet (27).

$LC_{50}$ (4 hr) inhalation rat 84.6 mg m$^{-3}$ (28).

$LD_{50}$ (24 hr) dermal rabbit 4100 mg kg$^{-1}$ (2).

$LD_{50}$ intraperitoneal rat, mouse 193, 250 mg kg$^{-1}$, respectively (29,30).

$LD_{50}$ subcutaneous mouse 221 mg kg$^{-1}$ (31).

$LD_{50}$ intravenous rat, mouse 50, 184 mg kg$^{-1}$, respectively (32,33).

The acute toxicity of malathion is reported to depend on its purity (34).

**Sub-acute data**

$LC_{50}$ oral (8 day) Japanese quail, bobwhite quail, ringnecked pheasant, mallard duck 2962->5000 mg kg$^{-1}$ diet (35).

Oral mice (20 day) 14 mg kg$^{-1}$ increased the primary response to sheep erythrocytes (sRBC) indicated by haemolysin titre and suppressed the secondary response to sRBC. The activity of α-naphthyl acetate esterase was also inhibited. These effects occurred with the inhibition of cholinesterase in the whole blood (36).

The uptake of $^{131}$I in rats given 44 mg kg$^{-1}$ for 12 wk (route not specified) was reduced by 6.25%. There was also a 37.5% reduction in serum protein bound iodine. Normal thyroid function was observed in rats whose malathion was discontinued for 2 wk before the animals were examined (37).

In rats given 90% technical malathion in the diet for 4-6 wk, at average daily intakes of 62 and 68 mg kg$^{-1}$ to ♂ and ♀, respectively, cholinesterase activity in the brain, plasma and erythrocytes was inhibited by ≈50%. No adverse effect was noted in the animals (38).

**Carcinogenicity and long-term effects**

In 21 month feeding trials, rats receiving 100 mg kg$^{-1}$ diet showed normal weight gain (2).

In a 2 yr feeding study in ♂ rats, a level of 100 mg kg$^{-1}$ 90% technical grade (6 mg kg$^{-1}$ bodyweight), the lowest dose tested, resulted in a 10-30% depression in brain, plasma and erythrocyte cholinesterase activity, with no effect on food intake or growth (38).

No adequate data for evaluation of carcinogenicity in humans, inadequate evidence for carcinogenicity in animals, IARC classification group 3 (39).

National Toxicology Program tested rats and mice orally via feed. No evidence for carcinogenicity in rats or mice (40).

There was no significant increase in the incidence of tumours in mice fed diets containing 8000 or 16,000 mg kg$^{-1}$ for 80 wk (41).

There was no significant increase in the incidence of tumours in rats fed diets containing 2000 or 4000 mg kg$^{-1}$ for 103 wk compared with controls (42).

There was no significant increase in the incidence of tumours in rats and mice fed diets containing metabolites of malathion when compared with controls (34).

**Teratogenicity and reproductive effects**

*Xenopus laevis* embryos were exposed to malathion or its metabolite malaoxon during the first 4 days of development. The following defects were observed in a dose-dependent manner: reduced size, abnormal pigmentation, abnormal gut, enlargement of the atria and aorta, bent notochord and lowered NAD$^{+}$ (43).

Malathion (technical grade, 95% pure) was fed to rats at a dietary concentration of 4000 mg kg$^{-1}$ (≈ daily intake 240 mg kg$^{-1}$ body weight) for 2 generations. ♂ and ♀ of 70-100 days of age were bred after 10 wk of test; survival of the progeny on days 7 and 21 after birth was found to be reduced, and the surviving offspring showed growth retardation and an increased incidence of ring-tail disease (44).

A single intraperitoneal injection of 600 or 900 mg kg$^{-1}$ on day 11 of pregnancy to rats produced maternal toxicity; foetal weight reduction but no malformations were induced with the high dose only (45).

Technical grade malathion (purity unspecified) administered daily at maternally tolerated doses of 50-300 mg kg$^{-1}$ via gavage to rats on days 6-15 of pregnancy induced no embryotoxicity (46).

Oral rabbit (7-12 days of gestation) 100 mg kg$^{-1}$ no detectable difference in the number of resorptions, foetal size or abnormalities compared with controls (47).

Pregnant humans exposed to malathion whilst their town was fumigated (over 23 days) showed no increase in congenital malformations or stillbirths when compared to the previous year (48).

**Metabolism and pharmacokinetics**

Following oral administration to animals, the major part of the dose was excreted in the urine and faeces within 24 hr. Degradation occurs by oxidative desulfurisation by liver microsomal enzymes, leading to the formation of malaoxon. Malathion and malaoxon are hydrolysed and thus detoxified by carboxylesterases. In insects, metabolism involves hydrolysis of the carboxylate and phosphorodithioate esters, and oxidation to malaoxon (2).

Rat intravenous $^{14}$C-malathion highest levels detected in liver and kidney which reached a peak 1-3 min after administration. After 24 hr low levels of radioactivity were detected into the liver, kidneys, intestines and the Harderian gland (49).

## Genotoxicity

*Salmonella typhimurium* TA98, TA1535, TA1537, with and without metabolic activation, negative (50).

*Escherichia coli* with metabolic activation negative (51).

Did not induce sex-linked recessive lethal mutations in *Drosophila melanogaster* (52).

*In vitro* Chinese hamster ovary cells with and without metabolic activation sister chromatid exchanges and chromosomal aberrations positive (53).

Induced an increase in sister chromatid exchanges in human foetal fibroblasts *in vitro* (54).

Induced chromosomal aberrations and sister chromatid exchanges in human lymphocytes *in vitro* (55).

Induced an increase in chromosomal aberrations in primary spermatocytes *in vivo* following oral administration to mice (56).

Threshold dose for cytogenetic toxicity in meristermatic cells of *Allium cepa* root tips 25 ppm (57).

## Any other adverse effects to man

♂ Pesticide applicators had a significant increase in the frequency of sister chromatid exchanges compared to controls. They also showed cell cycle delay and decrease in mitotic index (58).

## Any other adverse effects

Acute toxicity is due to acetylcholinesterase inhibition at nerve endings, leading to an accumulation of endogenous acetylcholine. The effects are manifested by muscarinic, nicotinic and control nervous system signs and symptoms: sweating, salivation, diarrhoea, bronchorrhoea, bradycardia, bronchoconstriction, muscle fasciculations and coma. The cause of death is primarily respiratory failure (59).

Causes dose related immunological changes (species unspecified) (60).

## Legislation

Limited under EC Directive on Drinking Water Quality 80/778/EEC. Pesticides: maximum admissible concentration 0.1 $\mu$g l$^{-1}$ (61).

Included in Schedule 6 (Release into Land: Prescribed Substances) Statutory Instrument No. 472, 1991 (62).

UK Department of Environment advisory value for drinking water 7 $\mu$g l$^{-1}$ (63).

## Any other comments

Toxic and carcinogenic potential and hazards reviewed (64-66).

Soil has some stabilising effect on malathion due to adsorption. Addition of humic material helps in the decay in soil (67).

Occupational exposure from pesticide application assessed (68).

WHO Class III. EPA Toxicity Class III. Tolerable daily intake (humans) 0.02 mg kg$^{-1}$ (2). LD$_{50}$ topical application to bees 0.71 $\mu$g bee$^{-1}$ (2).

## References

1. Hansch, C. et al *Medchem Project Issue No. 26* 1985, Pomona College, Claremont, CA
2. *The Agrochemicals Handbook* 3rd ed., 1991, RSC, London
3. Macek, K. J. et al *Trans. Am. Fish Soc.* 1970, **99**(1), 20-27
4. Reddy, P. M. *Environ. Ecol.* 1986, **6**(2), 488-490
5. Ram, R. N. et al *Environ. Pollut.* 1987, **44**(1), 49-60
6. Mikim, J. M. et al *Environ. Toxicol. Chem.* 1987, **6**(4), 313-328
7. Rao, K. S. et al *Environ. Ecol.* 1987, **5**(1), 203-204
8. Khalil, Z. et al *Isot. Radiat. Res.* 1987, **19**(2), 123-129
9. Tandon, R. S. et al *Environ. Pollut.* 1988, **52**(1), 1-9
10. Sanders, H. O. et al *Trans. Am. Fish Soc.* 1966, **95**(2), 165-169
11. Tsuda, T. et al *Comp. Biochem. Physiol., C: Comp. Pharmacol. Toxicol.* 1990, **96C**(1), 23-26
12. Walsh, A. H. et al *The Pathology of Fish* 1973, 515-541, Univ. Wisconsin Press, Madison, WI
13. Knoetze, C. *Water Report* 1979, **9**, 5-6
14. Zdybiecoska, M. W. *Przem. Chem.* 1988, **67**(8), 378-379 (Pol.) (*Chem. Abstr.* **109**, 196461n)
15. Walker, W. W. et al *J. Environ. Qual.* 1973, **2**, 229-232
16. Cotham, W. E. et al *J. Agric. Food Chem.* 1989, **37**(3), 824-828
17. Satpathy, G. R. et al *Pollut. Res.* 1990, **9**, 69-74
18. Huang, J. et al *Water Treat.* 1989, **4**(4), 441-447
19. Sharma, S. R. et al *Ecotoxicol. Environ. Saf.* 1987, **14**(1), 22-29
20. Dotson, D. A. et al *Report* NRL-MR-5781; AD-A16808-1/8/GAR, 1986, Natl. Tech. Inf. Ser., Springfield, VA
21. Chen, Z. M. et al *Ind. Eng. Chem. Prod. Res. Dev.* 1984, **23**, 5-11
22. Wang, T. C. et al *J. Assoc. Off. Anal. Chem.* 1991, **74**(5), 883-886
23. Sanchez, A. J. et al *Methodol. Aspects Study Pestic. Behav. Soil [Workshop Proc.]* 1988, 13-19 (Fr.) (*Chem. Abstr.* **111**, 148838k)
24. Jury, W. A. et al *J. Environ. Qual.* 1987, **16**(4), 422-428
25. Kahn, et al *Clay Res.* 1988, **7**(1-2), 5-10
26. Izmerov, N. F. et al *Toxicometric Parameters of Industrial Toxic Chemicals Under Single Exposure* 1982, 56, CIP, Moscow
27. Schafer, E. W. et al *Arch. Environ. Contam. Toxicol.* 1983, **12**, 355-382
28. *Gig. Sanit.* 1986, **51**(3), 73
29. *Arzneim.-Forsch.* 1972, **22**, 1926
30. *Proc. Soc. Exp. Biol. Med.* 1968, **129**, 699

31. *Osaka Tgaku Zasshi* 1959, **71**, 6099
32. *Arzneim.-Forsch.* 1955, **5**, 626
33. Rosical, L. et al *Bratisl. Lek. Listy* 1958, **38**(1), 151-160 (*Chem. Abstr.* **52**, 16618a)
34. *IARC Monograph* 1983, **30**, 103-129
35. Hill, E. F. et al *Lethal Dietary Toxicity of Environmental Pollutants to Birds* US Fish and Wildlife Service, Report Wildlife No. 191, 1975, Washington, DC
36. Li, S. et al *Tongji Yike Daxue Xuebao* 1987, **16**(2), 105-108 (Ch.) (*Chem. Abstr.* **107**, 170250b)
37. Balasubramanian, K. et al *IRCS Med. Sci* 1986, **14**(11), 1139-1140 (*Chem. Abstr.* **106**, 45437w)
38. Natl. Inst. Occ. Saf. Health *Criteria for Recommended Standard – Occ. Exp. to Malathion* 1976, US Dept. Health, Ed. & Welfare, Washington, DC
39. *IARC Monograph* 1987, **Suppl. 7**, 65
40. *National Toxicology Program Research and Testing Division* 1992, Report Nos. TR-0241, TR-192, NIEHS, Research Triangle Park, NC
41. Natl. Cancer Inst. *Bioassay of Malathion for Possible Carcinogenicity (Tech. Report Ser. No. 214: DHEW Publ. No. (NIH) 78-824)* 1978, US Dept. Health, Ed. & Welfare, Washington, DC
42. Natl. Cancer Inst. *Bioassay of Malathion for Possible Carcinogenicity (Tech. Report Ser. No. 192, DHEW Publ. No. (NIH) 79-1748)* 1979, US Dept. Health, Ed. & Welfare, Washington, DC
43. Snawdr, J. E. et al *Life Sci.* 1990, **46**(23), 11635-42
44. Halow, W. et al *Nature (London)* 1961, **192**, 464-465
45. Kimbrough, R. D. et al *Arch. Environ. Health* 1968, **16**, 805-808
46. Huang, C. C. *Proc. Soc. Exp. Biol. Med.* 1973, **142**, 36-40
47. Machin, M. G. A. et al *J. Toxicol. Environ. Health* 1989, **26**(3), 249-253
48. Arevalo, S. J. et al *Rev. Med. Chile* 1987, **115**(1), 37-46 (*Chem. Abstr.* **107**, 34826s)
49. Muan, B. et al *J. Agric. Food Chem.* 1989, **37**(1), 210-213
50. Wong, P. K. et al *Chemosphere* 1989, **18**(11-12), 2413-2422
51. Houk, O. S. et al *Mutat. Res.* 1987, **182**(4), 193-201
52. Velazquez, A. et al *Environ. Mutagen.* 1987, **9**(3), 343-348
53. Galloway, S. M. et al *Environ. Mol. Mutagen.* 1987, **10**(Suppl. 10), 1-175
54. Nicholas, A. H. et al *Mutat. Res.* 1979, **67**, 167-172
55. Herath, J. H. et al *Cytologia* 1989, **54**(1), 191-195
56. Bulsiewiez, H. et al *Folia Morphol. (Warsaw)* 1976, **35**, 361-368
57. Kumar, D. et al *Cytologia* 1989, **54**(3), 547-552
58. Rups, D. S. et al *Environ. Mol. Mutagen.* 1991, **18**(2), 136-138
59. Taylor, P. *Anticholinesterase Agents* in *Goodman and Gilman's The Pharmacological Basis of Therapeutics* 6th ed., 1980, 100-119, Macmillan, Oew York
60. Zhou, Y. et al *Zhonghua Yufang Yixue Zazhi* 1989, OP23(2), 74-76 (*Chem. Abstr.* **111**, 91799a).
61. *EC Directive Relating to the Quality of Water Intended for Human Consumption* 1982, 80/778/EEC, Office for Official Publications of the European Communities, 2 rue Mercier, L-2985 Luxembourg
62. *S. I. 1991 No. 472 The Environmental Protection (Prescribed Processes and Substances) Regulations* 1991, HMSO, London
63. *DoE Guidance on Safeguarding the Quality of Public Water Supplies* 1989, HMSO, London
64. Zimmerman, N. *Comments Toxicol.* 1990, **4**(1), 39-58
65. Santodonato, J. *Centre for Chemical Hazard Assessment – Report* 1985, SRC-TR-84-1049, Order No. PB86-147246/GAR, New York
66. *Dangerous Prop. Ind. Mater. Rep.* 1987, **7**(5), 63-75
67. Adnikari, M. et al *Clay Res.* 1986, **5**(1), 46-48
68. Ueda, S. et al *Joshi Eiyo Daigaku Kiyo* 1990, **21**, 176-186 (Japan.) (*Chem. Abstr.* **114**, 2423415d)

# M11  Maleic acid, dimethyl ester

$$H_3CO_2C \diagup\!\!\!= CO_2CH_3$$

**CAS Registry No.** 624-48-6
**Synonyms** 2-butenedioic acid (Z)-, dimethyl ester; dimethyl (Z)-butenedioate; dimethyl *cis*-ethylenedicarboxylate; dimethyl maleate; methyl maleate
**Mol. Formula** $C_6H_8O_4$          **Mol. Wt.** 144.13

## Physical properties

**M. Pt.** −17.5°C; **B. Pt.** 205°C; **Flash point** 91°C; **Specific gravity** 1.153; **Volatility** v.p. 1 mmHg at 45.7°C; v. den. 4.97.

## Mammalian and avian toxicity

**Acute data**
$LD_{Lo}$ oral rat 1410 mg $kg^{-1}$ (1).
$LD_{50}$ dermal rabbit 530 mg $kg^{-1}$ (1).

## References

1.  *Am. Ind. Hyg. Assoc. J.* 1962, **23**, 93

# M12  Maleic anhydride

$$O \diagdown\!\!= O \diagup\!\!\!\diagdown O$$

**CAS Registry No.** 108-31-6
**Synonyms** 2,5-furandione; *cis*-butenedioic anhydride; dihydro-2,5-dioxofuran; Toxilic anhydride
**Mol. Formula** $C_4H_2O_3$          **Mol. Wt.** 98.06
**Uses** Dye intermediate. Manufacture of copolymers, pharmaceuticals and pesticides. In Diels-Alder synthesis.

## Physical properties

**M. Pt.** 52.8°C; **B. Pt.** 202°C; **Flash point** 107°C (closed cup); **Specific gravity** $d_4^{20}$ 1.48; **Volatility** v.p. 1 mmHg at 44°C; v. den. 3.4.

**Solubility**
Water: miscible. Organic solvent: acetone, benzene, chloroform, ethanol, ethyl acetate, 1,4-dioxane, ligroin, toluene, xylene

## Occupational exposure

**US TLV (TWA)** 0.25 ppm (1 mg $m^{-3}$); **UK Long-term limit** 0.25 ppm (1 mg $m^{-3}$);

UN No. 2215; **HAZCHEM Code** 2X; **Conveyance classification** corrosive substance; **Supply classification** harmful(≥1%).

**Risk phrases** ≥25% – Harmful if swallowed – Irritating to eyes, respiratory system and skin – May cause sensitisation by inhalation – ≥15%<25% – Harmful if swallowed – Irritating to eyes, respiratory system and skin – May cause sensitisation by inhalation – ≥1%<10% – Harmful if swallowed – May cause sensitisation by inhalation (R22, R36/37/38, R42, R22, R36/37/38, R42, R22, R42)

**Safety phrases** Do not breathe dust – After contact with skin, wash immediately with plenty of water – Wear eye/face protection (S22, S28, S39)

## Ecotoxicity

**Fish toxicity**

$LC_{50}$ (96 hr) mosquito fish 240 mg $l^{-1}$ (1).
$LC_{50}$ (24 hr) bluegill sunfish 150 mg $l^{-1}$ (2).

**Invertebrate toxicity**

$EC_{50}$ (30 min) *Photobacterium phosphoreum* 44 ppm, Microtox test (3).

## Environmental fate

**Degradation studies**

99% removal by activated sludge in 4 hr. In view of the rapid hydrolysis, the reported biodegradation probably relates to maleic acid (1,4).

**Abiotic removal**

Hydrolyses rapidly to form maleic acid in water with a $t_{1/2}$ 0.37 min at 25°C (5).
In the vapour phase undergoes complete hydrolysis in 21 hr at 96% relative humidity; whereas, no hydrolysis occurs at 50% relative humidity (6).
Estimated $t_{1/2}$ 1.7 hr for reaction with photochemically produced hydroxyl radicals and ozone in the atmosphere (7).

## Mammalian and avian toxicity

**Acute data**

$LD_{50}$ oral rat, mouse, guinea pig, rabbit 390-875 mg $kg^{-1}$ (8-10).
$LD_{50}$ dermal rabbit 2620 mg $kg^{-1}$ (11).
$LD_{50}$ intraperitoneal rat 97 mg $kg^{-1}$ (8).

**Sub-acute data**

Inhalation rat, hamster, monkey (6 month) 0, 1.1, 3.3 or 9.8 mg $m^{-3}$ 6 hr $day^{-1}$ for 5 days $wk^{-1}$ caused dose related nasal and ocular irritation, and a reduction in body weight gain. No significant treatment related mortality was observed in any species (12).

**Carcinogenicity and long-term effects**

Oral rat (2 yr) 0, 10, 32 or 100 mg $kg^{-1}$ $day^{-1}$ caused a significant decrease in red blood cell count in all treated ♂ rats at 6 month and in high dose ♀ rats after 12 month. Haematocrit decreased in ♂ rat after 6 month, and thyroid clear cell adenomas and hypoplasia was observed in treated ♀ rats. There was no significant effect on body weight, mortality, neurology, opthalmology or urinalysis (13).

**Teratogenicity and reproductive effects**

Oral rat, 0, 30, 90 or 140 mg $kg^{-1}$ $day^{-1}$ on day 6-15 of gestation. No teratogenic effects were observed (14).

### Irritancy
1% solution instilled into rabbit eye caused severe irritation (exposure not specified) (15).

### Sensitisation
Asthma has been reported among exposed workers and in clinical studies (16,17).

## Genotoxicity
*Salmonella typhimurium* TA98, TA100, TA1535, TA1537, TA1538 with and without metabolic activation negative (18).
*In vivo* rat bone marrow, chromosomal aberrations negative (19).

## Any other adverse effects to man
Powerful irritant. Inhalation can cause pulmonary oedema (20).

## Any other comments
Toxicology and environment fate reviewed (21-24).
Human health effects, experimental toxicology, physico-chemical properties reviewed (25).
Autoignition temperature 477°C.

## References
1. Jones, H. R. *Environmental Control in the Organic and Petroleum Industries* 1971, Noyes Data Corporation
2. Turnbull, H. et al *Ind. Eng. Chem.* 1954, **46**(2), 324-333
3. Kaiser, K. L. E. et al *Water Pollut. Res. J. Can.* 1991, **26**(3), 361-431
4. Matsui, S. et al *Prog. Water Technol.* 1975, **7**, 645-659
5. Burton, C. A. et al *J. Chem. Soc.* 1963, 2918-2926
6. Rosenfeld, J. M. et al *Talenta* 1967, **14**, 91
7. Cupitt, L. T. *Fate of Toxic and Hazardous Materials in the Air Environment* 1980, US EPA-600/3-80-084
8. Ismerov, N. F. et al *Toxicometric Parameters of Industrial Toxic Chemicals Under Single Exposure* 1982, 79, CIP, Moscow
9. *Memorandum: Interagency Collaborative Group on Environmental Carcinogenesis* 17 June 1974, National Cancer Institute
10. *Gig. Tr. Prof. Zabol.* 1969, **13**, 42 (*Chem. Abstr.* **71**(17), 79284t)
11. *Toxicol. Appl. Pharmacol.* 1977, **42**, 417
12. Short, R. D. et al *Fundam. Appl. Toxicol.* 1988, **10**(3), 517-524
13. IIT Research Institute *Chronic Dietary Administration of Maleic Anhydride* 1983, **1**, EPA Document No. FYI-OTS-1283-0277
14. Short, R. D. et al *Fundam. Appl. Toxicol.* 1986, **7**(3), 359-366
15. *Am. J. Opthalmol.* 1946, **29**, 1363
16. Lee, H. S. et al *Br. J. Ind. Med.* 1991, **48**(4), 283-285
17. Graneek, B. J. et al *Clin. Respir. Physiol.* 1987, **23**(6), 577-581
18. Monsanto Co. *Mutagenicity Plate Assay: Maleic Anyhydride* 1977, EPA Document No. 878214770
19. Hazelton Laboratories America Inc. *In Vivo Bone Marrow Chromosome Study in Rats (Inhalation Exposure)* 1983, US EPA Document No. 878214783
20. *The Merck Index* 11th ed., 1989, 895-896, Merck & Co, Rahway, NJ
21. *Dangerous Prop. Ind. Mat. Rep.* 1990, **10**(4), 9-25
22. Randall, D. J. et al *Acute Toxicity Data* 1990, **1**(1), 75-76
23. *Chemical Safety Data Sheets* 1990, **3**, 152-155, RSC, London

24.  Howard, P. H. (Ed.) *Handbook of Environmental Fate and Exposure Data for Organic Chemicals* 1989, **1**, 382-385, Lewis Publishers, Chelsea, MI

25.  *EC Directive Relating to the Quality of Water Intended for Human Consumption* 1982, 80/778/EEC, Office for Official Publications of the European Communities, 2 rue Mercier, L-2985 Luxembourg

# M13   Maleic hydrazide

**CAS Registry No.** 123-33-1

**Synonyms** 1,2-dihydro-3,6-pyridazinedione; Antergon; 3,6-dihyroxypyridazine; 6-hydroxy-2*H*-pyridazin-3-one; 1,2-dihydropyridazine-3,6-one; 6-hydroxy-3(2*H*)-pyridazinone; MH; malic acid hydrazide; Malepin; Malzid

**Mol. Formula** $C_4H_4N_2O_2$                    **Mol. Wt.** 112.09

**Uses** Plant growth inhibitor and herbicide, principal use is in the control of sucker growth on tobacco, also tested to retard growth of trees, hedges and grass and sprouting of carrots, onions, beets, rutabagas and potatoes.

## Physical properties

**M. Pt.** 310-312°C; **B. Pt.** 260°C (decomp.); **Specific gravity** $d^{20}$ 1.60; **Partition coefficient** log $P_{ow}$ −1.959 (1); **Volatility** v.p. 0 mmHg at 50°C.

**Solubility**

Water: 6 g $l^{-1}$. Organic solvent: ethanol, dimethyl formamide

## Ecotoxicity

**Fish toxicity**

$LC_{50}$ (96 hr) rainbow trout, bluegill sunfish 1400-1600 mg $l^{-1}$ (1).

$LC_{50}$ (96 hr) harlequin fish 100 mg $l^{-1}$ (2).

Trout, bluegill sunfish, goldfish 5 ppm caused death in 8-12 hr. Test conditions: pH 7.0; dissolved oxygen content 7.5 ppm; total hardness (soap method) 300 ppm; free carbon dioxide 5 ppm; and temperature 12.8°C (3).

**Invertebrate toxicity**

$LC_{50}$ (96 hr) *Daphnia* sp. 107 mg $l^{-1}$ (4).

## Environmental fate

**Degradation studies**

In various soils from eastern, southern and midwestern USA, $t_{1/2}$ 14-100 days, may be related to the soils organic content. Mircrobial degradation appeared to be rapid (5).

**Abiotic removal**

Stable to hydrolysis at pH 3, 6, 9 (5).

Rapid photochemical degradation occurs in water (1).

### Absorption
Highly mobile in un-aged soils. Aerobic ageing of maleic hydrazide lowered the leaching potential (5).

## Mammalian and avian toxicity

### Acute data
$LD_{50}$ oral rat 3800-6800 mg kg$^{-1}$ (6).
$LD_{50}$ oral mallard duck, bobwhite quail >10,000 mg kg$^{-1}$ (1,4).
$LC_{50}$ (1 hr) inhalation rat 20 mg l$^{-1}$ (1,4).
$LD_{50}$ dermal rabbit >4000 mg kg$^{-1}$ (6).

### Sub-acute data
$LD_{50}$ (15 day) oral rat ♂ 6300 kg$^{-1}$, ♀ 6680 mg kg$^{-1}$ (7).

### Carcinogenicity and long-term effects
No adequate evidence for carcinogenicity to humans or animals, IARC classification group 3 (8).
Mouse (3 wk) gavage 1000 mg kg$^{-1}$ followed by (18 month) oral in feed *ad libitum* 3000 ppm. No evidence of increased tumour incidence compared to controls (9).
Oral (100 wk) rats 1% in diet no increase in number of tumours compared to the controls (10).
Oral (2 yr) rat 30 g kg$^{-1}$ diet (technical potassium salt) caused no adverse effects (4).
Subcutaneous (100 wk) mice, rats 500 mg kg$^{-1}$ wkly no difference in tumour incidence as compared to controls (10,11).

### Teratogenicity and reproductive effects
Four generation reproduction study in rats of feeding 5000, 10,000, 20,000, 50,000 ppm of sodium salt, showed no effects on fertility, lactation or other reproductive parameters (7).
Gavage rats (6-15 day of gestation) 0, 400, 800, 1200, 1600 mg kg$^{-1}$ rats killed on day 22 of gestation. No adverse maternal, foetotoxicity or teratogenic effects were observed (12).
Gavage rabbit (7-27 day of gestation) 0, 100, 300, 1000 mg kg$^{-1}$ day$^{-1}$. No maternal toxicity. 300 or 1000 mg kg$^{-1}$ caused malformed scapulae in foetuses indicating a no observed effect level of 100 mg kg$^{-1}$ day$^{-1}$ for developmental effects (7).

### Metabolism and pharmacokinetics
Oral rabbit 100 mg kg$^{-1}$ 43-62% excreted in urine unchanged within 48 hr (10).
Oral rat [$^{14}$C]-labelled maleic hydrazide, after 6 days only 12% had been excreted in faeces suggesting the remainder had been absorbed (13).
Pregnant rats, 0.5, 5.0 mg kg$^{-1}$, mothers killed on day 20 gestation and foetuses examined or mothers allowed to litter and feed pups before pups were killed and stomach contents examined. Maleic hydrazide can cross the placenta and is transmitted to pups via milk (7).
Oral rat (6 day) [$^{14}$C]-labelled maleic hydrazide , 0.2% recovered in expired carbon dioxide. 77% was recovered in urine which contained 92-94% of the unchanged compound and 6-8% in the form of conjugates of maleic hydrazide, and 12% was recovered in faeces. Only trace amounts were detected in the blood and tissues (13).

### Irritancy
Dermal rabbit (duration not specified) 0.5 ml caused mild irritation and 100 mg instilled into the eye caused no irritation (7).

---

## Genotoxicity

*Salmonella typhimurium* TA98, TA1535, TA1537 with and without metabolic activation negative (14).
*Escherichia coli* WP2*hcr* and *Salmonella typhimurium* TA98, TA100, TA1535, TA1537, TA1538 with and without metabolic activation negative (15).
*Bacillus subtilis* (metabolic activation unspecified) positive (16).
*In vitro* human lymphocytes and V97 Chinese hamster cells ≥100 ppm, dose related increases in sister chromatid exchanges (17).
*In vivo* intraperitoneal 100, 200 mg kg$^{-1}$ no effect on bone marrow erythrocyte micronuclei or the ratio of poly- to normo-chromatic erythrocytes (18).

## Any other adverse effects

Dermal (14 day) rabbit single dose 20 g kg$^{-1}$. 2 of 5 ♂ and 1 of 5 ♀ rabbits died on day 1. The rabbits that died were comatose and showed ataxia and shallow respiration (7).

## Legislation

Limited under EC Directive on Drinking Water Quality 80/778/EEC. Pesticides: maximum admissible concentration 0.1 μg l$^{-1}$ (19).
Included in Schedule 6 (Release into Land: Prescribed Substances) Statutory Instrument No. 472, 1991 (20).

## Any other comments

Mode of action in plants: inhibited meristemic cell division but had no effect on cell elongation (1).

## References

1. *The Agrochemicals Handbook* 3rd ed., 1991, RSC, London
2. Tooby, T. E. et al *Chem. Ind.* 1975, **21**, 523-525
3. Wood, E. M. *The Toxicity of 3400 Chemicals to Fish* 1987, EPA 560/6-89-001 PB 86-156 715, Washington DC
4. *The Pesticide Manual* 9th ed., 1991, British Crop Protection Council, Farnham
5. *Herbicide Handbook* 5th ed., Weed Society of America, Champaign, IL
6. Ben-Dyke, R. et al *World Rev. Pest Control* 1970, **9**, 119
7. *Drinking Water Health Advisory – Pesticides* 1989, 487-504, Lewis Publishers, Chelsea, MI
8. *IARC Monograph* 1987, **Suppl. 7**, 65
9. Innes, J. R. M. et al *J. Natl. Cancer Inst.* 1969, **42**, 1101
10. Barnes, J. M. et al *Nature* 1957, **180**, 62
11. Dickens, F. et al *Br. J. Cancer* 1965, **19**, 392
12. Khera, K. S. C. et al *J. Environ. Sci. Health, Part B* 1979, **14**, 563-577
13. Mays, D. L. et al *J. Agric. Food Chem.* 1968, **16**, 356-357
14. McCann, J. et al *Proc. Natl. Acad. Sci. U.S.A.* 1975, **72**(12), 1111-1117
15. Moriya, M. et al *Mutat. Res.* 1983, **116**, 185-216
16. Shiau, S. Y. et al *Mutat. Res.* 1980, **71**, 169-179
17. Sabharwal, P. S. et al *In Vitro* 1980, **16**(3), 205
18. Chanbey, R. C. et al *Mutat. Res.* 1978, **57**, 187-191
19. *EC Directive Relating to the Quality of Water Intended for Human Consumption* 1982, 80/778/EEC, Office for Official Publications of the European Communities, 2 rue Mercier, L-2985 Luxembourg
20. *S. I. 1991 No. 472 The Environmental Protection (Prescribed Processes and Substances) Regulations* 1991, HMSO, London

# M14   Malic acid

$$HO_2CCH_2CH(OH)CO_2H$$

**CAS Registry No.** 6915-15-7
**Synonyms** hydroxybutanedioic acid; deoxyteraric acid; hydroxysuccinic acid;
Pomalus Acid
**Mol. Formula** $C_4H_6O_5$                                  **Mol. Wt.** 134.09
**Occurrence** Trace amounts detected in forest soils of S.E. USA (1). Found in many
fruits.

## Physical properties

**M. Pt.** 131-132°C (D-, L-form); **B. Pt.** decomp.; **Specific gravity** 1.595;
**Partition coefficient** log $P_{ow}$ −1.26 (2).

## Solubility

Water: 558 g $l^{-1}$. Organic solvent: methanol, diethyl ether, ethanol, acetone

## Environmental fate

### Degradation studies

Activated sludge process treatment at 20°C 120 hr, 70-100% was removed (3).
$BOD_5$ 0.468 mg $l^{-1}O_2$, $BOD_5$ 0.34-0.57 mg $l^{-1}O_2$ Warburg sewage (2).
COD 0.68; 0.70 mg $l^{-1}O_2$ (2,4).
Activated sludge 6, 12, 24 hr (L-isomer)9.6%, 22.9%, 44,8% ThOD, (D-isomer)6.0%,
4.6%, 20.8% ThOD (5).

## Mammalian and avian toxicity

### Acute data

$LD_{Lo}$ oral rat 1600 mg $kg^{-1}$ (6).

### Irritancy

Dermal rabbit (24 hr) caused moderate irritation and 750 µg instilled into the eye
(24 hr) caused severe irritation (7).

## Genotoxicity

*Salmonella typhimurium* TA97, TA98, TA100, TA104 with and without metabolic
activation negative (8).

## Any other comments

Toxicity reviewed (9).

## References

1.  Fox, T. R. et al *Soil Sci. Soc. Am. J.* 1990, **54**, 1139-1144
2.  Verschueren, K. *Handbook of Environmental Data on Organic Chemicals* 2nd ed., 1983,
    Van Nostrand Reinhold, New York
3.  Muto, N. et al *Kenkyu Hokoku – Kanto Gakuin Daigaku Kogakubu* 1987, **31**(2), 257-66
4.  Lund, H. F. *Industrial Pollution Control Handbook* 1971, McGraw-Hill, New York
5.  Maloney, G. W. et al *J.- Water Pollut. Control Fed.* 1969, **41**(2), R-18-R33
6.  Patty, F. A. (Ed.) *Industrial Hygiene and Toxicology* 2nd rev. ed., 1953-63, Interscience
    Publishers, New York

7. Marhold, J. V. *Sbornik Vysledku Toxicologickeho Vysetreni Latek A Pripravku* 1972, Prague
8. Al-Ani, F. Y. et al *Mutat. Res.* 1988, **206**(4), 467-470
9. *BIBRA Toxicity Profile* 1991, British Industrial Biological Research Association, Carshalton

# M15   Malonic acid

## HO₂CCH₂CO₂H

**CAS Registry No.** 141-82-2
**Synonyms** propanedioic acid; carboxyacetic acid; dicarboxymethane; methanedicarboxylic acid
**Mol. Formula** $C_3H_4O_4$ **Mol. Wt.** 104.06
**Uses** Manufacture of barbiturates.

## Physical properties

**M. Pt.** ≈135°C (decomp.); **B. Pt.** sublimes *in vacuo*.

### Solubility

Water: 1538 g $l^{-1}$. Organic solvent: ethanol, methanol, propyl alcohol, diethyl ether, pyridine

## Environmental fate

### Degradation studies

$BOD_{25}$ 2, 5, 10, 20 days 0.31, 0.36, 0.52, 0.53 mg $l^{-1}O_2$, respectively, with a substance concentration of 13.5 mg $l^{-1}$ and an inoculum of soil microorganisms (1). Waste water treatment activated sludge 6, 12, 24 hr: 1.2% ThOD, toxic, 0.9% ThOD, respectively (2).

## Mammalian and avian toxicity

### Acute data

$LD_{50}$ oral rat, mouse 1310, 4000 mg $kg^{-1}$, respectively (3,4).
$LD_{50}$ intraperitoneal mouse 300 mg $kg^{-1}$ (5).

### Irritancy

Dermal (24 hr) rabbit 500 mg caused mild irritation and 100 mg caused severe eye irritation (duration unspecified) (3).

## Any other adverse effects

Salamander embryo at blastula stage 15-20 ppm induced abnormal development (6).

## References

1. Hammond, M. W. et al *Environ. Sci. Technol.* 1972, **6**(8), 732-735
2. Malaney, G. W. et al *J.- Water Pollut. Control Fed.* 1969, **41**(2), R18-R33
3. *BIOFAX, Data Sheets* 1971, 22-3, Industrial Bio-Test Laboratories, Inc., Northbrook, IL
4. *Biochem. J.* 1940, **34**, 1196
5. *NTIS Report AD277-689* Natl. Tech. Inf. Ser. Springfield, VA
6. Milkami, Y. *Gosei Senzai Kenkyukaishi* 1987, **10**(2), 43-4 (*Chem. Abstr.* **109**, 2110d)

# M16   Malononitrile

## CNCH2CN

**CAS Registry No.** 109-77-3
**Synonyms** propanedinitrile; cyanoacetonitrile; dicyanomethane; malonic acid
dinitrile; methylene cyanide; methylenedinitrile
**Mol. Formula** $C_3H_2N_2$                         **Mol. Wt.** 66.06
**Uses** Used in organic synthesis. Leaching gold from gold ores. Vitamin $B_1$ synthesis.

## Physical properties

**M. Pt.** 32°C; **B. Pt.** 218-219°C; **Flash point** 130°C (open cup); **Specific gravity**
$d_4^{20}$ 1.910.

### Solubility
Organic solvent: ethanol, diethyl ether, acetone, benzene

## Occupational exposure

**UN No.** 2647; **HAZCHEM Code** 2X; **Conveyance classification** toxic substance;
**Supply classification** toxic.
**Risk phrases** Toxic by inhalation, in contact with skin and if swallowed (R23/24/25)
**Safety phrases** Do not breathe vapour – Take off immediately all contaminated
clothing (S23, S27)

## Ecotoxicity

### Fish toxicity
$LC_{50}$ (12, 24, 48, 96 hr) rainbow trout 19.4, 6.2, 4.2, 1.6 mg $l^{-1}$, respectively (hard
water) (1).

## Mammalian and avian toxicity

### Acute data
$LD_{50}$ oral mouse, rat 19, 61 mg $kg^{-1}$, respectively (2,3).
$LD_{50}$ intravenous rabbit, mouse 28, 32 mg $kg^{-1}$, respectively (4,5).
$LD_{50}$ intraperitoneal mouse 12.9 mg $kg^{-1}$ (6).
$LD_{Lo}$ subcutaneous rabbit, dog, mouse 6-8 mg $kg^{-1}$ (7,8).

### Irritancy
Severe eye irritant and causes local skin irritation (species unspecified) (9).

## Genotoxicity

*Salmonella typhimurium* TA97, TA98, TA100, TA1535 with and without metabolic
activation negative (10).

## Any other adverse effects to man

Vapour inhalation can cause methaemoglobinaemia (9).

## Any other comments

EPA requiring and/or recommending testing for human health effects and for
chemical fate to establish levels at which waste is no longer hazardous (11).

---

May polymerise violently at 130°C. Prolonged exposure to the air and ultra-violet radiation will result in hydrogen cyanide release (9).

## References

1. Abram, F. S. H. et al *Water Res.* 1979, **13**, 631-635
2. *Khig. Zdraveopaz.* 1966, **39**, 50
3. Marhold, J. V. *Sbornik Vysledku Toxixologickeho Vysetreni Latek A Pripravku* 1972, Prague
4. *US Army Armament Research and Development Command* NX 07576, Chemical Systems Laboratory Aberdeen Proving Ground, MD
5. *Pol. J. Pharmacol. Pharm.* 1979, **31**, 563
6. Jones, G. N. R. *Nature* London 1970, **228**, 1315
7. *C. R. Seances Soc. Biol. Ses Fil.* 1927, **96**, 202
8. *Arch. Int. Pharmacodyn. Ther.* 1897, **3**, 77
9. *Chemical Safety Data Sheets* 1991, **4b**, 4-5 RSC, London
10. Zeiger, E. et al *Environ. Mol. Mutagen.* 1988, **11**(Suppl.12), 1-157
11. EPA *Fed. Regist.* 15 June 1988, **53**(115), 22300-22325

# M17   Mancozeb

**CAS Registry No.** 8018-01-7
**Synonyms** manganese, [[1,2-ethanediylbis[carbamodithioato]](Z-)]-, mixture with [[1,2-ethanediylbis[carbamodithioato]](2-)]zinc; Manzeb; dithane; Karamate; Zimanat
**Uses** Fungicide for fieldcrops, nuts, vegetables, fruits and ornamentals.

## Physical properties

**M. Pt.** 192-194°C (decomp. without melting); **Flash point** 137.8°C (open cup);
**Volatility** v.p. negligible at 20°C.

### Solubility
Water: 6-20 mg $l^{-1}$.

## Ecotoxicity

### Fish toxicity
$LC_{50}$ (48 hr) rainbow trout, guppy, brown bullhead, goldfish 1.9-7.7 mg $l^{-1}$ (1,2).
$LC_{50}$ (48 hr) carp 24 mg $l^{-1}$ (1).

## Environmental fate

### Anaerobic effects
*Methanosarcina barkeri* methanogenesis from glucose greatly inhibited by 100 mg $l^{-1}$ (3).

## Degradation studies

$t_{1/2}$ 15 days in unsterilised soil, initial concentration 3 ppm, after which the rate of loss slowed. In sterilised soil the rate of loss was slower for the first 15 days but subsequent rate of loss was similar in both cases. Hence micro-organisms play an important role in degradation but long term persistence is significant even in the presence of microflora (4).

### Abiotic removal

Decomposes on prolonged exposure to air or moisture. Rapidly hydrolysed in acidic media (5).

## Mammalian and avian toxicity

### Acute data

$LD_{50}$ oral rat 5-7.5 g $kg^{-1}$ (5,6).
$LD_{50}$ dermal rabbit, rat >5, >10 g $kg^{-1}$, respectively (2,5).

### Sub-acute data

Rat (17 day) 300 mg $kg^{-1}$ (route unspecified) caused hind limb paralysis indicating motor neuropathy (7).
Oral (12 wk) ♂ Wistar rats 0, 10, 50, 75, 113, 169, 253, 379 mg $kg^{-1}$ with feed. 379 mg $kg^{-1}$ caused the death of a third of the rats. ≥169 mg $kg^{-1}$ decreased growth and nutrient utilisation and the weight of kidneys, adrenals and testes were increased in the 2 highest dose groups. Liver detoxifying function was also reduced. ≥75 mg $kg^{-1}$ increased liver and thyroid weights. Even very low doses impaired thyroid function (8).
Oral (10 day) Japanese quail, mallard duck 3200-6400 mg $kg^{-1}$ $day^{-1}$ no mortalities (2,5).

### Carcinogenicity and long-term effects

Dermal (60 wk) ♀ mice 100 mg 3 × $wk^{-1}$. Development of tumours observed after 31 wk mainly benign squamous cell papillomas and keratoacanthomas. High mortality was observed after 54 wk due to mancozeb toxicity (9).
Dermal ♀ Swiss albino mice initiated with a single sub-carcinogenic dose of 7, 12-dimethylbenz[a]anthracene (52 μg). 7 days after initiation, mice were treated with mancozeb 100 mg $kg^{-1}$ 3 × $wk^{-1}$. 100% tumorigenisis was recorded after 17 wk of treatment, examination showed most of these tumours to be benign (10).

### Teratogenicity and reproductive effects

Gavage ♂, ♀ rats (8 wk) 70, 140, 280, 350, 700 mg $kg^{-1}$ 6 × $wk^{-1}$. High doses caused reduction in maternal body weight, paralysis of hind legs and dose-dependent mortality. Changes in fertility parameters were minimal. No difference in fertility parameters in $F_1$ generation (11).
High levels have caused birth defects in test animals (details unspecified) (2).
Vineyard workers (couples) exposed to pesticides including mancozeb showed an increased frequency of miscarriages and still births compared to controls, and an increase in the percentage of chromosomal aberrations (12).
Intragastric mice 100-1000 mg $kg^{-1}$ increased number of abnormal sperm, testes weight and histology were normal (13).

### Metabolism and pharmacokinetics

The major metabolite (species unspecified) is ethylenethiourea comprising almost 24% of the bioavailable dose in urine and bile. Ethylenethiourea residues were detected at levels of 1ppm in the thyroid and liver, by 24 hr residues were undetectable (14).

**Irritancy**

Pesticide patch test 1% (72 hr) on 200 volunteers (50 were agricultural workers) negative (15).

## Genotoxicity

*Salmonella typhimurium* TA92, TA98, TA100, TA102, TA1535, TA1537, TA2637 with and without metabolic activation negative (16,17).
*Aspergillus nidulans* (activation unspecified) induction of point mutation positive (18).
*In vitro* human peripheral blood lymphocytes with metabolic activation unscheduled DNA synthesis and sister chromatid exchanges negative, without activation unscheduled DNA synthesis positive and there was a dose-dependent inhibition of thymidine uptake (19).
Occupationally exposed workers *in vitro* short term peripheral lymphocyte cultures. Mancozeb exposure was associated with significant increases in the frequencies of structural chromosomal aberrations and sister chromatid exchanges (20).
*In vivo* mouse bone marrow cells chromosomal aberrations and physiological effects such as end-to-end chromosomal associations, clumping, uneven stretching of chromatin material, were observed (21).

## Any other adverse effects to man

Cytogenetic analysis of workers exposed during manufacture showed increased chromosomal aberrations especially in ♀ packers of the final product and increased immunoglobulins I and G and α-2-microglobulin (22).

## Legislation

Limited under EC Directive on Drinking Water Quality 80/778/EEC. Pesticides: maximum admissible concentration $0.1 \, \mu g \, l^{-1}$ (23).
Included in Schedule 6 (Release into Land: Prescribed Substances) Statutory Instrument No. 472, 1991 (24).

## Any other comments

Toxicity and occupational limit values reviewed (25).
Dissolves in powerful chelating agents but cannot be recovered from them (2).
$LD_{50}$ honeybee 0.193 mg bee$^{-1}$ (2).
Tolerable daily intake (TDI) human 0.05 mg kg$^{-1}$ body weight (2).

## References

1. Hejduk, J. et al *Acta. Vet. Brno* 1980, **49**, 251
2. *The Pesticide Manual* 9th ed., 1993, British Crop Protection Council, Farnham
3. Khalil, E. F. et al *Environ. Technol.* 1991, **12**(6), 471-6
4. Gennari, U. et al *Def. Veg.* 1988, **42**(252), 19-20 (Fr.) (*Chem. Abstr.* **110**, 10016g)
5. Lewis, M. A. et al *Registry of toxic effects of chemical substances* 1984, National Institute for Occupational Health and Safety No.83-107-4
6. *The Agrochemicals Handbook* 3rd ed., 1991, RSC, London
7. Csorba-Caszlo, E. et al *Egeszsegtudomany* 1989, **33**(2), 125-134 (Hung.) (*Chem. Abstr.* **112**, 153392k)
8. Szepoolgyi, J. et al *Food Chem. Toxicol.* 1989, **27**(8), 531-538

9. Shukla, Y. et al *Cancer Lett. (Shannon, Irel.)* 1990, **53**(2-3), 191-195
10. Shukla, Y. et al *Carcinogenesis (London)* 1988, **9**(8), 1511-1512
11. Fay, E. et al *Egeszsegtudomany* 1989, **33**(1), 51-57 (Hung.) (*Chem. Abstr.* **111**, 148444d)
12. Rita, P. et al *Environ. Res.* 1987, **44**(11), 1-5
13. Jablonicka, A. et al *Bratisl. Lek. Listy* 1988, **89**(8), 611-614 (Slo.) (*Chem. Abstr.* **110**, 35053m)
14. *National Pesticide Information Retrieval System, Mancozeb Fact Sheet* 1987, No.125, Purdue University
15. Lisi, P. et al *Contact Dermatitis* 1986, **15**(5), 266-269
16. Remondelli, P. et al *Med. Biol. Environ.* 1986, **14**(1), 377-386
17. Choi, E. J. et al *Environ. Mutagen. Carcinog.* 1985, **5**(1), 11-18
18. Martinez-Rossi, N. et al *Mutat. Res.* 1987, **176**(1), 29-35
19. Preocco, P. et al *Teratog., Carcinog., Mutagen.* 1989, **9**(2), 75-81
20. Jablonicka, A. et al *Mutat. Res.* 1989, **224**(2), 143-146
21. Gautman, D. C. et al *Experientia* 1991, **47**(3), 280-282
22. Vargova, M. et al *Prac. Lek.* 1988, **40**(10), 425-431 (Slo.) (*Chem. Abstr.* **111**, 63205b)
23. *EC Directive Relating to the Quality of Water Intended for Human Consumption* 1982, 80/778/EEC, Office for Official Publications of the European Communities, 2 rue Mercier, L-2985 Luxembourg
24. *S. I. 1991 No. 472 The Environmental Protection (Prescribed Processes and Substances) Regulations* 1991, HMSO, London
25. Chiesura, E. et al *Rapp. ISTISAN* 1990, **90/12** PE.2 (Ital.) (*Chem. Abstr.* **114**, 253236x)

# M18   D-Mandelic acid

**CAS Registry No.** 611-72-3
**Synonyms** (±)-α-hydroxybenzeneacetic acid; (±)-mandelic acid
**Mol. Formula** $C_8H_8O_3$                         **Mol. Wt.** 152.15

## Physical properties
**M. Pt.** 121-123°C; **Specific gravity** $d^{20}$ 1.30.

## Solubility
Organic solvent: ethanol, diethyl ether, chloroform

## Mammalian and avian toxicity

### Metabolism and pharmacokinetics
In rats the extent to which mandelic acid is metabolised to phenylglyoxylic acid is dependent on the enantiomeric composition of mandelic acid administered (1).

### Any other adverse effects to man
Occasional clinical side effects include giddiness, tinnitus, gastric disturbances, dysuria and haematuria. It should be avoided by patients with renal disfunction (2).

---

## Any other comments

Used in the treatment of urinary tract infections, usually as the ammonium or calcium salt (2).

## References

1. Drummond, L. et al *Xenobiotica* 1990, **20**(2), 159-168
2. *Martindale. The Extra Pharmacopoeia* 30th ed., 1993, 257, The Pharmaceutical Press, London

# M19    Maneb

$$Mn^{2+}-S$$

NHCH_2CH_2NHCS_2^-

**CAS Registry No.** 12427-38-2
**Synonyms** ethylene bis(dithiocarbamato)manganese; manganese ethylenebis(dithiocarbamate);[[1,2-ethanediylbis[carbamodithioato]](2-)] manganese; Amangan; Manzate; Nespor; Trimangol; MEB
**Mol. Formula** $(C_4H_6MnN_2S_4)_x$        **Mol. Wt.** $[265.29]_x$
**Uses** Fungicide.
**Occurrence** Residues have been isolated from crops (1).

## Physical properties

**M. Pt.** 192-204°C (decomp. without melting); **Specific gravity** $d^{20}$ 1.92; **Volatility** v.p. $<7.5 \times 10^{-8}$ mmHg at 20°C.

## Solubility

Organic solvent: chloroform, pyridine

## Occupational exposure

**UN No.** 2968 (stabilised); 2210; **Conveyance classification** substance which in contact with water emits flammable gas (stabilised); spontaneously combustible substance, substance which in contact with water emits flammable gas.

## Ecotoxicity

### Fish toxicity

Rainbow trout continuous exposure (dose unspecified) from fertilised egg to early fry state induced embryotoxicity and was teratogenic. Major effects, severe spinal and vertebral abnormalities mostly associated with retarded yolk sac resorption (2). $LC_{50}$ (96 hr) harlequin fish 0.53 mg l$^{-1}$ (3).

LC$_{50}$ (96 hr) guppy 3.7 mg l$^{-1}$ (4).
LC$_{50}$ (48 hr) carp 1.8 mg l$^{-1}$ (1).

**Invertebrate toxicity**
EC$_{50}$ (96 hr) reproduction *Chlorella pyrenoidosa* 3.2 mg l$^{-1}$ (5).
EC$_{50}$ (48 hr) *Daphnia magna*, brown shrimp 1-3.3 mg l$^{-1}$ (5,6).
LC$_{50}$ (21 day) *Daphnia magna* 0.11 mg l$^{-1}$ (5).
EC$_{50}$ (30 min) *Photobacterium phosphoreum* 3.31 ppm, Microtox test (7).

# Environmental fate

**Nitrification inhibition**
Minimum inhibitory concentration (3 hr) for *Microsomonas* and *Nitrobacter* 56 mg l$^{-1}$ (4).
*Nitrosomonas* in soil >960 mg l$^{-1}$ caused some inhibition of nitrification (8).

**Degradation studies**
Rapidly degraded in the environment by metabolism, hydrolysis, photolysis and oxidation (1).
t$_{1/2}$ in soil 6-36 day (9).

**Absorption**
Calculated K$_{oc}$ of >2000 indicates that maneb will adsorb to soil and sediments (10).

# Mammalian and avian toxicity

**Acute data**
LD$_{50}$ oral mouse, rat 2.6-3.0 g kg$^{-1}$ (11).
LD$_{50}$ oral mouse, rat 4-4.5 g kg$^{-1}$ (12).
LD$_{50}$ oral redwing blackbird >100 mg kg$^{-1}$ (13).
LC$_{50}$ inhalation (4 hr) rat >3.8 mg l$^{-1}$ (8).
LD$_{50}$ dermal rat, rabbit >5000 mg kg$^{-1}$ (8).

**Sub-acute data**
LC$_{50}$ (8 day) mallard duck, bobwhite quail >10,000 mg kg$^{-1}$ diet (1).
Inhalation (4.5 month) rat 2, 12, 135 mg m$^{-3}$ caused toxic and irritation in the trachea and lungs (14).

**Carcinogenicity and long-term effects**
No adequate evidence for carcinogenicity to humans or animals, IARC classification group 3 (15).
Gavage (78 wk) ♂ ♀ mouse, 46 mg kg$^{-1}$ at 7 days old and then the same amount not adjusted for body weight for up to 4 wk of age, for the rest of the period they were given 158 mg kg$^{-1}$ diet. No significant increase in tumour incidences recorded (16).
Gavage (9 month) mouse 6 wkly administrations of 500 mg kg$^{-1}$. 4 lung adenomas occurred compared to 0 in the untreated controls (17).
Gavage (22 month) rat 335 mg kg$^{-1}$ 2 × wk$^{-1}$ for life. 6/60 rats survived at 22 months. One rat developed a subcutaneous rhabdomyosarcoma and 1 developed a mammary carcinoma. 1/46 controls developed a fibrosarcoma (18,19).
Subcutaneous (78 wk) ♂, ♀ mouse 100 mg kg$^{-1}$ in 0.5% gelatine on day 28 of life. 70 mice survived to the end of the experiment (original number 72); no increased incidence of tumours observed (20).
Subcutaneous (22 month) mouse single injection of 12.5 mg kg$^{-1}$ in paraffin. Of 4/46 survivors at 22 months, 3 developed malignant tumours (2 fibrosarcomas and 1

thyroid carcinoma). One fibrosarcoma was seen in 46 controls which survived at 22 months (19).

**Teratogenicity and reproductive effects**
Doses of 0.002-0.1 $LD_{50}$ reduced fertility in both ♂ and ♀ mice characterised by disturbed spermatozoa maturation, low numbers or lack of differentiated spermatogenesis, autotrophic changes in Sertoli cells, suppressed ovopoiesis, and increased numbers of atresic follicles. No effect on thyroid activity was reported (14).
Chick embryos, eggs pre-incubation dipped in 0.2% or 1.2% aqueous dispersions for 30 sec, no evidence of teratogenic or embryotoxic effects (21).
Chick embryos, eggs pre-incubation dipped in 0.5, 1.5, 4.5 or 13.5 g $l^{-1}$ for 30 sec. All concentrations tested were teratogenic mainly inducing unilateral lower limb deformities (22).
Pregnant rats 40 mg $kg^{-1}$ caused embryo growth retardation. Also decreased activity of liver microsomal $N$-demethylase in nulliparous ♀ (23).
Gavage pregnant mice day 6-15 of gestation 1200 mg $kg^{-1}$ $day^{-1}$. Offspring examined 60, 65 days post-natally for skeletal variations. Major changes in frequences of parted frontals, abnormal metopic roots, reduced articular processes of thoracic vertebrae and carpal fusions occurred. Prenatal mortality was higher (20% compared to 7.5%) but litter size and litter weight were not altered (24).
Oral immature ♂, ♀ rat (1 month) 50 mg $kg^{-1}$ $day^{-1}$ after 10 wk control ♂ were mated with treated ♀ and control ♀ with treated ♂. A decline in fertility was observed in both sexes which was reversible after 3.5 months (25).
Gavage mouse, 2000 mg $kg^{-1}$ $day^{-1}$ on days 8-12 of gestation did not reduce the viabilities of foetuses or cause any carcinogenic effects (26).

**Metabolism and pharmacokinetics**
$^{14}$C-Maneb oral rat 55% excreted in urine and faeces within 3 days. Body organs contained 1.2%, 0.18% after 24 hr and 5 days, respectively, as metabolites ethylenediamine and ethylenebisthiuram monosulfide (27).
Acid hydrolysis in the stomach results in the production of ethylenediamine and carbon disulfide which is expired in air. The ethylenediamine is absorbed, some is oxidised to glycine and oxalic acid, and ultimately carbon dioxide. The remainder is excreted in urine together with inorganic sulfate. Maneb is also transformed into ethylene bisthiuram monosulfide, it then forms ethylenethiourea and then carbon disulfide which is expired in air (27).

**Irritancy**
1% skin patch test (72 hr) caused no irritation (28).

**Sensitisation**
In a 1% skin patch test (72 hr), 3 out of 125 agricultural workers experienced an allergic reaction (28).

## Genotoxicity
*Salmonella typhimurium* TA98, TA100, TA1537 with metabolic activation positive (29).
*Eschericha coli* WP2 *uvr A* reverse mutation with metabolic activation negative (29).
*Saccharomyces cerevisiae* D3 mitotic recombination with metabolic activation positive (29).

*Salmonella typhimurium* TA100, TA98 with and without metabolic activation negative (30).
*Saccharomyces cerevisiae* D$_7$ without metabolic activation negative (30).
*In vitro* human lung fibroblasts, unscheduled DNA synthesis positive (29).
*In vivo* chick embryo, eggs dipped in 0-27 g l$^{-1}$ for 30 sec, significantly increased sister chromatid exchanges at ≥13.5 g l$^{-1}$ (31).
*In vivo* mouse bone marrow cells slight increase in chromosomal aberrations (32).

## Any other adverse effects

Intraperitoneal ♂ mouse, single doses of 30, 60, 100, 200 or 1000 mg kg$^{-1}$ caused an inhibitory effect on locomotor activity and aggressiveness (33).
A trace contaminant and degradation product of maneb, ethylenethiourea has caused thyroid effects, tumours and birth defects in laboratory animals (34).

## Legislation

Limited under EC Directive on Drinking Water Quality 80/778/EEC. Pesticides: maximum admissible concentration 0.1 µg l$^{-1}$ (35).
Included in Schedule 6 (Release into Land: Prescribed Substances) Statutory Instrument No. 472, 1991 (36).

## Any other comments

WHO Class Table 5: EPA Toxicity Class IV (1).
Tolerable daily intake (human) 0.05 mg kg$^{-1}$ (1).
UK advisory value for drinking water 10 µg l$^{-1}$ (37).
Residues on vegetables degrade to ethylene thiourea (*qv*) during cooking (20).
Use, occurrence, analysis, carcinogenicity and mammalian toxicity reviewed (20).
Genetic toxicity reviewed (38).

## References

1. *The Agrochemicals Handbook* 3rd ed., 1991, RSC, London
2. van Leeuwen, C. J. et al *Aquat. Toxicol.* 1986, **9**(2-3), 129-145
3. Tooby, T. E. et al *Chem. Ind.* 1975, **21**, 523-525
4. van Leeuwen, C. J. et al *Aquat. Toxicol.* 1985, **7**(3), 145-164
5. *Environmental Properties of Chemicals* 1990, Research Report 91, Ministry of the Environment, Finland
6. Kemp, H. T. et al *Water Quality Data Book* Vol.5, 1973, EPA, Water Pollution Control Research Series 09/73
7. Kaiser, K. L. E. et al *Water Pollut. Res. J. Can.* 1991, **26**(3), 361-431
8. Parr, J. F. *Pest. Soil Water* 1974, 321-340
9. Nash, R. G. et al *J. Agric. Food Chem.* 1980, **28**, 322-330
10. Frietz, D. et al *Chemosphere* 1985, **14**, 1589-1616
11. *Hyg. Sanit. (USSR)* 1971, **36**(5), 22
12. Lewis, R. J. et al *Registry of toxic effects of chemical substances* National Institute for Occupational Safety and Health No.83-107-4
13. Schafer, E. W. et al *Arch. Environ. Contam. Toxicol.* 1983, **12**, 355-382
14. Kaloyanova, F. et al *J. Hyg. Epidemiol. Microbiol. Immunol.* 1989, **33**(1), 11-17
15. *IARC Monograph* 1987, **Suppl. 7**, 65
16. Innes, J. R. M. et al *J. Natl. Cancer Inst.* 1969, **42**, 1101-1114
17. Balin, P. N. *Vrach. Delo* 1970, **4**, 21-24
18. *IARC Monograph* 1976, **12**, 137-149
19. Andrianova, M. M. *Vopr. Pitan.* 1970, **29**, 71-74

20. NTIS *Evaluation of Carcinogenic, Teratogenic and Mutagenic Activities of Selected Pesticides and Industrial Chemicals* Vol.1, 1968, Washington, DC
21. Munk, R. et al *Ecotoxicol. Environ. Saf.* 1989, **17**(1), 112-118
22. Maci, R. et al *Ecotoxicol. Environ. Saf.* 1987, **13**(2), 169-173
23. Belyl, D. W. R. et al *Nahrung* 1988, **32**(10), 1007-1010
24. Beck, S. L. *Reprod. Toxicol.* 1990, **4**(4), 283-290
25. Muracon, L. V. *Farmacol Toksikol (Moscow)* 1969, **32**, 731-732
26. Kavlock, R. J. et al *Teratol., Carcinog., Mutagen.* 1987, **7**(1), 7-16
27. Seidler, H. et al *Nahrung* 1970, **14**, 363-373
28. Lisi, P. et al *Contact Dermatitis* 1986, **15**(5), 266-269
29. Garnett, N. E. et al *Mutat. Res.* 1986, **168**(3), 301-325
30. Choi, E. J. *Environ. Mutagen. Carcinog.* 1985, **5**(1), 11-18
31. Arias, E. *Mutat. Res.* 1988, **206**(2), 271-273
32. Antonovich, E. A. et al *Proceedings of the Symposium on Toxicology and Analytical Chemistry of Dithiocurbamates, Dubrovnic* 1970, 3-20, Beograd
33. Morato, G. S. et al *Neurotoxicol. Teratol.* 1989, **11**(5), 421-425
34. *The Pesticide Manual* 9th ed., 1991, British Crop Protection Council, Farnham
35. *EC Directive Relating to the Quality of Water Intended for Human Consumption* 1982, 80/778/EEC, Office for Official Publications of the European Communities, 2 rue Mercier, L-2985 Luxembourg
36. *S. I. 1991 No. 472 The Environmental Protection (Prescribed processes and Substances) Regulations* 1991, HMSO, London
37. *Guidance on Safeguarding the Quality of Public Water Supplies* 1989, 93-104, HMSO, London
38. Franekic, J. et al *International Reference Manual on Chemical Safety* 1994, Richardson, M.L. (ed.), 141–156, VCH Publishers, Weinheim

# M20　Manganese

## Mn

CAS Registry No. 7439-96-5
Synonyms Manganese
Mol. Formula Mn　　　　　　　　　　　　　Mol. Wt. 54.94
Uses In making steel and non-ferrous alloys.
Occurrence As oxide, sulfide, carbonate and silicate in the earth's crust, in soil, sediments, sea-floor nodules, plants and animals, but not as the free metal; constitutes 0.085% of earth's crust. Environmental pollutant from iron and steel works, mines, agrochemical production and use and production of dry-cell batteries and manganese oxide (1).

## Physical properties

**M. Pt.** 1244°C; **B. Pt.** 2095°C; **Specific gravity** $d^{20}$ 7.47 (α-form); $d^{20}$ 7.26 (β); $d^{1100}$ 6.37 (γ); $d^{1143}$ 6.28 (δ); **Volatility** v.p 1.0 mmHg at 1292°C.

## Occupational exposure

US TLV (TWA) 0.2 mg m$^{-3}$ (as Mn) intended change; UK Long-term limit 1 mg m$^{-3}$ (as Mn) (fume) 5 mg m$^{-3}$; UK Short-term limit 3 mg m$^{-3}$ (as Mn) (fume).

## Ecotoxicity

### Fish toxicity
$LC_{50}$ (28 day) rainbow trout 2.91 mg l$^{-1}$ (salt) (2).

### Invertebrate toxicity
$EC_{50}$ *Selenastrum capricornutum* 3.1 mg l$^{-1}$ (salt) (3).
$LC_{50}$ (48 hr) *Daphnia magna* 5.7 mg l$^{-1}$ (sulfide salt) (4).
$LC_{50}$ (48-96 hr) *Asellus aquaticus* 333-771 mg l$^{-1}$ for Mn(II) salt (5).

### Bioaccumulation
In chironomid larvae age and body weight affected the extent of retention. Among 4th instar larvae, younger individuals had higher concentrations than older instars (6). Levels may be concentrated up to a factor of 10, 100 and 100,000 by land mammals, fish and marine plants respectively (1).

## Environmental fate

### Degradation studies
Soil microorganisms in well aerated soils at >pH 5.5 can oxidize the divalent form rapidly. Oxidation is very slow in highly acid soils (7).

### Absorption
Radiolabel study using 2 whole soils and the natural aggregates of one of the soils. Waterlogging, air drying and incubation at 100 cm suction treatments were applied. Irrespective of treatment accumulation occurred in spots in the aggregates and throughout the whole soils. The manganese was converted rapidly to easily reducible forms by a microbe-dependent process in the soils and the spots were concluded to be areas where divalent manganese was oxidized microbially to insoluble oxides (8). Mobilization of manganese is favoured by low pH (9).
In river water enhanced mobility exhibited during denitrification phases, probably due to decreased redox potential and associated manganese oxyhydroxide reduction (10).

## Mammalian and avian toxicity

### Acute data
$LD_{50}$ injection chicken egg 765 µg egg$^{-1}$ (as chloride) (11).

### Sub-acute data
Intratracheal rat (1 month) 12.5 mg kg$^{-1}$ daily of welding dust containing 7.5-11% manganese caused changes to the cardiac-respiratory system including effects to metabolism and macrophage morphology and atelectasis and emphysema. No fibrogenic activity was observed (12).

### Carcinogenicity and long-term effects
Compounds found not to be carcinogenic (13).

### Teratogenicity and reproductive effects
Gross malformations to chick foetuses following injection to egg on day 2 of incubation (as chloride) (11).

### Metabolism and pharmacokinetics
The gastrointestinal and respiratory tracts are the major routes for absorbtion in humans reaching in the former <5% for healthy adults (14-16).
Absorption rate is influenced by dietary levels of iron and manganese, iron

deficiency, age and type of manganese compound (15,17).

Calcium may enhance absorption in the gastrointestinal tract (18).

Total body load in a 70 kg man $\approx$10-20 mg and is widely distributed (19,20).

Mitochondria-rich tissues (pancreas, liver, intestines and kidney) contain high concentrations (21,22).

Bound to $\beta_1$-globulin, probably transferrin, it is transported in the plasma (23,24).

It is able to pass through the blood-brain barrier and placenta (25).

$t_{1/2} \approx$37 days (15).

Inorganic forms are mainly excreted faecally via bile with only 0.1-1.3% of daily intake eliminated in urine (26-28).

**Irritancy**

500 mg (24 hr) instilled into rabbit eye or administered to skin caused mild irritation, and well defined erythema with slight oedema, respectively (29).

## Genotoxicity

*Salmonella typhimurium* TA98, TA102, TA1535, TA1537 with metabolic activation negative; TA1537 without metabolic activation positive (30).

Mouse polychromatic red blood cells in bone marrow and blood lymphocytes did not significantly induce micronuclei (31).

## Any other adverse effects to man

In an occupational exposure study significantly higher incidence of dyspnoea during exercise, cough in cold season and episodes of acute bronchitis were observed. Lung ventilatory parameters were slightly altered. Reaction time, hand tremor and audioverbal short-term memory were significantly altered. There was a slight increase in the amount of circulating neutrophils and in the values of certain serum parameters. A blood threshold concentration at $\approx$1 $\mu$g 100 ml whole blood$^{-1}$ is suggested by the response to the eye-hand coordination test. A time-weighted average exposure to airborne dust of $\approx$1 mg m$^{-3}$ for <20 yr may cause preclinical signs of poisoning (32). Lesions of the central nervous system have been shown to be most severe in the pallidum and striatum following autopsy of chronic poisoning cases. Chronic poisoning is a danger in the mining and processing of manganese ores, welding and in the manganese alloy and dry-cell battery industries. Psychological and neurological symptons characterise the condition (14,33,34).

Effects on communities in the vicinity of manganese production plants have been reported (35-37).

## Any other adverse effects

Caused depletion of dopamine, and probably serotonin, in the basal ganglia of rabbits, rats and monkeys (38-40).

Liver necrosis and reduced blood pressure effects reported (41,42).

## Legislation

Limited under EC Directive on Drinking Water Quality 80/778/EEC. Manganese: guide level 20 $\mu$g l$^{-1}$ and maximum admissible concentration 50 $\mu$g l$^{-1}$ (43).

Included in Schedule 4 (Release into Air: Prescribed Substances) Statutory Instrument No. 472, 1991 (44).

## Any other comments

Availability increased by fertilization of soil. Wheat production positively correlated to availability (45).

Toxicity reviewed (46).

Pathology, bioavailability, metabolism and biochemistry reviewed (47-51).

An essential trace element necessary for the formation of connective tissue and bone, lipid and carbohydrate metabolism, reproduction and growth. Daily requirement is 2-3 mg (19,20).

Conversion of sulfur dioxide to sulfuric acid is promoted by atmospheric manganese compounds (52).

The divalent form is 2.5-3 times more toxic than the trivalent form (1).

Toxicity data refer to bioavailable forms.

## References

1. *Environmental Health Criteria 17: Manganese* 1981, WHO, Geneva
2. Birge, W. J. et al *Aquatic toxicity tests on inorganic elements occurring in oil shale* 1980, EPA 600/9-80-022, NTIS, Springfield, VA, USA
3. Christensen, E. R. et al *Water Res.* 1979, **13**, 79-92
4. Biesinger, K. E. et al *J. Fish Res. Board Can.* 1972, **29**, 1691-1700
5. Martin, T. R. et al *Water Res.* 1986, **20**(9), 1137-1147
6. Krantzberg, G. *Hydrobiologia* 1989, **188-189**, 497-506
7. Zajic, J. E. *Microbial Biochemistry* 1969, Academic Press, New York
8. Uren, N. C. *Aust. J. Soil Res.* 1990, **28**(5), 677-683
9. Mitchell, R. L. *Trace elements in soil* 1971, Tech. Bull. No.20, Macaulay Institute for Soil Research, Aberdeen
10. Jacobs, L. A. et al *Geochim. Cosmochim. Acta* 1988, **52**(11), 2693-2706
11. Gilani, S. H. et al *J. Toxicol. Environ. Health* 1990, **30**(1), 23-31
12. Pokrovskaya, T. N. et al *Gig. Tr. Prof. Zabol.* 1990, (9), 37-40 (Russ.) (*Chem. Abstr.* **114**, 1907k)
13. Health and Safety Executive *Occupational Exposure Limits: Criteria Document Summaries* 1993, HMSO, London
14. Rodier, J. *Br. J. Ind. Med.* 1955, **12**, 21-35
15. Mena, I. et al *Neurology* 1969, **19**, 1000-1006
16. Pollack, S. et al *J. Clin. Invest.* 1965, **44**, 1470-1473
17. Thompson, A. B. R. et al *J. Lab. Clin. Med.* 1971, **78**(4), 642-655
18. Lassiter, J. W. et al *Trace Element Metabolism in Animals, Proceedings of WAAP/IBP International Symposium, Aberdeen, July 1969* 1970, E&S Livingstone, Edinburgh
19. Underwood, E. J. *Trace Elements in Human and Animal Nutrition* 3rd ed., 1971, Academic Press, New York
20. *WHO Technical Report Series, No.532* 1973
21. Maynard, L. S. et al *J. Biol. Chem.* 1955, **214**, 489-495
22. Thiers, R. E. et al *J. Biol. Chem.* 1957, **226**, 911-920
23. Mena, I. et al *J. Nucl. Med.* 1974, **15**, 516
24. Pauic, B. *Acta Vet. Scand.* 1967, **8**, 228-233
25. Widdowson, E. M. et al *Biol. Neonate* 1972, **20**(5-6), 360-367
26. Papavasiliou, P. S. et al *Am. J. Physiol.* 1966, **211**, 211-216
27. Tipton, I. H. et al *Health Phys.* 1969, **16**, 455-462
28. McLeod, B. E. et al *Br. J. Nutr.* 1972, **27**, 221-227
29. Marhold, J. V. *Sbornik Vysledku Toxixologickeho Vysetreni Latek A Pripravku* 1972, Prague
30. Wong, P. K. *Bull. Environ. Contam. Toxicol.* 1988, **40**(4), 597-603
31. Fan, L. et al *Gongye Weisheng Yu Zhiyebing* 1986, **12**(2), 77-80 (Ch.) (*Chem. Abstr.* **107**, 53901d)

32. Roels, H. et al *Am. J. Ind. Med.* 1987, **11**(3), 307-327
33. Mena, I. et al *Neurology* 1967, **17**, 128-136
34. Smyth, L. T. et al *J. Occup. Med.* 1973, **15**, 101-109
35. Elstad, D. *Nord. Med.* 1939, **3**, 2527-2533
36. Savic, M. *Biological Effects of Manganese* 1978, EPA-600/1-78-001, US EPA, Research Triangle Park, NC
37. Savic, M. et al *International Conference on Heavy Metals in the Environment Proceedings* 1975, 389-398, Institute for Environmental Studies, University of Toronto, Toronto
38. Neff, N. H. et al *Experientia* 1969, **25**, 1140-1141
39. Mustafa, S. J. et al *J. Neurochem.* 1971, **18**, 931-933
40. Bonilla, E. et al *J. Neurochem.* 1974, **22**, 297-299
41. Baxter, D. J. et al *Proc. Soc. Exp. Biol. Med.* 1965, **119**, 966-970
42. Kostial, K. et al *Br. J. Pharmacol.* 1974, **51**, 231-235
43. *EC Directive Relating to the Quality of Water Intended for Human Consumption* 1982, 80/778/EEC, Office for Official Publications of the European Communities, 2 rue Mercier, L-2985 Luxembourg
44. *S. I. 1991 No. 472 The Environmental Protection (Prescribed Processes and Substances) Regulations* 1991, HMSO, London
45. Qiu, F. et al *Turang Tongbao* 1986, **17**(7), 73-76
46. Anke, M. et al *Trace Elem. Anal. Chem. Med. Biol., Proc. Int. Workshop 4th* 1987, 201-236, Berlin
47. Nishida, M. et al *Saibo* 1990, **22**(1), 23-27, (Japan.) (*Chem. Abstr.* **113**, 92852g)
48. Brandt, M. et al *Manganese Metab. Enzyme Funct.* 1986, 3-16, Academic, Orlando, FL
49. Kies, C. *ACS Symp. Ser.* 1987, **354**, 1-8
50. Lonnerdal, B. et al *ACS Symp. Ser.* 1987, **354**, 9-20
51. Failla, M. L. *Manganese Metab. Enzyme Funct.* 1986, 93-105, Academic, Orlando, FL
52. McKay, H. A. C. *Atmos. Environ.* 1971, **5**, 7-14

# M21   Manganese cyclopentadienyl tricarbonyl

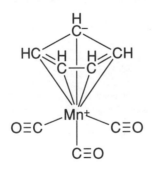

**CAS Registry No.** 12079-65-1
**Synonyms** tricarbonyl ($\eta^5$-2,4-cyclopentadien-1-yl)-manganese; tricarbonyl-π-cyclopentadienylmanganese; Cymantrene; cyclopentadienylmanganese tricarbonyl; MCT
**Mol. Formula** $C_8H_5MnO_3$        **Mol. Wt.** 204.07
**Uses** Additive in unleaded petrol.

## Physical properties

**M. Pt.** −1°C; **B. Pt.** 232-233°C.

## Occupational exposure

**US TLV (TWA)** 0.2 mg m$^{-3}$ (as Mn) intended change; **UK Long-term limit** 0.1 mg m$^{-3}$ (as Mn); **UK Short-term limit** 0.3 mg m$^{-3}$.

## Mammalian and avian toxicity

### Acute data

$LD_{50}$ oral rat, mouse 80-150 mg kg$^{-1}$ (1).
$LC_{50}$ (2 hr) inhalation rat 120 mg m$^{-3}$ (2).
$LD_{50}$ intraperitoneal rat 14 mg kg$^{-1}$ (3).
$LD_{50}$ intravenous mouse 710 μg kg$^{-1}$ (4).

### Sub-acute data

Subcutaneous ♂ rat 0.5, 1.0 or 2.5 mg Mn kg$^{-1}$ at 24 hr bronchoalveolar lavage albumin, protein content and lactate dehydrogenase activities increased. Lung manganese content was significantly raised (5).
Inhalation (10 month) rabbits, guinea pigs, rats 0.7-1 mg m$^{-3}$ caused central nervous system muscarinic effects, decreased resistance to infection and a decrease in diuresis with increase in urinary albumin content (1).

### Metabolism and pharmacokinetics

Following subcutaneous administration to rat manganese accumulated in the lung in a nonlipid-soluble form suggesting accumulation of metabolites rather than parent compound. A strong correlation between pulmonary manganese concentration and toxicity was observed. It was suggested that monooxygenase mediated metabolites were involved in lung toxicity and manganese accumulation (5).

## Legislation

Included in Schedule 4 (Release into Air: Prescribed Substances) Statutory Instrument No. 472, 1991 (6).

## References

1. Arhipova, O. G. *Gig. Tr. Prof. Zabol.* 1963, **7**, 43-49 (Russ.)
2. Arhipova, O. G. et al *Hyg. Sanit. (USSR)* 1963, **28**(4), 29-32
3. *Toxicology* 1985, **34**, 341
4. *Report NX No.11285* US Army Armament Research and Development Command, Chemical Systems Laboratory, NIOSH Exchange Chemicals, Aberdeen Proving Ground, MD 21010
5. Clay, R. J. et al *Toxicol. Appl. Pharmacol.* 1989, **98**(3), 434-443
6. *S. I. 1991 No. 472 The Environmental Protection (Prescribed Processes and Substances) Regulations* 1991, HMSO, London

# M22   Manganese dioxide

## $MnO_2$

**CAS Registry No.** 1313-13-9
**Synonyms** manganese oxide; battery manganese; manganese(IV) oxide
**Mol. Formula** $MnO_2$                                    **Mol. Wt.** 86.94
**Uses** In dry-cell batteries. Oxidising agent. Porcelain paint. Pigments. In therapeutics.
**Occurrence** As the mineral pyrolusite; and in seawater. Pollutant from mining, steel
casting, metallurgical processing and metal welding and cutting (1,2).

## Physical properties

**M. Pt.** 535°C (decomp.); **Specific gravity** 5.03.

## Occupational exposure

**US TLV (TWA)** 0.2 mg m$^{-3}$ (as Mn) intended change; **UK Long-term limit**
5 mg m$^{-3}$ (as Mn); **Supply classification** harmful.
**Risk phrases** Harmful by inhalation and if swallowed (R20/22)
**Safety phrases** Avoid contact with eyes (S25)

## Ecotoxicity

**Fish toxicity**
Common carp (140 day) 0.0137 mg g$^{-1}$ caused a measurable change in length and/or
weight (3).

## Mammalian and avian toxicity

**Acute data**
$LD_{100}$ subcutaneous mouse 550 mg kg$^{-1}$ (4).
$LD_{50}$ subcutaneous mouse 442 mg kg$^{-1}$ (5).
$LD_{Lo}$ intravenous rabbit 45 mg kg$^{-1}$ (6).

**Sub-acute data**
Inhalation (95 × 1 hr) monkey 0.6-3.0 mg m$^{-3}$ over 4 months caused central nervous
system effects (7).
Intratracheal (4 month) rat 0.3 mg m$^{-3}$ for 5-6 hr daily caused inflammatory changes (6).
Injection (14.5 month) monkey 2000 and 3500 mg with interval of 3 months caused
the proliferation of bizarre cells and extensive loss of neurones in the pallidum and
subthalamic nucleus (8).
Subcutaneous (3.5 month) squirrel monkey 400 mg divided into two doses
administered with a 5 wk interval caused muscular rigidity and tremor (9).

**Metabolism and pharmacokinetics**
Oral ♂ mouse (100 day) 2 g Mn kg$^{-1}$. Levels of manganese in liver, kidney and hair
were the highest followed by bone, pancreas and prostate gland. The leukocyte level
was significantly reduced (10).
Largely cleared from the lungs by ciliary action and a proportion of this is absorbed
from the gut (11).

## Genotoxicity

*Salmonella typhimurium* TA1535/psk1002 did not induce SOS reaction (12).

## Any other adverse effects

Inhalation of fine dust containing relatively low concentration causes pneumonitis. When induced in rat by intracheal administration shedding of bronchial and alveolar epithelium is seen with mononuclear cell infiltration of the alveolar walls and alveoli (13). Other effects include peribronchial and perivascular sclerosis, inflammatory changes, appearance of collagenic threads, accentuation of blood vessels and reduced resistance to infection (14-17).

Hepatic effects also reported (15).

Inhalation at 10-20 mg m$^{-3}$ for 4 hr daily during 3 to 6 months caused changes to lung tissue and reduced number of red blood cells (11).

## Legislation

Limited under EC Directive on Drinking Water Quality 80/778/EEC. Manganese: guide level 20 µg l$^{-1}$ and maximum admissible concentration 50 µg l$^{-1}$ (18).

Included in Schedule 4 (Release into Air: Prescribed Substances) Statutory Instrument No. 472, 1991 (19).

## Any other comments

Toxicity data refer to bioavailable forms.

Dissolves in HCl.

## References

1. Mihajlov, V. A. et al *Klinika patogenez i profilaktika profzabolevanij himiceskoj etiologii na predprijatijah cvefnoj i cernoj metallurgii, sverdlovsk Part II* 1969 (Russ.)
2. Erman, M. I. *Gig. Tr. Prof. Zabol.* 1972, **16**(5), 26-30 (Russ.)
3. Satoh, S. et al *Bull. Jpn. Soc. Sci. Fish.* 1987, **53**, 825
4. Date, S. *J. Kumamoto Med. Soc.* 1960, **34**, 159-182
5. *Zh. Vses. Khim. O-va. im. D. I. Mendeleeva* 1974, **19**, 186 (Russ.)
6. Spector, W. S. (Ed.) *Handbook of Toxicology* 1956, Volume 1, Saunders, Philadelphia, PA
7. *Evironmental Health Criteria 17: Manganese* 1981, WHO, Geneva
8. Pentoschew, A. et al *J. Neuropathol. Exp. Neurol.* 1963, **22**, 488-499
9. Neff, N. H. et al *Experientia* 1969, **25**, 1140-1141
10. Komura, J. et al *Hokuriku Koshu Eisei Gakkaishi* 1986, **13**(1), 75-78 (Japan.) (*Chem. Abstr.* **106**, 114869c)
11. Health and Safety Executive *Occupational Exposure Limits: Criteria Document Summaries* 1993, HMSO, London
12. Nakamura, S. et al *Sangyo Igaku* 1989, **31**(6), 430-431 (Japan.) (*Chem. Abstr.* **112**, 153457k)
13. Lloyd Davies, T. A. et al *Br. J. Ind. Med.* 1949, **6**, 82-90
14. Levina, E. N. et al *Hyg. Sanit. (USSR)* 1955, **20**(1), 25-28
15. Dokucaeva, V. F. et al *Biological Effects and Hygienic Significance of Atmospheric Concentrations* 1966, 173-185 (Russ.)
16. Nishiyama, K. et al *Jpn. J. Hyg.* 1975, **30**, 117 (Japan.)
17. Maigretter, R. Z. et al *Environ. Res.* 1976, **11**, 386-391
18. *EC Directive Relating to the Quality of Water Intended for Human Consumption* 1982, 80/778/EEC, Office for Official Publications of the European Communities, 2 rue Mercier, L-2985 Luxembourg
19. *S. I. 1991 No. 472 The Environmental Protection (Prescribed Processes and Substances) Regulations* 1991, HMSO, London

# M23 Manganese 2-methylcyclopentadienyl tricarbonyl

**CAS Registry No.** 12108-13-3

**Synonyms** manganese, tricarbonyl[(1,2,3,4,5-π)-1-methyl-2,4-cyclopentadien-1-yl]-; manganese, tricarbonyl(methyl-π-cyclopentadienyl)-; methylcymantrene; MMT; 2-methylcyclopentadienylmanganese tricarbonyl

**Mol. Formula** $C_9H_{17}MnO_3$                              **Mol. Wt.** 228.17

**Uses** Antiknock agent, used as an octane improver in unleaded gasoline.

## Physical properties

**M. Pt.** −1°C; **B. Pt.** 233°C; **Flash point** 96°C; **Specific gravity** 1.38; **Volatility** v.p. $4.7 \times 10^{-2}$ mmHg at 20°C.

### Solubility
Water: 70 ppm at 25°C.

## Occupational exposure

**US TLV (TWA)** 0.2 mg m$^{-3}$ (as Mn); **UK Long-term limit** 0.2 mg m$^{-3}$ (as Mn); **UK Short-term limit** 0.6 mg m$^{-3}$ (as Mn).

## Mammalian and avian toxicity

### Acute data
$LD_{50}$ oral mouse 230 mg kg$^{-1}$ (1).
$LD_{50}$ oral rat 50 mg kg$^{-1}$ (2).
$LD_{50}$ oral mice 152 mg kg$^{-1}$ (solvent propylene glycol) (3).
$LD_{50}$ oral mice 999 mg kg$^{-1}$ (solvent corn oil) (3).
$LC_{50}$ (4 hr) inhalation rat 76 mg m$^{-3}$ (1).
$LC_{50}$ (4 hr) inhalation mouse 58 mg m$^{-3}$ (4).
$LD_{50}$ dermal rabbit 140 mg kg$^{-1}$ (1).
$LD_{50}$ intraperitoneal mouse 152 mg kg$^{-1}$ (5).
$LD_{50}$ intraperitoneal rat 12.1 mg kg$^{-1}$ (3).
$LD_{50}$ intraperitoneal rat 23 mg kg$^{-1}$ (2).

### Metabolism and pharmacokinetics
Subcutaneous Sprague-Dawley rats 4 mg kg$^{-1}$, blood, lung, liver and kidney manganese levels increased between 1.5 and 96 hr, with peak organ levels occurring

---

at 3-6 hr. Manganese concentration in lung, liver and kidney averaged 13-, 4- and 4-fold higher, respectively, than in the blood indicating the accumulation and retention of it in these tissues. Maximal pulmonary toxicity occurred 24-48 hr after injection. No hepatic or renal injury occurred (6).

### Any other adverse effects

Subcutaneous (24 hr) ♂ rats 0.5, 1.0 or 2.5 mg manganese $kg^{-1}$. The pneumotoxic response was characterised by large increases in lavage albumin and protein content with smaller increases in lactate dehydrogenase activity. Lung manganese levels were significantly elevated after treatment (7).

Inhibits the binding of [$^{3}$H]-*tert*-butyl bicyclo-*o*-benzoate in mouse brain membranes with a median inhibitory concentration value of 240 µg $kg^{-1}$ (8).

### Legislation

Included in Schedule 6 (Release into Land: Prescribed Substances) Statutory Instrument No. 472, 1991 (9).

### Any other comments

Environmental effects, environmental toxicology, human health effects and workplace experience reviewed (10-12).

### References

1.   Hinderer, R. K. *Am. Ind. Hyg. Assoc. J.* 1979, **40**, 164
2.   *Toxicol. Appl. Pharmacol.* 1980, **56**, 353
3.   Cox, D. N. et al *Toxicol. Lett.* 1987, **39**(1), 1-5
4.   *Jpn. J. Ind. Health* 1978, **20**, 553
5.   *Shikoku Igaku Zasshi* 1978, **34**, 183
6.   McGinley, P. A. et al *Toxicol. Lett.* 1987, **36**(2), 137-145
7.   Clay, R. J. et al *Toxicol. Appl. Pharmacol.* 1989, **98**(3), 434-443
8.   Fishman, B. E. et al *Toxicology* 1987, **45**(2), 193-201
9.   *S. I. 1991 No. 472 The Environmental Protection (Prescribed Processes and Substances) Regulations* 1991, HMSO, London
10.   *BIBRA Toxicity Profile* 1991, British Industrial Biological Research Association, Carshalton
11.   Abbott, P. J. *Sci. Total Environ.* 1987, **67**(2-3), 247-255
12.   *ECETOC Technical Report No. 30(4)* 1991, European Chemical Industry Ecology and Toxicology Centre, B-1160 Brussels

# M24   Manganese nitrate

## $Mn(NO_3)_2$

**CAS Registry No.** 10377-66-9
**Synonyms** nitric acid, manganese (2+) salt; manganese dinitrate
**Mol. Formula** $MnN_2O_6$                                      **Mol. Wt.** 178.95
**Uses** Preparation of porcelain colourants. Intermediate in the manufacture of reagent grade $MnO_2$.

## Physical properties

**M. Pt.** 37.1°C (tetrahydrate); 25.0°C (hexahydrate); **Specific gravity**
2.13 (tetrahydrate); 1.8 (hexahydrate).

### Solubility
Water: freely soluble. Organic solvent: dioxane, tetrahydrofuran, acetonitrile

### Occupational exposure
**US TLV (TWA)** 0.2 mg m$^{-3}$ (as Mn) intended change; **UK Long-term limit**
5 mg m$^{-3}$ (as Mn); **UN No.** 2724; **Conveyance classification** oxidizing substance.

## Mammalian and avian toxicity

### Acute data
LD$_{50}$ intraperitoneal mouse 56 mg kg$^{-1}$ (1).

## Legislation
Limited under EC Directive on Drinking Water Quality 80/778/EEC. Manganese:
guide level 20 µg l$^{-1}$ and maximum admissible concentration 50 µg l$^{-1}$; nitrate: guide
level 25 mg l$^{-1}$ and maximum admissible concentration 50 mg l$^{-1}$ (2).
Included in Schedule 4 (Release into Air: Prescribed Substances) Statutory Instrument
No. 472, 1991 (3).

## References

1. Yamamoto, H. et al *Proceedings of the 42nd Annual Meeting of the Japan Association of Industrial Health, 28-31 March, 1959* 1969, Japan Association of Industrial Health, Fukuoka City (Japan.)
2. *EC Directive Relating to the Quality of Water Intended for Human Consumption* 1982, 80/778/EEC, Office for Official Publications of the European Communities, 2 rue Mercier, L-2985 Luxembourg
3. *S. I. 1991 No. 472 The Environmental Protection (Prescribed Processes and Substances) Regulations* 1991, HMSO, London

# M25   Manganese tetroxide

## Mn$_3$O$_4$

**CAS Registry No.** 1317-35-7
**Synonyms** manganese oxide (Mn$_3$O$_4$); manganese oxide; trimanganese tetraoxide;
M34
**Mol. Formula** Mn$_3$O$_4$                        **Mol. Wt.** 228.81
**Occurrence** As the mineral hausmannite.

## Physical properties
**Specific gravity** 4.80.

## Occupational exposure
**US TLV (TWA)** 0.2 mg m$^{-3}$ (as Mn) intended change; **UK Long-term limit**
1 mg m$^{-3}$ (as Mn).

## Mammalian and avian toxicity

**Sub-acute data**
Inhalation rat (2 month) (concentration unspecified) no noticeable histological changes in lung (1).

**Metabolism and pharmacokinetics**
Clearance $t_{1/2}$ from rat lung in 2 month inhalation study was several hours (1).

## Legislation
Included in Schedule 4 (Release into Air: Prescribed Substances) Statutory Instrument No. 472, 1991 (2).

## Any other comments
Soluble in HCl.

## References

1. Kyono, H. *Saibo* 1990, **22**(13), 502-507
2. *S. I. 1991 No. 472 The Environmental Protection (Prescribed Processes and Substances) Regulations* 1991, HMSO, London

# M26   D-Mannitol

## CH2OHCH(OH)CH(OH)CH(OH)CH(OH)CH2OH

**CAS Registry No.** 69-65-8
**Synonyms** cordycepic acid; Diosmol; Isotol; manna sugar; mannistol; mannitol; 1,2,3,4,5,6-hexanehexol
**Mol. Formula** $C_6H_{14}O_6$                          **Mol. Wt.** 182.17
**Uses** In making artificial resins and plasticizers. Pharmaceutical excipient and diluent. Anticaking and free-flow agent, flavouring agent and stabiliser in food industry. Diuretic. Renal function diagnostic aid. Used in reducing intracranial and intra-ocular pressure.
**Occurrence** Plants.

## Physical properties

**M. Pt.** 166-168°C; **B. Pt.** $_{3.5}$ 290-295°C; **Specific gravity** $d^{20}$ 1.52; **Partition coefficient** log $P_{ow}$ -3.10 (1).

**Solubility**
Water: 182 g $l^{-1}$. Organic solvent: ethanol, pyridine, aniline, glycerol

## Environmental fate

**Degradation studies**
BOD (5 day test), 59% ThOD; COD 87% ThOD (0.05 N $K_2Cr_2O_7$); and ThOD, 1.15 (2).
Utilised by microbes of industrial activated sludge (3).

---

## Mammalian and avian toxicity

### Acute data
$LD_{50}$ oral rat, mouse 13.5-22 g $kg^{-1}$ (4,5).
$LD_{50}$ intraperitoneal mouse 14 g $kg^{-1}$ (6).

### Carcinogenicity and long-term effects
Oral F344 rat and B6C3F$_1$ mouse (103 wk) 0, 2.5 or 5.0% in diet had no effect on
survival or mean body weight. Feed consumption was approximately the same as
controls. ♀ rats had increased incidence of gastric fundal gland dilation and a mild
nephrosis in the renal tubular epithelium of mice was observed. Retinopathy and
cataracts occurred at increased frequency in high-dose ♂ and low- and high-dose ♀
rats (7,8).
Non-carcinogenic to rat and mouse (9).
Oral rat, mouse up to highest tested dose of 50,000 ppm non-carcinogenic (10).
National Toxicology Program tested rats and mice orally in feed. No evidence of
carcinogenicity (8).

### Teratogenicity and reproductive effects
*In vitro* chick embryo neural retina cells not developmentally toxic; lowest observed
effect concentration >7.3 g $l^{-1}$ (11).
*In vivo* rat developmentally toxic intravenous dose 12.5 g $l^{-1}$ $kg^{-1}$ (12).

### Metabolism and pharmacokinetics
In man gastrointestinal tract absorbtion is minimal. Following intravenous injection
and before any significant metabolism by the liver can take place rapid excretion from
the kidneys occurs. Elimination $t_{1/2}$ 100 min. The eye is not penetrated and nor is the
blood-brain barrier crossed (13).

### Sensitisation
Hypersensitivity reactions have been reported (species unspecified) (13).

## Genotoxicity
*Salmonella typhimurium* (strain unspecified) with or without metabolic activation
negative (14).
*In vitro* mouse lymphoma L5178Y with and without metabolic activation negative (15).
*In vitro* Chinese hamster ovary cells with and without metabolic activation sister
chromatid exhanges and chromosomal aberrations negative (16).
*Drosophila melanogaster* wing spot test negative (17).

## Any other adverse effects to man
Fluid and electrolyte imbalance including circulatory overload and acidosis at high
doses. Expansion of extracellular volume can precipitate pulmonary oedema. Can
cause tissue dehydration. Dehydration of the brain may lead to central nervous system
symptoms. Causes diarrhoea when given orally. Headache, thirst, vomiting, dizziness,
chills, nausea and fever are some of the side effects when administered intravenously (13).
In renal failure patients central nervous system involvement, large osmolality gap,
fluid overload and severe hyponatraemia were seen following intravenous
administration over 1 to 3 days (18).
Following intravenous injection of 20% mannitol one patient suffered focal osmotic
nephrosis (19).
Oliguric renal failure reported (20,21).

## Any other adverse effects

Non-toxic to mouse lymphoma L5178Y cells for concentrations up to 5000 µg ml$^{-1}$ (15).

## References

1. Verschueren, K. *Handbook of Environmental Data on Organic Chemicals* 2nd ed., 1983, Van Nostrand Reinhold Co., New York
2. Wolters, N. *Unterschiedliche Bestimmungsmethoden zur Erfassung der Organischen Substanz in einer Verbindung* Lehrauftrag Wasserbiologie a.d. Technischen Hochschule, Darmstadt
3. Hallas, L. E. et al *J. Ind. Microbiol.* 1988, **3**(6), 377-385
4. *Yakkyoku* 1981, **32**, 1367
5. *FAO report 40* 1967, 161
6. *Proc. Soc. Exp. Biol. Med.* 1936, **35**, 98
7. Abdo, K. M. et al *Food Chem. Toxicol.* 1986, **24**(10-11), 1091-1097
8. *National Toxicology Program Research and Testing Division* 1992, Report No.TR-236, NIEHS, Research Triangle Park, NC
9. Ashby, J. et al *Mutat. Res.* 1991, **257**(3), 229-306
10. Haseman, J. K. et al *Environ. Mol. Mutagen.* 1990, **16**(Suppl.18), 15-31
11. Daston, G. P. et al *Toxicol. Appl. Pharmacol.* 1991, **109**(2), 352-366
12. Petter, C. *C. R. Soc. Biol.* 1967, **16**, 1010-1014
13. *Martindale. The Extra Pharmacopoeia* 30th ed., 1993, The Pharmaceutical Press, London
14. Buzzi, R. et al *Mutat. Res.* 1990, **234**(5), 269-288
15. Myhr, B. et al *Environ. Mol. Mutagen.* 1991, **18**(1), 51-83
16. Gulati, D. K. et al *Environ. Mol. Mutagen.* 1989, **13**(2), 133-193
17. Tripathy, N. K. et al *Mutat. Res.* 1990, **242**(3), 169-180
18. Borges, H. F. et al *Arch. Intern. Med.* 1982, **142**, 63-66
19. Goodwin, W. E. et al *J. Urol. (Baltimore)* 1970, **103**, 11-14
20. Whelan, T. V. et al *Arch. Intern. Med.* 1984, **144**, 2053-2055
21. Goldwasser, P. et al *Arch. Intern. Med.* 1984, 144, 2214-2216

# M27   D-Mannose

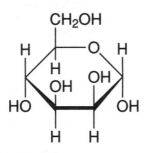

**CAS Registry No.** 3458-28-4
**Synonyms** Carubinose; mannose; D(+)-mannose; seminose
**Mol. Formula** $C_6H_{12}O_6$                    **Mol. Wt.** 180.16

## Physical properties

**M. Pt.** 133°C (α form); 133-140°C (mixture of anomers); **Specific gravity** d$^{20}$ 1.54 (β).

## Solubility
Water: 2500 g l$^{-1}$ ($\beta$). Organic solvent: absolute ethanol, methanol, pyridine

## Environmental fate

### Anaerobic effects
Utilised by *Blastocladiella ramosa* and *Blastocladiella pringsheimii* producing lactic, succinic, acetic and propionic acids (1).

### Degradation studies
Degraded in soil (2).

## Mammalian and avian toxicity

### Teratogenicity and reproductive effects
Infused (12 hr) pregnant rats during early neurulation, day 9.5-10 of development. The number of resorbed conceptions was increased. At term no foetal gross anomalies, but mean body weight and mean weight and protein content of livers, kidneys, hearts and brains were reduced. Skeletal development was significantly delayed (3).
*In vitro* rat embryos were cultured from 9.5 day of gestation. 3 or 6 mg ml$^{-1}$ (48 hr) caused inhibition of yolk sac expansion, smaller embryo size, delayed morphological development and abnormalities. Similar effects were seen in shorter-duration exposures, including delayed neural fold closure or an irregular neural groove (4).

## Any other comments
Mannose recognition has a role in lymphocyte entry to spleen and lymph node (5).

## References
1. Gleason, F. H. et al *Mycologia* 1989, **81**(5), 811-815
2. Izumori, K. et al *Kagawa Daigaku Nogakubu Gakujutsu Hokoku* 1987, **39**(1), 83-86 (Japan.) (*Chem. Abstr.* **111**, 170854j)
3. Buchanan, T. A. et al *Am. J. Obstet. Gynecol.* 1988, **158**(3, pt.1), 663-669
4. Moore, D. C. P. et al *Life Sci.* 1987, **41**(16), 1885-1893
5. Weston, S. A. et al *J. Immunol.* 1991, **146**(12), 4180-4186

# M28   MCPA-thioethyl

$Cl-\!\!\!\!\!\!\!\!\!\bigcirc\!\!\!\!\!\!-OCH_2C(O)SCH_2CH_3$ with $CH_3$ substituent

**CAS Registry No.** 25319-90-8
**Synonyms** (4-chloro-2-methylphenoxy)ethanethioic acid, *S*-ethyl ester; [(4-chloro-*o*-tolyl)oxy]thioacetic acid, *S*-ethyl ester; Herbit; HOK 7501; Phenothiol
**Mol. Formula** $C_{11}H_{13}ClO_2S$  **Mol. Wt.** 244.74
**Uses** Hormone-type herbicide.

## Physical properties

**M. Pt.** 41-42°C; **B. Pt.** $_7$ 165°C; **Volatility** v.p. $1.58 \times 10^{-4}$ mmHg at 20°C.

**Solubility**
Water: 2.3 mg $l^{-1}$ at 25°C. Organic solvent: acetone, xylene, n-hexane

## Ecotoxicity

**Fish toxicity**
$LC_{50}$ (48 hr) carp 2.5 mg $l^{-1}$ (1).

## Mammalian and avian toxicity

**Acute data**
$LD_{50}$ oral ♂, ♀ rat, ♂ ♀ mouse 749-877 mg $kg^{-1}$ (1).
$LD_{50}$ oral Japanese quail >1000 mg $kg^{-1}$ (1).
$LD_{50}$ dermal ♂ mouse >1500 mg $kg^{-1}$ (1).
$LD_{50}$ intraperitoneal ♂, ♀ rat 530-570 mg $kg^{-1}$ (1).

**Carcinogenicity and long-term effects**
Oral (2 yr) rat 100 mg $kg^{-1}$ in diet no effect level (1).
Oral (2 yr) mouse 20 mg $kg^{-1}$ in diet no effect level (1).

**Teratogenicity and reproductive effects**
Non-teratogenic and no reproductive effects to rats (1).

**Irritancy**
Non-irritating to rabbit skin and eyes (1).

## Legislation

Limited under EC Directive on Drinking Water Quality 80/778/EEC.Pesticides: maximum admissible concentration 0.1 μg $l^{-1}$ (2).
Included in Schedule 6 (Release into Land: Prescribed Substances) Statutory Instrument No. 472, 1991 (3).

## Any other comments

WHO Class III (1).

## References

1. *The Agrochemicals Handbook* 3rd ed., 1991, RSC, London
2. *EC Directive Relating to the Quality of Water Intended for Human Consumption* 1982, 80/778/EEC, Office for Official Publications of the European Communities, 2 rue Mercier, L-2985 Luxembourg
3. *S. I. 1991 No. 472 The Environmental Protection (Prescribed Processes and Substances) Regulations* 1991, HMSO, London

# M29  Mebeverine

**CAS Registry No.** 3625-06-7

**Synonyms** 3,4-dimethoxybenzoic acid, 4-[ethyl[2-(4-methoxyphenyl)-1-methylethyl]amino]butyl ester; veratric acid, 4-[ethyl(p-methoxy-α-methylphenethyl)amino]butyl ester

**Mol. Formula** $C_{25}H_{35}NO_5$                    **Mol. Wt.** 429.56

**Uses** Gastrointestinal antispasmodic.

**Occurrence** Has been detected at levels of 0.29 $\mu g\ l^{-1}$ in river water (1).

## Physical properties

**M. Pt.** 129-131°C.

## Solubility

Organic solvent: ethanol

## Any other adverse effects to man

Cystic fibrosis patient developed a perforated stercoral ulcer (2,3).

## Any other comments

Administered as the hydrochloride or embonate.

## References

1.  Richardson, M. L. et al *J. Pharm. Pharmacol.* 1985, **37**, 1-12
2.  Hassan, W. *Lancet* 1990, **335**, 1225
3.  Whitehead, A. M. *Lancet* 1990, **336**, 446

# M30  Mebhydrolin

CAS Registry No. 524-81-2
Synonyms 2,3,4,5-tetrahydro-2-methyl-5-(phenylmethyl)-1*H*-pyrido[4,3-*b*]indole;
5-benzyl-2,3,4,5-tetrahydro-2-methyl-1*H*-pyrido[4,3-b]indole;
3-methyl-9-benzyl-1,2,3,4-tetrahydro-γ-carboline; *N*-methyl-9-benzyl
tetrahydro-γ-carboline; Incidal
**Mol. Formula** $C_{19}H_{20}N_2$ **Mol. Wt.** 276.38
**Uses** Antimuscarinic and central sedative (antihistamine) used in the relief of
hypersensitivity reactions and in pruritic skin disorders.
**Occurrence** Has been detected in river water at levels of 0.15 μg l$^{-1}$ (1).

## Physical properties

**M. Pt.** 95°C; **B. Pt.** $_1$ 207-215°C.

## Solubility
Organic solvent: methanol, ethanol, acetone, chloroform

## Mammalian and avian toxicity

### Metabolism and pharmacokinetics
Absorbed slowly from the gastrointestinal tract. Following extensive metabolism only
small amounts of unchanged drug are detected in urine. Plasma $t_{1/2} \approx 4$ hr (2).

## Any other adverse effects to man

Granulocytopenia and agranulocytosis reported (2).

## Any other comments

Administered as the base or as the napadisylate salt.

## References

1. Richardson, M. L. et al *J. Pharm. Pharmacol.* 1985, **37**, 1-12
2. *Martindale. The Extra Pharmacopoeia* 30th ed., 1993, The Pharmaceutical Press, London

# M31   Mecarbam

$(CH_3CH_2O)_2P(S)SCH_2C(O)N(CH_3)CO_2CH_2CH_3$

**CAS Registry No.** 2595-54-2

**Synonyms** *S*-[*N*-ethoxycarbonyl-*N*- methylcarbamoylmethyl] *O,O*-diethyl phosphorodithioate; ethyl *N*-[diethoxythiophosphorylthio]acetyl-*N*-methylcarbamate; 6-ethoxy-2-methyl-3-oxo-7-oxa-5-thia-2-aza-6-phosphanonanoic acid, ethyl ester, 6-sulfide; (mercaptoacetyl)methylcarbamic acid, ethyl ester, *S*-ester with *O,O*-diethyl phosphorodithioate; Marfotoks; MS 1053; Murfotox; Murphotox; Pestan

**Mol. Formula** $C_{10}H_{20}NO_5PS_2$    **Mol. Wt.** 329.38

**Uses** Insecticide/acaricide.

## Physical properties

**B. Pt.** $_{0.02}$ 144°C; **Specific gravity** $d^{20}$ 1.222.

**Solubility**
Water: $<1$ g $l^{-1}$. Organic solvent: aliphatic hydrocarbons, miscible with alcohols, ketones, esters and aromatic and chlorinated hydrocarbons

## Occupational exposure

**Supply classification** toxic.
**Risk phrases** Toxic in contact with skin and if swallowed (R24/25)
**Safety phrases** Wear suitable protective clothing and gloves – If you feel unwell, seek medical advice (show label where possible) (S36/37, S44)

## Ecotoxicity

**Fish toxicity**
$LC_{50}$ (24, 48, 96 hr) harlequin fish 8, 7 and 4 μg $l^{-1}$ respectively (1).

## Environmental fate

**Degradation studies**
Persists in soil for 4 to 6 wk (2).

**Abiotic removal**
Below pH 3 subject to hydrolysis (2).

## Mammalian and avian toxicity

**Acute data**
$LD_{50}$ oral rat, mouse 36-106 mg $kg^{-1}$ (3,4).
$LC_{50}$ (6 hr) inhalation rat 0.7 mg l air$^{-1}$ (2).
$LD_{50}$ dermal rat 380->1220 mg $kg^{-1}$ (2,5).
$LD_{Lo}$ subcutaneous guinea pig 50 mg $kg^{-1}$ (6).

**Sub-acute data**
Oral (6 month) rat 1.6 mg $kg^{-1}$ day$^{-1}$ no effects, but 4.56 mg $kg^{-1}$ day$^{-1}$ caused slight depression in growth rate (2).

**Metabolism and pharmacokinetics**

Hydrolysis, oxidative desulfuration and degradation of the carbamoyl moiety was the principal metabolic pathway in the rat. A minor pathway was O-deethylation (2).

**Legislation**

Limited under EC Directive on Drinking Water Quality 80/778/EEC. Pesticides: maximum admissible concentration 0.1 µg l$^{-1}$ (7).
Included in Schedule 6 (Release into Land: Prescribed Substances) Statutory Instrument No. 472, 1991 (8).

**Any other comments**

WHO Class Ib; EPA Toxicity Class I; TDI (human) 0.002 mg kg$^{-1}$ (2).

**References**

1. Tooby, T. E. et al *Chemistry and Industry* 1975
2. *The Agrochemicals Handbook* 3rd ed., 1991, RSC, London
3. Frear, E. H. (Ed.) *Pesticide Index* 5th ed., 1976, College Science Publications, State College, PA
4. *Guide to the Chemicals Used in Crop Protection* 6th ed., 1973, Information Canada, Ottawa
5. *World Rev. Pest Control* 1970, **9**, 119
6. *J. Econ. Entomol.* 1969, **62**, 934
7. *EC Directive Relating to the Quality of Water Intended for Human Consumption* 1982, 80/778/EEC, Office for Official Publications of the European Communities, 2 rue Mercier, L-2985 Luxembourg
8. *S. I. 1991 No. 472 The Environmental Protection (Prescribed Processes and Substances) Regulations* 1991, HMSO, London

# M32   Mecoprop

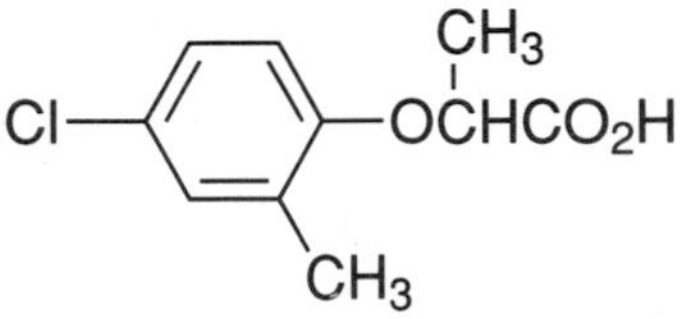

**CAS Registry No.** 7085-19-0 (racemate)
**Synonyms** 2-(4-chloro-2-methylphenoxy)-(±)- propanoic acid;
2-[(4-chloro-*O*-tolyl)oxy]-(±)-propionic acid; Anicon P; CMPP; Isocarnox; MCPP; Mechlorprop; [*RS*]-2-[4-chloro-*O*-tolyloxy]propionic acid
**Mol. Formula** $C_{10}H_{11}ClO_3$ **Mol. Wt.** 214.65
**Uses** Herbicide.

**Physical properties**

**M. Pt.** 93-94°C; **Partition coefficient** log $P_{ow}$ 0.1004 (1); **Volatility** v.p 2.33 × 10$^{-6}$ mmHg at 20°C.

## Solubility
Water: 620 mg $l^{-1}$ at 20°C. Organic solvent: acetone, chloroform, diethyl ether, ethanol, ethyl acetate

## Ecotoxicity

### Fish toxicity
$LC_{50}$ (96 hr) trout 150-220 mg $l^{-1}$ (1).
$LC_{50}$ (96 hr) bluegill sunfish >100 mg $l^{-1}$ (1).

## Environmental fate

### Anaerobic effects
Persistence was seen when incubated with primary digesting sludge at 35°C. The time required to achieve 10% theoretical reduction was >75 days and net gas production was −30% of the theoretical gas production. After 6 wk some detoxification or adaptation of the sludge was apparent (2).

### Degradation studies
The bacterium *Flavobacterium* sp. can grow using the compound as sole source of carbon and energy (3).
Utilized by mixed bacterial cultures enriched from soil with 2,4-dichlorophenol as an intermediate (4).
Incompletely degraded by 2 mixed bacterial cultures. The metabolites which had accumulated during the reaction were degraded upon further incubation (5).
Duration of soil residual activity is ≈2 months (1).

### Abiotic removal
Stable to heat and resistant to hydrolysis, reduction and atmospheric oxidation (6).

### Absorption
Salts have high soil mobility (1).

## Mammalian and avian toxicity

### Acute data
$LD_{50}$ oral mouse 650 mg $kg^{-1}$, rat 930-1166 mg $kg^{-1}$ (6).
$LD_{50}$ oral Japanese quail 740 mg $kg^{-1}$ (1).
$LC_{50}$ (4 hr) inhalation rat >12.5 mg $l^{-1}$ (6).
$LD_{50}$ dermal rat > 4000 mg $kg^{-1}$ (6).
$LD_{50}$ dermal rabbit 900 mg $kg^{-1}$ (1).

### Sub-acute data
Rat (21 day) 65 mg $kg^{-1}$ $day^{-1}$ gave no adverse effects observed (6).

### Carcinogenicity and long-term effects
Rat (2 yr) 1.1 mg $kg^{-1}$ $day^{-1}$ caused no noticeable effects. Rat (210 day) 100 mg $kg^{-1}$ in diet caused only a slight enlargement of the kidneys (6).

### Metabolism and pharmacokinetics
In an acute intoxication case plasma concentration was 298 mg $l^{-1}$, 3-4 hr after ingestion plasma $t_{1/2}$ ≈17 hr. Plasma elimination probably follows first-order kinetics (7).
Predominantly eliminated unchanged in the urine following oral administration in mammals (1).

### Irritancy

Slight skin irritant. Highly irritating to eyes (species unspecified) (1).

### Genotoxicity

*Escherichia coli* PQ37 SOS-Chromotest with and without metabolic activation negative (8).

### Any other adverse effects to man

Acute poisoning causes central nervous system involvement, loss of consciousness, inadequate respiration, rhabdomyolysis with renal failure and muscle cramps. Arterial blood pressure seriously decreased and in one case was shown to be caused by a reduction in peripheral vascular resistance (7).

### Any other adverse effects

In rats caused changes in the number of blood lymphocytes and granulocytes and structural changes in the spleen and thymus (9).

### Legislation

Limited under EC Directive on Drinking Water Quality 80/778/EEC.Pesticides: maximum admissible concentration 0.1 µg l$^{-1}$ (10).
Included in Schedule 6 (Release into Land: Prescribed Substances) Statutory Instrument No. 472, 1991 (11).

### Any other comments

The effect of pH on toxicity to toad studied (12).
WHO Class III; EPA Toxicity Class III (1).
Exists as mixture of 2 optically active isomers of which only the (*R*)-(+)-form is active as a herbicide (1).
Inhibited freshwater bacterial growth on agar plates; mean diameter of inhibition zone 17.0 mm (13).
Readily forms soluble salts, many of which are very soluble in water; 460 g mecoprop-sodium l$^{-1}$ (15˚C); 795 g mecoprop-potassium l$^{-1}$ (0˚C); 580 g mecoprop-bis[2-hydroxyethyl)ammonium l$^{-1}$ (20°C).

### References

1.  *The Agrochemicals Handbook* 3rd ed., 1991, RSC, London
2.  Battersby, N. S. et al *Appl. Environ. Microbiol.* 1989, **55**(2), 433-439
3.  Horvath, M. et al *Appl. Microbiol. Biotechnol.* 1990, **33**(2), 213-216
4.  Oh, K. H. et al *Environ. Contam. Toxicol.* 1991, **47**(2), 222-229
5.  Hallberg, K. B. et al *Curr. Microbiol.* 1991, **23**(2), 65-69
6.  *The Pesticide Manual* 9th ed., 1991, British Crop Protection Council, Farnham
7.  Meulenbelt, J. et al *Hum. Toxicol.* 1988, **7**(3), 289-292
8.  Mersch-Sundermann, V. et al *Zentralbl. Hyg. Umweltmed.* 1989, **189**(2), 135-146
9.  Moeller, T. et al *Z. Gesamte Hyg. Ihre Grenzgeb.* 1989, **35**(5), 258-260
10. *EC Directive Relating to the Quality of Water Intended for Human Consumption* 1982, 80/778/EEC, Office for Official Publications of the European Communities, 2 rue Mercier, L-2985 Luxembourg
11. *S. I. 1991 No. 472 The Environmental Protection (Prescribed Processes and Substances) Regulations* 1991, HMSO, London
12. Nishinchi, Y. *Seitai Kagaku* 1988, **9**(3), 19-26, (Japan.) (*Chem. Abstr.* **111**, 2387r)
13. Milner, C. R. et al *Bull. Environ. Contam. Toxicol.* 1986, **37**, 714-718

# M33   Mefenacet

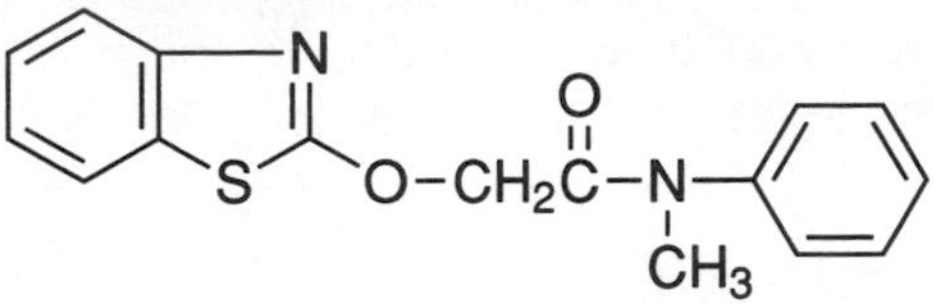

**CAS Registry No.** 73250-68-7

**Synonyms** 2-(2-benzothiazolyloxy)-*N*-methyl-*N*-phenyl acetamide; FOE 1976; NTN 801; 2-benzothiazol-2-yloxy-*N*-methylacetanilide

**Mol. Formula** $C_{16}H_{14}N_2O_2S$          **Mol. Wt.** 298.37

**Uses** Herbicide.

## Physical properties

**M. Pt.** 134.8°C; **Partition coefficient** log $P_{ow}$ 3.230 (1); **Volatility** v.p. $8.25 \times 10^5$ mmHg at 100°C.

## Solubility

Water: 4 mg $l^{-1}$ at 20°C. Organic solvent: dichloromethane, n-hexane, toluene, isopropanol

## Ecotoxicity

**Fish toxicity**

$LC_{50}$ (96 hr) trout, carp 6.8-8.0 mg $l^{-1}$ (1).

## Mammalian and avian toxicity

**Acute data**

$LD_{50}$ oral rat, mouse, dog >5000 mg $kg^{-1}$ (1).

$LC_{50}$ (4 hr) inhalation rat 0.02 mg $l^{-1}$ air (dust) (1).

$LD_{50}$ dermal rat, mouse >5000 mg $kg^{-1}$ (1).

**Carcinogenicity and long-term effects**

Oral (2 yr) rat 100 mg $kg^{-1}$ no-effect level (1).

## Legislation

Limited under EC Directive on Drinking Water Quality 80/778/EEC.Pesticides: maximum admissible concentration 0.1 µg $l^{-1}$ (2).

Included in Schedule 6 (Release into Land: Prescribed Substances) Statutory Instrument No. 472, 1991 (3).

The log $P_{ow}$ value exceeds the European Community recommended level 3.0 (6th and 7th amendments) (4).

## Any other comments

WHO Class Table 5 (1).

## References

1. *The Agrochemicals Handbook* 3rd ed., 1991, RSC, London
2. *EC Directive Relating to the Quality of Water Intended for Human Consumption* 1982, 80/778/EEC, Office for Official Publications of the European Communities, 2 rue Mercier, L-2985 Luxembourg

3.  *S. I. 1991 No. 472 The Environmental Protection (Prescribed Processes and Substances) Regulations 1991*, HMSO, London
4.  *1967 Directive on Classification, Packaging, and Labelling of Dangerous Substances 67/548/EEC; 6th Amendment EEC Directive 79/831/EEC; 7th Amendment EEC Directive 91/32/EEC 1991*, HMSO, London

# M34   Mefanamic acid

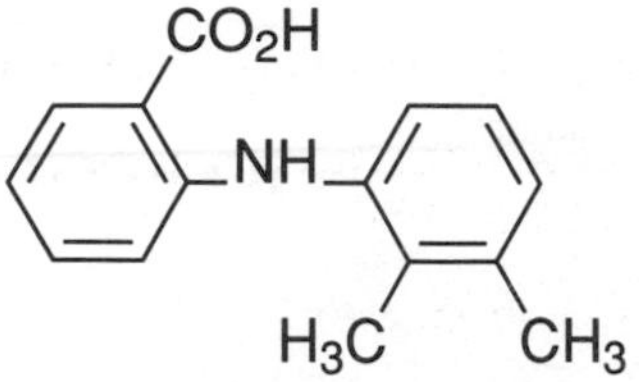

**CAS Registry No.** 61-68-7

**Synonyms** 2-[(2,3-dimethylphenyl)amino]benzoic acid; *N*-2,3-xylylanthranilic acid; C.I. 473; Coslan; Mefacit; mephenamic acid; Ponstan; Ponstil

**Mol. Formula** $C_{15}H_{15}NO_2$        **Mol. Wt.** 241.29

**Uses** Anti-inflammatory, analgesic and antipyretic.

**Occurrence** Detected at levels of 1.17 µg mefanamic $l^{-1}$ in river water (1).

## Physical properties

**M. Pt.** 230-231°C (effervescence).

**Solubility**

Water: 0.041 g $l^{-1}$ at 25°C and pH 7.1. Organic solvent: chloroform, ethanol, methanol

## Mammalian and avian toxicity

**Acute data**

$LD_{50}$ oral mouse, rat 525, 740 mg $kg^{-1}$, respectively (2,3).

$LD_{50}$ intravenous mouse, cat, rat 96-112 mg $kg^{-1}$ (4,5).

$LD_{50}$ intraperitoneal mouse 120 mg $kg^{-1}$ (6).

**Metabolism and pharmacokinetics**

Absorbed from the human gastrointestinal tract with peak plasma concentrations occurring 2–4 hr after ingestion. $t_{1/2}$ 2 to 4 hr. Bound to plasma proteins. Over 50% may be recovered as unchanged drug or conjugated metabolites in the urine (7). Has been detected in human breast milk (8).

**Sensitisation**

Skin rashes, urticaria and occasionally allergic glomerulonephritis. May precipitate asthma (7).

## Any other adverse effects to man

Gastrointestinal disturbances, particularly diarrhoea, are the most common side effects (7).

Nephrotoxicity had been reported as a problem in elderly patients (9).

Bullous pemphigoid and fixed eruptions of the skin (10-12).

Central nervous system toxicity, especially convulsions, and coma have occurred in overdose cases (13-15).

Haemolytic anaemia, pancytopenia, agranulocytosis and bone-marrow aplasia recorded (7).

## References

1.  Richardson, M. L. et al *J. Pharm. Pharmacol.* 1985, **37**, 1-12
2.  *J. Pharmacol.* 1971, **2**, 259
3.  *Toho Igakkai Zasshi* 1981, **28**, 99
4.  *Yakugaku Zasshi* 1969, **89**(10), 1392
5.  *Curr. Med. Res. Opin.* 1976, **4**, 17
6.  *Arzheim.–Forsch.* 1969, **19**, 36
7.  *Martindale. The Extra Pharmacopoeia* 30th ed., 1993, The Pharmaceutical Press, London
8.  Buchanan, R. A. et al *Curr. Ther. Res.* 1968, **10**, 592-596
9.  Taha, A. *Br. Med. J.* 1985, **291**, 661-662
10. Shepherd, A. N. et al *Postgrad. Med. J.* 1986, **62**, 67-68
11. Wilson, C. L. et al *Br. Med. J.* 1986, **293**, 1243
12. Long, C. C. et al *Br. J. Dermatol.* 1992, **126**, 409-411
13. Court, H. et al *Adverse Drug React. Acute Poisoning Rev.* 1984, **3**, 1-21
14. Gossinger, H. et al *Lancet* 1982, **ii**, 384
15. Hendrickse, M. T. *Lancet* 1988, **ii**, 1019

# M35   Mefluidide

**CAS Registry No.** 53780-34-0

**Synonyms** *N*-[2,4-dimethyl-5-[[(trifluoromethyl)sulfonyl]amino]phenyl]acetamide; Embark; MBR 12325; VEL 3973; 5′-[1,1,1-trifluoromethanesulfonamido]acet-2′,4′-xylidide

**Mol. Formula** $C_{11}H_{13}F_3N_2O_3S$    **Mol. Wt.** 310.30

**Uses** Plant growth regulator, herbicide.

## Physical properties

**M. Pt.** 183-185°C; **Volatility** v.p. <9.75 × $10^{-5}$ mmHg at 25°C.

## Solubility

Water: 0.18 g $l^{-1}$ at 25°C. Organic solvent: acetone, methanol, 1-octanol, dichloromethane, benzene

## Ecotoxicity

### Fish toxicity

$LC_{50}$ (96 hr) rainbow trout, bluegill sunfish > 100 mg $l^{-1}$ (1).

### Invertebrate toxicity
$LC_{50}$ (6 hr) *Daphnia magna* 40 mg l$^{-1}$ (1).

## Environmental fate

### Degradation studies
$t_{1/2}$ < 1 wk in soil. 5-Amino-2,4-dimethyltrifluoromethane sulfone anilide is a metabolite (2).

## Mammalian and avian toxicity

### Acute data
$LD_{50}$ oral mouse, rat 1920-4000 mg kg$^{-1}$ (3,4).
$LD_{50}$ oral mallard duck, bobwhite quail >4620 mg kg$^{-1}$ (2).
$LD_{50}$ dermal rabbit >4000 mg kg$^{-1}$ (2).

### Sub-acute data
$LC_{50}$ (8 day) oral mallard duck, bobwhite quail > 10,000 mg kg$^{-1}$ in diet (2).
Oral (90 day) rat 6000 mg kg$^{-1}$ in diet no-effect level (2).
Oral (90 day) dog 1000 mg kg$^{-1}$ in diet no effect level (2).

### Carcinogenicity and long-term effects
Oral rat (2 yr) 100 mg kg$^{-1}$ diet, no effect level (2).

### Metabolism and pharmacokinetics
Excreted unchanged when orally administered to mammals (2).

### Irritancy
Mild irritant to rabbit eye. Dermal rabbit non-irritating (2).

## Legislation
Limited under EC Directive on Drinking Water Quality 80/778/EEC. Pesticides: maximum admissible concentration 0.1 µg l$^{-1}$ (5).
Included in Schedule 6 (Release into Land: Prescribed Substances) Statutory Instrument No. 472, 1991 (6).

## Any other comments
Relation of pH to toxicity studied in the toad (7).
WHO Class Table 5 (2).

## References

1. *The Pesticide Manual* 9th ed., 1991, British Crop Protection Council, Farnham
2. *The Agrochemicals Handbook* 3rd ed., 1991, RSC, London
3. *Farm Chemicals Handbook* 1983, Meister Publishing, Willoughby, OH
4. *Agricultural Chemicals* 1976/77, Book 3, Thomson Publications, Fresno, CA
5. *EC Directive Relating to the Quality of Water Intended for Human Consumption* 1982, 80/778/EEC, Office for Official Publications of the European Communities, 2 rue Mercier, L-2985 Luxembourg
6. *S. I. 1991 No. 472 The Environmental Protection (Prescribed Processes and Substances) Regulations* 1991, HMSO, London
7. Nishiuchi, Y. *Seitai Kagaku* 1988, **9**(3), 19-26

# M36   Melamine

$$H_2N \diagdown \begin{array}{c} N \\ \diagup \end{array} NH_2$$

(structure of melamine: 1,3,5-triazine ring with three NH₂ groups)

**CAS Registry No.** 108-78-1

**Synonyms** 1,3,5-triazine-2,4,6-triamine; cyanuramide; cyanurotriamide; isomelamide; Theoharn; 2,4,6-triaminotriazine

**Mol. Formula** $C_3H_6N_6$          **Mol. Wt.** 126.12

**Uses** Forms synthetic resins with formaldehyde which are used in the synthesis of laminates, surface coating resins, plastics, paper products, textile resins.

## Physical properties

**M. Pt.** <250°C; **B. Pt.** sublimes; **Specific gravity** $d_4^{16}$ 1.573; **Volatility** v.p. 50 mmHg at 315°C; v. den. 4.34.

### Solubility
Organic solvent: very slightly soluble in hot ethanol

## Ecotoxicity

### Bioaccumulation
Non accumulative or low accumulative (1).

## Environmental fate

### Nitrification inhibition
*Nitrosomonas* sp. 100 mg $l^{-1}$ caused no inhibition of ammonium oxidation (2).
50 mg $l^{-1}$ (threshold concentration) caused no inhibition of nitrifying bacteria (3).

## Mammalian and avian toxicity

### Acute data
LD$_{50}$ oral rat, mouse 3160-3300 mg $kg^{-1}$ (4,5).
LD$_{100}$ intraperitoneal mouse 800 mg $kg^{-1}$ (5).
LD$_{100}$ intraperitoneal rat 3200 mg $kg^{-1}$ (5).

### Carcinogenicity and long-term effects
No adequate evidence for carcinogenicity to humans or animals, IARC classification group 3 (6).
National Toxicology Program tested (103 wk) rats and mice orally 2250 or 4500 ppm via food. No evidence of carcinogenicity in mice (♂ and ♀) or ♀ rats was reported. However, ♀ rats developed chronic inflammation of the kidneys and ♂ mice developed acute and chronic inflammation and epithelial hyperplasia of the urinary bladder and an increased incidence of bladder stones. ♂ rats developed transitional-cell carcinomas with incidences significantly higher in the high dose group. There was an association between bladder stones and bladder tumours (7,8).

**Teratogenicity and reproductive effects**
Intraperitoneal administration of 70 mg kg$^{-1}$ to rats on days 5 and 6, 8 and 9, or 12 and 13 of gestation caused no toxic effects or gross malformations of foetuses (9).

**Metabolism and pharmacokinetics**
50% of an oral dose of 250 mg kg$^{-1}$ recovered in urine of rats after 6 hr (10).
[$^{14}$C]-melamine ♂ rat 90% excreted in urine within 24 hr; analysis showed that the melamine was not metabolised. Negligible radioactivity was detected in exhaled air or faeces whilst it was found concentrated in the kidney and bladder. By 24 hr virtually no radioactivity was detected in tissues (11).

**Irritancy**
0.5% patch test negative for irritancy and allergic reactions (12).

## Genotoxicity

*Salmonella typhimurium* TA98, TA100, TA1535, TA1537, with and without metabolic activation negative (13).
*Escherichia coli* microscreen assay (metabolic activation unspecified) positive (14).
*In vitro* mouse lymphoma L5178Y tk$^+$/tk$^-$ with and without metabolic activation negative (15).
*In vitro* Chinese hamster ovary cells with and without metabolic activition sister chromatid exchanges equivocal and chromosomal aberrations negative (16).
*Drosophila melanogaster* fed on a diet containing melamine did not develop sex-linked recessive lethal mutations (17).
*In vitro* mouse bone marrow did not induce micronuclei (18).

## Any other adverse effects to man

Exposure to melamine-formaldehyde resins has caused dermatitis (19).

## Any other adverse effects

X-ray and infrared analysis of bladder stones obtained from melamine treated ♂ rats indicated the principal component was melamine (20).

## Legislation

Limited under EC Directive on Drinking Water Quality 80/778/EEC. Pesticides: maximum admissible concentration 0.1 µg l$^{-1}$ (21).
Included in Schedule 6 (Release into Land: Prescribed Substances) Statutory Instrument No. 472, 1991 (22).

## Any other comments

Toxicity and hazards reviewed (23,24).

## References

1.  *The list of the existing chemical substances tested on biodegradability by microorganisms or bioaccumulation in fish body* 1987, Chemicals Inspection and Testing Institute, Japan
2.  Hockenbury, M. R. et al *J.-Water Pollut. Control Fed.* 1977, **4a**(5), 768-777
3.  Zhadanova, N. Y. et al *Biokhim. Ochistka Stochnulzh Vod Predpr. Khim.* 1962, 73-127 (Russ.)
4.  *Toxicol. Appl. Pharmacol.* 1984, **72**, 292
5.  *Patty's Industrial Hygiene and Toxicology* 3rd rev. ed., 1981, **24**, 2769-2772, John Wiley & Sons, New York

6. *IARC Monograph* 1987, **Suppl. 7**, 65
7. *National Toxicology Program Research and Testing Division* 1992, Report No.TR-245, NIEHS, Research Triangle Park, NC
8. Haseman, J. K. et al *Environ. Mol. Mutagen.* 1990, **16**(Suppl. 18) 15-31
9. Thiersch, J. B. *Proc. Soc. Exp. Biol. Med.* 1957, **94**, 36-40
10. Lipscitz, W. L. et al *J. Pharmacol. Exp. Ther.* 1945, **83**, 235-401
11. Mast, R. W. et al *Food. Chem. Toxicol.* 1983, **21**, 807
12. Lisi, P. et al *Contact Dermatitis* 1987, **17**(4), 212-218
13. Haworth, S. et al *Environ. Mutagen.* 1983, (Suppl. 1), 3-142
14. Rossman, T. G. et al *Mutat. Res.* 1991, **260**(4), 349-367
15. McGregor, D. B. et al *Environ. Mol. Mutagen.* 1988, **12**(1), 85-154
16. Galloway, S. M. et al *Environ. Mol. Mutagen.* 1987, **10**(Suppl. 10), 1-175
17. Rohrborn, G. *Z. Verburyslehre* 1962, **93**, 1-6
18. Mast, R. W. et al *Environ. Mutagen.* 1982, **4**, 340-341
19. *IARC Monograph* 1986, **39**, 333-346
20. *Evaluation of Urolithiasis Induction by Melamine in male Weanling Fisher 344 Rats* 1982. American Cyanamid Co. Wayne, NJ
21. *EC Directive Relating to the Quality of Water Intended for Human Consumption* 1982, 80/778/EEC, Office for Official Publications of the European Communities, 2 rue Mercier, L-2985 Luxembourg
22. *S. I. 1991 No. 472 The Environmental Protection (Prescribed Processes and Substances) Regulations* 1991, HMSO, London
23. Vernon, P. A. et al *Acute Toxic. Data* 1990, **1**(2), 110
24. *Dangerous Prop. Ind. Mater. Res.* 1988, **8**(4), 40-41

# M37   Melatonin

**CAS Registry No.** 73-31-4
**Synonyms** *N*-[2-(5-methoxy-1*H*-indol-3-yl)ethyl]acetamide
**Mol. Formula** $C_{13}H_{16}N_2O_2$                    **Mol. Wt.** 232.28
**Uses** Potential to alleviate jet lag and delayed sleep phase syndrome.
**Occurrence** Hormone produced in the pineal gland.

## Physical properties
**M. Pt.** 116-118°C.

## Mammalian and avian toxicity

**Acute data**
$LD_{50}$ intravenous mouse 180 mg kg$^{-1}$ (1).

**Metabolism and pharmacokinetics**

Precursor is $N$-acetylserotonin (2).

The major metabolite is 6-hydroxymelatonin (3).

Radiolabelled ($^3$H) melatonin injected into spotted skunk jugular vein 1-3 hr before darkness. Relatively high amounts were found, after 22 min, in the pineal gland, liver, pituitary gland, ovary and kidney. Relatively small amounts were found in the brain, uterus, temporalis muscle and pancreas. The lung had the least (4).

## Genotoxicity

*Salmonella typhimurium* TA97, TA98, TA100 with and without metabolic activation negative (3).

*In vitro* mouse spleen lymphocytes DNA replication was unaffected by concentrations between 0.02 and $2 \times 10^{-5}$ g (2).

## Any other adverse effects

Daily administration of 25 µg to gerbil led to increased eye lens weight. Wistar/NIN rats increased lens weight was dependent on time of administration (noon and 5pm) (5). Intramuscular administration over 30 days to adult ♂ Indian finches inhibited thyroid activity, seasonal gonadal growth and activity and luteinizing hormone-dependent plumage pigmentation. Inhibition of body weight growth was dependent on time of administration and dose (6).

Intravenously or intracerebroventricularly to rats did not produce any cardiovascular or ventilatory effects (7).

## Any other comments

Physiology, pharmacology, actions in optic and central nervous systems, oncology interactions reviewed (8-11).

Animals studies indicate that concentrations of γ-aminobutyric acid and serotonin are increased in the hypothalalmus and midbrain and pyridoxal kinase activity is enhanced. Involved in oestrus, inhibition of gonadal development and protective changes in skin colouration. Secreted in a diurnal rhythm and may influence sleep patterns (12).

## References

1. *US Army Armament Research and Development Command NX02739* Chemical Systems Laboratory, Aberdeen Proving Ground, MD
2. Pawlikowski, M. et al *J. Neural Transm.* 1988, **73**(2), 161-166
3. Neville, S. et al *J. Pineal Res.* 1989, **6**(1), 73-76
4. Berria, M. et al *J Pineal Res.* 1990, **8**(2), 129-136
5. Joshi, B. N. et al *Adv. Pineal Res.* 1989, **3**, 253-357
6. Gupta, B. B. P. et al *Gen. Comp. Endocrinol.* 1987,**6**(3), 451-456
7. Ekholm, S. et al *Med. Biol.* 1986, **64**(4), 215-219
8. Claustrat, B. et al *Nucl. Med. Biol.* 1990, **17**(7), 625-632
9. Quay, W. B. *Adv. Pineal Res.* 1987, **22**, 141-154
10. Kvetnoi, I. M. et al *Eksp. Onkol.* 1986, **8**(4), 11-15 (Russ.) (*Chem. Abstr.* **106**, 173775s)
11. Blask, D. E. et al *Adv. Pineal Res.* 1989, **3**, 259-263
12. *Martindale. The Extra Pharmacopoeia* 30th ed., 1993, The Pharmaceutical Press, London

# M38   Melphalan

$$HO_2CHCH_2C\underset{NH_2}{}-\underset{}{\bigcirc}-N(CH_2CH_2Cl)_2$$

**CAS Registry No.** 148-82-3
**Synonyms** 4-[bis(2-chloroethyl)amino]-L-phenylalanine; Alkeran; 3025 C.B.;
levofalan; L-PAM; phenylalanine mustard; L-sarcolysine
**Mol. Formula** $C_{13}H_{18}Cl_2N_2O_2$          **Mol. Wt.** 305.21
**Uses** Antineoplastic chemotherapeutic agent.

## Physical properties
**M. Pt.** 182-183°C (decomp.).

**Solubility**
Organic solvent: ethanol, propylene glycol

## Mammalian and avian toxicity

### Acute data
$LD_{50}$ oral rat 11200 µg $kg^{-1}$ (1).
$LD_{Lo}$ intravenous man 8140 mg $kg^{-1}$ (2).
$LD_{50}$ intravenous mouse 23 mg $kg^{-1}$ (3).
Intravenous dog 0.75, 1.5 or 3.0 mg $kg^{-1}$. After 8-9 days post-administration lethargy,
vomiting, general muscle weakness, dehydration and diarrhoea was observed at high
dose. Necropsy revealed widespread lymphocytic depletion in all lymphoid tissues
and in the bone marrow and spleen marked hypoplasia of haematopoietic tissue. A
dose-related decrease in white blood cells occurred (4).

### Carcinogenicity and long-term effects
Sufficient evidence for carcinogenicity to humans and animals, IARC classification
group 1 (5).
Patients developed acute non-lymphocytic leukaemia following treatment (6-10).
Dose-related increase in lung tumours in mice and peritoneal sarcomas in rats
following intraperitoneal injection (11).
$TD_{50}$ for tumours in ♀ mouse, ♂ mouse, ♀ rat, ♂ rat were respectively 0.10, 0.11,
0.078 and 0.047 mg $kg^{-1}$ $day^{-1}$ (12).

### Metabolism and pharmacokinetics
600 µg $kg^{-1}$ body weight administered to cancer patients. When given orally systemic
availability was very variable. Mean plasma terminal phase $t_{1/2}$ 90 min and mean
24 hr excretion in urine was 10.9% of dose. Mean peak plasma level was 280 ng $ml^{-1}$
(13).
Following absorbtion it is quickly distributed throughout body water and is believed
to be inactivated mainly by spontaneous hydrolysis. 50-60% is protein bound initially
rising to 80-90% after 12 hr (14).

### Sensitisation
Skin rashes and hypersensitivity reactions may occur (species unspecified) (14).

## Genotoxicity

*Salmonella typhimurium* TA1535 with and without metabolic activation positive (15).
*Salmonella typhimurium* TA1537, TA1538 with and without metabolic activation negative (16).
*In vitro* human lymphocytes induced sister chromatid exchange in cells from heroin addicts and controls with incidence higher in the former (17).
*Drosophila melanogaster* wing spot test positive (18).
Mouse germ cell mutagenicity studies gave positive result in dominant lethal test in ♂, heritable translocations and postmeiotic and premeiotic specific locus mutations (19).
*In vivo* bone marrow cells of ♂ B6C3F mouse micronucleus test and chromosomal aberrations induction positive (20).

## Any other adverse effects to man

Neutropenia, thrombocytopenia, bone-marrow depression, gastrointestinal effects, haemolytic anaemia, vasculitis, pulmonary fibrosis, oedema, neurotoxicity, vesiculation of the skin and thrombophlebitis reported (14).

## Any other adverse effects

Ovarian failure in 3 young women given high doses (21).

## Legislation

Land disposal regulated under the Federal Resource Conservation and Recovery Act (22).

## References

1. *Iyakuhin Kenkyu* 1979, **10**, 710
2. *N.Z. Med. J.* 1984, **97**, 816
3. Battelle Columbus Laboratories *Report 1987* NTIS
4. *NTIS Report* PB 84-162486, Natl. Tech. Inf. Ser., Springfield, VA
5. *IARC Monograph* 1987, **Suppl. 7**
6. Greene, M. H. et al *New Engl. J. Med.* 1982, **307**, 1416-1421
7. Einhorn, N. et al *Cancer* 1982, **49**, 2234-2241
8. Law, I. P. et al *Oncology* 1977, **34**, 20-24
9. Gonzalez, F. et al *Ann. Intern. Med.* 1977, **86**, 440-443
10. Greene, M. H. et al *Ann. Intern. Med.* 1986, **105**, 360-367
11. *IARC Monograph* 1975, **9**, 167-180
12. Kaldor, J. M. et al *Eur. J. Cancer Clin. Oncol.* 1988, **24**(4), 703-711
13. Albert, D. S. et al *Clin. Pharmacol. Ther.* 1979, **26**, 737-745
14. *Martindale. The Extra Pharmacopoeia* 30th ed., 1993, The Pharmaceutical Press, London
15. Matheson, D. et al *Drug Chem. Toxicol.* 1978, **1**, 277-304
16. Minnich, V. et al *Cancer* 1976, **38**, 1253-1258
17. Vassiliades, N. et al *Forensic Sci. Int.* 1991, **50**(2), 269-276
18. Tripathy, N. K. et al *Mutat. Res.* 1990, **242**(3), 169-180
19. Shelby, M. D. et al *Environ. Health Perspect.* 1993, **100**, 283-291
20. Shelby, M. D. et al *Environ. Mol. Mutagen.* 1989, **13**(4), 339-342
21. Kellie, S. J. et al *Lancet* 1987, **i**, 1425
22. US EPA *Fed. Regist.* 1991, **56**(21), 3864-3928

# M39   Menazon

CAS Registry No. 78-57-9

**Synonyms** phosphorodithioic acid, *S*-[(4,6-diamino-1,3,5-triazin-2-yl)methyl]
*O,O*-dimethyl ester; phosphorodithioic acid, *S*-[(4,6-diamino-*s*-triazin-2-yl)methyl]
*O,O*-dimethyl ester; *S*-[(4,6-diamino-1,3,5-triazin-2-yl)methyl] *O,O*-dimethyl
phosphorodithioate; Azidithion; Sayfos

**Mol. Formula** $C_6H_{12}N_5O_2PS_2$                **Mol. Wt.** 281.30

**Uses** Aphicide. Fungicide.

**Occurrence** Residues have been found in meat and fish.

## Physical properties

**M. Pt.** 160-162°C (decomp.); **Volatility** v.p. $1 \times 10^{-6}$ mmHg at 25°C.

### Solubility

Water: 240 mg $l^{-1}$ at 20°C. Organic solvent: 2-ethoxyethanol, ethanol glycol,
2-methoxyethanol, tetrahydrofurfurol

## Occupational exposure

**Supply classification** harmful.

**Risk phrases** Harmful if swallowed (R22)

## Ecotoxicity

### Fish toxicity

$LC_{50}$ (48 hr) harlequin fish 220 mg $l^{-1}$ (1).

Carp, $\geq 0.1$ mg $l^{-1}$ decreased blood corpuscles, cardiovascular activity and growth rate (2).
Rainbow trout and stickleback 30 mg $l^{-1}$ caused death in 2-4 hr. Test conditions: total
hardness 0-17 mg $l^{-1}$; methyl orange alkalinity 14 mg $l^{-1}$; and pH 7.6 (3).

## Mammalian and avian toxicity

### Acute data

$LD_{50}$ oral chicken 487 mg $kg^{-1}$ (4).

$LD_{50}$ oral rat, mouse 427, 890 mg $kg^{-1}$, respectively (5,6).

$LD_{50}$ oral mouse 427 mg $kg^{-1}$ (6).

### Sub-acute data

Dermal rabbit (24 hr) 500-800 mg $kg^{-1}$ caused no local or systemic effects. Oral rat
(90 day) no-effect level 30 mg $kg^{-1}$ diet (1).

### Carcinogenicity and long-term effects

Oral rat (2 yr) 250, 1000, 4000 mg $kg^{-1}$ diet, showed no significant effect other than
inhibition of cholinesterase activity (1).

## Legislation

Limited under EC Directive on Drinking Water Quality 80/778/EEC. Pesticides: maximum admissible concentration 0.1 $\mu g\, l^{-1}$ (7).

Pesticides are included in Schedule 6 (Release into Land: Prescribed Substances) Statutory Instrument No. 472, 1991 (8).

## References

1. *The Pesticide Manual* 6th ed., 1979, 331, British Crop Protection Council, Croydon
2. Guseva, S. S. et al *Biol. Nauki (Moscow)* 1988, (1), 53-58 (Russ.) (*Chem. Abstr.* **100**, 145069n)
3. McPhee, C. et al *Lethal Effects of 2014 Chemicals to Fish* 1989, EPA 560/6-89-001, PB 89-156715, Washington, DC
4. *Toxicol. Appl. Pharmacol.* 1965, **7**, 606
5. *USDA Information Memorandum* 1966, **20**, 14, Agricultural Research Service, Beltsville, MD
6. *Spec. Publ.- Entomol. Soc. Am.* 1978, **78**(1), 48
7. *EC Directive Relating to the Quality of Water Intended for Human Consumption* 1982, 80/778/EEC, Office for Publications of the European Communities, 2 rue Mercier, L-2985 Luxembourg
8. *S. I. 1991 No.472, The Environmental Protection (Prescribed Processes and Substances) Regulations* 1991, HMSO, London

# M40   Menthol

**CAS Registry No.** 89-78-1

**Synonyms** 5-methyl-2-(1-methylethyl)cyclohexanol, (1α,2β,5α)-; *cis*-1,3-*trans*-1,4-menthol; menthacamphor; menthomenthol; peppermint camphor

**Mol. Formula** $C_{10}H_{20}O$                    **Mol. Wt.** 156.27

**Uses** In liqueurs, perfumery, confectionery, cigarettes, cough drops. A topical antipruritic. Veterinary local anaesthetic and antiseptic and internally as carminative and gastric sedative.

**Occurrence** *Mentha* spp.

## Physical properties

**M. Pt.** 41-43°C; **B. Pt.** 212°C; **Flash point** 110.6°C; **Specific gravity** 0.890; **Volatility** v.p. 1 mmHg at 56°C; v. den. 5.38.

## Solubility

Organic solvent: ethanol, chloroform, diethyl ether, petrol ether, glacial acetic acid, liquid petroleum

## Ecotoxicity

### Bioaccumulation
Confirmed to be non-accumulative or low accumulative (1).

## Mammalian and avian toxicity

### Acute data
$LD_{50}$ oral rat 3180 mg $kg^{-1}$ (2).
$LD_{Lo}$ oral cat 1500 mg $kg^{-1}$ (3).
$LD_{Lo}$ intraperitoneal rat, mouse 1500-1800 mg $kg^{-1}$ (4,5).
$LD_{50}$ intramuscular rat 10 g $kg^{-1}$ (6).
$LD_{Lo}$ intravenous cat $\approx$37 mg $kg^{-1}$ (5).

### Metabolism and pharmacokinetics
Excreted in urine and bile as a glucuronide (7).

### Irritancy
750 µg instilled into rabbit eye caused severe irritation (8).

### Sensitisation
May cause hypersensitivity reactions including contact dermatitis (7).

## Any other adverse effects to man

Ataxia, confusion, nystagmus, euphoria and diplopia were experienced when recommended dose of a preparation was exceeded (9).
Apnoea and collapse reported in infants following local application to nostrils. Taken orally, nausea, vomiting, vertigo, abdominal pain, ataxia, drowsiness and coma have been reported (7).

## Any other comments

Toxicity reviewed (10).

## References

1. *The list of the existing chemical substances tested on biodegradability by microorganisms or bioaccumulation in fish body* 1987, Chemicals Inspection and Testing Institute, Japan
2. *Food Cosmet. Toxicol.* 1964, **2**, 327
3. *Abdernalden's Handbuch der Biologischen Arbeitsmethoden* 1935, **4**, 1289
4. *Ann. Pharm. Fr.* 1952, **10**, 481
5. *Arch. Int. Pharmacodyn. Ther.* 1939, **63**, 43
6. *Naunyn-Schmiedeberg's Arch. Pharmakol.* 1954, **222**, 244
7. *Martindale. The Extra Pharmacopoeia* 30th ed., 1993, The Pharmaceutical Press, London
8. *Am. J. Ophthalmol.* 1946, **29**, 1363
9. O'Mullane, N. M. et al *Lancet* 1982, **i**, 1121
10. *BIBRA Toxicity Profile* 1991, British Industrial Biological Research Association, Carshalton

# M41   D,L-Menthol

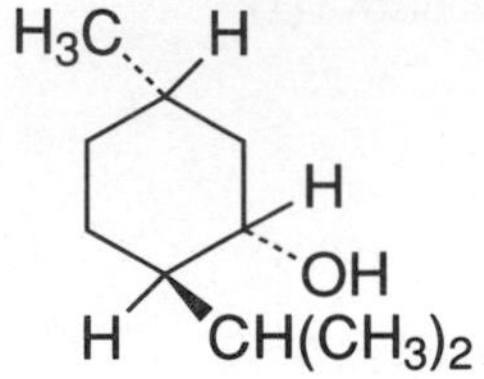

H3C  H

H

OH

H   CH(CH3)2

**CAS Registry No.** 15356-70-4
**Synonyms** 5-methyl-2-(1-methylethyl)cyclohexanol, $(1\alpha,2\beta,5\alpha)$-($\pm$)-;
($\pm$)-*cis*-1,3-, *trans*-1,4,-menthol; ($\pm$)-menthol
**Mol. Formula** $C_{10}H_{20}O$                                      **Mol. Wt.** 156.27
**Uses** Cigarettes, pharmaceutical rubs and liniments, nasal sprays, antipruritic lotions, expectorants, cough drops and foot powders.
**Occurrence** In the oils of *Mentha arvensis*.

## Physical properties

**M. Pt.** 43-44°C; **B. Pt.** 9 103-104°C; **Flash point** 91°C; **Specific gravity** $d_{15}^{15}$ 0.904.

**Solubility**
Organic solvent: ethanol, diethyl ether, acetone, benzene

## Mammalian and avian toxicity

**Acute data**
$LD_{50}$ oral rat, mouse 2900-3100 mg $kg^{-1}$ (1,2).
$LD_{Lo}$ subcutaneous rat 1 g $kg^{-1}$ (3).
$LD_{Lo}$ intraperitoneal cat, rabbit 1500-2000 mg $kg^{-1}$ (1,3).
$LD_{50}$ intraperitoneal guinea pig 865 mg $kg^{-1}$ (4).

**Carcinogenicity and long-term effects**
National Toxicology Program administered orally in feed 3750 or 7500 ppm to Fischer 344 rats and either 2000 or 4000 ppm to B6C3F1 mice for 103 wk. Mean body weights were slightly lower but no other effects were observed. ♀ rats showed the only dose-related trend in mortality. Non-carcinogenic (5,6).

**Irritancy**
Dermal rabbit (24 hr) 500 mg caused well defined erythema and slight oedema (7).

## Genotoxicity

*Salmonella typhimurium* TA97, TA98, TA100, TA1535 with and without metabolic activation negative (8).
*In vitro* mouse lymphoma L5178Y with and without metabolic activation negative (9).

## Any other adverse effects

Lethal to *in vitro* mouse lymphoma L5178Y cells at 200 µg $ml^{-1}$ with and without metabolic activation. Metabolic activation tended to reduce the toxicity of non-lethal concentrations (9).

---

## References

1. *Report 44A* 1967, 59, FAO UN, Washington, DC
2. *Q. J. Pharm. Pharmacol.* 1932, **5**, 233
3. *Muench. Med. Wochenschr.* 1926, **73**, 2011
4. *Ann. Pharm. Fr.* 1952, **10**, 481
5. *National Toxicology Program Research and Testing Division* 1992, Report No. TR-098, NIEHS, Research Triangle Park, NC
6. Haseman, J. K. et al *Environ. Mol. Mutagen.* 1990, **16**(Suppl. 18), 15-31
7. *Food Cosmet. Toxicol.* 1976, **14**, 443
8. Zeiger, E. et al *Environ. Mol. Mutagen.* 1988, **11**(Suppl. 12), 1-157
9. Myhr, B. et al *Environ. Mol. Mutagen.* 1991, **18**(1), 51-83

# M42   *p*-Menthane hydroperoxide

**CAS Registry No.** 80-47-7
**Synonyms** 1-methyl-1-(4-methylcyclohexyl)ethyl hydroperoxide;
*p*-menth-8-yl hydroperoxide
**Mol. Formula** $C_{10}H_{20}O_2$                    **Mol. Wt.** 172.27

## Physical properties

**Flash point** 71°C; **Specific gravity** $d_4^{15.5}$ 0.910-0.925.

## Occupational exposure

**UN No.** 2125; **HAZCHEM Code** 2W; **Conveyance classification** organic peroxide.

## Genotoxicity

*Salmonella typhimurium* TA97 with and without metabolic activation weakly positive (1).
*Salmonella typhimurium* TA98, TA100, TA1535, TA1537 with and without metabolic activation negative (1).

## References

1. Zeiger, E. et al *Environ. Mol. Mutagen.* 1988, **11**(Suppl.12), 1-157

# M43   Mephosfolan

**CAS Registry No.** 950-10-7
**Synonyms** diethyl 4-methyl-1,3-dithiolan-2-ylidenephosphoramidate;
(4-methyl-1,3-dithiolan-2-ylidene)phosphoramidic acid, diethyl ester;
phosphonadithioimidocarbonic acid, cyclic propylene *P,P*-diethyl ester; Cytrolane;
AC 47470; imidocarbonic acid, phosphonodithio-, cyclic propylene
**Mol. Formula** $C_8H_{16}NO_3PS_2$                     **Mol. Wt.** 269.32
**Uses** Insecticide. Acaricide.
**Occurrence** Environmental pollutant.

## Physical properties

**B. Pt.** $_{0.001}$ 120°C; **Specific gravity** $d^{26}$ 1.539; **Partition coefficient** log $P_{ow}$ 1.0414 (1).

**Solubility**
Water: 57 g $l^{-1}$ at 25°C. Organic solvent: acetone, benzene, dichloroethane, ethanol,
toluene, xylene

## Occupational exposure

**Supply classification** Very toxic.
**Risk phrases** Very toxic in contact with skin and if swallowed (R27/28)
**Safety phrases** Wear suitable protective clothing, gloves and eye/face protection – In
case of accident or if you feel unwell, seek medical advice immediately (show label
where possible) (S36/37/39, S45)

## Ecotoxicity

**Fish toxicity**
$LC_{50}$ (96 hr) rainbow trout 2 mg $l^{-1}$ (1).
$LC_{50}$ (96 hr) carp 55 mg $l^{-1}$ (1).

## Environmental fate

**Degradation studies**
$t_{1/2}$ in soil 7-13 day. Rate of degradation decreased with decreased soil pH (2).
Accelerated degradation by soil following as little as one exposure can reduce
efficiency as an insecticide; >95% degraded in previously treated hop yard soils after
8 wk at 15°C compared with 23-35% in untreated soil (3).

## Mammalian and avian toxicity

**Acute data**
$LD_{50}$ oral chicken 2.8 g $kg^{-1}$ (4).
$LD_{50}$ oral rat, mouse, quail 9-13 mg $kg^{-1}$ (1,5).
$LD_{50}$ dermal rabbit 10-29 mg $kg^{-1}$ (1,6).

## Sub-acute data

Oral rat (90 day) 15 mg kg$^{-1}$ diet caused a reduction in erythrocyte and brain cholinesterase activities, but no significant effect on weight gain (1).

## Legislation

Limited under EC Directive on Drinking Water Quality 80/778/EEC. Pesticides: maximum admissible concentration 0.1 µg l$^{-1}$ (7).
Included in Schedule 6 (Release into Land: Prescribed Substances) Statutory Instrument No. 472. 1991 (8).

## Any other comments

WHO Class 1a; EPA Toxicity Class 1 (1).
LD$_{50}$ topical application bee 3.5 µg bee$^{-1}$ (1).

## References

1. *The Agrochemicals Handbook* 3rd ed. 1991, RSC, London
2. Suett, D. L. et al *Crop. Prot.* 1990, **9**(1), 44-51
3. Suett, D. L. *Toxicol. Environ. Chem.* 1988, **18**(1), 37-49
4. *Experientia* 1974, **30**, 63
5. *Bull. Entom. Soc. Am.* 1969, **15**, 122
6. Spencer, E. Y. *Guide to Chemicals Used in Crop Protection* 7th ed. 1982, 367, Research Institute, Agriculture Canada Ottawa, Publication 1093
7. *EC Directive Relating to the Quality of Water Intended for Human Consumption* 1982, 80/778/EEC, Office for Official Publications of the European Communities, 2 rue Mercier, L-2985, Luxembourg
8. *S. I. 1991, No. 472 The Environmental Protection (Prescribed Process and Substances) Regulation* 1991, HMSO, London

# M44   Meprobamate

$$NH_2C(O)OCH_2C(CH_3)(CH_2CH_2CH_3)CH_2OC(O)NH_2$$

**CAS Registry No.** 57-53-4
**Synonyms** Miltown; 2-methyl-2-propyl-1,3-propanediol dicarbamate; carbamic acid, 2-methyl-2-propyltrimethylene ester
**Mol. Formula** $C_9H_{18}N_2O_4$         **Mol. Wt.** 218.25
**Uses** Anxiolytic with hypnotic, sedative and some muscle relaxant properties. Used to treat anxiety disorders and for short-term management of insomnia, although largely superseded by benzodiazepines.
**Occurrence** 2.6 µg l$^{-1}$ predicted in River Lee (1).

## Physical properties

**Solubility**
Water: 0.34% (w/w) at 20°C. Organic solvent: acetone, ethanol

## Mammalian and avian toxicity

**Acute data**
LD$_{50}$ oral starling 127 mg kg$^{-1}$ (2).

$LD_{50}$ oral rat, mouse 750, 1000 mg $kg^{-1}$, respectively (3,4).
$LD_{50}$ intravenous mouse, rat 230, 350 mg $kg^{-1}$, respectively (5,6).

### Teratogenicity and reproductive effects

Two studies reported significantly higher rates of severe congenital defects following administration to women during early pregnancy (7,8), while one did not (9). Teratogenic and embryotoxic in mice after subcutaneous administration of 0.75 mg $g^{-1}$ on day 10 or 12 of pregnancy. Principal abnormalities were club foot or affected fingers and toes (10).

Not teratogenic in rats or rabbits (dose, duration and route unspecified) (11).
1000 $\mu$g $ml^{-1}$ affected crown-rump length, differentiation, yolk sac size and vascularisation, and 300 $\mu$g $l^{-1}$ affected morphology in whole rat embryos cultured during early stages of organogenesis (12).

Rodent embryo limb bud cell culture *in vitro* teratogen screen positive (inhibited cartilage formation) (13).

### Metabolism and pharmacokinetics

Elimination from blood in rabbits unaffected by daily doses of 0.4 g $kg^{-1}$ ethanol and accelerated by 1.6 g $kg^{-1}$ (14).

Readily absorbed from human gut; plasma concentrations peak after 1-3 hr. Widely distributed; extensively metabolised in the liver and excreted in urine mainly as a hydroxylated metabolite and its glucuronide conjugate. <10% excreted unchanged. $t_{1/2}$ 6-16 hr, but may be prolonged after chronic administration. Crosses the placenta and appears in human milk at 4 × maternal plasma concentration (15).

### Sensitisation

Hypersensitivity may occur in patients, causing rashes, urticaria, purpura, angiodema, bronchospasm and anuria. Erythema multiforme and exfoliative or bullous dermatitis have also been reported (15).

### Any other adverse effects to man

Barbiturate-like profile of action. When administered at 600-3600 mg to men with previous histories of drug misuses, they reported mood and sleep sedation effects but not tranquilisation. Likelihood of abuse equals or exceeds that of Lorazepam, but is less likely to produce adverse behavioural effects (16).

Lowest lethal dose in humans (route unspecified) 441 mg $kg^{-1}$ (17).

Side effects include drowsiness, nausea, vomiting, diarrhoea, paraesthesia, weakness, headache, paradoxical excitement, dizziness, ataxia and visual disturbances. Hypotension, tachycardia and cardiac arrhythmias may occur. Blood disorders including agranulocytosis, eosinophilia, leucopenia, thrombocytopenia and aplastic anaemia have been reported. Caution is required for use in patients with impaired liver or kidney function, mental or respiratory function depression. It may cause convulsions in epileptics. It is unsafe for patients with acute porphyria and its effects are enhanced by alcohol and other central nervous system depressants (15).

### Legislation

Controlled substance (depressant) listed in US Code of Federal Regulations, TA621, Part 1308.14, 1987.

---

# References

1. Richardson, M. L. et al *J. Pharm. Pharmacol.* 1985, **37**, 1-12
2. Schafer, E. W. et al *Arch. Environ. Contam. Toxicol.* 1983, **12**, 355-382
3. *J. Med. Chem.* 1972, **15**, 998
4. Foye, W. O. (Ed.) *Principles of Medicinal Chemistry* 1974, Lea and Febiger, Philadelphia, PA
5. *Eur. J. Med. Chem.* 1977, **12**, 447
6. *Prensa Med. Argent.* 1957, **44**, 915
7. Milkovich, L. et al *New Engl. J. Med.* 1974, **291**, 1268-1271
8. Crombie, D. L. et al *New Engl. J. Med.* 1975, **293**, 198-199
9. Hartz, S. C. et al *New Engl. J. Med.* 1975, **292**, 726-728
10. Nishikawa, M. *Kaibogaku Zasshi* 1963, **38**, 258-263
11. Jelovsek, F. R. et al *Obstet. Gynecol.(N. Y.)* 1989, **74**(4), 624-636
12. Cicurel, L. et al *Experientia* 1988, **44**, 833-840
13. Renault, J-Y. et al *Teratog., Carcinog., Mutagen.* 1989, **9**, 83-96
14. Moriya, F. et al *Arukoru Kenkyu to Yakubutsu Izon* 1989, **24**(5), 391-396
15. *Martindale. The Extra Pharmacopoeia* 30th ed., 1993, The Pharmaceutical Press, London
16. Roache, J. D. et al *J. Pharmacol. Exp. Ther.* 1987, **243**(3), 978-988
17. Arena, J. M. *Poisoning; Toxicology, Symptoms, Treatments* 2nd ed., 1970, Charles C. Thomas, Springfield, IL

# M45   Mepronil

**CAS Registry No.** 55814-41-0
**Synonyms** 2-methyl-*N*-[3-(1-methylethoxy)phenyl]benzamide;
3'-isopropoxy-*o*-toluanilide; 3'-isopropoxy-2-methylbenzanilide
**Mol. Formula** $C_{12}H_{19}NO_2$                **Mol. Wt.** 209.29
**Uses** Systemic fungicide, mainly used to treat *Puccinia* and *Typhula* infection in wheat.

## Physical properties

**B. Pt.** 93.6-94.2°C; **Partition coefficient** log $P_{ow}$ 3.66; **Volatility** v.p. $4.2 \times 10^{-7}$ mmHg at 20°C.

## Solubility

Water: 12.7 mg $l^{-1}$ at 20°C. Organic solvent: acetone, methanol, acetonitrile, benzene, hexane

## Ecotoxicity

### Fish toxicity

$LC_{50}$ (96 hr) carp, rainbow trout 8-10 mg $l^{-1}$ (1).

## Mammalian and avian toxicity

### Acute data
$LD_{50}$ oral rat, mouse >10 g $kg^{-1}$ (1).
$LD_{50}$ oral hen >8 g $kg^{-1}$ (1).
$LC_{50}$ (6 hr) rat >1.32 mg $l^{-1}$ (1).
$LD_{50}$ dermal rabbit, rat >10 g $kg^{-1}$ (1).

### Carcinogenicity and long-term effects
No effect level in 2 yr feeding trial in ♂ rats 5.9 mg $kg^{-1}$; ♀ rats 72.9 mg $kg^{-1}$ (1).

### Irritancy
Non-irritating to eyes and skin of rabbits (1).

## Legislation

The log $P_{ow}$ value exceeds the European Community recommended level 3.0 (6th and 7th amendments) (2).

## Any other comments

WHO Class Table 5; EPA Toxicity Class IV (1).
$LD_{50}$ oral honey bee >0.1 mg $bee^{-1}$ (1).

## References

1. *The Agrochemicals Handbook* 3rd ed., 1991, RSC, London
2. *1967 Directive on Classification, Packaging and Labelling of Dangerous Substances 67/548/EEC; 6th Amendment EEC Directive 79/831/EEC; 7th Amendment EEC Directive 91/321/EEC* 1991,HMSO, London

# M46    2-Mercaptoacetic acid

## $HSCH_2CO_2H$

**CAS Registry No.** 68-11-1
**Synonyms** 2-mercaptoethanoic acid; thioglycolic acid; thioranic acid
**Mol. Formula** $C_2H_4O_2S$          **Mol. Wt.** 92.12
**Uses** Reagent for iron, tin, molybdenum and silver. In manufacture of thioglycolates.

## Physical properties

**M. Pt.** −16.5°C; **B. Pt.** $_{11}$ 104-106°C; **Specific gravity** $d_4^{20}$ 1.325; **Volatility** v.p. 10 mmHg at 18°C.

### Solubility
Water: miscible. Organic solvent: miscible with ethanol, diethyl ether, chloroform, benzene

## Occupational exposure

**Supply classification** toxic (≥2%), harmful.
**Risk phrases** ≥10% – Toxic by inhalation, in contact with skin and if swallowed – Causes burns – ≥5%<10% – Toxic by inhalation, in contact with skin and if

swallowed – Irritating to eyes and skin – ≥2%<5% – Toxic by inhalation, in contact with skin and if swallowed – Causes burns – ≥0.2%<2% – Harmful by inhalation and in contact with skin (R23/24/25, R34, R23/24/25, R36/38, R23/24/25, R34, R20/21)
**Safety phrases** Keep out of reach of children – Avoid contact with eyes – Take off immediately all contaminated clothing – After contact with skin, wash immediately with plenty of water (S2, S25, S27, S28)

## Ecotoxicity

**Fish toxicity**
5 ppm (24 hr) did not cause symptoms of sickness in rainbow trout, bluegill sunfish, yellow perch or goldfish. Test conditions: pH 7.0; dissolved oxygen content 7.5 ppm; total hardness (soap method) 300 ppm; methyl orange alkalinity 300 ppm; free carbon dioxide 5 ppm; and temperature 12.8°C (1).

## Environmental fate

**Degradation studies**
Activated sludge (special respirometer) BOD 20°C 1-5 days observed, feed 662 mg $l^{-1}$ acclimation: 1 day: no removal or oxidation (2).

## Mammalian and avian toxicity

**Acute data**
$LD_{50}$ oral mouse 242 mg $kg^{-1}$ (3).
$LD_{Lo}$ dermal rabbit 300 mg $kg^{-1}$ (4).
$LD_{50}$ intraperitoneal rat, mouse 70, 138 mg $kg^{-1}$, respectively (3,5).

**Irritancy**
Corrosive irritant to skin (6).

## References

1. Wood, E. M. *The toxicity of 3400 chemicals to fish* 1987, EPA Report 500/6-87-002, PB 87-200-275, Washington, DC
2. Ludzack, F. J. et al *J.- Water Pollut. Control Fed.* 1960, **32**, 1173
3. *Z. Gesamte Hyg. Ihre Grenzgeb.* 1974, **20**, 575
4. *Gig. Tr. Prof. Zabol.* 1969, **13**, 48
5. *Arzneim.- Forsch.* 1981, **31**, 1713
6. *Ind. Med.* 1946, **15**, 669

# M47   2-Mercaptobenzimidazole

**CAS Registry No.** 583-39-1
**Synonyms** 2-benzimidazolethiol; *o*-phenylenethiourea; autigene MB; USAF EK-6540
**Mol. Formula** $C_7H_6N_2S$             **Mol. Wt.** 150.20

## Ecotoxicity

### Invertebrate toxicity
$EC_{50}$ (5-30 min) *Photobacterium phosphoreum* 86.4 ppm Microtox test (1).

## Mammalian and avian toxicity

### Acute data
$LD_{50}$ oral rat, mouse 476, 1250 mg kg$^{-1}$, respectively (2,3).
$LD_{50}$ intravenous mouse 180 mg kg$^{-1}$ (4).
$LD_{50}$ intraperitoneal mouse 200 mg kg$^{-1}$ (5).

### Irritancy
500 mg applied to skin or eyes of rabbits for 24 hr caused mild irritation (2).
100 mg instilled into rabbits' eyes caused moderate to severe irritation with corneal involvement persisting >24 hr but with recovery within 21 days (6).

## Genotoxicity
*Salmonella typhimurium* TA98, TA100, TA1535, TA1537 with and without metabolic activation negative (7).

## Any other comments
Physical properties reviewed (8).
Human health effects, experimental toxicity and workplace experience reviewed (9).
Impurities in 2-mercaptobenzimidizole produced in Shanghai, China reported as responsible for mutagenicity in the Ames test (10).

## References
1. Kaiser, K. L. E. et al *Water Pollut. Res. J. Can.* 1991, **26**(3), 361-431
2. Marhold, J. V. *Prehled Prumyslove Toxikologie: Organicke Latky* 1986, Prague
3. *Gig. Tr. Prof. Zabol.* 1964, **8**(7), 39
4. *Report NX 04376* US Army Armament Res. Dev. Command, Aberdeen Proving Ground, MD
5. *NTIS Report* AD 277-689, Natl. Tech. Inf. Ser., Springfield, VA
6. Sugai, S. et al *J. Toxicol. Sci.* 1990, **15**, 245-262
7. Zeiger, E. et al *Environ. Mutagen.* 1987, **9**(Suppl.9), 1-109
8. Mitsuo, T. *Setchaku* 1988, **32**(2), 75-80
9. *ECETOC Technical Report No. 30(4)* 1991, European Chemical Industry Ecology and Toxicology Centre, B-1160 Brussels
10. Feng, B. et al *Zhonghua Laodong Weisheng Zhiyebing Zazhi* 1988, **6**(5), 261-264

# M48  2-Mercaptobenzothiazole

**CAS Registry No.** 149-30-4
**Synonyms** Captax; MBT; 2-benzothiazolethiol; Rotax; NCI-C56 519; USAF GY-3
**Mol. Formula** $C_7H_5NS_2$                    **Mol. Wt.** 167.25

**Uses** Rubber vulcanisation accelerator. Salts used as fungicide.

## Physical properties

**M. Pt.** 164-175°C.

**Solubility**
Organic solvent: ethanol, diethyl ether, benzene, acetone, carbon tetrachloride

## Ecotoxicity

**Fish toxicity**
Fatal concentration (48 hr) goldfish 2 mg $l^{-1}$ (1).

**Invertebrate toxicity**
$EC_{50}$ (5-30 min) *Photobacterium phosphoreum* 0.681 ppm Microtox test (2).
Killed microorganisms present in municipal treatment plant sludge in 2-3 days (3).

**Bioaccumulation**
No or low accumulation (4).

## Environmental fate

**Degradation studies**
Non-biodegradable (4).
Did not degrade during Pitter's kinetic test of biodegradation (3).
TOC, COD and $BOD_5$ 57.8, 230 and 5.5 mg $l^{-1}$ $O_2$ respectively; $\gamma$ irradiation did not improve biodegradation (3).

## Mammalian and avian toxicity

**Acute data**
$LD_{50}$ oral rat 1490-1560 mg $kg^{-1}$ (5,6).
$LD_{50}$ intraperitoneal mouse, rat 100, 300 mg $kg^{-1}$, respectively (7,8).

**Carcinogenicity and long-term effects**
National Toxicology Program tested rats and mice via gavage 5 days $wk^{-1}$ for 103 wk. No evidence of carcinogenicity in $\sigma$ mice dosed with 375 or 750 mg $kg^{-1}$ equivocal evidence in ♀ mice, some evidence of carcinogenicity in $\sigma$ and ♀ rats (9).
Increased tumour rates in NTP study were adrenal gland phaeochromocytoma in $\sigma$ and ♀ rats; pituitary gland adenoma in ♀ rats, leukaemia and adenoma of the pancreas and preputial gland in $\sigma$ rats; and liver adenoma and carcinoma in ♀ mice (10).
Reduced weight gain and renal histopathological changes reported in mice fed 1920 ppm in the diet for 20 months (5).

**Metabolism and pharmacokinetics**
Rats and guinea pigs absorbed 16.1-17.5% and 33.4% respectively of a topical radiolabelled dose. 72 hr after intravenous administration 90.9-100% appeared in urine of rats; results were similar after oral administration. No unchanged mercaptobenzothiazole was detected in urine, but two metabolites were identified; one was a thioglucuronide derivative and the other possibly a sulfonic acid derivative (11).

**Irritancy**
5 and 10% solutions caused skin irritation in guinea pigs (12).

### Sensitisation
Cross-sensitisation to morpholinylmercaptobenzothiazole reported in guinea pigs (12). Hypersensitivity reported in a miner, and cross sensitisation to the rubber additive 2,2′-dibenzothiazyl disulfide (13).

### Genotoxicity
*Salmonella typhimurium* TA98, TA100, TA1535, TA1537 with and without metabolic activation negative. Equivocal results also reported (14).
Mouse lymphoma L5178Y cell assay without metabolic activation negative, with metabolic activation positive; induced sister chromatid exchanges and chromosomal aberrations in Chinese hamster ovary cells with metabolic activation, but only sister chromatid exchanges without metabolic activation (15,16).

### Any other comments
Report compiled by the German Advisory Committee on Existing Chemicals under the German Chemical Act (17).
Human health effects, experimental toxicity, environmental effects, ecotoxicology and exposure levels reviewed (18).

### References
1.  McKee, J. E. et al *Water Quality Criteria* 1963, Resources Agency of California, State Water Quality Control Board, CA
2.  Kaiser, K. L. E. et al *Water Pollut. Res. J. Can.* 1991, **26**(3), 361-431
3.  Tolgyessy, P. et al *J. Radioanal. Nucl. Chem.* 1986, **107**(5), 315-320
4.  *List of existing chemical substances tested on biodegradability by microorganisms or bioaccumulation in fish body* 1987, Chemical Inspection and Testing Division, Japan
5.  Ogawa, Y. et al *Eisei Shikensho Hokoku* 1989, (107), 44-50 (Jap.) (*Chem. Abstr.* **113**, 110565x)
6.  Randall, D. J. et al *Acute Toxic. Data* 1990, **1**(1), 62
7.  *NTIS Report* AD 277-689, Natl. Tech. Inf. Ser., Springfield, VA
8.  *Med. Prac.* 1965, **16**, 35
9.  *National Toxicology Program Research and Testing Division*, 1992, Report No. TR-332, NIEHS, Research Triangle Park, NC 27709
10. Haseman, J. K. et al *Environ. Mol. Mutagen.* 1990, **16** (Suppl.16), 15-31
11. El Dareer, S. M. et al *J. Toxicol. Environ. Health* 1989, **27**(2), 65-84
12. Wang, X. et al *Contact Dermatitis* 1988, **19**(1), 11-15
13. Hegyi, E. *Prakt. Lek.* 1990, **42**(10), 436-437 (Slo.) (*Chem. Abstr.* **115**, 188972p)
14. Zeiger, E. et al *Environ. Mol. Mutagen.* 1987, **9**(Suppl. 9), 1-109
15. Zeiger, E. et al *Environ. Mol. Mutagen.* 1990, **16**(Suppl. 18), 1-14
16. Zeiger, E. et al *Environ. Mol. Mutagen.* 1990, **16**(Suppl. 18), 55-137
17. Haltrich, W. G. *Vom Wasser* 1989, **73**, 11024 (Ger.) (*Chem. Abstr.* **115**, 188972p)
18. *ECETOC Technical Report No. 30(4)* 1991, European Chemical Industry Ecology and Toxicology Centre, B-1160 Brussels

# M49    Mercaptobenzothiazole disulfide

**CAS Registry No.** 120-78-5
**Synonyms** benzothiazole disulfide; dibenzothiazyl disulfide; Altax; Vulkacit DM; MBTS; mercaptobenzthiazyl ether; 2,2′-dithiobis[benzothiazole]
**Mol. Formula** $C_{14}H_8N_2S_4$                    **Mol. Wt.** 332.49
**Uses** Accelerator in the rubber industry.

## Physical properties
**M. Pt.** 175°C; **Specific gravity** 1.5.

## Solubility
Water: slightly soluble. Organic solvent: methanol, ethanol

## Mammalian and avian toxicity

### Acute data
$LD_{50}$ oral mouse 7 g $kg^{-1}$ (1).
$LD_{50}$ intraperitoneal mouse, rat 100, 2600 mg $kg^{-1}$, respectively (1,2).
$LD_{50}$ intravenous mouse 180 mg $kg^{-1}$ (3).

### Carcinogenicity and long-term effects
Equivocal tumorigenic effects reported in rats after oral administration of a total of 172 g $kg^{-1}$ intermittently over 78 wk (4).

### Teratogenicity and reproductive effects
200 mg $kg^{-1}$ injected into the stomach of ♀ rats either on day 1 and 3 of oestrus (before pregancy) or on day 4 and 11 of pregnancy, and to ♂ rats twice with a 3 day interval resulted in a lengthened oestrus cycle, 30-46% reduced fertility and post-implantation embryomortality (5).

## Genotoxicity
*Salmonella typhimurium* TA98, TA100, TA1535, TA1537, TA1538 with and without metabolic activation negative. (6).
*Escherichia coli* $WP_2$ uvrA⁻ with or without metabolic activation equivocal (6).
Mouse lymphoma L5178Y cell assay with or without metabolic activation negative(6).
BALB/3T3 mouse cell transformation assay with or without metabolic activation negative(6).
Did not induce chromosomal aberrations in Chinese hamster ovary cells with or without metabolic activation (6).

## Any other comments
Human health effects, experimental toxicity and environmental effects reviewed (7).

## References

1. *Int. Polymer Sci. Technol.* 1976, **3**, 93
2. *NTIS Report* AD 277-689, Natl. Tech. Inf. Ser., Springfield, VA
3. *Report NX 02251* US Army Armament Res. Dev. Command, Chem. Systems Lab., NIOSH Exchange Chemicals, Aberdeen Proving Ground, MD
4. *NTIS Report* PB 223-159 Natl. Tech. Inf. Ser., Springfield, VA
5. Aleksandrov, S. E. *Bull. Exp. Biol. Med.* 1982, **93**, 107-109
6. Hinderer, R. K. et al *Environ. Mutagen.* 1983, **5**, 193-215
7. *ECETOC Technical Report No. 30(4)* 1991, European Chemical Industry Ecology and Toxicology Centre, B-1160 Brussels

# M50   6-Mercaptopurine

**CAS Registry No.** 50-44-2

**Synonyms** 6-purinethiol; purine-6-thiol; 1,7-dihydro- 6*H*-purine-6-thione; 6-thiopurine; Leukerin; Ismipur; Purinethol

**Mol. Formula** $C_5H_4N_4S$                                    **Mol. Wt.** 152.18

**Uses** Antineoplastic used for treatment chiefly in maintenance programmes, of acute myeloblastic leukaemia and chronic myelocytic leukaemia. It acts as an antimetabolite.

## Physical properties

**M. Pt.** 313-314°C (decomp.).

**Solubility**
Organic solvent: ethanol

## Mammalian and avian toxicity

**Acute data**
$LD_{50}$ oral mouse 260 mg $kg^{-1}$ (1).
$LD_{50}$ intraperitoneal ♂ mouse 240 mg $kg^{-1}$ (2).

**Carcinogenicity and long-term effects**
No evidence for carcinogenicity in limited studies in experimental animals, insufficient data to make an evaluation in humans (3).

**Teratogenicity and reproductive effects**
Transient profound oligospermia reported in a man treated with 150 mg $day^{-1}$ 6-mercaptopurine plus 80 mg $day^{-1}$ prednisone (4).
Of two babies born to a woman treated with 6-mercaptopurine and radiation during both pregnancies, one had severe multiple anomalies including microphthalmia,

corneal opacities, cleft palate and thyroid and ovarian hypoplasia. The mother received busulfan during the pregnancy which resulted in the anomalous baby (5).

Rats fed 1000 μg zinc $g^{-1}$ diet showed fewer adverse effects on reproduction and embryogenesis than those fed less zinc; the rats received 27.5 mg $kg^{-1}$ on day 11 of pregnancy (6).

*In utero* exposed mice were sterile or had small litters and more dead foetuses (7). 50% foetal death occurred in rats after oral administration of 5 mg $kg^{-1}$ on days 7 and 8 of pregnancy (8).

Malformations reported in rats and mice given 50 mg $kg^{-1}$ on days 5-9 or 6-12 (rats) and 6-8 (mice) of pregnancy; central nervous system anomalies occurred at 0.5-1 mg $kg^{-1}$ (9,10).

Gut, liver, limb, palate, mandible and tongue malformations reported in Syrian hamsters after intraperitoneal administration of 5-9 mg $kg^{-1}$ on day 9 of pregnancy; the teratogenic effect was less marked if dosed on day 8 or 10-11 (11).

**Metabolism and pharmacokinetics**
In humans variably and incompletely absorbed from the gut; 40-50% of oral doses have been reportedly absorbed (3,12).

Absolute bioavailability is somewhat lower and varies widely between individuals, which may be due to thiopurine methyltransferase activity (resulting in methylation and inactivation of mercaptopurine instead of formation of active nucleotides) (12). Distributed widely after absorption (12); after intravenous administration to mice concentrations were highest in the liver, gut and other organs but low in the brain (3). It crosses the blood-brain barrier to an extent; it was found in subtherapeutic levels (10% of plasma level) in cerebrospinal fluids (3,12).

Plasma $t_{1/2}$ is 10-90 min after intravenous injection (12) and 90 min after oral administration (3); it is not found in plasma after 8 hr (12).

Activated intracellularly by conversion to nucleotide derivatives (6-thionucleotides) (3,12).

Intracellular metabolites are 6-methylmercaptopurine, ribonucleoside mono-, di- and triphosphates (3).

Rapidly and extensively metabolised in the liver by methylation, oxidation and formation of inorganic sulfates. Considerable amounts are oxidised to thiouric acid by xanthine oxidase (12).

Excreted in urine unchanged (46% of an oral dose and 71% of an intravenous dose over 24 hr) and as 6-thiouric acid; 6-methylthio-2,8-dihydroxypurine; 6-methylthio-8-hydroxypurine glucuronide; inorganic sulfate; and 6-methylsulfinyl-8-hydroxypurine. The first two predominate after intravenous administration and the latter two after oral (3).

In mice given a single intraperitoneal dose of 10 mg $kg^{-1}$ 21.4% was excreted unchanged in urine, 18.9% as 6-thiouric acid and 29.5% as inorganic sulfate. Blood $t_{1/2}$ 14 min in mice and 9 min in rats (3).

# Genotoxicity
*Salmonella typhimurium* TA1535, G46 without metabolic activation positive (13). Induced sister chromatid exchanges in human peripheral lymphocytes and bone marrow cells *in vitro* (14,15).

*In vivo* mouse bone marrow micronucleus test positive (16,17).

Induced chromosomal aberrations in bone marrow cells of rats and Chinese hamster after intraperitoneal administration and in mice after oral or parental administration (18).

## Any other adverse effects to man

Major toxic effect is white cell depression. Higher doses used in the past (2.5-5 mg kg$^{-1}$ day$^{-1}$) caused reversible jaundice. Less common side effects include nausea, mucosal ulceration, skin rash and fever (3).

Crystalluria with haematuria, and skin hyperpigmentation have been observed rarely (12).

Its effects are enhanced by allopurinol. It may diminish warfarin activity. Its hepatoxicity may be enhanced by doxorubicin, and concurrent administration of methotrexate increases plasma concentrations of mercaptopurine (12).

## Any other adverse effects

It had pronounced effects on zinc, copper, calcium and magnesium metabolism in maternal and foetal rat tissue which may play a role in teratogenesis (6).

Predominant toxicity in rats and mice is bone marrow and gut epithelial damage and liver necrosis (3).

## Any other comments

Urine and faeces produced by patients up to 48 hr and 5 days respectively should be handled wearing protective clothing (12).

Destruction of mercaptopurine wastes to non-mutagenic residues by oxidation with potassium permanganate in sulfuric acid has been described (19).

## References

1. *Cesk. Farm.* 1965, **14**, 389
2. Scherf, H. R. et al *Recent Results Cancer Res.* 1975, **52**, 76-87
3. *IARC Monograph* 1981, **26**, 249-266
4. Hinkes, E. et al *J. Am. Med. Assoc.* 1973, **223**, 1490-1491
5. Diamond, I. et al *Pediatrics* 1960, **25**, 85-90
6. Amemiya, K. et al *Teratology* 1986, **34**(3), 321-334
7. Reimers, T. J. et al *Science (Washington, D.C., 1883-)* 1978, **201**, 65-67
8. Thiersch, J. B. *Ann. N. Y. Acad. Sci.* 1954, **60**, 220-227
9. Zunin, C. et al *Minerva Pediatr.* 1955, **7**, 66-71
10. Mercier-Parot, L. et al *C. R. Seances Soc. Biol. Ses Fil.* 1967, **161**, 762-768
11. Shah, R. M. et al *Can. J. Physiol. Pharmacol.* 1979, **57**, 53-58
12. *Martindale. The Extra Pharmacopoeia* 30th ed., 1993, The Pharmaceutical Press, London
13. Benedict, W. F. et al *Cancer Res.* 1977, **37**, 2209-2213
14. Fedortzeva, R. F. et al *Cytologia* 1973, **15**, 1339-1404
15. Zhang, S. et al *Cancer Genet. Cytogenet.* 1988, **31**(2), 157-163
16. *Mutat. Res.* 1988, **204**(2), 307-316
17. *Mutat. Res.* 1990, **234**(3-4), 205-222
18. Frohberg, H. et al *Arch. Toxicol.* 1975, **33**, 209-224
19. *IARC Scientific Publication 73* 1985

# M51   Mercury bisulfate

## HgSO$_4$

**CAS Registry No.** 7783-35-9
**Synonyms** mercuric sulfate
**Mol. Formula** HgO$_4$S                                   **Mol. Wt.** 296.65
**Uses** Battery electrolyte; with sodium chloride for extracting gold and silver from
roasted pyrites.
**Occurrence** Accumulation of mercury in the environment results from atmospheric
deposition. Both natural and anthropogenic sources contribute to atmospheric
deposition; these include volcanoes, land evasion, incinerators, landfills, hazardous
waste sites, sewage treatment plants, coal-combustion power plants and chlor-alkali
production plants. The 3 major mercury species are: elemental mercury Hg$^0$;
inorganic mercury Hg$^{2+}$; and methylmercury CH$_3$Hg$^+$. Refer to mercury and
methylmercury entries for additional toxicity data on mercury salts.

## Physical properties

**Specific gravity** 6.47.

**Solubility**
Water: decomp.

## Ecotoxicity

**Fish toxicity**
Intestinal transport rate for nutrients decreased in snakehead fish exposed to 3 μg l$^{-1}$
for 30 days (1).
Disrupted metabolism in *Notopterus notopterus* exposed to 44 μg l$^{-1}$ for 30 days
(Hg$^{2+}$) (1).
Teratogenic to rainbow trout after exposure of eggs to 0.12-0.21 μg l$^{-1}$ 4 days post
hatch, and after parental exposure to 0.70-0.79 μg l$^{-1}$ for 400 days (Hg$^{2+}$) (1).

**Invertebrate toxicity**
LC$_{50}$ (48 hr) *Daphnia magna* 9.3 μg l$^{-1}$ (Hg$^{2+}$) (1).
EC$_{50}$ (48 hr) *Daphnia magna* 5.2 μg l$^{-1}$ (Hg$^{2+}$) (1).

## Mammalian and avian toxicity

**Acute data**
LD$_{50}$ oral mouse, rat 25, 67 mg kg$^{-1}$, respectively (2,3).
LD$_{50}$ dermal rat 625 mg kg$^{-1}$ (2).
LD$_{50}$ intraperitoneal mouse 6300 μg kg$^{-1}$ (2).

## Legislation

Community Right-To-Know List.
Limited under EC Directive on Drinking Water Quality 80/778/EEC. Mercury:
maximum admissible concentration 1 μg l$^{-1}$ (4).
Included in Schedule 4 (Release into Air: Prescribed Substances) Statutory Instrument
No. 472, 1991 (5).
Quality objectives under EC Directives 82/176/EEC and 84/156/EEC 0.3 mg kg$^{-1}$

(wet weight) in a representative sample of fish flesh; 1 µg l$^{-1}$ (annual mean) total mercury in inland surface waters; 0.5 µg l$^{-1}$ (annual mean) dissolved mercury in estuarine waters; 0.3 µg l$^{-1}$ (annual mean) dissolved mercury in marine waters. A 'standstill' provision applies to concentrations in shellfish or sediments. Limit values under EC Directive 84/156/EEC 0.05 mg l$^{-1}$ effluent and 0.1 g l$^{-1}$ vinyl chloride production capacity for chemical industries using mercury catalysts in vinyl chloride production; 0.05 mg l$^{-1}$ effluent and 5 g kg$^{-1}$ mercury processed for chemical industries using mercury catalysts in other processes; 0.05 mg l$^{-1}$ effluent and 0.7 g kg$^{-1}$ mercury processed for manufacture of mercury catalysts used in vinyl chloride production; 0.05 mg l$^{-1}$ effluent and 0.05 g kg$^{-1}$ mercury processed for manufacture of organic and non-organic mercury compounds (other than mercury catalysts for vinyl chloride production); 0.05 mg l$^{-1}$ effluent and 0.03 g kg$^{-1}$ mercury processed for manufacture of primary batteries containing mercury; 0.05 mg l$^{-1}$ effluent for mercury recovery plants and extraction and refining of non-ferrous metals; 0.05 mg l$^{-1}$ effluent for plants treating toxic wastes containing mercury (6).

## Any other comments

Physico-chemical properties, human health effects, experimental toxicity, environmental effects, ecotoxicology, exposure levels and workplace experience reviewed (7).
Physico-chemical properties, hazards and legislation in France reviewed (8).
Toxicity of inorganic mercury and environmental effects reviewed (9,10).

## References

1. Zillioux, E. J. et al *Environ. Toxicol. Chem.* 1993, **12**, 2245-2264
2. *NTIS Report PB214-270* Natl. Tech. Inf. Ser. Springfield, VA
3. *Gig. Tr. Prof. Zabol.* 1981, **25**(7), 27
4. *EC Directive Relating to the Quality of Water Intended for Human Consumption* 1982, 80/778/EEC, Office for Official Publications of the European Communities, 2 rue Mercier, L-2985 Luxembourg
5. *S. I. 1991 No. 472 The Environmental Protection (Prescribed Processes and Substances) Regulations* 1991, HMSO, London
6. *DoE Circular 7/89: Water and the environment. The implementation of EC directives on pollution caused by certain dangerous substances discharged into the aquatic environment* 1989, HMSO, London
7. *ECETOC Technical Report No. 30(4)* 1991, European Chemical Industry Ecology and Toxicology Centre, B-1160 Brussels
8. *Cah. Notes Doc.* 1989, **137**, 711-716 (Fr.) (Chem. Abstr. **112**, 12422ly)
9. *Environmental Health Criteria 118: Inorganic Mercury* 1991, WHO/IPCS, Geneva
10. *Environmental Health Criteria 86: Mercury-Environmental Aspects* 1989, WHO/IPCS, Geneva

# M52 Mercury

## Hg

CAS Registry No. 7439-97-6
**Mol. Formula** Hg                                          **Mol. Wt.** 200.59
**Uses** Cathode in sodium chloride electrolysis; in the electrical industry, in control instruments in the home and industry and in laboratory and medical instruments. In gold extraction. In dental amalgam.
**Occurrence** Natural mercury arises from the Earth's crust through volcanic gases and evaporation from oceans and amounts to between 25,000 and 125,000 tonnes per yr. Anthropogenic sources also contribute to atmospheric disposition, sources include waste incinerators, landfills, hazardous waste sites, sewage treatment plants, coal-combustion power plants and chlor-alkali production plants. The 3 major mercury species are: elemental mercury $Hg^0$; inorganic mercury $Hg^{2+}$; and methylmercury $CH_3Hg^+$. 99% of atmospheric mercury exists as $Hg^0$. However, particulate mercury accounts for all of the mercury wet and dry deposition (1,2).
The biological effects of mercury depend on dose-response relationships for methylmercury and the organism. Assessment of risks to the health of humans and other organisms who consume organisms that accumulate mercury from aquatic ecosystems requires consideration of the atmosphere as a potentially important source (3).

## Physical properties

**M. Pt.** −38.87°C; **B. Pt.** 356.72°C; **Specific gravity** $d^{25}$ 13.534; **Volatility** v.p. 0.0012 mmHg at 20°C.

## Occupational exposure

**US TLV (TWA)** 0.025 mg m$^{-3}$ (as Hg) intended change; **UK Long-term limit** 0.05 mg m$^{-3}$ (as Hg) under review; **UK Short-term limit** 0.15 mg m$^{-3}$ (as Hg) under review; **UN No.** 2809; **Supply classification** toxic.
**Risk phrases** Toxic by inhalation – Danger of cumulative effects (R23, R33)
**Safety phrases** Keep container tightly closed – If you feel unwell, seek medical advice (show label where possible) (S7, S44)

## Ecotoxicity

**Fish toxicity**
LC$_{50}$ (96 hr) bluegill sunfish, rainbow trout, snakehead fish 0.16-0.9 mg l$^{-1}$ (4-6).
LC$_{50}$ (28 day) rainbow trout 0.005 mg l$^{-1}$ (7).

**Invertebrate toxicity**
EC$_{50}$ (48 hr) *Daphnia magna* 0.0052 mg l$^{-1}$ (8).
LC$_{50}$ (96 hr) *Lymnaea acuminata, Nais communis, Ilyodrilus frantzi, Aplexa hypnorum* 0.023-0.36 mg l$^{-1}$ (4,9,10).

## Environmental fate

**Degradation studies**
BOD, COD and MnO$_4^-$ values showed 40-60% reduction of organic matter when

mercury content in feed solution was ≥400 µg dm$^{-3}$ in bench-scale experiments of mercury toxicity in biological aerobic filters treating synthetic wastewater (11).

**Abiotic removal**
Limited measurements of methylmercury (2–14% of total mercury) indicated that disturbed wetland environments produce more methylmercury than undisturbed (3).

**Absorption**
Colloidal manganese oxides may sorb inorganic mercury and thereby affect cycling in lake waters (3).
High organic matter content increases the transport of mercury from watersheds, therefore increasing the available supply to fish (12,13).

## Mammalian and avian toxicity

**Acute data**
$LC_{50}$ (30 hr) inhalation rabbit 29 mg m$^{-3}$ (14).

**Carcinogenicity and long-term effects**
The WHO reported no evidence that inorganic mercury is carcinogenic, and the EPA classify it as group O (not classifiable as to human carcinogenicity); mercury has not been reviewed by IARC (15).

**Teratogenicity and reproductive effects**
Increased resorptions reported in rats exposed by inhalation to 500 or 1000 µg m$^{-3}$ on days 10-15 of pregnancy. Exposure throughout gestation caused cranial defects at 500 µg m$^{-3}$ and reduced maternal and foetal weights at 1000 µg m$^{-3}$ (16).
Elemental mercury vapour easily penetrates the placental barrier and, after oxidation, accumulates in rodents' foetal tissue. In guinea pigs exposed to 200-300 µg m$^{-3}$, 2 hr day$^{-1}$ (duration unspecified) or a single 150 min exposure to 8-11 mg m$^{-3}$, mercury concentrations in foetal lungs, brain, heart, kidneys and blood were much lower (5-100 ×) than in maternal tissues; mercury concentration in foetal liver was 2 × higher than in maternal liver (15).

**Metabolism and pharmacokinetics**
In humans 80% of the vapour is retained by the body, but the liquid is poorly absorbed via the gut (<1%). Most is then deposited in the kidney. Metabolism can be via oxidation of metallic to divalent mercury; reduction of divalent to metallic mercury; methylation of inorganic mercury and conversion of methylmercury to divalent inorganic mercury. Oxidation of metallic mercury vapour is too slow to prevent passage through the blood-brain barrier, placenta and other tissues.
Elimination in humans is mainly via urine and faeces, although some is exhaled.
$t_{1/2}$ is a few days or weeks for most of the absorbed mercury but in years for a fraction; possibly due to formation of a selenium complex (15).
After short-term human exposure, the first phase of elimination from blood had $t_{1/2}$ 2-4 days, accounting for 90% of the mercury. The second phase $t_{1/2}$ was 15-30 days (15).

**Irritancy**
Metallic mercury may cause contact dermatitis; mercurial pharmaceuticals cause Pink disease in children and mercury vapour may cause Kawasaki disease (15).

## Genotoxicity
HeLa cells *in vitro* have shown inhibition of DNA, RNA, and protein synthesis (hydroxide) (17).

## Any other adverse effects to man

The standard of epidemiological evidence is such that the effect of mercury vapour on the menstrual cycle and foetal development in the absence of signs of intoxication is open to question (15).

Acute inhalation of the vapour may cause chest pains, dyspnoea, cough, haemoptysis, pneumonitis and may be fatal. The central nervous system is the critical organ of vapour exposure. Subacute exposure may cause psychotic reactions; occupational exposure may cause erethism and, after continuing exposure, fine tremor (initially of the hands). Occupational exposure is also associated with proteinuria, and kidney effects have been reported at exposure levels below those causing central nervous system effects (15).

Skin uptake of metallic mercury vapour in human volunteers is only 1% of inhalation uptake (15).

Acute, prolonged exposure to elemental mercury and its vapour caused acute, inorganic mercury toxicity and long-term, probably irreversible, neurological sequelae in 53 construction workers. Their earliest symptoms were rapidly resolving metal fume fever. Central nervous system symptoms and abnormal performance on neuropsychological tests persisted for the prolonged period of follow-up (570 days). Serial mercury determination in the blood and urine confirmed the long $t_{1/2}$ and large volume of distribution of mercury. Estimated $t_{1/2}$, assuming linear first-order kinetics, 42 days. Blood mercury levels did not correlate with severity of symptoms, but there were significant correlations between neuropsychological tests and indices of mercury exposure (18).

## Any other adverse effects

Mercury is a potent neurotoxin in vertebrates because they lack external barriers or internal detoxifying systems (19,20).

## Legislation

Community Right-To-Know List.

Limited under EC Directive on Drinking Water Quality 80/778/EEC. Mercury: maximum admissible concentration 1 µg l$^{-1}$ (21).

Included in Schedule 4 (Release into Air: Prescribed Substances) Statutory Instrument No. 472, 1991 (22).

Quality objectives under EC Directives 82/176/EEC and 84/156/EEC 0.3 mg kg$^{-1}$ (wet weight) in a representative sample of fish flesh; 1 µg l$^{-1}$ (annual mean) total mercury in inland surface waters; 0.5 µg l$^{-1}$ (annual mean) dissolved mercury in estuarine waters; 0.3 µg l$^{-1}$ (annual mean) dissolved mercury in marine waters. A 'standstill' provision applies to concentrations in shellfish or sediments. Limit values under EC Directive 84/156/EEC 0.05 mg l$^{-1}$ effluent and 0.1 g l$^{-1}$ vinyl chloride production capacity for chemical industries using mercury catalysts in vinyl chloride production; 0.05 mg l$^{-1}$ effluent and 5 g kg$^{-1}$ mercury processed for chemical industries using mercury catalysts in other processes; 0.05 mg l$^{-1}$ effluent and 0.7 g kg$^{-1}$ mercury processed for manufacture of mercury catalysts used in vinyl chloride production; 0.05 mg l$^{-1}$ effluent and 0.05 g kg$^{-1}$ mercury processed for manufacture of organic and non-organic mercury compounds (other than mercury catalysts for vinyl chloride production); 0.05 mg l$^{-1}$ effluent and 0.03 g kg$^{-1}$ mercury processed for manufacture of primary batteries containing mercury; 0.05 mg l$^{-1}$

effluent for mercury recovery plants and extraction and refining of non-ferrous metals; 0.05 mg l$^{-1}$ effluent for plants treating toxic wastes containing mercury (23).

## Any other comments

Physico-chemical properties, human health effects, experimental toxicity, environmental effects, ecotoxicology, exposure levels and workplace experience reviewed (24). Toxicity (25-27), neurotoxicity (28,29), teratology (30), mutagenicity (31-33) and health hazards from use in dental amalgam reviewed (34,35).

Aquatic pollution (36), biotransformation (37), bioavailability (38,39) and bioaccumulation reviewed (40-42).

The toxicity and environmental fate of organomercury compounds has been comprehensively reviewed (43).

Few data are available on the biological effects of mercury that actually occur in the aquatic environment. Pre mid-1980s laboratory studies of mercury toxicity to aquatic species were undertaken using unrealistically high water concentrations. The bias resulted from field measurements on contaminated sites. Post mid-1980s improvements in analytical techniques have allowed more accurate determinations and this accounts for the decrease in enivronmental concentrations reported (44-46). Recent concern about the use of mercury amalgams in dental work and the potential toxic effects to humans has prompted some countries including the USA, Canada, Sweden and Germany to review their existing legislation on permitted use.

## References

1. Fitzgerald, W. F. et al *Water Air Soil Pollut.* 1991, **56** 745–768
2. Watras, C. J. et al *Mercury as a Global Pollutant* 1994, Lewis Publ., Chelsea, MI
3. Porcella, D. B. et al *Environ. Toxicol. Chem.* 1993, **12**, 2245-2264
4. Holcombe, G. W. et al *Ecotoxicol. Environ. Saf.* 1983, **8**(2), 106-117
5. Snarski, V. M. et al *Aquat. Toxicol.* 1982, **2**, 143-156
6. Saxena, O. P. et al *J. Environ. Biol.* 1983, **4**, 91
7. Birge, W. J. et al *Aquatic toxicity tests on inorganic elements occurring in oil shale: EPA 600/9-80-022* 1980, Natl. Tech. Inf. Ser. Springfield, VA
8. Khangarot, B. S. et al *Bull. Environ. Contam. Toxicol.* 1987, **38,** 722-726
9. Khangarot, B. S. et al *Acta Hydrochim. Hydrobiol.* 1982, **10**, 367
10. Chapman, P. M. et al *Hydrobiologia* 1986, **137**, 61-64
11. Nenov, V. et al *Environ. Prot. Eng.* 1986, **12**(2), 51-60
12. Lee, Y. H. et al *Environ. Toxicol. Chem.* 1990, **9**, 833-842
13. Miskimmen, B. M. *Bull. Environ. Contam. Toxicol.* 1991, **47**, 743-750
14. Lewis, R. J. et al *RTECS* 1984, NIOSH No.83-107-4
15. *Environmental Health Criteria No.118: Inorganic Mercury* 1991, WHO/IPCS, Geneva
16. Steffek, A. J. et al *J. Dent. Res.* 1987, **66,** 239
17. Schuster, G. S. *Exp. Cell. Res.* 1970, **59**, 163-169
18. Bluhm, R. E. et al *Hum. Exp. Toxicol.* 1992, **11**, 201-210
19. Baatrup, E. *Comp. Biochem. Physiol.* 1991, **100**C, 253-257
20. Sutterlin, A. M. *Chemical Senses and Flavour* 1974, 1, 167-178
21. *EC Directive Relating to the Quality of Water Intended for Human Consumption* 1982, 80/778/EEC, Office for Official Publications of the European Communities, 2 rue Mercier, L-2985 Luxembourg
22. *S. I. 1991 No. 472 The Environmental Protection (Prescribed Processes and Substances) Regulations* 1991, HMSO, London
23. *DoE Circular 7/89: Water and the environment. The implementation of EC directives on pollution caused by certain dangerous substances discharged into the aquatic environment* 1989, HMSO, London

24. *ECETOC Technical Report No. 30(4)* 1991, European Chemical Industry Ecology and Toxicology Centre, B-1160 Brussels
25. Gonzalez, F. E. *Med. Segur. Trab.* 1987, **34**(134), 30-41 (Span.) (*Chem. Abstr.* **107**, 213045h)
26. Clarkson, T. W. *Curr. Top. Nutr. Dis.* 1986, **18**(Essent. Toxic Trace Elem./ Hum. Health Dis.), 631-643
27. *Cah. Notes Doc.* 1989, **137**, 711-716 (Fr.) (*Chem. Abstr.* **112**, 124221y)
28. Clarkson, T. W. *Environ. Health Perspect.* 1987, **75**, 56-64
29. Weiss, B. *Trace Subst. Environ. Health* 1989, **22**, 77-89
30. Inouye, M. *Congenital Anomalies* 1989, **29**(4), 333-344
31. Moszczynski, P. et al *Wiad. Lek.* 1988, **41**(3), 177-184 (Pol.) (*Chem. Abstr.* **109**, 105902t)
32. Fan, L. et al *Gongye Weisheng Yu Zhiyebing* 1986, **12**(2), 77-80 (Ch.) (*Chem. Abstr.* **107**, 53901d)
33. Wong, P. K. *Bull. Environ. Contam. Toxicol.* 1988, **40**(4), 597-603
34. Enwonwu, C. O. *Environ. Res.* 1987, **42**(1), 257-274
35. Eley, B. M. et al *J. Dent.* 1988, **16**(2), 90-95
36. Bhatia, S. et al *J. Indian Counc. Chem.* 1988, **4**(1), 59-70
37. Moszczynski, P. et al *Postepy Hig. Med. Dosw.* 1990, **44**(1-3), 153-180 (Pol.) (*Chem. Abstr.* **115**, 249627d)
38. Bjoernberg, A. et al *Environ. Pollut.* 1988, **49**(1), 53-61
39. Stokes, P. M. et al *SCOPE*, 1987, **31** (Lead, Mercury, Cadmium, Arsenic Environ.), 255-277
40. Wiener, J. G. et al *Environ. Toxicol. Chem.* 1990, **9**(7), 821-823
41. Scheuhammer, A. M. *Environ. Pollut.* 1991, **71**(2-4), 329-375
42. Thriene, B. et al *Forum Staedte-Hyg.* 1989, **40**(6), 350-355 (Ger.) (*Chem. Abstr.* **112**, 135647t)
43. Trakhtenburg, I. M. *Scientific Reviews of Soviet Literature of Toxicity and Hazards of Chemicals. Organomercury Compounds 117* 1991, 1-74, Eng. Trans. (Ed.) Richardson, M. L., UNEP/IRPTC, Geneva, CIP, Moscow
44. Gill, G. A. et al *Environ. Sci. Technol.* 1990, **24**, 1392-1400
45. Fitzgerald, W. F. *The Role of Air-Sea Exchange in Geochemical Cycling* 1986, 363-408, C. Reidel, Dordrect, The Netherlands
46. Fitzgerald, W. F. *Chemical Oceanography* 1989, **10**, 152-185, Academic Press, New York

# M53   Mercury(II) acetate

## $(CH_3CO_2)_2Hg$

**CAS Registry No.** 1600-27-7
**Synonyms** acetic acid, mercury(2+)salt; diacetoxymercury; mercuric diacetate; mercuryl acetate
**Mol. Formula** $C_4H_6HgO_4$                    **Mol. Wt.** 318.68
**Uses** Chemical intermediate and catalyst, with fungicidal properties (1).

## Physical properties

**M. Pt.** 178-180°C; **Specific gravity** 3.28.

**Solubility**
Water: 400 g $l^{-1}$ at 20°C. Organic solvent: ethanol

---

## Occupational exposure

**US TLV (TWA)** 0.025 mg m$^{-3}$ (as Hg) intended change; **UK Long-term limit** 0.05 mg m$^{-3}$ (as Hg) under review; **UK Short-term limit** 0.15 mg m$^{-3}$ (as Hg) under review; **UN No.** 1629; **HAZCHEM Code** 2X; **Conveyance classification** toxic substance.

## Ecotoxicity

**Fish toxicity**

Intestinal transport rate for nutrients decreased in *Channa punctatus* exposed to 3 μg l$^{-1}$ (Hg$^{2+}$) for 30 days. Disrupted metabolism in *Notopterus notopterus* exposed to 44 μg l$^{-1}$ (Hg$^{2+}$) for 30 days (2).

Teratogenic to rainbow trout after exposure of eggs to 0.12-0.21 μg l$^{-1}$ (Hg$^{2+}$) 4 days post hatch, and after parental exposure to 0.70-0.79 μg l$^{-1}$ (Hg$^{2+}$) for 400 days (2).

$LD_{Lo}$ (1 hr) trout 5 ppm (3).

$LD_{Lo}$ (2 hr) bluegill sunfish 5 ppm (3).

$LD_{Lo}$ (3 hr) goldfish 5 ppm (3).

**Invertebrate toxicity**

$LC_{50}$ (48 hr) *Daphnia magna* 9.3 μg l$^{-1}$; $EC_{50}$ (48 hr) *Daphnia magna* 5.2 μg l$^{-1}$ (Hg$^{2+}$) (2).

$IC_{50}$ *Pseudomonas fluorescens* 7.65 μg of Hg at pH 6. Toxicity reduces with increasing pH and may be linked with the ability of the microorganism to methylate inorganic mercury (4,5).

Toxicity is also reduced by presence of cysteine in medium due to binding of mercury to thiol groups (5).

## Mammalian and avian toxicity

**Acute data**

$LD_{50}$ oral mouse, rat 24, 41 mg kg$^{-1}$, respectively (6).

$LD_{50}$ dermal rat 570 mg kg$^{-1}$ (7).

$LD_{50}$ intraperitoneal mouse 6.5 mg kg$^{-1}$ (7).

$LD_{50}$ subcutaneous mouse 20 mg kg$^{-1}$ (8).

**Teratogenicity and reproductive effects**

Embryotoxicity was studied in golden hamsters given 4-100 mg kg$^{-1}$ orally, 4-50 mg kg$^{-1}$ subcutaneously, 2-8 mg kg$^{-1}$ intraperitoneally or 4 mg kg$^{-1}$ intravenously on day 8 of pregnancy. Major manifestations of embryo damage were increased resorption rate and small, retarded and oedematous embryos. Embryotoxicity based on resorption rate decreased in the order intraperitoneal >subcutaneous >oral (9).

**Metabolism and pharmacokinetics**

Rats receiving the compound in diet, are reported to have absorbed 50% of dose (10).

Uptake into isolated rat hepatocytes is concentration dependent and causes inhibition of alcohol dehydrogenase and glutathione reductase activities (11).

## Any other adverse effects

In synaptosomal fractions of rat brain cerebral cortex, 5HT release can be induced at 34 mg l$^{-1}$ and 5HT binding at 3.4 mg l$^{-1}$ (12).

## Legislation

Limited under EC Directive on Drinking Water Quality 80/778/EEC. Mercury: maximum admissible concentration $1 \mu g \, l^{-1}$ (13).

Included in Schedule 4 (Release into Air: Prescribed Substances) Statutory Instrument No. 472, 1991 (14).

Quality objectives under EC Directives 82/176/EEC and 84/156/EEC 0.3 mg kg$^{-1}$ (wet weight) in a representative sample of fish flesh; $1 \mu g \, l^{-1}$ (annual mean) total mercury in inland surface waters; $0.5 \mu g \, l^{-1}$ (annual mean) dissolved mercury in estuarine waters; $0.3 \mu g \, l^{-1}$ (annual mean) dissolved mercury in marine waters. A 'standstill' provision applies to concentrations in shellfish or sediments. Limit values under EC Directive 84/156/EEC 0.05 mg l$^{-1}$ effluent and 0.1 g l$^{-1}$ vinyl chloride production capacity for chemical industries using mercury catalysts in vinyl chloride production; 0.05 mg l$^{-1}$ effluent and 5 g kg$^{-1}$ mercury processed for chemical industries using mercury catalysts in other processes; 0.05 mg l$^{-1}$ effluent and 0.7 g kg$^{-1}$ mercury processed for manufacture of mercury catalysts used in vinyl chloride production; 0.05 mg l$^{-1}$ effluent and 0.05 g kg$^{-1}$ mercury processed for manufacture of organic and non-organic mercury compounds (other than mercury catalysts for vinyl chloride production); 0.05 mg l$^{-1}$ effluent and 0.03 g kg$^{-1}$ mercury processed for manufacture of primary batteries containing mercury; 0.05 mg l$^{-1}$ effluent for mercury recovery plants and extraction and refining of non-ferrous metals; 0.05 mg l$^{-1}$ effluent for plants treating toxic wastes containing mercury (15).

## Any other comments

The compound is generally classified as an inorganic mercury compound. The toxicology, absorption, distribution and metabolism of inorganic mercury compounds has been reviewed (16).

The toxicity and environmental fate of organomercury compounds has been comprehensively reviewed (17).

## References

1. Freers, C. R. *US Patent Applic.* 1984, 580, 829 16 Feb 1984
2. Zillioux, E. J. et al *Environ. Toxicol. Chem.* 1993, **12**, 2245-2264
3. The Toxicity of 3400 Chemicals to Fish 1987, EPA 560/6-87-002 PB 87-200-275, Washington, DC
4. Farrell, R. E. *Appl. Environ. Microbiol.* 1990, **56**(10), 3006-3016
5. Ribo, J. M. *Hydrobiologia* 1989, 188-189
6. Grins, N. et al *Gig. Sanit.* 1981, **46**(8), 12-14
7. *Gig. Tr. Prof. Zabol.* 1981, **25**(7), 27
8. *Monatsschr. Ohrenheilkd. Laryngo-Rhinol.* 1939, **73**, 751
9. Gale, T. F. *Environ. Res.* 1974, **8**, 207-213
10. Fitzhugh, O. G. et al *Arch. Ind. Hyg. Occup. Med.* 1950, **2**, 433-442
11. Hellstroem-Lindahl, E. et al *Toxicol. Lett.* 1989, **49**(1), 87-98
12. Oudar, P. et al *Pharmacol. Toxicol. (Copenhagen)* 1989, **65**(4), 245-248
13. *EC Directive Relating to the Quality of Water Intended for Human Consumption* 1982, 80/778/EEC, Office for Official Publications of the European Communities, 2 rue Mercier, L-2985 Luxembourg
14. *S. I. 1991 No. 472 The Environmental Protection (Prescribed Processes and Substances) Regulations* 1991, HMSO, London
15. *DoE Circular 7/89: Water and the environment. The implementation of EC directives on pollution caused by certain dangerous substances discharged into the aquatic environment* 1989, HMSO, London

16.  Oehme, F. W. (Ed.) *Toxicity of Heavy Metals in the Environment* Part 1, 1978, Marcel Dekker Inc., New York
17.  Trakhtenburg, I. M. *Scientific Reviews of Soviet Literature of Toxicity and Hazards of Chemicals. Organomercury Compounds 117* 1991, 1-74, Eng. Trans. (Ed.) Richardson, M. L., UNEP/IRPTC Geneva, CIP Moscow

# M54   Mercury ammonium chloride

## Hg(Cl)NH₂

**CAS Registry No.** 10124-48-8
**Synonyms** mercury amide chloride; aminomercuric chloride
**Mol. Formula** HgNH₂Cl                                      **Mol. Wt.** 252.07
**Uses** Topical anti-infective.

## Physical properties
**Specific gravity** 5.38.

## Occupational exposure
**US TLV (TWA)** 0.025 mg m$^{-3}$ (as Hg) intended change; **UK Long-term limit** 0.05 mg m$^{-3}$ (as Hg) under review; **UK Short-term limit** 0.15 mg m$^{-3}$ (as Hg) under review; **UN No.** 1630; **Conveyance classification** toxic substance.

## Any other adverse effects to man

Allergic dermatitis reported (1).
Prolonged use may cause local pigmentation of skin or eyelids, ptyalism. Oral ingestion causes epigastric pain, nausea and diarrhoea (1).

## Legislation

Community Right-To-Know List.
Limited under EC Directive on Drinking Water Quality 80/778/EEC. Mercury: maximum admissible concentration 1 μg l$^{-1}$ (2).
Included in Schedule 4 (Release into Air: Prescribed Substances) Statutory Instrument No. 472, 1991 (3).
Quality objectives under EC Directives 82/176/EEC and 84/156/EEC 0.3 mg kg$^{-1}$ (wet weight) in a representative sample of fish flesh; 1 μg l$^{-1}$ (annual mean) total mercury in inland surface waters; 0.5 μg l$^{-1}$ (annual mean) dissolved mercury in estuarine waters; 0.3 μg l$^{-1}$ (annual mean) dissolved mercury in marine waters. A 'standstill' provision applies to concentrations in shellfish or sediments. Limit values under EC Directive 84/156/EEC 0.05 mg l$^{-1}$ effluent and 0.1 g l$^{-1}$ vinyl chloride production capacity for chemical industries using mercury catalysts in vinyl chloride production; 0.05 mg l$^{-1}$ effluent and 5 g kg$^{-1}$ mercury processed for chemical industries using mercury catalysts in other processes; 0.05 mg l$^{-1}$ effluent and 0.7 g kg$^{-1}$ mercury processed for manufacture of mercury catalysts used in vinyl

chloride production; 0.05 mg l$^{-1}$ effluent and 0.05 g kg$^{-1}$ mercury processed for manufacture of organic and non-organic mercury compounds (other than mercury catalysts for vinyl chloride production); 0.05 mg l$^{-1}$ effluent and 0.03 g kg$^{-1}$ mercury processed for manufacture of primary batteries containing mercury; 0.05 mg l$^{-1}$ effluent for mercury recovery plants and extraction and refining of non-ferrous metals; 0.05 mg l$^{-1}$ effluent for plants treating toxic wastes containing mercury (4).

## Any other comments

Should not be used therapeutically in conjunction with sulfur or iodine (1).
Physico-chemical properties, human health effects, ecotoxicology, experimental toxicology, environmental effects, exposure levels, and workplace experience reviewed (5).
Aquatic toxicology reviewed (6).
Soluble in acetic acid, ammonium carbonate and sodium thiosulfate.

## References

1.  *The Merck Index* 11th ed., Merck & Co., Rahway, NJ
2.  *EC Directive Relating to the Quality of Water Intended for Human Consumption* 1982, 80/778/EEC, Office for Official Publications of the European Communities, 2 rue Mercier, L-2985 Luxembourg
3.  *S. I. 1991 No. 472 The Environmental Protection (Prescribed Processes and Substances) Regulations* 1991, HMSO, London
4.  *DoE Circular 7/89: Water and the environment. The implementation of EC directives on pollution caused by certain dangerous substances discharged into the aquatic environment* 1989, HMSO, London
5.  *ECETOC Technical Report No. 30(4)* 1991, European Chemical Industry Ecology and Toxicology Centre, B-1160 Brussels
6.  Zilloux, E. J. et al *Environ. Toxicol. Chem.* 1993, **12**, 2245-2264.

# M55   Mercury benzoate

**CAS Registry No.** 583-15-3
**Synonyms** mercuric benzoate
**Mol. Formula** $C_{14}H_{10}O_4Hg$                     **Mol. Wt.** 442.82
**Uses** Formerly used as an antisyphilitic.

## Physical properties

**M. Pt.** 165°C.

## Solubility

Water: soluble in 90 parts cold water. Organic solvent: ethanol

## Occupational exposure

**UN No.** 1631; **Conveyance classification** toxic substance.

## Legislation

Community Right-To-Know List.

Limited under EC Directive on Drinking Water Quality 80/778/EEC. Mercury: maximum admissible concentration 1 μg l$^{-1}$ (1).

Included in Schedule 4 and 6 (Release into Air/Land: Prescribed Substances) Statutory Instrument No. 472, 1991 (2).

Quality objectives under EC Directives 82/176/EEC and 84/156/EEC 0.3 mg kg$^{-1}$ (wet weight) in a representative sample of fish flesh; 1 μg l$^{-1}$ (annual mean) total mercury in inland surface waters; 0.5 μg l$^{-1}$ (annual mean) dissolved mercury in estuarine waters; 0.3 μg l$^{-1}$ (annual mean) dissolved mercury in marine waters. A 'standstill' provision applies to concentrations in shellfish or sediments. Limit values under EC Directive 84/156/EEC 0.05 mg l$^{-1}$ effluent and 0.1 g l$^{-1}$ vinyl chloride production capacity for chemical industries using mercury catalysts in vinyl chloride production; 0.05 mg l$^{-1}$ effluent and 5 g kg$^{-1}$ mercury processed for chemical industries using mercury catalysts in other processes; 0.05 mg l$^{-1}$ effluent and 0.7 g kg$^{-1}$ mercury processed for manufacture of mercury catalysts used in vinyl chloride production; 0.05 mg l$^{-1}$ effluent and 0.05 g kg$^{-1}$ mercury processed for manufacture of organic and non-organic mercury compounds (other than mercury catalysts for vinyl chloride production); 0.05 mg l$^{-1}$ effluent and 0.03 g kg$^{-1}$ mercury processed for manufacture of primary batteries containing mercury; 0.05 mg l$^{-1}$ effluent for mercury recovery plants and extraction and refining of non-ferrous metals; 0.05 mg l$^{-1}$ effluent for plants treating toxic wastes containing mercury (3).

## Any other comments

Physico-chemical properties, human health effects, experimental toxicity, environmental effects, ecotoxicology, exposure levels and workplace experience reviewed (4).

Aquatic toxicology reviewed (5).

Protect from light.

## References

1. *EC Directive Relating to the Quality of Water Intended for Human Consumption* 1982, 80/778/EEC, Office for Official Publications of the European Communities, 2 rue Mercier, L-2985 Luxembourg
2. *S. I. 1991 No. 472 The Environmental Protection (Prescribed Processes and Substances) Regulations* 1991, HMSO, London
3. *DoE Circular 7/89: Water and the environment. The implementation of EC directives on pollution caused by certain dangerous substances discharged into the aquatic environment* 1989, HMSO, London
4. *ECETOC Technical Report No. 30(4)* 1991, European Chemical Industry Ecology and Toxicology Centre, B-1160 Brussels
5. Zillioux, E. J. et al *Environ. Toxicol. Chem.* 1993, **12**, 2245-2264

# M56  Mercury(I) bromide

Hg₂Br₂

CAS Registry No. 10031-18-2
**Synonyms** mercurous bromide
**Mol. Formula** HgBr                                    **Mol. Wt.** 280.50

## Physical properties

**B. Pt.** 390°C (subl.); **Specific gravity** 7.307; **Volatility** v. den. 19.3.

## Legislation

Community Right-To-Know List.

Limited under EC Directive on Drinking Water Quality 80/778/EEC. Mercury: maximum admissible concentration 1 μg $l^{-1}$ (1).

Included in Schedule 4 (Release into Air: Prescribed Substances) Statutory Instrument No. 472, 1991 (2).

Quality objectives under EC Directives 82/176/EEC and 84/156/EEC 0.3 mg $kg^{-1}$ (wet weight) in a representative sample of fish flesh; 1 μg $l^{-1}$ (annual mean) total mercury in inland surface waters; 0.5 μg $l^{-1}$ (annual mean) dissolved mercury in estuarine waters; 0.3 μg $l^{-1}$ (annual mean) dissolved mercury in marine waters. A 'standstill' provision applies. Limit values under EC Directive 84/156/EEC 0.05 mg $l^{-1}$ effluent and 0.1 g $l^{-1}$ vinyl chloride production capacity for chemical industries using mercury catalysts in vinyl chloride production; 0.05 mg $l^{-1}$ effluent and 5 g $kg^{-1}$ mercury processed for chemical industries using mercury catalysts in other processes; 0.05 mg $l^{-1}$ effluent and 0.7 g $kg^{-1}$ mercury processed for manufacture of mercury catalysts used in vinyl chloride production; 0.05 mg $l^{-1}$ effluent and 0.05 g $kg^{-1}$ mercury processed for manufacture of organic and non-organic mercury compounds (other than mercury catalysts for vinyl chloride production); 0.05 mg $l^{-1}$ effluent and 0.03 g $kg^{-1}$ mercury processed for manufacture of primary batteries containing mercury; 0.05 mg $l^{-1}$ effluent for mercury recovery plants and extraction and refining of non-ferrous metals; 0.05 mg $l^{-1}$ effluent for plants treating toxic wastes containing mercury (3).

## Any other comments

Physico-chemical properties, human health effects, experimental toxicity, environmental effects, ecotoxicology, exposure levels and workplace experience reviewed (4).

Toxicity of inorganic mercury and environmental effects reviewed (5,6).

Aquatic toxicology reviewed (7).

## References

1. *EC Directive Relating to the Quality of Water Intended for Human Consumption* 1982, 80/778/EEC, Office for Official Publications of the European Communities, 2 rue Mercier, L-2985 Luxembourg
2. *S. I. 1991 No. 472 The Environmental Protection (Prescribed Processes and Substances) Regulations* 1991, HMSO, London
3. *ECETOC Technical Report No. 30(4)* 1991, European Chemical Industry Ecology and Toxicology Centre, B-1160 Brussels

4.  *DoE Circular 7/89: Water and the environment. The implementation of EC directives on pollution caused by certain dangerous substances discharged into the aquatic environment* 1989, HMSO, London
5.  *Environmental Health Criteria 118: Inorganic Mercury* 1991, WHO/IPCS, Geneva
6.  *Environmental Health Criteria 86: Mercury-Environmental Aspects* 1989, WHO/IPCS, Geneva
7.  Zillioux, E. J. et al *Environ. Toxicol. Chem.* 1993, **12**, 2245-2264

# M57   Mercury(II) bromide

## HgBr₂

**CAS Registry No.** 7789-47-1
**Synonyms** mercuric bromide
**Mol. Formula** HgBr$_2$                **Mol. Wt.** 360.41

### Physical properties

**M. Pt.** 237°C; **B. Pt.** 322°C (subl.); **Specific gravity** d$^{25}$ 6.109;
**Volatility** v.p. 1 mmHg at 136.5°C.

### Solubility
Water: soluble in 200 parts water. Organic solvent: ethanol, methanol

### Ecotoxicity

**Fish toxicity**
Intestinal transport rate for nutrients decreased in snakehead fish exposed to 3 µg l$^{-1}$ (Hg$^{2+}$) for 30 days (1).
Disrupted metabolism in knifefish exposed to 44 µg l$^{-1}$ (Hg$^{2+}$) for 30 days (1).
Teratogenic to rainbow trout after egg exposure to 0.12-0.21 µg l$^{-1}$ (Hg$^{2+}$) 4 days post hatch, and after parental exposure to 0.70-0.79 µg l$^{-1}$ (Hg$^{2+}$) for 400 days (1).

**Invertebrate toxicity**
LC$_{50}$ (48 hr) *Daphnia magna* 9.3 µg l$^{-1}$ (Hg$^{2+}$) (1).
EC$_{50}$ (48 hr) *Daphnia magna* 5.2 µg l$^{-1}$ (Hg$^{2+}$) (1).

### Mammalian and avian toxicity

**Acute data**
LD$_{50}$ oral mouse, rat 35-40 mg kg$^{-1}$ (2).
LD$_{50}$ dermal rat 100 mg kg$^{-1}$ (2).
LD$_{50}$ intraperitoneal mouse 5 mg kg$^{-1}$ (2).

### Legislation
Community Right-To-Know List.
Limited under EC Directive on Drinking Water Quality 80/778/EEC. Mercury: maximum admissible concentration 1 µg l$^{-1}$ (3).
Included in Schedule 4 (Release into Air: Prescribed Substances) Statutory Instrument No. 472, 1991 (4).
Quality objectives under EC Directives 82/176/EEC and 84/156/EEC 0.3 mg kg$^{-1}$

(wet weight) in a representative sample of fish flesh; $1 \mu g \, l^{-1}$ (annual mean) total mercury in inland surface waters; $0.5 \mu g \, l^{-1}$ (annual mean) dissolved mercury in estuarine waters; $0.3 \mu g \, l^{-1}$ (annual mean) dissolved mercury in marine waters. A 'standstill' provision applies to concentrations in shellfish or sediments. Limit values under EC Directive 84/156/EEC 0.05 $mg \, l^{-1}$ effluent and $0.1 \, g \, l^{-1}$ vinyl chloride production capacity for chemical industries using mercury catalysts in vinyl chloride production; $0.05 \, mg \, l^{-1}$ effluent and $5 \, g \, kg^{-1}$ mercury processed for chemical industries using mercury catalysts in other processes; $0.05 \, mg \, l^{-1}$ effluent and $0.7 \, g \, kg^{-1}$ mercury processed for manufacture of mercury catalysts used in vinyl chloride production; $0.05 \, mg \, l^{-1}$ effluent and $0.05 \, g \, kg^{-1}$ mercury processed for manufacture of organic and non-organic mercury compounds (other than mercury catalysts for vinyl chloride production); $0.05 \, mg \, l^{-1}$ effluent and $0.03 \, g \, kg^{-1}$ mercury processed for manufacture of primary batteries containing mercury; $0.05 \, mg \, l^{-1}$ effluent for mercury recovery plants and extraction and refining of non-ferrous metals; $0.05 \, mg \, l^{-1}$ effluent for plants treating toxic wastes containing mercury (5).

## Any other comments

Physico-chemical properties, human health effects, experimental toxicity, environmental effects, ecotoxicology, exposure levels and workplace experience reviewed (6).
Toxicity of inorganic mercury and environmental effects reviewed (7,8).
Toxicity of mercuric compounds discussed (9).

## References

1. Zillioux, E. J. et al *Environ. Toxicol. Chem.* 1993, **12**, 2245-2264
2. *Gig. Tr. Prof. Zabol.* 1981, **25**(7), 27
3. *EC Directive Relating to the Quality of Water Intended for Human Consumption* 1982, 80/778/EEC, Office for Official Publications of the European Communities, 2 rue Mercier, L-2985 Luxembourg
4. *S. I. 1991 No. 472 The Environmental Protection (Prescribed Processes and Substances) Regulations* 1991, HMSO, London
5. *DoE Circular 7/89: Water and the environment. The implementation of EC directives on pollution caused by certain dangerous substances discharged into the aquatic environment* 1989, HMSO, London
6. *ECETOC Technical Report No. 30(4)* 1991, European Chemical Industry Ecology and Toxicology Centre, B-1160 Brussels
7. *Environmental Health Criteria 118: Inorganic Mercury* 1991, WHO/IPCS, Geneva
8. *Environmental Health Criteria 86: Mercury-Environmental Aspects* 1989, WHO/IPCS, Geneva
9. Korshun, M. N. *Gig. Sanit.* 1989, (1), 69-70 (Russ.) (*Chem. Abstr.* **110**, 149418q)

# M58   Mercury(I) chloride

## Hg₂Cl₂

**CAS Registry No.** 7546-30-7
**Synonyms** mercurous chloride; Cyclosan; subchloride of mercury; mercury monochloride; Precipite blanc; C.I. 77764
**Mol. Formula** HgCl                                      **Mol. Wt.** 236.04
**Uses** Used as a purgative, and in teething powders for children, now banned. Also previously used in ointments for the treatment of syphilis, and as an insecticide.

## Physical properties
**M. Pt.** 400-500°C (sublimes); **Specific gravity** 7.15.

## Solubility
Water: 2 mg $l^{-1}$ at 25°C. Organic solvent: ethanol

## Occupational exposure
**US TLV (TWA)** 0.025 mg m$^{-3}$ (as Hg) intended change; **UK Long-term limit** 0.05 mg m$^{-3}$ (as Hg) under review; **UK Short-term limit** 0.15 mg m$^{-3}$ (as Hg) under review.

## Ecotoxicity

### Fish toxicity
Intestinal transport rate for nutrients decreased in *Channa punctatus* exposed to 3 µg $l^{-1}$ (Hg$^{2+}$) for 30 days. Disrupted metabolism in *Notopterus notopterus* exposed to 44 µg $l^{-1}$ (Hg$^{2+}$) for 30 days (1).
Teratogenic to rainbow trout after exposure of eggs to 0.12-0.21 µg $l^{-1}$ (Hg$^{2+}$) 4 days post hatch, and after parental exposure to 0.70-0.79 µg $l^{-1}$ (Hg$^{2+}$) for 400 days (1).

### Invertebrate toxicity
LC$_{50}$ (48 hr) *Daphnia magna* 9.3 µg $l^{-1}$; EC$_{50}$ (48 hr) *Daphnia magna* 5.2 µg $l^{-1}$ (Hg$^{2+}$) (1).

## Mammalian and avian toxicity

### Acute data
LD$_{50}$ oral rat 166-210 mg kg$^{-1}$ (2,3).
LD$_{50}$ oral mouse 180 mg kg$^{-1}$ (2).
LD$_{Lo}$ oral dog 1-2 g dog$^{-1}$ (3).
LD$_{Lo}$ oral horse 12-16 g horse$^{-1}$ (3).
LD$_{50}$ dermal rat 1.5 g kg$^{-1}$ (2).
LD$_{50}$ intraperitoneal mouse 10 mg kg$^{-1}$ (2).

### Teratogenicity and reproductive effects
Injection into chick eggs caused death of all embryos at 0.5 mg and an 80% reduction in hatchability at 0.25 mg (4,5).

### Metabolism and pharmacokinetics
After oral ingestion, most of the compound passes out of the stomach unchanged, and

---

into the small intestine where some is converted to various irritant compounds. Most is eliminated in faeces within a week of ingestion (6).

## Genotoxicity

*Rec* assay with *Bacillus subtilis* positive (7).
Induced sister chromatid exchanges in Chinese hamster cells *in vitro* (8).
Mouse glioma cells *in vitro* inhibited DNA synthesis and transport systems for DNA precursors (9).

## Any other adverse effects to man

When used in teething powders for infants, the compound has caused acrodynia. The swelling, rawness and pink colouration of limbs was due principally to renal and cardiovascular effects associated with salt and water loss (6,10).
Repeated use of the compound as a purgative has led to dementia, erethism, colitis, renal failure and death due to absorption of mercury from the gastrointestinal tract (11).

## Legislation

Limited under EC Directive on Drinking Water Quality 80/778/EEC. Mercury: maximum admissible concentration 0.1 $\mu$g l$^{-1}$ (12).
Included in Schedule 4 (Release into Air: Prescribed Substances) and Schedule 5 (Release into Water: Prescribed Substances) Statutory Instrument No. 472, 1991 (13).
Quality objectives under EC Directives 82/176/EEC and 84/156/EEC 0.3 mg kg$^{-1}$ (wet weight) in a representative sample of fish flesh; 1 $\mu$g l$^{-1}$ (annual mean) total mercury in inland surface waters; 0.5 $\mu$g l$^{-1}$ (annual mean) dissolved mercury in estuarine waters; 0.3 $\mu$g l$^{-1}$ (annual mean) dissolved mercury in marine waters. A 'standstill' provision applies to concentrations in shellfish or sediments. Limit values under EC Directive 84/156/EEC 0.05 mg l$^{-1}$ effluent and 0.1 g l$^{-1}$ vinyl chloride production capacity for chemical industries using mercury catalysts in vinyl chloride production; 0.05 mg l$^{-1}$ effluent and 5 g kg$^{-1}$ mercury processed for chemical industries using mercury catalysts in other processes; 0.05 mg l$^{-1}$ effluent and 0.7 g kg$^{-1}$ mercury processed for manufacture of mercury catalysts used in vinyl chloride production; 0.05 mg l$^{-1}$ effluent and 0.05 g kg$^{-1}$ mercury processed for manufacture of organic and non-organic mercury compounds (other than mercury catalysts for vinyl chloride production); 0.05 mg l$^{-1}$ effluent and 0.03 g kg$^{-1}$ mercury processed for manufacture of primary batteries containing mercury; 0.05 mg l$^{-1}$ effluent for mercury recovery plants and extraction and refining of non-ferrous metals; 0.05 mg l$^{-1}$ effluent for plants treating toxic wastes containing mercury (14).

## Any other comments

Absorption, excretion and distribution of inorganic mercurial salts has been reviewed (3,15).
Human health effects, experimental toxicology, physico-chemical properties reviewed (16).
The molecular formula is sometimes written as $Hg_2Cl_2$ with the synonym dimercury dichloride.

## References

1. Zillioux, E. J. et al *Environ. Toxicol. Chem.* 1993, **12**, 2245-2264
2. *Gig. Tr. Prof. Zabol.* 1981, **25**(7), 27

3. Oehme, F. W. (Ed.) *Toxicology of Heavy Metals in the Environment* Part 1, 1978, Marcel Dekker Inc., New York
4. McLaughlin, J. et al *Toxicol. Appl. Pharmacol.* 1963, **5**, 760
5. Kuahara, S. et al *Med. Soc.* 1970, *44*, 81
6. *Martindale. The Extra Pharmacopoeia* 30th ed., 1993, The Pharmaceutical Press, London
7. Kanematsu, N. et al *Mutat. Res.* 1980, **77**, 109-116
8. Montaldi, A. et al *Environ. Mutagen.* 1985, **7**, 381-390
9. Nakada, S. et al *Toxicol. Appl. Pharmacol.* 1980, **53**, 24
10. Harvey, C. G. *Br. Med. J.* 1965, **1**, 1181
11. *J. Am. Med. Assoc.* 1974, **228**, 1446
12. *EC Directive Relating to the Quality of Water Intended for Human Consumption* 1982, 80/778/EEC, Office for Official Publications of the European Communities, 2 rue Mercier, L-2985 Luxembourg
13. *S. I. 1991 No. 472 The Environmental Protection (Prescribed Processes and Substances) Regulations* 1991, HMSO, London
14. *DoE Circular 7/89: Water and the environment. The implementation of EC directives on pollution caused by certain dangerous substances discharged into the aquatic environment* 1989, HMSO, London
15. Petering, H. G. et al *Pharmacol. Ther.* 1976, **1**, 131
16. *ECETOC Technical Report No. 30(4)* 1991, European Chemical Industry Ecology and Toxicology Centre, B-1160 Brussels

# M59   Mercury(II) chloride

## HgCl$_2$

**CAS Registry No.** 7487-94-7

**Synonyms** mercury bichloride; mercury perchloride; calochlor; corrosive sublimate

**Mol. Formula** HgCl$_2$                          **Mol. Wt.** 271.50

**Uses** Preserving wood and anatomical specimens. In electroplating and other metal processes. Topical antiseptic and disinfectant.

## Physical properties

**M. Pt.** 277°C; **Specific gravity** 5.4; **Volatility** v.p. $1.4 \times 10^{-4}$ mmHg at 34°C.

**Solubility**
Water: 69 g l$^{-1}$. Organic solvent: benzene, glycerol, diethyl ether

## Occupational exposure

**US TLV (TWA)** 0.025 mg m$^{-3}$ (as Hg) intended change; **UK Long-term limit** 0.05 mg m$^{-3}$ (as Hg) under review; **UK Short-term limit** 0.15 mg m$^{-3}$ (as Hg) under review; **UN No.** 1624; **HAZCHEM Code** 2X; **Conveyance classification** toxic substance; **Supply classification** very toxic.

**Risk phrases** Very toxic if swallowed – Causes burns – Toxic: danger of serious damage to health by prolonged exposure in contact with skin and if swallowed (R28, R34, R48/24/25)

**Safety phrases** Wear suitable protective clothing, gloves and eye/face protection – In case of accident or if you feel unwell, seek medical advice immediately (show label where possible) (S36/37/39, S45)

# Ecotoxicity

## Fish toxicity

Intestinal transport rate for nutrients decreased in *Channa punctatus* exposed to
3 µg l$^{-1}$ (Hg$^{2+}$) for 30 days. Disrupted metabolism in *Notopterus notopterus* exposed
to 44 µg l$^{-1}$ (Hg$^{2+}$) for 30 days (1).
Teratogenic to rainbow trout after exposure of eggs to 0.12-0.21 µg l$^{-1}$ (Hg$^{2+}$) 4 days
post hatch, and after parental exposure to 0.70-0.79 µg l$^{-1}$ (Hg$^{2+}$) for 400 days (1).
Trout, bluegill sunfish and goldfish exposed to 5 ppm exhibited signs of distress in
1, 3 and 4 hr, respectively. Test conditions: pH 7.0; dissolved oxygen content 7.5
ppm; total hardness (soap method) 300 ppm; methyl orange alkalinity 310 ppm; free
carbon dioxide 5 ppm; and temperature 12.8°C (2).

## Invertebrate toxicity

LC$_{50}$ (48 hr) *Daphnia magna* 9.3 µg l$^{-1}$; EC$_{50}$ (48 hr) *Daphnia magna*
5.2 µg l$^{-1}$ (Hg$^{2+}$) (1).
IC$_{50}$ *Pseudomonas fluorescens* 4.85 µg Hg at pH 6. Toxicity reduces with increasing
pH and may be linked with the ability of the microorganism to methylate inorganic
mercury (3,4).
Toxicity is also reduced by presence of cysteine in medium due to binding of mercury
to thiol groups (4).

## Bioaccumulation

Rainbow trout were exposed for 96 hr to 50 µg l$^{-1}$ (Hg$^{2+}$) at temperatures of 5, 10 and
20°C; bioconcentration factors were 5, 12 and 26, respectively (5).
*Acartia clausia* 24 hr exposure to 0.05-0.5 µg l$^{-1}$; bioconcentration factor 7500 (6).

# Mammalian and avian toxicity

## Acute data

LD$_{50}$ oral rat 1.5 mg kg$^{-1}$ (7).
LD$_{Lo}$ oral dog 0.1-0.3 g dog$^{-1}$ (8).
LD$_{Lo}$ oral cattle 4-8 g animal$^{-1}$ (8).
LD$_{50}$ dermal mouse 41 mg kg$^{-1}$ (9).
LD$_{50}$ subcutaneous mouse 4.5 mg kg$^{-1}$ (10).

## Carcinogenicity and long-term effects

National Toxicology Program tested rats and mice via gavage. Some evidence for
carcinogenicity in ♂ rats, equivocal evidence for carcinogenicity in ♀ rats and ♂ mice
and no evidence for carcinogenicity in ♀ mice. A Post Peer Review of data is in
progress (11).
Mercury(II) chloride in drinking water was not significantly tumorigenic to mice in
lifetime studies (12).

## Teratogenicity and reproductive effects

Increased incidence of post-implantation loss reported in rats given 16 or 24 mg kg$^{-1}$
day$^{-1}$ on days 6-15 of pregnancy. Foetal weight was reduced at 12 mg kg$^{-1}$. Delayed
ossification and a range of major malformations occurred at 24 mg kg$^{-1}$ (13).

## Metabolism and pharmacokinetics

Rapidly absorbed from the gastrointestinal tract with effects being visible after 10-15
min (14).

Mercury is distributed throughout soft tissue including the brain and spinal cord in which tissues there is some localisation (15-17).
It can be detected in liver, erythrocytes and blood plasma as well as the kidney where particularly high levels are found in the renal cortex (15,18).
Complexes are formed with proteins at many sites (8,14).
Mercury is secreted in saliva, bile, gastrointestinal juices and breast milk, and ultimately excreted in urine and faeces (8,14,19).
Covalent binding to thiol groups leads to interference with function of many enzymes and co-enzymes, and binding to the thiol groups of cysteine leads to accumulation in the keratin of hair, nails and skin (14,20).
Absorption across the rat small intestine is pH dependent, and the moiety absorbed is thought to be $Hg(OH)Cl$ or $Hg(OH)_2$ (21,22).

**Irritancy**
Severe eye and skin irritant (species unspecified) (23).

**Sensitisation**
Allergies to mercury(II) chloride have been reported as quite common in dentists and surgeons (24).
An auto-immune reaction may be involved in the nephritis induced by the compound (18,25).

## Genotoxicity
*Bacillus subtilis* rec assay positive (26).
*In vitro* mouse lymphoma forward mutation assay L5178 $tk^+/tk^-$ positive (27).
*In vitro* induced sister chromatid exchanges in Chinese hamster ovary cells (28).
DNA crosslinking in Novikoff Ascites hepatoma cells *in vitro* was induced by concentrations of 135 mg $l^{-1}$ (29).

## Any other adverse effects to man
Chronic poisoning with mercury can induce tremor and neuropsychiatric disturbances. These are probably due to a variety of actions on the nervous system including that on 5HT mediated pathways (30).
Single oral doses of 21 and 29 mg $kg^{-1}$ have been reported as fatal to humans, with damage to nervous system, gastrointestinal tract and kidney (31,32).

## Any other adverse effects
Inorganic mercury compounds are highly reactive neurotoxins altering nerve function by interference with $Ca^{2+}$ transport, whereas kidney damage may be related to induction of abnormal protein production (33-36).

## Legislation
Limited under EC Directive on Drinking Water Quality 80/778/EEC. Mercury: maximum admissible concentration 1 µg $l^{-1}$ (37).
Included in Schedule 4 (Release into Air: Prescribed Substances) and Schedule 5 (Release into Water: Prescribed Substances) Statutory Instrument No. 472, 1991 (38).
Quality objectives under EC Directives 82/176/EEC and 84/156/EEC 0.3 mg $kg^{-1}$ (wet weight) in a representative sample of fish flesh; 1 µg $l^{-1}$ (annual mean) total mercury in inland surface waters; 0.5 µg $l^{-1}$ (annual mean) dissolved mercury in estuarine waters; 0.3 µg $l^{-1}$ (annual mean) dissolved mercury in marine waters. A

'standstill' provision applies to concentrations in shellfish or sediments. Limit values under EC Directive 84/156/EEC 0.05 mg $l^{-1}$ effluent and 0.1 g $l^{-1}$ vinyl chloride production capacity for chemical industries using mercury catalysts in vinyl chloride production; 0.05 mg $l^{-1}$ effluent and 5 g $kg^{-1}$ mercury processed for chemical industries using mercury catalysts in other processes; 0.05 mg $l^{-1}$ effluent and 0.7 g $kg^{-1}$ mercury processed for manufacture of mercury catalysts used in vinyl chloride production; 0.05 mg $l^{-1}$ effluent and 0.05 g $kg^{-1}$ mercury processed for manufacture of organic and non-organic mercury compounds (other than mercury catalysts for vinyl chloride production); 0.05 mg $l^{-1}$ effluent and 0.03 g $kg^{-1}$ mercury processed for manufacture of primary batteries containing mercury; 0.05 mg $l^{-1}$ effluent for mercury recovery plants and extraction and refining of non-ferrous metals; 0.05 mg $l^{-1}$ effluent for plants treating toxic wastes containing mercury (39).

# References

1.  Zillioux, E. J. et al *Environ. Toxicol. Chem.* 1993, **12**, 2254-2264
2.  *The Toxicity of 3400 Chemicals to Fish* 1987, EPA 560/6-87-002 PB 87-200-275, Washington, DC
3.  Farrell, R. E. *Appl. Environ. Microbiol.* 1990, **56**(10), 3006-3016
4.  Ribo, J. M. *Hydrobiologia* 1989, 188-189
5.  MacLeod, J. C. et al *J. Fish Res. Board Can.* 1973, **30**, 485-492
6.  Hirota, R. et al *Bull. Jpn. Soc. Sci. Fish* 1983, **49**, 1249-1251
7.  *The Pesticide Manual* 9th ed., 1991, British Crop Protection Council, Farnham
8.  Oehme, F. W. (Ed.) *Toxicity of Heavy Metals in the Environment* Part 1, 1978, Marcel Dekker Inc., New York
9.  Lenga, R. E. *The Sigma-Aldrich Library of Chemical Safety Data* 2nd ed., 1988, Sigma-Aldrich, Milwaukee, WI
10. *Nippon Eiseigaku Zasshi* 1979, **34**, 193
11. *National Toxicology Program Research and Testing Division* 1992, Report No. TR-418, NIEHS, Research Triangle Park, NC
12. *IPCS Environmental Health Criteria 118: Inorganic Mercury*, 1991, WHO, Geneva
13. McAnulty, P. A. et al *Teratology* 1982, **25**(1), 26A
14. *Martindale. The Extra Pharmacopoeia* 30th ed., 1993, The Pharmaceutical Press, London
15. Houser, M. T. *Toxicol. Appl. Pharmacol.* 1988, **93**, 187-194
16. Moller-Madsen, B. et al *Toxicol. Appl. Pharmacol.* 1991, **108**, 457-473
17. Moller-Madsen, B. et al *Toxicol. Appl. Pharmacol.* 1990, **103**(2), 303-323
18. Kanfer, A. et al *Lab. Invest.* 1987, **57**(2), 138-143
19. Rothstein, A. et al *J. Pharmacol. Exp. Ther.* 1960, **130**, 166
20. Bowman, W. C. et al *Textbook of Pharmacology* 2nd ed., 1980, Blackwell, Oxford
21. Endo, T. et al *Pharmacol. Toxicol. (Copenhagen)* 1988, **63**(5), 361-368
22. Endo, T. et al *Toxicol. Appl. Pharmacol.* 1990, **66**(5), 347-353
23. Keith, L. H. et al (Eds.) *Compendium of Safety Data Sheets for Research and Industrial Chemicals* 1987, VCH, Deerfield Park
24. Kleinewska, P. *Br. J. Dermatol.* 1970, **83**, 543
25. Aten, J. *Am. J. Pathol.* 1988, **133**(1), 127-138
26. Kanematsu, N. et al *Mutat. Res.* 1980, **77**, 109-116
27. McGregor, D. B. et al *Environ. Mol. Mutagen.* 1988, **12**(1), 88-154
28. Montaldi, A. et al *Environ. Mutagen.* 1985, **7**, 381-390
29. Wedrychowski, A. *J. Biol. Chem.* 1986, **261**, 3370
30. Oudar, P. et al *Pharmacol. Toxicol. (Copenhagen)* 1989, **65**(4), 245-248
31. *New Eng. J. Med.* 1951, **244**, 459
32. *Med. J. Australia* 1978, **2**, 125
33. Hewett, S. J. et al *Toxicol. Appl. Pharmacol.* 1992, **113**, 267-273

34. Nachshen, D. A. *J. Gen. Physiol.* 1984, **83**, 941-967
35. Juang, M. S. *Toxicol. Appl. Pharmacol.* 1976, **35**, 339-348
36. Goering, P. L. *Toxicol. Appl. Pharmacol.* 1992, **113**, 184-191
37. *EC Directive Relating to the Quality of Water Intended for Human Consumption* 1982, 80/778/EEC, Office for Official Publications of the European Communities, 2 rue Mercier, L-2985 Luxembourg
38. *S. I. 1991 No. 472 The Environmental Protection (Prescribed Processes and Substances) Regulations* 1991, HMSO, London
39. *DoE Circular 7/89: Water and the environment. The implementation of EC directives on pollution caused by certain dangerous substances discharged into the aquatic environment* 1989, HMSO, London

# M60　Mercury(II) cyanide

## $Hg(CN)_2$

**CAS Registry No.** 592-04-1
**Synonyms** mercuric cyanide
**Mol. Formula** $C_2HgN_2$　　　　　　　**Mol. Wt.** 252.63
**Uses** Topical antiseptic for human and animal use. Chemical intermediate.

## Physical properties

**M. Pt.** 20°C (decomp.); **Specific gravity** 3.996.

### Solubility
Organic solvent: ethanol, methanol, glycerol

## Occupational exposure

**US TLV (TWA)** 0.025 mg m$^{-3}$ (as Hg) intended change; **UK Long-term limit** 0.05 mg m$^{-3}$ (as Hg) under review; **UK Short-term limit** 0.15 mg m$^{-3}$ (as Hg) under review; **UN No.** 1636; **Conveyance classification** toxic substance.

## Ecotoxicity

### Fish toxicity
Intestinal transport rate for nutrients decreased in *Channa punctatus* exposed to 3 µg l$^{-1}$ (Hg$^{2+}$) for 30 days. Disrupted metabolism in *Notopterus notopterus* exposed to 44 µg l$^{-1}$ (Hg$^{2+}$) for 30 days (1).
Teratogenic to rainbow trout after exposure of eggs to 0.12-0.21 µg l$^{-1}$ (Hg$^{2+}$) 4 days post hatch, and after parental exposure to 0.70-0.79 µg l$^{-1}$ (Hg$^{2+}$) for 400 days (1).

### Invertebrate toxicity
LC$_{50}$ (48 hr) *Daphnia magna* 9.3 µg l$^{-1}$; EC$_{50}$ (48 hr) *Daphnia magna* 5.2 µg l$^{-1}$ (Hg$^{2+}$) (1).

## Mammalian and avian toxicity

### Acute data
LD$_{50}$ oral rat, mouse 26, 33 mg kg$^{-1}$, respectively (2).

LD$_{Lo}$ intraperitoneal rat 7.5 mg kg$^{-1}$ (3).
LD$_{50}$ subcutaneous dog 2.7 mg kg$^{-1}$ (4).
LD$_{50}$ intravenous rabbit 2 mg kg$^{-1}$ (5).
LD$_{Lo}$ in ♀ humans has been reported as 10 mg kg$^{-1}$ (6).

**Metabolism and pharmacokinetics**
Absorption from rat small intestine is pH dependent in the range 5.4-7.4. The higher pH favours absorption, which may involve hydroxylation of the compound (7).

**Irritancy**
May cause eye and skin irritation (species unspecified) (8,9).

**Any other adverse effects to man**

The compound can be fatal by inhalation (8).
Acute poisoning may produce symptoms of cough, choking, diarrhoea, dizziness, rapid breath, headache and unconsciousness (9).
Symptoms of both mercury and cyanide poisoning are seen in humans, with those of cyanide being seen first (9).
Chronic poisoning may cause tremor and neuropsychiatric disturbances (10).

**Legislation**

Limited under EC Directive on Drinking Water Quality 80/778/EEC. Mercury: maximum admissible concentration 1 μg l$^{-1}$ (11).
Included in Schedule 4 (Release into Air: Prescribed Substances) and Schedule 5 (Release into Water: Prescribed Substances) Statutory Instrument No. 472, 1991 (12).
Quality objectives under EC Directives 82/176/EEC and 84/156/EEC 0.3 mg kg$^{-1}$ (wet weight) in a representative sample of fish flesh; 1 μg l$^{-1}$ (annual mean) total mercury in inland surface waters; 0.5 μg l$^{-1}$ (annual mean) dissolved mercury in estuarine waters; 0.3 μg l$^{-1}$ (annual mean) dissolved mercury in marine waters. A 'standstill' provision applies to concentrations in shellfish or sediments. Limit values under EC Directive 84/156/EEC 0.05 mg l$^{-1}$ effluent and 0.1 g l$^{-1}$ vinyl chloride production capacity for chemical industries using mercury catalysts in vinyl chloride production; 0.05 mg l$^{-1}$ effluent and 5 g kg$^{-1}$ mercury processed for chemical industries using mercury catalysts in other processes; 0.05 mg l$^{-1}$ effluent and 0.7 g kg$^{-1}$ mercury processed for manufacture of mercury catalysts used in vinyl chloride production; 0.05 mg l$^{-1}$ effluent and 0.05 g kg$^{-1}$ mercury processed for manufacture of organic and non-organic mercury compounds (other than mercury catalysts for vinyl chloride production); 0.05 mg l$^{-1}$ effluent and 0.03 g kg$^{-1}$ mercury processed for manufacture of primary batteries containing mercury; 0.05 mg l$^{-1}$ effluent for mercury recovery plants and extraction and refining of non-ferrous metals; 0.05 mg l$^{-1}$ effluent for plants treating toxic wastes containing mercury (13).

**Any other comments**
Toxicology of inorganic mercury compounds reviewed (10,14).

**References**

1. Zillioux, E. J. et al *Environ. Toxicol. Chem.* 1993, **12**, 2245-2264
2. Vernot, E. H. et al *Toxicol. Appl. Pharmacol.* 1977, **42**(2), 417-423
3. *Natl. Acad. Sci.* 1953, **65**, 28
4. *Proc. Soc. Exp. Biol.* 1964, **116**, 371
5. *J. Pharmacol. Exp. Ther.* 1931, **41**, 21

6. *J. Am. Med. Assoc.* 1916, **66**, 1694
7. Endo, T. et al *Pharmacol. Toxicol. (Copenhagen)* 1990, **66**(5), 347-353
8. Lenga, R. E. *The Sigma-Aldrich Library of Chemical Safety Data* 2nd ed., 1988, Sigma-Aldrich, Milwaukee, WI
9. *Dangerous Prop. Ind. Mater. Rep.* 1986, **6**(1), 68-72
10. Oehme, F. W. (Ed.) *Toxicity of Heavy Metals in the Environment* Part 1, 1978, Marcel Dekker Inc., New York
11. *EC Directive Relating to the Quality of Water Intended for Human Consumption* 1982, 80/778/EEC, Office for Official Publications of the European Communities, 2 rue Mercier, L-2985 Luxembourg
12. *S. I. 1991 No. 472 The Environmental Protection (Prescribed Processes and Substances) Regulations* 1991, HMSO, London
13. *DoE Circular 7/89: Water and the environment. The implementation of EC directives on pollution caused by certain dangerous substances discharged into the aquatic environment* 1989, HMSO, London
14. Korshun, M. N. *Gig. Sanit.* 1989, (1), 69-70 (Russ.) (*Chem. Abstr.* **110**, 149418q)

# M61   Mercury, diethyl-

## $Hg(C_2H_5)_2$

**CAS Registry No.** 627-44-1
**Synonyms** diethylmercury; ethylmercury
**Mol. Formula** $C_4H_{10}Hg$                    **Mol. Wt.** 258.71

## Physical properties

**B. Pt.** 159°C; **Specific gravity** $d^{20}$ 2.4660.

## Occupational exposure

**US TLV (TWA)** 0.01 mg m$^{-3}$ (as Hg); **US TLV (STEL)** 0.03 mg m$^{-3}$ (as Hg); **UK Long-term limit** 0.01 mg m$^{-3}$ (as Hg) under review; **UK Short-term limit** 0.03 mg m$^{-3}$ (as Hg) under review; **UN No.** 2024.

## Ecotoxicity

### Invertebrate toxicity

Significantly reduced emergence and hatching of the brine shrimp *Artemia franciscana* at 2.59 mg l$^{-1}$, the lowest concentration tested (1).

### Degradation

Threshold concentration in water reservoirs (in former USSR) BOD 5 µg l$^{-1}$ (2).

## Mammalian and avian toxicity

### Acute data

$LD_{50}$ oral mouse, rat 44-51 mg kg$^{-1}$ (3).
$LC_{50}$ (duration unspecified) inhalation mouse, rat 90-260 mg m$^{-3}$ (4).
$LD_{50}$ intraperitoneal mouse 45 mg kg$^{-1}$ (5).

---

## Teratogenicity and reproductive effects

Continuous exposure to the vapour at 6.43 µg mg$^{-3}$ for 115 days caused reduced sperm count and motility in ♂ rats. In ♀ rats 5.92 µg mg$^{-3}$ prolonged oestrus and reduced conception rate. Rats born to exposed parents had reduced life span, reduced development, and morphofunctional and biochemical gonad changes (6).

## Any other adverse effects to man

Two fatal cases of poisoning occurred in two women stenographers working in a warehouse in which diethyl mercury was stored (7).

In persons occupationally exposed, the circulation of mercury and urinary clearance was observed for weeks and sometimes months after exposure had ceased (8).

## Legislation

Community Right-To-Know List.

Limited under EC Directive on Drinking Water Quality 80/778/EEC. Mercury: maximum admissible concentration 1 µg l$^{-1}$ (9).

Included in Schedule 4 and 6 (Release into Air/Land: Prescribed Substances) Statutory Instrument No. 472, 1991 (10).

Quality objectives under EC Directives 82/176/EEC and 84/156/EEC 0.3 mg kg$^{-1}$ (wet weight) in a representative sample of fish flesh; 1 µg l$^{-1}$ (annual mean) total mercury in inland surface waters; 0.5 µg l$^{-1}$ (annual mean) dissolved mercury in estuarine waters; 0.3 µg l$^{-1}$ (annual mean) dissolved mercury in marine waters. A 'standstill' provision applies to concentrations in shellfish or sediments. Limit values under EC Directive 84/156/EEC 0.05 mg l$^{-1}$ effluent and 0.1 g l$^{-1}$ vinyl chloride production capacity for chemical industries using mercury catalysts in vinyl chloride production; 0.05 mg l$^{-1}$ effluent and 5 g kg$^{-1}$ mercury processed for chemical industries using mercury catalysts in other processes; 0.05 mg l$^{-1}$ effluent and 0.7 g kg$^{-1}$ mercury processed for manufacture of mercury catalysts used in vinyl chloride production; 0.05 mg l$^{-1}$ effluent and 0.05 g kg$^{-1}$ mercury processed for manufacture of organic and non-organic mercury compounds (other than mercury catalysts for vinyl chloride production); 0.05 mg l$^{-1}$ effluent and 0.03 g kg$^{-1}$ mercury processed for manufacture of primary batteries containing mercury; 0.05 mg l$^{-1}$ effluent for mercury recovery plants and extraction and refining of non-ferrous metals; 0.05 mg l$^{-1}$ effluent for plants treating toxic wastes containing mercury (11). Maximum admissible concentration in working zone atmosphere (in former USSR) 0.005 mg m$^{-3}$. Hazard class 1: dangerous by absorption through skin (8).

## Any other comments

Physico-chemical properties, human health effects, experimental toxicity, environmental effects, ecotoxicology, exposure levels and workplace experience reviewed (12).

Toxicity and hazards reviewed (8).

Aquatic toxicology reviewed (13).

Saturation concentration 6250 mg m$^{-3}$.

## References

1. Pandey, A. S. et al *Ecotoxicol. Environ. Saf.* 1991, **21**, 68-79
2. Samyslova, S. D. et al *Sanitary Protection of Water Reservoirs against Pollution with Industrial Effluents* 1965, 113–128, Ser. 7, Meditsina Publ., Moscow

3.  Izmerov, N. F. et al *Toxicometric Parameters of Industrial Toxic Chemicals under Single Exposure* 1982, CIP, Moscow
4.  *Hyg. Sanit. (USSR)*. 1973, **38**(1), 100
5.  *Yaku Kenkyu* 1959, **79**, 579
6.  Ignatyev, V. N. *Gig. Sanit.* 1980, **6**, 24-27
7.  Hill, W. H. *Can. J. Public Health* 1943, **34**, 158-160
8.  Trakhtenburg, I. M. *Scientific Reviews of Soviet Literature of Toxicity and Hazards of Chemicals. Organomercury Compounds 117* 1991, Eng. Trans. (Ed.) Richardson, M. L., UNEP/IRPTC Geneva, CIP Moscow
9   *EC Directive Relating to the Quality of Water Intended for Human Consumption* 1982, 80/778/EEC, Office for Official Publications of the European Communities, 2 rue Mercier, L-2985 Luxembourg
10. *S. I. 1991 No. 472 The Environmental Protection (Prescribed Processes and Substances) Regulations* 1991, HMSO, London
11. *DoE Circular 7/89: Water and the environment. The implementation of EC directives on pollution caused by certain dangerous substances discharged into the aquatic environment* 1989, HMSO, London
12. *ECETOC Technical Report No. 30(4)* 1991, European Chemical Industry Ecology and Toxicology Centre, B-1160 Brussels
13. Zillioux, E. J. et al *Environ. Toxicol. Chem.* 1993, **12**, 2245-2264

# M62   Mercury, dimethyl-

## $Hg(CH_3)_2$

**CAS Registry No.** 593-74-8
**Synonyms** dimethylmercury
**Mol. Formula** $C_2H_6Hg$                           **Mol. Wt.** 230.66
**Uses** Inorganic reagent.
**Occurrence** Formed in oceans by methylation of inorganic mercury. Environmental contaminant found in fish and birds.

## Physical properties

**B. Pt.** $_{740}$ 92°C; **Specific gravity** $d^{20}$ 3.1874.

## Solubility
Organic solvent: ethanol, diethyl ether

## Occupational exposure

**US TLV (TWA)** 0.01 mg m$^{-3}$ (as Hg); **US TLV (STEL)** 0.03 mg m$^{-3}$ (as Hg); **UK Long-term limit** 0.01 mg m$^{-3}$ (as Hg) under review; **UK Short-term limit** 0.03 mg m$^{-3}$ (as Hg) under review; **UN No.** 2024.

## Ecotoxicity

### Invertebrate toxicity
Significantly reduced emergence and hatching of the brine shrimp *Artemia franciscana* at 2.59 mg l$^{-1}$, the lowest concentration tested (1).

---

## Legislation

Community Right-To-Know List.

Limited under EC Directive on Drinking Water Quality 80/778/EEC. Mercury: maximum admissible concentration 1 μg l$^{-1}$ (2).

Included in Schedule 4 and 6 (Release into Air/Land: Prescribed Substances) Statutory Instrument No. 472, 1991 (3).

Quality objectives under EC Directives 82/176/EEC and 84/156/EEC 0.3 mg kg$^{-1}$ (wet weight) in a representative sample of fish flesh; 1 μg l$^{-1}$ (annual mean) total mercury in inland surface waters; 0.5 μg l$^{-1}$ (annual mean) dissolved mercury in estuarine waters; 0.3 μg l$^{-1}$ (annual mean) dissolved mercury in marine waters. A 'standstill' provision applies to concentrations in shellfish or sediments. Limit values under EC Directive 84/156/EEC 0.05 mg l$^{-1}$ effluent and 0.1 g l$^{-1}$ vinyl chloride production capacity for chemical industries using mercury catalysts in vinyl chloride production; 0.05 mg l$^{-1}$ effluent and 5 g kg$^{-1}$ mercury processed for chemical industries using mercury catalysts in other processes; 0.05 mg l$^{-1}$ effluent and 0.7 g kg$^{-1}$ mercury processed for manufacture of mercury catalysts used in vinyl chloride production; 0.05 mg l$^{-1}$ effluent and 0.05 g kg$^{-1}$ mercury processed for manufacture of organic and non-organic mercury compounds (other than mercury catalysts for vinyl chloride production); 0.05 mg l$^{-1}$ effluent and 0.03 g kg$^{-1}$ mercury processed for manufacture of primary batteries containing mercury; 0.05 mg l$^{-1}$ effluent for mercury recovery plants and extraction and refining of non-ferrous metals; 0.05 mg l$^{-1}$ effluent for plants treating toxic wastes containing mercury (4).

## Any other comments

Physico-chemical properties, human health effects, experimental toxicity, environmental effects, ecotoxicology, exposure levels and workplace experience reviewed (5).

Toxicity and hazards reviewed (6).

Aquatic toxicology reviewed (7).

At one time dimethylmercury was considered as a possibly important mercurial chemical species, but so far it has only been observed in marine environments (8).

## References

1. Pandey, A. S. et al *Ecotoxicol. Environ. Saf.* 1991, **21**, 68-79
2. *EC Directive Relating to the Quality of Water Intended for Human Consumption* 1982, 80/778/EEC, Office for Official Publications of the European Communities, 2 rue Mercier, L-2985 Luxembourg
3. *S. I. 1991 No. 472 The Environmental Protection (Prescribed Processes and Substances) Regulations* 1991, HMSO, London
4. *DoE Circular 7/89: Water and the environment. The implementation of EC directives on pollution caused by certain dangerous substances discharged into the aquatic environment* 1989, HMSO, London
5. *ECETOC Technical Report No. 30(4)* 1991, European Chemical Industry Ecology and Toxicology Centre, B-1160 Brussels
6. Trakhtenburg, I. M. *Scientific Reviews of Soviet Literature of Toxicity and Hazards of Chemicals. Organomercury Compounds 117* 1990, Eng. Trans. (Ed.) Richardson, M. L., UNEP/IRPTC, Geneva, CIP, Moscow
7. Zillioux, E. J. et al *Environ. Toxicol. Chem.* 1993, **12**, 2245-2264
8. Mason, R.P. et al *Nature* 1990, **347**, 457–459

# M63   Mercury gluconate

**CAS Registry No.** 63937-14-4
**Synonyms** mercurous gluconate
**Mol. Formula** $C_6H_{11}O_7Hg$ $\qquad\qquad\qquad$ **Mol. Wt.** 395.74

## Occupational exposure

**US TLV (TWA)** 0.025 mg m$^{-3}$ (as Hg) intended change; **UK Long-term limit**
0.05 mg m$^{-3}$ (as Hg) under review; **UK Short-term limit** 0.15 mg m$^{-3}$ (as Hg) under
review; **UN No.** 1637; **Conveyance classification** toxic substance.

## Legislation

Community Right-To-Know List.
Limited under EC Directive on Drinking Water Quality 80/778/EEC. Mercury:
maximum admissible concentration 1 µg l$^{-1}$ (1).
Included in Schedule 4 and 6 (Release into Air/Land: Prescribed Substances)
Statutory Instrument No. 472, 1991 (2).
Quality objectives under EC Directives 82/176/EEC and 84/156/EEC 0.3 mg kg$^{-1}$
(wet weight) in a representative sample of fish flesh; 1 µg l$^{-1}$ (annual mean) total
mercury in inland surface waters; 0.5 µg l$^{-1}$ (annual mean) dissolved mercury in
estuarine waters; 0.3 µg l$^{-1}$ (annual mean) dissolved mercury in marine waters. A
'standstill' provision applies to concentrations in shellfish or sediments. Limit values
under EC Directive 84/156/EEC 0.05 mg l$^{-1}$ effluent and 0.1 g l$^{-1}$ vinyl chloride
production capacity for chemical industries using mercury catalysts in vinyl chloride
production; 0.05 mg l$^{-1}$ effluent and 5 g kg$^{-1}$ mercury processed for chemical
industries using mercury catalysts in other processes; 0.05 mg l$^{-1}$ effluent and
0.7 g kg$^{-1}$ mercury processed for manufacture of mercury catalysts used in vinyl
chloride production; 0.05 mg l$^{-1}$ effluent and 0.05 g kg$^{-1}$ mercury processed for
manufacture of organic and non-organic mercury compounds (other than mercury
catalysts for vinyl chloride production); 0.05 mg l$^{-1}$ effluent and 0.03 g kg$^{-1}$ mercury
processed for manufacture of primary batteries containing mercury; 0.05 mg l$^{-1}$
effluent for mercury recovery plants and extraction and refining of non-ferrous
metals; 0.05 mg l$^{-1}$ effluent for plants treating toxic wastes containing mercury (3).

## Any other comments

Physico-chemical properties, human health effects, experimental toxicity,
environmental effects, ecotoxicology, exposure levels and workplace experience
reviewed (4).
Aquatic toxicology reviewed (5).

## References

1. *EC Directive Relating to the Quality of Water Intended for Human Consumption* 1982,
   80/778/EEC, Office for Official Publications of the European Communities, 2 rue Mercier,
   L-2985 Luxembourg
2. *S. I. 1991 No. 472 The Environmental Protection (Prescribed Processes and Substances)
   Regulations* 1991, HMSO, London
3. *DoE Circular 7/89: Water and the environment. The implementation of EC directives on
   pollution caused by certain dangerous substances discharged into the aquatic environment*
   1989, HMSO, London

4.  *ECETOC Technical Report No. 30(4)* 1991, European Chemical Industry Ecology and Toxicology Centre, B-1160 Brussels
5.  Zillioux, E. J. et al *Environ. Toxicol. Chem.* 1993, **12**, 2245-2264

# M64 Mercury(I) iodide

$Hg_2I_2$

**CAS Registry No.** 7783-30-4
**Synonyms** mercurous iodide; mercury protoiodide; yellow mercury iodide
**Mol. Formula** HgI                                **Mol. Wt.** 327.49
**Uses** Antibacterial.

## Physical properties

**M. Pt.** 290°C; **Specific gravity** 7.70.

## Occupational exposure

US **TLV (TWA)** 0.025 mg m$^{-3}$ (as Hg) intended change; **UK Long-term limit** 0.05 mg m$^{-3}$ (as Hg) under review; **UK Short-term limit** 0.15 mg m$^{-3}$ (as Hg) under review; **UN No.** 1638; **Conveyance classification** toxic substance.

## Mammalian and avian toxicity

### Acute data
$LD_{50}$ oral mouse 110 mg kg$^{-1}$ (1).
$LD_{50}$ intraperitoneal mouse 50 mg kg$^{-1}$ (1).

## Legislation

Community Right-To-Know List.
Limited under EC Directive on Drinking Water Quality 80/778/EEC. Mercury: maximum admissible concentration 1 µg l$^{-1}$ (2).
Included in Schedule 4 (Release into Air. Prescribed Substances) Statutory Instrument No. 472, 1991 (3).
Quality objectives under EC Directives 82/176/EEC and 84/156/EEC 0.3 mg kg$^{-1}$ (wet weight) in a representative sample of fish flesh; 1 µg l$^{-1}$ (annual mean) total mercury in inland surface waters; 0.5 µg l$^{-1}$ (annual mean) dissolved mercury in estuarine waters; 0.3 µg l$^{-1}$ (annual mean) dissolved mercury in marine waters. A 'standstill' provision applies to concentrations in shellfish or sediments. Limit values under EC Directive 87/156/EEC 0.05 mg l$^{-1}$ effluent and 0.1 g l$^{-1}$ vinyl chloride production capacity for chemical industries using mercury catalysts in vinyl chloride production; 0.05 mg l$^{-1}$ effluent and 5 g kg$^{-1}$ mercury processed for chemical industries using mercury catalysts in other processes; 0.05 mg l$^{-1}$ effluent and 0.7 g kg$^{-1}$ mercury processed for manufacture of mercury catalysts used in vinyl chloride production; 0.05 mg l$^{-1}$ effluent and 0.05 g kg$^{-1}$ mercury processed for manufacture of organic and non-organic mercury compounds (other than mercury catalysts for vinyl chloride production); 0.05 mg l$^{-1}$ effluent and 0.03 g kg$^{-1}$ mercury

processed for manufacture of primary batteries containing mercury; 0.05 mg l$^{-1}$ effluent for mercury recovery plants and extraction and refining of non-ferrous metals; 0.05 mg l$^{-1}$ effluent for plants treating toxic wastes containing mercury (4).

## Any other comments

Physico-chemical properties, human health effects, experimental toxicity, environmental effects, ecotoxicity, exposure levels and workplace experience reviewed (5).
Toxicity of inorganic mercury and environmental effects revieved (6,7).
Aquatic toxicology reviewed (8).

## References

1. *Arch. Toxikol.* 1964, **20**, 226
2. *EC Directive Relating to the Quality of Water Intended for Human Consumption* 1982, 80/778/EEC, Office for Official Publications of the European Communities, 2 rue Mercier, L-2985 Luxembourg
3. *S. I. 1991 No. 472 The Environmental Protection (Prescribed Processes and Substances) Regulations* 1991, HMSO, London
4. *DoE Circular 7/89: Water and the environment. The implementation of EC directives on pollution caused by certain dangerous substances discharged into the aquatic environment* 1989, HMSO, London
5. *ECETOC Technical Report No. 30(4)* 1991, European Chemical Industry Ecology and Toxicology Centre, B-1160 Brussels
6. *Environmental Health Criteria 118: Inorganic Mercury* 1991, WHO/IPCS, Geneva
7. *Environmental Health Criteria 86: Mercury-Environmental Aspects* 1989, WHO/IPCS, Geneva
8. Zillioux, E. J. et al *Environ. Toxicol. Chem.* 1993, **12**, 2245-2264

# M65   Mercury(II) iodide

## HgI$_2$

**CAS Registry No.** 7774-29-0
**Synonyms** mercuric iodide; mercury biniodide; red mercuric iodide
**Mol. Formula** HgI$_2$                    **Mol. Wt.** 454.40
**Uses** Topical antiseptic.

## Physical properties

**M. Pt.** 259°C; **B. Pt.** 350°C (sublimes); **Specific gravity** 6.28.

**Solubility**
Water: 0.006 g 100 g$^{-1}$ at 25°C.

## Occupational exposure

**US TLV (TWA)** 0.025 mg m$^{-3}$ (as Hg) intended change; **UK Long-term limit** 0.05 mg m$^{-3}$ (as Hg) under review; **UK Short-term limit** 0.15 mg m$^{-3}$ (as Hg) under review; **UN No.** 1634; **Conveyance classification** toxic substance.

## Ecotoxicity

### Fish toxicity

Intestinal transport rate for nutrients decreased in snakehead fish exposed to 3 µg l$^{-1}$ (Hg$^{2+}$) for 30 days (1).

Disrupted metabolism in knifefish exposed to 44 µg l$^{-1}$ (Hg$^{2+}$) for 30 days (1).

Teratogenic to rainbow trout after exposure of eggs to 0.12-0.21 µg l$^{-1}$ (Hg$^{2+}$) 4 days post hatch, and after parental exposure to 0.70-0.79 µg l$^{-1}$ (Hg$^{2+}$) for 400 days (1).

### Invertebrate toxicity

LC$_{50}$ (48 hr) *Daphnia magna* 9.3 µg l$^{-1}$; EC$_{50}$ (48 hr) *Daphnia magna* 5.2 µg l$^{-1}$ (Hg$^{2+}$) (1).

## Mammalian and avian toxicity

### Acute data

LD$_{50}$ oral mouse, rat 17-18 mg kg$^{-1}$ (2).

LD$_{Lo}$ oral human 357 mg kg$^{-1}$ (3).

LD$_{50}$ dermal rat 75 mg kg$^{-1}$ (2).

LD$_{50}$ intraperitoneal mouse 4200 µg kg$^{-1}$ (2).

### Teratogenicity and reproductive effects

Teratogenic and reproductive effects reported in mice exposed to 4870 ng m$^{-3}$ for 24 hr by inhalation on days 1-22 of pregnancy (4).

## Legislation

Community Right-To-Know List.

Limited under EC Directive on Drinking Water Quality 80/778/EEC. Mercury: maximum admissible concentration 1 µg l$^{-1}$ (5).

Included in Schedule 4 (Release into Air: Prescribed Substances) Statutory Instrument No. 472, 1991 (6).

Quality objectives under EC Directives 82/176/EEC and 84/156/EEC 0.3 mg kg$^{-1}$ (wet weight) in a representative sample of fish flesh; 1 µg l$^{-1}$ (annual mean) total mercury in inland surface waters; 0.5 µg l$^{-1}$ (annual mean) dissolved mercury in estuarine waters; 0.3 µg l$^{-1}$ (annual mean) dissolved mercury in marine waters. A 'standstill' provision applies to concentrations in shellfish or sediments. Limit values under EC Directive 84/156/EEC 0.05 mg l$^{-1}$ effluent and 0.1 g l$^{-1}$ vinyl chloride production capacity for chemical industries using mercury catalysts in vinyl chloride production; 0.05 mg l$^{-1}$ effluent and 5 g kg$^{-1}$ mercury processed for chemical industries using mercury catalysts in other processes; 0.05 mg l$^{-1}$ effluent and 0.7 g kg$^{-1}$ mercury processed for manufacture of mercury catalysts used in vinyl chloride production; 0.05 mg l$^{-1}$ effluent and 0.05 g kg$^{-1}$ mercury processed for manufacture of organic and non-organic mercury compounds (other than mercury catalysts for vinyl chloride production); 0.05 mg l$^{-1}$ effluent and 0.03 g kg$^{-1}$ mercury processed for manufacture of primary batteries containing mercury; 0.05 mg l$^{-1}$ effluent for mercury; 0.05 mg l$^{-1}$ effluent for mercury recovery plants and extraction and refining of non-ferrous metals; 0.05 mg l$^{-1}$ effluent for plants treating toxic wastes containing mercury (7).

**Any other comments**

Physico-chemical properties, human health effects, experimental toxicity, environmental effects, ecotoxicology, exposure levels and workplace experience reviewed (8).

Toxicity of inorganic mercury and environmental effects reviewed (9,10).

Toxicity of mercuric compounds discussed (11).

**References**

1. Zillioux, E. J. et al *Environ. Toxicol. Chem.* 1993, **12**, 2245-2264
2. *Gig. Tr. Prof. Zabol.* 1981, **25**(7), 27
3. *Z. Klin. Med. (1879)* 1927, **106**, 783
4. *Gig. Sanit.* 1981, **46**(5), 73
5. *EC Directive Relating to the Quality of Water Intended for Human Consumption* 1982, 80/778/EEC, Office for Official Publications of the European Communities, 2 rue Mercier, L-2985 Luxembourg
6. *S. I. 1991 No. 472 The Environmental Protection (Prescribed Processes and Substances) Regulations* 1991, HMSO, London
7. *DoE Circular 7/89: Water and the environment. The implementation of EC directives on pollution caused by certain dangerous substances discharged into the aquatic environment* 1989, HMSO, London
8. *ECETOC Technical Report No. 30(4)* 1991, European Chemical Industry Ecology and Toxicology Centre, B-1160 Brussels
9. *Environmental Health Criteria 118: Inorganic Mercury* 1991, WHO/IPCS, Geneva
10. *Environmental Health Criteria 86: Mercury-Environmental Aspects* 1989, WHO/IPCS, Geneva
11. Korshun, M. N. *Gig. Sanit.* 1989, (1), 69-70 (Russ.) (*Chem. Abstr.* **110**, 149418q)

# M66   Mercury(II), methoxyethyl-, acetate

## $CH_3OCH_2CH_2HgOC(O)CH_3$

**CAS Registry No.** 151-38-2

**Synonyms** Landisan; Mema; Mercuran; Panogen; Radosan; methoxyethylmercury acetate

**Mol. Formula** $C_5H_{10}HgO_3$                    **Mol. Wt.** 318.72

**Uses** Fungicide used in seed dressings.

**Physical properties**

**M. Pt.** 40-42°C. **Volatility** v.p. $1.3 \times 10^{-5}$ mmHg at 20°C

**Solubility**

Water: miscible. Organic solvent: ethylene glycol, methanol

**Occupational exposure**

**US TLV (TWA)** 0.01 mg m$^{-3}$ (as Hg); **UK Long-term limit** 0.01 mg m$^{-3}$ (as Hg); **UK Short-term limit** 0.03 mg m$^{-3}$ (as Hg).

## Environmental fate

### Degradation studies

In soil degraded to inorganic mercury salts or metallic mercury. Metallic mercury is ultimately converted to mercury sulfide by the reaction with hydrogen sulfide liberated by soil microorganisms (1).

## Mammalian and avian toxicity

### Acute data

$LD_{50}$ oral rat, mouse 25, 45 mg $kg^{-1}$, respectively (2,3).
$LD_{50}$ intragastric mouse, rat 60-70 mg $kg^{-1}$ (Radosan) (4).

### Metabolism and pharmacokinetics

In single doses alkoxyalkyl mercury compounds are rapidly converted to inorganic mercury and do not appear to cross the blood-brain barrier appreciably in mammals. Mercury accumulates principally in the kidneys and liver. With repeated exposure there is a slow but progressive increase in brain levels of mercury (5).

## Any other adverse effects to man

Symptoms of chronic organomercury compound toxicity include disorders of the central nervous system (general depression, weakness, disturbed co-ordination of movements, increased reflexes, excitability, tremors, paralysis, convulsions) and blood disorders (advancing anaemia, leucopenia, eosinopenia, declining lymphocyte count, onset of granules in neutrophils). The number of young cells and mitosis decline in the bone marrow (6).

## Legislation

Limited under EC Directive on drinking water quality 80/778/EEC. Mercury maximum admissible concentration 1 μg $l^{-1}$; individual pesticides maximum admissible concentration 0.1 μg $l^{-1}$ (7).
Included in Schedule 5 (Release into Water: Prescribed Substances) and Schedule 6 (Release into Land: Prescribed Substances) Statutory Instrument No. 472, 1991 (8).

## Any other comments

Organomercury compounds comprehensively reviewed (9).

## References

1. White-Stevens, R. (Ed.) *Pesticides in the Environment* 1971, **1**(1), 22, Marcel Dekker, New York
2. *Occup. Health Rev.* 1963, **15**, 5
3. *Eisei Kagaku* 1972, **18**, 248
4. Borisenko, N.F. *Hyg. Toxicol.* 1967, 97-100, Zdorov'ya Publ, Kiev
5. Gosselin, R. E. et al *Clinical Toxicology of Commercial Products* 5th ed., 1984, **3**, 267, Williams & Wilkins, Baltimore
6. Nazaretyan, K. L. *Domestic Poisoning with Grenosan* 1971, 178, Yerevan, Ayastan
7. *EC Directive Relating to the Quality of Water Intended for Human Consumption* 1982, 80/778/EEC, Office for Official Publications of the European Communities, 2 rue Mercier, L-2985 Luxembourg
8. *S. I. 1991 No. 472 The Environmental Protection (Prescribed Processes and Substances) Regulations* 1991, HMSO, London

9.  Trakhtenberg, I. M. et al *Scientific Reviews of Soviet Literature on Toxicity and Hazards of Chemicals Organomercury Compounds 117* 1993, Eng. Trans. (Ed.) Richardson, M.L., UNEP/IRPTC, Geneva, CIP Moscow

# M67   Mercury, methyl

## $CH_3Hg^+$

**CAS Registry No.** 22967-92-6
**Synonyms** Mercury(1+), methyl-; methylmercury ion(1+); methylmercury(1+); methylmercury
**Mol. Formula** $CH_3Hg$                        **Mol. Wt.** 215.63
**Occurrence** A pollutant particularly in sea and lake water and in aquatic and land animals (1-3).
Widely detected in trace amounts, in human tissue (4,5).
Formed from inorganic and organic mercury by many microorganisms (6-9).

## Occupational exposure

**US TLV (TWA)** 0.01 mg m$^{-3}$ (as Hg); **US TLV (STEL)** 0.03 mg m$^{-3}$ (as Hg); **UK Long-term limit** 0.01 mg m$^{-3}$ (as Hg) under review; **UK Short-term limit** 0.03 mg m$^{-3}$ (as Hg) under review.

## Ecotoxicity

### Fish toxicity
95% of the mercury in fish is methylmercury (10,11).
Guppy and medaka exposed to 1.8 µg l$^{-1}$ in water for 3 months caused impaired spermatogenesis in both species (12).
Brook trout (embryo) exposed to 0.88 µg l$^{-1}$ in water for 17 days suffered enzyme disruption, while rainbow trout (adult) exposed to 0.04 µg l$^{-1}$ for 64 days showed reduced growth (13).

### Invertebrate toxicity
*Rana pipiens* exposed to 1.0 µg l$^{-1}$ in water for 4 months showed arrested metamorphosis (13).

### Bioaccumulation
Calc. bioaccumulation factor $\log_{10}$ 6.5 in freshwater lake, using ratios of total fish to mercury and aqueous methylmercury measurements. Calc. methylmercury accumulation factor 3 million in fish (14-16).

## Environmental Fate

### Anaerobic effects
Methylmercury content was significantly higher in water and sediment in anaerobic model laboratory experiments with eutrophic sediments than similar systems maintained aerobically, whereas sediment accumulation of total mercury was faster in aerobic systems (17-19).

Sulfate-reducing bacteria and methanogens appear to be involved in anaerobic demethylation of freshwater sediments (20).

**Degradation**
Aerobic demethylation occurs in estuarine sediments but appears to be relatively unimportant in fresh water (20).

**Abiotic removal**
Abiotic methylation of $Hg^{2+}$ readily occurs in aquatic systems. It requires the presence of metals which act as catalysts and organic matter for the transformation process (21).
The net of methylation and demethylation processes produces the mercury available for bioaccumulation. Environmental factors affect these processes to increase or decrease net methylation (22).

**Absorption**
In a freshwater lake, the concentration of methylmercury in the sediment was ≈1% of total mercury, indicating substantial sediment flux (14).

## Mammalian and avian toxicity

**Acute data**
$LD_{50}$ oral mouse 53 mg kg$^{-1}$ (23).
$LD_{50}$ intraperitoneal mouse 22 mg kg$^{-1}$ (23).
Toxicity is manifest by initial excitation followed by depression and finally convulsions, paralysis and death (24).

**Teratogenicity and reproductive effects**
Positive effects on reproduction have been observed in mice receiving 6.5 g kg$^{-1}$ on day 9 subcutaneously (25).

**Metabolism and pharmacokinetics**
Detected in brain, liver, kidney, fatty tissue and hair of majority of humans examined. Also present in human breast milk (4).
Being lipophilic it is selectively accumulated in lipid-rich tissues. More than 50% is frequently found as metallothionines (26).
Metabolism can occur by dealkylation in hepatic microsomal tissue (27).
Methylmercury can be absorbed from the gastrointestinal tract or inhaled (5,23).
Biotic methylation, the transformation of $Hg^{2+}$ to $CH_3Hg^+$ is not well understood, but appears to be a co-metabolic reaction with no known specific gene control.
Demethylation, $CH_3Hg^+$ transformed to $Hg^{2+}$ and then $Hg^0$ is primarily enzyme mediated, taking place within a single cell. The demethylation process is controlled by 2 genes (28).

## Any other adverse effects to man

Studies of sections of human brain, liver, kidneys, fatty tissue and hair, have indicated that methylmercury is detectable in the majority of human organs and tissues, the percentage of positive results being as high as 87-97% (29).

## Any other adverse effects

Methylmercury passes the blood brain barrier and nuclear membranes to react directly with both cellular and nuclear components. Accumulation of mercury in the brain, compared to blood and muscle, is much less in fish than in mammals (30).

Significant toxic effects in waterbirds are associated with liver-Hg levels of 11 ppm or above, however lower toxicity thresholds have been reported (31).

## Legislation

Limited under EC Directive on Drinking Water Quality 80/778/EEC. Mercury: maximum admissible concentration 1 μg l$^{-1}$ (32).

Included in Schedule 4 (Release into Air: Prescribed Substances) and Schedule 5 (Release into Water: Prescribed Substances) Statutory Instrument No. 472, 1991 (33). Quality objectives under EC Directives 82/176/EEC and 84/156/EEC 0.3 mg kg$^{-1}$ (wet weight) in a representative sample of fish flesh; 1 μg l$^{-1}$ (annual mean) total mercury in inland surface waters; 0.5 μg l$^{-1}$ (annual mean) dissolved mercury in estuarine waters; 0.3 μg l$^{-1}$ (annual mean) dissolved mercury in marine waters. A 'standstill' provision applies to concentrations in shellfish or sediments. Limit values under EC Directive 84/156/EEC 0.05 mg l$^{-1}$ effluent and 0.1 g l$^{-1}$ vinyl chloride production capacity for chemical industries using mercury catalysts in vinyl chloride production; 0.05 mg l$^{-1}$ effluent and 5 g kg$^{-1}$ mercury processed for chemical industries using mercury catalysts in other processes; 0.05 mg l$^{-1}$ effluent and 0.7 g kg$^{-1}$ mercury processed for manufacture of mercury catalysts used in vinyl chloride production; 0.05 mg l$^{-1}$ effluent and 0.05 g kg$^{-1}$ mercury processed for manufacture of organic and non-organic mercury compounds (other than mercury catalysts for vinyl chloride production); 0.05 mg l$^{-1}$ effluent and 0.03 g kg$^{-1}$ mercury processed for manufacture of primary batteries containing mercury; 0.05 mg l$^{-1}$ effluent for mercury recovery plants and extraction and refining of non-ferrous metals; 0.05 mg l$^{-1}$ effluent for plants treating toxic wastes containing mercury (34).

## Any other comments

Human health effects, experimental toxicology, physico-chemical properties reviewed (35).

Toxic effects of organomercuric compounds reviewed (29).

*Pseudomonas putidas* FB1, *Candida albicans* and *Saccharomyces cerevisiae* strains can all produce methylmercury from elemental mercury or inorganic salts (6,9).

Methylmercury formed in lake and sea water accumulates in aquatic animals and enters the food chain (2,5).

Low pH of lake water favours the transformation of inorganic mercury to the methylated form (1-3,36).

Methylation and demethylation transformations in aquatic systems reviewed (22,37,38).

## References

1. Bloom, N. S. et al *Water Air Soil Pollut.* 1990, **53**(3-4), 251-265
2. Winfrey, M. R. *Environ. Toxicol. Chem.* 1990, **9**(7), 853-869
3. Shepot'ko, A. O. et al *Dokl. Akad. Nauk SSSR* 1990, **311**(1), 204-207 (Russ.) (*Chem. Abstr.* **113**, 103032t)
4. Bogomolova, Z. N. et al *Metals, Hygienic Aspects of Assessment and Enhancement of the Environment* Kasparov, A. A. et al (Ed.) 1983, USSR Acad. Sci., Moscow
5. Stenberg, A. I. *VIIIth Intern. Symp. on Sea Medicine* Odessa, Sept. 23-30 1976, 96-97, Moscow.
6. Baldi, F. et al *Microbios* 1988, **54**(218), 7-13
7. Baldi, F. *Microbiol. Ecol.* 1989, **17**(3), 263-274
8. Illyin, I. E. *Gig. Sanit.* 1985, (3), 7-11

9. Yannai, S. et al *Appl. Environ. Microbiol.* 1991, **57**(1), 245-247

10. Grieb, T. M. et al *Environ. Toxicol. Chem.* 1990, **9**, 919-930

11. Bloom, N. S. *Can. J. Fish Aquat. Sci.* 1992, **49**, 1010-1017

12. Wester, P. W. *Comp. Biochem. Physiol.* 1991, **100c**, 237-239

13. *Ambient Water Quality Criteria for Mercury* 1980, US EPA 440/5-80-058. Natl. Tech. Inf. Ser., Springfield, VA

14. Watras, C. J. et al *Mercury as a Global Pollutant* 1994, Lewis Publ., Chelsea, MI

15. Gill, G. A. et al *Environ. Sci. Technol.* 1990, **24** 1392-1400

16. Bloom, N. S. et al *Water Air Soil Pollut.* 1991, **56** 477-491

17. Regnell, O. et al *Appl. Environ. Microb.* 1991, **57** 789-795

18. Regnell, O. *Can. J. Fish Aquat. Sci.* 1990, **47**, 548-553

19. Hammer, U. T. et al *Arch. Environ. Contam. Toxicol.* 1988, **17**, 257-262

20. Oremland, R. S. et al *Appl. Environ. Microbiol.* 1991, **57**, 130-137

21. Lee, Y. H. et al *Water Air Soil Pollut.* 1985, **25** 391-400

22. Winfrey, M. R. et al. *Environ, Toxicol. Chem.* 1990, **9**, 853-869

23. *Issues of Industrial Hygiene, Occupational Pathology, Industrial Toxicology and Sanitary Chemistry, Gark* 1963, 12-15

24. Trakhtenberg, I. M. *Farmakol. Toxsikol. (Moscow)* 1951, (1), 48-51

25. *Teratology* 1982, **26**(1), 14A

26. *Problems of Hygiene and Toxicology of Pesticides Abst. Reps VIth Natl. Sci. Conf., Part 2.* 1982, Sc-Res. Inst. Toxicol., Kiev

27. *Prevention and Measures of Control of Fish Diseases in Intensive Breeding Technologies* Krasnodar, 1978 pp. 58-60

28. Summers, A. O. *Annu. Rev Microbiol.* 1986, **40**, 607-634

29. Trakhtenberg, I. M. et al *Scientific Reviews of Soviet Literature on Toxicity and Hazards of Chemicals: Organomercury compounds 117* 1993, Richardson, M. L. (Ed.), UNEP/IPRTC, Geneva, CIP, Moscow

30. Giblin, F. J. *Toxicol. Appl. Pharmacol.* 1973, **24**, 81-91

31. Scheuhammer, A. M. *Environ. Pollut.* 1991, **71** 329-375

32. *EC Directive Relating to the Quality of Water Intended for Human Consumption* 1982, 80/778/EEC, Office for Official Publications of the European Communities, 2 rue Mercier, L-2985 Luxembourg

33. *S. I. 1991 No. 472 The Environmental Protection (Prescribed Processes and Substances) Regulations* 1991, HMSO, London

34. *DoE Circular 7/89: Water and the environment. The implementation of EC directives on pollution caused by certain dangerous substances discharged into the aquatic environment* 1989, HMSO, London

35. *ECETOC Technical Report No. 30(4)* 1991, European Chemical Industry Ecology and Toxicology Centre, B-1160 Brussels

36. Spry, D. J. et al *Environ. Pollut.* 1991, **71**(2-4), 243-304

37. Gilmour, C. C. et al *Environ. Pollut.* 1991, **71**, 131-169

38. Olson, B. H. *Environ. Sci. Technol.* 1991, **25**, 604-611

# M68    Mercury(II), methyl-, acetate

## $CH_3HgOC(O)CH_3$

**CAS Registry No.** 108-07-6
**Synonyms** mercury, (acetato-$O$)methyl-; mercury, acetoxymethyl-; methylmercuric acetate
**Mol. Formula** $C_3H_6HgO_2$                                      **Mol. Wt.** 274.67

### Physical properties
**M. Pt.** 128-129°C.

### Occupational exposure
**US TLV (TWA)** 0.01 mg m$^{-3}$ (as Hg); **UK Long-term limit** 0.01 mg m$^{-3}$ (as Hg);
**UK Short-term limit** 0.03 mg m$^{-3}$ (as Hg).

### Any other adverse effects
*In vitro* rat forebrain, 34 mg kg$^{-1}$ significantly reduced synaptosomal $^{45}Ca^{2+}$ uptake measured during 1 (fast uptake) or 10 (total uptake) sec of incubation. 14 mg kg$^{-1}$ reduced total uptake of $^{45}Ca^{2+}$ by ≥70% and reduced fast uptake by 20-60%. It affects $Ca^{2+}$ uptake in the absence of depolarisation (1).
Acute bath application of micromolar concentrations blocks the nerve evoked release of acetylcholine at the neuromuscular junction by presynaptic effects (2).

### Legislation
Limited under EC Directive on Drinking Water Quality 80/778/EEC, maximum admissible concentration 1 μg Hg l$^{-1}$ (3).
Included in Schedule 5 (Release into Water: Prescribed Substances) Statutory Instrument No. 472, 1991 (4).

### References
1. Schafer, T. J. et al *J. Pharmacol. Exp. Ther*. 1989, **248**(2), 696-702
2. Traxinger, D. L. et al *Toxicol. Appl. Pharmacol*. 1987, **90**(1), 23-33
3. *EC Directive Relating to the Quality of Water Intended for Human Consumption* 1982, 80/778/EEC, Office for Official Publications of the European Communities, 2 rue Mercier, L-2985 Luxembourg
4. *S. I. 1991 No. 472 The Environmental Protection (Prescribed Processes and Substances) Regulations* 1991, HMSO, London

# M69   Mercury(II), methyl-, dicyanamide

## NCNHC(=NH)NHHgCH₃

**CAS Registry No.** 502-39-6
**Synonyms** mercury,(cyanoguanidinato-$N'$)methyl-; Agrosol; methylmercuric cyanoguanidine; Morsodren; Mort on Soil Drench; Pandrinox; Panogen PX
**Mol. Formula** $C_3H_6HgN_4$                    **Mol. Wt.** 298.70

## Physical properties
**M. Pt.** 156°C; **Volatility** v.p. $6.5 \times 10^{-5}$ mmHg at 35°C.

### Solubility
Water: 21.7 g $l^{-1}$. Organic solvent: acetone, ethanol, ethylene glycol

## Occupational exposure
**US TLV (TWA)** 0.01 mg m$^{-3}$ (as Hg); **UK Long-term limit** 0.01 mg m$^{-3}$ (as Hg); **UK Short-term limit** 0.03 mg m$^{-3}$ (as Hg).

### Abiotic removal
Readily absorbed by plants from soil; concentrations of 1 and 10 mg kg$^{-1}$ were added to 3 soil types: silty loam; coarse sandstone; and humic. Mercury content accumulated in potato tubers was: 83 and 53 µg kg$^{-1}$, respectively (silt); 100 and 327 µg kg$^{-1}$, respectively (sandstone); and 67 and 196 µg kg$^{-1}$, respectively (humic) (1).

## Mammalian and avian toxicity

### Acute data
LD$_{50}$ oral rat 68 mg kg$^{-1}$ (2).
LD$_{50}$ oral mouse 20 mg kg$^{-1}$ (3).
LD$_{50}$ intraperitoneal rat 13 mg kg$^{-1}$ (4).
LD$_{50}$ intraperitoneal mouse 20 mg kg$^{-1}$ (5).

### Teratogenicity and reproductive effects
TD$_{Lo}$ intraperitoneal (9-13 day pregnancy) mouse 8 mg kg$^{-1}$, reproductive effects (6).
TD$_{Lo}$ intraperitoneal (10 day pregnancy) mouse 4 mg kg$^{-1}$, teratogenic – nontransmissible changes produced in the offspring (7).

## Legislation
Limited under EC Directive on Drinking Water Quality 80/778/EEC, maximum admissible concentration 1 µg Hg $l^{-1}$ (8).
Included in Schedule 5 (Release into Water: Prescribed Substances) Statutory Instrument No. 472, 1991 (9).

## References
1. Mel'nikov, N. N. et al *Pesticides and Environment* 1977, 240, Khimiya Publ., Moscow
2. *Acta Pol. Pharm.* 1972, **29**, 623
3. *Pesticide Chemicals Official Compendium* 1966, 735, Association of the American Pesticide Control Officials, Inc.
4. *Toxicol. Appl. Pharmacol.* 1972, **23**, 197
5. *Acta Med. Scand.* 1952, **143**, 365
6. *Terat. J. Abnorm. Devel.* 1972, **5**, 181

7. *Annual Research Report of the Institute of Environmental Medicine, Nagoya University* 1972, **19**, 61
8. *EC Directive Relating to the Quality of Water Intended for Human Consumption* 1982, 80/778/EEC, Office for Official Publications of the European Communities, 2 rue Mercier, L-2985 Luxembourg
9. *S. I. 1991 No. 472 The Environmental Protection (Prescribed Processes and Substances) Regulations* 1991, HMSO, London

# M70   Mercury(I) nitrate

## $Hg_2(NO_3)_2$

**CAS Registry No.** 10415-75-5
**Synonyms** nitric acid mercury (1+) salt; mercurous nitrate; mercury protonitrate
**Mol. Formula** $HgNO_3$                                    **Mol. Wt.** 262.59
**Uses** For fire gilding and blackening brass.

## Occupational exposure

**US TLV (TWA)** 0.025 mg m$^{-3}$ (as Hg) intended change; **UK Long-term limit** 0.05 mg m$^{-3}$ (as Hg) under review; **UK Short-term limit** 0.15 mg m$^{-3}$ (as Hg) under review; **UN No.** 1627; **Conveyance classification** toxic substance.

## Ecotoxicity

### Fish toxicity

Intestinal transport rate for nutrients decreased in *Channa punctatus* exposed to 3 µg l$^{-1}$ ($Hg^{2+}$) for 30 days. Disrupted metabolism in *Notopterus notopterus* exposed to 44 µg l$^{-1}$ ($Hg^{2+}$) for 30 days (1).
Teratogenic to rainbow trout after exposure of eggs to 0.12-0.21 µg l$^{-1}$ ($Hg^{2+}$) 4 days post hatch, and after parental exposure to 0.70-0.79 µg l$^{-1}$ ($Hg^{2+}$) for 400 days (1).

### Invertebrate toxicity

$LC_{50}$ (48 hr) *Daphnia magna* 9.3 µg l$^{-1}$; $EC_{50}$ (48 hr) *Daphnia magna* 5.2 µg l$^{-1}$ ($Hg^{2+}$) (1).

## Mammalian and avian toxicity

### Acute data

$LD_{50}$ oral rat 170 mg kg$^{-1}$ (2).
$LD_{50}$ oral mouse 49 mg kg$^{-1}$ (2).
$LD_{50}$ dermal rat 2.33 g kg$^{-1}$ (3).
$LD_{50}$ intraperitoneal mouse 5 mg kg$^{-1}$ (4).

## Legislation

Limited under EC Directive on Drinking Water Quality 80/778/EEC. Mercury: maximum admissible concentration 1 µg l$^{-1}$ (5).
Included in Schedule 4 (Release into Air: Prescribed Substances) and Schedule 5 (Release into Water: Prescribed Substances) Statutory Instrument No. 472, 1991 (6).
Quality objectives under EC Directives 82/176/EEC and 84/156/EEC 0.3 mg kg$^{-1}$

---

(wet weight) in a representative sample of fish flesh; 1 μg l$^{-1}$ (annual mean) total mercury in inland surface waters; 0.5 μg l$^{-1}$ (annual mean) dissolved mercury in estuarine waters; 0.3 μg l$^{-1}$ (annual mean) dissolved mercury in marine waters. A 'standstill' provision applies to concentrations in shellfish or sediments. Limit values under EC Directive 84/156/EEC 0.05 mg l$^{-1}$ effluent and 0.1 g l$^{-1}$ vinyl chloride production capacity for chemical industries using mercury catalysts in vinyl chloride production; 0.05 mg l$^{-1}$ effluent and 5 g kg$^{-1}$ mercury processed for chemical industries using mercury catalysts in other processes; 0.05 mg l$^{-1}$ effluent and 0.7 g kg$^{-1}$ mercury processed for manufacture of mercury catalysts used in vinyl chloride production; 0.05 mg l$^{-1}$ effluent and 0.05 g kg$^{-1}$ mercury processed for manufacture of organic and non-organic mercury compounds (other than mercury catalysts for vinyl chloride production); 0.05 mg l$^{-1}$ effluent and 0.03 g kg$^{-1}$ mercury processed for manufacture of primary batteries containing mercury; 0.05 mg l$^{-1}$ effluent for mercury recovery plants and extraction and refining of non-ferrous metals; 0.05 mg l$^{-1}$ effluent for plants treating toxic wastes containing mercury (7).

## Any other comments

Toxicology, absorption, distribution and metabolism of inorganic mercury compounds has been reviewed (8).
Normally exists as dihydrate, melting point 70°C.

## References

1. Zillioux, E. J. et al *Environ. Toxicol. Chem.* 1993, **12**, 2245-2264
2. *Hyg. Sanit. (USSR)* 1981, **46**(8), 12
3. *Gig. Tr. Prof. Zabol.* 1981, **25**(7), 27
4. *Archiv. Toxikol.* 1964, **20**, 226
5. *EC Directive Relating to the Quality of Water Intended for Human Consumption* 1982, 80/778/EEC, Office for Official Publications of the European Communities, 2 rue Mercier, L-2985 Luxembourg
6. *S. I. 1991 No. 472 The Environmental Protection (Prescribed Processes and Substances) Regulations* 1991, HMSO, London
7. *DoE Circular 7/89: Water and the environment. The implementation of EC directives on pollution caused by certain dangerous substances discharged into the aquatic environment* 1989, HMSO, London
8. Oehme, F. W. (Ed.) *Toxicity of Heavy Metals in the Environment* Part 1, 1978, Marcel Dekker Inc., New York.

# M71    Mercury(II) nitrate

## Hg(NO$_3$)$_2$

**CAS Registry No.** 10045-94-0
**Synonyms** nitric acid, mercury(2+) salt; mercuric nitrate; mercury pernitrate
**Mol. Formula** HgN$_2$O$_6$                    **Mol. Wt.** 324.60
**Uses** Chemical intermediate. Used in the manufacture of felt. Fungicide.

## Physical properties

**Specific gravity** 4.3.

## Occupational exposure

**US TLV (TWA)** 0.025 mg m$^{-3}$ (as Hg); **UK Long-term limit** 0.05 mg m$^{-3}$ (as Hg) under review; **UK Short-term limit** 0.15 mg m$^{-3}$ (as Hg) under review; **UN No.** 1625; **Conveyance classification** toxic substance.

## Ecotoxicity

### Fish toxicity

Intestinal transport rate for nutrients decreased in *Channa punctatus* exposed to 3 µg l$^{-1}$ (Hg$^{2+}$) for 30 days. Disrupted metabolism in *Notopterus notopterus* exposed to 44 µg l$^{-1}$ (Hg$^{2+}$) for 30 days (1).
Teratogenic to rainbow trout after exposure of eggs to 0.12-0.21 µg l$^{-1}$ (Hg$^{2+}$) 4 days post hatch, and after parental exposure to 0.70-0.79 µg l$^{-1}$ (Hg$^{2+}$) for 400 days (1).

### Invertebrate toxicity

LC$_{50}$ (48 hr) *Daphnia magna* 9.3 µg l$^{-1}$; EC$_{50}$ (48 hr) *Daphnia magna* 5.2 µg l$^{-1}$ (Hg$^{2+}$) (1).
IC$_{50}$ *Pseudomonas fluorescens* 8.03 µg l$^{-1}$ Hg at pH 6. Toxicity reduces with increasing pH and may be linked with the ability of the microorganism to methylate inorganic mercury (2,3).
Toxicity is also reduced by presence of cysteine in medium, due to binding of mercury to thiol groups (3).

### Abiotic removal

The compound can be stabilized in industrial waste by the formation of insoluble complexes with pulverised fuel ash and Portland cement. Chemical and physical reactions are involved (4).

## Mammalian and avian toxicity

### Acute data

LD$_{50}$ oral rat, mouse 25-26 mg kg$^{-1}$ (5).
LD$_{50}$ intraperitoneal mouse 7.2 mg kg$^{-1}$ (6).
LD$_{50}$ subcutaneous mouse 20 mg kg$^{-1}$ (7).

### Teratogenicity and reproductive effects

Chronic mercury poisoning in ♀ mice following administration of the compound caused disturbances of the oestrus cycle (8).

### Irritancy

Can cause eye irritation, conjunctival and corneal ulceration (9).
Causes skin irritation and dermatitis (9).

## Any other adverse effects to man

Acute poisoning in humans produces symptoms including, gastrointestinal disturbance, anuria and anaemia (9).
Chronic poisoning with mercury can induce tremors and neuropsychiatric disturbances (9).
In humans symptoms are probably due to a variety of actions on the nervous system including those on 5HT mediated pathways (10).

## Any other adverse effects

After ingestion in rats and mice diarrhoea is seen (5).

# Legislation

Limited under EC Directive on Drinking Water Quality 80/778/EEC. Mercury: maximum admissible concentration 1 $\mu$g $l^{-1}$ (11).

Included in Schedule 4 (Release into Air: Prescribed Substances) and Schedule 5 (Release into Water: Prescribed Substances) Statutory Instrument No. 472, 1991 (12). Quality objectives under EC Directives 82/176/EEC and 84/156/EEC 0.3 mg $kg^{-1}$ (wet weight) in a representative sample of fish flesh; 1 $\mu$g $l^{-1}$ (annual mean) total mercury in inland surface waters; 0.5 $\mu$g $l^{-1}$ (annual mean) dissolved mercury in estuarine waters; 0.3 $\mu$g $l^{-1}$ (annual mean) dissolved mercury in marine waters. A 'standstill' provision applies to concentrations in shellfish or sediments. Limit values under EC Directive 84/156/EEC 0.05 mg $l^{-1}$ effluent and 0.1 g $l^{-1}$ vinyl chloride production capacity for chemical industries using mercury catalysts in vinyl chloride production; 0.05 mg $l^{-1}$ effluent and 5 g $kg^{-1}$ mercury processed for chemical industries using mercury catalysts in other processes; 0.05 mg $l^{-1}$ effluent and 0.7 g $kg^{-1}$ mercury processed for manufacture of mercury catalysts used in vinyl chloride production; 0.05 mg $l^{-1}$ effluent and 0.05 g $kg^{-1}$ mercury processed for manufacture of organic and non-organic mercury compounds (other than mercury catalysts for vinyl chloride production); 0.05 mg $l^{-1}$ effluent and 0.03 g $kg^{-1}$ mercury processed for manufacture of primary batteries containing mercury; 0.05 mg $l^{-1}$ effluent for mercury recovery plants and extraction and refining of non-ferrous metals; 0.05 mg $l^{-1}$ effluent for plants treating toxic wastes containing mercury (13).

# Any other comments

Toxicology, absorption, distribution and metabolism of inorganic mercury compounds reviewed (14).

# References

1. Zilliox, E. J. et al *Environ. Toxicol. Chem.* 1993, **12**, 2245-2264
2. Farrell, R. E. *Appl. Environ. Microbiol.* 1990, **56**(10), 3006-3016
3. Ribo, J. M. *Hydrobiologia* 1989, 188-189
4. Clark, A. J. et al *Chem. Environ. Proc. Int. Conf.* 1986, 680-689
5. Grins, N. et al *Hyg. Sanit. (USSR)* 1981, **46**(8), 12-14
6. *Gig. Tr. Prof. Zabol.* 1981, **25**(7), 27
7. *Monatsschr. Ohrenheilkd. Laryngo-Rhinol.* 1939, **73**, 751
8. Lach, H. *Srebryo. Z. Acta. Biol. Cracov., Ser. Zool.* 1972, **15**(1), 121-130
9. *Dangerous Prop. Ind. Mater. Rep.* 1988, **8**(4), 42-49
10. Oudar, P. et al *Pharmacol. Toxicol. (Copenhagen)* 1989, **65**(4), 245-248
11. *EC Directive Relating to the Quality of Water Intended for Human Consumption* 1982, 80/778/EEC, Office for Official Publications of the European Communities, 2 rue Mercier, L-2985 Luxembourg
12. *S. I. 1991 No. 472 The Environmental Protection (Prescribed Processes and Substances) Regulations* 1991, HMSO, London
13. *DoE Circular 7/89: Water and the environment. The implementation of EC directives on pollution caused by certain dangerous substances discharged into the aquatic environment* 1989, HMSO, London
14. Oehme, F. W. (Ed.) *Toxicity of Heavy Metals in the Environment Part 1* 1978, Marcel Dekker Inc., New York

# M72   Mercury nucleate

**CAS Registry No.** 12002-19-6
**Synonyms** mercurol

## Occupational exposure

**US TLV (TWA)** 0.025 mg m$^{-3}$ (as Hg) intended change; **UK Long-term limit**
0.05 mg m$^{-3}$ (as Hg) under review; **UK Short-term limit** 0.15 mg m$^{-3}$ (as Hg) under
review; **UN No.** 1639; **Conveyance classification** toxic substance.

## Legislation

Community Right-To-Know List.
Limited under EC Directive on Drinking Water Quality 80/778/EEC. Mercury:
maximum admissible concentration 1 µg l$^{-1}$ (1).
Included in Schedule 4 and 6 (Release into Air/Land: Prescribed Substances)
Statutory Instrument No. 472, 1991 (2).
Quality objectives under EC Directives 82/176/EEC and 84/156/EEC 0.3 mg kg$^{-1}$
(wet weight) in a representative sample of fish flesh; 1 µg l$^{-1}$ (annual mean) total
mercury in inland surface waters; 0.5 µg l$^{-1}$ (annual mean) dissolved mercury in
estuarine waters; 0.3 µg l$^{-1}$ (annual mean) dissolved mercury in marine waters. A
'standstill' provision applies to concentrations in shellfish or sediments. Limit values
under EC Directive 84/156/EEC 0.05 mg l$^{-1}$ effluent and 0.1 g l$^{-1}$ vinyl chloride
production capacity for chemical industries using mercury catalysts in vinyl chloride
production; 0.05 mg l$^{-1}$ effluent and 5 g kg$^{-1}$ mercury processed for chemical
industries using mercury catalysts in other processes; 0.05 mg l$^{-1}$ effluent and
0.7 g kg$^{-1}$ mercury processed for manufacture of mercury catalysts used in vinyl
chloride production; 0.05 mg l$^{-1}$ effluent and 0.05 g kg$^{-1}$ mercury processed for
manufacture of organic and non-organic mercury compounds (other than mercury
catalysts for vinyl chloride production); 0.05 mg l$^{-1}$ effluent and 0.03 g kg$^{-1}$ mercury
processed for manufacture of primary batteries containing mercury; 0.05 mg l$^{-1}$
effluent for mercury recovery plants and extraction and refining of non-ferrous
metals; 0.05 mg l$^{-1}$ effluent for plants treating toxic wastes containing mercury (3).

## Any other comments

Physico-chemical properties, human health effects, experimental toxicity,
environmental effects, ecotoxicology, exposure levels and workplace experience
reviewed (4).
Aquatic toxicology reviewed (5).

## References

1. *EC Directive Relating to the Quality of Water Intended for Human Consumption* 1982,
   80/778/EEC, Office for Official Publications of the European Communities, 2 rue Mercier,
   L-2985 Luxembourg
2. *S. I. 1991 No. 472 The Environmental Protection (Prescribed Processes and Substances)
   Regulations* 1991, HMSO, London
3. *DoE Circular 7/89: Water and the environment. The implementation of EC directives on
   pollution caused by certain dangerous substances discharged into the aquatic environment*
   1989, HMSO, London

4. *ECETOC Technical Report No. 30(4)* 1991, European Chemical Industry Ecology and Toxicology Centre, B-1160 Brussels
5. Zillioux, E. J. et al *Environ. Toxicol. Chem.* 1993, **12**, 2245-2264

# M73   Mercury oleate

$$H_3C(H_2C)_7 \quad (CH_2)_7C{-}O{-}Hg{-}O{-}C(CH_2)_7 \quad (CH_2)_7CH_3$$

**CAS Registry No.** 1191-80-6
**Synonyms** 9-octadecenoic acid, (Z)-, mercury salt; oleate of mercury.
**Mol. Formula** $C_{36}H_{66}O_4Hg$ **Mol. Wt.** 763.52
**Uses** Has been used as an ectoparasiticide, and in antifouling paints.

## Physical properties

**M. Pt.** 115°C

**Solubility**
Organic solvent: ethanol, diethyl ether, oils

## Occupational exposure

**US TLV (TWA)** 0.025 mg $m^{-3}$ (as Hg) intended change; **UK Long-term limit** 0.05 mg $m^{-3}$ (as Hg) under review; **UK Short-term limit** 0.15 mg $m^{-3}$ (as Hg) under review; **UN No.** 1640; **Conveyance classification** toxic substance.

## Ecotoxicity

**Fish toxicity**
5 ppm was lethal to rainbow trout, yellow perch and bluegill sunfish in 1 hr, 5 hr and 20 hr respectively. Test conditions: pH 7.0; dissolved oxygen content 7.5 ppm; total hardness (soap method) 300 ppm; methyl orange alkalinity 310 ppm; free carbon dioxide 5 ppm; and temperature 12.8°C (1).

## Legislation

Community Right-To-Know List.
Limited under EC Directive on Drinking Water Quality 80/778/EEC. Mercury: maximum admissible concentration 1 µg $l^{-1}$ (2).
Included in Schedule 4 (Release into Air: Prescribed Substances) Statutory Instrument No. 472, 1991 (3).
Quality objectives under EC Directives 82/176/EEC and 84/156/EEC 0.3 mg $kg^{-1}$ (wet weight) in a representative sample of fish flesh; 1 µg $l^{-1}$ (annual mean) total mercury in inland surface waters; 0.5 µg $l^{-1}$ (annual mean) dissolved mercury in estuarine waters; 0.3 µg $l^{-1}$ (annual mean) dissolved mercury in marine waters. A 'standstill' provision applies to concentrations in shellfish or sediments. Limit values under EC Directive 84/156/EEC 0.05 mg $l^{-1}$ effluent and 0.1 g $l^{-1}$ vinyl chloride production capacity for chemical industries using mercury catalysts in vinyl chloride production; 0.05 mg $l^{-1}$ effluent and 5 g $kg^{-1}$ mercury processed for chemical

industries using mercury catalysts in other processes; 0.05 mg l$^{-1}$ effluent and
0.7 g kg$^{-1}$ mercury processed for manufacture of mercury catalysts used in vinyl
chloride production; 0.05 mg l$^{-1}$ effluent and 0.05 g kg$^{-1}$ mercury processed for
manufacture of organic and non-organic mercury compounds (other than mercury
catalysts for vinyl chloride production); 0.05 mg l$^{-1}$ effluent and 0.03 g kg$^{-1}$
mercury processed for manufacture of primary batteries containing mercury;
0.05 mg l$^{-1}$ effluent for mercury recovery plants and extraction and refining of non-ferrous
metals; 0.05 mg l$^{-1}$ effluent for plants treating toxic wastes containing mercury (4).

## Any other comments

Physico-chemical properties, human health effects, experimental toxicity, environmental
effects, ecotoxicology, exposure levels and workplace experience reviewed (5).
Aquatic toxicology reviewed (6).

## References

1. Wood, E. M. *The toxicity of 3400 chemicals to fish* 1987, EPA Report 500/6-87-002
2. *EC Directive Relating to the Quality of Water Intended for Human Consumption* 1982, 80/778/EEC, Office for Official Publications of the European Communities, 2 rue Mercier, L-2985 Luxembourg
3. *S. I. 1991 No. 472 The Environmental Protection (Prescribed Processes and Substances) Regulations* 1991, HMSO, London
4. *DoE Circular 7/89: Water and the environment. The implementation of EC directives on pollution caused by certain dangerous substances discharged into the aquatic environment* 1989, HMSO, London
5. *ECETOC Technical Report No. 30(4)* 1991, European Chemical Industry Ecology and Toxicology Centre, B-1160 Brussels
6. Zillioux, E. J. et al *Environ. Toxicol. Chem.* 1993, **12**, 2245-2264

# M74   Mercury oxide

## Hg$_2$O

**CAS Registry No.** 12653-71-3
**Synonyms** Santar
**Mol. Formula** Hg$_2$O                          **Mol. Wt.** 417.18
**Uses** Treatment of apple canker and pruning cuts on fruit trees.

## Physical properties

**B. Pt.** 500°C (decomp.).

## Occupational exposure

**US TLV (TWA)** 0.025 mg m$^{-3}$ (as Hg) intended change; **UK Long-term limit**
0.05 mg m$^{-3}$ (as Hg) under review; **UK Short-term limit** 0.15 mg m$^{-3}$ (as Hg) under
review; **UN No.** 1641; **Conveyance classification** toxic substance.

## Legislation

Community Right-To-Know List.
Limited under EC Directive on Drinking Water Quality 80/778/EEC. Mercury:
maximum admissible concentration 1 μg l$^{-1}$ (1).

Included in Schedule 4 (Release into Air: Prescribed Substances) Statutory Instrument No. 472, 1991 (2).

Quality objectives under EC Directives 82/176/EEC and 84/156/EEC 0.3 mg kg$^{-1}$ (wet weight) in a representative sample of fish flesh; 1 µg l$^{-1}$ (annual mean) total mercury in inland surface waters; 0.5 µg l$^{-1}$ (annual mean) dissolved mercury in estuarine waters; 0.3 µg l$^{-1}$ (annual mean) dissolved mercury in marine waters. A 'standstill' provision applies to concentrations in shellfish or sediments. Limit values under EC Directive 84/156/EEC 0.05 mg l$^{-1}$ effluent and 0.1 g l$^{-1}$ vinyl chloride production capacity for chemical industries using mercury catalysts in vinyl chloride production; 0.05 mg l$^{-1}$ effluent and 5 g kg$^{-1}$ mercury processed for chemical industries using mercury catalysts in other processes; 0.05 mg l$^{-1}$ effluent and 0.7 g kg$^{-1}$ mercury processed for manufacture of mercury catalysts used in vinyl chloride production; 0.05 mg l$^{-1}$ effluent and 0.05 g kg$^{-1}$ mercury processed for manufacture of organic and non-organic mercury compounds (other than mercury catalysts for vinyl chloride production); 0.05 mg l$^{-1}$ effluent and 0.03 g kg$^{-1}$ mercury processed for manufacture of primary batteries containing mercury; 0.05 mg l$^{-1}$ effluent for mercury recovery plants and extraction and refining of non-ferrous metals; 0.05 mg l$^{-1}$ effluent for plants treating toxic wastes containing mercury (3).

**Any other comments**

Physico-chemical properties, human health effects, experimental toxicity, environmental effects, ecotoxicology, exposure levels and workplace experience reviewed (4).

Toxicity of inorganic mercury and environmental effects reviewed (5,6).

Aquatic toxicology reviewed (7).

**References**

1. *EC Directive Relating to the Quality of Water Intended for Human Consumption* 1982, 80/778/EEC, Office for Official Publications of the European Communities, 2 rue Mercier, L-2985 Luxembourg
2. *S. I. 1991 No. 472 The Environmental Protection (Prescribed Processes and Substances) Regulations* 1991, HMSO, London
3. *DoE Circular 7/89: Water and the environment. The implementation of EC directives on pollution caused by certain dangerous substances discharged into the aquatic environment* 1989, HMSO, London
4. *ECETOC Technical Report No. 30(4)* 1991, European Chemical Industry Ecology and Toxicology Centre, B-1160 Brussels
5. *Environmental Health Criteria 118: Inorganic Mercury* 1991, WHO/IPCS, Geneva
6. *Environmental Health Criteria 86: Mercury-Environmental Aspects* 1989, WHO/IPCS, Geneva
7. Zillioux, E. J. et al *Environ. Toxicol. Chem.* 1993, **12**, 2245-2264

# M75  Mercury(II) oxide

## HgO

**CAS Registry No.** 21908-53-2

**Synonyms** mercuric oxide red; mercuric oxide yellow; red oxide of mercury; Santar M; C.I. 77760

**Mol. Formula** HgO                    **Mol. Wt.** 216.59

**Uses** Chemical intermediate. Fungicide. Topical antiseptic. Pigment in marine bottom paints and porcelain paints (1).

## Physical properties

**M. Pt.** 500°C (decomp.); **Specific gravity** 11.14.

**Solubility**
Water: 50 mg $l^{-1}$.

## Ecotoxicity

**Fish toxicity**
Toxic to fish (2).
Intestinal transport rate for nutrients decreased in *Channa punctatus* exposed to 3 μg $l^{-1}$ ($Hg^{2+}$) for 30 days. Disrupted metabolism in *Notopterus notopterus* exposed to 44 μg $l^{-1}$ ($Hg^{2+}$) for 30 days (3).
Teratogenic to rainbow trout after exposure of eggs to 0.12- 0.21 μg $l^{-1}$ ($Hg^{2+}$) 4 days post hatch, and after parental exposure to 0.70-0.79 μg $l^{-1}$ ($Hg^{2+}$) for 400 days (3).

**Invertebrate toxicity**
$LC_{50}$ (48 hr) *Daphnia magna* 9.3 μg $l^{-1}$; $EC_{50}$ (48 hr) *Daphnia magna* 5.2 μg $l^{-1}$ ($Hg^{2+}$) (3).

**Abiotic removal**
Compound can be removed from wastewaters by flotation at pH 3-4.5 with potassium salts of saturated fatty acids (4).

## Mammalian and avian toxicity

**Acute data**
$LD_{50}$ oral mouse, rat 16-18 mg $kg^{-1}$ (5).
$LD_{50}$ dermal mouse 315 mg $kg^{-1}$ (6).
$LD_{50}$ intraperitoneal mouse 4.5 mg $kg^{-1}$ (6).

**Teratogenicity and reproductive effects**
Oral administration of 2.16 mg to ♀ rats on day 12 or 19 of pregnancy retarded foetal growth and inhibited eye formation (7).

**Metabolism and pharmacokinetics**
When introduced into rat duodenum, absorption and distribution through the body were rapid. Absorption showed different characteristics to that of many other inorganic mercury compounds such as mercury(II) chloride (8).

**Irritancy**
Strongly irritant (9).

**Sensitisation**
Strong allergen (9).

## Legislation

Limited under EC Directive on Drinking Water Quality 80/778/EEC. Mercury: maximum admissible concentration 1 μg $l^{-1}$ (10).
Included in Schedule 4 (Release into Air: Prescribed Substances) and Schedule 5 (Release into Water: Prescribed Substances) Statutory Instrument No. 472, 1991 (11).

---

Quality objectives under EC Directives 82/176/EEC and 84/156/EEC 0.3 mg kg$^{-1}$ (wet weight) in a representative sample of fish flesh; 1 μg l$^{-1}$ (annual mean) total mercury in inland surface waters; 0.5 μg l$^{-1}$ (annual mean) dissolved mercury in estuarine waters; 0.3 μg l$^{-1}$ (annual mean) dissolved mercury in marine waters. A 'standstill' provision applies to concentrations in shellfish or sediments. Limit values under EC Directive 84/156/EEC 0.05 mg l$^{-1}$ effluent and 0.1 g l$^{-1}$ vinyl chloride production capacity for chemical industries using mercury catalysts in vinyl chloride production; 0.05 mg l$^{-1}$ effluent and 5 g kg$^{-1}$ mercury processed for chemical industries using mercury catalysts in other processes; 0.05 mg l$^{-1}$ effluent and 0.7 g kg$^{-1}$ mercury processed for manufacture of mercury catalysts used in vinyl chloride production; 0.05 mg l$^{-1}$ effluent and 0.05 g kg$^{-1}$ mercury processed for manufacture of organic and non-organic mercury compounds (other than mercury catalysts for vinyl chloride production); 0.05 mg l$^{-1}$ effluent and 0.03 g kg$^{-1}$ mercury processed for manufacture of primary batteries containing mercury; 0.05 mg l$^{-1}$ effluent for mercury recovery plants and extraction and refining of non-ferrous metals; 0.05 mg l$^{-1}$ effluent for plants treating toxic wastes containing mercury (12).

## Any other comments

Toxicity reviewed (9).
WHO Class 1b; EPA Toxicity Class 1 (2).

## References

1. Bezo, M. et al *Hung. Teljes* HU46, 240 28 Oct 1988
2. *The Agrochemicals Handbook* 3rd ed., 1991, RSC, London
3. Zillioux, E. J. et al *Environ. Toxicol. Chem.* 1993, **12**, 2245-2264
4. Scrylev, L. D. *Izv. Vyssh. Uchebn. Zaved., Tsvetn. Metall.* 1988, (6), 13-18 (Russ.) (*Chem. Abstr.* **110**, 218410v)
5. *IPCS Environmental Health Criteria 118: Inorganic Mercury* 1991, World Health Organisation, Geneva
6. *Gig. Tr. Prof. Zabol.* 1981, **25**(7), 27
7. Rizzo, A. M. et al *Proc. Western Pharm. Soc.* 1972, **15**, 52-54
8. Endo, T. et al *Pharmacol. Toxicol. (Copenhagen)* 1990, **67**(5), 431-435
9. *Dangerous Prop. Ind. Mater. Rep.* 1989, **9**(3), 49-57
10. *EC Directive Relating to the Quality of Water Intended for Human Consumption* 1982, 80/778/EEC, Office for Official Publications of the European Communities, 2 rue Mercier, L-2985 Luxembourg
11. *S. I. 1991 No. 472 The Environmental Protection (Prescribed Processes and Substances) Regulations* 1991, HMSO, London
12. *DoE Circular 7/89: Water and the environment. The implementation of EC directives on pollution caused by certain dangerous substances discharged into the aquatic environment* 1989, HMSO, London

# M76   Mercury oxycyanide

HgO·Hg(CN)$_2$

**CAS Registry No.** 1335-31-5
**Synonyms** mercury cyanide oxide; mercuric oxycyanide
**Mol. Formula** $C_2Hg_2N_2O$ **Mol. Wt.** 469.22

**Uses** Topical antiseptic.

## Physical properties
**Specific gravity** 4.44.

## Occupational exposure
**US TLV (TWA)** 0.025 mg m$^{-3}$ (as Hg) intended change; **UK Long-term limit** 0.05 mg m$^{-3}$ (as Hg) under review; **UK Short-term limit** 0.15 mg m$^{-3}$ (as Hg) under review; **UN No.** 1642; **Conveyance classification** toxic substance; **Supply classification** explosive, toxic.

**Risk phrases** Toxic by inhalation, in contact with skin and if swallowed – Danger of cumulative effects (R23/24/25, R33)

**Safety phrases** After contact with skin, wash immediately with plenty of soap and water – This material and its container must be disposed of in a safe way – If you feel unwell, seek medical advice (show label where possible) (S28, S35, S44)

## Mammalian and avian toxicity

**Acute data**
LD$_{Lo}$ intravenous rabbit 2500 µg kg$^{-1}$ (1).

## Legislation
Community Right-To-Know List.
Limited under EC Directive on Drinking Water Quality 80/778/EEC. Mercury: maximum admissible concentration 1 µg l$^{-1}$ (2).
Included in Schedule 4 (Release into Air: Prescribed Substances) Statutory Instrument No. 472, 1991 (3).
Quality objectives under EC Directives 82/176/EEC and 84/156/EEC 0.3 mg kg$^{-1}$ (wet weight) in a representative sample of fish flesh; 1 µg l$^{-1}$ (annual mean) total mercury in inland surface waters; 0.5 µg l$^{-1}$ (annual mean) dissolved mercury in estuarine waters; 0.3 µg l$^{-1}$ (annual mean) dissolved mercury in marine waters. A 'standstill' provision applies to concentrations in shellfish or sediments. Limit values under EC Directive 84/156/EEC 0.05 mg l$^{-1}$ effluent and 0.1 g l$^{-1}$ vinyl chloride production capacity for chemical industries using mercury catalysts in vinyl chloride production; 0.05 mg l$^{-1}$ effluent and 5 g kg$^{-1}$ mercury processed for chemical industries using mercury catalysts in other processes; 0.05 mg l$^{-1}$ effluent and 0.7 g kg$^{-1}$ mercury processed for manufacture of mercury catalysts used in vinyl chloride production; 0.05 mg l$^{-1}$ effluent and 0.05 g kg$^{-1}$ mercury processed for manufacture of organic and non-organic mercury compounds (other than mercury catalysts for vinyl chloride production); 0.05 mg l$^{-1}$ effluent and 0.03 g kg$^{-1}$ mercury processed for manufacture of primary batteries containing mercury; 0.05 mg l$^{-1}$ effluent for mercury recovery plants and extraction and refining of non-ferrous metals; 0.05 mg l$^{-1}$ effluent for plants treating toxic wastes containing mercury (4).

## Any other comments
Physico-chemical properties, human health effects and experimental toxicity reviewed (5).
Aquatic toxicology reviewed (6).
Violent poison. Explodes on impact or when exposed to flame.

---

## References

1. *J. Pharmacol. Exp. Ther.* 1931, **41**, 21
2. *EC Directive Relating to the Quality of Water Intended for Human Consumption* 1982, 80/778/EEC, Office for Official Publications of the European Communities, 2 rue Mercier, L-2985 Luxembourg
3. *S. I. 1991 No. 472 The Environmental Protection (Prescribed Processes and Substances) Regulations* 1991, HMSO, London
4. *DoE Circular 7/89: Water and the environment. The implementation of EC directives on pollution caused by certain dangerous substances discharged into the aquatic environment* 1989, HMSO, London
5. *ECETOC Technical Report No. 30(4)* 1991, European Chemical Industry Ecology and Toxicology Centre, B-1160 Brussels
6. Zillioux, E. J. et al *Environ. Toxicol. Chem.* 1993, **12**, 2245-2264

# M77   Mercury potassium iodide

## $K_2HgI_4$

**CAS Registry No.** 7783-33-7

**Synonyms** Channing's solution; mercuric potassium iodide; potassium iodohydragyrate; potassium mercuric iodide; potassium tetraiodomercurate(II); dipotassium tetraiodate

**Mol. Formula** $HgI_4K_2$         **Mol. Wt.** 786.41

**Uses** Topical anti-infective. Disinfectant. Ammonia analysis.

## Physical properties

**Solubility**
Organic solvent: ethanol, diethyl ether, acetone

## Occupational exposure

**US TLV (TWA)** 0.025 mg m$^{-3}$ (as Hg) intended change; **UK Long-term limit** 0.05 mg m$^{-3}$ (as Hg) under review; **UK Short-term limit** 0.15 mg m$^{-3}$ (as Hg) under review; **UN No.** 1643; **Conveyance classification** toxic substance.

## Ecotoxicity

**Fish toxicity**
Intestinal transport rate for nutrients decreased in snakehead fish exposed to 3 µg l$^{-1}$ ($Hg^{2+}$) for 30 days (1).
Disrupted metabolism in knifefish exposed to 44 µg l$^{-1}$ ($Hg^{2+}$) for 30 days (1).
Teratogenic to rainbow trout after exposure of eggs to 0.12-0.21 µg l$^{-1}$ ($Hg^{2+}$) 4 days post hatch, and after parental exposure to 0.70-0.79 µg l$^{-1}$ ($Hg^{2+}$) for 400 days (1).

**Invertebrate toxicity**
$LC_{50}$ (48 hr) *Daphnia magna* 9.3 µg l$^{-1}$; $EC_{50}$ (48 hr) *Daphnia magna* 5.2 µg l$^{-1}$ ($Hg^{2+}$) (1).

## Mammalian and avian toxicity

### Acute data
$LD_{Lo}$ dermal guinea pig 1000 mg $kg^{-1}$ (2).
$LD_{Lo}$ intraperitoneal guinea pig 1000 mg $kg^{-1}$ (2).

## Legislation

Community Right-To-Know List.

Limited under EC Directive on Drinking Water Quality 80/778/EEC. Mercury: maximum admissible concentration 1 µg $l^{-1}$ (3).

Included in Schedule 4 (Release into Air: Prescribed Substances) Statutory Instrument No. 472, 1991 (4).

Quality objectives under EC Directives 82/176/EEC and 84/156/EEC 0.3 mg $kg^{-1}$ (wet weight) in a representative sample of fish flesh; 1 µg $l^{-1}$ (annual mean) total mercury in inland surface waters; 0.5 µg $l^{-1}$ (annual mean) dissolved mercury in estuarine waters; 0.3 µg $l^{-1}$ (annual mean) dissolved mercury in marine waters. A 'standstill' provision applies to concentrations in shellfish or sediments. Limit values under EC Directive 84/156/EEC 0.05 mg $l^{-1}$ effluent and 0.1 g $l^{-1}$ vinyl chloride production capacity for chemical industries using mercury catalysts in vinyl chloride production; 0.05 mg $l^{-1}$ effluent and 5 g $kg^{-1}$ mercury processed for chemical industries using mercury catalysts in other processes; 0.05 mg $l^{-1}$ effluent and 0.7 g $kg^{-1}$ mercury processed for manufacture of mercury catalysts used in vinyl chloride production; 0.05 mg $l^{-1}$ effluent and 0.05 g $kg^{-1}$ mercury processed for manufacture of organic and non-organic mercury compounds (other than mercury catalysts for vinyl chloride production); 0.05 mg $l^{-1}$ effluent and 0.03 g $kg^{-1}$ mercury processed for manufacture of primary batteries containing mercury; 0.05 mg $l^{-1}$ effluent for mercury recovery plants and extraction and refining of non-ferrous metals; 0.05 mg $l^{-1}$ effluent for plants treating toxic wastes containing mercury (5).

## Any other comments

Physico-chemical properties, human health effects, experimental toxicity, environmental effects, ecotoxicology, exposure levels and workplace experience reviewed (6).

Toxicity of inorganic mercury and environmental effects reviewed (7,8).

## References

1.   Zillioux, E. J. et al *Environ. Toxicol. Chem.* 1993, **12**, 2245-2264
2.   *Arch. Environ. Health* 1965, **11**, 201
3.   *EC Directive Relating to the Quality of Water Intended for Human Consumption* 1982, 80/778/EEC, Office for Official Publications of the European Communities, 2 rue Mercier, L-2985 Luxembourg
4.   *S. I. 1991 No. 472 The Environmental Protection (Prescribed Processes and Substances) Regulations* 1991, HMSO, London
5.   *DoE Circular 7/89: Water and the environment. The implementation of EC directives on pollution caused by certain dangerous substances discharged into the aquatic environment* 1989, HMSO, London
6.   *ECETOC Technical Report No. 30(4)* 1991, European Chemical Industry Ecology and Toxicology Centre, B-1160 Brussels
7.   *Environmental Health Criteria 118: Inorganic Mercury* 1991, WHO/IPCS, Geneva
8.   *Environmental Health Criteria 86: Mercury-Environmental Aspects* 1989, WHO/IPCS, Geneva

# M78   Mercury salicylate

**CAS Registry No.** 5970-32-1
**Synonyms** mercuric salicylate; mercurisalicylic acid; salicylic acid,
(hydroxymercuri)-, cyclic anhydride; benzoic acid, 2-hydroxy-, mercury complex
**Mol. Formula** $C_7H_4HgO_3$                **Mol. Wt.** 336.70
**Uses** Topical antiseptic.

## Occupational exposure

**US TLV (TWA)** 0.025 mg m$^{-3}$ (as Hg) intended change; **UK Long-term limit**
0.05 mg m$^{-3}$ (as Hg) under review; **UK Short-term limit** 0.15 mg m$^{-3}$ (as Hg) under
review; **UN No.** 1644; **Conveyance classification** toxic substance.

## Mammalian and avian toxicity

### Acute data
$LD_{Lo}$ subcutaneous mouse 10 mg kg$^{-1}$ (1).
$LD_{Lo}$ intramuscular rabbit 40 mg kg$^{-1}$ (2).

## Legislation

Community Right-To-Know List.
Limited under EC Directive on Drinking Water Quality 80/778/EEC. Mercury:
maximum admissible concentration 1 μg l$^{-1}$ (3).
Included in Schedule 4 and 6 (Release into Air/Land: Prescribed Substances)
Statutory Instrument No. 472, 1991 (4).
Quality objectives under EC Directives 82/176/EEC and 84/156/EEC 0.3 mg kg$^{-1}$
(wet weight) in a representative sample of fish flesh; 1 μg l$^{-1}$ (annual mean) total
mercury in inland surface waters; 0.5 μg l$^{-1}$ (annual mean) dissolved mercury in
estuarine waters; 0.3 μg l$^{-1}$ (annual mean) dissolved mercury in marine waters. A
'standstill' provision applies to concentrations in shellfish or sediments. Limit values
under EC Directive 84/156/EEC 0.05 mg l$^{-1}$ effluent and 0.1 g l$^{-1}$ vinyl chloride
production capacity for chemical industries using mercury catalysts in vinyl chloride
production; 0.05 mg l$^{-1}$ effluent and 5 g kg$^{-1}$ mercury processed for chemical
industries using mercury catalysts in other processes; 0.05 mg l$^{-1}$ effluent and
0.7 g kg$^{-1}$ mercury processed for manufacture of mercury catalysts used in vinyl
chloride production; 0.05 mg l$^{-1}$ effluent and 0.05 g kg$^{-1}$ mercury processed for
manufacture of organic and non-organic mercury compounds (other than mercury
catalysts for vinyl chloride production); 0.05 mg l$^{-1}$ effluent and 0.03 g kg$^{-1}$ mercury
processed for manufacture of primary batteries containing mercury; 0.05 mg l$^{-1}$
effluent for mercury recovery plants and extraction and refining of non-ferrous
metals; 0.05 mg l$^{-1}$ effluent for plants treating toxic wastes containing mercury (5).

## Any other comments

Physico-chemical properties, human health effects, experimental toxicity, environmental

effects, ecotoxicology, exposure levels and workplace experience reviewed (6).
Aquatic toxicology reviewed (7).
Poison. Incompatible with alkali iodides.

### References

1. *Monat. Bhrenheilk. Laryngo-Rhinol.* 1939, **73**, 751
2. *J. Pharmacol. Exp. Ther.* 1926, **27**, 385
3. *EC Directive Relating to the Quality of Water Intended for Human Consumption* 1982, 80/778/EEC, Office for Official Publications of the European Communities, 2 rue Mercier, L-2985 Luxembourg
4. *S. I. 1991 No. 472 The Environmental Protection (Prescribed Processes and Substances) Regulations* 1991, HMSO, London
5. *DoE Circular 7/89: Water and the environment. The implementation of EC directives on pollution caused by certain dangerous substances discharged into the aquatic environment* 1989, HMSO, London
6. *ECETOC Technical Report No. 30(4)* 1991, European Chemical Industry Ecology and Toxicology Centre, B-1160 Brussels
7. Zillioux, E. J. et al *Environ. Toxicol. Chem.* 1993, **12**, 2245-2264

# M79   Mercury(I) sulfate

## $Hg_2SO_4$

**CAS Registry No.** 7783-36-0
**Synonyms** sulfuric acid, dimercury(1+) salt; mercurous sulphate
**Mol. Formula** $Hg_2O_4S$                    **Mol. Wt.** 497.24
**Uses** Used in the manufacture of electrical batteries.

### Physical properties

**M. Pt.** decomp. with heat; **Specific gravity** 7.56.

**Solubility**
Water: 600 mg $l^{-1}$ at 25°C.

### Occupational exposure

**US TLV (TWA)** 0.025 mg $m^{-3}$ (as Hg) intended change; **UK Long-term limit** 0.05 mg $m^{-3}$ (as Hg) under review; **UK Short-term limit** 0.15 mg $m^{-3}$ (as Hg) under review; **UN No.** 1645; **Conveyance classification** toxic substance.

### Ecotoxicity

**Fish toxicity**
Intestinal transport rate for nutrients decreased in *Channa punctatus* exposed to 3 µg $l^{-1}$ ($Hg^{2+}$) for 30 days. Disrupted metabolism in *Notopterus notopterus* exposed to 44 µg $l^{-1}$ ($Hg^{2+}$) for 30 days (1).
Teratogenic to rainbow trout after exposure of eggs to 0.12-0.21 µg $l^{-1}$ ($Hg^{2+}$) 4 days post hatch, and after parental exposure to 0.70-0.79 µg $l^{-1}$ ($Hg^{2+}$) for 400 days (1).

**Invertebrate toxicity**
$LC_{50}$ (48 hr) *Daphnia magna* 9.3 µg $l^{-1}$; $EC_{50}$ (48 hr) *Daphnia magna* 5.2 µg $l^{-1}$ ($Hg^{2+}$) (1).

## Mammalian and avian toxicity

### Acute data
$LD_{50}$ oral rat 20.5 mg $kg^{-1}$ (2).
$LD_{50}$ oral mouse 152 mg $kg^{-1}$ (2).
$LD_{50}$ dermal rat 1.175 g $kg^{-1}$ (3).
$LD_{50}$ intraperitoneal mouse 11.5 mg $kg^{-1}$ (4).

## Legislation

Limited under EC Directive on Drinking Water Quality 80/778/EEC. Mercury: maximum admissible concentration 1 µg $l^{-1}$ (5).
Included in Schedule 4 (Release into Air: Prescribed Substances) and Schedule 5 (Release into Water: Prescribed Substances) Statutory Instrument No. 472, 1991 (6). Quality objectives under EC Directives 82/176/EEC and 84/156/EEC 0.3 mg $kg^{-1}$ (wet weight) in a representative sample of fish flesh; 1 µg $l^{-1}$ (annual mean) total mercury in inland surface waters; 0.5 µg $l^{-1}$ (annual mean) dissolved mercury in estuarine waters; 0.3 µg $l^{-1}$ (annual mean) dissolved mercury in marine waters. A 'standstill' provision applies to concentrations in shellfish or sediments. Limit values under EC Directive 84/156/EEC 0.05 mg $l^{-1}$ effluent and 0.1 g $l^{-1}$ vinyl chloride production capacity for chemical industries using mercury catalysts in vinyl chloride production; 0.05 mg $l^{-1}$ effluent and 5 g $kg^{-1}$ mercury processed for chemical industries using mercury catalysts in other processes; 0.05 mg $l^{-1}$ effluent and 0.7 g $kg^{-1}$ mercury processed for manufacture of mercury catalysts used in vinyl chloride production; 0.05 mg $l^{-1}$ effluent and 0.05 g $kg^{-1}$ mercury processed for manufacture of organic and non-organic mercury compounds (other than mercury catalysts for vinyl chloride production); 0.05 mg $l^{-1}$ effluent and 0.03 g $kg^{-1}$ mercury processed for manufacture of primary batteries containing mercury; 0.05 mg $l^{-1}$ effluent for mercury recovery plants and extraction and refining of non-ferrous metals; 0.05 mg $l^{-1}$ effluent for plants treating toxic wastes containing mercury (7).

## Any other comments

Toxicology, absorption, distribution and metabolism of inorganic mercury compounds reviewed (8).

## References

1. Zillioux, E. J. et al *Environ. Toxicol. Chem.* 1993, **12**, 2245-2264
2. Grins, N. et al *Hyg. Sanit. (USSR)* 1981, **46**(8), 12-14
3. *Gig. Tr. Prof. Zabol.* 1981, **25**(7), 27
4. *Arch. Toxikol.* 1964, **20**, 226
5. *EC Directive Relating to the Quality of Water Intended for Human Consumption* 1982, 80/778/EEC, Office for Official Publications of the European Communities, 2 rue Mercier, L-2985 Luxembourg
6. *S. I. 1991 No. 472 The Environmental Protection (Prescribed Processes and Substances) Regulations* 1991, HMSO, London
7. *DoE Circular 7/89: Water and the environment. The implementation of EC directives on pollution caused by certain dangerous substances discharged into the aquatic environment* 1989, HMSO, London
8. Oehme, F. W. (Ed.) *Toxicity of Heavy Metals in the Environment* Part 1, 1978, Marcel Dekker Inc., New York

# M80    Mercury(II) thiocyanate

## Hg(SCN)₂

**CAS Registry No.** 592-85-8

**Synonyms** thiocyanic acid mercury(2+) salt; mercuric thiocyanate; mercuric sulfocyanate; mercuric sulfocyanide; mercury dithiocyanate

**Mol. Formula** $C_2HgN_2S_2$                                     **Mol. Wt.** 316.75

**Uses** Manufacture of fireworks and in photography.

## Physical properties

**M. Pt.** 165°C.

### Solubility
Water: 0.69 g $l^{-1}$ at 25°C.

## Occupational exposure

**US TLV (TWA)** 0.025 mg m$^{-3}$ (as Hg) intended change; **UK Long-term limit** 0.05 mg m$^{-3}$ (as Hg) under review; **UK Short-term limit** 0.15 mg m$^{-3}$ (as Hg) under review; **UN No.** 1646.

## Ecotoxicity

### Fish toxicity
Intestinal transport rate for nutrients decreased in *Channa punctatus* exposed to 3 µg $l^{-1}$ (Hg$^{2+}$) for 30 days. Disrupted metabolism in *Notopterus notopterus* exposed to 44 µg $l^{-1}$ (Hg$^{2+}$) for 30 days (1).

Teratogenic to rainbow trout after exposure of eggs to 0.12-0.21 µg $l^{-1}$ (Hg$^{2+}$) 4 days post hatch, and after parental exposure to 0.70-0.79 µg $l^{-1}$ (Hg$^{2+}$) for 400 days (1).

LC$_{50}$ (24 hr) fathead minnow 0.39 mg $l^{-1}$ (2).

### Invertebrate toxicity
LC$_{50}$ (48 hr) *Daphnia magna* 9.3 µg $l^{-1}$; EC$_{50}$ (48 hr) *Daphnia magna* 5.2 µg $l^{-1}$ (Hg$^{2+}$) (1).

LC$_{50}$ (24 hr) grass shrimp 0.09 mg $l^{-1}$ (2).

## Mammalian and avian toxicity

### Acute data
LD$_{50}$ oral rat 46 mg kg$^{-1}$ (3).
LD$_{50}$ oral mouse 24.5 mg kg$^{-1}$ (3).
LD$_{50}$ dermal rat 685 mg kg$^{-1}$ (3).
LD$_{50}$ intraperitoneal 3.5 mg kg$^{-1}$ (3).

### Metabolism and pharmacokinetics
Absorption from rat small intestine is pH dependent over the range 5.5-7.4, with higher pHs favouring absorption. Possibly hydroxylation of the compound may be involved (4).

## Legislation

Limited under EC Directive on Drinking Water Quality 80/778/EEC. Mercury: maximum admissible concentration 1 µg $l^{-1}$ (5).

---

Included in Schedule 4 (Release into Air: Prescribed Substances) and Schedule 5 (Release into Water: Prescribed Substances) Statutory Instrument No. 472, 1991 (6). Quality objectives under EC Directives 82/176/EEC and 84/156/EEC 0.3 mg kg$^{-1}$ (wet weight) in a representative sample of fish flesh; 1 µg l$^{-1}$ (annual mean) total mercury in inland surface waters; 0.5 µg l$^{-1}$ (annual mean) dissolved mercury in estuarine waters; 0.3 µg l$^{-1}$ (annual mean) dissolved mercury in marine waters. A 'standstill' provision applies to concentrations in shellfish or sediments. Limit values under EC Directive 84/156/EEC 0.05 mg l$^{-1}$ effluent and 0.1 g l$^{-1}$ vinyl chloride production capacity for chemical industries using mercury catalysts in vinyl chloride production; 0.05 mg l$^{-1}$ effluent and 5 g kg$^{-1}$ mercury processed for chemical industries using mercury catalysts in other processes; 0.05 mg l$^{-1}$ effluent and 0.7 g kg$^{-1}$ mercury processed for manufacture of mercury catalysts used in vinyl chloride production; 0.05 mg l$^{-1}$ effluent and 0.05 g kg$^{-1}$ mercury processed for manufacture of organic and non-organic mercury compounds (other than mercury catalysts for vinyl chloride production); 0.05 mg l$^{-1}$ effluent and 0.03 g kg$^{-1}$ mercury processed for manufacture of primary batteries containing mercury; 0.05 mg l$^{-1}$ effluent for mercury recovery plants and extraction and refining of non-ferrous metals; 0.05 mg l$^{-1}$ effluent for plants treating toxic wastes containing mercury (7).

## Any other comments

Toxicology of inorganic mercury compounds reviewed (8,9).

## References

1. Zillioux, E. J. et al *Environ. Toxicol. Chem.* 1993, **12**, 2245-2264
2. Vershueren, K. *Handbook of Environmental Data on Organic Chemicals* 2nd ed., 1983, Van Nostrand Reinhold, New York
3. *Gig. Tr. Prof. Zabol.* 1981, **25**(7), 27
4. Endo, T. et al *Pharmacol. Toxicol. (Copenhagen)* 1990, **66**(5), 347-353
5. *EC Directive Relating to the Quality of Water Intended for Human Consumption* 1982, 80/778/EEC, Office for Official Publications of the European Communities, 2 rue Mercier, L-2985 Luxembourg
6. *S. I. 1991 No. 472 The Environmental Protection (Prescribed Processes and Substances) Regulations* 1991, HMSO, London
7. *DoE Circular 7/89: Water and the environment. The implementation of EC directives on pollution caused by certain dangerous substances discharged into the aquatic environment* 1989, HMSO, London
8. Korshun, M. N. *Gig. Sanit.* 1989, (1), 69-70 (Russ.) (*Chem. Abstr.* **110** 149418q)
9. Oehme, F. W. (Ed.) *Toxicity of Heavy Metals in the Environment Parts 1 and 2* 1978, Marcel Dekker Inc., New York

# M81   Merphalan

$$(ClH_2CH_2C)_2N-\!\!\left\langle\phantom{x}\right\rangle\!\!-CH_2\overset{\overset{\displaystyle NH_2}{|}}{C}HCO_2H$$

**CAS Registry No.** 531-76-0
**Synonyms** DL-phenylalanine, 4-[bis(2- chloroethyl)amino]-; DL-3-[[p-bis(2-chloroethyl)amino]phenyl]alanine; DL-phenylalanine mustard; DL-sarcolysin; CB-3307
**Mol. Formula** $C_{13}H_{18}Cl_2N_2O_2$                **Mol. Wt.** 305.21
**Uses** Antineoplastic agent particularly for breast and ovarian cancer.

## Physical properties

**M. Pt.** 172-174°C.

## Mammalian and avian toxicity

### Acute data

$LD_{50}$ intraperitoneal rat 23 mg $kg^{-1}$ (1).

### Carcinogenicity and long-term effects

No adequate evidence for carcinogenicity to humans, sufficient evidence for carcinogenicity to animals, IARC classification group 2B (2).
♂ mice receiving 1.25-2.5 mg $kg^{-1}$ intraperitoneally three times per wk for 6 months developed a significantly increased number of tumours, particularly lymphosarcomas.
♂ rats receiving 0.75 mg $kg^{-1}$ intraperitoneally showed an increased incidence of sarcomas particularly of peritoneal and reticulum cells while ♀ rats receiving the same dose showed an increased incidence of sarcomas, particularly of peritoneal and breast cells (3).
Virgin ♀ rats given a single intraperitoneal injection of 10 mg $kg^{-1}$ developed an increased incidence of mammaryfibroadenomas within 17 months (4).

### Teratogenicity and reproductive effects

Merphalan has been shown to be teratogenic in rats, when given during the first 10 days of pregnancy, causing termination of pregnancy, and various types of malformations (5).

### Metabolism and pharmacokinetics

Distribution throughout tissues in humans can be variable and lead to variable responses to treatment (6).
After administration by perfusion to cancer patients, 60% of dose can be removed within 45 min (7).
After intraperitoneal administration to rats of β-$^{14}$C-merphalan, there was selective binding of radioactivity to the soluble protein fraction of kidney after 2 days (1).
Merphalan reacts rapidly with heparinised blood *in vitro* (7).

## Genotoxicity

*Escherichia coli* sd-4-73, streptomycin independent reversion induction positive (8).
Cytotoxicity is associated with the bis(chloroethyl)amino group (9).

### Any other adverse effects to man

Can cause neutropenia and thrombocytopenia (10).

### Any other adverse effects

The compound exerts immunosuppressive effects in mice (11).

### Any other comments

Merphalan is a mixture of the two isomers medphalan and melphalan. The toxicology of the two isomers and merphalan itself have been reviewed (12).

### References

1.  Cohn, P. et al *Br. J. Cancer* 1957, **11**, 258-267
2.  *IARC Monograph* 1987, **Suppl. 7**, 65
3.  Weisburger, E. K. *Cancer* 1977, **40**, 1935-1949
4.  Presnov, M. A. et al *Vopr. Onkol.* 1964, **40**, 66-72
5.  Aleksandrov, V. A. *Dokl. Akad. Nauk. SSSR* 1966, **171**, 146-149
6.  Albert, D. S. et al *Clin. Pharmacol. Ther.* 1979, **26**, 737
7.  Klatt, O. et al *Proc. Soc. Exp. Biol. (N. Y.)* 1960, **104**, 629-631
8.  Szybalski, W. *Ann. N. Y. Acad. Sci.* 1958, **76**, 475-489
9.  Pantokiene, L. et al *Liet. TSR Mokslu Akad. Darb. Ser. B* 1986, (6), 141 (Russ.) (*Chem. Abstr.* **107**, 154719f)
10. Kazaryan, K. A. *Byull. Eksp. Biol. Med.* 1970, **69**, 87-91
11. *Martindale. The Extra Pharmacopoeia* 30th ed., 1993, The Pharmaceutical Press, London
12. *IARC Monograph* 1975, **9**, 169

# M82   Merphos

## P(SCH₂CH₂CH₂CH₃)₃

**CAS Registry No.** 150-50-5
**Synonyms** phosphorotrithious acid, tributyl ester; tributyl phosphorotrithioate; *S,S,S*-tributyl trithiophosphite; Folex; Deleaf defoliant; Butiphos
**Mol. Formula** $C_{12}H_{27}PS_3$ **Mol. Wt.** 298.51
**Uses** Herbicide (discontinued). Chemical intermediate.

### Physical properties

**B. Pt.** 150-152°C; **Flash point** 60-63°C; **Specific gravity** $d_4^{20}$ 0.987.

### Solubility

Organic solvent: ethanol, chloroform

### Mammalian and avian toxicity

#### Acute data

$LD_{50}$ oral ♂, ♀ rat 1475 mg kg$^{-1}$ (1).
$LD_{50}$ dermal rat ♂ 690 mg kg$^{-1}$; ♀ rat 615 mg kg$^{-1}$ (1).
$LD_{50}$ intraperitoneal rat 150 mg kg$^{-1}$ (2).
$LD_{Lo}$ subcutaneous atropinised chicken 600 mg kg$^{-1}$ (1).

**Sub-acute data**

In 90 day feeding trials, dogs and cats fed 750 mg kg$^{-1}$ diet showed depression of cholinesterase activity, but no other pathological or histological effects (3).

**Metabolism and pharmacokinetics**

Merphos can be absorbed from the gastrointestinal tract or through skin. The compound is more toxic in rats via the dermal route (1).

**Any other adverse effects**

Subcutaneous single injections of ≥600 mg kg$^{-1}$ to atropinised chickens caused muscle weakness or paralysis for more than 90 days, onset time 3-21 days (1).

**Legislation**

Limited under EC Directive on Drinking Water Quality 80/778/EEC. Pesticides: maximum admissible concentration 0.1 μg l$^{-1}$ (4).

Included in Schedule 6 (Release into Land: Prescribed Substances) Statutory Instrument No. 472, 1991 (5).

**Any other comments**

Toxicity and environmental fate reviewed (6).

**References**

1. Gaines, T. B. *Toxicol. Appl. Pharmacol.* 1969, **14**(3), 515-534
2. *Proc. Soc. Exp. Biol. Med.* 1963, **114**, 509
3. *Agrochemicals Handbook* 3rd ed., 1991, RSC, London
4. *EC Directive Relating to the Quality of Water Intended for Human Consumption* 1982, 80/778/EEC, Office for Official Publications of the European Communities, 2 rue Mercier, L-2985 Luxembourg
5. *S. I. 1991 No. 472 The Environmental Protection (Prescribed Processes and Substances) Regulations* 1991, HMSO, London
6. Izmerov, N. F. *Reviews of Scientific Literature in Russian on Selected Hazardous Chemicals: Butiphos 113* 1993, Richardson, M. L. (Ed.), UNEP/IRPTC, Geneva, CIP, Moscow

# M83   Mesitylene

**CAS Registry No.** 108-67-8

**Synonyms** benzene, 1,3,5-trimethyl-; *s*-trimethylbenzene; trimethylbenzol; Fleet-X

**Mol. Formula** C$_9$H$_{12}$                    **Mol. Wt.** 120.20

**Uses** Chemical intermediate.

**Occurrence** Ground, air and water pollutant. Detected in gasoline and in diesel exhaust. Occurs in the aroma of some cooked foods.

---

## Physical properties

**M. Pt.** −44.8°C; **B. Pt.** 164.7°C; **Flash point** 44°C; **Specific gravity** $d_4^{20}$ 0.8637; **Partition coefficient** log $P_{ow}$ 3.42.

### Solubility

Water: 20 mg $l^{-1}$. Organic solvent: ethanol, benzene, diethyl ether

## Occupational exposure

**UN No.** 2325; **HAZCHEM Code** 3⊠; **Conveyance classification** flammable liquid; **Supply classification** irritant.
**Risk phrases** ≥25% – Flammable – Irritating to respiratory system – <25% – Flammable (R10, R37, R10)

## Ecotoxicity

### Fish toxicity

$LC_{Lo}$ (96 hr) goldfish 13 mg $l^{-1}$ (1).

### Invertebrate toxicity

$EC_{50}$ (24 hr) *Daphnia magna* ≈50 mg $l^{-1}$. NOEC (21 day) *Daphnia* reproduction test 0.4 mg $l^{-1}$ (minimum value); 2 mg $l^{-1}$ (nominal value) (2).
$EC_{50}$ (48 hr) *Scenedesmus subspicatus* 25-53 mg $l^{-1}$ (3).

## Mammalian and avian toxicity

### Acute data

$LC_{50}$ (4 hr) inhalation rat 24 mg $m^{-3}$ (4).
$LD_{Lo}$ intraperitoneal guinea pig 1-3 mg $kg^{-1}$ (5).

### Irritancy

Skin and respiratory irritant (species unspecified) (6).

## References

1. Brenniman, G. et al *Water Res.* 1976, **10**, 165
2. Kuehn, R. et al *Water Res.* 1989, **23**(4), 501-510
3. Kuehn, R. et al *Water Res.* 1990, **24**(1), 31-38
4. *Gig. Sanit.* 1979, **44**(5)
5. *Arch. Ind. Hyg. Occup. Med.* 1954, **9**, 227
6. Luxton, S. G. (Ed.) *Hazards in the Chemical Laboratory* 5th ed., 1992, RSC, London

# M84 Mestranol

CAS Registry No. 72-33-3
Synonyms 19-norpregna-1,3,5(10)-trien-20-yn-17-ol, 3-methoxy-, (17α)-;
17α-ethynyl estradiol-3-methyl ether; 17α-ethynyl-3-methoxy-
1,3,5(10)-estratrien-17β-ol; Menophase; Norquen
Mol. Formula $C_{21}H_{26}O_2$                                    Mol. Wt. 310.44
Uses Oestrogen in oral contraceptives and for oestrogen deficiency conditions, active
by oral route.

## Physical properties

M. Pt. 150-151°C.

## Solubility
Organic solvent: ethanol, diethyl ether, acetone

## Mammalian and avian toxicity

### Acute data
$LD_{50}$ oral blackbird 1 g $kg^{-1}$ (1).
$LD_{50}$ oral coturnix 1 g $kg^{-1}$ (1).
$LD_{50}$ intraperitoneal mouse 3.5 g $kg^{-1}$ (2).

### Carcinogenicity and long-term effects
Sufficient evidence for carcinogenicity to humans and animals, IARC classification
group 1 (3).
♀ mice given 20 mg $kg^{-1}$ diet for their lifespan showed an increased incidence of
chromophobe adenomas and an increase in some non-metastasizing epithelial
tumorous lesions (4).
Castrated ♂ mice receiving 0.1 or 1.0 mg $kg^{-1}$ diet developed an increased incidence
of mammary tumours (5).
♀ rats given 3 mg orally on 6 days of the wk for 50 wk developed no mammary
tumours (6).
♀ monkeys receiving up to 50 times the human dose orally for 7 yr, developed some
slight ductal epithelial hyperplasia in mammary glands and some palpable nodules (7).
No significant increase in mammary tumour occurrence was seen in dogs dosed orally
with mestranol (8,9).
Mice receiving 0.1 mg subcutaneously twice wkly from 1 to 21 months of age
showed a significant increase in incidence of mammary tumours (10).

---

**Teratogenicity and reproductive effects**

Oral administration of 0.05-0.2 mg kg$^{-1}$ to mice on days 4-8 increased the number of resorptions, while on days 7-11, abortions but no teratogenic effects were seen (11). In rats, subcutaneous administration of 0.1 mg kg$^{-1}$ on days 2-4 terminated pregnancy (12).

Prenatal administration to rats by oral administration to mothers of 0.1 mg kg$^{-1}$ on days 14.5-19.5 influenced testosterone production of ♂ offspring both in foetal and adult stages (13).

**Metabolism and pharmacokinetics**

Readily absorbed from gastrointestinal tract of animals and humans with $t_{1/2}$ 50 hr. Enterohepatic circulation of metabolites is a significant feature of metabolism with several metabolites being eliminated in urine (14-16).

In humans, 54% of dose can be demethylated to ethynylestradiol, with this proportion varying in other species (17).

The demethylated product then follows the pathways for ethynylestradiol metabolism (14,18).

Mestranol is well distributed within the body and is thought able to cross into the placenta (13,19).

Found in secretions such as breast milk, the main compound found in plasma after intraveneous administration is ethynylestradiol 3-sulphate (14,19,20).

## Genotoxicity

*Salmonella typhimurium* TA100, TA1535, TA1537, TA1538 with and without metabolic activation negative (21).

*Drosophila melanogaster* lethal mutagenicity negative (22).

*In vivo* mouse bone marrow cells chromosomal aberrations negative (23).

Effects on cell proliferation *in vitro* at 10$^{-6}$ M have been found using hepato-carcinoma HepG2 cells (24).

## Any other comments

Toxicology and genotoxic reviewed (25,26).

## References

1. Schafer, E. W. et al *Arch. Environ. Contam. Toxicol.* 1983, **12**, 355-382
2. Gosselin, R. L. *Clinical Toxicology of Commercial Products* 3rd ed., 1968, Williams & Williams, Baltimore
3. *IARC Monograph* 1987, **Suppl. 7**, 288
4. Heston, W. E. et al *J. Natl. Cancer Inst.* 1973, **51**, 209-224
5. Rudali, G. et al *Rev. Eur. Etud. Clin. Biol.* 1971, **16**, 425-429
6. Gruenstein, M. et al *Cancer Res.* 1964, **24**, 1656-1658
7. Geil, R. G. et al *J. Toxicol. Environ. Health* 1977, **3**, 179-193
8. Etreby, E. et al *Pharmacol. Ther.* 1979, **5**, 369-402
9. Kwapien, R. P. et al *J. Natl. Cancer Inst.* 1980, **65**, 137-144
10. Welsch, C. W. et al *Br. J. Cancer* 1977, **35**, 322-328
11. Heinecke, H. et al *Pharmazie* 1975, **30**, 53-56
12. Saunders, F. J. et al *Toxicol. Appl. Pharmacol.* 1967, **11**, 229-244
13. Varma, S. K. *Acta Endocrinol. (Copenhagen)* 1987, **116**(2), 193-199
14. *Martindale. The Extra Pharmacopoeia* 30th ed., 1993, The Pharmaceutical Press, London
15. Goldzieher, J. W. et al *Am. J. Obstet. Gynecol.* 1990, **163**(6, Pt. 2), 2114-2119
16. Brewster, D. et al *Biochem. Pharmacol.* 1977, **26**, 943-946

17. Bolt, H. M. et al *Eur. J. Clin. Pharmacol.* 1974, **7**, 295-305
18. Mills, T. M. et al *J. Obstet. Gynaecol.* 1976, **126**, 987
19. *Contraception* 1977, **15**, 255
20. Bird, C. E. et al *J. Clin. Endocrinol. Metab.* 1973, **36**, 296-302
21. Rao, M. S. et al *Toxicol. Appl. Pharmacol.* 1983, **69**, 48-54
22. Paradi, E. *Mutat. Res.* 1981, **38**, 175-178
23. Ansari, G. A. S. *Indian. J. Hered.* 1977, **9**, 7-9
24. Coezy, E. *Endocrinology (Baltimore)* 1987, **120**(1), 133-141
25. *IARC Monograph* 1979, **21**, 257
26. *IARC Monograph* 1987, **Suppl. 6**, 44

# M85  Metalaxyl

**CAS Registry No.** 57837-19-1
**Synonyms** DL-alanine *N*-(2,6-dimethylphenyl)-*N*-(methoxyacetyl)-, methyl ester; Metaxanin; methyl (*N*-(2-methoxyacetyl)-*N*-(2,6-xylyl)-DL-alaninate; Ridomil; Subdue
**Mol. Formula** $C_{15}H_{21}NO_4$          **Mol. Wt.** 279.34
**Uses** Fungicide.
**Occurrence** Contaminant of food and water (1,2).

## Physical properties

**M. Pt.** 71.8-72°C; **Specific gravity** $d^{20}$ 1.21; **Volatility** v.p. $2.2 \times 10^{-6}$ mmHg at 20°C.

**Solubility**
Water: 7.1 g $l^{-1}$ at 20°C. Organic solvent: benzene, methanol

## Ecotoxicity

**Fish toxicity**
$LC_{50}$ (96 hr) rainbow trout, carp, bluegill sunfish >100 mg $l^{-1}$ (3).

**Invertebrate toxicity**
$EC_{50}$ (5 min) *Photobacterium phosphoreum* 119 ppm Microtox test (4).

## Environmental fate

**Degradation studies**
Residual activity in soil can be detected for 70-90 days (1).

## Mammalian and avian toxicity

### Acute data

$LD_{50}$ oral rat 566 mg $kg^{-1}$ (5).
$LD_{50}$ dermal rat >3100 mg $kg^{-1}$ (6).

### Sub-acute data

90 day feeding trial rats, NOEC 17 mg $kg^{-1}$ (diet) (6).
6 month feeding trial dogs, NOEC 7.6 mg $kg^{-1}$ (diet) (6).

### Metabolism and pharmacokinetics

In mammals, after oral administration the ether bond is hydrolysed and the methyl
ether bond oxidatively cleaved (3).

### Irritancy

Slight irritant to rabbit eyes and skin (dose/duration unspecified) (6).

### Sensitisation

No skin sensitisation in guinea pigs (3).

## Any other adverse effects

Acute toxicity after oral dosing with formulation Apron 35WS was higher in ♀ than ♂
rats. Mild cytoplasmic vacuolar lesions, cellular infiltration and necrotic patches were
seen in liver, along with some adrenal cortex toxicity. $LD_{50}$ oral ♀ rat 2.8 g $kg^{-1}$ (7).

## Legislation

Limited under EC Directive on Drinking Water Quality 80/778/EEC. Pesticides:
maximum admissible concentration 0.1 μg $l^{-1}$ (8).
Included in Schedule 6 (Release into Land: Prescribed Substances) Statutory
Instrument No. 472, 1991 (9).

## Any other comments

TDI (human) 0.03 mg $kg^{-1}$; WHO Class III; EPA Toxicity Class III (3).
Non-toxic to bees (3).
Transformation in soil and water has been reviewed (10).

## References

1. Frank, R. et al *Arch. Environ. Contam. Toxicol.* 1987, **16**(1), 9-22
2. Lute, M. et al *J. Assoc. Off. Anal. Chem.* 1988, **71**(2), 415-433
3. *The Agrochemicals Handbook* 3rd ed., 1991, RSC, London
4. Kaiser, K. L. E. et al *Water Pollut. Res. J. Can.* 1991, **26**(3), 361-431
5. *J. Hyg. Epidem. Microbiol. Immunol.* 1991, **35**, 375
6. *The Pesticide Manual* 9th ed., 1991, British Crop Protection Council, Farnham
7. Reddy, M. V. *Comp. Physiol. Ecol.* 1990, **15**(2), 51-57
8. *EC Directive Relating to the Quality of Water Intended for Human Consumption* 1982, 80/778/EEC, Office for Official Publications of the European Communities, 2 rue Mercier, L-2985 Luxembourg
9. *S. I. 1991 No. 472 The Environmental Protection (Prescribed Processes and Substances) Regulations* 1991, HMSO, London
10. Anan'eva, N. D. et al *Agrokhimiya* 1988, (9), 128-133 (Russ.) (*Chem. Abstr.* **109**, 206616h)

# M86   Metaldehyde

$$CH_3$$

**CAS Registry No.** 108-62-3
**Synonyms** 1,3,5,7-tetroxocane, 2,4,6,8-tetramethyl-; metacetaldehyde;
1'-2,C-4,C-6,C-8-tetramethyl-1,3,5,7-tetroxocane; acetaldehyde tetramer
**Uses** Molluscicide. In compressed form as a fuel.

## Physical properties

**M. Pt.** 246°C (in sealed tube); **B. Pt.** 112-115°C (sublimes with partial
depolymerisation); **Flash point** 36-40°C.

### Solubility
Water: 200 mg $l^{-1}$ at 17°C. Organic solvent: benzene, chloroform

## Occupational exposure

**UN No.** 1332; **Conveyance classification** flammable solid;
**Supply classification** harmful.
**Risk phrases** Flammable – Harmful if swallowed (R10, R22)
**Safety phrases** Keep out of reach of children – Keep away from food, drink and
animal feeding stuffs – Avoid contact with eyes – If swallowed seek medical advice
immediately and show this container or label (S2, S13, S25, S46)

## Ecotoxicity

**Fish toxicity**
Non-toxic to fish (1).

**Invertebrate toxicity**
Metaldehyde is orally active in molluscs on land and water and is thought to affect
nervous system activity (2,3).

## Environmental fate

**Abiotic removal**
Gradual degradation by depolymerisation to acetaldehyde is seen in the environment,
followed by oxidation to acetic acid (1).

## Mammalian and avian toxicity

**Acute data**
$LD_{50}$ oral guinea pig, dog, rat 600-630 mg $kg^{-1}$ (1,4).
$LD_{50}$ oral dog 1000 mg $kg^{-1}$ (4).
$LD_{Lo}$ oral child 100 mg $kg^{-1}$ (5).

LD$_{Lo}$ oral adult human 43 mg kg$^{-1}$ (6).
LD$_{50}$ dermal rat >5 g kg$^{-1}$ (1).

## Sensitisation
No significant dermatitis has been seen in patch tests (7).

## Legislation
Limited under EC Directive on Drinking Water Quality 80/778/EEC. Pesticides: maximum admissible concentration 0.1 μg l$^{-1}$ (8).
Included in Schedule 6 (Release into Land: Prescribed Substances) Statutory Instrument No. 472, 1991 (9).

## Any other comments
WHO Class III; EPA Toxicity Class II (1).

## References
1. *The Agrochemicals Handbook* 3rd ed., 1991, RSC, London
2. Briggs, G. G. et al *Crop. Prot.* 1987, **6**(5), 341-346
3. Mills, J. et al *Pestic. Sci.* 1990, **28**(1), 89-99
4. *The Pesticide Manual* 9th ed. 1991, British Crop Protection Council, Farnham
5. Klimmer, O. R. *Pflanzenschutz-und Schaedlings Bekaempfungsmittel* 2nd ed., 1971, Hundt-Verlag, Hattingen
6. Dreishach, R. H. *Handbook of Poisoning Diagnosis and Treatment* 8th ed., 1971, Lange Medical Publishing, CA
7. Lisi, P. *Contact Dermatitis* 1986, **15**(5), 266-269
8. *EC Directive Relating to the Quality of Water Intended for Human Consumption* 1982, 80/778/EEC, Office for Official Publications of the European Communities, 2 rue Mercier, L-2985 Luxembourg
9. *S. I. 1991 No. 472 The Environmental Protection (Prescribed Processes and Substances) Regulations* 1991, HMSO, London

# M87   Metamitron

**CAS Registry No.** 41394-05-2
**Synonyms** 1,2,4-triazin-5(4*H*)-one, 4-amino-3-methyl-6-phenyl-; 4-amino-4,5-dihydro-3-methyl-6-phenyl-1,2,4-triazin-5-one; BAY DRW1139; methiamitron; Metamitrone
**Mol. Formula** C$_{10}$H$_{10}$N$_4$O                    **Mol. Wt.** 202.22
**Uses** Herbicide.

## Physical properties
**M. Pt.** 166.6°C; **Partition coefficient** log P$_{ow}$ 0.833; **Volatility** v.p. $9.8 \times 10^{-5}$ mmHg.

## Solubility

Water: 1.82 g $l^{-1}$ at 20°C. Organic solvent: cyclohexanone, dichloromethane

## Ecotoxicity

### Fish toxicity

$LC_{50}$ (96 hr) goldfish 400-500 mg $l^{-1}$ (70% wet powder formulation) (1).
$LC_{50}$ (96 hr) carp 500 mg $l^{-1}$ (1).
$LC_{50}$ (96 hr) orfe 240-300 mg $l^{-1}$ (70% wet powder formulation) (1).

### Invertebrate toxicity

Metamitron is toxic to starfish at concentrations >10 ppm. Exposure to light, increases toxicity (2).

## Environmental fate

### Degradation studies

In soil, 20% of compound applied may be detectable after 4-6 wk but can be as low as 4% by 8 wk (1,3).
The major metabolite is the active desamino-metamitron, but neither this nor the parent compound appear to leach below the top 20 cm of soil (3).
Formation of bound residues occurs (4).

## Mammalian and avian toxicity

### Acute data

$LD_{50}$ oral ♂ ♀ rat 3343, 1832 mg $kg^{-1}$, respectively (5).
$LD_{50}$ oral mouse 1450-1463 mg $kg^{-1}$ (5).
$LD_{50}$ oral canary >1 g $kg^{-1}$ (5).
$LD_{50}$ dermal rat >1 g $kg^{-1}$ (5).

### Sub-acute data

A 90 day feeding trial in dogs established, NOEC 500 mg $kg^{-1}$ diet (5).

### Carcinogenicity and long-term effects

A 2 yr feeding trial in rats established, NOEC 250 mg $kg^{-1}$ diet (5).

### Metabolism and pharmacokinetics

After oral administration to mammals 50% of dose is eliminated in urine and 50% in faeces within 48 hr (1).

## Legislation

Limited under EC Directive on Drinking Water Quality 80/778/EEC. Pesticides: maximum admissible concentration 0.1 μg $l^{-1}$ (6).
Included in Schedule 6 (Release into Land: Prescribed Substances) Statutory Instrument No. 472, 1991 (7).

## References

1. *The Agrochemicals Handbook* 3rd ed., 1991, RSC, London
2. Gadkari, D. et al *Biol. Fertil. Soils* 1987, **3**(3), 171-177
3. Kubiak, R. *Weed Sci.* 1988, **36**(4), 514-518
4. Kubiak, R. et al *Int. J. Environ. Anal. Chem.* 1990, **39**(1), 47-57
5. *The Pesticide Manual* 9th ed., 1991, British Crop Protection Council, Farnham

6. *EC Directive Relating to the Quality of Water Intended for Human Consumption* 1982, 80/778/EEC, Office for Official Publications of the European Communities, 2 rue Mercier, L-2985 Luxembourg
7. *S. I. 1991 No. 472 The Environmental Protection (Prescribed Processes and Substances) Regulations* 1991, HMSO, London

# M88  Metazachlor

**CAS Registry No.** 67129-08-2
**Synonyms** acetamide, 2-chloro-*N*-(2,6-dimethylphenyl)-*N*-(1*H*-pyrazol-1-ylmethyl)-; 2-chloro-*N*-(pyrazol-1-ylmethyl)acet-2′,6′-xylidine; BAS47900H; Butisan S
**Mol. Formula** $C_{14}H_{16}ClN_3O$ **Mol. Wt.** 277.76
**Uses** Herbicide.
**Occurrence** Water pollutant. Food contaminant.

## Physical properties

**M. Pt.** 85°C; **Volatility** v.p. $3.7 \times 10^{-7}$ mmHg at 20°C.

**Solubility**
Water: 17 mg $l^{-1}$ at 20°C. Organic solvent: acetone, chloroform, ethanol

## Ecotoxicity

**Fish toxicity**
Toxic to trout, moderately toxic to carp (1).

## Environmental fate

**Degradation studies**
Degradation in soil is reduced or inhibited by antimicrobial agents and protein synthesis inhibitors (2).
Degradation products in soil include:
2-hydroxy-*N*-(2,6-dimethylphenyl)-*N*-(1*H*-pyrazol-1-ylmethyl)acetamide; 2-chloro-N-(2,6-dimethylphenyl)acetamide; and 4,4′-methylenebis(2,6-dimethylbenzeneamine). The latter product is seen only in low concentrations. Metazachlor $t_{1/2}$ 2 months, and the metabolites $t_{1/2}$ 3-3.5 months (3).

**Abiotic removal**
Does not show competitive adsorption with humic substances during water treatment with activated carbon, but does show carbon fouling through pre-adsorption (4).

## Mammalian and avian toxicity

### Acute data

$LD_{50}$ oral rat 2.15 g $kg^{-1}$ (5).
$LD_{50}$ dermal rat >6.8 g $kg^{-1}$ (5).

### Irritancy

Non-irritant to mucous membrane of rabbit (1).

## Legislation

Limited under EC Directive on Drinking Water Quality 80/778/EEC. Pesticides: maximum admissible concentration 0.1 µg $l^{-1}$ (6).
Included in Schedule 6 (Release into Land: Prescribed Substances) Statutory Instrument No. 472, 1991 (7).

## Any other comments

Non-toxic to bees (1).

## References

1. *The Agrochemicals Handbook* 3rd ed., 1991, RSC, London
2. Allen, R. et al *Pestic. Sci.* 1988, **22**(4), 297-305
3. Rouchaud, J. et al *Meded. Fac. Landbouwwet., Rijksuniv., Gent* 1989, **54**(20, 257-261
4. Haist-Gulde, B. et al *GWF, Gas-Wasserfach: Wasser/Abwasser* 1991, **132**(1), 8-15 (Ger.) (*Chem. Abstr.* **115** 56840j)
5. *The Pesticide Manual* 9th ed. 1991, British Crop Protection Council, Farnham
6. *EC Directive Relating to the Quality of Water Intended for Human Consumption* 1982, 80/778/EEC, Office for Official Publications of the European Communities, 2 rue Mercier, L-2985 Luxembourg
7. *S. I. 1991 No. 472 The Environmental Protection (Prescribed Processes and Substances) Regulations* 1991, HMSO, London

# M89   Metepa

**CAS Registry No.** 60057-39-6
**Synonyms** aziridine, 1,1′,1″-phosphinylidynetris[2-methyl]-;
1,1′,1′-phosphinylidynetris[2-methylaziridine];tris(2-methyl-1-azirinyl)phosphine oxide; methylapoxide
**Mol. Formula** $C_9H_{18}N_3OP$                         **Mol. Wt.** 215.24
**Uses** Chemosterilant. Also used in crease proofing and flameproofing textiles.

---

## Physical properties
**B. Pt.** 90-92°C (0.15-0.3 mmHg); **Specific gravity** $d_{25}^{25}$ 1.079.

## Solubility
Water: miscible. Organic solvent: miscible with ethanol, diethyl ether, chloroform

## Ecotoxicity

### Invertebrate toxicity
Pupae of mosquito flies exposed to ≤10,000 ppm for 2 hr or ♀ flies fed ≤0.5% in diet yielded no effect levels of 1000 ppm and 0.05% respectively. Higher doses to pupae and adults caused deaths (1).

## Mammalian and avian toxicity

### Acute data
$LD_{50}$ oral ♂ rat 136 mg $kg^{-1}$; ♀ rat 213 mg $kg^{-1}$ (2).
$LD_{50}$ oral mouse 292 mg $kg^{-1}$ (3).
$LD_{50}$ dermal mouse 375 mg $kg^{-1}$ (3).
$LD_{50}$ subcutaneous mouse 140 mg $kg^{-1}$ (3).

### Carcinogenicity and long-term effects
No adequate evidence for carcinogenicity to humans or animals, IARC classification group 3 (4).
Rats administered 0.625 mg $kg^{-1}$ orally for ≤422 days did not develop tumours. Lymphatic leukaemias were seen at a low incidence in animals given ≥2.5 mg $kg^{-1}$ but not in a dose-related manner (5).

### Teratogenicity and reproductive effects
Rats ♀ injected intraperitoneally with 30 mg $kg^{-1}$ on day 12 of pregnancy showed teratogenic and embryotoxic effects. All neonates had malformations including ectrodactylia. Daily injections of 1.25 mg $kg^{-1}$ on days 7-13 of pregnancy caused reduction in foetal and placental weight. In ♂ rats a sterilising effect was seen (5).

### Metabolism and pharmacokinetics
♂ mice injected intraperitoneally with 100 mg $kg^{-1}$ [$^{32}$P]-metepa eliminated 50% of the dose in urine within 12 hr, mostly in unchanged form or as phosphoric acid. Unchanged metepa was the major radioactive substance in blood at 2 hr but had almost disappeared by 6 hr (6).

## Genotoxicity
*Salmonella typhimurium* G-46 host mediated assay positive (7).
Mice dominant lethal assay positive (8).
Chronic treatment of ♂ mice 40 mg $kg^{-1}$ orally or intraperitoneally induced micronuclei in bone marrow cells and chromosomal aberrations in testicular germ cells (9).

## Any other comments
Toxicology reviewed (10).

## References
1. Mittal, P. K. *Res. Bull. Punjab. Univ. Sci.* 1987, **38**(3-4), 119-125
2. Gaines, T. B. *Toxicol. Appl. Pharmacol.* 1969, **14**, 515

3. Maehashi, H. *Ind. Health. (Jpn.)*, 1970, **84**, 54-65
4. *IARC Monograph* 1987, **Suppl. 7**, 73
5. Gaines, T. B. et al *Bull. W.H.O.* 1966, **34**, 317-320
6. Plapp, F. W. et al *J. Econ. Entomol.* 1962, **55**, 607-613
7. Devi, K. R. *Med. Sci. Res.* 1987, **15**(19), 1175-1176
8. Epstein, S. S. et al *Toxicol. Appl. Pharmacol.* 1970, **17**, 23-40
9. Devi, K. R. *Cell Chromosome Res.* 1986, **9**(2), 39-41
10. *IARC Monograph* 1975, **9**, 107

# M90    Metformin hydrochloride

## $(H_3C)_2NC(=NH)NHC(=NH)NH_2 \cdot HCl$

**CAS Registry No.** 1115-70-4
**Synonyms** 1,1-dimethylbiguanide hydrochloride; *N,N*-dimethylimidodicarbonimidic diamide monohydrochloride; Glucophage; Diabefagos; Meguan; metformin HCl
**Mol. Formula** $C_4H_{12}ClN_5$                    **Mol. Wt.** 165.63
**Uses** Hypoglycaemic drug used in oral treatment of maturity onset diabetes.

## Physical properties

**M. Pt.** 232°C.

**Solubility**
Water: 500 g $l^{-1}$. Organic solvent: 95% ethanol

## Mammalian and avian toxicity

### Acute data

$LD_{50}$ oral rat, mouse 1000, 1450 mg $kg^{-1}$, respectively (1).
$LD_{50}$ intraperitoneal mouse 420 mg $kg^{-1}$ (1).
$LD_{50}$ subcutaneous rat 300 mg $kg^{-1}$ (1).

### Metabolism and pharmacokinetics

After oral administration to humans, there is incomplete absorption from the gastrointestinal tract, but that part which is absorbed is excreted unchanged in urine (2,3). There is no binding to plasma protein and $t_{1/2}$ is 1.5-3 hr (3,4).
There is little or no effect on blood sugar of normal human subjects, rats, guinea pigs or rabbits but dogs are more sensitive with an oral dose of 100 mg $kg^{-1}$ reducing blood sugar significantly (5,6).
Some circulating insulin must be present for the hypoglycaemic effect (2).

### Irritancy

100 mg instilled into rabbit eye for 2 sec caused mild irritation and 500 mg applied to rabbit skin caused mild irritation (7,8).

## Any other adverse effects to man

The compound can cause lactic acid acidosis, and care has to be taken when using the compound in conditions where such acidosis might result. Other side effects include anorexia, nausea, vomiting and diarrhoea (4).
Some drug interactions occur including that with verapamil (4,9).

## Any other comments

The compound has been shown to be of use in treatment of ischaemia and anoxia (10).

## References

1. *The Merck Index* 11th ed., 1989, Merck & Co. Inc., Rahway, NJ
2. *Martindale. The Extra Pharmacopoeia* 30th ed., 1993, The Pharmaceutical Press, London
3. Sirtori, C. R. et al *Clin. Pharmacol. Ther.* 1978, **24**, 683
4. Marchetti, P. et al *Clin. Pharmacokinet.* 1989, **16**, 100-128
5. Kochergin, P. et al *Khim-Farm Zh.* 1987, **21**(12), 1517-1518 (Russ.) (*Chem. Abstr.* **108**, 106293m)
6. Tucker, G. T. et al *Br. J. Clin. Pharmacol.* 1981, **12**, 235
7. *Food Cosmet. Toxicol* 1982, **20**, 563
8. *Food Cosmet. Toxicol* 1982, **20**, 573
9. El-Bauomy, A. M. *Egypt. J. Vet. Sci.* 1987, **24**(2), 191-198
10. Wernsperger, N. et al *European Patent Application* 1988, EP283, 369. 21 Sep 88

# M91   Methabenzthiazuron

**CAS Registry No.** 18691-97-9
**Synonyms** 1-(1,3-benzothiazol-2-yl)-1,3-dimethylurea;
1-benzothiazol-2-yl-1,3-dimethylurea; *N*-2-benzothiazolyl-*N*,*N*'-dimethylurea;
1-(2-benzothiazolyl)-1,3-dimethylurea; methibenzuron; Tribunil
**Mol. Formula** $C_{10}H_{11}N_3OS$               **Mol. Wt.** 221.28
**Uses** Herbicide.

## Physical properties

**M. Pt.** 119-121°C; **Partition coefficient** log $P_{ow}$ 2.640 (1); **Volatility**
v.p. $4.4 \times 10^6$ mmHg at 20°C.

## Solubility
Water: 59 mg $l^{-1}$ at 20°C. Organic solvent: acetone, dichloromethane, dimethylformamide, hexane, isopropanol, toluene

## Ecotoxicity

### Fish toxicity
$LC_{50}$ (96 hr) rainbow trout, golden orfe 16-29 mg $l^{-1}$ (1).

## Environmental fate

### Degradation studies
60% loss from soil after 127 days (2).

## Mammalian and avian toxicity

### Acute data
$LD_{50}$ oral rat, mouse, rabbit, cat, dog >1000 mg $kg^{-1}$ (1,3).
$LD_{50}$ oral guinea pig >2500 mg $kg^{-1}$ (1).
$LC_{50}$ (4 hr) inhalation rat >500 mg $m^{-3}$ (1).
$LD_{50}$ dermal rat >500 mg $kg^{-1}$ (1).
$LD_{50}$ intraperitoneal rat 315-540 mg $kg^{-1}$ (3).

### Carcinogenicity and long-term effects
Oral rat (2 yr) no adverse effect level 150 mg $kg^{-1}$ diet (1).

### Legislation
Limited under EC Directive on Drinking Water Quality 80/778/EEC. Pesticides:
maximum admissible concentration 0.1 $\mu g\, l^{-1}$ (4).

### Any other comments
WHO Class Table 5; EPA Toxicity Class IV (1).
In plants metabolised to $N$-hydroxymethyl-$N$-(2-benzothiazolyl)urea and
$N$-methyl-$N$-(2-benzothiazolyl)urea as the water soluble glucosides (1).

### References
1. *The Agrochemicals Handbook* 3rd ed., 1991, RSC, London
2. Kubiak, R. et al *Weed Sci.* 1988, **36**(4), 514-518
3. Kimmerle, E. et al *Pflanzenschutz-Nachr.* 1969, **22**, 351
4. *EC Directive Relating to the Quality of Water Intended for Human Consumption* 1982, 80/778/EEC, Office for Official Publications of the European Communities, 2 rue Mercier, L-2985 Luxemburg

# M92   Methacrifos

**CAS Registry No.** 62610-77-9 ((*E*)-isomer)
**Synonyms** methyl (*E*)-3-(dimethoxyphosphinothioyloxy)-2 –methylacrylate;
(*E*)-*O*-2-methoxycarbonylprop-1-enyl *O,O*-dimethyl phosphorothioate; (*E*)-methyl
3-[(dimethoxyphosphinothioyl)oxy]-2-methyl-2-propenoate; OMS 2005; Damfin
**Mol. Formula** $C_7H_{13}O_5PS$ **Mol. Wt.** 240.22
**Uses** Acaricide. Insecticide.

## Physical properties
**B. Pt.** $_{0.01}$ 90°C; **Volatility** v.p. $1.2 \times 10^{-3}$ mmHg at 20°C.

### Solubility
Water: 400 mg l$^{-1}$ at 20°C. Organic solvent: diethyl ether, ethanol, benzene, dichloromethene, hexane, methanol

### Abiotic removal
Calculated hydrolysis $t_{1/2}$ 66 days at pH 1; 29 days at pH 7; and 9.5 days at pH 9 (1).

## Mammalian and avian toxicity

### Acute data
LD$_{50}$ oral rat 680 mg kg$^{-1}$ (1,2).
LC$_{50}$ (6 hr) 2200 mg m$^{-3}$ (1,2).
LD$_{50}$ dermal rat >3100 mg kg$^{-1}$ (1-3).

### Irritancy
Reported to be a mild skin irritant, but not irritating to the eyes of rabbits (1).

## Any other adverse effects
Inhibits cholinesterase activity (1,2).

## Legislation
Limited under EC Directive on Drinking Water Quality 80/778/EEC. Pesticides: maximum admissible concentration 0.1 μg l$^{-1}$ (4).

## Any other comments
WHO Class II (1,2).
TDI (human) 0.003 mg kg$^{-1}$ (temporary 1988-1990) (1,2).

## References
1. *The Pesticide Manual* 9th ed., 1991, 562, British Crop Protection Council, Farnham
2. *The Agrochemicals Handbook* 3rd ed., 1993, RSC, London
3. Wyniger, R. et al *Proc. – Br. Crop. Prot. Conf. – Pests Dis.* 1977, 1033
4. *EC Directive Relating to the Quality of Water Intended for Human Consumption* 1982, 80/778/EEC, Office for Official Publications of the European Communities, 2 rue Mercier, L-2985 Luxembourg

# M93    Methacrolein

## CH2=C(CH3)CHO

**CAS Registry No.** 78-85-3
**Synonyms** isobutenal; methenylaldehyde; 2-methylacrolein ; 2-methylpropenal; 2-methyl-2-propenal; methacrylaldehyde
**Mol. Formula** C$_4$H$_6$O                **Mol. Wt.** 70.09
**Uses** Manufacture of copolymers and resins.

## Physical properties
**M. Pt.** –81°C; **B. Pt.** 69°C; **Flash point** –15°C; **Specific gravity** d$_4^{20}$ 0.847; **Volatility** v.p. 120 mmHg at 20°C; v. den. 2.42.

**Solubility**

Water: 64 g l$^{-1}$. Organic solvent: diethyl ether, ethanol

## Occupational exposure

UN No. 2396; **HAZCHEM Code** 3WE.

## Ecotoxicity

**Fish toxicity**

Fatal to brown trout, bluegill sunfish, yellow perch and goldfish after 22 hr at 5 ppm. Test conditions: pH 7.0; dissolved oxygen content 7.5 ppm; total hardness (soap method) 300 ppm; methyl orange alkalinity 310 ppm; free carbon dioxide 5 ppm; temperature 12.8°C (1).

## Environmental fate

**Degradation studies**

65% ThOD removed by acclimatised sewage in 10 days at 20°C (2).

## Mammalian and avian toxicity

**Acute data**

$LD_{50}$ oral rat 110-140 mg kg$^{-1}$ (3).
$LC_{Lo}$ (4 hr) inhalation rat 125 ppm (4).
$LD_{50}$ dermal rabbit 360 mg kg$^{-1}$ (5).

**Irritancy**

Dermal rabbit (24 hr) 500 mg caused severe irritation. 50 µg instilled into rabbit eye caused severe irritation (3).

## Genotoxicity

*Salmonella typhimurium* TA104 without metabolic activation positive (6).

## Any other comments

Human health effects, experimental toxicology, physico-chemical properties reviewed (7).

## References

1. Wood, E. M. *The Toxicity of 3400 Chemicals to Fish* 1987, EPA 560/6-87-002; PB 87-200-275, Washington, DC
2. Verschueren, K. *Handbook of Environmental Data on Organic Chemicals* 2nd ed., 1983, 815, Van Nostrand Reinhold, New York
3. Marhold, J. V. *Sbornik Vysledku Toxixologickeho Vysetreni Latek A Pripravku* 1972, Prague
4. *J. Ind. Hyg. Toxicol.* 1941, **31**, 343
5. *J. Ind. Hyg. Toxicol.* 1941, **31**, 60
6. Marnett, L. J. et al *Mutat. Res.* 1985, **148**(1), 25-34
7. *ECETOC Technical Report No. 30(4)* 1991, European Chemical Industry Ecology and Toxicology Centre, B-1160 Brussels

# M94　Methacrolein diacetate

$$CH_2=C(CH_3)CH(O_2CCH_3)_2$$

**CAS Registry No.** 10476-95-6
**Synonyms** acetic acid, 2-methyl-2-propene-1,1-diol diester; methallylidene diacetate
**Mol. Formula** $C_8H_{12}O_4$　　　　　　　　　**Mol. Wt.** 172.18

## Physical properties

**M. Pt.** −15°C; **B. Pt.** 191°C; **Flash point** 83°C; **Specific gravity** $d_{20}^{20}$ 1.039; **Volatility**
v.p. 760 mmHg at 191°C.

## Mammalian and avian toxicity

### Acute data
$LD_{50}$ oral rat 440 mg $kg^{-1}$ (1).
$LC_{Lo}$ (1 hr) inhalation rat 62 ppm (1).
$LD_{50}$ dermal rabbit 44 mg $kg^{-1}$ (1).
$LD_{Lo}$ intraperitoneal mouse 250 mg $kg^{-1}$ (2).

### Irritancy
Causes severe irritation. High concentrations are extremely destructive to tissues of
the mucous membranes, upper respiratory tract, eyes and skin (species unspecified) (3).
Rabbit eye studies demonstrated severe irritation and corneal damage, being rated 9
on a scale of 10 (4).

## References

1.　*Am. Ind. Hyg. Assoc. J.* 1969, **30**, 470
2.　*Summary Tables of Biological Tests* 1953, **5**, 61
3.　Lenga, R. E. (Ed.) *Sigma-Aldrich Library of Chemical Safety Data* 2nd ed., 1988, **2**, 2227, Sigma-Aldrich, Milwaukee, WI
4.　*Patty's Industrial Hygiene and Toxicology* 3rd ed., 1982, **2c**, 4018, John Wiley & Sons, New York

# M95　Methacrylic acid

$$CH_2=C(CH_3)CO_2H$$

**CAS Registry No.** 79-41-4
**Synonyms** α-methacrylic acid; 2-methylpropenoic acid; 2-methyl-2-propenoic acid
**Mol. Formula** $C_4H_6O_2$　　　　　　　　　**Mol. Wt.** 86.09
**Uses** Manufacture of methacrylate resins and plastics.
**Occurrence** In oil from Roman chamomile.

## Physical properties

**M. Pt.** 16°C; **B. Pt.** 163°C; **Flash point** 77°C (open cup); **Specific gravity** $d_4^{20}$ 1.0153;
**Partition coefficient** log $P_{ow}$ 0.93; **Volatility** v.p. 1.0 mmHg at 25.5°C; v. den. 2.97.

## Solubility
Water: 89 g l$^{-1}$ at 25°C. Organic solvent: diethyl ether, ethanol, methanol

## Occupational exposure
**US TLV (TWA)** 20 ppm (70 mg m$^{-3}$); **UK Long-term limit** 20 ppm (70 mg m$^{-3}$); **UK Short-term limit** 40 ppm (140 mg m$^{-3}$); **UN No.** 2531; **HAZCHEM Code** 3X; **Conveyance classification** corrosive substance; **Supply classification** corrosive.
**Risk phrases** ≥25% – Causes burns – ≥2%<25% – Irritating to eyes and skin (R34, R36/38)
**Safety phrases** Keep away from heat – In case of contact with eyes, rinse immediately with plenty of water and seek medical advice (S15, S26)

## Ecotoxicity

### Fish toxicity
No toxic effects to stickleback and rainbow trout at 1 mg l$^{-1}$ for 24 hr (1).

### Bioaccumulation
Calculated bioconcentration factor of 3.0 indicates that environmental accumulation is unlikely (2).

## Environmental fate

### Degradation studies
ThOD 1.67 mg l$^{-1}$ O$_2$; BOD$_5$ 0.89 mg l$^{-1}$O$_2$ in standard dilute sewage (3,4).

### Abiotic removal
Methacrylic acid may undergo photolysis, based upon its slight absorption of light at wavelengths >290 nm in methanol (5).
Reaction with photochemically produced hydroxyl radicals in the atmosphere, estimated t$_{1/2}$ 6.12 hr (6).
Volatilisation t$_{1/2}$ 27.5 days in model river water and 300 days in model pond water (2,7).

### Absorption
Calculated K$_{oc}$ of 76 indicates that adsorption to soil and sediments would be insignificant (2).

## Mammalian and avian toxicity

### Acute data
LD$_{50}$ oral rat, mouse 60, 1600 mg kg$^{-1}$, respectively (8,9).
LD$_{50}$ oral redwing blackbird >100 mg kg$^{-1}$ (10).
LD$_{50}$ dermal rabbit 500 mg kg$^{-1}$ (11).
LD$_{50}$ intraperitoneal mouse 48 mg kg$^{-1}$ (12).

### Sub-acute data
Gavage rat (10 days) 5 or 10 mg kg$^{-1}$ day$^{-1}$ caused slight to moderate alveolar haemorrhage and lipid granuloma in the lungs and moderate to severe granularity of liver cytoplasm (13).
Gavage rat (5 days) 50, 100 or 1000 mg kg$^{-1}$ day$^{-1}$ caused sufficient reduction in feed intake and weight loss to terminate the study (14).
Inhalation rat (20 days) 300 ppm 6 hr day$^{-1}$ caused no clinical symptoms.
Histopathological examination revealed slight renal congestion (15).

## Irritancy

Dermal guinea pig (24 hr) 1 ml caused severe irritation. Application for 10 days caused necrosis (13).

## Sensitisation

Did not cause sensitisation in guinea pig skin tests (13).

## Genotoxicity

*Escherichia coli* DNA-cell-binding assay positive (16).

## Any other comments

Physical properties, mammalian toxicity and health hazards reviewed (17,18).

## References

1. McPhee, C. et al *Lethal Effects of 2014 Chemicals to Fish* 1989, EPA 560/6-89-001; PB 89-156-715, Washington, DC
2. Lyman, W. K. et al *Handbook of Chemical Property Estimation Methods Environmental Behaviour of Organic Compounds* 1982, McGraw-Hill, New York
3. Stanair, W. *Plant Engineering Handbook* 1950, McGraw-Hill, New York
4. Lund, H. F. *Industrial Pollution Control Handbook* 1971, McGraw-Hill, New York
5. Sadtler, N. A. *Sadtler Standard Spectra UV No. 1464*
6. Atkinson, R. *Int. J. Chem. Kinetics* 1987, **19**, 799-828
7. *EXAMS II Computer Simulation* 1987, US EPA, Washington, DC
8. Izmerov, N. F. et al *Toxicometric Parameters of Industrial Toxic Chemicals under Single Exposure* 1982, CIP, Moscow
9. *Hyg. Sanit. (USSR)* 1984, **49**(10), 64
10. Schafer, E. W. et al *Arch. Environ. Contam. Toxicol.* 1983, **12**, 355-382
11. *Documentation of Threshold Limit Values for Substances in Workroom Air* 1980, **4**, 257, American Conference of Governmental Industrial Hygienists Inc., Cincinnati, OH
12. *J. Pharm. Sci.* 1973, **62**, 778
13. *Health, Safety and Human Factors Laboratory Report* 1980, Eastman Kodak Co., Rochester, NY
14. *Report NX 00371* U.S. Army Armament Research and Development Command, Clinical Systems Laboratory, NIOSH Exchange Chemicals, Aberdeen Proving Ground, MD
15. Cage, J. C. *Br. J. Ind. Med.* 1970, **27**, 1-18
16. Kubinski, H. et al *Mutat. Res.* 1981, **89**, 95-136
17. *Chemical Safety Data Sheets* 1990, **3**, 156-158, RSC, London
18. *ECETOC Technical Report No. 30(4)* 1991, European Chemical Industry Ecology and Toxicology Centre, B-1160 Brussels

# M96    Methacrylic anhydride

[CH$_2$=C(CH$_3$)C(O)]$_2$O

**CAS Registry No.** 760-93-0
**Synonyms** methacrolyl anhydride; 2-methyl-2-propenoic acid anhydride
**Mol. Formula** C$_8$H$_{10}$O$_3$    **Mol. Wt.** 154.17
**Uses** Manufacture of polymers.

## Physical properties

**B. Pt.** $_{13}$ 87°C; **Flash point** 84°C; **Specific gravity** $d_{20}$ 1.035.

**Solubility**

Organic solvent: diethyl ether, ethanol

## Mammalian and avian toxicity

**Acute data**

$LC_{50}$ (2 hr) inhalation mouse 450 mg m$^{-3}$ (1).

## Any other adverse effects to man

Extremely destructive to tissue of the mucous membrane and upper respiratory tract, eyes and skin. Inhalation may be fatal as a result of spasm, inflammation and oedema of the larynx and bronchi, chemical pneumonitis and pulmonary oedema (2).

## References

1. Izmerov, N. F. et al *Toxicometric Parameters of Industrial Toxic Chemicals under Single Exposure* 1982, CIP, Moscow
2. Lenga, R. E. (Ed.) *Sigma-Aldrich Library of Chemical Safety Data* 2nd ed., 1988, **2**, 2226, Sigma-Aldrich, Milwaukee, WI

# M97   Methacrylonitrile

CH$_2$=C(CH$_3$)CN

**CAS Registry No.** 126-98-7

**Synonyms** 2-cyano-1-propene; isopropene cyanide; isopropenylnitrile; α-methacrylonitrile; 2-methylpropenenitrile

**Mol. Formula** C$_4$H$_5$N                              **Mol. Wt.** 67.09

**Uses** Manufacture of homo- and copolymers. Chemical intermediate.

## Physical properties

**M. Pt.** −35.8°C; **B. Pt.** 90-92°C; **Flash point** 13°C (open cup); **Specific gravity** $d_4^{20}$ 0.80; **Partition coefficient** log $P_{ow}$ 0.68;

**Volatility** v.p. 40 mmHg at 12.8°C; v. den. 2.31.

## Solubility

Water: 2.57 g 100 ml$^{-1}$ at 20°C. Organic solvent: miscible with acetone, octane, toluene, soluble in petroleum ether, ethanol, diethyl ether

## Occupational exposure

**US TLV (TWA)** 1 ppm (2.7 mg m$^{-3}$); **UK Long-term limit** 1 ppm (3 mg m$^{-3}$); **UN No.** 3079; **HAZCHEM Code** 3WE; **Conveyance classification** flammable liquid, toxic substance; **Supply classification** highly flammable, toxic (≥1%).

**Risk phrases** ≥1% – Highly flammable – Toxic by inhalation, in contact with skin and if swallowed – May cause sensitisation by skin contact – ≥0.2%<1% – Highly flammable – Harmful by inhalation, in contact with skin and if swallowed – May

cause sensitisation by skin contact (R11, R23/24/25, R43, R11, R20/21/22, R43)
**Safety phrases** Keep container in a well ventilated place – Keep away from sources of ignition – No Smoking – Handle and open container with care – Do not empty into drains – In case of accident or if you feel unwell, seek medical advice immediately (show label where possible) (S9, S16, S18, S29, S45)

## Ecotoxicity

### Bioaccumulation
Calculated bioconcentration factor of 2 indicates that environmental accumulation is unlikely (1).

## Environmental fate

### Degradation studies
Utilised as sole nitrogen source by *Klebsiella pneumoniae* culture isolated from sewage sludge. Metabolites included ammonia and acrylamide, which was further hydrolysed to acrylic acid. Optimum pH 8.0 and temperature 40-55°C (2).

### Abiotic removal
Reaction with photochemically produced hydroxyl radicals in the atmosphere, estimated $t_{1/2} \approx 2$ days (3).
Estimated volatilisation $t_{1/2}$ 5.3 hr from model river water and 60 hr from model pond water (1,4).

### Absorption
Calculated $K_{oc}$ of 18 indicates that adsorption to soil and sediments would be insignificant (1).

## Mammalian and avian toxicity

### Acute data
$LD_{50}$ oral mouse, rat 17, 250 mg $kg^{-1}$, respectively (5,6).
$LD_{Lo}$ oral mouse 15 mg $kg^{-1}$ (7).
$LC_{50}$ (4 hr) inhalation rat 330 ppm (8).
$LC_{50}$ (4 hr) inhalation mouse, rabbit 36-37 ppm (8).
$LD_{50}$ dermal rabbit 320 mg $kg^{-1}$ (8).
$LD_{Lo}$ intraperitoneal mouse 15 mg $kg^{-1}$ (9).

### Sub-acute data
Inhalation rat (9 day) 20, 50 or 100 ppm 7 hr $day^{-1}$ 5 days $wk^{-1}$ induced sickness by day 1 and a 20% weight loss by day 8. No gross or microscopic lesions were observed (10).
Inhalation rat (91 days) 50 or 100 ppm. Some ♂ rats died by day 1 in the high dose group. Weight gains were decreased and liver weight increased in all groups (10).
Inhalation dog (90 days) 9 or 14 ppm 7 hr $day^{-1}$ 5 days $wk^{-1}$. Central nervous system effects were observed, manifested by convulsions and loss of motor control in the hind limbs in the high dose group. Microscopic brain lesions were also observed. Elevated serum transaminase values were recorded on day 21 (10).

### Teratogenicity and reproductive effects
Oral rat, 50 mg $kg^{-1}$ $day^{-1}$ during the first or second wk of gestation and 100 mg $kg^{-1}$ $day^{-1}$ only during the second wk of gestation caused a dose dependent reduction in

maternal body weight gain. All treated rats aborted and were found to have developed dose-dependent oedema in the Fallopian tubes (11).

**Metabolism and pharmacokinetics**

Following oral administration of 100 mg kg$^{-1}$ $^{14}$C-labelled methacrylonitrile, 43% of $^{14}$C was excreted in the urine, 15% in the faeces and 2% expired as $CO_2$ in 5 days. Hydrogen cyanide was not detectable. The red blood cells retained significant levels of radioactivity for more than 5 days after administration whereas plasma levels declined sharply. >50% of the radioactivity in erythrocytes was detected covalently bound to cytoplasmic haemoglobin and membrane proteins. ≈13% of the administered dose was recovered as thiocyanate in the plasma and urine. The authors concluded that toxicity of methacrylonitrile may be attributable to the whole molecule and not to *in vivo* liberation of cyanide (12).

**Irritancy**

Dermal rabbit (24 hr) 500 mg caused mild irritation. 500 mg instilled into rabbit eye for 24 hr caused mild irritation (13).

## Genotoxicity

*Salmonella typhimurium* TA97, TA98, TA100, TA1535, TA1537 with and without metabolic activation negative (14).
*Drosophila melanogaster* sex-linked recessive lethal assay negative (15).

## Any other adverse effects to man

Whole body exposure to 24 ppm for 1 min caused nose, throat and eye irritation among 6-22% of exposed subjects. Concentrations of ≤14 ppm caused no effects (16).

## Any other adverse effects

Inhalation exposure of laboratory animals to 0.2-22.5 mg m$^{-3}$ was toxic to the kidney and central nervous system (17).

## Any other comments

Physical properties, use, mammalian toxicity and health precautions reviewed (18-20).

## References

1. Lyman, W. K. et al *Handbook of Chemical Property Estimation Methods Environmental Behaviour of Organic Compounds* 1982, McGraw-Hill, New York
2. Nawaz, M. S. et al *Arch. Microbiol.* 1991, **56**(3), 231-238
3. Atkinson, R. *J. Int. Chem. Kinet.* 1987, **19**, 799-828
4. *EXAMS II Computer Simulation* 1987, USEPA, Washington, DC
5. Smyth, M. D. et al *Am. Ind. Hyg. Assoc. J.* 1962, **23**, 95
6. *Juzen Igakkai Zasshi* 1985, **94**, 664
7. *Gig. Tr. Prof. Zabol.* 1985, **29**(5), 37
8. *Am. Ind. Hyg. Assoc. J.* 1968, **29**, 202
9. *J. Ind. Hyg. Toxicol.* 1949, **31**, 113
10. *Documentation of Threshold Limit Values and Biological Exposure Indices* 5th ed., 1986, 370, American Conference of Governmental Industrial Hygienists, Cincinnati, OH
11. Willhite, V. H. et al *Teratology* 1981, **23**, 317
12. Cavazos, R. et al *J. Appl. Toxicol.* 1989, **9**(1), 53-57
13. Marhold, J. V. *Prehled Prumyslove Toxikologie: Organicke Latky* 1986, Prague
14. Zeiger, E. et al *Environ. Mutagen.* 1987, **9**(Suppl. 9), 1-109
15. Zimmering, S. et al *Environ. Mol. Mutagen.* 1989, **14**(4), 245-251

16. McOmie, W. A. *J. Ind. Hyg. Toxicol.* 1949, **31**, 113
17. Kurzaliev, S. A. et al *Gig. Tr. Prof. Zabol.* 1988, (4), 51-52
18. *ECETOC Technical Report No. 30(4)* 1991, European Chemical Industry Ecology and Toxicology Centre, B-1160 Brussels
19. Farooqui, M. Y. H. et al *Toxicology* 1991, **65**(3), 239-250
20. *Chemical Safety Data Sheets* 1991, **46**, 32-34, RSC, London

# M98  Methacryloyl chloride

## $CH_2=C(CH_3)C(O)Cl$

**CAS Registry No.** 920-46-7
**Synonyms** methacrylic chloride; 2-methyl-2-propenoyl chloride
**Mol. Formula** $C_4H_5ClO$            **Mol. Wt.** 104.54
**Uses** Acylating agent. Organic synthesis. Manufacture of polymers used for contact lenses.

## Physical properties

**B. Pt.** 95-96°C; **Flash point** 2°C; **Specific gravity** $d_4^{20}$ 1.0871; **Volatility** v.p. 78 mmHg at 20°C.

## Solubility
Organic solvent: acetone, chloroform, diethyl ether

## Occupational exposure
**UK Long-term limit** 1 ppm (3 mg m$^{-3}$).

## Mammalian and avian toxicity

### Acute data
$LC_{50}$ (4 hr) inhalation rat 60 mg m$^{-3}$ (1).
$LC_{50}$ (2 hr) inhalation mouse 115 mg m$^{-3}$ (1).

### Any other adverse effects to man
Extremely destructive to tissue of mucous membranes and upper respiratory tract, eyes and skin. Inhalation may be fatal as a result of spasm, inflammation and oedema of the larynx and bronchi, chemical pneumonitis and oedema (2).

## References
1. Izmerov, N. F. et al *Toxicometric Parameters of Industrial Toxic Chemicals under Single Exposure* 1982, CIP, Moscow
2. Lenga, R. E. (Ed.) *Sigma-Aldrich Library of Chemical Safety Data* 2nd ed., 1988, **2**, 2227, Sigma-Aldrich, Milwaukee, WI

# M99    Methacryloyloxyethyl isocyanate

$$CH_2=C(CH_3)CO_2CH_2CH_2NCO$$

**CAS Registry No.** 30674-80-7
**Synonyms** 2-isocyanatoethyl methacrylate
**Mol. Formula** $C_7H_9NO_3$                              **Mol. Wt.** 155.15
**Uses** Preparation of polymers.

## Physical properties
**B. Pt.** $_{10}$ 87-89°C.

## Mammalian and avian toxicity

**Acute data**
$LD_{50}$ oral rat 670 mg $kg^{-1}$ (1).
$LC_{50}$ (6 hr) inhalation rat 4 ppm (1).

**Sub-acute data**
Inhalation guinea pig (2 wk) 0.01-0.6 ppm caused a significant increase in respiratory rate at 0.1-0.4 ppm; 0.5-0.6 ppm caused irritation responses (2).

**Teratogenicity and reproductive effects**
Inhalation ♂ rat, lowest toxic concentration, 80 ppb for 6 hr $day^{-1}$ for 49 days, reproductive effects (1).

## Genotoxicity
*In vivo* rat, dominant lethal assay negative (1).

## References
1.  Murray, J. S. et al *Drug Chem. Toxicol.* 1981, **3**, 381
2.  Mullin, L. S. et al *Toxicol. Appl. Pharmacol.* 1983, **71**(1), 113-122

# M100    Methallyl alcohol

$$CH_2=C(CH_3)CH_2OH$$

**CAS Registry No.** 513-42-8
**Synonyms** isopropenyl carbinol; 2-methylallyl alcohol; 2-methyl-2-propen-1-ol; 3-hydroxy-2-methylpropene
**Mol. Formula** $C_4H_8O$                              **Mol. Wt.** 72.11
**Uses** Organic synthesis.

## Physical properties
**B. Pt.** 113-115°C; **Flash point** 33°C; **Specific gravity** $d_4^{20}$ 0.852; **Volatility** v. den. 2.5.

## Solubility

Water: miscible. Organic solvent: diethyl ether, ethanol

## Occupational exposure

**UN No.** 2614; **HAZCHEM Code** 2P; **Conveyance classification** flammable liquid.

## Mammalian and avian toxicity

### Acute data

$LD_{Lo}$ oral mouse 500 mg kg$^{-1}$ (1).
$LC_{Lo}$ (2 hr) inhalation mouse 2900 ppm (1).
$LD_{Lo}$ dermal rabbit 2000 mg kg$^{-1}$ (1).

### Irritancy

Dermal rabbit, 500 mg caused moderate irritation (exposure not specified) (1).

## References

1.  Shell Chemical Co. *Report* 1961, 6

# M101   Methamidophos

## H2NP(O)(OCH3)SCH3

**CAS Registry No.** 10265-92-6
**Synonyms** *O,S*-dimethyl phosphoramidothioate; Aceptate-met; Filitox; Monitor;
methyl phosphoramidothioate; Tamaron
**Mol. Formula** $C_2H_8NO_2PS$                    **Mol. Wt.** 141.13
**Uses** Insecticide.

## Physical properties

**M. Pt.** 46.1°C; **Specific gravity** d$^{44.5}$ 1.31; **Volatility** v.p. $1.7 \times 10^{-5}$ mmHg at 20°C.

## Solubility

Water: >2 kg l$^{-1}$. Organic solvent: benzene, dichloromethane, ethanol, diethyl ether,
*n*-hexane, isopropanol, kerosene, xylene

## Occupational exposure

**Supply classification** very toxic.
**Risk phrases** Toxic in contact with skin – Very toxic if swallowed – Irritating to eyes
(R24, R28, R36)
**Safety phrases** Do not breathe dust – After contact with skin, wash immediately with
plenty of soap and water – Wear suitable protective clothing and gloves – In case of
accident or if you feel unwell, seek medical advice immediately (show label where
possible) (S22, S28, S36/37, S45)

## Ecotoxicity

### Fish toxicity

$LC_{50}$ (96 hr) guppy, rainbow trout, carp, goldfish 46-100 mg l$^{-1}$ (1).

**Invertebrate toxicity**
$IC_{50}$ (24 hr) *Daphnia magna* 1.88 mg $l^{-1}$; $LC_{50}$ (24 hr) *Daphnia magna* 56 mg $l^{-1}$ (2).

**Abiotic removal**
Hydrolysis $t_{1/2}$ 120 hr at 37°C, pH 9.0; 140 hr at 40°C, pH 2.0 (1).

## Mammalian and avian toxicity

**Acute data**
$LD_{50}$ oral rat, mouse, dog, rabbit, bobwhite quail 10-60 mg $kg^{-1}$ (1,3-5).
$LC_{50}$ (4 hr) inhalation rat 200 mg $m^{-3}$ (aerosol) (1).
$LD_{50}$ dermal rat, rabbit 118-130 mg $kg^{-1}$ (1, 6).
$LD_{50}$ dermal rat 50 mg $kg^{-1}$ (7).
$LD_{50}$ intraperitoneal rat 15 mg $kg^{-1}$ (8).

**Sub-acute data**
Oral rat (12 wk) the maximal dose which did not inhibit blood cholinesterase activity was 0.17 mg $kg^{-1}$ $day^{-1}$ (9).

**Carcinogenicity and long-term effects**
Oral rat and dog (2 yr) no adverse effect level 2 mg $kg^{-1}$ diet (1).

**Teratogenicity and reproductive effects**
Oral, single dose of 1 or 2 mg $kg^{-1}$ to pregnant rats caused embryolethality, growth retardation, encephaly and anotia. Visceral organs were not affected. The high dose caused cyanosis (time of administration not specified) (3).
Gavage ♂ mouse, 0.2, 0.8 or 3.2 mg $kg^{-1}$ $day^{-1}$ for 5 days. On day 35 sperm motility decreased and abnormal sperm rate increased markedly, the structure of mitochondria and smooth endoplasmic reticulum in Leydig cells changed, cells in convoluted tubules degenerated and interstitial tissue in the testes appeared oedematous. These effects were observed only in the mid and high dose groups (10).
Chick embryo 0.125-2.0 mg $egg^{-1}$ on day 4 of incubation caused embryo lethality, growth retardation and a low incidence of developmental anomalies, open umbelicus and thin debilitated toes and feet (11).

## Genotoxicity
*Salmonella typhimurium* TA98, TA100 with and without metabolic activation negative (12).
*In vivo* mouse bone marrow cells, micronucleus test and chromosomal aberrations positive (13).
*Vicia Faba* root tip cells, mitotic index significantly decreased and chromosomal aberrations induced (14).

## Any other adverse effects to man
Occupational exposure to ≤0.13 mg $m^{-3}$ during greenhouse spraying caused an 11% decrease in erythrocyte acetylcholinesterase activity (15).

## Legislation
EEC maximum residue limits for pome, stone and citrus fruit 0.3 ppm (1).
Limited under EC Directive on Drinking Water Quality 80/778/EEC. Pesticides: maximum admissible concentration 0.1 µg $l^{-1}$ (16).

Included in Schedule 6 (Release into Land: Prescribed Substances) Statutory Instrument No. 472, 1991 (17).

**Any other comments**

WHO Class Ib; EPA Toxicity Class I (1).
TDI (human) 0.6 µg kg$^{-1}$ (1).

**References**

1. *The Agrochemicals Handbook* 3rd ed., 1993, RSC, London
2. Xiu, R. et al *Zhonguo Huanjing Kexue* 1988, **8**(6), 38-41, (Ch.) (*Chem. Abstr.* **111**, 72800u)
3. Henafy, M. S. M. et al *Vet. Med. J.* 1986, **34**(3), 357-363
4. *USDA Inf. Mem.* 1966, **20**, 7
5. Gaines, T. B. et al *Fundam. Appl. Toxicol.* 1986, **7**, 299
6. *Guide to Chemicals Used in Crop Protection* 1973, **6**, 333
7. *Pesticide Index* 1976, **5**, 149
8. *Pestic. Biochem. Physiol.* 1980, **13**, 267
9. Guo, L. et al *Zhonghua Laodong Weisheng Zhiyebing Zazhi* 1986, **4**(5), 257-260 (Ch.) (*Chem. Abstr.* **108**, 133427y)
10. Lun, X. et al *Zhongguo Yaolixue Yu Dulixe Zazhi* 1988, **2**(2), 142-147, (Ch.) (*Chem. Abstr.* **109**, 68525z)
11. Hanafy, M. S. M. et al *Vet. Med. J.* 1986, **34**(3), 347-356
12. Guo, L. et al *Zhonghua Laodong Weisheng Zhiyebing Zazhi* 1986, **4**(5), 261-263, (Ch.) (*Chem. Abstr.* **106**, 133533e)
13. Amer, S. M. et al *Z. Naturforsch. C: Biosci.* 1987, **42**(1-2), 21-30 (Eng.) (*Chem. Abstr.* **106**, 170758c)
14. Adam, Z. M. et al *Cytologia* 1990, **55**(3), 349-355
15. Wagner, R. et al *Z. Gesammte Hyg. Ihre Grenzgeb.* 1987, **33**(5), 255-257, (Ger.) (*Chem. Abstr.* **106** 170758c)
16. *EC Directive Relating to the Quality of Water Intended for Human Consumption* 1982, 80/778/EEC, Office for Official Publications of the European Communities, 2 rue Mercier, L-2985 Luxembourg
17. *S. I. 1991 No. 472 The Environmental Protection (Prescribed Processes and Substances) Regulations* 1991, HMSO, London

# M102   Methane

$CH_4$

**CAS Registry No.** 74-82-8
**Synonyms** fire damp; marsh gas; methyl hydride
**Mol. Formula** $CH_4$                          **Mol. Wt.** 16.04
**Uses** Fuel. Organic synthesis.
**Occurrence** Found during anaerobic degradation of organic matter. Principal constituent of natural gas.

**Physical properties**

**M. Pt.** −183°C; **B. Pt.** −161°C; **Flash point** −183.2°C; **Specific gravity** $d_4^0$ 0.554;
**Partition coefficient** log $P_{ow}$ 1.09 (1); **Volatility** v.p. 40 mmHg at −86.3°C; v. den. 0.55.

**Solubility**
Water: 25 mg $l^{-1}$. Organic solvent: diethyl ether, ethanol

## Occupational exposure

**UN No.** 1971 (compressed); 1972 (refrigerated liquid); **HAZCHEM Code** 2SE comp; 2WE rliq; **Conveyance classification** flammable gas; **Supply classification** extremely flammable.
**Risk phrases** Extremely flammable (R12)
**Safety phrases** Keep container in a well ventilated place – Keep away from sources of ignition – No Smoking – Take precautionary measures against static discharges (S9, S16, S33)

## Environmental fate

### Degradation studies
ThOD 3.99 mg $l^{-1}$ $O_2$; $BOD_{35}$ 3.04 mg $l^{-1}$ $O_2$ at 25°C (2).
Utilised as sole carbon source by *Methylococcus sp.* (3,4).
Microbial degradation in soils, estimated $t_{1/2}$ 70 days (3).

### Abiotic removal
Reaction with photochemically produced hydroxyl radicals in the atmosphere $t_{1/2}$ 1900 days (5).
Estimated volatilisation $t_{1/2}$ 1.17 hr for model river water and 14 hr for model pond water (6,7).

### Absorption
Calculated $K_{oc}$ of 753 indicates that adsorption to soil may occur (6).

## Mammalian and avian toxicity

### Teratogenicity and reproductive effects
Inhalation mouse, 5-8% fuel gas (containing 85% methane with small amounts of ethane, propane and butane) for 1 hr on day 8 of gestation. Abnormalities in the foetal brain were found to result in brain hernia and hydrocephalus (8).

### Metabolism and pharmacokinetics
Absorbed via the lungs in mammals. When inhaled, the major proportion is exhaled unchanged. A small amount is metabolised to methanol (1,3).

## Any other adverse effects to man

Acts as a simple asphyxiant (1).
Liquified methane gas causes frost bite on skin contact (1).

## Any other comments

Physical properties, use, mammalian toxicity and health hazards reviewed (1,9, 10).
Autoignition temperature 650°C.

## References

1. *Ethel Browning's Toxicity and Metabolism of Industrial Solvents* 2nd ed., 1987, **1**, 257, Elsevier, New York
2. Gatallier, C. R. *Chimie et Industrie-Genie Chimique* 1971, **104**
3. Laurence, A. J. et al *Biochem. J.* 1970, **116**, 631
4. Fuerst, R. et al *Dev. Ind. Microbiol.* 1970, **11**, 301

5. Atkinson, R. *Chem. Rev. (Washington, D. C.)* 1985, **85**, 69-201
6. Lyman, W. K. et al *Handbook of Chemical Property Estimation Methods Environmental Behaviour of Organic Compounds* 1982, McGraw-Hill, New York
7. *EXAMS II Computer Simulation* 1987, US EPA, Washington, DC
8. Shepard, T. H. *Catalog of Teratogenic Agents* 5th ed., 1986, 371, Johns Hopkins University Press, Baltimore, MD
9. *Chemical Safety Data Sheets* 1992, **5**, 175-178, RSC, London
10. *ECETOC Technical Report No. 30(4)* 1991, European Chemical Industry Ecology and Toxicology Centre, B-1160 Brussels

# M103   Methanesulfonic acid

## $CH_3SO_3H$

**CAS Registry No.** 75-75-2
**Synonyms** methylsulfonic acid
**Mol. Formula** $CH_4O_3S$                    **Mol. Wt.** 96.11
**Uses** Catalyst. Organic synthesis. Solvent.

## Physical properties

**M. Pt.** 20°C; **B. Pt.** $_{10}$ 167°C; **Flash point** >110°C; **Specific gravity** $d_4^{10}$ 1.4812.

## Solubility

Water: miscible. Organic solvent: diethyl ether, ethanol, benzene, ethyl disulfide, chlorotoluene, toluene

## Occupational exposure

**Supply classification** corrosive.
**Risk phrases** Causes burns (R34)
**Safety phrases** In case of contact with eyes, rinse immediately with plenty of water and seek medical advice – Wear suitable protective clothing (S26, S36)

## Mammalian and avian toxicity

**Acute data**
$LD_{Lo}$ oral rat 200 mg kg$^{-1}$ (1).
$LD_{50}$ oral quail 1000 mg kg$^{-1}$ (2).
$LD_{Lo}$ intraperitoneal rat 50 mg kg$^{-1}$ (1).

## Genotoxicity

*Salmonella typhimurium* TA100, TA1535 with and without metabolic activation negative (3).

## Any other adverse effects to man

Extremely destructive to tissues of the mucous membranes and upper respiratory tract. Inhalation may be fatal as a result of spasm, inflammation and oedema of the larynx and bronchi, chemical pneumonitis and pulmonary oedema (4).

## Legislation

Included in Schedule 6 (Release into Land: Prescribed Substances) Statutory Instrument No. 472, 1991 (5).

## Any other comments

Human health effects, experimental toxicology, physico-chemical properties reviewed (6).

## References

1. *Kodak Company Report* 21 May 1971
2. Schafer, E. W. et al *Arch. Environ. Contam. Toxicol.* 1983, **12**, 355-382
3. Zeiger, E. et al *Environ. Mol. Mutagen.* 1989, **13**(4), 343-346
4. Lenga, R. E. (Ed.) *Sigma-Aldrich Library of Chemical Safety Data* 2nd ed., 1988, **2**, 2229, Sigma-Aldrich, Milwaukee, WI
5. *S. I. 1991 No. 472 The Environmental Protection (Prescribed Processes and Substances) Regulations* 1991, HMSO, London
6. *ECETOC Technical Report No. 30(4)* 1991, European Chemical Industry Ecology and Toxicology Centre, B-1160 Brussels

# M104    Methanesulfonyl fluoride

## $CH_3SO_2F$

**CAS Registry No.** 558-25-8
**Synonyms** Fumette; mesyl fluoride; MSF; chloromethylsulfone
**Mol. Formula** $CH_3FO_2S$                                   **Mol. Wt.** 98.10
**Uses** Formerly used as a fumigant insecticide.

## Physical properties

**B. Pt.** 123-124°C; **Specific gravity** $d_4^{18}$ 1.4805.

**Solubility**
Organic solvent: diethyl ether, ethanol

## Mammalian and avian toxicity

### Acute data

$LD_{50}$ oral rat 2 mg $kg^{-1}$ (1).
$LD_{Lo}$ subcutaneous rat, mouse, dog, rabbit 3.5 mg $kg^{-1}$ (2,3).
$LD_{50}$ intraperitoneal rat 3 mg $kg^{-1}$ (4).
$LD_{50}$ intravenous mouse, rabbit, dog 0.3-1.0 mg $kg^{-1}$ (1).

## Any other adverse effects to man

Extremely destructive to tissues of the mucous membranes and upper respiratory tract, eyes and skin. Inhalation may be fatal as a result of spasm, inflammation and oedema of the larynx and bronchi, chemical pneumonitis and pulmonary oedema (5).

---

### Any other adverse effects

Inhibited central nervous system cholinesterase activity selectively in monkeys (6).

### Legislation

Limited under EC Directive on Drinking Water Quality 80/778/EEC. Pesticides: maximum admissible concentration 0.1 $\mu g\ l^{-1}$ (7).
Included in Schedule 6 (Release into Land: Prescribed Substances) Statutory Instrument No. 472, 1991 (8).

### References

1. *Interagency Collaborative Group on Environmental Carcinogenesis, Memorandum* 17 June 1974, Natl. Cancer Institute
2. *Pesticide Index* 1969, **4**, 271
3. *The Pesticide Manual* 1968, **1**, 287, British Crop Protection Council, Worcestershire
4. *Nature (London)* 1954, **173**, 33
5. Lenga, R. E. (Ed.) *Sigma-Aldrich Library of Chemical Safety Data* 2nd ed., 1988, **2**, 2230, Sigma-Aldrich, Milwaukee, WI
6. Moss, D. E. et al *Curr. Res. Alzheimer Ther. Cholinesterase Inhib.* 1988, 305-314, Taylor and Francis, New York
7. *EC Directive Relating to the Quality of Water Intended for Human Consumption* 1982, 80/778/EEC, Office for Official Publications of the European Communities, 2 rue Mercier, L-2985 Luxembourg
8. *S. I. 1991 No. 472 The Environmental Protection (Prescribed Processes and Substances) Regulations* 1991, HMSO, London

# M105　Methanethiol

## $CH_3SH$

**CAS Registry No.** 74-93-1
**Synonyms** mercaptomethane; methyl mercaptan; methyl sulfhydrate; thiomethyl alcohol
**Mol. Formula** $CH_4S$           **Mol. Wt.** 48.11
**Uses** Organic synthesis. Odourant for hazardous gases.
**Occurrence** Aroma component of cooked meat, fish and cheeses. Degradation product in spoilt foods. Food metabolite in mammals. Occurs in fossil fuels.

### Physical properties

**M. Pt.** $-123°C$; **B. Pt.** $6°C$; **Flash point** $-18°C$ (open cup); **Specific gravity** $d_4^{20}$ 0.8665; **Volatility** v.p. 1520 mmHg at 26.1°C; v. den. 1.66.

### Solubility

Water: 23.3 g $l^{-1}$ at 20°C. Organic solvent: diethyl ether, ethanol

### Occupational exposure

**US TLV (TWA)** 0.5 ppm (0.98 mg $m^{-3}$); **UK Long-term limit** 0.5 ppm (1.0 mg $m^{-3}$);
**Supply classification** highly flammable, harmful.
**Risk phrases** Extremely flammable liquefied gas – Harmful by inhalation (R13, R20)

**Safety phrases** Keep away from sources of ignition – No Smoking – Avoid contact with eyes (S16, S25)

## Environmental fate

### Degradation studies
Degraded by *Thiobacillus thioparus* when isolated from peat (1).
Readily metabolised in estuarine and fresh water sediments (2).

### Abiotic removal
Reaction with photochemically produced hydroxyl radicals in the atmosphere, estimated $t_{1/2}$ 11.6 hr (3).
Reaction under photochemical smog conditions, forming formaldehyde, sulfur dioxide, methyl nitrate and inorganic sulfate, $t_{1/2}$ 2 hr (4,5).
Estimated volatilisation $t_{1/2}$ 2 hr in model river water (6).

## Mammalian and avian toxicity

### Acute data
$LC_{50}$ (2 hr) inhalation mouse 6.5 mg m$^{-3}$ (7).

### Sub-acute data
Inhalation rat, (3 month) 2, 20 or 60 ppm. Histopathological examination revealed liver damage. A dose-related reduction in body weight was reported (8).

### Metabolism and pharmacokinetics
Metabolised in rats to carbon dioxide and sulfate. The sulfate was excreted in the urine and 94% of the sulfur from methanethiol was eliminated in 24 hr (9).

## Any other adverse effects to man

Extremely destructive to tissues of the mucous membranes and upper respiratory tract, eyes and skin. Inhalation may be fatal as a result of spasm, inflammation and oedema of the larynx and bronchi, chemical pneumonitis and pulmonary oedema (10).

## Any other adverse effects

Methanethiol strongly inhibited rat liver mitochondrial respiration, apparently by reacting with cytochrome *c* oxidase (11).

## Any other comments

Physical properties, use, mammalian toxicity and health precautions reviewed (12,13).

## References

1. Cho, K. S. et al *J. Ferment. Bioeng.* 1991, **71**(6), 384-389
2. Keine, R. P. et al *Appl. Environ. Microbiol.* 1986, **52**, 1037-1045
3. Atkinson, R. *Chem. Rev. (Washington, D. C.)* 1985, **85**, 69-201
4. Sickles, J. E. et al *Proc. Air Pollut. Central Assoc.* 1985, **4**, 1-15
5. Grosjean, D. *Environ. Sci. Technol.* 1984, **18**, 460-480
6. Lyman, W. K. et al *Handbook of Chemical Property Estimation Methods Environmental Behaviour of Organic Compounds* 1982, McGraw-Hill, New York
7. *Gig. Tr. Prof. Zabol.* 1972, **16**(6), 46
8. Tansy, M. F. et al *J. Toxicol. Environ. Health* 1981, **8**(1-2), 71-88
9. Derr, R. F. et al *Res. Comm. Chem. Pathol. Pharmacol.* 1983, **39**(3), 503-506
10. Lenga, R. E. (Ed.) *Sigma-Aldrich Library of Chemical Safety Data* 2nd ed., 1988, **2**, 2230, Sigma-Aldrich, Milwaukee, WI

11.   Waller, R. L. *Toxicol. Appl. Pharmacol.* 1977, **42**(1), 111-118
12.   *Chemical Safety Data Sheets* 1992, **5**, 179-181, RSC, London
13.   *ECETOC Technical Report No. 30(4)* 1991, European Chemical Industry Ecology and Toxicology Centre, B-1160 Brussels

# M106   Methanol

## $CH_3OH$

**CAS Registry No.** 67-56-1
**Synonyms** carbinol; colonial spirit; methyl alcohol; methyl hydroxide; methylol; pyroxylic spirit; wood naphtha; wood spirit; wood alcohol
**Mol. Formula** $CH_4O$                                      **Mol. Wt.** 32.04
**Uses** Solvent. Antifreeze. Fuel and fuel additive. Organic synthesis.
**Occurrence** Occurs naturally in some woods. Traces have been identified in engine exhausts and cigarette smoke (1).

## Physical properties

**M. Pt.** −98°C; **B. Pt.** 64.6°C; **Flash point** 12°C (closed cup); **Specific gravity** $d_4^{20}$ 0.7915; **Partition coefficient** log $P_{ow}$ -0.77 (2); **Volatility** v.p. 100 mmHg at 21.2°C; v. den. 1.11.

### Solubility

Water: miscible. Organic solvent: acetone, diethyl ether, ethanol, benzene, chloroform

## Occupational exposure

**US TLV (TWA)** 200 ppm (262 mg m$^{-3}$); **US TLV (STEL)** 250 ppm (328 mg m$^{-3}$); **UK Long-term limit** 200 ppm (260 mg m$^{-3}$); **UK Short-term limit** 250 pm (310 mg m$^{-3}$); **UN No.** 1230; **HAZCHEM Code** 2PE; **Conveyance classification** flammable gas, toxic substance; **Supply classification** highly flammable, toxic (≥20%).
**Risk phrases** ≥20% – Highly flammable – Toxic by inhalation and if swallowed – ≥3%<20% – Highly flammable – Harmful by inhalation and if swallowed (R11, R23/25, R11, R20/22)
**Safety phrases** Keep out of reach of children – Keep container tightly closed – Keep away from sources of ignition – No Smoking – Avoid contact with skin (S2, S7, S16, S24)

## Ecotoxicity

### Fish toxicity

$LC_{50}$ (48 hr) trout 8000 mg l$^{-1}$ (3).

### Invertebrate toxicity

$LC_{50}$ (24 hr) brine shrimp 10,000 mg l$^{-1}$ (3).
$EC_{50}$ (30 min) *Photobacterium phosphoreum* 51,000-320,000 ppm, Microtox test (4).

### Bioaccumulation

Bioconcentration factor for golden ide <10 (5).

## Environmental fate

### Nitrification inhibition
$IC_{50}$ ammonic oxidation by *Nitrosomonas* 160 mg l$^{-1}$ (exposure not specified) (6).
Metabolised by the marine ammonia oxidising bacterium *Nitrococcus oceanus* with the liberation of $CO_2$ (7).

### Degradation studies
ThOD 1.5 mg l$^{-1}$ $O_2$; $BOD_5$ 76% $O_2$ of ThOD (4).
Under anaerobic conditions traces of carbon monoxide were formed together with methane by activated sludge inoculum (8).
Biodegradation in anaerobic groundwater systems was enhanced in the presence of nitrate at pH ≥7.0 (9).

### Abiotic removal
Reaction with photochemically produced hydroxyl radicals in the atmosphere, forming formaldehyde, estimated $t_{1/2}$ 18 days (10).
Volatilisation from model river water $t_{1/2}$ 5.3 hr, and from pond $t_{1/2}$ 2.6 days (11, 12).
Adsorption by activated carbon 7 mg g$^{-1}$ carbon (13).

## Mammalian and avian toxicity

### Acute data
$LD_{50}$ oral rat, mouse 5600, 7300 mg kg$^{-1}$, respectively (14,15).
$LC_{50}$ (4 hr) inhalation rat 64,000 ppm (16).
$LD_{50}$ intraperitoneal rat 7500 mg kg$^{-1}$ (17).
$LD_{50}$ intravenous rat, mouse 2100, 4700 mg kg$^{-1}$, respectively (17).

### Carcinogenicity and long-term effects
Inhalation rat (24 months) 1300 mg m$^{-3}$ and mouse (18 months) 13,000 mg m$^{-3}$. A preliminary report indicated no evidence of carcinogenicity (18).

### Teratogenicity and reproductive effects
Inhalation rat, lowest toxic concentration 20,000 ppm for 7 hr day$^{-1}$ on days 1-22 of gestation, teratogenic effects (musculoskeletal system, cardiovascular system and urogenital system) (19).
Inhalation rat, lowest toxic concentration 10,000 ppm for 7 hr day$^{-1}$ on days 7-18 of gestation, foetotoxic effects (20).

### Metabolism and pharmacokinetics
Readily absorbed from the gastrointestinal tract. It may also be absorbed by inhalation and through the skin. Oxidation by alcohol dehydrogenase with the formation of formaldehyde and formic acid takes place mainly in the liver and in the kidneys. These metabolites may be excreted in the urine or further metabolised to carbon dioxide and exhaled by the lungs (21).

### Irritancy
Dermal rabbit (24 hr) 500 mg caused moderate irritation (22).
100 mg instilled into rabbit eye for 24 hr caused moderate irritation (23).

## Genotoxicity
*Salmonella typhimurium* TA98, TA100, TA1535, TA1537, TA1538, with and without metabolic activation negative (24).

---

*Escherichia coli* WP2, WP67, CM871, with and without metabolic activation positive (24).
*In vitro* mouse lymphoma L5178Y, tk$^+$/tk$^-$ forward mutation assay positive (25).
*In vitro* Chinese hamster V79 lung fibroblasts, induction of micronuclei negative (26).
*In vivo* mouse, increase in micronuclei in blood and lung cells, sister chromatid exchanges and chromosomal aberrations in lung cells, synaptinemal complex damage in spermatocytes negative (27).

## Any other adverse effects to man

The outstanding features of methanol poisoning are metabolic acidosis with rapid, shallow breathing, visual disturbances which may lead to irreversible blindness, and severe abdominal pain. Ingestion of ≈30 ml may be fatal (28).
Epidemiological studies have demonstrated no increased incidence of cancer among exposed workers (29).

## Legislation

Included in Schedule 6 (Release into Land: Prescribed Substances) Statutory Instrument No. 472, 1991 (30).

## Any other comments

Environmental fate reviewed (1).
The metabolites of methanol (formaldehyde and formic acid) are believed to be responsible for the symptoms of poisoning (28).
Physical properties, use, mammalian toxicity and health precautions reviewed (31-36).
Autoignition temperature 470°C.

## References

1. Howard, P. H. *Handbook of Environmental Fate and Exposure Data for Organic Chemicals* 1991, **2**, 310-316, Lewis Publishers, Chelsea, MI
2. Hansch, C. et al *Crit. Rev. Toxicol.* 1989, **19**, 185-226
3. Price, K. S. et al *J.- Water Pollut. Control Fed.* 1974, **46**, 1
4. Kaiser, K. L. E. et al *Water Pollut. Res. J. Can.* 1991, **26**(3), 361-431
5. Freitag, D. et al *Chemosphere* 1985, **14**, 1589-1616
6. Hooper, A. *J. Bacteriol.* 1973, **115**, 480
7. Ward, B. B. *Arch. Microbiol.* 1987, **147**(2), 126-133
8. Hickey, R. F. et al *Biotechnol. Lett.* 1987, **9**(1), 63-66
9. Wilson, W. G. et al *Proc. Ind. Waste Conf.* 1987, **42nd**, 197-205
10. Atkinson, R. A. *Chem. Rev. (Washington, D. C.)* 1985, **85**, 60-201
11. Lyman, W. K. et al *Handbook of Chemical Property Estimation Methods Environmental Behaviour of Organic Compounds* 1982, McGraw-Hill, New York
12. *EXAMS II Computer Simulations* 1987, US EPA, Washington, DC
13. Guisti, D. M. et al *J.- Water Pollut. Control Fed.* 1974, **46**(5), 947-965
14. *Gig. Tr. Prof. Zabol.* 1975, **19**(11), 27
15. *Toxicology* 1982, **25**, 271
16. *Raw Material Data Handbook* 1974, **1**, 74
17. *Environ. Health Perspect.* 1985, **61**, 321
18. Katah, M. *Methanol Vapours and Health Effects, Workshop Summary Report* 1989, A7, ILSI Risk Science Institute, Washington, DC
19. *Teratology* 1984, **29**(2), 48A
20. *Fundam. Appl. Toxicol.* 1985, **5**, 727
21. Eells, J. T. et al *J. Pharmacol. Exp. Ther.* 1983, **227**, 349-353

22. Marhold, J. V. *Sbornik Vysledku Toxixologickeho Vysetreni Latek A Pripravku* 1972, Prague
23. Marhold, J. V. *Prehled Prumyslove Toxikologie: Organicke Latky* 1986, Prague
24. De Flora, S. et al *Mutat. Res.* 1984, **133**(3), 161-198
25. McGregor, D. B. et al *Environ. Mutagen.* 1985, **7**(Suppl. 3), 10
26. Gu, Z. et al *Waisheng Dulixne Zazhi* 1988, **2**(1), 1-4. (Ch.) (*Chem. Abstr.* **112**, 93613w)
27. Campbell, J. A. et al *Mutat. Res.* 1991, **260**(3), 257-264
28. *Martindale. The Extra Pharmacopoeia* 30th ed., 1103-1104, Pharmaceutical Press, London
29. Liemiatyeki, J. *Teratog., Carcinog., Mutagen.* 1982, **2**(2), 169-177
30. *S. I. 1991 No. 472 The Environmental Protection (Prescribed Processes and Substances) Regulations* 1991, HMSO, London
31. *Toxicological Data Sheets No. 5, Cah. Notes Doc.* 1990, **138**, 217-221
32. *Proc. APCA Ann. Meet.* 1988, **81st**(2), Paper 88/41.2
33. *Comm. Eur. Communities [Rep.] Methanol* 1988, EUR11553, 157-186
34. *Chemical Safety Data Sheets* 1989, **1**, 208-210, RSC, London
35. *Ethel Browning's Toxicology and Metabolism of Industrial Solvents* 2nd ed., 1992, **3**, 3-20
36. *ECETOC Technical Report No. 30(4)* 1991, European Chemical Industry Ecology and Toxicology Centre, B-1160 Brussels

# M107    Methapyrilene

**CAS Registry No.** 91-80-5

**Synonyms** 2-[(2-dimethylaminoethyl)-2-thenylamino]pyridine; *N,N*-dimethyl-*N'*-2-pyridinyl-*N'*-(2-thienylmethyl)-1,2-ethanediamide; Dormin; Histadyl; Lullamin; paradormalene; pyrathin; pyrinistol; Restinyl; Semikon; Tenalin; thenylpyramine; Thionylan

**Mol. Formula** $C_{14}H_{19}N_3S$                **Mol. Wt.** 261.39

**Uses** Antihistamine drug.

## Physical properties

**B. Pt.** $_3$ 173-175°C; **Flash point** 880 mg l$^{-1}$; **Specific gravity** log $P_{ow}$ 2.74 (calc.) (1); **Volatility** v.p. $7.0 \times 10^{-4}$ mmHg at 25°C.

## Solubility

Water: 879 mg l$^{-1}$.

## Ecotoxicity

### Bioaccumulation

Calculated bioaccumulation factor of 71 (2).

## Environmental fate

### Absorption
Estimated $K_{oc}$ of 740 indicates that adsorption to soil would be significant (2).

## Mammalian and avian toxicity

### Acute data
$LD_{50}$ oral mouse, guinea pig 180-380 mg $kg^{-1}$ (3).
$LD_{50}$ intravenous mouse 20 mg $kg^{-1}$ (3).
$LD_{50}$ intraperitoneal mouse 77 mg $kg^{-1}$ (4).
$LD_{50}$ subcutaneous rat 150 mg $kg^{-1}$ (5).

### Metabolism and pharmacokinetics
Following intravenous administration of 0.7 or 3.5 mg $kg^{-1}$ to ♂ rats, 40 and 35% of the administered dose was excreted in the urine, respectively and 38 and 44% was eliminated in the faeces, respectively within 24 hr. The major urinary metabolites were methapyrilene N-oxide; mono-N-desmethylmethapyrilene; and unchanged methapyrilene (6).
The major urinary metabolite identified in rats was (5-hydroxypyridyl) methapyrilene. After 4 wks treatment rats also excreted the 3- and 6-isomers.
N'-(2-pyridyl)-N,N-dimethylethylenediamine and its metabolite
N'-2-(5-hydroxypyridyl)-N,N-dimethylethylenediamine were also identified (7).

## Genotoxicity
*Salmonella typhimurium* TA1535, without metabolic activation, weakly positive (8).
*In vitro* Chinese hamster ovary cells, HGPRT assay negative (9).
*In vitro* mouse lymphoma $tk^+/tk^-$ forward mutation assay negative (10).
*In vitro* calf thymus, DNA binding occurred only after metabolic activation by rat liver microsomes and NADPH (11).

## Any other adverse effects to man
Overdoses produce excitement, convulsions, hyperpyrexia, cerebral oedema, depression and occasionally renal tubular necrosis. Death has been reported from an oral dose of 12 mg $kg^{-1}$ but others have survived 80 mg $kg^{-1}$ (12).

## Any other comments
Administered as the fumarate and hydrochloride (13).
Human health effects, experimental toxicology, physico-chemical properties reviewed (14).

## References
1. *GEMS: Graphical Exposure Modeling System. Fate of Atmospheric Pollutants* 1986, USEPA, Washington, DC
2. Lyman, W. K. et al *Handbook of Chemical Property Estimation Methods Environmental Behaviour of Organic Compounds* 1982, McGraw-Hill, New York
3. *Proc. Soc. Exp. Biol. Med.* 1952, **80**, 458
4. *J. Pharmacol. Exp. Ther.* 1949, **96**, 388
5. *C. R. Sceances Soc. Biol. Ses Fil.* 1950, **144**, 887
6. Kelly, D. W. et al *Drug Metab. Dispos.* 1990, **18**(6), 1018-1024
7. Kammer, R. C. et al *Xenobiotica* 1988, **18**(7), 869-881
8. Ashby, J. et al *Environ. Mol. Mutagen.* 1988, **12**(2), 243-252

9. Caseino, D. A. et al *Cancer Lett. (Shannon, Irel.)* 1984, **21**(5), 337-341
10. McGregor, D. B. et al *Environ. Mol. Mutagen.* 1991, **17**(3), 196-219
11. Lampe, M. A. et al *Carcinogenesis (London)* 1987, **8**(10), 1525-1529
12. Gosselin, R. E. et al *Clinical Toxicology of Commercial Products* 5th ed., 1984, **II**, 379, Williams & Wilkins, Baltimore, MD
13. *Martindale. The Extra Pharmacopoeia* 30th ed., 1993, 942, Pharmaceutical Press, London
14. *ECETOC Technical Report No. 30(4)* 1991, European Chemical Industry Ecology and Toxicology Centre, B-1160 Brussels

# M108   Methapyrilene hydrochloride

**CAS Registry No.** 135-23-9

**Synonyms** 2-[(2-(dimethylamino)ethyl)-2-thenylamino] pyridine hydrochloride; Barhist; Histidyl; Dozar; methacon; methoxylene; Semikon hydrochloride; thenylene hydrochloride; thenylpyramine hydrochloride

**Mol. Formula** $C_{14}H_{20}ClN_3S$          **Mol. Wt.** 297.85

**Uses** Antihistamine drug.

## Physical properties

**M. Pt.** 162°C.

**Solubility**
Water: 200% w/v. Organic solvent: chloroform, ethanol

## Mammalian and avian toxicity

### Acute data

$LD_{50}$ oral rat, mouse 180, 520 mg $kg^{-1}$, respectively (1, 2).
$LD_{50}$ subcutaneous rat, mouse 75, 150 mg $kg^{-1}$, respectively (2, 3).
$LD_{50}$ intravenous mouse, guinea pig 15-18 mg $kg^{-1}$ (3, 4).

### Carcinogenicity and long-term effects

Gavage rat single dose of 30, 100, 200 or 300 mg $kg^{-1}$. Rats were then fed 0.05% phenobarbital in diet for 3, 6 or 9 months. The number of altered hepatic foci was increased 2- to 4-fold for the highest dose, indicating that methapyrilene may act as a weak initiator of hepatocarcinogenesis (5).

Oral rat, 125 or 250 mg $kg^{-1}$ diet (exposure unspecified). The high dose induced liver carcinomas or neoplastic nodules in almost all rats, whereas the low dose induced neoplastic nodules in the liver of 4% animals (6).

### Metabolism and pharmacokinetics

Metabolites identified in mouse hepatocytes were methapyrilene glucuronide and desmethylmethapyrilene glucuronide (7).

## Genotoxicity

*Salmonella typhimurium* TA1535 without metabolic activation weakly positive (8).
*In vitro* mouse lymphoma L5178Y tk$^+$/tk$^-$ with metabolic activation positive (9).

## Any other comments

Human health effects, experimental toxicology, physico-chemical properties
reviewed (10).

## References

1. *J. Pharmacol. Exp. Ther.* 1947, **90**, 83
2. *Handbook of Analytical Toxicology* 1969, 73, Chemical Rubber Co. Cleveland, OH
3. *J. Pharmacol. Exp. Ther.* 1948, **93**, 210
4. *Arch. Int. Pharmacodyn. Ther.* 1958, **113**, 313
5. Glauert, H. P. et al *Cancer Lett. (Shannon, Irel.)* 1988, **46**(3), 189-194
6. Lijinsky, W. *Food Chem. Toxicol.* 1984, **22**(1), 27-30
7. Lay, J. O. et al *Rapid Commun. Mass Spectrom.* 1989, **3**(3), 72-75
8. Ashby, J. et al *Environ. Mol. Mutagen.* 1988, **12**(2), 243-252
9. Turner, N. T. et al *Mutat. Res.* 1987, **189**(3), 285-297
10. Lyman, W. K. et al *Handbook of Chemical Property Estimation Methods Environmental Behaviour of Organic Compounds* 1982, McGraw-Hill, New York

# M109   Methaqualone

**CAS Registry No.** 72-44-6

**Synonyms** Citexal; 3,4-dihydro-2-methyl-4-oxo-3-*o*-tolylquinazoline; Dormigan;
Halodorm; Hyminal; Mequin ; 2-methyl-3-(2-methylphenyl)-4-quinazolinone;
2-methyl-3-(2-methylphenyl)-4(3*H*)-quinazolinone; Metolquizalone; Orthonal;
Revonal; Tuazole; Quaalude

**Mol. Formula** C$_{16}$H$_{14}$N$_2$O                    **Mol. Wt.** 250.30

**Uses** Sedative. Hypnotic.

## Physical properties

**M. Pt.** 114-116°C.

## Solubility

Organic solvent: chloroform, ethanol, diethyl ether

## Environmental fate

### Absorption

Adsorption capacity of activated carbon 183 mg g$^{-1}$ carbon (1).

## Mammalian and avian toxicity

### Acute data
$LD_{Lo}$ oral man 114 mg kg$^{-1}$ (2).
$LD_{50}$ oral rat, mouse 230, 420 mg kg$^{-1}$, respectively (3-5).
$LD_{50}$ intraperitoneal rat, mouse 125, 180 mg kg$^{-1}$, respectively (4, 6).
$LD_{50}$ intravenous mouse 100 mg kg$^{-1}$ (2).

### Sub-acute data
Oral rat, (6 days) 100 mg kg$^{-1}$ day$^{-1}$ evoked a 2.5 fold increase in *p*-nitrophenol glucuronidation by hepatic microsomes (7).

### Teratogenicity and reproductive effects
Oral rabbit, lowest toxic dose, 900 mg kg$^{-1}$ day$^{-1}$ on days 8-16 of gestation, teratogenic effects (foetal death and musculoskeletal abnormalities) (8).

### Metabolism and pharmacokinetics
Major metabolite of the drug is 2-methyl-3-(2'-hydroxymethylphenyl)-4(3*H*)-quinozolinone (species unspecified) (9).

## Any other adverse effects to man
Administration of 2.9 or 5.9 mg kg$^{-1}$ increased mean reaction time to stimuli (10). Coma has occurred after taking 2.4 g and death after 8 g (11).

## Any other adverse effects
Studies in rats indicated that physical dependence on the drug may be similar in nature to that of benzodiazipines rather than barbiturates and alcohol (12).

## Legislation
Listed as a Controlled Substance (Depressant) in US Code of Federal Regulations (13).

## References
1. Schmiedel, R. et al *Zentralbl. Pharm. Pharmakother. Lab.* 1988, **127**(12), 789-788 (Ger.) (*Chem. Abstr.* **110**, 121335m)
2. *Arch. Toxikol.* 1963, **20**, 31
3. *Toxicol. Appl. Pharmacol.* 1959, **1**, 42
4. *Arzneim.-Forsch.* 1967, **17**, 229
5. Goldenthal, E. I. *Toxicol. Appl. Pharmacol.* 1971, **18**, 185
6. *Ind. J. Med. Res.* 1979, **69**, 1008
7. Kaur, S. et al *Pest. Comm. Chem. Pathol. Pharmacol.* 1982, **38**(2), 343
8. *Toxicol. Appl. Pharmacol.* 1967, **10**, 244
9. Poropatich, A. et al *Spectrosc. Lett.* 1990, **23**(1), 29-43
10. Smith, L. T. et al *Psychopharmacology* 1988, **94**(1), 126-132
11. *Goodman and Gillman's The Pharmacological Basis of Therapeutics* 6th ed., 1980, 367, MacMillan, New York
12. Suzuki, T. et al *Jpn. J. Pharmacol.* 1988, **46**(4), 403-410
13. *US Code of Federal Regulations* 1985, Title 21, Part 1308.11

# M110   Methazole

**CAS Registry No.** 20354-26-1
**Synonyms** Bioxone; 2-(3,4-dichlorophenyl)-4-methyl-
1,2,4-oxadiazolidine-3,5-dione; Oxydiazol; Mezapur; Paxilon; Probe; Tunic
**Mol. Formula** $C_9H_6Cl_2N_2O_3$                    **Mol. Wt.** 261.07
**Uses** Herbicide.

## Physical properties

**M. Pt.** 123-124°C; **Specific gravity** $d^{25}$ 1.24; **Partition coefficient** log $P_{ow}$ 2.587
(25°C) (1); **Volatility** v.p. $1.0 \times 10^{-6}$ mmHg at 25°C.

**Solubility**
Water: 1.5 mg $l^{-1}$ at 25°C. Organic solvent: acetone, cyclohexanone,
dichloromethane, dimethylformamide, methanol, xylene

## Occupational exposure

**Supply classification** harmful.
**Risk phrases** Harmful in contact with skin and if swallowed – Irritating to eyes and
skin (R21/22, R36/38)
**Safety phrases** Wear suitable protective clothing and gloves (S36/37)

## Ecotoxicity

**Fish toxicity**
$LC_{50}$ (96 hr) bluegill sunfish, rainbow trout 4-5 mg $l^{-1}$ (1).

## Environmental fate

**Degradation studies**
Microbial degradation in soil $t_{1/2}$ <30 days. Degradation product for plants and
microbes, 3,4-dichloroaniline (1).

**Abiotic removal**
Decomposed by UV light in methanol but suspension in water exposed to sunlight is
more stable (2).

## Mammalian and avian toxicity

**Acute data**
$LD_{50}$ oral rat 780-2500 mg $kg^{-1}$ (1,3,4).
$LC_{50}$ (4 hr) inhalation rat >200 g $m^{-3}$ (dust) (1).
$LD_{50}$ dermal rabbit >12,500 mg $kg^{-1}$ (1).
$LD_{50}$ intraperitoneal mouse 600 mg $kg^{-1}$ (2).

**Sub-acute data**

$LC_{50}$ (8 day) oral mallard duck, bobwhite quail 1800, 11,200 mg $kg^{-1}$ diet (1).

**Carcinogenicity and long-term effects**

Oral rat, mouse (2 yr) a yellow/brown pigmentation of liver or spleen tissue occurred at >100 mg $kg^{-1}$ diet (1).

**Teratogenicity and reproductive effects**

Not mutagenic and not teratogenic in rabbits at ≤60 mg $kg^{-1}$ $day^{-1}$, but foetotoxic at ≥30 mg $kg^{-1}$ $day^{-1}$ (route of administration and exposure not specified) (5).
Oral rat, 50 mg $kg^{-1}$ diet for 3 generations, cataracts were observed at ≥100 mg $kg^{-1}$ diet. No other adverse effect was observed (5).

**Metabolism and pharmacokinetics**

Metabolism in mammals involves decarboxylation to 1-(3,4-dichlorophenyl)-3-methylurea which undergoes $N$-demethylation to 3,4-dichlorophenylurea, and then to 3,4-dichloroaniline (1).

**Irritancy**

Mild skin and eye irritant (species unspecified) (4).

**Sensitisation**

Negative in skin sensitisation test on guinea pigs (4).

## Any other adverse effects

Chloracne has been reported in rabbit in studies and among some exposed workers at manufacturing plants (4).

## Legislation

Limited under EC Directive on Drinking Water Quality 80/778/EEC. Pesticides: maximum admissible concentration 0.1 μg $l^{-1}$ (6).
Included in Schedule 6 (Release into Land: Prescribed Substances) Statutory Instrument No. 472, 1991 (7).

## References

1. *The Agrochemicals Handbook* 3rd ed., 1991, RSC, London
2. Ivie, G. W. et al *J. Agric. Food Chem.* 1973, **21**, 386
3. *Farm Chemicals Handbook* 1983, C195, Meister Publishing Co., Willoughby, OH
4. Gaines, T. B. et al *Fundam. Appl. Toxicol.* 1986, **7**, 299
5. *The Pesticide Manual* 9th ed., 1993, 565, British Crop Protection Council, Farnham
6. *EC Directive Relating to the Quality of Water Intended for Human Consumption* 1982, 80/778/EEC, Office for Official Publications of the European Communities, 2 rue Mercier, L-2985 Luxembourg
7. *S. I. 1991 No. 472 The Environmental Protection (Prescribed Processes and Substances) Regulations* 1991, HMSO, London

# M111   Methidathion

**CAS Registry No.** 950-37-8
**Synonyms** $S$-2,3-dihydro-5-methoxy-2-oxo-1,3,4-thiadiazol-3-ylmethyl
$O,O$-dimethyl phosphorodithioate; 3,3-dimethoxyphosphinothioyl-
thiomethyl-5-methoxy-1,3,4-thiadiazol-2($3H$)-one;
$S$-[(5-methoxy-2-oxo-1,3,4-thiadiazol-3($2H$)-yl)methyl] $O,O$-dimethyl
phosphorodithioate; Supracide; Suprathim; Ultracide
**Mol. Formula** $C_6H_{11}N_2O_4PS_3$                    **Mol. Wt.** 302.33
**Uses** Insecticide. Acaricide.

## Physical properties

**M. Pt.** 39-40°C; **Partition coefficient** log $P_{ow}$ 2.22 (1); **Volatility** v.p. $1.4 \times 10^{-3}$
mmHg at 20°C.

### Solubility
Water: 240 mg $l^{-1}$ at 20°C. Organic solvent: acetone, benzene, cyclohexanone,
ethanol, methanol, octanol, xylene

## Occupational exposure

**Supply classification** very toxic.
**Risk phrases** Harmful in contact with skin – Very toxic if swallowed (R21, R28)
**Safety phrases** Do not breathe dust – After contact with skin, wash immediately with
plenty of soap and water – Wear suitable protective clothing and gloves – In case of
accident or if you feel unwell, seek medical advice immediately (show label where
possible) (S22, S28, S36/37, S45)

## Ecotoxicity

### Fish toxicity
$LC_{50}$ (96 hr) bluegill sunfish, rainbow trout 2-10 μg $l^{-1}$ (2).

## Environmental fate

### Degradation studies
Metabolised by the soil bacterium *Bacillus coagulans*. The major metabolite is
desmethyl methidathion (3).

### Abiotic removal
Hydrolysis $t_{1/2}$ 30 min at pH 13, 25°C (4).
≈20% loss by evaporation from glass beads and in field trials after 24 hr (5).

### Absorption
Sorption coefficient for organic matter ($K_{oc}$) 19.5. Sorption coefficient for pond
sediments ($K_d$) 0.63-3.45 (1).

## Mammalian and avian toxicity

### Acute data

$LD_{50}$ oral rat, mouse, guinea pig, rabbit 20-80 mg $kg^{-1}$ (2,6,7).
$LC_{50}$ (4 hr) inhalation rat 50 mg $m^{-3}$ (8).
$LD_{50}$ dermal rat 25 mg $kg^{-1}$ (6).
$LD_{50}$ dermal rabbit 200 mg $kg^{-1}$ (8).

### Carcinogenicity and long-term effects

Oral rat, dog (2 yr) no adverse effect level 0.15 mg $kg^{-1}$ $day^{-1}$ for rats, 0.25 mg $kg^{-1}$ $day^{-1}$ for dogs (2).

### Metabolism and pharmacokinetics

Following oral administration to rats of radiolabelled methidathion 48% of radiolabel was excreted in the urine and 38% in expired air within 24 hr (9).

Readily absorbed from the lumen of lactating cows. Milk contained the sulfone and sulfoxide (10).

### Irritancy

34 mg instilled into rabbit eye caused severe irritation (exposure not specified) (11).

## Genotoxicity

*In vitro* hamster cells, sister chromatid exchanges negative (12).

## Any other adverse effects to man

Reduced serum cholinesterase activity in exposed agricultural workers (13).
Human volunteers tolerated oral doses of 0.11 mg $kg^{-1}$ $day^{-1}$ for >42 days without reaction (4).

## Legislation

EEC maximum residue levels: citrus fruit 2 ppm; pome fruit 0.5 ppm; other fruit and vegetables 0.2 ppm (2).

Limited under EC Directive on Drinking Water Quality 80/778/EEC. Pesticides: maximum admissible concentration 0.1 μg $l^{-1}$ (14).

Included in Schedule 6 (Release into Land: Prescribed Substances) Statutory Instrument No. 472, 1991 (15).

## Any other comments

TDI (human) 0.005 mg $kg^{-1}$ (4).

## References

1. Froehe, Z. et al *Toxicol. Environ. Chem.* 1989, **19**(1-2), 69-82
2. *The Agrochemicals Handbook* 3rd ed., 1991, RSC, London
3. Gauthier, M. J. et al *Pestic. Biochem. Physiol.* 1988, **31**(1), 61-66
4. *The Pesticide Manual* 9th ed., 1993, 567-568, British Crop Protection Council, Farnham
5. Neuruner, H. et al *Bodenkultur* 1991, **42**(1), 57-70 (Ger.) (*Chem. Abstr.* **115**, 34704r)
6. *World Rev. Pest Control* 1970, **9**, 119
7. Gaines, T. B. et al *Fundam. Appl. Toxicol.* 1986, **7**, 299
8. *Farm Chemicals Handbook* 1983, C224, Meister Publishing Co., Willoughby, OH
9. *Foreign Compound Metabolism in Mammals* 1972, **2**, 144
10. Menzie, C. M. *Metabolism of Pesticides, An Update* 1974, 326, US Dept. of Interior, Fish, Wildlife Service, Report Wildlife No. 184, Washington, DC

11.  *Ciba-Geigy Toxicology Data/Indexes* 1 August 1973, Ardsley, NY

12.  Chen, H. H. et al *Mutat. Res.* 1988, **88**(3), 307-316

13.  Drevenkar, V. et al *Arch. Environ. Contam. Toxicol.* 1991, **20**(3), 417-422

14.  *EC Directive Relating to the Quality of Water Intended for Human Consumption* 1982, 80/778/EEC, Office for Official Publications of the European Communities, 2 rue Mercier, L-2985 Luxembourg

15.  *S. I. 1991 No. 472 The Environmental Protection (Prescribed Processes and Substances) Regulations* 1991, HMSO, London

# M112   Methimazole

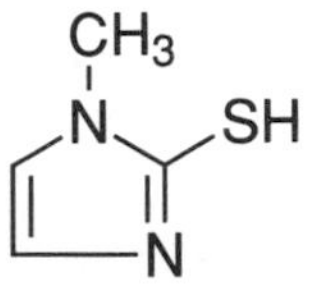

**CAS Registry No.** 60-56-0

**Synonyms** 1,3-dihydro-1-methyl-2*H*-imidazole-2-thione; Danantizal; Favistan; mercaptazole; 2-mercapto-1-methylimidazole; Mercazolyl; Thiamazole; Thymidazole; methyl mercaptoimidazole; 1-methylimidazole-2-thiol

**Mol. Formula** $C_4H_6N_2S$                          **Mol. Wt.** 114.17

**Uses** Antithyroid agent.

## Physical properties

**M. Pt.** 146-148°C; **B. Pt.** 280°C (decomp.).

**Solubility**

Water: ≈20%. Organic solvent: benzene, chloroform, diethyl ether, ethanol, petroleum ether

## Ecotoxicity

**Invertebrate toxicity**

$EC_{50}$ (30 min) *Photobacterium phosphoreum* 240-550 ppm, Microtox test (1).

## Mammalian and avian toxicity

**Acute data**

$LD_{50}$ oral rat 2250 mg $kg^{-1}$ (2).

$LD_{50}$ oral mouse 860 mg $kg^{-1}$ (2).

$LD_{50}$ subcutaneous mouse 345 mg $kg^{-1}$ (2).

$LD_{50}$ subcutaneous rat 1050 mg $kg^{-1}$ (3).

$LD_{50}$ intraperitoneal mouse 500 mg $kg^{-1}$ (4).

**Teratogenicity and reproductive effects**

Oral mouse, 0.1 mg $ml^{-1}$ in drinking water from day 16 of gestation to postpartum day 10. Mean body weight of exposed offspring were reduced and development of behaviour patterns were delayed (5).

## Metabolism and pharmacokinetics

In rats, biliary excretion accounted for 80-90% of injected dose. These metabolites were not glucuronides but labile conjugates (6).

## Genotoxicity

*In vivo* mouse bone marrow induction of micronuclei and chromosomal aberrations negative, mouse spermatocytes and spermatogonia, chromosomal aberrations negative (7).

## Any other adverse effects

*In vitro* studies indicate that immunosuppressive activity of methimazole does not involve the B- or T-lymphocytes (8, 9).

## Any other comments

Methimazole is a potent scavenger of free oxygen radicals. This could explain the suppression of the inflammatory response to skin exposed to UV irradiation. Use as a radioprotectant during radiotherapy has been suggested (10).

## References

1. Kaiser, K. L. E. et al *Water Pollut. Res. J. Can.* 1991, **26**(3), 361-431
2. *Drugs in Japan: Ethical Drugs* 6th ed., 1982, 447, Jakugyo Jihr Co. Tokyo
3. *Farmaco, Ed. Sci.* 1959, **14**, 54
4. *NTIS Report* AD 277-689, Natl. Tech. Inf. Ser., Springfield, VA
5. Rice, S. A. et al *Fundam. Appl. Toxicol.* 1987, **8**(4), 531-540
6. Taurog, A. et al *Endocrinology (Baltimore)* 1988, **122**(2), 592-601
7. Hashimoto, T. et al *J. Toxicol. Sci.* 1987, **12**(1), 23-32
8. Wilson, R. et al *Clin. Endocrinol. (Oxford)* 1988, **28**(4), 389-397
9. Bagnasw, M. et al *J. Endocrinol. Invest.* 1990, **13**(6), 493-499
10. Ferguson, M. M. et al *Lancet* 1985, *ii*, 325

# M113   Methiocarb

**CAS Registry No.** 2032-65-7
**Synonyms** methylcarbamic acid, 4-(methylthio)-3,5-xylyl ester; Bay 37344; Mercaptodimethur; Mesurol; Mesurol phenol; 3,5-dimethyl-4-methylthiophenyl methylcarbamate; Metmercapturon
**Mol. Formula** $C_{11}H_{15}NO_2S$                          **Mol. Wt.** 225.31
**Uses** Insecticide. Molluscide. Acaride. Bird repellent.
**Occurrence** Residues have been isolated from crops (1).

## Physical properties

**M. Pt.** 119°C; **Partition coefficient** log $P_{ow}$ 2.92 (2); **Volatility** v.p. $1.1 \times 10^{-5}$ mmHg at 20°C.

## Solubility

Water: 27 mg $l^{-1}$ at 20°C. Organic solvent: ethanol, diethyl ether, acetone, dichloromethane, *n*-hexane, toluene, propan-2-ol

## Occupational exposure

**Supply classification** toxic.
**Risk phrases** Toxic if swallowed (R25)
**Safety phrases** Do not breathe dust – Wear suitable gloves – If you feel unwell, seek medical advice (show label where possible) (S22, S37, S44)

## Ecotoxicity

### Fish toxicity

$LC_{50}$ (96 hr) bluegill sunfish, golden orfe, carp and rainbow trout 0.21-3.8 mg $l^{-1}$ (3).

### Invertebrate toxicity

$LC_{50}$ (24 hr) crayfish and grass shrimp 0.4-3.7 mg $l^{-1}$ (4).

### Bioaccumulation

Calculated bioconcentration factors of 91-98 indicate that significant environmental accumulation is unlikely (2).

## Environmental fate

### Abiotic removal

Effectively removed from waste-waters in bench scale powdered activated carbon tests (5).
$t_{1/2}$ for hydrolysis at 20°C is >1 yr at pH 4, <35 day at pH 7 and 6 hr at pH 9. The hydrolysis product is 4-methylthio-3,5-dimethylphenol (3,6).
$t_{1/2}$ for reactions with photochemically produced hydroxyl radicals in the atmosphere 7.92 hr (7).

### Absorption

$K_{oc}$ in silt loam soil 70 (8).

## Mammalian and avian toxicity

### Acute data

$LD_{50}$ oral rat, mouse, guinea pig 15-40 mg $kg^{-1}$ (9-11).
$LD_{50}$ oral redwing blackbird, starling, Japanese quail 5-50 mg $kg^{-1}$ (3,12).
$LD_{50}$ oral chicken 175-190 mg $kg^{-1}$ (3).
$LC_{50}$ (4 hr) inhalation rat >0.3 mg $l^{-1}$ (air) aerosol (9).
$LD_{50}$ dermal rat 350 mg $kg^{-1}$ (13).
$LD_{50}$ intraperitoneal mouse 16 mg $kg^{-1}$ (14).

### Carcinogenicity and long-term effects

Oral rat (2 yr) no adverse effect level 67 mg $kg^{-1}$ diet (3).

### Metabolism and pharmacokinetics

Following oral administration to dogs and mice (unspecified dose and duration) rapid absorption and excretion occurs. Metabolised via hydrolysis, oxidation, hydroxylation with principle excretion in urine in free or conjugated form. Only minor amounts excreted in faeces (3).

## Any other adverse effects

Inhibits cholinesterase activity (3).

When applied at the recommended application rate of 3 kg ha$^{-1}$ (for slug control) or at an excess rate 30 kg ha$^{-1}$ caused a reduction in the number of surface dwelling fauna (staphylinids and carabid beetles) (5).

## Legislation

Limited under EC Directive on Drinking Water Quality 80/778/EEC. Pesticides: maximum admissible concentration 0.1 µg l$^{-1}$ (15).

Included in Schedule 6 (Release into Land: Prescribed Substances) Statutory Instrument No. 472, 1991 (16).

EEC maximum admissible residue levels – salad 0.2 ppm; other vegetables and berries 0.1 ppm (3).

## Any other comments

WHO Class II; EPA Toxicity Class II (3).

Tolerable daily intake (human) (TDI) 0.001 mg kg$^{-1}$ (3).

Toxicity and environmental impact reviewed (17,18).

## References

1. Yess, N. J. *J.- Assoc. Off. Anal. Chem.* 1991, **74**(5), 121A-142A
2. Lyman, W. K. et al *Handbook of Chemical Property Estimation Methods Environmental Behaviour of Organic Compounds* 1982, McGraw-Hill, New York
3. *The Agrochemicals Handbook* 3rd ed., 1991, RSC, London
4. Marking, L. L. et al *Bull. Environ. Contam. Toxicol.* 1981, **26**(6), 705-716
5. Dietrich, M. J. et al *Environ. Prog.* 1988, **7**(2), 143-149
6. Eichelberger, J. W. et al *Environ. Sci. Technol.* 1971, **5**, 541
7. Eisenreich, S. J. et al *Environ. Sci. Technol.* 1981, **15**, 30
8. Swann, R. L. et al *Res. Rev.* 1983, **85**, 16
9. *Farm Chemicals Handbook* 1983, C150, Meister Publishing Co., Willoughby, OH
10. *Farmakol. Toksikol. (Moscow)* 1972, **35**, 356
11. Gaines, T. B. *Toxicol. Appl. Pharmacol.* 1969, **14**, 515
12. Schafer, E. W. et al *Arch. Environ. Contam. Toxicol.* 1983, **12**, 355-382
13. *Pesticide Chemicals Official Compendium* 1966, 105, Association of the American Pesticide Control Officials, Topeka, KS
14. *Toxicol. Appl. Pharmacol.* 1964, **6**, 402
15. *EC Directive Relating to the Quality of Water Intended for Human Consumption* 1982, 80/778/EEC, Office for Official Publications of the European Communities, 2 rue Mercier, L-2985 Luxembourg
16. *S. I. 1991 No. 472 The Environmental Protection (Prescribed Processes and Substances) Regulations* 1991, HMSO, London
17. *Dangerous Prop. Ind. Mater. Rep.* 1987, **7**(1), 66-69
18. *IPCS Environmental Health Criteria No. 64* 1986, WHO, Geneva

# M114  D-Methionine

CH3SCH2CH2CH(NH2)CO2H

**CAS Registry No.** 348-67-4
**Synonyms** (*R*)-(−)-methionine
**Mol. Formula** $C_5H_{11}NO_2S$           **Mol. Wt.** 149.21
**Uses** Lipotropic agent.
**Occurrence** Formed by *Candida*, *Pseudomonas*, *Hansenula* and *Rhodococcus* species.

## Physical properties
**M. Pt.** 273°C (decomp.).

**Solubility**
Water: miscible.

## Environmental fate

**Degradation studies**
Metabolised by *Halobacterium ralobium* by conversion to the L-form, possibly via keto acids (1).

## Mammalian and avian toxicity

**Acute data**
$LD_{50}$ intraperitoneal rat 5200 mg kg$^{-1}$ (2).

**Metabolism and pharmacokinetics**
Metabolised in the rat liver to form methyl sarcosine via oxidative deamination and reamination to give the L-form (3).
Readily absorbed through the skin (species unspecified) (4).

**Irritancy**
Causes skin and eye irritation (species unspecified) (4).

## Any other adverse effects
Intraluminal administration to rabbits initially enhanced intestinal myoelectric activity, followed by a dose-dependent inhibitory phase (5).
Dietary administration reduced body weight gain in rats. Hepatic methionine transsulfuration enzymes activities, spleen weight, Fe content and methionine levels were all increased (6).

## Any other comments
Essential amino acid for protein synthesis (7).

## References
1.  Tenaka, M. et al *Viva Origino* 1991, **18**(3), 109-117. (Japan.) (*Chem. Abstr.* **115**, 154686m)
2.  *Annals Appl. Biol.* 1956, **64**, 319
3.  London, R. E. et al *Biochemistry* 1988, **27**(20), 7864-7869
4.  Lenga, R. E. (Ed.) *Sigma-Aldrich Library of Chemical Safety Data* 2nd ed., 1988, **2**, 2233, Sigma-Aldrich, Milwaukee, WI
5.  Ko, S. K. et al *Tongmul Hakhoechi* 1988, **31**(2), 81-86 (Korean) (*Chem. Abstr.* **109**, 188063u)

6. Sugiyama, K. et al *Agric. Biol. Chem.* 1987, **51**(12), 3411-3413
7. *The Merck Index* 11th ed., 1989, 943, Merck & Co., Rahway, NJ

# M115   Methocarbamol

$$OCH_2CHCH_2OCNH_2$$

with OH above the CH and O below the C; H₃CO attached to benzene ring.

OH
|
OCH₂CHCH₂OCNH₂
‖
O

H₃CO

**CAS Registry No.** 532-03-6
**Synonyms** glycerylguaiacolate carbamate; guaiacol glyceryl ether carbamate;
2-hydroxy-3-(*o*-methoxyphenoxy)propyl-1-carbamate;
3-(2-methoxyphenoxy)-1-glyceryl carbamate;
3-(*o*-methoxyphenoxy)-1,2-propanediol 1-carbamate

**Mol. Formula** $C_{11}H_{15}NO_5$            **Mol. Wt.** 241.25
**Uses** Muscle relaxant. Antispasmodic drug.

## Physical properties

**M. Pt.** 92-94°C.

### Solubility
Water: ≈2.5% at 20°C. Organic solvent: benzene, chloroform, ethanol, propylene glycol

## Mammalian and avian toxicity

### Acute data
$LD_{50}$ oral rat, mouse, dog, hamster 1410-2000 mg $kg^{-1}$ (1,2).
$LD_{50}$ subcutaneous mouse 780 mg $kg^{-1}$ (3).
$LD_{50}$ intraperitoneal rat, mouse, hamster 820-1050 mg $kg^{-1}$ (1, 4).
$LD_{50}$ intravenous rabbit 680 mg $kg^{-1}$ (5).

### Metabolism and pharmacokinetics
Following an oral dose of 150 mg $kg^{-1}$ to rats, peak plasma levels were reached after 150 min. It is excreted in the urine primarily as the glucuronide and sulfate conjugates. A small amount is excreted in faeces (6,7).

### Any other adverse effects to man
Adverse effects include drowsiness, dizziness, blurred vision, gastrointestinal effects and hypersensitivity reactions including rashes, pruritus, urticaria, and conjunctivitis with nasal congestion (7).

## References

1. *J. Pharmacol. Exp. Ther.* 1960, **129**, 75

2.  Usdin, E. et al *Psychotropic Drugs and Related Compounds* 2nd ed., 1972, 398, Washington, DC
3.  *Acta Pharmacol. Toxicol.* 1962, **19**, 247
4.  *Arzneim.-Forsch.* 1967, **17**, 242
5.  *Int. J. Neuropharmacol.* 1966, **5**, 305
6.  Oback, R. et al *Biopharm. Drug Dispos.* 1988, **9**(5), 501-511
7.  *Martindale. The Extra Pharmacopoeia* 30th ed., 1993, 1206, The Pharmaceutical Press, London

# M116   Methomyl

$$CH_3C(SCH_3)=NOC(=O)NHCH_3$$

**CAS Registry No.** 16752-77-5
**Synonyms** *N*-[[(methylamino)carbonyl]oxy]ethanimidothioic acid, methyl ester; *S*-methyl *N*-[methylcarbamoyl)oxy]thioacetimidate; Lannate; Nudrin; *N*-[(methylcarbamoyl)oxy]thioacetimidic acid, methyl ester; Du Pont 1179
**Mol. Formula** $C_5H_{10}N_2O_2S$                    **Mol. Wt.** 162.21
**Uses** Insecticide. Acaricide.
**Occurrence** Traces have been isolated from natural waters and on crops (1,2).

## Physical properties
**M. Pt.** 78-79°C; **Specific gravity** $d_4^{24}$ 1.29; **Partition coefficient** log $P_{ow}$ −3.56 (1);
**Volatility** v.p. $5 \times 10^{-5}$ mmHg at 25°C.

## Solubility
Water: 57.9 g $l^{-1}$. Organic solvent: methanol, ethanol, acetone, propan-2-ol, toluene

## Occupational exposure
**US TLV (TWA)** 2.5 mg m$^{-3}$; **UK Long-term limit** 2.5 mg m$^{-3}$; **Supply classification** very toxic.
**Risk phrases** Very toxic if swallowed (R28)
**Safety phrases** Do not breathe dust – Wear suitable protective clothing and gloves – In case of accident or if you feel unwell, seek medical advice immediately (show label where possible) (S22, S36/37, S45)

## Ecotoxicity

**Fish toxicity**
$LC_{50}$ (96 hr) bluegill sunfish, rainbow trout 0.9-3.4 mg $l^{-1}$ (3,4).
$LC_{50}$ (48 hr) carp, goldfish, killifish, guppy 0.9-2.8 mg $l^{-1}$ (5).

**Invertebrate toxicity**
$EC_{50}$ (48 hr) *Daphnia magna* 9 µg $l^{-1}$ (4).

**Bioaccumulation**
Calculated bioconcentration factors of 1.3-1.7 indicate that the potential for environmental accumulation is unlikely (6).

## Environmental fate

### Nitrification inhibition
Nitrogen fixation ability of some bacteria was reduced by up to 80% when methomyl was applied at 20-160 mg kg$^{-1}$ soil (7).

### Degradation studies
Degraded by the soil fungi *Alternaria brassicola, Penicillium notatum, Aspergillus* and *Helminthosporium* species and *Verticillium agaricinum* (8,9).

Degradation $t_{1/2}$ 30-42 days in soil. Major degradation product was carbon dioxide. A minor degradation product, *S*-methyl-*N*-hydroxythioacetimidate, was a possible hydrolysis product (10).

Hydrolysis $t_{1/2}$ 56 wk at pH 8.0; 54 wk at pH 7.0; 38 wk at pH 6.0; 20 wk at pH 4.5, in sterile 1% ethanol (11).

### Abiotic removal
Degrades rapidly in chlorinated water. Rate increases with decreasing pH, increasing temperature and increasing chlorine concentration. Reaction rate is 1000 × faster with free chlorine than with chloramine. Methomyl sulfoxide and *N*-chloromethomyl are formed before degrading to acetic acid, methanesulfonic acid and dichloromethylamine (12).

### Absorption
Calculated $K_{oc}$ 10-500 indicates low to moderate adsorption to soil and sediments is likely (6,13).

## Mammalian and avian toxicity

### Acute data
$LD_{50}$ oral rat, mouse, mallard duck, pheasant, quail, starling 10-35 mg kg$^{-1}$ (3,14-17). $LC_{50}$ (4 hr) inhalation rat 300 mg m$^{-3}$. Animals showed characteristic signs of cholinesterase inhibition, including salivation, lachrymation and tremors (18). $LD_{50}$ dermal rabbit 5800 mg kg$^{-1}$ (19).

### Sub-acute data
Inhalation ♂ rat, 15 mg m$^{-3}$ 4 hr day$^{-1}$ 5 days wk$^{-1}$ for 3 months caused no change in plasma and red cell cholinesterase activity, histopathology or lipid concentration (20). $LD_{50}$ (5 days) bobwhite quail, Japanese quail, ring-necked pheasant, mallard duck 1100-2900 mg kg$^{-1}$ diet (21).

Oral rat (13 wk) 0, 1, 3, 10 or 30 mg kg$^{-1}$ day$^{-1}$ caused no fatality or clinical signs of toxicity. Body weight gain was reduced in treated ♀ but not ♂ at all dose levels. Kidney/body weight ratio, but not absolute kidney weights were increased at the 2 high dose levels. Red blood cell cholinesterase activity was elevated at the high dose levels but plasma and brain cholinesterase levels were unaffected (21).

### Carcinogenicity and long-term effects
Oral rat (22 month) 1, 2.5, 5, 10 or 20 mg kg$^{-1}$ day$^{-1}$. Autospy revealed kidney tubular hypertrophy, vacuolation of epithelial cells of the proximal convoluted tubules and histological alterations in the spleen at the high dose level. No effects were seen on plasma or red blood cell cholinesterase levels (14).

Oral rat (2 yr) 0, 2.5, 5 or 20 mg kg$^{-1}$ day$^{-1}$. Effects observed in the high dose group were reduced weight gain in both sexes, and in ♀ rats, lower erythrocyte counts, haemoglobin values and haemotocrits. Blood and brain cholinesterase activity levels and other clinical chemistry parameters were not significantly altered (22).

**Teratogenicity and reproductive effects**
Oral rat 0, 2.5 or 5.0 mg kg$^{-1}$ day$^{-1}$ for 3 generations. No adverse effects were reported on reproduction or lactation, and no pathological changes were found in the weanling pups of the $F_3$ generation (14).
Oral rabbit 0, 2, 6 or 16 mg kg$^{-1}$ day$^{-1}$ on days 7-19 of gestation. 1/5 animals in the high dose group died, exhibiting characteristic signs of cholinesterase inhibition. No teratogenic or embryotoxic effects were observed (1).

**Metabolism and pharmacokinetics**
Almost completely absorbed from the gastrointestinal tract following oral administration to rats. After 72 hr, 15-25% of the dose was expired as carbon dioxide, 33-50% was expired as acetonitrile, and ≈25% as metabolites in the urine. Metabolism was believed to involve partial isomerisation followed by hydrolysis of the 2 isomeric forms to yield 2 isomeric oximes that then break down to carbon dioxide and acetonitrile at different rates. No other metabolites were identified (23,24).

**Irritancy**
Eye irritant in rabbits. No irritation occurred after application to guinea pig skin (dose and duration unspecified) (3).

## Genotoxicity

*Salmonella typhimurium* TA98, TA100, TA1535, TA1537, TA1538 with and without metabolic activation negative (25,26).
*Drosophila melanogaster* sex-linked recessive lethal assay positive (27).
*In vivo* mouse sperm morphological abnormalities and chromosomal aberrations positive (28).

## Any other adverse effects to man

Accidental doses of 12-15 mg kg$^{-1}$ have been fatal to man (29,30).

## Any other adverse effects

Oral ♂ rat, single dose of 5, 10 or 15 mg kg$^{-1}$. Serum cholinesterase activity was inhibited in all treated groups. No haematological changes were observed. Biochemical studies revealed disorders of liver and kidney function in rats receiving the low dose. Histopathological alterations were seen in the lungs at all doses, and in the liver and kidneys at the high dose. All treated rats showed a decrease in superoxide dismutase activity, an increase in serum lipoperoxide level and a decrease in serum α-tocopherol level (31).

## Legislation

Limited under EC Directive on Drinking Water Quality 80/778/EEC. Pesticides: maximum admissible concentration 0.1 μg l$^{-1}$ (32).
Included in Schedule 6 (Release into Land: Prescribed Substances) Statutory Instrument No. 472, 1991 (33).
US maximum residue on hops 12 ppm (34).
EC maximum residue levels vegetables, citrus fruit 0.5 ppm; pome and stone fruit, grapes 1 ppm; salads 2 ppm (3).

## Any other comments

Physical properties, uses, occurrence, analysis, environmental fate, metabolism, mammalian toxicity, teratogenicity, mutagenicity, carcinogenicity and health advice reviewed (1,2,35,36).
Tolerable daily intake (TDI) human 0.001 mg kg$^{-1}$ (37).
WHO class II; EPA Toxicity Class II (3).

## References

1. *Drinking Water Health Advisory: Pesticides* 1989, 521-539, Lewis Publishers, Chelsea, MI
2. Howard, P. H. *Handbook of Environmental Fate and Exposure Data for Organic Chemicals* 1991, **3**, 495-501, Lewis Publishers, Chelsea, MI
3. *The Agrochemicals Handbook* 3rd ed., 1991, RSC, London
4. Johnson, W. E. et al *Handbook of Acute Toxicity of Chemicals in Fish and Aquatic Invertebrates* 1980, Resource Publication No. 137, US Fish and Wildlife Service, Washington, DC
5. Nishiuchi, T. *Aquiculture* 1974, **22**, 16-18
6. Lyman, W. K. et al *Handbook of Chemical Property Estimation Methods Environmental Behaviour of Organic Compounds* 1982, McGraw-Hill, New York
7. Huang, C. Y. *Botanical Bull. Acad. Sinica* 1978, **19**(1), 41-52
8. Adam, T, M. et al *Isot. Radiat. Res.* 1989, **21**(1), 59-64
9. Mostafa, I. Y. et al *Isot. Radiat. Res.* 1990, **22**(1), 39-48
10. Harvey, J. *Report* 1977, Du Pont de Nemours and Co. Wilmington, DE
11. Chapman, R. A. et al *J. Environ. Sci. Health, Part B* 1982, **B17**, 487
12. Miles, C. J. et al *Environ. Toxicol. Chem.* 1990, **9**(5), 535-540
13. Laskowski, D. A. et al *Residue Rev.* 1983, **85**, 139-158
14. Kaplan, A. M. et al *Toxicol. Appl. Pharmacol.* 1977, **40**, 1-17
15. Schafer, E. W. et al *Arch. Environ. Contam. Toxicol.* 1983, **12**, 355-382
16. Dashiell, O. L. et al *J. Appl. Toxicol.* 1984, **4**(5), 320-325
17. El-Sebae, A. H. et al *Proc. Br. Crop. Prot. Conf. Pests Diseases* 1970, 731-736
18. *Documentation of Threshold Limit Values for Substances in Workroom Air* 3rd ed., 1984, American Conference of Governmental Industrial Hygienists, Cincinnati, OH
19. *Farm Chemicals Handbook* 1980, D197, Meister Publishing Co., Willoughby, OH
20. Tanaka, I. et al *Am. Ind. Hyg. Assoc. J.* 1987, **48**(4), 330-334
21. Hill, E. F. et al *Lethal Dietary Toxicities of Environmental Pollutants to Birds* 1975, US Fish & Wildlife Service, Report Wildlife No. 191, Washington, DC
22. Haskell Laboratories *Long-term Feeding Study in Rats with Methomyl INX-1179 Final Report* 1981, Report No. 235-281
23. Baron, R. L. *Int. Symp. Pestic. Terminal Res.* 1971, 185-197
24. Dorough, H. W. *Metabolism of Carbamate Insecticides* 1977, NTIS, Springfield, VA, PB-266 233
25. Moriya, M. et al *Mutat. Res.* 1983, **116**, 185-216
26. Blevins, R. D. et al *Mutat. Res.* 1977, **56**, 1-6
27. Hemavathy, K. C. et al *Mutat. Res.* 1987, **191**(1), 41-43
28. Hemavathy, K. C. et al *Environ. Res.* 1987, **42**(2), 362-365
29. Araki, M. et al *Nippon Hoigaku Zashi* 1982, **36**, 584-588
30. Liddle, J. A. et al *Clin. Toxicol.* 1979, **15**, 159-167
31. Higashihara, E. *Yamaguchi Igaku* 1987, 36(4), 233-251, (Japan.) (*Chem. Abstr.* **107**, 192625e)
32. *EC Directive Relating to the Quality of Water Intended for Human Consumption* 1982, 80/778/EEC, Office for Official Publications of the European Communities, 2 rue Mercier, L-2985 Luxembourg

33. *S. I. 1991 No. 472 The Environmental Protection (Prescribed Processes and Substances) Regulations* 1991, HMSO, London
34. US EPA *Fed. Regist.* 14 Feb. 1990, **55**(31), 5219-5220
35. Martinez-Chuecos, J. et al *Hum. Exp. Toxicol.* 1990, **9**, 251-254
36. *ECETOC Technical Report No. 30(4)* 1991, European Chemical Industry Ecology and Toxicology Centre, B-1160 Brussels
37. *The Pesticide Manual* 9th ed., 1993, 570-571, British Crop Protection Council, Farnham

# M117   Methoprene

**CAS Registry No.** 40596-69-8
**Synonyms** isopropyl (*E,E*)-(*RS*)-11-methoxy-3,7,11-trimethyldodeca-2,4-dienoate; (*E,E*)-1-methylethyl 11-methoxy-3,7,11-trimethyl-2,4-dodecadienoate; Altosid; Diacon; Kabat; Minex; Pharorid; Precor
**Mol. Formula** $C_{19}H_{34}O_3$                            **Mol. Wt.** 310.48
**Uses** Insect growth regulator.

## Physical properties

**B. Pt.** $_{0.05}$ 100°C; **Flash point** 187°C (open cup); **Specific gravity** $d^{20}$ 0.926; **Volatility** v.p. $2.4 \times 10^{-5}$ mmHg at 25°C.

## Solubility

Water: 1.4 mg $l^{-1}$. Organic solvent: diethyl ether, ethanol

## Ecotoxicity

**Fish toxicity**
$LC_{50}$ (96 hr) bluegill sunfish, trout 4.4-4.6 mg $l^{-1}$ (1).
$LC_{50}$ (96 hr) mummichog 125 mg $l^{-1}$ (2).

**Abiotic removal**
Degraded by UV irradiation (3).

## Mammalian and avian toxicity

**Acute data**
$LD_{50}$ oral rat >34,600 mg $kg^{-1}$, dog >5000 mg $kg^{-1}$ (1,4).
$LD_{50}$ dermal rabbit 3000-3500 mg $kg^{-1}$ (1, 4).

**Sub-acute data**
$LD_{50}$ (8 day) oral chicken >4640 mg $kg^{-1}$ (1).

### Carcinogenicity and long-term effects
Oral (2 yr) rats receiving 5000 mg kg$^{-1}$ diet and mice receiving 2500 mg kg$^{-1}$ diet showed no ill-effects (1).

### Teratogenicity and reproductive effects
Oral rat, 3-generation study, 2500 mg kg$^{-1}$ diet caused no reproductive adverse effects (1).

### Metabolism and pharmacokinetics
In mammals, the secondary metabolite cholesterol has been identified (1).

## Genotoxicity
*Drosophila melanogaster* wing spot test, weakly positive (5).

## Any other adverse effects
Acts as an insect growth regulator by mimicking the action of insect juvenile hormones, causing death by preventing the transformation of larva to pupa (6).

## Legislation
Limited under EC Directive on Drinking Water Quality 80/778/EEC. Pesticides: maximum admissible concentration 0.1 μg l$^{-1}$ (7).
Included in Schedule 6 (Release into Land: Prescribed Substances) Statutory Instrument No. 472, 1991 (8).

## Any other comments
WHO Class Table 5; EPA Toxicity Class IV (1).
TDI (human) 0.1 mg kg$^{-1}$ (3).

## References
1. *The Agrochemicals Handbook* 3rd ed., 1991, RSC, London
2. Lee, B. M. et al *Bull. Environ. Contam. Toxicol.* 1989, **43**(6), 827-832
3. *The Pesticide Manual* 9th ed., 1993, 572, British Crop Protection Council, Farnham
4. *Environ. Health Perspect.* 1976, **14**, 119
5. Manec, F. et al *Mutat. Res.* 1987, **188**(3), 209-214
6. *Martindale. The Extra Pharmacopoeia* 30th ed., 1993, 1129, The Pharmaceutical Press, London
7. *EC Directive Relating to the Quality of Water Intended for Human Consumption* 1982, 80/778/EEC, Office for Official Publications of the European Communities, 2 rue Mercier, L-2985 Luxembourg
8. *S. I. 1991 No. 472 The Environmental Protection (Prescribed Processes and Substances) Regulations* 1991, HMSO, London

# M118   Methotrexate

$$\text{HO}_2\text{CH}_2\text{CH}_2\text{CH}_2\overset{\displaystyle \text{CO}_2\text{H}}{\underset{}{\text{C}}}\text{H}\overset{\displaystyle \text{O}}{\underset{\text{H}}{\text{C}}}-\text{N}-\overset{}{\underset{\text{CH}_3}{\text{N}}}-\text{CH}_2$$

**CAS Registry No.** 59-05-2
**Synonyms** L-(+)-amethopterin; N-(4-[[(2,4-diamino-6-pteridinyl)methyl]-methylamino]benzoyl-L-glutamic acid; amethopterin; 4 amino-10-methylfolic acid; 4-amino-N-methylpteroylglutamic acid; α-methopterin; methylaminopterin
**Mol. Formula** $C_{20}H_{22}N_8O_5$                          **Mol. Wt.** 454.45
**Uses** Antineoplastic agent. Antifungal agent. Treatment of psoriasis.

## Physical properties

**M. Pt.** 108-204°C.

## Solubility

Organic solvent: dilute alkaline hydroxides and carbonates, dilute hydrochloric acid

## Mammalian and avian toxicity

### Acute data

$LD_{50}$ oral rat, mouse 135-180 mg $kg^{-1}$ (1,2).
$LD_{50}$ intraperitoneal rat, mouse 6 mg $kg^{-1}$ (1).
$LD_{50}$ subcutaneous mouse 250 mg $kg^{-1}$ (3).
$LD_{50}$ intravenous rat, mouse 14, 65 mg $kg^{-1}$, respectively (2,4).

### Sub-acute data

$LD_{50}$ (5 day) intraperitoneal rat, mouse 1.1-2.0 mg $kg^{-1}$ $day^{-1}$ (1).
Intraperitoneal rat 0.1-0.4 mg $kg^{-1}$ $day^{-1}$ and dog 1 mg $kg^{-1}$ $day^{-1}$ caused colonic ulceration, sometimes associated with an ileitis and/or intestinal haemorrhage (5).
Oral rat (6 wk) 100, 150, 200 or 300 μg $kg^{-1}$ $day^{-1}$. The high dose caused systemic toxicity although hepatotoxicity was not observed. The lower doses were tolerated for longer periods and were associated with hepatotoxicity, ranging from focal to confluent necrosis (6).

### Carcinogenicity and long-term effects

No adequate evidence for carcinogenicity to humans or animals, IARC classification group 3 (7).
Oral hamster (2 yr) 5, 10 or 20 mg $kg^{-1}$ diet for life. Median survival time for the high dose group was 40-50 wk. Survival in the other treated groups was similar to that of controls. There was no significant difference in the incidence of tumours compared to controls (8).
Oral mouse (120 wk) 5, 8 or 10 mg $kg^{-1}$ diet $wk^{-1}$ on alternate wk for life. Median survival time was 80-90 wk. The incidence of tumours was not significantly different from those in controls (8).
Oral mouse, 0.1 mg $kg^{-1}$ $day^{-1}$ in drinking water for 18 months. 5/32 treated animals developed lung adenomas and 16/32 developed lung carcinomas (overall incidence

66%). This compared with incidence of 5.1% lung tumours in controls. One treated mouse developed a hepatoma (9).

Intraperitoneal rat, mouse (22 month) 0.15-1.0 mg kg$^{-1}$ 3 × wk$^{-1}$ for 6 months. The incidence of tumours was not significantly different from controls. Median survival times were significantly reduced in all treated groups (10).

Intravenous rat (2 yr) 1 mg kg$^{-1}$ wk$^{-1}$ for 52 wk. The incidence of tumours was not significantly different from that of controls. Survival rates were ≈40% that of controls (4).

### Teratogenicity and reproductive effects

Intraperitoneal mouse 20 mg kg$^{-1}$ on day 9 of gestation induced a high incidence of median facial clefts in offspring (11).

Oral cat, 0.5 mg kg$^{-1}$ day$^{-1}$ on days 11-14, 14-17 or 17-20 of gestation. Maternal toxicity was observed in ≈18% treated animals. Visceral abnormalities were seen only in some offspring of cats treated on days 17-20 of gestation (umbilical hernia, cleft palate, hydrocephalus, spina bifida and malformed limbs) (12).

Intraperitoneal rat, 0.2 and 0.3 mg kg$^{-1}$ on day 9 of gestation was teratogenic to 35 and 75% of foetuses and lethal to 63 and 84% of embryos, respectively.

Malformations were also induced by 0.25 mg kg$^{-1}$ administered on day 5. This dose was lethal to all embryos when administered on days 6-9 of gestation (13).

### Metabolism and pharmacokinetics

Methotrexate was detected in the plasma after 4 hr following dermal application to shaved mice (14).

Following infusion rats, metabolised in the liver to 7-hydroxymethotrexate which was excreted in the bile. Plasma $t_{1/2}$ 15 hr for methotrexate and 7-hydroxymethotrexate (15,16).

Other metabolites formed in rats and mice are 4-amino-4-deoxy-$N^{10}$-methylpteroic acid and various free pteridines which may be formed by the intestinal flora (17,18). Methotrexate is retained in the liver, to a degree which is in part dependent upon the dose. It was excreted in the urine and bile. Partial reabsorption occurred in the gut. Within 24-48 hr after administration 67-91% of the dose was eliminated by dogs and monkeys (19).

## Genotoxicity

*Salmonella typhimurium* TA97, TA98, TA100, TA1535, TA1537, TA1538 with and without metabolic activation marginally positive (20).

*Drosophila melanogaster* wing spot test positive (21).

*In vitro* human fibroblasts, sister chromatid exchanges negative (22).

*In vitro* mouse lymphoma L5178Y, tk$^+$/tk$^-$ mutation assay positive (23).

*In vitro* human bone marrow cells, sister chromatid exchanges positive (24).

*In vitro* hamster A(T$^1$)C1-3 cells, chromosomal damage positive (25).

*In vivo* mouse bone marrow micronucleus test positive (26).

*In vivo* mouse, dominant lethal assay positive (27).

## Any other adverse effects to man

A woman who took 2.5 mg day$^{-1}$ for 5 days during the 9th wk of pregnancy in an attempt to abort, gave birth to an infant with no frontal bone, craniosynostosis of the coronal and lambdoid sutures, limb reduction defects, hypertelorism and a flat nasal bridge. A similar case was reported in a woman who took 5 mg day$^{-1}$ for the first 2 months of pregnancy for the treatment of psoriasis (28,29).

Women treated with methotrexate prior to pregnancy do not appear to have an increased risk for spontaneous abortion or for offspring with congenital malformations (30,31).

Chromosomal damage was seen in bone marrow cells, but not in peripheral lymphocytes in methotrexate-treated patients (32,33).

Case reports of malignancy in patients treated for psoriasis include skin cancers, cervical cancer, non-Hodgkin's lymphoma, leukaemia, breast cancer, renal cancer and nasopharyngeal cancer (34-42).

Adverse effects are mainly associated with tissues with a rapid cell turnover, particularly the bone marrow, the alimentary tract epithelium, the epidermis, foetal tissue and germinal cells (43).

## Any other adverse effects

Intraperitoneal rat, single dose of $\leq 50$ mg kg$^{-1}$. The highest dose caused the bone marrow to become hypocellular within 15 hr the reduction in the erythroid series being the most pronounced. Recticulocytopenia and panleucopenia were also evident in the peripheral blood. These effects progressed in severity up to 72 hr. Lower doses caused similar effects, with particularly significant anaemia and lymphocytopenia. In addition the thymus, spleen and lymph nodes showed marked atrophy. Increased susceptibility to infection was also reported due to a decrease in the production of antibodies (1,5,44).

Administration into the cerebrospinal fluid of cats caused segmental axonal degeneration. None of the animals showed disseminated necrotising leukoencephalopathy. This suggested a direct toxic effect on the axon (45).

## Any other comments

Physical properties, use analysis, carcinogenicity, mammalian toxicity, metabolism, health effects and mutagenicity reviewed (46,47).

## References

1. Scherf, H. R. et al *Recent Results Cancer Res.* 1975, **52**, 76-87
2. *Drugs in Japan: Ethical Drugs* 6th ed., 1982, 841, Jakugyo Jiho, Tokyo
3. *Natl. Cancer Inst. Screening Program Data Summary: Developmental Therapeutics Program* Jan 1986, Bethesda MD
4. Schmaehl, D. et al *Arzneim.-Forsch.* 1970, **20**, 1461-1467
5. Ferguson, F. C. et al *J. Pharmacol. Exp. Ther.* 1950, **98**, 293-299
6. Hall, P. M. et al *Hepatology (Baltimore)* 1991, **14**(5), 906-910
7. *IARC Monograph* 1987, **Suppl. 7**, 241-242
8. Rustin, M. et al *Toxicol. Appl. Pharmacol.* 1973, **26**, 329-338
9. Roschlau, G. et al *Dtsch. Gesundheitswes.* 1971, **26**, 219-222
10. Weisburger, E. K. *Cancer* 1977, **40**, 1977
11. Darab, D. J. et al *Teratology* 1987, **36**(1), 77-86
12. Khera, K. S. et al *Teratology* 1976, **14**, 21-28
13. Jorden, R. L. et al *Teratology* 1977, **15**, 73-80
14. Bailey, D. N. et al *J. Toxicol., Cutaneous Ocul. Toxicol.* 1987, **6**(1), 9-12
15. Bremmes, R. M. et al *Cancer Res.* 1989, **49**(22), 6359-6364
16. Bore, P. et al *Cancer Drug Delivery* 1987, **4**(3), 177-183
17. Valerino, D. M. *Res. Commun. Chem. Pathol. Pharmacol.* 1972, **4**, 529-542
18. Valerino, D. M. et al *Biochem. Pharmacol.* 1972, **21**, 821-831
19. Henderson, E. S. et al *Cancer Res.* 1965, **25**, 1008-1017

20. Klopman, G. et al *Mutat. Res.* 1990, **228**(1), 1-58
21. Clements, J. et al *Mutat. Res.* 1988, **203**(2), 117-123
22. Littlefield, L. G. et al *Mutat. Res.* 1979, **67**, 259-269
23. Clive, D. et al *Mutagenesis* 1990, **5**(2), 191-197
24. Zheng, S. et al *Cancer Genet. Cytogenet.* 1988, **31**(2), 157-163
25. Benedict, W. F. et al *Cancer Res.* 1977, **37**, 2202-2208
26. Maier, P. et al *Mutat. Res.* 1976, **40**, 325-388
27. Epstein, S. S. et al *Toxicol. Appl. Pharmacol.* 1972, **23**, 288-325
28. Powell, H. R. et al *Med. J. Aust.* 1971, **2**, 1076-1077
29. Milunsky, A. et al *J. Pediatr.* 1968, **72**, 790-795
30. Ross, G. T. *Cancer* 1976, **37**, 1043-1047
31. Walden, P. A. M. et al *Am. J. Obstet. Gynecol.* 1976, **125**, 1108-1114
32. Melnyk, J. et al *Clin. Genet.* 1971, **2**, 28-31
33. Krogh Jensen, M. et al *Mutat. Res.* 1979, **64**, 339-343
34. Ringrose, C. A. D. *Am. J. Obstet. Gynecol.* 1974, **119**, 1132-1133
35. Schroeter, R. et al *Dermatologica* 1971, **143**, 131-136
36. Rees, R. B. et al *Arch. Dermatol.* 1967, **95**, 2-11
37. Craig, R. S. et al *Arch. Dermatol.* 1971, **103**, 505-506
38. Molin, L. et al *Arch. Dermatol.* 1972, **105**, 292
39. Rustin, G. J. S. et al *New Engl. J. Med.* 1983, **308**, 473-476
40. Nyfors, A. et al *Dermatologica* 1983, **167**, 260-261
41. Stern, R. S. et al *Cancer* 1982, **50**, 869-872
42. Bailin, P. L. et al *J. Am. Med. Assoc.* 1975, **232**, 359-362
43. Muller, S. A. et al *Arch. Dermatol.* 1969, **100**, 523-530
44. Custer, R. P. et al *J. Natl. Cancer Inst.* 1977, **58**, 1011-1017
45. Shibutani, M. et al *Acta Neuropathol.* 1989, **78**(3), 291-300
46. *IARC Monograph* 1981, **26**, 267-292
47. *Martindale. The Extra Pharmacopoeia* 30th ed., 1993, 488-492, The Pharmaceutical Press, London

# M119   Methoxsalen

**CAS Registry No.** 298-81-7
**Synonyms** 9-methoxy-7*H*-furo[3,2-*g*][1]benzopyran-7-one;
6-hydroxy-7-methoxy-5-benzofuranacrylic acid δ-lactone; 8-methoxypsoralen;
Ammoidin; Meladinine ; Meloxine; oxypsoralen; Puvalen; Puvamet; Xanthotoxin
**Mol. Formula** $C_{12}H_8O_4$                     **Mol. Wt.** 216.20
**Uses** Used with long wave UV light in the treatment of vitiligo and psoriasis and as a pigmentation agent.
**Occurrence** Occurs in *Angelica* and *Ammi* species. It is also produced by the fungus *Sclerotinia sclerotiorum* which causes pink rot in celery.

## Physical properties
**M. Pt.** 148-150°C.

### Solubility
Organic solvent: acetic acid, acetone, benzene, chloroform, ethanol, petroleum ether, propylene glycol

## Environmental fate

### Degradation studies
Readily hydrolysed, whereby the lactone ring is opened (1).

## Mammalian and avian toxicity

### Acute data
$LD_{50}$ oral rat, mouse 420, 790 mg $kg^{-1}$, respectively (2,3).
$LD_{50}$ intraperitoneal rat, mouse 160, 470 mg $kg^{-1}$, respectively (2,3).
$LD_{50}$ subcutaneous mouse 860 mg $kg^{-1}$ (4).

### Carcinogenicity and long-term effects
Sufficient evidence for carcinogenicity to humans and animals, IARC classification group 1 (with UV radiation) (5).
Intraperitoneal mouse 12 mg $kg^{-1}$ $day^{-1}$ for 1 yr induced no detectable toxic effects. However, administration of 4 mg $kg^{-1}$ followed by exposure to UV irradiation at 320-400 nm resulted in severe toxic effects including erythema, burns and liver damage (3).
Oral mouse (12 months) 0.6-40 mg $kg^{-1}$ $day^{-1}$ caused a significant increase in skin tumours or tumours of internal organs either alone or in combination with UV radiation (PUVA) (6,7).
Dermal mouse, 115 applications of 15 μg $cm^{-2}$ to the ear, followed by irradiation with $1.68 \times 10^4$ J $m^{-2}$ of 365 nm UV light. 37/40 treated animals developed skin tumours compared to 0/20 irradiated controls (8).
Dermal mouse (60 wk) 5 μg 2 × $wk^{-1}$ with subsequent exposure to UV light (300-400 nm) for 15-60 min. Subcutaneous malignant tumours (mammary adenocarcinomas, skin carcinomas and carcino-mixo-sarcomas) and lymphomas were seen in 43/100 treated animals compared with 11/25 controls exposed to UV light alone (9).
Intraperitoneal mouse (11 month) 0.4 mg $day^{-1}$ 6 days $wk^{-1}$ for 10 months. In 1 group of 20 mice, each injection was followed 1 hr later by a 20-60 min UV exposure at 250 nm. A 3rd group was exposed to UV irradiation alone. No epidermal tumours were observed in the 2nd group that received methoxsalen alone. The epidermal tumour incidence in the 2 groups exposed to UV light with and without methoxsalen were 10-20% and 20% respectively. In a concomitant experiment, mice were given intraperitoneal administration of 0.4 mg $day^{-1}$ 6 days $wk^{-1}$ for 6 wk 1 hr before 10 min exposure to UV light (>320 nm). All treated animals developed epidermal tumours (fibrosarcomas and squamous carcinomas of the ears and eye region) whereas none was observed in controls exposed only to UV light (10).
National Toxicology Program tested rats via gavage. No evidence of carcinogenicity in ♀ rats, clear evidence of carcinogenicity in ♂ rats (11).

### Metabolism and pharmacokinetics
Following oral administration of 10 mg $kg^{-1}$ of $^{14}$C-labelled methoxsalen to dogs, radioactivity recovered in urine represented 4-25% of the dose and in the faeces

47-95% of the dose within 4 days. In pigs 26-58% of the dose was recovered in urine indicating greater absorption (12).

Following intravenous administration of 2 mg kg$^{-1}$ to dogs, plasma t$_{1/2}$ 2.17 hr (13). Following intravenous administration to dogs <2% was excreted unchanged in the urine. Four urinary metabolites were isolated, 3 of which resulted from opening of the furan ring: 7-hydroxy-8-methoxy-2-oxo-2$H$-1-benzopyran-6-acetic acid; α,7-dihydroxy-8-methoxy-2-oxo-2$H$-1-benzopyran-6-acetic acid; and an unknown conjugate of the former at the 7-hydroxy position. The 4th metabolite, formed by opening of the pyrone ring, was an unknown conjugate of Z-3-(6-hydroxy-7-methoxybenzofuran-5-yl)-2-propenoic acid (14).

## Genotoxicity

*Salmonella typhimurium* TA97, TA98, TA100, TA1535, TA1537, TA1538, with and without metabolic activation marginally positive (15).

*Saccharomyces cerevisiae* forward mutation assay positive (16).

*In vitro* mouse lymphoma L5178Y cells, tk$^+$/tk$^-$ positive (17).

*In vitro* human fibroblasts, DNA damage positive. Pre-treatment with UV light increased the rate of DNA damage by 50% (18).

*In vivo* rodent bone marrow, induction of chromosomal aberrations and micronuclei negative (PVVA) (19).

*In vitro* human lymphocytes, sister chromatid exchanges negative (20).

## Any other adverse effects to man

Basal and squamous cell skin cancers have been reported in patients treated with methoxsalen and long wave UV light (PUVA) for psoriasis or mycosis fungoides (21-23).

3 cases of malignant melanoma of the skin have been reported in patients treated with methoxsalen and UV light (24,25).

In a follow up study of 1380 PUVA treated patients, the standardised incidence ratio for squamous cell carcinoma increased from 4.1 at low doses to 22.3 at medium doses and 56.8 at high doses. This effect was reported to be independent of possible confounding effects of therapy with ionising radiation and topical tar. The effect on basal-cell cancer incidence was much weaker (standardised incidence ratio 4.5 for high doses) (26).

## Any other adverse effects

Intraperitoneal ♂ rat, single injection of 5 or 10 mg kg$^{-1}$ at the end of a 14 hr light phase. After 2 hr when the normal nocturnal surge of $N$-acetyltransferase activity and melatonin content in the pineal gland had begun in controls, $N$-acetyltransferase was increased. Melatonin content was unaffected (27).

Methoxsalen forms cyclobutane mono- and di-adducts with pyrimide bases of DNA by photoaddition under UV irradiation. The regions with alternate sequence of A-T appear to be the best reception sites for the formation of mono-adducts, while the regions containing an alternate sequence of A-T and G-C appear to be the preferential sites for cross-linking (28).

## Any other comments

Use, occurrence, analysis, carcinogenicity, mammalian toxicity, metabolism and mutagenicity reviewed (1,29,30).

# References

1. *IARC Monograph* 1980, **24**, 101-124
2. Apostolou, A. et al *Drug Chem. Toxicol.* 1979, **2**, 309
3. Hakim, et al *J. Pharmacol. Exp. Ther.* 1961, **131**, 394-399
4. *Drugs in Japan: Ethical Drugs* 6th ed., 1982, 837, Jakugyo Jiho, Tokyo
5. *IARC Monograph* 1987, **Suppl. 7**, 243-244
6. Pathak, M. A. et al *Clin. Res.* 1978, **26**, 300A
7. Langner, A. et al *J. Invest. Dermatol.* 1977, **69**, 451-457
8. Dubertnet, L. et al *Br. J. Dermatol.* 1979, **101**, 379-389
9. Santamaria, L. et al *Bull. Chim. Farm.* 1979, **118**, 356-362
10. Griffin, A. C. et al *J. Invest. Dermatol.* 1958, **31**, 289-295
11. *National Toxicology Program Research and Testing Division* 1992, Report No. TR-359, NIEHS, Research Triangle Park, NC
12. Ikeda, G. J. et al *Food Chem. Toxicol.* 1990, **28**(5), 333-338
13. Monbaliu, J. G. et al *Biopharm. Drug Dispos.* 1988, **9**(1), 9-17
14. Kolis, S. J. *Drug Metab. Dispos.* 1979, **7**, 220-225
15. Klopman, G. et al *Mutat. Res.* 1990, **228**(1), 1-50
16. Averbeck, D. et al *Mutat. Res.* 1985, **148**, 47-57
17. Arlett, C. F. et al *Clin. Exp. Dermatol.* 1980, **5**, 147-150
18. Ljungman, M. *Carcinogenesis (London)* 1989, **10**(3), 447-451
19. Shelby, M. D. *Mutat. Res.* 1988, **204**(1), 3-15
20. Lambut, B. et al *Acta Derm.-Venereol.* 1978, **58**, 13-16
21. Roenigk, H. H. et al *J. Am. Acad. Dermatol.* 1981, **4**, 319-322
22. Abel, E. A. et al *J. Am. Acad. Dermatol.* 1981, **4**, 423-429
23. Stern, R. et al *J. Invest. Dermatol.* 1982, **78**, 147-149
24. Frenk, E. *Dermatologica* 1983, **167**, 152-154
25. Marx, J. L. et al *J. Am. Acad. Dermatol.* 1983, **9**, 904-911
26. Li, K. et al *Endocr. Res.* 1987, **13**(1), 43-48
27. Stern, R. S. et al *New Engl. J. Med.* 1984, **310**, 1156-1161
28. Dall'Acqua, F. et al *J. Invest. Dermatol.* 1979, **73**, 191-197
29. *Martindale. The Extra Pharmacopoeia* 30th ed., 1993, 763-765, The Pharmaceutical Press, London
30. *ECETOC Technical Report No. 30(4)* 1991, European Chemical Industry Ecology and Toxicology Centre, B-1160 Brussels

# M120    2-Methoxyacetic acid

$CH_2(OCH_3)CO_2H$

**CAS Registry No.** 625-45-6
**Synonyms** methoxyacetic acid; methylglycolic acid
**Mol. Formula** $C_3H_6O_3$                    **Mol. Wt.** 90.08
**Uses** Catalyst.

## Physical properties

**B. Pt.** 202-204°C; **Flash point** >110°C; **Specific gravity** $d_4^{20}$ 1.174.

**Solubility**
Water: miscible. Organic solvent: diethyl ether, ethanol

## Mammalian and avian toxicity

### Acute data
$LD_{Lo}$ oral rat 2000 mg $kg^{-1}$ (1).

### Teratogenicity and reproductive effects
Intraperitoneal rat lowest toxic dose 225 mg $kg^{-1}$ on day 8 of gestation, foetotoxic effects (2).
Oral rat, lowest toxic dose single dose of 590 mg $kg^{-1}$ induced abnormal sperm (3).

### Genotoxicity
*In vitro* mouse embryo, inhibition of DNA synthesis (thymidine incorporation) positive (4).
*In vivo* mouse exposed orally to 50, 100, 300, 600 or 900 mg $kg^{-1}$ (single dose). Results were evaluated 2, 7, 14, 28 and 45 day after treatment. Induced cytotoxic damage on primary spermatocytes. Affected nucleic acid synthesis and spermatid morphology (5).

### Any other adverse effects to man
Extremely destructive to tissue of the mucous membranes and upper respiratory tract, eyes and skin. Inhalation may be fatal as a result of spasm, inflammation and oedema of the larynx and bronchi, chemical pneumonitis and pulmonary oedema (6).

### Any other comments
Human health effects, experimental toxicology, physico-chemical properties reviewed (7).

### References
1. *Fundam. Appl. Toxicol.* 1982, **2**, 158
2. Brown, N. A. et al *Toxicol. Lett.* 1984, **22**, 93-100
3. *Teratog., Carcinog., Mutagen.* 1987, **7**, 141
4. Stediman, D. B. et al *Toxicol. Lett.* 1989, **45**, 111-117
5. Spano, M. et al *J. Toxicol. Environ. Health.* 1991, **34**(1), 157-176
6. Lenga, R. E. (Ed.) *Sigma-Aldrich Library of Chemical Safety Data* 2nd ed., 1988, **2**, 2237, Sigma-Aldrich, Milwaukee, WI
7. *ECETOC Technical Report No. 30(4)* 1991, European Chemical Industry Ecology and Toxicology Centre, B-1160 Brussels

# M121   1-Methoxyacetone

## $CH_3C(O)CH_2OCH_3$

**CAS Registry No.** 5878-19-3
**Synonyms** acetonyl methyl ether; 1-methoxy-2-propanone; methoxyacetone
**Mol. Formula** $C_4H_8O_2$                    **Mol. Wt.** 88.11
**Uses** Organic synthesis.

## Physical properties
**B. Pt.** 118°C; **Flash point** 25°C; **Specific gravity** $d^{20}$ 0.957.

## Solubility
Water: miscible. Organic solvent: diethyl ether, ethanol

## Mammalian and avian toxicity

### Acute data
$LD_{50}$ oral rat 9 g $kg^{-1}$ (1).
$LD_{50}$ dermal rabbit >20 g $kg^{-1}$ (1).

### Irritancy
Dermal rabbit (24 hr) 500 mg caused mild irritation. 500 mg instilled into rabbit eye for 24 hr caused mild irritation (2).

## References
1. *Toxicol. Appl. Pharmacol.* 1974, **28**, 313
2. Marhold, J. V. *Prehled Prumyslove Toxikologie: Organicke Latky* 1986, Prague

# M122   Methoxychlor

**CAS Registry No.** 72-43-5
**Synonyms** 1,1'-(2,2,2-trichloroethylidene)bis(4-methoxybenzene);
1,1-bis(*p*-methoxyphenyl)-2,2,2-trichloroethane; 2,2-bis(*p*-methoxyphenyl)-
1,1,1-trichlorethane; dimethoxy-DDT; di(*p*-methoxyphenyl)trichloromethylmethane;
DMDT; *p,p'*-DMDT ; Maralate; Metox; 1,1,1-trichloro-2,2-di(4-methoxyphenyl)ethane
**Mol. Formula** $C_{16}H_{15}Cl_3O_2$                    **Mol. Wt.** 345.66
**Uses** Insecticide. Ectoparasiticide.
**Occurrence** Residues have been isolated from water, sediments, soil, crops and in animal tissues (1-3).

## Physical properties
**M. Pt.** 89°C; **Specific gravity** $d^{25}$ 1.41; **Partition coefficient** log $P_{ow}$ 4.68-5.08 (4);
**Volatility** v.p. $1.43 \times 10^{-6}$ mmHg at 25°C; v. den. 12.

## Solubility
Water: 0.1 mg $l^{-1}$ at 25°C. Organic solvent: acetone, ethanol, benzene, chloroform, toluene, methanol, petroleum oils, xylene

## Occupational exposure
**US TLV (TWA)** 10 mg $m^{-3}$; **UK Long-term limit** 10 mg $m^{-3}$.

## Ecotoxicity

### Fish toxicity
$LC_{50}$ (96 hr) fathead minnow, bluegill sunfish, rainbow trout, brown trout, coho salmon, king salmon, yellow perch 7.5-67 $\mu g\ l^{-1}$ (4-8).

### Invertebrate toxicity
$LC_{50}$ (48 hr) *Daphnia* 0.8 $\mu g\ l^{-1}$ (8).
$LC_{50}$ (96 hr) *Asellus brevicaudus, Gammarus lacustris, Gammarus fasciatus, Palaemonetes kadiakensis, Orconectes nais* 0.5-3.2 $\mu g\ l^{-1}$ (9, 10).
$EC_{50}$ (96 hr) *Chlorella vulgaris* 49 $\mu g\ l^{-1}$ (11).

### Bioaccumulation
Bioconcentration factor for fathead minnow 8300, sheepshead minnow 140 (12, 13).

## Environmental fate

### Degradation studies
Biodegradation by estuarine sediment/water system, $t_{1/2}$ <2 wk. In aerobic sediments $t_{1/2}$ >100 days. Under anaerobic conditions the major degradation products are dechlorinated methoxychlor and mono- and dihydroxy derivatives (14, 15).

### Abiotic removal
Degraded in water by UV irradiation at 254 nm and hydrogen peroxide treatment (16). Hydrolysis $t_{1/2}$ 367 days at pH 3-7 and 270 days at pH 9. At pH 7 ethanone, 2-hydroxy-1,2-bis(methoxyphenyl) and ethanedione, bis(methoxyphenyl) are the major hydrolysis products. At pH 10 1,1-bis(*p*-methoxyphenyl)-2,2-dichloroethylene is the major product. This is also the major product of photolysis, with 4-methoxybenzaldehyde a minor photolysis product (17-19).
Volatilisation $t_{1/2}$ 4.5 days in model river water. This neglects the effect of adsorption which may increase the volatilisation $t_{1/2}$ to 2 yr (20,21).
Reaction with photochemically produced hydroxyl radicals in the atmosphere, estimated $t_{1/2}$ 6.8 hr (22).

### Absorption
Soil $K_{oc}$ range from 9700 in clay soil to 100,000 in silt (23).

## Mammalian and avian toxicity

### Acute data
$LD_{50}$ oral rat, mouse 1000, 7000 mg $kg^{-1}$, respectively (24, 25).
$LD_{50}$ dermal rabbit >2000 mg $kg^{-1}$ (8).
$LD_{50}$ intraperitoneal hamster 500 mg $kg^{-1}$ (26).

### Sub-acute data
Oral rat, monkey 400, 1000 or 2500 mg $kg^{-1}$ $day^{-1}$ for 3 or 6 months caused damage to the liver and small intestine (27).
$LD_{50}$ (5 day) oral bobwhite quail, Japanese quail, ring necked pheasant, mallard duck >5000 mg $kg^{-1}$ diet (28).
Dermal rabbit (13 wk) 2-3 ml of 30% solution 5 days $wk^{-1}$ produced paralysis of the forelimbs, some fatty degeneration of the liver and lesions of the central nervous system (29).

---

## Carcinogenicity and long-term effects

No adequate evidence for carcinogenicity to humans or animals, IARC classification group 3 (30).

Oral dog, pig (2 yr) 1000, 2000 or 4000 mg kg$^{-1}$ diet. In dogs the 2 higher doses produced nervousness, apprehension, excess salivation, tremors and convulsions. Nephritis and mammary hyperplasia were observed in the pigs at autopsy (28).

Oral mouse (92 wk) 750-2800 mg kg$^{-1}$ diet for 1 wk, when doses were increased 1000-3500 mg kg$^{-1}$ diet for 77 wk. There was no significant increase in the incidence of benign and malignant neoplasms compared with controls (31).

Oral rat (2 yr) 360-1500 mg kg$^{-1}$ diet. Haemangiosarcomas of the spleen in ♂ rats were the only tumours that showed an increased incidence (31).

Dermal mouse, 0.1 or 10 mg animal$^{-1}$. The mean survival time ranged from 342 days for low dose ♀ mice to 450 days in the other group. No skin tumours were observed (32).

## Teratogenicity and reproductive effects

Oral rat, single dose of 50, 100, 200 or 400 mg kg$^{-1}$. 200 and 400 mg kg$^{-1}$ doses were foetotoxic. A dose-related increase in the incidence of wavy ribs was induced by doses of 100, 200 and 400 mg kg$^{-1}$ (33).

Gavage ♀ mouse (2-4 wk) 1.25, 2.5 or 5.0 mg animal$^{-1}$ day$^{-1}$ 5 days wk$^{-1}$. A dose-related induction of persistent vaginal oestrus, and an increase in the number of atretic large follicles, indicating potential loss in fertility, were observed. These effects were very similar to those induced by oestrogens (34).

Intraperitoneal ♂ mouse 0.1 or 1.0 mg animal$^{-1}$ day$^{-1}$ on days 1-9 of age. Body weight and mortality were unaffected. Serum T concentrations were reduced. The high dose significantly decreased DNA content of the seminal vesicles, bulbo-urethral glands and ventral prostrate, but not of the testes, epididymides and efferent ductules. The lower dose decreased DNA content only of the bulbo-urethral glands and seminal vesicles. These effects were similar to those induced by 17β-oestradiol (35).

## Metabolism and pharmacokinetics

Metabolised by human and rat hepatic cytochrome P$_{450}$ monooxygenases by sequential demethylations to mono- and bis-didemethylated phenolic derivatives and a trihydroxy derivative 1,1,1-trichloro-2-(4-hydroxyphenyl)-2-(3,4-dihydroxyphenyl)ethane (36).

In mice 98% of orally administered methoxychlor was eliminated in the urine as conjugated metabolites within 24 hr (37).

Weanling rats fed 500 mg kg$^{-1}$ diet for 4-18 wk stored 14-36 mg kg$^{-1}$ in fat. Equilibrium was reached within 4 wk and methoxychlor was cleared from the fatty tissue within 2 wk after the end of exposure. Rats fed 100 mg kg$^{-1}$ diet stored 1-7 mg kg$^{-1}$ fat. None was stored in rats fed 25 mg kg$^{-1}$ diet. No sex differences were observed (38).

Residues were found in milk after cows were sprayed with aqueous suspensions of methoxychlor. A maximum level of 0.1 mg l$^{-1}$ was found after 1 day, and detectable levels persisted for 1 wk (39).

# Genotoxicity

*Salmonella typhimurium* TA97, TA98, TA100, TA1535, TA1537, TA1538, with and without metabolic activation negative (40).

*Escherichia coli* WP2, *Saccharomyces cerevisiae* D3 mutagenicity assays negative (41).

*Drosophila melanogaster* sex-linked recessive lethal assay negative (42).

*In vitro* mouse lymphoma L5178Y cells, tk$^+$/tk$^-$ with metabolic activation positive (43).

*In vitro* Chinese hamster ovary cells, chromosomal aberrations negative, sister chromatid exchanges positive (metabolic activation unspecified) (43).

*In vivo* mouse bone marrow cells, chromosome damage negative (44).

*In vivo* mouse, a slight increase in chromosome breakage as observed in spermatogonia, but there was no increase in sperm abnormalities and spermatocyte metaphases revealed no significant cytological damage (44).

*In vivo* mouse dominant lethal assay negative (44).

### Any other adverse effects to man

Oral man, 100 mg kg$^{-1}$ diet for 2 yr produced no toxic symptoms. 500 mg kg$^{-1}$ diet produced unspecified tissue changes (45).

### Legislation

EEC maximum residue limit for fruit and vegetables 10 ppm (8).

Limited under EC Directive on Drinking Water Quality 80/778/EEC. Pesticides: maximum admissible concentration 0.1 μg l$^{-1}$ (46).

Included in Schedule 6 (Release into Land: Prescribed Substances) Statutory Instrument No. 472, 1991 (47).

The log $P_{ow}$ value exceeds the European Community recommended level 3.0 (6th and 7th amendments) (48).

### Any other comments

Environmental fate reviewed (3).

WHO Guideline Value for drinking water 20 μg l$^{-1}$ (49).

UK Advisory Value for drinking water 20 μg l$^{-1}$ (50).

Use, occurrence, physical properties, analysis, carcinogenicity, mammalian toxicity, teratogenicity, metabolism and mutagenicity reviewed (1,51-55).

### References

1. *IARC Monograph* 1979, **20**, 259-281
2. Verschueren, K. *Handbook of Environmental Data on Organic Chemicals* 2nd ed., 1983, 821-824, Van Nostrand Reinhold, NY
3. Howard, P. H. *Handbook of Environmental Fate and Exposure Data for Organic Chemicals* 1991, **3**, 502-514, Lewis Publishers, Chelsea, MA
4. Harsch, C. et al *Medchem Project Issue No. 26*, 1985, Pomona College, Claremont, CA
5. Merna, J. W. *Report* 1971, Institute for Fisheries Research, Michigan Dept. of Natural Resources, Ann Arbor, MI
6. Henderson, C. et al *Trans. Am. Fish Soc.* 1959, **88**(1), 23-32
7. Oseid, D. M. et al *Water Res.* 1974, **8**, 781-788
8. *The Agrochemicals Handbook* 3rd ed., 1991, RSC, London
9. Sanders, H. O. *Toxicity of Pesticides to the Crustacean, Gammarus lacustris* 1969, Bureau of Sport Fisheries and Wildlife Technical Paper 25, Washington, DC
10. Sanders, H. O. *The Toxicity of Some Insecticides to Four Species of Malocostracan Crustacea* 1972, Fish Pesticide Research Laboratory, Columbia, MO
11. *ECETOC Technical Report No. 56* 1993, European Chemical Industry Ecology and Toxicology Centre, B-1160 Brussels
12. Parrish, R. R. et al *Chronic Toxicity of Methoxychlor, Malathion and Carbofuran to Sheephead Minnows* 1977, NTIS PB-272101

13. Veith, G. D. et al *J. Fish Res. Board Can.* 1979, **36**, 1040-1048
14. Walker, W. W. et al *Chemosphere* 1988, **17**(12), 2255-2270
15. Muir, D. C. G. et al *J. Environ. Sci. Health, Part B* 1984, **B19**, 271
16. Thiemann, W. et al *DVGW-Schriftenr., Wasser* 1988, **107**, 129-146
17. Wolfe, N. L. et al *Environ. Sci. Technol.* 1977, **11**, 1077
18. Kollig, H. P. et al *Hydrolysis Rate Constants, Partition Coefficients and Water Solubilities for 129 Chemicals* 1987, USEPA, Washington, DC
19. Zepp, R. G. et al *J. Agric. Food Chem.* 1976, **24**, 727
20. *EXAMS II Exposure Analysis Modeling System* 1986, USEPA, Athens, G.A
21. Lyman, W. K. et al *Handbook of Chemical Property Estimation Methods Environmental Behaviour of Organic Compounds* 1982, McGraw-Hill, New York
22. Atkinson, R. L. et al *Environ. Pollut., Ser. A* 1980, **2**, 111
23. Karickhoff, S. W. et al *Water Res.* 1979, **13**, 241
24. Hodge, H. C. et al *J. Pharmacol. Exp. Ther.* 1950, **99**, 140
25. *J. Agric. Food Chem.* 1977, **25**, 859
26. *Arch. Toxicol.* 1985, **58**, 152
27. Stein, A. A. *Ind. Med. Surg.* 1968, **37**, 540-541
28. Hill, E. F. et al *Lethal Dietary Toxicities of Environmental Pollutants to Birds* 1975, US Fish and Wildlife Service, Report Wildlife No. 191, Washington, DC
29. Hoag, H. B. et al *Arch. Int. Pharmacodyn.* 1950, **83**, 491-504
30. *IARC Monograph* 1987, **Suppl. 7**, 66
31. National Cancer Institute *Bioassay of Methoxychlor for Possible Carcinogenicity* 1978, TR-35, NIH-78-835, Washington, DC
32. Hodge, H. C. et al *J. Pharmacol. Exp. Ther.* 1966, **9**, 583-596
33. Khera, K. S. et al *Toxicol. Appl. Pharmacol.* 1978, **45**, 435-444
34. Martinez, E. M. et al *Reprod. Toxicol.* 1991, **5**(2), 139-147
35. Cooke, P. S. et al *Biol. Reprod.* 1990, **42**(3), 585-596
36. Kupfer, D. et al *Chem. Res. Toxicol.* 1990, **3**(1), 8-16
37. Kapoor, I. P. et al *J. Agric. Food Chem.* 1970, **18**, 1145-1152
38. Kunze, F. M. et al *Proc. Soc. Exp. Biol. (N.Y.)* 1975, 415-416
39. Cluett, M. L. et al *J. Agric. Food Chem.* 1960, **8**, 277-281
40. Klopman, G. et al *Mutat. Res.* 1990, **228**, 1-50
41. Poole, D. C. et al *Toxicol. Appl. Pharmacol.* 1977, **41**, 196
42. Benes, V. et al *Ind. Med.* 1969, **38**, 50-52
43. Caspary, W. J. et al *Mutat. Res.* 1988, **196**(1), 61-81
44. Degreave, N. et al *Mutat. Res.* 1977, **46**, 204
45. *Documentation of Threshold Limit Values for Substances in Air* 1974, 152, American Conference of Government Industrial Hygienists, Cincinnati, OH
46. *EC Directive Relating to the Quality of Water Intended for Human Consumption* 1982, 80/778/EEC, Office for Official Publications of the European Communities, 2 rue Mercier, L-2985 Luxembourg
47. *S. I. 1991 No. 472 The Environmental Protection (Prescribed Processes and Substances) Regulations* 1991, HMSO, London
48. *1967 Directive on Classification, Packaging, and Labelling of Dangerous Substances 67/548/EEC; 6th Amendment EEC Directive 79/831/EEC; 7th Amendment EEC Directive 91/32/EEC* 1991, HMSO, London
49. *Guidelines for Drinking Water Quality* 2nd ed., 1993, **1**, WHO, Geneva.
50. *Guidance on Safeguarding the Quality of Public Water Supplies* 1989, 93-104, HMSO, London
51. Wilber, S. B. et al *Drinking Water Criteria Document for Methoxychlor* 1990, ECAO-CIN-425, Order No. PB91-143461
52. *US EPA Report* 1987, Order No. PB87-200176
53. Kupfer, D. et al *Rev. Biochem. Toxicol.* 1987, **8**, 183-215

54. *Dangerous Prop. Ind. Mater. Rep.* 1987, **7**(5), 79-87
55. *ECETOC Technical Report No. 30(4)* 1991, European Chemical Industry Ecology and Toxicology Centre, B-1160 Brussels

# M123   7-Methoxycoumarin

**CAS Registry No.** 531-59-9
**Synonyms** herniarin; 7-methoxy-2*H*-1-benzopyran-2-one; ayapanin; methylumbelliferone
**Mol. Formula** $C_{10}H_8O_3$                    **Mol. Wt.** 176.17
**Occurrence** Present in citrus plant oils.

## Physical properties
**M. Pt.** 118-120°C.

## Solubility
Water: miscible. Organic solvent: methanol

## Mammalian and avian toxicity

### Acute data
$LD_{50}$ oral rat 4300 mg kg$^{-1}$ (1).
$LD_{50}$ dermal guinea pig >5000 mg kg$^{-1}$ (1).

### Carcinogenicity and long-term effects
A CASE study designated 7-methoxycoumarin as a marginal carcinogen (2).

### Sensitisation
Negative when investigated by 2 different methods for identifying sensitising capacity (details unspecified) (3).

## References
1. *Food Chem. Toxicol.* 1988, **26**, 375
2. Rosenkranz, H. S. *Carcinogenesis (London)* 1990, **11**(2), 349-353
3. Hanson, B. M. et al *Contact Dermatitis* 1986, **15**(3), 157-163

# M124   2-(2-Methoxyethoxy)ethanol

$$CH_3OCH_2CH_2OCH_2CH_2OH$$

**CAS Registry No.** 111-77-3
**Synonyms** diethylene glycol monomethyl ether; diglycol monomethyl ether; Dowanol DM; Jeffosol DM; methoxydiglycol; Methyl carbitol; Poly-solv DM
**Mol. Formula** $C_5H_{12}O_3$                    **Mol. Wt.** 120.15

**Uses** Adsorbant for hydrogen sulfide from natural gas and coal gas. Corrosion inhibitor for fuels and lubricating oils. Insect repellent. Solvent.
**Occurrence** Residues have been isolated from effluent and drinking waters (1).

## Physical properties

**M. Pt.** –70°C; **B. Pt.** 194.2°C; **Flash point** 93°C (open cup); **Specific gravity** $d_4^{20}$ 1.0354; **Partition coefficient** log $P_{ow}$ -0.68 (calc) (2); **Volatility** v.p. 0.2 mmHg at 20°C; v. den. 4.14.

### Solubility

Water: miscible. Organic solvent: acetone, diethyl ether, dimethyl formamide, ethanol, glycerol

## Ecotoxicity

### Fish toxicity

$LC_{50}$ (96 hr) bluegill sunfish 7500 mg $l^{-1}$ static bioassay at 23°C (3).

### Bioaccumulation

Calculated bioconcentration factor of 6 indicates that environmental accumulation is unlikely (4).

## Environmental fate

### Degradation studies

$BOD_5$ 5% of ThOD (5).

### Abiotic removal

Reaction with photochemically produced hydroxyl radicals in the atmosphere, estimated $t_{1/2} \approx 6$ hr (6).

### Absorption

Calculated $K_{oc}$ of 10 indicates that adsorption to soil and sediments is not significant (4).

## Mammalian and avian toxicity

### Acute data

$LD_{50}$ oral rat, mouse, guinea pig 4200-9200 mg $kg^{-1}$ (7, 8).
$LD_{50}$ dermal rabbit 6500 mg $kg^{-1}$ (9).
$LD_{Lo}$ intraperitoneal rat, mouse 2600-3000 mg $kg^{-1}$ (8, 10).

### Sub-acute data

Oral ♂ rat (20 days) 500, 1000 or 2000 mg $kg^{-1}$ for 1, 2, 5 or 20 days. The highest doses caused a reduction in liver weight. An increase in hepatic microsomal protein, and cytochrome $P_{450}$ were induced. Cytochrome $b_5$ and the reduced form of NADP cytochrome $c$ reductase were unaffected. The activity of cytosolic ADH was also unaffected (11).
Dermal guinea pig (13 wk) 0, 40, 200 or 1000 mg $kg^{-1}$ $day^{-1}$ 5 days $wk^{-1}$. The 2 high doses caused a decrease in spleen weight. Testicular weight was unaffected. The high dose induced an increase in lactate dehydrogenase activity, while all doses caused an increase in urinary calcium excretion (12).

### Teratogenicity and reproductive effects

Oral mouse, 500 mg $kg^{-1}$ on day 11 of gestation was embryotoxic (13).

Oral mouse, 720-5200 mg kg$^{-1}$ day$^{-1}$ on days 7-16 of gestation. At the high dose 2/9 dams died, all litters were resorbed and maternal body weight gain was reduced. At a dose of 3400 mg kg$^{-1}$ day$^{-1}$ 6/9 litters were resorbed. A dose-related increase in malformations, primarily of the ribs and cardiovascular system were recorded (14).

**Metabolism and pharmacokinetics**

An oral dose of 500 mg kg$^{-1}$ to pregnant mice was metabolised predominantly by O-demethylation to 2-(2-methoxyethoxy)ethanol with subsequent oxidation to (2-methoxyethoxy)acetic acid. Urinary excretion of this metabolite over 48 hr accounted for 63% of the administered dose. A smaller percentage of the administered dose was metabolised at the central ether linkage to produce 2-methoxyethanol which was further metabolised by alcohol dehydrogenase to methoxyacetic acid, a powerful developmental toxicant. Urinary excretion of this metabolite accounted for 26% of the administered dose within 48 hr. Unchanged 2-(2-methoxyethoxy)ethanol and methoxyacetic acid were detected in embryonic tissues (13).

**Irritancy**

500 mg instilled into rabbit eye for 24 hr caused moderate irritation (15).

## Legislation

Included in Schedule 6 (Release into Land: Prescribed Substances) Statutory Instrument No. 472, 1991 (16).

## Any other comments

Environmental fate reviewed (1).
Human health effects, experimental toxicology, physico-chemical properties reviewed (17).

## References

1.  Howard, P. H. *Handbook of Environmental Fate and Exposure Data for Organic Chemicals* 1993, **4**, 214-220, Lewis Publishers, Chelsea, MA
2.  *GEMS: Graphical Exposure Modeling System: Fate of Atmospheric Pollutants*, 1986, Washington, DC
3.  Geynor, W. D. et al *J. Hazard. Mater.* 1975/77, **1**
4.  Lyman, W. K. et al *Handbook of Chemical Property Estimation Methods Environmental Behaviour of Organic Compounds* 1982, McGraw-Hill, New York
5.  Bridie, A. L. et al *Water Res.* 1979, **13**, 627-630
6.  Atkinson, R. *Int. J. Chem. Kinet.* 1987, **19**, 799-828
7.  Smyth, E. M. et al *J. Ind. Hyg. Toxicol.* 1941, **23**, 259
8.  *Gig. Naselennykh Mest.* 1990, **29**, 57
9.  *Union Carbide Data Sheet* 21 Apr. 1967, Union Carbide Corp., New York
10.  *J. Pharm. Pharmacol.* 1959, **11**, 150
11.  Kawamoto, T. et al *Toxicology* 1990, **62**(3), 265-274
12.  Hobson, D. W. et al *Fundam. Appl. Toxicol.* 1986, **6**(2), 339-348
13.  Daniel, F. B. et al *Fundam. Appl. Toxicol.* 1991, **16**(3), 567-575
14.  Hardin, B. D. et al *Fundam. Appl. Toxicol.* 1986, **6**(3), 430-439
15.  Marhold, J. V. *Prehled Prumyslove Toxikologie: Organicke Latky* 1986, Prague
16.  *S. I. 1991 No. 472 The Environmental Protection (Prescribed Processes and Substances) Regulations* 1991, HMSO, London
17.  *ECETOC Technical Report No. 30(4)* 1991, European Chemical Industry Ecology and Toxicology Centre, B-1160 Brussels

# M125   2-[2-(2-Methyoxyethoxy)ethoxy]ethanol

$$CH_3OCH_2CH_2OCH_2CH_2OCH_2CH_2OH$$

**CAS Registry No.** 112-35-6
**Synonyms** Dowanol TMAT; methoxytriethylene glycol; methoxtriglycol; triethylene glycol monomethyl ether; triglycol monomethyl ether; 3,6,9-trioxadecanol
**Mol. Formula** $C_7H_{16}O_4$                                **Mol. Wt.** 164.20
**Uses** Solvent.

## Physical properties

**M. Pt.** –44°C; **B. Pt.** 249°C; **Flash point** 118°C (open cup);
**Specific gravity** $d^{20}$ 1.0494.

### Solubility
Water: miscible.

## Mammalian and avian toxicity

### Acute data
$LD_{50}$ oral rat 11,300 mg $kg^{-1}$ (1).
$LD_{50}$ dermal rabbit 7100 mg $kg^{-1}$ (1).

### Sub-acute data
Dermal rabbit, (21 days) 1000 mg $kg^{-1}$ $day^{-1}$ did not produce systemic toxicity. The low rate of dermal adsorption may have played a role in this outcome (2).

### Teratogenicity and reproductive effects
Oral rat, 100 mg $kg^{-1}$ $day^{-1}$ on days 6-15 of gestation did not cause maternal or foetal toxicity or any teratogenic effects (2).

### Irritancy
Dermal rabbit (24 hr) 10 mg caused mild irritation (1).
500 mg instilled into rabbit eye for 24 hr caused mild irritation (3).

## Legislation

Included in Schedule 6 (Release into Land: Prescribed Substances) Statutory Instrument No. 472, 1991 (4).

## Any other comments

Human health effects, experimental toxicology, physico-chemical properties reviewed (5).

## References

1. *Am. Ind. Hyg. Assoc. J.* 1962, **23**, 95
2. Leber, A. P. et al *J. Am. Coll. Toxicol.* 1990, **9**(5), 507-515
3. Marhold, J. V. *Prehled Prumyslove Toxikologie: Organicke Latky* 1986, Prague
4. *S. I. 1991 No. 472 The Environmental Protection (Prescribed Processes and Substances) Regulations* 1991, HMSO, London
5. *ECETOC Technical Report No. 30(4)* 1991, European Chemical Industry Ecology and Toxicology Centre, B-1160 Brussels

# M126  6-Methoxyguanine

**CAS Registry No.** 20535-83-5
**Synonyms** 2-amino-6-methoxypurine; 6-methoxy-1*H*-purin-2-amine;
$O^6$-methylguanine
**Mol. Formula** $C_6H_7N_5O$                **Mol. Wt.** 165.16
**Uses** Alkylating agent. Antineoplastic agent.

## Genotoxicity

M13mp8 phage, incorporation of a single residue of 6-methoxyguanine in the phage DNA induced G to A transitions (1).

*In vitro* human lymphocytes, chromosomal aberrations positive (metabolic activation unspecified) (2).

*In vivo* rat bone marrow, kidney, lung, spleen and intestine modulated the activity of the DNA repair enzyme $O^6$-alkylguanine-DNA alkyltransferase (3).

## Any other comments

*In vitro* Chinese hamster ovary and human cells, treatment with several alkylating agents indicated that the formation of 6-methoxyguanine is the major mutagenic lesion (4, 5).

## References

1.  Essingmann, J. M. et al *IARC Sci. Publ.* 1986, **70**, 393-399
2.  Abbondanalolo, A. et al *Prog. Clin. Biol. Res.* 1989, **318**, 335-343
3.  Dexter, E. Y. et al *Cancer Res.* 1989, **49**(13), 3520-3524
4.  Dunn, W. C. et al *Carcinogenesis (London)* 1991, **12**(1), 83-89
5.  Day, R. S. et al *J. Cell Sci., Suppl.* 1987, **6**, 333-353

# M127  Methoxymethylbenzene

**CAS Registry No.** 538-86-3
**Synonyms** benzyl methyl ether; methyl benzyl ether; α-methoxytoluene
**Mol. Formula** $C_8H_{10}O$                **Mol. Wt.** 122.17

**Occurrence** Aroma and taste components in plants.

**Physical properties**

**B. Pt.** 174°C; **Specific gravity** $d^{20}$ 0.987; **Partition coefficient** log $P_{ow}$ 1.35.

**Solubility**
Organic solvent: diethyl ether, ethanol, methanol

**Mammalian and avian toxicity**

**Acute data**
$LD_{50}$ oral rat 9800 mg $kg^{-1}$ (1).

**References**

1. *Am. Ind. Hyg. Assoc. J.* 1969, **30**, 470

# M128    4-Methoxy-4-methylpentan-2-one

## $CH_3C(O)CH_2C(CH_3)(OCH_3)CH_3$

**CAS Registry No.** 107-70-0
**Synonyms** 4-methoxy-4-methyl-2-pentanone; methoxyhexanone; Pentoxane; ME-6K
**Mol. Formula** $C_7H_{14}O_2$                              **Mol. Wt.** 130.19
**Uses** Azeotropic agent. Solvent.

**Physical properties**

**B. Pt.** 159.1°C; **Flash point** 60.5°C; **Specific gravity** $d_{25}^{25}$ 0.899; **Volatility** v.p. 3.16 mmHg at 25°C.

**Solubility**
Water: 280 g $l^{-1}$ at 25°C. Organic solvent: acetone, ethanol, *n*-hexane, allyl alcohol

**Occupational exposure**

**UN No.** 2293; **HAZCHEM Code** 3▨; **Conveyance classification** flammable liquid.
**Risk phrases** Flammable (R10)
**Safety phrases** Do not breathe vapour or spray (S23)

**Ecotoxicity**

**Fish toxicity**
$LC_{50}$ (24 hr) goldfish 3800 mg $l^{-1}$ (1).

**Bioaccumulation**
Calculated bioconcentration factor 0.5 indicates that environmental accumulation is unlikely (2).

**Environmental fate**

**Degradation studies**
$BOD_5$ 0.11 mg $l^{-1}$ $O_2$; COD 2.24 mg $l^{-1}$ $O_2$ (1).

**Abiotic removal**

Reaction with photochemically produced hydroxyl radicals, estimated $t_{1/2}$ 3.1 days (3). Laboratory studies demonstrated direct photolysis occurs at 313 nm in hexane, ethanol and allyl alcohol. Photolysis products included mesityl oxide; methanol; a hydroxyfuran derivative; and small amounts of acetone and methyl isoprenyl ether (4). Estimated volatilisation $t_{1/2}$ 22 days for model river water and 236 days for model pond water (2,5).

**Absorption**

Estimated $K_{oc}$ of 4.4 indicates that adsorption to soil and sediments is unlikely (6).

## Mammalian and avian toxicity

**Acute data**

$LD_{50}$ oral mouse 2100 mg $kg^{-1}$ (1).
$LC_{Lo}$ (15 hr) inhalation mouse 2300 ppm (1).
$LD_{Lo}$ dermal rabbit 3000 mg $kg^{-1}$ (1).

**Irritancy**

Dermal rabbit, 500 mg caused moderate irritation (exposure unspecified) (1).

## Genotoxicity

*Salmonella typhimurium* TA98, TA100, TA1535, TA1537 with and without metabolic activation negative (7).
*Saccharomyces cerevisiae* mitotic gene conversion negative (8).

## Legislation

Included in Schedule 6 (Release into Land: Prescribed Substances) Statutory Instrument No. 472, 1991 (9).

## Any other comments

Human health effects, experimental toxicology, physico-chemical properties reviewed (10).

## References

1. Shell Chemical Company *Report* 1961, 6
2. Lyman, W. K. et al *Handbook of Chemical Property Estimation Methods Environmental Behaviour of Organic Compounds* 1982, McGraw-Hill, New York
3. Atkinson, R. *J. Int. Chem. Kinet.* 1987, **19**, 799-828
4. Coyle, D. J. et al *J. Am. Chem. Soc.* 1964, **86**, 3850-3854
5. *EXAMS II Computer Simulation* 1987, US EPA, Washington, DC
6. Swann, R. L. et al *Residue. Rev.* 1983, **85**, 23
7. Zeiger, E. et al *Environ. Mutagen.* 1987, **9**(Suppl. 9), 1-110
8. Brooks, T. M. et al *Mutagenesis* 1988, **3**(3), 227-232
9. *S. I. 1991 No. 472 The Environmental Protection (Prescribed Processes and Substances) Regulations* 1991, HMSO, London
10. *ECETOC Technical Report No. 30(4)* 1991, European Chemical Industry Ecology and Toxicology Centre, B-1160 Brussels

# M129   2-Methoxy-4-nitroaniline

CAS Registry No. 97-52-9
**Synonyms** 2-methoxy-4-nitrobenzenamine; 4-nitro-*o*-anisidine;
2-amino-5-nitroanisole; Amorthol Fast Red B Base; C.I. 37125; Diazo Fast Red B;
Naphthenil Red B Base; Sanyo Fast Red B Base
**Mol. Formula** $C_7H_8N_2O_3$                    **Mol. Wt.** 168.15
**Uses** Organic synthesis. Manufacture of dyestuffs.

## Physical properties
**M. Pt.** 140-142°C; **Specific gravity** $d^{156}$ 1.211.

## Solubility
Organic solvent: dimethyl sulfoxide

## Mammalian and avian toxicity

**Acute data**
$LD_{50}$ oral rat 1000 mg $kg^{-1}$ (1).

**Irritancy**
Irritating to the eyes, skin, mucous membranes and upper respiratory tract (species
unspecified) (2).

## Genotoxicity
*Salmonella typhimurium* TA98, with metabolic activation positive (3).

## Any other adverse effects to man
Absorption into the body leads to the formation of methaemoglobin which in
sufficient amounts causes cyanosis (2).

## Any other comments
Human health effects, experimental toxicology, physico-chemical properties
reviewed (4).

## References
1.  Sax, N. I. et al *Dangerous Properties of Industrial Materials*, 8th ed., 1992, Van Nostrand
    Reinhold, New York
2.  Lenga, R. E. (Ed.) *Sigma-Aldrich Library of Chemical Safety Data* 2nd ed., 1988, **2**, 2258,
    Sigma-Aldrich, Milwaukee, WI
3.  Koovi, D. G. et al *Ann. Falsif. Expert. Clim. Toxicol.* 1987, **80**(854), 25-39
4.  *ECETOC Technical Report No. 30(4)* 1991, European Chemical Industry Ecology and
    Toxicology Centre, B-1160 Brussels

# M130   4-Methoxy-2-nitroaniline

CAS Registry No. 96-96-8
Synonyms 4-methoxy-2-nitrobenzenamine; 2-nitro-*p*-anisidine;
4-amino-3-nitroanisole
Mol. Formula $C_7H_{18}N_2O_3$                    Mol. Wt. 178.23
Uses Organic synthesis. Manufacture of dyestuffs.

## Physical properties

M. Pt. 123-126°C.

Solubility
Water: miscible. Organic solvent: acetone, diethyl ether, ethanol

## Occupational exposure

Supply classification very toxic.
Risk phrases Very toxic by inhalation, in contact with skin and if swallowed –
Danger of cumulative effects (R26/27/28, R33)
Safety phrases After contact with skin, wash immediately with plenty of soap and
water – Wear suitable protective clothing and gloves – In case of accident or if you
feel unwell, seek medical advice immediately (show label where possible) (S28,
S36/37, S45)

## Mammalian and avian toxicity

Acute data
$LD_{50}$ oral rat 14100 mg $kg^{-1}$ (1).
$LD_{50}$ oral redwing blackbird, starling >100 mg $kg^{-1}$ (2).

Irritancy
Irritating to the eyes, skin, mucous membranes and upper respiratory tract (species
unspecified) (3).

## Any other adverse effects to man

Absorption into the body leads to the formation of methaemoglobin which in
sufficient amounts causes cyanosis (3).

## Any other comments

Human health effects, experimental toxicology, physico-chemical properties
reviewed (4).
All reasonable efforts have been taken to find information on isomers of this
compound but no relevant data are available.

## References

1. Marhold, J. V. *Sbornik Vysledku Toxixologickeho Vysetreni Latek A Pripravku* 1972, Prague
2. Schafer, E. W. et al *Arch. Environ. Contam. Toxicol.* 1983, **12**, 355-382
3. Lenga, R. E. (Ed.) *Sigma-Aldrich Library of Chemical Safety Data* 2nd ed., 1988, **2**, 2259, Sigma-Aldrich, Milwaukee, WI
4. *ECETOC Technical Report No. 30(4)* 1991, European Chemical Industry Ecology and Toxicology Centre, B-1160 Brussels

# M131   4-Methoxyphenol

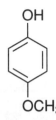

**CAS Registry No.** 150-76-5
**Synonyms** *p*-methoxyphenol; *p*-guaiacol; hydroquinone methyl ether; Leucobasal; Mequinol; PMF (antioxidant)
**Mol. Formula** $C_7H_8O_2$                    **Mol. Wt.** 124.14
**Uses** Antioxidant. Corrosion inhibitor. Organic synthesis. Polymerisation inhibitor. Manufacture of photographic compounds. Treatment of skin hyperpigmentation.

## Physical properties

**M. Pt.** 55-57°C; **B. Pt.** 243°C; **Flash point** 132°C (open cup); **Specific gravity** $d_{20}^{20}$ 1.55; **Partition coefficient** log $P_{ow}$ 1.34 (1).

### Solubility
Water: 40 g $l^{-1}$ at 25°C. Organic solvent: acetone, benzene, diethyl ether, ethanol, ethyl acetate

## Occupational exposure

**US TLV (TWA)** 5 mg $m^{-3}$; **UK Long-term limit** 5 mg $m^{-3}$.

## Ecotoxicity

### Fish toxicity
$LC_{50}$ (48 hr) goldfish $\approx$200 mg $l^{-1}$ (2).
$LC_{50}$ (96 hr) fathead minnow 140 mg $l^{-1}$, flowthrough bioassay (3).

## Environmental fate

### Degradation studies
Complete degradation by soil microflora in 8 days (4).

## Mammalian and avian toxicity

**Acute data**

$LD_{50}$ oral rat 1600 mg $kg^{-1}$ (5).
$LD_{50}$ intraperitoneal mouse 250 mg $kg^{-1}$ (6).

**Sub-acute data**

Oral rat (7 wk) 0.1-5% diet caused a dose related decrease in body weight gain (5).
Oral rat (duration not specified) 2% diet induced deep ulceration parallel to the
limiting ridge of the forestomach mucosa, with hyperplasia and mild hyperkeratosis in
the adjoining mucosa. These effects were similar to those caused by
3-*tert*-butyl-4-hydroxyanisole after 2 days administration (7).
Injection into the substantia nigra of rats led to a dose-related destruction of dopamine
neurons as indicated by a reduction in dopamine and its metabolites in the ipsilateral
striatum and loss of tyrosine hydroxylase immunoreactivity (8).

**Irritancy**

Dermal rabbit (12 days) 6000 mg $day^{-1}$ caused mild irritation (9).

## Genotoxicity

*In vitro* human peripheral lymphocytes inhibited DNA repair and semiconservative
DNA synthesis (10).

## Any other adverse effects to man

Severe reversible irregular hypopigmentation was reported in a West Indian woman
(11).

## Any other adverse effects

*In vitro* $IC_{50}$ (4 hr) mouse hepatocytes 32 mg $l^{-1}$ (12).

## Any other comments

Autoignition temperature 421°C.
All reasonable efforts have been taken to find information on isomers of this
compound but no relevant data are available.

## References

1. Camilleri, P. et al *J. Chem. Soc. Perkin Trans. II* 1988, (2), 1699-1707
2. McKee, J. E. et al *Water Quality Criteria* 1963, Resources Agency of California, State
   Water Quality Control Board
3. Schultz, T. W. et al *Ecotoxicol. Environ. Saf.* 1986, **12**(2), 146-153
4. Alexander, M. et al *J. Agric. Food Chem.* 1966, **14**, 410
5. *Patty's Industrial Hygiene and Toxicology* 2nd ed., 1963, **2**, 1688, John Wiley & Sons,
   New York
6. *NTIS Report AD691-490*, Natl. Tech. Inf. Serv., Springfield, VA
7. Altmann, H.-J. et al *Food Chem. Toxicol.* 1986, **24**(10/11), 1183-1158
8. Cumming, P. et al *Neuropharmacology* 1987, **26**(12), 1795-1797
9. *J. Ind. Hyg. Toxicol.* 1949, **31**, 79
10. Daugherty, J. P. et al *Res. Commun. Chem. Pathol. Pharmacol.* 1986, **54**(1), 133-136
11. Boyle, J. et al *Br. Med. J.* 1984, **288**, 1998-1999
12. Schiller, C. D. et al *Eur. J. Cancer* 1991, **27**(8), 1017-1022

# M132   Methoxyphenone

CAS Registry No. 41295-28-7
Synonyms (4-methoxy-3-methylphenyl)(3-methylphenyl)methanone; 4-methoxy-3,3′-dimethylbenzophenone; Kayametone; NK049
Mol. Formula $C_{16}H_{16}O_2$                     Mol. Wt. 240.30
Uses Catalyst.

## Physical properties
M. Pt. 62.0-62.5°C.

## Solubility
Water: 2 mg $l^{-1}$ at 20°C. Organic solvent: ethanol, diethyl ether

## Ecotoxicity

### Fish toxicity
$LC_{50}$ (48 hr) carp, goldfish 3.2-10 mg $l^{-1}$ (1).

### Abiotic removal
Slowly decomposed by sunlight (1).

## Mammalian and avian toxicity

### Acute data
$LD_{50}$ oral rat, mouse >4000 mg $kg^{-1}$ (2).
$LD_{50}$ dermal rat >4000 mg $kg^{-1}$ (2).

### Sub-acute data
Oral rat, mouse (90 day) no adverse effect level 1000-1500 mg $kg^{-2}$ (1).

## References
1.  *The Pesticide Manual* 8th ed., 1987, 556, British Crop Protection Council, Thornton Heath
2.  *Jpn. Pestic. Inf.* 1976, (27), 11

# M133   1-Methoxypropane

$$CH_3CH_2CH_2OCH_3$$

CAS Registry No. 557-17-5
Synonyms Metopryl; methyl propyl ether; Neothyl
Mol. Formula $C_4H_{10}O$                     Mol. Wt. 74.12

Uses Inhalation anaesthetic.

**Physical properties**

**B. Pt.** $_{761}$ 38.8°C; **Flash point** < −20°C; **Specific gravity** $d_4^{13}$ 0.7356.

**Solubility**
Water: 50 ml l$^{-1}$ at 25°C.

**Mammalian and avian toxicity**

**Acute data**
$LC_{50}$ (15 min) inhalation mouse 260 mg m$^{-3}$ (1).

**References**

1. *Anesthesiology* 1950, **11**, 455

# M134   1-Methoxy-2-propanol

## CH3CH(OH)CH2OCH3

**CAS Registry No.** 107-98-2
**Synonyms** Dowtherm 209; (±)-1-methoxypropan-2-ol; monopropylene glycol methyl ether; Propasol; propylene glycol monomethyl ether; Solvent M
**Mol. Formula** $C_4H_{10}O_2$                                                **Mol. Wt.** 90.12
**Uses** Solvent.

**Physical properties**

**M. Pt.** −96.7°C; **B. Pt.** 118-119°C; **Flash point** 33°C; **Specific gravity** $d_{25}^{25}$ 0.919;
**Volatility** v.p. 11.8 mmHg at 25°C; v. den. 3.11.

**Solubility**
Water: miscible. Organic solvent: diethyl ether, methanol

**Occupational exposure**

**UK Long-term limit** 100 ppm (360 mg m$^{-3}$); **UK Short-term limit** 300 ppm (1080 mg m$^{-3}$); **UN No.** 3092; **Conveyance classification** flammable liquid.
**Risk phrases** Flammable (R10)
**Safety phrases** Avoid contact with skin (S24)

**Ecotoxicity**

**Fish toxicity**
Not toxic to brown trout, bluegill sunfish, yellow perch or goldfish at 5 ppm after 24 hr exposure (1).

**Mammalian and avian toxicity**

**Acute data**
$LD_{50}$ oral rat, mouse 5700, 12,000 mg kg$^{-1}$, respectively (2, 3).
$LC_{Lo}$ (4 hr) inhalation rat 7000 ppm (4).

$LD_{50}$ subcutaneous rat 7800 mg $kg^{-1}$ (3).
$LD_{50}$ intravenous rat, rabbit 4200, 8000 mg $kg^{-1}$, respectively (3, 5).
$LD_{50}$ intraperitoneal rat 3700 mg $kg^{-1}$ (5).

**Sub-acute data**
Oral rat (13 wk) 0.5, 0.9, 1.8 or 3.6 mg $kg^{-1}$ $day^{-1}$ 5 days $wk^{-1}$ caused dose-related central nervous system depression, reduced food intake and growth depression. Liver enlargement was accompanied by cell necrosis mainly in the peripheral parts of the lobules. The high dose caused appreciable mortality. The high doses also caused kidney injury (3).
Oral rat (35 day) 3000 mg $kg^{-1}$ $day^{-1}$ on 26 days induced mild histopathological changes in the liver and kidneys (4).
Inhalation rat (13 wk) 300, 1000 or 3000 ppm 6 hr $day^{-1}$, 5 days $wk^{-1}$. The high dose group appeared sedated during the first 2 wk. Pathological changes were observed in the liver of the high dose group (6).
Dermal rabbit (3 month) 2, 4, 6 or 9 g $kg^{-1}$ $day^{-1}$. All the high dose and 8/9 of the 6 g $kg^{-1}$ $day^{-1}$ group died within 6 wk. Deaths were associated with loss of body weight and narcosis. The stomachs of these animals were distended with food indicating gastric retention. Renal necrosis or slight granular degeneration of the tubules was also observed (4).
Subcutaneous rat (4 wk) 0.5, 0.9, 1.8 or 3.7 g $kg^{-1}$ $day^{-1}$. Some fatalities occurred in the high dose group. Ataxia, prostration and liver lesions were seen at $\geq 1.8$ g $kg^{-1}$ $day^{-1}$, dyspnoea at $\geq 0.9$ g $kg^{-1}$ $day^{-1}$. Dose-related bodyweight loss and reduced urinary volume were observed (3).

**Teratogenicity and reproductive effects**
Oral and subcutaneous rat, mouse, rabbit 0.03-2.0 mg $kg^{-1}$ $day^{-1}$ on days 18-21 of gestation. Only rat foetuses showed any effect, a delayed ossification of the skull at the highest dose (0.7 mg $kg^{-1}$) (3).
Inhalation rat, lowest toxic concentration 3000 ppm 6 hr $day^{-1}$ on days 6-15 of gestation, teratogenic effects (7).

**Metabolism and pharmacokinetics**
Following oral administration of $^{14}C$-labelled substance, 11-25% of the radioactivity was found in urine, 0.7-1.5% in faeces and 57-63% was exhaled as carbon dioxide after 2 days (8).

**Irritancy**
Dermal rabbit 500 mg caused mild irritation. 230 mg instilled into rabbit eye caused mild irritation (exposure not specified) (5).

## Genotoxicity
*Salmonella typhimurium* TA98, TA100, TA1535, TA1537, TA1538 with and without metabolic activation negative (9).
*In vitro* Chinese hamster ovary cells, chromosomal aberrations negative (9).
*In vitro* primary rat hepatocytes, unscheduled DNA synthesis negative (9).

## Legislation
Included in Schedule 6 (Release into Land: Prescribed Substances) Statutory Instrument No. 472, 1991 (10).

## Any other comments

Physical properties, mammalian toxicity, teratogenicity and mutagenicity reviewed (11,12).

All reasonable efforts have been taken to find information on isomers of this compound but no relevant data are available.

## References

1. Wood, E. M. *The Toxicity of 3400 Chemicals to Fish* 1987, EPA 560/6-87-002; PB 87-200-275, Washington, DC
2. *Am. Ind. Hyg. Assoc. J.* 1962, **23**, 95
3. Sterger, E. G. et al *Arzneim.-Forsch.* 1972, **22**, 569-574
4. Rowe, V. K. et al *Arch. Ind. Hyg. Occup. Med.* 1954, **9**, 509
5. *Union Carbide Data Sheet* 15 Nov. 1971, Union Carbide Corp., New York
6. Lendry, T. D. et al *Fundam. Appl. Toxicol.* 1983, **3**, 627-630
7. *Fundam. Appl. Toxicol.* 1984, **4**, 784
8. Miller, R. R. et al *Toxicol. Appl. Pharmacol.* 1983, **67**, 229-237
9. McGregor, D. B. *Environ. Health. Perspect.* 1984, **57**, 97-103
10. *S. I. 1991 No. 472 The Environmental Protection (Prescribed Processes and Substances) Regulations* 1991, HMSO, London
11. *Toxicity Review No. 10: Glycol Ethers* 1985, HMSO, London
12. *ECETOC Technical Report No. 30(4)* 1991, European Chemical Industry Ecology and Toxicology Centre, B-1160 Brussels

# M135  1-Methoxy-2-propanol acetate

## $CH_3CH(OC(O)CH_3)CH_2OCH_3$

**CAS Registry No.** 108-65-6
**Synonyms** 2-acetoxy-1-methoxypropane
**Mol. Formula** $C_6H_{12}O_3$                       **Mol. Wt.** 132.16
**Uses** Solvent.

## Physical properties

**B. Pt.** 147°C.

## Ecotoxicity

**Fish toxicity**
Fatal to brown trout after 21 hr and to yellow perch after 24 hr at 5 ppm. Not toxic to bluegill sunfish or goldfish after 24 hr at 5 ppm. Test conditions: pH 7, dissolved oxygen content 7.5 ppm, total hardness (soap method) 300 ppm, methyl orange alkalinity 310 ppm, free carbon dioxide 5 ppm and temperature 12.8°C (1).

## Mammalian and avian toxicity

**Acute data**
$LD_{50}$ oral rat 8500 mg kg$^{-1}$ (2).
$LD_{50}$ dermal rat >5000 mg kg$^{-1}$ (2).
$LD_{50}$ intraperitoneal mouse 750 mg kg$^{-1}$ (3).

## Legislation

Included in Schedule 6 (Release into Land: Prescribed Substances) Statutory Instrument No. 472, 1991 (4).

## Any other comments

Human health effects, experimental toxicology, physico-chemical properties reviewed (5).

## References

1. Wood, E. M. *The Toxicity of 3400 Chemicals to Fish* 1987, EPA 560/6-87-002; PB 87-200-275
2. *Dow Chemical Company Report* MSD-1582
3. *NTIS Report* AD 691-490, Natl. Tech Inf. Ser., Springfield, VA
4. *S. I. 1991 No. 472 The Environmental Protection (Prescribed Processes and Substances) Regulations* 1991, HMSO, London
5. *ECETOC Technical Report No. 30(4)* 1991, European Chemical Industry Ecology and Toxicology Centre, B-1160 Brussels

# M136    2-Methoxy-1-propanol acetate

## $CH_3CH(OCH_3)CH_2OC(O)CH_3$

**CAS Registry No.** 70657-70-4
**Synonyms** 2-methoxypropyl acetate
**Mol. Formula** $C_6H_{12}O_3$          **Mol. Wt.** 132.16
**Uses** Solvent.

## Mammalian and avian toxicity

### Teratogenicity and reproductive effects

Inhalation rat 0, 0.6, 3.0 or 15 mg $l^{-1}$ for 6 hr day$^{-1}$ on days 6-15 of gestation. The 2 high doses caused a degree of maternal toxicity. Skeletal anomalies of the thoracic vertebrae were observed in foetuses of the high dose group (1).
Inhalation rabbit 0, 0.2, 0.8 or 3.0 mg $l^{-1}$ for 6 hr day$^{-1}$ on days 6-18 of gestation. Severe developmental malformations, without maternal toxicity, were observed in the high dose group (1).
Dermal rabbit, 1000 or 2000 mg kg$^{-1}$ day$^{-1}$ on days 6-18 of gestation caused no maternal or foetal toxicity (1).

## Legislation

Included in Schedule 6 (Release into Land: Prescribed Substances) Statutory Instrument No. 472, 1991 (2).

## Any other comments

Human health effects, experimental toxicology, physico-chemical properties reviewed (3).

## References

1. Merkle, J. et al *Fundam. Appl. Toxicol.* 1987, **8**(1), 71-79
2. *S. I. 1991 No. 472 The Environmental Protection (Prescribed Processes and Substances) Regulations* 1991, HMSO, London
3. *ECETOC Technical Report No. 30(4)* 1991, European Chemical Industry Ecology and Toxicology Centre, B-1160 Brussels

# M137   Methylacetamide

## CH3C(O)NHCH3

**CAS Registry No.** 79-16-3
**Synonyms** *N*-methylacetamide; monomethylacetamide
**Mol. Formula** $C_3H_7NO$                                **Mol. Wt.** 73.10

## Physical properties

**M. Pt.** 26-28°C; **B. Pt.** 204-206°C; **Flash point** >107°C (closed cup);
**Specific gravity** $d_4^{25}$ 0.9571.

## Solubility

Organic solvent: ethanol, diethyl ether, acetone, benzene

## Mammalian and avian toxicity

### Acute data

$LD_{50}$ oral rat 5000 mg kg$^{-1}$ (1).
$LD_{50}$ subcutaneous rat 3600 mg kg$^{-1}$ (2).
$LD_{50}$ intraperitoneal rat, mouse 2750, 4380 mg kg$^{-1}$, respectively (1,3).
$LD_{Lo}$ intravenous rabbit 16,940 mg kg$^{-1}$ (4).
$LD_{50}$ intravenous mouse 4015 mg kg$^{-1}$ (3).

## References

1. *J. Reprod. Fertil.* 1962, **4**, 219
2. *C. R. Hebd. Seances Acad. Sci.* 1960, **251**, 1937
3. *J. Pharm. Pharmacol.* 1964, **16**, 472
4. *Arzneim.-Forsch.* 1970, **20**, 1242

# M138   Methyl acetoacetate

## CH3C(O)CH2CO2CH3

**CAS Registry No.** 105-45-3
**Synonyms** methyl acetyl acetate; methyl- 3-oxobutyrate
**Mol. Formula** $C_5H_8O_3$                                **Mol. Wt.** 116.12
**Uses** Cross-linking catalyst. Chelating agent. Organic synthesis.

## Physical properties

**M. Pt.** 27.5°C; **B. Pt.** 169-170°C; **Flash point** 70°C; **Specific gravity** $d_4^{20}$ 1.0762;
**Partition coefficient** log $P_{ow}$ -0.264 (calc.) (1).; **Volatility** v.p. 0.7 mmHg at 20°C;
v. den. 4.00.

## Solubility

Water: ≈38%. Organic solvent: diethyl ether, ethanol

## Occupational exposure

**Supply classification** irritant.
**Risk phrases** Irritating to eyes (R36)
**Safety phrases** In case of contact with eyes, rinse immediately with plenty of water
and seek medical advice (S26)

## Ecotoxicity

**Bioaccumulation**
Calculated bioconcentration factors of 0.37-0.48 indicate that environmental
accumulation is unlikely (2).

## Environmental fate

**Absorption**
Calculated $K_{oc}$ of ≈17 indicates that adsorption to soil and sediments will not be
significant (2).

## Mammalian and avian toxicity

**Acute data**
$LD_{50}$ oral rat, mouse, rabbit 1250-3500 mg kg$^{-1}$ (3,4).
Inhalation rat, 8 hr exposure to saturated vapour was not lethal (4).
$LD_{50}$ dermal rabbit >10 ml kg$^{-1}$ (4).

**Irritancy**
Dermal rabbit (24 hr) 500 mg caused mild irritation (5).
2 mg instilled into rabbit eye caused severe irritation (exposure not specified) (6).

## Any other comments

Human health effects, experimental toxicology, physico-chemical properties
reviewed (7).
Autoignition temperature 280°C.

## References

1. *GEMS: Graphical Exposure Modeling System: Fate of Atmospheric Pollutants* 1987,
   Washington, DC
2. Lyman, W. K. et al *Handbook of Chemical Property Estimation Methods Environmental
   Behaviour of Organic Compounds* 1982, McGraw-Hill, New York
3. Romashov, P. G. et al *Gig. Sanit.* 1991, (3), 20-22 (Russ.) (*Chem. Abstr.* **115**, 200734x)
4. Smyth, H. F. et al *J. Ind. Hyg. Toxicol.* 1948, **30**, 63
5. *Food Cosmet. Toxicol.* 1978, **16**, 637
6. *Am. J. Ophthalmol.* 1946, **29**, 1363
7. *ECETOC Technical Report No. 30(4)* 1991, European Chemical Industry Ecology and
   Toxicology Centre, B-1160 Brussels

# M139   Methyl acetylene

$$CH_3C{\equiv}CH$$

**CAS Registry No.** 74-99-7
**Synonyms** allylene; 1-propyne; propine
**Mol. Formula** $C_3H_4$            **Mol. Wt.** 40.07
**Uses** Organic synthesis. Welding torch fuel.
**Occurrence** In exhaust gases of diesel engines and in cigarette smoke (1,2).

## Physical properties

**M. Pt.** –104°C; **B. Pt.** –23.3°C; **Specific gravity** $d^0$ 1.787; **Volatility**
v.p. 3876 mmHg at 20°C; v. den. 1.38.

## Solubility
Water: 3640 mg $l^{-1}$ at 20°C. Organic solvent: benzene, chloroform, ethanol

## Occupational exposure
**US TLV (TWA)** 1000 ppm (1640 mg $m^{-3}$).

## Mammalian and avian toxicity

### Sub-acute data
Inhalation rat (6 month) 29,000 ppm 6 hr $day^{-1}$ 5 days $wk^{-1}$. 8/20 rats died. Signs of
toxicity included excitement, ataxia, salivation, mydriasis and tremors (3).

## Any other comments
Human health effects, experimental toxicology, physico-chemical properties
reviewed (4).

## References
1. Weigert, W. et al *Chem.-Ztg.* 1973, **97**
2. Conkle, J. P. et al *Arch. Environ. Health* 1975, **30**(6), 290
3. *Documentation of Threshold Limit Values* 4th ed., 1980, American Conference of
   Governmental Industrial Hygienists, Cincinnati, OH
4. *ECETOC Technical Report No. 30(4)* 1991, European Chemical Industry Ecology and
   Toxicology Centre, B-1160 Brussels

# M140   2-Methylacrylamide

$$CH_2{=}C(CH_3)C(O)NH_2$$

**CAS Registry No.** 79-39-0
**Synonyms** 2-methyl-2-propenamide; methacrylamide; methacrylic acid amide;
2-methylpropenamide
**Mol. Formula** $C_4H_7NO$            **Mol. Wt.** 85.11

## Physical properties
M. Pt. 108-109°C; B. Pt. 215°C; Flash point 103°C; Specific gravity $d^{20}$ 1.053.

## Mammalian and avian toxicity

### Acute data
$LD_{50}$ oral mouse, rat 451, 459 mg $kg^{-1}$ respectively (1,2).
$LD_{50}$ intraperitoneal mouse 200 mg $kg^{-1}$ (3).

## References
1. *Arch. Toxicol.* 1981, **47**, 179
2. *Hyg. Sanit. (USSR)* 1980, **45**(10), 74
3. *NTIS Report* AD277-689 Natl. Tech. Inf. Ser., Springfield, VA

# M141   Methylal

## CH2(OCH3)2

CAS Registry No. 109-87-5
**Synonyms** dimethoxymethane; Anesthenyl; dimethyl formal; 2,4-dioxapentane;
formal; formaldehyde dimethyl acetal
**Mol. Formula** $C_3H_8O_2$                                    **Mol. Wt.** 76.10
**Uses** Special-purpose fuel and as a solvent for perfumes, coatings, adhesives and
resins.
**Occurrence** Detected in the drinking water of 2 cities (1).

## Physical properties
**M. Pt.** −105°C; **B. Pt.** $_{754}$ 41.6°C; **Flash point** −18°C (closed cup); **Specific gravity**
$d_4^{20}$ 0.8593; **Partition coefficient** log $P_{ow}$ 0.0 (2); **Volatility** v.p. 300 mmHg at 20°C,
400 mmHg at 25°C; v. den. 2.6.

## Solubility
Water: 330 g $l^{-1}$. Organic solvent: miscible with ethanol, diethyl ether

## Occupational exposure
**US TLV (TWA)** 1000 ppm (3110 mg $m^{-3}$); **UK Long-term limit** 1000 ppm (3100
mg $m^{-3}$); **UK Short-term limit** 1250 ppm (3880 mg $m^{-3}$); **UN No.** 1234;
**HAZCHEM Code** 2ME; **Conveyance classification** flammmable liquid.

## Ecotoxicity

### Fish toxicity
Non-toxic at 5 ppm (24 hr) to salmon, bluegill sunfish, yellow perch and goldfish.
Test conditions: pH, 7.0; dissolved oxygen, 7.5 ppm; total hardness (soap method),
300 ppm; methyl orange alkalinity, 310 ppm; phenolphthalein alkalinity, 0; free
carbon dioxide, 5 ppm; temperature, 12.8°C (3).

## Environmental fate

### Degradation studies

Wastewater treatment: bench scale activated sludge, fill and draw operations, COD, feed 333 mg $l^{-1}$ $O_2$, at 20°C, 30 days acclimation: 88% removed (4).

## Mammalian and avian toxicity

### Acute data

$LD_{50}$ oral rabbit 5708 mg $kg^{-1}$ (5).
$LC_{50}$ (7 hr) inhalation mouse 18,000 ppm (6).
$LD_{50}$ subcutaneous guinea pig 5 g $kg^{-1}$ (6).
$LD_{Lo}$ subcutaneous guinea pig 3013 mg $kg^{-1}$ (7).

### Sub-acute data

Inhalation (7 hr) guinea pig for 5 days at 45,000 ppm showed no significant histopathological effects (8).
Inhalation (7 hr) mouse for 15 exposures up to 14,000 ppm (individual dose) showed occasional slight fatty changes in the kidneys and pulmonary oedema (8).

### Carcinogenicity and long-term effects

If metabolised to formaldehyde it has been suggested that it may pose a carcinogenic risk. Studies examining the link between inhalation and carcinogenicity have not confirmed the relationship (9,10).

### Metabolism and pharmacokinetics

Metabolised to methanol (9).

### Irritancy

Causes irritation to the eyes and skin (species unspecified) (9).

### Sensitisation

Dermatitis, defatting and necrosis of the skin, but does not cause sensitisation (species unspecified) (9).

## Any other adverse effects to man

As air pollutant attributed as the cause of stomach troubles in workers manufacturing anion exchangers (11).
Likely to cause vomiting, diarrhoea, gastrointestinal upsets and narcosis with systemic toxicity to kidneys and liver if swallowed (9).

## Any other adverse effects

Absorbed through the skin of rabbits and guinea pigs. Anaesthesia and respiratory tract irritation, with systemic toxicity to the kidney, heart and liver at high concentrations (9).
Inhalation by laboratory animals at concentrations above 1400 ppm caused anaesthesia. Side effects include blood in the urine and variations in blood glucose levels (12).
No adverse effects were seen at 4000 ppm in acute inhalation toxicity studies (species and duration unspecified) (13).
Mice died at inhaled levels 18,000 to 153,000 ppm with eye irritancy, lung congestion, respiratory tract irritancy and bronchopneumonia effects. Liver, heart, lungs and kidney showed severe fatty degeneration (8).
Absorbed readily through the skin; a more toxic route than vapour inhalation (9).

## Legislation
Included in Schedule 6 (Release into Land: Prescribed Substances) Statutory Instrument No. 472, 1991 (14).

## Any other comments
Experimental toxicology, epidemiology, human health effects and workplace experience reviewed (15).
Hazards reviewed (16).

## References
1. US EPA *Preliminary Assessment of Suspected Carcinogens in Drinking Water* 1975, Office of Toxic Substances
2. Verschueren, K. *Handbook of Environmental Data on Organic Chemicals* 2nd ed., 1983, Van Nostrand Reinhold, New York
3. Wood, E. M. *The Toxicity of 3400 Chemicals to Fish* 1987, US EPA, EPA 560/6-87-002, PB87-200-275, Washington, DC
4. Ludzack, F. J. et al *J.- Water Pollut. Control Fed.* 1960, **32**, 1173
5. *Proc. Soc. Exp. Biol. Med.* 1932, **29**, 730
6. *Patty's Industrial Hygiene and Toxicology* 3rd rev. ed. 1981, **2a**, John Wiley & Sons, New York
7. *Br. J. Ind. Med.* 1951, **8**, 279
8. Weaver, F. L. *Br. J. Ind. Med.* 1951, **8**, 279
9. *Chemical Safety Data Sheets, Volume 1, Solvents* 1988, 215-217, RSC, London
10. Dahl, A. R. *Toxicol. Appl. Pharmacol.* 1983, **67**(2), 200-205
11. Kazeka, E. K. *Uch. Zap. – Mosk. Nauchno-Issled. Inst. Gig. im. F. F. Erismana* 1974, **21**, 6-9
12. Virtue, R. W. *Anesthesiology* 1951, **12**, 100-105
13. Gage, J. C. *Br. J. Ind. Med.* 1970, **27**(1), 1-18
14. *S. I. 1991 No. 472 The Environmental Protection (Prescribed Processes and Substances) Regulations* 1991, HMSO, London
15. *ECETOC Technical Report No. 30(4)* 1991, European Chemical Industry Ecology and Toxicology Centre, B-1160 Brussels
16. *Dangerous Prop. Ind. Mater. Rep.* 1987, **7**(6), 76-80

# M142   *N*-Methyl-2-aminoethanol

CH₃NHCH₂CH₂OH

**CAS Registry No.** 109-83-1
**Synonyms** (2-hydroxyethyl)methylamine; 2-(methylamino)ethanol; *N*-methylethanolamine; *N*-methyl-2-hydroxyethylamine; monomethylaminoethanol
**Mol. Formula** $C_3H_9NO$                    **Mol. Wt.** 75.11
**Uses** Absorbent for removal of carbon dioxide from natural gas. Amidation reagent. Catalyst. Corrosion inhibitor. Organic synthesis.
**Occurrence** *In vivo* precursor of choline in mammals.

## Physical properties
**M. Pt.** −4.5°C; **B. Pt.** 159°C; **Flash point** 72°C; **Specific gravity** $d_{20}^{20}$ 0.9414;
**Volatility** v.p. 0.7 mmHg at 20°C; v. den. 2.9.

## Solubility

Water: miscible. Organic solvent: acetone, benzene, diethyl ether, ethanol

## Occupational exposure

**Supply classification** corrosive.
**Risk phrases** Causes burns (R34)
**Safety phrases** Do not breathe vapour – In case of contact with eyes, rinse immediately with plenty of water and seek medical advice – Wear suitable protective clothing (S23, S26, S36)

## Mammalian and avian toxicity

**Acute data**
$LD_{50}$ oral rat 2300 mg kg$^{-1}$ (1).
$LD_{50}$ subcutaneous mouse 1800 mg kg$^{-1}$ (2).
$LD_{50}$ intraperitoneal rat 1300 mg kg$^{-1}$ (3).
$LD_{50}$ intraperitoneal mouse 125 mg kg$^{-1}$ (4).

**Teratogenicity and reproductive effects**
Oral rat, choline deficient diet supplemented with 1% for 15 days pre-delivery to 15 days post-partum. All pups died within 36 hr. Levels of choline and acetylcholine were elevated (5).

**Irritancy**
Dermal rabbit (24 hr) 10 mg caused mild irritation. 250 µg instilled into rabbit eye caused severe irritation (exposure unspecified) (1).

## Genotoxicity

*Salmonella typhimurium* TA97, TA98, TA100, TA1535, TA1537 with and without metabolic activation negative (6).

## Any other comments

Human health effects, experimental toxicology, physico-chemical properties reviewed (7).

## References

1. *AMA Arch. Ind. Hyg. Occup. Med.* 1954, **10**, 61
2. *Naunyn-Schmiedeberg's Arch. Exp. Pathol. Pharmakol.* 1955, **225**, 428
3. *Toxicol. Appl. Pharmacol.* 1968, **12**, 486
4. *NTIS Report No.2 AD277-689*, Natl. Tech. Inf. Ser., Springfield, VA
5. Zahriser, N. R. et al *J. Neurochem.* 1978, **30**(6), 1245
6. Zeiger, E. et al *Environ. Mutagen.* 1987, **9**(Suppl. 9), 1-109
7. *ECETOC Technical Report No. 30(4)* 1991, European Chemical Industry Ecology and Toxicology Centre, B-1160 Brussels

# M143   *N*-Methylaniline

CAS Registry No. 100-61-8
**Synonyms** anilinomethane; (methylamino)benzene; *N*-methylbenzenamine;
methylphenylamine; monomethylaniline; *N*-phenylmethylamine
**Mol. Formula** $C_7H_9N$                               **Mol. Wt.** 107.16
**Uses** Organic synthesis. Solvent. Acid acceptor.

## Physical properties
**M. Pt.** –57°C; **B. Pt.** 196°C; **Flash point** 79°C (closed cup); **Specific gravity** $d_4^{20}$ 0.989;
**Partition coefficient** log $P_{ow}$ 1.66 (1); **Volatility** v.p. 0.3 mmHg at 20°C; v. den. 3.70.

## Solubility
Water: ≈5 mg $l^{-1}$ at 25°C. Organic solvent: acetone, diethyl ether, ethanol, chloroform

## Occupational exposure
**US TLV (TWA)** 0.5 ppm (2.2 mg $m^{-3}$); **UK Long-term limit** 0.5 ppm (2 mg $m^{-3}$);
**UN No.** 2294; **HAZCHEM Code** 3X; **Conveyance classification** harmful substance;
**Supply classification** toxic.
**Risk phrases** Toxic by inhalation, in contact with skin and if swallowed – Danger of
cumulative effects (R23/24/25, R33)
**Safety phrases** After contact with skin, wash immediately with plenty of soap and
water – Wear suitable gloves – If you feel unwell, seek medical advice (show label
where possible) (S28, S37, S44)

## Ecotoxicity

### Invertebrate toxicity
$EC_{50}$ (30 min) *Photobacterium phosphoreum* 14 ppm, Microtox test (2).

## Environmental fate

### Nitrification inhibition
Inhibition of ammonia oxidation by *Nitrosomonas* 90% at 100 mg $l^{-1}$; 83% at
50 mg $l^{-1}$; 71% at 10 mg $l^{-1}$; 50% at 1 mg $l^{-1}$ (1).

### Degradation studies
Metabolised by *Mycobacterium* to yield *N,N*-dimethylaniline (3).
42% degradation by acclimated petrochemical plant wastewater at an initial
concentration of 100 mg $l^{-1}$ (4).

## Mammalian and avian toxicity

### Acute data
$LD_{Lo}$ oral rabbit, guinea pig 280, 1200 mg $kg^{-1}$, respectively (5, 6).
$LD_{Lo}$ intravenous rabbit, cat 24 mg $kg^{-1}$ (5).
$LD_{Lo}$ subcutaneous guinea pig 1200 mg $kg^{-1}$ (6).

### Carcinogenicity and long-term effects

Oral rat, mouse reported to induce cancer of the oesophagus when administered with sodium nitrite (details not given) (7,8).

### Metabolism and pharmacokinetics

Metabolised to aniline and 4-(methylamino)phenol in rabbits (9,10).

### Irritancy

Irritating to the skin. Vapour or mist is irritating to the eyes, mucous membranes and upper respiratory tract (11).

## Genotoxicity

*Salmonella typhimurium* TA97, TA98, TA100, TA1535 with and without metabolic activation negative (12).
*In vitro* primary rat hepatocytes, DNA repair assay negative (13).

## Any other adverse effects to man

Absorption into the body leads to the formation of methaemoglobin which in sufficient concentration causes cyanosis (11).

## Legislation

Included in Schedule 6 (Release into Land: Prescribed Substances) Statutory Instrument No. 472, 1991 (14).

## Any other comments

Physical properties, use, toxicity and safety precautions reviewed (15).
Human health effects, experimental toxicology, physico-chemical properties reviewed (16)
All reasonable efforts have been taken to find information on isomers of this compound but no relevant data are available.

## References

1. Verschueren, K. *Handbook of Environmental Data on Organic Chemicals* 2nd ed., 1983, 831-832, Van Nostrand Reinhold, New York
2. Kaiser, K. L. E. et al *Water Pollut. Res. J. Can.* 1991, **26**(3), 361-431
3. Lenfant, M. et al *Biochem. Biophys. Acta* 1970, **201**, 82
4. Matsui, S. et al *Water Sci. Technol.* 1989, **20**(10), 201-210
5. *J. Ind. Hyg. Toxicol.* 1949, **31**, 1
6. US Public Health Service *Public Health Bulletin* 1941, **271**, 16
7. Minnish, S. S. *Top. Chem. Carcin. Proc. Int. Symp. 2nd* 1971, 279-295
8. Sander, J. et al *N-Nitroso Comp. Anal. Form. Proc. Work. Conf.* 1971, 97-103
9. Goudette, L. E. et al *Biochem. Pharmacol.* 1959, **2**, 793
10. Ichikawa, Y. et al *Biochem. Biophys. Acta* 1969, **171**, 32
11. Lenga, R. E. (Ed.) *Sigma-Aldrich Library of Chemical Safety Data* 2nd ed., 1988, **2**, 2293, Sigma-Aldrich, Milwaukee, WI
12. Zeiger, E. et al *Environ. Mol. Mutagen.* 1988, **11**(Suppl. 12), 1-158
13. Yoshimi, M. et al *Mutat. Res.* 1988, **206**(2), 183-191
14. *S. I. 1991 No. 472 The Environmental Protection (Prescribed Processes and Substances) Regulations* 1991, HMSO, London
15. *Chemical Safety Data Sheets* 1991, **4b**, 35-37, RSC, London
16. *ECETOC Technical Report No. 30(4)* 1991, European Chemical Industry Ecology and Toxicology Centre, B-1160 Brussels

# M144    4-Methylanisole

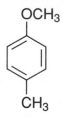

CAS Registry No. 104-93-8

Synonyms *p*-cresol methyl ether; *p*-methoxytoluene; 1-methoxy-4-methylbenzene;
*p*-methylanisole; 4-methylphenol methyl ether; *p*-tolyl methyl ether

**Mol. Formula** $C_8H_{10}O$                **Mol. Wt.** 122.17

**Uses** Organic synthesis. Flavour and fragrance agent.

**Occurrence** Aroma component of plant oils, cooked meats and dairy products.

## Physical properties

**B. Pt.** 176.5°C; **Flash point** 62°C; **Specific gravity** $d_{25}^{25}$ 0.9689; **Partition coefficient**
log $P_{ow}$ 2.81.

## Solubility
Organic solvent: diethyl ether, ethanol, chloroform

## Ecotoxicity

**Invertebrate toxicity**
$EC_{50}$ (30 min) *Photobacterium phosphoreum* 3.5 ppm, Microtox test (1).

## Mammalian and avian toxicity

**Acute data**
$LD_{50}$ oral rat 1920 mg kg$^{-1}$ (2).
$LD_{50}$ dermal rabbit >5000 mg kg$^{-1}$ (3).

**Metabolism and pharmacokinetics**
Metabolites identified in rabbits include 3-methyl-4-hydroxyanisole, anisic acid and
*p*-cresol (4, 5).

**Irritancy**
Dermal rabbit (24 hr) 500 mg caused moderate irritation (2).

## Any other comments

All reasonable efforts have been taken to find information on isomers of this
compound, but no relevant data are available.

## References

1. Kaiser, K. L. E. et al *Water Pollut. Res. J. Can.* 1991, **26**(3), 361-431
2. *Food Cosmet. Toxicol.* 1974, **12**, 385
3. Opdyke, D. L. J. (Ed.) *Monographs on Fragrance Raw Materials* 1979, 269, Pergamon Press, New York
4. Bray, A. G. et al *Biochem. J.* 1954, **60**, 255

5.  Testa, B. et al *Drug Metabolism: Chemical and Biochemical Aspects* 1976, 50, Marcel Dekker, New York

# M145    Methylazoxymethanol

$$CH_3N(O)=NCH_2OH$$

**CAS Registry No.** 590-96-5
**Synonyms** (methyl-*ONN*-azoxy)methanol; MAM
**Mol. Formula** $C_2H_6N_2O_2$                          **Mol. Wt.** 90.08
**Occurrence** As a glucoside (cycasin) in seeds of Cycadaceae from Guam.

## Mammalian and avian toxicity

### Carcinogenicity and long-term effects
No adequate evidence for carcinogenicity to humans, sufficient evidence for carcinogenicity to animals, IARC classification group 2B (1).
Intraperitoneal (237-244 day) ♀ Fisher rat, 2 mg × 10 injections, 4 mg × 4-12 injections or 6 mg × 2 injections to give a total dose of 12-48 mg (a total of 6 rats were used 1 for each injection regime). 4/6 rats developed tumours of the intestinal tract with the duodenum being the primary site of mucinous adenocarcinomas in 3 rats. The rats without intestinal tumours developed neoplasms at other sites such as the liver and kidneys (2).

### Teratogenicity and reproductive effects
Pregnant rats treated on gestation day 15 with 25 mg $kg^{-1}$ gave birth to hyperactive offspring that showed clear impairments in acquisition of instrumental learning in mazes. Elevated levels of noradrenaline and dopamine were observed in several regions of the brain (3).
Crj:CD(SD) pregnant rats treated with 40 mg $kg^{-1}$ on gestational days 12, 13, 14 or 15 had microencephalic offspring. Neurobehavioural ontogeny was retarded (4).
Single dose 25 mg $kg^{-1}$ injected into pregnant rats did not affect gestational and litter parameters. Degrees of altered physical and behavioural development were dependent on the time of administration (5).
The histochemical reactivity of the enzyme $NADH_2$-tetrazolium reductase at the level of frontal and occipital cortex, neostriatum and hippocampus was reduced in the offspring of mice treated on gestational day 15. Cholinacetyltransferase immunoreactivity within nerve cell bodies of the pontine tegmentum was also decreased, but acetylcholinesterase reactivity was increased in most brain areas. Nisal reactivity was modified (6).

### Metabolism and pharmacokinetics
Metabolised by rat liver cytochrome $P_{450}$ to methanol and formic acid. (7).
Metabolised by F344 rat liver microsomes, in presence of an NADPH-generating system, to methanol and formic acid. Spontaneous decomposition yields methanol and formaldehyde. Different enzymes are responsible for activation of the carcinogen in the liver and the colon (8).

### Genotoxicity
*Salmonella typhimurium* C50, D130, G46 reversion to histidine independence of

histidine-requiring mutants without metabolic activation positive (9).
*Drosophila melanogaster* sex-linked recessive lethal mutations positive (10).

### References

1. *IARC Monograph* 1987, **Suppl. 7**, 66
2. Laqueur, G. L. *Virchows Arch. A: Pathol. Anat.* 1965, **340**, 151-163
3. Archer, T. et al *Neurotoxicology* 1988, **10**(4), 341-347
4. Hashimoto, Y. et al *J. Vet. Med. Sci.* 1991, **53**(4), 643-649
5. Balduini, W. et al *Neurotoxicology* 1991, **12**(2), 179-188
6. Amenta, F. et al *Int. J. Tissue React.* 1986, **8**(6), 513-526
7. Sohn, O. S. et al *Carcinogenesis (London)* 1991, **12**(1), 127-131
8. Fiala, E. S. et al *J. Cancer Res. Clin. Oncol.* 1987, **113**(2), 145-150
9. Smith, D. W. E. *Science (Washington, D. C. 1883-)* 1966, **152**, 1273-1274
10. Teas, H. J. et al *Proc. Soc. Exp. Biol. Med.* 1967, **125**(3), 988-990

# M146   Methylazoxymethanol, acetate

$$CH_3N(O)=NCH_2OC(O)CH_3$$

**CAS Registry No.** 592-62-1
**Synonyms** (methyl-*ONN*-azoxy)methanol, acetate (ester); methylazoxymethyl acetate; MAM acetate
**Mol. Formula** $C_4H_8N_2O_3$                **Mol. Wt.** 132.12

## Physical properties

**B. Pt.** 191°C; **Flash point** 98°C; **Specific gravity** 1.172.

## Occupational exposure

**Supply classification** toxic.
**Risk phrases** May cause cancer – May cause birth defects (R45, R47)
**Safety phrases** Avoid exposure-obtain special instruction before use – If you feel unwell, seek medical advice (show label where possible) (S53, S44)

## Ecotoxicity

**Fish toxicity**
Medaka 0.3 ppm for 3 days or 0.1 ppm for 2 wk caused liver cell adenomas and carcinomas. Morphological alterations appeared from the second wk in the first group and at the fourth or fifth wk in the second (1).
Guppy exposed to $\leq$100 mg l$^{-1}$ for 2 hr developed pancreatic neoplasms in $\approx$9% of animals. Neoplasms included adenomas, acinoar cell carcinomas and adenocarcinomas (2).

## Mammalian and avian toxicity

**Acute data**
$LD_{Lo}$ oral mouse 35 mg kg$^{-1}$ (3,4).
$LD_{50}$ intraperitoneal rat, mouse 90-105 mg kg$^{-1}$ (5).
$LD_{50}$ intravenous mouse 10 mg kg$^{-1}$ (6).

## Carcinogenicity and long-term effects

No adequate evidence for carcinogenicity to humans, sufficient evidence for carcinogenicity to animals, IARC classification group 2B (7).

Oral rat 100 mg $kg^{-1}$ of diet for 2 wk, cumulative dose 13.4-13.7 mg $animal^{-1}$, produced 3 carcinomas of the colon, 31 kidney tumours, 4 liver-cell adenomas, 8 bile-duct adenomas and 1 hepatoma in the 26 tested animals (8).

Intravenous CD1 ♂ mouse given single dose 1-25 mg kg body $weight^{-1}$ showed no tumours within 14 months in 19 individuals (9).

Intravenous rat 35 mg kg body $weight^{-1}$ single dose produced intestinal, liver and kidney tumours after 6-7 months (4, 9, 10).

Offspring of prenatally intraperitoneally or intravenously treated Fischer rat 20 mg $kg^{-1}$ body weight developed a range of tumours within 356-637 days of birth (11).

## Teratogenicity and reproductive effects

Intraperitoneal rat 30 mg $kg^{-1}$ on gestation day 13 impaired motor behaviour of offspring, most severe in ♂ (12).

Intraperitoneal rat 1-25 mg $kg^{-1}$ on gestation day 15 caused dose-dependent reductions in cerebral hemisphere basal ganglia, diencephalon and mesencephalon in the offspring. The larger doses elevated levels of noradrenaline, dopamine and 5-HT in parts of the brain of the young (13).

When 20 mg $kg^{-1}$ was administered to rats on gestation days 14 and 16 massive reductions in brain weight of the offspring occurred when adult. The earlier application offspring included a small number of dwarves that had very small pituitary glands and an immature pattern of somatotrope distribution. This group generally had a significant, selective reduction in growth hormone releasing factor neurones, but an increased number of periventricular somatotropin release inhibiting factor neurones. Day 16 treated offspring showed accelerated postnatal growth which was significant in ♂. These animals had very large pituitary glands with some hypertrophy of somatotropes (14).

Young rats exposed prenatally showed a reduced ultrasonic vocalization and in ♀ reduced locomotor activity (15).

The pattern of cell death in prenatally treated rats described (16).

Intraperitoneal rat 30 mg $kg^{-1}$ on day 13 of gestation led to delays in cliff avoidance reflex and negative geotaxis, delays in achieving control of body in swimming and a prolonged time to complete a T-maze (17).

## Genotoxicity

*Salmonella typhimurium* his G46 without metabolic activation positive. Mutagenicity was higher at lower pH. Lignin, pectin and hemicellulose inhibited mutagenicity, cellulose did not effectively inhibit mutagenicity (18).

*Escherichia coli* HV with metabolic activation positive (19).

*In vivo* B6D2F$_1$ mouse bone marrow, alveolar macrophages, liver, kidney sister chromatid exchange negative (20).

## Any other adverse effects

Reduced DNA synthesis in rat liver and kidney (4).

Inhibited nucleolar RNA synthesis in the rat liver (9).

Single strand breaks in the DNA from rat liver remained unrepaired after 14 days (21).

## Any other comments

Human health effects, experimental toxicology, physico-chemical properties reviewed (22).

---

# References

1.  Harada, T. et al *J. Comp. Pathol.* 1988, **98**(4), 441-452
2.  Fournie, J. W. et al *J. Natl. Cancer Inst.* 1987, **78**, 715
3.  Spatz, M. *Ann. N. Y. Acad. Sci.* 1969, **163**, 848-859
4.  Zedeck, M. S. et al *Cancer Res.* 1970, **30**, 801-812
5.  *J. Natl. Cancer Inst.* 1979, **62**, 911
6.  *NX No. 04566* US Army Armament Research and Development Command, Chemical Systems Laboratory, NIOSH Exchange Chemicals, Aberdeen Proving Ground, MD
7.  *IARC Monograph 1987,* **Suppl. 7,** 66
8.  Laqueur, G. L. et al *J. Natl. Cancer Inst.* 1967, **39**, 355-371
9.  Zedeck, M. S. et al *Fed. Proc.* 1972, **31**, 1485-1492
10. Zedeck, M. S. et al *J. Natl. Cancer Inst.* 1974, **53**, 1419-1421
11. Laqueur, G. L. et al in *Transplacental Carcinogenesis* 1973, Tomatis, L. et al (Ed.), IARC Scientific Publications, No. 4, IARC, Lyon
12. Yamamoto, Y. et al *Congenital Anomalies* 1989, **29**(2), 51-58
13. Tamaru, M. et al *Teratology* 1988, **37**(2), 149-157
14. Rodier, P. M. et al *Teratology* 1991, **43**(3), 241-251
15. Cagiano, R. et al *Arch. Toxicol., Suppl. 1986* 1987, **11**, 148-151
16. Ashwell, K. W. S. et al *Neurotoxicol. Teratol.* 1988, **10**(1), 65-73
17. Yamamoto, Y. et al *Acta Med. Kinki Univ.* 1991, **16**(1), 51-56
18. Christie, G. L. et al *J. Food Prot.* 1989, **52**(6), 416-418
19. Zeilmaker, M. J. et al *Mol. Carcinog.* 1991, **4**(3), 180-188
20. Neft, R. E. et al *Teratog., Carcinog., Mutagen.* 1989, **9**(4), 219-237
21. Damjanov, I. et al *Cancer Res.* 1973, **33**, 2122-2128
22. *ECETOC Technical Report No. 30(4)* 1991, European Chemical Industry Ecology and Toxicology Centre, B-1160 Brussels

# M147   α-Methylbenzenemethanol

**CAS Registry No.** 98-85-1
**Synonyms** α-methylbenzyl alcohol; α-hydroxyethylbenzene; methylphenyl carbinol; α-phenethyl alcohol; 1-phenylethanol; styralyl alcohol; 1-phenylethyl alcohol
**Mol. Formula** $C_8H_{10}O$        **Mol. Wt.** 122.17
**Uses** Cosmetic ingredient. Food flavouring agent.

## Physical properties

**B. Pt.** 204°C; **Specific gravity** $d_{20}^{20}$ 1.015; **Volatility** v.p. 0.1 mmHg at 20°C; v. den. 4.21.

## Solubility

Organic solvent: fixed oils, propylene glycol, glycerin

## Mammalian and avian toxicity

### Acute data

$LD_{50}$ oral rat 400 mg $kg^{-1}$ (1).
$LD_{50}$ dermal rabbit 2500 mg $kg^{-1}$ (2).
$LD_{50}$ subcutaneous mouse 250 mg $kg^{-1}$ (3).
$LD_{Lo}$ intravenous dog 200 mg $kg^{-1}$ (4).

### Sub-acute data

Body weight gain in F344/N rats was reduced at 1500 mg $kg^{-1}$ gavage in 13 wk study. The only effects were ataxia, laboured breathing and lethargy for 30 min after dosing in rats and B6C3F1 mice and increases in liver weight to body weight ratios for rats (5).

### Carcinogenicity and long-term effects

National Toxicology Program tested gavage F344/N rat and B6C3F1 mouse 375 or 750 mg $kg^{-1}$ 5 days $wk^{-1}$ for 103 wk. Some evidence of carcinogenic activity (renal tubular cell adenomas and adenomas or adenocarcinomas (combined)) for ♂ rat. No evidence of carcinogenic activity in ♀ rat or mice (both sexes). In a 2 yr study significant reduction in body weight gain commenced at weeks 20-30 in high dose rats. Renal nephropathy was enhanced, particularly in ♂ rats, as ageing occurred. Non-neoplastic lesions (parathyroid hyperplasia, fibrous osteodystrophy of bone and calcification of the heart and glandular stomach) were found in ♂ rats. Mice only showed a reduction in body weight gain at the highest dose (5).

### Irritancy

Dermal rabbit (24 hr) 500 mg caused moderate to severe erythema and moderate oedema (2).
2 mg instilled into rabbit eye caused severe irritation (6).

## Genotoxicity

*Salmonella typhimurium* TA98, TA100, TA1535, TA1537 with and without metabolic activation negative (7).
*In vitro* mouse lymphoma L5178Y $tk^+/tk^-$ without metabolic activation positive for induction of trifluorothymidine resistance (5).
*In vitro* Chinese hamster ovary cells with metabolic activation chromosomal aberrations positive, without metabolic activation negative. Sister chromatid exchange was not observed (5).

## Any other comments

Experimental toxicology and human health effects reviewed (8).

## References

1. *J. Ind. Hyg. Toxicol.* 1944, **26**, 269
2. *Food Cosmet. Toxicol.* 1974, **12**, 995
3. *Arch. Int. Pharmacodyn. Ther.* 1958, **116**, 154
4. *J. Pharmacol. Exp. Ther.* 1920, **15**, 129
5. Dieter, M. P. *National Toxicology Program Report* 1990, NTP-TR-369, NIH/PUB-89-2824, NIEHS, Research Triangle Park, NC
6. *Am. J. Ophthalmol.* 1946, **29**, 1363
7. Zeiger, E. et al *Environ. Mutagen.* 1987, **9**(Suppl. 9), 1-109
8. *BIBRA Toxicity Profile* 1991, British Industrial Biological Research Association, Carshalton

# M148  2-Methylbenzothiazole

CAS Registry No. 120-75-2
Synonyms USAF EK-1853
Mol. Formula $C_8H_7NS$                        Mol. Wt. 149.22

## Physical properties
M. Pt. 12-14°C; B. Pt. 238°C; Flash point 102°C; Specific gravity 1.173.

## Mammalian and avian toxicity

### Acute data
$LD_{50}$ intravenous mouse 105 mg $kg^{-1}$ (1).
$LD_{50}$ intraperitoneal mouse 300 mg $kg^{-1}$ (2).

## References

1. *J. Pharmacol. Exp. Ther.* 1952, **105**, 486
2. *NTIS Report* AD277-689, Natl. Tech. Inf. Ser., Springfield, VA

# M149  4-(6-Methyl-2-benzothiazolyl)benzenamine

CAS Registry No. 92-36-4
Synonyms 2-(*p*-aminophenyl)-6-methylbenzothiazole; dehydrothio-*p*-toluidine;
DHPT; *p*-(6-methylbenzothiazol-2-yl)aniline
Mol. Formula $C_{14}H_{12}N_2S$                Mol. Wt. 240.33
Uses Manufacture of dyestuffs.

## Physical properties
M. Pt. 191°C; B. Pt. 434°C.

## Solubility
Water: <1 g $l^{-1}$ at 20°C. Organic solvent: acetone, dimethyl sulfoxide, ethanol

## Mammalian and avian toxicity

### Acute data
$LC_{Lo}$ (4 hr) inhalation rat 3000 mg $m^{-3}$ (1).

## Irritancy
Irritating to the eyes, skin, mucous membranes and upper respiratory tract (species unspecified) (2).

### References
1. *Food Chem. Toxicol.* 1984, **22**, 289
2. Lenga, R. E. (Ed.) *Sigma-Aldrich Library of Chemical Safety Data* 2nd ed., 1988, **1**, 210, Sigma-Aldrich, Milwaukee, WI

# M150    Methyl-1*H*-benzotriazole

**CAS Registry No.** 29385-43-1
**Synonyms** 4(or 5)-methyl-1*H*-benzotriazole; methylbenzotriazole; Cobratec TT100; tolyltriazole; TTZ
**Mol. Formula** $C_7H_7N_3$                    **Mol. Wt.** 133.15
**Uses** Corrosion inhibitor. Antioxidant. In photographic developers.

## Physical properties
**M. Pt.** 83°C; **B. Pt.** $_2$ 160°C; **Flash point** 182.2°C; **Specific gravity** $d_{25}^{100}$ 1.13; **Volatility** v.p. $3.0 \times 10^{-2}$ mmHg at 50°C; v. den. 4.6.

## Solubility
Water: <1 g l$^{-1}$ at 18°C. Organic solvent: acetone, dimethyl sulfoxide, ethanol

## Mammalian and avian toxicity

### Acute data
$LD_{50}$ oral rat 680 mg kg$^{-1}$ (1).

## Genotoxicity
*Salmonella typhimurium* TA97, TA98, TA100, TA1535 without metabolic activation negative, with metabolic activation weakly positive (2).

## Any other comments
Human health effects, experimental toxicology, physico-chemical properties reviewed (3).

### References
1. *Huntingdon Research Centre Report* 1972
2. Zeiger, E. et al *Environ. Mol. Mutagen.* 1988, **11**(Suppl. 12), 1-158
3. *ECETOC Technical Report No. 30(4)* 1991, European Chemical Industry Ecology and Toxicology Centre, B-1160 Brussels

# M151    1-Methyl-1*H*-benzotriazole

**CAS Registry No.** 13351-73-0
**Synonyms** 1-methylbenzotriazole
**Mol. Formula** $C_7H_7N_3$                    **Mol. Wt.** 133.15
**Uses** Corrosion inhibitor.

## Physical properties
**M. Pt.** 64-65°C; **B. Pt.** 270-271°C.

## Solubility
Organic solvent: benzene, petroleum ether

## Mammalian and avian toxicity

### Acute data
$LD_{50}$ intravenous mouse 375 mg $kg^{-1}$ (1).

## References
1.   *J. Pharmacol. Exp. Ther.* 1952, **105**, 486

# M152    5-Methyl-1*H*-benzotriazole

**CAS Registry No.** 136-85-6
**Synonyms** 5-methylbenzotriazole; 5-methyl-1,2,3-benzotriazole
**Mol. Formula** $C_7H_7N_3$                    **Mol. Wt.** 133.15
**Uses** Corrosion inhibitor. Preparation of antifogging photographic reagents.

## Physical properties
**M. Pt.** 80-82°C; **B. Pt.** $_{12}$ 210-212°C.

## Solubility
Organic solvent: acetone, ethanol

## Mammalian and avian toxicity

**Acute data**

$LD_{50}$ oral rat 1600 mg kg$^{-1}$ (1).

**Irritancy**

Irritating to the eyes, skin, mucous membranes and upper respiratory tract (species unspecified) (2).

### References

1. *Kodak Company Report* 21 May, 1971
2. Lenga, R. E. (Ed.) *Sigma-Aldrich Library of Chemical Safety Data* 2nd ed., 1988, **2**, 2298, Sigma-Aldrich, Milwaukee, WI

# M153 α-Methylbenzyl 3-hydroxycrotonate, dimethyl phosphate

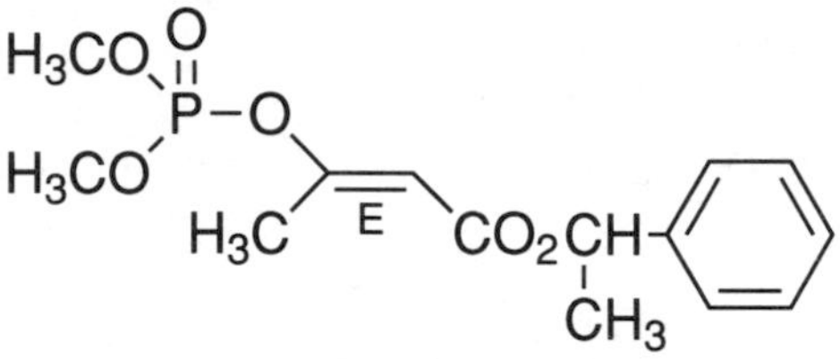

**CAS Registry No.** 7700-17-6

**Synonyms** 2-butenoic acid, 3-[(dimethoxyphosphinyl)oxy]-, 1-phenylethyl ester, (*E*)-; crotonic acid, 3-hydroxy-, α-methylbenzyl ester, dimethyl phosphate, (*E*)-; Ciodrin; Crotoxyphos; Cyodrin; Volfazol

**Mol. Formula** $C_{14}H_{19}O_6P$ **Mol. Wt.** 314.28

**Uses** Livestock insecticide (superseded).

## Physical properties

**B. Pt.** $_{0.03}$ 135°C; **Flash point** 79.4°C; **Specific gravity** d$^{25}$ 1.19; **Partition coefficient** log $P_{ow}$ 0.82 (1); **Volatility** v.p. $1.4 \times 10^{-5}$ mmHg at 20°C.

**Solubility**

Water: 1200 mg l$^{-1}$. Organic solvent: ethanol, highly chlorinated hydrocarbons, kerosene, acetone, chloroform.

## Occupational exposure

**Supply classification** toxic.

**Risk phrases** Toxic in contact with skin and if swallowed (R24/25)

**Safety phrases** After contact with skin, wash immediately with plenty of soap and water – Wear suitable protective clothing and gloves – If you feel unwell, seek medical advice (show label where possible) (S28, S36/37, S44)

## Ecotoxicity

### Fish toxicity
$LC_{50}$ (96 hr) largemouth bass, channel catfish 1100-2600 µg $l^{-1}$ (2).
$LC_{50}$ (96 hr) bluegill sunfish 152 µg $l^{-1}$ (2).
$LC_{50}$ (96 hr) fathead minnow 12 mg $l^{-1}$ (2).

### Invertebrate toxicity
$LC_{50}$ (24 hr) *Gammarus lacustris* 49 µg $l^{-1}$ at 15°C (2).
$LC_{50}$ (96 hr) *Gammarus fasciatus* 11 µg $l^{-1}$ (3).

## Environmental fate

### Degradation studies
$t_{1/2}$ ranged from 2 hr in silty clay loam soil to 71 hr in loamy sand. In aqueous systems at pH 9, 6 and 2, $t_{1/2}$ were respectively 180, 410 and 540 hr (4).
An enzyme isolated from clay loam hydrolysed the compound to dimethyl phosphate and α-methylbenzyl 3-hydroxycrotonate in 16 hr at 37°C (5).

### Abiotic removal
Slowly hydrolysed by water. 50% decomposition occurs in 35 hr at pH 9 and 87 hr at pH 1 at 38°C (6).

## Mammalian and avian toxicity

### Acute data
$LD_{50}$ oral rat, mouse 38, 40 mg $kg^{-1}$, respectively (7).
$LD_{50}$ oral ♂, ♀ rat 74, 110 mg $kg^{-1}$, respectively (8).
$LD_{50}$ oral chicken 111 mg $kg^{-1}$ (9).
$LD_{50}$ oral redwing blackbird 56.2 mg $kg^{-1}$ (10).
$LD_{50}$ dermal rabbit 385 mg $kg^{-1}$ (11).
$LD_{50}$ dermal ♀, ♂ rat 202, 375 mg $kg^{-1}$, respectively (8).
$LD_{50}$ intravenous mouse 4.5 mg $kg^{-1}$ (12).
$LD_{50}$ subcutaneous rat 47 mg $kg^{-1}$ (13).

### Sub-acute data
Oral rat 300-900 mg $kg^{-1}$ in diet for 90 days caused no effects to growth and no pathological changes. Blood cholinesterase activity was inhibited at oral doses of 20 mg $kg^{-1}$ in feed (6).

### Metabolism and pharmacokinetics
Radio-label studies showed in lactating ewes and goats that hydrolytic fission to the corresponding monomethyl phosphate and dimethyl phosphate occurred. Excretion in urine was the major eliminatory route. The very small amount of unmodified compound found in milk consisted solely of the β-isomer (14).
61-90% of radiolabel in the urine of orally administered ewe was associated with dimethyl phosphoric acid. There was rapid elimination with moderate absorption (4).
3-(Dimethoxyphosphinyloxy)crotonic acid was an important urinary metabolite appearing as 11% of metabolites after 3 hr and falling to 3% after 6 hr (15).

## Genotoxicity
*Saccharomyces cerevisiae* with and without metabolic activation mitotic recombination negative, but toxic (16).

---

*In vivo* mouse bone marrow dose- and administration route (intragastrically and inhalation)- dependent increase in chromosomal aberrations (17).

## Legislation

Regulated under the US Clean Water Act and Federal Insecticide, Fungicide and Rodenticide Act (18).

Limited under EC Directive on Drinking Water Quality 80/778/EEC. Pesticides: maximum admissible concentration 0.1 $\mu$g l$^{-1}$ (19).

Included in Schedule 6 (Release into Land: Prescribed Substances) Statutory Instrument No. 472, 1991 (20).

## References

1. Leo, A. J. *Log P Values Calculated Using the CLOGP Program for Compounds in ISHOW Files* Seaver Chemistry Laboratory, Claremont, CA
2. Johnson, W. W. et al *Handbook of Acute Toxicity of Chemicals to Fish and Aquatic Invertebrates* 1980, Fish and Wildlife Service, USDI, Washington, DC
3. Sanders, H. O. *The toxicities of some insecticides to four species of Malocostracan crustacea* 1972, Fish Pesticide Research Laboratory, Bureau of Sport Fisheries and Wildlife, Columbia, MO
4. Menzie, C. M. *Metabolism of Pesticides* 1969, Special Scientific Report No. 127, Fish and Wildlife Service, USDI, Washington, DC
5. Getzin, L. N. et al *Arch. Environ. Contam. Toxicol.* 1979, **8**(6), 661-672
6. *The Agrochemicals Handbook* 1st ed., 1983, RSC, London
7. *Gig. Sanit.* 1973, **38**(6), 30 (Russ.)
8. Gaines, T. B. *Toxicol. Appl. Pharmacol.* 1969, **14**, 515-534
9. *Toxicol. Appl. Pharmacol.* 1965, **7**, 606
10. Schaefer, E. W. et al *Arch. Environ. Contam. Toxicol.* 1983, **12**, 355-382
11. *Pesticide Chemicals Official Compendium* 1966, Association of the American Pesticide Control Officials, Inc., Topeka, KS
12. *J. Pharm. Pharmacol.* 1967, **19**, 612
13. *Br. J. Pharmacol.* 1970, **40**, 124
14. *Foreign Compound Metabolism in Mammals* 1970, Volume 1, 276
15. White-Stevens *Pesticides in Environment* 1971, Volume 1
16. L'vova, T. S. *Tsitol. Genet.* 1989, **23**(3), 68-70 (Russ.) (*Chem. Abstr.* **111**, 110812j)
17. German, I. V. *Gig. Sanit.* 1990, (5), 72-74 (Russ.) (*Chem. Abstr.* **113**, 93130g)
18. *Dangerous Prop. Ind. Mater. Rep.* 1993, **13**(1), Van Nostrand Reinhold, New York
19. *EC Directive Relating to the Quality of Water Intended for Human Consumption* 1982, 80/778/EEC, Office for Official Publications of the European Communities, 2 rue Mercier, L-2985 Luxembourg
20. *S. I. 1991 No. 472 The Environmental Protection (Prescribed Processes and Substances) Regulations* 1991, HMSO, London

# M154   Methyl-1,1'-biphenyl

CAS Registry No. 28652-72-4
Synonyms methylbiphenyl (mixed isomers); phenyltoluene
Mol. Formula $C_{13}H_{12}$                                    Mol. Wt. 168.24
Uses Solvent. Heat transfer fluid.
Occurrence In the essential oil of rue (*Ruta graveolens*). In fossil fuels. In diesel
fumes. Residues have been isolated from sediments and natural waters and in fish
tissues (1-6).

## Physical properties

M. Pt. –2 to 50°C; B. Pt. 255-272°C; Volatility v.p. 0.3 mmHg at 20°C.

### Solubility
Organic solvent: diethyl ether, dimethyl sulfoxide, ethanol

## Ecotoxicity

### Fish toxicity
$LC_{50}$ (96 hr) fathead minnow 1.5 mg $l^{-1}$, static bioassay (Sure-Sol 170) (7).

### Bioaccumulation
Bioconcentration factor in *Rengia cuneata* and *Crassostrea virginica* 300-1700 (3).

## Environmental fate

### Degradation studies
Degraded by the marine bacteria *Alcaligenes* and *Acinetobacter* species to yield
α-hydroxymuconic semialdehydes which were further degraded (8).

## Genotoxicity

*Salmonella typhimurium* TA97, TA98, TA100, TA1535 with and without metabolic
activation negative (9).

## Legislation

Included in Schedule 6 (Release into Land: Prescribed Substances) Statutory
Instrument No. 472, 1991 (10).

## Any other comments

Methyl-1,1'-biphenyl is the major component of the solvent Sure-Sol. Other
components include biphenyl and other aromatic hydrocarbons (11).

## References

1.   Williams, R. et al *Int. J. Environ. Anal. Chem.* 1986, **26**(1), 27-49
2.   Tattje, D. H. E. et al *Pharm. Weekbl.* 1978, **113**(45), 1169-1174

3. Fielding, M. et al *Organic Micro-Pollutants in Drinking Water* 1981, TR159, Water Research Centre, Medmenham
4. Shirohara, R. et al *Environ. Int.* 1980, **4**(2), 163-174
5. Veith, G. D. et al *Pestic. Monit. J.* 1981, **15**(1), 1-8
6. Neff, J. M. et al *Mar. Biol.* 1976, **38**(3), 279-289
7. *Toxicity and Environmental Data* 1976, Sun Oil Co
8. Fedorak, P. M. et al *Can. J. Microbiol.* 1983, **29**(5), 497-503
9. Zeiger, E. et al *Environ. Mol. Mutagen.* 1992, **19**(Suppl. 21), 2-141
10. *S. I. 1991 No. 472 The Environmental Protection (Prescribed Processes and Substances) Regulations* 1991, HMSO, London
11. *Product Information Sheet* 1982, Kveh Chemical Co

# M155   2-Methylbiphenyl

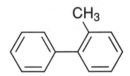

**CAS Registry No.** 643-58-3
**Synonyms** 2-methyl-1,1'-biphenyl; 2-phenyltoluene
**Mol. Formula** $C_{13}H_{12}$                    **Mol. Wt.** 168.24
**Occurrence** Airborne coal tar emissions, coal tar and wood preservative sludge (1). Biphenyls are formed by the anaerobic reduction of polychlorinated biphenyls (2).

## Physical properties

**B. Pt.** 255°C; **Flash point** 137°C (open cup) (3); **Specific gravity** 1.011.

## Environmental fate

### Degradation studies

Surfactants at low concentrations stimulate biodegradation of sorbed hydrocarbons in samples of aquifer sand and oil (4).

## Any other comments

Autoignition temperature: 495 and 502 °C reported.

## References

1. Lao, R. C. et al *J. Chromatogr.* 1975, **112**, 681-700
2. Rhee-Yull, G. et al *Environ. Toxicol. Chem.* 1993, **12**, 1033-1039
3. Bond, J. *Sources of Ignition. Flammability Characteristics of Chemicals and Products* 1991, 108, Butterworth-Heinemann, Oxford
4. Aronstein, B. N. et al *Environ. Toxicol. Chem.* 1992, **11**, 1227-1233.

# M156   4-Methylbiphenyl

CAS Registry No. 644-08-6
Synonyms 4-methyl-1,1'-biphenyl; 4-methyldiphenyl; 4-phenyltoluene
Mol. Formula $C_{13}H_{12}$                                    Mol. Wt. 168.24
Occurrence Pollutant of groundwater (1).
Airborne coal tar emissions, coal tar and wood preservative sludge (2).

## Physical properties

M. Pt. 44-47°C; B. Pt. 267-268°C; Flash point 110°C; Partition coefficient
log $P_{ow}$ 4.63.

## Ecotoxicity

Invertebrate toxicity
$EC_{50}$ (30 min) *Photobacterium phosphoreum* 2.22 ppm Microtox test (3).

Acute data
$LD_{50}$ rat 2570 mg $kg^{-1}$ (4).

## References

1. Tu, J. et al *Huanjing Huaxue* 1986, **5**(5), 60-74 (Ch.) (*Chem. Abstr.* **106**, 22946x)
2. Lao, R. C. et al *J. Chromatogr.* 1975, **112**, 681-700
3. Kaiser, K. L. E. et al *Water Pollut. Res. J. Can.* 1991, **26**(3), 361-431
4. *Food Cosmet. Toxicol.* 1975, **13**, 487

# M157   Methyl bromoacetate

$BrCH_2CO_2CH_3$

CAS Registry No. 96-32-2
Synonyms bromoacetic acid, methyl ester
Mol. Formula $C_3H_5BrO_2$                                    Mol. Wt. 152.98
Uses Alkylating agent. Chemical intermediate.

## Physical properties

B. Pt. $_{750}$ 130-133°C; Flash point 62°C; Specific gravity $d_4^{20}$ 1.6350.

## Solubility

Organic solvent: acetone, benzene, diethyl ether, ethanol

## Occupational exposure

UN No. 2643; HAZCHEM Code 2X; Conveyance classification toxic substance.

## Mammalian and avian toxicity

**Acute data**

$LD_{Lo}$ intravenous mouse 16 mg $kg^{-1}$ (1).

### Any other adverse effects to man

Extremely destructive to tissue of the mucous membranes and upper respiratory tract, eyes and skin. Inhalation may be fatal as a result of spasm, inflammation and oedema of the larynx and bronchi, chemical pneumonitis and pulmonary oedema (2).

### References

1. *Summary Tables Biological Tests* 1954, **6**, 138
2. Lenga, R. E. (Ed.) *Sigma-Aldrich Library of Chemical Safety Data* 2nd ed., 1988, **2**, 2304, Sigma-Aldrich, Milwaukee, WI

# M158   1-Methylbutadiene

CH₂=CHCH=CHCH₃

**CAS Registry No.** 504-60-9
**Synonyms** 1,3-pentadiene
**Mol. Formula** $C_5H_8$                    **Mol. Wt.** 68.12
**Uses** Monomer in plastics. Maleic acid intermediate.
**Occurrence** Released to the atmosphere by biomass combustion, synthetic rubber manufacture and tobacco smoke (1).
In car exhaust (2).
In waste incinerator emissions (3).

## Physical properties

**M. Pt.** –87.5°C; **B. Pt.** 42°C; **Specific gravity** $d_4^{20}$ 0.6760.

**Solubility**
Water: >341 mg $l^{-1}$ at room temperature. Organic solvent: ethanol, diethyl ether, acetone, benzene

## Ecotoxicity

**Bioaccumulation**
Not thought to bioconcentrate in aquatic organisms; based on water solubility and vapour pressure. Bioconcentration factors of <23 to 18 have been obtained by regression analysis (4).
Detected in clams and oysters at concentrations of 3.2 and 1.4 ppb wet weight, respectively, in a Louisiana lake (5).

## Environmental fate

**Abiotic removal**
The estimated Henry's Law constant ($6.89 \times 10^{-2}$ atm $mmole^{-3}$ at 25°C) and known vapour pressure (380-411 mmHg at 25°C) suggest rapid volatilisation from soil (6,7).

---

In model river 1 m deep flowing at 1 m sec$^{-1}$ with a wind speed of 3 m sec$^{-1}$ the estimate $t_{1/2}$ for volatilisation is 2.4 hr (4,8).
In distilled water direct photochemical degradation did not occur (9,10).
In the atmosphere expected to undergo a rapid gas-phase reaction with photochemically produced hydroxyl radicals and ozone. $t_{1/2}$ for the former 3.8 hr and the latter 5.2 hr (8,11,12).
Gas-phase reaction with nitrate radicals may be important in urban areas or in polluted atmospheres as night time degradation (13).
May be removed from the atmosphere by moisture deposition, however it would be expected to be rapidly revolatilised (14).

**Absorption**
Expected to be high to moderately mobile in soil based on the estimated soil adsorption coefficients ranging from <177 to 140 (4,15).

## Mammalian and avian toxicity

**Acute data**
$LD_{50}$ intravenous mouse 18 mg kg$^{-1}$ (16).

## Any other comments
Occupational exposure is likely to be via inhalation. Cigarette smoke or wood fire smoke inhalation may be a route of exposure for the general public (1).
Uses reviewed (17).

## References

1. Graedel, T. E. et al *Atmospheric Chemical Compounds* 1986, Academic Press, London
2. Hampton, C. V. et al *Environ. Sci. Technol.* 1982, **16**, 287-298
3. Junk, G. A. et al *Chemosphere* 1980, **9**, 187-230
4. Lyman, W. K. et al *Handbook of Chemical Property Estimation Methods: Environmental Behavior of Organic Compounds* 1982, McGraw-Hill, New York
5. Ferrario, J. B. et al *Bull. Environ. Contam. Toxicol.* 1985, **34**, 246-255
6. Hine, J. et al *J. Org. Chem.* 1975, **40**, 292-298
7. Daubert, T. E. et al *Physical and Thermodynamic Properties of Pure Chemicals, Data Compilation* 1989, Design Inst. Phys. Prop. Data, Am. Inst. Chem. Eng., Hemisphere Publishing Corp., New York
8. Atkinson, R. et al *Chem. Rev. (Washington, D. C.)* 1984, **84**, 437-470
9. Zepp, R. C. et al *Chemosphere* 1981, **10**, 109-117
10. Zepp, R. C. et al *Environ. Sci. Technol.* 1983, **17**, 462-468
11. Emsten, H. et al *J. Atmos. Chem.* 1984, **2**, 83-96
12. Atkinson, R. *Chem. Rev. (Washington, D. C.)* 1985, **85**, 69-201
13. Atkinson, R. et al *J. Phys. Chem.* 1988, 1205-1210
14. Zepp, R. G. et al *Environ. Sci. Technol.* 1985, **19**, 74-81
15. Swann, R. L. et al *Residue Rev.* 1983, **85**, 17-28
16. *NX No.04179* US Army Armament Research and Development Command, Chemical Systems Laboratory, NIOSH Exchange Chemicals, Aberdeen Proving Ground, MD
17. Zhang, W. et al *Shiyou Huagong* 1989, **18**(11), 777-782 (Ch.) (*Chem. Abstr.* **113**, 9068n)

# M159   2-Methylbutanal

## CH₃CH₂CH(CH₃)CHO

$CH_3CH_2CH(CH_3)CHO$

**CAS Registry No.** 96-17-3
**Synonyms** 2-ethylpropanal; 2-formylbutane; methylbutyraldehyde; α-methylbutanal; 2-methylbutyraldehyde; 2-methylbutyric aldehyde
**Mol. Formula** $C_5H_{10}O$             **Mol. Wt.** 86.13
**Uses** Organic synthesis.
**Occurrence** Aroma component of fruits, vegetables and cooked meat.

## Physical properties

**B. Pt.** 90-92°C; **Flash point** 4°C; **Specific gravity** $d^{20}$ 0.804.

**Solubility**
Organic solvent: methanol, diethyl ether, ethanol

## Ecotoxicity

**Fish toxicity**
$LC_{50}$ (96 hr) fathead minnow 740 mg $l^{-1}$ (1).

## Mammalian and avian toxicity

**Acute data**
$LD_{50}$ oral rat, mouse 6400, 3200 mg $kg^{-1}$, respectively (2, 3).
$LC_{50}$ (4 hr) inhalation rat 14,000 ppm (4).
$LD_{50}$ dermal rabbit 5700 mg $kg^{-1}$ (4).
$LD_{50}$ dermal guinea pig > 20,000 mg $kg^{-1}$ (2).

**Metabolism and pharmacokinetics**
*In vitro* rabbit liver undergoes oxidative cleavage with olefin formation catalysed by cytochrome $P_{450}$ (5).

**Irritancy**
500 mg instilled into rabbit eye for 24 hr caused severe irritation (2).
Dermal rabbit, 500 mg caused mild irritation (exposure not specified) (4).

## Genotoxicity

*Salmonella typhimurium* TA98, TA100, TA102 with metabolic activation positive (6).

## References

1. Protic, M. et al *Aquat. Toxicol.* 1989, **14**(1), 47-64
2. *Food Chem. Toxicol.* 1982, **20**(Suppl.) 739
3. *Patty's Industrial Hygiene and Toxicology* 2nd ed., 1963, **2**, 1968, John Wiley & Sons, New York
4. *Union Carbide Data Sheet* 5 October 1972
5. Roberts, E. S. et al *Proc. Natl. Acad. Sci. U.S.A.* 1991, **88**(20), 8963-8966
6. Aeschbacher, H. V. et al *Food Chem. Toxicol.* 1989, **27**(4), 227-232

# M160    3-Methylbutanenitrile

CH₃CH(CH₃)CH₂CN

**CAS Registry No.** 625-28-5
**Synonyms** isoamyl cyanide; isovaleronitrile; 3-methylbutyronitrile;
2-methylbutane-*sec*-mononitrile
**Mol. Formula** $C_5H_9N$                          **Mol. Wt.** 83.13
**Uses** Organic synthesis.
**Occurrence** Aroma component in tomatoes.

## Physical properties

**M. Pt.** −100.9°C; **B. Pt.** 130.3°C; **Specific gravity** $d_4^{15}$ 0.795; **Partition coefficient**
log $P_{ow}$ 0.85.

## Mammalian and avian toxicity

### Acute data
$LD_{50}$ oral mouse 230 mg kg$^{-1}$ (1).
$LD_{Lo}$ subcutaneous rabbit 43 mg kg$^{-1}$ (2).

## References

1. *Nippon Eiseigaku Zasshi* 1984, **39**, 423
2. *Arch. Int. Pharmacodyn. Ther.* 1899, **5**, 161

# M161    2-Methylbutan-2-ol

CH₃CH₂C(OH)(CH₃)₂

**CAS Registry No.** 75-85-4
**Synonyms** *tert*-pentyl alcohol; *tert*-amyl alcohol; amylene hydrate;
dimethylethylcarbinol; ethyldimethylcarbinol; *tert*-pentanol
**Mol. Formula** $C_5H_{12}O$                          **Mol. Wt.** 88.15
**Uses** Hypnotic.

## Physical properties

**M. Pt.** −9.0°C; **B. Pt.** $_{765}$ 102.5°C; **Flash point** 19°C (closed cup), 21°C (open cup);
**Specific gravity** $d^{20}$ 0.8084; **Partition coefficient** log $P_{ow}$ 0.89; **Volatility**
v.p. 10 mmHg at 17.2°C; v. den 3.03.

## Solubility
Water: 1 in 8 parts. Organic solvent: miscible with ethanol, benzene, diethyl ether,
glycerol, chloroform

## Occupational exposure
**US TLV (TWA)** 100 ppm (361 mg m$^{-3}$); **US TLV (STEL)** 125 ppm (452 mg m$^{-3}$);
**UK Long-term limit** 100 ppm (361 mg m$^{-3}$); **UK Short-term limit** 125 ppm (450
mg m$^{-3}$); **Supply classification** highly flammable, harmful (≥25%).

**Risk phrases** ≥25% – Highly flammable – Harmful by inhalation – ≥25% – Highly flammable (R11, R20, R11)

**Safety phrases** Keep container in a well ventilated place – Keep away from sources of ignition – No Smoking – Avoid contact with skin and eyes (S9, S16, S24/25)

## Ecotoxicity

**Fish toxicity**
NOEC (24 hr) Creek chub 1300 mg $l^{-1}$ (1).
$LC_{100}$ (24 hrs) Creek chub 2000 mg $l^{-1}$ (1).

**Invertebrate toxicity**
LOEC reproduction *Microcystis aeruginosa* 105 mg $l^{-1}$ (2).
Cell multiplication inhibition test *Pseudomonas putida* 410 mg $l^{-1}$, *Scenedesmus quadricauda* 1250 mg $l^{-1}$, and *Entosiphon sulcatum* 680 mg $l^{-1}$ (3).
Cell multiplication inhibition test *Uronema parduzi* Chatton-Lwoff 859 mg $l^{-1}$ (4).

## Environmental fate

**Degradation studies**
Using bench scale activated sludge, fill and draw operations after 6, 12 and 24 hr respectively 1.3, 1.7 and 3.7% ThOD (5).

## Mammalian and avian toxicity

**Acute data**
$LD_{50}$ oral rat 1-2 g $kg^{-1}$ (6,7).
$LD_{50}$ subcutaneous mouse 2100 mg $kg^{-1}$ (8).
$LD_{Lo}$ subcutaneous rat 1400 mg $kg^{-1}$ (9).
$LD_{Lo}$ intraperitoneal rat 1530 mg $kg^{-1}$ (10).
$LD_{Lo}$ rectal rat 1400 mg $kg^{-1}$ (9).

## Any other comments

Hazards and properties reviewed (11).
Experimental toxicology, human health effects and physico-chemical properties reviewed (12).
Included in QSAR on toxicity and narcosis to the tadpole (13).
Autoignition temperature, 435/437°C.

## References

1. Gillette, L. A. et al *Sewage Ind. Wastes* 1952, **24**(11), 1397-1401
2. Bringmann, G. et al *GWF, Gas Wasserfach: Wasser/Abwasser* 1976, **117**(9), 410-413
3. Bringmann, G. et al *Water Res.* 1980, **14**, 231-241
4. Bringmann, G. et al *Z. Wasser/Abwasser Forsch.* 1980, (1), 26-31
5. Gerhold, R. M. et al *J. – Water Pollut. Control. Fed.* 1966, **38**(4), 562
6. Schaffarzick et al *Science (Washington, D. C. 1883-)* 1952, **116**, 663
7. *Material Safety Data Sheet* 1978, Dow Chemical Co.
8. *Arzneim.-Forsch.* 1955, **5**, 161
9. *J. Pharmacol. Exp. Ther.* 1933, **49**, 36
10. *J. Ind. Hyg. Toxicol.* 1945, **27**, 1
11. Sejbl, J. *Rudy* 1986, **34**(12), 364-367 (Czech.) (*Chem. Abstr.* **106**, 181925f)
12. *ECETOC Technical Report No. 30(4)* 1991, European Chemical Industry Ecology and Toxicology Centre, B-1160 Brussels
13. Lipnick, R. L. *ASTM Spec. Tech. Publ.* 1988, **1007**, 468-489

# M162   3-Methylbutan-1-ol

CH₃CH(CH₃)CH₂CH₂OH

**CAS Registry No.** 123-51-3

**Synonyms** isopentyl alcohol; *iso*-amyl alcohol; fermentation amyl alcohol; isoamyl alcohol; isoamylol; isobutylcarbinol; isopentanol

**Mol. Formula** $C_5H_{12}O$ **Mol. Wt.** 88.15

**Uses** Solvent for fats, alkaloids and resins. In microscopy. For determining fat in milk. In manufacture of isoamyl compounds, isovaleric acid, artificial silk, lacquers, mercury fulminate and pyroxylin.

## Physical properties

**M. Pt.** −117.2°C; **B. Pt.** 131-132°C; **Flash point** 43°C (closed cup), 55°C (open cup); **Specific gravity** $d_4^{15}$ 0.813; **Volatility** v.p. 2.3 mmHg at 20°C, 4.8 mmHg at 30°C; v. den. 3.04.

### Solubility

Water: 20 g $l^{-1}$ at 14°C. Organic solvent: miscible with ethanol, diethyl ether, benzene, chloroform, glacial acetic acid and petroleum ether

## Occupational exposure

**US TLV (TWA)** 100 ppm (361 mg m⁻³); **US TLV (STEL)** 125 ppm (452 mg m⁻³); **UK Long-term limit** 100 ppm (360 mg m⁻³); **UK Short-term limit** 125 ppm (450 mg m⁻³).

## Ecotoxicity

### Invertebrate toxicity

Ratio of chemical water solubility to concentration causing immobilisation in 50% of marine barnacle larvae 0.038 (1).

## Environmental fate

### Degradation studies

BOD₅ determined using acclimated mixed microbial cultures 4.46 mg $l^{-1}$ (2).

Waste water treatment; bench scale activated sludge, fill and draw operations, 20°C, 1-5 days observed, feed 333 mg $l^{-1}$, 30 days acclimation, 79% removed (3).

BOD₅ 0.162 mg $l^{-1}O_2$ with standard dilution technique with sewage as seed material (4,5).

BOD₅: 59% ThOD; COD 77% ThOD (0.05 N $K_2Cr_2O_7$) (6).

ThOD: 2.740 mg $l^{-1} O_2$ (5).

First-order biodegradation rate constants (per day) and $t_{1/2}$: ground water, 0.045 and 15 days; for river water, 0.064 and 11 days; and fresh water-lake harbour water, 0.113 and 6 days (7).

## Mammalian and avian toxicity

### Acute data

LD₅₀ oral rat 1300 mg kg⁻¹ (8).

LD_Lo oral rabbit 4250 mg kg⁻¹ (9).

$LD_{50}$ dermal rabbit 3212 mg kg$^{-1}$ (10).
$LD_{Lo}$ intraperitoneal mouse, rat 233, 813 mg kg$^{-1}$, respectively (11,12).
$LD_{Lo}$ subcutaneous mouse 7480 mg kg$^{-1}$ (11).
$LD_{50}$ intravenous mouse 234 mg kg$^{-1}$ (12).

**Irritancy**

Dermal rabbit (24 hr) 500 mg caused moderate to severe erythema and moderate oedema and 20 mg instilled into rabbit eye (24 hr) caused moderate irritation (13). Acute patch testing 12/12 volunteers produced erythema. The reaction is provoked by the corresponding aldehyde (14).

**Legislation**

Included in Schedule 6 (Release into Land: Prescribed Substances) Statutory Instrument No. 472, 1991 (15).

**Any other comments**

Included in QSAR for upper respiratory tract irritation in mice (16).
Experimental toxicology and human health effects reviewed (17).
Experimental toxicology, human health effects, epidemiology and workplace experience reviewed (18).
Autoignition temperature, 340/350°C.

**References**

1. Vaishnav, D. D. et al *Arch. Environ. Contam. Toxicol.* 1990, **19**, 624-628
2. Baben, L. et al *J. Ind. Microbiol.* 1987, **2**, 107-115
3. Ludzack, F. J. et al *J. – Water Pollut. Control Fed.* 1960, **32**, 1173
4. Lund, H. F. *Industrial Pollution Control Handbook* 1971, McGraw-Hill, New York
5. Meinck, F. et al *Les eaux residuaires industrielles* 1970
6. Wolters, N. *Unterschiedliche Bestimmungsmethoden zur Erfassung der organischen Substanz in einer Verbindung* Lehrauftrag Wasserbiologie a.d. Technischen Hochschule, Darmstadt
7. Vaishnav, D. D. et al *Bull. Environ. Contam. Toxicol.* 1987, **39**, 237-244
8. *S. Afr. Med. J.* 1969, **43**, 795
9. *J. Lab. Clin. Med.* 1925, **10**, 985
10. *Am. Ind. Hyg. Assoc. J.* 1969, **30**, 470
11. *Food Cosmet. Toxicol.* 1978, **16**, 785
12. *Naunyn-Schmiedeberg's Arch. Pharmakol.* 1928, **132**, 214
13. Marhold, J. V. *Sbornik Vysledku Toxixologickeho Vysetreni Latek A Pripravku* 1972, Prague
14. Wilkin, J. K. et al *J. Invest. Dermatol.* 1987, **88**(4), 452-454
15. *S. I. 1991 No. 472 The Environmental Protection (Prescribed Processes and Substances) Regulations* 1991, HMSO, London
16. Abraham, M. H. et al *Quant. Struct.-Act. Relat.* 1990, **9**(1), 6-10
17. *BIBRA Toxicity Profile* 1991, British Industrial Biological Research Association, Carshalton
18. *ECETOC Technical Report No. 30(4)* 1991, European Chemical Industry Ecology and Toxicology Centre, B-1160 Brussels

# M163   3-Methylbutan-2-one

## $(CH_3)_2CHC(O)CH_3$

**CAS Registry No.** 563-80-4
**Synonyms** 2-acetylpropane; isopropyl methyl ketone; methylbutanone;
methyl isopropyl ketone
**Mol. Formula** $C_5H_{10}O$                                **Mol. Wt.** 86.13
**Uses** Perfumery.

## Physical properties

**M. Pt.** −92°C; **B. Pt.** 93°C; **Flash point** 6°C; **Specific gravity** 0.805.

## Occupational exposure

**US TLV (TWA)** 200 ppm (705 mg m$^{-3}$); **UN No.** 2397; **HAZCHEM Code** 3ME;
**Conveyance classification** flammable liquid; **Supply classification** highly flammable.
**Risk phrases** Highly flammable (R11)
**Safety phrases** Keep container in a well ventilated place − Keep away from sources
of ignition − No Smoking − Take precautionary measures against static discharges
(S9, S16, S33)

## Ecotoxicity

**Fish toxicity**
$LC_{50}$ (96 hr) fathead minnow 0.86 g l$^{-1}$ (1).

**Invertebrate toxicity**
$EC_{50}$ mixed bacterial culture 12.4 g l$^{-1}$ (2).

## Environmental fate

**Degradation studies**
$BOD_{15}$ using acclimated inocula of wastewater origin 4.60 mg l$^{-1}$ $O_2$, %ThOD 37.4 (3).
Using bench scale activated sludge, fill and draw operations at 6, 12 and 24 hr,
respectively, 1.2, 1.3 and 2.2% ThOD (4).

## Mammalian and avian toxicity

**Acute data**
$LC_{50}$ oral rat 148 mg kg$^{-1}$ (5).
$LD_{Lo}$ (4 hr) inhalation rat 5700 ppm (5).
$LD_{50}$ dermal rabbit 6350 mg kg$^{-1}$ (6).

**Irritancy**
Dermal rabbit (24 hr) 500 mg caused moderate to severe erythema and moderate
oedemas and 100 mg instilled into rabbit eye (24 hr) caused mild irritation (6).
Dermal rabbit 500 mg, open to atmosphere caused well defined erythema and slight
oedema (7).

## Genotoxicity

*Salmonella typhimurium* TA1535/psk1002 with and without metabolic activation,
*umu* test, negative (8).

---

*Bacillus subtilis* microsome rec-assay without metabolic activation negative, with metabolic activation DNA damaging potential (9).
*Saccharomyces cerevisiae* D61.M point mutation and/or mitotic recombination positive, but only weakly induced mitotic aneuploidy at levels of high toxicity (10).

**Any other adverse effects to man**
The threshold concentration for the eye (exposure to gas) was ≈1.5 × that for the lungs (11).

**Any other comments**
Subjective reactions by volunteers to compound as an indoor air pollutant studied (12). Inhibits acetylcholinesterase activity (13).

## References

1. Vaishnav, D. D. et al *Arch. Environ. Contam. Toxicol.* 1990, **19**, 624-628
2. Vaishnav, D. D. *Toxic. Assess.* 1986, **1**(2), 227-240
3. Vaishnav, D. D. et al *Chemosphere* 1987, **16**(4), 695-703
4. Gerhold, R. M. et al *J. – Water Pollut. Control Fed.* 1966, **38**(4), 562
5. Kennedy, G. L. et al *Toxicol. Lett.* 1991, **56**, 317-326
6. *Food Cosmet. Toxicol.* 1978, **16**, 819
7. *Food Cosmet. Toxicol.* 1978, **16**, 637
8. Ono, Y. et al *Water Sci. Technol.* 1991, **23**(1-3), 329-338
9. Matsui, S. et al *Water Sci. Technol.* 1989, **21**(8-9), 875-887
10. Zimmerman, F. K. et al *Mutat. Res.* 1985, **149**, 339-351
11. Douglas, R. B. et al *Ann. Occup. Hyg.* 1987, **31**(2), 265-267
12. Moelhave, L. et al *Atmos. Environ., Part A* 1991, **25A**(7), 1283-1293
13. Dafforn, A. et al *Biochem. Biophys. Acta* 1979, **569**, 23

# M164    3-Methyl-1-butene

$(CH_3)_2CHCH=CH_2$

**CAS Registry No.** 563-45-1
**Synonyms** α-isoamylene; isopentene; isopropylethylene
**Mol. Formula** $C_5H_{10}$                                         **Mol. Wt.** 70.14
**Uses** Organic synthesis. Manufacture of polymers.
**Occurrence** In fossil fuels. Detected in natural and drinking waters (1).

## Physical properties

**M. Pt.** –168°C; **B. Pt.** 20°C; **Flash point** –62°C; **Specific gravity** $d_4^{20}$ 0.6272; **Volatility** v. den. 2.4.

## Solubility

Water: 130 mg l$^{-1}$ at 20°C. Organic solvent: ethanol, diethyl ether, benzene

## Occupational exposure

**UN No.** 2561; **HAZCHEM Code** 3☒E; **Conveyance classification** flammable liquid.

## Environmental fate

### Degradation studies

Degradation by activated sludge, 0.8% of ThOD after 12 hr (initial concentration not specified) (2).

## Mammalian and avian toxicity

### Irritancy

Vapour or mist is irritating to the eyes, mucous membranes and upper respiratory tract (species unspecified) (3).

### Any other adverse effects to man

Symptoms of exposure include burning sensation, coughing, wheezing, laryngitis, shortness of breath, headache, nausea and vomiting (3).

### Any other comments

Autoignition temperature 365°C.

### References

1. Smith, R. L. *Trace Subst. Environ. Health* 1989, **22**, 215-232
2. Gerhold, R. M. et al *J. Water Pollut. Control Fed.* 1966, **38**(4), 562
3. Lenga, R. E. (Ed.) *Sigma-Aldrich Library of Chemical Safety Data* 2nd ed., 1988, 2309, Sigma-Aldrich, Milwaukee, WI

# M165   3-Methyl-2-buten-1-ol

$(CH_3)_2C=CHCH_2OH$

**CAS Registry No.** 556-82-1
**Synonyms** dimethylallyl alcohol; 3,3-dimethylallyl alcohol; 3-methylbut-2-en-1-ol; Prenol; prenyl alcohol
**Mol. Formula** $C_5H_{10}O$            **Mol. Wt.** 86.13
**Uses** Organic synthesis.
**Occurrence** In plant oils.

## Physical properties

**B. Pt.** 140°C; **Flash point** 43°C; **Specific gravity** $d^{20}$ 0.848; **Volatility** v.p. 1.4 mmHg at 20°C.

### Solubility

Organic solvent: vegetable oils

## Mammalian and avian toxicity

### Acute data

$LD_{50}$ oral rat 810 mg kg$^{-1}$ (1).
$LD_{50}$ dermal rabbit 3900 mg kg$^{-1}$ (1).

**Irritancy**

Dermal rabbit (24 hr) 500 mg caused moderate irritation (1).
Irritating to the eyes, upper respiratory tract and mucous membranes (species unspecified) (2).

**References**

1. *Food Cosmet. Toxicol.* 1979, **17**, 895
2. Lenga, R. E. (Ed.) *Sigma-Aldrich Library of Chemical Safety Data* 2nd ed., 1988, **2**, 2311, Sigma-Aldrich, Milwaukee, WI

# M166   3-Methyl-3-buten-1-ol

$CH_2=C(CH_3)CH_2CH_2OH$

**CAS Registry No.** 763-32-6
**Synonyms** Isobutenylcarbinol; 3-isopentenyl alcohol; isopropenylethyl alcohol; methallylcarbinol
**Mol. Formula** $C_5H_{10}O$                                   **Mol. Wt.** 86.13

**Physical properties**

**B. Pt.** $_{760}$ 129.80°C; **Flash point** 36°C; **Specific gravity** $d^{20}$ 0.853.

**Legislation**

Included in Schedule 4 (Release into Air: Prescribed Substances) Statutory Instrument No. 472, 1991 (1).

**Any other comments**

It is an attractant to houseflies and European engraver beetles (2,3).

**References**

1. *S. I. 1991 No. 472 The Environmental Protection (Prescribed Processes and Substances) Regulations* 1991, HMSO, London
2. Wilson, R. A. et al *Use of C8-tert-alkanols and C5-11-ω-alken-1-ols in attracting houseflies* 1988, **U.S. US 4, 764, 367**, *Cl.424-84;A01N25/00), US. Appl. 879,426, 27 Jun 1986, 54*
3. Kohnle, U. et al *Entomol. Exp. Appl.* 1988, **49**(1-2), 43-53

# M167   2-Methyl-3-buten-2-ol

$(CH_3)_2C(OH)CH=CH_2$

**CAS Registry No.** 115-18-4
**Synonyms** α,α-dimethylallyl alcohol; dimethylvinylcarbinol; dimethylvinylmethanol
**Mol. Formula** $C_5H_{10}O$                                   **Mol. Wt.** 86.13

**Physical properties**

**B. Pt.** 98-99°C; **Flash point** 13°C; **Specific gravity** $d^{20}$ 0.824; **Volatility** v.p. 517 mmHg at 25°C.

## Mammalian and avian toxicity

### Acute data
LD$_{50}$ oral rat 810 mg kg$^{-1}$ (1).
LD$_{50}$ dermal rabbit 3900 mg kg$^{-1}$ (1).

### Irritancy
Dermal rabbit 500 mg caused moderate irritation (1).

### Legislation
Included in Schedule 4 (Release into Air: Prescribed Substances) Statutory Instrument
No. 472, 1991 (2).

### Any other comments
It is an attractant to engraver beetle, bark beetle and *Pityogenes chalcographus* (3-5).

### References
1. *Food Cosmet. Toxicol.* 1979, **17**, 895
2. *S. I. 1991 No. 472 The Environmental Protection (Prescribed Processes and Substances) Regulations* 1991, HMSO, London
3. Konecny, K. et al *Verbenol- and alkynol-containing attractant for the engraver beetle Ips typographues.* 1987, **Czech CS 244,985**, (Cl.A01N31/02), Appl. 84/8,215, 29 Oct 1984, 5
4. Vite, J. P. et al *Attractant for the bark beetle* 1987, **Ger. Offen. DE 3,541,467**(Cl.Ao1N43/90), Appl. 23 Nov 1985, 3
5. Lofqvist, J. et al *Attractant for Pityogenes chalcographus* containing methyl-2,4-decadienoate and chalcogran 1987, **Eur. Pat. Appl. EP233,164**, (Cl. A01N43/90), SE Appl. 86/448, 7

# M168   3-Methyl-3-buten-2-one

CH$_3$C(O)C(CH$_3$)=CH$_2$

CAS Registry No. 814-78-8
**Synonyms** Isopropenyl methyl ketone; methyl isopropenyl ketone
**Mol. Formula** C$_5$H$_8$O                     **Mol. Wt.** 84.12

## Physical properties
**B. Pt.** 98°C; **Flash point** 21°C; **Volatility** v. den. 2.9.

## Mammalian and avian toxicity

### Acute data
LD$_{50}$ oral rat 180 mg kg$^{-1}$ (1).
LD$_{Lo}$ oral guinea pig 60 mg kg$^{-1}$ (2).
LC$_{Lo}$ (4 hr) inhalation rat 125 ppm (1).
LD$_{50}$ dermal rabbit 230 mg kg$^{-1}$ (1).

### Legislation
Included in Schedule 4 (Release into Air: Prescribed Substances) Statutory Instrument
No. 472, 1991 (3).

### References
1. *Arch. Ind. Hyg. Occup. Med.* 1951, **4**, 119
2. Patty, F. A., (Ed.) *Industrial Hygiene and Toxicology* 1963, **2**, 1754, John Wiley & Son, New York
3. *S. I. 1991 No. 472 The Environmental Protection (Prescribed Processes and Substances) Regulations* 1991, HMSO, London

# M169   *N*-Methylbutylamine

## CH₃CH₂CH₂CH₂NHCH₃

**CAS Registry No.** 110-68-9
**Synonyms** butylamine, *N*-methyl-; butylmethylamine; methylbutylamine; 1-butanamine, *N*-methyl-; *N*-methylbutamine
**Mol. Formula** $C_5H_{13}N$                    **Mol. Wt.** 87.17
**Uses** Used in the preparation of pharmaceuticals. Catalyst. Used in co-surfactants. Chemical intermediate.

### Physical properties

**M. Pt.** –75°C; **B. Pt.** 90.5-91.5°C; **Flash point** 1°C; **Specific gravity** 0.736; **Partition coefficient** log $P_{ow}$ 1.33; **Volatility** v. den. 3.0.

### Occupational exposure

**UN No.** 2945; **HAZCHEM Code** 2PE; **Conveyance classification** flammable liquid.

### Mammalian and avian toxicity

**Acute data**
$LD_{50}$ oral rat 420 mg kg$^{-1}$ (1).
$LC_{Lo}$ (4 hr) inhalation rat 2000 ppm (1).
$LD_{50}$ dermal rabbit 1260 mg kg$^{-1}$ (2).
$LD_{50}$ intravenous mouse 120 mg kg$^{-1}$ (3).
$LD_{50}$ intraperitoneal mouse 470 mg kg$^{-1}$ (3).

**Irritancy**
74 mg instilled into rabbit eye caused severe irritation (unspecified duration) (1).

### Any other adverse effects to man

Very damaging to the tissue of the upper respiratory tract and mucous membranes. Inhalation can cause spasm, inflammation and oedema of the larynx and bronchi, chemical pneumonitis and pulmonary oedema, these effects may be severe enough to cause death (4).

### Legislation

Included in Schedule 4 (Release into Air: Prescribed Processes) Statutory Instrument No. 472, 1991 (5).

### References
1. *Union Carbide Data Sheet* 1970, **7**, 6
2. *J. Pharmacol. Exp. Ther.* 1946, **88**, 82

3. *Am. Ind. Hyg. Assoc. J.* 1962, **23**, 95
4. Lenga, R. E. (Ed.) *Sigma-Aldrich Library of Chemical Safety Data* 2nd ed., 1988, Sigma-Aldrich, Milwaukee, WI
5. *S. I. 1991 No. 472 The Environmental Protection (Prescribed Processes and Substances) Regulations* 1991, HMSO, London

# M170   Methyl *tert*-butyl ether

## $CH_3OC(CH_3)_3$

**CAS Registry No.** 1634-04-4
**Synonyms** *tert*-butyl methyl ether; 2-methoxy-2-methylpropane; methyl 1,1-dimethylethyl ether; MTBE; 2-oxa-3,3-dimethylbutane
**Mol. Formula** $C_5H_{12}O$  **Mol. Wt.** 88.15
**Uses** Fuel additive especially as an alternative for lead in gasoline. Catalyst. Solvent for cholesterol gallstones. Chromatographic eluent. Cholelitholytic agent.
**Occurrence** Residues have been detected in groundwaters (1).

## Physical properties

**M. Pt.** −109°C; **B. Pt.** 55.2°C; **Flash point** −10°C; **Specific gravity** $d_4^{20}$ 0.7404;
**Partition coefficient** log $P_{ow}$ 0.94; **Volatility** v.p. 249 mmHg at 25°C.

## Solubility

Water: 51 g $l^{-1}$ at 25°C. Organic solvent: diethyl ether, ethanol

## Ecotoxicity

**Fish toxicity**
$LC_{50}$ (96 hr) fathead minnow 110 mg $l^{-1}$ (2).

**Invertebrate toxicity**
$EC_{50}$ (30 min) *Photobacterium phosphoreum* 11.4 ppm, Microtox test (3).

**Bioaccumulation**
Bioconcentration factor for Japanese carp 1-5 (4).

## Environmental fate

**Degradation studies**
Degradation by activated sludge 1% of ThOD in 21 days (5).

**Abiotic removal**
Reaction with photochemically produced hydroxyl radicals in the atmosphere $t_{1/2}$ 5-6 days (6,7).
Volatilisation $t_{1/2}$ 4.1 hr from model river water at 25°C, and 2.0 days from model pond water (8,9).

**Absorption**
Estimated $K_{oc}$ of 11.2 indicates that adsorption to soil and sediments would not be significant (8).

---

## Mammalian and avian toxicity

### Acute data

$LD_{50}$ oral rat, mouse 4000 mg $kg^{-1}$ (9,10).
$LC_{50}$ (4 hr) inhalation rat 24,000 ppm (10).
$LC_{50}$ (15 min) inhalation mouse 140 mg $m^{-3}$ (11).
$LD_{50}$ dermal rat >7.5 mg $g^{-1}$ (12).
$LD_{50}$ intraperitoneal mouse 1500 mg $kg^{-1}$ (9).
$LD_{50}$ intravenous mouse 230 mg $kg^{-1}$ (9).

### Teratogenicity and reproductive effects

Inhalation ♂ rats, 300, 1300, or 3400 ppm 6 hr $day^{-1}$, 5 days $wk^{-1}$ for 12 wk, mated with ♀ rats exposed to the same concentrations for 3 wk. Exposure continued throughout the mating period, and ♀ rats continued exposure during gestation and from day 5-21 of lactation. No parental toxicity was observed and fertility indices were unaltered (13).

### Metabolism and pharmacokinetics

Metabolised by rat hepatic microsomes, yielding equimolar amounts of *tert*-butanol and formaldehyde (14).

### Irritancy

Vapour or mist is irritating to the eyes, mucous membranes and upper respiratory tract (dose, species unspecified) (15).

## Any other adverse effects to man

Presence in the lungs may cause chemical pneumonia which can be fatal (15).
Skin irritant and may cause central nervous system depression (16).
A case of renal failure was reported in 1/8 patients infused with methyl *tert*-butyl ether for the removal of gallstones. Haemolysis due to extravasation after leakage alongside the catheter was suspected to be the cause of renal failure. Renal function recovered after 18 days dialysis (17).

## Any other adverse effects

Toxic effects in dogs, rabbits and mice include haemolysis, nerve paralysis, acute cholecystitis, hepatocyte cloudy swelling and focal necrosis, and acute duodenitis (9).

## Legislation

Included in Schedule 6 (Release into Land: Prescribed Substances) Statutory Instrument No. 472, 1991 (18).

## Any other comments

Autoignition temperature 224°C.
Environmental fate reviewed (1).
Human health effects, experimental toxicology, physico-chemical properties reviewed (19).
$LC_{50}$ *Rana temporaria* tadpole 2500 mg $l^{-1}$ (exposure not specified). Exposure to 100 mg $l^{-1}$ caused a marked increase in body weight and accelerated tadpole development, metamorphosis occurring 2 days earlier than in controls (20).

## References

1. Howard, P. H. *Handbook of Environmental Fate and Exposure Data for Organic Chemicals* 1993, **4**, 71-76
2. Protic, M. et al *Aquat. Toxicol.* 1989, **14**(1), 47-64
3. Kaiser, K. L. E. et al *Water Pollut. Res. J. Can.* 1991, **26**(3), 361-431
4. Fujiwara, Y. et al *Yukagaku* 1984, **33**, 111-114
5. Alexander, M. *Biotechnol. Bioeng.* 1973, **15**, 611-647
6. Bennett, P. J. et al *J. Atmos. Chem.* 1990, **10**, 29-38
7. Wallington, et al *Environ. Sci. Technol.* 1988, **22**, 842-844
8. Lyman, W. K. et al *Handbook of Chemical Property Estimation Methods: Environmental Behaviour of Organic Compounds* 1982, McGraw-Hill, New York
9. Guo, S. et al *Shanghai Yixue* 1989, **12**(8), 472-476 (Ch.) (*Chem. Abstr.* **112**, 91572h)
10. *NTIS Report PB87-17603*, Natl. Tech. Inf. Ser., Springfield, VA
11. Marsh, D. L. et al *Anesthesiology* 1950, **11**, 455
12. Smyth, A. F. et al *J. Ind. Hyg. Toxicol.* 1948, **30**, 63
13. Biles, R. W. et al *Toxicol. Ind. Health* 1987, **3**(4), 519-534
14. Brady, J. F. et al *Arch. Toxicol.* 1990, **64**(2), 157-160
15. Lenga, R. E. (Ed.) *Sigma-Aldrich Library of Chemical Safety Data* 2nd ed., 1988, **1**, 641, Sigma-Aldrich, Milwaukee, WI
16. *Martindale. The Extra Pharmacopoeia* 30th ed., 1993, 1104, The Pharmaceutical Press, London
17. Ponchon, T. et al *Lancet* 1988, 276-277
18. *S. I. 1991 No. 472 The Environmental Protection (Prescribed Processes and Substances) Regulations* 1991, HMSO, London
19. *ECETOC Technical Report No. 30(4)* 1991, European Chemical Industry Ecology and Toxicology Centre, B-1160 Brussels
20. Paulov, S. *Biologia (Bratislava)* 1987, **42**(2), 185-189 (Slo.) (*Chem. Abstr.* **107**, 2142z)

# M171   2-Methyl-3-(4-*tert*-butylphenyl)-propionaldehyde

**CAS Registry No.** 80-54-6

**Synonyms** benzenepropenal, 4-(1,1-dimethylethyl)-α-methyl-; hydrocinnamaldehyde, *p-tert*-butyl-α-methyl-; Lilial; Lilyal

**Mol. Formula** $C_{14}H_{20}O$                **Mol. Wt.** 204.31

## Physical properties

**B. Pt.** $_{10}$ 150°C; **Specific gravity** $d^{25}$ 0.939.

## Mammalian and avian toxicity

### Acute data

$LD_{50}$ oral rat 3700 mg $kg^{-1}$ (1).

---

**Irritancy**

Dermal rabbit (24 hr) 500 mg caused moderate irritation (1).

**Legislation**

Included in Schedule 4 (Release into Air: Prescribed Substances) Statutory Instrument No. 472, 1991 (2).

**References**

1. *Food Cosmet. Toxicol.* 1978, **16**, 637
2. *S. I. 1991 No. 472 The Environmental Protection (Prescribed Processes and Substances) Regulations* 1991, HMSO, London

# M172   2-Methyl-3-butyn-2-ol

$HC\equiv CC(OH)(CH_3)_2$

**CAS Registry No.** 115-19-5

**Synonyms** 2-Methylbutyn-3-ol-2; 1,1-dimethyl-2-propyn-1-ol; dimethylacetylenecarbinol; dimethylacetylenylcarbinol; α,α-dimethylpropargyl alcohol; 1,1-dimethylpropynol; ethynyldimethylcarbinol

**Mol. Formula** $C_5H_8O$                          **Mol. Wt.** 84.12

**Physical properties**

**M. Pt.** 2.6°C; **B. Pt.** $_{760}$ 104-105°C; **Flash point** 25°C; **Specific gravity** $d_{20}^{20}$ 0.8672; **Volatility** v. den. 2.49.

**Solubility**

Water: miscible. Organic solvent: miscible with acetone, carbon tetrachloride, acetone, benzene, mineral spirits, petroleum ether

**Ecotoxicity**

**Fish toxicity**

$LC_{50}$ fathead minnow 3290 mg l$^{-1}$ (1).

**Mammalian and avian toxicity**

**Acute data**

$LD_{50}$ oral mouse, rat 1800-1950 mg kg$^{-1}$ (2,3).
$LD_{50}$ intraperitoneal mouse 3600 mg kg$^{-1}$ (4).
$LD_{50}$ subcutaneous mouse 2340 mg kg$^{-1}$ (5).

**References**

1. Veith, G. D. et al *Xenobiotica* 1989, **19**(5), 555-565
2. Soehring, K. *Arzneim.-Forsch.* 1954, **4**, 477
3. *J. Pharmacol. Exp. Ther.* 1955, **115**, 230
4. *Merck Index* 1983, **10**, Merck and Co. Inc., Rahway, NJ
5. *Arzneim.-Forsch.* 1955, **5**, 161

# M173   Methyl carbamate

$NH_2CO_2CH_3$

**CAS Registry No.** 598-55-0
**Synonyms** methylurethane; NCI-C55594; urethylane
**Mol. Formula** $C_2H_5NO_2$                                        **Mol. Wt.** 75.07
**Uses** Manufacture of resins and pharmaceuticals.
**Occurrence** Isolated from *Salsola* species (1).

## Physical properties

**M. Pt.** 56-58°C; **B. Pt.** 176-177°C; **Specific gravity** $d_4^{56}$ 1.1361.

## Solubility
Water: 2170 g $l^{-1}$ at 11°C. Organic solvent: acetone, diethyl ether, ethanol

## Environmental fate

### Absorption
Adsorbs strongly to Na-, Mg-, Al- and Cu-montmorillonites by interaction between the C=O group and the exchangeable cations (2).

## Mammalian and avian toxicity

### Acute data
$LD_{50}$ oral mouse 6200 mg $kg^{-1}$ (3).
$LD_{50}$ subcutaneous mouse 4500 mg $kg^{-1}$ (4).
$LD_{Lo}$ intraperitoneal mouse 200 mg $kg^{-1}$ (5).

### Sub-acute data
Gavage rat, mouse (13 wk) 50-1000 mg $kg^{-1}$ $day^{-1}$ 5 days $wk^{-1}$ in rats,
100-2000 mg $kg^{-1}$ $day^{-1}$ 5 days $wk^{-1}$ in mice. Treatment in rats resulted in dose-related lesions in the liver, weight loss, testicular hypoplasia, bone marrow hyperplasia and pigmentation of the spleen. The highest dose reduced the mean survival time. Only weight loss and inflammatory changes of the liver were observed in mice. The proliferative nature of hepatic lesions in rats suggest that methyl carbamate is potentially hepatocarcinogenic (6).

### Carcinogenicity and long-term effects
No adequate evidence for carcinogenicity to humans or animals, IARC classification group 3 (7).
Subcutaneous mouse (6 month) $3 \times 0.1$ mg $kg^{-1}$ at 2 day intervals. After 3 months, 3/29 killed mice had lung adenomas compared with 3/22 controls and with 23/27 mice given 1 mg $kg^{-1}$ urethane plus 1 mg $kg^{-1}$ methyl carbamate. At 6 months 2/26 killed mice had lung adenomas compared with 0/26 controls and 6/28 mice administered urethane plus methyl carbamate (8).
Intraperitoneal mouse (4 month) 0, 1000, 2000 or 3000 mg $kg^{-1}$ once $wk^{-1}$ for 13 wk. Lung adenomas occurred in 17% untreated controls, 16% of the lower dose group, 9% of the 1000 mg $kg^{-1}$ treated group and 22% of the 2000 mg $kg^{-1}$ treated group. The high dose caused early deaths (9).
Intraperitoneal mouse (24 wk) $12 \times 5$ mg $animal^{-1}$ over 4 wk, lung adenomas occurred in 1/16 treated animals, in 6/31 vehicle controls and in 2/31 mice which

received no treatment (10).

Dermal mouse (18 wk) 75 mg animal$^{-1}$ wk$^{-1}$ for 15 wk. 3 days after the start of methyl carbamate treatment, 18 wkly applications of croton oil (0.3 ml of a 0.5% solution in acetone) were given. 1/18 treated mice developed a skin tumour compared with 1/20 controls given croton oil only, and 7/18 treated mice had a total of 12 lung adenomas (incidence of lung adenomas in controls was not reported) (11).

National Toxicology Program tested rats and mice via gavage. No evidence of carcinogenicity in ♂ and ♀ mice, clear evidence for carcinogenicity in ♂ and ♀ rats (12).

**Metabolism and pharmacokinetics**

Rats injected intraperitoneally with methyl carbamate or the corresponding N-hydroxycarbamate excreted methyl carbamate and N-hydroxycarbamate in the urine (13).

## Genotoxicity

*Salmonella typhimurium* TA97, TA98, TA100, TA1535, TA1537, TA1538 with and without metabolic activation negative (14).

*Bacillus subtilis* reverse mutation without metabolic activation positive (15).

*In vitro* Chinese hamster ovary cells, chromosomal aberrations and sister chromatid exchanges negative (16).

*In vitro* mouse lymphoma L5178Y cells, tk$^+$/tk$^-$ with and without metabolic activation negative (16).

Did not bind significantly to rat liver or kidney DNA *in vivo*, while ethyl carbamate was found to bind to rat liver DNA to a significant extent (17).

## Any other comments

Physical properties, use, occurrence, analysis, carcinogenicity, mammalian toxicity and mutagenicity reviewed (1,18).

## References

1. *IARC Monograph* 1976, **12**, 151-159
2. Fusi, P. et al *Appl. Clay Sci.* 1989, **4**(5-6), 403-409
3. Srivalova, T. I. *Toksikol. Nov. Prom. Khim. Veshctestv* 1973, **13**, 86-91
4. *Aust. J. Exp. Biol. Med. Sci.* 1967, **45**, 507
5. *Toxicol. Appl. Pharmacol.* 1972, **23**, 288
6. Quest, J. A. et al *Fundam. Appl. Toxicol.* 1987, **8**(3), 389-399
7. *IARC Monograph* 1987, **Suppl. 7**, 66
8. Yagubov, A. S. et al *Gig. Tr. Prof. Zabol.* 1973, **8**, 19-22
9. Larsen, C. D. *J. Natl. Cancer Inst.* 1947, **8**, 99-101
10. Shimkin, M. B. et al *Cancer Res.* 1969, **29**, 2184-2190
11. Roe, F. J. C. et al *Br. J. Cancer* 1955, **9**, 177-203
12. *National Toxicology Program Research and Testing Division* 1992, Report No. TR-328, NIEHS, Research Triangle Park, NC
13. Boylant, E. et al *Biochem. J.* 1965, **94**, 198-208
14. Klopman, G. et al *Mutat. Res.* 1990, **228**, 1-50
15. De Giovanni-Donelly, R. et al *Mutat. Res.* 1967, **4**, 543-551
16. Anderson, B. E. et al *Environ. Mol. Mutagen.* 1990, **16**(Suppl. 18), 55-137
17. Lawson, T. A. et al *Chem.-Biol. Interact.* 1973, **6**, 99-105
18. *ECETOC Technical Report No. 30(4)* 1991, European Chemical Industry Ecology and Toxicology Centre, B-1160 Brussels

# M174  Methyl 2-chloroacrylate

$$CH_2=C(Cl)CO_2CH_3$$

**CAS Registry No.** 80-63-7
**Synonyms** methyl 2-chloro-2-propenoate
**Mol. Formula** $C_4H_5ClO_2$                **Mol. Wt.** 120.54
**Uses** Manufacture of polymers.

**Physical properties**
**B. Pt.** $_{51}$ 52°C; **Specific gravity** $d_4^{20}$ 1.189.

**Solubility**
Organic solvent: diethyl ether

## Mammalian and avian toxicity

**Acute data**
$LC_{50}$ (2 hr) inhalation rat, mouse, rabbit, guinea pig, cat 500 mg m$^{-3}$ (1).

**Irritancy**
Dermal rabbit, 500 mg caused severe irritation (exposure not specified) (2).
Vapours are irritating to the eyes (humans) at 5-10 ppm (3).

**Genotoxicity**
*Salmonella typhimurium* TA100 with metabolic activation positive (4).

**Any other adverse effects to man**
Inhalation exposure has been reported to cause pulmonary oedema (3).
Skin contact causes painful keratitis and dermatitis with vesiculation (5).

## References

1. *Farmakol. Toksikol. (Moscow)* 1956, **19**, 60
2. Shell Chemical Co. *Report* 1961, 6
3. Lefaux, R. *Practical Toxicology of Plastics* 1968, 90, CRC Press, Cleveland, OH
4. Rosen, J. D. et al *Mutat. Res.* 1980, **78**, 113-119
5. Grant, W. M. *Toxicology of the Eye* 3rd ed., 1986, 615, Charles C. Thomas, Springfield, IL

# M175  Methyl chloroformate

$$ClCO_2CH_3$$

**CAS Registry No.** 79-22-1
**Synonyms** MCF; methoxycarbonyl chloride; methyl carbonochloridate; methyl chlorocarbonate
**Mol. Formula** $C_2H_3ClO_2$               **Mol. Wt.** 94.50
**Uses** Acylating agent. Organic synthesis. Insecticide.

## Physical properties
**M. Pt.** < –81°C; **B. Pt.** 70-72°C; **Flash point** 17°C; **Specific gravity** $d_4^{20}$ 1.223; **Volatility** v.p. 127 mmHg at 20°C; v. den. 3.26.

### Solubility
Water: miscible. Organic solvent: benzene, chloroform, diethyl ether, ethanol

## Occupational exposure
**UN No.** 1238; **HAZCHEM Code** 3WE; **Conveyance classification** toxic substance, flammable liquid, corrosive substance; **Supply classification** highly flammable, toxic.
**Risk phrases** Highly flammable – Toxic by inhalation – Irritating to eyes, respiratory system and skin (R11, R23, R36/37/38)
**Safety phrases** Keep container in a well ventilated place – Keep away from sources of ignition – No Smoking – Take precautionary measures against static discharges – If you feel unwell, seek medical advice (show label where possible) (S9, S16, S33, S44)

### Abiotic removal
Hydrolysis $t_{1/2}$ 35 min at 19.6°C (1).
Reaction with photochemically produced hydroxyl radicals in the atmosphere, estimated $t_{1/2}$ 74 days (2).

## Mammalian and avian toxicity

### Acute data
$LD_{50}$ oral rat, mouse, guinea pig 60-140 mg $kg^{-1}$ (3).
$LC_{50}$ (1 hr) inhalation rat 88 ppm (4).
$LD_{50}$ dermal mouse, rabbit 1750, 7100 mg $kg^{-1}$, respectively (3,4).
$LD_{50}$ intraperitoneal mouse 40 mg $kg^{-1}$ (5).

### Irritancy
Vapour is strongly irritating to the eyes (6).

## Any other adverse effects to man
Extremely destructive to tissue of the mucous membranes and upper respiratory tract, eyes and skin. Inhalation may be fatal as a result of spasm, inflammation and oedema of the larynx and bronchi, chemical pneumonitis and pulmonary oedema (7).

## Any other adverse effects
Chronic inhalation exposure causes changes in neuromuscular excitability, body temperature, respiration rate and organ weights (species not specified) (3).

## Legislation
Limited under EC Directive on Drinking Water Quality 80/778/EEC. Pesticides: maximum admissible concentration 0.1 μg $l^{-1}$ (8).
Included in Schedule 6 (Release into Land: Prescribed Substances) Statutory Instrument No. 472, 1991 (9).

## Any other comments
Physical properties, use, toxicity and safety precautions reviewed (10,11).
Autoignition temperature 505°C.

## References

1. Queen, A. *Can. J. Chem.* 1967, **45**, 1619-1629
2. Atkinson, R. *Int. J. Chem. Kinet.* 1987, **19**, 799-828
3. Gurova, A. L. et al *Gig. Sanit.* 1977, (5), 97-99
4. *Toxicol. Appl. Pharmacol.* 1977, **42**, 417
5. *NTIS Report AD691-490,* Natl. Tech. Inf. Ser., Springfield, VA
6. *The Merck Index* 11th ed., 1989, 952, Merck & Co., Rahway, NJ
7. Lenga, R. E. (Ed.) *Sigma-Aldrich Library of Chemical Safety Data* 2nd ed., 1988, **2**, 2318, Sigma-Aldrich, Milwaukee, WI
8. *EC Directive Relating to the Quality of Water Intended for Human Consumption* 1982, 80/778/EEC, Office for Official Publications of the European Communities, 2 rue Mercier, L-2985 Luxembourg
9. *S. I. 1991 No. 472 The Environmental Protection (Prescribed Processes and Substances) Regulations* 1991, HMSO, London
10. *Chemical Safety Data Sheets* 1991, **4b**, 41-43, RSC, London
11. *ECETOC Technical Report No. 30(4)* 1991, European Chemical Industry Ecology and Toxicology Centre, B-1160 Brussels

# M176  Methyl 2-chloropropionate

## $CH_3CHClCO_2CH_3$

**CAS Registry No.** 17639-93-9
**Synonyms** propanoic acid, 2-chloro-, methyl ester; methyl $\alpha$-chloropropionate
**Mol. Formula** $C_4H_7ClO_2$ **Mol. Wt.** 122.55

### Physical properties

**B. Pt.** 132-133°C; **Flash point** 38°C; **Specific gravity** $d^{20}$ 1.075.

### Occupational exposure

**UN No.** 2933; **HAZCHEM Code** 2Y; **Conveyance classification** flammable liquid.

### Mammalian and avian toxicity

**Acute data**
$LD_{Lo}$ intraperitoneal mouse 250 mg kg$^{-1}$ (1).

### Legislation

Included in Schedule 4 (Release into Air: Prescribed Substances) Statutory Instrument No. 472, 1991 (2).

### Any other comments

Lachrymator.

### References

1. *Summary Tables of Biological Tests* 1954, **6**, 228, National Research Council Chemical-Biological Coordination Centre
2. *S. I. 1991 No. 472 The Environmental Protection (Prescribed Processes and Substances) Regulations* 1991, HMSO, London

# M177    3-Methylcholanthrene

CAS Registry No. 56-49-5
**Synonyms** benz[*j*]aceanthrylene, 1,2-dihydro-3-methyl-; cholanthrene, 3-methyl-;
3-MC; 20-MC; 20-methylcholanthrene
**Mol. Formula** $C_{21}H_{16}$                                    **Mol. Wt.** 268.36
**Uses** Experimentally in cancer research.

## Physical properties

**M. Pt.** 180°C; **B. Pt.** $_{80}$ 280°C; **Specific gravity** 1.28; **Partition coefficient**
log $P_{ow}$ 6.75.

**Solubility**
Organic solvent: benzene, xylene, toluene

## Ecotoxicity

**Fish toxicity**
Carp, tench (48 hr) 10 mg $kg^{-1}$ induced changes in the RNA and DNA of the cells (1).

## Environmental fate

**Degradation studies**
A Gram-positive, rod-shaped bacterium mineralised 3-methylcholanthrene, 1.6% of
the original amount, to $CO_2$ when grown for 2 wk in pure culture with organic
nutrients (2).

**Absorption**
Log $K_{oc}$ 6.25 (3).

## Mammalian and avian toxicity

**Acute data**
$LD_{Lo}$ intraperitoneal mouse 100 mg $kg^{-1}$ (4).

**Carcinogenicity and long-term effects**
Pregnant rats exposed to 7, 21 or 63 mg $kg^{-1}$ on days 15-17 of gestation via gavage.
Adult offspring developed neoplastic changes in the lungs 6 months after *in utero*
exposure (5).
Displayed potent carcinogenic activity in mouse skin and rat mammary gland (6).
*In vitro* slices of mouse lung 0.5 or 1 mg, dose-dependent carcinogenesis
(proliferation) and toxic (dystrophy and necrosis) effects were observed (7).
Incidence of liver and lung tumours in mice exposed transplacentally were
significantly influenced by the sensitivity of both mothers and foetuses to induction of

cytochrome P450 by polycyclic aromatic hydrocarbons (8).
Oral (3 month) 6 ♂ rats 0.03%, report of study in preparation (9).

**Teratogenicity and reproductive effects**
In pregnant mice caused lung and liver tumours in the offspring, the incidences of
which were greatly influenced by the *Ah* locus-regulated induction phenotype for
arylhydrocarbon hydroxylase activity in the mother and foetuses (10).

**Metabolism and pharmacokinetics**
Metabolic products vary with the type of enzyme induction (10).
In foetal rat liver, 1- or 2-hydroxy-, *cis*- and *trans*-1,2-dihydroxy-, 11,12-dihydroxy-,
11,12-dihydro-, and 1- and 2-keto-3-cholanthrene were produced (11).
When incubated with human bone marrow preparations in air for 60 min at 37°C;
major metabolites were: 1-hydroxy-3- methylcholanthrene;
1-keto-3-methylcholanthrene; and cholanthrene. It can undergo biochemical reactions
in preparations of human bone marrow, giving rise to the formation of metabolites
which are known to be carcinogenic in rats and mice (12).
Peyer's patches have an importance in the absorption from the gut and subsequent
retention and hence may be a likely target organ for lymphoid carcinogenesis
following oral exposure to carcinogenic polycyclic aromatic hydrocarbons (13).

## Genotoxicity

*In vitro* Chinese hamster cells chromosomal aberrations and sister chromatid
exchanges marginally positive (metabolic activation unspecified) (14).
*In vitro* human cells with or without metabolic activation negative (details
unspecified) (15).
*In vitro* Syrian hamster embryo cells morphological transformation positive
(metabolic activation unspecified) (16).
Rat primary lung cells *in vitro* without metabolic activation and *in vivo* caused high
frequencies of sister chromatid exchanges (17).
Human peripheral blood lymphocytes from non-smoking, non-drug taking individuals
were cultured in 10% foetal or calf serum. The level of *ss*-DNA increased but not of
*ds*-DNA on exposure to 3-methyl cholanthrene. The time-dependent changes
increased for ≥4 day of exposure, indicating that the repair enzymes were not able to
compensate for the DNA damage (18).
*In vitro* carp kidney cells, chromosomal aberrations and micronucleated erythrocytes
positive (19).
BALB/c 3T3 cells simultaneous cell transformation and mutation assay, positive (20).

## Any other adverse effects

Promotes the induction or activation of cytochrome $P_{450}$ of the aryl hydrocarbon
hydroxylase system (21).

## Legislation

The log $P_{ow}$ value exceeds the European Community recommended level 3.0 (6th and
7th amendments) (22).

## References

1.  Al-Sabti, K. *Cytobios* 1986, **47**, 147
2.  Heitkamp, M. A. et al *Appl. Environ. Microbiol.* 1988, **54**(6), 1612-1614

3.  Sabljic, A. *QSAR, Environ. Toxicol., Proc. Int. Workshop, 2nd* 1986, 309-332
4.  *Toxicol. Appl. Pharmacol.* 1972, **23**, 288
5.  *J. Natl. Cancer Inst.* 1972, **48**, 185
6.  Cavalieri, E. *J. Cancer Res. Clin. Oncol.* 1988, **114**(1), 16-22
7.  Paternain, J. L. et al *Rev. Esp. Fisiol.* 1987, **43**(2), 223-227
8.  Miller, M. S. et al *Carcinogenesis (London)* 1989, **10**(5), 875-883
9.  *Directory of Agents Being Tested for Carcinogenicity* 1992, No. 15, IARC, Lyon
10. Miller, M. S. et al *Carcinogenesis (London)* 1990, **11**(11), 1979-1984
11. Buerki, K. et al *Biochem. Biophys. Acta* 1972, **260**(1), 98
12. Myers, S. R. et al *Drug Metab. Dispos.* 1990, **18**(5), 664-669
13. Bost, K. L. et al *Carcinogenesis (London)* 1986, **7**(8), 1251-1256
14. Propescu, N. C. et al *J. Natl. Cancer Inst.* 1977, **59**(1), 289
15. Styles, J. A. *Br. J. Cancer* 1978, **37**(6), 931
16. Le Boeuf, R. A. et al *Mutat. Res.* 1989, **222**(3), 205-218
17. Wong, W. Z. et al *Mutat. Res.* 1990, **241**(1), 7-13
18. Claussen, J. et al *ALTA, Altern. Lab. Anim.* 1987, **14**(3), 168-171
19. Al-Sabti, K. *Comp. Biochem. Physiol., C: Comp. Pharmacol. Toxicol.* 1986, **85C**, 5-9
20. Fitzgerald, O. J. et al *Mutagenesis* 1989, **4**(4), 286-291
21. *Patty's Industrial Hygiene and Toxicology* 3rd rev. ed., 1981, **2B**, John Wiley & Sons, New York
22. *1967 Directive on Classification, Packaging, and Labelling of Dangerous Substances 67/548/EEC; 6th Amendment EEC Directive 79/831/EEC; 7th Amendment EEC Directive 91/32/EEC* 1991, HMSO, London

# M178   1-Methylchrysene

**CAS Registry No.** 3351-28-8
**Synonyms** chrysene, 1-methyl-
**Mol. Formula** $C_{19}H_{14}$                                    **Mol. Wt.** 242.32
**Occurrence** In fossil fuels. Has been identified in tobacco and marijuana smoke (1).

## Physical properties

**M. Pt.** 253-257°C; **B. Pt.** sublimes at 130-140°C (in vacuum).

**Solubility**
Organic solvent: acetone, benzene, ethanol, hexane, toluene

## Mammalian and avian toxicity

### Carcinogenicity and long-term effects
No adequate evidence for carcinogenicity to humans or animals, IARC classification group 3 (2).

Dermal mouse (72 wk) 100 µg animal$^{-1}$ 3 × wk$^{-1}$ for 72 wk, at which time 7/20 treated animals were still alive. No skin tumours were observed in treated animals or vehicle controls (3, 4).

Dermal mouse (24 wk) 100 µg animal$^{-1}$ on alternate days for 20 days, followed 10 days later by applications of 2.5 µg animal$^{-1}$ 12-O-tetradecanoylphorbol 13-acetate 3 × wk$^{-1}$ for 20 wk at which time 19/20 mice were still alive. In this group 6 animals each had a skin tumour, as compared to none in vehicle controls and 10 skin tumours in 6/20 surviving positive controls treated with benzo[a]pyrene (4).

### Genotoxicity

*Salmonella typhimurium* TA100 with metabolic activation positive (5).

### Any other comments

Physical properties, occurrence, carcinogenicity and metabolism of methylchrysenes reviewed (1, 6).

### References

1. *IARC Monograph* 1983, **32**, 379-397
2. *IARC Monograph* 1987, **Suppl. 7**, 66
3. Hoffman, D. et al *Science (Washington, D. C. 1883-)* 1974, **183**, 215-216
4. Hecht, S. S. et al *J. Natl. Cancer Inst.* 1974, **53**, 1121-1133
5. Coombs, M. M. et al *Cancer Res.* 1976, **36**, 4525-4529
6. *ECETOC Technical Report No. 30(4)* 1991, European Chemical Industry Ecology and Toxicology Centre, B-1160 Brussels

# M179   2-Methylchrysene

**CAS Registry No.** 3351-32-4

**Synonyms** chrysene, 2-methyl-

**Mol. Formula** $C_{19}H_{14}$                    **Mol. Wt.** 242.32

**Occurrence** In fossil fuels. Has been identified in tobacco smoke, marijuana smoke, engine exhausts and in vegetables (1).

### Physical properties

**M. Pt.** 225-230°C.

### Solubility

Organic solvent: acetic acid, acetone, benzene, ethanol, cyclohexane

## Mammalian and avian toxicity

### Carcinogenicity and long-term effects
No adequate evidence for carcinogenicity to humans or animals, IARC classification group 3 (2).

Dermal mouse (72 wk) 100 µg animal$^{-1}$ 3 × wk$^{-1}$ for 72 wk. The first skin tumour was reported at 40 wk at which time 18/20 animals were still alive. At 72 wk 10/20 treated mice were still alive; 21 skin tumours were observed in 11 tumour-bearing animals. Seven of these tumours were carcinomas. No skin tumours were reported in vehicle controls (3).

Dermal mouse (24 wk) 100 µg animal$^{-1}$ on alternate days for 20 days, followed 10 days later by application of 5 µg animal$^{-1}$ of 12-*O*-tetradecanoylphorbol 13-acetate 3 × wk$^{-1}$ for 20 wk. At this time 19/20 mice were still alive. In this group 13 skin tumours were present in 8 mice, as compared with none in vehicle controls and 10 skin tumours in 6/20 surviving positive controls treated with benzo[*a*]pyrene (3).

### Genotoxicity
*Salmonella typhimurium* TA100 with metabolic activation positive (4).

### Any other comments
Physical properties, occurrence, carcinogenicity and metabolism of methylchrysenes reviewed (1).

### References
1. *IARC Monograph* 1983, **32**, 379-397
2. *IARC Monograph* 1987, **Suppl. 7**, 66
3. Hecht, S. S. et al *J. Natl. Cancer Inst.* 1974, **53**, 1121-1133
4. Coombs, M. M. et al *Cancer Res.* 1976, **36**, 4525-4529

# M180   3-Methylchrysene

**CAS Registry No.** 3351-31-3
**Synonyms** chrysene, 3-methyl-
**Mol. Formula** $C_{19}H_{14}$                    **Mol. Wt.** 242.32
**Occurrence** In fossil fuels. Has been identified in tobacco smoke, marijuana smoke, engine exhausts and in vegetables (1).

### Physical properties
**M. Pt.** 170-174°C.

## Solubility

Organic solvent: acetone, benzene, cyclohexane, ethanol, petroleum ether

## Mammalian and avian toxicity

### Carcinogenicity and long-term effects

No adequate evidence for carcinogenicity to humans or animals, IARC classification group 3 (2).

Dermal mouse (72 wk) 100 µg animal$^{-1}$ 3 × wk$^{-1}$. The first skin tumour was observed at 15 wk at which time all 20 treated mice were still alive. At 72 wk 8/20 mice were still alive, 5 of which were bearing a total of 6 skin tumours. Four of these were carcinomas. No skin tumours occurred in vehicle controls (3,4).

Dermal mouse (24 wk) 10, 30 or 100 µg animal$^{-1}$ on alternate days for 20 days, followed 10 days later by application of 2.5 µg animal$^{-1}$ of 12-O-tetradecanoylphorbol 13-acetate 3 × wk$^{-1}$ for 20 wk. At this time survival rates were 17/20, 16/20 and 20/20 respectively. Skin tumour incidences were 3 in 3 tumour bearing mice, 8 in 4 tumour bearing mice, and 26 in 14 tumour bearing mice for each treated group respectively. No skin tumours occurred in vehicle controls, while 6/20 surviving positive controls treated with benzo[a]pyrene had a total of 10 skin tumours (3).

## Genotoxicity

*Salmonella typhimurium* TA100 with metabolic activation positive (5).

## Any other comments

Physical properties, occurrence, carcinogenicity and metabolism of methylchrysenes reviewed (1).

## References

1. *IARC Monograph* 1983, **32**, 379-397
2. *IARC Monograph* 1987, **Suppl. 7**, 66
3. Hecht, S. S. et al *J. Natl. Cancer Inst.* 1974, **53**, 1121-1133
4. Hoffmann, D. et al *Science* 1974, **183**, 215-216
5. Coombs, M. M. et al *Cancer Res.* 1976, **36**, 4525-4529

# M181   5-Methylchrysene

CAS Registry No. 3697-24-3
Synonyms chrysene, 5-methyl-
Mol. Formula C$_{19}$H$_{14}$          Mol. Wt. 242.32
Uses No commercial production or known use.

## Physical properties

**M. Pt.** 117-119°C.

### Solubility
Organic solvent: acetone

## Mammalian and avian toxicity

### Carcinogenicity and long-term effects
No adequate evidence for carcinogenicity to humans, sufficient evidence for carcinogenicity to animals, IARC classification group 2B (1).

Dermal 20 Swiss ♀ albino Ha/ICR/Mil (72 wk) mice 100 μg 0.1 ml$^{-1}$ acetone. All the mice developed skin tumours by 25 wk, by 35 wk all mice had died, 99 tumours had developed in 20/20 animals (2,3).

Dermal 20 Swiss ♀ Ha/ICR mice (62 wk), treated with concentrations of 0.01 or 0.005% (solvent unspecified). At 55 wk all 20 mice in 0.01% group had died with a total of 38 skin tumours. At the low dose 7/20 animals survived, 22 tumours (site unspecified) were found in 9 tumour-bearing mice (2).

Subcutaneous ♂ Swiss mice 2 mg animal$^{-1}$. No injection-site tumour was observed, but only 4/20 animals were alive 6 months after injection. First tumour developed at 114 day and average latency was 125 day (site unspecified) (4).

Subcutaneous ♂ C57B1 mice 0.05 mg animal$^{-1}$ fortnightly for 20 wk, 22/25 mice had 24 fibrosarcomas with an average latency period of 25 wk (site unspecified) (5).

### Metabolism and pharmacokinetics
*In vitro* rat liver cytosol and *in vivo* rat dorsal subcutaneous tissue. Undergoes dealkylation in *in vitro* rat liver cells to yield chrysene, and a bioxygenation to yield the corresponding hydroxyalkyl substituted chrysene. Substitution of a methyl group may be a necessary step in the metabolic activation and carcinogenicity of this compound (6).

## Genotoxicity
*Salmonella typhimurium* TA98, TA100 with metabolic activation positive (7).
Produced a DNA repair response in humans but not in rat hepatocytes (8).

## Legislation
Included in Schedule 4 (Release into Air: Prescribed Substances) Statutory Instrument No. 472, 1991 (9).

## Any other comments
Experimental toxicology and human health effects reviewed (10).

## References

1. *IARC Monograph* 1987, **Suppl. 7**, 66
2. Hecht, S. S. et al *J. Natl. Cancer Inst.* 1974, **53**, 1121-1133
3. Hoffman, D. et al *Science (Washington, DC. 1883-)* 1974, **183**, 215-216
4. Dunlap, C. E. et al *Cancer Res.* 1943, **3**, 606-607
5. Hecht, S. S. et al *Carcinogenesis, Vol. 1, Polynuclear Aromatic Hydrocarbons: Chemistry, Metabolism and Carcinogenesis* 1976, 325-340, Raven Press, New York
6. Myers, S. R. et al *Chem.-Biol. Interact.* 1991, **77**, 203
7. Lee-Ruff, E. et al *Environ. Mutagen.* 1987, **9**(2), 183-189

8. Butterworth, B. E. et al *Cancer Res.* 1989, **49**, 1075-1084
9. *S. I. 1991 No. 472 The Environmental Protection (Prescribed Processes and Substances) Regulations* 1991, HMSO, London
10. *ECETOC Technical Report No. 30(4)* 1991, European Chemical Industry Ecology and Toxicology Centre, B-1160 Brussels

# M182   6-Methylchrysene

CH₃

**CAS Registry No.** 1705-85-7
**Synonyms** chrysene, 6-methyl-
**Mol. Formula** $C_{19}H_{14}$            **Mol. Wt.** 242.32
**Occurrence** In fossil fuels. Has been identified in tobacco smoke, marijuana smoke, engine exhausts and in vegetables (1).

## Physical properties
**M. Pt.** 161°C.

**Solubility**
Organic solvent: acetone, ethanol, ethyl acetate

## Mammalian and avian toxicity

### Carcinogenicity and long-term effects
No adequate evidence for carcinogenicity to humans or animals, IARC classification group 3 (2).
Dermal mouse (72 wk) 100 μg animal$^{-1}$ 3 × wk$^{-1}$. The first skin tumour was reported after 20 wk at which time 19/20 animals were still alive. At 72 wk 12 mice were still alive, 3 of which had a total of 3 skin tumours, including 1 adenoma. No such tumours occurred in vehicle controls (3,4).
Dermal mouse (24 wk) 100 μg animal$^{-1}$ on alternate days for 20 days followed 10 days later by application of 2.5 μg animal of 12-*O*-tetradecanoylphorbol 13-acetate 3 × wk$^{-1}$ for 20 wk. At this time 19/20 mice were still alive, of which 7 animals were bearing 11 skin tumours. No skin tumours occurred in vehicle controls while 6/20 surviving positive controls treated with benzo[a]pyrene had a total of 10 skin tumours (3).

### Metabolism and pharmacokinetics
Metabolised in rat liver by demethylation to give chrysene, by alkylation to give dimethylchrysene, and by oxidation to give hydroxyalkyl substituted chrysene (5).

### Genotoxicity

*Salmonella typhimurium* TA100 with metabolic activation positive (6).

### Any other comments

Physical properties, occurrence, carcinogenicity and metabolism reviewed (1,7).

### References

1.  *IARC Monograph* 1983, **32**, 379-397
2.  *IARC Monograph* 1987, **Suppl. 7**, 66
3.  Hecht, S. S. et al *J. Natl. Cancer Inst.* 1974, **53**, 1121-1133
4.  Hoffmann, D. et al *Science (Washington, D. C. 1883-)* 1974, **183**, 215-216
5.  Myers, S. R. et al *Chem.-Biol. Interact.* 1991, **77**(2), 203-221
6.  Coombs, M. M. et al *Cancer Res.* 1976, **36**, 4525-4529
7.  *ECETOC Technical Report No. 30(4)* 1991, European Chemical Industry Ecology and Toxicology Centre, B-1160 Brussels

# M183   Methyl 2-cyanoacrylate

$$CH_2=C(CN)CO_2CH_3$$

**CAS Registry No.** 137-05-3
**Synonyms** Adhere; Coapt; cyanolyt; mecrylate; methyl 2-cyano-2-propenoate
**Mol. Formula** $C_5H_5NO_2$                    **Mol. Wt.** 111.10
**Uses** Manufacture of adhesives and polymers. Surgical tissue adhesive.

### Physical properties

**B. Pt.** $_{1.8}$ 47-49°C; **Specific gravity** $d_4^{27}$ 1.1044; **Partition coefficient** log $P_{ow}$ 0.030
(1); **Volatility** v.p. 0.18 mmHg at 25°C.

**Solubility**
Water: <1 g $l^{-1}$ at 22°C.

### Occupational exposure

**US TLV (TWA)** 2 ppm (9.1 mg m$^{-3}$); **US TLV (STEL)** 4 ppm (18 mg m$^{-3}$);
**UK Long-term limit** 2 ppm (8 mg m$^{-3}$) under review; **UK Short-term limit**
4 ppm (16 mg m$^{-3}$) under review.

### Ecotoxicity

**Bioaccumulation**
Calculated bioconcentration factor 0.62 indicates that environmental accumulation is unlikely (2).

### Environmental fate

**Degradation studies**
Biodegradation in screening tests using sewage seed, 28% ThOD after 5 days, and 66% ThOD after 5 days with acclimated sewage seed (3).

**Abiotic removal**

Reaction with photochemically produced hydroxyl radicals and ozone in the atmosphere $t_{1/2}$ 2.18 days (4).

95% removal from atmosphere containing 10 ppm by activated carbon up to 200 mg kg$^{-1}$ carbon (5).

**Absorption**

Calculated $K_{oc}$ 25 indicates that adsorption to soil and sediments would not be significant (2).

## Mammalian and avian toxicity

**Acute data**

LD$_{50}$ oral starling >100 mg kg$^{-1}$ (6).
LC$_{50}$ (6 hr) inhalation rat 100 ppm (7).
LD$_{50}$ dermal guinea pig >10 ml kg$^{-1}$ (7).

**Teratogenicity and reproductive effects**

Intratesticular monkey, lowest toxic dose 100 mg kg$^{-1}$ (effects on spermatogenesis, details unspecified) (8).

**Metabolism and pharmacokinetics**

Rapidly absorbed through the skin of guinea pigs and eliminated in the urine. Initial metabolites possessed the carbon skeleton of the monomer, whereas metabolites excreted from day 2 onward represented absorbed and degraded polymeric material (9).

**Irritancy**

Human volunteers exposed to 40-60 ppm suffered irritation of the eyes and blurred vision. After 2 hr exposure 5 and 20 ppm caused lachrymation and rhinorrhoea (10).

## Genotoxicity

*Salmonella typhimurium* TA100, with and without metabolic activation positive, TA97, TA98, TA1535, TA1537 with and without metabolic activation negative (11,12).

## Any other adverse effects to man

Instillation into the eyes may cause double vision and lachrymation. There is usually no residual damage (13).

## Any other adverse effects

*In vitro* rat polymorphonuclear leukocytes. Cell degranulation increased and migration decreased in a concentration-dependent manner. Cytotoxicity decreased in the presence of inhibitors of prostaglandin synthase (14).

## Any other comments

Human health effects, experimental toxicology, physico-chemical properties reviewed (15).

## References

1. *GEMS: Graphical Exposure Modeling System: Fate of Atmospheric Pollutants*, 1986, US EPA Office of Toxic Substances, Washington, DC
2. Lyman, W. K. et al *Handbook of Chemical Property Estimation Methods: Environmental Behavior of Organic Compounds* 1982, McGraw-Hill, New York

3. Price, K. S. et al *J. -Water Pollut. Control Fed.* 1974, **46**, 63-77
4. Atkinson, R. *Int. J. Chem. Kinet.* 1987, **19**, 799-828
5. Benson, L. A. *Am. Hyg. Ind. Assoc. J.* 1975, **36**(10), 741-744
6. Schafer, E. W. et al *Arch. Environ. Contam. Toxicol.* 1983, **12**, 355-382
7. *Documentation of Threshold Limit Values and Biological Exposure Indices* 5th ed., 1986, 383, American Conference of Governmental Industrial Hygienists, Cincinnati, OH
8. *Fert. Steril.* 1983, **39**(Suppl.) 441
9. Reynolds, R. C. et al *J. Surg. Res.* 1966, **6**(3), 132-136
10. Grant, W. M. *Toxicology of the Eye* 2nd ed., 1974, 338, Charles C. Thomas, Springfield, IL
11. Rietveld, E. C. et al *Mutat. Res.* 1987, **188**(2), 97-104
12. Zeiger, E. et al *Environ. Mutagen.* 1987, **9**(Suppl. 9), 1-109
13. *Martindale. The Extra Pharmacopoeia* 30th ed., 1993, 1359, The Pharmaceutical Press, London
14. Papatheofanis, F. J. *J. Biomed. Mater. Res.* 1989, **23**(6), 661-668
15. *ECETOC Technical Report No. 30(4)* 1991, European Chemical Industry Ecology and Toxicology Centre, B-1160 Brussels

# M184   Methylcyclohexane

**CAS Registry No.** 108-87-2
**Synonyms** cyclohexylmethane; hexahydrotoluene; Sextone B; toluene hexahydride
**Mol. Formula** $C_7H_{14}$ **Mol. Wt.** 98.19
**Uses** Solvent for cellulose ethers. Organic synthesis.

## Physical properties

**M. Pt.** –126°C; **B. Pt.** 101°C; **Flash point** –3.8°C; **Specific gravity** $d_4^{20}$ 0.770; **Partition coefficient** log $P_{ow}$ 3.88 (1); **Volatility** v.p. 144 mmHg at 20°C; v. den. 3.38.

## Solubility
Water: 14 mg $l^{-1}$ at 20°C. Organic solvent: diethyl ether, ethanol

## Occupational exposure

**US TLV (TWA)** 400 ppm (1610 mg m$^{-3}$); **UK Long-term limit** 400 ppm (1600 mg m$^{-3}$); **UK Short-term limit** 500 ppm (2000 mg m$^{-3}$); **UN No.** 2296; **HAZCHEM Code** 3ME; **Conveyance classification** flammable liquid; **Supply classification** highly flammable.
**Risk phrases** Highly flammable (R11)

**Safety phrases** Keep container in a well ventilated place – Keep away from sources of ignition – No Smoking – Take precautionary measures against static discharges (S9, S16, S33)

## Ecotoxicity

**Fish toxicity**
$LC_{50}$ (96 hr) golden shiner 72 mg $l^{-1}$ (emulsion) (2).

**Invertebrate toxicity**
$LC_{50}$ (96 hr) *Cyclops viridis* 865 mg $l^{-1}$ (3).
$LC_{50}$ (96 hr) *Thiara tuberculata* 1160 mg $l^{-1}$ (3).
$LC_{50}$ (96 hr) *Chironomus* larvae 1000 mg $l^{-1}$ (3).

**Bioaccumulation**
Confirmed to be non-accumulative or low accumulative in fish despite its high log $P_{ow}$ (4).

## Environmental fate

**Degradation studies**
*m*-Xylene-adapted microorganisms in an aquifer column were unable to metabolise the compound (5).
Biodegradation 75% after 192 hr at 13˚C, initial concentration 0.05 µg $l^{-1}$ (method unspecified) (6).

## Mammalian and avian toxicity

**Acute data**
$LD_{Lo}$ oral rabbit 4000 mg $kg^{-1}$ (7).
$LD_{50}$ oral mouse 2250 mg $kg^{-1}$ (7).
$LC_{50}$ (2 hr) inhalation mouse 41,500 mg $m^{-3}$ (8).

**Sub-acute data**
Inhalation (10 wk) rabbits 4600 mg $m^{-3}$, 6 hr $day^{-1}$, 5 day $wk^{-1}$, appeared to be non-toxic (9).

**Metabolism and pharmacokinetics**
In rats it is primarily excreted in the urine with ≈15% eliminated in exhaled air. The primary urinary metabolites are *cis*- and *trans*-isomers of methylcyclohexanols, which are further conjugated with glucuronic acid (10,11).
Oral ♂ rats, urinary metabolites included: cyclohexylmethanol, 3-methylcyclohexanol; *trans*-4-methylcyclohexanol; 2-*cis*-hydroxy-4-*cis*-methylcyclohexanol; 2-*cis*-hydroxy-4-*trans*-methylcyclohexanol; and 2-*trans*-hydroxy-4-*cis*-methylcyclohexanol. Metabolism of the ring is favoured (11).

## Any other adverse effects

Lethal concentrations cause mucous secretions, lacrimation, salivation, laboured breathing and diarrhoea (species unspecified) (9).
Histopathological examination revealed only very slight renal tissue damage following oral adiministration to rats (11).
Subtle liver and kidney damage and convulsions have been noted in inhalation tests on rabbits (12).

---

## Legislation

Included in Schedules 4 and 6 (Release into Air/Land: Prescribed Substances) Statutory Instrument No. 472, 1991 (13).

The log $P_{ow}$ value exceeds the European Community recommended level (6th and 7th amendments) (14).

## Any other comments

Experimental toxicology, human health effects, epidemiology and workplace experience reviewed (15,16).

Narcosis in aquatic species including fathead minnows, guppies, *Daphnia magna* and *Artemia* discussed (17).

Autoignition temperature 285°C.

## References

1. Sangster, J. *J. Phys. Ref. Data.* 1989, **18** (3), 1111-1129
2. Hartwell, J. *Survey of Compounds which have been Tested for Carcinogenic Activity* 2nd ed., 1951, National Cancer Institute U.S. Public Health Service, Washington, DC
3. Panigrahi, A. K. et al *Environ. Toxicol.* 1989, **7**(1), 44-49
4. *The List of the Existing Chemical Substances Tested on Biodegradability by Microorganisms or Bioaccumulation in Fish Body* 1987, Chemicals Inspection and Testing Institute, Japan
5. Kuhn, E. P. et al *Appl. Environ. Microbiol.* 1988, **54**(2), 490-496
6. Jamison, V. W. et al *Proc. 3rd. Int. Biodegrad. Symp.* 1976, Applied Science Publishers (Elsevier, Amsterdam)
7. Nikunen, E. *Environmental Properties of Chemicals* 1990, VAPK Publishing, Helsinki
8. Izmerov, N. F. et al *Toxicometric Parameters of Industrial Toxic Chemicals under Single Exposure* 1982, CIP, Moscow
9. Treon, J. F. *J. Ind. Hyg. Toxicol.* 1943, **25**(6), 323
10. Elliot, T. H. *Biochem. J.* 1965, **65**, 70-76
11. Parnell, M. J. et al *Chemosphere* 1988, **17**(7), 1321-1327
12. Treon, J. F. *J. Ind. Hyg. Toxicol.* 1943, **25**(6), 199
13. *S. I. 1991 No. 472 The Environmental Protection (Prescribed Processes and Substances) Regulations* 1991, HMSO, London
14. *1967 Directive on Classification, Packaging, and Labelling of Dangerous Substances 67/548/EEC; 6th Amendment EEC Directive 79/831/EEC; 7th Amendment EEC Directive 91/32/EEC* 1991, HMSO, London
15. *BIBRA Toxicity Profile* 1991, British Biological Industrial Research Association, Carshalton
16. *ECETOC Technical Report No. 30(4)* 1991, European Chemical Industry Ecology and Toxicology Centre, B-1160 Brussels
17. Abernethy, S. et al *QSAR, Environ. Toxicol., Proc. Int. Workshop, 2nd* 1986, 1-16

# M185   Methylcyclohexanol

OH
CH₃

[structure diagram of methylcyclohexanol]

**CAS Registry No.** 25639-42-3
**Synonyms** hexahydrocresol; hexahydromethylphenol; methyl adronal; methylanol; methylhexalin; Sextol
**Mol. Formula** $C_7H_{14}O$                          **Mol. Wt.** 114.19
**Uses** Solvent. Antioxidant in lubricants.

## Physical properties

**M. Pt.** −50°C; **B. Pt.** 155-180°C; **Flash point** 63°C (closed cup); **Specific gravity** $d_{15.5}^{15.5}$ 0.924; **Volatility** v.p. 1.5 mmHg at 30°C; v. den. 3.93.

## Solubility

Water: 35 g $l^{-1}$ at 20°C. Organic solvent: diethyl ether, ethanol

## Occupational exposure

**US TLV (TWA)** 50 ppm (234 mg $m^{-3}$); **UK Long-term limit** 50 ppm (235 mg $m^{-3}$); **UK Short-term limit** 75 pm (350 mg $m^{-3}$); **UN No.** 2617.

## Mammalian and avian toxicity

### Acute data

$LD_{50}$ oral rat, rabbit 1660-2000 mg $kg^{-1}$ (1,2).
Oral rabbit, single dose of 1750 mg $kg^{-1}$ caused rapidly developing anaesthesia with spasmodic jerking of the head and rhythmic movement of the forelegs. The only histopathological evidence observed were degenerative changes in the liver. No significant abnormalities of the blood were observed (2).
$LD_{Lo}$ dermal rabbit 6800 mg $kg^{-1}$ (3).
$LD_{50}$ subcutaneous rat 2900 mg $kg^{-1}$ (1).

### Sub-acute data

Inhalation rabbit (10 wk) 500 ppm 6 hr $day^{-1}$ 5 days $wk^{-1}$ induced salivation, conjunctival congestion and irritation, and lethargy (1).
Inhalation dog (6 days) exposure to saturated air (0.2%) 10 min $day^{-1}$ caused no signs of intoxication (4).
Dermal rabbit (6 days) 10 ml applied to intact skin 1 hr $day^{-1}$ was fatal (2).

### Metabolism and pharmacokinetics

Following inhalation exposure and dermal application, methylcyclohexanol is excreted in the urine as the glucuronide (species unspecified) (5).

### Irritancy

Rabbits exposed to 2300 mg $m^{-3}$ showed symptoms of eye irritation (exposure not specified) (6).

## Legislation

Included in Schedule 6 (Release into Land: Prescribed Substances) Statutory Instrument No. 472, 1991 (7).

## Any other comments

Human health effects, experimental toxicology, physico-chemical properties reviewed (8).

The registry number 25639-42-3 is the general registry number for methylcyclohexanols.

Autoignition temperature 296°C.

## References

1. *J. Ind. Hyg. Toxicol.* 1943, **25**, 415
2. Treon, J. F. et al *J. Ind. Hyg. Toxicol.* 1943, **25**, 189-213
3. *Handbook of Toxicology* 1956, **1**, 194, Saunders, Philadelphia
4. Pohl, J. *Gewerbehyg. Unfallverhuet.* 1925, **12**, 91
5. *Encyclopedia of Occupational Health and Safety* 1983, 111, International Labour Office, Geneva
6. Grent, W. M. *Toxicology of the Eye* 3rd ed., 1986, 614, Charles C. Thomas, Springfield, IL
7. *S. I. 1991 No. 472 The Environmental Protection (Prescribed Processes and Substances) Regulations* 1991, HMSO, London
8. *ECETOC Technical Report No. 30(4)* 1991, European Chemical Industry Ecology and Toxicology Centre, B-1160 Brussels

# M186    2-Methylcyclohexanol

**CAS Registry No.** 583-59-5
**Synonyms** cyclohexanol, 2-methyl-; 2-methylcyclohexyl alcohol
**Mol. Formula** $C_7H_{14}O$                    **Mol. Wt.** 114.19
**Uses** Solvent.

## Physical properties

**M. Pt.** –21°C; **B. Pt.** 163-166°C; **Flash point** 58°C; **Specific gravity** 0.930; **Volatility** v. den. 3.39.

## Solubility

Organic solvent: diethyl ether, ethanol

## Occupational exposure

**Conveyance classification** flammable liquid; **Supply classification** harmful.
**Risk phrases** ≥25% – Harmful by inhalation (R20)
**Safety phrases** Avoid contact with skin and eyes (S24/25)

## Ecotoxicity

**Invertebrate toxicity**

$LC_{50}$ (96 hr) algae 395.0 mg $l^{-1}$ (1).
$LC_{50}$ (48 hr) *Daphnia magna* 267 mg $l^{-1}$ (1).

**Mammalian and avian toxicity**

**Acute data**

$LD_{50}$ intramuscular mouse 1000 mg $kg^{-1}$ (2).

**Any other adverse effects to man**

Vapour irritating to eyes and respiratory system. Inhalation of high concentrations of vapour leads to signs of narcosis (3).

**Legislation**

Included in Schedules 4 and 6 (Release into Air/Land: Prescribed Substances) Statutory Instrument No. 472, 1991 (4).

**Any other comments**

Mixture of *cis-* and *trans*-isomers.

**References**

1. Haley, M. V. *Gov. Rep. Announce. Index (U. S.)* 1989, **89**(18), Abstr. No. 949,416
2. *J. Sci. Ind. Res., Sect. C: Biol. Sci.* 1962, **21**, 342
3. Henning, H. (Ed.) *Solvent Safety Sheets: A Compendium for the Working Chemist* 1993, 159, RSC, London
4. *S. I. 1991 No. 472 The Environmental Protection (Prescribed Processes and Substances) Regulations* 1991, HMSO, London

# M187   3-Methylcyclohexanol

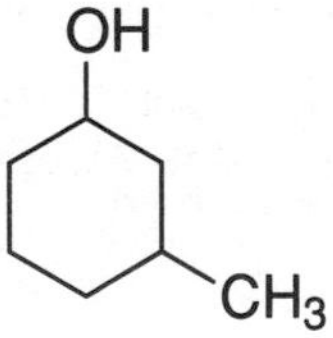

**CAS Registry No.** 591-23-1
**Synonyms** 3-methyl-1-cyclohexanol; *m*-methylcyclohexanol
**Mol. Formula** $C_7H_{14}O$                     **Mol. Wt.** 114.19
**Occurrence** Isolated from *Mentha polegium*.

**Physical properties**

**B. Pt.** 163°C; **Flash point** 62°C; **Specific gravity** $d^{20}$ 0.914.

**Solubility**
Organic solvent: diethyl ether, ethanol

**Mammalian and avian toxicity**

**Acute data**
LD$_{50}$ intramuscular mouse 1000 mg kg$^{-1}$ (1).

**Any other comments**
Autoignition temperature 295°C

**References**
1.  *J. Sci. Ind. Res. Sect. C: Biol. Sci.* 1962, **21**, 342

# M188  4-Methylcyclohexanol

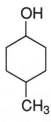

**CAS Registry No.** 589-91-3
**Synonyms** hexahydro-*p*-cresol; methyladronal; methylanol;
4-methyl-1-cyclohexanol; 4-methylcyclohexyl alcohol; *p*-methylcyclohexanol; Sextol
**Mol. Formula** C$_7$H$_{14}$O                    **Mol. Wt.** 114.19

**Physical properties**
**M. Pt.** –50°C; **B. Pt.** 171-173°C; **Flash point** 70°C; **Specific gravity** d$^{20}$ 0.914;
**Partition coefficient** log P$_{ow}$ 1.79; **Volatility** v.p. 1.5 mmHg at 30°C; v. den. 3.94.

**Solubility**
Organic solvent: diethyl ether, ethanol

**Environmental fate**

**Degradation studies**
Adapted activated sludge, 94% COD at 40 mg l$^{-1}$ COD g$^{-1}$ dry inoculum hr$^{-1}$ at 20°C,
when utilised as sole carbon source (1).

**Mammalian and avian toxicity**

**Acute data**
LD$_{50}$ oral rabbit 1750-2000 mg kg$^{-1}$ (2).

**Sub-acute data**
Inhalation rabbit (50 day) lowest lethal concentration 500 ppm for 6 hr day$^{-1}$; no
adverse effect level 230 ppm for 6 hr day$^{-1}$ (2).

## Any other comments

Autoignition temperature 295°C. A mixture of *cis*- and *trans*-isomers.

## References

1. Pitter, P. *Water Res.* 1976, **10**, 231-235
2. Treon, J. F. et al *J. Ind. Hyg. Toxicol.* 1943, **25**, 323

# M189   Methylcyclohexanone

**CAS Registry No.** 1331-22-2
**Synonyms** cyclohexanone, methyl-; methylcyclohexan-1-one
**Mol. Formula** $C_7H_{12}O$                    **Mol. Wt.** 112.17
**Uses** Solvent, manufacture of lacquers, varnishes and plastics; leather industry; rust remover.

## Physical properties

**M. Pt.** -14°C; **B. Pt.** 160-170°C; **Flash point** 48°C (closed cup); **Specific gravity** $d_5^{15}$ 0.925; **Volatility** v.den. 3.86.

## Solubility

Organic solvent: diethyl ether, ethanol

## Occupational exposure

**UN No.** 2297; **HAZCHEM Code** 3Y; **Conveyance classification** flammable liquid.

## Mammalian and avian toxicity

### Acute data

$LD_{50}$ oral rat 2140 mg $kg^{-1}$ (1).
$LD_{Lo}$ oral rabbit 1000 mg $kg^{-1}$ (2).
$LD_{Lo}$ dermal rabbit 4900 mg $kg^{-1}$ (2).

### Metabolism and pharmacokinetics

Metabolised via methylcyclohexanol and conjugated with glucuronic acid (species unspecified) (3).

## Any other adverse effects to man

Narcotic and causes eye and respiratory tract irritation at high concentrations (3).

## Legislation

Included in Schedules 4 and 6 (Release into Air/Land: Prescribed Substances) Statutory Instrument No. 472, 1991 (4).

## Any other comments
Physical properties, toxicity and safety precautions reviewed (1).

## References
1. *Chemical Safety Data Sheets: Solvents* 1989, **1**, 228, RSC, London
2. *J. Ind. Hyg. Toxicol.* 1943, **25**, 199
3. Tao, C. C. *Biochem. J.* 1962, **84**, 38P
4. *S. I. 1991 No. 472 The Environmental Protection (Prescribed Processes and Substances) Regulations* 1991, HMSO, London

# M190   2-Methylcyclohexanone

**CAS Registry No.** 583-60-8
**Synonyms** 1-methylcyclohexan-2-one; methylanone; methanon ; Sexton B
**Mol. Formula** $C_7H_{12}O$                **Mol. Wt.** 112.17
**Uses** Organic synthesis. Solvent.

## Physical properties
**M. Pt.** –14°C; **B. Pt.** 162-162.5°C; **Flash point** 46°C (closed cup); **Specific gravity** $d_4^{20}$ 0.924; **Volatility** v. den. 3.86.

## Solubility
Organic solvent: diethyl ether, ethanol

## Occupational exposure
**US TLV (TWA)** 50 ppm (229 mg $m^{-3}$); **US TLV (STEL)** 75 ppm (344 mg $m^{-3}$);
**UK Long-term limit** 50 ppm (230 mg $m^{-3}$); **UK Short-term limit** 75 ppm (345 mg $m^{-3}$);
**UN No.** 2297; **HAZCHEM Code** 3Y; **Conveyance classification** flammable liquid;
**Supply classification** harmful (>25%).
**Risk phrases** ≥25% – Flammable – Harmful by inhalation – <25% – Flammable
(R10, R20, R10)
**Safety phrases** Avoid contact with eyes (S25)

## Ecotoxicity

### Invertebrate toxicity
Toxicity threshold, cell multiplication test *Pseudomonas putida, Microcystis aeruginosa, Scenedesmus quadricauda,* and *Entosiphon sulcatum* 26-160 mg $l^{-1}$ (exposure not specified) (1, 2).

---

# Mammalian and avian toxicity

### Acute data

$LD_{50}$ oral rabbit, rat 1000, 2140 mg kg$^{-1}$, respectively (3, 4).

$LC_{50}$ (4 hr) inhalation rat 2500 ppm (3).

$LD_{50}$ dermal rabbit 1640 mg kg$^{-1}$ (3).

$LD_{50}$ intraperitoneal mouse 200 mg kg$^{-1}$ (5).

### Sub-acute data

Inhalation rabbit (50 day) no adverse effect level 180 ppm 6 hr day$^{-1}$. Higher concentrations caused lethargy, salivation, lachrymation and eye irritation (6).

### Metabolism and pharmacokinetics

*In vivo* undergoes reduction to the *cis-* and *trans*-methylcyclohexanols which are excreted in the urine as sulfuric and glucuronic acid conjugates (species unspecified) (7,8).

## Any other adverse effects to man

Reported to be narcotic at concentrations >100 ppm, and an eye and respiratory irritant at >50 ppm (4).

## Legislation

Included in Schedule 6 (Release into Land: Prescribed Substances) Statutory Instrument No. 472, 1991 (9).

## Any other comments

Human health effects, experimental toxicology, physico-chemical properties reviewed (10).

## References

1. Bringmann, G. et al *Z. Wasser/Abwasser Forsch.* 1980, (1), 26-31
2. Bringmann, G. et al *Water Res.* 1980, **14**, 231-241
3. *Documentation of Threshold Limit Values for Substances in Workroom Air* 1980, **4**, 272, American Conference of Governmental Industrial Hygienists Inc. Cincinnati, OH
4. Smyth, H. F. et al *Am. Ind. Hyg. Assoc. J.* 1969, **30**, 470-476
5. *NTIS Report AD691-490* Natl. Tech. Inf. Ser., Springfield, VA
6. Gerarde, H. W. *Arch. Environ. Health* 1963, **6**, 329
7. Testa, B. et al *Drug Metabolism: Chemical and Biochemical Aspects* 1976, 258, Marcel Dekker, New York
8. Tao, C. C. et al *Biochem. J.* 1962, **84**, 38-39
9. *S. I. 1991 No. 472 The Environmental Protection (Prescribed Processes and Substances) Regulations* 1991, HMSO, London
10. *ECETOC Technical Report No. 30(4)* 1991, European Chemical Industry Ecology and Toxicology Centre, B-1160 Brussels

# M191  Methylcyclopentane

CH₃

**CAS Registry No.** 96-37-7
**Mol. Formula** $C_6H_{12}$                                    **Mol. Wt.** 84.16
**Uses** Solvent.
**Occurrence** In fossil fuels. Has been detected in human respired air in smokers and traces in non-smokers (1).

## Physical properties

**M. Pt.** –142°C; **B. Pt.** 72°C; **Flash point** –23°C; **Specific gravity** $d_4^{20}$ 0.750; **Partition coefficient** log $P_{ow}$ 3.37; **Volatility** v.p. 100 mmHg at 17.9°C.

### Solubility

Organic solvent: acetone, benzene, carbon tetrachloride, diethyl ether, ethanol, petroleum ether

## Occupational exposure

**UN No.** 2298; **HAZCHEM Code** 3◪E; **Conveyance classification** flammable liquid.

## Environmental fate

### Degradation studies

Degraded by *Pseudomonas* sp., *Nocardia* sp. and *Micrococcus* sp. isolated from gasoline contaminated soil (2).
Biodegradation by natural flora in groundwater, in the presence of other components of high octane gasoline, 10% after 192 hr at 0.4 μg l⁻¹ at 13°C (3).

## Mammalian and avian toxicity

### Acute data

$LC_{Lo}$ inhalation (duration unspecified) mouse 95 g m⁻³ (4).

### Sub-acute data

Gavage rat (4 wk) 500 or 2000 mg kg⁻¹ day⁻¹ 5 days wk⁻¹ induced nephropathy. The low dose caused 10% mortality and the high dose 40% mortality (5).

### Irritancy

Irritating to the skin. Vapour or mist is irritating to the eyes, mucous membranes and upper respiratory tract (dose, duration unspecified) (6).

## Any other adverse effects to man

Polyneuropathy of the motor type was reported among exposed workers (7).

## Legislation

Included in Schedule 6 (Release into Land: Prescribed Substances) Statutory Instrument No. 472, 1991 (8).

---

The log $P_{ow}$ value exceeds the European Community recommended level of 3.0 (6th and 7th amendments) (9).

## References

1.  Conkle, J. P. *Arch. Environ. Health* 1975, **30**(6), 290
2.  Ridgeway, H. F. et al *Appl. Environ. Microbiol.* 1990, **56**(11), 3565-3575
3.  Jamison, V. W. et al *Proc. Third Int. Biodegrad. Symp.* 1976
4.  *Naunyn-Schmiedeberg's Arch. Exp. Pathol. Pharmakol.* 1930, **149**, 116
5.  Halder, C. A. et al *Toxicol. Ind. Health* 1985, **1**(3), 67-87
6.  Lenga, R. E. (Ed.) *Sigma-Aldrich Library of Chemical Safety Data* 2nd ed., 1988, **2**, 2330 Sigma-Aldrich, Milwaukee, WI
7.  Brugnone, F. et al *Int. Arch. Occup. Environ. Health* 1979, **42**(3-4), 355-363
8.  *S. I. 1991 No. 472 The Environmental Protection (Prescribed Processes and Substances) Regulations* 1991, HMSO, London
9.  *1967 Directive on Classification, Packaging and Labelling of Dangerous Substances 67/548/EEC; 6th Amendment EEC Directive 79/831/EEC; 7th Amendment EEC Directive 91/32/EEC* 1991, HMSO, London

# M192   Methyl demeton

## $CH_3CH_2SCH_2CH_2OP(S)(OCH_3)_2$

**CAS Registry No.** 8022-00-2

**Synonyms** phosphorothioic acid, *O,O*-dimethyl *O*- [2-(ethylthio)ethyl] ester; *O*-[2-(ethylthio)ethyl] *O,O*-dimethylphosphorothioate; Metasystox; Metasystox forte; methylmercaptophos; methyl systox

**Mol. Formula** $C_6H_{15}O_3PS_2$ **Mol. Wt.** 230.29

**Uses** Insecticide. Acaricide.

## Physical properties

**B. Pt.** $_{0.15}$ 89°C; **Specific gravity** $d_4^{20}$ 1.207; **Volatility** v.p. $3 \times 10^{-4}$ mmHg at 20°C.

## Solubility

Water: 3300 mg $l^{-1}$ at 20°C. Organic solvent: acetone, dichloromethane, diethyl ether, ethanol, propan-2-ol

## Occupational exposure

**US TLV (TWA)** 0.5 mg m$^{-3}$.

## Ecotoxicity

**Fish toxicity**
$LC_{50}$ (96 hr) mirror carp 6.3-9.1 mg $l^{-1}$ (1).
$LC_{50}$ (24 hr) harlequin fish 9 mg $l^{-1}$ (2).

**Invertebrate toxicity**
$LC_{50}$ (96 hr) marine wedge clam 6.4 µg $l^{-1}$ (3).
$LC_{50}$ (24 hr) *Macrobrachium lamerrei* 1.6 mg $l^{-1}$ (4).

## Environmental fate

**Degradation studies**
In plants the thioethyl group is oxidised to give the sulfoxide and sulfone (5).

**Abiotic removal**
Undergoes hydrolysis to give dimethyl phosphate (5).

## Mammalian and avian toxicity

**Acute data**
$LD_{50}$ oral rat, mouse, 46-65 mg $kg^{-1}$ (5,6).
$LC_{50}$ (4 hr) inhalation cat 20 mg $m^{-3}$ (6).
$LD_{50}$ dermal rat 300 mg $kg^{-1}$ (7).
$LD_{50}$ intravenous mouse 0.5-1.0 mg $kg^{-1}$ (8).
$LD_{50}$ intraperitoneal guinea pig 65 mg $kg^{-1}$ (9).

**Carcinogenicity and long-term effects**
In a 2 yr feeding trial, the no effect level for rats was 1 mg $kg^{-1}$ (demeton-*S*-methyl) (5).

## Genotoxicity

*Salmonella typhimurium* TA98, TA102, TA1535 with and without metabolic activation negative (10).
Onion and barley chromosomal aberrations positive (11).

## Any other adverse effects to man

Inhibits cholinesterase activity (5).
Reported to cause alterations in intraocular pressure in various eye layers. Symptoms of acute poisoning include nausea, headache, dizziness, vomiting and hyperaemia of the nasal mucosa and membranes of the respiratory organs. It also causes inner ear irritations (12).

## Legislation

Limited under EC Directive on Drinking Water Quality 80/778/EEC. Pesticides: maximum admissible concentration 0.1 μg $l^{-1}$ (13).
Included in Schedule 6 (Release into Land: Prescribed Substances) Statutory Instrument No. 472, 1991 (14).

## Any other comments

Physical properties, toxicity and safety precautions reviewed (15).
WHO Class 1b; EPA Toxicity Class 1. Tolerable daily intake (TDI) human 0.3 μg $kg^{-1}$.
EEC MRL Carrots 0 ppm; other fruits and vegetables 0.4 mg $kg^{-1}$ (demeton-*S*-methyl) (5).
Consists of an isomeric mixture of demeton-*O*-methyl and demeton-*S*-methyl.

## References

1.  Kulshrestha, S. K. et al *Ecotoxicol. Environ. Saf.* 1986, **12**(2), 114-119
2.  Pimental, D. *Ecological Effects of Pesticides on Non-Target Species* June 1971, Presidential Report, Office of Science and Technology
3.  Muley, D. V. et al *Fish Technol.* 1987, **24**(1), 27-30
4.  Avelin, M. et al *J. Environ. Biol.* 1986, **7**(3), 189-195
5.  *The Agrochemicals Handbook* 3rd ed., 1991, RSC, London
6.  Izmerov, N. F. et al *Toxicometric Parameters of Industrial Toxic Chemicals under Single Exposure* 1982, CIP, Moscow

7. *World Rev. Pest Control* 1970, **9**, 119
8. Spencer, E. Y. *Guide to the Chemicals Used in Crop Protection* 7th ed., 1982, 169, Publication 1093, Research Institute, Agriculture Canada, Ottawa
9. Hayes, U. J. *Pesticides Studied in Man* 1982, 341, Williams & Wilkins, Baltimore, MD
10. Ladhar, S. S. et al *Ind. J. Exp. Biol.* 1990, **28**(4), 390-391
11. Grover, I. S. et al *Prog. Clin. Biol. Res.* 1990, **340E**, 91-106
12. *Encyclopedia of Occupational Health and Safety* 1983, International Labour Office, Geneva
13. *EC Directive Relating to the Quality of Water Intended for Human Consumption* 1982, 80/778/EEC, Office for Official Publications of the European Communities, 2 rue Mercier, L-2985 Luxembourg
14. *S. I. No. 472 The Environmental Protection (Prescribed Processes and Substances) Regulations* 1991, HMSO, London
15. *Dangerous Prop. Ind. Mater. Rep.* 1987, **7**(5), 75-78

# M193   Methyldichlorosilane

## CH3SiHCl2

**CAS Registry No.** 75-54-7
**Synonyms** dichlorohydridomethylsilicon; dichloromethylsilane; monomethyldichlorosilane
**Mol. Formula** $CH_4Cl_2Si$                    **Mol. Wt.** 115.04

### Physical properties
**M. Pt.** –93°C; **B. Pt.** 41°C; **Flash point** –32°C; **Specific gravity** 1.1; **Volatility** v. den. 4.0.

### Occupational exposure
**UN No.** 1242; **HAZCHEM Code** 4WE; **Conveyance classification** substance which in contact with water emits flammable gas, flammable liquid, corrosive substance.

### Mammalian and avian toxicity

**Acute data**
$LC_{Lo}$ (4 hr) inhalation rat 300 ppm (1).

**Irritancy**
Dermal (24 hr) rabbit 500 mg caused severe irritation and 20 mg instilled into rabbit eye (24 hr) caused moderate irritation (1).

### Legislation
Included in Schedule 4 (Release into Air: Prescribed Substances) Statutory Instrument No. 472, 1991 (2).

### References
1. Marhold, J. V. *Sbornik Vysledku Toxixologickeho Vysetreni Latek A Pripravku* 1972, 216, Prague
2. *S. I. 1991 No. 472 The Environmental Protection (Prescribed Processes and Substances) Regulations* 1991, HMSO, London

# M194  *N*-Methyl-*N*,4-dinitrosoaniline

**CAS Registry No.** 99-80-9
**Synonyms** *N*-methyl-*N*,4-dinitrosobenzenamine; Elastopar; Heat Pre; Nitrosan K
**Mol. Formula** $C_7H_7N_3O_2$                              **Mol. Wt.** 165.15

## Physical properties

**M. Pt.** 101°C.

## Mammalian and avian toxicity

### Carcinogenicity and long-term effects

No adequate evidence for carcinogenicity to humans, limited evidence for
carcinogenicity to animals, IARC classification group 3 (1).

Gavage weanling rat 0.1, 0.3, 1, 3, 10 or 30 mg $rat^{-1}$, $5 \times wk^{-1}$ for 52 wk (commercial
preparation containing 33% of the compounds). The 2 highest concentrations showed
carcinogenic activity. Tumours affected breast, pituitary gland, testicles, thyroid gland
and lung (2,3).

Intraperitoneal rat 5 mg $rat^{-1}$ $wk^{-1}$ for 6 months (total dose, 600 mg kg body
$weight^{-1}$). The 24 rats showed 1 hepatoma, 1 thymoma, 2 local sarcomas, 1 pancreatic
adenoma and 1 pituitary tumour. 1 hepatoma was observed in the controls (4).

## References

1. *IARC Monograph* 1987, **Suppl. 7**, 66
2. Weistburger, J. H. et al *Naturwissenschaften* 1966, **53**, 508
3. Hadidian, Z. et al *J. Natl. Cancer Inst.* 1968, **41**, 985
4. Boyland, E. et al *Eur. J. Cancer* 1968, **4**, 233

# M195   Methyldopa

**CAS Registry No.** 555-30-6
**Synonyms** L-tyrosine, 3-hydroxy-α-methyl-; alanine, 3-(3,4-dihydroxyphenyl)-
2-methylaniline, L-;  Aldomet; Dopegyt; Medomet; Presinol
**Mol. Formula** $C_{10}H_{13}NO_4$                    **Mol. Wt.** 211.22
**Uses** Antihypertensive.

## Physical properties
**M. Pt.** >300°C; **Partition coefficient** log $P_{ow}$ –2.09 (1).

## Solubility
Water: 10 g $l^{-1}$ at 25°C.

## Mammalian and avian toxicity

### Acute data
$LD_{50}$ oral rat 5000 mg $kg^{-1}$ (2).
$LD_{50}$ oral rabbit 713 mg $kg^{-1}$ (3).
$LD_{50}$ intravenous rabbit 713 mg $kg^{-1}$ (2).
$LD_{50}$ intraperitoneal rat 300 mg $kg^{-1}$ (3).
$LD_{50}$ intraperitoneal mouse 150 mg $kg^{-1}$ (4).

### Sub-acute data
$TD_{Lo}$ (17 wk) oral woman 1830 mg $kg^{-1}$, peripheral nervous system effects (5).
$TD_{Lo}$ (3 yr) oral woman 44 g $kg^{-1}$, gastrointestinal tract effects (diarrhoea,
constipation, ulceration) (6).
$TD_{Lo}$ (22 wk) oral man 1070 mg $kg^{-1}$, skin effects (erythema, rash, sensitisation of
skin, petechial haemorrhage) (7).

### Teratogenicity and reproductive effects
Maternal mice (route, duration, dose unspecified), there were no adverse effects on
live litter size, survival and growth of live pups to postnatal day 4 or pup survival
and growth to postnatal day 21. *In utero* exposure to ≥ 250 mg $kg^{-1}$ $day^{-1}$ may result
in depressed brain growth relative to body growth (8).

### Metabolism and pharmacokinetics
Variably and incompletely absorbed by an amino-acid active transport system.
Bioavailability is 50%. Extensively metabolised and excreted in urine mainly as
unchanged drug and the *O*-sulfate conjugate. Crosses the blood-brain barrier and is
decarboxylated in the central nervous system to α-methylnoradrenaline. Crosses the
placenta and small amounts are excreted in breast milk (2).
Analysis of solution on both sides of the monolayer, used as a model of the
blood-brain barrier, failed to show any metabolites of methyldopa. May traverse the

blood-brain barrier by means of a carrier transport system, notably the large neutral amino acid transport (9).

Three volunteers were perfused with 21, 211 or 2112 mg kg$^{-1}$ at pH 6. At higher concentrations of drug in the perfusion solution, the free fraction in plasma samples was increased significantly. Although absorption is more efficient at lower concentrations, bioavailability may not be substantially enhanced due to increased sulfation in the gut wall (10).

### Sensitisation
Reported to cause skin sensitisation in humans (8).

## Genotoxicity
*Salmonella typhimurium* TA97, TA98, TA100, TA1535, with and without metabolic activation negative (11,12).

*In vitro* rat bone marrow cells, no clastogenic effect. *In vitro* human lymphocyte cultures, increased frequency of sister chromatid exchanges, without any effect on the frequency of chromosomal aberrations (metabolic activation unspecified in all cell types) (13).

## Any other adverse effects to man
Drowsiness, dizziness, light headedness, nausea, headache, weakness and fatigue, decreased libido and impotence have been reported frequently. Can impair concentration and memory; cause mild psychoses, depression, disturbed sleep and nightmares, paraesthesia, Bell's palsy, involuntary choreoathetotic movements and Parkinsonism (2).

Ischaemic heart disease was prevalent in treated group irrespective of blood pressure levels. Risk factors such as body mass index, skinfold thickness, serum cholesterol, albumin, creatinine, blood urea nitrogen, and uric acid at entry were elavated in the treated group (14).

## Any other adverse effects
Potency for protection was related to affinity for α-adrenergic binding sites labelled with [$^{3}$H]clonidine (15).

Intravenous dog 5.0 and 10 mg kg$^{-1}$ min$^{-1}$ for 30 min decreased the arterial blood pressure and renal vascular resistance by 27, 48, 63 and 79%, respectively (16).

## Any other comments
Human health effects reviewed (17).

## References
1. McCoy, G. D. et al *Carcinogenesis (London)* 1990, **11**(7), 1111-1117
2. *Martindale. The Extra Pharmacopoeia* 30th ed., 1993, The Pharmaceutical Press, London
3. Usdin, E. et al *Psychotropic Drugs and Related Compounds* 2nd ed., 1972, 348, Washington, DC
4. *J. Med. Chem.* 1977, **20**, 1378
5. *S. Afr. Med. J.* 1984, **65**, 194
6. *Am. Heart J.* 1983, **105**, 1037
7. *Cutan. Med. Practit.* 1986, **38**, 187
8. George, J. D. et al *Gov. Rep. Announce. Index (U. S.)* 1981, **87**(14), Abstr. No. 730,806
9. Chastain, J. E., et al *Neurosci. Res. Commun.* 1989, **4**(3), 147-152
10. Merfeld, A. E. et al *J. Pharm. Pharmacol.* 1986, **38**(11), 815-822

**M195. Methyldopa**

11.  Zeiger, E. et al *Environ. Mol. Mutagen.* 1988, **11**(Suppl. 12), 1-157
12.  Zeiger, E. *Environ. Mol. Mutagen.* 1990, **16**(Suppl. 18), 32-54
13.  Grisiolia, C. K. et al *Mutat. Res.* 1991, **259**(2), 127-132
14.  Kuramoto, K. et al *Jpn. Circ. J.* 1988, **52**(1), 1-8
15.  Buccafusco, J. J. et al *Toxicol. Lett.* 1987, **38**(1-2), 67-78
16.  Hassan, A. B. et al *Vet. Med. J.* 1986, **34**(1), 107-120
17.  Murphy, W. G. et al *Biochem. Soc. Trans.* 1991, **19**(1), 183-186

# M196   Methyldymron

**CAS Registry No.** 42609-73-4
**Synonyms** urea, *N*-methyl-*N'*-(1-methyl-1-phenylethyl)-*N*-phenylurea-;
Dimelon-methyl; K1441; methyldimuron
**Mol. Formula** $C_{17}H_{20}N_2O$                          **Mol. Wt.** 268.36
**Uses** Herbicide. Cell division inhibitor.

## Physical properties
**M. Pt.** 76°C; **Partition coefficient** log $P_{ow}$ 3.010 (1).

**Solubility**
Water: 120 mg $l^{-1}$ at 20°C. Organic solvent: acetone, n-hexane, methanol

## Ecotoxicity

**Fish toxicity**
$LC_{50}$ (48 hr) carp 14 mg $l^{-1}$ (2).

## Mammalian and avian toxicity

**Acute data**
$LD_{50}$ oral ♀ rats 3950 mg $kg^{-1}$; ♂ rats 5850 mg $kg^{-1}$ (1,2).
$LD_{50}$ oral ♀ mice 5270 mg $kg^{-1}$; oral ♂ mice 5000 mg $kg^{-1}$ (2).
$LD_{50}$ dermal rat 11.4 g $kg^{-1}$ (2).
$LD_{50}$ subcutaneous mouse 7600-7800 mg $kg^{-1}$ (3).

## Legislation
Included in Schedule 6 (Release into Land: Prescribed Substances) Statutory
Instrument No. 472, 1991 (4).
The log $P_{ow}$ value exceeds the European Community recommended level 3.0 (6th and
7th amendments) (5).

**Any other comments**

Rapidly metabolised in plants. Inhibits cell division (2).

**References**

1. *The Pesticide Manual* 9th ed., 1991, The British Crop Protection Council, Farnham
2. *The Agrochemicals Handbook* 3rd ed., 1991, RSC, London
3. *Shokubutsu Boeki* 1978, **32**, 488
4. *S. I. 1991 No. 472 The Environmental Protection (Prescribed Processes and Substances) Regulations* 1991, HMSO, London
5. *1967 Directive on Classification, Packaging, and Labelling of Dangerous Substances 67/548/EEC; 6th Amendment EEC Directive 79/831/EEC; 7th Amendment EEC Directive 91/32/EEC* 1991, HMSO, London

# M197   Methylenebis(4-cyclohexyl isocyanate)

$$OCN-\langle\rangle-CH_2-\langle\rangle-NCO$$

**CAS Registry No.** 5124-30-1

**Synonyms** 1,1'-methylenebis[4- isocyanatocyclohexane]; Nacconate H12; Isocyanic acid, methylenedi-4,1-cyclohexylene ester; bis(4-isocyanatocyclohexyl)methane

**Mol. Formula** $C_{15}H_{22}N_2O_2$          **Mol. Wt.** 262.35

**Occupational exposure**

**US TLV (TWA)** 0.005 ppm (0.054 mg m$^{-3}$); **UK Long-term limit** 0.02 mg m$^{-3}$; **UK Short-term limit** 0.07 mg m$^{-3}$; **Supply classification** toxic.

**Risk phrases** ≥20% – Toxic by inhalation – Irritating to eyes, respiratory system and skin – May cause sensitisation by inhalation and skin contact – ≥2%<20% – Toxic by inhalation – May cause sensitisation by inhalation and skin contact – ≥0.5%<2% – Harmful by inhalation – May cause sensitisation by inhalation and skin contact (R23, R36/37/38, R42/43, R23, R42/43, R20, R42/43)

**Safety phrases** In case of contact with eyes, rinse immediately with plenty of water and seek medical advice – After contact with skin, wash immediately with plenty of soap and water – In case of insufficient ventilation, wear suitable respiratory equipment – In case of accident or if you feel unwell, seek medical advice immediately (show label where possible) (S26, S28, S38, S45)

**Mammalian and avian toxicity**

**Acute data**

LD$_{50}$ oral rat 9900 mg kg$^{-1}$ (1).
LC$_{50}$ (5 hr) inhalation rabbit 20 ppm (1).

**Legislation**

Included in Schedule 4 (Release into Air: Prescribed Substances) Statutory Instrument No. 472, 1991 (2).

## References

1. *Documentation of the Threshold Limit Values and Biological Exposure Indices* 5th ed., 1986, 392, Cincinnati, OH
2. *S. I. 1991 No. 472 The Environmental Protection (Prescribed Processes and Substances) Regulations* 1991, HMSO, London

# M198    4,4′-Methylenebis[*N,N*- dimethyl-benzenamine]

$$(H_3C)_2N-\!\!\langle\ \rangle\!\!-CH_2-\!\!\langle\ \rangle\!\!-N(CH_3)_2$$

**CAS Registry No.** 101-61-1

**Synonyms** 4,4′-methylenebis[*N,N*- dimethylaniline]; Arnold's base; Michler's base; Michler's methane; tetrabase; tetramethyldiaminodiphenylmethane

**Mol. Formula** $C_{17}H_{22}N_2$                          **Mol. Wt.** 254.38

**Uses** Manufacture of dyes. Reagent for lead determination.

## Physical properties

**M. Pt.** 90-91°C; **B. Pt.** 390°C (sublimes without decomp.).

### Solubility

Organic solvent: benzene, carbon disulfide, diethyl ether

## Mammalian and avian toxicity

### Acute data

$LD_{50}$ oral mouse 3160 mg $kg^{-1}$ (1).

### Sub-acute data

Oral (4 wk) mice up to 11,380 mg $kg^{-1}$ in diet caused no death or growth inhibition, oral (4 wk) rats 3155 mg $kg^{-1}$ in diet did not affect survival, caused severe reduction in weight gain (2).

Oral (30 wk) mouse, rat 1250 or 2500 mg $kg^{-1}$ of diet (mouse), 375 or 750 mg $kg^{-1}$ of diet (rat) caused no reduction in body weight gain but increased compound dependent non-neoplastic proliferative lesions of the thyroid (2).

### Carcinogenicity and long-term effects

No adequate evidence for carcinogenicity to humans or animals, IARC classification group 3 (3).

Oral (59 wk) ♂, ♀ rats, 750, 375 ppm in feed, followed by 45 wk observation period. No significant association between administered dose and toxicity. Dose dependent increase in incidences of follicular-cell carcinomas of the thyroid. Non-neoplastic proliferative lesions of the thyroid were observed in dosed animals (4).

Oral (78 wk) ♂, ♀ mice 1250, 2500 ppm in food followed by 13 wk observation period. No significant associations between administered dose and toxicity. Liver neoplasms were observed in both ♂ and ♀ mice as well as increased incidences of hepatocellular adenomas. Incidences of hepatocellular carcinomas were not significantly increased above controls. Both sexes had a significant positive association for the incidence of a combination of hepatocellular adenomas and combination of hepatocellular adenomas and hepatocellular carcinomas; however, for ♂ this was not statistically significant. Non-neoplastic proliferative lesions of the thyroid were observed in dosed animals of both sexes (4).

## Genotoxicity

*Salmonella typhimurium* TA98, TA97, TA100, TA1535, TA1537 with and without metabolic activation negative (5).
*Saccharomyces cerevisiae* with and without metabolic activation mitotic recombination negative (6).
*In vitro* mouse lymphoma $tk^+/tk^-$ with and without metabolic activation positive (7).
*In vitro* mouse embryo cell line C3H/10T1/2 positive (8).

## Any other adverse effects

Following injection to rats [$^3$H]-4,4′- methylenebis[*N,N*-dimethylbenzenamine], radioactivity found bound to liver nucleic acids (9).

## References

1. Lewis, R. J. *Sax's Dangerous Properties of Industrial Materials* 8th ed., 1992, Van Nostrand Reinhold, New York
2. *National Cancer Institute* 1979, Tech. Rep. Ser. No.186; DHEW Publ. No. (NIH) 79-1762, Washington, DC
3. *IARC Monograph* 1987, **Suppl.7**, 66
4. *National Toxicology Program Research and Testing Division* 1979, Report No. TR-186, NIEHS, Research Triangle Park, NC
5. Ashby, J. et al *Mutat. Res.* 1988, 204(1), 17-115
6. Simmon, U. F. *J. Natl. Cancer. Inst.* 1979, **62** 901-909
7. Mitchell, A. D. et al *Environ. Mol. Mutagen.* 1988, **12**(Suppl. 13), 1-18
8. Dunkel, U. C. et al *Environ. Mol. Mutagen.* 1988, **12**, 21-31
9. Schribner, J. D. et al *Cancer Lett. (Shannon, Irel.)* 1980, **9**, 117-121.

# M199    4,4′-Methylenebis[2-methylaniline]

**CAS Registry No.** 838-88-0
**Synonyms** benzenamine, 4,4′-methylenebis[2- methylbenzenamine]; *o*-toluidine, 4,4′- methylenebis-
**Mol. Formula** $C_{15}H_{18}N_2$                                   **Mol. Wt.** 226.32
**Uses** In the production of dyes.

**Occurrence** May be present in waste streams from plants which produce it as an intermediate for further processing.

## Physical properties
**M. Pt.** 149°C.

**Solubility**
Water: miscible. Organic solvent: ethanol

## Occupational exposure
**Supply classification** toxic.
**Risk phrases** May cause cancer – Harmful if swallowed – May cause sensitisation by skin contact (R45, R22, R43)
**Safety phrases** Avoid exposure-obtain special instruction before use – If you feel unwell, seek medical advice (show label where possible) (S53, S44)

## Mammalian and avian toxicity

**Carcinogenicity and long-term effects**
No adequate evidence for carcinogenicity to humans, sufficient evidence for carcinogenicity to animals, IARC classification group 2B (1).
Oral (55 wk) rat, 1000 ppm induced liver tumours in both sexes and skin and mammary tumours in ♂ (2).
Oral (7 yr) ♀ dog 100 ng $3 \times$ wk$^{-1}$ for 6 wk; then 100 ng $5 \times$ wk for 5 wk$^{-1}$; and then 50 ng $5 \times$ wk$^{-1}$ for ≤7 yr. Treated dogs developed renal atrophy with elevated blood urea nitrogen ≈6 month prior to being killed *in extremis*. 3/6 dogs survived 5.2-7.0 yr, these developed hepatocellular carcinomas, and 2/3 developed primary lung tumours. No liver or lung tumours were seen in the controls which were kept for 8.3-9 yr (3).
Oral (16 month) 24 ♂ rats, total dose 10.2 g kg$^{-1}$ body weight. Over a 10 month period from the start of study, 18 malignant and 2 benign liver tumours and 12 subcutaneous tumours were detected (4).
Oral (1 yr) ♂, ♀ rats 200 ppm diet. Tumours developed in lung, liver and skin (5).
Oral (180 day) ♂ rats 50 mg kg$^{-1}$ body weight. Tumours developed in the liver, lung, mammary gland and skin (5).

**Irritancy**
100 mg instilled into rabbit eye for 24 hr caused moderate irritation (6).

## Legislation
Included in Schedule 4 (Release into Air: Prescribed Substances) Statutory Instrument No. 472, 1991 (7).

## Any other comments
Experimental toxicology and human health effects reviewed (8).

## References
1. *IARC Monograph* 1987, **Suppl. 7**, 66
2. *Toxicol. Appl. Pharmacol.* 1975, **31**, 159
3. Stula, E. F. et al *Toxicol. Appl. Pharmacol.* 1978, **1** (39), 339-356
4. Munn, A. *Bladder Cancer. A Symposium* 1967, 187, Birmingham
5. Stula, E. F. et al *Toxicol. Appl. Pharmacol.* 1971, **19**, 30

6. Marhold, J. V. *Sbornik Vysledku Toxixologickeho Vysetreni Latek A Pripravku* 1972, Prague
7. *S. I. 1991 No. 472 The Environmental Protection (Prescribed Processes and Substances) Regulations* 1991, HMSO, London
8. *ECETOC Technical Report No. 30(4)* 1991, European Chemical Industry Ecology and Toxicology Centre, B-1160 Brussels

# M200   Methylenebis(1-thiosemicarbazide)

## $H_2NC(S)NHNHCH_2NHNHC(S)NH_2$

**CAS Registry No.** 39603-48-0
**Synonyms** hydrazinecarbothioamide-2,2′- methylenebis-; Bisthiosemi; Kayanex; methylenebis(1-thiosemicarbazide)
**Mol. Formula** $C_3H_{10}N_6S_2$                          **Mol. Wt.** 194.28
**Uses** Rodenticide.

## Physical properties

**Solubility**
Organic solvent: dimethyl sulfoxide

## Mammalian and avian toxicity

**Acute data**
$LD_{50}$ oral mouse, guinea pig 30-32 mg $kg^{-1}$ (1).
$LD_{50}$ oral cat 150 mg $kg^{-1}$ (1).
$LD_{50}$ oral chicken 120 mg $kg^{-1}$ (1).

**Sub-acute data**
No observed effect level (NOEL) oral (90 day) mice 100 mg $kg^{-1}$ diet; rats 50 mg $kg^{-1}$ diet (1).

## Legislation

Included in Schedule 6 (Release into Land: Prescribed Substances) Statutory Instrument No. 472, 1991 (2).

## Any other comments

Toxicity studied (3).

## References

1. *The Pesticide Manual* 9th ed. 1991, British Crop Protection Council, Farnham
2. *S. I. 1991 No. 472 The Environmental Protection (Prescribed Processes and Substances) Regulations* 1991, HMSO, London
3. Nishiuchi, Y. *Seitai Kagaku* 1988, **9**(3), 19-26 (Japan.)

# M201  Methylene Blue

$$\left[ (H_3C)_2N - \underset{S^+}{\overset{N}{\bigcirc\bigcirc\bigcirc}} - N(CH_3)_2 \right] Cl^-$$

**CAS Registry No.** 61-73-4
**Synonyms** phenothiazin-5-ium, 3,7-bis(dimethylamino)-, chloride; C.I. Basic Blue 9;
methylenium ceruleum; Sandocryl Blue BRL; tetramethylthionine chloride;
Yamamoto Methylene Blue ZF; C.I. 52015
**Mol. Formula** $C_{16}H_{18}N_3SCl$        **Mol. Wt.** 319.86
**Uses** Stain in bacteriology. Analytical reagent. Oxidation-reduction indicator.
Antimethaemoglobinemic. Antidote for cyanide. Antiseptic.

## Physical properties

**M. Pt.** 190°C (decomp.); **Specific gravity** 0.908.

**Solubility**
Water: 40 g $l^{-1}$.

## Ecotoxicity

**Invertebrate toxicity**
$LC_{50}$ (96 hr) *Penaeus californiensis* 100 mg $l^{-1}$ (1).

## Environmental fate

**Nitrification inhibition**
At the highest concentration tested 100 mg $l^{-1}$, no inhibition of ammonia oxidation by
activated sludge observed (2).

**Abiotic removal**
Methylene blue-sensitised photolysis rates were highest at basic pH (3).

**Absorption**
Absorption onto sodium-montmorillonite was as much as 120% of the
cation-exchange capacity (4).

## Mammalian and avian toxicity

**Acute data**
$LD_{50}$ oral mouse 3500 mg $kg^{-1}$ (5).
$LD_{Lo}$ oral dog 500 mg $kg^{-1}$ (6).
$LD_{50}$ intravenous rat 1250 mg $kg^{-1}$ (7).
$LD_{50}$ intravenous mouse 77 mg $kg^{-1}$ (5).
$LD_{Lo}$ intravenous monkey, dog 10, 50 mg $kg^{-1}$ respectively (6).
$LD_{50}$ intraperitoneal mouse, rat 150, 180 mg $kg^{-1}$ respectively (8,9).

**Teratogenicity and reproductive effects**
$TD_{Lo}$ (1-22 day pregnancy) oral rat 2500 mg $kg^{-1}$, unspecified reproductive
effects (10).

There have been several reports of haemolytic anaemia and hyperbilirubinaemia in neonates exposed to methylene blue in the amniotic cavity (11-14).

**Metabolism and pharmacokinetics**

Absorbed from the gastrointestinal tract, and believed to be reduced in the tissue to the leucoform which is slowly excreted mainly in the urine, together with some unchanged dye (15).

74% of a 10 mg dose was recovered in the urine as unchanged dye or leucomethylene blue (16).

## Genotoxicity

*Salmonella typhimurium* TA98, TA100, TA1537 (metabolic activation unspecified) negative (17).

## Any other adverse effects to man

High doses may cause nausea, vomiting, abdominal and chest pain, headache, dizziness, mental confusion, profuse sweating, dyspnoea and hypertension. May cause necrotic abscesses if injected subcutaneously (15).

## Any other adverse effects

Mice, bearing Ehrlich ascites tumour, intraperitoneal (1-2 day) (dose unspecified) after tumour transplants, high toxicity was induced but no specific cytotactic effects (18).

## Legislation

Included in Schedule 4 (Release into Air: Prescribed Substances) Statutory Instrument No. 472, 1991 (19).

## Any other comments

Adsorbs completely on DNA-sepharose and DNA-polyacrylamide gels, but not on DNA-cellulose gels (20).

## References

1. Hanks, K. S. *Aquaculture* 1976, **7**, 293
2. Richardson, M. *Nitrification Inhibition in the Treatment of Sewage* 1985, 56, RSC, London
3. Watts, R. J. et al *Chemosphere* 1988, **17**(10), 2083-2091
4. Marguilies, L. et al *Clays Clay Miner.* 1988, **36**(3), 270-276
5. *Cesk. Farm.* 1963, **12**, 94
6. *Abern. Handb. Biol. Arbeitsmeth.* 1935, **4**, 1366
7. *Arzneim.-Forsch.* 1968, **18**, 678
8. *Naunyn-Schmiedeberg's Arch. Pharmacol.* 1947, **204**, 288
9. *NTIS AD691-490* Natl. Tech. Info. Ser., Springfield, VA
10. *Am. J. Anat.* 1962, **110**, 29
11. Cowett, R. M. et al *Obstet. Gynecol.* 1976, **48**(Suppl.), 74s-75s
12. Serota, F. T. et al *Lancet* 1979, **ii**, 1142-1143
13. Crooks, J. *Arch. Dis. Child* 1982, **57**, 872-873
14. Nicolini, V. et al *Lancet* 1990, **336**, 1258-1259
15. *Martindale. The Extra Pharmacopoeia* 30th ed., 1993, The Pharmaceutical Press, London
16. Disanto, A. R. et al *J. Pharm. Sci.* 1972, **61**, 1086-1089
17. Ferguson, L. R. et al *Mutat. Res.* 1988, **209**, 57-62

18. Mihail, N. et al *Rev. Chir., Oncol. Radiol, ORL, Oftalmol., Stomatol. Oncol.* 1989, **28**(2), 115-121. (Rom.)
19. *S. I. 1991 No. 472 The Environmental Protection (Prescribed Processes and Substances) Regulations* 1991, HMSO, London
20. Yamaguchi, T. *Agric. Biol. Chem.* 1988, **52**(3), 845-847

# M202    4,4'-Methylenedianiline dihydrochloride

$$\left[ \quad H_3\overset{+}{N}-\langle\!\!\!=\!\!\!\rangle-CH_2-\langle\!\!\!=\!\!\!\rangle-\overset{+}{N}H_3 \quad \right] \ 2Cl^-$$

**CAS Registry No.** 13552-44-8
**Synonyms** 4,4'-methylenedibenzenamine dihydrochloride
**Mol. Formula** $C_{13}Cl_2H_{16}N_2$                              **Mol. Wt.** 271.19

## Physical properties
**M. Pt.** 288°C.

## Mammalian and avian toxicity

### Sub-acute data
Oral (14 day) ♂ rat 1600, 3200 mg kg$^{-1}$ in drinking water, dose related body weight reductions observed (1).
Gavage (14 day) ovariectomised ♀ rats 150 mg kg$^{-1}$ body weight, increased weight of uterus, thyroid gland and adrenal gland (2).
Oral (13 wk) ♂, ♀ rat 400, 800 mg l$^{-1}$ in drinking water, a third of low dose and all of high dose animals developed bile-duct hyperplasia (1).

### Carcinogenicity and long-term effects
No adequate evidence for carcinogenicity to humans, sufficient evidence for carcinogenicity to animals, IARC classification group 2B (3).
Oral (103 wk) ♂, ♀ mice 150, 300 mg l$^{-1}$ in drinking water, an increased incidence of follicular-cell adenomas of the thyroid was observed in the high dose animals and a dose related incidence of thyroid-gland follicular-cell hyperplasia in both ♂ and ♀ with 2/50 ♀ developing thyroid follicular-cell carcinomas. ♀ also had an increased incidence of hepatocellular adenomas whilst both sexes had increased incidences of hepatocellular carcinomas. Also ♂ had significant increase in liver neoplastic nodules (1,4).
Gavage (30 days) ♀ rats (40 days old) 30 mg every 3 day in sesame oil. Observed for a further 9 month no increased incidence in mammary lesions were observed (5).

## References
1. *National Toxicology Program Research and Testing Division* 1983, Report No.248, Research Triangle Park, NC
2. Tullner, W. W. *Endocrinology* 1960, **66**, 470-974
3. *IARC Monograph.* 1987, **Suppl.7**, 66

4. Weisburger, E. K. et al *J. Natl. Cancer Inst.* 1984, **72**, 1457-1463
5. Griswold, D. P., *Cancer Res.* 1968, **28**, 924-933

# M203 Methyl cellulose

**CAS Registry No.** 9004-67-5
**Synonyms** methyl ether cellulose; Adulsin; Bulkaloid; Celacol M; cellulose methyl ether; cellulose methylate; cellumeth; Culminal MC; Morpolose M400; Methocel A; Methocel MC; methoxycellulose; methulose; *O*-methyl cellulose; Tylose SL400; Visosal
**Uses** Binding agent used in ceramics, mortar and cosmetics. Laxative. Bulk producer in the preparation of diabetic foods.

## Physical properties

**Solubility**
Water: miscible. Organic solvent: glacial acetic acid

## Environmental fate

**Degradation studies**
$LC_{100}$ *Ruminococcus flavefaciens* 0.1% (wt/vol) in a culture metabolising cellulose. However, methyl cellulose did not inhibit growth on cellobiose or cellulooligosaccharides (1).
Degraded by UV irradiation (2).

## Mammalian and avian toxicity

**Acute data**
$LD_{50}$ intraperitoneal mouse 280 g $kg^{-1}$ (3).
$LD_{Lo}$ intravenous mouse 1000 mg $kg^{-1}$ (4).

**Sub-acute data**
Oral rat (6 month) 6200 mg $kg^{-1}$ $day^{-1}$ caused no observable adverse effects (5).

**Teratogenicity and reproductive effects**
Oral rat (3 generations) 5% diet did not produce any adverse effects, including reproductive function (5).
Gavage rat 10 mg $kg^{-1}$ $day^{-1}$ on days 6-15 of gestation induced the formation of a thin central tendon in the diaphragm of all foetuses. In some weaned pups the liver protruded within this tendon (6).

**Metabolism and pharmacokinetics**
Oral doses of 5000 or 10,000 mg to humans were almost entirely eliminated in the faeces (5).

**Irritancy**
Reported to be non-irritating to the eye (species unspecified) (7).

## Genotoxicity

*Salmonella typhimurium* TA98, TA100, TA1535, TA1537, TA1538 with and without metabolic activation negative (8).

*Escherichia coli* WP2, with and without metabolic activation negative (8).
*Paramecium tetraurelia*, formation of macronuclear anlagen negative (9).

## Any other adverse effects

Intravenous administration of low viscosity solution (15 CP) to dogs caused a progressive decrease in the volume of urine. Necrotising vascular disease and death were due to renal failure (5).

Injection of 2% solution (volume not specified) into the anterior chamber of rabbit eye caused an increase in intraocular pressure which peaked in 3 hr, returning to normal in 9 hr. Corneal thickness was increased (peak at 12 hr) returning to normal after 6 days. Endothelial cell diameter decreased by 6% after 2 wk. The endothelial cells also showed decreased microvilli and enlargement of intercellular spaces (10).

## References

1. Rasmussen, M. A. et al *Appl. Environ. Microbiol.* 1988, **54**(4), 890-897
2. Yamada, H. et al *Shizuoka Daigaku Nogakubu Kenkyu Hokoku* 1986, (36), 65-68 (Japan.) (*Chem. Abstr.* **107**, 79730x)
3. *NTIS Report* AD628-313, Natl. Tech. Inf. Ser., Springfield, VA
4. *J. Am. Pharm. Assoc. Sci. Ed.* 1956, **45**, 685
5. *Patty's Industrial Hygiene and Toxicology* 3rd ed., 1981, **2b**, 2552-2553, John Wiley & Sons, New York
6. Lu, C. C. et al *Teratology* 1988, **37**(6), 571-576
7. *AMA Drug Evaluation* 3rd ed., 1977, American Medical Assoc., Dept. of Drugs, PSG Publishing Co., Littleton, MA
8. Prival, M. J. et al *Mutat. Res.* 1991, **260**(4), 321-329
9. Fukushima, S. et al *Kankyo Kagaku Kenkyusho Kenkyu Hokoku* (*Kinki Daigaku*) 1986, **14**, 1-6 (Japan.) (*Chem. Abstr.* **106**, 137093j)
10. Chang, K. H. et al *K'at'allik Taehak Uihakpu Nonmunjip* 1986, **39**(3), 907-919 (Korean) (*Chem. Abstr.* **106**, 188976g)

# M204    4-(1-Methylethyl)benzaldehyde

**CAS Registry No.** 122-03-2
**Synonyms** cumic aldehyde; *p*-isopropylbenzaldehyde
**Mol. Formula** $C_{10}H_{12}O$                    **Mol. Wt.** 148.21
**Uses** Perfumery.
**Occurrence** In essential oils of eucalyptus, myrrh and cassia.

## Physical properties

**B. Pt.** 235-236°C; **Flash point** 92-8°C; **Specific gravity** $d^{20}$ 0.978.

### Solubility

Organic solvent: ethanol, diethyl ether

### Ecotoxicity

**Invertebrate toxicity**

50% inhibitory growth concentration *Tetrahymena pyriformis* GL-C 32.5 g m$^{-3}$ (1).

### Mammalian and avian toxicity

**Acute data**

$LD_{50}$ oral rat, mouse 1390, 2440 mg kg$^{-1}$ respectively (2-4).
$LD_{50}$ dermal rabbit 2800 mg kg$^{-1}$ (5).

**Metabolism and pharmacokinetics**

Oral ♂ rabbits 12 g with 20 ml water, the rabbits were then allowed food and water
*ad. lib.*, urine was collected for 3 days. Metabolites in urine were *p*-cumyl alcohol;
*o*-cumyl alcohol; 8-hydroxycuminic acid; 9-hydroxycumnic acid; and
2-(*p*-carboxyphenyl) propinoic acid (6).

**Irritancy**

Dermal rabbit (24 hr) 500 mg caused irritation (5).

### References

1. Schultz, T. W. et al *ASTM Spec. Tech. Publ.* 1988, **1007**(Aquat. Toxicol. Environ. Fate:
   11th Vol.), 410-423
2. Madhyasha, X. M. et al *Indian J. Biochem.* 1968, **5**, 161
3. Rachwell, P. et al *Nutr. Cancer* 1979, **1**, 10
4. Jenner, P. M. et al *Food Cosmet. Toxicol.* 1964, **2**, 395
5. Opdyke, D. L. J. *Food Cosmet. Toxicol.* 1974, **12**, 395
6. Ishida, T. et al *Xenobiotica* 1989, **19**(8), 843-855

# M205   3-(1-Methylethyl)phenol methylcarbamate

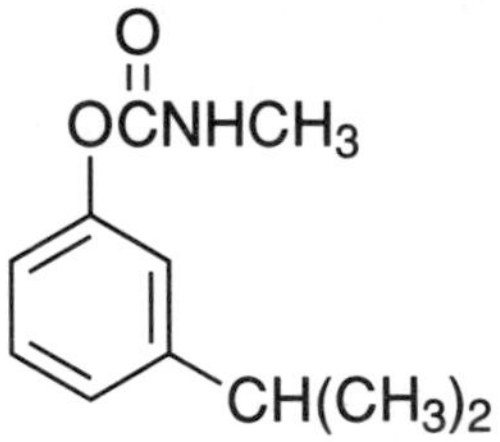

**CAS Registry No.** 64-00-6
**Synonyms** methylcarbamic acid, *m*-cumenyl ester; *m*-cumenol methylcarbamate;
*m*-cumenyl methyl carbamate; *m*-isopropylphenol methylcarbamate;
3-isopropylphenyl methylcarbamate
**Mol. Formula** $C_{11}H_{15}NO_2$                    **Mol. Wt.** 193.25

**Uses** Insecticide (not now in common use) (1).

## Physical properties
**M. Pt.** 72-74°C.

### Solubility
Water: 85 ppm at 30°C. Organic solvent: xylene, toluene, acetone, isopropanol, dimethylformamide

## Mammalian and avian toxicity

### Acute data
$LD_{50}$ oral rat, mouse, guinea pig, chicken 10-29 mg $kg^{-1}$ (2-5).
$LD_{50}$ oral redwing blackbird, 3.16-10.0 mg $kg^{-1}$, starling 17.0 mg $kg^{-1}$ (6).
$LD_{50}$ dermal rabbit, rat 40, 113 mg $kg^{-1}$ respectively (7,4).
$LD_{50}$ intraperitoneal rat 14.2 mg $kg^{-1}$ (8).
$LD_{50}$ intravenous mouse, rat 1.4-3.2 mg $kg^{-1}$ (9,10).
$LD_{50}$ intramuscular dog, rat 13-14 mg $kg^{-1}$ (10).

## Legislation
Limited under EC Directive on Drinking Water Quality 80/778/EEC. Pesticides: maximum admissible concentration 0.1 μg $l^{-1}$ (11).
Included in Schedule 6 (Release into Land: Prescribed Substances) Statutory Instrument No. 472, 1991 (12).

## References
1. *The Pesticide Manual* 9th ed., 1991, British Crop Protection Council, Farnham
2. *Toxicol. Appl. Pharmacol.* 1972, **21**, 315
3. *J. Agric. Food Chem.* 1970, **18**, 793
4. *The Pesticide Manual* 1968, British Crop Protection Council, Worcestershire
5. *Toxicol. Appl. Pharmacol.* 1967, **11**, 49
6. Schafer, E. W. et al *Arch. Environ. Contam. Toxicol.* 1983, **12**, 355382
7. *Farm Chemicals Handbook* 1980, Meister Publishing Co., Willoughby, OH
8. *Bull. WHO* 1971, **44**(1-3), 241
9. US Army Armament Research and Development Command, *Report NX02085*, Chemical Systems Laboratory, Aberdeen Proving Ground, MD
10. *Br. J. Ind. Med.* 1965, **22**, 317
11. *EC Directive Relating to the Quality of Water Intended for Human Consumption* 1982, 80/778/EEC, Office for Official Publications of the European Communities, 2 rue Mercier, L-2985 Luxembourg
12. *S. I. 1991 No. 472 The Environmental Protection (Prescribed Processes and Substances) Regulations* 1991, HMSO, London

# M206   2-Methyl-5-ethylpyridine

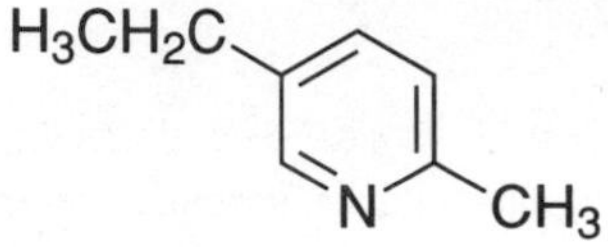

**CAS Registry No.** 104-90-5

**Synonyms** pyridine, 5-ethyl-2-methyl-; 2-picoline, 2-ethyl-; aldehydecollidine; aldehydine; MEP; 5-ethyl-2-methylpyridine

**Mol. Formula** $C_8H_{11}N$ **Mol. Wt.** 121.18

**Occurrence** Found at 24.4-73.5 $\mu g\ g^{-1}$ in coke oven emissions; 3.12 mg $g^{-1}$ in coal tar sample; and 0.37g $l^{-1}$ in raw wood preservative sludge sample (1).

## Physical properties

**B. Pt.** $_{20}$ 74-75°C; **Specific gravity** $d_4^{23}$ 0.9184.

## Solubility

Organic solvent: ethanol, diethyl ether, benzene

## Ecotoxicity

**Fish toxicity**

$LC_{50}$ (96 hr) fathead minnow 81.2 mg $l^{-1}$ (2).

## Mammalian and avian toxicity

**Acute data**

$LD_{50}$ oral mouse, rat 282-368 mg $kg^{-1}$ (3).

$LC_{Lo}$ (4 hr) inhalation rat 1000 ppm (4).

$LD_{50}$ dermal rabbit 1000 mg $kg^{-1}$ (5).

$LD_{50}$ subcutaneous mouse, rat 294, 826 mg $kg^{-1}$ respectively (3).

**Irritancy**

Dermal rabbit 10 mg (24 hr) caused severe irritation (4).

250 $\mu g$ instilled into rabbit eye caused severe irritation (4).

## Legislation

Included in Schedule 4 (Release into Air: Prescribed Substances) Statutory Instrument No. 472, 1991 (6).

## Any other comments

All reasonable efforts have been made to find information on isomers of this compound, but no relevant data are available.

## References

1. Lao, R. C. et al *J. Chromatogr.* 1975, **112**, 681-700
2. Schultz, T. W. et al *Chemosphere* 1989, **18**(11-12), 2283-2291
3. Izmerov, N. F. et al *Toxicometric Parameters of Industrial Toxic Chemicals under Single Exposure* 1982, CIP, Moscow
4. *AMA, Arch. Ind. Hyg. Occup. Med.* 1951, **4**, 119
5. *Union Carbide Data Sheet* 6/29/66, Industrial Medicine and Toxicology Department, Union Carbide Corp., 270 Park Ave., New York

6. *S. I. 1991 No. 472 The Environmental Protection (Prescribed Processes and Substances) Regulations* 1991, HMSO, London

# M207   1-(1-Methylethyl)-1*H*-pyrrole-2,5-dione

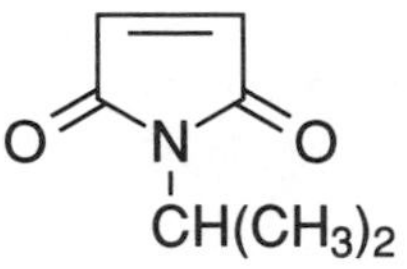

**CAS Registry No.** 1073-93-4
**Synonyms** *N*-isopropylmaleimide
**Mol. Formula** $C_7H_9NO_2$                    **Mol. Wt.** 139.16

**Physical properties**
**M. Pt.** 27.5°C; **B. Pt.** $_{17}$ 85°C.

**Mammalian and avian toxicity**

**Acute data**
$LD_{50}$ intravenous mouse 18 mg kg$^{-1}$ (1).

**References**

1. US Army Armament Research and Development Command, *Report*, Chemical Systems Laboratory, NIOSH Exchange Chemicals, Aberdeen Proving Ground, MD

# M208   Methyl fluorosulfonate ('Magic Methyl')

$FSO_3CH_3$

**CAS Registry No.** 421-20-5
**Synonyms** fluorosulfuric acid, methyl ester; methyl fluorosulfate
**Mol. Formula** $CH_3FO_3S$                    **Mol. Wt.** 114.10
**Uses** In organic synthesis as a methylating agent.

**Physical properties**
**M. Pt.** –95°C; **B. Pt.** 92-94°C; **Specific gravity** 1.412.

**Mammalian and avian toxicity**

**Acute data**
$LD_{50}$ oral mouse 112 mg kg$^{-1}$ (1).
$LC_{50}$ (1 hr) inhalation rat 5-6 ppm (1).
$LD_{Lo}$ dermal rabbit 455 mg kg$^{-1}$ (1).

**Irritancy**
100 mg instilled into rabbit eye for 4 sec caused severe irritation (1).

### Genotoxicity

*Salmonella typhimurium* TA1538 without metabolic activation positive, TA98 weakly positive, TA100 and TA1535 negative. BHK 21/C13 cell transformation assay positive (2).

### Any other adverse effects to man

Can cause fatal pulmonary oedema (3).
Acute exposure resulted in the death of a research worker (4).

### References

1. Hite, M. et al *Am. Ind. Hyg. Assoc. J.* 1979, **40**, 600
2. Ashby, J. et al *Mutat. Res.* 1978, **51**, 285-287
3. van den Ham, D. M. W. et al *Chem. Eng. News* 1976, **54**, 5
4. Admiraal, J. *Lancet* 1976, 854

# M209   *N*-Methylformamide

## HC(O)NHCH₃

**CAS Registry No.** 123-39-7
**Synonyms** Monomethylformamide
**Mol. Formula** $C_2H_5NO$                      **Mol. Wt.** 59.07

### Physical properties

**M. Pt.** –40°C; **B. Pt.** 120.83°C; **Flash point** 102.83°C; **Specific gravity** 1.011.

### Ecotoxicity

**Fish toxicity**
*In vitro* zebra fish embryo 0.01-1.5% v/v in incubation media during embryonic development, caused changes to vertebral column, neural tube, heart and blood vessels and growth retardation (1).

### Mammalian and avian toxicity

**Acute data**
$LD_{50}$ oral mouse, rat 2.6, 4.0 g kg$^{-1}$ respectively (2,3).
$LD_{50}$ subcutaneous mouse 3.1 g kg$^{-1}$ (4).
$LD_{50}$ intraperitoneal mouse, rat 802, 3500 mg kg$^{-1}$ respectively (5,3).
$LD_{50}$ intravenous mouse 1.6 mg kg$^{-1}$ (2).
$LD_{50}$ intramuscular mouse 2.7 g kg$^{-1}$ (2).

**Sub-acute data**
Inhalation (2 wk) ♂ rat 0, 50, 130, 400 ppm 6 hr day$^{-1}$, 5 days wk$^{-1}$ followed by 2 wks post exposure recovery period. 130, 400 ppm caused liver damage, with the changes being more marked in the 400 ppm group, the effects were partially reversible. No other organs were affected. No observed effect level was 50 ppm (6).

**Teratogenicity and reproductive effects**

Logistic regression and discriminant analysis used to predict developmental toxicity in humans, negative. Designated positive for developmental toxicity in rats, mice and negative in rabbits (7).

*In vitro* chick embryos 0.01-1.5% v/v in incubation media during embryonic development, caused changes to vertebral column, neural tube, heart, blood vessels, and growth retardation (1).

**Metabolism and pharmacokinetics**

*In vitro* suspensions of mouse hepatocytes metabolised to N-alkylcarbamoylating metabolites and depleted pools of glutathione. The amount of metabolism and glutathione depletion were dose dependent (8).

*In vivo* rat, mouse [$^{14}$C]-*N*-methylformamide metabolites detected in urine included an (unspecified) *N*-acetylcysteine conjugate, methylamine and *N*-hydroxymethylformamide. Metabolism was more extensive and faster in mice than in rats (9).

Metabolised *in vivo* (species not specified) to *N*-acetyl-*S*-(*N*-methylcarbamoyl)-cysteine via oxidation at the formyl carbon. *In vitro* mice microsomes, cytosol; the metabolite was generated by microsomes but not the cytosol (10).

**Irritancy**

Eye (24, 48, 72 hr) rabbit 100 µl did not cause irritation (11).

**Any other adverse effects**

Produced a dose-dependent zone 3 haemorrhagic necrosis in mice, the threshold dose was 100-200 mg kg$^{-1}$. 1000 mg kg$^{-1}$ in rats caused hepatic damage in some animals with a slight elevation in plasma transaminases. There was a recordable difference between the hepatoxic effects in mice and rats. Also liver nonprotein sulfhydryl was dose dependently depleted in mice but not rats (9).

**Any other comments**

*In vivo* murine TL × 5 lymphoma growth was inhibited by *N*-methylformamide.
*In vitro* murine TL × 5 lymphoma cells (72 hr) 0.25, 1% v/v growth rate and viability decreased dose-dependently (12).

**References**

1. Groth, G. et al *Cah. Notes Doc.* 1990, **140**, 712-715
2. *Toxicol. Appl. Pharmacol.* 1985, **34** 173
3. *J. Reprod. Fertil.* 1962, **4**, 219
4. Lewis, R. J. *Sax's Dangerous Properties of Industrial Materials* 8th ed., 1992, Van Nostrand Reinhold, New York
5. *National Cancer Institute Screening Program Data Summary* January 1986, Bethesda, MD
6. Kennedy, G. L., Jr. et al *Fundam. Appl. Toxicol.* 1990, **14**(4), 810-816
7. Jelovesk, F. R. et al *Obstet. Gynecol. (N. Y.)* 1989, **74**(4), 624-636
8. Shaw, A. J. et al *Toxicol. Appl. Pharmacol.* 1988, **95**(1), 162-170
9. Tulip, K. et al *Arch. Toxicol.* 1988, **62**(2-3), 167-176
10. Cross, H. et al *Chem. Res. Toxicol.* 1990, **3**(4), 357-362
11. Jacobs, G. A. et al *Food Chem. Toxicol.* 1989, **27**(4), 255-258
12. Bill, C. A. et al *Cancer Res.* 1988, **48**(12), 3389-3393

# M210 Methyl formate

H-C(=O)-OCH₃

**CAS Registry No.** 107-31-3

**Synonyms** formic acid, methyl ester; methyl methanoate

**Mol. Formula** $C_2H_4O_2$  **Mol. Wt.** 60.05

**Uses** Blowing agent. Organic synthesis. Resin hardener. Solvent. Fumigant and larvicide.

**Occurrence** Detected in cigarette smoke and gasoline engine exhaust (1,2). Present in wastewater from urea-formaldehyde resin manufacturing plant (3). Flavour component of some fruits (4).

## Physical properties

**M. Pt.** –100°C; **B. Pt.** 34°C; **Flash point** –19°C (closed cup); **Specific gravity** $d_4^{20}$ 0.975; **Partition coefficient** log $P_{ow}$ -0.264 (4); **Volatility** v.p. 400 mmHg at 16°C; v. den. 2.07.

## Solubility

Water: ≈ 33%. Organic solvent: ethanol, methanol

## Occupational exposure

**US TLV (TWA)** 100 ppm (246 mg m⁻³); **US TLV (STEL)** 150 ppm (368 mg m⁻³); **UK Long-term limit** 100 ppm (250 mg m⁻³); **UK Short-term limit** 150 ppm (375 mg m⁻³); **UN No.** 1243; **HAZCHEM Code** 2SE; **Conveyance classification** flammable liquid; **Supply classification** extremely flammable.

## Ecotoxicity

### Bioaccumulation

Calculated bioconcentration factor 0.6 indicates that environmental pollution is unlikely (5).

## Environmental fate

### Abiotic removal

$t_{1/2}$ for hydrolysis at 25°C: 22 days at pH 6; 2.2 days at pH 7; 9.1 hr at pH 8; and 0.91 hr at pH 9 (6).

Estimated $t_{1/2}$ for reaction with photochemically produced hydroxyl radicals in the atmosphere 74 days (7).

$t_{1/2}$ for volatilisation from model river water 5.3 hr and from pond water 60 hr (5,8).

### Absorption

Estimated $K_{oc}$ 5 indicates that methyl formate will leach readily from soil (5).

## Mammalian and avian toxicity

### Acute data

LD₅₀ oral rabbit 1620 mg kg⁻¹ (9).

LC₅₀ (1 hr) inhalation guinea pig ≈ 25,000 ppm (1).

## Irritancy

Causes severe irritation. High concentrations are extremely destructive to tissues of the mucous membranes and upper respiratory tract, eyes and skin (10).

## Any other adverse effects to man

Inhalation human (1 min) 1500 ppm caused no adverse effects (2).

## Legislation

Included in Schedule 6 (Release into Land: Prescribed Substances) Statutory Instrument No. 472, 1991 (11).

## Any other comments

Environmental fate reviewed (4).

Human health effects, experimental toxicology, physico-chemical properties reviewed (12).

Autoignition temperature 465°C.

## References

1. Seizeiger, E. D. et al *J. Air Pollut. Control Assoc.* 1972, **21**(1)
2. *Patty's Industrial Hygiene and Toxicology* 3rd ed., 1981, **2A**, 2263, John Wiley & Sons, New York
3. Barcelo, D. et al *Analysis of Volatile Components of Waste Water from a Urea-Formaldehyde Glue Plant.* 1980, 335-343, Pergamon Series Environmental Science, **3**, (Anal. Tech. Environ. Chem.), New York
4. Howard, P. H. *Handbook of Environmental Fate and Exposure Data for Organic Chemicals* 1993, **4**, 398-403, Lewis Publishers, Chelsea, MI
5. Lyman, W. J. et al *Handbook of Chemical Properties Estimation Methods* 1982, McGraw-Hill, New York
6. Mobey, W. et al *J. Phys. Chem. Ref. Data* 1978, **7**, 383-414
7. Atkinson, R. *J. Phys. Chem. Ref. Data* 1989, **Monograph 1**, 145
8. *EXAMS II Computer Simulation* 1987, USEPA, Office of Toxic Substances, Washington, DC
9. *Ind. Med. Surg.* 1972, **41**, 31
10. Lenga, R. E. *Sigma-Aldrich, Library of Chemical Safety Data* 2nd ed., 1988, **2**, 2352, Sigma-Aldrich, Milwaukee, WI
11. *S. I. 1991 No. 472 The Environmental Protection (Prescribed Processes and Substances) Regulations* 1991, HMSO, London
12. *ECETOC Technical Report No. 30(4)* 1991, European Chemical Industry Ecology and Toxicology Centre, B-1160 Brussels

# M211   *N*-Methyl-*N*-formylhydrazine

## $NH_2NH(CH_3)CHO$

**CAS Registry No.** 758-17-8

**Synonyms** formic acid, methylhydrazide; 1-formyl-1-methylhydrazine; MFH

**Mol. Formula** $C_2H_6N_2O$          **Mol. Wt.** 74.08

**Occurrence** Edible false morel mushroom (*Gyromitra esculenta*).

## Physical properties

**B. Pt.** $_{12}$ 95°C.

## Mammalian and avian toxicity

### Acute data
$LD_{50}$ oral mouse 118 mg kg$^{-1}$ (1).

### Carcinogenicity and long-term effects
Swiss mice given 0.0078% in drinking water for life developed tumours: benign hepatomas, adenomas and adenocarcinomas of the lungs, adenomas of the gall bladder, cholangiomas, liver cell carcinomas and cholangiocarcinomas (2). Administered as 40 wkly subcutaneous injections to Swiss mice at doses of 20 µg g$^{-1}$ body weight for ♀ and 10 µg g$^{-1}$ for ♂. Lung tumour incidence was significantly increased: 56% in ♀ and 40% in ♂. Other organs showed no detectable carcinogenic effect (3).

### Metabolism and pharmacokinetics
Hydrolysed non-enzymatically from acetaldehyde formylmethylhydrazone. Under physiological conditions hydrolyses to $N$-methylhydrazine and formic acid (4-6). Detected in the peritoneal fluid of mice after 3 hr oral administration of 9 mg acetaldehyde formylmethylhydrazone (7).
Oxidation to a hydroxylamine derivative is mediated by rat liver cytochrome $P_{450}$ according to a spectral study (8).

## Genotoxicity
*In vitro* rat, mouse hepatocytes DNA Repair Test negative (9).

## Any other adverse effects
Intragastric administration to Wistar rats had no effect on renal function (10). Following oral administration to rats there was a transient time- and dose-dependent decrease in cytochrome $P_{450}$ and an inhibition of cytochrome $P_{450}$-mediated metabolism in liver microsomes of $p$-nitroanisole and aminopyrine (11). Lowered the cytochrome $P_{450}$ concentration in rat liver microsomes following intragastric administration (8).

## Legislation
Included in Schedule 6 (Release into Land: Prescribed Substances) Statutory Instrument No. 472, 1991 (12).

## References
1. *Toxicol. Appl. Pharmacol.* 1978, **45**, 429
2. Toth, B. et al *J. Natl. Cancer Inst.* 1978, **60**, 201-204
3. Toth, B. et al *Neoplasma* 1983, **30**(4), 437-441
4. Nagel, D. et al *Proc. Am. Assoc. Cancer Res.* 1976, **17**, 76
5. Nagel, D. et al *Cancer Res.* 1977, **37**, 3458-3460
6. von Wright, A. et al *Toxicol. Lett.* 1978, **2**, 261-265
7. von Wright, A. et al *Mutat. Res.* 1978, **54**, 167-173
8. Braun, R. et al *Xenobiotica* 1980, **10**, 557-564
9. Mori, H. et al *Jpn. J. Cancer Res.* 1988, **79**, 204-211
10. Braun, R. et al *Toxicology* 1979, **13**, 187-196
11. Braun, R. et al *Toxicology* 1979, **12**, 155-163
12. *S. I. 1991 No. 472 The Environmental Protection (Prescribed Processes and Substances) Regulations* 1991, HMSO, London

# M212  2-Methylfuran

CAS Registry No. 534-22-5
Synonyms α-methylfuran; 5-methylfuran; Sylvan; Silvan
Mol. Formula $C_5H_6O$                                    Mol. Wt. 82.10
Uses Chemical intermediate.

## Physical properties

M. Pt. –88.7°C; B. Pt. 63°C; Flash point –30°C; Specific gravity $d_4^{20}$ 0.914;
Partition coefficient log $P_{ow}$ 0.85; Volatility v.p. 139 mmHg at 20°C; v. den. 2.8.

Solubility
Water: 3 g $l^{-1}$ at 20°C.

## Occupational exposure

UN No. 2301; HAZCHEM Code 3YE; Conveyance classification flammable liquid.

## Ecotoxicity

Invertebrate toxicity
LOEC reproduction (semi chronic) *Scenedesmus quadricauda* 40 mg $l^{-1}$ (1).
LOEC reproduction (semi chronic) *Microcystis aeruginosa* 40 mg $l^{-1}$ (2).
LOEC reproduction (semi chronic) *Uronema parduczi* 26 mg $l^{-1}$ (2).

## Environmental fate

Anaerobic effects
Disappearance in sulfate reducing and methanogenic aquifer slurries was measured
under anaerobic conditions. Sulfate reducing aquifer slurries 112, 29, 34% remaining
after 1, 3, 8 month respectively, methanogenic aquifer slurries 111, 106, 102 % of
substrate remaining after 1, 3, 8 months respectively (3).

## Mammalian and avian toxicity

Acute data
$LD_{50}$ oral rat 167 mg $kg^{-1}$ (4).
$LD_{50}$ oral redwing blackbird 98 mg $kg^{-1}$ (5).
$LC_{Lo}$ (4 hr) inhalation rat 377 ppm (4).
$LC_{50}$ (2 hr) inhalation rat 10 g $m^{-3}$ (6).

## Genotoxicity

*In vitro* Chinese hamster ovary cells without metabolic activation high frequency of
chromatid breaks and sister chromatid exchanges (7).

## Any other comments

Inhibited the mutagenicity of known mutagens Trp-P-1, B[*a*]P and 2-aminofluorene
towards *Salmonella typhimurium* TA98, TA100, with metabolic activation (8).

All reasonable efforts have been taken to find information on isomers of this compound, but no relevant data are available.

## References

1. Bringmann, G. et al *Water Res.* 1980, **14**, 231-41
2. Bringmann, G. et al *GWF, Gas-Wasserfach: Wasser/Abwasser* 1976, **117**(9)
3. Kuhn, E. P. et al *Environ. Toxicol. Chem.* 1989, **8**, 1149-1158
4. Marhold, J. V. *Sbornik Vysledku Toxixologickeho Vysetreni Latek A Pripravku* 1972, Prague
5. Schafer, E. W. et al *Arch. Environ. Contam. Toxicol.* 1983, **12**, 355-382
6. Izmerov, N. F. et al *Toxicometric Parameters of Industrial Toxic Chemicals under Single Exposure* 1982, CIP, Moscow
7. Stich, F. H. et al *Cancer Lett. (Shannon, Irel.)* 1981, **13**, 89-95
8. Kong, Z. L. et al *Agric. Biol. Chem.* 1989, **53**(8), 2073-2079

# M213   Methylheptenone

**CAS Registry No.** 409-02-9
**Synonyms** heptenone, methyl-
**Mol. Formula** $C_8H_{14}O$                        **Mol. Wt.** 126.20
**Uses** In perfumery.
**Occurrence** Occurs as 5-hepten-2-one, 6-methyl- in a variety of fruits, flowers and leaves.

## Physical properties

**M. Pt.** −67°C; **B. Pt.** 173-174°C; **Flash point** 50°C.

**Solubility**
Organic solvent: ethanol, diethyl ether, chloroform

## Environmental fate

**Degradation studies**
Epoxidation by fungi can occur (1).

## Mammalian and avian toxicity

**Acute data**
$LD_{50}$ oral rat 3.5 g kg$^{-1}$ (2).

## References

1. Abraham, W. R. *J. Essent. Oil Res.* 1990, **2**(5), 251-7
2. *Food Cosmet. Toxicol.* 1975, **13**, 859

# M214   6-Methyl-5-hepten-2-one

$(CH_3)_2C=CHCH_2CH_2C(O)CH_3$

**CAS Registry No.** 110-93-0
**Synonyms** 2-methyl-2-hepten-6-one; 2-methyl-6-oxo-2-heptene;
2-oxo-6-methylhept-5-ene; 6-methyl-$\Delta^5$-hepten-2-one; Sulcatone
**Mol. Formula** $C_8H_{14}O$                    **Mol. Wt.** 126.20

## Physical properties

**B. Pt.** 173-174°C **Flash point** 57°C (closed cup); **Specific gravity** 0.855;
**Volatility** v. den. 4.4.

## Ecotoxicity

**Fish toxicity**
Trout, bluegill sunfish, yellow perch and goldfish (24 hr) 5 ppm non-toxic. Test
conditions: pH, 7.0; dissolved oxygen, 7.5 ppm; total hardness (soap method), 300
ppm; methyl orange alkalinity, 310 ppm; phenolphthalein alkalinity, 0; free carbon
dioxide, 5 ppm; and temperature 12.8°C. (1)

**Invertebrate toxicity**
$EC_{50}$ (5 min) *Photobacterium phosphoreum* 17.4 ppm Microtox test (2).

## Mammalian and avian toxicity

**Acute data**
$LD_{50}$ (estimated) oral redwing blackbird >111 mg $kg^{-1}$ (3).
$LD_{50}$ intragastic mouse, rat 2.41, 4.25 g $kg^{-1}$ respectively (4).

**Irritancy**
Skin irritation in guinea pigs, mice and rabbits (dose and duration unspecified) (4).

**Sensitisation**
Caused dermatitis in rabbits, mice and guinea pigs (3).

## Any other comments

Recommended maximum permissible concentration for occupational inhalation
exposure, 6.75 mg $m^{-3}$.

## References

1.  US EPA *The Toxicity of 3400 Chemicals to Fish* 1987, EPA 560/6-87-002, US EPA,
    Washington DC
2.  Kaiser, K. L. E. *Water Pollut. Res. J. Can.* 1991, **26**(3), 361-431
3.  Schafer, E. W. et al *Arch. Environ. Contam. Toxicol.* 1983, **12**, 355-382
4.  Migukina, N. V. et al *Gig. Tr. Prof. Zabol.* 1989, (8), 52-53 (Russ.) (*Chem. Abstr.* **112**,
    50126x).

# M215   5-Methylhexan-2-one

## $CH_3CH(CH_3)CH_2CH_2C(O)CH_3$

**CAS Registry No.** 110-12-3
**Synonyms** isoamyl methyl ketone; isopentyl methyl ketone; methyl isoamyl ketone; methyl isopentyl ketone
**Mol. Formula** $C_7H_{14}O$                               **Mol. Wt.** 114.19

## Physical properties

**B. Pt.** 144°C; **Flash point** 35.98°C; **Specific gravity** $d_4^{20}$ 0.888; **Partition coefficient** log $P_{ow}$ 1.88.

### Solubility
Water: 5.4 g $l^{-1}$. Organic solvent: miscible with most organic solvents

## Occupational exposure

**US TLV (TWA)** 50 ppm (234 mg $m^{-3}$); **UK Long-term limit** 50 ppm (240 mg $m^{-3}$); **UK Short-term limit** 75 ppm (360 mg $m^{-3}$); **UN No.** 2302; **HAZCHEM Code** 3Y; **Conveyance classification** flammable liquid.
**Risk phrases** Flammable (R10)
**Safety phrases** Do not breathe vapour (S23)

## Ecotoxicity

### Fish toxicity
$LC_{50}$ (96 hr) fathead minnow 159 mg $l^{-1}$ (1).

### Invertebrate toxicity
LOEC reproduction *Microcystis aeruginosa* 90 mg $l^{-1}$ (2).
Toxicity threshold (cell multiplication inhibition test) *Pseudomonas putida, Scenedesmus quadricauda, Entosiphon sulcatum, and Uronema parduczi Chatton-Lwoff* 115-980 mg $l^{-1}$ (3,4).
$EC_{50}$ (5 min) *Photobacterium phosphoreum* 972-1438 ppm Microtox test (5).

### Abiotic removal
Activated carbon adsorbability (0.169 g $g^{-1}$ carbon), influent contained 986 mg $l^{-1}$ effluent contained 146 mg $l^{-1}$ a 85.2% reduction (6).

## Mammalian and avian toxicity

### Acute data
$LD_{50}$ oral rat 4.8 g $kg^{-1}$ (7,8).
$LC_{50}$ (4 hr) inhalation rat 4000 ppm (7).
$LD_{50}$ dermal rabbit 10 g $kg^{-1}$ (9).

## References

1. Veith, G. D. et al in *Aquatic Toxicology and Hazard Assessment: 6th Symposium* 1983, ASTM STP803, American Society of Testing Materials, Philadelphia, PA
2. Bringman, G. et al *GWF Gas-Wasserfach: Wasser/Abwasser* 1976, **117**(9) (Ger.)
3. Bringman, G. et al *Z. Wasser/Abwasser Forsch.* 1980, (1), 26-31
4. Bringman, G. et al *Water Res.* 1980, **14**(1), 231-241

5. Curtis, C. A. et al *Aquatic Toxicology and Hazard Assessment: 5th Conference* 1982, 170-178, J. G. Pearson Ed., ASTM STP 766, American Society for Testing and Materials, Philadelphia, PA
6. Guisti, D. M. et al *J. -Water Pollut. Control Fed.* 1971, **43**(8), 1716
7. *Am. Ind. Hyg. Assoc. J.* 1962, **23**, 95
8. Kennedy, G. L., et al *Toxicol. Lett.* 1991, **56**(3), 317-326
9. *Union Carbide Data Sheet* 7/8/63, Union Carbide Corp., New York

# M216  Methylhydrazine

## NH₂NHCH₃

**CAS Registry No.** 60-34-4
**Synonyms** Monomethylhydrazine; MMH
**Mol. Formula** $CH_6N_2$                                    **Mol. Wt.** 46.07
**Uses** Intermediate in chemical synthesis. In rocket fuel. Solvent.
**Occurrence** In the edible mushroom, *Gyromitia esculenta*.

## Physical properties

**M. Pt.** –52.4°C; **B. Pt.** 87°C; **Flash point** 21°C (closed cup); **Specific gravity** $d^{25}$ 0.874; **Volatility** v. den. 1.6.

### Solubility
Water: miscible. Organic solvent: ethanol, petroleum ether, diethyl ether

## Occupational exposure

**US TLV (TWA)** 0.01 ppm (0.019 mg m$^{-3}$) intended change; **UN No.** 1244;
**HAZCHEM Code** 2WE; **Conveyance classification** toxic substance, corrosive substance, flammable liquid.

## Ecotoxicity

### Fish toxicity
$LC_{50}$ (96 hr) guppy 2.58 mg l$^{-1}$ (1).

### Invertebrate toxicity
$EC_{50}$ (5, 10 or 15 min) *Photobacterium phosphoreum* 15.3 ppm Microtox test (2).

## Environmental fate

### Nitrification inhibition
*Nitromonas* sp., denitrifying bacteria, and *Nitrobacter* sp. 50% inhibition 1.4, 4.3, 12.3 mg l$^{-1}$ respectively (3).

### Degradation studies
*Achromobacter* sp., *Pseudomonas* sp. in culture media and soil samples could not utilise methylhydrazine as a sole carbon source, and did not promote degradation to its final oxidation products of carbon dioxide and water (4).

---

## Mammalian and avian toxicity

### Acute data

$LD_{50}$ oral mouse, rat 33, 70 mg $kg^{-1}$, respectively (5,6).

$LC_{50}$ (1-4 hr) inhalation rat, mouse, dog, 34-96 ppm (7–9).

$LC_{50}$ (1 hr) inhalation Rhesus monkey 162 ppm (10).

Intraperitoneal rat 25-60 mg $kg^{-1}$ caused proximal tubule damage in kidneys (11).

$LD_{50}$ dermal rabbit, rat 95, 183 mg $kg^{-1}$, respectively (12,13).

$LD_{50}$ subcutaneous rat 25 mg $kg^{-1}$ (14).

$LD_{50}$ intravenous rat, mouse 17, 33 mg $kg^{-1}$, respectively (13,15).

$LD_{50}$ intraperitoneal rat, mouse 15, 21 mg $kg^{-1}$, respectively (13,16).

### Carcinogenicity and long-term effects

Oral (lifetime study) Swiss mice 0.01% in drinking water. Enhanced the development of lung tumours by shortening their latent period (17).

## Genotoxicity

*Escherichia coli* PQ37 with and without metabolic activation SOS chromotest negative (18).

*Salmonella typhimurium* TA100, TA98, TA1535, TA1537, TA1538 with and without metabolic activation negative (19).

*Salmonella typhimurium* TA100 positive for spot test, negative for plate test with and without metabolic activation and host-mediated assay (20).

## Any other adverse effects to man

Inhalation (10 min) ♂ 90 ppm, primary effect lachrymation and bronchospasms although these were not excessive. Clinical chemistry was normal, the only haematologic abnormality was Heinz body formation in 3-5% of erythrocytes on day 7 post-exposure, which disappeared by day 60 (10).

## Any other comments

Human health effects, experimental toxicology and workplace experience reviewed (21).

Strong reducing agent. Will ignite spontaneously in contact with strong oxidising agents. Autoignition temperature 194°C.

## References

1. Slonium, A. R. *Water Res.* 1977, **11**, 889-895
2. Kaiser, K. L. E. et al *Water Pollut. Res. J. Can.* 1991, **26**(3), 361-431
3. Kane, D. A. et al *Arch. Environ. Contam. Toxicol.* 1983, 447-453
4. Ou, T. L. *Bull. Environ. Contam. Toxicol.* 1988, **41**(6), 851-857
5. Witkin *Arch. Ind. Health* 1956, **13**, 34
6. Gregory, et al *Clin. Toxicol.* 1971, **4**, 435
7. *Aerospace Medical Research Laboratory Report* 1967, TR-67-137, Air Force Systems Command, Wright-Patterson Air Force Base, OH
8. *AMA Arch. Ind. Hyg. Assoc.* 1970, **31**, 667
9. *Am. Ind. Hyg. Assoc.* 1970, **31**, 667
10. *Patty's Industrial Hygiene and Toxicology* 3rd ed., 1981, 2A, John Wiley & Sons, New York
11. Hong, W. et al *Zhongguo Yaolixue Yu Dulixe Zazhi* 1987, **1**(5), 366-370 (Ch.) (*Chem. Abstr.* **108**, 144927d)

12. *Proc. Soc. Exp. Biol. Med.* 1969, **131**, 226
13. *Clin. Toxicol* 1971, **435**, 4
14. *Br. J. Cancer* 1974, **30**, 429
15. *Med. Prac.* 1973, **24**, 71
16. *Proc. Soc. Exp. Biol. Med.* 1967, **124**, 172
17. Toth, B. *Int. J. Cancer* 1972, **9**, 109-118
18. Von der Hude, W. et al *Mutat. Res.* 1988, **203**(2), 81-84
19. Mortelmans, K. et al *Environ. Mutagen.* 1986, **7**, 1-119
20. von Wright, A. et al *Mutat. Res.* 1978, **54**, 167-173
21. *ECETOC Technical Report No. 30(4)* 1991, European Chemical Industry Ecology and Toxicology Centre, B-1160 Brussels

# M217   Methylhydrazine sulfate

## CH₃NHNH₂·H₂SO₄

**CAS Registry No.** 302-15-8
**Synonyms** methyl hydrazine monosulfate
**Mol. Formula** $CH_8N_2O_4S$                 **Mol. Wt.** 144.15

## Physical properties
**M. Pt.** 142°C.

**Solubility**
Organic solvent: slightly soluble ethanol

## Mammalian and avian toxicity

**Acute data**
$LD_{50}$ intraperitoneal mouse 160 mg $kg^{-1}$ (1).

**Carcinogenicity and long-term effects**
Oral (lifetime study) Swiss mice 0.01% in drinking water. 46% developed lung tumours, 20% of ♀ and 16% of ♂ developed malignant lymphomas and 6% of ♀ developed breast tumours (2).

## Genotoxicity
*Salmonella typhimurium* TA100, TA1537, TA98 with and without metabolic activation negative (3).
*In vitro* rat, mouse hepatocyte/DNA repair test positive (4).

## References
1. *Russ. Pharmacol. Toxicol.* 1973, **36**, 27
2. Toth, B. *Int. J. Cancer* 1972, **9**, 109-118
3. Shimizu, H. et al *Jpn. J. Hyg.* 1978, **33**, 474-485 (Japan.)
4. Mori, H. et al *Jpn. J. Cancer. Res.* 1988, **79**(2), 204-211

# M218　Methyl *p*-hydroxybenzoate

CO$_2$CH$_3$

OH

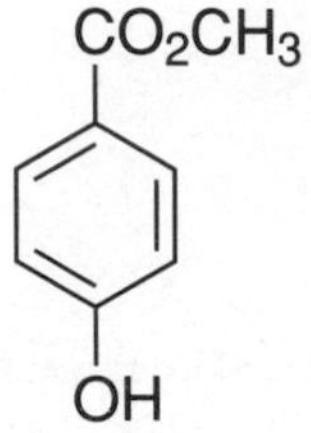

**CAS Registry No.** 99-76-3
**Synonyms** methylparaben; 4-hydroxybenzoic acid, methyl ester; Nipagin M;
Tegosept M; Methyl Chemosept; Methyl Parasept
**Mol. Formula** C$_8$H$_8$O$_3$　　　　　　　　　　　　**Mol. Wt.** 152.15
**Uses** Preservative in foods, beverages and cosmetics. Has antibacterial and antifungal
properties.
**Occurrence** In several animal species as a volatile aromatic component of scent (1,2).

## Physical properties
**M. Pt.** 131°C; **B. Pt.** 270-280°C (decomp.).

**Solubility**
Water: 2.5 g l$^{-1}$. Organic solvent: ethanol, acetone, diethyl ether

## Ecotoxicity

**Invertebrate toxicity**
EC$_{50}$ (30 min) *Photobacterium phosphoreum* 6.34 ppm Microtox test (3).

## Environmental fate

**Degradation studies**
*Pseudomonas cepacia* can grow in the presence of methyl *p*-hydroxybenzoate and
can contribute to its degradation (4).

**Mammalian and avian toxicity**

**Acute data**
LD$_{50}$ oral guinea pig 3 g kg$^{-1}$ (5).
LD$_{50}$ oral dog 3 g kg$^{-1}$ (6).
LD$_{50}$ intraperitoneal mouse 960 mg kg$^{-1}$ (7).
LD$_{50}$ subcutaneous mouse 1200 mg kg$^{-1}$ (8).

**Metabolism and pharmacokinetics**
Can be absorbed from mammalian intestines and through skin and body of goldfish (9).
Present in an injectable preparation of gentamicin and was excreted in the urine of
preterm infants, following intramuscular injection, mainly in the conjugated form.
*p*-Hydroxybenzoic acid  was detected as the urinary metabolite (10).

**Irritancy**
Can cause stinging of skin, particularly the face, possibly through calcium channel
activation (11).

## Genotoxicity

*Salmonella typhimurium* TA100, TA1537, TA98 and TA1535 with and without metabolic activation negative (12).
Induced chromosomal aberrations in Chinese hamster cells without, but not with, metabolic activation (13).

## Any other adverse effects to man

Hypersensitivity reactions occur with hydroxybenzoates, generally delayed reaction contact dermatitis. Immediate reactions with urticaria and bronchospasm have been reported. Activity can be adversely affected by the presence of other excipients or active ingredients (14-16).

## Any other comments

Human health effects, experimental toxicology, physico-chemical properties reviewed (17,18).
Contact dermatitis in humans reviewed (19-22).
All reasonable efforts have been taken to find information on isomers of this compound, but no relevant data are available.

## References

1. Goodwin, M. et al *Science (Washington, D. C., 1883-)* 1979, **203**, 559
2. Nishida, R. et al *Appl. Entomol. Zool.* 1990, **25**(1), 105-112
3. Kaiser, K. L. E. et al *Water Pollut. Res. J. Can.* 1991, **26**(3), 361
4. Suemitsu, R. *Bokin Bobai* 1990, **18**(12), 579-582
5. *Food Agric. Organisation* 1967, **40**, 23
6. *Patty's Industrial Hygiene and Toxicology* 3rd rev. ed., 1981, **2A,2B**, John Wiley & Sons, New York
7. *J. Am. Pharm. Assoc.* 1956, **45**, 260
8. *Arch. Int. Pharmacodyn. Ther.* 1960, **128**, 135
9. Sakiya, Y. et al *Int. J. Pharm.* 1988, **47**(1-3), 185-196
10. Hindmarsh, K. W. *J. Pharm. Sci.* 1983, **72**, 1039-1041
11. Sone, T. et al *Nippon Koshohin Kagakkaishi* 1990, **14**(1), 8-16 (Japan.) (*Chem. Abstr.* **113**, 197637u)
12. Prival, M. J. *Food Chem. Toxicol.* 1982, **20**, 427-432
13. Matsuoka, A. et al *Mutat. Res.* 1979, **66**, 277-290
14. Michaelsson, G. et al *Br. J. Dermatol.* 1973, **88**, 525-532
15. Warin, R. P. et al *Br. J. Dermatol.* 1976, **94**, 401-406
16. Kaminer, Y. et al *Clin. Pharm.* 1982, **1**, 469-470
17. *ECETOC Technical Report No. 30(4)* 1991, European Chemical Industry Ecology and Toxicology Centre, B-1160 Brussels
18. *WHO Tech. Rep. Ser.* 1974, 539, FAO/WHO Expert Committee on Food Additives
19. *Arch. Dermatol.* 1973, **108**, 537-540
20. Moole, J. *J. Am. Coll. Toxicol.* 1984, **3**, 147-209
21. Fisher, A. A. *J. Am. Acad. Dermatol.* 1982, **6**, 116-117
22. Lederman, D. A. *Oral Surg.* 1980, **49**, 28-33

# M219   1-Methylimidazole

CAS Registry No. 616-47-7
Synonyms 1-methyl-1*H*-imidazole
Mol. Formula $C_4H_6N_2$                           Mol. Wt. 82.11

## Physical properties

M. Pt. –60°C; B. Pt. 198°C; Flash point 92°C (closed cup); Specific gravity 1.030.

## Occupational exposure

Supply classification corrosive.
Risk phrases Harmful in contact with skin and if swallowed – Causes burns
(R21/22, R34)
Safety phrases In case of contact with eyes, rinse immediately with plenty of water
and seek medical advice – Wear suitable protective clothing (S26, S36)

## Mammalian and avian toxicity

Acute data
$LD_{50}$ oral mouse 1400 mg $kg^{-1}$ (1).
$LD_{50}$ intraperitoneal mouse 380 mg $kg^{-1}$ (1).

## Any other comments

Human health effects, experimental toxicology, physico-chemical properties
reviewed (2).

## References

1.  *Toxicol. Appl. Pharmacol.* 1969, **14**, 301
2.  *ECETOC Technical Report No. 30(4)* 1991, European Chemical Industry Ecology and
    Toxicology Centre, B-1160 Brussels

# M220   2,2′-(Methylimino)diethanol

$CH_3N(CH_2CH_2OH)_2$

CAS Registry No. 105-59-9
Synonyms 2,2′-(methylimino)bisethanol; diethanol methylamine; MDEA;
methylenediethanolamine; methyliminodiethanol; bis(2-hydroxyethyl)methylamine
Mol. Formula $C_5H_{13}NO_2$                           Mol. Wt. 119.16

## Physical properties
**B. Pt.** $_{747}$ 246-248°C; **Specific gravity** $d^{20}$ 1.0377.

## Occupational exposure
**Supply classification** irritant.
**Risk phrases** Irritating to eyes (R36)
**Safety phrases** Avoid contact with skin (S24)

## Environmental fate
### Degradation studies
Analysis of partially degraded aqueous 2,2′-(methylimino)diethanol solutions, most important degradation products were methanol; ethylene oxide; trimethylamine; ethylene glycol; 2-(dimethylamino)ethanol; 1,4-dimethylpiperazine; triethanolamine; N-(hydroxyethyl)- methylpiperazine; and N,N-bis(hydroxyethyl)piperazine (1).

## Mammalian and avian toxicity
### Acute data
$LD_{50}$ oral mouse, rat 500, 4780 mg $kg^{-1}$, respectively (2,3).

### Irritancy
Dermal rabbit (24 hr) 10 and 502 mg caused mild irritation (2,4).

## Genotoxicity
*Salmonella typhimurium* TA98, TA100, TA1535, TA1537 with and without metabolic activation negative (5).

## References
1. Chakma, A. et al *J. Chromatogr.* 1988, **457**, 287-297
2. *Arch. Ind. Health* 1954, **10**, 61
3. *NTIS Report* AD277-689, Natl. Tech. Inf. Ser., Springfield, V.A
4. *Union Carbide Data Sheet.* 13/7/71, Union Carbide Corp., New York
5. Zeiger, E. et al *Environ. Mol. Mutagen.* 1987, **9** (Suppl.9), 1-109

# M221    5-Methylindan

**CAS Registry No.** 874-35-1
**Synonyms** 1*H*-indene, 2,3-dihydro-5-methyl-
**Mol. Formula** $C_{10}H_{12}$                    **Mol. Wt.** 132.21

## Ecotoxicity
### Invertebrate toxicity
*Phaeodactylum tricornutum* (24 hr) 1-2 mg $l^{-1}$ reduced photosynthesis by 50%. 300 µg $l^{-1}$ modified the feeding of *Trigriopus brevicornis* (1).

## Legislation

Included in Schedule 4 (Release into Air: Prescribed Substances) Statutory Instrument No. 472, 1991 (2).

## Any other comments

All reasonable efforts have been taken to find information on isomers of this compound, but no relevant data are available.

## References

1. Lacaze, J. C. et al *Sci. Eau* 1987, **6**(4), 415-433
2. *S. I. 1991 No. 472 The Environmental Protection (Prescribed Processes and Substances) Regulations* 1991, HMSO, London

# M222   3-Methylindole

**CAS Registry No.** 83-34-1
**Synonyms** 1*H*-indole, 3-methyl-; β-methylindole; Scatole; Skatol; Skatole
**Mol. Formula** $C_9H_9N$                    **Mol. Wt.** 131.18
**Occurrence** Present in cigarette smoke. Constituent of faeces, beetroot, nectandra woodland and coal tar.

## Physical properties

**M. Pt.** 95°C; **B. Pt.** $_{755}$ 265-266°C; **Partition coefficient** log $P_{ow}$ 2.60.

**Solubility**
Organic solvent: ethanol, chloroform, diethyl ether and benzene

## Environmental fate

**Nitrification inhibition**
At 7 mg $l^{-1}$ 75% inhibition of ammonia oxidation (activated sludge) (1).

## Mammalian and avian toxicity

**Acute data**

$LD_{50}$ oral rat 3450 mg $kg^{-1}$ (2).
$LD_{Lo}$ oral mouse 470 mg $kg^{-1}$ (3).
$LD_{Lo}$ oral cattle 200 mg $kg^{-1}$ (2).
$LD_{Lo}$ intravenous cattle 60 mg $kg^{-1}$ (2).
$LD_{50}$ intraperitoneal mouse 175 mg $kg^{-1}$ (2).
Intraperitoneal (24 hr) rat 50-300 mg $kg^{-1}$. A dose-dependent decrease in splenic weight 24-75% and nucleated splenic cell number 22-68% was observed (4).

**Sub-acute data**

Intraperitoneal (28 day) mice 400 mg kg$^{-1}$, cellular swelling was apparent in the olfactory epithelium by 6 hr. Necrosis of the olfactory epithelium and subepithelial glands were diffuse by 48 hr. Subsequent ulceration resulted in epithelial hyperplasia, squamous metaplasia, fibroplasia and ossification (5).

Infusion (72 hr) goats 35 mg kg$^{-1}$ body weight, prostaglandin concentrations in lungs were unaffected. Plasma and lung prostaglandin and thromboxane B$_2$ concentrations did not appear to be altered in 3-methylindole-induced lung disease (6).

**Metabolism and pharmacokinetics**

It requires activation by cytochrome P$_{450}$ to be cytotoxic in rabbit pulmonary cells (7).

## Genotoxicity

*Salmonella typhimurium* TA100 without metabolic activation, strong mutagenic activity (8).

## Any other adverse effects

A dose of 550 mg kg$^{-1}$ specifically damaged pulmonary tissue in Swiss-Webster mice without causing any hepatic or renal necrosis. When a glutathione depleter was administered to mice 3 hr before a low dose of 3-methylindene (75 mg kg$^{-1}$), histopathalogical examination after 4 hr showed renal damage. The production of a toxic metabolite in the livers of glutathione-depleted mice that is circulated to susceptible renal cells may be the mechanism of this toxicity (9).

The electrophilic imine methide may be the intermediate which binds with and depletes glutathione. An imine methide is the primary reactive intermediate in 3-methylindole mediated pneumotoxicity (10).

## Legislation

Included in Schedule 4 (Release into Air: Prescribed Substances) Statutory Instrument No. 472, 1991 (11).

## Any other comments

Toxicity reviewed (12-14).

It is a pneumotoxic metabolite of L-tryptophan that forms in the digestive tract of humans and ruminants (15).

Soluble in hot water.

All reasonable efforts have been taken to find information on isomers of this compound, but no relevant data are available.

## References

1. Richardson, M. *Nitrification Inhibition in the Treatment of Sewage* 1985, RSC, London
2. *Food Cosmet. Toxicol.* 1976, **14**, 863
3. *Arch. Environ. Contam. Toxicol.* 1985, **14**, 111
4. Updyke, L. W. et al *Toxicol. Appl. Pharmacol.* 1991, **109**(3), 379-390
5. Turk, M. A. M. et al *Vet. Pathol.* 1987, **24**(5), 400-403
6. Acton, K. S. et al *Comp. Biochem. Physiol., A: Comp. Physiol.* 1989, **94A**(4), 677-681
7. Appleton, M. L. et al *Adv. Exp. Med. Biol.* 1991, **283** (Biol. React. Intermed. 4), 245-248
8. Nagao, M. et al *Proc. Int. Symp. Princess Takamatsu Cancer Res. Fund 1985* 1986, **16**, 77-86
9. Yost, G. S. et al *Toxicol. Appl. Pharmacol.* 1990, **103**(1), 40-51
10. Huijzer, J. C. et al *Toxicol. Appl. Pharmacol.* 1987, **90**(1), 60-68

11. *S. I. 1991 No. 472 The Environmental Protection (Prescribed Processes and Substances) Regulations* 1991, HMSO, London

12. Yost, G. S. *Chem. Res. Toxicol.* 1989, **2**(5), 273-279

13. Carlson, J. R. et al *J. Toxicol., Toxin Rev.* 1986, **5**(2), 217-227

14. *Dangerous Prop. Ind. Mater. Rep.* 1987, **7**(6), 84-87

15. Updyke, L. W. et al *Toxicol. Appl. Pharmacol.* 1991, **109**(3), 391-398

# M223 Methyl isobutyrate

## $CH_3CH(CH_3)CO_2CH_3$

**CAS Registry No.** 547-63-7

**Synonyms** propanoic acid, 2-methyl-, methyl ester; 2-methylpropanoic acid, methyl ester; isobutyric acid, methyl ester

**Mol. Formula** $C_5H_{10}O_2$          **Mol. Wt.** 102.13

**Uses** Chemical intermediate.

**Occurrence** In aromas of fruits such as blackberry and herbs such as dill. Pollutant in drinking water (1).

## Physical properties

**M. Pt.** -84 to -85°C; **B. Pt.** 93°C; **Flash point** 3°C; **Specific gravity** $d^{20}$ 0.891.

**Solubility**

Organic solvent: miscible ethanol, diethyl ether

## Mammalian and avian toxicity

**Acute data**

$LD_{50}$ oral rat 16 g $kg^{-1}$ (2).

$LD_{50}$ intraperitoneal rat 3.2 g $kg^{-1}$ (2).

## References

1. Li, Z. et al *Xi'an Yike Daxue Xuebao* 1988, **9**(1), 53-8 (Ch.) (*Chem. Abstr.* **114**, 176718e)

2. *Patty's Industrial Hygiene and Toxicology* 3rd rev. ed., 1981, **2A,2B**, John Wiley & Sons, New York

# M224 Methyl isocyanate

## $CH_3NCO$

**CAS Registry No.** 624-83-9

**Synonyms** isocyanatomethane; isocyanic acid, methyl ester; MIC

**Mol. Formula** $C_2H_3NO$          **Mol. Wt.** 57.05

**Uses** Chemical intermediate in syntheses particularly of insecticides and herbicides.

## Physical properties

**M. Pt.** –17°C; **B. Pt.** 37-39°C; **Flash point** –7°C; **Specific gravity** $d^{20}$ 0.967;

**Volatility** v.p. 400 mmHg at 20.6°C.

## Occupational exposure

**US TLV (TWA)** 0.02 ppm (0.047 mg m$^{-3}$); **UK Long-term limit** MEL 0.02 mg m$^{-3}$ (as –NCO); **UK Short-term limit** MEL 0.07 mg m$^{-3}$ (as –NCO); **UN No.** 2480; **Conveyance classification** toxic substance, flammable liquid; **Supply classification** extremely flammable, toxic.

**Risk phrases** Extremely flammable – Toxic by inhalation, in contact with skin and if swallowed – Irritating to eyes, respiratory system and skin (R12, R23/24/25, R36/37/38)

**Safety phrases** Keep container in a well ventilated place – Never add water to this product – In case of fire, use dry powder or carbon dioxide – If you feel unwell, seek medical advice (show label where possible) (S9, S30, S43, S44)

## Environmental fate

### Nitrification inhibition

Examination of forest nursery soils demonstrated reduced ability of bacterial and fungal populations as ammonifiers and denitrifiers after exposure during the 1984 Bhopal accident (1).

## Mammalian and avian toxicity

### Acute data

LD$_{50}$ oral rat 51.5 mg kg$^{-1}$ (2).
LC$_{50}$ (6 hr) inhalation guinea pig, rat 5400, 6100 ppb, respectively (3).
LD$_{50}$ subcutaneous rat 261 mg kg$^{-1}$ (2).
LD$_{50}$ subcutaneous mouse 82-85 mg kg$^{-1}$ (2).
Absence of a cyanide-like action in acute toxicity has been established (2).

### Carcinogenicity and long-term effects

A battery carcinogen prediction model has indicated a significant potential for carcinogenicity in rodents, but judged the potency of the compound to be low (4).

### Teratogenicity and reproductive effects

Pregnant mice exposed to 9-15 ppm for 3 hr demonstrated >75% complete resorption (5).
Foetal toxicity in mice and rats is thought to be partly independent of maternal toxicity and thought to result from actions of the compound on foetal tissue (6).

### Metabolism and pharmacokinetics

Guinea pigs inhaling $^{14}$C-methyl isocyanate for 1-6 hr at concentrations of 0.5-15 ppm, retained some of the compound in the upper respiratory passage. Clearance of $^{14}$C from tissues was gradual over 3 days with quantities in bile and urine being parallel with concentrations in blood. In those animals that were pregnant, the compound passed into the foetuses (7).
The compound can be absorbed through the skin (8).

### Irritancy

Sensory and pulmonary irritation have been established in mouse and guinea pig (9). In humans lachrymation and irritation of the throat are produced along with skin irritation (8,10).

### Sensitisation

Methyl isocyanate causes sensitisation (species unspecified) (8).

## Genotoxicity

*Salmonella typhimurium* with and without metabolic activation negative (11).
*Drosophila melanogaster* sex-linked recessive lethal assay negative (11).
*In vitro* Chinese hamster ovary cells with and without metabolic activation, sister chromatid exchanges positive and chromosomal aberrations positive (11).
Mouse micronucleus test weak positive (12).
Sister chromatid exchanges *in vivo* mouse, weak positive (13).

## Any other adverse effects to man

The major effect on Bhopal survivors was lung damage. Inhalation initially caused vomiting and coughing with damage to lungs resulting in oedema, permanent fibrosis, emphysema and bronchitis. Lingering respiratory illness has been the main long-term effect on survivors of the accident, with some evidence of neuromuscular dysfunction (14-16).
Of women who were pregnant and living near Bhopal at the time of the 1984 accident, 43% failed to give rise to live births (5).

## Any other comments

Human health effects, experimental toxicology, physico-chemical properties reviewed (17).
The genotoxicity of the compound has been reviewed (18).
An industrial accident in Bhopal, India in 1984 at a chemical plant manufacturing carbaryl, resulted in over 2000 people dying and approximately 200,000 people being exposed to the vapour. The original incident has been well documented (19,20) and follow-up studies continue to be reported (1,5,14,15,16,21).

## References

1. Mohammed, G. et al *J. Trop. For.* 1986, **2**(2), 168-169
2. Vijayarahavan, R. et al *Indian J. Exp. Biol.* 1987, **25**(8), 531-534
3. *Fundam. Appl. Toxicol.* 1986, **6**, 747
4. Ennever, F. K. *Toxicol. Appl. Pharmacol.* 1987, **91**(3), 502-505
5. Varma, D. R. *Environ. Health Perspect.* 1987, **72**, 153-157
6. Varma, D. R. et al *J. Toxicol. Environ. Health* 1990, **30**(1), 1-14
7. Ferguson, J. S. *Toxicol. Appl. Pharmacol.* 1988, **94**(1), 104-117
8. Mellon Inst. *Special Report 26-75* to Union Carbide Chem. Co. 1963
9. Alarie, Y. et al *Environ. Health Perspect.* 1987, **72**, 159-167
10. Kimmerle, G. et al *Arch. Toxicol.* 1960, **20**, 235
11. Mason, J. M. et al *Environ. Mutagen.* 1987, **9**, 19-28
12. Tice, R. et al *Environ. Mutagen.* 1987, **9**, 37-58
13. Klingerman, A. D. *Environ. Mutagen.* 1987, **9**, 29-36
14. Anderson, N. et al *Br. J. Ind. Med.* 1988, **45**(7), 469-475
15. *Chem. Eng. News.* 1986, **64**(9), 4
16. *Occup. Saf. Health* 1986, **16**(4), 4
17. *ECETOC Technical Report No. 30(4)* 1991, European Chemical Industry Ecology and Toxicology Centre, B-1160 Brussels
18. Shelby, M. D. et al *Environ. Health Perspect.* 1987, **72**, 183-187
19. *Chem. Eng. News* 1985, **63**, 14
20. *Chem. Eng. News* 1984, **62**, 6
21. *Indian J. Exp. Biol.* 1988, **26**, 149-176

# M225　Methyl isothiocyanate

## CH3NCS

**CAS Registry No.** 556-61-6
**Synonyms** isothiocyanatomethane; methyl mustard oil; Trapex
**Mol. Formula** $C_2H_3NS$　　　　　　　　　**Mol. Wt.** 73.12
**Uses** Pesticide used for soil fumigation.
**Occurrence** Pollutant of air, soil and water.

## Physical properties

**M. Pt.** 35-36°C; **B. Pt.** 119°C; **Flash point** 32°; **Specific gravity** $d^{37}$ 1.069;
**Partition coefficient** log $P_{ow}$ 1.374; **Volatility** v.p. $2.03 \times 10^{-5}$ at 20°C.

## Solubility

Water: 8.2 g $l^{-1}$ at 20°C. Organic solvent: ethanol, diethyl ether

## Occupational exposure

**UN No.** 2477; **HAZCHEM Code** 2XE; **Conveyance classification** flammable
liquid, toxic substance; **Supply classification** toxic.
**Risk phrases** Toxic by inhalation and if swallowed – Causes burns – May cause
sensitisation by skin contact (R23/25, R34, R43)
**Safety phrases** Wear suitable protective clothing and gloves – In case of insufficient
ventilation, wear suitable respiratory equipment – If you feel unwell, seek medical
advice (show label where possible) (S36/37, S38, S44)

## Ecotoxicity

**Fish toxicity**
$LC_{50}$ (96 hr) bluegill sunfish 0.13 mg $l^{-1}$ (1).
$LC_{50}$ (96 hr) rainbow trout, mirror carp 0.37-0.57 mg $l^{-1}$ (1).

## Environmental fate

**Degradation studies**
Disappearance from damp soil by degradation and evaporation is temperature
dependent. Most disappears within 3 wk at 20°C or by 8 wk at 0°C (1).

**Abiotic removal**
Rapidly hydrolysed by alkalis; sensitive to oxygen and sunlight (1).

## Mammalian and avian toxicity

**Acute data**
$LD_{50}$ oral ♂ rat 175-220 mg $kg^{-1}$ (1, 2).
$LD_{50}$ oral ♂ mouse 90 mg $kg^{-1}$ (1).
$LC_{50}$ (1 hr) inhalation rat 1.3-1.9 mg $l^{-1}$ (1, 2).
$LD_{50}$ dermal mouse 1.87 g $kg^{-1}$ (1).
$LD_{50}$ dermal rabbit 33-263 mg $kg^{-1}$ (1,2).

$LD_{Lo}$ oral human 1 g kg$^{-1}$ with accompanying central nervous system effects (3).
In the rat exposure to vapour for 1 hr causes reversible damage to epithelium of
proximal bronchioles and upper airways (4).
$LD_{50}$ oral mallard duck 136 mg kg$^{-1}$ (1).

### Sub-acute data
$LC_{50}$ (5 day) mallard duck 10.9 g kg$^{-1}$ diet (1).
$LC_{50}$ (5 day) pheasant >5 g kg$^{-1}$ diet (1).

### Carcinogenicity and long-term effects
2 yr feeding trials in the rat established a no observed effect concentration for rat of
10 mg l$^{-1}$ in drinking water and for mouse of 20 mg l$^{-1}$ in drinking water (1).
A 1 yr feeding trial with dogs established a no observed effect concentration of
0.4 mg kg$^{-1}$ daily (1).

### Irritancy
Strong irritant of rabbit skin and eye (1).

## Legislation

Limited under EC Directive on Drinking Water Quality 80/778/EEC. Pesticides:
maximum admissible concentration 0.1 μg l$^{-1}$ (5).
Included in Schedule 6 (Release into Land: Prescribed Substances) Statutory
Instrument No. 472, 1991 (6).

## Any other comments

Human health effects, experimental toxicology, physico-chemical properties
reviewed (7).
Embryos of the clawed frog exposed to methyl isothiocyanate during development
showed disturbances of collagen formation and of the notochordal sheath (8).
Not toxic to bees when used as directed (1).

## References

1. *The Agrochemicals Handbook* 3rd ed., 1991, RSC, London
2. Vernot, E. H. et al *Toxicol. Appl. Pharmacol.* 1977, **42**, 417
3. *Br. Med. J.* 1981, **283**, 18
4. Dinsdale, D. et al *Arch. Toxicol.* 1987, **59**(6), 385-390
5. *EC Directive Relating to the Quality of Water Intended for Human Consumption* 1982, 80/778/EEC, Office for Official Publications of the European Communities, 2 rue Mercier, L-2985 Luxembourg
6. *S. I. 1991 No. 472 The Environmental Protection (Prescribed Processes and Substances) Regulations* 1991, HMSO, London
7. *ECETOC Technical Report No. 30(4)* 1991, European Chemical Industry Ecology and Toxicology Centre, B-1160 Brussels
8. Birch, W. X. et al *Cytobios* 1986, **48**(194-195), 175-184

# M226   Methyl isovalerate

## $(CH_3)_2CHCH_2CO_2CH_3$

**CAS Registry No.** 556-24-1
**Synonyms** 3-methylbutanoic acid, methyl ester; isovaleric acid, methyl ester; methyl isopentanoate
**Mol. Formula** $C_6H_{12}O_2$                     **Mol. Wt.** 116.16
**Occurrence** In aromas of fruits and prepared meat foods. Produced by *Pseudomonas* strains on beef during cold storage (1).
Pollutant of drinking water (2).

## Physical properties
**B. Pt.** 116-117°C; **Specific gravity** $d_4^{20}$ 0.881.

## Solubility
Organic solvent: miscible ethanol, diethyl ether

## Occupational exposure
**UN No.** 2400; **HAZCHEM Code** 3ME; **Conveyance classification** flammable liquid.

## Mammalian and avian toxicity

### Acute data
$LD_{50}$ oral rabbit 5.7 g $kg^{-1}$ (3).
$LC_{50}$ (2 hr) inhalation mouse 2 g $m^{-3}$ (4).

## References

1. Edwards, R. A. et al *J. Appl. Bacteriol.* 1987, **62**(5), 403-412
2. Li, Z. et al *Xi'an Yike Daxue Xuebao* 1988, **9**(1), 53-8 (Ch.) (*Chem. Abstr.* **114**, 170718e)
3. *Ind. Med. Surg.* 1972, **41**, 31
4. Izmerov, N. F. et al *Toxicometric Parameters of Industrial Toxic Chemicals under Single Exposure* 1982, CIP, Moscow

# M227   Methyl lactate

## $CH_3CH(OH)CO_2CH_3$

**CAS Registry No.** 547-64-8
**Synonyms** 2-hydroxypropanoic acid, methyl ester
**Mol. Formula** $C_4H_8O_3$                     **Mol. Wt.** 104.11
**Uses** Solvent particularly for cellulose acetate.
**Occurrence** In aromas of a variety of plant leaves, in molasses from sugar mills (1).
In pickles (2).

## Physical properties
**B. Pt.** 144-145°C; **Specific gravity** $d^{19}$ 1.09.

### Solubility
Water: (decomp.). Organic solvent: ethanol

### Occupational exposure
**Risk phrases** Flammable (R10)
**Safety phrases** Do not breathe vapour (S23)

### Mammalian and avian toxicity
**Irritancy**
Compound is a mild irritant (species unspecified) (3).

### Any other comments
Human health effects, experimental toxicology, physico-chemical properties reviewed (4).

### References
1. Enriquez, M. et al *Sobre Deriv. Cana Azucar* 1989, **23**(1), 27-33 (Span.) (*Chem. Abstr.* **112**, 219097b)
2. Habashi, M. et al *Nippon Shokuhin Kogyo Gakkaishi* 1990, **37**(1), 15-19 (Japan.) (*Chem. Abstr.* **113**, 76827x)
3. *The Merck Index* 11th ed., 1989, Merck & Co. Inc., Rahway, NJ
4. *ECETOC Technical Report No. 30(4)* 1991, European Chemical Industry Ecology and Toxicology Centre, B-1160 Brussels

# M228   Methyl mercaptoacetate

## HSCH₂CO₂CH₃

**CAS Registry No.** 2365-48-2
**Synonyms** mercaptoacetic acid, methyl ester; methyl thioglycolate; thiglycolic acid, methyl ester; USAF EK-7119
**Mol. Formula** $C_3H_6O_2S$                    **Mol. Wt.** 106.14
**Uses** Chemical intermediate for pharmaceuticals and agrochemicals.
**Occurrence** Produced by *Pseudomonas* strains growing on stored, chilled beef (1).

### Physical properties
**B. Pt.** $_{10}$ 42-43°C; **Flash point** 30°C; **Specific gravity** 1.187.

### Mammalian and avian toxicity
**Acute data**
$LD_{50}$ oral rat 209 mg kg$^{-1}$ (2).
$LD_{50}$ intraperitoneal rat 252 mg kg$^{-1}$ (2).
$LD_{50}$ intraperitoneal mouse 100 mg kg$^{-1}$ (3).

### References
1. Edwards, R. A. et al *J. Appl. Bacteriol.* 1987, **62**(5), 403-412
2. *NTIS Report* AD277-689, Natl. Tech. Inf. Ser., Springfield, VA
3. *Z. Gesamte Hyg. Ihre Grenzgeb.* 1974, **20**, 575

# M229    Methyl methacrylate

$$CH_2=C(CH_3)CO_2CH_3$$

**CAS Registry No.** 80-62-6

**Synonyms** 2-propenoic acid, 2-methyl-, methyl ester; methacrylic acid, methyl ester; methyl 2-methylpropenoate; MME

**Mol. Formula** $C_5H_8O_2$                                    **Mol. Wt.** 100.12

**Uses** In production of acrylic polymers.

**Occurrence** Environmental pollutant.

## Physical properties

**M. Pt.** –48°C; **B. Pt.** 100-101°C; **Flash point** 10°C; **Specific gravity** $d_4^{20}$ 0.936;

**Partition coefficient** log $P_{ow}$ 1.38; **Volatility** v.p. 40 mmHg at 25.5°C.

### Solubility

Organic solvent: miscible ethanol, diethyl ether, acetone

## Occupational exposure

**US TLV (TWA)** 100 ppm (410 mg m$^{-3}$); **UK Long-term limit** 100 ppm (410 mg m$^{-3}$) under review; **UK Short-term limit** 125 ppm (510 mg m$^{-3}$) under review;

**UN No.** 1247; **HAZCHEM Code** 3☒E; **Conveyance classification** flammable liquid; **Supply classification** highly flammable, irritant (>1%).

**Risk phrases** ≥20% – Highly flammable – Irritating to eyes, respiratory system and skin – May cause sensitisation by skin contact – ≥1%<20% – Highly flammable – May cause sensitisation by skin contact (R11, R36/37/38, R43, R11, R43)

**Safety phrases** Keep container in a well ventilated place – Keep away from sources of ignition – No Smoking – Do not empty into drains – Take precautionary measures against static discharges (S9, S16, S29, S33)

### Abiotic removal

Use of ozone to treat gaseous discharge turns the compound into low-toxicity non-aromatic compounds (1).

## Mammalian and avian toxicity

### Acute data

LD$_{50}$ oral rat 8.4 ml kg$^{-1}$ (2).

LD$_{Lo}$ oral rabbit 7 ml kg$^{-1}$ (2).

LC$_{Lo}$ (5 hr) inhalation rabbit, guinea pig, rat <15 mg l$^{-1}$ (2).

Inhalation (duration unspecified) occupationally exposed workers >63.3 mg m$^{-3}$ is toxic. Effects include damage to nervous and cardiovascular systems (3).

LD$_{50}$ dermal rabbit >40 ml kg$^{-1}$ (4).

LD$_{50}$ intraperitoneal mouse 1.2 ml kg$^{-1}$ (5).

Intravenous doses of 1.25 g l$^{-1}$ infused into dogs caused rapid fall in blood pressure and respiratory arrest (6).

### Sub-acute data

Guinea pigs exposed for 3 hr daily for 15 days to 39 mg l$^{-1}$ methyl methacrylate vapour developed degenerative changes in liver (4).

### Carcinogenicity and long-term effects

No adequate evidence for carcinogenicity to humans or animals, IARC classification group 3 (7).

National Toxicology Program tested mice and rats by inhalation and food no evidence of carcinogenicity in either sex of either species but non-neoplastic lesions in nasal cavity were seen in both species (8,9).

Rats receiving 6 to 2000 mg $l^{-1}$ in drinking water for 2 yr developed no treatment-related tumours (10).

Rats painted on the back of the neck $3 \times wk^{-1}$ for 4 months developed no local tumours during their lifespan (11).

### Teratogenicity and reproductive effects

Rats receiving 0.13-0.44 mg $kg^{-1}$ on days 5, 10 and 15 of gestation showed no teratogenic effects but reduced foetal weight was seen at all doses (5,12,13).

Pregnant rats inhaling 100 mg $l^{-1}$ on days 6-15 of gestation for 54 or 18 min $day^{-1}$ showed poor maternal weight gain and food consumption. Foetal weight and size were reduced and some skeletal malformations were seen (14).

### Metabolism and pharmacokinetics

In rats up to 88% of a single dose of 5.7 mg $kg^{-1}$ methyl $^{14}C$-methacrylate has been found to be excreted as $^{14}CO_2$ in 10 days, irrespective of route of administration. Excreted metabolites included: $^{14}C$-methyl malonate; $^{14}C$-succinate; and possibly $^{14}C$-β-hydroxyisobutyrate and 2-formylpropionate (10).

### Sensitisation

Allergic responses in man are reported (15,16).

Allergic stomatitis has been reported in a patient wearing a denture containing methyl methacrylate monomer (17).

## Genotoxicity

*Salmonella typhimurium* TA97, TA98, TA100, TA1535, TA1537 with or without metabolic activation negative (18).

L5178Y $tk^+/tk^-$ mouse lymphoma assay without metabolic activation negative, with metabolic activation positive (19).

Workers exposed to high doses of methyl methacrylate showed an increased frequency of sister chromatid exchanges in lymphocytes (20).

## Any other adverse effects to man

Some evidence that workers occupationally exposed to methyl methacrylate develop chronic cough and mild airways obstruction (21).

## Any other adverse effects

Exposure of mice to 164 mg $l^{-1}$ in air for 14 min increased sleeping time induced by sodium pentobarbital (22).

## Any other comments

Human health effects, experimental toxicology, physico-chemical properties reviewed (23).

The carcinogenicity and toxicology of the compound has been assessed (24).

---

# References

1. Nosterov, N. G. et al *Plast. Massy* 1991, (2), 20-1 (Russ.) (*Chem. Abstr.* **114**, 253163w)
2. Deichmann, W. *J. Ind. Hyg. Toxicol.* 1941, **23**, 313-351
3. Ouyang, G. et al *Zhonghua Laodong Weisheng Zhiyebing Zazhi* 1990, **8**(1), 5-9 (Ch.) (*Chem. Abstr.* **113**, 19063w)
4. Spealman, C. R. et al *Ind. Med.* 1945, **14**, 292-298
5. Autian, J. *Environ. Health Perspect.* 1975, **11**, 141-152
6. Homsy, C. A. et al *Clin. Orthop. Relat. Res.* 1972, **83**, 317-328
7. *IARC Monograph* 1987, **Suppl. 7**, 66
8. *National Toxicology Program Research and Testing Division* 1992, Report No. TR-314, NIEHS, Research Triangle Park, NC
9. Chan, P. et al *Toxicology* 1988, **52**(3), 237-252
10. Borzelleca, J. F. et al *Toxicol. Appl. Pharmacol.* 1964, **6**, 29-36
11. Oppenheimer, B. S. et al *Cancer Res.* 1955, **15**, 333-340
12. Singh, A. R. et al *J. Dent. Res.* 1972, **51**, 1632-1638
13. Singh, A. R. et al *Toxicol. Appl. Pharmacol.* 1972, **22**, 314-315
14. Nicholas, C. A. et al *Toxicol. Appl. Pharmacol.* 1979, **50**, 451
15. Fisher, A. A. *J. Am. Med. Assoc.* 1954, **156**, 238-242
16. Kanerva, L. et al *Arch. Toxicol., Suppl.* 1986, **9**, 456-459
17. Kanzaki, T. et al *Hifu* 1989, **31**(Suppl. 7), 11-16 (Japan.) (*Chem. Abstr.* **112**, 104825v)
18. Zeiger, E. et al *Environ. Mutagen.* 1987, **9**(Suppl. 9), 1-109
19. Amtower, A. L. et al *Environ. Mutagen.* 1986, **8**(Suppl. 6), 4
20. Marez, T. et al *Mutagenesis* 1991, **6**(2), 127-129
21. Marez, T. et al *Br. J. Ind. Med.* 1993, **50**(10), 894-897
22. Lawrence, W. H. et al *J. Dent. Res.* 1972, **51**, 878
23. *ECETOC Technical Report No. 30(4)* 1991, European Chemical Industry Ecology and Toxicology Centre, B-1160 Brussels
24. *IARC Monograph* 1979, **19**, 187

# M230   Methyl methanesulfonate

$CH_3SO_3CH_3$

**CAS Registry No.** 66-27-3
**Synonyms** methanesulfonic acid, methyl ester; methyl methanesulfonic acid; methyl mesylate; MMS
**Mol. Formula** $C_2H_6O_3S$          **Mol. Wt.** 110.13
**Uses** Potential use as male chemosterilant for insects and mammalian pests.

## Physical properties

**B. Pt.** $_{753}$ 203°C; **Flash point** 104°C; **Specific gravity** $d_4^{20}$ 1.2943.

## Solubility

Water: 200 g $l^{-1}$. Organic solvent: ethanol, diethyl ether

## Ecotoxicity

### Fish toxicity

Embryotoxicity and teratogenicity were observed in Japanese medaka fish embryos exposed for 2 hr to 0.08-13.2 g $l^{-1}$ (1).

## Mammalian and avian toxicity

### Acute data
$LD_{50}$ oral blackbird 56.2 mg $kg^{-1}$ (2).
$LD_{50}$ oral quail 75 mg $kg^{-1}$ (2).

### Carcinogenicity and long-term effects
No adequate evidence for carcinogenicity to humans, sufficient evidence for
carcinogenicity to animals, IARC classification group 2B (3).
♂ mice receiving the equivalent of about 30 mg $kg^{-1}$ via drinking water for up to
20 months developed an increased incidence of lung tumours compared to controls (4).
♂ rats receiving the compound subcutaneously for 48 wk at doses upwards of 4 mg $kg^{-1}$
wkly developed an increased incidence of tumours particularly at the injection site (5).
♂ and ♀ rats receiving a single intraperitoneal injection of upwards of 72 mg $kg^{-1}$
developed tumours of tissues within the nervous system including malignant
neurofibromas and astrocytomas (6).
♀ mice receiving 2.2 mg $l^{-1}$ subcutaneously once wkly for 64 wk developed some
sarcomas at the injection site (7).
♀ mice receiving a dermal application of 4.4 mg $l^{-1}$ once wkly for 64 wk did not
develop any tumours (7).

### Teratogenicity and reproductive effects
Single intravenous injections of 20-68 mg $kg^{-1}$ to pregnant rats on day 15 or 21 of
gestation resulted in the development of neurogenic tumours in ≈ 20% of offspring.
No neurogenic tumours were found in offspring of pregnant ♀ treated on day 9 of
gestation (8).
Single intraperitoneal doses of 50 mg $kg^{-1}$ to ♂ rats resulted in infertility during the
second or third wk, which lasted for 28 days (9).

### Metabolism and pharmacokinetics
In mice, the compound is rapidly distributed throughout the body, including the
nervous system (10).
In rats the compound has been shown able to cross the placental barrier (11).
Rats receiving 100 mg $kg^{-1}$ intravenously yielded metabolites including:
methylmercapturic acid sulphoxide; 2-hydroxy-3-(methylsulphinyl)propionic acid;
(methylsulphinyl)acetic acid; methylmercapturic acid; and
N-(methylthioacetyl)glycine (12).
Conjugation with glutathione in rat liver has been demonstrated (13).

## Genotoxicity
*Salmonella typhimurium* TA100, TA1535 without metabolic activation positive (14).
*Salmonella typhimurium* TA100 with metabolic activation positive (15).
*Escherichia coli* PQ37 SOS chromotest positive (16).
*Salmonella typhimurium* SV50 Ara^r assay positive (17).
Unscheduled DNA synthesis in cultured hepatocytes positive (18).
DNA repair synthesis induced in HeLa cells (19).
Caused DNA damage but not strand-break repair in human diploid fibroblasts *in vitro* (20).

## Any other adverse effects to man
Therapeutic doses of 2.8 to 800 mg $kg^{-1}$ for up to 350 days led to significant
gastrointestinal and hepatotoxic effects in cancer patients (21).

---

## Any other adverse effects
Methylation of nucleic acid in liver, brain and foetal tissue of rats has been observed, and *in vitro* alkylation of Chinese hamster ovary cells has been reported (22-24).

## Any other comments
Human health effects, experimental toxicology, physico-chemical properties reviewed (25). The toxicology and carcinogenicity of the compound has been reviewed (26).

## References
1. Solomon, F. P. et al *Environ. Toxicol. Chem.* 1987, **6**(10), 747-753
2. Schafer, E. W. *Arch. Environ. Contam. Toxicol.* 1983, **12**, 355-382
3. *IARC Monograph* 1987, **Suppl. 7**, 66
4. Clapp, N. K. et al *Science (Washington, D. C. 1883-)* 1968, **161**, 913-914
5. Druckrey, H. et al *Z. Krebsforsch.* 1970, **74**, 241-270
6. Swann, P. F. et al *Biochem. J.* 1968, **110**, 39-47
7. Segal, A. et al *Cancer Res.* 1987, **47**, 3402
8. Kleihues, P. et al *Eur. J. Cancer* 1972, **8**, 641-645
9. Jackson, H. et al *J. Reprod. Fertil.* 1961, **2**, 447-465
10. Cumming, R. B. et al *Mutat. Res.* 1970, **10**, 365-377
11. Kleihues, P. et al *Z. Krebsforsch.* 1974, **81**, 273-283
12. Barnsley, E. A. *Biochem. J.* 1968, **106**, 18P-19P
13. Pillinger, D. J. et al in Roth, L. J. (Ed.) *Isotopes in Experimental Pharmacology* 1965, 83-89, University of Chicago Press, Chicago
14. McCann, J. et al *Proc. Natl. Acad. Sci. U. S. A.* 1975, **72**, 5135-5139
15. Gajcy, H. et al *Pol. J. Pharmacol. Pharm.* 1987, **39**(1), 27-32
16. Von der Hude, W. *Mutat. Res.* 1988, **203**, 81-94
17. Xu, J. et al *Mutat. Res.* 1984, **130**, 79-86
18. Fautz, R. et al *Mutat. Res.* 1991, **253**(2), 173-179
19. Jong, K. et al *Cell Biol. Toxicol.* 1991, **7**(1), 49-58
20. Snyder, R. D. et al *Environ. Mutagen.* 1985, **7**, 267-279
21. Bateman, J. R. et al *Cancer Chemother. Rep.* 1966, **50**, 675-682
22. Kleihues, P. et al *J. Neurochem.* 1973, **20**, 595-606
23. Sellakumar, A. R. et al *J. Natl. Cancer Inst.* 1987, **79**(2), 285-289
24. Dunn, W. C. et al *Carcinogenesis (London)* 1991, **12**(1), 83-89
25. *IARC Monograph* 1974, **7**, 253
26. *ECETOC Technical Report No 30*(4) 1991, European Chemical Industry Ecology and Toxicology Centre, B-1160 Brussels

# M231    7-Methyl-3-methylene-1,6-octadiene

$$CH_2=CHC(=CH_2)CH_2CH_2CH=C(CH_3)_2$$

**CAS Registry No.** 123-35-3
**Synonyms** 1,6-octadiene, 7-methyl-3-methylene-; Myrcene; β-Myrcene
**Mol. Formula** $C_{10}H_{16}$                     **Mol. Wt.** 136.24
**Uses** Intermediate in the manufacture of perfume chemicals.

## Physical properties
**M. Pt.** 163-178°C; **B. Pt.** 120°C; **Flash point** 37°C; **Specific gravity** 0.789.

## Solubility
Organic solvent: chloroform, diethyl ether, glacial acetic acid

## Ecotoxicity

### Invertebrate toxicity

$LD_{50}$ (24 hr) ♀ *Musca domestica* 360 µg insect$^{-1}$ (1).

$LD_{50}$ (24 hr) ♂ *Blatella germanica* >1580 µg insect$^{-1}$ (1).

$LC_{50}$ (24 hr) *Sitophilus oryzae* and *Blatella germanica* >100 ppm (1).

Embryotoxic effect in *Blatella germanica* Ooethecae at 0, 198, 395 and 790 µg mean % of Ooethecae producing offspring 90, 80, 63.3, 46.7%, respectively (1).

## Environmental fate

### Nitrification inhibition

At 25°C it did not significantly affect nitrification in soil (2).

Inhibits activity of ammonium monooxygenase (3).

Inhibition of net mineralisation at low additions progressing to net immobilisation with high additions, apparent inhibition of nitrification was observed (4).

### Degradation studies

*Aspergillus niger* 45% of original substrate was present on the 1st day to 2% by day 7. Main products after 120 hr were: 2-methyl-6-methylen-7-octene-2,3-diol; 6-methyl-2-vinyl-5-heptene-1,2-diol; and 7-methyl-3-methylen-6-octene-1,2-diol (5). Recovery of 30 µl added to soil was 5.28% after 24 days (4).

## Mammalian and avian toxicity

### Carcinogenicity and long-term effects

CASE prediction of the carcinogenicity in rodents gave a negative result (6).

### Metabolism and pharmacokinetics

Oral rats, metabolites found in urine were 10-hydroxylinalool; 7-methyl-3-methylenoct-6-ene-1,2-diol; 1-(hydroxymethyl)-4-isopropenylcyclohexanol; 10-carboxylinalool; and 2-hydroxy-7-methyl-3-methylenoct-6-enoic acid. Rat liver microsomes *in vitro* studies convertion to 10-hydroxylinalool in the presence of NADPH and oxygen (7). Rats (4 day) no significant effect on the hepatic drug-metabolising enzymes (8).

### Irritancy

Dermal rabbit (24 hr) 500 mg caused moderate irritation (8).

## Genotoxicity

*In vitro* human lymphocytes with or without metabolic activation, induced neither chromosomal aberrations nor sister chromatid exchanges (9).

*In vitro* V-79 cells, with or without metabolic activation did not cause increased mutation frequencies at the *hprt* locus (9).

## Any other adverse effects

Naturally occurring terpenoids induce cytochrome $P_{450}$ (species unspecified) (10).

## Any other comments

Not found in nature (11).

## References

1. Coats, J. R. et al *ACS Symp. Ser.* 1991, **449**(Nat. Occuring Pest Bioregul.), 305-316

2.   Bremner, J. M. et al *Soil Sci. Soc. Am. J.* 1988, **52**(6), 1630-1633
3.   White, C. S. *Ecology* 1988, **69**(5), 1631-1633
4.   White, C. S. *Biogeochemistry* 1991, **12**, 43-68
5.   Yamazaki, Y. et al *Agric. Biol. Chem.* 1988, **52**(11), 2921-2922
6.   Rosenkrantz, H. S. et al *Carcinogenesis (London)* 1990, **11**(2), 349-353
7.   Madyastha, K. M. et al *Xenobiotica* 1987, **17**(5), 539-549
8.   *Food Cosmet. Toxicol* 1976, **14**, 615
9.   Kauderer, B. et al *Environ. Mol. Mutagen.* 1991, **18**(1), 28-34
10.   Austin, C. A. *Biochem. Pharmacol.* 1988, **37**(11), 2223-2229
11.   Budvari, S. *Merck Index* 11th ed., 1989, Merck & Co. Inc., Rahway, N.J.

# M232   3-Methyl-4-(methylthio)phenol

**CAS Registry No.** 3120-74-9
**Synonyms** phenol, 3-methyl-4-(methylthio)-; *m*-cresol, 4-(methylthio)-; MMTP
**Mol. Formula** $C_8H_{10}OS$                               **Mol. Wt.** 154.23

## Physical properties

**M. Pt.** 52-54°C; **B. Pt.** 224-228°C.

## Mammalian and avian toxicity

### Acute data

$LD_{50}$ oral mouse, rat, guinea pig 1000, 3400, 3500 mg $kg^{-1}$, respectively (1).
$LD_{Lo}$ intraperitoneal mouse 100 mg $kg^{-1}$ (2).

## Legislation

Limited under EC Directive on Drinking Water Quality 80/778/EEC. Maximum
admissible concentration 0.5 µg ethanol $l^{-1}$ excluding natural phenols which do not
react to chlorine (3).
Included in Schedule 4 (Release Into Air: Prescribed Substances) Statutory
Instrument No. 472, 1991 (4).

## Any other comments

All reasonable efforts have been taken to find information on isomers of this
compound, but no relevant data are available.

## References

1.   *Gig. Sanit.* 1978, **43**(8), 10 (Russ.)
2.   *NTIS Report* AD438-895, Natl. Tech. Inf. Ser., Springfield, VA

3. *EC Directive Relating to the Quality of Water Intended for Human Consumption* 1982, 80/778/EEC, Office for Official Publications of the European Communities, 2 rue Mercier, L-2985 Luxembourg
4. *S. I. 1991 No. 472 The Environmental Protection (Prescribed Processes and Substances) Regulations* 1991, HMSO, London

# M233   *N*-Methylmorpholine

**CAS Registry No.** 109-02-4
**Synonyms** 4-methylmorpholine
**Mol. Formula** $C_5H_{11}NO$         **Mol. Wt.** 101.15

## Physical properties

**M. Pt.** –65°C; **B. Pt.** $_{750}$ 115-116°C; **Flash point** 24°C; **Specific gravity** 0.920; **Volatility** v. den. 3.5.

## Occupational exposure

**UN No.** 2535; **HAZCHEM Code** 2PE; **Conveyance classification** flammable liquid, corrosive substance.

## Mammalian and avian toxicity

### Acute data
$LD_{50}$ oral rat 2720 mg kg$^{-1}$ (1).
$LC_{Lo}$ (4 hr) inhalation rat 2000 ppm (2).
$LC_{50}$ (2 hr) inhalation mouse 25,200 mg m$^{-3}$ (3).
$LD_{50}$ dermal rabbit 1242 mg kg$^{-1}$ (1).

### Irritancy
Dermal rabbit 460 mg, open to atmosphere, caused mild irritation (4).
Eye rabbit 920 µg caused severe irritation (exposure unspecified) (4).

## Legislation

Included in Schedule 4 (Release into Air: Prescribed Substances) Statutory Instrument No. 472, 1991 (5).

## References

1. *J. Ind. Hyg. Toxicol.* 1949, **31**, 60
2. *J. Ind. Hyg. Toxicol.* 1949, **30**, 60
3. *Toksikol. Nov. Prom. Khim. Veshchestv* 1979, **15**, 116
4. *Union Carbide Data Sheet* 1959, **6**, 30, Union Carbide Corp., New York
5. *S. I. 1991 No. 472 The Environmental Protection (Prescribed Processes and Substances) Regulations* 1991, HMSO, London

# M234　Methylnaphthalene

**CAS Registry No.** 1321-94-4
**Synonyms** methylnaftalen; naphthalene, methyl-
**Mol. Formula** $C_{11}H_{10}$　　　　　　　　　**Mol. Wt.** 142.20
**Occurrence** In the aroma of a variety of foods including fruits and cheeses. Also in coal tar, exhaust fumes, in seawater and in fish (1).

## Ecotoxicity

**Invertebrate toxicity**
Toxicity and detrimental actions to *Selenastrum capricornutum* have been demonstrated (2).

**Bioaccumulation**
Compound has been detected in white croaker in sea water close to Los Angeles, probably as a result of the compound entering the food chain (1).
Accumulation and retention have been demonstrated in blue crab, with the hepatopancreas and gill tissue accumulating the highest concentrations (3).

## Environmental fate

**Degradation studies**
Following a diesel spillage, contaminated soil containing methylnaphthalene was restored to near normality in 13 wk by liming, fertilising and tilling of the soil (4).

## Mammalian and avian toxicity

**Acute data**
$LD_{50}$ oral rat 4.36 g $kg^{-1}$ (5).

**Irritancy**
Moderate irritant of rabbit skin at 500 mg for 24 hr (5).

## Any other comments

Human health effects, experimental toxicology, physico-chemical properties reviewed (6).
The compound occurs as a mixture of isomers.

## References

1. Malins, D. C. et al *Environ. Sci. Technol.* 1987, **21**(8), 765-770
2. Gaw, J. P. *Acta Hydrochim. Hydrobiol.* 1988, **16**(6), 617-620
3. Melzian, B. D. *Oil Chem. Pollut.* 1987, **3**(5), 367-399
4. Wang, X. et al *Environ. Sci. Technol.* 1990, **24**(7), 1086-1089
5. Marhold, J. V. *Sbornik Vysledku Toxixologickeho Vysetreni Latek A Pripravku* 1972, Prague
6. *ECETOC Technical Report No. 30(4)* 1991, European Chemical Industry Ecology and Toxicology Centre, B-1160 Brussels

# M235   1-Methylnaphthalene

CH₃

**CAS Registry No.** 90-12-0
**Synonyms** α-methylnaphthalene; naphthalene, 1-methyl-
**Mol. Formula** $C_{11}H_{10}$          **Mol. Wt.** 142.20
**Uses** Insecticide manufacture. Solvent.
**Occurrence** Asphalt, naphtha, coal and petroleum. Trace contaminant in some water samples (1-3).

## Physical properties

**M. Pt.** –22°C; **B. Pt.** 240-243°C; **Flash point** 82°C; **Specific gravity** 1.001;
**Partition coefficient** log $P_{ow}$ 3.87; **Volatility** v. den. 4.91.

### Solubility
Water: 26 mg l$^{-1}$ at 25°C. Organic solvent: ethanol, diethyl ether, benzene

## Ecotoxicity

### Fish toxicity
$LC_{50}$ (48 hr) (static bioassay) brown trout yearlings 8.4 mg l$^{-1}$ (4).
Exposure to a low concentration decreased feeding. At a higher level all fish stopped feeding for 3 days and one stopped for 10 days (5).
$LD_{50}$ intraperitoneal rainbow trout 2.92 g kg$^{-1}$ (6).
$LC_{50}$ (96 hr) fathead minnow (18-22°C) 9 mg l$^{-1}$ (7).

### Bioaccumulation
Uptake and depuration by oysters from oil treated enclosures; concentration in oysters was 36 μg g$^{-1}$. The concentration in the water was 3 μg l$^{-1}$, the accumulation factor was 12,000 (8).

## Environmental fate

### Degradation studies
Degradation by free living bacteria at 4 sites ranged from 37-106 hr to a concentration of 1 μg l$^{-1}$ (9).
The $t_{1/2}$ in two different sandy loam soils ranged from 1.4-2.1 days and 1.6-3.2 days respectively (10).
*Alcaligenes denitrificans* WW1 can utilise it as a sole carbon source (11).
Pollution with 100-240 μg l$^{-1}$ showed that the bacterial communities in the heavily polluted water have a higher degree of adaptation to hydrocarbon degradation than communities from slightly polluted water. Bacteria from the unpolluted water were considered as being unadapted (12).
When it was added to a fluoranthene bacteria community it was not detected (<10 ng l$^{-1}$) after 3 days. Recovery in killed cells was 85.2%. Recovery from *Pseudomonas putida* Pp67 cells was 64.5% (13).

Biodegradation at 0.1 mg l$^{-1}$ after 135 hr; 0% in normal sewage and 95% in adapted sewage (14).

## Mammalian and avian toxicity

### Acute data
LD$_{50}$ oral rat 1840 mg l$^{-1}$ (15).
Intraperitoneal (24 hr) rats 71 mg kg$^{-1}$, microscopy of lungs showed no lesions (16).

### Carcinogenicity and long-term effects
Dermal mouse bioassay at 100 µl 3 × wk$^{-1}$ inhibition of benzo[a]pyrene carcinogenesis in mouse skin was observed (17).

### Metabolism and pharmacokinetics
In vitro mammalian microsomes principal metabolite 1-methylnaphthoic acid (18).

## Genotoxicity

Salmonella typhimurium TA98, TA100 with and without metabolic activation negative (19).
Salmonella typhimurium TM677 with metabolic activation, positive (20).

## Any other adverse effects

Oral rats 1.5-2 g kg$^{-1}$, liver function was impaired but restored to normal after 3-5 days (21).
In vitro ascites sarcoma BP8 cells growth inhibited at 142 mg l$^{-1}$ (22).

## Legislation

The log P$_{ow}$ value exceeds the European Community recommended level 3.0 (6th and 7th amendments) (23).

## Any other comments

Ecotoxicology, environmental effects, experimental toxicity, exposure levels and human health effects reviewed (24).

## References

1. Kopfler, F. L. et al *Human Exposure to Water Pollutants* Suffet, I. R. (Ed.) *HEEP* **79**, 90906
2. Drost, G. *Groundwater Pollut. Eur. Proc. Conf.* 1972, 126, (*Chem. Abstr.* **84**, 169394b)
3. Grob, K. et al *J. Chromatogr.* 1974, **90**, 303
4. Woodiwiss, F. S. et al *Water Pollut. Control* 1974, **73**, 396
5. Purdy, J. E. *J. Fish Biol.* 1989, **34**(4), 621-629
6. Hodson, P. V. et al *Environ. Toxicol. Chem.* 1988, **7**, 443-454
7. Mattosov, V. R. et al *EPA-600/3-76-097* Oct 1976
8. Lee, R. F. et al *Environ. Sci. Technol.* 1978, **12**, 832-838
9. Arvin, E. et al *Water Sci. Technol.* 1988, **20**(3), 109-118
10. Park, K. S. et al *Environ. Toxicol. Chem.* 1990, **9**, 187-195
11. Weissenfels, W. D. et al *Appl. Microbiol. Biotechnol.* 1991, **34**(4), 528-535
12. Aamand, J. et al *J. Contam. Hydrol.* 1989, **4**(4), 299-312
13. Mueller, J. G. et al *Appl. Environ. Microbiol.* 1989, **55**(12), 3085-3090
14. Gaffney, P. E. *J. Water Purif. Control. Fed.* 1976, **48**(12), 2731-2737
15. Izmerov, N. F. et al *Toxicometric Parameters of Industrial Toxic Chemicals under Single Exposure* 1982, 85, CIP, Moscow
16. Dinsdale, D. et al *Arch. Toxicol., Suppl. 1986* 1987, **11** (Mech. Models Toxicol.), 288-291

17. Schemeltz, I. et al *Carcinog. - Compr. Surv.* 1978, **3**, 47
18. Snyder, R. (Ed.) *Ethel Browning's Toxicity and Metabolism of Industrial Solvents* 2nd ed., 1987, Elsevier, Amsterdam
19. Florin, I. et al *Toxicology* 1980, **18**, 219-232
20. Kaden, D. A. et al *Cancer. Res.* 1979, **39**, 4
21. Ott, H. et al *Eur. At. Energy Community EURATOM (Rep)* 1970, EUR-4558, 14 (*Chem. Abstr.* **75**, 86750c)
22. Pilotti, A. et al *Toxicology* 1975, **5**, 49
23. *1967 Directive on Classification, Packaging and Labelling of Dangerous Substances 67/548/EEC; 6th Amendment EEC Directive 79/831/EEC; 7th Amendment EEC Directive 91/32/EEC* 1991, HMSO, London
24. *ECETOC Technical Report No. 30(4)* 1991, European Chemical Industry Ecology and Toxicology Centre, B-1160 Brussels.

# M236   2-Methylnaphthalene

**CAS Registry No.** 91-57-6
**Synonyms** β-methylnaphthalene; naphthalene, 2- methyl-
**Mol. Formula** $C_{11}H_{10}$                    **Mol. Wt.** 142.20
**Occurrence** Trace contaminant in some water samples (1-3).
Coal tar pitch fumes.

## Physical properties

**M. Pt.** 34-36°C; **B. Pt.** 241-242°C; **Flash point** 97°C; **Specific gravity** $d^{20}$ 1.00; **Partition coefficient** log $P_{ow}$ 4.00.

## Ecotoxicity

**Fish toxicity**
Intraarterial rainbow trout 10 mg kg$^{-1}$, terminal $t_{1/2}$ was 9.9 hr, it is metabolised mainly to water soluble metabolites which were excreted into the urine and bile. The apparent bioavailability was 20% (4).

**Invertebrate toxicity**
*Meretrix casta* var *ovum* (96 hr) 10 mg l$^{-1}$ specific activities of Na-K-Mg-ATPase and Na-K-ATPase were elevated and decreased in the hepatopancreas and gill, respectively (5).

**Bioaccumulation**
Rainbow trout (4 wk) bioaccumulation factor 23,500 (6).
The bioconcentration factor in coho salmon was 28 (7).
Bioconcentration factor flowthrough method (species unspecified) 407 (8).

## Environmental fate

### Degradation studies

Microbial degradation to carbon dioxide, in seawater at 12°C in the dark after 24 hr incubation at 50 µg l$^{-1}$: 0.10 µg l$^{-1}$ day$^{-1}$ turnover time was 500 days (9).
In seawater with oil oxidising micro-organisms there was 17.1% breakdown after 21 days at 22°C in stoppered bottles (10).
Undergoes microbial oxidation by *Pseudomonas putida* 39D and *Pseudomonas putida* NCIB 9816 (11).
Microcosms inoculated with *Mycobacterium* sp. showed enhanced mineralisation of this compound (12).
*Alcaligenes denitrificans* WW1 can utilise it as sole carbon source (13).
Biodegradation by indigenous microorganisms, undetected after 8 days, initial concentration 9.5 µg m$^{-1}$ (14).
Biodegraded to carbon dioxide in 30-670 days (15).
In sea water at 12°C biodegraded within 500 days (16).

### Abiotic removal

Reaction with hydroxyl radicals is the dominant loss process (17).

## Mammalian and avian toxicity

### Acute data

LD$_{50}$ oral rat 1630 mg kg$^{-1}$ (18).
LD$_{Lo}$ intraperitoneal mouse 1000 mg kg$^{-1}$ (19).
Intraperitoneal (24 hr) mice 400 mg kg$^{-1}$, pulmonary damage was detected, depletion of pulmonary reduced glutathione was observed. Lipid peroxidation and phospholipid content in the lung were unaffected (20).

### Sub-acute data

Inhalation (6.5 month) (species unspecified) 100 mg m$^{-3}$ produced an increase in both the respiration rate and oxygen consumption (21).

### Carcinogenicity and long-term effects

Dermal mouse bioassay 100 µl 3 × wk$^{-1}$ inhibition of benzo[*a*]pyrene carcinogenesis in mouse skin was observed (22).

### Metabolism and pharmacokinetics

Oral rats, rabbits urinary metabolites included: 2-naphthoic acid; glycine conjugate of 2-naphthoic acid; 7-methyl-1- and 2-naphthols; and, 1,2-dihydro-1,2-dihydroxy-7-methylnaphthalene (23).
*In vitro* rat hepatic microsomes metabolised to 2-naphthoic acid and a methylnaphthalene dihydrodiol (24).
Metabolites produced by pulmonary and hepatic microsomes from DBA/2J mice were: 3 dihydrodiols; 2-naphthyl alcohol; and other unidentified metabolites (25).
Metabolites from rat liver microsomes were: 2-(hydroxymethyl)naphthalene; 3,4-dihydrodiol; 5,6-dihydrodiol; and, 7,8-dihydrodiol (26).
Metabolites in rat urine were 2-naphthoic acid and 2-naphthoylglycine (27).

## Genotoxicity

*Salmonella typhimurium* TM677 (metabolic activation unspecified) weakly mutagenic (28).

*In vitro* human lymphocytes with or without metabolic activation, weakly positive or negative, respectively. Sister chromatid exchange frequencies were significantly increased with metabolic activation (29).

## Any other adverse effects

*In vitro* ascites sarcoma BP8 cells, growth inhibited at 142 mg l$^{-1}$ (30).
Oral rat 1.5-2 g kg$^{-1}$, liver function was impaired but restored to normal after 3-5 days (31).

## Legislation

The log $P_{ow}$ value exceeds the European Community recommended level 3.0 (6th and 7th amendments) (32).

## Any other comments

Ecotoxicology, environmental effects, experimental toxicology, exposure levels and human health effects reviewed (33).
Metabolism reviewed (34).

## References

1. Kopfler, F. C. et al *Human Exposure to Water Pollutants* Suffet, I. R. (Ed.) *HEEP* **79**, 09006
2. Drost, G. *Groundwater Pollut. Eur. Proc. Conf.* 1972, (Publ. 1974), 126 (*Chem. Abstr.* **84**, 196394b)
3. Grob, K. et al *J. Chromatogr.* 1974, **90**, 303
4. Kennedy, C. J. et al *Environ. Toxicol. Chem.* 1990, **9**(2), 133-139
5. Kulkarni, B. G. *J. Environ. Biol.* 1990, **11**(3), 275-278
6. Melancon, M. J. et al *Arch. Environ. Contam. Toxicol.* 1978, **7**, 207
7. Raubul, W. T. et al, 1977, Oil Spill Conference, USA
8. Sabljic, A. *QSAR Environ. Toxicol., Proc. Int. Workshop, 2nd* 1986, 309-332
9. Richard, F. L. et al *Rapp. P. V. Reun. Comm. Int. Explor. Sci. Mer Mediterr.* 1977, **171**, 150-156
10. McKenzie, P. et al *Microbiology in Agriculture, Fisheries and Food* 1976 F. A. Skinner (Ed.), Academic Press
11. Deluca, M. E. et al *Tetrahedron Lett.* 1990, **31**(1), 13-16
12. Heitkamp, M. A. et al *Appl. Environ. Microbiol.* 1989, **55**(8), 1968-1973
13. Weissenfels, W. D. et al *Appl. Microbiol. Biotechnol.* 1991, **34**(4), 528-535
14. Mueller, J. G. et al *Appl. Environ. Microbiol.* 1991, **57**(5), 1277-1285
15. Lee, R. F., 1977, Oil Spill Conference, USA
16. Lee, R. F. et al *Rapp. P. V. Reun. Comm. Int. Explor. Sci. Mer Mediterr.* 1977, **171**, 150
17. Arey, J. et al *Int. J. Chem. Kinet.* 1989, **21**(9), 775-799
18. Izmerov, N. F. et al *Toxicometric Parameters of Industrial Toxic Chemicals under Single Exposure* 1982, CIP, Moscow
19. Griffing, K. A. et al *Toxicol. Appl. Pharmacol.* 1981, **61**, 185
20. Honda, T. et al *Chem. Pharm. Bull.* 1990, **38**(11), 3130-3135
21. Reshyetyuk, A. L. *Vopr. Gig. Tr. Prof. Patol. Ugol. Gornorud. Met. Prom.* 1968, **144** (Russ.) (*Chem. Abstr.* 73, 118667h)
22. Schemeltz, I. *Carcinog.- Comp. Surv.* 1978, **3**, 47
23. Grimes, A. J. et al *Biochem. J.* 1956, **62**, 11
24. Kaubisch, N. et al *Biochemistry* 1972, **11**, 3080-3088
25. Griffin, K. A. et al *Toxicology* 1983, **26**, 213
26. Breger, R. F. et al *Drug Metab. Dispos.* 1983, **11**, 319
27. Melancon, M. J. et al *Drug Metab. Dispos.* 1982, **10**, 128

28.  Kaden, D. A. et al *Cancer. Res.* 1979, **39**, 4152
29.  Kulka, U. et al *Mutat. Res.* 1988, **208**(3-4), 155-158
30.  Pilotti, A. et al *Toxicology* 1975, **5**, 49
31.  Ott, H. et al *Eur. At. Energy Community EURATOM (REP)* 1970, EUR-4558, 14 (*Chem. Abstr.* **75**, 86750c)
32.  *1967 Directive on Classification, Packaging and Labelling of Dangerous Substances 67/548/EEC; 6th Amendment EEC Directive 79/831/EEC; 7th Amendment EEC Directive 91/32/EEC* 1991, HMSO, London
33.  *ECETOC Technical Report No. 30(4)* 1991, European Chemical Industry Ecology and Toxicology Centre, B-1160 Brussels
34.  Buckpitt, A. R. et al *Pharmacol. Ther.* 1989, **41**(1-2), 393-410.

# M237  2-Methyl-4-nitroaniline

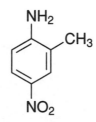

**CAS Registry No.** 99-52-5
**Synonyms** 2-methyl-4-nitrobenzenamine; 4-nitro-2-toluidine; C.I. 37100; Fast Red Base RL; Red RL Base; 2-methyl-*p*-nitroaniline
**Mol. Formula** $C_7H_8N_2O_2$  **Mol. Wt.** 152.15

## Physical properties
**M. Pt.** 131-133°C.

## Ecotoxicity

**Invertebrate toxicity**
$EC_{50}$ (30 min) *Photobacterium phosphoreum* 3.57 ppm Microtox test (1).

## Any other comments
Experimental toxicology, human health effects and workplace experience reviewed (2).

## References
1.  Kaiser, K. L. E. et al *Water Pollut. Res. J. Can.* 1991, **26**(3), 361-431
2.  *ECETOC Technical Report No. 30(4)* 1991, European Chemical Industry Ecology and Toxicology Centre, B-1160 Brussels

# M238  2-Methyl-5-nitroaniline

CAS Registry No. 99-55-8
**Synonyms** 2-methyl-5-nitrobenzenamine; 5-nitro-*o*-toluidine; C.I. 37105; C.I. azoic diazo component 12; Fast Scarlet G; Scarlet G Base
**Mol. Formula** $C_7H_8N_2O_2$                                **Mol. Wt.** 152.15
**Uses** Dye.

## Ecotoxicity

**Invertebrate toxicity**
$EC_{50}$ (30 min) *Photobacterium phosphoreum* 15.2 ppm Microtox test (1).

## Mammalian and avian toxicity

**Acute data**
$LD_{50}$ oral rat 574 mg kg$^{-1}$ (2).

**Carcinogenicity and long-term effects**
National Toxicology Program tested Fischer 344 rats and B6C3F1 mice via the feed. Carcinogenic in mice causing hepatocellular carcinomas in both sexes, an increase in the combined incidence of haemangiomas and haemangiosarcomas in ♂ mice and an increased incidence of haemangiosarcomas in ♀ mice. Not carcinogenic in the rats (3).

## References

1. Kaiser, K. L. E. et al. *Water Pollut. Res. J. Can.* 1991, **26**(3), 361-431
2. Progress Report *NIH-71-E-2144* Submitted to the National Cancer Institute 1973, Mason Research Institute, Worcester, MA
3. *National Toxicology Program Research and Testing Division* 1992, Report No. TR-107, NIEHS, Research Triangle Park, NC

# M239   4-Methyl-2-nitroaniline

CAS Registry No. 89-62-3
**Synonyms** benzenamine, 4-methyl-2-nitro-; *p*-toluidine, 2-nitro-; Amarthol Fast Red
GL Base; Fast Red Base GL; Lithosol Scarlet Base M; C.I. 37110
**Mol. Formula** $C_7H_8N_2O_2$                                   **Mol. Wt.** 152.15

## Physical properties

**M. Pt.** 115-116°C.

## Ecotoxicity

**Fish toxicity**
$LC_{50}$ (96 hr) fathead minnow 26.1 mg $l^{-1}$ (1).
$LC_{50}$ (30 min) rainbow trout 5 mg $l^{-1}$ (2).
$LC_{50}$ (3 hr) bluegill sunfish 5 mg $l^{-1}$ (2).
$LC_{50}$ (3 hr) goldfish 5 mg $l^{-1}$ (2).
$LC_{50}$ (4-7 hr) chinook salmon 10 mg $l^{-1}$ (2).
$LC_{50}$ (1-2 hr) coho salmon 10 mg $l^{-1}$ (2).

**Invertebrate toxicity**
$EC_{50}$ (48 hr) *Daphnia magna* 14.2 mg $l^{-1}$ (1).
$EC_{50}$ (30 min) *Photobacterium phosphoreum* 5.92 mg $l^{-1}$ Microtox test (3).

## Genotoxicity

Hepatocyte/DNA repair test with primary cultured rat hepatocytes 0.152 g $l^{-1}$
negative (4).

## Legislation

Included in Schedule 4 (Release into Air: Prescribed Substances) Statutory Instrument
No. 472, 1991 (5).

## References

1. Pearson, J. G. et al *Aquat. Toxicol.* 1979, 284-301 ASTM STP 667, Marking, L. L et al (Ed.)
2. Newsome, L. D. et al *QSAR, Environ. Toxicol. Proc. Int. Workshop, 2nd* 1987, 231-250
3. Kaiser, K. L. E. et al *Water Pollut. Res. J. Can.* 1991, **26**(3), 361-431
4. Yoshimi, N. et al *Mutat. Res.* 1988, **206**, 183-191
5. *S. I. 1991 No. 472 The Environmental Protection (Prescribed Processes and Substances) Regulations* 1991, HMSO, London

# M240    4-Methyl-3-nitroaniline

**CAS Registry No.** 119-32-4
**Synonyms** 4-methyl-3-nitrobenzenamine; 3-nitro-*p*-toluidine;
1-amino-3-nitro-4-methylbenzene; 5-nitro-4-toluidine; *m*-nitro-*p*-toluidine
**Mol. Formula** $C_7H_8N_2O_2$                     **Mol. Wt.** 152.15

## Physical properties

**M. Pt.** 116°C; **Flash point** 175°C (closed cup); **Specific gravity** 1.312; **Volatility**
v. den. 5.80.

## Solubility

Organic solvent: ethanol

## Ecotoxicity

**Fish toxicity**
$LC_{50}$ (96 hr) fathead minnow 34.0-71.3 mg $l^{-1}$ (1,2).

**Invertebrate toxicity**
$EC_{50}$ (30 min) *Photobacterium phosphoreum* 5.40 ppm Microtox test (3).
$EC_{50}$ (48 hr) *Daphnia magna* 22.5 mg $l^{-1}$ (2).

## Mammalian and avian toxicity

**Acute data**
$LD_{50}$ oral rat 6860 mg $kg^{-1}$ (4).
$LD_{50}$ oral redwing blackbird 3.16 mg $kg^{-1}$ (5).
$LD_{50}$ oral starling 31.6 mg $kg^{-1}$ (5).
$LD_{50}$ intravenous mouse 180 mg $kg^{-1}$ (6).

## References

1.  Holcombe, G. W. et al *Environ. Pollut., Ser. A* 1984, **35**, 367-381
2.  Pearson, J. G. et al *Aquat. Toxicol.* 1979, ASTM STP 667, Marking, L. L. et al (Ed.)
3.  Kaiser, K. L. E. et al *Water Pollut. Res. J. Can.* 1991, **26**(3), 361-431
4.  Marhold, J. V. *Sbornik Vysledku Toxixologickeho Vysetreni Latek A Pripravku* 1972, Prague
5.  Schafer, E. W. *Toxicol. Appl. Pharmacol.* 1972, **21**, 315
6.  *Report NX No.04522* US Army Armament Research and Development Command, Chemical Systems Laboratory, NIOSH Exchange Chemicals, Aberdeen Proving Ground, MD

# M241    *N*-Methyl-4-nitroaniline

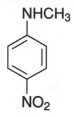

**CAS Registry No.** 100-15-2
**Synonyms** *N*-methyl-4-nitrobenzenamine; *p*-(methylamino)nitrobenzene
**Mol. Formula** $C_7H_8N_2O_2$                                   **Mol. Wt.** 152.15

## Physical properties
**M. Pt.** 152-154°C.

## Ecotoxicity
**Invertebrate toxicity**
$EC_{50}$ (30 min) *Photobacterium phosphoreum* 1.21 ppm Microtox test (1).

## References
1.   Kaiser, K. L. E. et al *Water Pollut. Res. J. Can.* 1991, **26**(3), 361-431

# M242    2-Methyl-1-nitroanthraquinone

**CAS Registry No.** 129-15-7
**Synonyms** 2-methyl-1-nitro-9,10-anthracenedione; NCI-CO1 923;
1-nitro-2-methylanthraquinone; 1-N-2-MA
**Mol. Formula** $C_{15}H_9NO_4$                                   **Mol. Wt.** 267.24
**Uses** Organic synthesis. Manufacture of dyes.

## Physical properties
**M. Pt.** 270-271°C; **Partition coefficient** log $P_{ow}$ 2.4894 (1).

## Solubility
Organic solvent: benzene, chloroform, diethyl ether, nitrobenzene

## Mammalian and avian toxicity

### Acute data
$LD_{50}$ oral rat >500 mg kg$^{-1}$ (2).
$LD_{50}$ oral redwing blackbird >113 mg kg$^{-1}$ (3).
$LD_{50}$ intraperitoneal rat 1100 mg kg$^{-1}$ (4).

### Sub-acute data
Oral rat (6 wk) 0.15% diet caused a reduction in body weight gain. 0.06 and 0.12% diet caused hyperplasia of the lymphoid tissue and of the stomach, as well as inflammatory changes of the stomach (5).

### Carcinogenicity and long-term effects
No adequate evidence for carcinogenicity to humans, sufficient evidence for carcinogenicity to animals, IARC classification group 2B (6).
Oral mouse 0, 300 or 600 mg kg$^{-1}$ diet. All treated mice died within 338 days. Subcutaneous haemangiosarcomas developed in 88/90 treated ♂ mice and 79/82 treated ♀ mice. 4 of these tumours metastasized to the lung. Mesenteric haemangiosarcomas were observed in 6 ♂ and 8 ♀ treated mice, compared with 1/49 ♂ and 0/48 ♀ controls (5).
Oral rat (109 wk) 0, 600 or 1200 mg kg$^{-1}$ diet for 78 wk. Survival rates were 29/48, 35/49 and 27/49 respectively for ♂ rats, and 22/50, 40/50 and 29/49 respectively for ♀ rats. An increase in the incidence of hepatocellular carcinoma was observed only in ♂ animals: 1/48 controls, 5/48 low dose and 9/49 high dose rats. Fibromas of the subcutaneous tissue were observed in 3/48 control, 10/49 low dose and 34/49 high dose ♂ rats, and in 1/50 control, 0/50 low dose and 13/49 high dose ♀ rats. Subcutaneous haemangiosarcomas occurred in 3/49 high dose ♂ rats (5).

## Genotoxicity
*Salmonella typhimurium* TA97, TA98, TA100 with metabolic activation positive (7).
*In vitro* Chinese hamster ovary cells, sister chromatid exchanges and chromosomal aberrations negative (8).
*In vitro* mouse lymphoma L5178Y cells, tk$^{+}$/tk$^{-}$ with metabolic activation, equivocal results (9).

## Any other comments
Physical properties, use, carcinogenicity and toxicity reviewed (10,11).

## References
1.  McCoy, G. D. et al *Carcinogenesis (London)* 1990, **11**(7), 1111-1117
2.  *Review* 1953, **5**, 33, Natl. Academy Sciences, Natl. Res. Council. Chem-Biol. Coordination Centre
3.  Schafer, E. W. et al *Arch. Environ. Contam. Toxicol.* 1983, **12**, 355-382
4.  *Gig. Tr. Prof. Zabol.* 1977, **21**(12), 27
5.  National Cancer Institute *Tech. Rep. Ser. No. 29* 1978, DHEW Publ. No. (NIH) 78-829, Govt. Printing Office, Washington, DC
6.  *IARC Monograph* 1987, **Suppl. 7**, 66
7.  Mortelmans, K. et al *Environ. Mutagen.* 1986, **8**(Suppl. 7), 1-119
8.  Rosenkranz, H. S. et al *Environ. Mol. Mutagen.* 1990, **16**(3), 149
9.  Harrington-Brock, K. et al *Mutagenesis* 1991, **6**(1), 35-46
10.  *IARC Monograph* 1982, **27**, 205-210

11. *ECETOC Technical Report No. 30(4)* 1991, European Chemical Industry Ecology and Toxicology Centre, B-1160 Brussels

# M243   Methylnitropropyl-4-nitrosoaniline

$$NHCH_2\underset{\underset{CH_3}{|}}{\overset{\overset{NO_2}{|}}{C}}-CH_3$$

**CAS Registry No.** 24458- 48-8

**Synonyms** benzenamine, *N*-(2-methyl-2-nitropropyl)-4- nitroso-; Nitrol; aniline, *N*-(2-methyl-2-nitropropyl)-*p*-nitroso-;

**Mol. Formula** $C_{10}H_{13}N_3O_3$              **Mol. Wt.** 223.23

## Physical properties

**M. Pt.** 131-132°C.

## Mammalian and avian toxicity

**Acute data**

$LD_{50}$ oral rat 2730 mg $kg^{-1}$ (1).

**Carcinogenicity and long-term effects**

$TD_{Lo}$ (route unspecified) (2 yr) rat 81,800 mg $kg^{-1}$ (total dose), carcinogenic effects (effects unspecified) (2).

## Legislation

Included in Schedule 4 (Release into Air: Prescribed Substances) Statutory Instrument No. 472, 1991 (3).

## Any other comments

Experimental toxicology, environmental effects and human health effects reviewed (4).

## References

1. *Chemical Hazard Information Profile Draft Report* 27/12/1983, U.S. EPA Office of Toxic Substances, Washington, DC
2. US EPA Office of Pesticides and Toxic Substances *8EHQ-0183-0165*
3. *S. I. 1991 No. 472 The Environmental Protection (Prescribed Processes and Substances) Regulations* 1991, HMSO, London
4. *ECETOC Technical Report No. 30(4)* 1991, European Chemical Industry Ecology and Toxicology Centre, B-1160 Brussels

# M244  *N*-Methyl-*N*-nitrosobenzamide

**CAS Registry No.** 63412-06-6
**Synonyms** MNB
**Mol. Formula** $C_8H_8N_2O_2$                        **Mol. Wt.** 164.17

## Mammalian and avian toxicity

**Acute data**
$LD_{50}$ intraperitoneal rat 70 mg $kg^{-1}$ (1).

## References

1.  *J. Natl. Cancer Inst.* 1979, **62**, 1523


# M245  *N*-(*N*-Methyl-*N*-nitrosocarbamoyl)-L-ornithine

$$HO_2C(NH_2)CH(CH_2)_3NHC(O)N(NO)CH_3$$

**CAS Registry No.** 63642-17-1
**Synonyms** *N*-(*N*-nitroso-*N*-methylcarbamoyl)-L-ornithine
**Mol. Formula** $C_7H_{14}N_4O_4$                        **Mol. Wt.** 218.21

## Mammalian and avian toxicity

**Carcinogenicity and long-term effects**
Induces acinar cell tumours in rat (1).
Induces duct-like pancreatic carcinomas in hamster (2).

## Genotoxicity

*In vivo* (alkaline elution analysis) and *in vitro* ♂ Syrian hamster, Lewis rat pancreatic acinar cell DNA damaged in a dose-related manner (3).

## References

1.  Longnecker, D. S. et al *J. Environ. Pathol. Toxicol.* 1980, **4**, 117-129
2.  Longnecker, D. S. et al *J. Natl. Cancer Inst.* 1983, **71**, 1327-1336
3.  Curphey, T. J. et al *Carcinogenesis (London)* 1987, **8**(8), 1033-1037

# M246  N-Methyl-N-nitroso-p-toluenesulfonamide

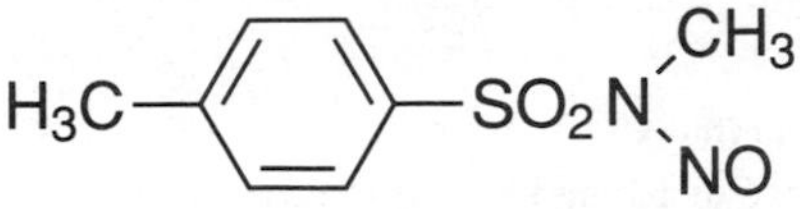

$$H_3C-\text{C}_6\text{H}_4-SO_2N(CH_3)(NO)$$

**CAS Registry No.** 80-11-5

**Synonyms** N,4-dimethyl-N-nitrosobenzenesulfonamide; Diazald; Diazale

**Mol. Formula** $C_8H_{10}N_2O_3S$      **Mol. Wt.** 214.24

**Uses** In preparation of diazomethane.

## Physical properties

**M. Pt.** 62°C.

**Solubility**

Organic solvent: diethyl ether, petroleum ether, benzene, chloroform, carbon tetrachloride

## Mammalian and avian toxicity

**Acute data**

$LD_{50}$ oral rat 2700 mg $kg^{-1}$ (1).

$LD_{50}$ oral redwing blackbird >100 mg $kg^{-1}$ (2).

$LD_{50}$ intraperitoneal mouse 19 mg $kg^{-1}$ (3).

**Carcinogenicity and long-term effects**

Reported to be non-carcinogenic (4).

## References

1. *Naturwissenschaften* 1961, **48**, 165
2. Schafer, E. W. et al *Arch. Environ. Contam. Toxicol.* 1983, **12**, 355-382
3. *Cancer Res.* 1970, **30**, 11
4. Fishbein, L. et al *Chemical Mutagens* 1970, Academic, London

# M247  N-Methyloloacrylamide

$$CH_2=CHC(O)NHCH_2OH$$

**CAS Registry No.** 924-42-5

**Synonyms** 2-propenamide, N-(hydroxymethyl)-; N-methanolacrylamide; monomethylolacrylamide; Uramine T80

**Mol. Formula** $C_4H_7NO_2$      **Mol. Wt.** 101.11

## Physical properties

**M. Pt.** 74°C.

## Mammalian and avian toxicity

### Acute data
$LD_{50}$ oral mouse, rat 420, 474 mg kg$^{-1}$, respectively (1,2).
$LD_{50}$ intraperitoneal rat 563 mg kg$^{-1}$ (3).

### Carcinogenicity and long-term effects
Gavage (2 yr) ♂/♀ rats and mice, 6 or 12 mg kg$^{-1}$ day$^{-1}$, no evidence of carcinogenicity in ♂/♀ rats, clear evidence of carcinogenicity in ♂/♀ mice based on increased incidence of neoplasms of the Harderian gland (4).

## Genotoxicity
*Salmonella typhimurium* TA97, TA98, TA100, TA1535 with or without metabolic activation, negative (4).
*In vitro* Chinese hamster ovary cells, sister chromatid exchanges and chromosomal aberrations were induced (metabolic activation unspecified) (4).
*In vitro* B6C3F1 mice bone marrow, no increase in micronucleated polychromatic erythrocytes was observed (4).

## Legislation
Included in Schedule 4 (Release into Air: Prescribed Substances) Statutory Instrument No. 472, 1991 (5).

## Any other comments
Toxicity reviewed (6).
Experimental toxicity and human health effects reviewed (7).

## References
1. *Jpn. J. Hyg.* 1979, **34**, 183
2. Grayson, M. et al (Eds.) *Kirk-Othmer Encyclopedia of Chemical Technology* 1978, 306, John Wiley & Sons, New York
3. *Biochem. Pharmacol.* 1970, **19**, 2591
4. *National Toxicology Program Research and Testing Division* 1992, Report No. TR-352, NIEHS, Research Triangle Park, NC
5. *S. I. 1991 No. 472 The Environmental Protection (Prescribed Processes and Substances) Regulations* 1991, HMSO, London
6. *BIBRA Toxicity Profile* 1991, British Industrial Biological Research Association, Carlshalton
7. *ECETOC Technical Report No. 30(4)* 1991, European Chemical Industry Ecology and Toxicology Centre, B-1160 Brussels

# M248   Z-Methyl oleate

$$\overset{\displaystyle Z}{\overset{\displaystyle \diagup\diagdown}{H_3CO_2C(H_2C)_7 \quad (CH_2)_7CH_3}}$$

**CAS Registry No.** 112-62-9

**Synonyms** 9-octadecenoic acid, methyl ester, (Z)-; *cis*-oleic acid, methyl ester; methyl *cis*-9-octadecenoate; Emerest 2301; Emerest 2801; Emery 2310

**Uses** Lubricant and chemical intermediate. In liposomes for experimental and therapeutic purposes (3).

**Occurrence** In a variety of fruits, flowers and plant extracts, including Eucalyptus oil (1).

In cellular membranes (2).

## Physical properties

**M. Pt.** −20°C; **B. Pt.** 168-170°C; **Flash point** 177°C; **Specific gravity** $d_4^{20}$ 0.874.

**Solubility**
Organic solvent: miscible with ethanol, diethyl ether

## Environmental fate

**Degradation studies**
Readily biodegraded by *Pseudomonas aeruginosa* strain (4).

## Mammalian and avian toxicity

**Carcinogenicity and long-term effects**
Dermal application of 0.05 ml 20% (v/v) in acetone $3 \times wk^{-1}$ for 1 yr had a promoting effect on skin tumours initiated by 7,12-dimethylbenz[*a*]anthracene; even without initiation it had some activity in lymphoma carcinogenesis (5).

## Any other comments

Human health effects, experimental toxicology, physico-chemical properties reviewed (6).

## References

1. Panda, R. et al *Indian For.* 1987, **113**(6), 434-440
2. Seifert, R. et al *Eur. J. Biochem.* 1987, **162**(3), 563-569
3. Bittman, R. et al *Biochem. Biophys. Acta* 1986, **863**(1), 115-120
4. Williams, J. et al *J. -Water Pollut. Control. Fed.* 1973, **45**(8), 1671-1681
5. Arffmann, E. et al *Acta Pathol. Microbiol. Scand.* 1974, **82**, 127-136
6. *ECETOC Technical Report No. 30(4)* 1991, European Chemical Industry Ecology and Toxicology Centre, B-1160 Brussels

# M249   2-Methylpentane

## CH3CH2CH2CH(CH3)2

**CAS Registry No.** 107-83-5
**Synonyms** isohexane
**Mol. Formula** $C_6H_{14}$                                      **Mol. Wt.** 86.18

### Physical properties

**M. Pt.** –154°C; **B. Pt.** 60°C; **Flash point** –23°C (closed cup); **Specific gravity** 0.653;
**Volatility** v. den. 3.00.

### Occupational exposure

**UN No.** 1208; **HAZCHEM Code** 3☒E; **Conveyance classification** flammable liquid.

### Environmental fate

#### Degradation studies

Incubation with natural flora in groundwater in the presence of other components of
high octane gasoline (100 $\mu l\, l^{-1}$). Biodegradation 6% after 192 hr at 13°C, initial
concentration 1.72 $\mu l\, l^{-1}$ (1).

### Mammalian and avian toxicity

#### Sub-acute data

Gavage (4 wk) ♂ F-344 rats (dose unspecified) 5 days $wk^{-1}$ nephrotoxicity was
observed (2).

#### Metabolism and pharmacokinetics

Pulmonary retention 15-18%. Absorption rate measured was 0.11 $\mu g\, cm^{-2}\, hr^{-1}$ in the
rat and is expected to be lower in humans (3).
*In vitro* rate of dermal absorption was found to be 0.11 $\mu g\, cm^{-2}\, hr^{-1}$ (4).

### Any Other Adverse Effects

An intraperitoneal 2 wk dose to ♀ rats of 1.5 g $kg^{-1}$ resulted in a 24 hr urinary
excretion of 2.64 $\mu g$ β-2-microglobulin and 352 $\mu g$ albumin (5).

### Legislation

Included in Schedule 4 (Release into Air: Prescribed Substances) Statutory Instrument
No. 472, 1991 (6).

### References

1. Jamison, V. W. et al *Proc. Third Int. Biodegrad. Symp.* 1976, Applied Science Publishers (Elsevier, Amsterdam)
2. Halder, C. A. et al *Toxicol. Ind. Health* 1985, **1**(3), 67-87
3. Deutsch-Wenzel, R. P. et al *Energy Res. Abstr.* 1988, **13**(5), Abstr. No. 9919
4. Tsuruta, H. *Ind. Health* 1982, **20**, 339-345
5. Bernard, A. M. et al *Toxicol. Lett.* 1989, **45**(2-3), 271-280
6. *S. I. 1991 No. 472 The Environmental Protection (Prescribed Processes and Substances) Regulations* 1991, HMSO, London

# M250   3-Methylpentane

CH3CH2CH(CH3)CH2CH3

**CAS Registry No.** 96-14-0
**Mol. Formula** $C_6H_{14}$                                       **Mol. Wt.** 86.18

## Physical properties

**B. Pt.** 63.3°C; **Flash point** –7°C; **Specific gravity** $d_4^{20}$ 0.664; **Partition coefficient** log $P_{ow}$ 3.60; **Volatility** v.p. 100 mmHg at 10.5°C; v. den. 2.97.

## Environmental fate

### Degradation studies

Incubation with natural flora in groundwater in the presence of the other components of high-octane gasoline (100 µl l$^{-1}$). Biodegradation 7% after 192 hr at 13°C initial concentration 1.30 µl l$^{-1}$ (1).

## Legislation

Included in Schedule 4 (Release into Air: Prescribed Substances) Statutory Instrument No. 472, 1991 (2).
The log $P_{ow}$ value exceeds the European Community recommended value 3.0 (6th and 7th amendments) (3).

## Any other comments

Physico-chemical properties, legislation and storage reviewed (4).

## References

1. Jamison, V. W. et al *Proc. Third Int. Biodegrad. Symp.* 1976, Applied Science Publishers
2. *S. I. 1991 No. 472 The Environmental Protection (Prescribed Processes and Substances) Regulations* 1991, HMSO, London
3. *1967 Directive on Classification, Packaging and Labelling of Dangerous Substances 67/548/EEC; 6th Amendment EEC Directive 79/831/EEC, 7th Amendment EEC Directive 91/32/EEC* 1991, HMSO, London
4. Institut National de Recherche et de Securite *Cah. Notes Doc.* 1989, **135**, 381-384.

# M251   2-Methylpentane-2,4-diol

CH3CH(OH)CH2C(OH)(CH3)CH3

**CAS Registry No.** 107-41-5
**Synonyms** 2,4-pentanediol, 2-methyl-; diolane; hexylene glycol; Isol; 1,1,3-trimethyltrimethylenediol
**Mol. Formula** $C_6H_{14}O_2$                              **Mol. Wt.** 118.18
**Uses** Cosmetics. Hydraulic brake fluid. Coupling agent for castor oil.

## Physical properties

**M. Pt.** −40°C; **B. Pt.** 197.1°C; **Flash point** 96°C; **Specific gravity** $d^{20}$ 0.9234; **Volatility** v.p. 0.05 mmHg at 20°C.

## Solubility

Organic solvent: diethyl ether

## Occupational exposure

**UK Long-term limit** 25 ppm (125 mg m$^{-3}$); **UK Short-term limit** 25 ppm (125 mg m$^{-3}$); **Supply classification** irritant.
**Risk phrases** ≥10% − Irritating to eyes and skin (R36/38)

## Ecotoxicity

**Fish toxicity**
$LC_{50}$ (96 hr) Mississippi silverside 10 g l$^{-1}$ (1).
$LC_{50}$ (24 hr) goldfish >5 g l$^{-1}$ (2).

**Invertebrate toxicity**
$EC_{50}$ (5 min) *Photobacterium phosphoreum* 3038 ppm Microtox test (3).

## Mammalian and avian toxicity

**Acute data**
$LD_{50}$ oral rat 3700 mg kg$^{-1}$ (4).
$LD_{50}$ oral mouse 3097 mg kg$^{-1}$ (5).
$LD_{50}$ oral rabbit 3200 mg kg$^{-1}$ (6).
$LD_{50}$ oral guinea pig 2800 mg kg$^{-1}$ (6).
$LD_{50}$ dermal rabbit 8560 mg kg$^{-1}$ (7).
$LD_{50}$ intraperitoneal rat 1500 mg kg$^{-1}$ (8).
$LD_{50}$ intraperitoneal mouse 1299 mg kg$^{-1}$ (9).

**Sub-acute data**
$TC_{Lo}$ inhalation human (15 month) 50 ppm, eye effects: irritation, diplopia, cataracts and eye ground. Pulmonary system effects: effects on respiration and respiratory pathway (7,10).

**Metabolism and pharmacokinetics**
Eliminated in the urine partly in conjugated forms (species unspecified) (11).

**Irritancy**
Dermal rabbit (24 hr) 465 mg, caused moderate irritation (12).
Eye rabbit 93 mg caused severe irritation (exposure unspecified) (13).

## Any other adverse effects to man

Eye irritant. Central nervous system depression produced by oral administration (14).

## Legislation

Included in Schedule 4 (Release into Air: Prescribed Substances) Statutory Instrument No. 472, 1991 (15).

## Any other comments

It has been defined as inert (16).

## References

1. Verschueren, K. *Handbook of Environmental Data of Organic Chemicals* 1983, Van Nostrand Reinhold Co. Inc., New York
2. Bridie, A. L. et al *Water Res.* 1979, **13**, 627-630
3. Kaiser, K. L. E. et al *Water Pollut. Res. J. Can.* 1991, **26**(3), 361-431
4. *Raw Material Data Handbook, Vol. 1 Organic Solvents* 1974, 68
5. *J. Am. Pharm. Assoc., Sci. Ed.* 1956, **45**, 669
6. *Fed. Proc.* 1945, **4**, 142
7. Deichmann, W. B. *Toxicology of Drugs and Chemicals* 1969, 312, 731, Academic Press, New York
8. *J. Pharm. Pharmacol.* 1959, **11**, 150
9. Lewis, R. J. (Ed.) *Sax's Dangerous Properties of Industrial Materials* 8th ed., 1992, Van Nostrand Reinhold, New York
10. *J. Ind. Hyg. Toxicol.* 1946, **28**, 262
11. Jacobsen, E. *Acta Pharmacol. Toxicol.* 1958, **14**, 207-213
12. *J. Pharmacol. Exp. Ther.* 1944, **82**, 377
13. *BIOFAX Industrial Bio-Test Laboratories, Inc., Data Sheets* 1970, 12-14
14. *Chemical Safety Data Sheets: Vol. 1 Solvents* 1989, RSC, London
15. *S. I. 1991 No. 472 The Environmental Protection (Prescribed Processes and Substances) Regulations* 1991, HMSO, London
16. Shell Chemical Corporation *Industrial Hygiene Bulletin* SC:57-101 and SC57-102, New York

# M252   3-Methyl-2,4-pentanedione

## $CH_3C(O)CH(CH_3)C(O)CH_3$

**CAS Registry No.** 815-57-6
**Synonyms** 3-methylacetoacetone
**Mol. Formula** $C_6H_{10}O_2$                 **Mol. Wt.** 114.15
**Uses** Organic synthesis.

## Physical properties

**B. Pt.** 172-174°C; **Flash point** 56°C; **Specific gravity** $d_4^{20}$ 0.981.

## Ecotoxicity

**Invertebrate toxicity**
$EC_{50}$ (15 min) *Photobacterium phosphoreum* 1400 ppm, Microtox test (1).

## Mammalian and avian toxicity

**Irritancy**
Irritating to the skin. Vapour or mist is irritating to the eyes, mucous membranes and upper respiratory tract (species unspecified) (2).

## References

1. Kaiser, K. L. E. et al *Water Pollut. Res. J. Can.* 1991, **26**(3), 361-431
2. Lenga, R. E. (Ed.) *Sigma-Aldrich Library of Chemical Safety Data* 2nd ed., 1988, **2**, 2402, Sigma-Aldrich, Milwaukee, WI

# M253  2-Methyl-1-pentanol

$$CH_3CH_2CH_2CH(CH_3)CH_2OH$$

**CAS Registry No.** 105-30-6
**Synonyms** amyl methyl alcohol; 1,3-dimethylbutanol; isohexyl alcohol; isopropyl dimethyl carbinol; methyl amyl alcohol; methyl isobutyl carbinol; 2-methylpent-1-ol; 2-methyl-2-propylethanol
**Mol. Formula** $C_6H_{14}O$                    **Mol. Wt.** 102.18
**Uses** Solvent. Organic synthesis.

## Physical properties

**B. Pt.** 148°C; **Flash point** 50°C; **Specific gravity** $d_4^{20}$ 0.8263; **Volatility** v.p. 1.1 mmHg at 20°C; v. den. 3.52.

## Solubility

Organic solvent: acetone, diethyl ether, ethanol

## Mammalian and avian toxicity

### Acute data

$LD_{50}$ dermal rat 1400 mg $kg^{-1}$ (1).
$LD_{50}$ dermal rabbit 3600 mg $kg^{-1}$ (1).

### Irritancy

Dermal rabbit (24 hr) 10 mg caused mild irritation (2).
750 µg instilled into rabbit eye for 24 hr caused severe irritation (3).

## Any other adverse effects to man

Inhalation human, 50 ppm caused lung irritation (exposure not specified) (4).

## Legislation

Included in Schedule 6 (Release into Land: Prescribed Substances) Statutory Instrument No. 472, 1991 (5).

## References

1. *AMA Arch. Ind. Hyg. Occup. Med.* 1954, **10**, 61
2. *Am. Ind. Hyg. Assoc. J.* 1962, **23**, 95
3. Marhold, J. V. *Prehled Prumyslove Toxikologie: Organicke Latky* 1986, Prague
4. *J. Ind. Hyg. Toxicol.* 1946, **28**, 262
5. *S. I. 1991 No. 472 The Environmental Protection (Prescribed Processes and Substances) Regulations* 1991, HMSO, London

# M254    2-Methyl-3-pentanol

## CH3CH2CH(OH)CH(CH3)2

**CAS Registry No.** 565-67-3
**Synonyms** ethyl isopropyl carbinol; 1-isopropylpropanol
**Mol. Formula** $C_6H_{14}O$                    **Mol. Wt.** 102.18
**Uses** Fuel.

### Physical properties
**B. Pt.** 128°C; **Flash point** 46°C; **Specific gravity** $d_4^{20}$ 0.8243; **Partition coefficient** log $P_{ow}$ 1.65 (1).

### Solubility
Organic solvent: miscible with diethyl ether, ethanol

### Mammalian and avian toxicity

#### Acute data
$LD_{50}$ intravenous mouse 320 mg $kg^{-1}$ (2).

### References
1. Leahy, D. E. *J. Pharm. Sci.* 1986, **75**(7), 629-636
2. *Report* US Army Armament Research and Development Command, Chemical Systems Laboratory, NIOSH Exchange Chemicals, Aberdeen Proving Ground, MD 21010

# M255    3-Methyl-3-pentanol

## CH3CH2C(CH3)(OH)CH2CH3

**CAS Registry No.** 77-74-7
**Synonyms** 3-pentanol, 3-methyl-; diethyl carbinol; 1-ethyl-1-methyl-1-propanol; 3-hydroxy-3-methylpentane; methyldiethyl carbinol
**Mol. Formula** $C_6H_{14}O$                    **Mol. Wt.** 102.18

### Physical properties
**M. Pt.** –38°C; **B. Pt.** 123°C; **Flash point** 46°C; **Specific gravity** 0.824.

### Solubility
Organic solvent: ethanol, diethyl ether

### Environmental fate

#### Degradation studies
In activated sludge, biodegradation resulted in formation of the corresponding fatty acid (1).

## Mammalian and avian toxicity

**Acute data**

$LD_{50}$ oral rat 710 mg kg$^{-1}$ (2).

$LD_{Lo}$ oral mouse 750 mg kg$^{-1}$ (3).

$LD_{50}$ subcutanous mouse 1100 mg kg$^{-1}$ (4).

## Legislation

Included in Schedule 4 (Release Into Air: Prescribed Substances) Statutory Instrument No. 472, 1991 (5).

## References

1. Nitsuma, T. et al *Tohoku Gakuin Daigaku Kogakubu Kenkyu Hokoku* 1988, **23**(1), 57-60 (Japan.)
2. *J. Pharmacol. Exp. Ther.* 1955, **115**, 230
3. Leube, F. *Narkoseversuche mit Hoheren Alkoholen und Stickstoffderivaten* 1931 (Ger.)
4. *Arzneim.-Forsch.* 1955, **5**, 161 (Ger.)
5. *S. I. 1991 No. 472 The Environmental Protection (Prescribed Processes and Substances) Regulations* 1991, HMSO, London

# M256   4-Methyl-2-pentanol

## (CH₃)₂CHCH₂CH(OH)CH₃

**CAS Registry No.** 108-11-2

**Synonyms** isobutyl methyl carbinol; isobutylmethylmethanol; methyl amyl alcohol; MAOH; MIBC; 3-MIC

**Mol. Formula** $C_6H_{14}O$                                  **Mol. Wt.** 102.18

**Uses** Solvent for dyes, oils, gums, resins, waxes, nitrocellulose and ethylcellulose. Organic synthesis. Froth flotation. Brake fluids.

**Occurrence** Trace amounts in some water samples.

## Physical properties

**M. Pt.** –90°C; **B. Pt.** 131.8°C; **Flash point** 41°C; **Specific gravity** $d^{20}$ 0.802; **Volatility** v.p. 2.8 mmHg at 20°C; v. den. 3.53.

**Solubility**

Water: 17 g l$^{-1}$ at 20°C. Organic solvent: ethanol, hydrocarbons, diethyl ether

## Occupational exposure

**UK Long-term limit** 25 ppm (100 mg m$^{-3}$); **UK Short-term limit** 40 ppm (160 mg m$^{-3}$); **UN No.** 2053; **HAZCHEM Code** 3▼; **Conveyance classification** flammable liquid; **Supply classification** irritant.

**Risk phrases** ≥25% – Flammable – Irritating to respiratory system – <25% – Flammable (R10, R37, R10)

**Safety phrases** Avoid contact with skin and eyes (S24/25)

## Ecotoxicity

**Fish toxicity**

$LC_{50}$ (24 hr) goldfish 360 mg l$^{-1}$ (1).

**Invertebrate toxicity**

$EC_{50}$ (duration and species unspecified) 2.45 g $l^{-1}$ (2).

**Environmental fate**

**Degradation studies**

Oxidation parameters, $BOD_5$ 2.12 mg $l^{-1}$ $O_2$ NEN3235 (Nederlandse norm)
5.4 mg $l^{-1}$ $O_2$; COD 2.60 mg $l^{-1}$ $O_2$ NEN3235 5.3 mg $l^{-1}$ $O_2$ (Nederlandse norm) (3).

**Mammalian and avian toxicity**

**Acute data**

$LD_{50}$ oral rat 2590 mg $kg^{-1}$ (4).
$LD_{Lo}$ oral mouse 1000 mg $kg^{-1}$ (5).
$LC_{Lo}$ (4 hr) inhalation rat 2000 ppm (6).
$LD_{50}$ intraperitoneal mouse 812 mg $kg^{-1}$ (7).
$LD_{50}$ dermal rabbit 3560 mg $kg^{-1}$ (4).

**Irritancy**

Dermal rabbit (24 hr) 10 mg caused mild irritation (4).
Eye rabbit 20 mg (uncovered) caused severe irritation (duration unspecified) (4).

**Genotoxicity**

*Salmonella typhimurium* and *Escherichia coli* with and without metabolic activation negative (8).

**Legislation**

Included in Schedule 4 (Release into Air: Prescribed Substances) Statutory Instrument No. 472, 1991 (9).

**Any other comments**

Metabolite of methyl isobutyl ketone in humans (10).

**References**

1. Bridie, A. L. et al *Water Res.* 1979, **13**, 623
2. Vaishnav, D. D. *Toxic Assess.* 1986, **1**(2), 227-240
3. Shell Chemie *Shell Industrie Chemicalien gids* 1/1/1975 Shell Nederland Chemie, Afd. Industrie-chemicalien, Wassenaarseweg 80, s-Gravenhage, Nederland
4. *Arch. Ind. Hyg. Occup. Med.* 1951, **4**, 119
5. *University of California, Pub. Pharmacol.* 1949, **2**, 217
6. Lewis, R. J. (Ed.) *Sax's Dangerous Properties of Industrial Materials* 8th ed., 1992, Van Nostrand Reinhold, New York
7. *J. Ind. Hyg. Toxicol.* 1949, **31**, 343
8. Shimizu, H. et al *Sangyo Igaku* 1985, **27**(6), 400-419
9. *S. I. 1991 No. 472 The Environmental Protection (Prescribed Processes and Substances) Regulations* 1991, HMSO, London
10. Hjelm, E. W. et al *Int. Arch. Occup. Environ. Health* 1990, **62**(1), 19-26

# M257   2-Methyl-1-pentene

$$CH_3CH_2CH_2C(CH_3)=CH_2$$

**CAS Registry No.** 763-29-1
**Synonyms** 2-methylpentene; 2-methylpent-1-ene; 1-methyl-1-propylethene
**Mol. Formula** $C_6H_{12}$                                  **Mol. Wt.** 84.16
**Uses** Organic synthesis.
**Occurrence** Has been detected in engine exhausts (1).

## Physical properties

**M. Pt.** –136°C; **B. Pt.** 62°C; **Flash point** –26°C; **Specific gravity** $d_{15.5}^{15.5}$ 0.684;
**Volatility** v.p. 326 mmHg at 37.3°C; v. den. 2.9.

## Solubility

Water: 78 mg $l^{-1}$ at 20°C. Organic solvent: benzene, chloroform, ethanol, petroleum ether

## Environmental fate

### Degradation studies

Degradation by activated sludge 1.0% of ThOD after 6 hr; 1.1% of ThOD after 12 hr; 1.7% of ThOD after 24 hr (initial concentration not specified) (2).

## Mammalian and avian toxicity

### Acute data

$LC_{50}$ (4 hr) inhalation rat 115 g m$^{-3}$ (3).
$LC_{50}$ (2 hr) inhalation mouse 130 g m$^{-3}$ (3).

### Irritancy

Vapour or mist is irritating to the eyes, mucous membranes and upper respiratory tract (species unspecified) (4).

## Any other adverse effects to man

Symptoms of exposure include burning sensation, coughing, wheezing, laryngitis, shortness of breath, headache, nausea and vomiting (3).

## Any other comments

Autoignition temperature 250°C.

## References

1. Lipari, F. *J. Chromatogr.* 1990, **503**(1), 51-68
2. Gerhold, R. M. et al *J. -Water Pollut. Control Fed.* 1966, **38**(4), 562
3. *Russ. Pharmacol. Toxicol. (Engl. Transl.)* 1968, **31**, 162
4. Lenga, R. E. (Ed.) *Sigma-Aldrich Library of Chemical Safety Data* 2nd ed., 1988, **2**, 2406, Sigma-Aldrich, Milwaukee, WI

# M258   2-Methyl-2-pentene

CH3CH2CH=C(CH3)CH3

**CAS Registry No.** 625-27-4
**Synonyms** 2-ethyl-1,1-dimethylethylene; 2-methylpent-2-ene
**Mol. Formula** $C_6H_{12}$                    **Mol. Wt.** 84.16
**Uses** Organic synthesis.
**Occurrence** In gasoline.

## Physical properties

**M. Pt.** −135°C; **B. Pt.** 67°C; **Flash point** −23°C; **Specific gravity** $d_4^{20}$ 0.690;
**Volatility** v. den. 2.9.

## Solubility

Organic solvent: benzene, carbon tetrachloride, chloroform, ethanol, petroleum ether

## Mammalian and avian toxicity

### Acute data

$LC_{50}$ (4 hr) inhalation rat 114 g m$^{-3}$ (1).
$LC_{50}$ (2 hr) inhalation mouse 130 g m$^{-3}$ (1).

### Sub-acute data

Gavage rat (4 wk) 500 or 2000 mg kg$^{-1}$ day$^{-1}$ 5 days wk$^{-1}$ for 4 wk caused fatality of
1/10 animals in each group (2).

### Irritancy

Irritating to the skin. Vapour or mist is irritating to the eyes, skin, mucous membranes
and upper respiratory tract (species unspecified) (3).

## References

1.  *Russ. Pharmacol. Toxicol. (Engl. Transl.)* 1968, **31**, 162
2.  Halder, C. A. et al *Toxicol. Ind. Health* 1985, **1**(3), 67-87
3.  Lenga, R. E. (Ed.) *Sigma-Aldrich Library of Chemical Safety Data* 2nd ed., 1988, **2**, 2406,
    Sigma-Aldrich, Milwaukee, WI

# M259   4-Methyl-1-pentene

CH3CH(CH3)CH2CH=CH2

**CAS Registry No.** 691-37-2
**Synonyms** isobutylethylene; 4-methylpent-1-ene
**Mol. Formula** $C_6H_{12}$                    **Mol. Wt.** 84.16
**Uses** Preparation of polymers.

## Physical properties

**M. Pt.** −153.6°C; **B. Pt.** 53-54°C; **Flash point** −31°C; **Specific gravity** $d^{20}$ 0.664;
**Volatility** v.p. 424 mmHg at 38°C; v. den. 2.9.

## Solubility

Water: 48 mg l$^{-1}$ at 20°C. Organic solvent: benzene, chloroform, ethanol, petroleum ether

## Environmental fate

### Degradation studies

Degradation by activated sludge 0.9% of ThOD after 6 hr; 1.4% of ThOD after 12 hr (initial concentration not specified) (1).

## Mammalian and avian toxicity

### Irritancy

Irritating to the skin. Vapour or mist is irritating to the eyes, skin, mucous membranes and upper respiratory tract (species unspecified) (2).

## Any other comments

Autoignition temperature 250°C.

## References

1. Gerhold, R. M. et al *J. -Water Pollut. Control Fed.* 1966, **38**(4), 562
2. Lenga, R. E. (Ed.) *Sigma-Aldrich Library of Chemical Safety Data* 2nd ed., 1988, **2**, 2407, Sigma-Aldrich, Milwaukee, WI

# M260    4-Methyl-2-pentene

## CH₃CH(CH₃)CH=CHCH₃

**CAS Registry No.** 4461-48-7
**Synonyms** 1-isopropyl-2-methylethylene; 4-methylpent-2-ene
**Mol. Formula** $C_6H_{12}$                    **Mol. Wt.** 84.16
**Uses** Organic synthesis.
**Occurrence** In fossil fuels.

## Physical properties

**M. Pt.** –134.4°C; **B. Pt.** 58°C; **Flash point** –32°C; **Specific gravity** $d_4^{20}$ 0.670; **Volatility** v. den. 2.90.

## Solubility

Organic solvent: benzene, chloroform, petroleum ether

## Environmental fate

### Degradation studies

Degradation by activated sludge 0.6% of ThOD after 6 hr; 1.3% of ThOD after 24 hr (initial concentration not specified) (1).

## References

1. Gerhold, R. M. et al *J. -Water Pollut. Control Fed.* 1966, **38**(4), 562

# M261   4-Methylpent-3-en-2-one

$$CH_3C(CH_3)=CHC(O)CH_3$$

**CAS Registry No.** 141-79-7

**Synonyms** 3-penten-2-one, 4-methyl-; isobutenyl methyl ketone; isopropylidene acetone; mesityl oxide; methyl 2,2-dimethylvinyl ketone; methyl isobutenyl ketone

**Mol. Formula** $C_6H_{10}O$                    **Mol. Wt.** 98.15

**Uses** Solvent for nitrocellulose, gums and resins. In lacquers, varnishes and enamels. In methyl isobutyl ketone synthesis.

## Physical properties

**M. Pt.** −59°C; **B. Pt.** 130°C; **Flash point** 30.5°C; **Specific gravity** $d_4^{20}$ 0.8539; **Volatility** 10 mmHg at 20°C; v. den. 3.38.

## Solubility

Water: 28 g $l^{-1}$ at 20°C. Organic solvent: ethanol, diethyl ether

## Occupational exposure

**US TLV (TWA)** 15 ppm (60 mg m$^{-3}$); **US TLV (STEL)** 25 ppm (100 mg m$^{-3}$); **UK Long-term limit** 15 ppm (60 mg m$^{-3}$); **UK Short-term limit** 25 ppm (100 mg m$^{-3}$); **UN No.** 1229; **HAZCHEM Code** 3W; **Conveyance classification** flammable liquid; **Supply classification** harmful.

**Risk phrases** ≥5% − Flammable − Harmful by inhalation, in contact with skin and if swallowed  <5% − Flammable (R10, R20/21/22, R10)

**Safety phrases** Avoid contact with eyes (S25)

## Ecotoxicity

**Fish toxicity**

LC$_{50}$ goldfish 540 mg $l^{-1}$ (1).

## Mammalian and avian toxicity

### Acute data

LD$_{50}$ oral rat 1120 mg kg$^{-1}$ (2).
LD$_{50}$ oral mouse 710 mg kg$^{-1}$ (3).
LD$_{50}$ oral rabbit 1000 mg kg$^{-1}$ (4).
LC$_{50}$ inhalation (4 hr) rat 9 g m$^{-3}$ (3).
LC$_{50}$ inhalation (2 hr) mouse 10 g m$^{-3}$ (3).
LD$_{50}$ dermal rabbit 5150 mg kg$^{-1}$ (5).
LD$_{50}$ intraperitoneal mouse 354 mg kg$^{-1}$ (6).

### Irritancy

Dermal rabbit 430 mg, open to atmosphere, caused mild irritation (7).
Eye rabbit 4325 µg caused severe irritation (duration unspecified) (8).
In human eye irritation was observed at 25 ppm and nasal irritation at 50 ppm (9).

### Any other adverse effects

Inhalation (4 hr) rats (dose unspecified), leucopenia without any change in differential or red blood cell counts was observed. Leucopenia was caused when exposure

---

reached irritant level (10).

Injury to lungs, liver and kidney were observed in animal experiments (11).

Inhalation studies in rats revealed hypertrophy of the liver, changes to the kidney and spleen, and anaemia and other changes to the blood (12).

### Legislation

Included in Schedule 4 (Release into Air: Prescribed Substances) Statutory Instrument No. 472, 1991 (13).

### Any other comments

Body temperature effects, effects on heart rate and convulsions have been reported (14,15). Hazardous properties reviewed (16).

### References

1.  Bridie, A. L. et al *Water Res.* 1979, **13**, 627-630
2.  Hann, W. *Water Quality Characteristics of Hazardous Materials* 1974, **4**, Civil Engineering Department, Texas A & M University
3.  Izmerov, N. F. et al *Toxicometric Parameters of Industrial Toxic Chemicals under Single Exposure* 1982, CIP, Moscow
4.  *Shell Chemical Company, Technical Data Bulletin* 1971
5.  *Raw Material Data Handbook* 1974, **1**, 71
6.  Lewis, R. J. (Ed.) *Sax's Dangerous Properties of Industrial Materials* 8th ed., 1992, Van Nostrand Reinhold, New York
7.  *Union Carbide Data Sheet* 11/3/1971
8.  *Am. J. Ophthalmol.* 1946, **29**, 1363
9.  Silverman, L. J. *J. Ind. Hyg. Toxicol.* 1946, **28**, 262-266
10. Brondeau, M. T. et al *J. Appl. Toxicol.* 1990, **10**(2), 83-86
11. Smyth, H. F. et al *J. Ind. Hyg. Toxicol.* 1942, **24**, 46
12. Ito, S. *Yokohama Igaku* 1969, **20**(3), 253-255
13. *S. I. 1991 No. 472 The Environmental Protection (Prescribed Processes and Substances) Regulations* 1991, HMSO, London
14. Specht, H. *U.S. Publ. Health Bull.* 1940, No. 176, 1-66
15. *Fed. Reg.* 20 Dec 1987, **50**(245), 51857-51867
16. *Dangerous Prop. Ind. Mater. Rep.* 1989, **9**(5), 58-65

# M262   Methyl phenylacetate

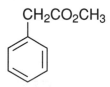

$CH_2CO_2CH_3$

**CAS Registry No.** 101-41-7

**Synonyms** methyl phenylethanoate; methyl benzeneacetate

**Mol. Formula** $C_9H_{10}O_2$ **Mol. Wt.** 150.18

**Uses** Organic synthesis. In perfumes.

**Occurrence** Aroma component of cooked meat and many plants.

**Physical properties**

**B. Pt.** 218°C; **Flash point** 89°C; **Specific gravity** $d^{20}$ 1.044; **Partition coefficient** log $P_{ow}$ 1.83; **Volatility** v. den. 5.18.

**Solubility**
Organic solvent: ethanol, fixed oils

**Mammalian and avian toxicity**

**Acute data**
$LD_{50}$ oral rat 2600 mg $kg^{-1}$ (1).
$LD_{50}$ dermal rabbit 2400 mg $kg^{-1}$ (1).

**Irritancy**
Dermal rabbit (24 hr) 500 mg caused irritation (1).

**References**

1. *Food Cosmet. Toxicol.* 1974, **12**, 807

# M263   Methyl phosphonic dichloride

## $CH_3P(O)Cl_2$

**CAS Registry No.** 676-97-1
**Synonyms** phosphonic dichloride, methyl-
**Mol. Formula** $CH_3Cl_2OP$                          **Mol. Wt.** 132.91
**Uses** Chemical intermediate and dehydrating agent.

**Physical properties**
**M. Pt.** 32°C; **B. Pt.** 162°C.

**Mammalian and avian toxicity**

**Acute data**
$LC_{50}$ (4 hr) inhalation rat 26 ppm (1).
Compound is an inhibitor of cholinesterases (2).

**References**

1. *Am. Ind. Hyg. Assoc. J.* 1964, **25**, 470
2. Ashani, Y. et al *Biochemistry* 1990, **29**(10), 2456-2463

# M264  1-Methylpiperidine

**CAS Registry No.** 626-67-5
**Synonyms** *N*-methylpiperidine
**Mol. Formula** $C_6H_{13}N$                    **Mol. Wt.** 99.18

## Physical properties
**B. Pt.** 107°C; **Flash point** 3°C; **Specific gravity** $d^{15}$ 0.821; **Partition coefficient** $\log P_{ow}$ 1.30.

## Occupational exposure
**UN No.** 2399; **HAZCHEM Code** 2WE; **Conveyance classification** flammable liquid.

## Mammalian and avian toxicity

**Acute data**
$LD_{Lo}$ subcutaneous rabbit 400 mg $kg^{-1}$ (1).

## Legislation
Included in Schedule 4 (Release into Air: Prescribed Substances) Statutory Instrument No. 472, 1991 (2).

## References

1.  *Ber. Dtsch. Chem. Ges., Abt. B: Abhand.* 1901, **32**, 2408
2.  *S. I. 1991 No. 472 The Environmental Protection (Prescribed Processes and Substances) Regulations* 1991, HMSO, London

# M265  2-Methylpropanoic acid, 2,2,4-trimethylpentyl ester

$$(CH_3)_2CHCO_2CH_2C(CH_3)_2CH_2CH(CH_3)_2$$

**CAS Registry No.** 36679-74-0
**Synonyms** 2,2,4-trimethylpentyl isobutyrate; 2,2,4-trimethylpentyl 2-methylpropanoate
**Mol. Formula** $C_{12}H_{24}O_2$                    **Mol. Wt.** 200.32

## Physical properties
**B. Pt.** $_{747}$ 199-202°C; **Specific gravity** $d^{20}$ 0.855.

## Mammalian and avian toxicity

### Acute data
$LD_{50}$ oral rat 75 ml $kg^{-1}$ (toxic effects included ataxia and effects on the eyes and salivary glands) (1).
$LD_{50}$ dermal rabbit >16 ml $kg^{-1}$ (1).

### Irritancy
Dermal rabbit, 500 mg caused irritation. 100 mg instilled into rabbit eye caused irritation (exposure unspecified) (1).

### References
1. *Acute Toxicity Data* 1992, **1**, 193

# M266   2-Methyl-2-propenoic acid, butyl ester

$$CH_2=C(CH_3)CO_2CH_2CH_2CH_2CH_3$$

**CAS Registry No.** 97-88-1
**Synonyms** methacrylic acid, butyl ester; butyl methacrylate
**Mol. Formula** $C_8H_{14}O_2$          **Mol. Wt.** 142.20

## Physical properties
**B. Pt.** 163°C; **Flash point** 50°C; **Specific gravity** $d_4^{20}$ 0.895; **Partition coefficient** log $P_{ow}$ 2.88; **Volatility** v.p. 4.9 mmHg at 20°C; v. den. 4.8.

## Occupational exposure
**Supply classification** irritant.
**Risk phrases** Flammable – Irritating to eyes, respiratory system and skin – May cause sensitisation by skin contact (R10, R36/37/38, R43)

## Ecotoxicity

### Fish toxicity
$LC_{50}$ (72 hr) goldfish 5.52 mg $l^{-1}$ (1).

### Invertebrate toxicity
$EC_{50}$ (30 min) *Photobacterium phosphoreum* 55.3 mg $l^{-1}$ Microtox test (2).

## Mammalian and avian toxicity

### Acute data
$LD_{50}$ oral rat 22,600 mg $kg^{-1}$ (3).
$LD_{50}$ oral mouse 13,500 mg $kg^{-1}$ (4).
$LD_{Lo}$ oral rabbit 6270 mg $kg^{-1}$ (5).
$LC_{50}$ (4 hr) inhalation rat 4910 ppm (6).
$LD_{50}$ dermal rabbit 11,300 mg $kg^{-1}$ (3).
$LD_{50}$ intraperitoneal rat 2304 mg $kg^{-1}$ (7).
$LD_{50}$ intraperitoneal mouse 1490 mg $kg^{-1}$ (8).

### Metabolism and pharmacokinetics

Intraperitoneal rat (concentration unspecified) quickly absorbed and accumulated in liver, kidney, blood, heart and brain (9).

### Genotoxicity

*Salmonella typhimurium* TA98, TA100, TA1535, TA1537, TA1538 with and without metabolic activation negative (10,11).

### Legislation

Included in Schedule 4 (Release into Air: Prescribed Substances) Statutory Instrument No. 472, 1991 (12).

### References

1. Reinert, K. H. *Regul. Toxicol. Pharmacol.* 1987, **7**, 384-389
2. Kaiser, K. L. E. et al *Water Pollut. Res. J. Can.* 1991, **26**(3), 361-431
3. *Am. Ind. Hyg. Assoc. J.* 1969, **30**, 470
4. *Gig. Tr. Prof. Zabol.* 1975, **19**(9), 57
5. *J. Ind. Hyg. Toxicol.* 1941, **23**, 343
6. *J. Toxicol. Environ. Health* 1985, **16**, 811.
7. *J. Dental Res.* 1972, **51**, 1632
8. *J. Pharm. Sci.* 1973, **62**, 778
9. Svetlakov, A. V. et al *Gig. Tr. Prof. Zabol.* 1989, (3), 51-52 (Russ.) (*Chem. Abstr.* **111**, 110595r)
10. Waegemakers, T. H. et al *Mutat. Res.* 1984, **137** (2-3), 95-102
11. Zeiger, E. et al *Environ. Mutagen.* 1987, **9**(Suppl. 9) 1-109
12. *S. I. 1991 No. 472 The Environmental Protection (Prescribed Processes and Substances) Regulations* 1991, HMSO, London

# M267    1-Methylpropyl acetate

$$CH_3CO_2CH(CH_3)CH_2CH_3$$

**CAS Registry No.** 105-46-4
**Synonyms** acetic acid, 1-methylpropyl ester; acetic acid, *sec*-butyl ester; 2-acetoxybutane; *sec*-butyl acetate; 2-butyl acetate; *sec*-butyl alcohol acetate
**Mol. Formula** $C_6H_{12}O_2$                    **Mol. Wt.** 116.16

### Physical properties

**B. Pt.** 112-113°C (*dl*-form) 116-117°C (*d*-form); **Flash point** 31°C (open cup) (*dl*-form); **Specific gravity** $d_4^{25}$ 0.865 (*dl*-form).

### Solubility

Organic solvent: ethanol, diethyl ether

### Occupational exposure

**Supply classification** highly flammable.
**Risk phrases** Highly flammable (R11)
Safety phrases Keep away from sources of ignition – No Smoking –
Do not breathe vapour – Do not empty into drains – Take precautionary measures against static discharges (S16, S23, S29, S33)

## Mammalian and avian toxicity

### Acute data
Exposure to 10,000 ppm for 5 hr caused irritation and death in guinea pigs (1).

### Legislation
Included in Schedule 4 (Release into Air: Prescribed Substances) Statutory Instrument No. 472, 1991 (2).

### Any other comments
Human health effects, experimental toxicology, physico-chemical properties, epidemiology and workplace experience reviewed (3).

### References

1. Tagaki, K. et al *J. Pharm. Pharmacol.* 1966, **18**(12), 795-800
2. *S. I. 1991 No. 472 The Environmental Protection (Prescribed Processes and Substances) Regulations* 1991, HMSO, London
3. *ECETOC Technical Report No. 30(4)* 1991, European Chemical Industry Ecology and Toxicology Centre, B-1160 Brussels

# M268  Methyl propyl ketone

$$CH_3C(O)CH_2CH_2CH_3$$

**CAS Registry No.** 107-87-9
**Synonyms** methyl isopropyl ketone; 2-pentanone; ethylacetone; pentan-2-one
**Mol. Formula** $C_5H_{10}O$                    **Mol. Wt.** 86.13
**Uses** Solvent.
**Occurrence** Volatile emission product from a variety of foods including smoked meat (1). Mammalian urinary product particularly during starvation (2).

## Physical properties

**M. Pt.** −78°C; **B. Pt.** 102°C; **Flash point** 7.22°C; **Specific gravity** $d_4^{20}$ 0.809; **Partition coefficient** log $P_{ow}$ 0.91.

## Solubility
Organic solvent: miscible with ethanol, diethyl ether

## Occupational exposure

**US TLV (TWA)** 200 ppm (705 mg m$^{-3}$); **US TLV (STEL)** 250 ppm (881 mg m$^{-3}$); **UK Long-term limit** 200 ppm (700 mg m$^{-3}$); **UK Short-term limit** 250 ppm (875 mg m$^{-3}$); **UN No.** 1249; **HAZCHEM Code** 3▪E; **Conveyance classification** flammable liquid.

## Mammalian and avian toxicity

### Acute data
LD$_{50}$ oral rat 3.7 g kg$^{-1}$ (3).
LC$_{50}$ (4 hr) inhalation rat 2000 ppm (3).

$LD_{50}$ dermal rabbit 6.5 g $kg^{-1}$ (4).
$LD_{50}$ intraperitoneal rat 1.25 mg $kg^{-1}$ (4).

**Irritancy**
Irritant to soft mucous tissues including the lungs and upper respiratory tract (5-7).
Irritant to eyes (5).

**Genotoxicity**
Weak inducer of aneuploidy in *Saccharomyces cerevisiae* D61.M (8).

**Any other adverse effects to man**
The defatting action on skin can lead to dermatitis (7).
Prolonged exposure can cause headache, drowsiness and ultimately death (7).

**Any other comments**
Human health effects, experimental toxicology, physico-chemical properties reviewed (9).

**References**

1. Wittkowski, R. et al *Food Chem.* 1990, **37**(2), 135-144
2. Yancey, M. et al *J. Chromatogr.* 1986, **382**, 3-18
3. Smyth, H. F. et al *Am. Ind. Hyg. Assoc. J.* 1962, **23**, 95
4. *Raw Material Data Handbook* 1974, **1**, 83
5. Douglas, R. B. et al *Ann. Occup. Hyg.* 1987, **31**(2), 265-267
6. Abraham, M. H. *Quant. Struct.-Act. Relat.* 1990, **9**(1), 6-10
7. *Material Safety Data Sheet* 1990, M and B Laboratory Products
8. Zimmerman, F. K. et al *Mutat. Res.* 1985, **149**, 339
9. *ECETOC Technical Report No. 30(4)* 1991, European Chemical Industry Ecology and Toxicology Centre, B-1160 Brussels

# M269    2-Methylpyrazine

**CAS Registry No.** 109-08-0
**Synonyms** pyrazine, 2-methyl-
**Mol. Formula** $C_5H_6N_2$                    **Mol. Wt.** 94.12
**Uses** Chemical intermediate.
**Occurrence** In the aroma of a variety of cooked foods of vegetable and animal origin including coffee (1) and boiled shrimps (2).
Formed by thermal pretreatment of sewage sludge (3).

**Physical properties**
**M. Pt.** −29°C; **B. Pt.** $_{737}$133°C; **Flash point** 50°C; **Specific gravity** $d^{25}$ 1.030.

**Solubility**
Water: miscible. Organic solvent: ethanol, acetone

## Ecotoxicity

### Invertebrate toxicity

$EC_{50}$ 30 min *Photobacterium phosphoreum* 430 ppm Microtox test (4).

### Mammalian and avian toxicity

#### Acute data

$LD_{50}$ oral rat 1.8 g kg$^{-1}$ (5).
$LD_{50}$ intraperitoneal mouse 1.82 g kg$^{-1}$ (6).
The compound has central depressant, weak hypnotic and anticonvulsant actions with hypnotic doses $\approx$ 1.25 g kg$^{-1}$ intraperitoneally in mice and anticonvulsant doses of 130-580 mg kg$^{-1}$ intraperitoneally (6).

### Genotoxicity

In mutagenicity tests with *Salmonella typhimurium* TA100 and TA102 methylpyrazine was judged not to contribute to the positive result seen with coffee (2). *Salmonella typhimurium* TA100, TA98, TA1537 with and without metabolic activation negative. Induced chromosomal aberrations in Chinese hamster ovary cells with or without metabolic activation (7).

### Any other comments

All reasonable efforts have been taken to find information on isomers of this compound, but no relevant data are available.

### References

1. Shenderyuk, V. V. et al *Rybn. Khoz. (Moscow)* 1990, (1), 86-89 (Russ.) (*Chem. Abstr.* **112**, 177182t)
2. Aeschbacher, H. V. et al *Food Chem. Toxicol.* 1989, **27**(4), 227-232
3. Pinnekamp, J. *Gewaesserschutz, Wasser, Abwasser* 1986, **85**, 331-358 (Ger.) (*Chem. Abstr.* **107**, 160853m)
4. Kaiser, K. L. E. et al *Water Pollut. Res. J. Can.* 1991, 26(3), 361-431
5. *Drug Chem. Toxicol. (1977)* 1980, **3**, 249
6. *Toxicol. Appl. Pharmacol.* 1970, **17**, 244
7. Stich, H. F. et al *Food Chem. Toxicol.* 1980, **18**, 581-584

# M270   1-Methylpyrene

**CAS Registry No.** 2381-21-7
**Synonyms** pyrene, 1-methyl-
**Mol. Formula** $C_{17}H_{12}$                    **Mol. Wt.** 216.29

**Occurrence** In diesel exhaust and as air and water pollutant. Present in sediment of some urban waters (1).

## Physical properties
**M. Pt.** 70-75°C; **B. Pt.** 410°C.

## Ecotoxicity
### Bioaccumulation
1-Methylpyrene enters the food chain, but is biotransformed. In the eider duck on the Baltic coast the distribution between tissues was gall bladder > adipose tissue > liver. Compound was also detected in seston and blue mussel, the latter a food source for the ducks (2).

## Environmental fate
### Absorption
1-Methylpyrene is retained by aquatic humic substances in lake water (3).

## Mammalian and avian toxicity
### Carcinogenicity and long-term effects
Newborn mouse bioassay, 42 mg kg$^{-1}$ intraperitoneally for 3 days induced liver neoplasms (4).

### Teratogenicity and reproductive effects
Compound is toxic to chick embryos (5).

## Genotoxicity
*Salmonella typhimurium* TM677 with metabolic activation positive (6).
*In vitro* primary hepatocyte culture unscheduled DNA synthesis positive (4).

## Legislation
Limited under EC Directive on Drinking Water Quality 80/778/EEC. Polycyclic aromatic hydrocarbons: maximum admissible concentration 0.2 µg l$^{-1}$ (7).

## References
1. Broman, D. et al *Bull. Environ. Contam. Toxicol.* 1987, **38**(6), 1020-1028
2. Broman, D. et al *Environ. Toxicol. Chem.* 1990, **9**(4), 429-442
3. Johnsen, S. et al *Sci. Total Environ.* 1987, **62**, 13-25
4. Rice, J. E. et al *J. Toxicol. Environ. Health* 1987, **21**(4), 525-532
5. Brunstroem, B. *Environ. Pollut.* 1990, **67**(2), 133-143
6. *Cancer Res.* 1979, **39**, 4152
7. *EC Directive Relating to the Quality of Water Intended for Human Consumption* 1982, 80/778/EEC, Office for Official Publications of the European Communities, 2 rue Mercier, L-2985 Luxembourg

# M271   2-Methylpyrene

**CAS Registry No.** 3442-78-2
**Synonyms** pyrene, 2-methyl-
**Mol. Formula** $C_{17}H_{12}$                    **Mol. Wt.** 216.29
**Occurrence** Pollutant of soil, water and atmosphere. Present in sediment of some urban waters (1).

## Physical properties

**M. Pt.** 142-144°C; **B. Pt.** 410°C.

## Ecotoxicity

### Bioaccumulation

2-Methylpyrene enters the food chain, but is biotransformed. It has been detected in eider ducks living on the Baltic coast and in seston and their food source the blue mussel. Tissue concentrations were gall bladder >adipose tissue >liver (2).

## Mammalian and avian toxicity

### Carcinogenicity and long-term effects

Compound has been included in a CASE-SAR analysis (3).

### Teratogenicity and reproductive effects

Compound is toxic to chick embryos (4).

## Legislation

Limited under EC Directive on Drinking Water Quality 80/778/EEC. Polycyclic aromatic hydrocarbons: maximum admissible concentration 0.2 µg $l^{-1}$ (5).

## References

1. Broman, D. et al *Bull. Environ. Contam. Toxicol.* 1987, **38**(6), 1020-1028
2. Broman, D. et al *Environ. Toxicol. Chem.* 1990, **9**(4), 429-442
3. Richard, A. M. et al *Mutat. Res.* 1990, **242**(4), 285-303
4. Brunstroem, B. et al *Environ. Pollut.* 1990, **67**(2), 133-143
5. *EC Directive Relating to the Quality of Water Intended for Human Consumption* 1982, 80/778/EEC, Office for Official Publications of the European Communities, 2 rue Mercier, L-2985 Luxembourg

# M272   4-Methylpyrene

**CAS Registry No.** 3553-12-6
**Synonyms** pyrene, 4-methyl-
**Mol. Formula** $C_{17}H_{12}$                    **Mol. Wt.** 216.29
**Occurrence** Environmental pollutant. Present in sediment of some urban waters (1), and in topsoils polluted by synthetic-rubber manufacture (2).

## Physical properties
**M. Pt.** 222-225°C.

## Mammalian and avian toxicity

### Carcinogenicity and long-term effects
Compound has been included in a CASE-SAR analysis of potential carcinogenicity (3).

## Legislation
Limited under EC Directive on Drinking Water Quality 80/778/EEC. Polycyclic aromatic hydrocarbons: maximum admissible concentration 0.2 µg l$^{-1}$ (4).

## References

1. Broman, D. et al *Bull. Environ. Contam. Toxicol.* 1987, 38(6), 1020-1028
2. Nikiforova, E. M. et al *Pochvovedenie* 1989, (2), 70-78 (Russ.) (*Chem. Abstr.* **111**, 38455a)
3. Richard, A. M. et al *Mutat. Res.* 1990, **242**(4), 285-303
4. *EC Directive Relating to the Quality of Water Intended for Human Consumption* 1982, 80/778/EEC, Office for Official Publications of the European Communities, 2 rue Mercier, L-2985 Luxembourg

# M273   2-Methylpyridine

**CAS Registry No.** 109-06-8
**Synonyms** pyridine, 2-methyl-; 2-picoline; α-methylpyridine; *o*-picoline; α-picoline
**Mol. Formula** $C_6H_7N$                    **Mol. Wt.** 93.13
**Uses** Solvent. Dye and resin intermediate.
**Occurrence** In coal tar and bone oil. Produced in coal and shale oil gasification. Identified in cigarette smoke (1).

## Physical properties

**M. Pt.** –70°C; **B. Pt.** 128-129°C; **Specific gravity** $d_4^{15}$ 0.950; **Partition coefficient** log $P_{ow}$ 1.06 (1); **Volatility** v.p. 8 mmHg at 20°C.

## Solubility

Water: miscible. Organic solvent: ethanol, diethyl ether

## Occupational exposure

**UN No.** 2313; **HAZCHEM Code** 2S; **Conveyance classification** flammable liquid; **Supply classification** harmful.
**Risk phrases** Flammable – Harmful by inhalation, in contact with skin and if swallowed – Irritating to eyes and respiratory system (R10, R20/21/22, R36/37)
**Safety phrases** In case of contact with eyes, rinse immediately with plenty of water and seek medical advice – Wear suitable protective clothing (S26, S36)

## Ecotoxicity

### Invertebrate toxicity

$LC_{100}$ (24 hr) *Tetrahymena pyriformis* 6.0 g $l^{-1}$ (2).
$EC_{50}$ (30 min) *Photobacterium phosphoreum* 109 ppm Microtox test (3).
Toxic to *Nitzchia closterium* at 93.7 mg $l^{-1}$. Reported in oil shale waste waters at 3.7-46.8 mg $l^{-1}$ (4).

### Bioaccumulation

Bioconcentration factor of 4 predicted (5).
Bioconcentration not significant in organisms (6).

## Environmental fate

### Nitrification inhibition

100 mg $l^{-1}$ (activated sludge) inhibited $NH_3$ oxidation by 40% (7).

### Carbonaceous inhibition

Completely removed in 8 months under methanogenic conditions in aquifer slurries (8).

### Anaerobic effects

Completely removed in 8 months under sulfate-reducing conditions in aquifer slurries (8).

### Degradation studies

ThOD 2.75 mg $l^{-1}$ $O_2$ (9).
Readily biodegradable (10).
93.7 mg with a fertile garden soil inoculum 100% degraded in 14-32 days (11).
187 µg $g^{-1}$ incubated in silt loam soil, 2.7% remained after 16 days (12).

### Abiotic removal

Reacts with hydroxyl radicals and scavenged by rain in the atmosphere $t_{1/2}$ 11.2 days (6).

## Mammalian and avian toxicity

### Acute data

$LD_{50}$ oral mouse, rat, guinea pig 674-900 mg $kg^{-1}$ (13).

LD$_{50}$ oral redwing blackbird, starling, quail >1000 mg kg$^{-1}$ (14).
LC$_{Lo}$ (4 hr) inhalation rat 4000 ppm (15).
LD$_{50}$ dermal rabbit 410 mg kg$^{-1}$ (5).
LD$_{50}$ intraperitoneal rat 200 mg kg$^{-1}$ (16).

**Carcinogenicity and long-term effects**
Carcinogenicity not determined, but expected to be non-carcinogenic from structure (17).

## Genotoxicity
*Salmonella typhimurium* TA97, TA98, TA100, TA102 with and without metabolic activation negative (18).

## Legislation
Included in Schedule 4 (Release into Air: Prescribed Substances) Statutory Instrument No. 472, 1991 (19).

## Any other comments
Human health effects, experimental toxicology, environmental effects and physico-chemical properties reviewed (20).
Characteristic sweet odour.

## References
1.  Graedel, T. E. *Chemical Compounds in the Atmosphere* 1978, Academic Press, New York
2.  Schultz, T. W. et al *Arch. Environ. Contam. Toxicol.* 1978, **7**, 457-463
3.  Kaiser, K. L. E. et al *Water Pollut. Res. J. Can.* 1991, **26**(3), 361-431
4.  Mann, K. et al *Fuel* 1987, **66**, 404-407
5.  Lyman, W. K. et al *Handbook of Chemical Property Estimation Methods: Environmental Behaviour of Organic Compounds* 1982, McGraw-Hill, New York
6.  Howard, P. H. (Ed.) *Handbook of Fate and Exposure Data for Organic Chemicals* 1990, **2**, Lewis Publishers, Chelsea, MI
7.  Stafford, D. A. *J. Appl. Bacteriol.* 1974, **37**, 75-82
8.  Kuhn, E. P. et al *Environ. Toxicol. Chem.* 1989, **8**, 1149-1158
9.  Meinck, F. et al *Les Eaux Residuaires Industrielles* 1970
10. *The list of the existing chemical substances tested on biodegradability by microorganisms or bioaccumulation in fish body* 1987, Chemicals Inspection and Testing Institute, Japan
11. Naile, M. N. et al *Soil Biol. Biochem.* 1972, **4**, 313-323
12. Sims, G. K. et al *J. Environ. Qual.* 1985, **14**, 480-484
13. *Hyg. Sanit.* 1968, **33**, 341
14. Schafer, E. W. et al *Arch. Environ. Contam. Toxicol.* 1983, **12**, 355-382
15. *AMA, Arch. Ind. Hyg. Occup. Med.* 1951, **4**, 119
16. *Fundam. Appl. Toxicol.* 1985, **5**, 920
17. Sakamoto, Y. et al *Bull. Chem. Soc. Jpn.* 1989, **62**, 330-332
18. Claxton, L. D. et al *Mutat. Res.* 1987, **176**(2), 185-188
19. *S. I. 1991 No. 472 The Environmental Protection (Prescribed Processes and Substances) Regulations* 1991, HMSO, London
20. *ECETOC Technical Report No. 30(4)* 1991, European Chemical Industry Ecology and Toxicology Centre, B-1160 Brussels

# M274   3-Methylpyridine

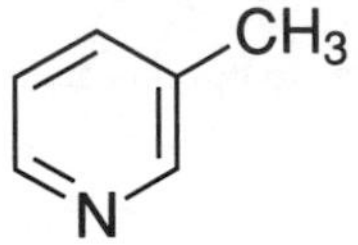

**CAS Registry No.** 108-99-6

**Synonyms** pyridine, 3-methyl-; 3-picoline, β-methylpyridine; *m*-picoline; β-picoline

**Mol. Formula** $C_6H_7N$                    **Mol. Wt.** 93.13

**Uses** Solvent. Dye and resin intermediate. In manufacture of waterproofing agents, insecticides, niacin and niacinamide.

## Physical properties

**B. Pt.** 143-144°C; **Specific gravity** $d_4^{15}$ 0.9613; **Partition coefficient** log $P_{ow}$ 1.20.

### Solubility
Water: miscible. Organic solvent: ethanol, acetone, diethyl ether (miscible)

## Occupational exposure

**UN No.** 2313; **HAZCHEM Code** 2S; **Conveyance classification** Flammable liquid.

## Ecotoxicity

### Invertebrate toxicity
$EC_{50}$ (30 min) *Photobacterium phosphoreum* 74.0 ppm Microtox test (1).
$EC_{50}$ (2.5 day) *Tetrahymena pyriformis* 862.4 mg $l^{-1}$ (2).

## Environmental fate

### Nitrification inhibition
100 mg $l^{-1}$ (activated sludge) did not inhibit $NH_3$ oxidation (3).

### Anaerobic effects
Only slightly degraded in methanogenic aquifer slurry (4).
Partially transformed (26%) in 8 months under sulfate-reducing conditions in aquifer slurries (4).

### Degradation studies
Adapted or fresh sludge is able to biodegrade wastewater containing ≥312 mg $l^{-1}$ pyridine bases (5).
ThOD, 2.75 mg $kg^{-1}$ $O_2$; COD, 4% ThOD (0.05 N $K_2Cr_2O_7$); $KMnO_4$, 2% ThOD (0.01 N $KMnO_4$) (6).
River water oxidation substrate, chemical analysis, 20% for 2 days. Observed feed, 1 mg $l^{-1}$ 100% removed after 16 day acclimation (7).

## Mammalian and avian toxicity

### Acute data
$LD_{50}$ oral redwing blackbird, quail, starling >1000 mg $kg^{-1}$ (8).

---

LD$_{50}$ intraperitoneal rat, mouse 150-596 mg kg$^{-1}$ (9, 10).
LD$_{50}$ intravenous mouse 298 mg kg$^{-1}$ (10).

**Carcinogenicity and long-term effects**

Carcinogenicity not determined, but from structure expected to be non-carcinogenic (11).

**Irritancy**

A man occupationally exposed to a number of chemicals, predominantly 3-methylpyridine, developed skin eruptions on his face (12).

## Genotoxicity

*Salmonella typhimurium* TA97, TA98, TA100, TA102 with and without metabolic activation negative (13).

## Legislation

Included in Schedule 4 (Release into Air: Prescribed Substances) Statutory Instrument No. 472, 1991 (14).

## Any other comments

Human health effects, experimental toxicology and environmental effects reviewed (15).

## References

1.  Kaiser, K. L. E. et al *Water Pollut. Res. J. Can.* 1991, **26**(3), 361-431
2.  Schultz, T. W. et al *Ecotoxicol. Environ. Saf.* 1987, **13**, 76
3.  Stafford, D. A. *J. Appl. Bacteriol.* 1974, **37**, 75-82
4.  Kuhn, E. P. et al *Environ. Toxicol. Chem.* 1989, **8**, 1149-1158
5.  Roubickova, J. *Vodni Hospod.: B* 1986, **3**(10), 271-277 *(Czech.) (Chem. Abstr.* **106**, 55222m)
6.  Wolters, N. *Unterschiedliche Bestimmungsmethoden zur Erfassung der organischen Substanz in einer Verbindung*, Lehrauftrag Wasser biologie a.d. Technischer Hochschule, Darmstadt, Germany
7.  Ludzak, F. J. et al *J. -Water Pollut. Control Fed.* 1960, **32**, 1173
8.  Schafer, E. W. et al *Arch. Environ. Contam. Toxicol.* 1983, **12**, 355-382
9.  *Fundam. Appl. Toxicol.* 1985, **5**, 920
10.  *J. Pharm. Exp. Ther.* 1946, **88**, 82
11.  Sahamoto, Y. et al *Bull. Chem. Soc. Jpn.* 1989, **62**, 330-332
12.  *BIBRA Toxicity Profiles* 1990, British Industrial Biological Research Association, Carshalton
13.  Claxton, L. D. et al *Mutat. Res.* 1987, **176**(2), 185-188
14.  *S. I. 1991 No. 472 The Environmental Protection (Prescribed Processes and Substances) Regulations* 1991, HMSO, London
15.  *ECETOC Technical Report No. 30(4)* 1991, European Chemical Industry Ecology and Toxicology Centre, B-1160 Brussels

# M275   4-Methylpyridine

CH₃

**CAS Registry No.** 108-89-4

**Synonyms** pyridine, 4-methyl-; 4-picoline; Ba 35846; γ-methylpyridine; *p*-picoline; γ-picoline

**Mol. Formula** $C_6H_7N$                          **Mol. Wt.** 93.13

**Uses** In manufacture of isonicotinic acid and derivatives. In waterproofing agents for fabrics. Solvent for resins.

**Occurrence** In coal tar, bone oil, urine of horses.

## Physical properties

**B. Pt.** 145°C; **Specific gravity** $d_4^{15}$ 0.9571; **Partition coefficient** log $P_{ow}$ 1.22.

### Solubility

Organic solvent: ethanol, diethyl ether, acetone

## Occupational exposure

**UN No.** 2313; **HAZCHEM Code** 2S; **Conveyance classification** flammable liquid; **Supply classification** toxic.

**Risk phrases** Flammable – Harmful by inhalation and if swallowed – Toxic in contact with skin – Irritating to eyes, respiratory system and skin (R10, R20/22, R24, R36/37/38)

**Safety phrases** In case of contact with eyes, rinse immediately with plenty of water and seek medical advice – Wear suitable protective clothing – If you feel unwell, seek medical advice (show label where possible) (S26, S36, S44)

## Ecotoxicity

### Fish toxicity

$LC_{50}$ (96 hr) fathead minnow 402.8 mg $l^{-1}$ (1).

### Invertebrate toxicity

$EC_{50}$ (30 min) *Photobacterium phosphoreum* 26.9 ppm Microtox test (2).

$EC_{50}$ (60 hr) *Tetrahymena pyriformis* 730 mg $l^{-1}$ (3).

## Environmental fate

### Nitrification inhibition

1.0 mg $l^{-1}$ inhibited nitrification by 50% (4).

100 mg $l^{-1}$ (activated sludge) inhibited $NH_3$ oxidation by 90% (5).

### Anaerobic effects

In methanogenic incubations 47-84% of 4-picoline was recovered as $CH_4$ (6).

Biotransformed within 3 months under sulfate-reducing conditions in aquifer slurries (6).

### Degradation studies
Adapted or fresh sludge is able to biodegrade wastewater containing $\leq$312 mg l$^{-1}$ pyridine bases (7).

## Mammalian and avian toxicity

### Acute data
LD$_{50}$ oral rat 1290 mg kg$^{-1}$ (8).
LD$_{50}$ oral redwing blackbird, starling, quail 100-1000 mg kg$^{-1}$ (9).
LC$_{Lo}$ (4 hr) inhalation rat 1000 ppm (10).
LD$_{50}$ dermal rabbit 270 mg kg$^{-1}$ (10).
LD$_{50}$ intraperitoneal rat 163 mg kg$^{-1}$ (11).

### Carcinogenicity and long-term effects
Carcinogenicity not investigated, but expected to be negative (12).

### Metabolism and pharmacokinetics
Unmetabolised 4-picoline found in expired air and urine, for 24 hr following exposure, of rats treated with 300 mg kg$^{-1}$ (13).

### Irritancy
Dermal rabbit 10 mg applied for 24 hr (open) caused severe irritation (10).
750 µg instilled into rabbit eye caused severe irritation (10).

## Genotoxicity
*Salmonella typhimurium* TA97, TA98, TA100, TA102 with and without metabolic activation negative (14).

## Legislation
Included in Schedule 4 (Release into Air: Prescribed Substances) Statutory Instrument No. 472, 1991 (15).

## Any other comments
Human health effects, experimental toxicology, physico-chemical properties and environmental effects reviewed (16).

## References

1. Schultz, T. W. et al *Chemosphere* 1989, **18**(11-12), 2283-2291
2. Kaiser, K. L. E. et al *Water Pollut. Res. J. Can.* 1991, **26**(3), 361-431
3. Schultz, T. W. et al *Ecotox. Environ. Saf.* 1985, **10**, 97-111
4. Greenfield, J. H. et al *36th Ind. Waste Conf.* Purdue UN, 772
5. Stafford, D. A. *J. Appl. Bacteriol.* 1974, **34**, 75-82
6. Kuhn, E. P. et al *Environ. Toxicol. Chem.* 1989, **8**(12), 1149-1158
7. Roubickova, J. *Vodni Hospod.: B* 1986, **36**(10), 271-277 (Czech.) (*Chem. Abstr.* **106**, 55222m)
8. *Union Carbide Data Sheet* No. 2/21/58, 1958, Industrial Medicine and Toxicology Dep., Union Carbide Corp., 270 Park Avenue, New York
9. Schafer, E. W. et al *Arch. Environ. Contam. Toxicol.* 1983, **12**, 355-382
10. *AMA, Arch. Ind. Hyg. Occup.* 1954, **10**, 61
11. *Fundam. Appl. Toxicol.* 1985, **5**, 920

12. Sahamoto, Y. et al *Bull. Chem. Soc. Jpn.* 1989, **62**, 330-332
13. Nguyen, P. L. et al *Arch. Toxicol., Suppl.* 1988, **12**(Target Organ Process), 308-312
14. Claxton, L. D. et al *Mutat. Res.* 1987, **176**(2), 185-188
15. *S. I. 1991 No. 472 The Environmental Protection (Prescribed Processes and Substances) Regulations* 1991, HMSO, London
16. *ECETOC Technical Report No. 30(4)* 1991, European Chemical Industry Ecology and Toxicology Centre, B-1160 Brussels

# M276   *N*-Methyl-3-pyridinecarboxamide

**CAS Registry No.** 114-33-0
**Synonyms** nicotinamide, *N*-methyl-; 3-(methylcarbamoyl)pyridine; nicotinic acid methylamide
**Mol. Formula** $C_7H_8N_2O$                **Mol. Wt.** 136.15
**Occurrence** Metabolite of nicotinamide and is a human metabolic product. Product of tobacco smoke.

## Physical properties

**M. Pt.** 104-105°C.

## Mammalian and avian toxicity

**Teratogenicity and reproductive effects**
Teratogenic in chickens at 5 and 10 mg $kg^{-1}$ (1).

**Metabolism and pharmacokinetics**
Catabolic metabolism is considered to occur constantly in humans consuming ordinary foods (2).

## Genotoxicity

*Salmonella typhimurium* TA98, TA100, TA1535, TA1537 with and without metabolic activation negative (3).
Weakly induced sister chromatid exchanges in UV irradiated human lymphocytes (4).

## Any other adverse effects

Inhibitor of poly(ADP-ribose) polymerase (5).

## Legislation

Included in Schedule 4 (Release into Air: Prescribed Substances) Statutory Instrument No. 472, 1991 (6).

## References

1. Landover, W. et al *J. Exp. Zool.* 1967, **164**, 499
2. Shibata, K. *Nippon Kasei Gakkaishi* 1990, **41**(10), 985-988
3. Florin, I. et al *Toxicology* 1980, **18**, 219
4. Honi, T. A. *Biochem. Biophys. Res. Commun.* 1981, **100**, 463
5. Miwa, M. et al *Biochem. Biophys. Res. Commun.* 1981, **100**, 470
6. *S. I. 1991 No. 472 The Environmental Protection (Prescribed Processes and Substances) Regulations* 1991, HMSO, London

# M277   *N*-Methylpyrrolidone

**CAS Registry No.** 872-50-4
**Synonyms** 2-pyrrolidinone, 1-methyl-; 1-methylpyrrolidone; M-pyrol
**Mol. Formula** $C_5H_9NO$                      **Mol. Wt.** 99.13
**Uses** Solvent.

## Physical properties

**M. Pt.** –24°C; **B. Pt.** 202°C; **Flash point** 96°C (open cup); **Specific gravity** $d_4^{25}$ 1.027.

### Solubility
Water: miscible. Organic solvent: aliphatic hydrocarbons, miscible with methanol, ethanol, diethyl ether, ethyl acetate, benzene, chloroform

## Occupational exposure

**UK Long-term limit** 100 ppm (400 mg m$^{-3}$); **Supply classification** irritant (≥10%).
**Risk phrases** ≥10% – Irritating to eyes and skin (R36/38)
**Safety phrases** In case of fire and/or explosion do not breathe fumes (S41)

## Mammalian and avian toxicity

### Acute data
$LD_{50}$ oral rat, mouse 3914, 5130 mg kg$^{-1}$ respectively (1, 2).
$LD_{50}$ dermal rabbit 8000 mg kg$^{-1}$ (3).
$LD_{50}$ intraperitoneal rat, mouse 2472, 3050 mg kg$^{-1}$ respectively (1, 2).
$LD_{50}$ intravenous mouse, rat 54,500, 80,500 μg kg$^{-1}$ respectively (4).

### Irritancy
Irritant to the eyes and corneal lesions and moderate conjunctivitis has been reported in humans (5).
100 mg instilled into rabbit eye (72 hr) caused moderate irritation (6).

### Genotoxicity

Found to induce aneuploidy in *Saccharomyces cerevisiae* (7).

### Legislation

Included in Schedule 4 (Release Into Air: Prescribed Substances) and Schedule 6 (Release Into Land: Prescribed Substances) Statutory Instrument No. 472, 1991 (8).

### Any other comments

Human health effects, experimental toxicology, physico-chemical properties, epidemiology and workplace experience reviewed (1).

### References

1. *Arzneim.-Forsch.* 1976, **26**, 1581
2. *US EPA Report* No. 8EHQ-1087-0695, Office of Pesticides and Toxic Substances, US EPA, Washington, D.C
3. *Raw Material Data Handbook, V. 1. Organic Solvents* 1974, 84, National Association of Printing Ink Research Institute, Francis McDonald Sinclair Memorial Laboratory, Lehigh University, Bethlehem, PA
4. *Iyakuhin Kenkyu* 1987, **18**, 922
5. Stasenara, K. P. et al *Toksikol Norykh. Prom. Khim. Veschester* 1965, **7**, 27-38
6. *Food Chem. Toxicol.* 1988, **26**, 475
7. Mayer, V. M. et al *Environ. Mol. Mutagen.* 1988, **11**, 31-40
8. *S. I. 1991 No. 472 The Environmental Protection (Prescribed Processes and Substances) Regulations* 1991, HMSO, London

# M278   2-Methylquinoline

**CAS Registry No.** 91-63-4
**Synonyms** quinoline, 2-methyl-; quinaldine; Khinaldin
**Mol. Formula** $C_{10}H_9N$                    **Mol. Wt.** 143.19
**Uses** Anaesthetic in transport and handling of fish.
**Occurrence** Occurs in coal tar.

### Physical properties

**M. Pt.** –2°C; **B. Pt.** 246-247°C; **Flash point** 110°C; **Specific gravity** 1.058; **Partition coefficient** log $P_{ow}$ 2.23 (calculated) (1).

### Solubility

Organic solvent: chloroform, diethyl ether

### Ecotoxicity

**Fish toxicity**
No toxic effect in fertilisation of rainbow trout at 0.05% (2).

**Invertebrate toxicity**

$LC_{10}$ (24 hr) *Tetrahymena pyriformis* 0.40 g l$^{-1}$ (3).
$LC_{50}$ (96 hr) *Xenopus laevis* embryo 26.4 mg l$^{-1}$ (4).

## Mammalian and avian toxicity

**Acute data**

$LD_{50}$ oral rat 1230 mg kg$^{-1}$ (5).
$LD_{50}$ dermal rabbit 1870 mg kg$^{-1}$ (5).

**Irritancy**

Dermal rabbit (24 hr) 10 mg caused mild irritation and 750 μg instilled into rabbit eye (72 hr) caused severe irritation (5).

## Genotoxicity

*Salmonella typhimurium* TA100 with metabolic activation negative (6).

## Legislation

Included in Schedule 4 (Release into Air: Prescribed Substances) Statutory Instrument No. 472, 1991 (7).

## Any other comments

Metabolite of tryptophan in rat intestine (8).

## References

1. Verschueren, K. *Handbook of Environmental Data on Organic Chemicals* 2nd ed., 1983, Van Nostrand Reinhold, New York
2. Marking, L. L. et al *Insect Fish Control* 1973, **47-50**, II/I/18
3. Schultz, T. W. et al *Arch. Environ. Contam. Toxicol.* 1978, **7**, 457-463
4. Eldridge, M. B. et al *J. Fish. Res. Board Can.* 1978, **35**, 1084-1088
5. *AMA, Arch. Ind. Hyg. Occup. Med.* 1951, **4**, 119
6. La Voie, E. J. et al *Carcinogenesis (London)* 1991, **12**(2), 217-220
7. *S. I. 1991 No. 472 The Environmental Protection (Prescribed Processes and Substances) Regulations* 1991, HMSO, London
8. Dumont, J. J. et al *Bull. Environ. Contam. Toxicol.* 1979, **22**, 159

# M279   4-Methylquinoline

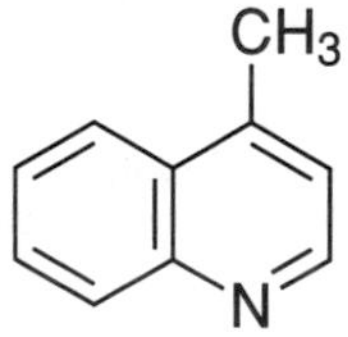

**CAS Registry No.** 491-35-0
**Synonyms** quinoline, 4-methyl-; lepidine; cincholepidine; 4-lepidin; γ-methylquinoline
**Mol. Formula** $C_{10}H_9N$                    **Mol. Wt.** 143.19

**Occurrence** In cigarette smoke (1).

**Physical properties**
**M. Pt.** 9-10°C; **B. Pt.** 261-263°C; **Flash point** >107°C; **Specific gravity** $d_4^{20}$ 1.0826.

**Solubility**
Organic solvent: miscible with ethanol, benzene, diethyl ether

**Ecotoxicity**

**Invertebrate toxicity**
$EC_{50}$ (30 min) *Photobacterium phosphoreum* 5.08 ppm Microtox test (2).
$EC_{50}$ (48 hr) *Daphnia magna* 11 mg $l^{-1}$ (3).

**Mammalian and avian toxicity**

**Carcinogenicity and long-term effects**
Structure-activity studies have predicted tumorigenic activity on mouse skin (4).
Assays in newborn ♂ mice show hepatocarcinogenicity (4).

**Genotoxicity**
*Salmonella typhimurium* TA100 with metabolic activation positive (4).

**Legislation**
Included in Schedule 4 (Release into Air: Prescribed Substances) Statutory Instrument No. 472, 1991 (5).

**References**

1. Dong, M. et al *Carcinog.-Compr. Survey* 1978, **37**, 97
2. Kaiser, K. L. E. et al *Water Pollut. Res. J. Can.* 1991, **26**(3), 361-431
3. Herbes, S. E. et al *Bull. Environ. Contam. Toxicol.* 1977, **27**(1)
4. La Voie, E. J. et al *Carcinogenesis (London)* 1983, **12**(2), 217-220
5. *S. I. 1991 No. 472 The Environmental Protection (Prescribed Processes and Substances) Regulations* 1991, HMSO, London

# M280   5-Methylquinoline

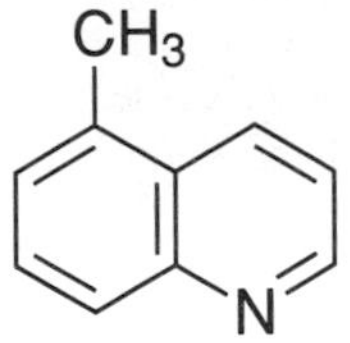

**CAS Registry No.** 7661-55-4
**Synonyms** quinoline, 5-methyl-

**Physical properties**
**M. Pt.** 19°C; **B. Pt.** 262.7°C; **Specific gravity** $d_4^{20}$ 1.0832.

## Solubility
Organic solvent: acetone, miscible with alcohol, diethyl ether

## Ecotoxicity

### Invertebrate toxicity
$EC_{50}$ (15 min) *Photobacterium phosphoreum* 0.991 ppm Microtox test (1).

## Legislation
Included in Schedule 4 (Release into Air: Prescribed Substances) Statutory Instrument No. 472, 1991 (2).

## References

1. Kaiser, K. L. E. et al *Water Pollut Res. J. Can.* 1991, **26**(3), 361-431
2. *S. I. 1991 No. 472 The Environmental Protection (Prescribed Processes and Substances) Regulations* 1991, HMSO, London

# M281   6-Methylquinoline

**CAS Registry No.** 91-62-3
**Synonyms** quinoline, 6-methyl-; *p*-methylquinoline; *p*-toluquinoline
**Mol. Formula** $C_{10}H_9N$                          **Mol. Wt.** 143.19
**Uses** Fragrance.
**Occurrence** In coal liquefaction waste water. In whisky and tea (1).

## Physical properties
**B. Pt.** 259°C; **Flash point** >107°C; **Specific gravity** 1.063; **Partition coefficient** log $P_{ow}$ 2.57 (2).

## Ecotoxicity

### Invertebrate toxicity
$EC_{50}$ (15 min) *Photobacterium phosphoreum* 2.79 ppm Microtox test (3).
$LC_{50}$ (48 hr) *Daphnia magna* 11 mg $l^{-1}$ in a mixture with resorcinol (4).
$LC_{100}$ (24 hr) *Tetrahymena pyriformis* 0.22 g $l^{-1}$ (5).

## Environmental fate

### Degradation studies
Hydroxylated, but not degraded by *Pseudomonas aeruginosa* QP and *Pseudomonas putida* QP (6).

## Mammalian and avian toxicity

**Acute data**
$LD_{50}$ oral rat 1260 mg $kg^{-1}$ (1).
$LD_{50}$ dermal rabbit 5000 mg $kg^{-1}$ (1).
$LD_{50}$ intraperitoneal mouse 386 mg $kg^{-1}$ (1).

**Carcinogenicity and long-term effects**
In feeding studies in F344 rats at 0.05-0.1% for 104 wk, no indication of carcinogenicity (7).

**Metabolism and pharmacokinetics**
Oxidised in dog to quinoline 6-carboxylic acid (1).

**Irritancy**
Dermal rabbit (24 hr) 500 mg caused moderate irritation (1).

**Sensitisation**
No sensitisation in humans (1).

## Genotoxicity

*Salmonella typhimurium* TA98, TA100 without metabolic activation negative, with metabolic activation positive (7,8).
*Salmonella typhimurium* TA100 with metabolic activation negative (9).

## Any other adverse effects

Increased rat liver aryl hydrocarbon hydroxylase activity when injected intraperitoneally (10).

## Legislation

Included in Schedule 4 (Release into Air: Prescribed Substances) Statutory Instrument No. 472, 1991 (11).

## References

1. Opdyke, D. L. J. *Food Cosmet. Toxicol* 1979, **17**, 871
2. Verschueren, K. *Handbook of Environmental Data on Organic Chemicals* 2nd ed., 1983, Van Nostrand Reinhold, New York
3. Kaiser, K. L. E. et al *Water Pollut. Res. J. Can.* 1991, **26**(3), 361-431
4. Herbes, S. E. et al *Bull. Environ. Contam. Toxicol.* 1977, **27**(1)
5. Schultz, W. T. et al *Arch. Environ. Contam. Toxicol.* 1978, **7**, 457-463
6. Aislabie, J. et al *Appl. Environ. Microbiol.. 1990,* **56**(2), 345-351
7. Fukyshima, G. et al *Cancer Lett.* 1981, **14**, 115
8. Nagao, M. et al *Mutat. Res.* 1977, **42**, 335
9. La Voie, E. J. et al *Carcinogenesis (London)* 1991, **12**(2), 217-220
10. *Acta Pharmacol. Toxicol.* 1980, **A105**, 80
11. *S. I. 1991 No. 472 The Environmental Protection (Prescribed Processes and Substances) Regulations* 1991, HMSO, London

# M282    7-Methylquinoline

**CAS Registry No.** 612-60-2
**Synonyms** quinoline, 7-methyl-; *m*-toluquinoline
**Mol. Formula** $C_{10}H_9N$                                          **Mol. Wt.** 143.19
**Occurrence** Detected in water samples.

## Physical properties

**M. Pt.** 35-37°C; **B. Pt.** 258°C; **Flash point** >110°C.

## Genotoxicity

*Salmonella typhimurium* TA98, TA100 without metabolic activation negative, with metabolic activation positive (1, 2).

## Legislation

Included in Schedule 4 (Release into Air: Prescribed Substances) Statutory Instrument No. 472, 1991 (3).

## References

1.   Epler, J. L. et al *Environ. Health Perspect.* 1979, **30**, 174
2.   Nagao, M. et al *Mutat. Res.* 1977, **42**, 335
3.   *S. I. 1991 No. 472 The Environmental Protection (Prescribed Processes and Substances) Regulations* 1991, HMSO, London

# M283    8-Methylquinoline

**CAS Registry No.** 611-32-5
**Synonyms** quinoline, 8-methyl-
**Mol. Formula** $C_{10}H_9N$                                          **Mol. Wt.** 143.19

## Physical properties

**M. Pt.** −80°C; **B. Pt.** $_{34}$ 143°C; **Flash point** 105°C; **Specific gravity** 1.052; **Partition coefficient** log $P_{ow}$ 2.60 (1).

---

## Ecotoxicity

**Fish toxicity**

Anaesthetic concentration in sea water for sharks 7.16-21.48 mg $l^{-1}$ (2).

**Invertebrate toxicity**

$EC_{50}$ (15 min) *Photobacterium phosphoreum* 9.03 ppm Microtox test (3).
$LC_{100}$ (24 hr) *Tetrahymena pyriformis* 0.22 g $l^{-1}$ (4).

## Mammalian and avian toxicity

**Acute data**

$LD_{50}$ intraperitoneal mouse ≈71.6-429.6 mg $kg^{-1}$ (2).

**Carcinogenicity and long-term effects**

35.7, 71.5, 143.0 µg $l^{-1}$ administered intraperitoneally to newborn CD-1 mice on day 1, 8, 15 of life for 1 yr caused no significant tumorigenic activity (4).
Newborn Sprague-Dawley rats administered 28.6 mg $kg^{-1}$ by injection on day 1 of life and then 14.3 mg $kg^{-1}$ wkly at wk 2-7 and 28.6 mg $kg^{-1}$ at wk 8 exhibited no significant difference in tumourigenic activity compared to controls (5).

## Genotoxicity

*Salmonella typhimurium* TA100 with metabolic activation positive (6).

## Legislation

Included in Schedule 4 (Release into Air: Prescribed Substances) Statutory Instrument No. 472, 1991 (7).

## Any other comments

All reasonable efforts have been taken to find information on isomers of this compound but no relevant data are available.

## References

1. Verschueren, K. *Handbook of Environmental Data on Organic Chemicals* 2nd ed., 1983, Van Nostrand Reinhold, New York
2. Brown, E. A. Comp. Biochem. Physiol., A: Comp. Physiol. 1972, **42**(1), 223-231
3. Kaiser, K. L. E. et al *Water Pollut. Res. J. Can.* 1991, **26**(3), 361-431
4. Schultz, T. W. et al *Arch. Environ. Contam. Toxicol.* 1978, **7**, 457-463
5. La Voie, E. J. et al *Food Chem. Toxicol.* 1988, **26**(7), 625-629
6. La Voie, E. J. et al *Carcinogenesis (London)* 1991, **12**(3), 217-220
7. *S. I. 1991 No. 472 The Environmental Protection (Prescribed Processes and Substances) Regulations* 1991, HMSO, London

# M284    Methyl salicylate

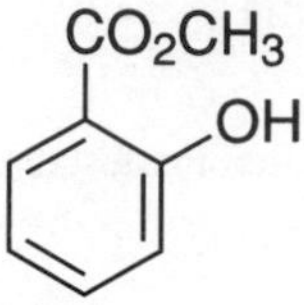

CO$_2$CH$_3$

OH

**CAS Registry No.** 119-36-8
**Synonyms** betula oil; methyl 2-hydroxybenzoate; sweet birch oil; teaberry oil; wintergreen oil
**Mol. Formula** C$_8$H$_8$O$_3$                    **Mol. Wt.** 152.15
**Uses** Anti-inflammatory drug. In perfumery. Flavouring agent. UV-absorber in sunburn lotions.
**Occurrence** In plant oils.

## Physical properties
**M. Pt.** –8 to –7°C; **B. Pt.** 222°C; **Flash point** 99°C (closed cup); **Specific gravity** $d_{25}^{25}$ 1.184; **Partition coefficient** log P$_{ow}$ 2.55 (1); **Volatility** v.p. 1 mmHg at 54°C; v. den. 5.24.

## Solubility
Water: ≈670 mg l$^{-1}$ at 20°C. Organic solvent: chloroform, diethyl ether, dimethyl sulfoxide, ethanol, glacial acetic acid

## Ecotoxicity

**Fish toxicity**
Not toxic to stickleback and rainbow trout at 10 mg l$^{-1}$ for 24 hr (sodium salt) (2).

## Environmental fate

**Degradation studies**
BOD$_5$ 55% ThOD (3).

**Abiotic removal**
Hydrolysis t$_{1/2}$ 12.1 hr at pH 9.2 and 25°C; and 3.2 hr at pH 11.26 and 24°C.
t$_{1/2}$ ≈22 days at pH 7.5 (4,5).
Photolysis t$_{1/2}$ 48 min in solution (absorption maximum 305 nm in methanol) (6,7).
Reaction with photochemically produced hydroxyl radicals in the atmosphere, estimated t$_{1/2}$ 5.7 day (8).

## Mammalian and avian toxicity

**Acute data**
LD$_{50}$ oral rat, guinea pig, rabbit, dog 700-2800 mg kg$^{-1}$ (9,10).
LD$_{Lo}$ subcutaneous guinea pig 1500 mg kg$^{-1}$ (11).

**Sub-acute data**
Inhalation rat (20 day) no adverse effect level 120 ppm for 7 hr day$^{-1}$ (12).

### Teratogenicity and reproductive effects
$TD_{Lo}$ oral hamster 1750 mg $kg^{-1}$ on day 7 of gestation (teratogenic effects on central nervous system) (13).
$TD_{Lo}$ dermal hamster 5250 mg $kg^{-1}$ on day 7 of gestation (teratogenic effects on central nervous system) (13).
$TD_{Lo}$ intraperitoneal rat 400 mg $kg^{-1}$ on day 12 of gestation (foetal mortality) (14).
$TD_{Lo}$ intraperitoneal rat 500 mg $kg^{-1}$ $day^{-1}$ on days 11-12 of gestation (developmental effects on urogenital system) (14).

### Metabolism and pharmacokinetics
Undergoes rapid hydrolysis to salicylic acid, mainly in the liver. In some species, including the rabbit, partially excreted as sulfate or glucuronide conjugates (10,15). Reported to cross the placental barrier (15).
Absorbed through the skin in humans (16).

### Irritancy
Dermal rabbit (24 hr) 500 mg caused moderate irritation (17).
500 mg instilled into rabbit eye for 24 hr caused mild irritation (18).

## Genotoxicity
*Salmonella typhimurium* TA98, TA100, TA1535, TA1537 with and without metabolic activation negative (19).

## Any other adverse effects to man
Lethal oral doses in man, 30 ml in adults, 10 ml in children. Symptoms of poisoning, which are similar to those for aspirin, include nausea, vomiting, acidosis, pulmonary oedema, pneumonia and convulsions (20).

## Any other comments
Autoignition temperature 454°C.

## References
1. Camilleri, P. et al *J. Chem. Soc., Perkin Trans. 2* 1988, (9), 1699-1707
2. McPhee, C. et al *Lethal Effects of 2014 Chemicals to Fish* 1989, EPA 560/6-89-001, PB 89-156-715, Washington, DC
3. Maggio, P. et al *Tinctoria* 1976, **73**, 15-20
4. Magid, L. J. et al *J. Org. Chem.* 1974, **39**, 3142-3144
5. Sennet, S. et al *An. Quim.* 1973, **69**, 13-23
6. Sadtler, S. *UV Spectrum* 1969, Sadtler Reg. Lab. Inc. Philadelphia, PA
7. Kondo, M. *Simulation Studies of Degradation of Chemicals in Water and Soils* 1978, Office of Health Studies, Environmental Agency, Japan
8. *GEMS: Graphical Exposure Modeling System. Fate of Atmospheric Pollutants* 1986, Washington, DC
9. Jenner, P. M. et al *Food Cosmet. Toxicol.* 1962, **2**, 327
10. *Patty's Industrial Hygiene and Toxicology* 2nd ed., 1978, **2**, John Wiley and Sons, New York
11. *FAO Nutrition Meetings Report Series* 1967, **44A**, 63
12. Cage, J. C. *Br. J. Ind. Med.* 1970, **27**
13. *Teratology* 1983, **28**, 421
14. *Fundam. Appl. Toxicol.* 1988, **11**, 381
15. Goodman, L. S. et al (Eds.) *The Pharmacological Basis of Therapeutics* 5th ed., 1975, 334, Macmillan, New York
16. *Martindale. The Extra Pharmacopoeia* 30th ed., 1993, The Pharmaceutical Press, London

17. *Food Cosmet. Toxicol.* 1978, **16**, 821
18. Marhold, J. V. *Prehled Prumyslove Toxikologie: Organicke Latky* 1986, Prague
19. Mortelmans, K. et al *Environ. Mutagen.* 1986, **8**(Suppl. 7), 1-119
20. Gosselin, R. E. et al (Ed.) *Clinical Toxicology of Commercial Products* 4th ed., 1976, 295-303, Williams & Wilkins, Baltimore, MD

# M285   Methylstyrene

**CAS Registry No.** 25013-15-4

**Synonyms** ethenylmethylbenzene; *ar*-methylstyrene; tolylethylene; vinyltoluene (mixed isomers)

**Mol. Formula** $C_9H_{10}$ **Mol. Wt.** 118.18

**Uses** Manufacture of polymers. Solvent.

**Occurrence** In essential oil of some plants. Residues have been identified in natural waters and in drinking water (1,2).

## Physical properties

**M. Pt.** –77°C; **B. Pt.** 170-171°C; **Flash point** 51.7°C; **Specific gravity** $d_{25}^{25}$ 0.890; **Volatility** v.p. 1.1 mmHg at 20°C; v. den. 4.1.

## Solubility

Organic solvent: acetone, carbon tetrachloride, diethyl ether, dimethyl sulfoxide, ethanol, methanol

## Occupational exposure

**UN No.** 2618; **HAZCHEM Code 3▣**; **Conveyance classification** flammable liquid.

## Ecotoxicity

### Bioaccumulation

Bioconcentration factor for goldfish 32-35 (3).

## Environmental fate

### Abiotic removal

Reaction with photochemically produced hydroxyl radicals and ozone in the atmosphere, estimated $t_{1/2}$ 6 hr (4,5).

Estimated volatilisation $t_{1/2}$ 10 days from model pond water and 3.5 hr in model river water (6, 7).

### Absorption

Estimated $K_{oc}$ 370 indicates the methylstyrene will adsorb moderately to soil and sediments (7).

## Mammalian and avian toxicity

### Acute data
$LD_{50}$ oral rat, mouse 2255, 3160 mg $kg^{-1}$ respectively (8,9)
$LC_{50}$ (4 hr) inhalation mouse 3000 mg $m^{-3}$ (exposure not specified) (9).
Inhalation rat (4 hr) 50 ppm induced leucopenia, without any change in differential or red blood cell counts (10).
$LD_{Lo}$ dermal rat, mouse 4500 mg $kg^{-1}$ (11).
$LD_{50}$ intraperitoneal rat 2300 mg $kg^{-1}$ (8).

### Carcinogenicity and long-term effects
Inhalation rat, (2 yr) 0, 100 or 300 ppm, 6 hr $day^{-1}$, 5 days $wk^{-1}$ for 103 wk. Mice were exposed to 0, 10 or 25 ppm under the same schedule. There was no evidence of carcinogenicity in ♂ and ♀ rats or mice (12).

### Teratogenicity and reproductive effects
Intraperitoneal rat, lowest toxic dose 3800 mg $kg^{-1}$ $day^{-1}$ on days 1-15 of gestation, teratogenic effects (13).

### Metabolism and pharmacokinetics
Following inhalation exposure of rats, methylstyrene was metabolised to glutathione conjugates via the formation of electrophilic intermediates (14).

### Irritancy
90 mg instilled into rabbit eye caused mild irritation (exposure not specified) (8).
Inhalation human, lowest irritant concentration 400 ppm (8).

## Genotoxicity
*Salmonella typhimurium* TA97, TA98, TA100, TA1535, TA1537 with and without metabolic activation negative (15).
*In vitro* mouse lymphoma L5178Y cells, $tk^+/tk^-$ without metabolic activation positive (16).
*In vivo* mouse, ratio of polychromatic to normachromatic erythrocytes was slightly decreased. There was no increase in normochromatic cells with micronuclei (17).

## Any other adverse effects to man
Causes drowsiness and central nervous system depression. At high concentrations causes dizziness, drunkenness and anaesthesia (18).
Human volunteers noted that odour was detectable at 50 ppm, odour was strong and tolerable at 200 ppm, odour was strong and objectionable at 300 ppm, eye and nasal irritation occurred at 400 ppm (19).

## Any other adverse effects
Intraperitoneal rat, mouse, Chinese hamster, single doses of up to 500 mg $kg^{-1}$ caused a dose-dependent decrease in glutathione content in the liver and kidneys. In mice, the highest dose decreased microsomal cytochrome $P_{450}$ content and 7-ethoxycoumarin *O*-deethylase activity acutely within 6 hr (20).

## Legislation
Included in Schedule 6 (Release into Land: Prescribed Substances) Statutory Instrument No. 472, 1991 (21).

## Any other comments

Human health effects, experimental toxicology, physico-chemical properties reviewed (22).

Autoignition temperature 490°C.

Usually occurs as a mixture of 50-70% 3-methylstyrene and 30-45% 4-methylstyrene.

## References

1. Suffet, I. H. et al *Water Res.* 1980, **14**, 853-867
2. Waggott, A. *Chem. Water Reuse* 1981, **2**, 55-99
3. Ogata, M. et al *Bull. Environ. Contam. Toxicol.* 1984, **33**, 561-567
4. Atkinson, R. *Int. J. Chem. Kinet.* 1987, **19**, 799-828
5. Atkinson, R. et al *Chem. Rev. (Washington, D. C.)* 1984, **84**, 437-470
6. *Exams II Computer Simulation* 1987, US EPA, Athens, GA
7. Lyman, W. K. et al *Handbook of Chemical Property Estimation Methods Environmental Behaviour of Organic Compounds* 1982, McGraw-Hill, New York
8. *Acute Toxicity Data* 1990, **1**, 77
9. *Hyg. Sanit. (USSR)* 1969, **34**(7-9), 334
10. Brondeau, M. T. et al *J. Appl. Toxicol.* 1990, **10**(2), 83-86
11. *Gig. Sanit.* 1969, **34**(9), 40
12. *National Toxicology Program Research and Testing Division* 1990, Report No. TR-375, NIEHS, Research Triangle Park, NC
13. *Scand. J. Work, Environ. Health* 1981, **7**(Suppl. 4), 66
14. Heinonen, T. et al *Acta Pharmacol. Toxicol.* 1982, **51**(1), 69-75
15. Zeiger, E. et al *Environ. Mutagen.* 1987, **9**(Suppl. 9), 1-109
16. McGregor, D. B. et al *Environ. Mol. Mutagen.* 1988, **12**(1), 85-154
17. Norppa, H. *Toxicol. Lett.* 1981, **8**(4-5), 247-252
18. Keith, L. H. et al (Ed.) *Compendium of Safety Data Sheets for Research and Industrial Chemicals* 1987, **6**, 1598, VCH, New York
19. *Documentation of Threshold Limit Values and Biological Exposure Indices* 5th ed., 1986, 630, American Conference of Governmental Industrial Hygienists, Cincinnati, OH
20. Heinonen, T. et al *Biochem, Pharmacol.* 1980, **29**(19), 2675-2679
21. *S. I. 1991 No. 472 The Environmental Protection (Prescribed Processes and Substances) Regulations* 1991, HMSO, London
22. *ECETOC Technical Report No. 30(4)* 1991, European Chemical Industry Ecology and Toxicology Centre, B-1160 Brussels

# M286   Methyl sulfoxide

## $(CH_3)_2SO$

**CAS Registry No.** 67-68-5

**Synonyms** Delton; Demasorb; dimethyl sulfoxide; dimexide; DMSO; Dolicur; Dromisol; methylsulfinylmethane; Syntexan; Topsym; sulfinyl bis[methane]

**Mol. Formula** $C_2H_6OS$           **Mol. Wt.** 78.13

**Uses** Solvent. Antifreeze. Anti-inflammatory drug. Organic synthesis. Paint remover.

**Occurrence** Has been detected in natural and drinking waters (1,2).

## Physical properties

**M. Pt.** 18.5°C; **B. Pt.** 189°C; **Flash point** 90°C (open cup); **Specific gravity** $d_4^{20}$ 1.100; **Partition coefficient** log $P_{ow}$ -1.35; **Volatility** v.p. 0.37 mmHg at 20°C; v. den. 2.71.

## Solubility

Water: miscible. Organic solvent: acetone, benzene, chloroform, ethanol

## Ecotoxicity

### Fish toxicity

Exposure of coho salmon to 1% v/v solution for 100 days caused no adverse effects (3).

### Invertebrate toxicity

$EC_{50}$ (5 min) *Photobacterium phosphoreum* 103,000 ppm, Microtox test (4).

## Environmental fate

### Degradation studies

< 20% degradation was observed in a screening test using activated sludge inoculum (5).
Transformed to methyl sulfide by anaerobic bacteria isolated from sediments (6).

### Abiotic removal

Undergoes disproportionation in water to methyl sulfide and methyl sulfone (7).
Reaction with photochemically produced hydroxyl radicals in the atmosphere $t_{1/2} \approx 7$ hr (8).

### Absorption

Methyl sulfoxide is reported to be adsorbed chemically and physically to clay minerals (9).

## Mammalian and avian toxicity

### Acute data

$LD_{50}$ oral rat 14.5 g $kg^{-1}$ (10).
$LD_{50}$ oral mouse 7900 mg $kg^{-1}$ (11).
$LD_{50}$ oral redwing blackbird, starling >100 mg $kg^{-1}$ (12).
$LD_{50}$ subcutaneous rat 12 g $kg^{-1}$ (13).
$LD_{50}$ intraperitoneal mouse, rat 2500, 8200 mg $kg^{-1}$, respectively (14, 15).
$LD_{50}$ intravenous dog, mouse, rat 2500, 3800, 5360 mg $kg^{-1}$, respectively (16-18).

### Sub-acute data

Gavage rat (45 day) 5000 mg $kg^{-1}$ $day^{-1}$ caused a slight loss in body weight, liver cell necrosis and inflammation with irritation of the portal spaces (19).

### Carcinogenicity and long-term effects

Gavage rat (18 month) 20 mg $kg^{-1}$ $day^{-1}$ 5 days $wk^{-1}$ for 18 months caused minimal changes in body weight, in haematological parameters and in the eyes (20).
♀ CH3/He mice were administered *N*-butyl-*N*-(4-hydroxybutyl)nitrosamine in drinking water for 8 wk. After this they were given wkly intravesicular instillations of 0.1 ml methyl sulfoxide for 10 wk. After 30 wk the animals were killed and their

urinary bladders were examined, 98.7% of the treated animals developed bladder carcinomas compared to 27.7% of controls (21).

In a similar study ♀ mice administered for 7 wk 0.05 ml intravesically after an initial 5 wk exposure to *N*-butyl-*N*-(4-hydroxybutyl)nitrosamine in drinking water, 25% developed bladder carcinomas compared to 0% in the controls (21).

**Teratogenicity and reproductive effects**

Intraperitoneal rat 10 g kg$^{-1}$ day$^{-1}$ on days 6-12 of gestation caused developmental effects to the central nervous system and musculoskeletal system, including anencephalia, malformed limbs and celosomia (19).

Intraperitoneal mouse, lowest toxic dose 8500 mg kg$^{-1}$ on day 10 of gestation caused foetal death (22).

Intraperitoneal mouse, lowest toxic dose 5500 mg kg$^{-1}$ on day 10 of gestation caused developmental effects to the musculo-skeletal system (22).

**Metabolism and pharmacokinetics**

Readily absorbed following administration by all routes. Metabolised by oxidation to dimethyl sulfone, and by reduction to dimethyl sulfide. Dimethyl sulfoxide and the sulfone metabolite are excreted in the urine and faeces. Dimethyl sulfide is excreted via the lungs and skin and is responsible for the characteristic garlic-like odour from patients (23).

**Irritancy**

Dermal rabbit (24 hr) 500 mg caused mild irritation, 500 mg instilled into rabbit eye for 24 hr caused mild irritation (24).

## Genotoxicity

*Salmonella typhimurium* TA97, TA98, TA100, TA1535, TA1537, TA1538 with and without metabolic activation negative (25).

*Escherichia coli* WP2$_S$(λ), Microscreen assay with and without metabolic activation positive (26).

*Saccharomyces cerevisiae* induction of mitotic chromosome loss, ambiguous results reported (27).

*Drosophila melanogaster* sex-linked recessive lethal assay negative (28).

## Any other adverse effects to man

Systemic effects, which may occur after administration by any route, include gastrointestinal disturbances, drowsiness, headache and hypersensitivity reactions (23).

Intravenous infusion induced transient haemolysis and haemoglobinuria. Infusion strengths >10% were associated with grossly discoloured urine but there was no evidence of kidney damage (29).

Two patients developed raised circulating levels of liver and muscle enzymes, mild jaundice, and evidence of haemolysis after intravenous administration for the treatment of arthritis. One patient also developed acute renal tubular necrosis, deterioration in the level of consciousness, and evidence of cerebral infarction (30).

## Legislation

Included in Schedule 6 (Release into Land: Prescribed Substances) Statutory Instrument No. 472, 1991 (31).

---

## Any other comments

Physical properties, use, toxicity and safety precautions reviewed (32-34).
In ♀ mice retarded the development of epidermal carcinomas which were chemically
induced with 9,10-dimethyl-1,2-benzanthracene and triamcinolone (35).
Autoignition temperature 215°C.

## References

1.  Kool, H. J. et al *Crit. Rev. Environ. Control* 1982, **12**, 307-357
2.  Andreae, M. O. *Anal. Chem.* 1980, **52**, 150-153
3.  Benville, P. E. et al *Toxicol. Appl. Pharmacol.* 1968, **12**, 156-178
4.  Kaiser, K. L. E. et al *Water Pollut. Res. J. Can.* 1991, **26**(3), 361-431
5.  Zahn, R. et al *Z. Wasser Abwasser Forsch.* 1980, **13**, 1-7
6.  Zinder, S. H. et al *Arch. Microbiol.* 1978, **116**, 35-40
7.  Harvey, G. R. et al *Geophys. Res. Lett.* 1986, **13**, 49-51
8.  Barnes, I. et al Comm. Eur. Communities, [Rep.] EUR, EUR 10832 1987, 327-337
9.  Morril, L. G. et al *Toxic Chemicals in Soil Environment* 1985, **2**, DPG-C-TA-85-02B,
    AD-A158, 215Z
10. Fishman, F. G. et al *Toxicol. Appl. Pharmacol.* 1969, **15**, 74-82
11. *Chim. Ther.* 1968, **3**, 10
12. Schafer, E. W. et al *Arch. Environ. Contam. Toxicol.* 1983, **12**, 355-382
13. *Arzneim.-Forsch.* 1964, **14**, 1050
14. *Russ. Pharmacol. Toxicol. (Engl. Transl.)* 1972, **35**, 300
15. *Food Chem. Toxicol.* 1984, **22**, 665
16. *Cancer Chemother. Rep.* 1963, **31**, 7
17. Deichmann, W. D. *Toxicology of Drugs and Chemicals* 1969, 656, Academic Press, NY
18. Wilson, J. E. et al *Toxicol. Appl. Pharmacol.* 1965, **7**, 104-112
19. Caujolle, F. M. E. et al *Ann. N. Y. Acad. Sci.* 1967, **141**, 110
20. Noel, P. R. B. *Toxicology* 1975, **3**, 143-169
21. Ohtani, M. et al *Nippon Hinyokika Gakkai Zasshi* 1992, **83**(a), 1422-1428
22. *Monographia* 1973, **61**, 131, Acta Universitatis Carolinae
23. *Martindale. The Extra Pharmacopoeia* 30th ed., 1993, The Pharmaceutical Press, London
24. Marhold, J. V. *Sbornik Vysledku Toxixologickeho Vysetreni Latek A Pripravku* 1972,
    Prague
25. Klopman, G. et al *Mutat. Res.* 1990, **228**(1), 1-50
26. De Merini, D. M. et al *Mutat. Res.* 1991, **263**(2), 107-113
27. Whittaker, S. G. et al *Mutat. Res.* 1989, **224**(1), 31-78
28. Mollet, P. *Mutat. Res.* 1978, **40**, 383-388
29. Muther, R. S. et al *J. Am. Med. Assoc.* 1980, **244**, 2081
30. Yellowlees, P. et al *Lancet* 1980, **ii**, 1004-1006
31. *S. I. 1991 No. 472 The Environmental Protection (Prescribed Processes and Substances)
    Regulations* 1991, HMSO, London
32. Willhite, C. C. et al *J. Appl. Toxicol.* 1984, **4**, 155-160
33. *Chemical Safety Data Sheets* 1989, **1**, 115-118
34. *ECETOC Technical Report No. 30(4)* 1991, European Chemical Industry Ecology and
    Toxicology Centre, B-1160 Brussels
35. Siegel, W. V. et al *Oral Surg., Oral Med., Oral Pathol.* 1969, **27**(6), 772-779

# M287    Methyl 2,3,5,6-tetrachloro-4-pyridyl sulfone

$$SO_2CH_3$$

**CAS Registry No.** 13108-52-6
**Synonyms** 2,3,5,6-tetrachloro-4-(methylsulfonyl)pyridine
**Mol. Formula** $C_6H_3Cl_4NO_2S$                    **Mol. Wt.** 294.97
**Uses** Fungicide. Disinfectant.

## Physical properties
**M. Pt.** 138-140°C.

## Occupational exposure
**Supply classification** harmful.
**Risk phrases** Harmful in contact with skin and if swallowed – Irritating to eyes –
May cause sensitisation by skin contact (R21/22, R36, R43)
**Safety phrases** In case of contact with eyes, rinse immediately with plenty of water
and seek medical advice – After contact with skin, wash immediately with plenty of
soap and water (S26, S28)

## Mammalian and avian toxicity

**Acute data**
$LD_{50}$ oral mouse 770 mg kg$^{-1}$ (1).

## Legislation
Limited under EC Directive on Drinking Water Quality 80/778/EEC. Pesticides:
maximum admissible concentration 0.1 µg l$^{-1}$ (2).
Included in Schedule 6 (Release into Land: Prescribed Substances) Statutory
Instrument No. 472, 1991 (3).

## Any other comments
Human health effects, experimental toxicology, physico-chemical properties
reviewed (4).

## References
1. *Nichidai Igaku Zasshi* 1981, **40**, 329
2. *EC Directive Relating to the Quality of Water Intended for Human Consumption* 1982,
   80/778/EEC, Office for Official Publications of the European Communities, 2 rue Mercier,
   L-2985 Luxembourg
3. *S. I. 1991 No. 472 The Environmental Protection (Prescribed Processes and Substances)
   Regulations* 1991, HMSO, London
4. *ECETOC Technical Report No. 30(4)* 1991, European Chemical Industry Ecology and
   Toxicology Centre, B-1160 Brussels

# M288  2-Methyltetrahydrofuran

CAS Registry No. 96-47-9
Synonyms furan, tetrahydro-2-methyl-; tetrahydro-2-methylfuran; tetrahydrosylvan
Mol. Formula $C_5H_{10}O$                          Mol. Wt. 86.13

## Physical properties

B. Pt. 80°C; Flash point –11/–12°C; Specific gravity $d_4^{20}$ 0.853.

## Mammalian and avian toxicity

### Acute data
$LD_{Lo}$ oral rat 5720 mg $kg^{-1}$ (1).
$LC_{50}$ (4 hr) inhalation rat 6000 ppm (2).
$LD_{50}$ dermal rabbit 4500 mg $kg^{-1}$ (2).

### Irritancy
500 mg instilled into rabbit eye (24 hr) caused mild irritation (1).

### Legislation
Included in Schedule 4 (Release into Air: Prescribed Substances) Statutory Instrument No. 472, 1991 (3).

### Any other comments
All reasonable efforts have been taken to find information on isomers of this compound but no relevant data are available.

### References
1. Marhold, J. V. *Sbornik Vysledku Toxixologickeho Vysetreni Latek A Pripravku* 1972, Prague
2. Deichmann, W. B. *Toxicology of Drugs and Chemicals* 1969, Academic Press, New York
3. *S. I. 1991 No. 472 The Environmental Protection (Prescribed Processes and Substances) Regulations* 1991, HMSO, London

# M289  2-(Methylthio)benzothiazole

CAS Registry No. 615-22-5
Synonyms benzothiazole, 2-(methylthio)-; 2-(methylmercapto)benzothiazole; methylcaptax
Mol. Formula $C_8H_7NS_2$                          Mol. Wt. 181.28
Uses Defoliant. Used in the vulcanisation of rubber goods. In automobile antifreeze.

## Physical properties
**M. Pt.** 43-46°C; **Partition coefficient** log $P_{ow}$ 5 (1).

### Solubility
Organic solvent: ethanol, chloroform

## Ecotoxicity
### Bioaccumulation
Average bioconcentration factors for the leeches *Dina dubia*, *Erpobdella punctata* and *Helobdella stagnalis* in a creek polluted industrially with 2-(methylthio)benzothiazole were 400, 200 and 100 × respectively (1).

## Mammalian and avian toxicity
### Metabolism and pharmacokinetics
2-(methylthio)benzothiazole incubated with $^{35}$S-labelled GSH and rat liver homogenate was oxidised to its corresponding methylsulfoxide and/or methylsulfone which becomes a substrate for GSH conjugation. The methylthio group is degraded to formaldehyde and sulfate (2).

## Any other adverse effects
Mutagenic (increased chromosomal aberrations) in *Gossypium barbadense* seeds (3).

## Legislation
Included in Schedule 4 (Release into Air: Prescribed Substances) Statutory Instrument No. 472, 1991 (4).
The log $P_{ow}$ value exceeds the European Community recomended value level 3.0 (6th and 7th amendments) (5).

## References
1. Metcalfe, J. L. et al *Environ. Monit. Assess.* 1988, **11**, 147-169
2. Larsen, G. L. et al *Xenobiotica* 1988, **18**(3), 313-322
3. Vesmanova, O. Y. et al *Tsitol. Genet.* 1988, **22**(2), 17-21 (Russ.) (*Chem. Abstr.* **109**, 106243r)
4. *S. I. 1991 No. 472 The Environmental Protection (Prescribed Processes and Substances) Regulations* 1991, HMSO, London
5. *1967 Directive on Classification, Packaging and Labelling of Dangerous Substances 67/548/EEC; 6th Amendment EEC Directive 79/831/EEC; 7th Amendment EEC Directive 91/32/EEC* 1991, HMSO, London

# M290   Methyl thiocyanate
## CH3NCS

**CAS Registry No.** 556-64-9
**Synonyms** methyl rhodanate; methyl sulfocyanate
**Mol. Formula** $C_2H_3NS$                    **Mol. Wt.** 73.12
**Uses** Insecticide.

## Physical properties
**M. Pt.** –5°C; **B. Pt.** 131°C; **Flash point** 38°C; **Specific gravity** $d_4^{25}$ 1.0678.

**Solubility**

Organic solvent: miscible with diethyl ether, ethanol

## Mammalian and avian toxicity

### Acute data

$LD_{50}$ oral rat 60 mg $kg^{-1}$ (1).
$LD_{50}$ intraperitoneal mouse 23 mg $kg^{-1}$ (2).
$LD_{50}$ intravenous mouse 18 mg $kg^{-1}$ (3).

### Sub-acute data

Oral chicken 0.33% diet (duration unspecified) had no effect on thyroid size but reduced egg iodine and reduced food intake (4).

### Metabolism and pharmacokinetics

Liver enzymes liberate cyanide (species unspecified) (5).

### Sensitisation

May cause allergic reaction (species unspecified) (6).

## Genotoxicity

*Salmonella typhimurium* TA100 without metabolic activation negative (7).

## Any other adverse effects to man

May be fatal if inhaled, swallowed or absorbed through the skin. High concentrations are extremely destructive to tissues of the mucous membranes and upper respiratory tract, eyes and skin. Prolonged exposure, may cause nausea, dizziness, headache, severe irritation or burns, lung irritation, chest pain and oedema which may be fatal (6).

## Legislation

Limited under EC Directive on Drinking Water Quality 80/778/EEC. Pesticides: maximum admissible concentration 0.1 μg $l^{-1}$ (8).
Included in Schedule 6 (Release into Land: Prescribed Substances) Statutory Instrument No. 472, 1991 (9).

## References

1.  Marhold, J. V. *Prehled Prumyslove Toxikologie: Organicke Latky* 1986, Prague
2.  *Pestic. Biochem. Physiol.* 1972, **2**, 95
3.  *Report NX#02864*, US Army Armament Research and Development Command, Chemical Systems Laboratory, NIOSH Exchange Chemicals, Aberdeen Proving Ground, MD21010
4.  Papas, A. et al *Can. J. Anim. Sci.* 1979, **59**(1), 119
5.  Gosselin, R. E. et al (Eds.) *Clinical Toxicology of Commercial Products* 4th ed., **3**,9, Williams & Wilkins, Baltimore, MD
6.  Lenga, R. E. (Ed.) *Sigma-Aldrich Library of Chemical Safety Data* 2nd ed., 1988, **2**, 2446, Sigma-Aldrich, Milwaukee, WI
7.  Yamaguchi, T. *Agric. Biol. Chem.* 1980, **44**(12), 3017
8.  *EC Directive Relating to the Quality of Water Intended for Human Consumption* 1982, 80/778/EEC, Office for Official Publications of the European Communities, 2 rue Mercier, L-2985 Luxembourg
9.  *S. I. 1991 No. 472 The Environmental Protection (Prescribed Processes and Substances) Regulations* 1991, HMSO, London

# M291  Methylthiouracil

CAS Registry No. 56-04-2

Synonyms 2,3-dihydro-6-methyl-2-thioxo-4(1$H$)pyrimidinone; 6-methyl-2-thiouracil; Alkiron; Metacil; Methiacil; MTU; Orcanon; Thiothymin

**Mol. Formula** $C_5H_6N_2OS$          **Mol. Wt.** 142.18

**Uses** Thyroid inhibitor.

## Physical properties

**M. Pt.** 326-331°C.

## Mammalian and avian toxicity

**Acute data**

$LD_{50}$ oral rat 2790 mg $kg^{-1}$ (1).

$LD_{Lo}$ oral rabbit 2500 mg $kg^{-1}$ (2).

$LD_{50}$ intraperitoneal mouse, rat 200-920 mg $kg^{-1}$ (3,4).

**Carcinogenicity and long-term effects**

Thyroid adenomas with metastases in the lungs have been reported in rats and mice treated with methylthiouracil. Malignancy and metastases have been questioned. In further work mice on an iodine-rich diet received methylthiouracil in drinking water as a 1% solution (1000 mg $l^{-1}$), and others on a low-iodine diet received 0.2-0.5% mixed in the food pellets. Thyroid adenomas and pulmonary nodules were seen but were not malignant. Hepatomas were also reported. 6 months after treatment was discontinued no adenomas were present and pulmonary nodules of thyroid tissue were demonstrated (5).

Thyroid adenomas and carcinomas reported in rats fed 2.5 mg $day^{-1}$ plus intraperitoneal injections of radioactive iodine ($^{131}$I) (6,7).

## Legislation

Land disposal prohibited under U.S. Federal Resource Conservation and Recovery Act (8).

## Any other comments

Thyroid hyperplasia produced by administration of methylthiouracil to infant rats can be prevented by concurrent administration of thyroid hormone (9).

Human health effects, experimental toxicology, physico-chemical properties reviewed (10).

## References

1.  *Farmaco, Ed. Sci.* 1952, **7**, 313

2.  *Merck Index* 11th ed., 1989, Merck & Co., Rahway, NJ
3.  *NTIS Report* AD277-689, Natl. Tech. Inf. Ser., Springfield, VA
4.  *Farmaco, Ed. Sci.* 1958, **13**, 882
5.  Jemec, B. *Cancer* 1977, **40**, 2188-2202
6.  Douiach, I. *Br. J. Cancer* 1953, **7**, 181-202
7.  Field, J. B. et al *Cancer Res.* 1959, **19**, 870-873
8.  *Fed. Regist.* 31 Jan 1991, **56**(21), 3864-3928
9.  Elphinstone, N. *Lancet* 27 Jun 1953, **1**, 1231-1233
10. *ECETOC Technical Report No. 30(4)* 1991, European Chemical Industry Ecology and Toxicology Centre, B-1160 Brussels

# M292   Methyl *p*-toluenesulfonate

$$H_3C-\langle\!\bigcirc\!\rangle-SO_3CH_3$$

**CAS Registry No.** 80-48-8

**Synonyms** 4-methylbenzenesulfonic acid, methyl ester; *p*-toluenesulfonic acid, methyl ester; methyl *p*-methylbenzenesulfonate; methyl toluene-4-sulfonate; methyl *p*-tosylate

**Mol. Formula** $C_8H_{10}O_3S$                    **Mol. Wt.** 186.23

**Uses** Methylating agent.

**Occurrence** Pollutant of drinking water wells in areas irrigated with sewage effluent (1).

## Physical properties

**M. Pt.** 27.5°C; **B. Pt.** $_5$ 144-145°C; **Flash point** 152°C (open cup) (2); **Specific gravity** $d_{25}^{25}$ 1.230.

## Solubility

Organic solvent: benzene, ethanol, diethyl ether

## Mammalian and avian toxicity

### Acute data

$LD_{50}$ oral rat 341 mg kg$^{-1}$ (3).
$LD_{50}$ subcutaneous rat 250 mg kg$^{-1}$ (4).

### Carcinogenicity and long-term effects

Subcutaneous BD-strain rat once a week developed local sarcomas. The number of tumours was dose dependent (5).
Subcutaneous BD rat single dose 50 mg kg$^{-1}$. 5/12 animals developed local tumours (6).

### Irritancy

Dermal rabbit (24 hr) 500 mg caused severe erythema to slight eschar formation and severe oedema. 500 mg instilled into rabbit eye (24 hr) caused mild irritation (3).

### Sensitisation

Positive allergic reactions, confirmed using patch tests, reported in patients reacting to dental impression materials (7).

### Genotoxicity

*Drosophila melanogaster* negative when orally administered, positive when injected in sex-linked recessive lethal test. Fed in combination with inhibitors of cytochrome $P_{450}$, sufficient mutagen reached the gonads to produce significant genetic damage (8).

### Any other comments

Experimental toxicology and human health effects reviewed (9).

### References

1. Tu, J. et al *Huanjing Huaxue* 1986, **5**(5), 60-74 (Ch.) (*Chem. Abstr.* **106**, 22946x)
2. Bond, J. *Sources of Ignition: Flammability Characterisitics of Chemicals and Products* 1991, p.113, Butterworth-Heinemann, Oxford
3. Marhold, J. V. *Sbornik Vysledku Toxixologickeho Vysetreni Latek A Pripravku* 1972, Prague
4. *Z. Kinderheilkd.* 1970, **74**, 241
5. Druckrey, H. et al *Z. Krebsforsch.* 1970, **74**(3), 241-273
6. Preussman, R. *Food Cosmet. Toxicol.* 1968, **6**, 576-577
7. van Groeningen et al *Contact Dermatitis* 1975, **1**, 373-376
8. Zijlstra, J. A. et al *Mutat. Res.* 1988, **198**(1), 73-83
9. *BIBRA Toxicity Profile* 1991, British Industrial Biological Research Association, Carshalton

# M293   Methyltrichlorosilane

## CH3SiCl3

**CAS Registry No.** 75-79-6
**Synonyms** trichloromethylsilane
**Mol. Formula** $CH_3Cl_3Si$ **Mol. Wt.** 149.48
**Uses** Has been used to impart water repellency and wet strength to paper.

### Physical properties

**B. Pt.** 66°C; **Flash point** −15°C; **Specific gravity** 1.273.

### Occupational exposure

**UN No.** 1250; **HAZCHEM Code** 4WE; **Conveyance classification** flammable liquid, corrosive substance; **Supply classification** highly flammable, irritant.
**Risk phrases** ≥1% – Highly flammable – Reacts violently with water – Irritating to eyes, respiratory system and skin – <1% – Highly flammable – Reacts violently with water (R11, R14, R36/37/38, R11, R14)
**Safety phrases** In case of contact with eyes, rinse immediately with plenty of water and seek medical advice – Wear eye/face protection (S26, S39)

## Mammalian and avian toxicity

**Acute data**
$LD_{Lo}$ oral rat 1000 mg $kg^{-1}$ (1).
$LC_{50}$ (4 hr) inhalation rat 450 ppm (2).
$LC_{50}$ (2 hr) inhalation mouse 180 mg $m^{-3}$ (3).
$LD_{Lo}$ intraperitoneal rat 30 mg $kg^{-1}$ (1).

**Irritancy**
Mild eye and severe skin irritant in rabbits (2).

## Any other comments

Human health effects, experimental toxicology, physico-chemical properties
reviewed (4).

## References

1. *J. Ind. Hyg. Toxicol.* 1948, **30**, 332
2. Marhold, J. V. *Prehled Prumyslove Toxikologie: Organicke Latky* 1986, Prague
3. *Toksikol. Nov. Prom. Khim. Veshchestv* 1961, **3**, 23
4. *ECETOC Technical Report No. 30(4)* 1991, European Chemical Industry Ecology and
   Toxicology Centre, B-1160 Brussels

# M294   1-Methylurea

## NH2C(O)NHCH3

**CAS Registry No.** 598-50-5
**Synonyms** *N*-methylurea; methylurea
**Mol. Formula** $C_2H_6N_2O$                    **Mol. Wt.** 74.08

## Physical properties

**M. Pt.** 101°C; **B. Pt.** decomp.; **Specific gravity** $d_{20}^{20}$ 1.205.

**Solubility**
Organic solvent: ethanol

## Mammalian and avian toxicity

**Acute data**
$LD_{Lo}$ oral rat 500 mg $kg^{-1}$ (1).

**Teratogenicity and reproductive effects**
Not teratogenic in rat or mouse *in vivo* (dose, duration and route unspecified). No
degenerative effect reported on neuronal or non-neuronal foetal rat brain cells *in
vitro* (2).

**Irritancy**
Non-irritating to skin (species unspecified) (3).

### Genotoxicity

*Salmonella typhimurium* TA98, TA100 with or without metabolic activation negative. *Bacillus subtilis* TKJ5211 his[+] reversion without metabolic activation positive; his[+] met[+] double reversion without metabolic activation negative (4).

### References

1.  *Natl. Acad. Sci. Natl. Res. Council* 1953, **5**, 47
2.  Khera, K. S. et al *Toxicol. in Vitro* 1988, **2**(4), 257-273
3.  Han, S. K. et al *Arch. Pharmacol. Res.* 1991, **14**(1), 12-18
4.  Tanooka, H. *Mutat. Res.* 1977, **42**, 19-32

# M295   2-Methylvaleraldehyde

## CH$_3$CH$_2$CH$_2$CH(CH$_3$)CHO

**CAS Registry No.** 123-15-9
**Synonyms** 2-formylpentane; 2-methylpentanal; 2-methylvaleric aldehyde
**Mol. Formula** C$_6$H$_{12}$O             **Mol. Wt.** 100.16
**Uses** Organic synthesis.
**Occurrence** In plant and animal oils.

### Physical properties

**M. Pt.** –100°C; **B. Pt.** 119-120°C; **Flash point** 16°C; **Specific gravity** d$^{20}$ 0.808; **Volatility** v. den. 3.5.

### Solubility

Water: miscible. Organic solvent: acetone, diethyl ether

### Occupational exposure

**UN No.** 2367; **HAZCHEM Code** 3YE; **Conveyance classification** flammable liquid.

### Ecotoxicity

**Fish toxicity**
LC$_{50}$ (96 hr) fathead minnow 190 mg l$^{-1}$ (1).

### Mammalian and avian toxicity

**Irritancy**
Irritating to the skin. Vapour or mist is irritating to the eyes, skin, mucous membranes and upper respiratory tract (species and dose unspecified) (2).

### Any other comments

Autoignition temperature 199°C.

### References

1.  Protic, A. et al *Aquat. Toxicol.* 1989, **14**(1), 47-64
2.  Lenga, R. E. (Ed.) *Sigma-Aldrich Library of Chemical Safety Data* 2nd ed., 1988, **2**, 2460, Sigma-Aldrich, Milwaukee, WI

# M296   Methyl vinyl ether

## $CH_3OCH=CH_2$

**CAS Registry No.** 107-25-5
**Synonyms** Agrisynth MVE; methoxyethene; methoxyethylene; vinyl methyl ether
**Mol. Formula** $C_3H_6O$                                     **Mol. Wt.** 58.08
**Uses** Manufacture of polymers.

## Physical properties

**M. Pt.** −123°C; **B. Pt.** 5-6°C; **Flash point** −51°C; **Specific gravity** $d_4^{20}$ 0.7725;
**Volatility** v.p. 1052 mmHg at 20°C; v. den. 2.0.

**Solubility**
Water: 15 g $l^{-1}$ at 20°C. Organic solvent: acetone, benzene, diethyl ether, ethanol

## Occupational exposure

**Supply classification** highly flammable.
**Risk phrases** Extremely flammable liquefied gas (R13)
**Safety phrases** Keep container in a well ventilated place – Keep away from sources
of ignition – No Smoking – Take precautionary measures against static discharges
(S9, S16, S33)

## Ecotoxicity

**Bioaccumulation**
Calculated bioconcentration factor 2.7 indicates that environmental accumulation is
unlikely (1).

## Environmental fate

**Abiotic removal**
Reaction with photochemically produced hydroxyl radicals and ozone in the
atmosphere, estimated $t_{1/2}$ 10 hr (2).
Estimated volatilisation $t_{1/2}$ 42 hr from model pond water and 3.3 hr in model river
water (1,3).

**Absorption**
Estimated soil $K_{oc}$ 22 indicates that methyl vinyl ether will not adsorb to soil and
sediments (1).

## Mammalian and avian toxicity

**Acute data**
$LD_{50}$ oral rat 4900 mg $kg^{-1}$ (4).
$LD_{50}$ dermal rabbit >8000 mg $kg^{-1}$ (5).

## Any other adverse effects to man

Inhalation exposure causes intoxication, blurred vision, headache, dizziness,
excitation and loss of consciousness. The liquid or concentrated vapour is irritating to
the eyes and causes frost bite to the skin. Aspiration of the liquid causes chemical
pneumonitis (6).

## Any other comments

Human health effects, experimental toxicology, physico-chemical properties reviewed (7).
Autoignition temperature 287°C.

## References

1. Lyman, W. K. et al *Handbook of Chemical Property Estimation Methods Environmental Behaviour of Organic Compounds* 1982, McGraw-Hill, New York
2. Atkinson, R. et al *Chem. Rev. (Washington, D. C.)* 1984, **84**, 437-470
3. *EXAMS II Computer Simulation* 1987, USEPA, Athens, GA
4. Deichmann, W. B. *Toxicology of Drugs and Chemicals* 1969, 395, Academic Press, New York
5. *GAF Material Safety Data Sheet*, GAF Chemicals Corp., Wayne, NJ
6. *CHRIS – Hazardous Chemical Data* 1978, **2**, US Coast Guard, Dept. of Transportation, Washington, DC
7. *ECETOC Technical Report No. 30(4)* 1991, European Chemical Industry Ecology and Toxicology Centre, B-1160 Brussels

# M297   Methyl vinyl ketone

$$CH_3C(O)CH=CH_2$$

**CAS Registry No.** 78-94-4
**Synonyms** acetylethylene; 3-buten-2-one; 1-buten-3-one; methylene acetone; $\delta$-oxo-$\alpha$-butylene
**Mol. Formula** $C_4H_6O$                    **Mol. Wt.** 70.09
**Uses** Alkylating agent. Michael-acceptor. Organic synthesis. Manufacture of polymers.
**Occurrence** In fulvic acids. In various plants.

## Physical properties

**M. Pt.** –7°C; **B. Pt.** 81.4°C; **Flash point** –6°C; **Specific gravity** $d_4^{20}$ 0.8636; **Partition coefficient** log $P_{ow}$ 0.117 (calc.) (1); **Volatility** v.p. 84 mmHg at 25°C; v. den. 2.41.

## Solubility

Water: 10%. Organic solvent: acetone, diethyl ether, dimethyl sulfoxide, ethanol, glacial acetic acid, methanol

## Occupational exposure

**UN No.** 1251; **HAZCHEM Code** 2PE; **Conveyance classification** flammable liquid.

## Ecotoxicity

**Invertebrate toxicity**
$EC_{50}$ *Selenastrum capricornutum* < 10 mg $l^{-1}$ (exposure not specified) (2).

**Bioaccumulation**
Calculated bioconcentration factor 0.72 indicates that environmental accumulation is unlikely (1).

# Environmental fate

### Degradation studies
$BOD_5$ 10% ThOD; COD 100% ThOD (3).

### Abiotic removal
Photooxidation by UV light in aqueous medium at 50°C, 16.8% degradation to carbon dioxide after 24 hr (4).
Reaction with photochemically produced hydroxyl radicals in the atmosphere, estimated $t_{1/2}$ 21 hr (5).
Estimated volatilisation $t_{1/2}$ 3.4 days from model river water and 37 days from model pond water (1,6).

### Absorption
Estimated $K_{oc}$ 28 indicates that adsorption to soil and sediments would not be significant (7).

## Mammalian and avian toxicity

### Acute data
$LD_{50}$ oral rat, mouse 30-33 mg $kg^{-1}$ (8,9).
$LC_{50}$ (4 hr) inhalation rat 7 mg $m^{-3}$ (9).

### Metabolism and pharmacokinetics
Readily absorbed through the skin (species unspecified) (10).
Binds to protein sulfhydryl groups and GSH (11).

### Sensitisation
Induced strong contact sensitivity in guinea pigs immunised with acrylate in Freund's complete adjuvant (12).

## Genotoxicity
*Salmonella typhimurium* TA98 with and without metabolic activation negative, TA100 with metabolic activation positive, and without metabolic activation negative (13,14).

## Any other adverse effects to man
Skin irritant and lachrymator (15).
Suspected of causing conjunctivitis and injury to the corneal epithelium in occupational exposure study in former Czechoslovakia (10).

## Any other comments
Human health effects, experimental toxicology, physico-chemical properties reviewed (16).
Autoignition temperature 491°C.

## References
1. Lyman, W. K. et al *Handbook of Chemical Property Estimation Methods Environmental Behaviour of Organic Compounds* 1982, McGraw-Hill, New York
2. Bollman, M. A. et al *Report on Algal Toxicity Tests, of Selected Office of Toxic Substances (OTS) Chemicals* 1989, EPA/600/3-90/041
3. Dore, M. et al *Trib. CEBEDEAU* 1975, **28**(379), 3-11
4. Knoevenagel, K. et al *Arch. Environ. Contam. Toxicol.* 1976, **4**, 324-333

5. Atkinson, R. *Chem. Rev. (Washington, D. C.)* 1985, **85**, 69-201
6. *EXAMS II Computer Simulation* 1987, US EPA, Athens, GA
7. Swann, R. L. et al *Residue Rev.* 1983, **85**, 17-28
8. *Nippon Eiseigaku Zasshi* 1979, **34**, 183
9. Izmerov, N. F. et al *Toxicometric Parameters of Industrial Toxic Chemicals under Single Exposure* 1982, CIP, Moscow
10. Grant, W. M. *Toxicology of the Eye* 3rd ed., 1986, 629, Charles C. Thomas, Springfield, IL
11. Lash, L. H. et al *Arch. Biochem. Biophys.* 1991, **286**(1), 46-56
12. Turk, J. L. et al *J. Allergy Clin. Immunol.* 1986, **78**(5, Pt. 2), 1082-1085
13. Neudecker, T. et al *Mutat. Res.* 1989, **227**(2), 131-134
14. Zeiger, E. et al *Environ. Mol. Mutagen.* 1992, **19**(Suppl. 21), 2-141
15. *The Merck Index* 11th ed., 1989, 963, Merck and Co.Inc, Rahway, NJ
16. *ECETOC Technical Report No. 30(4)* 1991, European Chemical Industry Ecology and Toxicology Centre, B-1160 Brussels

# M298   Metiram

**CAS Registry No.** 9006-42-2
**Synonyms** Polyram; zinc ammoniate ethylenebis(dithiocarbamate)-poly(ethylenethiuram disulfide)
**Mol. Formula** $C_{16}H_{33}N_{11}S_{16}Zn_3$ **Mol. Wt.** 1088.65
**Uses** Fungicide.

## Physical properties

**M. Pt.** 140°C (decomp.); **Volatility** v.p. $<7.25 \times 10^{-8}$ mmHg at 20°C.

## Solubility
Organic solvent: pyridine

## Ecotoxicity

**Fish toxicity**
$LC_{50}$ (48 hr) harlequin fish 17 mg l$^{-1}$ (1).
$LC_{50}$ (96 hr) rainbow trout, carp 1.1, 85 mg l$^{-1}$, respectively (1).

## Environmental fate

**Degradation studies**
Degraded to derivatives of thiourea, thiuram monosulfide, thiuram disulfide and sulfur (1).

## Mammalian and avian toxicity

**Acute data**
$LD_{50}$ oral rabbit, mouse, rat 620-2850 mg kg$^{-1}$ (2,3).
$LC_{50}$ (4 hr) inhalation rat >5.7 mg l$^{-1}$ (1).
$LD_{50}$ dermal rat >2000 mg kg$^{-1}$ (1).

**Carcinogenicity and long-term effects**
No effect level for rats in 2 yr feeding study 3.1 mg kg$^{-1}$ (1).

## Irritancy
Mild skin and eye irritant (species unspecified) (1).

## Genotoxicity
Induced mitotic gene conversion in *Saccharomyces cerevisiae* D4 (4).

## Legislation
Included in Schedule 4 (Release Into Air: Prescribed Substances) Statutory Instrument No. 472, 1991 (5).
Limited under EC Directive on Drinking Water Quality 80/778/EEC. Zinc: guide level 100 µg l$^{-1}$ (at outlets), 5000 µg l$^{-1}$ (for consumer) (6).

## Any other comments
LD$_{50}$ contact, oral bee >16->40 µg bee$^{-1}$ (1).
WHO class Table 5, EPA toxicity class v (1).
Community Right-To-Know List in the U.S.

## References
1. *The Agrochemicals Handbook* 3rd ed., 1991, RSC, London
2. *Veterinariya (Moscow)* 1986, **56**(6), 59
3. *Vrach. Delo* 1975, (9), 130
4. Siebert, D. et al *Mutat. Res.* 1970, **10**, 533-543
5. *S. I. 1991 No. 472 The Environmental Protection (Prescribed Processes and Substances) Regulations* 1991, HMSO, London
6. *EC Directive Relating to the Quality of Water Intended for Human Consumption* 1982, 80/778/EEC, Office for Official Publications of the European Communities, 2 rue Mercier, L-2985 Luxembourg

# M299   Metobromuron

**CAS Registry No.** 3060-89-7
**Synonyms** *N'*-(4-bromophenyl)-*N*-methoxy-*N*-methylurea; Patoran; 3-(4-bromophenyl)-1-methoxy-1-methylurea; Oatirab
**Mol. Formula** $C_9H_{11}BrN_2O_2$                 **Mol. Wt.** 259.11
**Uses** Selective phenylurea herbicide.

## Physical properties
**M. Pt.** 95-96°C; **Specific gravity** d$^{20}$ 1.60; **Partition coefficient** log P$_{ow}$ 2.41 (2); **Volatility** v.p. $3 \times 10^{-6}$ mmHg at 20°C.

## Solubility
Water: 300 ppm at 20°C. Organic solvent: methanol, ethanol, acetone, chloroform

## Ecotoxicity

**Fish toxicity**

$LC_{50}$ (96 hr) guppy 25.76 mg $l^{-1}$ (1).
$LC_{50}$ (96 hr) rainbow trout, crucian carp, bluegill sunfish, 36-40 mg $l^{-1}$ (2).

**Invertebrate toxicity**

$LC_{50}$ (96 hr) *Daphnia magna* 1.78 mg $l^{-1}$ (1).
Low concentrations (0.05 mg $l^{-1}$) promoted growth of *Scenedesmus* sp. as indicated by dry weight, chlorophyll a content and the ratio of chlorophyll a:b; growth was inhibited at 0.1, 0.5 and 1.0 mg $l^{-1}$ (3).

## Environmental fate

**Degradation studies**

In soil $t_{1/2}$ 30 days (2).

## Mammalian and avian toxicity

**Acute data**

$LD_{50}$ oral rat, mouse 2000-2100 mg $kg^{-1}$ (4,5).
$LC_{50}$ (4 hr) inhalation rat >1.1 mg $l^{-1}$ air (2).
$LD_{50}$ dermal rat, rabbit >3000, >10,200 mg $kg^{-1}$, respectively (2).
$LD_{50}$ intraperitoneal rat, mouse 430-850 mg $kg^{-1}$ (5).

**Carcinogenicity and long-term effects**

No effect level in dogs and rats in 2 yr feeding trials was 100 and 250 mg $kg^{-1}$ diet, respectively (2).

**Irritancy**

Mild skin and eye irritant in rabbits (2,6).

## Legislation

Limited under EC Directive on Drinking Water Quality 80/778/EEC. Pesticides: maximum admissible concentration 0.1 μg $l^{-1}$ (7).
Included in Schedule 6 (Release into Land: Prescribed Substances) Statutory Instrument No. 472, 1991 (8).

## Any other comments

WHO Class Table 5; EPA Toxicity Class III (2).

## References

1. Strateva, A. et al *Probl. Khig.* 1986, **11**, 32-37 (Bulg.) (*Chem. Abstr.* **107**, 72436x)
2. *The Agrochemicals Handbook* 3rd ed., 1991, RSC, London
3. El-Dib, M. A. et al *Water, Air Soil Pollut.* 1991, **55**, 295-303
4. *World Rev. Pest Control* 1970, **9**, 119
5. *Proc. Eur. Soc. Toxicol.* 1976, **17**, 351
6. *Ciba Geigy Toxicol. Index* 1977
7. *EC Directive Relating to the Quality of Water Intended for Human Consumption* 1982, 80/778/EEC, Office for Official Publications of the European Communities, 2 rue Mercier, L-2985 Luxembourg
8. *S. I. 1991 No. 472 The Environmental Protection (Prescribed Processes and Substances) Regulations* 1991, HMSO, London

# M300   Metolachlor

$$CH_3 \quad CH_3$$

(structural diagram: benzene ring with CH₃ groups at positions 2 and 6, H₃CH₂C group, and N-substituted acetamide)

CH₃ CH₃
  CHCH₂OCH₃
—N
  CCH₂Cl
  ‖
H₃CH₂C   O

**CAS Registry No.** 51218-45-2
**Synonyms** 2-chloro-6'-ethyl-N-(2-methoxy- 1-methylethyl)acet-o-toluidide;
2-chloro-N-(2- ethyl-6-methylphenyl)-N-(2-methoxy-1- methylethyl)acetamide
**Mol. Formula** $C_{15}H_{22}ClNO_2$                **Mol. Wt.** 283.80
**Uses** Herbicide.

## Physical properties

**B. Pt.** $_{0.001}$ 100°C; **Specific gravity** $d^{20}$ 1.12; **Partition coefficient** log $P_{ow}$ 3.450 (1);
**Volatility** v.p. $1.3 \times 10^{-5}$ mmHg at 20°C.

## Solubility

Water: 530 mg $l^{-1}$ at 20°C. Organic solvent: miscible with methanol, benzene, dimethylformamide, ethylene dichloride

## Ecotoxicity

### Fish toxicity

$LC_{50}$ (96 hr) rainbow trout, carp, bluegill sunfish 2-15 mg $l^{-1}$ (1).

## Environmental fate

### Degradation studies

In soil $t_{1/2}$  30 days (2).
Some (aerobic) metolachlor-degrading microbial population reported in sandy loam soil previously treated with metolachlor (3).
Microbial activity reported to be responsible for mineralisation of metolachlor in a soil perfusion system; degradation was enhanced in acclimated soils (4).
Metolachlor was transformed by *Streptomyces* sp. in a liquid medium and in $Na_2CO_3$ treated soil (5).
80% of added metolachlor (50 µg $ml^{-1}$) was absorbed and transformed by a stable bacterial community, suggesting seeding aquatic environments with a mixture of microorganisms rather than individual species may be advantageous in removal or detoxification of metolachlor (6).

## Mammalian and avian toxicity

### Acute data

$LD_{50}$ oral quail, rock dove, house sparrow, common grackle, starling 2.37, 5.6, 7.5, 23.7 and 31.6 mg $kg^{-1}$, respectively (7).
$LD_{50}$ oral rat 2780 mg $kg^{-1}$ (1).

$LD_{50}$ dermal rat >3170 mg kg$^{-1}$ (1).
$LC_{50}$ (6 hr) rat >1.75 mg l$^{-1}$ (1).

**Sub-acute data**
$LC_{50}$ (8 day) oral bobwhite quail, mallard duck >10 g kg$^{-1}$ diet (1).

**Carcinogenicity and long-term effects**
No effect level in dogs, rats in 90 day feeding trials 500-1000 mg kg$^{-1}$ diet (1).

**Metabolism and pharmacokinetics**
*In vitro* transformation by rat liver enzymes is via conjugation with GSH and oxidation to produce the I-GSH conjugate (8).

**Irritancy**
Mild skin irritant but non-irritating to eyes of rabbits (1).

## Genotoxicity

*Salmonella typhimurium* TA98, TA100, TA1535, TA1538 with and without metabolic activation negative (technical grade), positive (commercial grade) (9). *Saccharomyces cerevisiae* D4 with or without metabolic activation negative (technical grade); without metabolic activation negative, with metabolic activation positive (commercial grade) (9).

## Legislation

Limited under EC Directive on Drinking Water Quality 80/778/EEC. Pesticides: maximum admissible concentration 0.1 µg l$^{-1}$ (10).
Included in Schedule 6 (Release into Land: Prescribed Substances) Statutory Instrument No. 472, 1991 (11).
WHO guideline value for drinking water quality 10 µg l$^{-1}$ (12).
The log $P_{ow}$ value exceeds the European Community recommended level 3.0 (6th and 7th amendments) (13).

## Any other comments

WHO Class III; EPA toxicity class III (1).
Environmental fate reviewed (14).

## References

1. *The Agrochemicals Handbook* 3rd ed., 1991, RSC, London
2. Bowman, B. T. *Environ. Toxicol. Chem.* 1990, **9**(4), 453-461
3. Zhang, R. *J. Environ. Sci. (China)* 1989, **1**(2), 98-103
4. Liu, S. Y. et al *Biol. Fertil. Soils* 1988, **5**(4), 276-281
5. Liu, S. Y. et al *Biodegradation* 1990, **1**(1), 9-17
6. Liu, S. Y. et al *Appl. Environ. Microbiol.* 1989, **55**(3), 733-740
7. Schafer, E. W. et al *Arch. Environ. Contam. Toxicol.* 1983, **12**, 355-382
8. Feng, P. C. C. et al *J. Agric. Food Chem.* 1989, **37**(4), 1088-1093
9. Plewa, M. J. et al *Mutat. Res.* 1984, **136**, 233-245
10. *EC Directive Relating to the Quality of Water Intended for Human Consumption* 1982, 80/778/EEC, Office for Official Publications of the European Communities, 2 rue Mercier, L-2985 Luxembourg
11. *S. I. 1991 No. 472 The Environmental Protection (Prescribed Processes and Substances) Regulations* 1991, HMSO, London
12. *Guidelines for drinking-water quality* 2nd ed., 1993, WHO, Geneva

13. *1967 Directive on Classification, Packaging, and Labelling of Dangerous Substances 67/548/EEC; 6th Amendment EEC Directive 79/831/EEC; 7th Amendment EEC Directive 91/321/EEC* 1991, HMSO, London
14. Chesters, G. et al *Rev. Environ. Contam. Toxicol.* 1989, **110**, 1-74.

# M301   Metolcarb

**CAS Registry No.** 1129-41-5
**Synonyms** *m*-tolyl methyl carbamate; dicresyl; MTMC; 3-methylphenyl *N*-methylcarbamate
**Mol. Formula** $C_9H_{11}NO_2$                    **Mol. Wt.** 165.19
**Uses** Insecticide.

## Physical properties

**M. Pt.** 76-77°C; **Volatility** v.p. $1.09 \times 10^{-3}$ mmHg at 25°C.

**Solubility**
Water: 0.26% at 25°C.

## Occupational exposure

**Supply classification** harmful.
**Risk phrases** Harmful if swallowed (R22)

## Environmental fate

**Degradation studies**
Carbamates are readily degraded by soil microorganisms, and are photodecomposed in water by UV irradiation (1,2).

## Mammalian and avian toxicity

**Acute data**
$LD_{50}$ oral mouse 109-268 mg kg$^{-1}$ (3,4).
$LD_{50}$ inhalation rat 475 mg kg$^{-1}$ (5).
$LD_{50}$ dermal mouse 6000 mg kg$^{-1}$ (6).

## Genotoxicity

Induced chromosomal aberrations in Chinese hamster fibroblasts *in vitro* (metabolic activation unspecified) (7).

## Any other adverse effects to man

Most carbamates are active acetylcholinesterase activity inhibitors (2).

## Legislation

Limited under EC Directive on Drinking Water Quality 80/778/EEC. Pesticides: maximum admissible concentration 0.1 µg l$^{-1}$ (8).
Included in Schedule 6 (Release into Land: Prescribed Substances) Statutory Instrument No. 472, 1991 (9).

## References

1. *The list of the existing chemical substances tested on biodegradability by microorganisms or bioaccumulation in fish body* 1987, Chemicals Inspection & Testing Institute, Japan
2. *Environmental Health Criteria 64: Carbamate pesticides: a general introduction* 1986, WHO/IPCS, Geneva
3. *Farm Chemicals Handbook* 1983, Meister Publ. Co., Willoughby, OH
4. *Farm Chemicals Handbook* 1980, Meister Publ. Co., Willoughby, OH
5. *Environ. Qual. Saf.* 1975, **3**, 618
6. *Guide to the Chemicals used in Crop Protection* 1968, Information Canada, Ottawa, ON
7. Ishidate, M. et al *Mutat. Res.* 1977, **48**, 337-354
8. *EC Directive Relating to the Quality of Water Intended for Human Consumption* 1982, 80/778/EEC, Office for Official Publications of the European Communities, 2 rue Mercier, L-2985 Luxembourg
9. *S. I. 1991 No. 472 The Environmental Protection (Prescribed Processes and Substances) Regulations* 1991, HMSO, London

# M302   Metoxuron

**CAS Registry No.** 19937-59-8
**Synonyms** 3-(3-chloro-4-methoxyphenyl)-1,1-dimethylurea ; Dosanex; *N,N*-dimethyl-*N'*-(4-methoxy-3-chlorophenyl)urea
**Mol. Formula** $C_{10}H_{13}ClN_2O_2$          **Mol. Wt.** 228.68
**Uses** Herbicide.

## Physical properties

**M. Pt.** 126-127°C.

## Solubility

Water: 678 mg $l^{-1}$ at 24°C. Organic solvent: acetone, cyclohexanone, acetonitrile, ethanol

## Ecotoxicity

**Fish toxicity**
$LC_{50}$ (96 hr) rainbow trout, harlequin fish 20-40 mg $l^{-1}$ (1,2).

## Environmental fate

**Degradation studies**
Disappeared very quickly from loamy and clay loam soils via biodegradation, which was unaffected by supplying soil with ammoniacal nitrogen (3).

## Mammalian and avian toxicity

**Acute data**
$LD_{50}$ oral rat, rabbit, mouse 1600-2540 mg $kg^{-1}$ (4-6).

**Sub-acute data**
In 90 day feeding trials, rats receiving 1250 mg $kg^{-1}$ diet, and dogs receiving 2500 mg $kg^{-1}$ diet showed no ill effects (1).

## Genotoxicity

*Salmonella typhimurium* TA1535 with metabolic activation positive (7).
Inhibited testicular DNA synthesis (species unspecified) (7).
Micronucleus test positive (species unspecified) (7).

## Legislation

Limited under EC Directive on Drinking Water Quality 80/778/EEC. Pesticides: maximum admissible concentration 0.1 $\mu g\ l^{-1}$ (8).
Included in Schedule 6 (Release into Land: Prescribed Substances) Statutory Instrument No. 472, 1991 (9).

## References

1. Tooby, T. E. et al *Chem. Ind. (London)* 1975, **21**, 523-525
2. *Pesticide Manual* 1983, British Crop Protection Council, Farnham
3. Simon-Sylvestre, G. et al *J. Environ. Sci. Health, Part B* 1987, **B22**(5), 537-552 (Fr.) (*Chem. Abstr.* **107**, 231348p)
4. *World Rev. Pest Control* 1970, **9**, 119
5. *Hyg. Sanit. (USSR)* 1983, **48**(2), 85
6. *Hyg. Sanit. (USSR)* 1982, **47**(2), 13
7. Seiler, J. P. *Mutat. Res.* 1978, **58**, 353-359
8. *EC Directive Relating to the Quality of Water Intended for Human Consumption* 1982, 80/778/EEC, Office for Official Publications of the European Communities, 2 rue Mercier, L-2985 Luxembourg
9. *S. I. 1991 No. 472 The Environmental Protection (Prescribed Processes and Substances) Regulations* 1991, HMSO, London

# M303   Metribuzin

CAS Registry No. 21087-64-9
Synonyms 4-amino-6-*tert*-butyl-3-(methylthio)-1,2,4-triazin-5(4*H*)-one;
4-amino-6-(1,1-dimethylethyl)-3-(methylthio)-1,2,4-triazin-5(4*H*)-one
Mol. Formula $C_8H_{14}N_4OS$                     Mol. Wt. 214.29
Uses Herbicide.

## Physical properties
M. Pt. 125-126°C; Specific gravity $d_4^{20}$ 1.31; Partition coefficient log $P_{ow}$ 1.602 (1);
Volatility v.p. $4.35 \times 10^{-7}$ mmHg at 20°C.

## Solubility
Water: 1.05 g $l^{-1}$ at 20°C. Organic solvent: dimethyl formamide, cyclohexane,
dichloroform, acetone

## Occupational exposure
US TLV (TWA) 5 mg $m^{-3}$; Supply classification harmful.
Risk phrases Harmful if swallowed (R22)

## Ecotoxicity

### Fish toxicity
$LC_{50}$ (96 hr) harlequin fish 140 mg $l^{-1}$ (2).
$LC_{50}$ (96 hr) rainbow trout, bluegill sunfish 76-80 mg $l^{-1}$ (1).

## Environmental fate

### Degradation studies
Degraded in soil via deamination, with further degradation to water-soluble
conjugates. $t_{1/2}$ 1-2 months (1).
In pond water $t_{1/2}$ 7 days (1).
Surface accumulation of crop residue in no-tillage soil inhibited metribuzin
mineralisation compared to conventional tillage soils (3).

### Abiotic removal
Undergoes non-biological degradation in 4 dry Manitoba soils at 15°C, the rate law
describing this degradation was somewhat less than first order. $t_{1/2}$ 90-115 days at an
application rate of 1.8 ppm, $t_{1/2}$ longer at higher application rate (4).

## Mammalian and avian toxicity

### Acute data
$LD_{50}$ oral rat 1936-1986 mg $kg^{-1}$ (5).
$LD_{50}$ oral guinea pig, mouse 250, 711 mg $kg^{-1}$, respectively (1).
$LD_{50}$ oral bobwhite quail 164 mg $kg^{-1}$ (1).
$LD_{50}$ dermal rat 2000 mg $kg^{-1}$ (5).

**Sub-acute data**

$LC_{50}$ (5 day) diet bobwhite quail, mallard duck >4000 mg $kg^{-1}$ (1).
Liver damage in guinea pigs followed administration (dose unspecified) directly
into gastric lumen $6 \times wk^{-1}$ for 30 or 90 days (6).

**Carcinogenicity and long-term effects**

No effect level for rats and dogs in 2 yr feeding trials 100 mg $kg^{-1}$ (1).

**Metabolism and pharmacokinetics**

In mammals 90% of oral dose eliminated via urine and faeces, within 96 hr equally (1).

## Genotoxicity

*Escherichia coli* SOS chromotest negative (7).

## Legislation

Limited under EC Directive on Drinking Water Quality 80/778/EEC. Pesticides:
maximum admissible concentration 0.1 $\mu g \, l^{-1}$ (8).
Included in Schedule 6 (Release into Land: Prescribed Substances) Statutory
Instrument No. 472, 1991 (9).

## Any other comments

WHO Class Table 5, EPA toxicity class III (1).

## References

1. *The Agrochemicals Handbook* 3rd ed., 1991, RSC, London
2. Tooby, T. E. et al *Chem. Ind.* 1975, **21**, 523-525
3. Locke, M. A. et al *Pestic. Sci.* 1991, **31**(2), 239-247
4. Webster, G. R. B. et al *Bull. Environ. Contam. Toxicol.* 1978, **20**, 401-408
5. *Pesticide Dictionary* 1976, Meister Publ. Co., Willoughby, OH
6. Tomaszewski, J. et al *Acta Biol. Hung.* 1986, **37**(2), 121-126
7. Xu, H. H. et al *Toxic. Assess.* 1990, **5**(1), 1-14
8. *EC Directive Relating to the Quality of Water Intended for Human Consumption* 1982, 80/778/EEC, Office for Official Publications of the European Communities, 2 rue Mercier, L-2985 Luxembourg
9. *S. I. 1991 No. 472 The Environmental Protection (Prescribed Processes and Substances) Regulations* 1991, HMSO, London

# M304   Metronidazole

**CAS Registry No.** 443-48-1
**Synonyms** 2-methyl-5-nitroimidazole-1-ethanol; 1-(2-hydroxyethyl)-2-methyl-
5-nitroimidazole; 1-($\beta$-ethylol)-2-methyl-5-nitro-3-azapyrrole
**Mol. Formula** $C_6H_9N_3O_3$                     **Mol. Wt.** 171.16

**Uses** Antiprotozoal used in treatment of *Trichomonas vaginalis*, *Entamoeba histolytica* and *Giardia lamblia* infection.

## Physical properties
**M. Pt.** 158-160°C.

**Solubility**
Water: 10 g l$^{-1}$ at 20°C. Organic solvent: ethanol

## Mammalian and avian toxicity

**Acute data**
LD$_{50}$ oral rat, mouse 3.0-3.8 g kg$^{-1}$ (1,2).
LD$_{50}$ intraperitoneal, subcutaneous mouse 3-3.6 g kg$^{-1}$ (1).

**Carcinogenicity and long-term effects**
No adequate evidence for carcinogenicity to humans, sufficient evidence for carcinogenicity to animals, IARC classification group 2B (3).
Significantly increased incidence of lung tumours after oral administration to ♀ and ♂ mice, and of lymphomas in ♀ mice and of mammary, pituitary, testicular and liver tumours in rats. Increased incidence of colonic tumours induced in rats by subcutaneous administration of 1,2-dimethylhydrazine (3).
Administration by gavage of 2 mg day$^{-1}$ for 100 days caused a significant increase in lung tumours in ♂ mice and of lymphomas in ♀ mice (4).
Administration by gavage of 2 mg day$^{-1}$, 5 days wk$^{-1}$, every alternate wk for life resulted in a significant increase in overall tumour incidence in ♀ but not ♂ mice. Increased tumour incidence was observed in F$_1$ but not F$_2$ generation (5).
Excess lung cancer (but no significant increase overall in cancer-related morbidity or mortality) reported in a follow-up study of 771 patients in Rochester, Minnesota given metronidazole. No increase in lung cancer was observed in a follow-up of 2460 San Franciscan patients (6).
Some excess cervical cancer reported in women treated with metronidazole, although this neoplasm has risk factors in common with vaginal trichomoniasis, the main indication in women for treatment with the drug (3).
Studies in mice imply prolonged treatment with metronidazole may predispose to photocarcinogenesis (7).

**Teratogenicity and reproductive effects**
Crosses the placenta, is excreted in breast milk and its use in pregnancy is controversial. Its use during the first trimester is contraindicated by the manufacturer and the U.S. Centers for Disease Control. In the 2nd and 3rd trimester use for trichomoniasis may be acceptable if alternatives have failed, but it should not be given as a single dose (6).
Administration by gavage to Swiss strain mice of 2 mg day$^{-1}$, 5 days wk$^{-1}$, every alternate wk in a multigeneration study was not teratogenic (5).

**Metabolism and pharmacokinetics**
Readily absorbed after oral administration and bioavailability approaches 100%; peak plasma concentrations of 5-10 μg ml$^{-1}$ occur 1 hr after single doses of 250-500 mg. Bioavailability from rectal suppositories is 60-80%, with peak plasma levels about half that of oral doses occurring after 4 hr. It is widely distributed and appears in most body tissues and fluids in concentrations similar to plasma. Metabolised in liver by

side-chain oxidation and glucuronide formation. Principal oxidative metabolites are 1-(2-hydroxyethyl)-2-hydroxymethyl-5- nitroimidazole; and 2-methyl-5-nitroimidazole-1-acetic acid; with small amounts of acetamide; and $N$-(2-hydroxyethyl)oxamic acid. Excretion, mainly as metabolites, is via urine. Plasma elimination $t_{1/2}$ 8 hr, but longer in neonates and patients with liver disease (6).

## Genotoxicity

*Salmonella typhimurium* TA100, TA1535 with and without metabolic activation positive; TA1537, TA1538, TA98 with and without metabolic activation negative (8). *Escherichia coli* with metabolic activation positive (8).
Did not induce sex-linked recessives in *Drosophila melanogaster* (8).
Did not induce sister chromatid exchanges in human lymphocytes or in Chinese hamster cells *in vitro* without metabolic activation (8).
Did not induce chromosomal aberrations in human lymphocytes *in vitro* without metabolic activation (8).
Did not induce micronuclei in bone marrow cells of mice or rats (8).
No increased incidence of chromosomal aberrations in lymphocytes or bone marrow of treated patients (8).
*In vivo* mouse bone marrow micronucleus test positive (9).

## Any other adverse effects to man

Gastrointestinal disturbances are the most common adverse effects, including nausea and an unpleasant metallic taste. Headache, anorexia, vomiting, diarrhoea, dry mouth, furred tongue, glossitis and stomatitis may occur. High doses or prolonged treatment may cause peripheral neuropathy and epileptiform seizures. Leucopenia, skin rash, urethral discomfort, dark urine, raised liver enzyme values and thrombophlebitis may occur. Bone marrow aplasia, deafness, myopia, pancreatitis and gynaecomastia have been reported (6).

## Any other comments

Human health effects, experimental toxicology, physico-chemical properties reviewed (10).

## References

1.  *Oyo Yakuri* 1974, **8**, 1089
2.  *J. Med. Chem.* 1977, **20**, 1522
3.  *IARC Monograph* 1987, **Suppl. 7**, 250-251
4.  Cavaliere, A. et al *Tumori* 1983, **69**, 379-382
5.  Chacko, M. et al *J. Cancer Res. Clin. Oncol.* 1986, **112**, 135-140
6.  *Martindale. The Extra Pharmacopoeia* 30th ed., 1993, The Pharmaceutical Press, London
7.  Kelly, G. E. et al *Photochem. Photobiol.* 1989, **49**(1), 59-65
8.  *IARC Monograph* 1986, **Suppl. 6**, 399-401
9.  Paik, S. G. *Environ. Mutagen. Carcinog.* 1985, **5**(2), 61-72
10. *ECETOC Technical Report No. 30(4)* 1991, European Chemical Industry Ecology and Toxicology Centre, B-1160 Brussels

# M305   Metsulfuron-methyl

**CAS Registry No.** 74223-64-6

**Synonyms** benzoic acid, 2-[[[[(4-methoxy-6-methyl-1,3,5-triazin-2-yl)-amino]carbonylamino]sulfonyl]-, methyl ester; Ally; Brush-off; Escort; Granstar; Gropper

**Mol. Formula** $C_{14}H_{15}N_5O_6S$           **Mol. Wt.** 381.37

**Uses** Sulfonylurea herbicide.

## Physical properties

**M. Pt.** 158°C; **Partition coefficient** log $P_{ow}$ 1.0 at pH 5; −1.854 at pH 7 (1); **Volatility** v.p. $5.79 \times 10^{-5}$.

## Solubility

Water: 1.1 mg $l^{-1}$ at 25°C at pH 5; 9.5 g $l^{-1}$ at pH7. Organic solvent: ethanol, methanol, acetone, dichloromethane

## Ecotoxicity

### Fish toxicity

$LC_{50}$ (96 hr) rainbow trout, bluegill sunfish >12.5 mg $l^{-1}$ (1).

### Invertebrate toxicity

$LC_{50}$ (48 hr) *Daphnia magna* >150 mg $l^{-1}$ (1).

## Environmental fate

### Degradation studies

Degraded by non-chemical hydrolysis (especially at low soil pH) and by soil microorganisms. $t_{1/2}$ 20 to >150 days depending on soil type. Degradation decreased with increasing soil depth (2).

$t_{1/2}$ 1-4 wk. Breakdown is quicker at higher temperatures, higher levels of soil moisture and at lower soil pH (1).

### Absorption

Adsorption was negatively correlated with soil pH and positively correlated with soil organic matter content. Because adsorption is weak at high pH, leaching can occur after high rainfall (2).

## Mammalian and avian toxicity

### Acute data

$LD_{50}$ oral ♂, ♀ rat >5000 mg $kg^{-1}$ (1).
$LD_{50}$ oral mallard duck >5000 mg $kg^{-1}$ (1).
$LC_{50}$ (4 hr) inhalation ♂, ♀ rat >5 mg $l^{-1}$ air (1).
$LD_{50}$ dermal rabbit >2000 mg $kg^{-1}$ (1).

### Sub-acute data

NOEC for rats and dogs was 50 and 200 mg kg diet$^{-1}$, respectively, in 2 yr feeding study (1).

### Metabolism and pharmacokinetics

Excreted predominantly unchanged by mammals (species unspecified) following oral administration. The sulfonylurea and methoxy carbonyl groups are only partly degraded by *O*-demethylation and hydroxylation (1).

### Irritancy

Moderate rabbit eye irritant (reversible). Mild guinea pig skin irritant (1).

### Sensitisation

Not a sensitiser of guinea skin (1).

## Genotoxicity

Non-mutagenic in Ames test (details unspecified) (1).

## Legislation

Limited under EC Directive on Drinking Water Quality 80/778/EEC. Pesticides: maximum admissible concentration 0.1 µg l$^{-1}$ (3).
Included in Schedule 6 (Release into Land: Prescribed Substances) Statutory Instrument No. 472, 1991 (4).

### Any other comments

WHO Class Table 5; EPA Toxicity Class IV (1).
Non toxic to bees (1).

### References

1. *The Agrochemicals Handbook* 3rd ed., 1991, RSC, London
2. Walker, A. et al *Weed Res.* 1989, **29**(4), 281-287
3. *EC Directive Relating to the Quality of Water Intended for Human Consumption* 1982, 80/778/EEC, Office for Official Publications of the European Communities, 2 rue Mercier, L-2985 Luxembourg
4. *S. I. No. 472 The Environmental Protection (Prescribed Processes and Substances) Regulations* 1991, HMSO, London

# M306  Metyrapone

CAS Registry No. 54-36-4
**Synonyms** 2-methyl-1,2-di-3-pyridyl-1-propanone; methapyrapone;
2-methyl-1,2-bis(3-pyridyl)-1-propanone; 2-methyl-1,2-di-3-pyridinyl-1-propanone
**Mol. Formula** $C_{14}H_{14}N_2O$                   **Mol. Wt.** 226.28

**Uses** Diagnostic aid in pituitary function; an adrenal 11-β hydroxylase inhibitor which blocks endogenous glucocorticoid synthesis.

## Physical properties

**M. Pt.** 50-51°C.

## Mammalian and avian toxicity

### Acute data

$LD_{50}$ oral rat 520 mg $kg^{-1}$ (1).
$LD_{Lo}$ intraperitoneal mouse 300 mg $kg^{-1}$ (2).

### Teratogenicity and reproductive effects

Decreased lung tissue disaturated phosphatidylcholine, disaturated phosphatidylcholine/total phospholipids, superoxide dismutase, catalase and glutathione peroxidase reported in offspring of rats given 45 mg $kg^{-1}$ 2 × $day^{-1}$ for 3 days prior to delivery (3).

Not teratogenic in rats (dose, duration and route unspecified); thymidine/sulfate ratio 0.76 in mouse limb bud micromass assay (proposed preliminary screen for developmental toxins) (4).

## References

1. *Drugs in Japan. Ethical Drugs* 6th ed., 1982, 827
2. *J. Pharmacol. Exp. Ther.* 1964, **146**, 395
3. Sosenko, I. R. S. et al *Pediatr. Res.* 1986, **20**(7), 672-675
4. Wise, L. D. et al *Teratology* 1990, **41**, 341-351

# M307   Mevinphos

$$(CH_3O)_2P(O)OC(CH_3)=CHCO_2CH_3$$

**CAS Registry No.** 7786-34-7
**Synonyms** 2-methoxycarbonyl-1-methylvinyl dimethyl phosphate; 3-[(dimethoxyphosphinyl)oxy]-2-butenoic acid methyl ester; Phosdrin; methyl 3-(dimethoxyphosphinyloxy)crotonate
**Mol. Formula** $C_7H_{13}O_6P$          **Mol. Wt.** 224.15
**Uses** Acaricide and insecticide.

## Physical properties

**M. Pt.** *E*-isomer, 21°C; *Z*-isomer, 6.9°C; **B. Pt.** $_{0.3}$ 99-103°C; **Specific gravity** $d^{20}$ 1.24; **Volatility** v.p. $1.28 \times 10^{-4}$ mmHg at 20°C.

## Solubility

Organic solvent: miscible with acetone, benzene, chloroform, carbon tetrachloride, toluene

## Occupational exposure

**US TLV (TWA)** 0.01 ppm (0.092 mg $m^{-3}$); **US TLV (STEL)** 0.03 ppm

(0.27 mg m$^{-3}$); **UK Long-term limit** 0.01 ppm (0.1 mg m$^{-3}$); **UK Short-term limit**
0.03 pm (0.3 mg m$^{-3}$); **Supply classification** very toxic.
**Risk phrases** Very toxic in contact with skin and if swallowed (R27/28)
**Safety phrases** Do not breathe vapour – After contact with skin, wash immediately
with plenty of soap and water – Wear suitable protective clothing and gloves – In case
of accident or if you feel unwell, seek medical advice immediately (show label where
possible) (S23, S28, S36/37, S45)

## Ecotoxicity

### Fish toxicity
LC$_{50}$ (24 hr) rainbow trout, bluegill sunfish 0.034-0.041 mg l$^{-1}$ (1,2).
LC$_{50}$ (96 hr) American eel, mummichog, striped killifish, bluehead, striped mullet,
Atlantic silverside, northern puffer 65-800 µg l$^{-1}$ (3).
LC$_{50}$ (96 hr) bluegill sunfish, largemouth bass 70-100 µg l$^{-1}$ (2).

### Invertebrate toxicity
LC$_{50}$ (96 hr) *Daphnia pulex, Simocephalus serrulatus, Gammarus fasciatus*
0.16-2.80 µg l$^{-1}$ (4,5).
LC$_{50}$ (96 hr) sand shrimp, grass shrimp, hermit crab 11-69 µg l$^{-1}$ (6).
LC$_{50}$ (96 hr) *Palaemonetes kadiakensis, Asellus brevicaudus* 12-56 µg l$^{-1}$ (5).
LC$_{50}$ (96 hr) *Gammarus lacustris* 130 µg l$^{-1}$ (7).
LD$_{50}$ *Bufo arenarum* 880 mg kg$^{-1}$ (8).

## Environmental fate

### Degradation studies
t$_{1/2}$ 13 days in silty clay acid and sandy clay neutral soils (9).
Degradation rate in surface and groundwater samples from 50 to 0.1 µg l$^{-1}$,
164 days (10).

### Abiotic removal
50% hydrolysis occurs in 120 days at pH 6; 35 days at pH 7; 3 days at pH 9; and
1.4 hr at pH 11 (11).

## Mammalian and avian toxicity

### Acute data
LD$_{50}$ oral rat 3 mg kg$^{-1}$ (12).
LD$_{50}$ oral blackbird, duck 3-4.6 mg kg$^{-1}$ (12).
LD$_{50}$ oral house sparrow, starling, common grackle, rock dove, quail 1.8, 3.8, 4.2, 4.2
and 23.7 mg kg$^{-1}$, respectively (13).
LC$_{50}$ (1 hr) inhalation rat 14 ppm (14).
LD$_{50}$ dermal rat, duck 4.2 and 11 mg kg$^{-1}$, respectively (12,15).
LD$_{50}$ intraperitoneal rat, mouse 1.35-2 mg kg$^{-1}$ (16,17).
LD$_{50}$ intravenous, subcutaneous mouse 680 and 1180 µg kg$^{-1}$, respectively (18).

### Carcinogenicity and long-term effects
No adverse effect level in rats and dogs in 2 yr feeding trials 4-5 mg kg$^{-1}$ diet (11).

### Metabolism and pharmacokinetics
Eliminated as metabolites (unspecified) in urine and faeces in 3-4 days after oral
administration to mammals (11).

**Irritancy**

Mild irritant to skin and eyes of rabbits (11).

## Legislation

Limited under EC Directive on Drinking Water Quality 80/778/EEC. Pesticides: maximum admissible concentration 0.1 µg l$^{-1}$ (19).

Included in Schedule 6 (Release into Land: Prescribed Substances) Statutory Instrument No. 472, 1991 (20).

EEC maximum residue limit stone or citrus fruit 0.2 ppm, leaf vegetables 0.5 ppm, other fruit and vegetables 0.1 ppm (11).

Reportable quantity regulated under U.S. Federal Comprehensive Environmental Response, Compensation and Liability Act (21).

## Any other comments

WHO Class Ia; EPA toxicity class I; tolerable daily intake in humans 0.0015 mg kg$^{-1}$ (11). Toxicity and environmental effects of organophosphates reviewed (22).

Commercial product is a mixture of the *cis-* and *trans*-isomers, the former being 100 × more insecticidally active.

## References

1. Edwards, C. A. *Pesticides in Aquatic Environments* 1977, Khan, M. A. Q. (Ed.), Plenum Press, New York
2. Verschueren, K. *Handbook of Environmental Data on Organic Chemicals* 2nd ed., 1983, Van Nostrand Reinhold, New York
3. Eisler, R. *Bureau Sport Fisheries Wildlife Technical Paper 46* 1970, US Government Printing Office, Washington, DC
4. Sanders, H. O. et al *Trans. Am. Fish. Soc.* 1966, **95**, 165-169
5. Sanders, H. O. *The toxicities of some insecticides to four species of malocostracan crustacea* 1972, Fish Pesticide Research Laboratory, Bureau of Sport Fisheries and Wildlife, Columbia, MO
6. Eisler, R. *Crustaceana* 1969, **16**(3), 302-310
7. Sanders, H. O. *Bureau Sport Fisheries Wildlife Technical Paper 25* 1969, US Government Print Office, Washington, DC
8. Juarez, A. O. et al *Commun. Biol.* 198, **6**(3), 259-263
9. Sattar, M. A. *Chemosphere* 1990, **20**(3-4), 387-396
10. Frank, R. et al *Bull. Environ. Contam. Toxicol.* 1991, **47**(3), 374-380
11. *The Agrochemicals Handbook* 3rd ed., 1991, RSC, London
12. Lewis, R. J. et al *RTECS* 1984, No. 83-107-4, NIOSH
13. Schafer, E. W. et al *Arch. Environ. Contam. Toxicol.* 1983, **12**, 355-382
14. *Arch. Ind. Hyg. Occup. Med.* 1954, **9**, 45
15. *Toxicol. Appl. Pharmacol.* 1960, **2**, 88
16. *Proc. Soc. Exp. Biol. Med.* 1963, **114**, 509
17. *Can. J. Biochem. Physiol.* 1961, **39**, 1790
18. *J. Pharm. Pharmacol.* 1967, **19**, 612
19. *EC Directive Relating to the Quality of Water Intended for Human Consumption* 1982, 80/778/EEC, Office for Official Publications of the European Communities, 2 rue Mercier, L-2985 Luxembourg
20. *S. I. 1991 No. 472 The Environmental Protection (Prescribed Processes and Substances) Regulations* 1991, HMSO, London
21. *Fed. Regist.* 14 Aug 1989, **54**(155), 33426-33484
22. *Environmental Health Criteria 63: Organophosphorus Insecticides: a general introduction* 1986, WHO/IPCS, Geneva

# M308 Mexacarbate

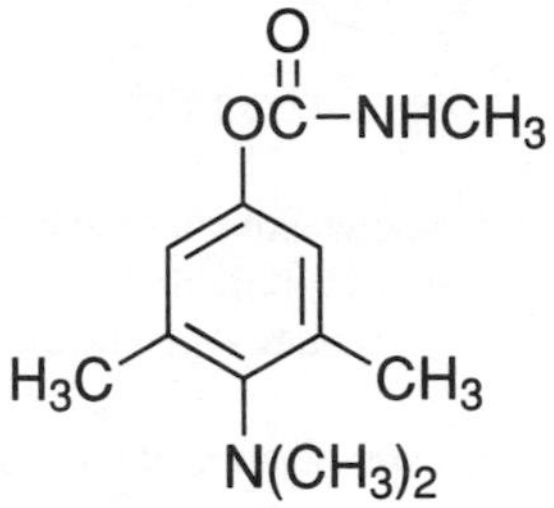

$$OC-NHCH_3$$

(structure)

O
‖
OC–NHCH₃

H₃C        CH₃

N(CH₃)₂

**CAS Registry No.** 315-18-4
**Synonyms** 4-(dimethylamino)-3,5-xylyl- *N*-methylcarbamate;
4-(dimethylamino)-3,5-dimethylphenol methylcarbamate (ester); Zectran
**Mol. Formula** $C_{12}H_{18}N_2O_2$        **Mol. Wt.** 222.29
**Uses** Insecticide and molluscicide.

## Physical properties

**M. Pt.** 85°C; **Volatility** v.p. <0.1 mmHg at 139°C.

### Solubility
Water: 0.01% at 25°C. Organic solvent: acetone, acetonitrile, methylene chloride, ethanol, benzene

## Occupational exposure

**Supply classification** very toxic.
**Risk phrases** Harmful in contact with skin – Very toxic if swallowed (R21, R28)
**Safety phrases** Wear suitable protective clothing and gloves – In case of accident or if you feel unwell, seek medical advice immediately (show label where possible) (S36/37, S45)

## Ecotoxicity

### Fish toxicity
$LC_{50}$ (96 hr) bluegill sunfish, coho salmon, yellow perch, brown trout, Atlantic salmon, fathead minnow 0.6-23.7 mg $l^{-1}$ (1).
$LC_{50}$ (96 hr) coho salmon, perch, brown trout, rainbow trout, yellow perch 1.73-20 mg $l^{-1}$ (2-4).

### Invertebrate toxicity
$EC_{50}$ (48 hr) *Daphnia pulex* 0.1 mg $l^{-1}$ (5).
$LC_{50}$ (96 hr) *Gammarus fasciatus* 0.04 mg $l^{-1}$ (6).
$LC_{50}$ (2 day) *Pycnopsyche* sp., *Simulium venustum*, *Ophiogomphus* sp. 0.099-0.492 mg $l^{-1}$ (7).
LOEC *Oscillatoria terebribormis, Synechoccus lividus, Navicula pelliculosa, Scenedesmus quadricauda* 1-10 mg $l^{-1}$ (8).
$LC_{50}$ (96 hr) *Lymnea acuminata* 1.7 mg $l^{-1}$ (9).
$LC_{50}$ (96 hr) crayfish 8.8 µg $ml^{-1}$; behavioural abnormalities occurred at levels >1.0 µg $ml^{-1}$ but as duration of exposure increased, animals recovered (10).

### Bioaccumulation

Bioconcentration factor in mosquito larvae 0-8 (at 5.5-10.8 ppb in water), in brine shrimp 18 (at 5 ppb in water), in silverside fish 45 (at 4.7 ppb in water) (11).

## Environmental fate

### Degradation studies

Presence of 4-methylamino and 4-amino-3,5-xylyl methylcarbamate, and 4-dimethylamino-3,5-xylenol in water and aquatic plants following aerial spraying showed demethylation and hydrolytic routes are major metabolic pathways for degradation of mexacarbate (12).

Carbamates are readily degraded by soil microorganisms, and are photodecomposed in water by UV radiation (13).

## Mammalian and avian toxicity

### Acute data

$LD_{50}$ oral mouse, rat, dog, rabbit 12-37 mg kg$^{-1}$ (14-17).
$LD_{50}$ oral duck, chicken, pigeon 3-5.6 mg kg$^{-1}$ (15).
$LD_{50}$ dermal mouse 107 mg kg$^{-1}$ (18).
$LD_{50}$ intraperitoneal mouse 7.8 mg kg$^{-1}$ (19).

### Carcinogenicity and long-term effects

No adequate evidence for carcinogenicity to humans or animals, IARC classification group 3 (20).

The USA National Toxicology Program tested mexacarbate via feed, negative in ♂ and ♀ rats and mice (21).

### Metabolism and pharmacokinetics

An oral dose in dogs was eliminated in urine as 4-dimethylamino-3,5-xylenol, predominantly in the conjugated form, as conjugated forms of the 2,6-dimethylhydroquinone and as small amounts of 2,6-dimethyl-$p$-benzoquinone (2).

Carbamates do not accumulate in mammals (13).

## Genotoxicity

*Bacillus subtilis* without metabolic activation positive (22).

Induction of sister chromatid exchanges in Chinese hamster ovary cells are predicted by a CASE study (23).

## Any other adverse effects to man

Most carbamates are active acetylcholinesterase inhibitors (13).

A fatal case of poisoning was reported 4 hr after ingestion of 55 g; the victim suffered bradycardia and heart failure (13).

Classical signs of cholinesterase poisoning, progressing to paralysis of the extremities was reported in an aerial spray plane pilot overexposed to mexacarbate via a leaking pump line (13).

## Legislation

Limited under EC Directive on Drinking Water Quality 80/778/EEC. Pesticides: maximum admissible concentration 0.1 μg l$^{-1}$ (24).

Included in Schedule 6 (Release into Land: Prescribed Substances) Statutory Instrument No. 472, 1991 (25).

## Any other comments

Metabolism reviewed (26).

## References

1. Mauk, W. L. et al *Arch. Environ. Contam. Toxicol.* 1977, **6**, 385
2. Macek, K. J. et al *Trans. Am. Fish. Soc.* 1970, **99**(1), 20-27
3. Mauck, W. L. et al *Arch. Environ. Contam. Toxicol.* 1977, **6**, 385
4. Kemp, H. T. et al *Water Pollut. Control Res. Series* 09/73, EPA
5. Sanders, H. O. et al *Trans. Am. Fish Soc.* 1966, **95**, 165-169
6. Sanders, H. O. *The toxicities of some insecticides to four species of Malocostracan crustacea* 1972, Fish Pesticide Research Laboratory, Bureau of Sport Fisheries and Wildlife, Columbia, OH
7. Poirier, D. G. et al *Can. Entomol.* 1987, **119**, 755
8. Snyder, C. E. et al *J. Phycol.* 1974, **10**, 137-139
9. Singh, D. K. et al *Arch. Environ. Contam. Toxicol.* 1983, **12**, 483
10. Sundaram, K. M. S. *ASTM Spec. Tech. Publ.* 1989, **1027** (Aquatic Toxicol. Hazard Assess. 12th vol.) 270-286
11. Khan, M. A. Q. (Ed.) *Pesticides in Aquatic Environments* 1977, Plenum Press, New York
12. Sundaram, K. M. S. et al *Rep. FPM-X-For Pest Manage. Inst.* 1986, FPM-X-73
13. *Environmental Health Criteria 64: Carbamate pesticides: a general introduction* 1986, WHO/IPCS, Geneva
14. *Pestic. Sci.* 1971, **2**, 10
15. *J. Econ. Entomol.* 1969, **62**, 1307
16. *Farm Chemicals Handbook* 1989, Meister Publ. Co., Willoughby, OH
17. *Sp. Publ. Entomol. Soc. Am.* 1978, **78-1**, 59
18. *J. Agric. Food Chem.* 1967, **15**, 479
19. *J. Agric. Food Chem.* 1968, **16**, 561
20. *IARC Monograph* 1987, **Suppl. 7**, 56
21. *National Toxicology Program Research and Testing Division* 1992, Report No. TR-147, NIEHS, Research Triangle Park NC 27709
22. *IARC Monograph* 1976, **12**, 237-243
23. Rosenkranz, H. S. et al *Mutagenesis* 1990, **5**(6), 559-571
24. *EC Directive Relating to the Quality of Water Intended for Human Consumption* 1982, 80/778/EEC, Office for Official Publications of the European Communities, 2 rue Mercier, L-2985 Luxembourg
25. *S. I. 1991 No. 472 The Environmental Protection (Prescribed Processes and Substances) Regulations* 1991, HMSO, London
26. Meikle, R. W. *Bull. Environ. Contam. Toxicol.* 1973, **10**, 29-36

# M309   Miconazole

**CAS Registry No.** 22916-47-8
**Synonyms** 1-[2-(2,4-dichlorophenyl)-2-[(2,4-dichlorophenyl)methoxy]ethyl]-
1*H*-imidazole
**Mol. Formula** $C_{18}H_{14}Cl_4N_2O$          **Mol. Wt.** 416.14
**Uses** Topical antifungal. Administered intravenously for severe fungal infections.

## Physical properties

**Solubility**
Organic solvent: acetone, diethyl ether, 2-propanol

## Mammalian and avian toxicity

**Acute data**
$LD_{50}$ oral mouse 872 mg kg$^{-1}$ (1).
$LD_{50}$ intraperitoneal rat, mouse 350-450 mg kg$^{-1}$ (1).

**Teratogenicity and reproductive effects**
No effect on mortality, growth or incidence of external, visceral or skeletal
abnormalities reported in foetuses of rats given 30 mg kg$^{-1}$ intravenously on days
6-18 of pregnancy (2).

**Metabolism and pharmacokinetics**
Incompletely absorbed from the human gut; peak plasma concentrations of 1 μg ml$^{-1}$ 4 hr
after a 1 g dose. Doses >9 mg kg$^{-1}$ by intravenous infusion usually produce plasma
concentrations >1 μg ml$^{-1}$. It diffuses well into infected joints but penetration into
cerebrospinal fluid and sputum is poor. >90% is bound to plasma proteins.
Metabolised in the liver to inactive metabolites. 10-20% of an oral or intravenous
dose is excreted in urine as metabolites. 50% of an oral dose is excreted mainly
unchanged in faeces. Elimination pharmacokinetics after intravenous infusion are
triphasic with $t_{1/2}$ of 0.4, 2.5 and 24 hr. Little is absorbed dermally (3).

**Sensitisation**
Sensitisation to miconazole nitrate reported in the guinea pig maximisation test (4).

## Genotoxicity

Disturbed normal cell division by altering chromosome distribution and lead to
mitotic non-disjunction in *Aspergillus nidulans* (5).
Micronucleus test in mice negative (6).

## References

1. *Arzneim.-Forsch.* 1981, **31**, 2145
2. Shibutani, Y. et al *Iyakuhin Kenkyu* 1986, **17**(5), 991-997 (Japan.) (*Chem. Abstr.* **106**, 60825q)
3. *Martindale. The Extra Pharmacopoeia* 30th ed., 1993, The Pharmaceutical Press, London
4. Nishioka, K. et al *Hifu* 1990, **32**(Suppl. 9), 155-160 (Japan.) (*Chem. Abstr.* **114**, 220809h)
5. Bellincampi, D. et al *Mutat. Res.* 1980, **79**, 169-172
6. Vanparys, P. *Mutat. Res.* 1990, **244**(2), 95-103

# M310    Mineral oil

**CAS Registry No.** 8012-95-1
**Synonyms** adepsine oil; crystosol; oil mist, mineral; parrafin oil
**Uses** Mainly as lubricating oils.
**Occurrence** Petroleum crude oil refinery process streams.

## Physical properties

**Solubility**
Organic solvent: benzene, diethyl ether, chloroform

## Mammalian and avian toxicity

**Acute data**
$LD_{50}$ oral mouse 22 g $kg^{-1}$ (1).

**Carcinogenicity and long-term effects**
Sufficient evidence for carcinogenicity of untreated or mildly treated mineral oils in humans and animals, IARC classification group 1. Inadequate evidence for carcinogenicity of highly refined mineral oils in humans and animals, IARC classification group 3 (2).
Mineral oils exposure has been associated strongly and consistently with some squamous cell cancers, especially of the scrotum; excesses of gastrointestinal, sinonasal, oral, bladder and lung cancers have been reported in many groups of workers exposed to mineral oils in many different industries including metal workers, toolmakers and printers (2).

**Teratogenicity and reproductive effects**
Used crankcase oil was embryolethal and teratogenic when applied at 1-15 μl to mallard duck and quail eggs (3).

**Metabolism and pharmacokinetics**
5 hr after oral administration of 0.66 ml $kg^{-1}$ to rats 1.5% was absorbed unchanged and 1.5% found as non-mineral oil substances. The liver, spleen, kidney, brain and fat contained mineral oil. Within 2 days 0.3% remained in the animals (3).
Mineral oil [class 5] is poorly absorbed from the gut, but was found in lungs, liver, lymph and spleen of a man who ingested considerable amounts over many years (3).

### Irritancy

Mild skin and eye irritant in rabbits (4,5).

Neither paraffinic nor naphthenic base stock caused primary eye irritation in rabbits (3).

Several light mineral oils caused hypertrophy, hyperplasia, hyperkeratosis and dilipidation after application to guinea pigs' skin; skin damage increased with increasing molecular size (3).

Occupational exposure causes eczematous dermatitis, contact dermatitis, oil acne, folliculitis, lipid granuloma and melanosis (3).

### Sensitisation

Neither paraffinic nor naphthenic base stock caused skin sensitisation in guinea pigs (3).

## Genotoxicity

*Salmonella typhimurium* TA98 with metabolic activation positive for vacuum distillates, hydrotreated oil and used steel-hardening oil, negative for white oil and unused steel-hardening oil (3).

*Salmonella typhimurium* TA1537, TA1538, TA98, TA100 with and without metabolic activation positive for used crankcase oils; both positive and negative results reported for unused crankcase oils (3).

Urine of workers occupationally exposed to mineral oils [class 8] and iron oxide was mutagenic in *Salmonella typhimurium* TA98, TA100 with metabolic activation (3).

Two insulation oils from highly-refined mineral-base oils induced transformation of Syrian hamster embryo cells and enhanced transformation of mouse cells; unused new, re-refined and used crankcase oils induced transformation of Syrian hamster embryo cells (6).

Increased frequency of chromosomal aberrations reported in peripheral lymphocytes of glass workers exposed to mineral oil mists (6).

## Any other adverse effects to man

Occupational ingestion, aspiration or inhalation exposure causes lipid pneumonia and lipid granuloma of the lung (3).

## Legislation

White mineral oil is approved by the US FDA for use as a direct additive to food for human consumption; mineral oil is approved, with some limitations, for use as an animal feed additive and in materials in contact with food (3).

UK Mineral Hydrocarbons in Food Regulations 1966 prohibits use in food composition or preparation (3).

## Any other comments

Human health effects, experimental toxicology, physico-chemical properties reviewed (7).

PAH content of petroleum-derived lubricants may increase during use, and used petrol engine oils can contain up to 1% lead (3).

## References

1. *Arch. Toxikol.* 1973, **30**, 243
2. *IARC Monograph* 1987, **Suppl. 7**, 66, 252

3. *IARC Monograph* 1984, **33**, 87-168
4. *Cosmet. Toiletries* 1979, **94**(8), 41
5. *Arch. Ind. Health* 1956, **14**, 265
6. *IARC Monograph* 1987, **Suppl. 6**, 403
7. *ECETOC Technical Report No. 30(4)* 1991, European Chemical Industry Ecology and Toxicology Centre, B-1160 Brussels

# M311   Mirex

**CAS Registry No.** 2385-85-5
**Synonyms** 1,1a,2,2,3,3a,4,5,5,5a,5b,6-dodecachlorooctahydro-1,3,4-metheno-1*H*-cyclobuta[*cd*]pentalene; dechlorane; hexachlorocyclopentadiene dimer

**Mol. Formula** $C_{10}Cl_{12}$                    **Mol. Wt.** 545.55
**Uses** Insecticide. In flame-retardant coatings.

## Physical properties

**M. Pt.** 485°C (decomp.); **Partition coefficient** log $P_{ow}$ 5.28;
**Volatility** v.p. $3 \times 10^{-7}$ mmHg at 25 °C.

## Solubility

Water: 0.2 mg $l^{-1}$ at 24°C. Organic solvent: tetrahydrofuran, carbon disulfide, chloroform, benzene

## Ecotoxicity

### Fish toxicity

$LC_{50}$ (96 hr) rainbow trout, goldfish, bluegill sunfish 0.2-30 mg $l^{-1}$ (1).

### Invertebrate toxicity

$LC_{50}$ (96 hr) *Gammarus fasciatus* 0.3 mg $l^{-1}$ (2).
$LC_{50}$ (96 hr) *Daphnia magna* 0.6 mg $l^{-1}$ (3).
$LC_{50}$ (120 hr) grass shrimp 190 µg $l^{-1}$ (4).
$LC_{65}$ (48 hr) juvenile crayfish 0.1 µg $l^{-1}$ (4).

### Bioaccumulation

Bioconcentration factor in mussels 34,200-73,700; crustaceans 16,860-71,400; algae 12,000 (5).

Tissue concentration in skunk and fox rose rapidly after spraying. Concentrations in aquatic ecosystems did not rise until 1 yr after spraying (6).

Bioaccumulates at all trophic levels and is biomagnified through food chains. Following application to land, residues were $<10\ \mu g\ l^{-1}$ in water and were 0 to 0.07 mg $kg^{-1}$ in sediment. Concentrations increased significantly up the food chain; birds, 0-0.17 mg $kg^{-1}$ and mammals 0-4.4 mg $kg^{-1}$. Accumulates to high levels in insectivorous birds with reported levels of 1-10 mg $kg^{-1}$ (4).

## Environmental fate

### Degradation studies

Microbial biodegradation does not occur except occasionally under anaerobic conditions, and even then slowly; dechlorination to a monohydro derivative by anaerobic microbial action in sewage sludge (4).

### Abiotic removal

Photodegraded to toxic products, chlordecone, monohydromirex and dihydromirex (6). Slow photodegradation under UV to photomirex (4).

Environmental $t_{1/2}$ is many years (4).

$t_{1/2}$ 48.4 hr in water under intense UV radiation at 90-95°C (4).

Photodegradation products include: 10-monohydromirex; 8-monohydromirex; 5,10-dihydromirex; chlordecone; and 2,8-dihydromirex (4).

## Mammalian and avian toxicity

### Acute data

$LD_{50}$ oral rat 235 mg $kg^{-1}$ (7).
$LD_{50}$ oral duck 2400 mg $kg^{-1}$ (7).
$LD_{50}$ oral Japanese quail 10,000 mg $kg^{-1}$ (4).
$LD_{50}$ oral starling, pheasant 562-1600 mg $kg^{-1}$ (4,8).
$LD_{50}$ dermal rabbit, rat 800-2000 mg $kg^{-1}$ (4,7).

### Sub-acute data

$LD_{50}$ (8 day) oral ring-necked pheasant, bobwhite quail, Japanese quail, mallard duck 1540->5000 mg $kg^{-1}$ (4).

### Carcinogenicity and long-term effects

No adequate data on carcinogenicity to humans, sufficient evidence for carcinogenicity to animals, IARC group 2B (9).

US National Toxicological Program evaluated mirex via feed, clear evidence of carcinogenicity in ♂, ♀ rat, inducing increased incidence of benign neoplastic nodules in liver in ♂ and ♀, as well as adrenal gland pheochromocytomas and transitional cell papillomas of the kidney in ♂, and mononuclear cell leukaemia in ♀ (10).

Increased incidence of hepatomas reported in mice fed 26 mg $kg^{-1}$ for 70 wk. Increased incidence of reticulum cell sarcomas in mice given 1000 mg $kg^{-1}$ subcutaneously (4).

### Teratogenicity and reproductive effects

Reduced litter size in mice fed 5 mg $kg^{-1}$ diet for 30 days before mating. Cessation of mating occurred in mice fed 17.8 mg $kg^{-1}$ for 3 months. Reduced litter size, neonatal viability and cataracts in neonates reported in rats fed 25 mg $kg^{-1}$. Rats fed 6 mg $kg^{-1}$ $day^{-1}$ on days 8-15 gestation gave birth to dead, oedematous foetuses. Visceral foetal anomalies, reduced foetal weight and survival reported at maternally toxic levels of

12.5 mg kg$^{-1}$ day$^{-1}$; administered orally on days 6-15 gestation in rats (4).
Reduced viability of neonates in an *in vivo* mouse teratology screen (11).

**Metabolism and pharmacokinetics**
Absorbed via the gut, by inhalation and skin. Not metabolised in any animal species
investigated; $t_{1/2}$ in the body is several months, being stored in adipose tissue because
it is lipophilic. It crosses the placenta and is excreted in breast milk. Excretion is via
faeces as unchanged mirex (4).

## Genotoxicity
*Salmonella typhimurium* (strain unspecified) with and without metabolic activation
negative (4).
Dominant lethal test in rats negative (4).
Microscreen phage-induction assay with or without metabolic activation positive (12).

## Any other adverse effects
Toxicity in oral short-term studies is characterised by reduced body weight,
hepatomegaly, induction of mixed function oxidases, morphological changes in liver
cells and occasionally death (4).

## Legislation
Regulated by US OSHA under its Air Contaminants Standard 1989 (13).
Limited under EC Directive on Drinking Water Quality 80/778/EEC. Pesticides:
maximum admissible concentration 0.1 µg l$^{-1}$ (14).
Included in Schedule 6 (Release into Land: Prescribed Substances) Statutory
Instrument No. 472, 1991 (15).
Included in the UN Banned and Severely Restricted list (16).
The log $P_{ow}$ value exceeds the European Community recommended level 3.0 (6th and
7th amendments) (17).

## Any other comments
Biodegradation of chlorinated organic compounds by *Phanerochaete chrysosporium*
reviewed (18).
Hazards reviewed (19).
Toxicity reviewed (20).
Aquatic toxicity reviewed (4).

## References
1.  *Technical Information A-10179 R2/72* 1972, Stauffer Chemical Company
2.  Sanders, H. O. *Toxicity of pesticides to the crustacean Gammarus lacustris* 1969, Bureau
    of Sport Fisheries and Wildlife, Government Printing Office Washington, DC
3.  Sanders, H. O. *J.- Water Pollut. Control. Fed.* 1970, **42**(8), 15544-15550
4.  *Environmental Health Criteria 44: Mirex* 1984, WHO/IPCS, Geneva
5.  Verschueren, K. *Handbook of Environmental data on Organic Chemicals* 2nd ed., 1983,
    Van Nostrand Reinhold, New York
6.  Carlson, D. A. et al *Science* 1976, **194**, 939-941
7.  Lewis, R. J. et al *RTECS* 1984, No. 83-107-4, NIOSH
8.  Schafer, E. W. et al *Arch. Environ. Contam. Toxicol.* 1983, **12**, 355-382
9.  *IARC Monograph* 1987, **Suppl. 7**, 66
10. *National Toxicology Program Research and Testing Division* 1992, Report No. TR-313,
    NIEHS, Research Triangle Park, NC

11. Kavlock, R. J. et al *Teratog., Carcinog., Mutagen.* 1987, **7**, 7-16
12. Houk, V. S. et al *Mutat. Res.* 1987, **182**(4), 193-201
13. Paxmau, D. G. et al *Regul. Toxicol. Pharmacol.* 1990, **12**(3, Pt. 1), 296-308
14. *EC Directive Relating to the Quality of Water Intended for Human Consumption* 1982, 80/778/EEC, Office for Official Publications of the European Communities, 2 rue Mercier, L-2985 Luxembourg
15. *S. I. 1991 No. 472 The Environmental Protection (Prescribed Processes and Substances) Regulations* 1991, HMSO, London
16. *Consolidated list of products whose consumption and/or sale has been banned, withdrawn, severely restricted or not approved by government: prepared in accordance with General Assembly resolutions 37/137,38/149* UN Environmental Program, New York
17. *1967 Directive on Classification, Packaging, and Labelling of Dangerous Substances 67/548/EEC; 6th Amendment EEC Directive 79/831/EEC; 7th Amendment EEC Directive 91/32/EEC* 1991, HMSO, London
18. Bumpus, J. A. et al *ACS Symp. Ser.* 1987, **338**, 340-349
19. *Dangerous Prop. Ind. Mater. Rep.* 1987, **7**(5), 88-91
20. US EPA Report 1987, EPA/600/8-88/046

# M312   Mirex, photo-

**CAS Registry No.** 39801-14-4
**Synonyms** 1,1a,2,2a,3,3a,4,5,5,5a,5b-undecachlorooctahydro-1,3,4-metheno-1*H*-cyclobuta[*cd*]pentalene; 8-hydromirex; 8-monohydromirex; photomirex
**Mol. Formula** $C_{10}HCl_{11}$ **Mol. Wt.** 511.10
**Occurrence** Photodegradation product of the insecticide Mirex.

## Ecotoxicity

### Bioaccumulation
Found to be the 4th highest organochlorine contaminant of herring-gull eggs and body lipids (after PCBs, DDE and Mirex). Also present at similar ratios in coho salmon muscle and liver, alewifes and smelt from Lake Ontario (1,2).

## Mammalian and avian toxicity

### Acute data
$LD_{50}$ oral rat 200 mg $kg^{-1}$ (1).
Oral quail 1000 mg $kg^{-1}$ caused moderate hepatic changes consisting of an increase in liver weight. No indication of severe liver damage (3).
Oral rat 100 or 200 mg $kg^{-1}$ showed mottled and congested livers and kidneys.
♀ developed haemorrhagic ovaries (1).

## Teratogenicity and reproductive effects
Designated positive for developmental toxicity in rats, negative in rabbits and predicted to be negative in humans (4).

## Genotoxicity
*Salmonella typhimurium* TA98, TA100, TA1535, TA1537, TA1538 with and without metabolic activation negative (1).

## References
1. Hallett, D. J. et al *J. Agric. Food Chem.* 1978, **26**, 388-391
2. Hallet, D. J. et al *J. Agric. Food Chem.* 1976, **24**, 1189
3. Strick, J. J. T. W. A. et al *Bull. Environ. Contam. Toxicol.* 1980, **24**, 350-355
4. Jelovsek, F. R. et al *Obstet. Gynecol. (N. Y.)* 1989, **74**(4), 624-636

# M313   Misonidazole

$$CH_2OCH_3$$
$$CH_2CHOH$$

**CAS Registry No.** 13551-87-6
**Synonyms** α-(methoxymethyl)-2-nitro-1*H*-imidazole-1-ethanol; α-(methoxymethyl)-2-nitroimidazole-1-thananol; SR 1354
**Mol. Formula** $C_7H_{11}N_3O_4$                    **Mol. Wt.** 201.18

## Physical properties
**M. Pt.** 110-111°C.

## Mammalian and avian toxicity

### Acute data
$LD_{50}$ oral mouse, rat 1869 and 2131 mg kg$^{-1}$, respectively (1,2).
$LD_{50}$ subcutaneous mouse 1414 mg kg$^{-1}$ (3).
$LD_{50}$ intraperitoneal mouse 1340 mg kg$^{-1}$ (4).

### Metabolism and pharmacokinetics
[$^3$H]-misonidazole was injected into ♂ and ♀ (lactating) mice at doses of 75 or 750 mg kg$^{-1}$. At the higher dose groups radiolabel was detected in meibomian gland ducts > oesphagus keratinized layer >liver periportalzone >hair bulb (5).
[$^3$H]-, [$^{14}$C]-misonidazole (24 hr) intravenous tumour-bearing mice. The 24 hr tissue retention (highest to lowest) was oesphageal epithelium, liver, foot pad, eyelid, lung, subcutaneous lung tumour (A110), oesphageal wall, uterus, eyeball, blood, salivary gland, spleen, voluntary muscle, pancreas and inguinal fat (6).
*In vitro* sciatic nerves of CH3/He mice show an apparent biological $t_{1/2}$ correlated to hydrophilicity (7).

## Genotoxicity

*Escherichia coli uvr*ABC excinuclease proficient and deficient strains grown under oxic and hypoxic conditions showed some induction of the SOS response (8).

## Any other adverse effects

Has increased cytotoxicity towards hypoxic compared to oxic cells. DNA is considered to be the target for cytotoxic activity (8).

Under anaerobic conditions both rat liver microsomes and cytosol catalysed the reductive metabolism and DNA binding of misonidazole (9).

Unspecified species, high-dose and short-term treatment induced oedematous or necrotic changes in the cerebellar and vestibular nuclei (10).

## References

1. *National Cancer Institute Screening Program Data Summary* January 1986, Bethesda, MD
2. *NITS Report PB81-121212*, Natl. Tech. Inf. Ser., Springfield, VA
3. *Antimicrob. Agents Chemother.* 1967, 513
4. *Radiat. Res.* 1982, **91**, 186
5. Cobb, L. M. et al *Int. J. Radiat. Oncol., Biol., Phys.* 1989, **16**(4), 953-956
6. Cobb, L. M. et al *Int. J. Radiat. Oncol., Biol., Phys.* 1990, **18**(2), 347-351
7. Sasai, K. et al *Int. J. Radiat. Biol.* 1990, **57**(5), 971-980
8. Widdick, D. A. et al *Mutat. Res.* 1991, **259**(1), 89-93
9. Djuric, Z. *Toxicol. Appl. Pharmacol.* 1989, **101**(1), 47-54
10. Yoshimura, S. et al *J. Toxicol. Pathol.* 1990, **3**(1), 65-89

# M314   Mitin N

**CAS Registry No.** 370-50-3

**Synonyms** *N,N'*-bis[4-chloro-3-(trifluoromethyl)phenyl]urea; 4,4'-dichloro-3,3'-bis(trifluoromethyl)carbanilide; Flucofuron

**Mol. Formula** $C_{15}H_8Cl_2F_6N_2O$                **Mol. Wt.** 417.14

**Uses** Insecticide used to control *Tineidae* sp. larvae, which attack cotton fabrics (moth-proofer).

## Physical properties

**M. Pt.** 241-242°C.

## Ecotoxicity

**Fish toxicity**

$LC_{50}$ (1 or 7 day) juvenile rainbow trout 150 and 96 µg $l^{-1}$ respectively (1).

Yearling rainbow trout (40 wk) at concentrations <10 μg l$^{-1}$ survived for 40 wk. At concentrations >10 μg l$^{-1}$ a dose-related decline in survival time was observed (1).

## Bioaccumulation
Will accumulate in rainbow trout tissues. Fish exposed to lethal concentrations accumulated average concentrations of 10 mg kg$^{-1}$ in muscle and 20 mg kg$^{-1}$ in viscera. Fish exposed to sub-lethal concentrations accumulated average concentrations of 1 mg kg$^{-1}$ in muscle and 3 mg kg$^{-1}$ in viscera (1).

## Mammalian and avian toxicity

### Acute data
LD$_{50}$ (unspecified route) rat 750 mg kg$^{-1}$ (1).

## Legislation
Limited under EC Directive on Drinking Water Quality 80/778/EEC. Pesticides: maximum admissible concentration 0.1 μg l$^{-1}$ (2).
Included in Schedule 6 (Release into Land: Prescribed Substances) Statutory Instrument No. 472, 1991 (3).

## References
1. Abram, F. S. H. et al *The Toxicities of Mitin N and Eulan WA New to Rainbow Trout* June 1981, 156-m, Water Research Centre Environmental Protection, Stevenage
2. *EC Directive Relating to the Quality of Water Intended for Human Consumption* 1982, 80/778/EEC, Office for Official Publications of the European Communities, 2 rue Mercier, L-2985 Luxembourg
3. *S. I. 1991 No. 472 The Environmental Protection (Prescribed Processes and Substances) Regulations* 1991, HMSO, London

# M315   Mitomycin C

CAS Registry No. 50-07-7
**Synonyms** azirino[2′,3′:3,4]pyrrolo[1,2-*a*]indole-4,7–dione, 6-amino-8-[[(aminocarbonyl)oxy]methyl]–1,1a,2,8,8a,8b-hexahydro-8a-methoxy-5-methyl–, [1a*S*-(1aα,8β,8aα,8bα)]-; azirino[2′,3′:3,4]pyrrolo[1,2-*a*]indole-4,7-dione, 6-amino-1,1a,2,8,8a,8b-hexahydro-8-(hydroxymethyl)-8a-methoxy-5-methyl, carbamate(ester); Ametycine; Mutamycin
**Mol. Formula** C$_{15}$H$_{18}$N$_4$O$_5$          **Mol. Wt.** 334.33

**Uses** Antibiotic, acting as a bioreductive alkylating agent, used in the treatment of malignant neoplasms.

**Occurrence** Produced by the growth of *Streptomyces caespitosus*.

## Physical properties

**M. Pt.** >360°C.

**Solubility**

Organic solvent: acetone, cyclohexanone, butyl acetate

## Ecotoxicity

**Invertebrate toxicity**

$EC_{50}$ (5, 10 or 20 min) *Photobacterium phosphoreum* >15.3 ppm Microtox test (1).

## Mammalian and avian toxicity

**Acute data**

$LD_{50}$ oral mouse, rat 23-30 mg $kg^{-1}$ (2).
$LD_{50}$ oral redwing blackbird, starling, quail 7.5->17.8 mg $kg^{-1}$ (3).
$LD_{50}$ subcutaneous rat, mouse 3250 and 7800 mg $kg^{-1}$, respectively (4).
$LD_{50}$ intraperitoneal rat, mouse 2-4 mg $kg^{-1}$ (5,6).
$LD_{50}$ intravenous dog, rat, mouse 1-4 mg $kg^{-1}$ (2,6,7).

**Sub-acute data**

Intravesical instillation rat wkly long-term experiment (dose and duration unspecified). Effects on the bladder included high percentage of pyuria and microscopic haematuria with submucosal/muscular fibrosis and severe dysplastic changes in the urothelium (8).
Intraperitoneal (5 wk) rat 1.7 mg $kg^{-1}$ wkly. Caused lung changes characterised by a decrease in aminopeptidase excretion, alveolar septal congestion, tubular damage with acute enzyme leakage from cells followed by enzyme depletion (9).

**Carcinogenicity and long-term effects**

No adequate evidence for carcinogenicity to humans, sufficient evidence for carcinogenicity to animals, IARC classification group 2B (10).
Dose estimated to induce tumours in 50% of a group of animals was calculated to be ♀ rat 1.1 mg $kg^{-1}$ $day^{-1}$, ♂ rat 0.73 mg $kg^{-1}$ $day^{-1}$ (11).
Subcutaneous (66 wk) mice (strains BTK, C57BL, C3H, DDO) 0.2 μg in saline 2 × wkly for 35 wk, followed by 31 wk observation. All BTK and 2/10 C57BL mice developed local sarcomas within 39-54 wk. No tumours were noted in the other 2 strains or any of the controls (12).
Intraperitoneal (18 month) Charles River CD rats 0.038 or 0.15 mg $kg^{-1}$ 3 × wkly for 6 month and then observed for a further 12 months. Peritoneal sarcomas developed in 30/31 ♀ and 27/29 ♂ (13).
Intravenous (lifetime) ♂ BR46 rats 0.52 mg $kg^{-1}$ 5 × within 2 wk. 79/96 survived to the time of the appearance of the 1st tumour (average time for tumour onset was 18 months). 34% developed malignant tumours which included lymphosarcomas, abdominal polymorphic-cell sarcomas, mammary carcinomas and sarcomas, subcutaneous fibrosarcomas and squamous-cell carcinomas. 3 developed benign tumours. Malignant tumours were observed in 6.2% of the controls (7).

**Teratogenicity and reproductive effects**
Intraperitoneal ICR mice day 1, 2 or 3 of pregnancy 1.5-5 mg kg$^{-1}$ (single dose) dose-dependent increase in foetal mortality and a decrease in the number of implants, foetal weight and the number of ossified sacral and caudal vertebrae. 5 mg kg$^{-1}$ caused an increase in the incidence of external malformation umbilical hernia (14).
*In vitro* morula-stage mouse embryo 0.004-0.5 μg ml$^{-1}$ caused a decrease in the number of inner cell mass cells at the blastocyst stage and a decrease in the tropectoderm population at the highest dosage only. Post-blastocyst development was retarded: fewer embryos formed trophoblastic outgrowths and the inner cell mass was poorly developed. Embryo transfer experiments showed that the reduction in inner cell mass cells diminished the potential of embryogenesis and successful implantation (15).

**Metabolism and pharmacokinetics**
Intravenous injection humans rapidly disappears from the blood, and is widely distributed but does not cross the blood-brain barrier. Metabolised mainly in the liver, about 10% is excreted unchanged in the urine, small amounts are also found in the bile and faeces (16).
Intravenous Wistar rat 2 mg kg$^{-1}$ 18% was collected unchanged in urine within 24 hr. 8 mg kg$^{-1}$ 35% was recovered in urine but none in faeces or tissues (17).
Intravenous guinea pig 8 mg kg$^{-1}$ after 30 min the drug was found concentrated in the kidneys and excreted in urine. Traces were detected in blood, it was not detected in the liver, spleen or brain (18).

**Sensitisation**
Patients receiving intravesical mitomycin C can develop severe eczematous symptoms which appear to be due to a hypersensitivity reaction (19).

## Genotoxicity
*Saccharomyces cerevisiae* D5 induction of mitotic crossing over positive (20).
*Escherichia coli* K-12 without metabolic activation negative (21).
*In vitro* human whole-blood and separated-lymphocyte cultures without metabolic activation micronuclei positive (22).
*In vitro* Chinese hamster ovary K1 cells with and without 1-β-D-arabinofuranosylcytosine induced sister-chromatid exchanges and chromosomal aberrations in s-phase synchronised cells (23).
*In vivo* rat liver and bone marrow cells chromosomal aberrations positive (24).
*In vivo* adult ♂ grasshopper injected with 0.2 ml of 0.01% mitomycin C caused an increase in chromosomal aberrations in the late spermatocytes. The induced aberration frequency was calculated to ≈20% (25).
*In vivo* intraperitoneal (5 day) ♂ rats 5-50 mg kg$^{-1}$ day$^{-1}$ percentage aberrant sperm heads increased 4-fold (26).

## Any other adverse effects to man
Main adverse effects are delayed cumulative bone-marrow suppression. Profound leucopaenia and thrombocytopenia occurs after 4 wk with recovery ≈8-10 wk after a dose (16).

## Any other adverse effects
Subconjunctival injection rabbit eye cornea 0.4 mg ml$^{-1}$, cytotoxic in intact and lesioned eye (27).

Intravesical instillation rats, 1 dose, short-term experiment (dose unspecified). Effects on the bladder included increased weight, oedematous changes in the muscle layer, congestion of the mucosa and infiltration of polymorphonuclear cells 3 days after instillation. These local changes had disappeared within 10 days (8).
Intraperitoneal rat 2.5 mg kg$^{-1}$, rats were examined after 5 days, alanine aminopeptidase:creatinine ratio increased compared with the control group (9). Forelimb buds from 14 day old rat foetuses were cut into pieces and transplanted subcutaneously into athymic (nude) mice. The mice were treated with mitomycin C on days 7, 9 and 11 after grafting and were examined on day 20, the differentiation of the grafts was inhibited compared to the controls (28).

## Any other comments

Pharmacokinetics, clinical pharmacology and antitumour mechanism reviewed (29,30).
Nephrotoxicity reviewed (31).

## References

1. Kaiser, K. L. E. et al *Water Pollut. Res. J. Can.* 1991, **26**(3), 361-431
2. *Cancer Res.* 1960, **20**, 1354
3. Schafer, E. W. et al *Arch. Environ. Contam. Toxicol.* 1983, **12**, 355-382
4. *Drugs in Japan. Ethical Drugs* 6th ed., 1982, Japan Pharmaceutical Centre, Tokyo
5. *Adv. Teratol.* 1968, **3**, 181
6. *J. Antibiot., Ser. A* 1960, **13**, 27
7. Schmal, D. et al *Arzneim.-Forsch.* 1970, **20**, 1461-1467
8. Chang, S. Y. et al *Br. J. Urol.* 1990, **66**(6), 623-627
9. Verweij, J. et al *J. Cancer Res. Clin. Oncol.* 1988, **114**(2), 137-141
10. *IARC Monograph* 1987, **Suppl. 7**, 67
11. Kaldor, J. M. et al *Eur. J. Cancer Clin. Oncol.* 1988, **24**(4), 703-711
12. Ikegmi, R. et al *Acta Pathol. Jpn.* 1967, **17**, 495-501
13. *IARC Monograph* 1976, **10**, 171-179
14. Nayao, T. et al *Senten Ijo* 1986, **26**(2), 93-101
15. Tam, P. P. L. *Teratology* 1988, **37**(3), 205-215
16. *Martindale. The Extra Pharmacopoeia* 30th ed., 1993, The Pharmaceutical Press, London
17. Schwartz, H. S. et al *J. Pharm. Exp. Ther.* 1961, **133**, 335-342
18. Fujita, H. *Jpn. J. Clin. Oncol.* 1971, **12**, 151-162
19. Clover, G. B. et al *Br. J. Dermatol.* 1990, **122**, 217-224
20. Ferguson, L. R. et al *Mutat. Res.* 1988, **204**, 239-249
21. Fram, R. J. et al *Mutat. Res.* 1986, **166**, 229-242
22. Migliore, L. et al *Mutat. Res.* 1989, **227**, 167-172
23. Um, K. H. et al *Environ. Mutagen. Carcinog.* 1987, **7**(1), 1-8
24. Rossi, A. M. et al *Mutagenesis* 1986, **1**(5), 335-338
25. Mohanta, B. K. et al *Environ. Ecol.* 1986, **4**(1), 124-129
26. Ye, W. et al *Zhongguo Yaolixue Yu Dulixue Zazhi* 1987, **1**(3), 229-230 (Ch.) (*Chem. Abstr.* **108**, 180288e)
27. Ishimaru, H. et al *Atarashii Ganka* 1988, **5**(6), 875-882, (Japan.) (*Chem. Abstr.* **109**, 183161q)
28. Shiota, K. et al *Reprod. Toxicol.* 1990, **4**(2), 95-103
29. Hagiwara, A. et al *Oncologia* 1987, **20**(2), 105-112, (Japan.) (*Chem. Abstr.* **107**, 168078h)
30. Taguchi, T. *Kagaku Ryoho no Ryoiki* 1987, **3**(10), 1649-1657 (Japan.) (*Chem. Abstr.* **108**, 593)
31. Ries, F. *Eur. J. Cancer Clin. Oncol.* 1988, **24**(6), 951-953

# M316   Molinate

$$\underset{\displaystyle \underset{N}{|}}{\overset{\displaystyle \overset{O}{\overset{||}{C}}}{}} - SCH_2CH_3$$

**CAS Registry No.** 2212-67-1
**Synonyms** 1*H*-azepine-1-carbothioic acid, hexahydro-, *S*-ethyl ester; Felan; Hydram; Jalan; Ordram; Yulan
**Mol. Formula** $C_9H_{17}NOS$                    **Mol. Wt.** 187.31
**Uses** Herbicide used primarily for weed control in rice culture.

## Physical properties

**B. Pt.** $_{10}$ 202°C; **Specific gravity** $d^{20}$ 1.063; **Partition coefficient** log $P_{ow}$ 2.8808 (1); **Volatility** v.p. $5.6 \times 10^{-3}$ mmHg at 25°C.

## Solubility

Water: 0.88 g $l^{-1}$. Organic solvent: miscible with acetone, methanol, ethanol, kerosene, 4-methylpentan-2-one, benzene, xylene

## Occupational exposure

**Supply classification** harmful.
**Risk phrases** Harmful if swallowed (R22)
**Safety phrases** Avoid contact with skin (S24)

## Ecotoxicity

**Fish toxicity**
$LC_{50}$ (48 hr) bluegill sunfish 0.48 mg $l^{-1}$ (2).
$LC_{50}$ (96 hr) mosquito fish 16.4 mg $l^{-1}$ (3).
$LC_{50}$ (21 day) carp 0.18 mg $l^{-1}$ (4).

**Invertebrate toxicity**
$LC_{50}$ (48 hr) *Asellus brevicaudus*, *Daphnia magna* 0.4-0.6 mg $l^{-1}$ (2).
$LC_{50}$ (96 hr) prawn, scud, crayfish 1.0-5.6 mg $l^{-1}$ (5,6).

## Environmental fate

**Degradation studies**
Degraded by *Streptomyces*, *Mycobacterium* and *Flavobacterium* species isolated from soil (7).
*Nocardia* sp., *Micrococcus* sp. isolated from garden soils and rice field drains degraded molinate completely to various hydroxy and oxidised products (8,9).
Metabolites of microbial breakdown in soil include ethyl mercaptan, carbon dioxide and dialkylamine (1).
In soil $t_{1/2} \approx 2$-5 wk (1).

**Abiotic removal**

$[^{14}C]$-molinate added to tap water decreased to 40% of original concentration in 14 days, this loss was primarily due to volatilisation. Major metabolites were molinate sulfoxide; 3- and 4-hydroxymolinate; 4-ketomolinate; and ketohexamethylene-imine (5).

## Mammalian and avian toxicity

### Acute data

$LD_{50}$ oral rat, mouse 369 and 530 mg $kg^{-1}$, respectively (10,11).
$LC_{Lo}$ inhalation rat, cat 200 mg $kg^{-1}$ (12).
$LD_{50}$ dermal rabbit 3536 mg $kg^{-1}$ (13).
$LD_{50}$ subcutaneous rat 1167 mg $kg^{-1}$ (11).

### Sub-acute data

$LC_{50}$ (5 day) mallard duckling >9300 ppm in diet (14).
$LC_{50}$ (9 wk) quail >1000 ppm in diet (14).

### Metabolism and pharmacokinetics

*In vitro* disposition and biotransformation by whole blood of the common carp. Accumulation by erythrocytes was nearly complete by 4 hr. Oxidised by erythrocytes to its sulfoxide and cleaved to form the mercapturic acid in both erythrocytes and plasma (15).

Oral rat rapidly metabolised within 72 hr; 50% eliminated as carbon dioxide, 25% is excreted in urine and 5-20% in faeces (1).

### Irritancy

Dermal (48, 72 hr) human 1% caused no irritation or sensitisation (16).
Moderate irritation to rabbit eye with iris and corneal dullness, all effects were gone by 5 days (5).

## Legislation

Limited under EC Directive on Drinking Water Quality 80/778/EEC. Pesticides: maximum admissible concentration 0.1 μg $l^{-1}$ (17).
Included in Schedule 6 (Release into Land: Prescribed Substances) Statutory Instrument No. 472, 1991 (18).

## Any other comments

Effect on fish reproduction disorders reviewed (19).
Cucumbers germinated and grown in soil which had previously been used to grow rice treated with molinate showed no adverse effects (20).

## References

1. *The Agrochemicals Handbook* 3rd ed., 1991, RSC, London
2. Kemp, H. T. et al *Water Quality Data Book* Vol.5, 1973, EPA, Water Pollution Control Research Series 09/73
3. Chaiyarach, S. et al *Bull. Environ. Contam. Toxicol.* 1975, **14**(3), 281-284
4. Kawastu, H. *Nippon Suisan Gakkaishi* 1977, **43**, 905
5. Verschueren, K. *Handbook of Environmental Data on Organic Chemicals* 2nd ed., 1983, 882-884, Van Nostrand Reinhold, New York
6. Sanders, H. O. *Toxicity of Pesticides to the Crustacean, Gammarus lacustris* 1969, Technical Paper 25, Bureau of Sport Fisheries and Wildlife, Washington, DC
7. Imai, Y. et al *Nippon Noyaku Gakkaishi* 1986, **11**(4), 563-572
8. Skryabin, G. K. et al *Dokl. Akad. Nauk SSSR* 1978, **239**(3), 717-720

9.  Golovleva, L. A. et al *4th International Congress of Pesticide Chemistry* 1978, Abstract V-610, Zurich, Switzerland
10. *The Pesticide Manual* 8th ed., 1987, British Crop Protection Council, Thornton Heath
11. *Vrach. Delo* 1969, (1), 119
12. *Hyg. Sanit. (USSR)* 1970, **35**(8), 35
13. *Farm Chemicals Handbook* 1983, C173, Meister Publishing Co., Willoughby, OH
14. Dietz, F. et al *GWF, Gas- Wasserfach: Wasser/Abwasser* 1978, **119**(6)
15. Tjeerdema, R. S. et al *Pestic. Biochem. Physiol.* 1988, **31**(1), 24-35
16. Lisi, P. et al *Contact Dermatitis* 1987, **17**(4), 212-218
17. *EC Directive Relating to the Quality of Water Intended for Human Consumption* 1982, 80/778/EEC, Office for Official Publications of the European Communities, 2 rue Mercier, L-2985 Luxembourg
18. *S. I. 1991 No. 472 The Environmental Protection (Prescribed Processes and Substances) Regulations* 1991, HMSO, London
19. Popova, G. U. et al *Eksp. Vodn. Tosksikol.* 1987, **12**, 192-201 (Russ.) (*Chem. Abstr.* **110**, 2498g
20. Krishnasamy, S. M. et al *Indian J. Agron.* 1988, **33**(4), 450-451

# M317   Molybdenum

## Mo

**CAS Registry No.** 7439-98-7
**Synonyms** MChVL; TsM1
**Mol. Formula** Mo                                              **Mol. Wt.** 95.94
**Uses** Tungsten production. Lubricant additive in colloidal form. In the ferromolybdenum form for manufacturing special steels used in tools, rifle barrels, propeller shafts etc.
**Occurrence** In the ores molybdenite and Wulfenite. Occurence in earth's crust 1-5 ppm.

## Physical properties

**M. Pt.** 2622°C; **B. Pt.** 4825°C; **Specific gravity** 10.28.

## Occupational exposure

**US TLV (TWA)** 10 mg m$^{-3}$ (as Mo); **UK Long-term limit** 10 mg m$^{-3}$ (as Mo); **UK Short-term limit** 20 mg m$^{-3}$ (as Mo).

## Ecotoxicity

**Fish toxicity**
$LC_{50}$ (96 hr) barb 550 mg l$^{-1}$ (as ammonium molybdate) (1).
$LC_{100}$ (96 hr) barb 650 mg l$^{-1}$ (as ammonium molybdate) (1).
Symptoms produced in barb include fin erosion, impaired swimming ability, dark colour and surfacing behaviour (1).
$LC_{50}$ (96 hr) swim-up, advanced fry of chinook salmon, coho salmon >100 mg l$^{-1}$ (form unspecified) (2).

### Bioaccumulation

In analysis of plankton, molluscs and marine plants molybdenum was only detected in marine plant tissues (3).

Found in *Calyptogena* sp., *Mytilus edulis*, *Ridgeia pisceae*, *Riftia pachyptila*, *Paravinella palmiformis*, *Munidopsis* sp. collected from hydrothermal springs in Juna de Fuca and the Gulf of California (4).

Deep-sea macroinvertebrates (mussels and limpets) from submarine thermal springs accumulated molybdenum at levels corresponding to the levels discharged from the springs (5).

Three microorganism species were able to biosorb molybdenum from dilute solutions (6).

## Mammalian and avian toxicity

### Acute data

$LD_{50}$ (route unspecified) mouse, rabbit 300 and 700 mg $kg^{-1}$, respectively (as ammonium molybdate) (7).

$LD_{100}$ (route unspecified) sheep 1000 mg $kg^{-1}$ (as ammonium molybdate) (7).

$LD_{Lo}$ intraperitoneal rat 114 mg $kg^{-1}$ (form unspecified) (8).

$LD_{Lo}$ intraperitoneal rabbit 70 mg $kg^{-1}$ (form unspecified) (9).

Acute symptoms in mouse, rabbit and sheep included general depression anaemia, decreased blood haemoglobin levels and alkaline phosphatase activity and increased xanthine oxidase in the liver and uric acid in the serum. Accumulation was highest in the kidney, liver and lungs (7).

Oral sheep 1000 mg $kg^{-1}$ (form unspecified) developed symptoms of poisoning 1-15 hr after administration and death 5-6 hr later. Levels: kidneys, 530-550 mg $kg^{-1}$; lungs, 480-500 mg $kg^{-1}$; liver, 250-370 mg $kg^{-1}$; and muscles, 100-105 mg $kg^{-1}$ (10).

### Sub-acute data

Sheep, rabbit chronic studies (duration, dose unspecified) resulted in weight reduction of liver and spleen, decreased haemoglobin levels, change in hepatic xanthine oxidase activity and uric acid level in the serum. Accumulation was highest in the liver (7).

Oral (3 month) rabbits, 1-10 mg $kg^{-1}$ (as ammonium molybdate) did not cause any changes to health or physiological indexes. Levels accumulated in the tissues and organs were dose dependent and the majority of accumulation was in the liver (10).

Oral (duration unspecified) rats 0.1% as ammonium molybdate in diet caused significant anaemia and marked cardiac hypertrophy and copper deficiency (11).

### Teratogenicity and reproductive effects

The air sacs of chick eggs were injected with various doses of a molybdenum salt on day 2 of incubation. On day 14 the developing embryos were examined. $LD_{50}$ 333 µg $egg^{-1}$. Teratogenic effects were observed (12).

## Genotoxicity

*Escherichia coli* PQ37, PQ35 SOS Chromotest with and without metabolic activation negative (as molybdenum chloride) (13).

*Salmonella typhimurium* TA1535, TA1537, a correlation was observed between toxicity and mutagenicity (14).

*In vivo* rats administered 0.0125 mg $kg^{-1}$ (maximum level allowed in water drinking water) not mutagenic (form unspecified) (15).

## Any other adverse effects to man

There is a correlation between levels in soil and age-adjusted incidences of heart disease and cancer death rates (16).

Concentrations in soil and food relate inversely to the significantly higher than national average rates of human stomach and oesophageal cancers in China (17).

## Legislation

Included in Schedule 4 (Release into Air: Prescribed Substances) Statutory Instrument No. 472, 1991 (18).

## Any other comments

Toxicity, biokinetics (including tissue distribution and metabolism), environmental effects reviewed (19-21).

Heavy metal toxicity to *Azotobacter chroococcum* described (22).

Has been detected in roe deer and mallard kidney and liver tissues (23).

Did not have a significant genotoxic effect on *Drosophila melanogaster* cells (form unspecified) (24).

Until 2 nitrogenases were isolated from *Azobacter* that lacked molybdenum, the element was considered indispensible for catalytic function in nitrogen fixation (25).

Analysis of lung tissue from 21 workers previously employed in metal refining showed levels of 0.037 $\mu$g g$^{-1}$ wet weight lung whilst control groups showed 0.03 $\mu$g g$^{-1}$ (26).

Human health effects, workplace experience, epidemiology and experimental toxicology reviewed (27).

## References

1. Pundir, P. et al *Pollut. Res.* 1990, **9**(1-4), 101-105
2. Hamilton, S. J. et al *Arch. Environ. Contam. Toxicol.* 1990, **19**(3), 366-373
3. Saenko, U. N. *Dokl. Akad. Nauk SSSR* 1989, **306**(3), 759-763 (Russ.) (*Chem. Abstr.* **111**, 131032v)
4. Lakashin, V. N. et al *Geokhimiya* 1990, (2), 279-285 (Russ.) (*Chem. Abstr.* **112**, 240085t)
5. Smith, D. R. et al *Mar. Biol. (Berlin)* 1989, **102**(1), 127-133
6. Prytogel, P. A. et al *Report* 1989, EGG-M-88376; Order no. DE90002073, available from NTIS, Natl. Tech. Inf. Ser., Springfield, VA
7. Arzumanyan, I. Z. *Veterinariya (Moscow)* 1987, (3), 74-76, (Russ.) (*Chem. Abstr.* **107**, 2230b)
8. Browning, E. *Toxicity of Industrial Metals* 1961, 214, Butterworths, London
9. *NTIS Report PB249-458*, Natl. Tech. Inf. Ser., Springfield, VA
10. Arzumanyan, I. Z. *Dokl. Vses. Akad. S-kh. Nauk im V. I. Lenina* 1986, (9), 46-47, (Russ.) (*Chem. Abstr.* **106**, 14412e)
11. Wang, Y. et al *Baiqiuen Yike Daxue Xuebao* 1988, **14**(3), 218-220 (Ch.) (*Chem. Abstr.* **110**, 149426r)
12. Galani, S. H. et al *J. Toxicol. Environ. Health* 1990, **30**(1), 23-31
13. Olivier, P. et al *Mutat. Res.* 1987, **189**(3), 263-269
14. D'yachenko, O. Z. et al *Gig. Sanit.* 1990, (8), 83-85, (Russ.) (*Chem. Abstr.* **113**, 186421t)
15. Saichenko, S. P. *Probl. Gig. Tr., Profpatol. Toksikol. Gornodobyvayushchei Metall. Prom-sti* 1985, 75-80 (Russ.) (*Chem. Abstr.* **106**, 209072b)
16. Jackson, M. L. *Appl. Geochem.* 1986, **1**(2), 172-180
17. Jackson, M. L. *Appl. Geochem.* 1986, **1**(4), 487-492
18. *S. I. 1991 No. 472 The Environmental Protection (Prescribed Processes and Substances) Regulations* 1991, HMSO, London

19. Anke, M. et al *Trace Elem. Anal. Chem. Med. Biol., Proc. Int. Workshop, 4th* 1986, 201-236, Braetter, P. et al (Ed.), de Gruyter, Berlin
20. Erzberger, A. *ISH-Heft* 1988, **124** (Ger.) (*Chem. Abstr.* **110**, 35399d)
21. Eisler, R. *Report* 1989, Biological-85(1.19), Contaminant Hazard Reviews-19; Order No. PB89-235998 (*Chem. Abstr.* **115**, 55776t)
22. Gulyas, F. et al *Hazard Waste: Detect., Control, Treat., Proc. World Conf.* 1987, Abbon, R. (Ed.), Elsevier, Amsterdam
23. Holm, J. et al *Fleischwirtschaft* 1987, **67**(9), 1145-1149 (Ger.) (*Chem. Abstr.* **108**, 1726b)
24. Chopikashvili, L. V. et al *Tsitol. Genet.* 1989, **23**(3), 35-38 (Russ.) (*Chem. Abstr.* **111**, 148548r)
25. Pau, R. N. *Trends Biochem. Sci.* 1989, **14**(5), 183-186
26. Hewitt, P. J. *Environ. Geochem. Health* 1988, **10**(3-4), 113-116
27. *ECETOC Technical Report No. 30(4)* 1991, European Chemical Industry Ecology and Toxicology Centre, B-1160 Brussels

# M318   Molybdenum trioxide

## $MoO_3$

**CAS Registry No.** 1313-27-5
**Synonyms** molybdenum oxide ($MoO_3$); molybdena; molybdenum(VI) oxide; molybdenum(VI) trioxide; molybdic acid anhydride; molybdic anhydride
**Mol. Formula** $MoO_3$                                **Mol. Wt.** 143.94
**Uses** Reagent for chemical analysis.

## Physical properties

**M. Pt.** 795°C; **B. Pt.** 1155°C; **Specific gravity** $d_4^{26}$ 4.696.

## Solubility
Water: 0.49 g $l^{-1}$ at 28°C.

## Occupational exposure

**US TLV (TWA)** 5 mg m$^{-3}$ (soluble compounds as Mo); **UK Long-term limit** 5 mg m$^{-3}$ (soluble compounds as Mo); **UK Short-term limit** 10 mg m$^{-3}$ (soluble compounds as Mo).

## Mammalian and avian toxicity

### Acute data
$LD_{50}$ oral rat 125 mg kg$^{-1}$ (1).
$LD_{50}$ subcutaneous mouse 4 mg kg$^{-1}$ (2).
$LD_{Lo}$ intraperitoneal guinea pig 400 mg kg$^{-1}$ (3).

### Carcinogenicity and long-term effects
Inhalation rat, mouse 1, 33, 100 mg m$^{-3}$ under investigation by The National Toxicology Program, National Institute of Environmental Health Sciences (4).

### Teratogenicity and reproductive effects
Inhalation (13 wk) ♂ B6C3F$_1$ mice 10-100 ppm, no change in reproductive organ weight or sperm motility, density or morphology (5).

Inhalation (13 wk) ♂ F344 rat 10-100 mg m$^{-3}$, statistically significant reduction in epididymis weight and an increase in the number of abnormal sperm was observed. Testis and cauda weight and sperm motility and sperm density were not significantly different from controls (5).

## Legislation

Included in Schedule 4 (Release into Air: Prescribed Substances) Statutory Instrument No. 472, 1991 (6).

## Any other comments

Hazards reviewed (7).
Human health effects, epidemiology, workplace experience and experimental toxicology reviewed (8).

## References

1. Browning, E. *Toxicology of Industrial Metals* 1961, 214, Butterworths, London
2. *Zh. Vses. Khim. O-va. im. D. I. Mendeleeva* 1974, **19**, 186
3. *Environ. Qual. Saf., Suppl.* 1975, **1**, 1
4. *IARC Directory of Agents Being Tested for Carcinogenicity* 1992, No.15, Lyon
5. Morrissey, R. E. et al *Fundam. Appl. Toxicol.* 1988, **11**(2), 343-358
6. *S. I. 1991 No. 472 The Environmental Protection (Prescribed Processes and Substances) Regulations* 1991, HMSO, London
7. *Dangerous Prop. Ind. Mater. Rep.* 1988, **8**(3), 73-78
8. *ECETOC Technical Report No. 30(4)* 1991, European Chemical Industry Ecology and Toxicology Centre, B-1160 Brussels

# M319   Monalide

**CAS Registry No.** 7287-36-7
**Synonyms** *N*-(4-chlorophenyl)-2,2-dimethylpentanamide;
4'-chloro-2,2-dimethylvaleranilide; Potablan
**Mol. Formula** C$_{13}$H$_{18}$ClNO                    **Mol. Wt.** 239.75
**Uses** Herbicide for post-emergence control of broad-leaved weeds.

## Physical properties

**M. Pt.** 87-88°C; **Volatility** v.p. $1.8 \times 10^{-6}$ mmHg at 25°C.

## Solubility

Water: 22.8 mg l$^{-1}$ at 23°C. Organic solvent: cyclohexanone, xylene, petroleum ether

## Ecotoxicity

**Fish toxicity**
$LC_{50}$ (duration unspecified) guppy >100 mg $l^{-1}$ (1,2).

## Environmental fate

**Degradation studies**
In compost soil at pH 4.85, 5.2, 10.8; $t_{1/2}$ 726, 1161, 1412 hr, respectively (3).
The decomposition product 4-chloroaniline was isolated from soil (3).

**Abiotic removal**
Decomposition in buffer solutions using 2 methods with slightly different pH ranges;
pH 1.1-11.0 and pH 1.8-10.9. Regardless of the method used decomposition was
fastest in acid media, with decomposition in alkaline media being slower and slowest
under neutral conditions. At pH 7, $t_{1/2}$ 80-372 wk depending on the method of
determination used (3).

## Mammalian and avian toxicity

**Acute data**
$LD_{50}$ oral rat 2600 mg $kg^{-1}$ (4).
$LD_{50}$ dermal rabbit 2600 mg $kg^{-1}$ (5).
$LD_{50}$ dermal rat, rabbit >800 mg $kg^{-1}$ (1,2).

**Sub-acute data**
NOEL (28 day) rat 150 mg $kg^{-1}$ $day^{-1}$ (1,2).

## Legislation

Limited under EC Directive on Drinking Water Quality 80/778/EEC. Pesticides:
maximum admissible concentration 0.1 µg $l^{-1}$ (6).
Included in Schedule 6 (Release into Land: Prescribed Substances) Statutory
Instrument No. 472, 1991 (7).

## Any other comments

WHO Class Table 5; EPA Toxicity Class III (1).

## References

1. *The Agrochemicals Handbook* 3rd ed., 1991, RSC, London
2. *The Pesticide Manual* 9th ed., 1991, British Crop Protection Council, Farnham
3. Duske, J. et al *Cesk. Farm.* 1988, **37**(4), 178-183 (*Chem. Abstr.* **10**, 337713)
4. Thomson, W. T. *Agric. Chem.* 1977, **32**, 220
5. *Pesticide Manual* 1968, 307, British Crop Protection Council, Worcestershire
6. *EC Directive Relating to the Quality of Water Intended for Human Consumption* 1982,
   80/778/EEC, Office for Official Publications of the European Communities, 2 rue Mercier,
   L-2985 Luxembourg
7. *S. I. 1991 No. 472 The Environmental Protection (Prescribed Processes and Substances)
   Regulations* 1991, HMSO, London

# M320   Monobutyl phthalate

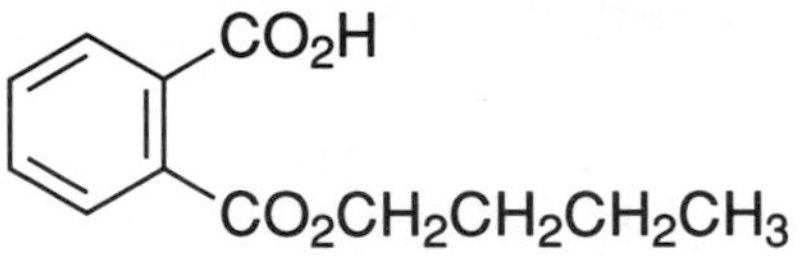

$$CO_2H$$
$$CO_2CH_2CH_2CH_2CH_3$$

**CAS Registry No.** 131-70-4

**Synonyms** 1,2-benzenedicarboxylic acid, monobutyl ester; phthalic acid, monobutyl ester; phthalic acid, butyl ester; butyl hydrogen phthalate; mono-$n$-butyl phthalate

**Mol. Formula** $C_{12}H_{14}O_4$                     **Mol. Wt.** 222.24

## Physical properties

**M. Pt.** 90°C

**Solubility**
Organic solvent: ethanol

## Mammalian and avian toxicity

**Acute data**
$LD_{50}$ intraperitoneal mouse 1 g $kg^{-1}$ (1).

**Sub-acute data**
Oral (34-36 day) ♂ Wistar rat 0.5 or 5% in diet. The higher dose group showed growth depression, liver enlargement, testicular atrophy, decrease of succinate and pyruvate dehyrogenase activities in liver mitochondria and changes in liver and testis biochemistry and histology (2).
Oral (1 wk) ♂ rat 0.8 g $kg^{-1}$ $day^{-1}$ caused severe testicular injury (3).

## Any other comments

Human health effects and experimental toxicology reviewed (4).

## References

1. *C. R. Hebd. Seances Acad. Sci. Ser. D.* 1971, **273**, 2165
2. Murakumi, K. et al *Nippon Eiseigaku Zasshi* 1986, **41**(4), 775-781
3. Mikuriya, H. et al *Jikeikai Med. J.* 1988, **35**(3), 403-409
4. *ECETOC Technical Report No. 30(4)* 1991, European Chemical Industry Ecology and Toxicology Centre, B-1160 Brussels

# M321   Monocrotaline

**CAS Registry No.** 315-22-0
**Synonyms** 20-norcrotolanan-11,15-dione, 14,19-dihydro-12,13-dihydroxy-, (13α,14α)-; monocrotolin
**Mol. Formula** $C_{16}H_{23}NO_6$                              **Mol. Wt.** 325.36
**Uses** In Africa and India crushed roots of *Crotalaria* sp. are used for a colic remedy, fever relief, remedy for haemoptysis and in the treatment of childhood malaria.
**Occurrence** Pyrrolizidine alkaloid which is the major toxic constituent of *Crotalaria spectabilis* Roth. Leguminosae (1).

## Physical properties
**M. Pt.** 197-198°C (decomp.).

**Solubility**
Water: 1.2%. Organic solvent: ethanol, chloroform

## Ecotoxicity

**Fish toxicity**
Non-toxic to trout, bluegill sunfish, yellow perch, goldfish (24 hr) at 5 ppm. Test conditions; pH 7.0; dissolved oxygen content 7.5 ppm; total hardness (soap method) 300 ppm; methyl orange alkalinity 310 ppm; free carbon dioxide 5 ppm; and temperature 12.8°C (2).
Three-spine stickleback, steelhead trout, sockeye salmon (24 hr) no fish death or loss of equilibrium at 10 mg l$^{-1}$. Test conditions: artesian well water total hardness 67-120 mg l$^{-1}$; methyl orange alkalinity 151-183 mg l$^{-1}$; total dissolved solids 160-175 mg l$^{-1}$; pH 7.1 (3).

## Mammalian and avian toxicity

**Acute data**
LD$_{50}$ oral mouse 166-261 mg kg$^{-1}$ (4,5).
LD$_{50}$ oral rat 66-71 mg kg$^{-1}$ (6).
LD$_{Lo}$ subcutaneous rat 60 mg kg$^{-1}$ (7).
LD$_{50}$ intravenous rat 92 mg kg$^{-1}$ (8).
LD$_{50}$ intravenous mouse 261 mg kg$^{-1}$ (9).
LD$_{Lo}$ intraperitoneal rat 130 mg kg$^{-1}$ (10).
LD$_{50}$ intraperitoneal mouse 259 mg kg$^{-1}$ (11).
LD$_{50}$ intraperitoneal ♂, ♀ rat 178-189 mg kg$^{-1}$ (12).

Guinea pig (route unspecified) 240 mg kg$^{-1}$ (4 times LD$_{50}$ for rats) showed no clinical or pathological effects (13).

## Sub-acute data

Gavage (14 day) ♀ C57B1/6 mice 0-150 mg kg$^{-1}$. Dose-dependent suppression in the antibody response to sheep red blood cells, cytotoxic T-lymphocyte response was decreased (38% of control), the number of cytotoxic T-lymphocytes spleen$^{-1}$ was reduced to 12% of control and the antibody titers were dose dependently suppressed (14). Oral (6 wk) ♂ mice 2.4, 4.8 or 24.0 mg kg$^{-1}$ day$^{-1}$ in drinking water continuously. Pulmonary endothelial function was investigated. A dose-dependent decrease in lung angiotensin converting enzyme and plasminogen activator activity, indicative of endothelial dysfunction, was found. However, these responses were only significant at the highest dose. Microscopy revealed dose-dependent pulmonary inflammation and exudative reactions (15).

Oral (6 month) rat 0.1 LD$_{50}$ 2 × wkly in drinking water, lesions in the lungs, liver, kidneys and heart. 2 ♀ rats developed preneoplastic lesions in the liver (16). Subcutaneous infant stumptail monkeys 30 mg kg$^{-1}$ followed by 60 mg kg$^{-1}$ during the 2, 4, 6 months of the experiment produced severe lung lesions and cardiac hypertrophy, but little liver damage. The same doses given to adolescent monkeys caused severe hepatic veno-occlusive lesions (17).

## Carcinogenicity and long-term effects

No adequate evidence for carcinogenicity to humans, sufficient evidence for carcinogenicity to animals, IARC classification group 2B (18).

Gavage (72 wk) ♂ Sprague-Dawley (CD) rat 25 mg kg$^{-1}$ wk$^{-1}$ for 4 wk then 8 mg kg$^{-1}$ wk$^{-1}$ for 38 wk followed by observation until 72 wk. 42/72 rats survived until the appearance of the 1st tumour (55 wk) of these 10 developed liver cell carcinomas. Lung metastases were also observed (19).

Gavage (72 wk) ♂ Sprague-Dawley (CD) rats fed a diet marginally deficient in lipotrophes 25 mg kg$^{-1}$ wk$^{-1}$ for 4 wk and then 8 mg kg$^{-1}$ wk$^{-1}$ for 38 wk followed by observation up to 72 wk. 35/50 survived until the appearance of the 1st tumour at 46 wk and of these 14 developed liver cell carcinomas. Lung metastases were also observed (19).

## Teratogenicity and reproductive effects

Oral rat 0.5 LD$_{50}$ in feed during pregnancy and lactation, toxic effects studied in newborn rats for up to 18 months after lactation. Toxic effects detected in lungs, liver, spleen and kidneys of newborns during early stages. These toxic effects were more severe in rats exposed during pregnancy and lactation than in rats only exposed during lactation (20).

## Metabolism and pharmacokinetics

The major metabolite excreted in rat urine is an $N$-acetylcysteine conjugate of (±)-6,7-dihydro-7-hydroxy-1-hydroxymethyl-5$H$-pyrrolizine (21).

[$^{14}$C]-monocrotaline (60 mg kg$^{-1}$, 200 μCi kg$^{-1}$) applied to rats subcutaneously. At 4 hr the distribution in tissues was 27.7, 24.1, 21.8, 11.7, 2.6 μg g$^{-1}$ of tissue for red blood cells, liver, kidney, lung and plasma, respectively. At 24 hr the distribution in tissues was 15.9, 8.1, 2.9, 3.3, 0.7 μg g$^{-1}$ of tissue for red blood cells, liver, kidney, lung and plasma, respectively (22).

Kinetic studies of [$^{14}$C]-monocrotaline (60 mg kg$^{-1}$, 10 μCi kg$^{-1}$) intravenously to rats demonstrated rapid elimination of radioactivity with ≈ 90% recovery of injected radioactivity in urine and bile by 7 hr. The plasma level of radioactivity dropped from 36.76 μg g$^{-1}$ to 3.58 μg g$^{-1}$ by 7 hr, red blood cell levels decreased from

46.85 µg g$^{-1}$ to 26.35 µg g$^{-1}$ by 7 hr. Intravenous rat with cannulated bile duct [$^{14}$C]-monocrotaline; 83% of the dose was excreted in urine within 7 hr and 12% of the dose was found in bile (22).

## Genotoxicity

*In vitro* human embryo kidney cell cultures irreversibly inhibited DNA synthesis (23).
*In vivo* ♂ *Drosophila melanogaster* injected into abdomen sex-linked recessive lethals positive (24).
*In vivo* after intraperitoneal injection of 0-30 mg kg$^{-1}$ caused DNA-DNA interstrand crosslinking in a dose-dependent manner in rat liver cells (25).

## Any other adverse effects to man

*In vitro* human embryo kidney cell culture inhibited carbohydrate synthesis and the cellular utilisation of carbohydrates. Formation of lactate was decreased and the intracellular pH markedly increased (23).
Consumption of *Crotalaria*. sp. containing monocrotaline can lead to veno-occlusive disease (26).

## Any other comments

Pulmonary effects reviewed (27).
Human health effects and experimental toxicology reviewed (28).

## References

1.  *The Merck Index* 11th ed., 1989, Merck & Co. Inc., Rahway, NJ
2.  *Lethal Effects of 3400 Chemicals to Fish* 1987, EPA560/6-87-002, PB87-200-275, Washington, DC
3.  *Lethal Effects of 2014 Chemicals to Fish* 1989, EPA560/6-89-001, PB89-156715, Washington, DC
4.  Goldenthal, E. L. et al *Toxicol. Appl. Pharmacol.* 1964, **6**, 434-441
5.  Harris, P. N. et al *J. Pharmacol. Exp. Ther.* 1942, **75**, 78-82
6.  Newberne, P. M. et al *Toxicol. Appl. Pharmacol.* 1971, **18**, 387
7.  *Toxicol. Appl. Pharmacol.* 1972, **23**, 470
8.  *J. Pharmacol. Exp. Ther.* 1945, **83**, 265
9.  *J. Pharmacol. Exp. Ther.* 1942, **75**, 78
10. *Chem.-Biol. Interact.* 1976, **12**, 299
11. *Toxicol. Appl. Pharmacol.* 1981, **59**, 424
12. Bull, L. B. et al *The Pyrrolizidine Alkaloids* 1968, North Holland, Amsterdam
13. Chesney, C. F. et al *Toxicol. Appl. Pharmacol.* 1973, **26**, 385-392
14. Deyo, J. A. et al *Fundam. Appl. Toxicol.* 1990, **14**(4), 842-849
15. Molteni, A. et al *Virchows Arch., B* 1989, **57**(3), 149-155
16. Sriraman, P. K. et al *Indian J. Anim. Sci.* 1987, **57**(10), 1060-1068
17. Allen, J. R. et al *Exp. Mol. Pathol.* 1972, **17**, 220-232
18. *IARC Monograph* 1987, **Suppl.7**, 67
19. Newberne, P. N. et al *Plant Foods for Man* 1973, 23-31, Newman, P. N. (ed.)
20. Sriraman, P. K. et al *Indian J. Anim. Sci.* 1988, **58**(11), 1292-1295
21. Estep, J. E. et al *Toxicol. Lett.* 1990, **54**(1), 61-69
22. Estep, J. E. et al *Drug Metab. Dispos.* 1991, **19**, 135-139
23. Yin, L. et al *Zhongguo Zhongyao Zazhi* 1990, **15**(6), 364-366 (Ch.) (*Chem. Abstr.* **113**, 165075t)
24. Clark, A. M. *Z. Vererbungsl.* 1960, **91**, 74-80
25. Petry, T. W. et al *Carcinogenesis (London)* 1987, **8**(3), 415-419
26. *IARC Monograph* 1976, **10**, 291-301

27. Roth, R. A. et al *Comments Toxicol.* 1989, **3**(2), 131-144
28. *ECETOC Technical Report No. 30(4)* 1991, European Chemical Industry Ecology and Toxicology Centre, B-1160 Brussels

# M322   Monocrotophos

**CAS Registry No.** 6923-22-4
**Synonyms** phosphoric acid, dimethyl 1-methyl-3-(methylamino)-3-oxo-1-propenyl ester, (E)-; phosphoric acid, dimethyl ester, ester with 3-hydroxy-N-methyl crotonamide, (E)-; Azodrin; Nuvacron
**Mol. Formula** $C_7H_{14}NO_5P$                    **Mol. Wt.** 223.17
**Uses** Systemic insecticide. Acaricide.

## Physical properties

**M. Pt.** 25-30°C (commercial solid product), 54-55°C (crystals); **B. Pt.** $_{0.0005}$ 125°C; **Specific gravity** $d^{20}$ 1.33; **Volatility** v.p. $7 \times 10^{-6}$ mmHg at 20°C.

## Solubility

Water: 1 kg $kg^{-1}$ at 20°C. Organic solvent: acetone, ethanol

## Occupational exposure

**US TLV (TWA)** 0.25 mg $m^{-3}$; **Supply classification** very toxic.
**Risk phrases** Toxic in contact with skin – Very toxic if swallowed (R24, R28)
**Safety phrases** Do not breathe vapour – Wear suitable protective clothing and gloves – In case of accident or if you feel unwell, seek medical advice immediately (show label where possible) (S23, S36/37, S45)

## Ecotoxicity

### Fish toxicity

$LC_{50}$ (96 hr) snakehead fish 10 ppm (1).
Snakehead fish (15-120 day) 1 ppm. Fish at 15 day were hypoglycaemic and hypolactaemic, after 30 days anaerobic metabolism prevailed over aerobic metabolism, after 60 and 120 days both aerobic and anaerobic pathways were impaired (1).
$LC_{50}$ (acute exposure) rainbow trout 4.9 mg $l^{-1}$ (2).
$LC_{50}$ (96 hr) harlequin fish 450 mg $l^{-1}$ (3).
$LC_{100}$ (96 hr) cichlid 18.6 mg $l^{-1}$ (4).

### Invertebrate toxicity

$LC_{50}$ (96 hr) marine edible crab 0.577 ppm (5).

Clawed frog embryos, caused dose-dependent developmental defects (dose unspecified) including abnormal pigmentation, abnormal gut development, notochordal defects and reduced growth (6).
LC$_{50}$ (48, 96 hr) shrimp 4.46, 1.59 mg l$^{-1}$, respectively (7,8).
LC$_{50}$ (96 hr) oarfooted crustacean 0.24 mg l$^{-1}$ (9).

## Environmental fate

### Nitrification inhibition
$\geq$5 μg ml$^{-1}$ inhibited nitrogen fixation activity of *Nostoc linckia* (10).

### Anaerobic effects
Two successive applications at field doses to flooded rice soil decreased significantly the population size of soil algae and also altered the species composition of the native algal flora (11).

### Degradation studies
20 day incubation in flooded rice soil decreased levels to trace amounts. In flooded rice soil that had been autoclaved monocrotophos persisted indicating microbiological action was important in degradation (11).
Soils treated 5 × 15 day intervals. Isolated algae were then incubated with the monocrotophos 5-50 ppm for 30 days all 5 algal species degraded the insecticide. Degradation was almost complete by 30 days (12).
$t_{1/2}$ in aqueous environment at 25°C; pH 3 131 day; pH 9 26 day. Hydrolysis followed 1st-order kinetics and the major hydrolytic degradation products were *N*-methylacetoacetamide and *O*-demethylmonocrotophos (13).
Soil metabolism studies showed rapid and extensive decomposition eventually to carbon dioxide and unextractable residues. The intermediate degradation products were *N*-methylacetoacetamide, *N*-(hydroxymethyl)monocrotophos and 3-hydroxy-*N*-methylbutyramide (13).

### Abiotic removal
Degradation did not occur in the absence of light (14).
Photodegradation was greater on soil surfaces than on glass. Photodegradation: alluvia <black <red loamy >laterite soil. Photolysis was greater on flooded moist soils than dry loam soil (14).
UV light was more effective at degradation than sunlight. Rate of photodegradation in tap water was twice that in distilled water (14).
River water in a sealed jar under sunlight and artificial light 0.01 mg l$^{-1}$ initial concentration, 100% remained after 8 wk (15).

### Absorption
Mobile in soil under test conditions (13).

## Mammalian and avian toxicity

### Acute data
LD$_{50}$ oral rat, mouse 8-15 mg kg$^{-1}$ (16,17).
LD$_{50}$ oral rat ♂ 17, ♀ 20 mg kg$^{-1}$ (18).
LD$_{50}$ oral northern bobwhite, quail, duck 0.8-3.4 mg kg$^{-1}$ (19,20).
LD$_{50}$ oral redwing blackbird, house sparrow, redbilled quelea, common grackle, common pigeon, quail, starling 1-5.62 mg kg$^{-1}$ (21).
LD$_{50}$ oral mallard 42.2 mg kg$^{-1}$ (21).
LC$_{50}$ (4 hr) inhalation rat 63 mg m$^{-3}$ (22).

$LD_{50}$ dermal rat ♂ 126, ♀ 112 mg kg$^{-1}$ (18).
$LD_{50}$ dermal duck 30 mg kg$^{-1}$ (23).
$LD_{50}$ subcutaneous rat, mouse 6964-8710 μg kg$^{-1}$ (24,25).
$LD_{Lo}$ intraperitoneal rat 20 mg kg$^{-1}$ (26).
$LD_{50}$ intraperitoneal mouse 3800 μg kg$^{-1}$ (27).
$LD_{50}$ intravenous rat, mouse 9200 μg kg$^{-1}$ (25,28).

**Sub-acute data**
Intragastric (2 wk) rat 0.3-2.4 mg kg$^{-1}$ day$^{-1}$ showed dose-, time-, sex-related changes
in blood chemistry and body and organ weights (29).
Oral (13 wk) ♂, ♀ Wistar-derived rats 0, 0.1, 0.25, 0.5, 2.0, 8.0 ppm in diet to animals
in each treatment group. Half the animals in each treatment group were killed after 8
wk and the remainder were killed after 13 wk. No clinical symptoms or deaths due to
the treatment were observed. There was a slight reduction of body weight in both
sexes at the 8 ppm dose level. At all doses there was a dose-related decrease of
plasma, erythrocyte and brain cholinesterase activities. The inhibition of brain
cholinesterase was biologically significant in the 2.0 and 8.0 ppm dose groups.
Cholinesterase activity was almost completely recovered by 5 wk post-treatment (30).

**Carcinogenicity and long-term effects**
Oral (104 wk) ♂, ♀ CD mice 1, 2, 5 or 10 ppm in diet. No evidence of
treatment-related oncogenic effects at any of the dose levels (30).
Oral (24 month) ♂, ♀ Wistar-derived rats 0.01, 0.03, 0.1, 1.0 or 10 ppm in diet. No
evidence of treatment-related oncogenic effects at any of the dose levels (30).

**Teratogenicity and reproductive effects**
Oral ♂, ♀ rats (5 wk old at start of study) 0, 0.1, 1, 3 or 10 ppm in diet for at least
15 wk ($F_0$ generation) these were mated and the progeny ($F_1$) were fed the same diet
for 18 wk prior to mating to produce an $F_2$ generation. Significantly lower (6-9%
lower) body weights were seen in the ♂ $F_0$ and $F_1$. No effects on sperm count were
seen in the ♂ $F_0$ or ♂ $F_1$. Mating performance, fertility index and gestation index were
not different among the $F_0$ groups. 10 ppm $F_1$ ♂ mating index was lower and fewer
litters were produced compared to controls. Mean litter size, viability index and
lactation index were significantly reduced at 10 ppm for both $F_1$ and $F_2$ generations
and the viability index of 3 ppm $F_2$ was also reduced. Mean pup weights were
reduced for the $F_1$ at 10 ppm and the $F_2$ at 10 and 3 ppm. There were 3 total litter
losses at $F_1$, $F_2$ at 10 ppm and 1 at $F_2$ at 3 ppm. The no-effect-level in this
reproduction study was 1 ppm (30).
Gavage (days 6 to 15 of gestation) ♀ rat 0, 0.3, 1.0 or 2.0 mg kg$^{-1}$ day$^{-1}$. At the
2.0 mg kg$^{-1}$ mean body weight and crown rump length of the foetuses were
significantly lower. At 1.0, 2.0 mg kg$^{-1}$ the mean percentage of runt foetuses was
increased. At 2 mg kg$^{-1}$ the percentage of foetuses with non-ossified sternebrae was
doubled compared to the controls. Malformed and/or misshapen brain was observed
in foetuses at 0.3, 1 and 2 mg kg$^{-1}$; this type of malformation is uncommon to this
strain of rat (30).

**Metabolism and pharmacokinetics**
Oral mammals 60-65% excreted within 24 hr, mostly in urine (31).

## Genotoxicity
*Escherichia coli* SOS Chromotest with and without metabolic activation negative (32).

*Salmonella typhimurium* TA98, TA100, TA102, TA1535, TA1537, TA1538 with and without metabolic activation TA100 positive, all others negative (30,33).
*Saccharomyces cerevisiae* D3, D7 with and without metabolic activation positive (30).
*In vitro* primary rat tracheal epithelial cells, Chinese hamster ovary cells with and without metabolic activation sister chromatid exchanges positive (34).
*In vitro* Chinese hamster ovary cells with and without metabolic activation clastogenesis positive (35).
*In vitro* human lymphocytes time- and concentration-dependent increases of chromosome damage and sister chromatid exchanges (36).
*In vivo* rat bone marrow cells chromosomal aberrations equivocal (37).
*In vivo* mice 0.9, 1.8, 3.6 mg kg$^{-1}$ increased the number of abnormal sperm to 2.12, 3.20, 5.36%, respectively, control 2.09% (38).

## Any other adverse effects to man

Occupationally exposed pesticide sprayers had significantly increased numbers of sister chromatid exchanges and chromosomal abberations, at all durations of exposure, compared to unexposed controls. They also showed cell cycle delay and a decrease in the mitotic index (39,40).

## Legislation

Limited under EC Directive on Drinking Water Quality 80/778/EEC.Pesticides: maximum admissible concentration 0.1 μg l$^{-1}$ (41).
Included in Schedule 6 (Release into Land: Prescribed Substances) Statutory Instrument No. 472, 1991 (42).

## Any other comments

Pesticide Suitability Rating (calculated to indicate suitability for use in the household) designates monocrotophos to be just above the acceptable limit (43).

## References

1.  Samuel, M. et al *Pestic. Biochem. Physiol.* 1989, **34**(1), 1-8
2.  Kenaga, E. E. *Down Earth* 1979, **35**(2), 25-31
3.  Tooby, T. E. et al *Chem. Ind. (London)* 1975, **21**, 523-525
4.  Mustafa, M. et al *Int. Pest Control* 1982, **24**, 90
5.  Rao, K. S. et al *Environ. Ecol.* 1987, **5**(1), 181-182
6.  Snawder, J. E. et al *J. Environ. Sci. Health, Part B* 1989, **B24**(3), 205-218
7.  Nagabhushanam, R. et al *Indian J. Comp. Physiol.* 1983, **1**, 71
8.  Omkar et al *Crustaceana* 1985, **48**, 1
9.  Khattat, F. H. et al *EPA Final Report* 1976, EPA-68/01/0151
10. Megharaj, M. et al *Bull. Environ. Contam. Toxicol.*, 1988, **41**(2), 277-281
11. Megharaj, M. et al *Chemosphere* 1988, **17**(5), 1033-1039
12. Megharaj, M. et al *Bull. Environ. Contam. Toxicol.* 1987, **39**(2), 251-256
13. Lee, P. W. et al *J. Agric. Food Chem.* 1990, **38**(2), 267-273
14. Dureja, P. *Bull. Environ. Contam. Toxicol.* 1989, **43**(2), 239-245
15. Eichelberger, J. W. et al *Environ. Sci. Technol.* 1971, **5**(6), 541-544
16. *Farm Chemicals Handbook* 1983, C161, Meister Publishing Co., Willoughby, OH
17. *Agricultural Research Service* 1966, **20**, 21, USDA, Beltsville, MD
18. Gaines, T. B. *Toxicol. Appl. Pharmacol.* 1969, **14**, 515
19. Wiemeyer, S. N. et al *Environ. Toxicol. Chem.* 1991, **10**(9), 1134-1148
20. *Toxicol. Appl. Pharmacol.* 1972, **22**, 556
21. Schafer, E. W. et al *Arch. Environ. Contam. Toxicol.* 1983, **12**, 355-382
22. *Egeszsegtudomany* 1980, **24**, 173

23. *Toxicol. Appl. Pharmacol.* 1979, **47**, 451
24. *Br. J. Pharmacol.* 1970, **40**, 124
25. *J. Pharm. Pharmacol.* 1967, **19**, 612
26. *Bull. Environ. Contam. Toxicol.* 1978, **19**, 47
27. *Toxicol. Appl. Pharmacol.* 1968, **13**, 37
28. *NTIS Report PB277-077*, Natl. Tech. Inf. Ser., Springfield, VA
29. Janardhan, A. et al *Bull. Environ. Contam. Toxicol.* 1989, **44**(2), 230-239
30. *IPCS Pesticide Residues in Food – 1991 Evaluations* 1991, Part II, 287-300 Toxicology, World Health Organization, Geneva
31. *The Agrochemicals Handbook* 3rd ed., 1991, RSC, London
32. Xu, H. H. et al *Toxic. Assess.* 1990, **5**(1), 1-14
33. Moriya, M. et al *Mutat. Res.* 1983, **116**, 185-216
34. Wang, T. C. et al *Mutat. Res.* 1987, **188**(4), 311-321
35. Lin, M. F. et al *Mutat. Res.* 1988, **188**(3), 241-250
36. Rupa, D. S. et al *Bull. Environ. Contam. Toxicol.* 1988, **41**(5), 737-741
37. Adhikari, N. et al *Environ. Mol. Mutagen.* 1988, **12**(2), 235-242
38. Kumar, D. V. et al *Bull. Environ. Contam. Toxicol.* 1988, **41**(2), 189-194
39. Rupa, D. S. et al *Environ. Mol. Mutagen.* 1991, **18**(2), 136-138
40. Rupa, D. S. et al *Environ. Res.* 1989, **49**, (1), 1-6
41. *EC Directive Relating to the Quality of Water Intended for Human Consumption* 1982, 80/778/EEC, Office for Official Publications of the European Communities, 2 rue Mercier, L-2985 Luxembourg
42. *S. I. 1991 No. 472 The Environmental Protection (Prescribed Processes and Substances) Regulations* 1991, HMSO, London
43. Deo, P. G. et al *Int. Pest. Control* 1988, **30**(5), 118-121

# M323   Mono(2-ethylhexyl) phthalate

**CAS Registry No.** 4376-20-9

**Synonyms** 1,2-benzenedicarboxylic acid, mono(2-ethylhexyl) ester; phthalic acid, mono(2-ethylhexyl) ester; 2-ethylhexyl hydrogen phthalate

**Mol. Formula** $C_{16}H_{22}O_4$                     **Mol. Wt.** 278.35

## Physical properties

**M. Pt.** 14-16°C.

## Mammalian and avian toxicity

### Acute data

$LD_{50}$ oral rat 1340 mg $kg^{-1}$ (1).

$LD_{50}$ intraperitoneal mouse, rat 240 and 415 mg $kg^{-1}$, respectively (2).

$LD_{50}$ intravenous rat, mouse 150 and 208 mg $kg^{-1}$, respectively (2).

### Sub-acute data

Oral ♂ Chinese hamster, ♂ Syrian hamster (14 day) 0.5% in diet caused an increase in liver size and an induction of peroxisomal enzyme activity. The effects were more pronounced in Chinese hamsters (3).

Oral ♂ rat (>21 day) various doses, doses ≥1000 ppm caused peroxisomal proliferation (4).

Gavage ♂ cyanomologus monkeys (>21 day) up to 500 mg kg$^{-1}$ day$^{-1}$ did not cause peroxisomal proliferation (4).

**Teratogenicity and reproductive effects**

♂ rat single dose (route unspecified) 0.8 g kg$^{-1}$ induced testicular atrophy which was age-dependent with only prepubertal rats being susceptible. Testicular zinc levels were affected to varying degrees dependent on treatment (5).

Oral mice (6-13 day gestation) 545 mg kg$^{-1}$ day$^{-1}$. Mice were allowed to deliver litters and the pups were studied for 3 days. The number of viable litters was significantly reduced, 2/33 pregnancies compared to the control of 34/38. Of the viable litters there was no significant decrease in live born per litter, percentage survival, birth weight or weight gain (6).

Designated negative for developmental toxicity in rabbits, positive for mice, with a predicted negative developmental toxicity for humans (7).

**Metabolism and pharmacokinetics**

*In vitro* cultures of ♂ rat hepatocytes, Sertoli cells, Leydig cells, incubated with [$^{14}$C]-mono(2-ethylhexyl) phthalate for up to 24 hr caused no significant reduction in viability. Hepatocytes extensively metabolised [$^{14}$C]mono(2-ethylhexyl) phthalate to a variety of products within 1 hr; Sertoli cells and Leydig cells showed no significant metabolism in 24 hr. Hepatocytes were much more efficient at uptake than Sertoli or Leydig cells (8).

## Genotoxicity

*Salmonella typhimurium* TA97, TA98, TA100, TA102 with and without metabolic activation negative (9,10).

*In vitro* Chinese hamster cells DNA amplification negative (9).

*In vitro* rat, hamster hepatocytes DNA damage negative (9).

*In vitro* rat hepatocytes unscheduled DNA synthesis negative (11).

## Any other adverse effects

*In vitro* rat, human hepatocyte cell cultures induced enzymes indicative of peroxisomal proliferation (12).

## Any other comments

Human health effects and experimental toxicology reviewed (13).

## References

1. *Toxicol. Appl. Pharmacol.* 1978, **45**, 250
2. *NTIS Report PB250-102*, Natl. Tech. Inf. Ser., Springfield, VA
3. Lake, B. G. et al *Food Chem. Toxicol.* 1986, **24**(6-7), 573-575
4. Short, R. D. et al *Toxicol. Ind. Health* 1987, **3**(2), 185-195
5. Teirlynck, O. et al *Toxicol. Lett.* 1988, **40**(1), 85-91
6. Hardin, B. D. et al *Teratog., Carcinog., Mutagen.* 1987, **7**(1), 29-48
7. Jelovsek, F. R. et al *Obstet. Gynecol. (N. Y.)* 1989, **74**(4), 624-636
8. Albro, P. W. et al *Toxicol. Appl. Pharmacol.* 1989, **100**(2), 193-200
9. Schmezer, P. et al *Carcinogenesis (London)* 1988, **9**(1), 37-43

10. Dirven, H. A. A. M. et al *Mutat. Res.* 1991, **260**(1), 121-130
11. Cattley, R. C. et al *Cancer Lett. (Shannon, Irel.)* 1986, **33**(3), 269-277
12. Butterworth, B. E. et al *Cancer Res.* 1989, **49**(5), 1075-1084
13. *ECETOC Technical Report No. 30(4)* 1991, European Chemical Industry Ecology and Toxicology Centre, B-1160 Brussels

# M324   Monolinuron

**CAS Registry No.** 1746-81-2
**Synonyms** $N'$-(4-chlorophenyl)-$N$-methoxy-$N$-methylurea;
3-(*p*-chlorophenyl)-1-methoxy-1-methylurea; Aresin; Arezin; Arezine; Monorotox
**Mol. Formula** $C_9H_{11}ClN_2O_2$                                   **Mol. Wt.** 214.65
**Uses** Herbicide.

## Physical properties

**M. Pt.** 80-83°C; **Partition coefficient** log $P_{ow}$ 2.2041 (1); **Volatility**
v.p. $1.5 \times 10^{-4}$ mmHg.

**Solubility**
Water: 0.735 g $l^{-1}$ at 25°C. Organic solvent: acetone, dioxane, xylene, chloroform, diethyl ether

## Occupational exposure

**Supply classification** harmful.
**Risk phrases** Harmful if swallowed (R22)
**Safety phrases** Do not breathe dust (S22)

## Ecotoxicity

**Fish toxicity**
$LC_{50}$ (96 hr) brown trout (non-migratory), common carp, cichlid, snakehead fish, 3.1, 12.9, 54, 105 mg $l^{-1}$, respectively (2,3).

**Invertebrate toxicity**
$LC_{50}$ (96 hr) *Daphnia magna* 1.3 mg $l^{-1}$ (2).
LOEC (duration unspecified) *Microcystis aeruginosa* 0.14 mg $l^{-1}$ (4).
$LC_{50}$ (96 hr) mosquito larvae 24.2 mg $l^{-1}$ (2).

## Environmental fate

### Degradation studies

In soils $t_{1/2} \approx 45\text{-}60$ day (1).
Breakdown in soil involves cleavage of the methyl and methoxy groups on the terminal nitrogen atom, with simultaneous ring hydroxylation and formation of 3-(2-hydroxy-4-chlorophenyl)urea and the corresponding 3-hydroxy compound (1).

### Abiotic removal

Major decomposition product after photochemical exposure was 3-phenyl-1-methylurea (5).

### Absorption

Soil adsorption $K_{oc}$ 250-500 (6).

## Mammalian and avian toxicity

### Acute data

$LD_{50}$ oral dog, rat 500, 1800 mg $kg^{-1}$, respectively (7,8).
$LD_{50}$ oral mallard duck >500 mg $kg^{-1}$ (1,6).
$LD_{50}$ oral bobwhite quail, Japanese quail 1260, >1690 mg $kg^{-1}$, respectively (1,6).
$LD_{50}$ intragastric duck, mouse, chicken, rabbit 0.38-1.80 g $kg^{-1}$. Affected the nervous system, heart function and blood composition. Accumulation was mainly in the liver, kidneys, lungs and heart (9).

### Carcinogenicity and long-term effects

In a 2 yr feeding study, no effect level for rats was 250 mg $kg^{-1}$ (1).

## Legislation

Limited under EC Directive on Drinking Water Quality 80/778/EEC. Pesticides: maximum admissible concentration 0.1 $\mu g \, l^{-1}$ (10).
Included in Schedule 6 (Release into Land: Prescribed Substances) Statutory Instrument No. 472, 1991 (11).

## Any other comments

Behaviour in soils predicted (12).

## References

1. *The Agrochemicals Handbook* 3rd ed., 1991, RSC, London
2. *Environmental Properties of Chemicals* 1990, Research Report 91, Ministry of the Environment, Finland
3. Rao, K. S. et al *J. Fish Biol.* 1979, **14**, 517
4. Bringmann, G. et al *GWF, Gas- Wasserfach: Wasser/Abwasser* 1976, **117**(9)
5. Welther-Sandor, M. et al *Magy. Kem. Foly.* 1987, **93**(3), 129-132 (Hung.) (*Chem. Abstr.* **107**, 54051p)
6. *The Pesticide Manual* 9th ed., 1991, British Crop Protection Council, Farnham
7. Frear, E. H. (Ed.) *The Pesticide Index* 5th ed., 1976, 158, College Science Publications, PA
8. *World Rev. Pest Control* 1970, **9**, 119
9. Ermolin, A. V. et al *Veterinariya (Moscow)* 1990, (9), 59-60 (Russ.) (*Chem. Abstr.* **114**, 116619k)
10. *EC Directive Relating to the Quality of Water Intended for Human Consumption* 1982, 80/778/EEC, Office for Official Publications of the European Communities, 2 rue Mercier, L-2985 Luxembourg

11.  *S. I. 1991 No. 472 The Environmental Protection (Prescribed Processes and Substances) Regulations* 1991, HMSO, London
12.  Blume, H. P. et al *Landwirtsch. Forsch.* 1987, **40**(1), 41-50 (Ger.) (*Chem. Abstr.* **108**, 33490m)

# M325   Monooctyl phthalate

**CAS Registry No.** 5393-19-1

**Synonyms** 1,2-benzenedicarboxylic acid, monooctyl ester; phthalic acid, monooctyl ester; octyl hydrogen phthalate

**Mol. Formula** $C_{16}H_{22}O_4$                    **Mol. Wt.** 278.35

## Physical properties

**M. Pt.** 21.5-22.5°C in petroleum ether.

## Any other adverse effects

Inhibited the state 3 oxygen consumption in mitochondrial function of rat testis, at a concentration as low as 18 mg $l^{-1}$ (1).

## Any other comments

Included in Schedule 4 (Release Into Air: Prescribed Substances) Statutory Instrument No. 472, 1991 (2).

## References

1.  Oishi, S. *Arch. Toxicol.* 1990, **64**(2), 143-147
2.  *S. I. 1991 No. 472 The Environmental Protection (Prescribed Processes and Substances) Regulations* 1991, HMSO, London

# M326   Monopentyl phthalate

**CAS Registry No.** 24539-56-8

**Synonyms** 1,2-benzenedicarboxylic acid, monopentyl ester; phthalic acid, monopentyl ester; monoamyl phthalate; mono-*n*-pentyl phthalate

**Mol. Formula** $C_{13}H_{16}O_4$                    **Mol. Wt.** 236.27

**Physical properties**

M. Pt. 75.4-75.6°C in petroleum ether.

**Any other adverse effects**

Produced ultrastructural changes in Sertoli cells in primary co-cultures of rat seminiferous tubules; changes were in the configuration of the plasma membrane, changes in microfilament distribution and an increased density of ribosomes, smooth endoplasmic reticulum and the Golgi body (1).

**Legislation**

Included in Schedule 4 (Release Into Air: Prescribed Substances) Statutory Instrument No. 472, 1991 (2).

**References**

1. Creasy, D. M. et al *Toxicol. in Vitro* 1988, **2**(2), 83-95
2. *S. I. 1991 No. 472 The Environmental Protection (Prescribed Processes and Substances) Regulations* 1991, HMSO, London

# M327    Monopropylene glycol methyl ether

CH₃CH(OH/CH₃)CH₂(OH/CH₃)

**CAS Registry No.** 1320-67-8

**Synonyms** propanol, methoxy-; Dowanol PM; methyl ether of propylene glycol; propylene glycol monomethyl ether

**Mol. Formula** $C_4H_{10}O_2$                                    **Mol. Wt.** 90.12

**Physical properties**

**M. Pt.** –96.7°C; **B. Pt.** 120°C; **Flash point** 168.3°C; **Specific gravity** $d_{25}^{25}$ 0.919.

**Mammalian and avian toxicity**

**Metabolism and pharmacokinetics**

Inhalation (6 hr) rats 300 and 3000 ppm 1 or 10 × daily, blood levels failed to plateau, indicating that absorption was limited by respiration. Monopropylene glycol methyl ether blood levels were higher in ♂ than in ♀ rats receiving a single 3000 ppm exposure (1).

**Irritancy**

Eye rabbit 100 mg caused severe irritation (2).

**Legislation**

Included in Schedule 4 (Release Into Air: Prescribed Substances) Statutory Instrument No. 472, 1991 (3).

**References**

1. Morgott, D. A. et al *Toxicol. Appl. Pharmacol.* 1987, **89**(1), 19-28
2. *Am. J. Ophthalmol.* 1946, **29**, 1363
3. *S. I. 1991 No. 472 The Environmental Protection (Prescribed Processes and Substances) Regulations* 1991, HMSO, London

# M328   Monothiuram

$(CH_3)_2NC(S)SC(S)N(CH_3)_2$

**CAS Registry No.** 97-74-5
**Synonyms** carbamic acid, dimethyldithioanhydrosulfide; thiodicarbonic diamide
$([(H_2(N)C(S)]_2S)$, tetramethyl-; tetramethylthiurammonium sulfide;
1,1'-thiobis(*N,N*-dimethylthio)formamide; bis(dimethylthiocarbamoyl) sulfide
**Mol. Formula** $C_6H_{12}N_2S_3$                     **Mol. Wt.** 208.37
**Uses** Component of bird repellent (1). Antioxidant in resin manufacture.
Vulcanisation accelerator. Fungicide.

## Physical properties
**M. Pt.** 104°C; **Partition coefficient** log $P_{ow}$ 1.17 (2).

**Solubility**
Organic solvent: ethanol, chloroform

## Ecotoxicity

**Fish toxicity**
$LC_{50}$ (96 hr) guppy 5.3 mg $l^{-1}$ (2).

**Invertebrate toxicity**
$EC_{50}$ (48 hr) *Daphnia magna* 2.9 mg $l^{-1}$ (2).

**Bioaccumulation**
$EC_{50}$ (96 hr) *Chlorella pyrenoidosa* 1.0 μg $l^{-1}$ (2).

## Environmental fate

**Nitrification inhibition**
The minimum inhibiting concentration (3 hr) for *Nitrosomonas* and *Nitrobacter*
is 32 mg $l^{-1}$ (2).

## Mammalian and avian toxicity

**Acute data**
$LD_{Lo}$ oral rat 500 mg $kg^{-1}$ (3).
$LD_{50}$ oral mouse 818 mg $kg^{-1}$ (4).
$LD_{Lo}$ oral cat 100 mg $kg^{-1}$ (5).
$LD_{Lo}$ oral guinea pig 10 mg $kg^{-1}$ (6).
$LD_{50}$ intraperitoneal mouse 300 mg $kg^{-1}$ (7).

**Sub-acute data**
In rats, pretreated by gavage with 2 doses of 3.3 mg $kg^{-1}$ or 53 mg $kg^{-1}$ at 90 min
or 18 hr before intraperitoneal administration of 2 g $kg^{-1}$ in ethanol showed significant
increases in blood acetaldehyde. Oral 3.3 mg $kg^{-1}$ dose could be detected up to
48 hr after administration (8).

**Carcinogenicity and long-term effects**
$TD_{Lo}$ subcutaneous mouse 100 mg $kg^{-1}$ reported equivocal tumorigenic effects (9).

### Teratogenicity and reproductive effects

$TD_{Lo}$ (6-14 days) subcutaneous pregnant mouse 900 mg $kg^{-1}$ reported reproductive effects (10).

Day 3 chicken embryos were injected and testing continued to day 14 of incubation, embryotoxicity was observed, most common malformations were eye defects and open coeloms (11).

### Irritancy

1% concentration caused a positive patch test reaction in 11-34% of cases evaluated (12).

### Genotoxicity

*Salmonella typhimurium* TA98, TA100 with metabolic activation positive (13).

### Any other adverse effects

Administration oral ♀ rat 26 mg $kg^{-1}$ caused prolongation of the hexobarbital sleeping time, related to inhibition of microsomal oxygenases. Decrease in erythrocytes and Hb count in blood (4).

Inhibited lipid peroxidation in rat hepatic microsomes (14).

### References

1. Yonezawa Chemical Industry *Jpn. Kokai Tokkyo Koho JP* 58109401 [83109401] (*Chem. Abstr.* **99**, 117875z)
2. van Leuwen, C. J. *Aquat. Toxicol.* 1985, **7**, 145-164
3. *J. Pharm. Exp. Ther.* 1947, **90**, 260
4. Alanis, O. T. et al *Environ. Res.* 1982, **28**(1), 199-221
5. *Rubber Chem. Technol.* 1971, **44**, 513
6. *NTIS Report* PB223-159, Natl. Tech. Inf. Ser. Springfield, VA
7. *NTIS Report* AO-277-689, Natl. Tech. Inf. Ser. Springfield, VA
8. Garcia de Torres, G. et al *Drug Chem. Toxicol. (1977)* 1983, **6**(4), 317-328
9. *NTIS Report* PB 223-160, Natl. Tech. Inf. Ser., Springfield, VA
10. Kanto, H. et al *Hifu* 1985, **27**(3), 501-509 (Japan.) (*Chem. Abstr.* **104**, 46819)
11. Korhonen, A. et al *Scand. J. Work, Environ. Health* 1982, **8**(1), 63-69
12. *Mutat. Res.* 1979, **68**, 313
13. You, X. et al *Huanjing Kexue* 1982, **3**(6), 39-42 (Ch.) (*Chem. Abstr.* **98**, 84705t)
14. Feundt, K. J. et al *G. Ital. Med. Lav.* 1981, **3**(2-3), 117-119

# M329   Montmorillonite

**CAS Registry No.** 1318-93-0
**Synonyms** montmorillonite; Arcillite; Bentolite; Deriton; Walkerde Flygtol GA
**Uses** In industrial chromatograph technique. In the petroleum industry. Catalyst carrier.
**Occurrence** A clay forming the principal constituent of bentonite and Fuller's Earth.

## Environmental fate

### Nitrification inhibition
Increasing the concentration of montmorillonite in a soil enhanced the rate of nitrification (1).

### Any other adverse effects
Potential health risk due to exposure to montmorillonite dust can be excluded. Scanning electron microscopic investigations did not reveal fibrous particles (2).

### Any other comments
A constituent of Fuller's Earth used in the treatment of paraquat poisoning (3). Montmorillonite clay is a good adsorber of ammonium but a poor absorber of potassium (4).

### References
1. James, R. A. et al *Appl. Environ. Microbiol.* 1991, **57**(11), 3212-3219
2. Werner, I. et al *Staub - Reinhalt. Luft* 1988, **48**(4), 169-171
3. *Martindale The Extra Pharmacopoeia* 30th ed., 1993, The Pharmaceutical Press, London
4. Bajwa, M. I. *J. Agron. Crop Sci.* 1987, **158**(1), 65-68

# M330   Monuron

**CAS Registry No.** 150-68-5
**Synonyms** $N'$-(4-chlorophenyl)-$N,N$-dimethylurea; 3-($p$-chlorophenyl)-1,1-dimethylurea; Telvar; Karmex Monuron Herbicide; Telvar Monuron Weed Killer; Karmex W. monuron herbicide
**Mol. Formula** $C_9H_{11}ClN_2O$                    **Mol. Wt.** 198.65
**Uses** Herbicide.

## Physical properties
**M. Pt.** 170.5-171.5°C; **Specific gravity** $d_{20}^{20}$ 1.27; **Partition coefficient** log $P_{ow}$ 2.13 (1); **Volatility** v.p $5 \times 10^{-7}$ mmHg at 25°C; $1.78 \times 10^{-3}$ mmHg at 100°C.

## Solubility
Water: 230 ppm at 25°C. Organic solvent: methanol, ethanol, acetone

## Occupational exposure

**Supply classification** harmful.

**Risk phrases** Harmful if swallowed – Possible risk of irreversible effects (R22, R40)

**Safety phrases** Wear suitable protective clothing and gloves (S36/37)

## Ecotoxicity

**Fish toxicity**

$LC_{50}$ (48 hr) coho salmon 110 mg$^{-1}$ (2).

**Invertebrate toxicity**

Microalgae, cyanobacteria minimum inhibitory concentration to growth >17.1 μg ml$^{-1}$ (3).

$EC_{50}$ (5 min) *Photobacterium phosphoreum* 228 ppm Microtox test (4).

$EC_{50}$ (duration unspecified) *Phaeodactylum tricornutum* 90 mg l$^{-1}$ affected photosynthesis (5).

LOEC (duration unspecified) *Dunaliella euchlora, Phaeodactylum tricornutum* 1.0 mg l$^{-1}$ affected reproduction (6).

## Environmental fate

**Nitrification inhibition**

In soil 40.0 ppm did not inhibit nitrification (7).

**Degradation studies**

In soils 75-100% disappeared in 10 months (8).

**Abiotic removal**

Acetone solutions of monuron injected into water samples and exposed to natural and artificial light at room temperature. 40, 30, 20 and 0% remained after 1, 2, 4 and 8 wk, respectively (9).

160 mg l$^{-1}$ activated carbon will reduce 5 mg l$^{-1}$ to 0.1 mg l$^{-1}$; and 29 mg l$^{-1}$ activated carbon will reduce 1 mg l$^{-1}$ to 0.1 mg l$^{-1}$ (10).

## Mammalian and avian toxicity

**Acute data**

$LD_{50}$ oral rat 1053-3700 mg kg$^{-1}$ (11,12).

$LD_{Lo}$ oral guinea pig 670 mg kg$^{-1}$ (13).

$LD_{50}$ intraperitoneal mouse 1000 mg kg$^{-1}$ (14).

**Carcinogenicity and long-term effects**

No adequate evidence for carcinogenicity to humans, limited evidence for carcinogenicity to animals, IARC classification group 3 (15).

Oral (2 yr) rat 25 mg kg$^{-1}$ day$^{-1}$ no toxic effects, 250 mg kg$^{-1}$ day$^{-1}$ caused slight toxic effects (16).

National Toxicology Program tested rats and mice via feed (maximum dose rats 1500 ppm, mice 10,000 ppm) no evidence for carcinogenicity in ♂, ♀ mice and ♀ rats. Positive evidence for carcinogenicity in ♂ mice: tumours found were kidney tubular cell adenocarcinoma (1/50), kidney tubular cell adenoma (2/50) and liver neoplastic nodules/carcinoma (6/49) (17).

Oral (78 wk) ♂, ♀ mice (strains (C57BL/6XC3H/Anf) F$_1$, (C57BL/6XAKR) F$_1$) 215 mg kg$^{-1}$ body weight in 0.5% gelatin via gavage at 7 days of age daily for up to

4 wk and then 517 mg kg$^{-1}$ in diet for the rest of the experimental period. A significant tumour incidence was seen for lung adenomas in ♂ of the 2nd strain only (18,19).
Gavage (13 month) random bred and C57BL mice 6 mg animal$^{-1}$. 13/23 random bred mice developed tumours and 7/25 C57BL mice developed tumours compared to the controls of 0 and 1 respectively. Tumour sites included: stomach, lung and kidney in mice; and intestine, liver, lung and kidney in rats. Survival of the controls was unspecified (20).

**Metabolism and pharmacokinetics**
Oral rat 875 mg kg$^{-1}$ peak blood concentration 2 hr after administration. Distribution was even throughout the body. Monuron-related products were secreted in milk of lactating animals and excreted in urine. Oral rat 175 mg kg$^{-1}$ day$^{-1}$ for 60 days or 0.1-0.2 mg kg$^{-1}$ day$^{-1}$ for 6 months showed tissue retention of monuron-related substances of lungs >heart >liver, brain, kidneys >milk, bone marrow, thyroid gland (21).
In mammals metabolised by oxidative $N$-demethylation, hydroxylation of the aromatic nucleus and fission of the urea residue to give chloroaniline derivatives. The principal urinary metabolites were: $N$-(4-chlorophenyl)urea; $N$-(2-hydroxy-4-chlorophenyl)-$N'$-methylurea; $N$-(2-hydroxy-4-chlorophenyl)urea; and, $N$-(3-hydroxy-4-chlorophenyl)urea accounting for 14.5, 1.5, 6.5 and 2.2% of the initial dose, respectively (22,23).

## Genotoxicity
*Salmonella typhimurium* TA98, TA100, TA1535, TA1537 with and without metabolic activation negative (24).
*In vitro* Chinese hamster ovary cells with and without metabolic activation sister chromatid exchanges positive, chromosomal aberrations negative without metabolic activation and positive with metabolic activation (25).

## Legislation
As of July 1973 no longer registered for use on agricultural crops in the US (26).
Limited under EC Directive on Drinking Water Quality 80/778/EEC. Pesticides: maximum admissible concentration 0.1 µg l$^{-1}$ (27).
Included in Schedule 6 (Release into Land: Prescribed Substances) Statutory Instrument No. 472, 1991 (28).

## Any other comments
Hazards reviewed (29).
Accumulation of pesticides and their association with fish fertility disorders reviewed (30).

## References
1. McCoy, G. D. et al *Carcinogenesis (London)* 1990, **11**(7), 1111-1117
2. *Environmental Properties of Chemicals* 1991, Research Report 91, 384, Ministry of the Environment, Finland
3. Paterson, M. D. et al *Lett. Appl. Microbiol.* 1988, **7**(4), 87-90
4. Kaiser, K. L. E. et al *Water Pollut. Res. J. Can.* 1991, **26**(3), 341-431
5. Walsh, G. E. *Hyacinth Control J.* 1972, **10**, 45-48
6. Ukeles, R. *Appl. Microbiol.* 1962, **10**(6), 532-537

7. Parr, J. F. *Pestic. Soil Water* 1974, 321-340
8. Edwards, C. A. *Residue Rev.* 1966, **13**, 83
9. Eichelberger, J. W. et al *Environ. Sci. Technol.* 1971, **5**(6). 541-544
10. Mohamed, A. E.-D. et al *Water Res.* 1977, **11**, 617-620
11. *Fundam. Appl. Toxicol.* 1986, **7**, 299
12. Bailey, G. W. et al *Residue Rev.* 1965, **10**, 97
13. *Pharmacol. Rev.* 1962, **14**, 225
14. *NTIS Report no. AD277-689*, Natl. Tech. Inf. Serv., Springfield, VA
15. *IARC Monograph* 1987, **Suppl.7**, 67
16. Hodge, H. C. et al *Arch. Ind. Health* 1958, **17**, 45-47
17. Haseman, J. K. et al *Environ. Mol. Mutagen.* 1990, **16**(Suppl.18), 15-31
18. Innes, J. R. M. et al *J. Natl. Cancer Inst.* 1969, **42**, 1101-1114
19. NTIS *Evaluation of Carcinogenic, Teratogenic and Mutagenic Activities of Selected Pesticides and Industrial Chemicals* 1968, Vol. 1, Washington, DC
20. Rubenchik, B. L. et al *Vopr. Onkol.* 1970, **16**, 51-53
21. Fridman, E. B. *Zdravookhr. Turkm.* 1968, **12**, 25-27
22. Ernst, W. *J. S. Afr. Chem. Inst.* 1969, **22**, 579-588
23. Ernst, W. et al *Food Cosmet. Toxicol.* 1965, **3**, 789-796
24. Haworth, S. et al *Environ. Mutagen.* 1983, **1**(Suppl.) 3-142
25. Galloway, S. M. et al *Environ. Mol. Mutagen.* 1987, **10**(Suppl.10), 1-175
26. *Initial Scientific and Mini-economic Review of Monuron. Substitute Chemical Program* 1975, EPA-540/1-75-028, US EPA, Washington, DC
27. *EC Directive Relating to the Quality of Water Intended for Human Consumption* 1982, 80/778/EEC, Office for Official Publications of the European Communities, 2 rue Mercier, L-2985 Luxembourg
28. *S. I. 1991 No. 472 The Environmental Protection (Prescribed Processes and Substances) Regulations* 1991, HMSO, London
29. *Dangerous Prop. Ind. Mater. Rep.* 1989, **9**(1), 44-50
30. Popova, G. U. et al *Eksp. Vodn. Toksikol.* 1987, **12**, 191-201 (Russ.) (*Chem. Abstr.* **110**, 2498g)

# M331   Monuron TCA

**CAS Registry No.** 140-41-0
**Synonyms** acetic acid, trichloro-, compound with
$N'$-(4-chlorophenyl)-$N,N$-dimethylurea (1:1); GC-2996; Urox, Urox 379
**Mol. Formula** $C_{11}H_{12}Cl_4N_2O_3$                     **Mol. Wt.** 362.04
**Uses** Herbicide. Inhibitor of photosynthesis.

## Physical properties

**M. Pt.** 78-81°C.

---

## Mammalian and avian toxicity

### Acute data
LD$_{50}$ oral rat 2300 mg kg$^{-1}$ (1).
LD$_{50}$ subcutaneous rabbit 1000 mg kg$^{-1}$ (2).

### Legislation
Included in Schedule 6 (Release into Land: Prescribed Substances) Statutory Instrument No. 472, 1991 (3).
Limited under EC Directive on Drinking Water Quality 80/778/EEC. Pesticides: maximum admissible concentration 0.1 µg l$^{-1}$ (4).

### Any other comments
It is toxic to beans for 5 or 12 months, respectively, after application in March at 20 or 50 kg ha$^{-1}$, and for 12 months (at both application levels), when applied in July (5).

### References
1. Frear, E. H. (Ed.) *Pesticide Index* 4th ed., 1969, 292, College Science Publications, State College, PA
2. *Farm Chem. Handbook* 1983, C249, Meister Publishing Co., Willoughby, OH
3. *S. I. 1991 No. 472 The Environmental Protection (Prescribed Processes and Substances) Regulations* 1991, HMSO, London
4. *EC Directive Relating to the Quality of Water Intended for Human Consumption* 1982, 80/778/EEC, Office for Official Publications of the European Communities, 2 rue Mercier, L-2985 Luxembourg
5. Alan, P.B. *Tr. Sukhum. Opyt. Sta. Efiromaslich. Kul't.* 1968, **7**, 87-94 (*Chem. Abstr.* **70**, 86496w)

# M332  Morphine

CAS Registry No. 57-27-2
**Synonyms** morphinan-3,6-diol, 7,8-didehydro-4,5-epoxy-17-methyl-, (5α,6α)-; Duromorph; Morphia; (–)-morphine; Morphium; Ospalivina
**Mol. Formula** C$_{17}$H$_{19}$NO$_3$          **Mol. Wt.** 285.35
**Uses** Narcotic analgesic.
**Occurrence** Principal alkaloid of opium.

## Physical properties
**M. Pt.** 197°C; **B. Pt.** 254°C.

## Mammalian and avian toxicity

### Acute data
$LD_{50}$ oral rat, mouse 335, 524 mg kg$^{-1}$, respectively (1,2).
$LD_{50}$ subcutaneous mouse 220 mg kg$^{-1}$ (3).
$LD_{50}$ intravenous rat 140 mg kg$^{-1}$ (4).
$LD_{50}$ intraperitoneal rat, mouse 160, 140 mg kg$^{-1}$, respectively (5,6).

### Teratogenicity and reproductive effects
Oral rat 10, 35 or 70 mg kg$^{-1}$ day$^{-1}$. The pregnancy rate was reduced at the intermediate and high doses to 57 and 6%, respectively. No teratogenic effects were observed at any dosage, but growth retardation was present at the intermediate dose (7). Infusion lambs (10 day) 3 mg hr$^{-1}$ did not affect foetal survival or the response of foetal breathing movements to hypercapria. Respiratory effects were observed which may be related to accumulation of morphine-3-$\beta$-D-glucuronide. Higher doses 10 and 30 mg hr$^{-1}$ caused seizures and decreased foetal survival (8).

### Metabolism and pharmacokinetics
Morphine salts are well absorbed from the gastro-intestinal tract in humans but have poor oral bioavailability since they undergo first-pass metabolism in the liver and gut. It is conjugated with glucuronic acid in the liver and gut to produce morphine-3-glucuronide which is inactive, whereas the active metabolite is morphine-6-glucuronide. Other active metabolites are normorphine, codeine and morphine ethereal sulphate. It is distributed throughout the body but mainly in the kidneys, liver, lungs and spleen with lower concentrations in the brain and muscles. Mean plasma elimination $t_{1/2}$ 1.7 hr. Up to 10% of a dose may eventually be excreted, as conjugates, through the bile into the faeces. The remainder is excreted in the urine mainly as conjugates. $\approx$90% is excreted in 24 hr with traces in urine $\geq$48 hr (9).
It was well absorbed from rat gastrointestinal tract in the order jejunum>duodenum>ileum>middle intestine>rectum, but it was poorly absorbed from the stomach (10).
Microsomal cytochrome $P_{450}$ linked metabolism plays a minor role in the hepatic toxicity of morphine in rats, whilst morphine-6-dehydrogenase plays a major part in this toxicity (11).

## Any other adverse effects to man
Nausea, vomiting, constipation, drowsiness and confusion occur with normal doses. Tolerance generally develops with long term use. Larger doses produce respiratory depression and hypotension with circulatory failure and deepening coma. Pulmonary oedema after overdosage is a common cause of fatalities among opiate addicts.
Rhabdomyolysis progressing to renal failure has been reported in overdosage (9).
A report of severe rectovaginal spasm following intrathecal administration. The spasms were successfully controlled with midazolan (12).
In acute oral poisoning the stomach should be emptied. A laxative may be given to aid peristalsis (13).
Adult human hepatocytes incubated with morphine resulted in cytotoxic effects which were observed at $\approx$100 times the plasma concentration required to produce analgesia. 285 mg l$^{-1}$ reduced the glycogen content by 50% and 228 mg l$^{-1}$ inhibited albumen synthesis by $\approx$50% after 24 hrs of pretreatment. Intracellular glutathione was reduced by 50% after 2-3 hrs of incubation with 570 mg l$^{-1}$, opiate doses during tolerance or abuse may be a cause of liver dysfunction (14).

## Any other adverse effects

0.002-1.0 mg injected into the cerebral ventricles of unanaesthetized cats evoked vomiting, lasting for 1-7 minutes, number of vomitings induced ranged from 2-5 (15). Intraperitoneal 5-, 10- and 20-day old rats 10 mg kg$^{-1}$, overt sedation in all 3 age groups and induced catalepsy which was particularly apparent in the 5- and 10-day old animals (16).

It has no direct peripheral effects on heart rate or blood vessel tone nor has it any effect on norepinephrine and epinephrine release from the sympathetic nerves and the adrenal medulla in the rat (17).

Subcutaneous mice 20 mg kg$^{-1}$ showed renal elimination of phenol red in mice, tolerance was more readily induced to the effects of narcotics on venal blood flow and tubular function than the reduction of glomerular function (18).

♂/♀ deer mice 10 mg kg$^{-1}$ produced maximum analgesic responses in adults. There are significant sex and population differences in opiate-induced analgesia in young and adult deer mice (19).

## Legislation

Included in Schedule 4 (Release Into Air: Prescribed Substances) Statutory Instrument No. 472, 1991 (20).

Restricted drug (9).

## Any other comments

Metabolism reviewed (21).

Abuse leads to habituation or addiction (9).

## References

1. *Drugs of the Future* 1977, **2**, 39
2. *Arzneim.-Forsch.* 1974, **24**, 200
3. *Arzneim.-Forsch.* 1958, **8**, 1958
4. *Br. J. Pharmacol. Chemother.* 1952, **7**, 196
5. *Pharmazie.* 1976, **31**, 655, (Ger.)
6. *Jpn. J. Pharmacol.* 1957, **53**, 565
7. Fujinaga, M. et al *Teratology* 1988, **38**(5), 401-410
8. Olsen, G. D. et al *J. Pharmacol. Exp. Ther.* 1988, **247**(1), 162-168
9. *Martindale. The Extra Pharmacopoeia* 30th ed., 1993, The Pharmaceutical Press, London
10. Tan, T. et al *Chem. Pharm. Bull.* 1989, **37**(1), 168-173
11. Ohno, Y. et al *Biochem. Pharmacol.* 1988, **37**(14), 2862-2863
12. Littrell, R. A. et al *Clin. Pharm.* 1992, **11**, 57-59
13. Henry, J. et al *Br. Med. J.* 1984, **289**, 990-993
14. Gomez-Lechion, M. J. et al *Mol. Toxicol.* 1987, **1**(4), 453-463
15. Beleslin, D. B. et al *Period. Biol.* 1986, **88**(2), 102-103
16. Jackson, H. C. et al *Psychopharmacology (Berlin)* 1989, **97**(3), 404-409
17. Feuerstein, G. et al *Neuropeptides (Edinburgh)* 1987, **9**(2), 139-150
18. Garty, M. et al *J. Pharmacol. Exp. Ther.* 1986, **239**(2), 346-350
19. Kavaliers, M. et al *Brain Res.* 1990, **516**(2), 326-331
20. *S. I. 1991 No. 472 The Environmental Protection (Prescribed Processes and Substances) Regulations* 1991, HMSO, London
21. Oguri, K. et al *Eisei Kagaku* 1986, **32**(6), 413-426

# M333 Morpholine

**CAS Registry No.** 110-91-8
**Synonyms** diethylene imidoxide; diethylene oximide; Drewamine;
tetrahydro-*p*-oxazine; tetrahydro-1,4-oxazine
**Mol. Formula** $C_4H_9NO$                    **Mol. Wt.** 87.12
**Uses** Rubber accelerator. Solvent. Organic synthesis. Additive to boiler water. Waxes
and polishes. Corrosion inhibitor. Optical brightener for detergents.

## Physical properties

**M. Pt.** –7 to –5°C; **B. Pt.** 128.9°C; **Flash point** 35°C (open cup); **Specific gravity**
$d_4^{20}$ 1.007; **Partition coefficient** log $P_{ow}$ –1.08 (1); **Volatility** v.p. 10 mmHg at 23°C;
v. den 3.00.

## Solubility

Water: miscible. Organic solvent: acetone, benzene, diethyl ether, ethanol, 2-hexanone

## Occupational exposure

**US TLV (TWA)** 20 ppm (71 mg m$^{-3}$); **UK Long-term limit** 20 ppm (70 mg m$^{-3}$);
**UK Short-term limit** 30 ppm (105 mg m$^{-3}$); **UN No.** 2054; **HAZCHEM Code** 2P;
**Conveyance classification** flammable liquid; **Supply classification** corrosive.
**Risk phrases** ≥25% – Flammable – Harmful by inhalation, in contact with skin and
if swallowed – Causes burns – ≥10%<25% – Flammable – Causes burns – ≥1%<10%
– Flammable – Irritating to eyes and skin (R10, R20/21/22, R34, R10, R34, R10, R36/38)
**Safety phrases** Do not breathe vapour – Wear suitable protective clothing (S23, S36)

## Ecotoxicity

**Fish toxicity**
$LC_{50}$ (96 hr) bluegill sunfish, static bioassay in fresh water at 23°C, 350 ppm (2).
$LC_{50}$ (96 hr) bluegill sunfish, static bioassay in synthetic seawater at 23°C, 400 ppm (2).

**Invertebrate toxicity**
$EC_{50}$ (5, 15 or 30 min) *Photobacterium phosphoreum* concentration 60, 51 or
57 mg l$^{-1}$, respectively, Microtox test (3).
Cell multiplication inhibition test *Pseudomonas putida* 310 mg l$^{-1}$, *Microcystis
aeruginosa* 1.7 mg l$^{-1}$, *Entosiphon sulcatum* 12 mg l$^{-1}$ (4,5).
$LC_{50}$ (24 hr) *Daphnia magna* 119 mg l$^{-1}$ (6).

## Environmental fate

**Degradation studies**
$BOD_5$: 0.9% ThOD (7).
$BOD_{15}$: 4.0% ThOD (7).
$BOD_{20}$: 5.1% ThOD (8).

No biodegradation of morpholine 10, 50 and 100 mg $l^{-1}$ was observed after 14 days incubation at 20°C in media inoculated with river mud (9).

*Mycobacterium* MorG, enzymes for ethanolamine, glycollate and pyrrolidine catabolism were strongly induced. Catabolised initially by an analogous route to pyrrolidine, producing 2-(2-aminoethoxy)acetate which can be oxidatively cleaved to give rise directly to glycolate and indirectly to ethanolamine (10).

### Abiotic removal
Computer estimated $t_{1/2}$ 4 hr for the reaction of morpholine with hydroxyl radicals in the atmosphere (11).

### Absorption
The estimated soil sorption coefficient is 8. This indicates that morpholine will not absorb strongly to soil (12).

## Mammalian and avian toxicity

### Acute data
$LD_{50}$ oral ♀ rat 1.05 g $kg^{-1}$ (13).
$LD_{50}$ oral guinea pig 0.9 g $kg^{-1}$ (14).
$LC_{50}$ (1 hr) inhalation rat 22.2 mg $l^{-1}$ (15).
Inhalation lethality (8 hr) rats >8000 ppm (16).
$LD_{50}$ dermal rabbit 0.5 ml $kg^{-1}$ (15).

### Sub-acute data
Oral (4 wk) ♂ rats 323 mg $kg^{-1}$ $day^{-1}$ indicated an increase in weight of adrenal glands and a lower mean body weight gain. Lower concentrations 27.6 and 93.1 mg $kg^{-1}$ $day^{-1}$ had no apparent effects. Histopathology was not mentioned but no gross lesions were found (17).
Dermal (30 day) guinea pigs 0.9, 0.18 or 0.27 g $kg^{-1}$ no lesions were observed, the skin was thickened at the application site (18).

### Carcinogenicity and long-term effects
In the stomach the reaction of morpholine with sodium nitrite produces *N*-nitrosomorpholine which is a known carcinogen, producing stomach, liver and lung tumours (19-21).

### Metabolism and pharmacokinetics
It is not readily metabolised in rat, dog and rabbit (22–27).
It is metabolised extensively in guinea pigs via *N*-methylation and *N*-oxidation (24).
87% of an administered dose was excreted in the urine of rats within 24 hr (23).
Intraperitoneal rat, hamster and guinea pig 125 mg $kg^{-1}$ blood plasma $t_{1/2}$ 115, 120 and 30 min, respectively. In all 3 species ≈80% was excreted in urine within 24 hr (24).

### Irritancy
Eyes rabbit, it has caused corneal conjuctivitis, clouding and general conjunctivitis (dose and duration unspecified) (28).
Eye rabbits 40% solution caused severe corneal necrosis (duration unspecified) (29).
Rabbits, undiluted morpholine caused necrosis within 24 hr (29).
In humans it is a skin, eye and mucous membrane irritant and a skin sensitiser (30).

### Sensitisation
10% did not cause sensitisation in guinea pigs (31-33).

## Genotoxicity

*Salmonella typhimurium* (activation and strain unspecified), negative (34).
*Salmonella typhimurium* TA98, TA100 and TA1530 with or without metabolic activation negative (35).
No increase in the number of chromosomal aberrations in the blood cells of workers exposed to 0.54-0.93 or 0.74-2.14 mg m$^{-3}$ (36).

## Any other adverse effects

Inhalation rabbit 250 ppm induced enzyme changes in pulmonary lavage fluids that may be related to pulmonary damage (37).
Inhalation rabbit 250 ppm for 33 exposures, maximum induction of α-mannosidase and acid phosphatase activities in ♀ rabbit was 1.7-fold and 2-fold, respectively, and in ♂ rabbit, 3-fold and unchanged, respectively (38).
In rats, high concentrations of vapours have caused death due to lung congestion, with degeneration, fatty changes and cellular necrosis of the liver and kidneys (39).
Subchronic inhalation studies at 25, 100 and 250 ppm for 6 hr day$^{-1}$, 5 day wk$^{-1}$ for 13 wk indicated salivation and nasal damage at the 2 higher concentrations although no haematological or organ weight changes were noted (40).

## Legislation

Included in Schedule 4 (Release Into Air: Prescribed Substances) Statutory Instrument No. 472, 1991 (41).

## Any other comments

Morpholine concentration in cigarette smoke condensate was 0.08 μg cigarette (42).
Human health effects, experimental toxicity, environmental effects, ecotoxicology, exposure levels, epidemiology and workplace exposure reviewed (43).
Autoignition temperature 310°C.

## References

1. Verschueren, K. *Handbook of Environmental Data on Organic Chemicals* 2nd ed., 1983, Van Nostrand Reinhold, New York
2. Dawson, G. W. et al *J. Hazard. Mater.* 1975/77, **1**, 303-318
3. Kaiser, K. L. E. et al *Water Pollut. Res. J. Can.* 1991, **26**(3), 361-431
4. Bringmann, G. et al *Water Res.* 1980, **14**, 231-241
5. Bringmann, G. et al *GWF- Wasserfach: Wasser/Abwasser* 1976, **117**(9)
6. Calamari, D. et al *Princ. Interp. Results Test. Proc. Ecotox.* 1982, Report EUR 7549 EN/FR
7. Lamb, C. et al *Proc. 7th Ind. Waste Conf. Purdue Eng. Bull.* 1953, 79
8. Lund, H. F. *Industrial Pollution Control Handbook* 1971, 14-21, McGraw-Hill, New York
9. Callamari, D. et al *Chemosphere* 1980, **9**, 753-762
10. Swain, A. et al *Appl. Microbiol. Biotechnol.* 1991, **35**(1), 110-114
11. Howard, P. H. (Ed.) *Handbook of Fate & Exposure for Organic Chemicals* 1990, **2**, 357, Lewis Publishers, Chelsea, MI
12. Kenaga, E. E. *Ecotoxicol Environ Saf.* 1980, **4**, 26-38
13. Smyth, et al *Ind. Hyg. Occup. Med.* 1954, **10**, 61
14. Patty, F. H. *Industrial Hygiene and Toxicology* 1967, **2**, Interscience Publishers, New York
15. *Toxic and Hazardous Industrial Safety Manual* 1977, 350, International Technical Information Institute, Japan
16. Kennedy, G. L. et al *Toxicol. Lett.* 1991, **56**(3), 317-326
17. *BIO.FAX Data Sheet 10-4/70* 1970, Industrial Bio Test Laboratories, Northbrook, IL
18. Shea, T. E. Jr. *J. Ind. Hyg. Toxicol.* 1939, **21**(7), 236-245
19. Newberne, P. M. et al *Food Cosmet. Toxicol* 1973, **11**(5), 819-825

20. Mirvish, S. S. *J. Natl. Cancer Inst.* 1983, **71**(1), 81-85
21. Shank, R. C. et al *Food Cosmet. Toxicol* 1976, **14**, 1-8
22. Tanaka, A. et al *J. Food Hyg. Soc.* 1978, **19**(3), 329-334
23. Maller, R. K. et al *Cancer Res.* 1957, **17**, 284
24. Sohn, O. S. et al *Toxicol. Appl. Pharmacol.* 1982, **64**(3), 486-491
25. Ohnishi, T. *Japan. J. Hyg.* 1984, **39**(4), 729-748
26. Rhodes, C. et al *Xenobiotica* 1977, **7**(1-2), 112
27. Van Stee, E. W. et al *Toxicology* 1981, **20**(1), 53-60
28. Grant, W. M. *Toxicology of the Eye: Drugs, Chemicals, Plants, Venoms* 2nd ed., 1974, C. C. Thomas, Springfield, IL
29. Smyth, H. F. et al *V. Arch. Ind. Hyg. Occup. Med.* 1954, **10**, 61-68
30. Cosmetic, Toiletry and Fragrance Assoc. *J. Am. Coll. Toxicol.* 1989, **8**(4), 707-748
31. Wang, X. et al *Shangai Yike Daxue Xuebao* 1986, **13**(3), 209-212, (Ch.)
32. Wang, X. et al *Shangai Yike Daxue Xuebao* 1986, **13**(5), 376-382, (Ch.)
33. Wang, X. et al *Contact Dermititis* 1988, **19**(1), 11-15
34. Zeiger, E. *Mutat. Res.* 1971, **12**(4), 472-474
35. Khudoley, V. V. et al *Arch. Geschwulstforsch.* 1987, **57**(6), 453-462
36. Katosova, L. D. et al *Gig. Tr. Prof. Zabol.* 1991, **6**, 35-36, (Russ.)
37. Tombropoulos, E. G. et al *Fed. Proc.* 1979, 38/3, No. 1873
38. Tombropoulos, E. G. et al *Toxicol. Appl. Pharmacol.* 1983, **70**(1), 1-6
39. Shea, T. E. *J. Ind. Hyg. Toxicol.* 1939, **21**, 236
40. Conaway, C. C. et al *Fundam. Appl. Toxicol.* 1984, **4**(3 part 1), 465-472
41. *S. I. 1991 No. 472 The Environmental Protection (Prescribed Processes and Substances) Regulations* 1991, HMSO, London
42. Singer, G. M. et al *J. Agric. Food Chem.* 1976, **24**, 553-555
43. *ECETOC Technical Report No. 30(4)* 1991, European Chemical Industry Ecology and Toxicology Centre, B-1160 Brussels

# M334    5-(Morpholinomethyl)-3-[(5-nitro-furfurylidene) amino]-2-oxazolidinone

**CAS Registry No.** 139-91-3
**Synonyms** 2-oxazolidinone, 5-(4-morpholinylmethyl)-3-[[(5-nitro-2-furanyl)-methylene]amino]-; Altabactina; Furaladone; Furzoline; Furmethanol; Ibifur; Sepsinol
**Mol. Formula** $C_{13}H_{16}N_4O_6$                    **Mol. Wt.** 324.30
**Uses** Antibacterial agent.

## Physical properties

**B. Pt.** 206°C (decomp.).

**Solubility**
Water: 750 mg $l^{-1}$ at 25°C.

## Mammalian and avian toxicity

### Acute data

$LD_{50}$ intraperitoneal mouse 1000 mg $kg^{-1}$ (1).
$LD_{50}$ (route unspecified) mice 525 mg $kg^{-1}$ (2).

### Metabolism and pharmacokinetics

Oral rat 100 mg $kg^{-1}$ body weight, plasma levels of 3.2 mg $l^{-1}$ were observed after 4 hr. When given 138 mg $kg^{-1}$ ≈3.4% was recovered from urine within 48 hr (3).

### Any other adverse effects

Oral chicken therapeutic levels, blebs of an electron dense substance developed in adrenal cortical mitochondria and aspartate transaminase activity was reduced (4).

### Legislation

Included in Schedule 4 (Release Into Air: Prescribed Substances) Statutory Instrument No. 472, 1991 (5).

### References

1. *J. Pharm. Pharmacol.* 1964, **16**, 663
2. Paul, H. E. et al *Exp. Chemother.* 1964, **2**(1), 307-370
3. Paul, M. F. et al *Antibiot. Chemother.* 1960, **10**, 287-302
4. Barltet, A. L. et al *J. Vet. Pharmacol. Ther.* 1990, **13**(2), 206-216
5. *S. I. 1991 No. 472 The Environmental Protection (Prescribed Processes and Substances) Regulations* 1991, HMSO, London.

# M335   Morphothion

**CAS Registry No.** 144-41-2

**Synonyms** phosphorodithioic acid, *O,O*-dimethyl *S*-[2-(4-morpholinyl)-2-oxoethyl] ester; Ekatin F; Ekatin M; Morphotox; *O,O*-dimethyl *S*-(morpholino-carbonylmethyl) phosphorodithioate; *O,O*-dimethyl *S*-[2-(4-morpholinyl)- 2-oxoethyl] phosphorodithioate; phosphorodithioic acid, *O,O*-dimethyl ester, *S*-ester with 4-(mercaptoacetyl)morpholine

**Mol. Formula** $C_8H_{16}NO_4PS_2$                      **Mol. Wt.** 285.32

**Uses** Superseded pesticide.

## Physical properties

**M. Pt.** 65°C.

### Solubility

Organic solvent: acetone, 1,4-dioxane, acetonitrile

## Occupational exposure

**Supply classification** toxic.

**Risk phrases** Toxic by inhalation, in contact with skin and if swallowed (R23/24/25)

**Safety phrases** Keep out of reach of children – Keep away from food, drink and animal feeding stuffs – If you feel unwell, seek medical advice (show label where possible) (S2, S13, S44)

## Mammalian and avian toxicity

### Acute data

$LD_{50}$ oral mouse, rat, rabbit 130-190 mg $kg^{-1}$ (1,2).
$LD_{50}$ dermal rat 283 mg $kg^{-1}$ (3).

## Legislation

Included in Schedule 4 (Release Into Air: Prescribed Substances) Statutory Instrument No. 472, 1991 (4).
Limited under EC Directive on Drinking Water Quality 80/778/EEC. Pesticides: maximum admissible concentration 0.1 µg $l^{-1}$ (5).

## Any other comments

Environmental aspects reviewed (6).

## References

1. Frear, E. H. (Ed.) *Pesticide Index* 5th ed., 1976, 159, College Science Publications, State College, PA
2. *Bull. Entomol. Soc. Am.* 1966, **12**, 161
3. *World Rev. Pest Control* 170, **9**, 119
4. *S. I. 1991 No. 472 The Environmental Protection (Prescribed Processes and Substances) Regulations* 1991, HMSO, London
5. *EC Directive Relating to the Quality of Water Intended for Human Consumption* 1982, 80/778/EEC, Office for Official Publications of the European Communities, 2 rue Mercier, L-2985 Luxembourg
6. U.K. Dept. of the Environment *Methods Exam. Waters Assoc. Mater.* 1986, (Organo-Phosporous Pestic. Sewage Sludge; Organo-Phosphorous Pestic. River Drinking Water, Addit. 1985), 1-20

# M336  MSMA

## CH3As(=O)(OH)ONa

**CAS Registry No.** 2163-80-6

**Synonyms** arsonic acid, methyl-, monosodium salt; Ansar 170; Daconate; Gepiron; monosodium acid methanearsonate

**Mol. Formula** $CH_4AsO_3Na$           **Mol. Wt.** 161.95

**Uses** Herbicide.

## Physical properties

**M. Pt.** 113-116°C (sesquihydrate);132-139°C(hexahydrate).

**Solubility**
Water: 1.4 kg kg$^{-1}$ at 20°C. Organic solvent: methanol

## Ecotoxicity

**Fish toxicity**
LC$_{50}$ (48 hr) bluegill sunfish >1000 mg l$^{-1}$ (1).

**Invertebrate toxicity**
Mixed culture of *Citrobacter freundii*, *Aeromonas* sp. and *Klebsiella* sp. isolated from soil LC$_{50}$ (96 hr) 220 mg l$^{-1}$, LC$_{50}$ (48 hr) 60 mg l$^{-1}$ and LC$_{50}$ (24 hr) 27 mg l$^{-1}$ (2). LC$_{50}$ juvenile cray fish and adult crayfish 101 and 1019 ppm, respectively (3).

## Environmental fate

**Degradation studies**
Degradation via oxidative demethylation by microbial population. Degradation rate is reduced at high clay contents, high adsorptive capacity for monosodium methanearsonate, and at low clay content, lower microbial population. 30°C at 20 and 150% field capacity the t$_{1/2}$ ranged from 88-178 and 25-178 days, respectively (4).

**Absorption**
It is adsorbed by clay soils (5).
It is fixed by iron and aluminium in the soil (5).

## Mammalian and avian toxicity

**Acute data**
LD$_{50}$ oral rat 700 mg kg$^{-1}$ (6).
LD$_{50}$ oral mouse 1800 mg kg$^{-1}$ (7).
LD$_{50}$ oral rabbit 102 mg kg$^{-1}$ caused constipation, diarrhoea, oligourea, generalised weakness and the loss of appetite (8).
LD$_{50}$ (route unspecified) mammal 50 mg kg$^{-1}$ (9).
LD$_{50}$ (route unspecified) snowshoe hare 173 mg kg$^{-1}$ (10).

**Metabolism and pharmacokinetics**
Oral rabbit (12 wk) 54% was eliminated in urine 46% in faeces (10).
Dermal ♀ rat young and adult (dose unspecified) significantly reduced skin penetration in young animals; parallel dose-absorption curves in young and adult rats indicated a lack of significant dose effect (11).

**Irritancy**
Dermal rabbit 54 mg, uncovered, caused mild irritation and eye rabbit 34 mg caused mild irritation (12).

## Genotoxicity

*Salmonella typhimurium* TA97, TA98, TA100 and TA1535 with or without metabolic activation, negative (13).
*Allium cepa* root meristems (1 hr), a marked clastogenic effect (14).

## Legislation

Included in Schedule 4 (Release Into Air: Prescribed Substances) Statutory Instrument No. 472, 1991 (15).
Limited under EC Directive on Drinking Water Quality 80/778/EEC. Maximum admissible concentration 50 μg As l$^{-1}$ (16).

# References

1. *The Agrochemicals Handbook* 3rd ed., 1991, RSC, London
2. Anderson, A. C. et al *Bull. Environ. Contam. Toxicol.* 1980, **24**, 124-127
3. Naqvi, S. M. et al *Environ. Pollut.* 1987, **48**(4), 275-283
4. Akram, K. H. et al *Weed Sci.* 1986, **34**, 781-787
5. Hitbold, A. E. *Arsenical Pesticides (ACS Symp. Series 7)* 1975, 53-69, American Chemical Society, Washington, DC
6. National Acadamy of Sciences *Arsenic, Medical and Biological Effects of Environmental Pollutants* 1977, Assembly of Life Sciences, National Research Council, Washington, DC
7. *Farm Chemicals Handbook* 1983, C163, Meister Publishing Co., Willoughby, OH
8. Jaghabir, M. T. W. et al *Bull. Environ. Contam. Toxicol.* 1988, **40**(1), 119-122
9. Melnikov, N. N. *Chemistry of Pesticides* 1971, 260, Springer-Verlag, New York
10. Exon, J. H. et al *Nutr. Rep. Int.* 1974, **9**, 351-357
11. Shah, P. V. et al *J. Toxicol. Environ. Health* 1987, **21**, 353-366
12. *Ciba-Geigy Toxicol. Data/Indexes* 1977, Ciba-Geigy Corp., Ardsley, New York
13. Rao, B. V. et al *Cytologia* 1988, **53**(2), 255-261
14. Zeiger, E. et al *Environ. Mol. Mutagen.* 1988, **11**(Suppl. 12), 1-157
15. *S. I. 1991 No. 472 The Environmental Protection (Prescribed Processes and Substances) Regulations* 1991, HMSO, London
16. *EC Directive Relating to the Quality of Water Intended for Human Consumption* 1982, 80/778/EEC, Office for Official Publications of the European Communities, 2 rue Mercier, L-2985 Luxembourg

# M337    Muscimol

**CAS Registry No.** 2763-96-4
**Synonyms** 3(2*H*)-isoxazolone, 5-(aminomethyl)-; Agarin; Agarine; Pantherine
**Mol. Formula** $C_4H_6N_2O_2$          **Mol. Wt.** 114.10
**Uses** Molecular probe to study γ-aminobutyric acid receptors. Sedative. Antiemetic.
**Occurrence** Hallucinogenic agent found in poisonous mushrooms.

## Physical properties

**M. Pt.** 175°C (decomp.).

## Mammalian and avian toxicity

### Acute data

$LD_{50}$ oral mouse, rat 17, 45 mg $kg^{-1}$, respectively (1,2).
$LD_{50}$ subcutaneous mouse 3800 μg $kg^{-1}$ (1).
$LD_{50}$ intravenous rat 4500 μg $kg^{-1}$ (1).
$LD_{50}$ intravenous mouse 5620 mg $kg^{-1}$ (3).

---

$LD_{Lo}$ intravenous rabbit 10 mg $kg^{-1}$ (1).
$LD_{50}$ intraperitoneal mouse 2500 μg $kg^{-1}$ (2).

**Teratogenicity and reproductive effects**

Muscimol binding by membranes of neocortex synaptosomes from 2 month-old rats exposed in utero (days 5-20 of pregnancy), was 27% higher than in controls. Impaired γ-aminobutyric acid system may be the cause of behavioural teratogenicity in offspring of rats with induced alcoholism (4).

**Any other adverse effects to man**

Adverse effects usually occur within 2 hr of ingestion, symptoms include ataxia, euphoria, delirium and hallucinations. Fatalities are rare (5).

**Legislation**

Included in Schedule 4 (Release Into Air: Prescribed Substances) Statutory Instrument No. 472, 1991 (6).

**References**

1. *Arzneim.-Forsch.* 1968, **18**, 311
2. *J. Med. Chem.* 1980, **23**, 702
3. *U.S. Army Armament Research Development Med. NX#11824* Chemical Systems Laboratory, Aberdeen Proving Ground, MD
4. Zhulin, V. V. et al *Byull. Eksp. Biol. Med.* 1988, **106**(10), 460-462
5. *Martindale. The Extra Pharmacopoeia* 30th ed., 1993, The Pharmaceutical Press, London
6. *S. I. 1991 No. 472 The Environmental Protection (Prescribed Processes and Substances) Regulations* 1991, HMSO, London

# M338   Musk ambrette

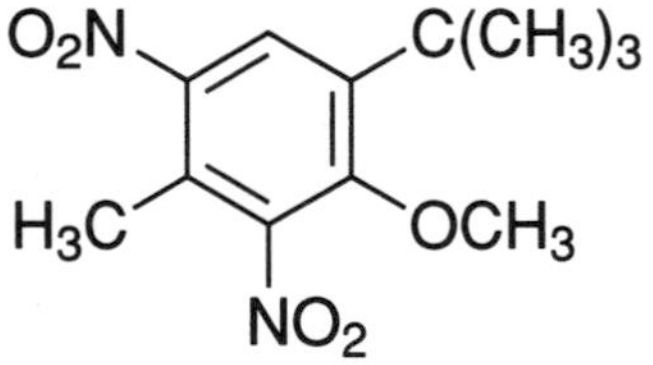

**CAS Registry No.** 83-66-9

**Synonyms** benzene, 1-(1,1-dimethylethyl)-2-methoxy-4-methyl-3,5-dinitro-; Amber musk; anisole, 6-*tert*-butyl-3-methyl-2,4-dinitro-

**Mol. Formula** $C_{12}H_{16}N_2O_5$          **Mol. Wt.** 268.27

**Uses** Perfumery.

**Occurrence** *Moschus moschiferus.*

**Physical properties**

**M. Pt.** 84-85°C; **B. Pt.** $_{16}$ 185°C; **Volatility** v.p. $5.9 \times 10^{-5}$ to $7.3 \times 10^{-3}$ mmHg at 30.3 to 72.3 °C.

## Mammalian and avian toxicity

**Acute data**

$LD_{50}$ oral rat 339 mg $kg^{-1}$ (1).

**Irritancy**

Dermal rabbit (24 hr) 500 mg caused moderate irritation (2).

**Sensitisation**

Guinea pig photoallergy model, 6 exposures over 2 wk period, contact photoallergy was detected (3).

## Genotoxicity

*Salmonella typhimurium* TA100 with metabolic activation, positive (4).

*Salmonella typhimurium* TA98, TA100, TA1535, TA1537 with or without metabolic activation, negative (5).

Induced a significant number of micronuclei in polychromatic erythrocytes of mouse bone marrow, and also induced chromosomal aberrations (4).

## Any other adverse effects to man

Causes photoallergic reactions in humans (6).

## Any other adverse effects

Neurotoxic and causes testicular atrophy in rats (7).

## Legislation

Included in Schedule 4 (Release Into Air: Prescribed Substances) Statutory Instrument No. 472, 1991 (8).

## Any other comments

Toxicity reviewed (9).

## References

1. *Food Cosmet. Toxicol.* 1964, **2**, 327
2. *Food Cosmet. Toxicol.* 1975, **13**, 681
3. Gerberick, G. F. et al *Contact Dermatitis* 1989, **20**(4), 251-259
4. Kayal, J. J. et al *N. Nitroso. Compd. Pap. Int. Symp. 1989* 1990, 131-138
5. Zeiger, E. et al *Environ. Mutagen.* 1987, **9**(Suppl. 9), 1-110
6. Lovell, W. W. et al *Int. J. Cosmet. Sci.* 1988, **10**(6), 271-279
7. Ford, R. A. et al *Food Chem. Toxicol.* 1990, **28**(1), 55-61
8. *S. I. 1991 No. 472 The Environmental Protection (Prescribed Processes and Substances) Regulations* 1991, HMSO, London
9. *BIBRA Toxicity Profile* 1991, British Biological Industrial Research Association, Carshalton

# M339   Mustard gas

$$S(CH_2CH_2Cl)_2$$

**CAS Registry No.** 505-60-2
**Synonyms** ethane, 1,1'-thiobis[2-chloroethane]; bis($\beta$-chloroethyl) sulfide; Kampstaff
Lost; Senfgas; sulfur mustard; Yperite
**Mol. Formula** $C_4H_8Cl_2S$                     **Mol. Wt.** 159.08
**Uses** In chemical warfare.

## Physical properties

**M. Pt.** 13-14°C; **B. Pt.** 215-217°C; **Flash point** 105°C; **Specific gravity** 1.2741;
**Volatility** v.p. $9.0 \times 10^{-2}$ mmHg at 30°C; v. den. 5.4.

## Solubility

Water: 0.68 g $l^{-1}$ at 25°C. Organic solvent: soluble in fat solvents

## Mammalian and avian toxicity

### Acute data

$LC_{50}$ (10 min) inhalation rat, mouse 100, 120 mg $m^{-3}$, respectively (1).
$LC_{50}$ (10 min) inhalation dog, monkey 70, 80 mg $m^{-3}$, respectively (1).
$LC_{Lo}$ (10 min) inhalation human 1500 mg $m^{-3}$ (2).
$LD_{Lo}$ dermal human 64 mg $kg^{-1}$ (3).
$LD_{50}$ dermal rat 5 mg $kg^{-1}$ (4).
$LD_{50}$ dermal rabbit 40 mg $kg^{-1}$ (1).
$LD_{50}$ subcutaneous rat 1500 $\mu$g $kg^{-1}$ (4).
$LD_{50}$ subcutaneous mouse 20 mg $kg^{-1}$ (1).
$LD_{50}$ intravenous dog 200 $\mu$g $kg^{-1}$ (1).

### Carcinogenicity and long-term effects

Sufficient evidence for carcinogenicity to humans, limited evidence for
carcinogenicity to animals, IARC classification group 1 (5).
The estimated potency factor for mustard gas is 1.3 relative to benzo[a]pyrene (6).
Inhalation 80 ♂/♀ mice 12.5 $cm^3$ $l^{-1}$ for 15 min. After 11 month all animals were
killed, 33 of mice had lung tumours. Number of tumours $mouse^{-1}$ was 0.66 (7).
Subcutaneous 40 mice 0.05 ml $wk^{-1}$ for 5.6 wks. At natural death 4 fibrosarcomas
were found at site of injection. Mammary tumours were found in 10 animals (8).
Intravenous 30 ♂/♀ mice 4 injections of 0.25 ml. All survivors killed at the age of 6
month 14/15 mice had pulmonary tumours (9).
Mustard gas inhibited the carcinogenic action of coal tar on mouse skin (10).
3354 humans who had worked in manufacture of mustard gas were traced for
mortality to the end of 1984. Large and highly significant excesses were observed as
compared with national death rates for deaths from cancer of the larynx, pharynx and
all other buccal cavity and upper respiratory sites combined. For lung cancer, a highly
significant but more moderate excess was observed (11).

### Teratogenicity and reproductive effects

Intragastric intubation rats and rabbits. Rats were dosed 6-15 days of gestation with
0, 0.5, 1.0 or 2.0 mg $kg^{-1}$, rabbits were dosed 6-19 days of gestation with 0, 0.4, 0.6 or
0.8 mg $kg^{-1}$. At necropsy in rats, reduction in body weight was observed in maternal

animals and their ♀ foetuses at the lowest dose, incidence of foetal malformations were not increased. In rabbits the highest dose induced maternal mortality and depressed body weight measures, but did not affect foetal developments. It is not teratogenic in rats and rabbits since foetal effects were observed at maternally toxic doses only (12).

Oral ♂/♀ rat 0.5 mg kg$^{-1}$, then mating between treated and untreated rats. An increase in early foetal resorptions and preimplantation losses and decrease of total live embryo implants; a significant increase in the percentage of abnormal sperm was detected in ♂ exposed to 0.5 mg kg$^{-1}$ (13).

**Metabolism and pharmacokinetics**
Intravenous rabbit 5 mg kg$^{-1}$ body weight of $^{35}$S-labelled mustard gas, it was rapidly diffused throughout the body, 20% radioactivity was excreted in the urine within 12 hr, excretion via the bile was noted, main organs of retention were the liver, lungs and kidneys (14).

The main urinary metabolites were thiodiglycol and conjugates (15%), glutathione-bis(β-chloroethyl) sulfide conjugates (45%), glutathione-bis(β-chloroethyl) sulfone conjugates (7%) and bis(β-chloroethyl) sulfone and conjugates (8%) (15).

Mustard gas reacts *in vivo* with proteins and nucleic acids of the lung, liver and kidneys of A/J mice (16).

The perfusion of lungs isolated from dogs showed that equilibrium between the blood and tissues was reached after 5 min and that 14% of the radioactivity was retained in the lung (17).

**Irritancy**
10 μl of a 0.01 to 1.00% dilution was applied to an epidermis explant for 18 hr at 36°C, it increased amounts of histamine, plasminogen-activating activity and prostaglandin $E_2$ (18).

Dermal (1 hr) pigs high dose after 72 hr the pigs exhibited microvesicle formations of varying intensities (19).

## Genotoxicity
*Salmonella typhimurium uvrB$^+$G46*, (metabolic activation unspecified) positive (19).
*In vitro* mouse bone marrow, induced micronucleii (20).
At low concentrations inhibited DNA synthesis in *Escherichia coli*, in L-cells and in HeLa and Chinese hamster cells (21-25).
It was the first chemical reported to induce mutation and chromosomal rearrangement in *Drosophila melanogaster* (26).
Induced chromosomal aberrations in cultured rat lymphosarcoma cell lines (27).
Dominant lethal mutations in adult ♂ virgin rat were induced after exposure to 0.1 mg m$^{-3}$ for 52 wk (28).
Of 1700 mustard gas factory workers examined, variants were detected in 85 and 6 cases by electrophoresis and enzyme activity measurement, respectively. All of the 66 cases in which parents could also be examined were genetic variants. In 62,747 equivalent locus tests made by electrophoresis, the mutation rate was 0-4.77 × 10$^{-5}$ locus$^{-1}$ generation$^{-1}$ showing no significant difference from the spontaneous mutation rate (29).

## Any other adverse effects to man

Causes conjunctivitis, blindness within 12 hr, cough, oedema of eyelids, erythema of skin, severe pruritus. May cause oedema, ulceration, necrosis of respiratory tract and exposed skin. Ingestion may cause nausea and vomiting (30).

A report of 11 cases of exposure in fishermen who retrieved leaking gas shells from underwater dumps, all 11 had very inflamed skin, in the axillary and genitofemoral regions, yellow blisters on the hands and legs, painful irritation of the eyes and transient blindness. Two developed pulmonary oedema. There was evidence of a mutagenic effect and in view of the increased risk of lung cancer in soldiers and workers exposed to the gas, it can be assumed that the fishermen heavily exposed also had an increased cancer risk (31).

Most patients exposed to mustard gas recover completely and only a small proportion will have long term eye and lung damage (32), although death from respiratory, renal and bone-marrow failure may occur (33).

## Legislation

Included in Schedule 4 (Release Into Air: Prescribed Substances) Statutory Instrument No. 472, 1991 (34).

## Any other comments

Pharmacokinetics and toxicology reviewed (35,36).

Human health effects, experimental toxicology, environmental effects, epidemiology and workplace experience reviewed (37).

## References

1. *NTIS Report PB158-507*, Natl. Tech. Inf. Serv., Springfield VA
2. *NTIS Report PB214-270*, Natl. Tech. Inf. Serv., Springfield, VA
3. *WHO, Tech. Rep. Series* 1970, 24
4. *Can. J. Res., Section E, Med. Sci.* 1947, **25**, 141
5. *IARC Monograph* 1987, **Suppl. 7**, 67
6. Watson, A. P. et al *Regul. Toxicol. Pharmacol.* 1989, **10**(1), 1-25
7. Heston, W. E. et al *Proc. Soc. Exp. Biol. (N.Y.)* 1953, **82**, 457-460
8. Heston, W. E. *J. Natl. Cancer Inst.* 1953, **14**, 131-140
9. Heston, W. E. *J. Natl. Cancer Inst.* 1950, **11**, 415-423
10. Berenblum, I. *J. Path. Bact.* 1935, **40**, 549-558
11. Easton, D. F. et al *Br. J. Ind. Med.* 1988, **45**(10), 652-659
12. Hackett, P. L. et al *Gov. Rep. Announce. Index (U.S.)* 1988, **88**(8), Abstr. No. 819,732
13. Sasser, L. B. et al *Gov. Rep. Announce. Index (U.S.)* 1990, **90**(6), Abstr. No. 012,515
14. Boursnell, J. C. et al *Biochem. J.* 1946, **40**, 756-764
15. Davison, C. et al *Biochem. Pharmacol.* 1961, **7**, 65-74
16. Abell, C. W. *Proc. Am. Assoc. Cancer Res.* 1964, **5**, 1
17. Pierpont, H. et al *J. Am. Med. Assoc.* 1962, **179**, 421-423
18. Rikimaru, T. et al *J. Invest. Dermatol.* 1991, **96**(8), 888-897
19. Mitcheltree, L. W. et al *J. Toxicol., Cutaneous Ocul. Toxicol.* 1989, **8**(3), 307-317
20. Ashby, J. et al *Mutat. Res.* 1991, **257**(3), 307-311
21. Lawley, P. D. et al *Nature (London)* 1965, **206**, 480-483
22. Crathorn, A. R. et al *Nature (London)* 1966, **211**, 150-153
23. Roberts, J. J. et al *Eur. J. Cancer* 1971, **7**, 515-524
24. Reid, B. D. et al *Biochem. Biophys. Acta* 1969, **179**, 179-188
25. Roberts, J. J. et al *Chem.-Biol. Interact.* 1971, **3**, 49-68
26. Averbach, C. et al *Nature (London)* 1946, **157**, 302

27. Scott, D. et al *Mutat. Res.* 1974, **22**, 207-221
28. Rozmiarek, H. et al *Mutat. Res.* 1973, **21**, 13-14
29. Fujita, M. *Hiroshima Daigaku Igaku Zasshi* 1987, **35**(3), 305-339 (Japan.)
30. Budvari, S. (Ed.) *The Merck Index* 11th ed., 1989, Merck & Co. Inc., Rahway, NJ
31. Wulf, H. C. et al *Lancet* 1985, **i**, 690-691
32. Murray, V. S. G. et al *Br. Med. J.* 1991, **302**, 129-130
33. Rees, J. et al *Lancet* 1991, **337**, 430
34. *S. I. 1991 No. 472 The Environmental Protection (Prescribed Processes and Substances) Regulations* 1991, HMSO, London
35. Wormser, V. *Trends Pharmacol. Sci.* 1991, **12**(4), 164-167
36. Gray, P. J. *From. Gov. Rep. Announce. Index (U. S.)* 1990, **90**(7), Abstr. No. 015,117
37. *ECETOC Technical Report No. 30(4)* 1991, European Chemical Industry Ecology and Toxicology Centre, B-1160 Brussels

# M340  Myclobutanil

**CAS Registry No.** 88671-89-0
**Synonyms** 1*H*-1,2,4-triazole-1-propanenitrile, α-butyl-α-(4-chlorophenyl)-; Nova; Nu-Flo w M; Rally; Systhane
**Mol. Formula** $C_{15}H_{17}ClN_4$     **Mol. Wt.** 288.78
**Uses** Fungicide.

## Physical properties

**M. Pt.** 63-68°C; **B. Pt.** $_{1.0}$ 202-208°C; **Partition coefficient** log $P_{ow}$ 2.94 at pH 7-8 and 25°C; **Volatility** v.p. $1.6 \times 10^{-6}$ mmHg at 25°C.

## Solubility
Water: 142 ppm at 25°C. Organic solvent: ketones, esters, alcohols, aromatic hydrocarbons

## Ecotoxicity

**Fish toxicity**
LC$_{50}$ (96 hr) bluegill sunfish 2.4 mg l$^{-1}$, LC$_{50}$ rainbow trout 4.2 mg l$^{-1}$ (1).

**Invertebrate toxicity**
LC$_{50}$ *Daphnia* sp. 11 mg l$^{-1}$ (1).

## Environmental fate

**Degradation studies**
No degradation under anaerobic conditions (1).
In soil, decomposition is through formation of highly polar triazole compounds with further degradation by ring splitting (2).

**Abiotic removal**

Aqueous solutions degraded on exposure to light, $t_{1/2}$ 222 days in sterile water; 0.8 days in sensitised sterile water; 25 days in pond water; soil $t_{1/2}$ 66 days in silt loam (1).

## Mammalian and avian toxicity

### Acute data

$LD_{50}$ oral ♂ rats 1600 mg kg$^{-1}$ for ♀ rats 2290 mg kg$^{-1}$ (1).
$LD_{50}$ oral mouse 1.36->4.42 g kg$^{-1}$ (3).
$LD_{50}$ oral rat 1.6-2.7 g kg$^{-1}$ (3).
$LD_{50}$ dermal rabbit >5000 mg kg$^{-1}$ (1).
$LD_{50}$ bobwhite quail 510 mg kg$^{-1}$ (1).
$LD_{50}$ grey partridge 1635 mg kg$^{-1}$ (1).

### Sub-acute data

Oral (90 day) rat, dog no effect at 100 mg kg$^{-1}$ diet (1).
Oral (13 wk) rat 0, 6.2, 18.8 or 192 mg kg$^{-1}$ body weight day$^{-1}$ in ♂ and 0, 6.9, 19.6 or 225 mg kg$^{-1}$ body weight day$^{-1}$ in ♀. Histomorphological changes in the liver, kidney and adrenal glands were observed (3).
Oral (12 month) 12 ♂ beagle dogs 0, 0.3, 3.1, 14.3 or 54.2 mg kg$^{-1}$ body weight day$^{-1}$, ♀ beagle dogs 0, 0.4, 3.8, 15.7 or 58.2 mg kg$^{-1}$ body weight day$^{-1}$, respectively. There was increased hepatocellular hypertrophy, increased liver weights and increased serum alkaline phosphatase levels in the liver (3).

### Carcinogenicity and long-term effects

Oral mice (2 yr) up to 500 ppm, no oncogenic activity noted (3).
Oral rats (1 yr) 800 ppm, no evidence of carcinogenic activity (3).

### Teratogenicity and reproductive effects

Oral 25 CRI:CD(5D)BR rats up to 1000 ppm, it decreased number of ♀ delivering litters, increased number of stillborn pups. In the $F_1$ ♂ diffuse atrophy of the testes, decreased spermatozoa and/or necrotic spermatocytes of the epididymis and atrophy of the prostrate were observed (3).
Oral 8 ♀ rats 464 or 700 mg kg$^{-1}$ resulted in mortality (25 and 100% respectively), decreased body weights, scant faeces, chromodactyorrheoa, red exudate around mouth, rough and urine stained hair coat, and salivation (3).

### Metabolism and pharmacokinetics

Oral gavage ♂/♀ CH:CD1 mouse 2, 20 or 200 mg $^{14}$C myclobutanil kg$^{-1}$ body weight. After 96 hr ≈81% excreted in urine and faeces. Comparable amounts were found in urine (41-57%) and faeces (31-52%). Elimination was biphasic with $t_{1/2}$ of 0.6 and 6 to 30.1 hr, respectively (3).
Oral 4 Sprague-Dawley ♂/♀ rats 150 mg $^{14}$C-compound kg$^{-1}$ body weight. Eliminated via urine (48% ♂, 37% ♀) and faeces (51% ♂, 63% ♀) (3).
In mice it is extensively metabolised to more polar compounds. The major polar metabolites in rats are the lactone, ketone, alcohol, carboxylate, dialcohol and sulfate conjugates (3).

### Irritancy

Eye rabbit (dose, duration unspecified) 91.9% purity caused corneal and conjunctival effects suggestive of moderate to severe irritating potential. Dermal (4 hr) rabbit 0.5 ml was practically non-irritating (3).

## Genotoxicity
*Bacillus subtilis* DNA repair test, negative (3).
*Salmonella typhimurium* TA98, TA100, TA1535 and TA1537 with or without metabolic activation, negative (3).
*In vitro* rat hepatocytes, did not induce unscheduled DNA synthesis (3).
*In vitro* Chinese hamster ovary cells with or without metabolic activation negative (3).

## Any other adverse effects
Mixed function oxidase activity, $g^{-1}$ of liver as estimated by *N*-demethylation of benzphetamine was significantly increased in ♂ (2.1-3.3 fold at ≥1000 ppm) and ♀ (1.7-2.2 fold at 3000 ppm). There was no increase in peroxisomal [$^{14}$C]-palmitoyl-CoA oxidase activity indicative of peroxisomal proliferation (3).

## Legislation
Included in Schedule 4 (Release Into Air: Prescribed Substances) Statutory Instrument No. 472, 1991 (4).

## References
1. *The Pesticide Manual* 9th ed., 1991, British Crop Protection Council, Farnham
2. *The Agrochemicals Handbook* 3rd ed., 1991, RSC, London
3. *IPCS Pesticide Residues in Food – 1992 Evaluations 1992* 1992, Part II Toxicology, WHO, Geneva
4. *S. I. 1991 No. 472 The Environmental Protection (Prescribed Processes and Substances) Regulations* 1991, HMSO, London

# M341   Myristic acid

## $CH_3(CH_2)_{12}CO_2H$

**CAS Registry No.** 544-63-8
**Synonyms** tetradecanoic acid; *neo*-fat 14; *n*-tetradecanoic acid; 1-tridecanecarboxylic acid; Univol U 316S
**Mol. Formula** $C_{14}H_{28}O_2$                    **Mol. Wt.** 228.38

## Physical properties
**M. Pt.** 58.5°C; **B. Pt.** $_{100}$ 250.5°C; **Flash point** 110°C; **Specific gravity** $d_4^{54}$ 0.8622; **Partition coefficient** log $P_{ow}$ 6.1.

## Solubility
Organic solvent: ethanol, methanol, diethyl ether, benzene, chloroform

## Ecotoxicity

### Fish toxicity
Lethal dose goldfish 8 mg $l^{-1}$ (1).

## Environmental fate

### Anaerobic effects
At 5 g COD $m^{-3}$ $day^{-1}$ it inhibited *Methanotrix* sp. bacteria (2).

### Degradation studies
$BOD_5$: 2% ThOD (3).
COD: 30% ThOD (3).

## Mammalian and avian toxicity

### Acute data
$LD_{50}$ intravenous mouse 43 mg $kg^{-1}$ (4).

### Irritancy
Dermal human (3 day) 75 mg caused moderate irritation (5).

## Genotoxicity

*Salmonella typhimurium* TA97, TA98, TA100, TA1535 and TA1537 with or without metabolic activation, negative (6).

## Legislation

Included in Schedule 4 (Release Into Air: Prescribed Substances) Statutory Instrument No. 472, 1991 (7).
The log $P_{ow}$ value exceeds the European Communities recommended level 3.0 (6th and 7th amendments) (8).

## Any other comments

It is a nonirritant and is safe in present practice of use and concentration in cosmetics (9).
It is present in most species of marine brown and red algae (10).
After a chemical leak into groundwater in the municipality of Les Franqueses del Valle as, Catalonia, Spain, the levels of myristic acid in well water during a 7 month period ranged from 0.6-19.8 µg $l^{-1}$ (11).

## References

1. Bock, K. J. *Muench. Beitr. Abwass.-, Fischerei-und Flussbiol.* 1967, **9**(2)
2. Kamei, M. et al *Mizu Shori Gijutsu* 1989, **30**(12), 709-717. (Japan.)
3. Dore, M. et al *La tribune du Cebedeau* 1975, **28**, 374
4. Drill, V. A. (Ed.) *Cutaneous Toxicity* 1977, 127, Academic Press, New York
5. *Acta Pharmacol. Toxicol.* 1961, **18**, 141
6. Zeiger, E. et al *Environ. Mol. Mutagen.* 1988, **11**(Suppl. 12), 1-157
7. *S. I. 1991 No. 472 The Environmental Protection (Prescribed Processes and Substances) Regulations* 1991, HMSO, London
8. *1967 Directive on Classification, Packaging, and Labelling of Dangerous Substances 67/548/EEC; 6th Amendment EEC Directive 79/831/EEC; 7th Amendment EEC Directive 91/32/EEC* 1991, HMSO, London
9. Cosmetic, Toiletry and Fragrance Assoc. *J. Am. Coll. Toxicol.* 1987, **6**(3), 321-401
10. Sakagami, H. et al *Nippon Suisan Gakkaishi* 1990, **56**(6), 973-983, (Japan.)
11. Galceran, M. T. et al *Waste Management* 1990, **10**, 261-268

# M342    Myristyltrimethylammonium bromide

$$CH_3(CH_2)_{13}N^+(CH_3)_3\ Br^-$$

**CAS Registry No.** 1119-97-7
**Synonyms** 1-tetradecanaminium, *N,N,N*-trimethyl-, bromide; ammonium,
trimethyltetradecyl-, bromide; Morpan T; myristyltrimethylammonium bromide;
Mytab; Quaternium 13
**Mol. Formula** $C_{17}H_{38}NBr$                         **Mol. Wt.** 336.41
**Uses** Disinfectant. Deodorant. Laboratory reagent. Cationic detergent.

## Physical properties

**M. Pt.** 245-250°C.

## Mammalian and avian toxicity

### Acute data

$LD_{50}$ intravenous mouse, rat 12, 15 mg $kg^{-1}$, respectively (1).

### Irritancy

Draize (albino rabbit) eye irritation test scores: IO max (maximal ocular irritation
score; obtained after 1 or 24 hr); the ocular irritation score obtained 7 days later was
42.66, 25.83. Classified as strongly irritant (2).

### Any other adverse effects

*In vitro* V79 Chinese hamster cells $LC_{50}$ with, without metabolic activation 26.29,
4.08 mg $l^{-1}$ respectively (2).
Growth inhibition test $IC_{50}$ (50% reduction in cell protein) 4.51 mg $l^{-1}$ (2).
$IC_{50}$ in cultures of rat sublingual mucosa and mouse embryo fibroblasts were 5.7 and
0.46 mg $l^{-1}$, repectively (3).

## Legislation

Included in Schedule 4 (Release Into Air: Prescribed Substances) Statutory
Instrument No. 472, 1991 (4).

## References

1. *The Merck Index* 11th ed., 1989, 1000, Merck & Co. Inc., Rahway, NJ
2. Tachon, P. et al *Int. J. Cosmet. Sci.* 1989, **11**, 233-243
3. Gajjar, L. *Toxicol. in Vitro* 1990, **4**(4/5), 280-283
4. *S. I. 1991 No. 472 The Environmental Protection (Prescribed Processes and Substances) Regulations* 1991, HMSO, London

# M343　Myrtenal

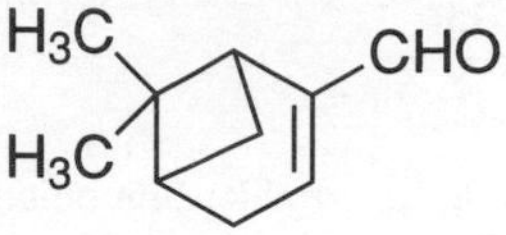

**CAS Registry No.** 564-94-3
**Synonyms** bicyclo[3.1.1]hept-2-ene-2-carboxaldehyde, 6,6-dimethyl-;
2-norpinene-2-carboxaldehyde, 6,6-dimethyl-
**Mol. Formula** $C_{10}H_{14}O$             **Mol. Wt.** 150.22

## Physical properties

**B. Pt.** 220-221°C; **Flash point** 78°C; **Specific gravity** 0.987.

## Environmental fate

**Degradation studies**

*Euglena gracillis* Z. reduced unsaturated terpene aldehyde to the corresponding alcohol (1).

## Mammalian and avian toxicity

**Acute data**

$LD_{50}$ oral rat 2300 mg kg$^{-1}$ (2).
$LD_{50}$ intravenous mouse 170 mg kg$^{-1}$ (2).
$LD_{50}$ dermal rabbit >5 g kg$^{-1}$ (2).

## Legislation

Included in Schedule 4 (Release Into Air: Prescribed Substances) Statutory Instrument No. 472, 1991 (3).

## References

1. Noma, Y. et al *Phytochemistry* 1991, **30**(4), 1147-1151
2. *Food Chem. Toxicol.* 1988, **26**, 329
3. *S. I. 1991 No. 472 The Environmental Protection (Prescribed Processes and Substances) Regulations* 1991, HMSO, London

# GLOSSARY OF NAMES FOR EXPERIMENTAL ORGANISMS

**Latin – Common**

| | |
|---|---|
| *Abramis brama* | bream |
| *Acanthurus spp.* | surgeon fish |
| *Acartia sp.* | oar-footed crustacean |
| *Acartia tonsa* | oar-footed crustacean |
| *Accipiter nisus* | sparrow hawk |
| *Accipiter cooperi* | Cooper's hawk |
| *Achromobacter sp.* | bacteria |
| *Acinetobacter sp.* | bacteria |
| *Acmaea testudinalis* | Barents sea limpet |
| *Acroneuria* | stonefly |
| *Acroneuria pacifica* | stonefly |
| *Acronychia savaera* | shrub |
| *Aedes sp.* | mosquito fly |
| *Aerobacter sp.* | bacteria |
| *Aequipecten gibbus* | calico scallop |
| *Aequipecten (Pecten) irradians* | bay scallop |
| *Aeromonas sp.* | microorganisms |
| *Aeromonas hydrophila* | bacteria |
| *Agelaius tricolor* | tricolored blackbird |
| *Agelaius phoeniceus* | redwing blackbird |
| *Agrobacterium sp.* | bacteria |
| *Alburnus alburnus* | bleak |
| *Alcaligenes sp.* | bacteria |
| *Alcaligenes denitrificans* | bacteria |
| *Alcaligenes faecalis* | bacteria |
| *Allorchestes compressa* | amphipod |
| *Alosa pseudoharengus* | alewife |
| *Ambystoma mexicanum* | Mexican axolotl |
| *Ameiurus melas* | black bullhead |
| *Amycolata autoptopica* | bacteria |
| *Anabaena sp.* | blue-green algae |
| *Anabaena oryzae* | blue-green algae |
| *Anagasta kuehniella* | Mediterranean flour moth |
| *Anarchichas lupus* | wolf-fish |
| *Anas acuta* | common pintail |
| *Anas platyrhynchos* | mallard |
| *Anas discors* | blue-winged teal |
| *Anchoa mitchilly* | bay anchovy |

| | |
|---|---|
| *Anguilla anguilla* | eel |
| *Anguilla japonica* | Japanese eel |
| *Anguilla rostrata* | American eel |
| *Anguilla vulgaris* | eel |
| *Ankistrodesmus falcatus* | green algae |
| *Annelida spp.* | segmented worms |
| *Anodonta anatina* | mollusc |
| *Anodonta cygnea* | freshwater clam |
| *Anolis carolinensis* | anole |
| *Anopheles spp.* | mosquito fly |
| *Anser anser* | greylag goose |
| *Anthopleura elegantissima* | sea anemone |
| *Aphelocoma coerulescens* | scrub jay |
| *Apis fabae* | bean aphid |
| *Apis melliferra* | honey bee |
| *Apis spp.* | honey bee |
| *Aplexa* | snail |
| *Aplexa hypnorum* | snail |
| *Aposemus sylvaticus* | field mouse |
| *Aquila chrysaetos* | golden eagle |
| *Aratinga pertinax* | brown-throated conure |
| *Aratinga canicularis* | orange-fronted conure |
| *Arbacia punctulata* | sea urchin |
| *Ardea cinerea* | heron |
| *Ardeola spp.* | egret |
| *Artemia franciscana* | brine shrimp |
| *Artemia spp.* | shrimp |
| *Artemia salina* | brine shrimp |
| *Artemisia dracunculus* | tarragon |
| *Arthrobacter sp.* | bacteria |
| *Arthrobacter simplex* | bacteria |
| *Aschelminthes spp.* | round worms |
| *Asellus* | sowbug |
| *Asellus aquaticus* | sowbug |
| *Asellus brevicaudus* | sowbug |
| *Aspergillus flavus* | fungi |
| *Aspergillus fumigatus* | fungi |
| *Aspergillus nidulans* | fungi |
| *Aspergillus niger* | fungi |
| *Aspergillus oryzae* | fungi |
| *Aspergillus parasiticus* | fungi |
| *Asperula sp.* | woodruff |
| *Atropa belladonna* | belladonna |
| *Aulosira fertilissima* | cyanobacteria |
| *Aureobasidium sp.* | yeast |
| *Azobacter chroococcum* | bacterium |
| | |
| *Bacillus firmus* | bacteria |
| *Bacillus licheniformis* | bacteria |

| | |
|---|---|
| *Bacillus macerans* | bacteria |
| *Bacillus megaterium* | bacteria |
| *Bacillus pumilus* | bacteria |
| *Bacillus subtilis* | bacteria |
| *Bacteroides sp.* | bacteria |
| *Baetis* | mayfly |
| *Balanus spp.* | barnacle |
| *Barbus conchonius* | red barb |
| *Barbus stigma* | barb |
| *Barbus ticto ticto* | barb |
| *Beta vulgaris* | red beetroot |
| *Biomphalaria glabrata* | freshwater snail |
| *Biston betularia* | peppered moth |
| *Blarina brevicauda* | short-tailed shrew |
| *Blastocladiella ramosa* | aquatic fungus |
| *Blastocladiella pringsheimii* | aquatic fungus |
| *Bombus spp.* | bumble bee |
| *Bombycilla cedrorum* | cedar waxwing |
| *Brachydanio rerio* | zebra fish |
| *Brachidontes recurvis* | mussel |
| *Branta canadensis* | Canada goose |
| *Brassica nigra* | black mustard seed |
| *Brevoortia tyrannus* | Atlantic menhaden |
| *Brevoortia patronus* | Gulf menhaden |
| *Brevibacterium sp.* | bacteria |
| *Bufo americanus* | American toad |
| *Bufo fowleri* | Fowler's toad |
| *Bufo bufo* | common toad |
| *Bufo regularis* | Eygptian toad |
| *Bullio rhodostoma* | mollusc |
| *Buteo swainsoni* | Swainson's hawk |
| | |
| *Cadra cautella* | almond moth |
| *Calamospiza melanocory* | lark bunting |
| *Calanoida* | oar-footed crustacean |
| *Callinectes sapidus* | blue crab |
| *Callithrix jacchus* | marmoset monkey |
| *Cancer magister* | Dungeness crab |
| *Caranx spp.* | pompano, jack cravally |
| *Carassius auratus* | goldfish |
| *Carassius carassius* | Crucian carp |
| *Carcinus maenas* | green crab |
| *Carpodacus cassinii* | cassins finch |
| *Carpodacus mexicanus* | house finch |
| *Cassidix major* | boat-tailed grackle |
| *Catostomus catostomus* | longnose sucker |
| *Catostomus commersoni* | white sucker |
| *Catostomus spp.* | suckers |
| *Cemiscus nitilus* | roach |

| | |
|---|---|
| *Ceratophyllum demersum* | coontail |
| *Ceriodaphnia dubia* | water flea |
| *Chandrus crispus* | sea moss |
| *Channa striatus* | snakehead fish |
| *Chanos chanos* | milkfish |
| *Chaoborus* | phantom midges |
| *Chelon labrosus* | grey mullet |
| *Chelydra serpentina* | snapping turtle |
| *Chilomonas paramecium* | protozoa, flagellate |
| *Chironomus* | midge |
| *Chironomus plumosus* | midge |
| *Chironomus riparius* | midge |
| *Chironomus tentans* | midge |
| *Chlamydomonas sp.* | green algae |
| *Chlamydomonas reinhardti* | green algae |
| *Chlamydomonas variabilis* | green algae |
| *Chlorella sp.* | green algae |
| *Chlorella fusca* | green algae |
| *Chlorella kesslerii* | green algae |
| *Chlorella pyrenoidosa* | green algae |
| *Chloristoneura occidentalis* | budworm |
| *Chlorococcum sp.* | green algae |
| *Chromobacterium sp.* | bacteria |
| *Chthamalus spp.* | barnacles |
| *Circus cyaneus* | northern harrier, marsh hawk |
| *Cirrhina mrigala* | freshwater carp |
| *Citrobacter freundii* | bacteria |
| *Claassenia sabulosa* | stonefly |
| *Cladocera* | water flea |
| *Clarias anguillaris* | catfish |
| *Clarias batrachus* | walking catfish |
| *Clarias lazera* | catfish |
| *Clarias macrocephalus* | catfish |
| *Clostridium difficile* | bacteria |
| *Clupea harengus* | herring |
| *Clupea harengus harengus* | Altantic herring |
| *Clupea sprattus* | sprat |
| *Clupea pallasi* | Pacific herring |
| *Coleoptera* | beetle |
| *Colinus virginianus* | common bobwhite, bobwhite quail |
| *Colisa fasciata* | giant gourami |
| *Colpidium* | protozoa |
| *Columba livia* | common pigeon, rock dove |
| *Columbina talpacoti* | ruddy ground dove |
| *Columbina passerina* | common or ground dove |
| *Conus spp.* | cone shells |

| | |
|---|---|
| *Coregonus albula* | vendace |
| *Coregonus lavaretus* | rowan, lavaret |
| *Corvus corax* | northern raven |
| *Corvus brachyrhynchos* | American crow |
| *Coturnix coturnix* | quail |
| *Coturnix coturnix japonica* | Japanese quail |
| *Crangon spp.* | snapping shrimp |
| *Crangon crangon* | brown shrimp |
| *Crangon septemspinosa* | sand shrimp |
| *Crangonyx pseudogracilis* | crustacean |
| *Crassostrea virginica* | eastern oyster |
| *Crassostrea gigas* | Pacific oyster |
| *Cryptotis parva* | least shrew |
| *Ctenopharyngodon* | carp |
| *Ctenopharyngodon idella* | white amur grass carp |
| *Culex* | mosquito fly |
| *Culex pipiens* | mosquito fly |
| *Culex theileri* | mosquito fly |
| *Cyanocitta cristata* | blue jay |
| *Cyanocorax yncas* | green jay |
| *Cycas sp.* | palm |
| *Cycas circinalis* | false sagopalm |
| *Cycas revoluta* | Japanese cycad |
| *Cymatogaster aggregata* | shiner perch |
| *Cypridopsis vidua* | seed shrimp |
| *Cyprinodon variegatus* | sheepshead minnow |
| *Cyprinus carpio* | carp |
| *Cytophaga* | bacteria |
| | |
| *Daphnia cucullata* | water flea |
| *Daphnia laevis* | water flea |
| *Daphnia longispina* | water flea |
| *Daphnia/Daphnia magna/Daphnia pulex* | water flea |
| *Dasypus novemcinctus* | armadillo |
| *Datura stramonium* | jimsonweed |
| *Delphinus delphis* | common dolphin |
| *Desulfovibrio vulgaris* | bacteria |
| *Dichapetalum cymosum* | woody plant |
| *Didelphis virginiana* | opossum |
| *Dipodomys sp.* | kangaroo rat |
| *Diphtheria bacillus* | bacteria |
| *Dorax cureatus* | marine wedge clam |
| *Drosophila melanogaster* | fruit fly |
| *Dunaliella sp.* | green algae |
| *Dunaliella bioculata* | green algae |
| *Dunaliella euchlora* | green algae |
| | |
| *Echinogammarus pirloti* | amphipod |
| *Echinometra spp.* | sea urchin |

| | |
|---|---|
| *Eisenia sp.* | kelp |
| *Eisenia foetida* | kelp |
| *Elaphe sp.* | rat snake |
| *Elminius modestus* | barnacle |
| *Engraulis engrasicholus* | European anchovy |
| *Engraulis mordax* | northern anchovy |
| *Entosiphon* | protozoa |
| *Entosiphon sulcatum* | protozoa, flagellate |
| *Ephemerella* | mayfly |
| *Eptesicus capensis* | bat |
| *Eptesicus fuscus* | big brown bat |
| *Eremophila alpestris* | horned lark |
| *Escherichia coli* | bacteria |
| *Esox lucius* | northern pike |
| *Euplectes orix* | red bishop |
| *Eutamias sp.* | western chipmunk |
| *Euthynnus pelamis* | skipjack tuna |
| | |
| *Falco peregrinus* | peregrine falcon |
| *Falco sparverius* | American kestrel |
| *Felis catus* | cat |
| *Fundulus* | killifish |
| *Fundulus heteroclitus* | mummichog |
| *Fusarium merismoides* | ascomycetes |
| *Fusarium oxysporum* | ascomycetes |
| | |
| *Gadus merlangus* | whiting |
| *Gadus morhua* | Atlantic cod |
| *Gallus gallus* | domestic chicken |
| *Gambusia affinis* | mosquito fish |
| *Gammarus* | scud |
| *Gammarus fasciatus* | scud |
| *Gammarus lacustris* | scud |
| *Gammarus pseudolimnaeus* | scud |
| *Gammarus pulex* | scud |
| *Gasterosteus aculeatus* | three spined stickleback |
| *Glossina* | tsetse fly |
| *Glycera dibranchiata* | bloodworm |
| *Gnathopogon caerulescens* | willow shiner |
| *Gobio gobio* | gudgeon |
| *Gobius minutus* | goby |
| *Gracilaria verrucosa* | red seaweed |
| *Gracilaria foliifera* | red seaweed |
| *Grus canadensis* | sandhill crane |
| *Gryllus pennsylvanicus* | field cricket |
| *Gymnodinium breve* | unicellar biflagellate algae |
| | |
| *Haematococcus pluvialis* | algae |
| *Haliotis spp.* | abalone |

| | |
|---|---|
| *Hansenula anomala* | bacteria |
| *Helcioniscus argentatus* | saltwater limpet |
| *Helcioniscus exaratus* | saltwater limpet |
| *Helix sp.* | land snail |
| *Hemigrapsus spp.* | shore crabs |
| *Heteropneustes fossilis* | airsac catfish |
| *Hexagenia* | mayfly |
| *Hippoglossus hippoglossus* | halibut |
| *Hirundinidae* | swallows |
| *Homarus americanus* | northern lobster |
| *Hyalella* | shrimp |
| *Hyalella azteca* | shrimp |
| *Hydra oligactis* | protozoa |
| *Hydrilla verticillata* | aquatic macrophyte |
| *Hyla versicolor* | treefrog |
| *Hyla crucifer* | spring peeper |
| *Hypophthalmichthys molitrix* | silver carp |
| *Hypophthalmichthys nobilis* | bighead |
| | |
| *Ictalurus melas* | black bullhead |
| *Ictalurus nebulosus* | brown bullhead |
| *Ictalurus punctatus* | channel catfish |
| *Ictalurus natalis* | yellow bullhead |
| *Ictalurus nebulosus* | brown bullhead |
| *Ictalurus bebulosus* | American catfish |
| *Ictalurus* | catfish |
| *Ictiobus cyprinellus* | bigmouth buffalo |
| *Ictiobus* | suckers |
| *Ischnura verticalis* | damselfly |
| *Isoperla* | stonefly |
| | |
| *Jordanella floridae* | American flagfish |
| | |
| *Klebsiella pneumoniae* | bacterium |
| *Klebsiella planticola* | bacterium |
| *Kuhlia sandvicensis* | mountain bass |
| | |
| *Labeo bicolor* | red tailed black |
| *Labeo rohita* | carp |
| *Lagodon* | common sunfish |
| *Lagodon rhomboides* | marine pin perch |
| *Laminaria digibacta* | Atlantic kelp |
| *Laminaria agardhii* | Atlantic kelp |
| *Larus delawarensis* | ring-billed gull |
| *Lasiurus borealis* | red bat |
| *Lebistes reticulatus* | guppy |
| *Lepomis cyanellus* | green sunfish |
| *Lepomis macrochirus* | bluegill sunfish |
| *Lepomis megalotis* | longear sunfish |

| | |
|---|---|
| *Lepomis microlophus* | redear sunfish |
| *Lepomis gibbosus* | pumpkinseed |
| *Lepomis humilis* | common sunfish |
| *Leporidae* | hare |
| *Leptocottus armatus* | staghorn sculpin |
| *Leptotila verreauxi* | white-fronted dove |
| *Lepus europaeus* | European hare |
| *Lestes congener* | damselfly |
| *Leuciscus cephalus* | chub |
| *Leuciscus idus* | ide |
| *Leuciscus leuciscus* | dace |
| *Ligularia clivorum* | plant |
| *Limanda limanda* | common dab |
| *Lophortyx californica* | California quail |
| *Lota lota* | burbot |
| *Lucania parva* | rainwater killifish |
| *Lumbricus terrestris* | common earthworm |
| *Lutjanus campechanus* | red snapper |
| *Lymnea acuminata* | mollusc |
| *Lymnaea stagnalis* | freshwater snail |
| *Lytechinus spp.* | sea urchin |
| | |
| *Macaca fascicularis* | macaque |
| *Macaca mulatta* | rhesus monkey |
| *Macoma balthica* | mollusc |
| *Macrobrachium lamarrii* | shrimp |
| *Macrocystis pyrifera* | kelp |
| *Macromia* | dragonfly |
| *Macropodus cupanus* | spike-tailed paradise fish |
| *Macropodus opercularis* | paradisefish |
| *Mallotus villosus* | capelin |
| *Meleagris gallopavo* | wild turkey |
| *Meleagris* | turkey |
| *Melicope leptococca* | citrus tree from New Caledonia |
| *Melilotus* | sweet clover |
| *Melopsittacus undulatus* | budgerigar |
| *Menidia audens* | Mississippi silverside |
| *Menidia menidia* | Atlantic silverside |
| *Menidia beryllina* | inland silverside |
| *Menidia peninsulae* | tidewater silverside |
| *Mercenaria mercenaria* | hard clam |
| *Metapenaeus monoceros* | shrimp |
| *Methanobacterium sp.* | bacteria |
| *Methanobrevibacter sp.* | bacteria |
| *Methanococcus sp.* | bacteria |
| *Methanosarcina sp.* | bacteria |
| *Micrococcus sp.* | bacteria |
| *Microcystis aeruginosa* | blue-green algae |
| *Microcystis sp.* | blue-green algae |

| | |
|---|---|
| *Micromesistius poutassou* | blue whiting |
| *Micropogon undulatus* | croaker |
| *Micropterus dolomieui* | smallmouth bass |
| *Micropterus salmoides* | largemouth bass |
| *Microtus arvalis* | common vole |
| *Microtus longicaudus* | vole |
| *Microtus montanus* | vole |
| *Microtus pennsylvanicus* | meadow vole |
| *Mollienesia latipinna* | sailfin molley |
| *Molothrus ater* | brown-headed cowbird |
| *Molothrus bonariensis* | shiny cowbird |
| *Monodonta turbinata* | marine snails |
| *Monoraphidium* | green algae |
| *Monoraphidium griffithii* | diatom |
| *Morone labrax* | sea bass |
| *Morone saxatilis* | striped bass |
| *Mugil cephalus* | striped mullet |
| *Mulloidichthys spp.* | goatfish |
| *Muntiacus muntjac* | Indian deer, Indian muntjac |
| *Mus musculus* | house mouse |
| *Musca domestica* | housefly |
| *Mustela vison* | mink |
| *Mya arenaria* | soft shell clam |
| *Mycobacterium tuberculosis* | bacteria |
| *Myiopsitta monacha* | monk parakeet |
| *Myocastor coypus* | nutria |
| *Myotis grisescens* | gray bat |
| *Myotis lucifugus* | little brown bat |
| *Mysidopsis* | shrimp |
| *Mysidopsis bahia* | mysid shrimp |
| *Mystus vittatus* | striped catfish |
| *Mytilus californianus* | California sea mussel |
| *Mytilus edulis* | bay mussel |
| *Myzus persicae* | peach-potato aphid |
| *Navicula pelliculosa* | diatom |
| *Neanthes arenaceodentata* | ragworm |
| *Nemoria esthamus* | freshwater fish |
| *Neotoma cinerea* | pack rat |
| *Nereis diversicolor* | sandworm |
| *Nereis vexillosa* | sandworm |
| *Nereis virens* | sandworm |
| *Neurospora crassa* | mould |
| *Nitrobacter agilis* | nitrogen-fixing bacteria |
| *Nitrosococcus oceanus* | nitrogen-fixing bacteria |
| *Nitrosomonas europaea* | nitrogen-fixing bacteria |
| *Nitrosomonas sp.* | nitrogen-fixing bacteria |
| *Nitocra* | shrimp |

| | |
|---|---|
| *Nitocra spinipes* | shrimp |
| *Nocardia sp.* | bacteria |
| *Nocardia rhodochrous* | bacteria |
| *Nostoc linckia* | starfish |
| *Nostoc muscorum* | blue-green algae |
| *Notopterus notopterus* | knifefish |
| *Nycticeius schlieffeni* | bat |
| | |
| *Olyzias latipes* | medaka |
| *Oncorhynchus gorbuscha* | pink salmon |
| *Oncorhynchus keta* | chum salmon |
| *Oncorhynchus masou* | masu salmon |
| *Oncorhynchus nerka* | sockeye salmon, blueback salmon |
| *Oncorhynchus nerka kennerlyi* | kokanee |
| *Oncorhynchus tschawytscha* | king salmon |
| *Oncorhynchus kisutch* | coho salmon |
| *Ophicephalus punctatus* | snakehead fish |
| *Ophiogomphus sp.* | aquatic invertebrate |
| *Oreochromis niloticus* | catfish, hybrid tilapia |
| *Orconectes nais* | crayfish |
| *Ortalis vetula* | plain chachalaca |
| *Oryctolagus cuniculus* | domestic New Zealand white rabbit |
| *Oryzias latipes* | rice fish |
| *Oscillatoria* | blue-green algae |
| *Osmerus* | smelt |
| *Ostracoda* | shrimp |
| *Otospermophilus beecheyi* | California ground squirrel |
| *Ovalipes ocellatus* | calico crab |
| *Ovis sp.* | sheep |
| | |
| *Palaemon elegans* | decapod |
| *Palaemonon macrodactylus* | grass shrimp |
| *Palaemonetes pugio* | crustacean |
| *Pandalus montagui* | shrimp |
| *Panulirus argus* | spiny lobster |
| *Panulirus japonicus* | Pacific lobster |
| *Panulirus pencillatus* | lobster |
| *Paracentrotus lividus* | sea urchin |
| *Paracoccus sp.* | bacteria |
| *Paralichthys lethostigma* | southern flounder |
| *Paralichthys dentatus* | summer flounder |
| *Paralichthys olivaceus* | flatfish |
| *Paralithodes camchatica* | king crab |
| *Paramecium caudatum* | ciliate |
| *Paramecium tetraurelia* | ciliate |
| *Paraphrys vetulus* | English sole |
| *Passer luteus* | golden sparrow |

| | |
|---|---|
| *Passer domesticus* | house sparrow |
| *Patella spp.* | limpets |
| *Pecten maximus* | scallop |
| *Penaeus monodon* | shrimp |
| *Penaeus setiferus* | white shrimp |
| *Penaeus duorarum* | pink shrimp |
| *Penaeus aztecus* | brown shrimp |
| *Penaeus orientalis* | prawn |
| *Penaeus stylirostris* | shrimp |
| *Pendaeus spp.* | shrimp |
| *Penicillium notatum* | bacteria |
| *Perca fluviatilis* | perch |
| *Perca flavescens* | yellow perch |
| *Periplaneta americana* | common cockroach |
| *Perna viridis* | green mussel |
| *Peromyscus gossypinus* | cotton mouse |
| *Peromyscus leucopus* | white-footed mouse |
| *Peromyscus maniculatus* | deer mouse |
| *Peromyscus polionotus* | old-field mouse |
| *Petromyzon marinus* | sea lamprey |
| *Phaeodactylum tricornutum* | diatom |
| *Phanerochaete chrysosporium* | fungi |
| *Phasianus colchicus* | ring-necked pheasant |
| *Phebalium dentatum* | mollusc |
| *Phormidium fragile* | blue-green algae |
| *Photobacterium phosphoreum* | bacteria |
| *Phoxinus phoxinus* | minnow |
| *Phrynosoma cornutum* | horned lizard |
| *Physa fontinalis* | freshwater snail |
| *Pica nuttalli* | yellow-billed magpie |
| *Pica pica* | black-billed magpie |
| *Pichia membranaefaciens* | bacteria |
| *Pimephales promelas* | fathead minnow |
| *Plasmodium berghei* | protozoa |
| *Plasmodium falciparum* | bacteria |
| *Platichthys flesus* | flounder |
| *Platichthys stellatus* | starry flounder |
| *Platyhelminthes* | flatworms |
| *Pleuronectes platessa* | plaice |
| *Pleurotus ostreatus* | mushroom |
| *Ploceus taeniopterus* | northern masked weaver |
| *Ploceus cucullatus* | village weaver |
| *Podophthalmus vigil* | crab |
| *Poecilia latipinna* | sailfin molley |
| *Poecilia reticulata* | guppy |
| *Pollachius pollachius* | pollack |
| *Pollachius virens* | saithe, coalfish |
| *Polyodon spathula* | paddlefish |
| *Pomoxis annularis* | white crappie |

| | |
|---|---|
| *Pomoxis nigromaculatus* | black crappie |
| *Porphyra spp.* | red algae |
| *Porthetria dispar* | gypsy moth |
| *Portunus sanguinolentus* | crab |
| *Procambarus clarki* | crayfish |
| *Proteus morganii* | bacteria |
| *Proteus vulgaris* | bacteria |
| *Protococcus* | green algae |
| *Prototheca staminea* | little neck clam |
| *Psetta maxima, Scophthalmus maximus* | turbot |
| *Pseudocalanus minutus* | copepod |
| *Pseudomonas acidovorans* | bacteria |
| *Pseudomonas aeruginosa* | bacteria |
| *Pseudomonas cepacia* | bacteria |
| *Pseudomonas fluorescens* | bacteria |
| *Pseudomonas pancimobilis* | bacteria |
| *Pseudomonas phaseolicola* | bacteria |
| *Pseudomonas putida* | bacteria |
| *Pseudomonas striata* | bacteria |
| *Pseudopleuronectes americanus* | winter flounder |
| *Pteronarcella* | stonefly |
| *Pteronarcella badia* | stonefly |
| *Pteronarcys* | stonefly |
| *Pteronarcys californica* | stonefly |
| *Pteronarcys dorsata* | stonefly |
| *Puntius conchonius* | rosy barb |
| *Puntius sophore* | barb |
| *Puntius ticto* | barb |
| *Pycnopsyche sp.* | aquatic insect |
| *Pygosteus pungitius* | stickleback (12-spined) |
| *Pylodictis olivaris* | flathead catfish |
| *Pyrethrum cinerariaefolium* | chrysanthemum |
| | |
| *Quelea quelea* | red-billed quelea |
| *Quiscalus quiscula* | common grackle |
| | |
| *Rana catesbeiana* | bullfrog |
| *Rana clamitans* | green frog |
| *Rana esculenta* | edible frog |
| *Rana palustris* | pickerel frog |
| *Rana temporaria* | common frog |
| *Rana sp.* | frog |
| *Rana sylvatica* | woodfrog |
| *Rana pipiens* | leopard frog |
| *Ranina serrata* | crab |
| *Rasbora daniconius* | slender rasbora |
| *Rasbora heteromorpha* | harlequin fish |
| *Rattus norvegicus* | old world rat |
| *Rhabdosargus holubi* | an estuarine fish |

| | |
|---|---|
| *Rhizobium trifolli* | subterranean clover |
| *Rhodococcus rhodochrous* | bacteria |
| *Rhodotorula rubra* | bacteria |
| | |
| *Saccobranchus fossilis* | water flea |
| *Saccharomyces nounii* | yeast |
| *Saccharomyces cerevisiae* | yeast |
| *Saimiri sciureus* | squirrel monkey |
| *Salmo trutta* | brown trout |
| *Salmo salar* | Atlantic salmon |
| *Salmo gairdneri* | rainbow trout |
| *Salmo clarki* | cutthroat trout |
| *Salmonella typhimurium* | bacteria |
| *Salmo trutta lacustris* | brown trout |
| *Salmo trutta m. trutta* | sea trout |
| *Salvelinus alpinus* | arctic char |
| *Salvelinus fontinalis* | American char or brook trout |
| *Salvelinus namaycush* | lake trout |
| *Sardina pilchardus* | European pilchard |
| *Sardinops caerula* | Pacific sardine |
| *Sardinops melanosticta* | Japanese pilchard |
| *Sardinops sagax* | Chilean pilchard |
| *Sargassum fluitans* | brown algae |
| *Sarotheredon aureus* | cichlid |
| *Sarotheredon galilaeus* | cichlid |
| *Sarotheredon mossambicus* | cichlid |
| *Scardafella inca* | inca dove |
| *Scardafella squammata* | scaly dove |
| *Scenedesmus obliquus* | green algae |
| *Scenedesmes pannonicus* | green algae |
| *Scenedesmus quadricauda* | green algae |
| *Scenedesmus sp.* | green algae |
| *Schizosaccharomyces octosporus* | yeast |
| *Scomber japonicus* | Pacific mackerel |
| *Scomber scombrus* | Atlantic mackerel |
| *Scophthalmus maximus* | European turbot |
| *Scophthalmus rhombus* | turbot |
| *Scylla serrata* | crab |
| *Selenastrum* | green algae |
| *Selenastrum capricornutum* | green algae |
| *Selenastrum costatum* | green algae |
| *Semotilus atromaculatus* | creek chub |
| *Senecio jacobaea* | ragwort |
| *Seriola lalandei* | yellowtail |
| *Serratia marcescens* | bacteria |
| *Shigella* | bacteria |
| *Sigmodon hispidus* | cotton rat |
| *Siliqua patula* | razor clam |
| *Silurus glanis* | sheatfish |

| | |
|---|---|
| *Simocephalus* | water flea |
| *Simocephalus serrulatus* | water flea |
| *Simulium venustum* | black flies |
| *Siphonaria normalis* | ribbed limpet |
| *Skeletonema costatum* | algae |
| *Solea solea* | common sole, Dover sole |
| *Sorex araneus* | common shrew |
| *Sorex cinereus* | shrew |
| *Sorex vagrans* | shrew |
| *Sphaerium stratinum* | fingernail clam |
| *Sphaerechinus granularis* | sea urchin |
| *Spisula solidissima* | surf clam |
| *Spiza americana* | dickcissel |
| *Sporophila minuta* | ruddy-breasted seedeater |
| *Sporobolomyces salmonicolor* | yeast |
| *Squalus cephalus* | chub |
| *Staphylococcus* | bacteria |
| *Staphylococcus aureus* | bacteria |
| *Stemphylium loti* | cyanogenic plant |
| *Stenotomus chrysops* | scup |
| *Stibiobacter senarmontii* | bacteria |
| *Stizostedion lucioperca* | pikeperch, sander |
| *Stizostedion vitreum* | walleye |
| *Stolephorus purpureus* | anchovy |
| *Stomoxys calcitrans* | stable fly |
| *Streptococcus* | bacteria |
| *Streptomyces antiboticus* | bacteria |
| *Streptomyces aureus* | bacteria |
| *Streptomyces chrysomallus* | bacteria |
| *Streptomyces lincolnensis* | bacteria |
| *Streptomyces parvullus* | bacteria |
| *Streptomyces peucetius* | bacteria |
| *Streptomyces sanguis* | bacteria |
| *Streptomyces sp.* | bacteria |
| *Streptomyces violaceoruber* | bacteria |
| *Streptoverticillium ladakanus* | bacteria |
| *Strix aluco* | tawny owl |
| *Strongylocentrotus purpuratus* | purple sea urchin |
| *Sturnus vulgaris* | European starling |
| *Sylvilagus sp.* | rabbit |
| *Sus scrofa* | wild boar |
| *Sylvilagus floridanus* | cottontail rabbit |
| *Synechococcus sp.* | algae |
| | |
| *Talpa europaea* | mole |
| *Tamias striatus* | eastern chipmunk |
| *Tangavious aeneus* | red-eyed cowbird, bronzed cowbird |
| *Tenebrio sp.* | mealworm |

| | |
|---|---|
| *Terrapene sp.* | box turtle |
| *Tetrahymena* | protozoa, ciliata |
| *Tetrahymena elliotti* | ciliata |
| *Tetrahymena pyriformis* | ciliata |
| *Thamnophis sirtalis* | garter snake |
| *Theragra chalcogramma* | Alaska pollack |
| *Thiobacillus thiooxidans* | bacteria |
| *Thunnus albacares* | yellowfin tuna |
| *Thymallus thymallus* | grayling |
| *Tilapia mossambica* | cichlid |
| *Tilapia nilotica* | cichlid |
| *Tilapia rendalli* | cichlid |
| *Tinca tinca* | tench |
| *Tivela stultorum* | pismo clam |
| *Toxostoma curvirostre* | curve-billed thrasher |
| *Toxostoma rufum* | brown thrasher |
| *Trachinotus carolinus* | pompano |
| *Trachurus capensis* | Cape horse mackerel |
| *Trachurus murphyi* | Chilean jack mackerel |
| *Tribolium spp.* | flour beetles |
| *Trichiurus lepturus* | largehead hairtail |
| *Trichogaster trichopterus* | three spot gourami |
| *Trifolium subterraneaum* | clover |
| *Trigriopus brevicornis* | copepod |
| *Tubifex tubifex* | worm |
| *Turdus migratorius* | American robin |
| *Turdus ericetorum* | song thrush |
| *Tyria jacobaeae* | cinnabar moth |
| *Tyto alba* | barn owl |
| | |
| *Ulva spp.* | sea lettuce |
| *Umbra pygmaea* | mud minnow |
| *Unio pictorium* | mollusc |
| *Uronema parduczi* | protozoa, ciliata |
| | |
| *Valerianella locusta* | herb |
| *Vicia faba* | broad bean |
| *Viviparus bengalensis* | freshwater snail |
| *Volatia jacarina* | blue-black grassquit |
| *Volsella demissa* | Atlantic ribbed mussel |
| *Vorticella campanula* | ciliate |
| | |
| *Westiellopsis prolifica* | cyanobacteria |
| | |
| *Xanthocephalus xanthocephalus* | yellow-headed blackbird |
| *Xenopus laevis* | clawed frog |
| *Xeroderma pigmentosum* | human pigment cell line |
| | |
| *Zenaida auriculata* | eared dove |

| | |
|---|---|
| *Zenaida macroura* | mourning dove |
| *Zenaida asiatica* | white-winged dove |
| *Zonotrichia leucophrys* | white-crowned sparrow |
| *Zonotrichia atricapilla* | golden-crowned sparrow |

**Common – Latin**

| | |
|---|---|
| abalone | *Haliotis sp.* |
| airsac catfish | *Heteropneustes fossilis* |
| Alaska pollack | *Theragra chalcogramma* |
| alewife | *Alosa pseudoharengus* |
| algae | *Haematococcus pluvialis* |
| algae | *Lenastrum costatum* |
| algae | *Skeletonema costatum* |
| algae | *Synechoccoccus sp.* |
| almond moth | *Cadra cautella* |
| American catfish | *Ictalurus nebulosus* |
| American char | *Salvelinus fontinalis* |
| American crow | *Corvus brachyrhynchos* |
| American eel | *Anguilla rostrata* |
| American flagfish | *Jordanella floridae* |
| American kestrel | *Falco sparverius* |
| American robin | *Turdus migratorius* |
| American toad | *Bufo americanus* |
| amphipod | *Allorchestes compressa* |
| amphipod | *Echinogammarus pirloti* |
| anchovy | *Stolephorus purpureus* |
| anole | *Anolis carolinensis* |
| aquatic fungus | *Blastocladiella ramosa* |
| aquatic fungus | *Blastocladiella pringsheimii* |
| aquatic invertebrate | *Ophiogomphus sp.* |
| aquatic insect | *Pycnopsyche sp.* |
| aquatic macrophyte | *Hydrilla verticillata* |
| Arctic char | *Salvelinus alpinus* |
| armadillo | *Dasypus novemcinctus* |
| ascomycetes | *Fasarium merismoides* |
| ascomycetes | *Fusarium oxysporum* |
| Atlantic cod | *Gadus morhua* |
| Atlantic herring | *Clupea harengus harengus* |
| Atlantic kelp | *Laminaria digibacta, L. agardhii* |
| Atlantic mackerel | *Scomber scombrus* |
| Atlantic menhaden | *Brevoortia tyrannus* |
| Atlantic ribbed mussel | *Volsella demissa* |
| Atlantic salmon | *Salmo salar* |
| Atlantic silverside | *Menidia menidia* |
| bacteria | *Achromobacter sp.* |
| bacteria | *Acinetobacter sp.* |
| bacteria | *Aeromonas hydrophila* |

| | |
|---|---|
| bacteria | *Alcaligenes denitrificans* |
| bacteria | *Alcaligenes faecalis* |
| bacteria | *Agrobacterium sp.* |
| bacteria | *Amycolata autoptophica* |
| bacteria | *Arthrobacter sp.* |
| bacteria | *Arthrobacter simplex* |
| bacteria | *Bacillus licheniformis* |
| bacteria | *Bacillus megaterium* |
| bacteria | *Brevibacterium sp.* |
| bacteria | *Chromobacterium sp.* |
| bacteria | *Citrobacter freundii* |
| bacteria | *Clostridium difficile* |
| bacteria | *Cytophaga* |
| bacteria | *Desulfovibrio vulgaris* |
| bacteria | *Hansenula anomala* |
| bacteria | *Mycobacterium tuberculosis* |
| bacteria | *Paracoccus sp.* |
| bacteria | *Penicillium notatum* |
| bacteria | *Photobacterium phosphoreum* |
| bacteria | *Pichia membranaefaciens* |
| bacteria | *Proteus morganii* |
| bacteria | *Proteus vulgaris* |
| bacteria | *Pseudomonas acidovorans* |
| bacteria | *Pseudomonas aeruginosa* |
| bacteria | *Pseudomonas cepacia* |
| bacteria | *Pseudomonas fluorescens* |
| bacteria | *Pseudomonas pancimobilis* |
| bacteria | *Pseudomonas phaseo* |
| bacteria | *Pseudomonas striata* |
| bacteria | *Rhabdosargus holubi* |
| bacteria | *Rhodococcus* |
| bacteria | *Rhodococcus rhodochrous* |
| bacteria | *Rhodotarula rubra* |
| bacteria | *Serratia marcescens* |
| bacteria | *Shigella* |
| bacteria | *Staphylococcus aureus* |
| bacteria | *Stibiobacter senarmontii* |
| bacteria | *Streptococcus* |
| bacteria | *Streptomyces antiobotius* |
| bacteria | *Streptomyces chrysomallus* |
| bacteria | *Streptomyces ladakanus* |
| bacteria | *Streptomyces lincolnensis* |
| bacteria | *Streptomyces parvullus* |
| bacteria | *Streptomyces peucetius* |
| bacteria | *Streptococcus aureus* |
| bacteria | *Streptococcus sanguis* |
| bacteria | *Streptomyces violaceoruber* |
| bacteria | *Streptoverticillium ladakanus* |
| bacteria | *Thiobacillus thiooxidans* |

| | |
|---|---|
| bacteria | *Azobacter chroococcum* |
| barb | *Puntius conchonius* |
| barb | *Puntius sophore* |
| barb | *Puntius ticto* |
| Barents sea limpet | *Acmaea testudinalis* |
| barn owl | *Tyto alba* |
| barnacle | *Balanus sp.* |
| barnacle | *Elminius modestus* |
| barnacles | *Chthamalus sp.* |
| bass | *Roccus saxatilis, Morone* |
| bat | *Eptesicus capensis* |
| bat | *Nycticeius schlieffeni* |
| bay anchovy | *Anchoa mitchilli* |
| bay mussel | *Mytilus edulis* |
| bay scallop | *Aequipecten (Pecten) irradians* |
| bean aphid | *Apis fabae* |
| beetle | *Coleoptera* |
| belladonna | *Atropa belladonna* |
| big brown bat | *Eptesicus fuscus* |
| bighead | *Hypophthalmichthys nobilis* |
| bigmouth buffalo | *Ictiobus cyprinellus* |
| black-billed magpie | *Pica pica* |
| black bullhead | *Ameiurus melas,* |
| | *Ictalurus melas* |
| black crappie | *Pomoxis nigromaculatus* |
| black flies | *Simulium venustum* |
| bleak | *Alburnus alburnus* |
| bloodworm | *Glycera dibranchiata* |
| blue crab | *Callinectes sapidus* |
| blue-green algae | *Anabaena oryzae* |
| blue-green algae | *Anabaena sp.* |
| blue-green algae | *Phormidium fragile* |
| blue jay | *Cyanocitta cristata* |
| blue whiting | *Micromesistius poutassou* |
| blue-black grassquit | *Volatia jacarina* |
| blue-green algae | *Anabaena oryzae* |
| blue-green algae | *Microcystis aeruginosa* |
| blue-green algae | *Microcystis sp.* |
| blue-green algae | *Nostoc muscorum* |
| blue-green algae | *Oscillatoria* |
| blue-green algae | *Phormidium fragile* |
| blue-winged teal | *Anas discors* |
| blueback salmon | *Oncorhynchus nerka* |
| bluegill sunfish | *Lepomis macrochirus* |
| boat-tailed grackle | *Cassidix major* |
| bobwhite quail | *Colinus virginianus* |
| box turtle | *Terrapene sp.* |
| bream | *Abramis brama* |
| brine shrimp | *Artemia franciscana* |

| | |
|---|---|
| brine shrimp | *Artemia salina* |
| broad bean | *Vicia faba* |
| bronzed cowbird | *Tangavious aeneus* |
| brook trout | *Salvelinus fontinalis* |
| brown bullhead | *Ictalurus nebulosus* |
| brown shrimp | *Crangon crangon* |
| brown shrimp | *Penaeus aztecus* |
| brown thrasher | *Toxostoma rufum* |
| brown trout | *Salmo trutta* |
| brown trout, non-migratory | *Salmo trutta m. fario* |
| brown trout | *Salmo trutta m. lacustris* |
| brown-headed cowbird | *Molothrus ater* |
| brown-throated conure | *Aratinga pertinax* |
| budgerigar | *Melopsittacus undulatus* |
| budworm | *Choristoneura occidentalis* |
| bullfrog | *Rana catesbeiana* |
| bumble bee | *Bombus sp.* |
| burbot | *Lota lota* |
| | |
| calico crab | *Ovalipes ocellatus* |
| calico scallop | *Aequipecten gibbus* |
| California ground squirrel | *Otospermophilus beecheyi* |
| California quail | *Loportyx californica* |
| California sea mussel | *Mytilus californianus* |
| Canada goose | *Branta canadensis* |
| Cape horse mackerel | *Trachurus capensis* |
| capelin | *Mallotus villosus* |
| carp | *Ctenopharyngodon* |
| | *Cyprinus carpio* |
| cassins finch | *Carpodacus cassinii* |
| cat | *Felis catus* |
| catfish | *Clarias anguillaris* |
| catfish | *Clarias lazera* |
| catfish | *Clarias macrocephalus* |
| catfish | *Ictalurus* |
| cedar waxwing | *Bombycilla cedrorum* |
| channel catfish | *Ameiurus nebulosus* |
| channel catfish | *Ictalurus punctatus* |
| Chilean jack mackerel | *Trachurus murphyi* |
| Chilean pilchard | *Sardinops sagax* |
| chub | *Leuciscus cephalus* |
| chub | *Squalius cephalus* |
| chum salmon | *Oncorhynchus keta* |
| cichlid | *Sarotheredon aureus* |
| cichlid | *Sarotheredon galilaeus* |
| cichlid | *Sarotheredon mossambicus* |
| cichlid | *Tilapia rendalli* |
| ciliate | *Paramecium caudatum* |
| ciliate | *Paramecium tetraurelia* |

| | |
|---|---|
| cinnabar moth | *Tyria jacobaeae* |
| citrus tree | *Melicope leptococca* |
| clawed frog | *Xenopus laevis* |
| clover | *Trifolium subterraneaum* |
| coalfish | *Pollachius virens* |
| cod | *Gadus morhua* |
| coho salmon | *Oncorhynchus kisutch* |
| common bobwhite | *Colinus virginianus* |
| common cockroach | *Periplaneta americana* |
| common dab | *Limanda limanda* |
| common dolphin | *Delphinus delphis* |
| common earthworm | *Lumbricus terrestris* |
| common frog | *Rana temporaria* |
| common grackle | *Quiscalus quiscula* |
| common or ground dove | *Columbina passerina* |
| common pigeon | *Columba livia* |
| common pintail | *Anas acuta* |
| common shrew | *Sorex araneus* |
| common sole | *Solea solea* |
| common sunfish | *Lepomis humilis* |
| common sunfish | *Lagodon* |
| common toad | *Bufo bufo* |
| common vole | *Microtus arvalis* |
| cone shells | *Conus sp.* |
| Cooper's hawk | *Accipiter cooperi* |
| coontail | *Ceratophyllum demersum* |
| copepod | *Pseudocalanus minutus* |
| copepod | *Trigriopus brevicornis* |
| cotton mouse | *Peromyscus gossypinus* |
| cotton rat | *Sigmodon hispidus* |
| cottontail rabbit | *Sylvilagus floridanus* |
| crab | *Podophthalmus vigil* |
| crab | *Portunus sanguinolentus* |
| crab | *Ranina serrata* |
| crayfish | *Orconectes nais* |
| crayfish | *Procambarus clarki* |
| creek chub | *Semolitus atromaculatus* |
| croaker | *Micropogon undulatus* |
| Crucian carp | *Carassius carassius* |
| crustacean | *Crangonyx pseudogracilis* |
| crustacean | *Palaemonetes pugio* |
| curve-billed thrasher | *Toxostoma curvirostre* |
| cutthroat trout | *Salmo clarki* |
| cyanobacteria | *Aulosira fertilissima* |
| cyanobacteria | *Westiellopsis prolifica* |
| cyanogenic plant | *Stemphylium loti* |
| | |
| dace | *Leuciscus leuciscus* |
| damselfly | *Ischnura verticalis* |

| | |
|---|---|
| damselfly | *Lestes congener* |
| decapod | *Palaemon elegans* |
| deer mouse | *Peromyscus maniculatus* |
| diatom | *Monoraphidium graffithii* |
| diatom | *Navicula pelliculosa* |
| diatom | *Phaeodactylum tricornutum* |
| dickcissel | *Spiza americana* |
| domestic New Zealand white rabbit | *Oryctolagus cuniculus* |
| domestic chicken | *Gallus gallus* |
| dragonfly | *Macromia* |
| Dungeness crab | *Cancer magister* |
| | |
| eared dove | *Zenaida auriculata* |
| eastern chipmunk | *Tamias striatus* |
| eastern oyster | *Crassostrea virginica* |
| edible frog | *Rana esculenta* |
| eel | *Anguilla anguilla* |
| eel | *Anguilla vulgaris* |
| egret | *Ardeola sp.* |
| Egyptian toad | *Bufo regularis* |
| English sole | *Paraphrys vetulus* |
| European anchovy | *Engraulis engrasicholus* |
| European hare | *Lepus europaeus* |
| European pilchard | *Sardina pilchardus* |
| European starling | *Sturnus vulgaris* |
| European turbot | *Scophthalmus maximus, Psetta maxima* |
| | |
| fathead minnow | *Pimephales promelas* |
| field cricket | *Gryllus pennsylvanicus* |
| field mouse | *Aposemus sylvaticus* |
| flatfish | *Paralichthys divaceus* |
| flathead catfish | *Pylodictis olivaris* |
| flatworms | *Platyhelminthes* |
| flour beetles | *Tribolium sp.* |
| Fowler's toad | *Bufo fowleri* |
| freshwater carp | *Cirrhina mrigala* |
| freshwater clam | *Anodonta cygnea* |
| freshwater fish | *Nemoria esthamus* |
| freshwater snail | *Biomphalaria glabrata* |
| freshwater snail | *Lymnaea stagnolis* |
| freshwater snail | *Physa fontinalis* |
| freshwater snail | *Viviparus bengalensis* |
| frog | *Rana sp.* |
| fruit fly | *Drosophila melanogaster* |
| fungi | *Aspergillus flavus* |
| fungi | *Aspergillus fumigatus* |
| fungi | *Aspergillus nidulans* |
| fungi | *Aspergillus niger* |
| fungi | *Aspergillus oryzae* |

| | |
|---|---|
| fungi | *Aspergillus parasiticus* |
| fungi | *Phanerorochaete chrysosporium* |
| | |
| garden pea | *Pisum sativum* |
| garter snake | *Thamophis sirtalis* |
| giant gourami | *Colisa fasciata* |
| grass shrimp | *Palaemonon macrodactylus* |
| goatfish | *Mulloidichthys sp.* |
| goby | *Gobius minutus* |
| golden eagle | *Aquila chrysaetos* |
| golden sparrow | *Passer luteus* |
| golden-crowned sparrow | *Zonotrichia atricapilla* |
| goldfish | *Carassius auratus* |
| gray bat | *Myotis grisescens* |
| grayling | *Thymallus thymallus* |
| green algae | *Ankistrodesmus falcatus* |
| green algae | *Chlamydonas sp.* |
| green algae | *Chlamydmonas reinhardti* |
| green algae | *Chlamydomonas variabilis* |
| green algae | *Chlorella fusca* |
| green algae | *Chlorella kesslerii* |
| green algae | *Chlorella pyrenoidosa* |
| green algae | *Chlorella sp.* |
| green algae | *Chlorococcum sp.* |
| green algae | *Dunaliella euchlora* |
| green algae | *Dunaliella sp.* |
| green algae | *Monoraphidium* |
| green algae | *Protococcus* |
| green algae | *Scenedesmus pannonicus* |
| green algae | *Scenedesmus quadricauda* |
| green algae | *Scenedesmus sp.* |
| green algae | *Selenastrum* |
| green algae | *Selenastrum capricornutum* |
| green algae | *Selenastrum costatum* |
| green crab | *Carcinus maenas* |
| green frog | *Rana clamitans* |
| green frog | *Rana clamitans* |
| green jay | *Cyanocorax yncas* |
| green mussel | *Perna viridis* |
| green sunfish | *Lepomis cyanellus* |
| grey mullet | *Chelon labrosus* |
| greylag goose | *Anser anser* |
| gudgeon | *Gobio gobio* |
| Gulf menhaden | *Brevoortia patronus* |
| guppy | *Lebistes reticulatus* |
| guppy | *Poecilia reticulata* |
| gypsy moth | *Porthetria dispar* |
| | |
| halibut | *Hippoglossus hippoglossus* |

| | |
|---|---|
| hard clam | *Mercenaria mercenaria* |
| hare | *Leporidae* |
| harlequin fish | *Rasbora heteromorpha* |
| heron | *Ardea cinerea* |
| herring | *Clupea harengus* |
| honey bee | *Apis melliferra* |
| honey bee | *Apis sp.* |
| horned lark | *Eremophila alpestris* |
| horned lizard | *Phrynosoma cornutum* |
| house finch | *Carpodacus mexicanus* |
| house mouse | *Mus musculus* |
| house sparrow | *Passer domesticus* |
| housefly | *Musca domestica* |
| | |
| ide | *Leuciscus idus* |
| inca dove | *Scardafella inca* |
| Indian deer | *Muntiacus muntjac* |
| Indian muntjac | *Muntiacus muntjac* |
| inland silverside | *Menidia beryllina* |
| | |
| Japanese eel | *Anguilla japonica* |
| Japanese pilchard | *Sardinops melanosticta* |
| Japanese quail | *Coturnix coturnix japonica* |
| jimsonweed | *Datura stramonium* |
| | |
| kangaroo rat | *Dipodomys sp.* |
| kelp | *Eisenia sp.* |
| kelp | *Eisenia foetida* |
| kelp | *Macrocystis pyrifera* |
| killifish | *Fundulus* |
| king crab | *Paralithodes camchatica* |
| king salmon | *Oncorhynchus tschawytscha* |
| knifefish | *Notopterus notopterus* |
| kokanee | *Oncorhynchus nerka kennerlyi* |
| | |
| lake trout | *Salvelinus namaycush* |
| land snail | *Helix sp.* |
| largehead hairtail | *Trichiurus lepturus* |
| largemouth bass | *Micropterus salmoides* |
| lark bunting | *Calamospiza melanocory* |
| sailfin molly | *Poecilia latipinna* |
| least shrew | *Cryptotis parva* |
| leopard frog | *Rana pipiens* |
| limpets | *Patella sp.* |
| little brown bat | *Myotis lucifugus* |
| little neck clam | *Prototheca staminea* |
| lobster | *Panulirus pencillatus* |
| longear sunfish | *Lepomis megalotis* |
| longnose sucker | *Catostomus catostomus* |

| | |
|---|---|
| macaque | *Macaca fascicularis* |
| mallard | *Anas platyrhynchos* |
| marine pin perch | *Lagodon rhomboides* |
| marine snails | *Monodonta turbinata* |
| marine wedge clam | *Dorax cureatus* |
| marsh hawk | *Circus cyaneus* |
| masu salmon | *Oncorhynchus masou* |
| mayfly | *Baetis* |
| mayfly | *Ephemerella* |
| mayfly | *Hexagenia* |
| meadow vole | *Microtus pennsylvanicus* |
| mealworm | *Tenebrio sp.* |
| Mediterranean flour moth | *Anagasta kuehniella* |
| Mexican axolotl | *Ambystoma mexicanum* |
| microorganism | *Aeromonas sp.* |
| midge | *Chironomus* |
| midge | *Chironomus plumosus* |
| midge | *Chironomus riparius* |
| midge | *Chironomus tentans* |
| milkfish | *Chanos chanos* |
| mink | *Mustela vison* |
| minnow | *Phoxinus phoxinus* |
| Mississippi silverside | *Menidia audens* |
| mole | *Talpa europaea* |
| monk parakeet | *Myiopsitta monacha* |
| mollusc | *Anodonta anatina* |
| mollusc | *Lymnea acuminata* |
| mollusc | *Macoma balthica* |
| mollusc | *Phebalium dentalum* |
| mosquito fish | *Gambusia affinis* |
| mosquito fly | *Aedes sp.* |
| mosquito fly | *Anopheles sp.* |
| mosquito fly | *Culex* |
| mosquito fly | *Culex pipiens* |
| mosquito fly | *Culex theileri* |
| mould | *Neanthes arenaceodentata* |
| mould | *Neurospora crassa* |
| mountain bass | *Kuhlia sandvicensis* |
| mourning dove | *Zenaida macroura* |
| mud minnow | *Umbra pygmaea* |
| mummichog | *Fundulus heteroclitus* |
| mushroom | *Pleurotus ostreatus* |
| mussel | *Brachidontes recurvus* |
| mysid shrimp | *Mysidopsis bahia* |
| | |
| nitrogen-fixing bacteria | *Nitrobacter agilis* |
| nitrogen-fixing bacteria | *Nitrosomonas europaea* |
| nitrogen-fixing bacteria | *Nitrosococcus oceanus* |
| northern anchovy | *Engraulis mordax* |

| | |
|---|---|
| northern harrier | *Circus cyaneus* |
| northern lobster | *Homarus americanus* |
| northern masked weaver | *Ploceus taeniopterus* |
| northern pike | *Esox lucius* |
| northern raven | *Corvus corax* |
| nutria | *Myocastor coypus* |
| | |
| oar-footed crustacean | *Acartia sp.* |
| oar-footed crustacean | *Acartia tonsa* |
| oar-footed crustacean | *Calanoida* |
| old world rat | *Rattus norwegicus* |
| old-field mouse | *Peromyscus polionotus* |
| opossum | *Didelphis virginiana* |
| orange-fronted conure | *Aratinga canicularis* |
| | |
| Pacific herring | *Clupea pallasi* |
| Pacific lobster | *Panuliris japonicus* |
| Pacific mackerel | *Scomber japonicus* |
| Pacific oyster | *Crassostrea gigas* |
| Pacific sardine | *Sardinops caerula* |
| pack rat | *Neotoma cinerea* |
| paddlefish | *Polyodon spathula* |
| paradisefish | *Macropodus opercularis* |
| peach-potato aphid | *Myzus persicae* |
| peppered moth | *Biston betularis* |
| perch | *Perca fluviatilis* |
| peregrine falcon | *Falco peregrinus* |
| phantom midges | *Chaoborus* |
| pickerel frog | *Rana palustris* |
| pikeperch, sander | *Stizostedion lucioperca* |
| pink salmon | *Oncorhynchus gorbuscha* |
| pink shrimp | *Penaeus duorarum* |
| pismo clam | *Tivela stultorum* |
| plaice | *Pleuronectes platessa* |
| plain chachalaca | *Ortalis vetula* |
| pollack | *Pollachius pollachius* |
| pompano | *Trachinotus carolinus* |
| pompano, jack cravally | *Caranx sp.* |
| powan, lavaret | *Coregonus lavaretus* |
| prawn | *Penaeus orientalis* |
| protozoa | *Colpidium* |
| protozoa | *Entosiphon* |
| protozoa | *Hydra oligaltis* |
| protozoa, ciliata | *Tetrahymena* |
| protozoa, ciliata | *Uronema parduczi* |
| protozoa, flagellate | *Chilomonas paramecium* |
| protozoa, flagellate | *Entosiphon sulcatum* |
| pumpkinseed | *Lepomis gibbosus* |
| purple sea urchin | *Strongylocentrotus purpuratus* |

| | |
|---|---|
| quail | *Coturnix coturnix* |
| rabbit | *Sylvilagus sp.* |
| ragworm | *Neanthes arenaceodentata* |
| ragwort | *Senecio jacobaea* |
| rainbow trout | *Salmo gairdneri* |
| rainwater killifish | *Lucania parva* |
| rat snake | *Elaphe sp.* |
| razor clam | *Siliqua patula* |
| red algae | *Porphyra sp.* |
| red bat | *Lasiurus borealis* |
| red beetroot | *Beta vulgaris* |
| red bishop | *Euplectes orix* |
| red seaweed | *Gracilaria verucosa, G. foliifera* |
| red snapper | *Lutjanus campechanus* |
| red tailed black | *Labeo bicolor* |
| red-billed quelea | *Quelea quelea* |
| red-eyed cowbird or | *Tangavious aeneus* |
| red-winged blackbird | *Agelaieus phoeniceus* |
| redear sunfish | *Lepomis microlophus* |
| rhesus monkey | *Macaca mulatta* |
| ribbed limpet | *Siphonaria normalis* |
| rice fish | *Oryzias latipes* |
| ring-billed gull | *Larus delawarensis* |
| ring-necked pheasant | *Phasianus colchicus* |
| roach | *Cemiscus nitilus* |
| rock dove | *Columba livia* |
| rosy barb, redbarb | *Puntius conchonius, Barbus conchonius* |
| round worms | *Aschelminthes sp.* |
| ruddy ground dove | *Columbina talpacoti* |
| ruddy-breasted seedeater | *Sporophila minuta* |
| sailfin molley | *Poecilia latipinna, Mollienesia* |
| saithe, pollack | *Pollachius virens* |
| saltwater limpet | *Helcioniscus argentatus* |
| saltwater limpet | *Helcioniscus exaratus* |
| sand shrimp | *Crangon septemspinosa* |
| sandhill crane | *Grus canadensis* |
| sandworm | *Nereis diversicolor* |
| sandworm | *Nereis virens, Nereis vexillosa* |
| scaly dove | *Scardafella squammata* |
| scrub jay | *Aphelocoma coerulescens* |
| scud | *Gammarus* |
| scud | *Gammarus fasciatus* |
| scud | *Gammarus lacustris* |
| scud | *Gammarus pseudolimnaeus* |
| scud | *Gammarus pulex* |
| scud | *Stenotomus chrysops* |
| sea anemone | *Anthopleura elegantissima* |

| | |
|---|---|
| sea bass | *Morone labrax* |
| sea lamprey | *Petromyzon marinus* |
| sea lettuce | *Ulva sp.* |
| sea moss | *Chandrus crispus* |
| sea trout | *Salmo trutta m. trutta* |
| sea urchin | *Arbacia puntulata* |
| sea urchin | *Echinometra sp.* |
| sea urchin | *Lytechinus sp.* |
| sea urchin | *Paracentrotus lividus* |
| segmented worms | *Annelida sp.* |
| sheatfish | *Silurus glanis* |
| sheep | *Ovis sp.* |
| sheepshead minnow | *Cyprinodon variegatus* |
| shiner perch | *Cymatogaster aggregata* |
| shiny cowbird | *Molothrus bonariensis* |
| shore crabs | *Hemigrapsus sp.* |
| short-tailed shrew | *Blarina brevicauda* |
| shrew | *Sorex cinereus* |
| shrew | *Sorex vagrans* |
| shrimp | *Artemia spp.* |
| shrimp | *Hyalella* |
| shrimp | *Hyalella aztera* |
| shrimp | *Macrobrachium lamarrii* |
| shrimp | *Metapenaeus monoceros* |
| shrimp | *Mysidopsis* |
| shrimp | *Nitocra* |
| shrimp | *Nitocra spinipes* |
| shrimp | *Ostracoda* |
| shrimp | *Penaeus monodon* |
| shrimp | *Pendaeus sp.* |
| shrub | *Acronychia savaera* |
| silver carp | *Hypophthalmichthys molitrix* |
| skipjack tuna | *Euthynnus pelamis* |
| slender rasbora | *Rasbora daniconius* |
| smallmouth bass | *Micropterus dolomieui* |
| smelt | *Osmerus* |
| snail | *Aplexa* |
| snail | *Aplexa hypnorum* |
| snakehead fish | *Channa striatus* |
| snakehead fish | *Ophiocephalus punctatus* |
| snapping shrimp | *Crangon sp.* |
| snapping turtle | *Chelydra serpentina* |
| sockeye salmon | *Oncorhynchus nerka* |
| soft shell clam | *Mya arenaria* |
| song thrush | *Turdus ericetorum* |
| southern flounder | *Paralichthys lethostigma* |
| sowbug | *Asellus* |
| sowbug | *Asellus brevicaudus* |
| sparrowhawk | *Accipiter nisus* |

| | |
|---|---|
| spike-tailed paradisefish | *Macropodus cupanus* |
| spiny lobster | *Panuliris argus* |
| sprat | *Clupea sprattus* |
| spring peeper | *Hyla crucifer* |
| squirrel monkey | *Saimiri sciureus* |
| stable fly | *Stomoxys calcitrans* |
| staghorn sculpin | *Leptocottus armatus* |
| starfish | *Nostoc linckia* |
| starry flounder | *Platichthys stellatus* |
| stickleback, three-spined | *Gasterosteus aculeatus* |
| stickleback, twelve-spined | *Pygosteus pungitius* |
| stonefly | *Acroneuria* |
| stonefly | *Acroneuria pacifica* |
| stonefly | *Claassenia sabulosa* |
| stonefly | *Isoperla* |
| stonefly | *Pteronarcys dorsata* |
| stonefly | *Pteronarcella* |
| stonefly | *Pteronarcella badia* |
| stonefly | *Pteronarcys* |
| stonefly | *Pteronarcys californica* |
| striped bass | *Morone saxatilis* |
| striped bass | *Roccus saxatilis* |
| striped catfish | *Mystus vittatus* |
| striped mullet | *Mugil cephalus* |
| subterranean clover | *Rhizobium trifolli* |
| suckers | *Catostomus sp.* |
| suckers | *Ictiobus* |
| summer flounder | *Paralichthys dentatus* |
| surf clam | *Spirula solidissima* |
| surgeon fish | *Acanthurus sp.* |
| Swainson's hawk | *Buteo swainsoni* |
| swallows | *Hirundinidae* |
| | |
| tarragon | *Artemisia dracunculus* |
| tawny owl | *Strix aluco* |
| tench | *Tinca tinca* |
| three spot gourami | *Trichogaster trichopterus* |
| tidewater silverside | *Menidia peninsulae* |
| treefrog | *Hyla versicolor* |
| tricolored blackbird | *Agelaeus tricolor* |
| tsetse fly | *Glossina* |
| turbot | *Pseta maxima* |
| turbot | *Scophthalmus rhombus* |
| turkey | *Meleagris* |
| | |
| unicellular biflagellate algae | *Gymnodinium breve* |
| | |
| vendace | *Coregonus albula* |
| village weaver | *Ploceus cucullatus* |

| | |
|---|---|
| vole | *Microtus longicaudus* |
| vole | *Microtus montanus* |
| | |
| walking catfish | *Clarias batrachus* |
| walleye | *Stizostedion vitreum* |
| water flea | *Ceriodaphnia dubia* |
| water flea | *Cladocera* |
| water flea | *Daphnia cucullata* |
| water flea | *Daphnia laevis* |
| water flea | *Daphnia longispina* |
| water flea | *Daphnia sp.* |
| water flea | *Saccobranchus fossilis* |
| water flea | *Simocephalus* |
| water flea | *Simocephalus serrulatus* |
| watermint | *Mentha aquatica* |
| western chipmunk | *Eutamias sp.* |
| white amur grass carp | *Ctenopharyngodon idella* |
| white crappie | *Pomoxis annularis* |
| white shrimp | *Penaeus setiferus* |
| white sucker | *Catostomus commersoni* |
| white-crowned sparrow | *Zonotrichia leucophrys* |
| white-footed mouse | *Peromyscus leucopus* |
| white-fronted dove | *Leptotila verreauxi* |
| white-winged dove | *Zenaida asiatica* |
| whiting | *Gadus merlangus* |
| wild boar | *Sus scrofa* |
| wild turkey | *Meleagris gallopavo* |
| willow shiner | *Gnathopogon caerulescens* |
| winter flounder | *Pseudopleuronectes americanus* |
| wolf-fish | *Anarchichas lupus* |
| woodfrog | *Rana sylvatica* |
| worm | *Tubifex tubifex* |
| | |
| yeast | *Aureobasidium sp.* |
| yeast | *Saccharomyces cerevisiae* |
| yeast | *Saccharomyces nounii* |
| yeast | *Schizosaccharomyces octosporus* |
| yeast | *Sporobolomyces salmonicolor* |
| yellow bullhead | *Ictalurus natalis* |
| yellow perch | *Perca flavescens* |
| yellow-billed magpie | *Pica nuttalli* |
| yellow-headed blackbird | *Xanthocephalus xanthocephalus* |
| yellowfin tuna | *Thunnus albacares* |
| yellowtail | *Seriola lalandei* |
| | |
| zebra fish | *Brachydanio rerio* |

# ABBREVIATIONS

| | |
|---|---|
| ACGIH | American Conference of Governmental Industrial Hygienists |
| ADI | acceptable daily intake |
| AFNOR | Association Francaise de Normalization |
| AIDS | Acquired Immune-Deficiency Syndrome |
| ATP | adenosine triphosphate |
| BAN | British Approved Name |
| BOD | Biological Oxygen Demand |
| Bq | Becquerel |
| BrdU | bromodeoxyuridine |
| BSI | British Standards Institute |
| c | centi |
| ca. | circa |
| CAS | Chemical Abstracts Service |
| CAS RN | Chemical Abstracts Service Registry Number |
| C.I. | Colour Index |
| Ci | Curie |
| CIP | Centre for International Projects |
| COD | Chemical Oxygen Demand |
| $d_4^{20}$ | density (temperature stated where known) |
| decomp. | decomposition |
| °C | degree centigrade |
| DDE | 2,2-bis(4-chlorophenyl)-1,1-dichloroethylene |
| DMSO | dimethyl sulfoxide |
| DNA | deoxyribonucleic acid |
| DO | dissolved oxygen |
| DoE | Department of the Environment (UK) |
| DPTA | diethylenetriaminepentaacetic acid |
| EC | effective concentration |
| EC | European Community |
| ECETOC | European Chemical Industry Ecology and Toxicology Centre |
| ECG | electrocardiogram |
| ED | effective dose |
| EDTA | (ethylenedinitrilo)tetraacetic acid |
| Ed | editor |
| ed | edition |
| EEC | European Economic Community |
| EEG | electroencephalogram |

| | |
|---|---|
| ELISA | Enzyme-linked immunosorbent assay |
| EPA | Environmental Protection Agency (US) |
| et al | and others (authors) |
| FAO | Food and Agriculture Organisation |
| FDA | Food and Drug Administration (US) |
| FIX ZESTE | *Drosophila melanogaster* test for aneuploidy |
| | FIX – ♀ free heterozygous inverted X chromosome |
| | ZESTE – X-linked sexual mutation |
| FMN | flavin mononucleotide |
| GC-MS | gas chromatography-mass spectrometry |
| g | gram |
| G-6-PD | glucose-6-phosphate dehydrogenase |
| GSH | reduced glutathione |
| ha | hectare |
| Hb | haemoglobin |
| HGPRT | hypoxanthine-guanine phosphorisobutyltransferase |
| HMSO | Her Majesty's Stationery Office |
| HPLC | high performance liquid chromatography |
| hr | hour |
| HSC | Health & Safety Commission |
| HSE | Health & Safety Executive |
| 5-HT | 5-hydroxytryptamine |
| Hz | Hertz |
| IARC | International Agency for Research on Cancer |
| ILO | International Labour Office |
| IR | infrared |
| IRPTC | International Register of Potentially Toxic Chemicals |
| ISO | International Standards Organisation |
| i.u. | International Unit |
| IUPAC | International Union of Pure and Applied Chemistry |
| J | Joule |
| JETOC | Japan Chemical Industry Ecology and Toxicology Information Center |
| k | kilo |
| $K_a$ | dissociation constant, acids |
| kg | kilogram |
| $K_{oc}$ | soil sorption coefficient |
| $K_{ow}$ | dissociation constant, octanol:water |
| $LC_{50}$ | lethal concentration – 50 |
| $LD_{50}$ | lethal dose – 50 |
| $LD_{Lo}$ | lowest lethal dose |
| LOEC | lowest observed effect concentration |
| Ltd | limited |
| l | litre |
| $\log P_{ow}$ | $\log_{10}$ of partition of coefficient – octanol:water |
| μ | micro |

| | |
|---|---|
| m | milli/metre |
| MAC | maximum admissible concentration |
| MEL | maximum exposure limit |
| MetHb | methaemoglobin |
| MITI | Ministry of International Trade and Industry, Japan |
| mm | millimetre |
| mmHg | millimetres mercury |
| min | minute |
| ml | millilitre |
| M | Molar |
| n | nano |
| N | Newton |
| n | normal |
| N | Normality |
| $NAD^+$ | Nicotinamide-adenine dinucleotide (oxidised form) |
| NADH | Nicotinamide-adenine dinucleotide (reduced form) |
| NAPQI | $N$-acetyl-$p$-benzoquinone imine |
| NIOSH | National Institute of Occupational Safety and Health |
| NOEC | no observed effect concentration |
| NTA | nitrilotriacetic acid |
| NTIS | National Technical Information Service |
| OECD | Organisation for Economic Cooperation and Development |
| p | pico |
| PAH | polycyclic aromatic hydrocarbon |
| PCB | polychlorinated biphenyl |
| PEG | polyethylene glycol |
| pKa | log Ka |
| ppb | parts per billion ($10^9$) |
| pph | parts per hundred ($10^2$) |
| ppm | parts per million ($10^6$) |
| ppt | parts per thousand ($10^3$) |
| pH | $\log_{10}$ hydrogen ion concentration |
| psi | pound force – square inch |
| PTFE | polytetrafluoroethylene |
| QSAR | Quantitative Structure – Activity Relationships |
| $R_{50}$ | repellency factor |
| RAD | Radiation absorbed dose 1 Erg $g^{-1}$ |
| $RD_{50}$ | 50% decrease in respiratory rate |
| RN | Registry Number |
| RNA | ribonucleic acid |
| RSC | The Royal Society of Chemistry |
| RTECS | Registry of Toxic Effects of Chemical Substances |
| sec | second |
| SMART | Somatic Mutation and Recombination Test (in *Drosophila melanogaster*) |
| s/v | surface/volume |
| $t_{1/2}$ | half life |

| | |
|---|---|
| TCA | tricarboxylic acid |
| ThOD | Theoretical Oxygen Demand |
| TLV | Threshold Limit Value |
| TOC | Total Organic Carbon |
| TPA | 12-*O*-tetradecanoylphorbol-13-acetate |
| TSH | thyrotropin |
| TWA | Time-Weighted Average |
| UK | United Kingdom |
| UNEP | United Nations Environment Programme |
| USA | United States of America |
| UV | ultraviolet |
| v. den. | vapour density |
| vol | volume |
| v.p. | vapour pressure |
| v/v | volume/volume |
| w/v | weight/volume |
| W | Watt |
| wk | week |
| WHO | World Health Organization |
| wt | weight |
| w/w | weight/weight |
| yr | year |

# INDEX OF NAMES AND SYNONYMS

3-aminoiodobenzene I38
4-aminoiodobenzene I39
*m*-aminoiodobenzene I38
aminomercuric chloride M54
2-amino-6-methoxypurine M126
4-amino-10-methylfolic acid M118
*d*-2-amino-3-methylpentanoic acid I99
4-amino-*N*-methylpteroylglutamic acid M118
*d*-α-amino-β-methylvaleric acid I99
2-amino-5-nitroanisole M129
4-amino-3-nitroanisole M130
1-amino-3-nitro-4-methylbenzene M240
2-(*p*-aminophenyl)-6-methylbenzothiazole
   M149
Ammoidin M119
ammonium ferrous sulfate I61
ammonium, trimethyltetradcyl-, bromide M342
Amorthol Fast Red B Base M129
*tert*-amyl alcohol M161
amylene hydrate M161
amyl methyl alcohol M253
Anestacon L42
anesthenyl M141
Anglislite L28
anhydrone M7
anhydrous magnesium perchlorate M7
Anicon P M32
DL-alanine *N*-(2,6-dimethylphenyl)-
   *N*-(methoxyacetyl)-, methyl ester M85
aniline, *N*-(2-methyl-2-nitropropyl)-
   *p*-nitroso- M243
anilinomethane M143
animag M6
anisole, 6-*tert*-butyl-3-methyl-2,4-dinitro-
   M338
Ansar 170 M336
Antergon M13
Antideprin I13
Arcillite M329
Arelon I127
Aresin M324
Arezin M324
Arezine M324
Argiflex K2
Armyl L61
Arnold's base M198
Arsenal I5
arsenic acid (H₃AsO₄), lead(2+) salt L11
arsenic acid, lead salt L11
arsenious acid, lead salt L12
arsenious acid (HAsO₂), lead(2+) salt L12
arsonic acid, methyl-, monosodium salt M336
arsonous dichloride, (2-chloroethenyl)- L40
Assault I5

Assert I4
Atorel I32
Australol I125
autigene MB M47
Avantin I115
ayapanin M123
1-azaindene I26
2-azaindole I18
2-azanaphthalene I129
1*H*-azepine-1-carbothioic acid, hexahydro-,
   *S*-ethyl ester M316
Azidithion M39
aziridine, 1,1′,1″-phosphinylidynetris-
   [2-methyl]- M89
azirino[2′,3′:3,4]pyrrolo[1,2-*a*]indole-4,7-dione,
   6-amino-1,1a,2,8,8a,8b-hexahydro-
   8-(hydroxymethyl)-8a-methoxy-5-methyl-,
   carbamate (ester) M315
azirino[2′,3′:3,4]pyrrolo[1,2-*a*]indole-
   4,7-dione, 6-amino-8-[[(aminocarbonyl)-
   oxy]methyl]-1,1a,2,8,8a,8b-hexahydro-
   8a-methoxy-5-methyl-,
   [1a*S*-(1aα,8β,8aα,8bα)]- M315
Azodrin M322
Ba 35846 M275
banana oil I104
Barhist M108
BAS 47900H M88
bauxite residue I68
Bay 37344 M113
BAY DRW1139 M87
BBCE I8
Belustine L54
Bentolite M329
Bentrol I53
benz[*j*]aceanthrylene, 1,2-dihydro-3-methyl-
   M177
2-benzazine I129
benzenamine,
   4-[(4-aminophenyl)(4-imino-3-methyl-
   2,5-cyclohexadien-1-ylidene)methyl]-
   2-methyl-, monohydrochloride M1
benzenamine, 4,4′-methylenebis-[2-methyl]-
   M199
benzenamine, 4-methyl-2-nitro- M239
benzenamine, *N*-(2-methyl-2-nitropropyl)-
   4-nitroso- M243
benzeneacetic acid, 3-benzoyl-α-methyl- K10
1,3-benzenedicarbonitrile I109
*m*-benzenedicarboxylic acid I108
1,2-benzenedicarboxylic acid, monobutyl
   ester M320
1,2-benzenedicarboxylic acid,
   mono(2-ethylhexyl) ester M323

1,2-benzenedicarboxylic acid, monooctyl
ester M325
1,2-benzenedicarboxylic acid, monopentyl
ester M326
benzene, 1-(1,1-dimethylethyl)-2-methoxy-
4-methyl-3,5-dinitro- M338
benzeneethanol, α-[2-(dimethylamino)-
1-methylethyl]-α-phenyl-, propanoate
(ester), [R-(R'S')]- L38
benzene hexachloride L45
ε-benzene hexachloride L46
benzene iodide I40
benzene, iodo- I40
benzenepropenal, 4-(1,1-dimethylethyl)-
α-methyl- M171
benzene, 1,3,5-trimethyl- M83
2-benzimidazolethiol M47
benzocyclopentane I15
benzoic acid, 2-[[[(4-methoxy-6-methyl-
1,3,5-triazin-2-yl)amino]carbonylamino]-
sulfonyl]-, methyl ester M305
4,5-benzoindotricarbocyanine I25
benzo[c]pyridine I129
1H-benzopyrazole I18
benzopyridine I129
benzothiazole disulfide M49
benzothiazole, 2-(methylthio)- M289
2-benzothiazolethiol M48
1-benzothiazol-2-yl-1,3-dimethylurea M91
1-(1,3-benzothiazol-2-yl)-1,3-dimethylurea
M91
1-(2-benzothiazolyl)-1,3-dimethylurea M91
N-2-benzothiazolyl-N,N'-dimethylurea M91
2-benzothiazol-2-yloxy-N-methylacetanilide
M33
2-(2-benzothiazolyloxy)-N-methyl-
N-phenylacetamide M33
3-benzoylhydratropic acid K10
benzylmethyl ether M127
Berkomine I14
betula oil M284
ε-BHC L46
γ-BHC L45
bicyclo[3.1.1]hept-2-ene-2-carboxaldehyde,
6,6-dimethyl- M343
Bioxone M110
bis(acetato)dihydroxytrilead L27
4-[bis(2-chloroethyl)amino]-L-phenylalanine
M38
DL-3-[p-[bis(2-chloroethyl)amino]phenyl]-
alanine M81
N,N-bis(β-chloroethyl)amino-N',O-propyl-
enephosporic acid ester diamide I107
bis(β-chloroethyl) sulfide M339

N,3-bis(2-chloroethyl)tetrahydro-
2H-1,3,2-oxazaphosphorin-2-amine,
2-oxide I107
N,N'-bis[4-chloro-3-(trifluoromethyl)-
phenyl]urea M314
bis(cyanoethyl)amine I8
bis(dimethyldithiocarbamato)lead L16
bis(dimethylthiocarbamoyl) sulfide M328
S-[1,2-bis(ethoxycarbonyl)ethyl]
O,O-dimethyl phosphorodithioate M10
bis(8-guanidinooctyl)amine I10
bis(2-hydroxyethyl)methylamine M220
bis(4-isocyanatocyclohexyl)methane M197
bis(2-methoxyethyl) ether M124
2,2-bis(p-methoxyphenyl)-1,1,1-trichlorethane
M122
1,1-bis(p-methoxyphenyl)-2,2,2-trichloro-
ethane M122
2-isocyanatoethyl methacrylate M99
bisphenol A I120
borate(1-), tetrafluoro-, lead(II) L18
bromoacetic acid, methyl ester M157
O-(4-bromo-2,5-dichlorophenyl) O,O-methyl
phenylphosphonothioate L35
3-(4-bromophenyl)-1-methoxy-1-methylurea
M299
N'-(4-bromophenyl)-N-methoxy-N-methylurea
M299
Brufen I2
Brush-off M305
Bulkaloid M203
Burstane L55
1-butanamine, N-methyl- M169
iso-butane I75
butanedioic acid,
[(dimethoxyphosphinyl)thio]-, diethyl
ester M9
butane, 1-iodo- I41
butane, 2-iodo- I42
butanoic acid, 1-methylethyl ester I116
2-butanol, 4-(dimethylethylamino)-
3-methyl-1,2-diphenyl-, propionate (ester),
(−) L38
2-butenedioic acid, (Z)-, dimethyl ester M11
2-butenedioic acid, (E)-, iron(2+) salt (1:1) I66
cis-butenedioic anhydride M12
2-butenoic acid,
3-[(dimethoxyphosphinyl)oxy]-,
1-phenylethyl ester, (E)- M153
2-butenoic acid, 2-methyl-7-[[2,3-dihydroxy-
2-(1-methoxyethyl)-3-methyl-1-oxobutoxy]-
methyl]-2,3,5,7a-tetrahydro-1H-pyrrolizin-
1-yl ester, [1S-[1a(Z),7(2S*,3R*),7aa]]- L7
1-buten-3-one M297

3-buten-2-one M297
3-buten-2-one, 4-(2,6,6-trimethyl-
2-cyclohexen-l-yl-, (*E*)- I52
Butiphos M82
Butisan S M88
2-butyl acetate M267
*sec*-butyl acetate M267
*sec*-butyl alcohol, acetate M267
butylamine, *N*-methyl- M169
γ-butylene I77
butyl hydrogen phthalate M320
butyl α-hydroxypropionate L2
butyl iodide I41
*n*-butyl iodide I41
2-butyl iodide I42
*sec*-butyl iodide I42
*tert*-butyl iodide I49
butyl lactate L2
*n*-butyl lactate L2
butyl methacrylate M266
butylmethylamine M169
*tert*-butyl methyl ether M170
butyric acid, isopropyl ester I116
butyrinase L48
3025 C.B M38
C.I. 473 M34
C.I. 37100 M237
C.I. 37105 M238
C.I. 37110 M239
C.I. 37125 M129
C.I. 52015 M201
C.I. 73000 I21
C.I. 73015 I22
C.I. 75125 L60
C.I. 75781 I22
C.I. 77575 L10
C.I. 77580 L17
C.I. 77610 L15
C.I. 77613 L21
C.I. 77622 L24
C.I. 77630 L28
C.I. 77640 L29
C.I. 77760 M75
C.I. 77764 M58
C.I. azoic diazo component 12 M238
C.I. Basic Blue 9 M201
C.I. Basic Violet 14 monohydrochloride M1
C.I. Natural Yellow 27 L60
C.I. Pigment Metal 4 L10
C.I. Pigment Red I68
C.I. Pigment White 3 L28
C.I. Pigment Yellow 48 L15
CA69-15 I53
calochlor M59

Captax M48
carbamic acid, dimethyldithioanhydrosulfide
M328
Carbamic acid, (1,1-dimethylethyl)-,
3-[[(dimethylamino)carbonyl]amino]phenyl
ester K4
carbamic acid, 2-methyl-2-propyltrimethylene
ester M44
carbinol M106
carbofos M10
carbomethene K8
carbonochloridic acid, 1-methylethyl ester I117
carbon triiodide I46
carboxyacetic acid M15
*N*-(carboxymethyl)glycine I11
Cardio Green I25
carob gum L53
ψ,ψ-carotene L60
Cartagena ipecacuanha I54
Carubinose M27
Carvene L44
CB-3307 M81
1'-2,C-4,C-6,C-8-tetramethyl-
1,3,5,7-tetroxocane M86
Celacol M M203
cellulose methylate M203
cellulose methyl ether M203
cellumeth M203
*Cephaelis ipecacuanha* I54
Certrol I53
Channing's solution M77
Chemiazid I100
Chinofer I64
Chipco 26019 I56
chloric acid, magnesium salt M3
chlorine iodide I36
chlorine monoiodide I36
1-(4-chlorobenzoyl)-5-methoxy-2-methyl-
1*H*-indole-3-acetic acid I31
2-chloro-2-(difluoromethoxy)-1,1,1-trifluoro-
ethane I96
4'-chloro-2,2-dimethylvaleranilide M319
3-(2-chloroethyl)-2-[(2-chloroethyl)amino]-
tetrahydro-2*H*-1,3,2-oxazaphosphorin
2-oxide I107
1-(2-chloroethyl)-3-cyclohexyl-1-nitrosourea
L54
chloroethylcyclohexylnitrosourea L54
2-chloro-6'-ethyl-*N*-(2-methoxy-1-methylethyl)
acet-*o*-toluidide M300
2-chloro-*N*-(2-ethyl-6-methylphenyl)-
*N*-(2-methoxy-1-methylethyl)acetamide
M300
chloroformic acid, isopropyl ester I117

3-(3-chloro-4-methoxyphenyl)-1,1-dimethyl-
urea M302

*O*-[5-chloro-1(1-methylethyl)-1*H*-1,2,4-triazol-
3-yl]-*O*,*O*-diethyl phosphorothioic acid
ester I72

(4-chloro-2-methylphenoxy)ethanethioic acid,
*S*-ethyl ester M28

(±)-2-(4-chloro-2-methylphenoxy)propanoic
acid M32

chloromethylsulfone M104

*N*-(4-chlorophenyl)-2,2-dimethylpentanamide
M319

3-(*p*-chlorophenyl)-1,1-dimethylurea M330

*N*′-(4-chlorophenyl)-*N*,*N*-dimethylurea M330

3-(*p*-chlorophenyl)-1-methoxy-1-methylurea
M324

*N*′-(4-chlorophenyl)-*N*-methoxy-*N*-methylurea
M324

2-chloro-*N*-(pyrazol-1-ylmethyl)acet-
2′,6′-xylidine M88

(±)-2-[(4-chloro-*o*-tolyl)oxy]propionic acid
M32

[*RS*]-2- [4-chloro-*o*-tolyloxy]propionic acid
M32

[(4-chloro-*o*-tolyl)oxy]thioacetic acid, *S*-ethyl
ester M28

1-chloro-2,2,2-trifluoroethyl difluoromethyl
ether I96

chlorovinylarsine dichloride L40

5β-cholan-24-oic acid, 3α-hydroxy- L52

cholan-24-oic acid, 3-hydroxy-, (3α,5β)- L52

cholanthrene, 3-methyl- M177

Chopper I5

Chromic acid (H$_2$CrO$_4$), lead(2+) salt(1:1) L14

chrysene, 1-methyl- M178

chrysene, 2-methyl- M179

chrysene, 3-methyl- M180

chrysene, 5-methyl- M181

chrysene, 6-methyl- M182

Ciba 9491 I45

Ciba-Geigy A 12223 I72

Ciclisin L61

Cidolysal L61

cincholepidine M279

Ciodrin M153

Citexal M109

CMPP M32

Coapt M183

Cobratec TT100 M150

colonial spirit M106

Compound 469 I96

Compound 711 I92

Compound 864 I9

COP1 I51

cordycepic acid M26

corrosive sublimate M59

Coslan M34

Cpiron I66

*p*-cresol methyl ether M144

*m*-cresol, 4-(methylthio)- M232

crotonic acid, 3-hydroxy-, α-methylbenzyl
ester, dimethyl phosphate, (*E*)- M153

Crotoxyphos M153

crystosol M310

Culminal MC M203

*p*-cumenol I125

*m*-cumenol, methylcarbamate M205

*m*-cumenyl methyl carbamate M205

*o*-cumenyl methylcarbamate I111

cumic aldehyde M204

cyanoacetonitrile M16

cyanolyt M183

3-(cyanomethyl)indole I27

cyano(methylmercuro)guanidine M69

2-cyanopropane I90

2-cyano-1-propene M97

cyanuramide M36

cyanurotriamide M36

α-cyclocitrylideneacetone I52

cyclohexane 1,2,3,4,5,6-hexachloro-,
(1α,2α,3β,4α,5α,β)- L45

*cis*-1,2,3,5-*trans*-4,6-cyclohexanehexol I33

cyclohexane, isopropyl- I118

cyclohexane, 1-methylethyl- I118

*trans*-cyclohexane,
1-methyl-4-(1-methylethyl)-, I122

cyclohexanol, 2-methyl- M186

cyclohexanone, methyl- M189

cyclohexene, 1-methyl-4-(1-methylethenyl)-,
(*R*)- L44

cyclohexylmethane M184

cyclopentadienylmanganese tricarbonyl M21

1*H*-cyclopentapyrimidine-2,4(3*H*,5*H*)-dione,
3-cyclohexyl-6,7-dihydro- L34

cyclopropanecarboxylic acid,
3-[(dihydro-2-oxo-3(2*H*)-thienylidene)-
methyl]-2,2-dimethyl-, [5-(phenylmethyl)-
3-furanyl]methyl ester, [1*R*-[1α,3α(*E*)]]- K1

cyclopropanecarboxylic acid,
2,2-dimethyl-3-(2-methyl-1-propenyl)-,
2-methyl-4-oxo-3-(2-pentenyl)-
2-cyclopenten-1-yl ester,
[1*R*-1α[*S*\*(*Z*)]*,*3β](*E*)]- J1

cyclopropanecarboxylic acid, 3-(3-methoxy-
2-methyl-3-oxo-1-propenyl)-2,2-dimethyl-,
2-methyl-4-oxo-3-(2-pentenyl)-
2-cyclopenten-1-yl ester,
[1*R*-[1α[*S*\*(*Z*)],3β(*E*)]]- J2

Cyclosan M58
Cymantrene M21
Cyodrin M153
Cytrolane M43
Daconate M336
Dagger I4
dambose I33
Damfin M92
Danantizal M112
dechlorane M311
dehydrite M7
dehydrothio-*p*-toluidine M149
Dehyquart C L9
Deleaf defoliant M82
Delton M286
Demasorb M286
Dendrid I43
2′-deoxy-5-iodouridine I43
3-deoxynorlutin L62
1-(2-deoxy-β-D-ribofuranosyl)-5-iodouracil I43
deoxyteraric acid M14
Deprinol I14
Deriton M329
Despirol K6
Devicorun I131
DHPT M149
Diabefagos M90
diacetoxymercury M53
Diacon M117
*N*-(4-[[(2,4-diamino-6-pteridinyl)methyl]
    methylamino]benzoyl-L-glutamic acid M118
*S*-[(4,6-diamino-1,3,5-triazin-2-yl)methyl]
    *O,O*-dimethyl phosphorodithioate M39
Dian I120
1,4:3,6-dianhydro-D-glucitol I131
1,2-diazaindene I18
Diazald M246
Diazale M246
Diazo Fast Red B M129
1,3-diazole I7
dibasic lead phosphite L25
dibenzothiazyl disulfide M49
dicarboxymethane M15
4,4′-dichloro-3,3′-bis(trifluoromethyl)-
    carbanilide M314
*O*-(2,5-dichloro-4-bromophenyl) *O*-methyl
    phenylthiophosphonate L35
dichloro(2-chlorovinyl)arsine L40
dichlorohydridomethylsilicon M193
dichloromethylsilane M193
1-[2-(2,4-dichlorophenyl)-
    2-[(2,4-dichlorophenyl)methoxy]ethyl]-
    1*H*-imidazole M309
3,4-dichlorophenyl isocyanate I91

2-(3,4-dichlorophenyl)-4-methyl-
    1,2,4-oxadiazolidine-3,5-dione M110
1-[2-(2,4-dichlorophenyl)-2-(2-propenyloxy)-
    ethyl]-1*H*-imidazole I3
dicresyl M301
1,3-dicyanobenzene I109
*m*-dicyanobenzene I109
dicyanomethane M16
diethanol methylamine M220
diethyl carbinol M255
diethyl (dimethoxythiophosphorylthio)-
    succinate M10
diethylene glycol monomethyl ether M124
diethylene imidoxide M333
diethylenimide M333
diethylmercury M61
diethyl 4-methyl-1,3-dithiolan-2-ylidene-
    phosphoramidate M43
diglycine I11
diglycol monomethyl ether M124
diglykokoll I11
10,11-dihydro-5*H*-dibenz[*b,f*]azepine I12
10,11-dihydro-*N,N*-dimethyl-5*H*-dibenz[*b,f*]-
    azepine-5-propanamine I13
2-[7-[1,3-dihydro-1,1-dimethyl-3-(4-sulfobutyl)-
    2*H*-benz[*e*]indol-2-ylidene]-
    1,3,5-heptatrienyl]-1,1-dimethyl-
    3-(4-sulfobutyl)-1*H*-benz[*e*]indolium,
    hydroxide, inner salt, sodium salt I25
dihydro-2,5-dioxofuran M12
2,3-dihydro-1*H*-indene I15
2,3-dihydro-1*H*-inden-5-ol I17
*S*-2,3-dihydro-5-methoxy-2-oxo-
    1,3,4-thiadiazol-3-ylmethyl *O,O*-dimethyl
    phosphorodithioate M111
1,3-dihydro-1-methyl-2*H*-imidazole-2-thione
    M112
(±)-2-[4,5-dihydro-4-methyl-4-(1-methylethyl)-
    5-oxo-1*H*-imidazol-2-yl]-5-ethyl-
    3-pyridinecarboxylic acid I6
2-[4,5-dihydro-4-methyl-4-(1-methylethyl)-
    5-oxo-1*H*-imidazol-2-yl]-4(or 5)-methyl-
    benzoic acid, methyl ester I4
2-[4,5-dihydro-4-methyl-4-(1-methylethyl)-
    5-oxo-1*H*-imidazol-2-yl]-
    3-pyridinecarboxylic acid I5
3,7-dihydro-1-methyl-3-(2-methylpropyl)-
    1*H*-purine-2,6-dione I85
3,4-dihydro-2-methyl-4-oxo-3-*o*-tolyl-
    quinazoline M109
2,3-dihydro-6-methyl-2-thioxo-
    4(1*H*)pyrimidinone M291
2-(1,3-dihydro-3-oxo-2*H*-indol-2-ylidene)-
    1,2-dihydro-3*H*-indol-3-one I21

2-(1,3-dihydro-3-oxo-5-sulfo-2*H*-indol-
2-ylidene)-2,3-dihydro-3-oxo-1*H*-indole-
5-sulfonic acid, disodium salt I22
1,7-dihydro-6*H*-purine-6-thione M50
1,2-dihydro-3,6-pyridazinedione M13
1,2-dihydropyridazine-3,6-one M13
*p,p*'-dihydroxydiphenylpropane I120
3,6-dihyroxypyridazine M13
Diiodine I35
2,6-diiodo-4-cyanophenol I53
diiron trisulfate I71
diisopropoxylphosphoryl fluoride I95
diisopropyl fluorophosphate I95
diisopropyl fluorophosphonate I95
diisopropyl phosphofluoridate I95
1,3-diketohydrindene I16
Dilcit I34
Dimelon-methyl M196
3-[(dimethooxyphosphinyl)oxy]-2-butenoic
acid methyl ester M307
3,4-dimethoxybenzoic acid,
4-[ethyl[2-(4-methoxyphenyl)-1-methyl-
ethyl]amino]butyl ester M29
dimethoxy-DDT M122
dimethoxymethane M141
di(*p*-methoxyphenyl)trichloromethylmethane
M122
[(dimethoxyphosphinothioyl)thio]butanedioic
acid, diethyl ester M10
3,3-dimethoxyphosphinothioyl
thiomethyl-5-methoxy-
1,3,4-thiadiazol-2(3*H*)-one M111
dimethylacetic acid I89
dimethyl acetonitrile I90
dimethylacetylenecarbinol M172
dimethylallyl alcohol M165
3,3-dimethylallyl alcohol M165
α,α-dimethylallyl alcohol M167
4-(dimethylamino)-3,5-dimethylphenol
methylcarbamate (ester) M308
2-[(2-(dimethylamino)ethyl)-2-thenyl-
amino]pyridine hydrochloride M108
2-[(2-dimethylaminoethyl)-2-thenyl-
amino]pyridine M107
5-[3-(dimethylamino)propyl]-10,11-dihydro-
5*H*-dibenz[*b,f*]azepine I13
4-(dimethylamino)-3,5-xylyl
*N*-methylcarbamate M308
1,1-dimethylbiguanide hydrochloride
M90
1,3-dimethylbutanol M253
dimethyl (*Z*)-butenedioate M11
1,1-dimethylethane I75
dimethylethylcarbinol M161

dimethylcarbamic acid,
3-methyl-1-(methylethyl)-1*H*-pyrazol-5-yl
ester I98
dimethyl *cis*-ethylenedicarboxylate M11
*N*'-[5-(1,1-dimethylethyl)-3-isoxazolyl]-
*N,N*-dimethylurea I133
dimethyl formal M141
*N,N*-dimethylimidodicarbonimidio diamide
monohydrochloride M90
dimethyl maleate M11
dimethylmercury M62
*N,N*-dimethyl-*N*'-(4-methoxy-3-chloro-
phenyl)urea M302
*N,N*-dimethyl-*N*'-[4-(1-methylethyl)phenyl]-
urea I127
3,5-dimethyl-4-methylthiophenyl
methylcarbamate M113
*O,O*-dimethyl *S*-(morpholinocarbonylmethyl)
phosphorodithioate M335
*O,O*-dimethyl *S*-[2-(4-morpholinyl)-
2-oxoethyl] phosphorodithioate M335
*N*,4-dimethyl-*N*-nitrosobenzenesulfonamide
M246
2-[(2,3-dimethylphenyl)amino]benzoic acid
M34
*O,S*-dimethyl phosphoramidothioate M101
α,α-dimethylpropargyl alcohol M172
1,1-dimethylpropynol M172
1,1-dimethyl-2-propyn-1-ol M172
2,3-dimethylpyridine L57
2,4-dimethylpyridine L58
2,5-dimethylpyridine L59
α,γ-dimethylpyridine L58
*N,N*-dimethyl-*N*'-2-pyridinyl-*N*'-(2-thienyl-
methyl)-1,2-ethanediamide M107
dimethyl sulfoxide M286
*N*-[2,4-dimethyl-5-[[(trifluoromethyl)-
sulfonyl]amino]phenyl]acetamide M35
dimethylvinylcarbinol M167
dimethylvinylmethanol M167
dimexide M286
2,6-dinitro-*N,N*-dipropylcumidine I112
18,19-dinorpregn-4-en-20-yn-3-one,
13-ethyl-17-hydroxy-, (17α)- L37
diolane M251
Diosmol M26
di-*o*-toluidine M199
2,4-dioxapentane M141
2,3-dioxoindoline I28
Dioxolan I121
1,3-dioxolane-4-methanol, 2,2-dimethyl- I121
Dipanol L43
diphenylolpropane I120
dipotassium tetraiodate M77

ferric trifluoride I65
ferroglucin I64
Ferrone I66
ferrous ammonium sulfate I61
ferrous chloride I62
ferrous diammonium disulfate I61
ferrous dichloride I62
ferrous sulphate I70
Filitox M101
fire damp M102
Fleet-X M83
Florasan I3
flores martis I63
Flucofuron M314
fluorosulfuric acid, methyl ester M208
FOE 1976 M33
Folex M82
formal M141
formaldehyde dimethyl acetal M141
formic acid, chloroisopropyl ester I117
formic acid, isobutyl ester I81
formic acid, isopropyl ester I119
formic acid, methyl ester M210
formic acid, 1-methylethyl ester I119
formic acid, methylhydrazide M211
2-formylbutane M159
1-formyl-1-methylhydrazine M211
2-formylpentane M295
Freemans white lead L28
fructoline L53
Fuji-one I126
fumaric acid, iron(2+) salt (1:1) I66
Fumette M104
Fungaflor I3
Fungazil I3
Furaladone M334
2,5-furandione M12
furan, tetrahydro-2-methyl- M288
N-furfuryladenine K11
6-(furfurylamino)purine K11
Furmethanol M334
Furzoline M334
GC-2996 M331
Gepiron M336
Glucophage M90
β-D-glucopyranose,
    4-O-β-D-galactopyranosyl- L4
glycerol acetonide I121
glycerol dimethylketal I121
glycerol ester hydrolase L48
glycerylguaiacolate carbamate M115
Glycophen I56
Glyoxaline I7
Graminon I127

Granstar M305
green vitriol I70
Gropper M305
p-guaiacol M131
guaiacol glyceryl ether carbamate M115
Halodorm M109
ε-HCH L46
γ-HCH L45
Heat Pre M194
1-heptanol, 6-methyl- I102
heptenone, methyl- M213
Herbicide 634 L34
Herbit M28
herniarin M123
Herplex I43
heteroauxin I29
$(1\alpha,2\alpha,3\alpha,4\beta,5\beta,6\beta)$-1,2,3,4,5,6-hexachloro-
    cyclohexane L46
ε-1,2,3,4,5,6-hexachlorocyclohexane L46
ε-hexachlorocyclohexane L46
hexachlorocyclopentadiene dimer M311
1,2,3,4,10,10-hexachloro-1,4,4a,5,8,8a-hexa-
    hydro-1,4,5,8-*endo,endo*-
    dimethanonaphthalene I92
Hexadimethrine bromide I51
hexafluoromagnesium (1:1) silicate (2-) M4
hexahydrocresol M185
hexahydro-p-cresol M188
hexahydromethylphenol M185
hexahydrotoluene M184
1,2,3,4,5,6-hexanehexol M26
Hexaniat I34
Hexilure L34
hexone I84
hexylene glycol M251
Histadyl M107
Histidyl M108
HOK 7501 M28
Hormex I30
Hormodiu I30
Hydram M316
hydratopic acid, m-benzoyl- K10
hydrazinecarbothioamide, 2,2'-methylenebis- M200
hydrindene I15
hydrindonaphthene I15
hydriodic ether I44
hydrocinnamaldehyde,
    p-tert-butyl-α-methyl- M171
Hydrogloss K2
8-hydromirex M312
Hydronol I131
hydroquinone methyl ether M131
(±)-α-hydroxybenzeneacetic acid M18
4-hydroxybenzoic acid, methyl ester M218

Isobide, 1,4:3,6-dianhydrosorbital I131
isoborneol acetate I74
Isobornyl acetate I74
isobutanal I88
Isobutane I75
isobutanoic acid I89
Isobutanol I76
isobutenal M93
Isobutene I77
Isobutenylcarbinol M166
isobutenyl methyl ketone M261
2-isobutoxynaphthalene I86
isobutyl alcohol I76
Isobutyl acetate I78
Isobutyl acrylate I79
Isobutylamine I80
isobutylcarbinol M162
isobutylene I77
isobutylethylene M259
Isobutyl formate I81
p-isobutylhydratropic acid I2
isobutyl iodide I48
Isobutyl isobutanoate I82
Isobutyl isobutyrate I82
Isobutyl methacrylate I83
isobutyl α-methylacrylate I83
isobutyl methyl carbinol M256
Isobutyl methyl ketone I84
isobutyl methylmethanol M256
Isobutylmethylxanthine I85
3-isobutyl-1-methylxanthine I85
Isobutyl 2-naphthyl ether I86
2-(4-isobutylphenyl)propionic acid I2
isobutyl propanoate I87
isobutyl 2-propenoate I79
Isobutyl propionate I87
isobutyltrimethylethane I101
Isobutyraldehyde I88
Isobutyric acid I89
isobutyric acid, isobutyl ester I82
isobutyric acid, methyl ester M223
Isobutyronitrile I90
Isocarnox M32
2-isocyanatoethyl methacrylate M99
isocyanatomethane M224
3-isocyanatomethyl-3,5,5-trimethylcyclohexyl
    isocyanate I106
Isocyanic acid, 3,4-dichlorophenyl
    ester I91
Isocyanic acid,
    methylenedi-4,1-cyclohexylene ester
    M197
isocyanic acid, methyl ester M224
Isodrin I92

Isoeugenol I93
Isofenphos I94
Isofluorphate I95
Isoflurane I96
isohexane M249
Isohexanol I97
isohexyl alcohol M253
isoindazole I18
Isol M251
Isolan I98
D-Isoleucine I99
isomelamide M36
Isoniazid I100
isonicotinic acid hydrazide I100
isonicotinic acid, 2-isopropylhydrazide,
    phosphate I57
isonicotinoyl hydrazide I100
Isonin I100
Isooctane I101
isooctanol I102
isooctaphenone I105
Isooctyl alcohol I102
isopentanal I134
Isopentane I103
isopentanol M162
isopentene M164
3-isopentenyl alcohol M166
Isopentyl acetate I104
isopentyl alcohol M162
isopentyl alcohol acetate I104
isopentyl methyl ketone M215
isophenphos I94
Isophorone I105
Isophorone diamine diisocyanate I106
Isophorone diisocyanate I106
Isophosphamide I107
Isophthalic acid I108
Isophthalonitrile I109
Isoprene I110
Isoprocarb I111
Isopropalin I112
isopropanol I115
isopropene cyanide M97
Isopropenyl acetate I113
Isopropenylbenzene I114
isopropenyl carbinol M100
isopropenylethyl alcohol M166
Isopropenyl methyl ketone M168
isopropenylnitrile M97
3′-isopropoxy-2-methylbenzanilide M45
3′-isopropoxy-o-toluanilide M45
isopropylacetone I84
Isopropyl alcohol I115
isopropyl aldehyde I88

lanthanum chloride (La$_2$Cl$_6$) L6
Lanthanum L5
lanthanum chloride (LaCl$_3$) L6
Lanthanum trichloride L6
Lasiocarpine L7
Lauric acid L8
Laurostearic acid L8
Lauryl pyridinium chloride L9
Lead L10
lead acetate L30
lead acetate, basic L27
Lead arsenate L11
Lead arsenite L12
lead, bis(acetato-$O$)tetrahydroxytri- L27
lead, bis(dimethylcarbamodithioato-$S,S'$)-,
　(T-4)- L16
lead bis(thiocyanate) L33
lead boron fluoride L18
Lead Bottoms L28
Lead chloride L13
lead chloride (PbCl$_2$) L13
Lead chromate L14
lead chromium oxide L14
Lead cyanide L15
lead dichloride L13
lead difluoride L19
lead diiodide L21
Lead dimethyldithiocarbamate L16
lead dinitrate L22
Lead dioxide L17
lead diperchlorate L23
lead diphosphate L24
lead distearate L26
lead dithiocyanate L33
Lead Flake L10
Lead fluoborate L18
Lead fluoride L19
lead fluoride (PbF$_2$) L19
lead fluorosilicate L20
Lead hexafluorosilicate L20
lead hydrogen arsenate L11
lead(II) thiocyanate L33
lead(II) iodide L21
Lead iodide L21
lead(IV) acetate L30
Lead nitrate L22
lead orthophosphate L24
lead oxide brown L17
lead oxide phosphonate L25
lead(2+) perchlorate L23
Lead perchlorate L23
lead peroxide L17
Lead phosphate L24
Lead phosphite dibasic L25

lead silicon fluoride L20
Lead stearate L26
Lead subacetate L27
Lead sulfate L28
Lead sulfide L29
lead superoxide L17
Lead tetraacetate L30
Lead, tetraethyl- L31
lead tetrafluoroborate L18
Lead, tetramethyl- L32
Lead thiocyanate L33
Ledate L16
Lenacil L34
4-lepidin M279
lepidine M279
Leptophos L35
Leropropoxyphene L38
D-Leucine L36
Leucobasal M131
Leucoline I129
Leukerin M50
levofalan M38
Levonorgestrel L37
Levopropoxyphene L38
levulic acid L39
Levulinic acid L39
Lewisite L40
Lidocaine L42
Lignin alkali L41
ignocaine L42
Lilial M171
Lilyal M171
Limonene L43
(+)-limonene L44
$d$-limonene L44
$R$-Limonene L44
α-limonene L43
Lindane L45
ε-Lindane L46
Linodil I34
Linoleic acid L47
linolic acid L47
Lipase L48
Liromat M9
Lithium L49
lithium chloride oxide L51
Lithium hydride L50
Lithium hypochlorite L51
lithium monohydride L50
lithium oxychloride L51
Lithocholic Acid L52
Lithosol Deep Blue B I21
Lithosol Scarlet Base M M239
Locust Bean Gum L53

Mercury(II) bromide  M57
Mercury(I) chloride  M58
Mercury(II) chloride  M59
Mercury(II) cyanide  M60
mercury cyanide oxide  M76
mercury, (cyanoguanidinato-$N'$)methyl-  M69
Mercury, diethyl-  M61
Mercury, dimethyl-  M62
mercury dithiocyanate  M80
Mercury gluconate  M63
Mercury(I) iodide  M64
Mercury(II) iodide  M65
mercuryl acetate  M53
Mercury(II), methoxyethyl-, acetate  M66
Mercury(1+), methyl-  M67
Mercury, methyl-  M67
Mercury(II), methyl-, acetate  M68
Mercury(II), methyl-, dicyanamide  M69
mercury monochloride  M58
Mercury(I) nitrate  M70
Mercury(II) nitrate  M71
Mercury nucleate  M72
Mercury oleate  M73
Mercury oxide  M74
Mercury(II) oxide  M75
Mercury oxycyanide  M76
mercury perchloride  M59
mercury pernitrate  M71
Mercury potassium iodide  M77
mercury protoiodide  M64
mercury protonitrate  M70
Mercury salicylate  M78
Mercury(I) sulfate  M79
Mercury(II) thiocyanate  M80
Merphalan  M81
Merphos  M82
Mesitylene  M83
mesityl oxide  M261
mesoinosite  I33
Mesonex  I34
Mestranol  M84
Mesurol  M113
Mesurol phenol  M113
mesyl fluoride  M104
Metacen  I31
metacetaldehyde  M86
Metacil  M291
Metalaxyl  M85
Metaldehyde  M86
Metamitron  M87
Metamitrone  M87
Metasystox  M192
Metasystox forte  M192
Metaxanin  M85

Metazachlor  M88
Metepa  M89
metformin HCl  M90
Metformin hydrochloride  M90
Methabenzthiazuron  M91
methacon  M108
Methacrifos  M92
Methacrolein  M93
Methacrolein diacetate  M94
methacrolyl anhydride  M96
methacrylaldehyde  M93
methacrylamide  M140
Methacrylic acid  M95
α-methacrylic acid  M95
methacrylic acid amide  M140
methacrylic acid, butyl ester  M266
methacrylic acid, isobutyl ester  I83
methacrylic acid, methyl ester  M229
Methacrylic anhydride  M96
methacrylic chloride  M98
Methacrylonitrile  M97
α-methacrylonitrile  M97
Methacryloyl chloride  M98
Methacryloyloxyethyl isocyanate  M99
methenylaldehyde  M93
Methallyl alcohol  M100
methallylcarbinol  M166
methallylidene diacetate  M94
Methamidophos  M101
Methane  M102
methanedicarboxylic acid  M15
methane, iodo-  I47
Methanesulfonic acid  M103
methanesulfonic acid, methyl ester  M230
Methanesulfonyl fluoride  M104
Methanethiol  M105
methane, triiodo-  I46
Methanol  M106
$N$-methanolacrylamide  M247
methanon  M190
Methanone, (4-methoxy-3-methylphenyl)
    (3-methylphenyl)-  M132
methapyrapone  M306
Methapyrilene  M107
Methapyrilene hydrochloride  M108
Methaqualone  M109
Methazole  M110
1,3,4-metheno-1$H$-cyclobuta[$cd$]pentalene-
    2-pentanoic acid, 1,1a,3,3a,4,5,5,5a,5b,
    6-deca- chlorooctahydro-2-hydroxy-γ-oxo-,
    ethyl ester  K6
Methiacil  M291
methiamitron  M87
methibenzuron  M91

Methidathion M111
Methimazole M112
Methiocarb M113
(R)-(–)-methionine M114
D-Methionine M114
Methocarbamol M115
Methocel A M203
Methocel MC M203
Methomyl M116
Methoprene M117
α-methopterin M118
Methotrexate M118
Methoxsalen M119
methoxtriglycol M125
methoxyacetic acid M120
2-Methoxyacetic acid M120
methoxyacetone M121
1-Methoxyacetone M121
7-methoxy-2H-1-benzopyran-2-one M123
2-methoxycarbon-1-methylvinyl dimethyl
  phosphate M307
methoxycarbonyl chloride M175
(E)-O-2-(methoxycarbonyl)prop-1-enyl
  O,O-dimethyl phosphorothioate M92
methoxycellulose M203
Methoxychlor M122
7-Methoxycoumarin M123
methoxydiglycol M124
4-methoxy-3,3'-dimethylbenzophenone M132
methoxyethene M296
2-(2-Methoxyethoxy)ethanol M124
2-[2-(2-Methoxyethoxy)ethoxy]ethanol M125
methoxyethylene M296
9-methoxy-7H-furo[3,2-g][1]benzopyran-
  7-one M119
6-Methoxyguanine M126
methoxyhexanone M128
N-[2-(5-methoxy-1H-indol-3-yl)ethyl]-
  acetamide M37
methoxylene M108
Methoxymethylbenzene M127
1-methoxy-4-methylbenzene M144
methoxyethylmercury acetate M66
α-(methoxymethyl)-2-nitro-1H-imidazole-
  1-ethanol M313
α-(methoxymethyl)-2-nitroimidazole-
  1-ethanol M313
4-methoxy-4-methyl-2-pentanone M128
4-Methoxy-4-methylpentan-2-one M128
2-methoxy-2-methylpropane M170
2-Methoxy-4-nitroaniline M129
4-Methoxy-2-nitroaniline M130
2-methoxy-4-nitrobenzenamine M129
4-methoxy-2-nitrobenzenamine M130

S-[(5-methoxy-2-oxo-1,3,4-thiadiazol-
  3(2H)-yl)methyl] O,O-dimethyl
  phosphorodithioate M111
4-Methoxyphenol M131
p-methoxyphenol M131
Methoxyphenone M132
3-(2-methoxyphenoxy)-1-glyceryl carbamate
  M115
3-(o-methoxyphenoxy)-1,2-propanediol
  1-carbamate M115
1-Methoxypropane M133
(±)-1-methoxypropan-2-ol M134
1-Methoxy-2-propanol M134
1-Methoxy-2-propanol, acetate M135
2-Methoxy-1-propanol, acetate M136
1-methoxy-2-propanone M121
2-methoxy-4-propenylphenol I93
2-methoxypropyl 1-acetate M136
8-methoxypsoralen M119
6-methoxy-1H-purin-2-amine M126
p-methoxytoluene M144
α-methoxytoluene M127
methoxytriethyene glycol M125
methulose M203
methylbutyraldehyde M159
Methylacetamide M137
N-methylacetamide M137
Methyl acetoacetate M138
3-methylacetoacetone M252
methyl acetylacetate M138
Methyl acetylene M139
2-methylacrolein M93
methacrylaldehyde M93
2-Methylacrylamide M140
methyl adronal M185
methyladronal M188
Methylal M141
methyl alcohol M106
2-methylallyl alcohol M100
(methylamino)benzene M143
N-[[(methylamino)carbonyl]oxy]-
  ethanimidothioic acid, methyl ester M116
2-(methylamino)ethanol M142
N-Methyl-2-aminoethanol M142
p-(methylamino)nitrobenzene M241
methylaminopterin M118
methyl amyl alcohol M253
methyl amyl alcohol M256
N-Methylaniline M143
4-Methylanisole M144
p-methylanisole M144
methylanol M185
methylanol M188
methylanone M190

methylapoxide M89
Methylazoxymethanol M145
(methyl-*ONN*-azoxy)methanol M145
Methylazoxymethanol, acetate M146
(methyl-*ONN*-azoxy)methanol, acetate (ester) M146
methylazoxymethyl acetate M146
*N*-methylbenzenamine M143
methyl benzeneacetate M262
α-Methylbenzenemethanol M147
4-methylbenzenesulfonic acid, methyl ester M292
2-Methylbenzothiazole M148
*p*-(6-methylbenzothiazol-2-yl)aniline M149
4-(6-Methyl-2-benzothiazolyl)benzenamine M149
methylbenzotriazole M150
1-methylbenzotriazole M151
1-Methyl-1*H*-benzotriazole M151
4(or 5)-methyl-1*H*-benzotriazole M150
5-methyl-1,2,3-benzotriazole M152
5-Methyl-1*H*-benzotriazole M152
5-methylbenzotriazole M152
Methyl-1*H*-benzotriazole M150
α-methylbenzyl alcohol M147
methyl benzyl ether M127
α-Methylbenzyl 3-hydroxycrotonate, dimethyl phosphate M153
*N*-methyl-9-benzyltetrahydro-γ-carboline M30
3-methyl-9-benzyl-1,2,3,4-tetrahydro-γ-carboline M30
Methyl-1,1'-biphenyl M154
2-Methylbiphenyl M155
2-methyl-1,1'-biphenyl M155
4-Methylbiphenyl M156
4-methyl-1,1'-biphenyl M156
methylbiphenyl (mixed isomers) M154
2-methyl-1,2-bis(3-pyridyl)-1-propanone M306
2-methylbivinyl I110
Methyl bromoacetate M157
1-Methylbutadiene M158
2-methyl-1,3-butadiene I110
2-methylbutadiene I110
*N*-methylbutamine M169
2-Methylbutanal M159
3-methylbutanal I134
α-methylbutanal M159
2-methylbutane I103
2-methylbutane-*sec*-mononitrile M160
3-Methylbutanenitrile M160
3-methylbutanoic acid, methyl ester M226
2-Methylbutan-2-ol M161

3-Methylbutan-1-ol M162
methylbutanone M163
3-Methylbutan-2-one M163
3-Methyl-1-butene M164
2-Methyl-3-buten-2-ol M167
3-Methyl-2-buten-1-ol M165
3-Methyl-3-buten-1-ol M166
3-methylbut-2-en-1-ol M165
3-Methyl-3-buten-2-one M168
2-Methylbutyn-3-ol-2 M172
3-methylbutyl acetate I104
methylbutylamine M169
*N*-Methylbutylamine M169
Methyl *tert*-butyl ether M170
2-Methyl-3-(4-*tert*-butylphenyl) propionaldehyde M171
2-methylbutryonitrile M160
2-Methyl-3-butyn-2-ol M172
2-methylbutyraldehyde M159
3-methylbutyraldehyde I134
2-methylbutyric aldehyde M159
3-methylbutyronitrile M160
methylcaptax M289
methylcarbamic acid, *m*-cumenyl ester M205
Methyl carbamate M173
methylcarbamic acid, 4-(methylthio)-3,5-xylyl ester M113
*N*-[(methylcarbamoyl)oxy]thioacetimidic acid, methyl ester M116
3-(methylcarbamoyl)pyridine M276
Methyl carbitol M124
methyl carbonochloridate M175
Methyl cellulose M203
*O*-methyl cellulose M203
Methyl Chemosept M218
Methyl 2-chloroacrylate M174
methyl chlorocarbonate M175
Methyl chloroformate M175
methyl 2-chloro-2-propenoate M174
methyl α-chloropropionate M176
Methyl 2-chloropropionate M176
20-methylcholanthrene M177
3-Methylcholanthrene M177
1-Methylchrysene M178
2-Methylchrysene M179
3-Methylchrysene M180
5-Methylchrysene M181
6-Methylchrysene M182
Methyl 2-cyanoacrylate M183
methyl 2-cyano-2-propenoate M183
Methylcyclohexane M184
Methylcyclohexanol M185
2-Methylcyclohexanol M186
3-Methylcyclohexanol M187

3-methyl-1-cyclohexanol  M187
4-methyl-1-cyclohexanol  M188
4-Methylcyclohexanol  M188
*m*-methylcyclohexanol  M187
*p*-methylcyclohexanol  M188
methylcyclohexan-1-one  M189
Methylcyclohexanone  M189
1-methylcyclohexan-2-one  M190
2-Methylcyclohexanone  M190
2-methylcyclohexyl alcohol  M186
4-methylcyclohexyl alcohol  M188
2-methylcyclopentadienylmanganese
    tricarbonyl  M23
Methylcyclopentane  M191
methylcymantrene  M23
Methyl demeton  M192
Methyldichlorosilane  M193
methyldiethylcarbinol  M255
(*E*)-methyl 3-[(dimethoxyphosphino
    thioyl)oxy]-2-methyl-2-propenoate  M92
(*E*)-methyl 3-(dimethoxyphosphino
    thioyl oxy)-2-methylacrylate  M92
methyl 3-(dimethoxyphosphinyl
    oxy)crotonate  M307
methyl 1,1-dimethylethyl ether  M170
methyl 2,2-dimethylvinyl ketone  M261
methyldimuron  M196
*N*-Methyl-*N*,4-dinitrosoaniline  M194
*N*-methyl-*N*,4-dinitrosobenzenamine  M194
4-methyldiphenyl  M156
2-methyl-1,2-di-3-pyridinyl-1-propanone
    M306
2-methyl-1,2-di-3-pyridyl-1-propanone
    M306
(4-methyl-1,3-dithiolan-2-ylidene)-
    phosphoramidic acid, diethyl ester  M43
Methyldopa  M195
Methyldymron  M196
methylene acetone  M297
Methylenebis[4-cyclohexyl isocyanate]  M197
4,4′-methylenebis[*N*,*N*-dimethylaniline]
    M198
1,1′-methylenebis[(4-isocyanatocyclohexane)]
    M197
4,4′-Methylenebis[2-methylaniline]  M199
4,4′-Methylenebis[(N,N-dimethyl)
    benzenamine]  M198
Methylenebis[1-thiosemicarbazide]  M200
Methylene Blue  M201
methylene cyanide  M16
4,4′-Methylenedianiline dihydrochloride
    M202
4,4′-methylenedibenzeneamine
    dihydrochloride  M202

methylenediethanolamine  M220
methylenedinitrile  M16
1,2-(methylenedioxy)-4-propenylbenzene
    I130
methylenium ceruleum  M201
*N*-methylethanolamine  M142
(1-methylethenyl)benzene  I114
methyl ether cellulose  M203
methyl ether of propylene glycol  M327
4-(1-Methylethyl)benzaldehyde  M204
4-(1-methylethyl)-2,6-dinitro-
    *N*,*N*-dipropybezenamine  I112
4,4′-(1-methylethylidene)bisphenol  I120
(*E*,*E*)-1-methylethyl-11-methoxy-
    3,7,11-trimethyl-2,4-dodecadienoate
    M117
*N*-(1-methylethyl)-4-[(2-methylhydrazino)-
    methyl]benzamide  I1
2-(1-methylethyl)naphthalene  I123
4-(1-methylethyl)phenol  I125
3-(1-Methylethyl)phenol, methylcarbamate
    M205
2-(1-methylethyl)phenyl methylcarbamate
    I111
2-Methyl-5-ethylpyridine  M206
methyl fluorosulfate  M208
Methyl fluorosulfonate  M208
*N*-Methylformamide  M209
Methyl formate  M210
*N*-Methyl-*N*-formylhydrazine  M211
2-Methylfuran  M212
5-methylfuran  M212
α-methylfuran  M212
methylglycolic acid  M120
$O^6$-methylguanine  M126
Methylheptenone  M213
2-methyl-2-hepten-6-one  M214
6-methyl-$\Delta^5$-hepten-2-one  M214
6-Methyl-5-hepten-2-one  M214
methylhexalin  M185
5-Methylhexan-2-one  M215
Methylhydrazine  M216
methyl hydrazine monosulfate  M217
Methyl hydrazine sulfate  M217
methyl hydride  M102
methyl hydroxide  M106
Methyl *p*-hydroxybenzoate  M218
methyl 2-hydroxybenzoate  M284
*N*-methyl-2-hydroxyethylamine  M142
1-methyl-1*H*-imidazole  M219
1-Methylimidazole  M219
1-methylimidazole-2-thiol  M112
2,2′-(methylimino)bis[ethanol]  M220
methyliminodiethanol  M220

2,2'-(Methylimino)diethanol M220
5-Methylindan M221
3-Methylindole M222
β-methylindole M222
methyl iodide I47
methyl isoamyl ketone M215
methyl isobutenyl ketone M261
methyl isobutyl carbinol M253
methyl isobutyl ketone I84
methylisobutylxanthine I85
Methyl isobutyrate M223
Methyl isocyanate M224
methyl isopentanoate M226
methyl isopentyl ketone M215
methyl isopropenyl ketone M168
methyl isopropyl ketone M163
Methyl isothiocyanate M225
Methyl isovalerate M226
Methyl lactate M227
methyl maleate M11
methyl mercaptan M105
Methyl mercaptoacetate M228
2-(methylmercapto)benzothiazole M289
methyl mercaptoimidazole M112
methylmercaptophos M192
methylmercuric acetate M68
methylmercury M67
methylmercury(II) cation M67
methylmercury ion(1+) M67
methyl mesylate M230
Methyl methacrylate M229
Methyl methanesulfonate M230
methyl methanesulfonic acid M230
methyl methanoate M210
methyl (N-(2-methoxyacetyl)-
    N-(2,6-xylyl)-DL-alaninate M85
methyl p-methylbenzenesulfonate M292
3-methyl-9-benzyl-1,2,3,4-tetrahydro-
    γ-carboline M30
S-methyl N-[(methylcarbamoyl)-
    oxy]thioacetimidate M116
1-methyl-1-(4-methylcyclohexyl)ethyl
    hydroperoxide M42
7-Methyl-3-methylene-1,6-octadiene M231
5-methyl-2-(1-methylethenyl)cyclohexanol,
    [1R-(1α,2β,5α)]- I128
1-methyl-4-(1-methylethenyl)cyclohexene
    L43
2-methyl-N-[3-(1-methylethoxy)phenyl]-
    benzamide M45
5-methyl-2-(1-methylethyl)cyclohexanol,
    (1α,2β,5α)- M40
5-methyl-2-(1-methylethyl)cyclohexanol,
    (1α,2β,5α), (±)- M41

2-methyl-3-(2-methylphenyl)-4(3H)-
    quinazolinone M109
2-methyl-3-(2-methylphenyl)-4-quinazolinone
    M109
methyl 2-methylpropenoate M229
α-methyl-4-(2-methylpropyl)benzeneacetic
    acid I2
3-Methyl-4-(methylthio)phenol M232
4-methylmorpholine M233
N-Methylmorpholine M233
methyl mustard oil M225
methylnaftalen M234
Methylnaphthalene M234
1-Methylnaphthalene M235
2-Methylnaphthalene M236
α-methylnaphthalene M235
β-methylnaphthalene M236
2-Methyl-4-nitroaniline M237
2-Methyl-5-nitroaniline M238
2-methyl-p-nitroaniline M237
4-Methyl-2-nitroaniline M239
4-Methyl-3-nitroaniline M240
N-Methyl-4-nitroaniline M241
2-methyl-1-nitro-9,10-anthracenedione M242
2-Methyl-1-nitroanthraquinone M242
2-methyl-4-nitrobenzenamine M237
2-methyl-5-nitro-benzenamine M238
4-methyl-3-nitrobenzenamine M240
N-methyl-4-nitrobenzenamine M241
2-methyl-5-nitroimidazole-1-ethanol M304
Methylnitropropyl-4-nitrosoaniline M243
N-Methyl-N-nitrosobenzamide M244
N-(N-Methyl-N-nitrosocarbamoyl)-L-ornithine
    M245
N-Methyl-N-nitroso-p-toluenesulfonamide
    M246
methyl cis-9-octadecenoate M248
methylol M106
Z-Methyl oleate M248
N-Methyloloacrylamide M247
methyl3-oxobutyrate M138
2-methyl-6-oxo-2-heptene M214
methylparaben M218
Methyl Parasept M218
2-methylpentanal M295
2-Methylpentane M249
3-Methylpentane M250
2-Methylpentane-2,4-diol M251
3-Methyl-2,4-pentanedione M252
2-Methyl-1-pentanol M253
2-Methyl-3-pentanol M254
4-Methyl-2-pentanol M256
3-Methyl-3-pentanol, 3-methyl- M255
4-methyl-2-pentanone I84

2-methylpentene M257
2-Methyl-1-pentene M257
2-Methyl-2-pentene M258
2-methylpent-1-ene M257
2-methylpent-2-ene M258
4-Methyl-1-pentene M259
4-methylpent-1-ene M259
4-Methyl-2-pentene M260
4-methylpent-2-ene M260
4-Methylpent-3-en-2-one M261
3-methyl-2-(2-pentenyl)-(Z)-
   2-cyclopenten-1-one J3
2-methylpent-1-ol M253
4-methylphenol methyl ether M144
Methyl phenylacetate M262
methylphenylamine M143
methylphenylcarbinol M147
methyl phenylethanoate M262
3-methylphenyl N-methylcarbamate M301
Methyl phosphonic dichloride M263
methyl phosphoramidothioate M101
1-Methylpiperidine M264
N-methylpiperidine M264
methylpropanal I88
2-methylpropanal I88
2-methyl-1-propanamine I80
2-methyl-propane I75
2-methylpropanitrile I90
2-methylpropanoic acid I89
2-methylpropanoic acid, methyl ester M223
2-methylpropanoic acid, 2-methylpropyl ester
   I82
2-methylpropanoic acid, 2,2,4-trimethylpentyl
   ester M262
2-methyl-1-propanol I76
2-methyl-2-propenal M93
2-methylpropenal M93
2-methyl-2-propenamide M140
2-methylpropenamide M140
2-methylpropene I77
2-methylpropenenitrile M97
2-methyl-2-propenoic acid M95
2-methylpropenoic acid M95
2-methyl-2-propenoic acid anhydride M96
2-Methyl-2-propenoic acid, butyl ester M266
2-methyl-2-propenoic acid, 2-methylpropyl
   ester I83
2-methyl-2-propen-1-ol M100
2-methyl-2-propenoyl chloride M98
2-methylpropionaldehyde I88
2-methylpropionic acid I89
2-methylpropenoic acid, 2-methylpropyl
   ester I82
2-(2-methylpropoxy)naphthalene I86

1-Methylpropyl acetate M267
2-methylpropyl acetate I78
2-methylpropyl acetic acid ester I78
2-methylpropyl alcohol I76
2-methylpropylamine I80
β-methylpropyl ethanoate I78
2-methyl-2-propylethanol M253
1-methyl-1-propylethene M257
methyl propyl ether M133
2-methylpropyl formate I81
2-methylpropyl formic acid ester I81
2-methylpropyl isobutyrate I82
Methyl propyl ketone M268
2-methylpropyl methacrylate I83
2-methylpropyl methyl ketone I84
2-methyl-2-propyl-1,3-propanediol
   dicarbamate M44
2-methylpropyl propanoate I87
2-methylpropyl propanoic acid ester I87
2-methylpropyl2-propenoic acid ester I79
2-methylpropyl propinoate I87
2-Methylpyrazine M269
1-Methylpyrene M270
2-Methylpyrene M271
4-Methylpyrene M272
2-Methylpyridine M273
3-Methylpyridine M274
4-Methylpyridine M275
α-methylpyridine M273
β–methylpyridine M274
γ-methylpyridine M275
N-Methyl-3-pyridinecarboxamide M276
1-(1-Methyl)-1H-pyrrole-2,5-dione M207
1-methylpyrrolidinone M277
N-Methylpyrrolidone M277
2-Methylquinoline M278
4-Methylquinoline M279
5-Methylquinoline M280
6-Methylquinoline M281
7-Methylquinoline M282
8-Methylquinoline M283
p-methylquinoline M281
γ-methylquinoline M279
methyl rhodanate M290
Methyl salicylate M284
Methylstyrene M285
ar-methylstyrene M285
α-methylstyrene I114
α-methylstyrol I114
methyl sulfhydrate M105
methylsulfinylmethane M286
methyl sulfocyanate M290
methylsulfonic acid M103
Methyl sulfoxide M286

methyl systox M192
Methyl 2,3,5,6-tetrachloro-4-pyridyl sulfone
   M287
2-Methyltetrahydrofuran M288
2-(Methylthio)benzothiazole M289
Methyl thiocyanate M290
methyl thioglycolate M228
Methylthiouracil M291
6-methyl-2-thiouracil M291
Methyl *p*-toluenesulfonate M292
methyl toluene-4-sulfonate M292
methyl *p*-tosylate M292
Methyltrichlorosilane M293
methylumbelliferone M123
methylurea M294
1-Methylurea M294
*N*-methylurea M294
methylurethane M173
2-Methylvaleraldehyde M295
2-methylvaleric aldehyde M295
1-methylvinyl acetate I113
Methyl vinyl ether M296
Methyl vinyl ketone M297
methylmercury(1+) M67
2-[2-(2-Methyoxyethoxy)ethoxy]ethanol
   M125
Metiram M298
Metmercapturon M113
Metobromuron M299
Metolachlor M300
Metolcarb M301
Metolquizalone M109
Metopryl M133
Metox M122
Metoxuron M302
Metribuzin M303
Metronidazole M304
Metsulfuron-methyl M305
Metyrapone M306
Mevinphos M307
Mexacarbate M308
Mezapur M110
MF H M211
MH M13
MIA I37
Miazole I7
MIBC M256
MIC M224
3-MIC M256
Michler's base M198
Michler's methane M198
Miconazole M309
Milk acid L1
Miltown M44

Mineral oil M310
Minex M117
MIPC I111
MIPCIN I111
MIPSIN I111
Miral I72
Mirex M311
Mirex, photo- M312
Misonidazole M313
Mitin N M314
Mitomycin C M315
Mitrol I10
MME M229
MMH M216
MMS M230
MMT M23
MMTP M232
MNB M244
Mohr's Salt I61
Moldison M10
Molinate M316
molybdena M318
Molybdenum M317
molybdenum oxide (MoO3) M318
Molybdenum trioxide M318
molybdenum(VI) oxide M318
molybdenum(VI) trioxide M318
molybdic acid anhydride M318
molybdic anhydride M318
Monalide M319
Monitor M101
monoamyl phthalate M326
monobasic lead acetate L27
Monobutyl phthalate M320
mono-*n*-butyl phthalate M320
Monocron 9491 I45
monocrotalin M321
Monocrotaline M321
Monocrotophos M322
Mono(2-ethylhexyl) phthalate M323
8-monohydromirex M312
monoiodoacetic acid I37
monoiodoethane I44
monoisobutylamine I80
Monolinuron M324
monomethylacetamide M137
monomethylaminoethanol M142
monomethylaniline M143
monomethyldichlorosilane M193
Monomethylformamide M209
Monomethylhydrazine M216
monomethylolacrylamide M247
Monooctyl phthalate M325
Monopentyl phthalate M326

mono-*n*-pentyl phthalate  M326
monopropylene glycol methyl ether  M134
Monopropylene glycol methyl ether  M327
Monorotox  M324
monosodium acid methanearsonate  M336
Monothiuram  M328
Montmorillonite  M329
Monuron  M330
Monuron TCA  M331
Morpan T  M342
Morphia  M332
morphinan-3,6-diol,
    7,8-didehydro-4,5-epoxy-17-methyl-,
    (5α,6α)-  M332
Morphine  M332
(–)-morphine  M332
Morphium  M332
Morpholine  M333
5-(Morpholinomethyl)-3-[(5-nitro-
    furfurylidene)amino]-2-oxazolidinone
    M334
Morphothion  M335
Morphotox  M335
Morpolose M400  M203
Morsodren  M69
Morton Soil Drench  M69
Motrin  I2
M-pyrol  M277
MS 1053  M31
MSF  M104
MSMA  M336
MTBE  M170
MTMC  M301
MTU  M291
Muccira  K3
Murfotox  M31
Murphotox  M31
Muscimol  M337
Musk ambrette  M338
Mustard gas  M339
Mutamycin  M315
Myclobutanil  M340
Myofer M  I64
Myrcene  M231
β-Myrcene  M231
Myristic acid  M341
myristyltrimethylammonium bromide  M342
Myrtenal  M343
Mytab  M342
Nacconate H12  M197
naphthalene, methyl-  M234
naphthalene, 1-methyl-  M235
naphthalene, 2-methyl-  M236
Naphthenil Red B Base  M129

Natulan  I1
Natural lead sulfide  L29
*n*-butyl iodide  I41
NCI-C55594  M173
NCI-C56 519  M48
NCI-C6 0979  I93
NCI-CO1 923  M242
Neo-fat 12  L8
Neo-fat 14  M341
Neothyl  M133
neovitamin A acid  I132
Nespor  M19
nicotinamide, *N*-methyl-  M276
nicotinic acid methylamide  M276
Nipagin M  M218
nitric acid, iron(3+) salt  I67
nitric acid, lead (2+) salt  L22
nitric acid, manganese(2+) salt  M24
nitric acid, mercury(1+) salt  M70
nitric acid, mercury(2+) salt  M71
nitric acid, 1-methylethyl ester  I124
2-nitro-*p*-anisidine  M130
4-nitro-*o*-anisidine  M129
Nitrol  M243
1-nitro-2-methylanthraquinone  M242
Nitrosan K  M194
*N*-(*N*-nitroso-*N*-methylcarbamoyl)-L-ornithine
    M245
3-nitro-*p*-toluidine  M240
4-nitro-2-toluidine  M237
5-nitro-4-toluidine  M240
5-nitro-*o*-toluidine  M238
*m*-nitro-*p*-toluidine  M240
NK 711  L35
NK049  M132
1-N-2-MA  M242
20-norcrotolanan-11,15-dione,
    14,19-dihydro-12,13-dihydroxy-,
    (13α,14α)-  M321
normal lead stearate  L26
2-norpinene-2-carboxaldehyde, 6,6-dimethyl-
    M343
Norplant  L37
19-norpregna-1,3,5(10)-trien-20-yn-17-ol,
    3-methoxy-, (17α)–  M84
19-nor-17α-pregn-4-en-20-yn-17-ol  L62
19-norpregn-4-en-20-yn-17-ol, (17α)-  L62
Norquen  M84
Nova  M340
NTN 801  M33
Nudrin  M116
Nu-Flow M  M340
Nuodox PM0 10  M73
Nuvacron  M322

Nuvanol N I45
Oatirab M299
1,3,4,5,6,7,8,8-octachloro-1,3,3a,4,7,7a-
hexahydro-4,7-methanoisobenzofuran
I73
1,3,4,5,6,7,8,8-octachloro-3a,4,7,7a-tetrahydro-
4,7-methanoisobenzofuran I73
9,12-octadecadienoic acid, (Z,Z)- L47
octadecanoic acid, lead(2+) salt L26
9-octadecenoic acid, (Z)-, mercury salt M73
9-octadecenoic acid, methyl ester, (Z)-
M248
1,6-octadiene, 7-methyl-3-methylene- M231
octyl hydrogen phthalate M325
Oftanol I94
oil mist, mineral M310
oleate of mercury M73
cis-oleic acid, methyl ester M248
OMS 1211 I45
OMS 2005 M92
Ophthalmadine L43
Orcanon M291
Ordram M316
Orgametril L62
Orthonal M109
Orudis K10
Ospalivina M332
2-oxa-3,3-dimethylbutane M170
2-oxazolidinone,
5-(4-morpholinylmethyl)-3-[[(5-nitro-
2-furanyl)methylene]amino]- M334
δ-oxo-α-butylene M297
2-oxo-6-methylhept-5-ene M214
4-oxovaleric acid L39
oxycarbophos M9
Oxydiazol M110
oxypsoralen M119
P128 L29
P37 L29
Palohex I34
L-PAM M38
Pandrinox M69
Panoctin I10
Panoctine I10
Panogen PX M69
Pantherine M337
paradormalene M107
Parogen M66
parrafin oil M310
Patoran M299
Paxilon M110
Pb-S 100 L10
pear oil I104
pentacarbonyliron I69

1,3-pentadiene M158
2,4-pentanediol, 2-methyl- M251
pentanoic acid, 4-oxo- L39
tert-pentanol M161
1-pentanol, 4-methyl- I97
3-pentanol, 3-methyl- M255
2-pentanone M268
pentan-2-one M268
3-penten-2-one, 4-methyl- M261
Pentoxane M128
tert-pentyl alcohol M161
peppermint camphor M40
perchloric acid, lead(2+) salt L23
Percin I100
Perlex Paste 500 L24
Pestan M31
petroleum gases, liquified L55
Pharorid M117
α-phenethyl alcohol M147
phenol, 3-methyl-4-(methylthio)- M232
phenothiazin-5-ium, 3,7-bis(dimethylamino)-,
chloride M201
Phenothiol M28
DL-phenylalanine,
4-[bis(2-chloroethyl)amino]- M81
phenylalanine mustard M38
DL-phenylalanine mustard M81
1,10-(o-phenylene)pyrene I20
o-phenylenepyrene I20
o-phenylenethiourea M47
1-phenylethanol M147
1-phenylethyl alcohol M147
phenyl iodide I40
N-phenylmethylamine M143
2-phenylpropene I114
phenyltoluene M154
2-phenyltoluene M155
4-phenyltoluene M156
Phosdrin M307
1,1',1'-phosphinylidynetris[2-methylaziridine]
M89
phosphonadithioimidocarbonic acid, cyclic
propylene P,P-diethyl ester M43
phosphonic dichloride, methyl- M263
phosphoric acid, dimethyl ester, ester with
3-hydroxy-N-methylcrotonamide, (E)-
M322
phosphoric acid, dimethyl
1-methyl-3-(methylamino)-3-oxo-
1-propenyl ester, (E)- M322
phosphoric acid, lead(2+) salt (2:3) L24
phosphorodithioic acid, S-[(4,6-diamino-
1,3,5-triazin-2-yl)methyl] O,O-dimethyl
ester M39

phosphorodithioic acid, S-[(4,6-diamino-
s-triazin-2-yl)methyl] O,O-dimethyl ester
M39

phosphorodithioic acid, O,O-dimethyl ester,
S-ester with 4-(mercaptoacetyl)morpholine
M335

phosphorodithioic acid, O,O-dimethyl
S-[2-(4-morpholinyl)-2-oxoethyl] ester
M335

phosphorothioic acid, S-benzyl
O,O-diisopropyl ester I55

phosphorothioic acid, O,O-bis(1-methylethyl)
S-(phenylmethyl) ester I55

phosphorothioic acid,
O-(2,5-dichloro-4-iodophenyl)
O,O-dimethyl ester I45

phosphorothioic acid, O,O-diethyl-
O-(5-phenyl-3-isoxazolyl) ester I135

phosphorothioic acid, O,O-dimethyl
O-[2(ethylthio)ethyl] ester M192

phosphorotrithious acid, tributyl ester M82

phosphothion M10

Phosvel L35

photomirex M312

m-phthalic acid. I108

phthalic acid, butyl ester M320

phthalic acid, monobutyl ester M320

phthalic acid, mono(2-ethylhexyl) ester M323

phthalic acid, monooctyl ester M325

phthalic acid, monopentyl ester M326

m-phthalodinitrile I109

Pichtosin I74

Pichtosine I74

2-picoline M273

4-picoline M275

m-picoline M274

o-picoline M273

p-picoline M275

α-picoline M273

β-picoline M274

γ-picoline M275

2-picoline, 2-ethyl- M206

3-picoline M274

Piperazine, 1-acetyl-4-[4-[[2-(2,4-dichloro-
phenyl)-2-(1H-imidazol-1-yl-methyl)-
1,3-dioxolan-4-yl]methoxy]phenyl]-, cis-
K9

p-Isopropylphenol I125

Pivot I5

Pivot I6

plumbane, tetraethyl- L31

plumbane, tetramethyl- L32

plumbic acetate L30

plumbic oxide L17

plumbous chloride L13

plumbous chromate L14

Plumbous fluoride L19

plumbous iodide L21

plumbous nitrate L22

plumbous sulfide L29

PMF (antioxidant) M131

Polybrene I51

poly(dimethyliminio) hexamethylene
(dimethylimino) trimethylene dibromide
I51

Polylin 515 L47

Polyram M298

Poly-solv DM M124

pomalus acid M14

Ponstan M34

Ponstil M34

Porcelain clay K2

Potablan M319

potassium iodohydragyrate M77

potassium mercuric iodide M77

potassium tetraiodomercurate(II) M77

Prazepine I13

Precipite blanc M58

Precor M117

Prenol M165

prenyl alcohol M165

Presinol M195

Probe M110

Procarbazine hydrochloride I1

Profenid K10

Promidione I56

propanedinitrile M16

propanedioic acid M15

propane, 1-iodo-2-methyl- I48

propane, 2-iodo-2-methyl- I49

propane-2-nitrate I124

propanenitrile, 2-hydroxy- L3

propanoic acid, 2-chloro-, methyl ester M176

propanoic acid, 2-hydroxy- L1

propanoic acid, 2-methyl-, methyl ester M223

2-propanol I115

propanol, methoxy- M327

Propasol M134

2-propenamide, N-(hydroxymethyl)- M247

1-propene-2-methyl I77

2-propenoic acid, 2-methyl, methyl ester
M229

1-propen-2-ol acetate I113

propen-2-yl acetate I113

5-(1-propenyl)-1,3-benzodioxole I130

4-propenylguaiacol I93

propine M139

propionic acid, isobutyl ester I87

sulfur mustard M339
Supracide M111
Suprathim M111
Suprex clay K2
sweet birch oil M284
Sylvan M212
Syntexan M286
Systhane M340
Tab gum K3
Tamaron M101
Tandex K4
TAP85 L45
teaberry oil M284
Tegosept M M218
TEL L31
Telfairic acid L47
Telodrin I73
Telvar M330
Telvar Monuron Weed Killer M330
Tenalin M107
tetrabase M198
2,3,5,6-tetrachloro-4-(methylsulfonyl)pyridine
  M287
tetracycline methylenelysine L61
1-tetradecanaminium, *N,N,N*-trimethyl-,
  bromide M342
tetradecanoic acid M341
*n*-tetradecanoic acid M341
tetraethylplumbane L31
tetrahydro-2-methylfuran M288
2,3,4,5-tetrahydro-2-methyl-5-(phenylmethyl)-
  1*H*-pyrido[4,3-*b*]indole M30
tetrahydro-1,4-oxazine M333
tetrahydro-*p*-oxazine M333
tetrahydrosylvan M288
tetramethyldiaminodiphenylmethane M198
tetramethyllead L32
tetramethylplumbane L32
tetramethylthionine chloride M201
tetramethylthiurammonium sulfide M328
1,3,5,7-tetroxocane, 2,4,6,8-tetramethyl- M86
tetryl formate I81
thenylene hydrochloride M108
thenylpyramine M107
thenylpyramine hydrochloride M108
Theoharn M36
Thiamazole M112
thiglycolic acid, methyl ester M228
1,1'-thiobis(*N,N*-dimethylthio)formamide
  M328
thiocyanic acid, lead(2+) salt L33
thiocyanic acid mercury(2+) salt M80
thiodicarbonic diamide ([(H2(N)C(S)]2S),
  tetramethyl- M328

thioglycolic acid M46
thiomethyl alcohol M105
Thionylan M107
6-thiopurine M50
thioranic acid M46
Thiothymin M291
Thymidazole M112
Tofranil I14
Toleron I66
Tolkan I127
toluene hexahydride M184
*p*-toluenesulfonic acid, methyl ester M292
*o*-toluidine, 4,4'-methylene- M199
*p*-toluidine, 2-nitro- M239
*m*-toluquinoline M282
*p*-toluquinoline M281
tolyl ethylene M285
*m*-tolyl methyl carbamate M301
*p*-tolyl methyl ether M144
tolyltriazole M150
Tomlinite L41
Topsym M286
Toxilic anhydride M12
tragacol L53
Trapex M225
triacylglycerol hydrolase L48
2,4,6-triaminotriazine M36
1,3,5-triazine-2,4,6-triamine M36
1,2,4-triazin-5(4*H*)-one,
  4-amino-3-methyl-6-phenyl- M87
1*H*-1,2,4-triazole-1-propanenitrile,
  α-butyl-α-(4-chlorophenyl)- M340
tribunil M91
tributyl phosphorotrithioate M82
*S,S,S*-tributyl trithiophosphite M82
tricarbonyl (η$^5$-2,4-cyclopentadien-
  1-yl)manganese M21
tricarbonyl-π-cyclopentadienylmanganese
  M21
1,1,1-trichloro-2,2-di(4-methoxyphenyl)ethane
  M122
1,1'-(2,2,2-trichloroethylidene)bis-
  (4-methoxybenzene) M122
trichloroindium I24
trichloromethylsilane M293
1-tridecanecarboxylic acid M341
triethylene glycol monomethyl ether M125
5'-[1,1,1-trifluoromethanesulfonamido]
  acet-2',4'-xylidide M35
triglyceridase L48
triglycol monomethyl ether M125
triiodomethane I46
trilead phosphate L24
trimanganese tetraoxide M25

# INDEX OF CAS REGISTRY
# NUMBERS

91-80-5    Methapyrilene  M107
92-36-4    4-(6-Methyl-2-benzothiazolyl)-
           benzenamine  M149
95-13-6    1H-Indene  I19
96-14-0    3-Methylpentane  M250
96-17-3    2-Methylbutanal  M159
96-32-2    Methyl bromoacetate  M157
96-37-7    Methylcyclopentane  M191
96-47-9    2-Methyltetrahydrofuran  M288
96-96-8    4-Methoxy-2-nitroaniline  M130
97-52-9    2-Methoxy-4-nitroaniline  M129
97-54-1    Isoeugenol  I93
97-74-5    Monothiuram  M328
97-85-8    Isobutyl isobutyrate  I82
97-86-9    Isobutyl methacrylate  I83
97-88-1    2-Methyl-2-propenoic acid,
           butyl ester  M266
98-83-9    Isopropenylbenzene  I114
98-85-1    α-Methylbenzenemethanol  M147
99-52-5    2-Methyl-4-nitroaniline  M237
99-55-8    2-Methyl-5-nitroaniline  M238
99-76-3    Methyl p-hydroxybenzoate
           M218
99-80-9    N-Methyl-N,4-dinitrosoaniline
           M194
99-89-8    p-Isopropylphenol  I125
100-15-2   N-Methyl-4-nitroaniline  M241
100-61-8   N-Methylaniline  M143
100-79-8   Isopropylidene glycerol  I121
101-41-7   Methyl phenylacetate  M262
101-61-1   4,4'-Methylenbis[N,N-dimethyl
           benzenamine]  M198
102-36-3   Isocyanic acid, 3,4-dichlorophenyl
           ester  I91
104-74-5   Lauryl pyridinium chloride  L9
104-90-5   2-Methyl-5-ethylpyridine  M206
104-93-8   4-Methylanisole  M144
105-30-6   2-Methyl-1-pentanol  M253
105-45-3   Methyl acetoacetate  M138
105-46-4   1-Methylpropyl acetate  M267
105-59-9   2,2'-(Methylimino)diethanol
           M220
106-63-8   Isobutyl acrylate  I79
107-25-5   Methyl vinyl ether  M296
107-31-3   Methyl formate  M210
107-41-5   2-Methylpentane-2,4-diol  M251
107-70-0   4-Methoxy-4-methylpentan-
           2-one  M128
107-83-5   2-Methylpentane  M249
107-87-9   Methyl propyl ketone  M268
107-98-2   1-Methoxy-2-propanol  M134
108-07-6   Mercury (II), methyl-, acetate
           M68
108-10-1   Isobutyl methyl ketone  I84

108-11-2   4-Methyl-2-pentanol  M256
108-22-5   Isopropenyl acetate  I113
108-23-6   Isopropyl chloroformate  I117
108-31-6   Maleic anhydride  M12
108-47-4   2,4-Lutidine  L58
108-62-3   Metaldehyde  M86
108-65-6   1-Methoxy-2-propanol, acetate
           M135
108-67-8   Mesitylene  M83
108-78-1   Melamine  M36
108-87-2   Methylcyclohexane  M184
108-89-4   4-Methylpyridine  M275
108-99-6   3-Methylpyridine  M274
109-02-4   N-Methylmorpholine  M233
109-06-8   2-Methylpyridine  M273
109-08-0   2-Methylpyrazine  M269
109-77-3   Malononitrile  M16
109-83-1   N-Methyl-2-aminoethanol  M142
109-87-5   Methylal  M141
110-12-3   5-Methylhexan-2-one  M215
110-19-0   Isobutyl acetate  I78
110-68-9   N-Methylbutylamine  M169
110-91-8   Morpholine  M333
110-93-0   6-Methyl-5-hepten-2-one  M214
111-77-3   2-(2-Methoxyethoxy)ethanol
           M124
111-94-4   3,3'-Iminobis[propanenitrile]  I8
112-35-6   2-[2-(2-Methoxyethoxy)ethoxy]-
           ethanol  M125
112-62-9   Z-Methyl oleate  M248
113-52-0   Imipramine hydrochloride  I14
114-33-0   N-Methyl-3-pyridinecarboxamide
           M276
115-11-7   Isobutene  I77
115-18-4   2-Methyl-3-buten-2-ol  M167
115-19-5   2-Methyl-3-butyn-2-ol  M172
119-32-4   4-Methyl-3-nitroaniline  M240
119-36-8   Methyl salicylate  M284
119-38-0   Isolan  I98
119-65-3   Isoquinoline  I129
120-58-1   Isosafrole  I130
120-72-9   Indole  I26
120-75-2   2-Methylbenzothiazole  M148
120-78-5   Mercaptobenzothiazole disulfide
           M49
121-75-5   Malathion  M10
121-91-5   Isophthalic acid  I108
122-03-2   4-(1-Methylethyl)benzaldehyde
           M204
123-15-9   2-Methylvaleraldehyde  M295
123-33-1   Maleic hydrazide  M13
123-35-3   7-Methyl-3-methylene-
           1,6-octadiene  M231
123-39-7   N-Methylformamide  M209

| | | | |
|---|---|---|---|
| 123-51-3 | 3-Methylbutan-1-ol M162 | 443-48-1 | Metronidazole M304 |
| 123-76-2 | Levulinic acid L39 | 463-51-4 | Ketene K8 |
| 123-92-2 | Isopentyl acetate I104 | 465-73-6 | Isodrin I92 |
| 125-12-2 | Isobornyl acetate I74 | 482-89-3 | Indigo I21 |
| 126-98-7 | Methacrylonitrile M97 | 488-10-8 | Jasmone J3 |
| 127-41-3 | α-Ionone I52 | 491-35-0 | 4-Methylquinoline M279 |
| 129-15-7 | 2-Methyl-1-nitroanthraquinone M242 | 494-19-9 | Iminodibenzyl I12 |
| 131-70-4 | Monobutyl phthalate M320 | 496-11-7 | Indan I15 |
| 133-32-4 | 4-Indol-3-ylbutyric acid I30 | 502-39-6 | Mercury (II), methyl-, dicyanamide M69 |
| 135-23-9 | Methapyrilene hydrochloride M108 | 502-65-8 | Lycopene C.I. L60 |
| 136-85-6 | 5-Methyl-1H-benzotriazole M152 | 504-60-9 | 1-Methylbutadiene M158 |
| 137-05-3 | Methyl 2-cyanoacrylate M183 | 505-60-2 | Mustard gas M339 |
| 137-58-6 | Lignocaine L42 | 513-38-2 | 1-Iodo-2-methylpropane I48 |
| 138-22-7 | Lactic acid, butyl ester L2 | 513-42-8 | Methallyl alcohol M100 |
| 138-86-3 | Limonene L43 | 513-48-4 | 2-Iodobutane I42 |
| 139-91-3 | 5-(Morpholinomethyl)-3-[(5-nitrofurfurylidene)-amino]-2-oxazolidinone M334 | 524-81-2 | Mebhydrolin M30 |
| | | 525-79-1 | Kinetin K11 |
| | | 531-59-9 | 7-Methoxycoumarin M123 |
| | | 531-76-0 | Merphalan M81 |
| 140-41-0 | Monuron TCA M331 | 532-03-6 | Methocarbamol M115 |
| 141-01-5 | Iron(II) fumarate I66 | 534-22-5 | 2-Methylfuran M212 |
| 141-79-7 | 4-Methylpent-3-en-2-one M261 | 538-86-3 | (Methoxymethyl)benzene M127 |
| 141-82-2 | Malonic acid M15 | 540-37-4 | 4-Iodoaniline I39 |
| 142-73-4 | Iminodiacetic acid I11 | 540-38-5 | 4-Iodophenol I50 |
| 143-07-7 | Lauric acid L8 | 540-42-1 | Isobutyl propionate I87 |
| 144-41-2 | Morphothion M335 | 540-84-1 | Isooctane I101 |
| 148-82-3 | Melphalan M38 | 541-25-3 | Lewisite L40 |
| 149-30-4 | 2-Mercaptobenzothiazole M48 | 542-55-2 | Isobutyl formate I81 |
| 150-50-5 | Merphos M82 | 542-69-8 | 1-Iodobutane I41 |
| 150-68-5 | Monuron M330 | 544-63-8 | Myristic acid M341 |
| 150-76-5 | 4-Methoxyphenol M131 | 546-67-8 | Lead tetraacetate L30 |
| 151-38-2 | Mercury(II), methoxyethyl-, acetate M66 | 547-63-7 | Methyl isobutyrate M223 |
| | | 547-64-8 | Methyl lactate M227 |
| 193-39-5 | Indeno[1,2,3-cd]pyrene I20 | 555-30-6 | Methyldopa M195 |
| 271-44-3 | Indazole I18 | 556-24-1 | Methyl isovalerate M226 |
| 288-32-4 | Imidazole I7 | 556-61-6 | Methyl isothiocyanate M225 |
| 297-78-9 | Isobenzan I73 | 556-64-9 | Methyl thiocyanate M290 |
| 298-81-7 | Methoxsalen M119 | 556-82-1 | 3-Methyl-2-buten-1-ol M165 |
| 302-15-8 | Methylhydrazine sulfate M217 | 557-17-5 | 1-Methoxypropane M133 |
| 303-34-4 | Lasiocarpine L7 | 558-17-8 | 2-Iodo-2-methylpropane I49 |
| 305-33-9 | Iproniazid phoshate I57 | 558-25-8 | Methanesulfonyl fluoride M104 |
| 315-18-4 | Mexacarbate M308 | 563-45-1 | 3-Methyl-1-butene M164 |
| 315-22-0 | Monocrotaline M321 | 563-80-4 | 3-Methylbutan-2-one M163 |
| 319-78-8 | D-Isoleucine I99 | 564-94-3 | Myrtenal M343 |
| 328-38-1 | D-Leucine L36 | 565-67-3 | 2-Methyl-3-pentanol M254 |
| 348-67-4 | D-Methionine M114 | 583-15-3 | Mercury benzoate M55 |
| 366-70-1 | Ibenzmethyzin hydrochloride I1 | 583-39-1 | 2-Mercaptobenzimidazole M47 |
| 370-50-3 | Mitin N M314 | 583-59-5 | 2-Methylcyclohexanol M186 |
| 409-02-9 | Methylheptenone M213 | 583-60-8 | 2-Methylcyclohexanone M190 |
| 421-20-5 | Methyl fluorosulfonate ('Magic Methyl') M208 | 583-61-9 | 2,3-Lutidine L57 |
| | | 589-91-3 | 4-Methylcyclohexanol M188 |
| | | 589-93-5 | 2,5-Lutidine L59 |
| 434-13-9 | Lithocholic Acid L52 | 590-86-3 | Isovaleraldehyde I134 |

| | |
|---|---|
| 590-96-5 | Methylazoxymethanol M145 |
| 591-23-1 | 3-Methylcyclohexanol M187 |
| 591-50-4 | Iodobenzene I40 |
| 592-04-1 | Mercury(II) cyanide M60 |
| 592-05-2 | Lead cyanide L15 |
| 592-62-1 | Methylazoxymethanol, acetate M146 |
| 592-85-8 | Mercury(II) thiocyanate M80 |
| 592-87-0 | Lead thiocyanate L33 |
| 593-74-8 | Mercury, dimethyl- M62 |
| 598-50-5 | 1-Methylurea M294 |
| 598-55-0 | Methyl carbamate M173 |
| 606-23-5 | 1,3-Indandione I16 |
| 611-32-5 | 8-Methylquinoline M283 |
| 611-72-3 | D-Mandelic acid M18 |
| 612-60-2 | 7-Methylquinoline M282 |
| 615-22-5 | 2-(Methylthio)benzothiazole M289 |
| 616-47-7 | 1-Methylimidazole M219 |
| 624-48-6 | Maleic acid, dimethyl ester M11 |
| 624-83-9 | Methyl isocyanate M224 |
| 625-27-4 | 2-Methyl-2-pentene M258 |
| 625-28-5 | 3-Methylbutanenitrile M160 |
| 625-45-6 | 2-Methoxyacetic acid M120 |
| 625-55-8 | Isopropyl formate I119 |
| 626-01-7 | 3-Iodoaniline I38 |
| 626-17-5 | Isophthalonitrile I109 |
| 626-67-5 | 1-Methylpiperidine M264 |
| 626-89-1 | Isohexanol I97 |
| 627-44-1 | Mercury, diethyl- M61 |
| 632-99-5 | Magenta I M1 |
| 638-11-9 | Isopropyl butyrate I116 |
| 643-58-3 | 2-Methylbiphenyl M155 |
| 644-08-6 | 4-Methylbiphenyl M156 |
| 652-67-5 | Isosorbide I131 |
| 676-97-1 | Methyl phosphonic dichloride M263 |
| 691-37-2 | 4-Methyl-1-pentene M259 |
| 696-29-7 | Isopropylcyclohexane I118 |
| 758-17-8 | N-Methyl-N-formylhydrazine M211 |
| 760-93-0 | Methacrylic anhydride M96 |
| 763-29-1 | 2-Methyl-1-pentene M257 |
| 763-32-6 | 3-Methyl-3-buten-1-ol M166 |
| 771-51-7 | Indole-3-acetonitrile I27 |
| 797-63-7 | Levonorgestrel L37 |
| 814-78-8 | 3-Methyl-3-buten-2-one M168 |
| 815-57-6 | 3-Methyl-2,4-pentanedione M252 |
| 838-88-0 | 4,4'-Methylenebis[2-methyl-aniline] M199 |
| 860-22-0 | Indigo carmine I22 |
| 872-50-4 | N-Methylpyrrolidone M277 |
| 874-35-1 | 5-Methylindan M221 |

| | |
|---|---|
| 920-46-7 | Methacryloyl chloride M98 |
| 924-42-5 | N-Methyloloacrylamide M247 |
| 950-10-7 | Mephosfolan M43 |
| 950-37-8 | Methidathion M111 |
| 992-21-2 | Lymecycline L61 |
| 1072-35-1 | Lead stearate L26 |
| 1073-93-4 | 1-(1-Methylethyl)-1H-pyrrole-2,5-dione M207 |
| 1114-34-7 | D-Lyxose L63 |
| 1115-70-4 | Metformin hydrochloride M90 |
| 1119-97-7 | Myristyltrimethylammonium bromide M342 |
| 1129-41-5 | Metolcarb M301 |
| 1172-63-0 | Jasmolin II J2 |
| 1191-80-6 | Mercury oleate M73 |
| 1309-37-1 | Iron oxide I68 |
| 1309-48-4 | Magnesium oxide M6 |
| 1309-60-0 | Lead dioxide L17 |
| 1313-13-9 | Manganese dioxide M22 |
| 1313-27-5 | Molybdenum trioxide M318 |
| 1314-87-0 | Lead sulfide L29 |
| 1317-35-7 | Manganese tetroxide M25 |
| 1318-93-0 | Montmorillonite M329 |
| 1320-67-8 | Monopropylene glycol methyl ether M327 |
| 1321-94-4 | Methylnaphthalene M234 |
| 1331-22-2 | Methylcyclohexanone M189 |
| 1332-58-7 | Kaolin K2 |
| 1335-31-5 | Mercury oxycyanide M76 |
| 1335-32-6 | Lead subacetate L27 |
| 1470-94-6 | 5-Indanol I17 |
| 1600-27-7 | Mercury(II) acetate M53 |
| 1634-04-4 | Methyl tert-butyl ether M170 |
| 1634-78-2 | Malaoxon M9 |
| 1678-82-6 | 1-Isopropyl-4-methylcyclohexane I122 |
| 1689-83-4 | Ioxynil I53 |
| 1705-85-7 | 6-Methylchrysene M182 |
| 1712-64-7 | Isopropyl nitrate I124 |
| 1746-81-2 | Monolinuron M324 |
| 2027-17-0 | 2-Isopropylnaphthalene I123 |
| 2032-65-7 | Methiocarb M113 |
| 2163-80-6 | MSMA M336 |
| 2164-08-1 | Lenacil L34 |
| 2173-57-1 | Isobutyl 2-naphthyl ether I86 |
| 2212-67-1 | Molinate M316 |
| 2338-37-6 | Levopropoxyphene L38 |
| 2365-48-2 | Methyl mercaptoacetate M228 |
| 2381-21-7 | 1-Methylpyrene M270 |
| 2385-85-5 | Mirex M311 |
| 2595-54-2 | Mecarbam M31 |
| 2631-40-5 | Isoprocarb I111 |
| 2763-96-4 | Muscimol M337 |
| 3060-89-7 | Metobromuron M299 |

# INDEX OF MOLECULAR
# FORMULAE

| | | | |
|---|---|---|---|
| C4H6O | Methyl vinyl ketone M297 | C5H8O3 | Methyl acetoacetate M138 |
| C4H6O | Methacrolein M93 | C5H9N | 3-Methylbutanenitrile M160 |
| C4H6O2 | Methacrylic acid M95 | C5H9NO | N-Methylpyrrolidone M277 |
| C4H6O5 | Malic acid M14 | C5H10 | 3-Methyl-1-butene M164 |
| C4H7ClO2 | Isopropyl chloroformate I117 | C5H10HgO3 | Mercury(II), methoxyethyl-, |
| C4H7ClO2 | Methyl 2-chloropropionate | | acetate M66 |
| | M176 | C5H10N2O2S | Methomyl M116 |
| C4H7N | Isobutyronitrile I90 | C5H10O | 2-Methyl-3-buten-2-ol M167 |
| C4H7NO | 2-Methylacrylamide M140 | C5H10O | 2-Methyltetrahydrofuran |
| C4H7NO2 | N-Methyloloacrylamide | | M288 |
| | M247 | C5H10O | Isovaleraldehyde I134 |
| C4H7NO4 | Iminodiacetic acid I11 | C5H10O | Methyl propyl ketone M268 |
| C4H8 | Isobutene I77 | C5H10O | 3-Methyl-3-buten-1-ol M166 |
| C4H8Cl2S | Mustard gas M339 | C5H10O | 3-Methylbutan-2-one M163 |
| C4H8N2O3 | Methylazoxymethanol, | C5H10O | 3-Methyl-2-buten-1-ol M165 |
| | acetate M146 | C5H10O | 2-Methyl butanal M159 |
| C4H8O | Methallyl alcohol M100 | C5H10O2 | Methyl isobutyrate M223 |
| C4H8O | Isobutyraldehyde I88 | C5H10O2 | Isobutyl formate I81 |
| C4H8O2 | 1-Methoxyacetone M121 | C5H10O5 | D-Lyxose L63 |
| C4H8O2 | Isobutyric acid I89 | C5H11NO | N-Methylmorpholine M233 |
| C4H8O2 | Isopropyl formate I119 | C5H11NO2S | D-Methionine M114 |
| C4H8O3 | Methyl lactate M227 | C5H12 | Isopentane I103 |
| C4H9I | 2-Iodobutane I42 | C5H12O | 2-Methylbutan-2-ol M161 |
| C4H9I | 2-Iodo-2-methylpropane I49 | C5H12O | 3-Methylbutan-1-ol M162 |
| C4H9I | 1-Iodobutane I41 | C5H12O | Methyl tert-butyl ether M170 |
| C4H9I | 1-Iodo-2-methylpropane I48 | C5H12O3 | 2-(2-Methoxyethoxy)ethanol |
| C4H9NO | Morpholine M333 | | M124 |
| C4H10 | Isobutane I75 | C5H13N | N-Methylbutylamine M169 |
| C4H10Hg | Mercury, diethyl- M61 | C5H13NO3 | 2,2′-(Methylimino)diethanol |
| C4H10O | 1-Methoxypropane M133 | | M220 |
| C4H10O | Isobutanol I76 | C6H3Cl4NO2S | Methyl 2,3,5,6-tetrachloro- |
| C4H10O2 | Monopropylene glycol | | 4-pyridyl sulfone M287 |
| | methyl ether M327 | C6H5I | Iodobenzene I40 |
| C4H10O2 | 1-Methoxy-2-propanol M134 | C6H5IO | 4-Iodophenol I50 |
| C4H10O8Pb3 | Lead subacetate L27 | C6H6Cl6 | ε-Lindane L46 |
| C4H11N | Isobutylamine I80 | C6H6Cl6 | Lindane L45 |
| C4H12ClN5 | Metformin hydrochloride | C6H6IN | 4-Iodoaniline I39 |
| | M90 | C6H6IN | 3-Iodoaniline I38 |
| C4H12Pb | Lead, tetramethyl- L32 | C6H7N | 4-Methylpyridine M275 |
| C5FeO5 | Iron pentacarbonyl I69 | C6H7N | 2-Methylpyridine M273 |
| C5H4N4S | 6-Mercaptopurine M50 | C6H7N | 3-Methylpyridine M274 |
| C5H5NO2 | Methyl 2-cyanoacrylate | C6H7N3O | Isoniazid I100 |
| | M183 | C6H7N5O | 6-Methoxyguanine M126 |
| C5H6N2 | 2-Methylpyrazine M269 | C6H8O4 | Maleic acid, dimethyl ester |
| C5H6N2OS | Methylthiouracil M291 | | M11 |
| C5H6O | 2-Methylfuran M212 | C6H9N3 | 3,3′-Iminobis[propanenitrile] |
| C5H8 | 1-Methylbutadiene M158 | | I8 |
| C5H8 | Isoprene I110 | C6H9N3O3 | Metronidazole M304 |
| C5H8O | 2-Methyl-3-butyn-2-ol M172 | C6H10O | 4-Methylpent-3-en-2-one |
| C5H8O | 3-Methyl-3-buten-2-one | | M261 |
| | M168 | C6H10O2 | 3-Methyl-2,4-pentanedione |
| C5H8O2 | Isopropenyl acetate I113 | | M252 |
| C5H8O2 | Methyl methacrylate M229 | C6H10O4 | Isosorbide I131 |
| C5H8O3 | Levulinic acid L39 | C6H11N2O4PS3 | Methidathion M111 |

| | | | |
|---|---|---|---|
| C$_6$H$_{11}$O$_7$Hg | Mercury gluconate M63 | C$_7$H$_7$N$_3$O$_2$ | *N*-Methyl-*N*,4-dinitrosoaniline M194 |
| C$_6$H$_{12}$ | Methylcyclopentane M191 | | |
| C$_6$H$_{12}$ | 2-Methyl-2-pentene M258 | C$_7$H$_8$N$_2$O | *N*-Methyl-3-pyridine- |
| C$_6$H$_{12}$ | 4-Methyl-1-pentene M259 | | carboxamide M276 |
| C$_6$H$_{12}$ | 4-Methyl-2-pentene M260 | C$_7$H$_8$N$_2$O$_2$ | 2-Methyl-4-nitroaniline |
| C$_6$H$_{12}$ | 2-Methyl-1-pentene M257 | | M237 |
| C$_6$H$_{12}$N$_2$PbS$_4$ | Lead dimethyldithiocarbamate L16 | C$_7$H$_8$N$_2$O$_2$ | *N*-Methyl-4-nitroaniline M241 |
| C$_6$H$_{12}$N$_2$S$_3$ | Monothiuram M328 | C$_7$H$_8$N$_2$O$_2$ | 4-Methyl-2-nitroaniline |
| C$_6$H$_{12}$N$_5$O$_2$PS$_2$ | Menazon M39 | | M239 |
| C$_6$H$_{12}$O | 2-Methylvaleraldehyde M295 | C$_7$H$_8$N$_2$O$_2$ | 4-Methyl-3-nitroaniline M240 |
| C$_6$H$_{12}$O | Isobutyl methyl ketone I84 | C$_7$H$_8$N$_2$O$_2$ | 2-Methyl-5-nitroaniline |
| C$_6$H$_{12}$O$_2$ | 1-Methylpropyl acetate M267 | | M238 |
| C$_6$H$_{12}$O$_2$ | Methyl isovalerate M226 | C$_7$H$_8$N$_2$O$_3$ | 2-Methoxy-4-nitroaniline |
| C$_6$H$_{12}$O$_2$ | Isobutyl acetate I78 | | M129 |
| C$_6$H$_{12}$O$_3$ | 1-Methoxy-2-propanol, acetate M135 | C$_7$H$_8$O$_2$ | 4-Methoxyphenol M131 |
| | | C$_7$H$_9$N | 2,4-Lutidine L58 |
| C$_6$H$_{12}$O$_3$ | Isopropylidene glycerol I121 | C$_7$H$_9$N | 2,5-Lutidine L59 |
| C$_6$H$_{12}$O$_3$ | 2-Methoxy-1-propanol, acetate M136 | C$_7$H$_9$N | *N*-Methylaniline M143 |
| | | C$_7$H$_9$N | 2,3-Lutidine L57 |
| C$_6$H$_{12}$O$_6$ | D-Mannose M27 | C$_7$H$_9$NO$_2$ | 1-(1-Methyl)-1*H*-pyrrole- |
| C$_6$H$_{12}$O$_6$ | Inositol I33 | | 2,5-dione M207 |
| C$_6$H$_{13}$N | 1-Methylpiperidine M264 | C$_7$H$_9$NO$_3$ | Methacryloyloxyethyl |
| C$_6$H$_{13}$NO$_2$ | D-Leucine L36 | | isocyanate M99 |
| C$_6$H$_{13}$NO$_2$ | D-Isoleucine I99 | C$_7$H$_{11}$N$_3$O$_4$ | Misonidazole M313 |
| C$_6$H$_{14}$ | 2-Methylpentane M249 | C$_7$H$_{12}$O | 2-Methylcyclohexanone |
| C$_6$H$_{14}$ | 3-Methylpentane M250 | | M190 |
| C$_6$H$_{14}$FO$_3$P | Isofluorphate I95 | C$_7$H$_{12}$O | Methylcyclohexanone |
| C$_6$H$_{14}$O | 4-Methyl-2-pentanol M256 | | M189 |
| C$_6$H$_{14}$O | 2-Methyl-1-pentanol M253 | C$_7$H$_{12}$O$_2$ | Isobutyl acrylate I79 |
| C$_6$H$_{14}$O | 2-Methyl-3-pentanol M254 | C$_7$H$_{13}$O$_5$PS | Methacrifos M92 |
| C$_6$H$_{14}$O | 3-Methyl-3-pentanol M255 | C$_7$H$_{13}$O$_6$P | Mevinphos M307 |
| C$_6$H$_{14}$O | Isohexanol I97 | C$_7$H$_{14}$ | Methylcyclohexane M184 |
| C$_6$H$_{14}$O$_2$ | 2-Methylpentane-2,4-diol M251 | C$_7$H$_{14}$NO$_5$P | Monocrotophos M322 |
| | | C$_7$H$_{14}$N$_4$O$_4$ | *N*-(*N*-Methyl-*N*-nitroso- |
| C$_6$H$_{14}$O$_6$ | D-Mannitol M26 | | carbamoyl)-L-ornithine |
| C$_6$H$_{15}$O$_3$PS$_2$ | Methyl demeton M192 | | M245 |
| C$_7$H$_3$Cl$_2$NO | Isocyanic acid, | C$_7$H$_{14}$O | 2-Methylcyclohexanol M186 |
| | 3,4-dichlorophenyl ester | C$_7$H$_{14}$O | 3-Methylcyclohexanol M187 |
| | I91 | C$_7$H$_{14}$O | 5-Methylhexan-2-one M215 |
| C$_7$H$_3$I$_2$NO | Ioxynil I53 | C$_7$H$_{14}$O | Methylcyclohexanol M185 |
| C$_7$H$_4$HgO$_3$ | Mercury salicylate M78 | C$_7$H$_{14}$O | 4-Methylcyclohexanol M188 |
| C$_7$H$_5$NS$_2$ | 2-Mercaptobenzothiazole M48 | C$_7$H$_{14}$O$_2$ | 4-Methoxy-4-methylpentan-2-one M128 |
| C$_7$H$_6$N$_2$ | Indazole I18 | C$_7$H$_{14}$O$_2$ | Isopentyl acetate I104 |
| C$_7$H$_6$N$_2$S | 2-Mercaptobenzimidazole M47 | C$_7$H$_{14}$O$_2$ | Isobutyl propionate I87 |
| | | C$_7$H$_{14}$O$_2$ | Isopropyl butyrate I116 |
| C$_7$H$_7$N$_3$ | Methyl-1*H*-benzotriazole M150 | C$_7$H$_{14}$O$_3$ | Lactic acid, butyl ester L2 |
| | | C$_7$H$_{15}$Cl$_2$N$_2$O$_2$P | Isophosphamide I107 |
| C$_7$H$_7$N$_3$ | 5-Methyl-1*H*-benzotriazole M152 | C$_7$H$_{16}$O$_4$ | 2-(2-(2-Methyoxyethoxy)-ethoxy)ethanol M125 |
| C$_7$H$_7$N$_3$ | 1-Methyl-1*H*-benzotriazole M151 | C$_7$H$_{18}$N$_2$O$_3$ | 4-Methoxy-2-nitroaniline M130 |

| | |
|---|---|
| C$_{10}$H$_{16}$ | Limonene L43 |
| C$_{10}$H$_{17}$N$_3$O$_2$ | Isouron I133 |
| C$_{10}$H$_{17}$N$_3$O$_2$ | Isolan I98 |
| C$_{10}$H$_{18}$O | Isopulegol I128 |
| C$_{10}$H$_{19}$O$_6$PS$_2$ | Malathion M10 |
| C$_{10}$H$_{19}$O$_7$PS | Malaoxon M9 |
| C$_{10}$H$_{20}$ | 1-Isopropyl-4-methyl-cyclohexane I122 |
| C$_{10}$H$_{20}$NO$_5$PS$_2$ | Mecarbam M31 |
| C$_{10}$H$_{20}$O | 2-Methyl-3-(4-*tert*-butyl-phenyl)propionaldehyde M171 |
| C$_{10}$H$_{20}$O | Menthol M40 |
| C$_{10}$H$_{20}$O | D,L-Menthol M41 |
| C$_{10}$H$_{20}$O$_2$ | *p*-Menthane hydroperoxide M42 |
| C$_{11}$H$_{10}$ | 1-Methylnaphthalene M235 |
| C$_{11}$H$_{10}$ | 2-Methylnaphthalene M236 |
| C$_{11}$H$_{10}$ | Methylnaphthalene M234 |
| C$_{11}$H$_{12}$Cl$_4$N$_2$O$_3$ | Monuron TCA M331 |
| C$_{11}$H$_{13}$ClO$_2$S | MCPA-thioethyl M28 |
| C$_{11}$H$_{13}$F$_3$N$_2$O$_3$S | Mefluidide M35 |
| C$_{11}$H$_{15}$NO$_2$ | 3-(1-Methylethyl)phenol methylcarbamate M205 |
| C$_{11}$H$_{15}$NO$_2$ | Isoprocarb I111 |
| C$_{11}$H$_{15}$NO$_2$S | Methiocarb M113 |
| C$_{11}$H$_{15}$NO$_5$ | Methocarbamol M115 |
| C$_{11}$H$_{16}$O | Jasmone J3 |
| C$_{12}$H$_2$MgO$_8$ | Magnesium perchlorate M7 |
| C$_{12}$H$_8$Cl$_6$ | Isodrin I92 |
| C$_{12}$H$_8$O$_4$ | Methoxsalen M119 |
| C$_{12}$H$_{13}$NO$_2$ | 4-Indol-3-ylbutyric acid I30 |
| C$_{12}$H$_{14}$O$_4$ | Monobutyl phthalate M320 |
| C$_{12}$H$_{16}$N$_2$O$_5$ | Musk ambrette M338 |
| C$_{12}$H$_{18}$N$_2$O | Isoproturon I127 |
| C$_{12}$H$_{18}$N$_2$O$_2$ | Isophorone diisocyanate I106 |
| C$_{12}$H$_{18}$N$_2$O$_2$ | Mexacarbate M308 |
| C$_{12}$H$_{18}$O$_4$S$_2$ | Isoprothiolane I126 |
| C$_{12}$H$_{19}$NO$_2$ | Mepronil M45 |
| C$_{12}$H$_{20}$ClN$_3$O | Ibenzmethyzin hydrochloride I1 |
| C$_{12}$H$_{20}$O$_2$ | Isobornyl acetate I74 |
| C$_{12}$H$_{20}$O$_{11}$ | β-Lactose L4 |
| C$_{12}$H$_{24}$O$_2$ | (2-Methyl-2,2,4-trimethyl-pentyl)propanoic acid M265 |
| C$_{12}$H$_{24}$O$_2$ | Lauric acid L8 |
| C$_{12}$H$_{27}$PS$_3$ | Merphos M82 |
| C$_{13}$H$_{10}$BrCl$_2$O$_2$PS | Leptophos L35 |
| C$_{13}$H$_{12}$ | Methyl-1,1'-biphenyl M154 |
| C$_{13}$H$_{12}$ | 2-Methylbiphenyl M155 |
| C$_{13}$H$_{12}$ | 4-Methylbiphenyl M156 |
| C$_{13}$H$_{13}$Cl$_2$N$_3$O$_3$ | Iprodione I56 |
| C$_{13}$H$_{14}$ | 2-Isopropylnaphthalene I123 |
| C$_{13}$H$_{15}$Cl$_2$N$_2$ | 4,4'-Methylenedianiline dihydrochloride M202 |
| C$_{13}$H$_{15}$N$_3$O$_3$ | Imazapyr I5 |
| C$_{13}$H$_{16}$N$_2$O$_2$ | Melatonin M37 |
| C$_{13}$H$_{16}$N$_4$O$_6$ | 5-Morpholinomethyl)-3-[(5-nitrofurfurylidene)-amino]-2-oxazolidinone M334 |
| C$_{13}$H$_{16}$NO$_4$PS | Isoxathion I135 |
| C$_{13}$H$_{16}$O$_4$ | Monopentyl phthalate M326 |
| C$_{13}$H$_{18}$ClNO | Monalide M319 |
| C$_{13}$H$_{18}$Cl$_2$N$_2$O$_2$ | Merphalan M81 |
| C$_{13}$H$_{18}$Cl$_2$N$_2$O$_2$ | Melphalan M38 |
| C$_{13}$H$_{18}$N$_2$O$_2$ | Lenacil L34 |
| C$_{13}$H$_{18}$O$_2$ | Ibuprofen I2 |
| C$_{13}$H$_{21}$O$_3$PS | Iprobenfos I55 |
| C$_{14}$H$_8$N$_2$S$_4$ | Mercaptobenzothiazole disulfide M49 |
| C$_{14}$H$_{10}$O$_4$Hg | Mercury benzoate M55 |
| C$_{14}$H$_{12}$N$_2$S | 4-(6-Methyl-2-benzothiazolyl)-benzenamine M149 |
| C$_{14}$H$_{13}$N | Iminodibenzyl I12 |
| C$_{14}$H$_{14}$Cl$_2$N$_2$O | Imazalil I3 |
| C$_{14}$H$_{14}$N$_2$O | Metyrapone M306 |
| C$_{14}$H$_{15}$N$_5$O$_6$S | Metsulfuron-methyl M305 |
| C$_{14}$H$_{16}$ClN$_3$O | Metazachlor M88 |
| C$_{14}$H$_{16}$O | Isobutyl 2-naphthyl ether I86 |
| C$_{14}$H$_{19}$N$_3$S | Methapyrilene M107 |
| C$_{14}$H$_{19}$O$_6$P | α-Methylbenzyl 3-hydroxycrotonate, dimethyl phosphate M153 |
| C$_{14}$H$_{20}$ClN$_3$S | Methapyrilene hydrochloride M108 |
| C$_{14}$H$_{21}$N$_3$O$_3$ | Karbutilate K4 |
| C$_{14}$H$_{22}$N$_2$O | Lignocaine L42 |
| C$_{14}$H$_{25}$N$_3$O$_9$ | Kasugamycin K5 |
| C$_{14}$H$_{28}$O$_2$ | Myristic acid M341 |
| C$_{15}$H$_8$C$_{12}$F$_6$N$_2$O | Mitin N M314 |
| C$_{15}$H$_9$NO$_4$ | 2-Methyl-1-nitro anthraquinone M242 |
| C$_{15}$H$_{15}$NO$_2$ | Mefanamic acid M34 |
| C$_{15}$H$_{16}$O$_2$ | 4,4'-Isopropylidenediphenol I120 |
| C$_{15}$H$_{17}$ClN$_4$ | Myclobutanil M340 |
| C$_{15}$H$_{18}$N$_2$ | 4,4'-Methylene bis[2-methylaniline] M199 |
| C$_{15}$H$_{18}$N$_4$O$_5$ | Mitomycin C M315 |
| C$_{15}$H$_{19}$N$_3$O$_3$ | Imazethapyr I6 |
| C$_{15}$H$_{21}$NO$_4$ | Metalaxyl M85 |
| C$_{15}$H$_{22}$ClNO$_2$ | Metolachlor M300 |
| C$_{15}$H$_{22}$N$_2$O$_2$ | Methylene bis(4-cyclohexyl- |

| | | | |
|---|---|---|---|
| | isocyanate) M197 | C$_{21}$H$_{30}$O$_3$ | Jasmolin I J1 |
| C$_{15}$H$_{23}$N$_3$O$_4$ | Isopropalin I112 | C$_{21}$H$_{33}$NO$_7$ | Lasiocarpine L7 |
| C$_{15}$H$_{24}$NO$_4$PS | Isofenphos I94 | C$_{22}$H$_{12}$ | Indeno[1,2,3-$cd$]pyrene I20 |
| C$_{16}$H$_{10}$N$_2$Na$_2$O$_8$S$_2$ | | C$_{22}$H$_{29}$NO$_2$ | Levopropoxyphene L38 |
| | Indigo carmine I22 | C$_{22}$H$_{30}$O$_5$ | Jasmolin II J2 |
| C$_{16}$H$_{10}$N$_2$O$_2$ | Indigo I21 | C$_{24}$H$_{26}$O$_3$S | Kadethrin K1 |
| C$_{16}$H$_{14}$N$_2$O | Methaqualone M109 | C$_{24}$H$_{40}$O$_3$ | Lithocholic Acid L52 |
| C$_{16}$H$_{14}$N$_2$O$_2$S | Mefenacet M33 | C$_{25}$H$_{35}$NO$_5$ | Mebeverine M29 |
| C$_{16}$H$_{14}$O$_3$ | Ketoprofen K10 | C$_{26}$H$_{28}$C$_{12}$N$_4$O$_4$ | Ketoconazole K9 |
| C$_{16}$H$_{15}$Cl$_3$O$_2$ | Methoxychlor M122 | C$_{29}$H$_{38}$N$_4$O$_{10}$ | Lymecycline L61 |
| C$_{16}$H$_{16}$O$_2$ | Methoxyphenone M132 | C$_{36}$H$_{66}$O$_4$Hg | Mercury oleate M73 |
| C$_{16}$H$_{18}$N$_3$SCl | Methylene Blue M201 | C$_{36}$H$_{72}$O$_4$Pb | Lead stearate L26 |
| C$_{16}$H$_{20}$N$_2$O$_3$ | Imazamethabenz-methyl I4 | C$_{40}$H$_{56}$ | Lycopene C.I. L60 |
| C$_{16}$H$_{22}$O$_4$ | Mono(2-ethylhexyl) phthalate M323 | C$_{42}$H$_{30}$N$_6$O$_{12}$ | Inositol niacinate I34 |
| | | C$_{43}$H$_{48}$N$_2$NaO$_6$S$_2$ | |
| C$_{16}$H$_{22}$O$_4$ | Monooctyl phthalate M325 | | Indocyanine green I25 |
| C$_{16}$H$_{23}$NO$_6$ | Monocrotaline M321 | ClI | Iodine monochloride I36 |
| C$_{16}$H$_{33}$N$_{11}$S$_{16}$Zn$_3$ | | ClLiO | Lithium hypochlorite L51 |
| | Metiram M298 | Cl$_2$Fe | Iron(II) chloride I62 |
| C$_{17}$H$_{12}$ | 1-Methylpyrene M270 | Cl$_2$H$_2$O$_8$Pb | Lead perchlorate L23 |
| C$_{17}$H$_{12}$ | 2-Methylpyrene M271 | Cl$_2$MgO$_6$ | Magnesium chlorate M3 |
| C$_{17}$H$_{12}$ | 4-Methylpyrene M272 | Cl$_2$Pb | Lead chloride L13 |
| C$_{17}$H$_{12}$Cl$_{10}$O$_4$ | Kelevan K6 | Cl$_3$Fe | Iron(III) chloride I63 |
| C$_{17}$H$_{19}$NO$_3$ | Morphine M332 | Cl$_3$H$_{20}$O | α-Ionone I52 |
| C$_{17}$H$_{20}$N$_2$O | Methyldymron M196 | Cl$_3$In | Indium trichloride I24 |
| C$_{17}$H$_{22}$N$_2$ | 4,4′-Methylenebis-($N$,$N$-dimethyl)benzenamine M198 | Cl$_3$La | Lanthanum trichloride L6 |
| | | Cl$_4$Ir | Iridium tetrachloride I59 |
| | | C$_r$O$_4$Pb | Lead chromate L14 |
| C$_{17}$H$_{30}$ClN | Lauryl pyridinium chloride L9 | F$_2$Pb | Lead fluoride L19 |
| | | F$_6$SiMg | Magnesium fluorosilicate M4 |
| C$_{17}$H$_{38}$NBr | Myristyltrimethylammonium bromide M342 | | |
| | | F$_6$SiPb | Lead hexafluorosilicate L20 |
| C$_{18}$H$_{14}$Cl$_4$N$_2$O | Miconazole M309 | Fe | Iron I60 |
| C$_{18}$H$_{32}$O$_2$ | Linoleic acid L47 | FeF$_3$ | Iron(III) fluoride I65 |
| C$_{18}$H$_{36}$O$_2$ | Z-Methyl oleate M248 | FeH$_8$N$_2$O$_8$S$_2$ | Iron(II) ammonium sulfate I61 |
| C$_{18}$H$_{41}$N$_7$ | Iminoctadine I10 | | |
| C$_{19}$H$_{14}$ | 1-Methylchrysene M178 | FeN$_3$O$_9$ | Iron(III) nitrate I67 |
| C$_{19}$H$_{14}$ | 2-Methylchrysene M179 | FeO$_4$S | Iron(II) sulfate I70 |
| C$_{19}$H$_{14}$ | 3-Methylchrysene M180 | Fe$_2$O$_3$ | Iron oxide I68 |
| C$_{19}$H$_{14}$ | 5-Methylchrysene M181 | Fe$_2$O$_{12}$S$_3$ | Iron(III) sulfate I71 |
| C$_{19}$H$_{14}$ | 6-Methylchrysene M182 | HO$_5$PPb$_3$ | Lead phosphite dibasic L25 |
| C$_{19}$H$_{16}$ClNO$_4$ | Indomethacin I31 | Hg | Mercury M52 |
| C$_{19}$H$_{20}$N$_2$ | Mebhydrolin M30 | HgBr | Mercury(I) bromide M56 |
| C$_{19}$H$_{24}$N$_2$ | Imipramine I13 | HgBr$_2$ | Mercury(II) bromide M57 |
| C$_{19}$H$_{25}$ClN$_2$ | Imipramine hydrochloride I14 | HgCl | Mercury(I) chloride M58 |
| | | HgCl$_2$ | Mercury(II) chloride M59 |
| C$_{19}$H$_{34}$O$_3$ | Methoprene M117 | HgI | Mercury(I) iodide M64 |
| C$_{20}$H$_{19}$N$_3$ClH | Magenta I M1 | HgI$_2$ | Mercury(II) iodide M65 |
| C$_{20}$H$_{22}$N$_8$O$_5$ | Methotrexate M118 | HgI$_4$K$_2$ | Mercury potassium iodide M77 |
| C$_{20}$H$_{28}$O | Lynestrenol L62 | | |
| C$_{20}$H$_{28}$O$_2$ | Isotretinoin I132 | HgNO$_3$ | Mercury(I) nitrate M70 |
| C$_{21}$H$_{16}$ | 3-Methylcholanthrene M177 | HgN$_2$O$_6$ | Mercury(II) nitrate M71 |
| C$_{21}$H$_{26}$O$_2$ | Mestranol M84 | HgNH$_2$Cl | Mercury ammonium chloride M54 |
| C$_{21}$H$_{28}$O$_2$ | Levonorgestrel L37 | | |

| | | | |
|---|---|---|---|
| HgO | Mercury(II) oxide M75 | Mg₃P₂ | Magnesium phosphide M8 |
| HgO₄S | Mercury bisulfate M51 | Mn | Manganese M20 |
| Hg₂O | Mercury oxide M74 | MnN₂O₆ | Manganese nitrate M24 |
| Hg₂O₄S | Mercury(I) sulfate M79 | MnO₂ | Manganese dioxide M22 |
| I₂ | Iodine I35 | Mn₃O₄ | Manganese tetroxide M25 |
| I₂Pb | Lead iodide L21 | Mo | Molybdenum M317 |
| In | Indium I23 | MoO₃ | Molybdenum trioxide M318 |
| Ir | Iridium black I58 | N₂H₆Pb | Lead nitrate L22 |
| La | Lanthanum L5 | O₂Pb | Lead dioxide L17 |
| Li | Lithium L49 | O₄P₂Pb₃ | Lead phosphate L24 |
| LiH | Lithium hydride L50 | O₄SPb | Lead sulfate L28 |
| Lu | Lutetium L56 | Pb | Lead L10 |
| MgN₂O₆ | Magnesium nitrate M5 | PbS | Lead sulfide L29 |
| MgO | Magnesium oxide M6 | | |

| | | | |
|---|---|---|---|
| $HgO$ | Mercury(II) oxide M75 | $Mg_3P_2$ | Magnesium phosphide M8 |
| $HgO_4S$ | Mercury bisulfate M51 | $Mn$ | Manganese M20 |
| $Hg_2O$ | Mercury oxide M74 | $MnN_2O_6$ | Manganese nitrate M24 |
| $Hg_2O_4S$ | Mercury(I) sulfate M79 | $MnO_2$ | Manganese dioxide M22 |
| $I_2$ | Iodine I35 | $Mn_3O_4$ | Manganese tetroxide M25 |
| $I_2Pb$ | Lead iodide L21 | $Mo$ | Molybdenum M317 |
| $In$ | Indium I23 | $MoO_3$ | Molybdenum trioxide M318 |
| $Ir$ | Iridium black I58 | $N_2H_6Pb$ | Lead nitrate L22 |
| $La$ | Lanthanum L5 | $O_2Pb$ | Lead dioxide L17 |
| $Li$ | Lithium L49 | $O_4P_2Pb_3$ | Lead phosphate L24 |
| $LiH$ | Lithium hydride L50 | $O_4SPb$ | Lead sulfate L28 |
| $Lu$ | Lutetium L56 | $Pb$ | Lead L10 |
| $MgN_2O_6$ | Magnesium nitrate M5 | $PbS$ | Lead sulfide L29 |
| $MgO$ | Magnesium oxide M6 | | |